AA002394

2012 Proceedings of the 19th International Conference Mixed Design of Integrated Circuits and Systems

(MIXDES 2012)

Warsaw, Poland
24-26 May 2012

IEEE Catalog Number: CFP12MIX-PRT
ISBN: 978-1-4577-2092-5

Copyright © 2012, Department of Microelectronics
All Rights Reserved

***This publication is a representation of what appears in the IEEE
Digital Libraries. Some format issues inherent in the e-media version may
also appear in this print version.*

IEEE Catalog Number: CFP12MIX-PRT
ISBN 13:   978-1-4577-2092-5

Additional Copies of This Publication Are Available From:

Curran Associates, Inc
57 Morehouse Lane
Red Hook, NY 12571 USA
Phone: (845) 758-0400
Fax: (845) 758-2633
E-mail: curran@proceedings.com
Web: www.proceedings.com

2012 Proceedings of the 19th International Conference Mixed Design of Integrated Circuits and Systems

(MIXDES 2012)

Warsaw, Poland
24-26 May 2012

IEEE Catalog Number: CFP12MIX-PRT
ISBN: 978-1-4577-2092-5

Preface

The current edition of International Conference "Mixed Design of Integrated Circuits and Systems" is a 19th in series of events held yearly since 1994 in the most renowned places in Poland. This year we meet together in Warsaw in the frame of the 5th Microwave and Radar Week.

The MIXDES conference is one of the largest in the Central Europe in the field, encompassing interdisciplinary research in design, modelling, simulation, testing and manufacturing in various areas, such as micro- and nanoelectronics, semiconductors, sensors, actuators, biomedical applications and power devices. All submissions from 32 countries were reviewed and scored by members of the Programme Committee to put together a high quality technical programme of 117 papers organised in oral and poster presentations.

In addition to the regular programme, there will be five invited papers:

- *Ballistic Transport in Nanoscale Semiconducting Devices*
  Prof. Vijay K. Arora (Wilkes University, USA)

- *Simulation and Modeling of Nanoscale Multiple-gate SOI MOSFETs*
  Prof. Benjamin Iñiguez (Universitat Rovira i Virgili, Spain)

- *The Future of Nanoelectronics*
  Dr. Walter Riess (IBM Research GmbH, Switzerland)

- *Transistor and Interconnect Modeling for Design of Carbon Nanotube Integrated Circuits*
  Prof. Ashok Srivastava (Louisiana State University, USA)

- *Trends and Challenges in Micro- and Nanoelectronics for the Next Decade*
  Prof. Cor Claeys (IMEC, Belgium)

The program of MIXDES 2012 includes also four special sessions:

- *Compact Modeling Support for Nanoscaled IC Technology and Design*
  organised by Dr. Daniel Tomaszewski (Inst. of Electron Techn., Poland)
  and Dr. Władysław Grabiński (GMC Suisse)

- *Especial TRAMS (Terascale Reliable Adaptive Memory Systems) Project Session*
  organised by Prof. Antonio Rubio (Universitat Politècnica de Catalunya, Spain)

- *Technologies towards Cognitive Transceivers - Par4CR Project*
  organised by Dr. Martha Suarez, Ewa Kurjata-Pfitzner and Dr. Jerzy Synka
  (Institute of Electron Technology, Poland)

- *xTCA for Instrumentation*
  organised by Dr. Dariusz Makowski (Technical University of Łódź, Poland)
  and Dr. Stefan Simrock (ITER, France)

We would like to take this opportunity to thank many individuals, especially the reviewers, who have worked so hard to make this meeting happen. We hope that you will enjoy your visit to Warsaw and next year in June we will meet together in Gdynia at MIXDES 2013 to celebrate together 20 years of MIXDES Conferences.

Number of Accepted Papers and Authors by Country

Country	Number of		Country	Number of		Country	Number of	
	papers	co-authors		papers	co-authors		papers	co-authors
Algeria	1	4	Germany	10	31	Romania	2	4
Austria	2	5	Greece	2	7	Slovakia	1	5
Belarus	3	11	Iran	5	15	Spain	6	19
Belgium	2	9	Ireland	1	3	Sweden	1	1
Bulgaria	3	13	Italy	1	7	Switzerland	4	10
Canada	1	1	Lithuania	2	3	The Netherlands	5	11
China	3	8	Malaysia	1	1	Tunisia	1	2
Czech Republic	1	2	Mexico	1	3	UK	1	1
Estonia	2	5	Norway	1	1	Ukraine	1	3
Finland	1	2	Poland	42	98	USA	3	7
France	4	13	Portugal	3	10			
						Total:	**117**	**315**

As it can be seen from the above statistics, the total number of authors and co-authors is 315. It should be underlined that some papers are common for two or even three institutions involved in a joint project.

Number of Accepted Papers by Country

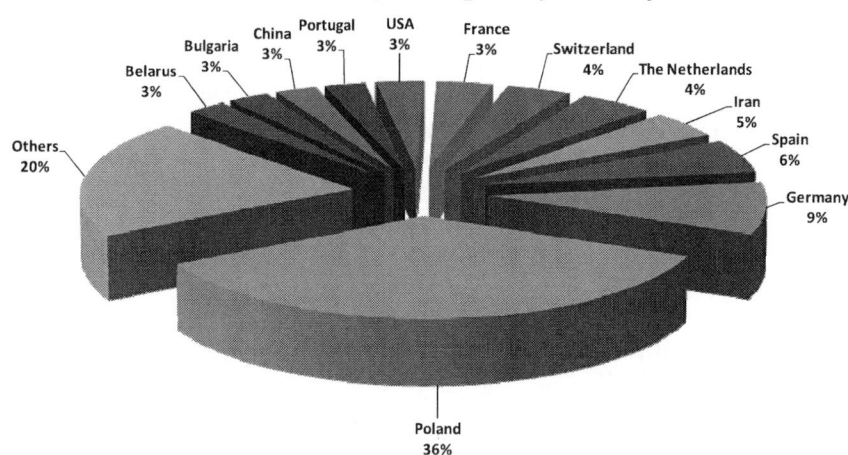

During the past years, the best papers chosen by the chairmen of all the sessions were printed by *Kluwer Academic Publishers*, in a special issue of the *Electron Technology Journal, Microelectronics Reliability Journal, Analog Integrated Circuits & Signal Processing Journal, VLSI Design Journal, Journal of Microelectronics and Computer Science* and the best papers from Poland in the *Elektronika Journal*. This year we continue this practice. Additionally, the authors of selected papers have received the *Best Paper Award* diplomas and the IEEE ED Poland Chapter has given the best student paper award.

We would like to thank all the participants for an inspiring discussion during the sessions and we hope that the conference will help to create and enhance links amongst the experts in different domains, and will help all the participating universities in finding the optimal teaching program for the students involved in the design of modern integrated circuits, devices, and microsystems.

Łódź, May 2012

Andrzej NAPIERALSKI
Department of Microelectronics and Computer Science
Technical University of Łódź, Poland
General Chairman of MIXDES 2012

International Programme Committee

Prof.	M. Adamski	Technical University of Zielona Góra, Poland
Prof.	M. Bucher	Technical University of Crete, Greece
Prof.	J. Cabestany	Universitat Politecnica de Catalunya, Spain
Prof.	J.-J. Charlot	Jean Jacques Charlot Consulting, France
Prof.	Z. Ciota	Technical University of Łódź, Poland
Prof.	J. Collet	LAAS - CNRS, Toulouse, France
Prof.	A. Dąbrowski	Poznań University of Technology, Poland
Prof.	G. De Mey	University of Ghent, Belgium **(Vice-Chairman)**
Prof.	A. De Vos	University of Ghent, Belgium
Prof.	J. Deen	McMaster University, Canada
Prof.	J.-M. Dorkel	LAAS - CNRS, Toulouse, France
Prof.	A. Filipkowski	Warsaw University of Technology, Poland
Dr.	D. Foty	Gilgamesh Associates, USA
Prof.	M. Glesner	Technische Hochschule Darmstadt, Germany
Prof.	L. Golonka	Wrocław University of Technology, Poland
Dr.	W. Grabiński	GMC, Switzerland
Prof.	A. Handkiewicz	Poznań University of Technology, Poland
Prof.	A. Hatzopoulos	Aristotle University of Thessaloniki, Greece
Dr.	S. Hausman	Technical University of Lodz, Poland
Prof.	A. Jakubowski	Warsaw University of Technology, Poland
Prof.	A. Kobus	Institute of Electron Technology, Warsaw, Poland
Prof.	A. Kos	AGH University of Science and Technology, Poland
Prof.	W. Kuźmicz	Warsaw University of Technology, Poland **(Programme Chairman)**
Prof.	M. Lobur	State University Lviv, Ukraine
Dr.	M.M. Louerat	Université Pierre et Marie Curie, Paris, France
Prof.	T. Łuba	Warsaw University of Technology, Poland
Prof.	B. Macukow	Warsaw University of Technology, Poland
Prof.	J. Madrenas	Universitat Politecnica de Catalunya, Spain
Prof.	A. Martinez	LAAS - CNRS, Toulouse, France **(Honorary Chairman)**
Prof.	A. Materka	Technical University of Łódź, Poland
Prof.	W. Mathis	University of Magdeburg, Germany
Prof.	H.S. Momose	Toshiba Corporation, Japan
Prof.	J.M. Moreno	Universitat Politecnica de Catalunya, Spain
Dr.	M. Napieralska	Technical University of Łódź, Poland
Prof.	A. Napieralski	Technical University of Łódź, Poland **(General Chairman)**
Prof.	J. Nishizawa	Semiconductor Research Institute, Japan
Dr.	J.L. Noullet	Chipyards, France
Prof.	M. Ogorzałek	AGH University of Science and Technology, Poland
Prof.	L. Opalski	Warsaw University of Technology, Poland
Prof.	A. Pfitzner	Warsaw University of Technology, Poland
Dr.	B.F. Romanowicz	Nano Science and Technology Institute, USA
Prof.	J.A. Rubio	Universitat Politecnica de Catalunya, Spain
Prof.	J. Rutkowski	Silesian University of Technology, Poland
Prof.	A. Rybarczyk	Poznań University of Technology, Poland
Dr.	J.-M. Sallese	Swiss Federal Institute of Technology, Switzerland
Prof.	D. Sankowski	Technical University of Łódź, Poland
Prof.	N. Stojadinović	University of Niš, Serbia
Prof.	V. Székely	Technical University of Budapest, Hungary
Prof.	T. Szmuc	AGH University of Science and Technology, Poland
Prof.	G. Szymański	Poznań University of Technology, Poland
Prof.	M. Tadeusiewicz	Technical University of Łódź, Poland
Dr.	D. Tomaszewski	Institute of Electron Technology, Warsaw, Poland
Dr.	P. Tounsi	INSA de Toulouse, France
Dr.	M. Turowski	CFD Research Corporation, USA
Prof.	R. Ubar	Tallinn Technical University, Estonia
Prof.	G. Wachutka	Technische Universitaet Muenchen, Germany
Prof.	K. Wawryn	Technical University of Koszalin, Poland
Prof.	B. Więcek	Technical University of Łódź, Poland
Prof.	S. Yoshitomi	Toshiba Corporation, Japan
Prof.	J. Zarębski	Gdynia Maritime Academy, Poland
Dr.	M. Zubert	Technical University of Łódź, Poland

Organising Committee

Prof.	A. Napieralski	(Chairman)
Dr.	M. Orlikowski	(Secretary)
Dr.	M. Napieralska	(Vice-Chairman)
Dr.	G. Jabłoński	
Dr.	M. Janicki	Department of Microelectronics and Computer Science,
Dr.	M. Piotrowicz	Technical University of Łódź, Poland
Prof.	W. Kuźmicz	Institute of Micro- and Optoelectronics, Warsaw University of Technology, Poland

Table of Contents

Preface 3

Table of Contents 7

General Invited Papers

Ballistic Transport in Nanoscale Devices 17
V.K. ARORA (Univ. Tekn. Malaysia, MALAYSIA and Wilkes Univ., USA)

Simulation and Modeling of Nanoscale Multiple-gate SOI MOSFETs 25
B. IÑIGUEZ (Univ. Rovira i Virgili, SPAIN), R. RITZENTHALER (IMEC, BELGIUM), F. LIME, B. NAE (Univ. Rovira i Virgili, SPAIN)

Transistor and Interconnect Modeling for Design of Carbon Nanotube Integrated Circuits 30
A. SRIVASTAVA (Louisiana State Univ., USA)

Trends and Challenges in Micro- and Nanoelectronics for the Next Decade 37
C. CLAEYS (Imec, BELGIUM)

Compact Modeling Support for Nanoscaled IC Technology and Design

Analog Circuits Sizing Using the Fixed Point Iteration Algorithm with Transistor Compact Models 45
F. JAVID, R. ISKANDER (Univ. Pierre and Marie Curie, FRANCE), F. DUSBIN (CEA DAM/DIF, FRANCE), M.-M. LOUËRAT (Univ. Pierre and Marie Curie, FRANCE)

Analytical Drain Current Core Model for Undoped Double Gate MOS Transistor 51
P. SAŁEK, L. ŁUKASIAK, A. JAKUBOWSKI (Warsaw Univ. of Techn., POLAND)

Analytical Model for Predicting Subthreshold Slope Improvement versus Negative Swing of S-shape Polarization in a Ferroelectric FET 55
A. RUSU, A.M. IONESCU (EPFL, SWITZERLAND)

Characterization of Parameter Variability and Correlations for FD SOI CMOS Technology 60
D. TOMASZEWSKI, G. GŁUSZKO, J. MALESIŃSKA, K. KUCHARSKI (Institute of Electron Techn., POLAND)

Impact Ionization Effect in Deep Submicron MOSFET Features Simulation 66
D. SPERANSKY, T.T. TRUNG (Belarusian State Univ., BELARUS)

KLU Sparse Direct Linear Solver Implementation into NGSPICE 69
F. LANNUTTI, P. NENZI, M. OLIVIERI (Sapienza - Univ. Roma, ITALY)

Simulation Study of Nanoscale Double-Gate CMOS Circuits Using Compact Advanced Transport Models 74
M. CHERALATHAN (Univ. Rovira i Virgili, SPAIN), E. CONTRERAS (CINVESTAV, MEXICO), J. ALVARADO (Benemérita Univ. Autónoma de Puebla, MEXICO), A. CERDEIRA (CINVESTAV, MEXICO), B. IÑIGUEZ (Univ. Rovira i Virgili, SPAIN)

Standardization of Multigate MOSFET Modeling 78
N. CHEVILLON, F. PRÉGALDINY, C. LALLEMENT (Univ. Strasbourg, FRANCE), J.-M. SALLESE (EPFL, SWITZERLAND)

Time-domain Waveform Based Extraction of FinFET Non-linear I-V Model 84
D. SCHREURS, G. AVOLIO (Univ. Leuven, BELGIUM), A. RAFFO, G. VANNINI (Univ. Ferrara, ITALY), G. CRUPI, A. CADDEMI (Univ. Messina, ITALY)

Two-dimensional Physics-based Modeling of Dopant-segregated Schottky Barrier UTB MOSFETs 88
M. SCHWARZ, T. HOLTIJ (Tech. Hochschule Mittelhessen, GERMANY and Univ. Rovira i Virgili, SPAIN), A. KLOES (Tech. Hochschule Mittelhessen, GERMANY), B. IÑIGUEZ (Univ. Rovira i Virgili, SPAIN)

Verilog-A Compact Semiconductor Device Modelling and Circuit Macromodelling with the QucsStuddio-ADMS "Turn-Key" Modelling System 94
M.E. BRINSON (London Metropolitan Univ., UK), M. MARGRAF (Qucs, GERMANY)

Especial TRAMS (Terascale Reliable Adaptive Memory Systems) Project

Enhancing 6T SRAM Cell Stability by Back Gate Biasing Techniques for 10nm SOI FinFETs under Process and Environmental Variations 103
Z. JAKŠIĆ, R. CANAL (Univ. Politècnica de Catalunya, SPAIN)

Mitigating Lower Layer Failures with Adaptive System Reconfiguration 109
T. RAMÍREZ, E. HERRERO, N. AXELOS, J. CARRETERO, N. FOUTRIS, D. SANCHEZ, X. VERA (Intel Labs, SPAIN)

SRAM Lifetime Improvement by Using Adaptive Proactive Reconfiguration 115
P. POUYAN, E. AMAT, A. RUBIO (Univ. Politècnica de Catalunya, SPAIN)

Strain Relevance on the Improvement of the 3T1D Cell Performance 120
E. AMAT, C.G. ALMUDÉVER, N. AYMERICH, R. CANAL, A. RUBIO (Univ. Politècnica de Catalunya, SPAIN)

Variability and Reliability Analysis of CNFET in the Presence of Carbon Nanotube Density Fluctuations 124
C.G. ALMUDÉVER, A. RUBIO (Univ. Politècnica de Catalunya, SPAIN)

Technologies towards Cognitive Transceivers - Par4CR Project

Digital Hardware Resources for Steering a Nonlinear Interference Suppressor 133
E.J.G. JANSSEN, H. HABIBI, D. MILOSEVIC, P.G.M. BALTUS, A.H.M. VAN ROERMUND (Eindhoven Univ. of Techn., THE NETHERLANDS)

LINC Architecture: A Discussion 139
E. HABEKOTTÉ, F. VAN DER WILT (Catena, THE NETHERLANDS)

xTCA for Instrumentation

Development of uTCA Hardware for BAM System at FLASH and XFEL 147
S.B. HABIB, D. SIKORA (Warsaw Univ. of Techn., POLAND), J. SZEWIŃSKI, S. KOROLCZUK (National Center for Nuclear Research, POLAND)

FMC-based Neutron and Gamma Radiation Monitoring Module for xTCA Applications 152
T. KOZAK, D. MAKOWSKI, A. NAPIERALSKI (Tech. Univ. Łódź, POLAND)

Image Acquisition Module for uTCA-based Systems 156
A. MIELCZAREK, D. MAKOWSKI, G. JABŁOŃSKI, P. PEREK, A. NAPIERALSKI (Tech. Univ. Łódź, POLAND)

Image Visualisation and Processing in DOOCS and EPICS 161
P. PEREK, J. WYCHOWANIAK, D. MAKOWSKI, M. ORLIKOWSKI, A. NAPIERALSKI (Tech. Univ. Łódź, POLAND)

1 Design of Integrated Circuits and Microsystems

A 10-bit 1.2V High Speed Current-steering DAC Using a Novel Switching Scheme .... N / A
H. WANG (Peking Univ., Shanghai and Peking Univ., Beijing, CHINA), Y. YAO, T. WANG, H. WANG (Peking Univ., Shanghai, CHINA), Y. CHENG (Peking Univ., Shanghai and Peking Univ., Beijing, CHINA)

A 6-bit 2GS/s 1.2V 78mW Flash ADC in 65nm CMOS .... N / A
H. WANG (Peking Univ., Shanghai and Peking Univ., Beijing, CHINA), T. WANG, Y. YAO, H. WANG (Peking Univ., Shanghai, CHINA), Y. CHENG (Peking Univ., Shanghai and Peking Univ., Beijing, CHINA)

A Balun Transimpedance Amplifier with Adjustable Gain for Integrated SPO_2 Optic Sensors 178
J. CARVALHO, L.B. OLIVEIRA, J.P. OLIVEIRA, J. GOES (Univ. Nova de Lisboa, PORTUGAL), M.M. SILVA (Tech. Univ. Lisbon, PORTUGAL)

A Calibratable Capacitance Array Based Approach for High Resolution CR SAR ADCs 183
J.H. MUELLER, S. STRACHE, L. BUSCH, R. WUNDERLICH, S. HEINEN (RWTH Aachen Univ., GERMANY)

A Charge Pump with Power-on Reset Circuit 189
A. GOŁDA, A. KOS (AGH Univ. of Science and Techn., POLAND)

A High Speed A/D Converter for Using in Low Power CMOS Image Sensors 192
M. TEYMOURI (Urmia Univ. of Techn., IRAN), B. ALIZADEH (Islamic Azad Univ., IRAN), A. DADASHI (Urmia Univ., IRAN),
A. MAHMOUDI (Islamic Azad Univ., IRAN)

A High Speed Two-Stage Dual-Path Operational Amplifier in 40nm Digital CMOS 198
H. CHEN, V. MILOVANOVIC, H. ZIMMERMANN (Vienna Univ. of Techn., AUSTRIA)

A Low Noise Low Offset Current Mode Instrumentation Amplifier 203
A. VOULKIDOU, S. SISKOS, T. LAOPOULOS (Aristotle Univ. Thessaloniki, GREECE)

A Novel Charge Recycling Approach to Low-Power Circuit Design 208
C. ULAGANATHAN, C.L. BRITTON, JR., J. HOLLEMAN, B.J. BLALOCK (The Univ. of Tennessee, USA)

A Switched-Capacitor Implementation of Short-Term Synaptic Dynamics 214
M. NOACK, C. MAYR, J. PARTZSCH, M. SCHULTZ, R. SCHÜFFNY (Tech. Univ. Dresden, GERMANY)

Area Efficient Front-end Readout Electronics for Pixel Detector Based on Inverter Amplifier 219
R. KŁECZEK, P. OTFINOWSKI, P. GRYBOŚ (AGH Univ. of Science and Techn., POLAND)

Area Efficient Low Power Neural Amplifiers Using MOS and MIM Capacitors in Submicron Technologies for Ultra Low Corner Frequencies 223
P. KMON, P. GRYBOŚ, R. SZCZYGIEŁ, M. ŻOŁĄDŹ (AGH Univ. of Science and Techn., POLAND)

Convex Combination Initialization Method for Kohonen Neural Network Implemented in the CMOS Technology 227
R. DŁUGOSZ (EPFL, SWITZERLAND and Univ. of Techn. and Life Sciences, POLAND), T. TALAŚKA (Univ. of Techn. and Life
Sciences, POLAND), P.-A. FARINE (EPFL, SWITZERLAND), W. PEDRYCZ (Univ. Alberta, CANADA)

Crystal Oscillator with Dual Amplitude Stabilization Feedback Loop 231
K. SIWIEC (Warsaw Univ. of Techn., POLAND)

Dc-Gain Enhanced Fast-settling Triple Folded Cascode Op-Amp 235
A. MAHMOUDI (Islamic Azad Univ., IRAN), A. DADASHI (Urmia Univ., IRAN), S. MASOUMI, M.K. KOUZEHKANAN (Islamic Azad
Univ., IRAN)

Design Criteria of Frequency Selection for the Internal Oscillator of UHF RFID Tag .... N / A
V. ZAITSAY, U. STEPANETS (Belarusian State Univ., BELARUS), M. AUDZEYEU (National Academy of Sciences, BELARUS)

Design of a Fully Programmable Analog Interval Type-2 Triangular/Trapezoidal Fuzzifier 243
H. YAZDANJOUEI, H. FEIZY, A. KHOEI, K. HADIDI (Urmia Univ., IRAN)

Design of an Analog Output Buffer for Active Matrix Displays Using Low-Temperature Polycrystalline Silicon Thin-Film Transistors 249
I. PAPPAS, S. SISKOS, A.A. HATZOPOULOS (Aristotle Univ. Thessaloniki, GREECE)

Fast-Settling Gain Stage Using Replica Amplification for High Performance Pipeline ADCs 255
M.K. KOUZEHKANAN (Islamic Azad Univ., IRAN), A. DADASHI (Urmia Univ., IRAN), M. TEYMOURI (Urmia Univ. of Techn., IRAN),
S. MASOUMI (Islamic Azad Univ., IRAN)

FPGA Implementation of Chaotic Pseudo-Random Bit Generators 260
P. DĄBAL, R. PEŁKA (Military Univ. of Techn., POLAND)

Fully Integratable 4-Phase Charge Pump Architecture for High Voltage Applications 265
L. SHEN, K. HOFMANN (Tech. Univ. Darmstadt, GERMANY)

High-resolution Hold-off Time Control Circuit for Geiger-mode Avalanche Photodiodes 269
S. DENG, A.P. MORRISON (Univ. College Cork, IRELAND)

Integrated Circuit for Wireless Inductive Powering Implemented in 180nm CMOS Process 273
M. ŻOŁĄDŹ, P. KMON, P. OTFINOWSKI, J. RAUZA, P. GRYBOŚ (AGH Univ. of Science and Techn., POLAND)

RF Varactor Design Based on Evolutionary Algorithms 277
P. PEREIRA, H. FINO, M. VENTIM-NEVES (Univ. Nova de Lisboa, PORTUGAL)

Temperature and Supply Voltage Compensated Biasing for Digitally Controlled Oscillators 283
S. HÖPPNER, S. HAENZSCHE, S. HARTMANN, S. SCHIEFER, R. SCHÜFFNY (Tech. Univ. Dresden, GERMANY)

Wideband Low-Noise RF Front-End for CNT-NEMS Sensors 289
C. KAUTH, M. PASTRE, M. KAYAL (EPFL, SWITZERLAND)

2 Thermal Issues in Microelectronics

1/f Noise Temperature Behaviour of Poly Resistors 297
W.C. PFLANZL, E. SEEBACHER (Austriamicrosystems, AUSTRIA)

Compensation of the Temperature Fluctuations in the Silicon Photomultiplier Measurement System 300
M. BASZCZYK, P. DOROSZ, S. GŁĄB, W. KUCEWICZ, Ł. MIK, M. SAPOR (AGH Univ. of Science and Techn., POLAND)

DC Measurements Method of the Thermal Resistance of Power MOSFETs 304
K. GÓRECKI, J. ZARĘBSKI (Gdynia Maritime Univ., POLAND)

Overheat Security System for High Speed Embedded Systems 309
M. FRANKIEWICZ, A. GOŁDA, A. KOS (AGH Univ. of Science and Techn., POLAND)

Paths of the Heat Flow from Semiconductor Devices to Surrounding 313
K. GÓRECKI, J. ZARĘBSKI (Gdynia Maritime Univ., POLAND)

Technology Migration and Thermal Coupling 319
M. JANICKI, P. ZAJĄC, M. SZERMER, A. NAPIERALSKI (Tech. Univ. Lodz, POLAND)

Thermal Models for Dynamic Clock Control 323
M. FRANKIEWICZ, A. KOS (AGH Univ. of Science and Techn., POLAND)

Thermographic Measurements of Planar Inductors 328
*I. PAPAGIANNOPOULOS, V. CHATZIATHANASIOU (Aristotle Univ. Thessaloniki, GREECE), G. DE MEY (Ghent Univ., BELGIUM),
B. WIĘCEK, M. KAŁUŻA (Tech. Univ. Lodz, POLAND)*

3 Analysis and Modelling of ICs and Microsystems

A Specific Parameters Analysis of CMOS Hall Effect Sensors with Various Geometries 335
M.-A. PAUN, J.-M. SALLESE, M. KAYAL (EPFL, SWITZERLAND)

Combined Hardware and Software Tracing of Real and Virtual Embedded System Parts 340
C. KOEHLER (Univ. Siegen, GERMANY), A. MAYER (Infineon Technologies AG, GERMANY), M. WURM (Univ. Siegen, GERMANY)

Conducted Emissions Susceptibility Study for an High Precision LDO 346
A. CREOSTEANU (Tech. Military Academy, ROMANIA), L. CREOSTEANU (Politehnica Univ., Bucharest, ROMANIA)

Device-Circuit Models for Extreme Environment Space Electronics 350
M. TUROWSKI, A. RAMAN (CFDRC, USA)

Estimating the Impact of Complete Analog Channel Selection on Zero-IF Multi-Standard Radio Receivers Power Consumption 356
*S. SPIRIDON ("POLITEHNICA" Univ. Bucharest, ROMANIA and Broadcom, THE NETHERLANDS), D. CLAUDIUS,
M. BODEA ("POLITEHNICA" Univ. Bucharest, ROMANIA)*

Extracting the Parameters of an EEHEMT Nonlinear Model for InP HEMT, Operating at G-band Frequency 360
S. ESKANDARI, F.T. HAMEDANI (Semnan Univ., IRAN)

Joint Simulation of Mixed-Signal Integrated Circuits and Printed Circuit Boards 364
L. CEDERSTRÖM, A. GRAUPNER (Zentrum Mikroelektronik Dresden AG, GERMANY)

Methodology for Development of LVS and LPE Rule Sets Adapted for MEMS Processes 370
*A. PASHEV, D. GAYDAZHIEV (Smartcom Bulgaria AD, BULGARIA), I. UZUNOV (Tech. Univ. Sofia, BULGARIA), D. PUKNEVA (Smartcom
Bulgaria AD, BULGARIA), E. MANOLOV (Tech. Univ. Sofia, BULGARIA)*

Modeling and Optimization of a Ker Charge Pump Loaded by a Resistive Circuit 376
J. HEITZ, N. DUMAS, V. FRICK, C. LALLEMENT, L. HÉBRARD (Univ. Strasbourg, FRANCE)

Offset Compensation for Voltage- and Current Amplifiers with CMOS Inverters 382
W. MACHOWSKI, J. JASIELSKI (AGH Univ. of Science and Techn., POLAND)

Surface Potential Model of a High-k HfO_2-Ta_2O_5 Capacitor 386
G. ANGELOV, N. BONEV, R. RUSEV, M. HRISTOV (Tech. Univ. Sofia, BULGARIA), A. PASKALEVA (Bulgarian Academy of Sciences, BULGARIA)

Symbolic Analysis in Gyrator-Capacitor Filters 392
P. KATARZYŃSKI, M. MELOSIK, M. NAUMOWICZ, S. SZCZĘSNY (Poznan Univ. of Techn., POLAND)

Technology and Device Design and Optimization for the MOSFET Hall Sensor on SOI Structure 398
L. DOLGIY, I. LOVSHENKO, V. NELAYEV, I. SHELIBAK (BSUIR, BELARUS), S. SHVEDOV, A. TURTSEVICH (R&D Center "BelMicroSystems", BELARUS)

Verilog-A Modeling of Electrical Circuit with Adding Element Based on Branched Hydrogen Bonding Network 401
E. GIEVA, R. RUSEV, G. ANGELOV, R. RADONOV, T. TAKOV, M. HRISTOV (Tech. Univ. Sofia, BULGARIA)

4 Microelectronics Technology and Packaging

Characterization of Test Devices for Development of Nanowire Sensor FETs 407
M. ZABOROWSKI, D. TOMASZEWSKI, A. PANAS, P. GRABIEC (Institute of Electron Techn., POLAND)

Design Model and Data Management for 3D Integration Technologies 412
A. GRÜNEWALD, K. HAHN, R. BRÜCK (Univ. Siegen, GERMANY)

Multisensor System for Monitoring Human Psychophysiologic State in Extreme Conditions with the Use of Microwave Sensor 417
K. RÓŻANOWSKI (Military Inst. of Aviation Medicine, POLAND), T. SONDEJ (Military Univ. of Techn., POLAND), J. LEWANDOWSKI (Military Inst. of Aviation Medicine, POLAND), M. ŁUSZCZYK, Z. SZCZEPANIAK (Przemysłowy Inst. Telekom. S.A., POLAND)

5 Testing and Reliability

Determining Effective Testability Degree of Analog Circuits 427
Z. KINCL, Z. KOLKA (Brno Univ. of Techn., CZECH REPUBLIC)

DFT for Analog and Mixed Signal IC Based on IDDQ Scanning 431
B. GUIBANE (Faculty of Sciences of Monastir, TUNISIA), B. HAMDI (Issats, TUNISIA)

GNU Radio and USRP2 as a Universal Platform for Verification of Wireless Communication Devices Used in Automotive Applications 436
M. SZELEST (Delphi Poland S.A., POLAND), W. UZDRZYCHOWSKI, D. GRZECHCA (Silesian Univ. of Techn., POLAND)

On-chip Parametric Test of Binary-weighted R-2R ladder D/A Converter and Its Efficiency 441
D. ARBET, G. GYEPES, J. BRENKUŠ, V. STOPJAKOVÁ, J. MIHÁLOV (Slovak Univ. of Techn., SLOVAKIA)

Reliable On-Chip Network Design Using an Agent-based Management Method 447
M. VALINATAJ (Babol Univ. of Techn., IRAN), P. LILJEBERG, J. PLOSILA (Univ. Turku, FINLAND)

6 Power Electronics

Survey and Analysis of the Design Issues of a Low Cost Micro Power DC-DC Step Up Converter for Indoor Light Energy Harvesting Applications 455
C. CARVALHO (Instituto Politécnico de Lisboa, PORTUGAL), J.P. OLIVEIRA, N. PAULINO (Univ. Nova de Lisboa, PORTUGAL)

The Effect of Thermal Inertia in Photovoltaic Module Simulation 461
M. PIOTROWICZ, W. MARAŃDA (Tech. Univ. Lodz, POLAND)

7 Signal Processing

A Hierarchical Algorithm for Moving Vehicle Identification Based on Acoustic Noise Analysis 467
S. ASTAPOV, A. RIID (Tallinn Univ. of Techn., ESTONIA)

Application of the Newton Method to First-order Implicit Fractional Transfer Function Approximation 473
A. TEPLJAKOV, E. PETLENKOV, J. BELIKOV (Tallinn Univ. of Techn., ESTONIA)

Combining Sound Source Tracking Algorithms Based on Microphone Array to Improve Real-Time Localization 478
C. IBALA (Univ. Limerick, IRELAND), J. VACHAUDEZ, G. FOURTOUNIS, P. POSSA, C. VALDERRAMA (Polytechnic Faculty of Mons, BELGIUM)

Design and Implementation of a Monopulse Radar Signal Processor 484
B. LIU, W. CHANG, X. LI (National Univ. of Defense Techn., CHINA)

Fractional Delay Filter Design with Extracted Window Offsetting 489
M. BLOK (Gdansk Univ.of Techn., POLAND)

Gearbox Fault Diagnosis Using a Hybrid De-noising Method Based on Ensemble Empirical Mode Decomposition and Wavelet Transform . . . . . N / A
M. HAFIDA, B.R. ELHADI, F. AHMED, A. MERABET (Univ. Ferhat Abbas, ALGERIA)

Means of Spatial Structures in the Problem of Finding Relationships between Objects . . . . N / A
G. CHETVERIKOV, V. LESHCHYNSKYI, I. VECHIRSKA (KHNURE, UKRAINE)

Measurement of Settling Time of High-speed D/A Converters 507
R. KVEDARAS, V. KVEDARAS, T. USTINAVIČIUS (Vilnius Gediminas Tech. Univ., LITHUANIA)

Realtime Physics Engine for Robots Movement: Enhanced Virtual Environment 511
A. BURDZIUK, J. POCHMARA, K. ŁAKOMY, P. SZABLATA, R. KOPPA (Poznan Univ. of Techn., POLAND)

8 Embedded Systems

Digital Hardware for Prime Numbers Generation 519
P.M. SZECÓWKA, W. BUSZKO (Wroclaw Univ. of Techn., POLAND)

Hierarchical UML Activity Diagrams into Control Interpreted Petri Nets Transformation 523
M. GROBELNY, I. GROBELNA, M. ADAMSKI (Univ. Zielona Góra, POLAND)

Wireless Communication Solutions for Distributed Strain Measure Systems in Mechanical Structures 527
A. ANDRZEJCZAK, P. PIETRZAK, M. MAKOWSKI, A. NAPIERALSKI (Tech. Univ. Lodz, POLAND)

9 Medical Applications

Automatisation of Computer-aided Burn Wounds Evaluation 535
W. TYLMAN, M. JANICKI, A. NAPIERALSKI (Tech. Univ. Lodz, POLAND)

Low Power System for Measurement of Skin Conductance and Temperature of Patient Body 539
M. MAJCHRZYCKI, I. KAROŃ, K. KOLANOWSKI, A. RYBARCZYK (Poznan Univ. of Techn., POLAND)

Prediction of Fatigue and Sleep Onset Using HRV Analysis 543
M. SZYPULSKA, Z. PIOTROWSKI (Military Univ. of Techn., POLAND)

10 Student Projects

A 6-bit 122 MS/s Digital-to-Analog Converter for Contactless Applications in CMOS 90 nm Technology 549
M. BRZEZIŃSKI (Warsaw Univ. of Techn., POLAND), T. POMORSKI (INSIDE Secure, POLAND), W.A. PLESKACZ (Warsaw Univ. of Techn., POLAND)

Application of the DLFSR Generators in Spread Spectrum Communication 555
R. STĘPIEŃ, J. WALCZAK (Silesian Univ. of Techn., POLAND)

FPGA Implementation of an Evolutionary Algorithm Based Charge Management for Electric Vehicles 559
M. MIELKE, S. HARDT, A. GRÜNEWALD, R. BRÜCK (Univ. Siegen, GERMANY)

Programmable Gain Amplifier for 13.56 MHz Radio Receiver in CMOS 90 nm Technology 564
M. TEODOROWSKI, W.A. PLESKACZ (Warsaw Univ. of Techn., POLAND), T. TAKESHIAN (INSIDE Secure, FRANCE),
T. POMORSKI (INSIDE Secure, POLAND)

Switched Capacitor Low Noise Voltage Converter Design Strategies in 90 nm CMOS Process 570
A. GRODZICKI (Warsaw Univ. of Techn., POLAND)

Index of Authors 575

14

General Invited Papers

Ballistic Transport in Nanoscale Devices

Scattering-limited to collision-free progression in a high electric field

Vijay K. Arora

Faculty of Electrical Engineering
Universiti Teknologi Malaysia
UTM Skudai 81310, Malaysia
and
Division of Engineering and Physics
Wilkes University
Wilkes-Barre, PA 18707, U. S. A.
E-Mail: vijay.arora@wilkes.edu

Abstract—**Ballistic transport from low-field to high-field regime is reviewed with transition from low-field ballistic mobility to high-field drift velocity limited to the intrinsic velocity for a given dimensionality. Equilibrium Fermi-Dirac to Boltzmann to nonequilibrium Arora distribution is delineated and applied. Ballistic injection from the contacts is shown to be of paramount importance as channels scale down to lengths below the scattering-limited mean free path (mfp). Mobility and drift velocity expressions covering the wide spectrum are obtained and compared with existing experimental data. The gamut spans low to high field transport, nondegenerate to degenerate statistics, and scattering-limited stochastic to unilateral streamlined regime.**

Index Terms—**ballistic mobility; nonequilibrium Arora Dsitribution; NEADF; intrinsic velocity; phonon emission; contacts injection.**

I. INTRODUCTION

Ballistic name is borrowed from military firing a projectile from a launcher hitting the target uninterrupted (unscattered) by the media through which the projectile travels. In the same spirit, ballistic transport is the movement of unscattered carriers (electrons or holes) in their journey from source to drain through the conducting channel. Injected electrons from contacts are expected to reach very high speeds in a ballistic channel as electric field is very high. This anticipation of enhanced drift characterized either by higher channel drift velocity or channel mobility has been thrashed by the published experimental data. The experiments are conclusive that the velocity saturates to a level of intrinsic velocity defined as the average of the magnitude of the stochastic velocity vectors. High-field induced, channel-length-limited, and scattering-enhanced low-dimensional (LD) channels result in reduced mobility. The conduction band is lifted from its normal bulk state by quantum confinement in LD nanostructures. This lifting of the conduction band edge leads to enhanced density of states and hence enhanced scattering resulting in reduction of channel mobility. A crucial question that remains unanswered in the present landscape of transport theories is this: Does high mobility leads to higher saturation velocity and hence superior transport behavior? It is easy to surmise that a higher mobility will lead to a higher saturation velocity, but facts point to the contrary.

In most cases, injected carriers thermalize through collisions on entering the channel. However, if the channel length is below the scattering-limited mean free path (mfp), the probability of carrier going collision free is greatly enhanced. The theories about stochastic nature of the carriers in random walk within a semiconductor are well established in the published literature. A continuing divergence of viewpoints is clearly apparent when dealing with either ballistic processes or transport in a high electric field with or without the presence of a high magnetic field. The most accepted format based on Monte Carlo experiments confirms that the carrier drift in response to an electric field is scattering-limited. In almost every prediction the involvement of optical phonon emission is invoked. It may as well be a photon emission with transition from higher quantized energy levels to a lower one in an LD nanostructure as carriers are accelerated to a higher quantum state. Energy balance theories make use of hot-electron temperature that is a function of electric field [1]. Myths and realities of hot electrons are reviewed by Arora [2]. Arora's paradigm [3] restricting saturation to the intrinsic velocity contradicts scattering-limited-saturation-velocity response to a towering electric field. Arora et al. [4] conclude that the drift saturation is in direct response to the reorientation of randomly oriented velocity vectors in equilibrium to unilateral ones in a towering electric field. This is in direct contrast to the theories where carriers are projected to be hot or randomness enhanced in a high electric field. Temperature (actually entropy) is closely connected with stochastic nature of the motion; randomness increases with the rising temperature. In Arora's framework, carriers respond to the forces of an applied electric field; streamlining the drift in a unilateral motion [3, 4] thereby resulting in reduced randomness. Yes, in a degenerate system, where the energy of carriers is higher than in a nondegenerate setup, the higher electron temperature is justified, but not in response to the electric field [5, 6]. An electric field in fact reorients the velocity vectors without changing their magnitude. Accelerating effect on electrons with velocity vector oriented antiparallel to an applied electric field is compensated by the decelerating effect on those electrons with velocity vector aligned in the parallel direction. This acceleration-deceleration effect on carriers oriented in and opposite to the direction of an electric field results in saturation velocity limited to the intrinsic velocity obtainable from the Fermi-Dirac statistics [5, 7-12].

The low-field characterization of nanoscale channels also gets modified as ballistic injection from the contacts takes on an increasing importance [13]. The carrier drift and hence mobility should increase as collisions are avoided in a ballistic channel with length below the scattering-limited mfp. This perception is contradicted by a number of experiments showing decreases in mobility as channels are scaled beneath the mfp [14]. Moreover magnetoresisance mobility in ballistic channels differs by a scattering-dependent magnetoresistance factor from the drift mobility [15, 16].

II. EQUILIBRIUM DISTRIBUTION FUNCTION

With large carrier population in a semiconductor, the distribution of energy and velocity takes on an increasing importance. In equilibrium, the carriers follow the Fermi-Dirac (FD) statistics. The probability $f(E)$ that a quantum state of energy E is occupied is given by [17, 18]

$$f(E) = \frac{1}{e^{\frac{E-E_F}{k_B T}} + 1} \qquad (2.1)$$

where E_F is the Fermi energy with the probability of occupation ½ at $E=E_F$. The placement of the Fermi energy with respect to the conduction band edge describes the degeneracy of the system. The degeneracy of an electron population with carrier concentration n_d depends on the ratio $u_d = (n/N_c)_d$ with N_{cd} the density of states for dimensionality d. d=3 is for a bulk semiconductor, d=2 for a nanolayer and d=1 for a nanowire. n_3 and N_{c3} have dimensions per unit volume, n_2 and N_{c2} per unit area, and n_1 and N_{c1} per unit length. $u_d = (n/N_c)_d$ is thus the effective probability of a quantum state filled. The effective density of states (DOS) or more correctly quantum states N_{cd} for conduction band is given by

$$N_{cd} = 2\left(\frac{m^* k_B T}{2\pi\hbar^2}\right)^{d/2} = \frac{2}{\left(2\sqrt{\pi}\lambda_D\right)^d} \qquad (2.2)$$

$$\lambda_D = \frac{\hbar}{\sqrt{2m^* k_B T}} \quad \lambda_D = \lambda_D / 2\pi \qquad (2.3)$$

λ_D is the de Broglie thermal wavelength of an electron with effective mass m*. The use of equilibrium distribution function along with the DOS yields reduced Fermi energy η_d [10, 11] extractable from

$$u_d = \mathfrak{I}_{\frac{d}{2}-1}(\eta_d) \quad d=3, 2, or 1 \qquad (2.4)$$

with

$$\eta_d = \frac{(E_F - E_c)_d}{k_B T} \qquad (2.5)$$

$E_c = E_{co} + \varepsilon_{od}$ with $\varepsilon_{o3} = 0$ is the modification in the conduction band energy as conduction band is lifted due to quantum effects by zero-point energy ε_{od} of the lowest quantized level. $\mathfrak{I}_j(\eta_d)$ is the Fermi-Dirac integral (FDI) discussed, for example, in [7]. In the Boltzmann statistics, an approximation of FD statsics, the 1 in the denominator of Equation (2.1) is neglected, converting FD distribution to an exponential Maxwell-Boltzmann (MB) distribution in the non-degenerate (ND) regime with $u_d < 1$. The FDI is always an exponential ($\mathfrak{I}_j(\eta_d) \approx e^{\eta_d}$) in MB/ND limit and does not depend on order j. This is the main attraction of using MB statistics even if it is not correct when one is in degenerate regime with $u_d > 1$.

The intrinsic velocity v_{id} [11] is the average magnitude of the stochastic velocity vectors as given by

$$v_{id} = v_{thd} \frac{\mathfrak{I}_{\frac{d-1}{2}}(\eta_d)}{(n/N_c)_d} = v_{thd} \frac{\mathfrak{I}_{\frac{d-1}{2}}(\eta_d)}{\mathfrak{I}_{\frac{d-2}{2}}(\eta_d)} \qquad (2.6)$$

with

$$v_{thd} = v_{th} \frac{\Gamma\left(\frac{d+1}{2}\right)}{\Gamma\left(\frac{d}{2}\right)} \quad v_{th} = \sqrt{\frac{2k_B T}{m^*}} \qquad (2.7)$$

In the ND limit, the intrinsic velocity is

$$v_{idND} = v_{thd} = v_{th} \frac{\Gamma\left(\frac{d+1}{2}\right)}{\Gamma\left(\frac{d}{2}\right)} \qquad (2.8)$$

ND velocity even though not correct is widely used even for strongly degenerate situations [19] for convenience. The reason for its widespread use arises from its independence on the carrier concentration as it depends on temperature only. In the other extreme of strong degenerate regime with $u_d \gg 1$, the intrinsic velocity is independent of temperature and depends strongly on carrier concentration as is given by

$$v_{idDeg} = 2\frac{d}{d+1}\frac{\hbar}{m^*}\sqrt{\pi}\left[\Gamma\left(\frac{d+2}{2}\right)\frac{n_d}{2}\right]^{\frac{1}{d}} \qquad (2.11)$$

Figure 1 shows normalized intrinsic velocity ratio v_{id} / v_{thd} of intrinsic velocity to its ND counterpart. As expected, the ratio is unity in the nondegenerate regime when $n_d < N_{cd}$ (or $u_d < 1$) increasing to much higher value in the degenerate regime. The rise is particularly striking for a 1D nanowire.

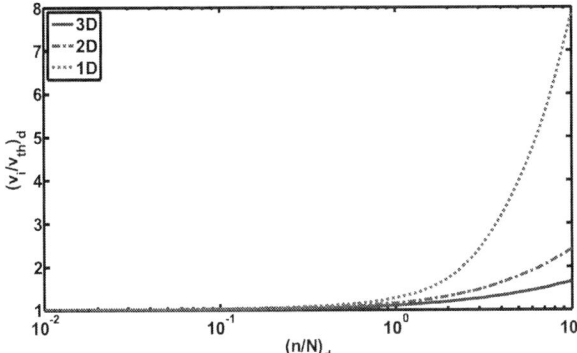

Figure 1. Normalized intrinsic velocity as a function of normalized concentration.

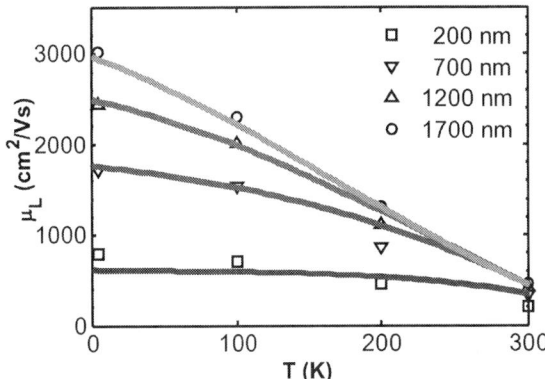

Figure 2. Ballistic mobility a function of temperature for L=200, 700, 1200, and 1700 nm. The solid lines are from nonstationary (transient) transport and markers are the experimental data.

III. BALLISTIC MOBILITY

The ballistic mobility originates from nonstationary (transient) transport in response to an ohmic electric field [20]. The source and drain reservoirs launch electrons into the channel with injection velocity v_{inj} of the contacts. $v_{inj} \simeq v_F$ is the Fermi velocity of the metal contacts at which probability of tunneling is the highest. Electrons so injected transit the channel with finite ballisticity $\exp(-L/\ell_B)$, defined as the probability of a collision-free flight. The ballistic mfp $\ell_B = \ell_{o\infty}(v_{inj}/v_{id})$ differs from the long-channel mfp $\ell_{o\infty}$ in a low electric field by a factor (v_{inj}/v_{id}), where v_{id} is the channel intrinsic velocity. Ballistic mfp is much larger than its counterpart in a long channel as $v_{inj} >> v_{id}$. Monte Carlo procedures and experiments authenticate this paradigm [7, 13, 14, 20]. The length-limited ohmic mobility μ_{oL} in this model is given by

$$\mu_{oL} = \mu_{o\infty}(1 - e^{-\frac{L}{\ell_B}}) \qquad (3.1)$$

The long channel mfp $\ell_{o\infty}$ is obtainable from experimentally measured $\mu_{o\infty}$

$$\ell_{o\infty} = \frac{m^* v_{md}}{q}\mu_{o\infty} \qquad (3.2)$$

with mobility velocity v_{md} intimately connected to v_{id} through

$$v_{md} = d\frac{1}{2}\left(\frac{\Gamma(d/2)}{\Gamma(d+1)}\right)^2 v_{id} \qquad (3.3)$$

The injection velocity for Aluminum contact is the Fermi velocity $v_{inj} \approx v_F = 2.02\times10^6 m/s$ related to Fermi energy of 11.63 eV. ND carrier statistics is certainly not applicable to Schottky metal contacts with high electron concentration.

Figure 2 shows the agreement with the experimental data over the wide range of temperatures [14]. The solid lines are from the ballistic model of Eq. (3.1). Lusakowski et al. [21] were unable to get agreement with experiments based on model proposed by Shur [22] and Wang and Lundstrom [23]. In fact, following the probabilistic nature of electronic motion, Shur [22] and Wang and Lundstrom [23] are correct in adding probabilities of an electron either going ballistic or undergoing collision. The difference is in their identification of the thermal velocity as the injection velocity from highly degenerate contacts. Additionally, Lusakowski et al. were not able to make a clear distinction between magnetoresistance and drift mobility. A poor agreement is attained if $\ell_{o\infty}$ from magnetoresistance mobiity is utilized. The data acquired by Lusakowski et al. [21] is for much reduced channel lengths as compared to that of Robertson and Dumin [24] obtained in 1985. However, Robertson and Dumin cover a wide spectrum of high electric fields with temperature ranging from cryogenic to room (300 K). The mobility degradation was conjectured to be due to high-field effects. In fact, the high-field and ballistic effects are intertwined. High-field effects are negligible for V=0.1 V in a ballistic channel. When mgantoresistance factor is phased in for Ref. [21], the agreement with the experimental data is as good as can be expected. Figure 3 shows the agreement of theory with the experimental data. $\ell_{o\infty}$ extracted from the drift mobility is 10.1 nm for $n_s = 5\times10^{10}cm^{-2}$, 9.8 nm for $5\times10^{11}cm^{-2}$, 9.5 nm for $1\times10^{12}cm^{-2}$, 9.0 nm for $2\times10^{12}cm^{-2}$, and 8.0 nm for $5\times10^{12}cm^{-2}$. The corresponding values of the ballistic mfp, $\ell_B = 108, 104, 99, 91,$ and 72 nm, are much higher than $\ell_{o\infty}$. The agreement with experiments is greatly improved over and above the agreement in Monte Carlo simulation of Huet et al. [20]. The real nature of ballisticity is apparent in Equation (3.1). $(1 - e^{-\frac{L}{\ell_B}})$ in Equation (3.1) is the probability of a collision and its complement $e^{-\frac{L}{\ell_B}}$ is the probability of ballisticity. The probability of ballistic transport approaches unity only when $L \to 0$. Probability of collision taking place is 100% in a long channel ($L \gg \ell_B$) reducing to zero only in $L \to 0$ limit. This probabilistic nature of ballisticity is a clear indication that no matter how small the channel, there is a finite non-zero probability of scattering.

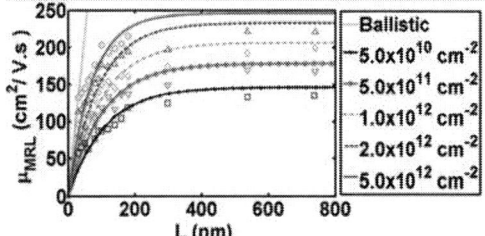

Figure 3. Length-limited magnetoresitance mobility μ_{MRL} versus size L of the conducting channel. Solid lines represent theory based on nonstationary model and markers are the experimental data obtained from Ref. 12.

IV. NONEQUILBRIUM DISTRIBUTION

Although above formalism is adequate for ballistic mobility in a low electric field, a careful scrutiny is needed for ballistic transport in a high electric field. Nonequilubrium Arora distribution function (NEADF) is a platform for ballistic transport in a high electric field. NEADF is an outgrowth of the equilibrium Fermi-Dirac distribution and is given by

$$f(E, \mathrm{E}, \theta) = \cfrac{1}{e^{\frac{E - (E_F - q\vec{\mathrm{E}} \cdot \vec{\ell})}{k_B T}} + 1}$$
$$= \frac{1}{e^{x - H(\theta)} + 1} \qquad (4.1)$$

with

$$H(\theta) = \eta - \delta_o \cos\theta \qquad (4.2)$$

$$x = \frac{E - E_c}{k_B T} \quad \eta = \frac{E_F - E_c}{k_B T} \quad \delta_o = \frac{\mathrm{E}}{\mathrm{E}_{co}} \qquad (4.3)$$

$$\mathrm{E}_{co} = \frac{k_B T}{q \ell_{o\infty}} = \frac{V_t}{\ell_{o\infty}}, \quad V_t = \frac{k_B T}{q} \qquad (4.4)$$

$q\vec{\mathrm{E}} \cdot \vec{\ell} = q_\mathrm{E} \ell \cos(\theta)$ is positive from $0 \le \theta \le \pi/2$ with extreme value $q\vec{\mathrm{E}} \cdot \vec{\ell} = +q_\mathrm{E}\ell$ at $\theta = 0$. In this range electrons are moving in the positive polar direction with electric field applied in the negative polar direction. $q\vec{\mathrm{E}} \cdot \vec{\ell} = q_\mathrm{E}\ell \cos(\theta)$ is negative from $\pi/2 \le \theta \le \pi$ with extreme value $q\vec{\mathrm{E}} \cdot \vec{\ell} = -q_\mathrm{E}\ell$ at $\theta = \pi$. In general, $\ell(E) = \tau(E)v$ is the energy-dependent mfp. An averaged constant value ℓ cuts down greatly the numerical work and brings out vividly the salient features of NEADF. The electrochemical potential E_F during the free flight of a carrier changes by $q_\mathrm{E}\ell$ as electrons tend to sink and holes tend to float on the tilted energy band diagram. This observation may suggest that an applied electric field tends to organize the otherwise complete random motion.

An electric dipole $q\ell$ due to the quasi-free motion of the carriers tends to organize in the direction of electric field for holes and in the opposite direction for electrons. The collisions tend to bring electrons closer to the conduction band edge. If the electric field is strong, this unidirectional motion gives carrier drift comparable to the intrinsic velocity. The normalized effective reduced Fermi energy $H(\theta)$ is now directional as compared to isotropic zero-field reduced Fermi energy η_o. NEADF is compatible with low-field Boltzmann transport framework. In the limit of low-field, the distribution function is linear in electric field

$$f(E, \mathrm{E}, \theta) = f_o(E) + \delta_o \cos\theta \frac{df_o}{dx} \qquad (4.5)$$

The ohmic mobility obtained from the linearized framework is the same as given by Equation (3.2)

$$\mu_{o\infty} = \frac{q \ell_{o\infty}}{m^* v_{md}} \qquad (4.6)$$

The carrier concentration and hence the Fermi energy in a high field originates from the product of NEADF and DOS resulting in an intractable integrals for all three dimensionalities

$$n_3 = (N_{c3}/2) \int_0^\pi d\theta \mathfrak{S}_{1/2}(H(\theta)) \sin(\theta)$$
$$= (N_{c3}/2) \int_{-1}^{+1} d[\cos(\theta)] \mathfrak{S}_{1/2}(H(\theta)) \qquad (4.7)$$

$$n_2 = (N_{c2}/2\pi) \int_0^{2\pi} d\theta \mathfrak{S}_o(H(\theta))$$
$$= (N_{c2}/2\pi) \int_0^{2\pi} d\theta \ln(1 + e^{H(\theta)}) \qquad (4.8)$$

$$n_1 = (N_{c1}/2)[\mathfrak{S}_{-1/2}(\eta_1 + \delta) + \mathfrak{S}_{-1/2}(\eta_1 - \delta)] \qquad (4.9)$$

In order to see the effect of random motion being transformed to a streamlined one, a simple substitution of average value of $\cos\theta = \pm 1/d$ in $H(\theta)$ simplifies the integral. The net number $\Delta n_d / n_d$ of electrons going in the positive direction (opposite to the electric field) is then obtained as

$$\Delta n_d / n_d = (n_{d+} - n_{d-})/(n_{d+} + n_{d-}) \qquad (4.10)$$

with

$$n_{d\pm} = (N_{cd}/2)\mathfrak{S}_{(d-2)/2}(\eta_d \pm \delta_o/d) \qquad (4.11)$$

A simple expression covering arbitrary degeneracy is obtainable from Eq. (4.10) by defining an equivalent degeneracy electron temperature given by [14]

$$\frac{T_{ed}}{T} = \frac{v_{id}}{v_{thd}} \frac{v_{ud}}{v_{thd}} = \frac{v_{id} v_{ud}}{v_{thd}^2} \qquad (4.12)$$

There are two aspects of T_{ed} being elevated over and above its ND counterpart depending on lattice temperature. One factor v_{id}/v_{thd} is the obvious rise in thermal velocity for a given dimensionality rising higher towards Fermi velocity. v_{id} is proportional to Fermi velocity depending on carrier concentration in extreme degeneracy. Since Fermi velocity is much higher than thermal velocity, the average energy and hence velocity appear elevated and can be connected to an enhanced electron temperature. The other factor v_{ud}/v_{thd} is much more subtle. In an ND regime, $T_e = T$. However, a distinction arises in the degenerate regime as electrons oriented in the direction parallel to the applied electric field change orientation opposite (antiparallel) to the electric filed. Pauli Exclusion Principle does not allow occupation of already occupied quantum states below the Fermi energy. The converted electrons increasingly unidirectional must go to the higher quantum states thereby elevating the Fermi energy to new level that is appropriate for $2n_d$ electrons filling half the k-space, as shown in Fig. 4.

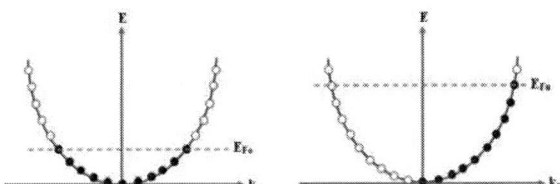

Figure 4. The stochastic carriers in equilibrium becoming unidirectional in high electric field thereby lifting the Fermi energy from E_{Fo} to E_{Fu} in extreme degeneracy to avoid conflict with Pauli Exclusion Principle.

The density of electrons does not increase as there are $2n_d$ antiparallel to the electric field and none in the parallel direction, giving average n_d. The unidirectional transformation thus raises the Fermi level appropriate to $2n_d$ electrons. In the ND regime, $v_{ud} = v_{id} = v_{thd}$. However as the degeneracy level rises, so does the transformation of v_{id} to v_{ud}. The normalized elevated electron temperature $(T_e/T)_d$ is thus the geometric mean of $(v_i/v_{th})_d$ and $(v_u/v_{th})_d$ [5, 6]. The ratio $(v_u/v_i)_d$ as a function of normalized carrier concentration $(n/N_c)_d$ for d=3, 2, and 1 is shown in Figure 5. $v_{u3} = 2^{1/3} v_{i3}$, $v_{u2} = 2^{1/2} v_{i2}$, and $v_{u1} = 2 v_{i2}$ in the extreme degenerate state.

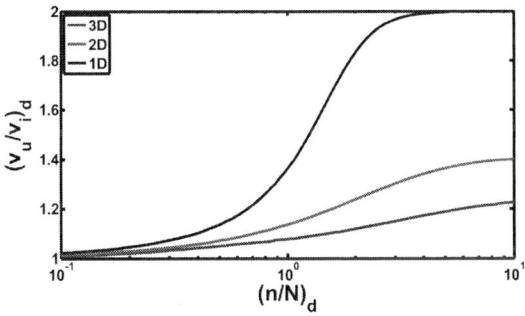

Figure 5. The ratio of unidirectional velocity to that of intrinsic velocity with increasing concentration.

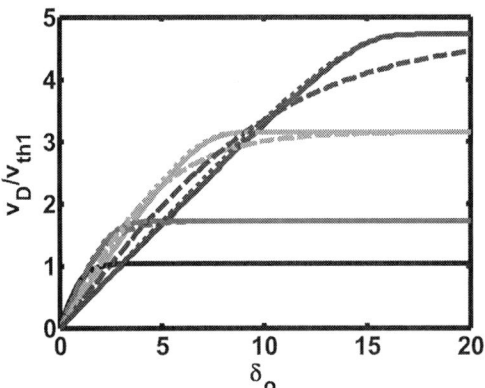

Figure 6. The normalized drift velocity as a function of normalized electric field $\delta_o = \mathrm{E}/\mathrm{E}_{co}$ for 1D electron gas as in a nanowire. Three set of curves are shown with $u_1 = 3$ (top), $u_1 = 2$, $u_1 = 1$, and $u_1 = 0.1$ (bottom). In each set, solid lines are exact values as obtained from the distribution function, dashed ones are from $v_{Dd} = v_{ud}\tanh(\mathcal{E}/\mathcal{E}_{cd})$ and dotted lines are from $(\Delta n/n)_1 (v_u/v_{th})_1$. The curves overlap for lower u values and are not visible distinctly.

$\Delta n_d/n_d$ of Eq. (4.8) is now greatly simplified to

$$\Delta n_d / n_d = \tanh(\delta) \qquad (4.13)$$

with

$$\delta = \frac{\mathrm{E}}{\mathrm{E}_{cd}} \qquad (4.14)$$

$$\mathrm{E}_{cd} = \mathrm{E}_{co} d \frac{T_{ed}}{T} = \mathrm{E}_{co} d \frac{v_{id}}{v_{thd}} \frac{v_{ud}}{v_{thd}} \qquad (4.15)$$

The drift response to an electric field is now straightforward

$$v_{Dd} = v_{ud} \tanh(\mathcal{E}/\mathcal{E}_{cd}) \qquad (4.16)$$

Figure 6 shows the drift response to an applied electric field in a 1D nanowire. There are five concentration values represented with $u_1 = (n/N_c)_1 = 3$ (top), 2, 1, and 0.1 (bottom). Solid lines are taken from exact calculation of the drift velocity from the distribution function. The dashed and dotted curves are from approximation of Equation (4.16) and $(\Delta n/n)_1 (v_u/v_{th})_1$ respectively. Similar analysis can be performed on 2D and 3D high-field transport. Reduction to tanh function either in Equation (4.13) or (4.16) makes this expression susceptible to wide variety of applications.

The emission of a quantum of energy in the form of a phonon (acoustic or optical) or a photon is always possible as electric field raises the energy of an accelerated electron in the antiparallel direction to an applied electric field [25]. The energy of an emitted photon in an LD nanostructure is equal to the difference of two lowest quantized levels. Phonons and photons are bosons following Bose-Einstein statistics. The probability of the emitted quantum is given by $N_o + 1$ with N_o given by

21

$$N_o = \frac{1}{\exp(\hbar\omega_o / k_B T) - 1} \tag{4.17}$$

The average energy of an emitted quantum is

$$E_Q = (N_o + 1)\hbar\omega_o. \tag{4.18}$$

The energy $q_E \ell_Q = E_Q$ triggering a quantum emission in an inelastic scattering length ℓ_Q gives

$$\ell_Q = E_Q / q_E \tag{4.19}$$

The effective mfp ℓ in the high-field limit, following the transient response model [5], is given by

$$\ell = \ell_{o\infty}\left(1 - e^{\frac{\ell_Q}{\ell_{oL}}}\right) \tag{4.20}$$

Equation (4.20) indicates that the inelastic-scattering quantum length ℓ_Q is infinite in a low electric field, making quantum emission highly improbable ($\ell \approx \ell_{o\infty}$). $\ell_{o\infty}$ in Equation (4.20) is replaced by ℓ_{oL} in a ballistic channel. In the presence of quantum emission, the velocity may saturate to a value lower than v_{ud} as tanh factor is always less than 1. v_{satd} so limited is given by

$$v_{satd} \approx v_{ud}\, tanh(E_Q / k_B T_{ed}) \tag{4.21}$$

$E_Q = k_B T$ for low energy quanta, e. g. acoustic phonons ($\hbar\omega_q \ll k_B T$) giving $v_{satd} \approx v_{ud}\, tanh(1)$. In this scenario, not only the optic phonon, but also the acoustic phonon will limit the saturation velocity. In the other extreme case, the emitted quantum having a large energy as compared to the thermal energy $(E_Q \gg k_B T)$, the saturation velocity is v_{ud} as $tanh(E_Q / k_B T) \approx 1$.

V. SCALED-DOWN CHANNELS

In a low (ohmic) electric field, the drift velocity v_D in response to an applied electric field E is linear: $v_D = -\mu_{o\infty}E$, –ve sign showing electron drift opposite to the electric field. $\mu_{o\infty}$ with subscripts is deliberately chosen to indicate low-field (o) and a long-channel (∞). Even when the field is high and velocity response sublinear, the mobility still exists, but is a function of electric field as velocity response is linear up to a critical electric field E_c. As drift is sublinear above E_c, the ratio $\mu = |v_D| / E$ decreases with the electric field. The incremental mobility $\mu = dv_D / dE$ for a signal floating over a dc bias degrades even further. The distinction is, therefore, needed between field-dependent direct and incremental (or signal) mobility.

The Boltzmann framework is still valid in a scaled-down channel so far the electric field E is below its critical value E_c ($E < E_c$) when $\mu_{o\infty}$ is replaced with μ_{oL} appropriate for a channel of finite length. In this case, the critical electric field is higher as ohmic mobility degrades because of ballistic effects. The drift response is sublinear in $E > E_c$ domain, ultimately reaching saturation of drift velocity limited to the intrinsic velocity v_{id}. $\mu(E) \approx v_{i2} / E$ in an extreme electric field or $\mu(V) = v_{i2}L / V$ as $E = V / L$. The application of NEADF is attractive to differentiate different domains of mobility and its degradation. High-field and ballistic effects are interweaved in scaled-down channels.

The length-limited mobility $\mu_L = v_D / E$ as compared to long-channel mobility μ_∞, following paradigm posed by NEADF, is given by [14]

$$\frac{\mu_L(V)}{\mu_{o\infty}} = \frac{v_D}{\mu_{o\infty}E} = \frac{\tanh(V / V_c)}{(V / V_{c\infty})} \tag{5.1}$$

with

$$V_c = V_{c\infty}\frac{1}{(1 - e^{-L/\ell_B})} \tag{5.2}$$

$\mu_L(V)$ is a function of an applied electric field or voltage as $E = V / L$. $V_{c\infty}$ is the critical voltage when $\ell_L = \ell_{o\infty}$. $\mu_L(V)$ approaches $\mu_{o\infty}$ in a long channel when $V < V_{c2\infty}$, as in top graph in Figure 7 as $\mu_L(V) / \mu_{o\infty} = 1$ when $V / V_{c2\infty} < 1$. Equation (5.1) transforms when factored into high-field and ballistic effects

$$\frac{\mu_L(V)}{\mu_{o\infty}} = \frac{\tanh(V / V_c)}{(V / V_c)}(1 - e^{-L/\ell_B}) \tag{5.3}$$

The subscript d is dropped as these general results apply to all dimensionalities. The first factor indicates mobility degradation due to high-field effects with modification in the critical voltage that is necessarily higher as mobility degrades due to ballistic effects. The second factor is the familiar ballistic factor as stated above. V_c plays a pivotal role in determining whether or not high-field effects are important. In the limit of small voltage $V \ll V_c$, $\mu_L - \mu_{oL}$ and the ratio simplifies to

$$\frac{\mu_{oL}}{\mu_{o\infty}} = 1 - e^{-L/\ell_B} \tag{5.4}$$

In the other extreme when $V \gg V_c$, μ_L decreases with voltage as follows

$$\frac{\mu_{oL}}{\mu_{o\infty}} \approx \frac{V_{c\infty}}{V} \tag{5.5}$$

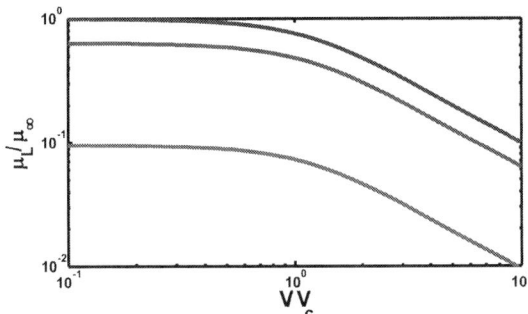

Figure 7. The normalized mobility as a function of normalized voltage (or field) for $L/\ell_B = \infty$ (top), 1, 0.1 (*bottom*).

The distinction between V_c and $V_{c\infty}$ holds key to high-field effects in scaled-down channels. In a long channel, $V_{c\infty}$ approach zero at low-temperatures. However, in a ballistic channel the critical voltage $V_c > V_{c\infty}$ not only is finite but is higher. High-field effects are then negligible for an applied voltage, as is clear from Figure 7. However, ballistic effects in $V < V_c$ regime show their existence. As applied voltage $V > V_c$, the high-field effects are bound to show their existence degrading the mobility even further. Top curve in Figure 7 is for a long channel when $L/\ell_B = \infty$. As channel length becomes comparable, the mobility drops by a factor $1 - e^{-1} = 0.63$. $1 - e^{-L/\ell_B} = L/\ell_B$ for a smaller length channel. The mobility is only 10% of the degrade value due to high-field effects when $L/\ell_B = 0.1$.

VI. CONCLUSIONS

The ballistic transport covering a wide spectrum is described and agreement obtained with the experimental observations. A number of experiments note a quantum resistance in 1D scaled-down channels where density of states spikes at the quantum level. The resistance of 1D wire of length L is [26, 27]

$$R = \rho_1 L = \frac{1}{n_1 q \mu_{oL}} L \qquad (6.1)$$

$\mu_{0L} \simeq \mu_{0\infty}(L/\ell_B) = qL/m_o v_F$ of a ballistic channel with $L << \ell_B$ and $v_{inj} = v_F$ of the contacts. The application of the Fermi velocity $v_F = \pi \hbar n_1 / 2m_o$ for a 1D conductor gives $R = h/4q^2$ which is quantum conductance for a single spike and will be scaled by the number of channels N filled up to the Fermi energy. Because of limited space, these ballistic effects could not be covered in this paper. However, as one can see these are well within the reach of the formalism presented. The results will change depending on the number of valleys in the band structure. If high-field effects are also invoked, the unilateral Fermi velocity will pay an active role. The rise in resistance in a scaled-down channel has been noted by a number of authors [27-29]. The enhancement of extrinsic contact resistance will change the channel behavior in a

distinct way. When Ohm's law is not obeyed and conduction becomes nonlinear even the division of voltage between contacts and channel may change and may change the transport behavior.

Quantum conductance is a rich and varied subject that goes beyond the simple interpretation given above as high-field effects may be predominant that are not included in the simple description given above. But, NEADF is in agreement with the Landaur-Buttiker formalism. However, measurement of quantum conduction and its interpretation in the light of formalism presented is a valuable method not only to study fundamental physics, but also to value the quality of contacts and even develop sensors.

In the fast changing landscape of graphene and traditional semiconductor devices with reduced dimensionality, it is hoped that these results will provide strong theoretical foundation on which to assess the performance of the emerging devices and methods to characterize these.

ACKNOWLEDGMENT

The author acknowledges the excellent hospitality of the Universiti Teknologi Malaysia (UTM) in providing environment conducive to creativity and innovation and award of a grant under Foreign Academic Visitors Fund (FAVF) grant number 4D037. The assistance of Michael Tan in completing this manuscript is gratefully acknowledged.

REFERENCES

[1] E. Conwell, *High Field transport in Semiconductors* vol. Supplement 9. New York: Academic Press, 1967.

[2] V. K. Arora, "Hot electrons : A myth or reality?," in *International Workshop on the Physics of Semiconductor Devices*, Delhi, India, 2002, pp. 563-569.

[3] V. K. Arora, "High-field distribution and mobility in semiconductors," *Japanese Journal of Applied Physics, Part 1: Regular Papers & Short Notes*, vol. 24, pp. 537-545, 1985.

[4] V. K. Arora, *et al.*, "Transition of equilibrium stochastic to unidirectional velocity vectors in a nanowire subjected to a towering electric field," *Journal of Applied Physics*, vol. 108, pp. 114314-8, 2010.

[5] V. K. Arora, *Quantum Nanoengineering*. Wilkes-Barre, PA: Wilkes University, 2012.

[6] V. K. Arora, "Quantum Transport in Nanowires and Nanographene," in *28th International Conference on Microelectronics (MIEL2012)*, Nis, Serbia, 2012.

[7] R. Qindeel, *et al.*, "Low-Dimensional Carrier Statistics in Nanostructures," *Current Nanoscience*, vol. 7, pp. 235-239, Apr 2011.

[8] R. Vidhi, *et al.*, "The Drift Response to a High-Electric-Field in Carbon Nanotubes," *Current Nanoscience*, vol. 6, pp. 492-495, Oct 2010.

[9] I. Saad, *et al.*, "The dependence of saturation velocity on temperature, inversion charge, and electric field in a nanoscale MOSFET," *International Journal of Nanoelectronics and Materials* vol. 3, pp. 17-34, 2010 2010.

[10] I. Saad, *et al.*, "Ballistic mobility and saturation velocity in low-dimensional nanostructures," *Microelectronics Journal*, vol. 40, pp. 540-542, Mar 2009.

[11] V. K. Arora, "Theory of Scattering-Limited and Ballistic Mobility and Saturation Velocity in Low-Dimensional Nanostructures," *Current Nanoscience*, vol. 5, pp. 227-231, May 2009.

[12] M. T. Ahmadi, *et al.*, "The high-field drift velocity in degenerately-doped silicon nanowires," *International Journal of Nanotechnology*, vol. 6, pp. 601-617, 2009.

[13] M. A. Riyadi and V. K. Arora, "The channel mobility degradation in a nanoscale MOSFET due to injection from the ballistic contacts," *Journal of Applied Physics*, vol. 109, p. 056103 2011.

[14] V. K. Arora, *et al.*, "Temperature-dependent ballistic transport in a channel with length below the scattering-limited mean free path," *Journal of Applied Physics,* vol. 111, p. 054301, 2012.

[15] V. K. Arora, *et al.*, "Concentration dependence of drift and magnetoresistance ballistic mobility in a scaled-down metal-oxide semiconductor field-effect transistor," *Appl. Phys. Lett.,* vol. 99, p. 063106, 2011.

[16] V. K. Arora and H. N. Spector, "Transition in magnetoresistance behavior from classical to quantum regime," *Physical Review B,* vol. 24, pp. 3616-3619, 1981.

[17] B. L. Anderson and R. L. Anderson, *Fundamentals of Semiconductor Devices.* New York, N. Y.: McGraw-Hill, 2005.

[18] B. G. Streetman and S. K. Banerjee, *Solid State Electronic Devices,* Sixth Edition ed. Upper Saddle River, N. J.: Prentice-Hall, 2006.

[19] M. Lundstrom and J. Guo, Eds., *Nanoscale Transistor: Device Physics, Modeling and Simulation.* Springer, 2006, p.^pp. Pages.

[20] K. Huet, *et al.*, "Monte Carlo study of apparent magnetoresistance mobility in nanometer scale metal oxide semiconductor field effect transistors," *Journal of Applied Physics,* vol. 104, p. 4504, Aug 15 2008.

[21] J. Lusakowski, *et al.*, "Ballistic and pocket limitations of mobility in nanometer Si metal-oxide semiconductor field-effect transistors," *Applied Physics Letters,* vol. 87, pp. -, Aug 1 2005.

[22] M. S. Shur, "Low ballistic mobility in submicron HEMTs," *Ieee Electron Device Letters,* vol. 23, pp. 511-513, Sep 2002.

[23] J. Wang and M. Lundstrom, "Ballistic transport in high electron mobility transistors," *IEEE Transactions on Electron Devices,* vol. 50, pp. 1604-1609, Jul 2003.

[24] P. Robertson and D. Dumin, "Ballistic transport and properties of submicrometer silicon MOSFET's from 300 to 4.2 K," *Electron Devices, IEEE Transactions on,* vol. 33, pp. 494-498, 1986.

[25] V. K. Arora, "Quantum engineering of nanoelectronic devices: the role of quantum emission in limiting drift velocity and diffusion coefficient," *Microelectronics Journal,* vol. 31, pp. 853–859, 2000.

[26] V. K. Arora, "High-electric-field initiated information processing in nanoelectronic devices," in *Nanotechnology for Telecommunications Handbook,* S. Anwar, Ed., ed Oxford, UK: CRC/Taylor and Francis Group, 2010, pp. pp. 309-334

[27] T. Saxena, *et al.*, "Microcircuit Modeling and Simulation Beyond Ohm's Law," *IEEE Transactions on Education,* vol. 54, pp. 34-40, Feb 2011.

[28] D. R. Greenberg and J. A. d. Alamo, "Velocity saturation in the extrinsic device: a fundamental limit in HFET's," *IEEE Trans. Electron Devices,* vol. 41, pp. 1334-1339, 1994.

[29] M. L. P. Tan, *et al.*, "Resistance blow-up effect in micro-circuit engineering," *Solid-State Electronics,* vol. 54, pp. 1617-1624, Dec 2010.

Simulation and Modeling of Nanoscale Multiple-Gate SOI MOSFETs

Benjamin. Iñiguez[1], Romain Ritzenthaler[2], François Lime[1], Bogdan Nae[1]

[1]Department of Electronic, Electric and Automatic Engineering, University Rovira i Virgili, Tarragona, Spain
[2]IMEC, Kapeldreef 75, Leuven, Belgium
benjamin.iniguez@urv.cat

Abstract—**This paper presents some insights into the modeling of different Multi-Gate SOI MOSFET structures, and in particular Tri-Gate MOSFETs (TGFETs). For long-channel case an electrostatic model can be developed from the solution of the 2D Poisson's equation in the section perpendicular to the channel. allowing it to be incorporated in quasi-2D compact models. For short-channel devices a model can be derived from a 3D electrostatic analysi. The subthreshold current model was successfully compared with experimental measurements, in terms of subthrehold slope, threshold voltage and DIBL.**

Index Terms—**modeling, simulation, SOI, MOSFET, multiple-gate**

I. Introduction

This paper presents some recent advancements in the area of modelling of Multi-Gate SOI MOSFETs (MuGFETs), and in particular Tri-Gate MOSFETs (TGFETs), applicable to its variants such as ΠFET and ΩFET structures by taking into account the electrostatics of the cross section perpendicular to the channel, and the BOX, which brings an additional electrostatic coupling component from the back-gate. For long-channel devices, the electrostatic modelling can be developed from the solution of the 2D Poisson's equation in the section perpendicular to the channel, and can be easily incorporated to quasi-2D compact drain current models [1] by means of a correction in the threshold voltage. For short channels, the electrostatic modelling must be developed from a solution of the 3D Poisson's equation, since the electrostatics in the cross section is affected by the lateral field from the drain. In all cases the Poisson's equation is solved in subthreshold, which is the regime where short-channel and interface coupling effects are dominant (above threshold these effects tend to vanish due to the screening of electric fields in the channel). The developed 3D electrostatic framework for Tri-Gate MOSFETs can be included in physically-based compact drain current models.

In Section II-III, the threshold voltage modelling of ΩFET transistors is presented [2]; the model can be extended to the case of ΠFETs, Triple-gate FETs and planar Fully Depleted SOI (planar FDSOI) structures. The back-interface parasitic activation is highlighted.

In Section IV, an electrostatic potential expression is derived for short-channel ΠFETs and TGFETs. The solution of the 3D analytical potential is presented and the approximations made in order to obtain the analytical subthreshold current are explained and validated. The subthreshold current model was successfully compared with experimental measurements, in terms of subthreshold slope, threshold voltage ('Roll-off') and DIBL [3].

II. Long Channel TGFETs

In long channel devices, the interface coupling between front- and back- gates can be modeled using the 2D Poisson's equation. In the subthreshold regime, the minority carrier concentration can be ignored; it is considered that the subthreshold approximation is valid up to the threshold voltage. Therefore, for undoped channels and subthreshold operation, the 2D Poisson's equation describing the electrostatics can be approximated by the Laplace's equation:

$$\frac{\partial^2 \psi(x,y)}{\partial x^2} + \frac{\partial^2 \psi(x,y)}{\partial y^2} = 0 \tag{1}$$

with the boundary conditions determined by the surface potential and the axis defined in Fig. 1.

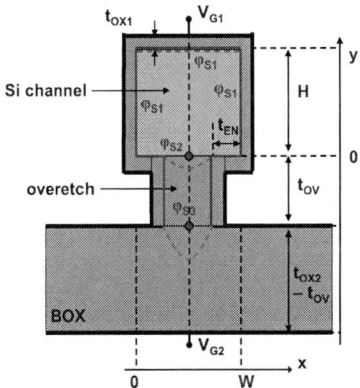

Fig. 1. Sketch of the transversal cross-section of an ΩFET. A TGFET corresponds to $t_{OV} = 0$ nm and $t_{EN} = 0$ nm, and a ΠFET to $t_{OV} \neq 0$ nm and $t_{EN} = 0$ nm.

Besides, the surface potential along the top and lateral gates is considered constant (φ_{S1}, see Fig. 1) and along the Triple-gate body/gate oxide. This approximation is not completely accurate since the surface potential is also affected by the back-gate (φ_{S1} shows variations from the bottom to the top of the lateral gates). However, as the threshold occurs when one point in the channel reaches the threshold value (φ_{ST}, which is a function of the channel doping, but is independent of the width and height of the channel), a constant φ_{S1} is an acceptable approximation to model the threshold voltage in undoped devices. This approximation is also consistent with neglecting corner effects, since we assume that the surface potential at the corners is equal to its value at x = W/2 (Fig. 1).

Considering the boundary conditions of the potential at the body/overetched region and overetched region/BOX interface, the potential will be assumed parabolic (with a value of φ_{S1} at the lateral interfaces or all along the penetration of the gate under the body, and a value of respectively φ_{S2} (x = W/2, y = 0) and φ_{S3} (x = W/2, y = -t_{OV}) at mid-channel (Fig. 1). The parabolic approximation for the back-interface boundary condition is very accurate for narrow fins, and acceptable for wide fins (where a constant back-interface potential would be more exact since the influence of the lateral gates is negligible); however, for the same reasons given for the front-surface potential, it can be considered that the first point in the back-channel to reach threshold gives a good estimation of the global threshold voltage of the whole channel. Considering channel widths W and heights H > 10 nm, the quantum confinement effects will also be neglected.

Using this modeling scheme, we obtain analytical expressions of the surface potentials at the interfaces (in terms of hyperbolic functions), from which threshold voltage expressions are derived. The calculation of the back-interface accumulation $V_{G2,ACC2}$ and inversion $V_{G2,INV2}$ voltages is made by considering limit values of the surface potentials φ_{S1} and φ_{S2} An expression front-gate threshold voltage V_{TH1} is derived according to the regime of the back-gate. The threshold voltage is constant when the back-interface is accumulated $V_{TH1,ACC2}$ or inverted $V_{TH1,INV2}$, and decreases linearly with V_{G2} when the back-interface is depleted (Fig. 2). Similar considerations as in the front-gate are applied to the back-gate in order to yield the back-gate threshold voltage.

Both the front and the back channels and their respective threshold voltages have to be merged in order to build a complete threshold voltage model of these devices. Four regimes can be defined: no channel inverted, one of the channels inverted, and both channels inverted. In an asymmetrical structure like an ΩFET, the threshold voltage of a channel is also a function of the bias applied on the other gate. If the back-gate bias V_{G2} is kept constant while varying the front-gate bias V_{G1}, the threshold voltage 'seen from measurements' (the global device threshold voltage) is a combination of the activation degrees of the front- and back-channel. For $V_{G2}<V_{G2,INV2}$, only one threshold voltage is noticeable, and is due to the inversion of the front-channel. Because of the coupling between the two channels, for $V_{G2}<V_{G2,INV2}$ the first channel to invert while ramping the front-gate bias is the back-channel. Using extraction methods like the constant current method or the maximum of transconductance method, only this threshold voltage is extracted. However, using other methods like the double derivative of the drain current the activation of the front-interface becomes apparent too (and accordingly two peaks can be seen).

In a transistor where the channel width W is much larger than the channel height H (*therefore* behaving as a planar Fully Depleted SOI (FDSOI) MOSFET), the threshold voltage is solely a function of the fin height. Compared to the FDSOI configuration (W =10 μm), we observe that in a TGFET transistor (Fig. 2) when the fin width W is decreased, the back-gate accumulation bias $V_{G2,ACC2}$ is shifted toward more negative biases and a smaller front-channel threshold voltage $V_{TH1,ACC2}$.

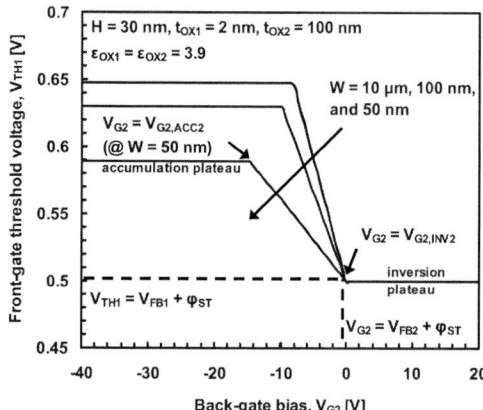

Fig. 2. Model of the *front-gate* threshold voltage V_{TH1} vs. back-gate bias V_{G2} for TGFET transistors for several fin widths W. Fin height H =30 nm, gate oxide thickness t_{OX1}=2 nm, BOX thickness t_{OX2}=100 nm.

As the back-gate has a smaller effect than for wider transistors, the electrostatic control of the channel by the front-gate on the channel is enhanced and the front-channel threshold voltage V_{TH1} is smaller than for wide transistors. As a consequence, the front-gate to back-gate coupling coefficient (defined as the slope of the $V_{TH1}(V_{G2})$ curve when the back-channel is depleted) decreases if the fin width is reduced: the lateral electrostatic coupling screens the vertical coupling induced by the back-gate. Similarly, the back-interface activation is effectively screened by the lateral coupling and takes place therefore for a voltage closer to the front-interface threshold. The threshold voltage model for ΩFETs can be extended to other structures, like ΠFETs, TGFETs, and planar FDSOI structures (Table I).

TABLE I
VARIATIONS OF THE CORE STRUCTURE

Structure	Features
ΩFET (core structure)	$t_{OV} \neq 0$, $t_{EN} \neq 0$
ΠFET	$t_{OV} \neq 0$, $t_{EN} \approx 0$
TGFET	$t_{OV} \approx 0$, $t_{EN} \approx 0$
Planar FDSOI	$t_{OV} \approx 0$, $t_{EN} \approx 0$, W>>H

In Fig. 3, numerical simulations (using Silvaco ATLAS) of the threshold voltage versus back-interface voltage curves are plotted for various gate widths and transistor configurations (Triple-gate, ΠFET and ΩFET geometries). The general agreement between the model and the numerical simulations is excellent. We observe that the threshold voltage of the Tri-gateFET is much more sensitive to back-gate bias than for both ΠFETs and ΩFETs, due to the better isolation of ΠFETs/ΩFETs to the back-gate bias thanks to the penetration of the front-gate in the BOX.

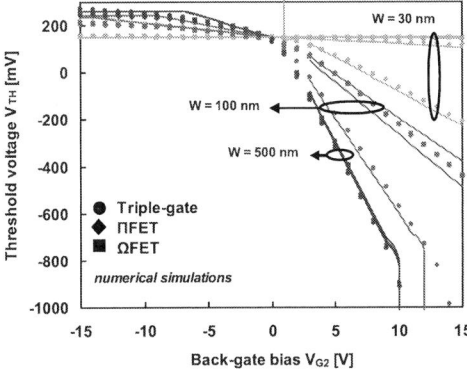

Fig. 3. Numerical simulations of the extracted threshold voltage (symbols, constant current method) versus substrate voltage V_{G2} for Tri-gate (circles), ΠFETs (triangles) and ΩFETs (squares), and comparison with the analytical model (solid lines). Gate widths W are 30, 100 and 500 nm. Drain voltage V_{DS} is 50 mV and gate length L_G is 10 μm. t_{OX1} = 2 nm, t_{OX2} = 100 nm, H = 26 nm, t_{OV} = 30 nm, t_{EN} = 5 nm.

III. EFFECT OF THE INTERFACE CONDUCTIONS

The model has been compared with experimental data from ΩFETs transistors (Fig. 4, process details and performance described in [4]). The agreement between the model calculations and the experimentally extracted threshold voltage values is very good. The invariant point predicted by the analytical model is clearly noticeable in the experimental measurements (Fig. 4) and occurs when $VB_{G2B} = VB_{FB2B} + \varphi B_{STB}$.

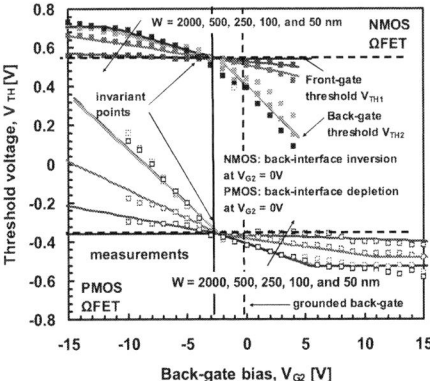

Fig. 4. Experimental threshold voltage values (constant current method, squares) as a function of back-gate bias V_{G2} and fin width W (from W = 2 μm down to W = 50 nm) for NMOS and PMOS and comparison with the analytical model (solid lines). L_G = 10 μm, H = 26 nm, t_{OX2} = 100 nm, and V_{DS} = 50 mV.

At this point corresponding to the beginning of back-interface inversion, the surface potential at the back-gate interface φB_{S2B} attains a value compensating the vertical electric field, leading to a flat potential in the channel at the front-gate threshold voltage.

For these transistors, the flat-band voltage of the back-channel V_{FB2} is extracted with the invariant point (Fig. 4) and is around -3V. Therefore, any measured characteristics with V_{G2} grounded (the usual case) correspond to the back-interface inversion for n-channel devices, and to the back-interface depletion for p-channel devices.

Therefore, using the model derived for long channels ΠFET, TGFETs and planar FDSOI transistors, we demonstrated that experimental fully depleted devices can operate in the 'back-interface inversion' regime even at V_{G2} grounded [5].

IV. SHORT CHANNELS TGFETS

In this section, the 3D analytical modelling of short-channels ΠFETs (and TGFETs transistors) is investigated. Based on a solution of the 3D Laplace's equation, the interface coupling in the structure is described and the potential calculated. Using the 'most leaky path' approach, the potential is then integrated and a simplified expression for the subthreshold current of ΠFET transistors can be found. The short-channel effects (subthreshold swing degradation, subthreshold slope, threshold voltage roll-off and DIBL) are inherently included in the model and were compared to experimental data.

In the subthreshold regime the 3DPoisson's equation can be approximated by the 3DnLaplace's equation:

$$\frac{\partial^2 \varphi(x,y,z)}{\partial x^2} + \frac{\partial^2 \varphi(x,y,z)}{\partial y^2} + \frac{\partial^2 \varphi(x,y,z)}{\partial z^2} \approx 0 \qquad (2)$$

with $\varphi(x,y,z)$ the electrostatic potential in the channel.

The following assumptions have been made to find an analytical solution:
- The channel doping is 10^{15} cm^{-3}, and the source and drain junctions are assumed to be abrupt.
- The influence of the field penetration from the drain and the overlapped regions of the gate through the BOX is neglected.
- Considering width W and height H of the channel above 10 nm, the quantum effects are neglected; numerical simulations have clearly shown that they do not induce significant variations of the threshold voltage and subthreshold slope down to W = H = 10 nm.
- The corner effects are also neglected, considering the use of undoped channels.

As far as the boundary conditions are concerned, the potential profile in the lateral direction is assumed parabolic both at the interfaces between channel and overetch region, and between overetch region and BOX (Fig. 5).

The equation that has to be solved is therefore a 3D Laplace's equation for the device configuration shown in Fig. 5. In order to calculate the potential in the structure, the influences of the six terminals corresponding to the external boundary conditions (namely source, drain, the three sides of the front-gate, and the back-gate) are considered separately ('superposition theorem'). For each terminal, the potential is developed in a series of hyperbolic functions (a similar approach as in [6]) while setting the other terminals to zero. This approach has the advantage of creating symmetries and simplifying the calculation of the hyperbolic function series coefficients. The series coefficients are only functions of the considered boundary conditions (Dirichlet boundary condition with a constant or parabolic value, and Neumann boundary condition).

Finally, the Gauss's theorem is applied at the back-interface in order to take into account the effects of the back-gate and the Π-shape of the transistor.

The comparison of the analytical formula with numerical simulations (using COMSOL Femlab to solve the Laplace's equation) shows an excellent agreement.

Calculating the minimum of the potential barrier and its location (where the subthreshold current is the weakest), the subthreshold current of the transistor can be derived. In TGFET and ΠFET devices where all the sides of the front-gate are connected, the minimum of potential is located at mid-channel ($x = W/2$, Fig. 5) due to obvious symmetry considerations and at the body/overetched region interface ($y = t_{OV}$, Fig. 5). Regarding the location of the minimum potential value along the source drain axis (z axis), the approximate expression derived in [6] for FinFETs has been used and shown to be applicable to TGFET devices in general.

Then, the subthreshold current is obtained analytically without the need of any fitting parameters, after a simplification of the integral of the potential along the section of the 'most leaky path'.

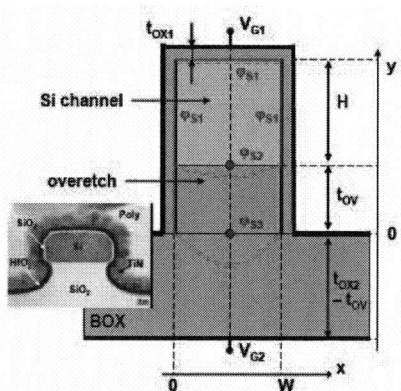

Fig. 5. Sketch of the transversal cross-section of a ΠFET with the notations used in this work. A TGFET corresponds to $t_{OV} = 0$ nm, and a ΠFET to $t_{OV} \neq 0$. Inset shows a TEM image of the devices. The gate encroachment under the Si film is small enough so the ΩFET can be considered as a ΠFET.

The threshold voltage in the linear regime and in saturation has been extracted using the constant current method (Fig. 6); a reasonable agreement is observed. It can be seen that the V_{TH} decreases with the channel length when the devices are operated in the saturation regime. The threshold voltage roll-off effect is accurately reproduced.

Similarly, the difference of the threshold voltage between low and high values of the drain voltage gives straightforwardly the DIBL effect. Similar trends as for the subthreshold slope are observed, and a good agreement is obtained too.

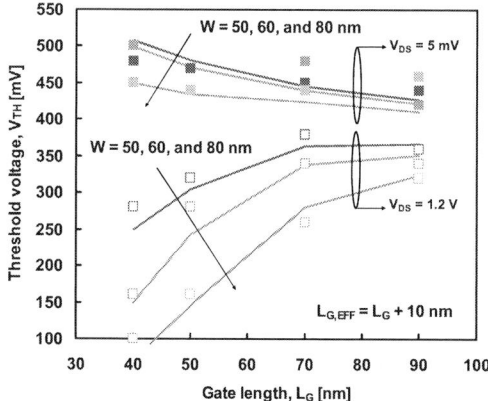

Fig. 6. Threshold voltage V_{TH} extracted with the constant current method (at 0.1 μA, symbols) and with the analytical model (solid lines) vs. channel length L_G. The channel width W varies from 50 to 80 nm, and V_{DS} from 5 mV (closed symbols) to 1.2 V (open symbols t_{OX1}=1.95 nm, t_{OX2} = 100 nm, H = 26 nm.

The ΠFET structure intrinsically contains all the other multi-gate transistors by simply changing the device parameters Therefore, the analytical expressions derived for ΠFETs can be used as a core model to depict the subthreshold slope SS vs. channel length L_G for a wide range of devices (Fig. 7). We found an excellent agreement wbetween the model and simulation data from [7] and our own experimental measurements for planar FDSOI wide devices.

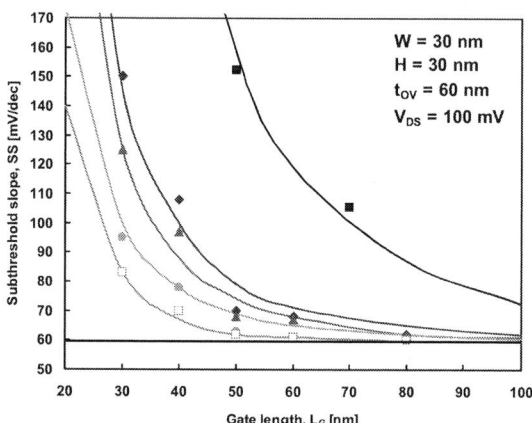

Fig. 7. Subthreshold slope SS vs. channel length L_G obtained with numerical simulations (symbols) and the analytical model (solid lines) for planar FDSOI FETs (squares), DGFET (diamonds), TGFET (triangles), ΠFET (circles), and GAA (open squares) transistors. All numerical simulation results are obtained from [7], except for planar FDSOI FETs (H = 26 nm, W = 10 μm), where our experimental measurements are used.

We remark that that without the use of any fitting parameter, the model can be extended to nearly all the MuGFETs devices (GAA /IIFETs /TGFETs/ FinFETs/ DGMOSFETs/ FDSOI planar devices), and therefore can be used as a core model for the scaling and calibration of a wide range of MuGFETs.

Our 3D electrostatic modelling can be incorporated into a physically-based compact Tri-Gate MOSFET drain current model (such as [8]) using the expressions we obtained for the threshold voltage and the subthreshold swing.

V. CONCLUSIONS

We have presented new modelling schemes for Tri-Gate SOI MOSFET (TGFET) structures, based on the solution of the cross-sectional 2D and the 3D Poisson's equation in subthreshold (I.e., Laplace's equation) using adequate techniques. The effect of the back gate is considered. Analytical solutions for the electrostatic potentials are obtained, which lead to expressions of the threshold voltage that include the interface coupling and short-channel effects and also of the subthreshold current. The developed electrostatic modelling can be extended to most Multi-Gate SOI MOS structures, and eventually incorporated into a compact drain current model by means of the threshold voltage and the subthreshold swing.

ACKNOWLEDGEMENTS

This work was supported by Ministerio de Ciencia e Innovación under project TEC2011-28357-C02-01 , by the European Commission under Contract FP7-PEOPLE-2007-3-

1-IAPP No. 218255 "Compact Modelling Network (COMON)", by the ICREA Academia Award and by the PGIR Grant from URV, and by the project 2010 CONE2 00061 from the Catalan Government.

REFERENCES

[1] A. Yesayan, F. Prégaldiny, N. Chevillon, C. Lallement, J. M. Sallese, "Physics-based compact model for ultra-scaled FinFETs," Solid-State *Electron.* 62, 2011, pp. 165-173..

[2] R. Ritzenthaler M. Tang, O. Faynot, F. Lime, F. Prégaldiny, C. Lallement, S. Cristoloveanu, and B. Iñiguez, "A 2D analytical model of threshold voltage for Pi-gate FinFET transistors," EUROSOI 2010, Grenoble, January 2010

[3] R. Ritzenthaler,F. Lime, O. Faynot, S. Cristoloveanu, B. Iñiguez, "3D analytical modelling of subthreshold characteristics in vertical Multiple-gate FinFET transistors," Solid-State Electronics, Volumes 65–66, Pages 1-262 (November–December 2011)

[4] C. Jahan et al., "10nm Ω FETs transistors with TiN metal gate and HfO2," Digest of Technical Papers, 2005, Symposium on VLSI Technology , pp. 112-113, 2005.

[5] R. Ritzenthaler, F. Lime, M. Ricoma, F. Martinez, O. Faynot, F. Pascal, M. Valenza, E. Miranda, S. Cristoloveanu, and B. Iñiguez, "Parasitic Back-Inferface Conduction in Planar and Triple-Gate SOI Transistors", IEEE 2010 International SOI conference, San Diego (USA), 2010

[6] G. Pei et al., "FinFET Design Considerations Based on 3D Simulation and Analytical Modeling," Electron Devices, IEEE Transactions on , vol. 49, no. 8, pp. 1411-1419, Aug. 2002.

[7] J.-T. Park, J.-P. Colinge, and C.H. Diaz, "Pi-Gate SOI MOSFET," IEEE Electron Device Letters , vol. 22, no. 8, pp. 405-406, Aug. 2001

[8] N. Chevillon, J. M. Sallese, C. Lallement, F. Prégaldiny, M. Madec, J. Sedlmeir, J. Aghassi, "Generalization of the concept of equivalent thickness and capacitance to multigate MOSFETs modeling," IEEE Trans. On Electron Devices, vol. 59, no. 1, pp.60-71

MIXDES 2012, 19th International Conference *"Mixed Design of Integrated Circuits and Systems"*, May 24-26, 2012, Warsaw, Poland

Transistor and Interconnect Modeling for Design of Carbon Nanotube Integrated Circuits[*]

Ashok Srivastava

Department of Electrical and Computer Engineering
Louisiana State University, Baton Rouge, LA 70803, U.S.A.
ashok@ece.lsu.edu

Abstract—The one-dimensional carbon nanotube (CNT) has excellent electrical, mechanical and thermal properties which have made the CNT one of the promising materials for applications in nanoelectronics and micro/nano-systems. Nanometer CMOS technology, especially in 22 nm and below, is plagued due to performance degradation of conventional Cu/low-k dielectric as an interconnect material for gigascale integration. In search for novel technologies, no such material has aroused so much interest other than carbon nanomaterials since the discovery of carbon nanotube. Recent work on analytical modeling equations describing the current transport in carbon nanotube field effect transistors and carbon nanotube interconnects will be presented for use in design of emerging logic devices similar to CMOS design style.

Index Terms—Carbon nanotubes; CNT-FETs; CNT interconnect; CNT integrated circuits; VLSI

I. INTRODUCTION

Carbon nanotubes are one-dimensional (1D) graphene sheets rolled into a tubular structure of nanometer size [1,2]. Their properties depend on the diameter and wrapping angle determined by the chiral vector which is characterized by the indices (n,m) of the graphene [2]. The one-dimensional carbon nanotube has excellent electrical, mechanical and thermal properties [1,3] which have made the CNT one of the promising materials for applications in nanoelectronics [4-7] and micro/nano-systems [8]. In nanoelectronics, CNT-FET is very promising in design of emerging logic devices for nano scale integration and there is a noticeable amount of published and on going research on understanding current transport in CNT-FETs and developing models for use in circuit design and simulators [9-18].

The structure of a CNT-FET is similar to the structure of a typical MOSFET [19,20], where a single walled CNT forms the channel between two electrodes, which work as the source and drain of the transistor. Both the n- and p-type CNT-FETs have been fabricated in the past decade [13,21] and multistage complementary logic gates have been demonstrated [11,14,22-24]. Efforts have also been made in modeling of the current transport behavior and models have been developed for the design of CNT-FETs based logic circuits [15,16,25-28].

Nanometer CMOS technology, especially in *22* nm and below, is plagued due to performance degradation of conventional Cu/low-k dielectric as an interconnect material for gigascale integration [29]. Thus the need for other materials possibly substituting Cu/low-k dielectric interconnections has brought forward other novel interconnect technologies for next generation VLSI interconnects. In search for novel technologies, no such material has aroused so much interest other than carbon nanomaterials since the discovery of carbon nanotube in 1991 by Iijima [30]. Carbon nanotube carries a current density of $\sim 10^{10}$ A/cm^2 which is higher by a two to three orders of magnitude in Cu. Its mean free path is in micrometer range compared to ~ 40 nm mean free path in Cu. The large mean fee path in CNT allows a ballistic transport over a wider range of micrometers resulting in reduced resistivity, and strong atomic bonds [31] provide tolerance to electromigration [29,32]. Higher thermal conductivity makes the CNT suitable for use in tall vias of 3D ICs [33,34,35].

In this paper, we present our work on analytical models characterizing the current transport in CNT-FETs for the analysis and design of integrated circuits. We also present our work on the carbon nanotube and its electrical modeling as interconnects in VLSI. Single walled, multiwalled and bundle of CNTs as interconnects have been considered.

In Section II, the current transport equation of a CNT-FET is obtained by relating the carbon nanotube potential to the terminal voltages. The charge inside the carbon nanotube is described from the electronic structure of the carbon nanotube. A model for the carbon nanotube potential is then derived and the current transport equation is obtained. Voltage transfer characteristics of logic gates are presented using complementary CNT-FETs. In Section III, equivalent circuit models of CNT interconnections have been discussed. Performance of CNT interconnects have been evaluated and compared with Cu interconnects. A complementary CNT-FET inverter pair has been analyzed with three types of CNT interconnections and simulated in Verilog-AMS. Section IV presents conclusion.

---

[*]Part of the author's work is reported in *physica status solidi (a)*, vol. 206, no. 7, pp. 1569-1578. 2009 and *Journal of Nanophotonics*, vol. 4, 041690 (17 May 2010), pp. 1-26, 2010 (online).

Part of the work is supported by the United States Air Force Research Laboratory under agreement number FA9453-10-1-0002. The U.S. Government is authorized to reproduce and distribute reprints for Government purposes notwithstanding any copyright notation thereon.

II. CURRENT TRANSPORT MODELING – CNT-FET

A. Charge and Potential Diustribution

Figure 1(a) shows the basic cross section of a CNT-FET including the charge distributions. Figure 1(b) shows the corresponding potential distributions between the gate and the substrate. In Fig. 1(a), charge distributions are explained as follows: the charge on the gate, Q_g, the charges in the oxide layers, Q_{01} and Q_{02}, the charge inside the CNT, Q_{cnt}, and the charge in the substrate, Q_{subs}. In Fig. 1(b), six different potential distributions are shown, which are also described as follows. The voltage between the gate and the substrate (back gate) is V_{gb}, the potential drop across the oxides are ψ_{ox1} and ψ_{ox2}, the surface potential in the substrate with respect to the back gate is ψ_{subs}, the potential across the CNT is ψ_{cnt} and the work function difference between the gate and the substrate materials is ϕ_{ms}.

Using Kirchoff's voltage law, the potential balance and charge neutrality condition, we can write:

$$V_{gb} = \phi_{ms} + \psi_{subs} + \psi_{ox2} + \psi_{cnt} + \psi_{ox1} , \quad (1)$$

$$Q_g' + Q_{01}' + Q_{cnt}' + Q_{02}' + Q_{subs}' = 0 . \quad (2)$$

The prime in Eq. (2) denotes the charge per unit area.

B. Current Equation

In a CNT-FET, both diffusion and drift carrier transport mechanisms contribute to the current. We have considered both diffusion and drift transport mechanism since fabricated CNT-FETs [23,24,36] have carbon nanotube lengths longer than the magnitude of the optical phonon mean free path (*100 nm*) [37,38]. In CNT-FETs, CNT is formed by rolling a grapheme sheet into a tubular structure. Current is confined mainly to its circumference which can be described by the following equation of the form [39,40], $|R|$ in Eq. (3) is the circumference of the nanotube and μ is the carrier mobility in a carbon nanotube. In Eq. (3), we have used the charge per

$$I_{ds} = I_{diff}(x) + I_{drift}(x) = \frac{|R|}{2L}\left[\int_{\psi_{cnt}(0)}^{\psi_{cnt}(L)} \mu(-Q_{cnt}')d\psi_{cnt} + \frac{kT}{q}\int_{Q_{cnt}'(0)}^{Q_{cnt}'(L)} \mu dQ_{cnt}' \right], \quad (3)$$

unit area, but only to show that we have considered a surface area, $|R|L/2$ for the carbon nanotube. The mobility can be replaced by $\mu_{eff} = \gamma\mu_{graphite}$, as each CNT (n,m) will have different values for the mobility, γ is a conversion factor for CNT from graphite with a value varying from 0 to 1. In addition, γ can also be used to represent how much surface area of the CNT is responsible for the charge flow. The charge, Q_{cnt} can be obtained as follows [41]:

$$Q_{cnt} = C_{ox1}\left(-V_{gb} + \psi_{cnt,s} + V_{fb}\right). \quad (4)$$

Equation (4) can be expressed in terms of the charge per unit area by diving it by $|R|L/2$ to obtain Q_{cnt}'. By substituting Eq. (4) in Eq. (3), the following expression for the current is obtained:

$$I_{ds} = \beta\left[f\left\{\psi_{cnt,s}(L),V_{gs}\right\} - f\left\{\psi_{cnt,s}(0),V_{gs}\right\} \right], \quad (5)$$

where

$$f\left\{\psi_{cnt,s}(x),V_{gs}\right\} = \left(V_{gs} + V_{sb} - V_{fb} + \frac{kT}{q}\right)\psi_{cnt,s}(x) - \tfrac{1}{2}\psi_{cnt,s}^2(x), \quad (6)$$

and

$$\beta = \gamma\frac{\mu\,C_{ox1}}{L^2}. \quad (7)$$

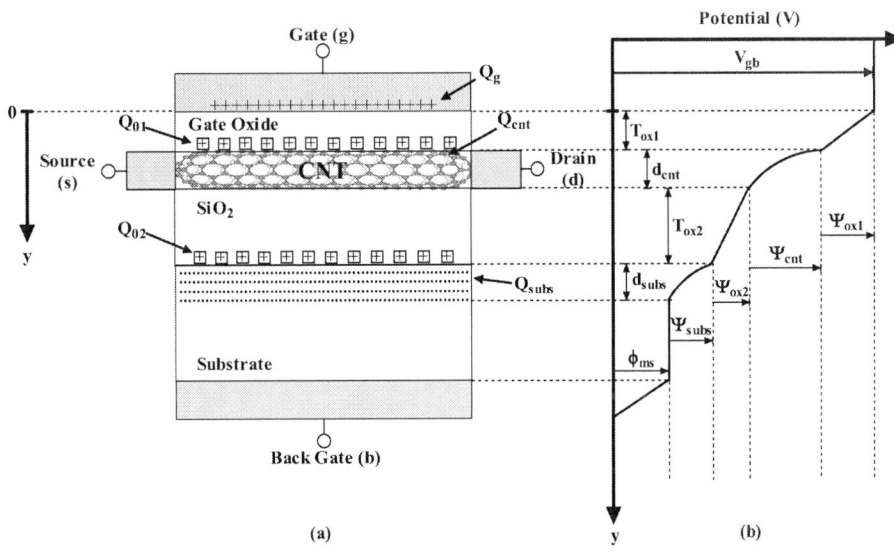

Figure 1. (a) Plot of the charges from the gate to the substrate and (b) plot of the potential distribution from the gate to the substrate in a CNT-FET.

In order for this assumption to be valid, the gate to substrate voltage must satisfy the condition: $V_{gb} \geq V_{sb} + V_{fb} + \phi_0 - \frac{\Delta E_F}{q} + \frac{E_c}{q} - \frac{kT}{q} - \frac{Ie^{-1}}{m}$. Two regions of operation can be defined as follows, a saturation region for $V_{ds} \geq V_{gs} - \left(V_{fb} + \phi_0 - \frac{\Delta E_F}{q} + \frac{E_c}{q} - \frac{kT}{q} - \frac{Ie^{-1}}{m}\right)$, and a linear region for $V_{ds} \leq V_{gs} - \left(V_{fb} + \phi_0 - \frac{\Delta E_F}{q} + \frac{E_c}{q} - \frac{kT}{q} - \frac{Ie^{-1}}{m}\right)$. The term under parenthesis of equation is the threshold voltage, V_{th} term. The saturation voltage, $V_{ds,sat}$ and V_{th} of a CNT-FET are described as follows:

$$V_{th} = V_{fb} + \phi_0 - \frac{\Delta E_F}{q} + \frac{E_c}{q} - \frac{kT}{q} - \frac{Ie^{-1}}{m}, \qquad (8)$$

$$V_{ds,sat} = V_{gs} - \left(V_{fb} + \phi_0 - \frac{\Delta E_F}{q} + \frac{E_c}{q} - \frac{kT}{q} - \frac{Ie^{-1}}{m}\right). \qquad (9)$$

Equation (5) can be easily modified to account for variation of current with V_{ds} in saturation region by introducing a parameter, λ which is equivalent to channel length modulation parameter in a MOSFET. The modified Eq. (5) is as follows:

$$I_{ds} = \beta \left[f\left\{\psi_{cnt,s}(L),V_{gs}\right\} - f\left\{\psi_{cnt,s}(0),V_{gs}\right\} \right] (1 + \lambda V_{ds}) \quad (10)$$

Figure 2 shows the I-V characteristics for a carbon nanotube with chiral vector (11,9) for different overdrive gate voltages obtained from Eqs. (5) and (10). In Fig. 2, experimentally measured data taken from [36] for chiral vector (11,9) are also plotted for the comparison which follow very closely the modeled behavior.

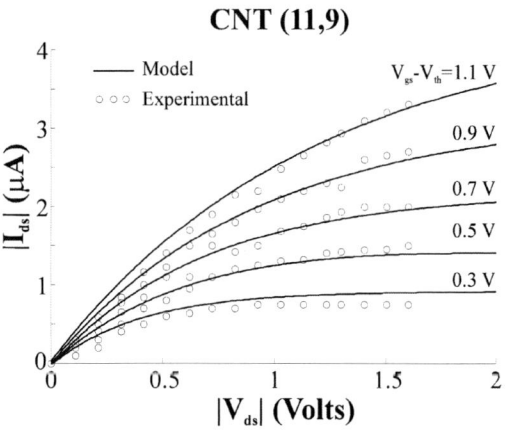

Figure 2. I-V characteristics of CNT-FET (11,9) with $V_{fb} = -0.79\ V$ and $\phi_0 = 0$. The device dimensions are $T_{ox1} = 15\ nm$, $T_{ox2} = 120\ nm$ and $L = 250\ nm$. In modeled curve, $Q_{01} = Q_{02} = 0$ and $\lambda = 0.1 V^{-1}$.

CNT-FETs can be made both n- and p-types as in CMOS [19], making possible the implementation of CNT-FETs as fully complementary logic, such as the inverters, NOR and NAND gates as shown in Fig. 3. The model equations characterizing the current voltage transport described in Section II B are for n-type CNT-FETs, can also be used for p-type CNT-FETs by changing the polarities of the voltages as

in a standard p-MOSFET. Using the current transport model we have modeled voltage transfer characteristics of basic gates using complementary CNT-FETs.

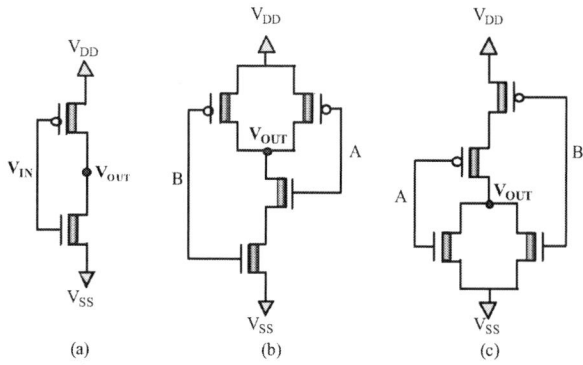

Figure 3. CNT-FET logic: (a) Inverter, (b) two input NAND gate and (c) two input NOR gate.

Figure 4 shows the voltage transfer characteristic of an inverter for a chiral vector (11,9). The dotted line in Fig. 4 is the experimental curve plotted from the work of Derycke et al., [23] and Martel et al., [24] for comparison. The two solid lines in Fig. 4 correspond to $\lambda = 0.1$ and $0\ V^{-1}$ with and without channel length modulation, respectively. It should be noted in Fig. 4 that modeled curves have been obtained for conditions of the experiment for comparison. It is seen from the Fig. 4 that the experimental voltage transfer characteristics closely follow the modeled voltage transfer characteristic corresponding to channel length modulation, $\lambda = 0.1\ V^{-1}$. Experimental transfer characteristic shows some deviation from the modeled behavior which may be attributed to influence of process variation on device parameters.

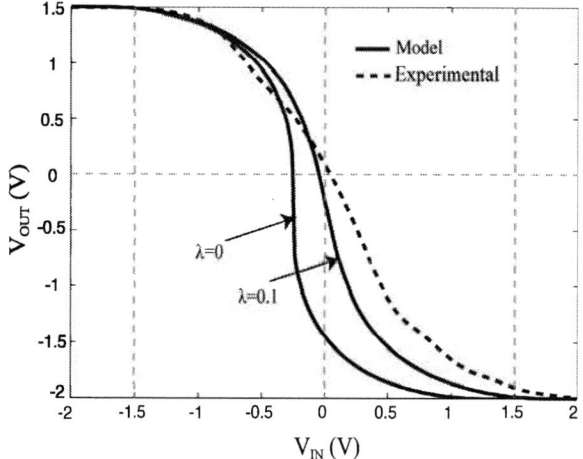

Figure 4. Voltage transfer characteristics of an inverter using CNT-FETs (11,9) with $V_{fb} = 0\ V$, $\lambda = 0, 0.1\ V^{-1}$ and $\phi_0 = 0$. The dimensions of both the n-type CNT-FET and p-type CNT-FET are: $T_{ox1} = 15\ nm$, $T_{ox2} = 120\ nm$ and $L = 250\ nm$.

Figure 5 shows the voltage transfer characteristics of an inverter and two input NAND gate for a chiral vector (11,9). Figure 6 shows the voltage transfer characteristics of an inverter and a NOR gate for a chiral vector (11,9). The voltage transfer characteristic of the inverter is included to show its full output voltage. The power supply voltage is *2 V*. It is seen that the inverter, NOR and NAND gates give full logic swing similar to inverter and gates designed in CMOS. The inverter switching threshold voltage is *1 V*. In NAND and NOR gates, the switching threshold voltage is dependent on input voltage conditions. The voltage transfer characteristics in these gates also exhibit sharp transition at switching thresholds and are similar to characteristics observed in corresponding gates implemental in CMOS.

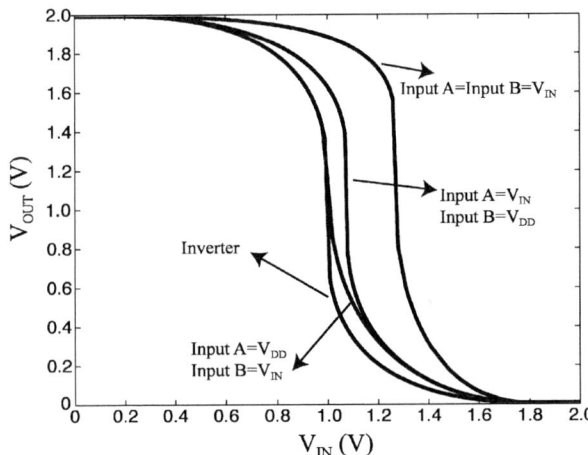

Figure 5. Voltage transfer characteristics of an inverter and a NAND gate using CNT-FETs (11,9) with $V_{fb} = 0\ V$ and $\phi_0 = 0$. The dimensions of both the n-type CNT-FET and p-type CNT-FET are: $T_{ox1} = 15\ nm$, $T_{ox2} = 120\ nm$ and $L = 250\ nm$.

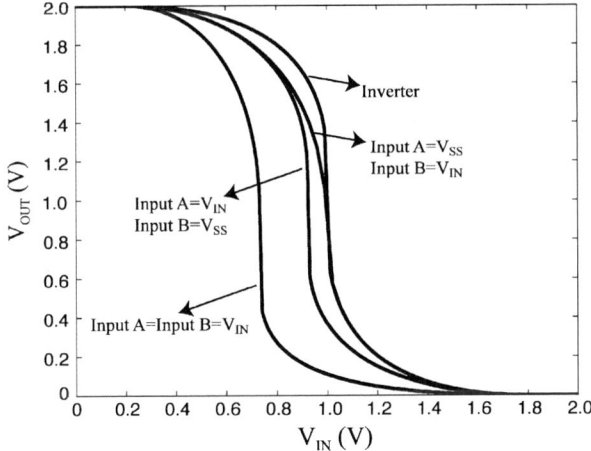

Figure 6. Voltage transfer characteristics of an inverter and a NOR gate using CNT-FETs (11,9) with $V_{fb} = 0\ V$ and $\phi_0 = 0$. The dimensions of both the n-type CNT-FET and p-type CNT-FET are: $T_{ox1} = 15\ nm$, $T_{ox2} = 120\ nm$ and $L = 250\ nm$.

III. CNT INTERCONNECTION

In a recent work [42], we have made modification in two-dimensional fluid model to include electron-electron repulsive interaction and built a semi-classical one-dimensional fluid model. In this model, metallic single-walled carbon nanotube (SWCNT) is considered and represented by a transmission line model. The SWCNT is regarded as a graphene sheet rolled to form a tube of infinitesimally thin layer. The conduction electrons are distributed on the lateral surface of the SWCNT cylindrical shell. The electrons are embedded in a rigid uniform positive charge background with a uniform surface carrier concentration. Besides its accuracy compared with two-dimensional fluid model and Lüttinger liquid theory, one-dimensional fluid model is simple in mathematical modeling and easier to extend for electronic transport modeling of multiwalled carbon nanotubes and single-walled carbon nanotube bundles as interconnections. In the following sub-sections, we will describe theoretical modeling of SWCNT, MWCNT and SWCNT bundles for interconnections.

A. One-Dimensional Fluid Model

If we regard the graphene sheet which is rolled to form a CNT is infinitesimally thin, then the conduction electrons are distributed on the lateral surface of the CNT cylinder shell and the electrons are embedded in a rigid uniform positive charge background with a uniform surface number density. Thus, the motion of electrons is confined to the surface. Furthermore, electrical charge neutrality requires that in equilibrium the conduction electron charge density precisely cancels with that of the background positive ions. According to this analysis, fluid model could be utilized to study the electron transport along the CNT. This model is shown in Fig. 7. The cylinder shell radius is *r* and length is *l*. The cylinder axis is oriented along the *z*-axis of the reference system. Two assumptions have been made to utilize this model. One is the electrons can only move along the *z*-axis; other is that all other fluid variables, such as the tangential component of the electric field to the lateral surface, *s'* of the nanotube, are almost uniform in the cross section plane of the CNT. These two assumptions are valid if both the nanotube length and the smallest wavelength of the electromagnetic field are much greater than the nanotube radius [43,44].

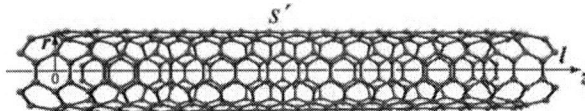

Figure 7. Geometry of a single-walled carbon nanotube.

If we neglect heat transfer and viscosity in the CNT, Euler's equation with Lorentz Force term [45,47] can be used to describe transport of electrons in a CNT as follows,

$$mN\left(\frac{\partial}{\partial t} + \vec{V} \cdot \nabla\right)\vec{V} = -\nabla \vec{P} - eN\vec{\mathcal{E}} - mN\nu\vec{V}, \quad (11)$$

where $N(\vec{R},t)$ is the electron three-dimension carrier density, $\vec{V}(\vec{R},t)$ is the electron mean velocity, $\vec{R}$ is the position vector,

$\bar{P}$ is the pressure, m is the electron mass, e is the electronic charge and $\vec{\mathcal{E}}$ is electric field. The last term on the right hand side represents the effect of scattering of electrons with the positive charge background and ν is the electron relaxation frequency.

It should be noted that Lorentz force term, which belongs to body force terms in fluid dynamics, is a source momentum [48]. The external electric field provides both the potential and kinetic energy to the fluid. As a result, one-dimensional fluid model can be expressed as follows [48],

$$mn\left(\frac{\partial}{\partial t}+v_z\frac{\partial}{\partial z}\right)v_z=-\frac{\partial p}{\partial z}-en\left\{(1-\alpha)\mathcal{E}_z\big|_{s'}\right\}-mn\nu v_z,\quad (12)$$

where n is the electron density in this one-dimensional system, v_z is the electron mean velocity in z direction, p is the pressure in one-dimensional system. $\mathcal{E}_z$ is electric field in z direction. In one-dimensional fluid model, α is defined as [42],

$$\alpha \equiv \frac{\mathcal{E}_{zP}}{\mathcal{E}_z}=\frac{E_P}{E}=\frac{E_P}{E_K+E_P},\quad (13)$$

where $\mathcal{E}_{zP}$ is the part of the electrical field which provides potential energy to electrons in z-direction. E is the total energy of electrons. E_P and E_K are the potential and kinetic energies of electrons, respectively.

The fluid model described by Eq. (12) assumes flow of one-dimensional electron fluid under the low external electric fields, $\mathcal{E}_z$. The difference between our one-dimensional fluid model described in Eq. (12) with two-dimensional fluid model [45-47] is the Lorentz force term. The external electric field provides both the potential and kinetic energies to the one-dimensional fluid.

Figures 8 (a) and (b) show S_{21} and S_{11} parameters of MWCNT, SWCNT bundle and Cu interconnects of lengths corresponding to ballistic transport (*1* μm), local interconnection (*10* μm and *100* μm) and global interconnection (*500* μm). For comparison, we choose $\beta = 1/3$ and $50\ \Omega$ terminal impedance, which is a typical impedance for high frequency transmission lines. For the MWCNT and SWCNT bundle, the electrostatic capacitance depends upon the geometry of the structure and is approximately equal to that of Cu interconnects [49-51].

Figure 8 (a) shows the *3*dB bandwidths for both the CNT and Cu interconnects. The transmission efficiency of both the CNT and Cu interconnects decreases with increasing lengths. However, Cu interconnect has a larger *3*dB bandwidth in comparison with CNT interconnects. It should also be noticed that the short length CNT interconnects still have over a *100* GHz *3*dB bandwidth. Figure 8 (a) also shows large S_{21} for SWCNT bundle and MWCNT interconnects than that of the Cu interconnect. In Fig. 8 (b) for S_{11} parameters at frequencies less than *100* GHz, Cu interconnect has the largest reflection losses while SWCNT bundle interconnect has the least reflection losses. The results show that SWCNT bundle interconnect has better performance than the MWCNT interconnect.

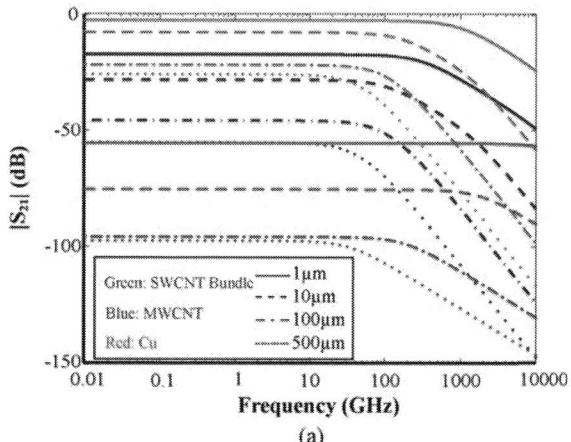

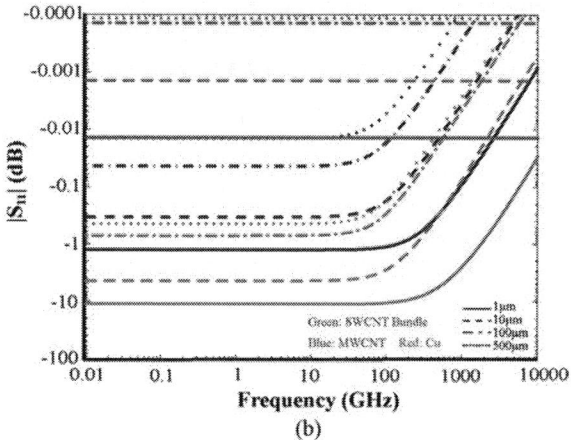

Figure 8. Calculated S-parameters of different interconnects: (a) S_{21} (amplitude) and (b) S_{11} (amplitude).

Figure 9 shows the CNT-FET inverter pair at *1*V supply voltage. The interconnection can be Cu or MWCNT or SWCNT bundle. The delay analysis includes the CNT-FET models developed by Srivastava *et al.* [41] and dynamic models reported in Ref. 52.

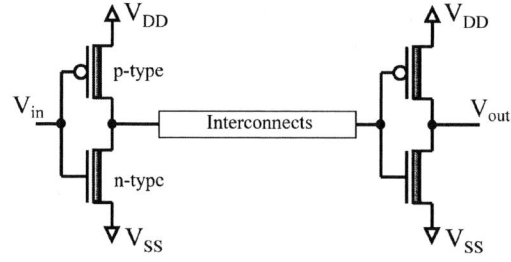

Figure 9. Inverter pair with interconnects.

We have utilized the process parameters from the 2016 node for *22* nm technology [53] assuming *22* nm diameter of a MWCNT, *22* nm width and *44* nm thickness of a SWCNT bundle. Relatively global interconnects have larger cross section and smaller resistivity. The lengths are on the order of hundred micrometers. We have utilized the process parameters from the 2016 node of *22* nm technology [53] assuming *33* nm diameter of a MWCNT, *33* nm width and *87* nm thickness of a SWCNT bundle. Simulations in cadence/Spectre (modeling through Verilog-AMS) are performed for different lengths of Cu, MWCNT and SWCNT bundle interconnects corresponding to ballistic transport length (*1* μm), local interconnects (*10* μm, *100* μm) to global interconnects (*500* μm). The results are shown in Fig. 10. Dependence of delay on interconnection length in Fig. 10 shows that the increase in delay for Cu interconnects is larger than that of MWCNT and SWCNT bundle interconnects. The delays of MWCNT interconnects ($\beta = 1$ and $\beta = 1/3$) are smaller than that of SWCNT bundle and Cu interconnects. The delays are smaller for $\beta = 1$ than for $\beta = 1/3$ for both MWCNT and SWCNT bundle interconnects and is due to more interconnect channels with increase in β.

Power dissipation is another challenge to next generation interconnects. We have simulated power dissipation for MWCNT and SWCNT bundle interconnects in *22* nm technology node and compared with the Cu wire interconnects [54]. We concluded that CNT interconnects dissipates less power and especially for local interconnections. Maximum power dissipation in CNTs interconnections is no more than the 8% of the Cu interconnections [54].

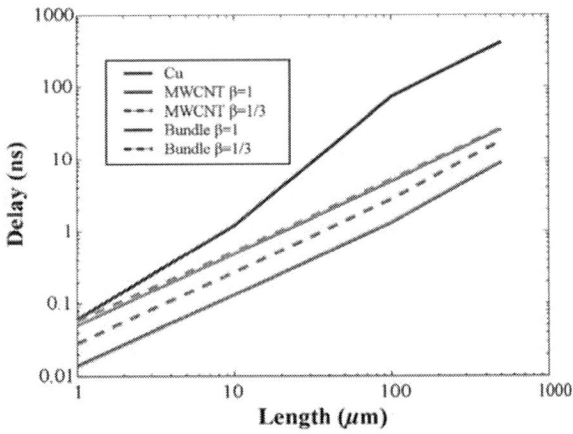

Figure 10. Propagation delays of interconnects of different lengths for *22* nm technology.

IV. CONCLUSIONS

In this work, the charge transport in a CNT-FET, based on our recently reported carrier concentration model, has been used to relate the carbon nanotube potential and the gate substrate voltage. Analytical solutions have been developed relating the carbon nanotube potential and the gate substrate voltage. These solutions are then used to model the current transport in a CNT-FET depending on the chiral vector and device geometry analytically. The analytical transport models have been used to generate CNT-FET I-V characteristics and compared with the recently reported experimental data for the chiral vector (11,9). A close agreement is obtained between the analytical models and the experimental observations in linear and saturation regions.

The analytical model equations for the current transport have been used to characterize voltage transfer characteristics of complementary logic devices such as the inverter, NAND and NOR gates. The voltage transfer characteristics of the CNT-FET gates are similar to the voltage transfer characteristics of typical CMOS gates and show a sharp transition at the inverter logic threshold voltage, 1.0 V for a 2 V operation.

The one-dimensional fluid model can be applied to CNT interconnects using low resistance contacts in current low-voltage nanometer CMOS technologies. The applicability of MWCNT and SWCNT bundle as interconnect wires for next generation design of integrated circuits has been explored theoretically and compared with Cu interconnects in *22* nm technology node. Results of the one-dimensional fluid theory for SWCNT interconnect extended to MWCNT and SWCNT bundle interconnects show that MWCNT and SWCNT bundle interconnects have better performance than the Cu interconnects. MWCNT and SWCNT bundle interconnects exhibit higher transmission efficiency and lower reflection losses, smaller delays and less power dissipations. For applications requiring small circuit delays MWCNT interconnects should be used due to smaller capacitances. Applications requiring large transmission efficiency and low reflection losses, CNT bundles should be used for interconnects since the numbers of conducting channels per shell are more in SWCNTs bundle than the number of conducting channels per shell in MWCNT of the same size. These findings suggest that MWCNT and SWCNT bundle can replace Cu as interconnection wires in next generation of VLSI integrated circuits.

ACKNOWLEDGMENT

The author acknowledges Drs. Y. Xu, J. Marulanda, A. K. Sharma and Mr. C. Mayberry for useful discussions and feedback in CNT-FETs, VLSI interconnects and emerging integrated circuits.

REFERENCES

[1] M. S. Dresselhaus, G. Dresselhaus, and P. Avouris, Carbon Nanotube: Synthesis, Properties, Structure, and Applications, Springer Verlag, 2001.

[2] J. Wildoer, L. Venema, A. Rinzler, R. Smalley, and C. Dekker, Nature, vol. 391, p. 59, 1998.

[3] R. Saito, M. S. Dresselhaus, and G. Dresselhaus, "Physical Properties of Carbon Nanotubes", Imperial London: College Press, 1998.

[4] M. Haselman and S. Hauck, "The future of integrated circuits: a survey of nanoelectronics," Proceedings of IEEE, vol. 98, no. 1, pp. 11-38, 2010.

[5] A. Maffucci, "Carbon nanotubes in nanopackaging applications," IEEE Nanotechnology Magazine, vo. 3, no. 3, pp. 22-25, 2009.

[6] N. Alam, A. K. Kureshi, M. Hasan, and T. Arslan, "Carbon nanotube interconnects for low-power high-speed applications," Proc. IEEE International Symposium on Circuits and Systems (ISCAS 2009), pp. 2273-2276, 2009.

[7] P. Avouris, J. Appenzeller, R. Martel, and S. L. Wind, "Carbon nanotube electronics," Proc. of IEEE, vo. 91, no. 11, pp. 1772-1784, 2003.

[8] T. S. Cho, K.-J. Lee, J. Kong, and A. P. Chandrakasan, "A low power carbon nanotube chemical sensor system," Proc. IEEE 2007 Custom Integrated Circuits Conference (CICC), pp. 181-184, 2007.

[9] S. J. Tans, A. R. M. Vershueren, and C. Dekker, "Room-temperature transistor based on a single carbon nanotube," Nature, vo. 393, pp. 49-52, 1998.

[10] R. Martel, T. Schmidt, H. R. Shea, T. Hertel, and P. Avouris, "Single- and multi-wall carbon nanotube field effect transistors," Appl. Phys. Letters, vo. 73, pp. 2447-2449, 1998.

[11] A. Bachtold, P. Hadley, T. Nakanishi, and C. Dekker, "Logic circuits with carbon nanotube transistors," Science, vol. 294, pp. 1317-1320, 2001.

[12] H. S. P. Wong, "Field effect transistors - from silicon MOSFETs to carbon nanotube FETs," Proc. 23th International Conference on Microelectronics, (MIEL), 103-107 (2002).

[13] A. Javey, Q. Wang, W. Kim, and H. Dai, "Advancements in complementary carbon nanotube field-effect transistors," IEDM Technical Digest, pp. 31.2.1-31.2.4, December 2003.

[14] A. Javey, Q. Wang, A. Ural, Y. Li, and H. Dai, "Carbon nanotube transistors arrays for multistage complementary logic and ring oscillators," Nano Letters, vol. 2, pp. 929-932, 2002.

[15] A. Raychowdhury, S. Mukhopadhyay, and K. Roy, "A circuit-compatible model of ballistic carbon nanotube field effect transistors," IEEE Transactions on Computer-Aided Design of Integrated Circuits and Systems, vol. 23, pp. 1411-1420, 2004.

[16] I. O'Connor, J. Liu, F. Gaffiot, F. Prégaldiny, C. Lallement, C. Maneux, J. Goguet, S. Frégonèse, T. Zimmer, L. Anghel, T.-T. Dang, and R. Leveugle, "CNTFET modeling and reconfigurable logic-circuit design," IEEE Transactions on Circuits and Systems, Part-1, vol. 54, pp. 2365-2379, 2007.

[17] S. Fregonese, C. Maneux, T. Zimmer, "Implementation of tunneling phenomena in a CNTFET compact model," IEEE Transactions on Electron Dev., vol. 56, no. 6, pp. 2224-2231, 2009.

[18] A. Srivastava, J. Marulanda, Y. Xu and A. K. Sharma, "Current transport modeling of carbon nanotube field effect transistors," physica status solidi (a), vol. 206, no. 7, pp. 1569-1578, 2009.

[19] H. S. P. Wong, in: Proceedings of 23th International Conference on Microelectronics, (MIEL), pp. 103-107, 2002.

[20] A. Javey, J. Guo, D. B. Farmer, Q. Wang, D. Wang, R. G. Gordon, M. Lundstrom, and H. Dai, Nano Letts. Vol. 4, p. 447, 2004.

[21] Y. Nosho, Y. Ohno, S. Kishimoto, and T. Mizutani, Appl. Phys. Letts. Vol. 86, p. 073105, 2005.

[22] A. Javey, H. Kim, M. Brink, Q. Wang, A. Ural, J. Guo, P. Mcientyre, P. McEuen, M. Lundstrom, and H. Dai, Nature Materials, vol. 1, p. 241, 2002.

[23] V. Derycke, R. Martel, J. Appenzeller, and P. Avouris, Nano Letts., vol. 1, p. 453, 2001.

[24] R. Martel, V. Derycke, J. Appenzeller, S. Wind, and P. Avouris, in: Proceedings 39th Design Automation Conference, pp. 94-98, 2002.

[25] D. L. John, L. C. Castro, J. P. Clifford, and D. L. Pulfrey, IEEE Trans. on Nanotechnology, vol. 2, p. 175, 2003.

[26] C. Dwyer, M. Cheung, and D. J. Sorin, in: Proceedings 4th IEEE Conference on Nanotechnology, pp. 386-388, 2004.

[27] A. Raychowdhury and K. Roy, IEEE Trans. on Nanotechnology, vol. 4, p. 168, 2005.

[28] A. Hazeghi, T. Krishnamohan, and H.-S. P. Wong, IEEE Trans. on Electron Devices, vol. 54, p. 439, 2007.

[29] K.-H. Koo, P. Kapur, and K. C. Saraswat, "Compact performance models and comparison for gigascale on-chip global interconnect technologies," IEEE Trans. Electron Dev., vol. 56, no. 9, pp. 1787-1798, September 2009.

[30] S. Iijima, "Helical microtubules of graphitic carbon," Nature, vol. 354, pp. 56-58, November 1991.

[31] J. W. G. Wilder, L. C. Venema, A. G. Rinzler, R. E. Smalley, and C. Dekker, "Electronic structure of atomically resolved carbon nanotubes," Nature, vol. 391, pp. 59-62, 1998.

[32] J.-H. Ting, C.-C. Chiu, and F.-Y. Huang, "Carbon nanotube array vias for interconnect applications," J. Vac. Sci. & Technol., vol. B 27, no. 3, pp. 1086-1092, 2009.

[33] H. Li, C. Xu, N. Srivastava and K. Banerjee, "Carbon nanomaterials for next-generation interconnects and passives: physics, status, and prospects," IEEE Trans. Electron Dev., vol. 56, no. 9, pp. 1799-1821, 2009.

[34] T. Xu, Z. Wang, J. Miao, X. Chen, and C. M. Tan, "Aligned carbon nanotubes for through-wafer interconnects," Appl. Phys. Lett., vol. 91, no. 4, p. 042108, July 2007.

[35] A. G. Chiariello, A. Maffucci, and G. Miano, "Signal integrity analysis of carbon nanotube on-chip interconnects," Proc. IEEE Workshop on Signal Propagation on Interconnects (SPI '09), pp. 1-4, 2009.

[36] S. J. Wind, J. Appenzeller, R. Martel, V. Derycke, and P. Avouris, Appl. Phys. Letts., vol. 80, p. 3817, 2002.

[37] Z. Yao, C. L. Kane, and C. Dekker, Phys. Rev. Letts., vol. 84, 2000.

[38] A. Javey, J. Guo, Q. Wang, M. Lundstron, and H. Dai, Nature, vol. 424, p. 654, 2003.

[39] Y. Tsividis, "Operation and Modeling of the MOS Transistor", Singapore: McGraw-Hill, 1999.

[40] B. G. Streetman, Solid State Electronic Devices, India: Prentice Hall, 2000.

[41] A. Srivastava, J. Marulanda, Y. Xu and A.K. Sharma, "Current transport modelling of carbon nanotube field effect transistors," physica status solidi (a), vol. 206, no. 7, pp. 1569-1578, 2009.

[42] Y. Xu, and A. Srivastava, "A model for carbon nanotube interconnects," Int. J. Circ. Theor. Appl., published online in Wiley InterScience, vol. 38, issue 6, pp. 559-575, 2010.

[43] G. Miano, and F. Villone, "An integral formulation for the electrodynamics of metallic carbon nanotubes based on a fluid model," IEEE Transactions on Antennas and Propagation, vol. 54, no. 10, pp. 2713-2724, 2006.

[44] A. G. Chiariello, A. Maffucci, G. Miano, F. Villone, and W. Zamboni, "Metallic carbon nanotube interconnects, part I: a fluid model and a 3D integral formulation," Proc. IEEE Workshop on Signal Propagation on Interconnects, pp. 181-184, 2006.

[45] A. L. Fetter, "Electrodynamics of a layered electron gas. I. single layer," Ann. of Physics, vol. 81, no. 2, pp. 367-393, 1973.

[46] A. L. Fetter, "Electrodynamics of a layered electron gas. II. Periodic array," Ann. of Physics, vol. 88, no. 1, pp. 1-25, 1974.

[47] A. Maffucci, G. Miano, and F. Villone, "A transmission line model for metallic carbon nanotube interconnects," Int. J. Circuit Theory Appl., vol. 36, no. 1, pp. 31-51, 2008.

[48] G. K. Batchelor, "An Introduction to Fluid Dynamics," Oxford: Cambridge University Press, 1967.

[49] N. Srivastava, Li Hong, F. Kreupl, and K. Banerjee, "On the applicability of single-walled carbon nanotubes as VLSI interconnects," IEEE Transactions on Nanotechnology, vol. 8, no. 4, pp. 542-559, 2009.

[50] H. Li, W.-Y. Yin, K. Banerjee, and J.-F. Mao, "Circuit modeling and performance analysis of multi-walled carbon nanotube interconnects," IEEE Trans Electron Dev., vol. 55, no. 6, pp. 1328-1337, 2008.

[51] A. Naeemi, and J. D. Meindl, "Design and performance modeling for single-walled carbon nanotubes as local, semiglobal, and global interconnects in gigascale integrated systems," IEEE Trans. on Electron Dev., vol. 54, no. 1, pp. 26-37, 2007.

[52] Y. Xu, and A. Srivastava, "Dynamic response of carbon nanotube field effect transistor circuits," Proc. 2009 NSTI Nanotechnology Conference and Expo, vol. 1, pp. 625-628, May 3-7, 2009.

[53] International Technology Roadmap for Semiconductors, (http://www.itrs.net/Links/ 2007ITRS/Home2007.htm).

[54] A. Srivastava, Y. Xu and A.K. Sharma, "Carbon nanotubes for next generation very large scale integration interconnects," J. Nanophotonics, (invited paper – online), special session on carbon nanotubes, vol. 4, 041690 (17 May 2010), pp. 1-26, 2010.

MIXDES 2012, 19th International Conference *"Mixed Design of Integrated Circuits and Systems"*, May 24-26, 2012, Warsaw, Poland

Trends and Challenges in Micro- and Nanoelectronics for the Next Decade

Cor Claeys

Imec

Leuven, Belgium

Abstract—The research related to micro- and nano-electronics can be grouped in three main directions, i.e., More Moore, Beyond CMOS and More than Moore. For each of them some general trends and challenges are addressed. The convergence of the top-down technology, with bottom-up methods derived from fundamental disciplines such as materials physics, chemistry and biotechnology is opening a totally new world of applications. The diversity of the 'More than Moore' domain, includes applications related to ambient intelligence, automotives, molecular electronics, nano-biotechnology, polymer electronics, sustainable energy based on photovoltaic cells and human healthcare.

Index Terms—Strain engineering; Scaling, Moore's law; heterogeneous integration; packaging; healthcare; automotives

I. INTRODUCTION

Since the discovery of the semiconductor transistor in 1948 by Shockley, Bardeen and Brattain [1], followed by the invention of the planar technology [2] enabling to manufacture integrated circuits, the microelectronics industry has been exponentially growing. Based on Moore's empirical law [3] that the packaging density should double every 18 months and the cost/function decrease, today advanced semiconductor chips are manufactured in a 28 nm technology on 300 mm diameter silicon wafers. Research is pushing towards 22 nm and below while the introduction of 450 mm wafers is at the horizon and expected to happen within two to three years.

In the last decade microelectronics devices have been a driving force for societal applications and are the cornerstone for a green sustainable world. Key fields such as security, energy, healthcare, transport, communication and infotainment are gaining more and more market so that microelectronics is becoming an inherent part of everyday life. Typical examples are smartphone, mobile internet devices, netbook, MEMS applications, global positioning systems, healthcare, etc. The increasing microelectronics content necessitates the introduction of advanced components with higher system functionality based on heterogeneous integration combined with advanced 3D packaging approaches.

As illustrated in Fig. 1, the microelectronics landscape can be divided into three basic research domains. First there is the drive towards a continuous performance enhancement by further scaling the device dimensions in line with Moore's law

and therefore referred to as 'More Moore'. As will be shown further, this requires the implementation of new materials and advanced process modules, and the switching over to non-planar or alternative device architectures. The ultimate scaling will be hampered by physical limitations as one enters the Si atomic scale.

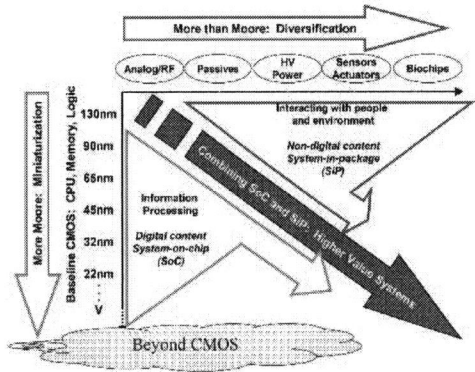

Figure 1. Schematic illustration of the main research domains in micro- and nanoelectronics [4].

At the end of the roadmap one will have to switch over to another operating principle of the devices to further enhance the electrical performance and/or reduce the power consumption. This research domain is referred to as Beyond CMOS. Extensive research is focusing on tunnelFETs (TFETs), based on tunneling instead of thermionic emission and characterized by a very low substhreshold swing (<60 mV/dec) [5-6]. Another exciting research field is carbon-based electronics using carbon nanotubes (CNT) and/or graphene nanoribbons (GNR) [7-8]. Ultimately, operating concepts such as e.g. quantum computing [9] and spintronics [10] could be used for future applications.

Increased functionality is driving the 'More than Moore' field enabling System-on-Chip (SOC) and System-in-Package (SiP) applications. The interaction of the device with the environment and with people is important. This implies research related to heterogeneous integration, 3D packaging, etc and has a large number of applications in fields such as automotive, ambient intelligence, healthcare, domotics, etc. A large variety of technologies are being investigated, such as e.g. analog/RF, on-chip integration of passive elements, sensors and actuators, biochips, etc.

II. DEVICE SCALING – MORE MOORE

Moore's law, first postulated in middle of the 60-ties and illustrated in Fig. 2, drives the scaling research. The minimum required feature size is different for logic, microprocessor and memory circuits.

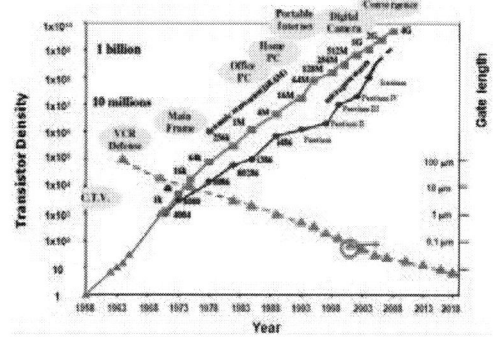

Figure 2. Moore's law showing the evolution of the transistor packing density (left axis) as a function of time, requiring the reduction of the device geometries (right axis) (after S. Deleonibus [11]).

The technological innovations necessary for a timely manufacturing introduction of the next technology generation, guided by the International Technology Roadmap for Semiconductors (ITRS), requires research on new processing modules and new materials such as e.g. silicides, ultra shallow junctions, alternative metallization systems, high-κ gate dielectrics, low-κ dielectrics for intermetal isolation, metal gates to replace polysilicon, Cu-based metallization systems, strain engineering, etc. Today, 28 nm technologies are appearing in manufacturing, while industrial research is oriented towards 22 nm and below. Some of the on-going processing/device efforts will be briefly discussed.

A. Strain Engineering

Since the 90 nm technology node it is common practice to introduce strain engineering, based on either global (wafer-level) or local (device-level) techniques, to enhance the carrier mobility and drive current of MOS devices. Several new materials and/or processing approaches to create strained channel regions are introduced. For CMOS both device types have to be optimized so that an uniaxial compressive stress for pMOSFETs and an uniaxial tensile stress for nMOSFETs is used. This can e.g. be achieved using an embedded source/drain (S/D) technology, i.e., a recess of the source and drain regions of the transistor is done before the selective epitaxial growth of $Si_{1-x}Ge_x$ [12-13] for pFETs and $Si_{1-y}C_y$ [14-15] for nFETs. Since SiGe/SiC has a larger/smaller lattice constant than the Si substrate, the embedded SiGe/SiC source-drain areas induce a lateral compressive/tensile stress in the Si channel, boosting up the device performance, as illustrated in Fig. 3. Another approach is using a SiN cap layer, referred to as Contact Etch Stop Layers (CESLs) [16]. Depending on the hydrogen content of the CESL either compressive (c-CESL) or tensile (t-CESL) caps are obtained, so that with a dual cap-layer approach performance optimization is achieved for both p- and n-channel devices.

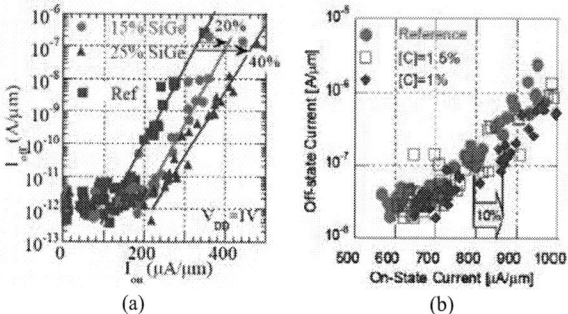

(a)   (b)

Figure 3. The use of SiGe [13] (a) and SiC [15] (b) source/drain regions to improve the carrier mobility (I_{on}) of p- and n-channel devices, respectively.

Strain engineering can also be based on the use of high-mobility substrates such as e.g. SiGe layers, strained Si (sSi) on a strain relaxed buffer layer and strained Silicon-on-Insulator (sSOI). Although the global approach enables to achieve higher stress levels than the use of stressors, the latter is more effective to translate strain into improved device performance for shorter channels [17].

B. Ge and III-V Materials

Ge has a higher hole and electron mobility than Si so that there is a strong interest in using Ge for high performing transistors. The state-of-the-art in Ge processing has been reviewed [18] and the feasibility to fabricate good performing p-channel MOSFETs reported [19]. Further performance enhancement can be achieved by implementing strain engineering, resulting in a 70% hole mobility increase for biaxially-strained low EOT pFETs [20]. Key challenges for device optimisation are the surface passivation in order to achieve a low interface state density, the formation of the Ge layer (e.g., epitaxially grown on Si, GeOI, Ge condensation technique etc.) to ensure a low defect density, the source/drain junction engineering for a low parasitic series resistance and good Ohmic contacts [21]. Figure 4 gives a TEM image of a 65 nm Ge transistor with high-κ dielectric and metal gate and shows the low field mobility versus gate length, corrected for r_{ext}, for a biaxially strained Ge technology compared to relaxed Ge. However, Ge n-FETs are suffering from both poorer interface quality (high interface defect density) leading to low channel mobility and poor n$^+$ doping control (too low activation level, in the order of 5-6×10^{19} cm^{-3}, and concentration enhanced diffusion).

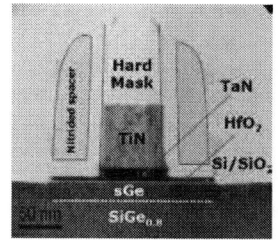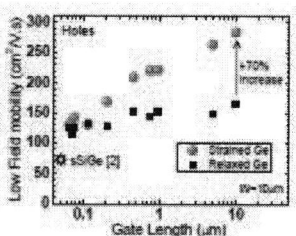

Figure 4. TEM image of a 65 nm pFET with high-κ dielectric and metal gate (left) and low field mobility vs gate length, corrected for r_{ext}, for a biaxially strained Ge technology compared to relaxed Ge (right) [20].

38

III-V compounds are considered strong potential candidates for n-channel devices. A recent review on Ge and III-V for advanced CMOS devices, outlining the progress and remaining difficulties is given in [22]. The main challenge will be to develop a high-reliable, high yielding full integrated technology enabling the on-chip co-integration of both Ge and III-V circuits on a silicon CMOS platform. Recently, a Ge/III-V CMOS process flow with n-channel InGaAs and p-channel Ge devices has been demonstrated, as illustrated in Fig. 5 [23]. First electrical data are promising, but scaling aspects have to be studied.

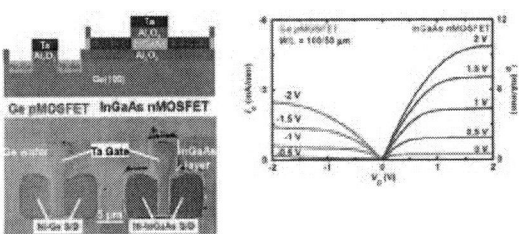

Figure 5. Schematic illustration and SEM picture of a Ge/III-V CMOS process flow with n-channel InGaAs and p-channel Ge device (left), showing promising eletrical performance date (right)[23].

C. Multi-Gate Devices(MuGFETs)

To enhance the device performance, alternative gate concepts, with an evolution from from planar single gate to double gate, tri-gate or FinFET, and gate-all-around (GAA) or nanowire concepts, have been extensively studied. Although manufacturing issues have delayed their introduction in productions lines, FinFET and MuGFET concepts are presently being used for 22 nm technologies.

FinFETs enable a better gate control, an ideal subthreshold slope and a lower leakage current. Due to the undoped channel region the mobility improves and the random dopant fluctuations reduce. SEM photographs of some fabricated Si FinFETs down to 15 nm width are shown in Fig. 6. For MugFETs both bulk and SOI approaches are investigated. Floating body bulk FinFETs are gaining interest as one transistor capacitor-less random access memory (1T-RAM) because they are easier to scale than planar devices, enable a higher charge storage area in the fin and avoid self-heating effects as occurring in SOI devices [24]. By optimizing the ground plane doping a 10 s retention time at 85°C is reported for a FinFET with a 90 nm gate length and 20 nm fin width [25].

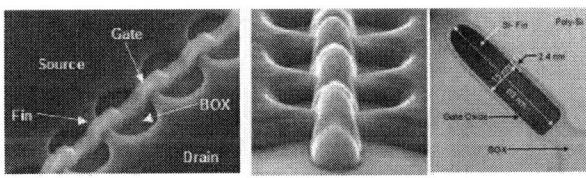

Figure 6. SEM photographs of processed FinFETs down to 15 nm width.

D. Tunnel Field Effect Transistors (TFETs)

For MOSFETs the thermionic nature of the turn-off mechanism limits the subthreshold slope to 60 mV/dec at 300 K and is forming a trade-off between leakage current and performance at low voltage operation. This can be overcome by switching over to tunnel FETs relying on band-to-band-tunneling modulated by the gate voltage. Basically a TFET is a gated p-i-n structure

The excellent subthreshold behavior of TFETs was first demonstrated by Appenzeller et al. [26] for carbon nanotubes (CNTs) and by Choi et al. [27] for Si devices. TFETs can be realized based by either a horizontal or vertical technology using either a planar, double gate or MuGFET approach. Each of these structures has advantages and drawbacks from a viewpoint of process complexity, achievable packing density, and a possible implementation of hetero-structures to tune the band gap to increase the tunneling efficiency.

The fabrication can be done either bottom up (i.e., growth of the Si nanowire using a catalyst such as Au) or top down (i.e., based on the etching of the Si). The nanowire TFET in Fig. 7 is an example of last technique, i.e., etching of an epitaxially grown Si stack [28]. In the other case it is important to select a catalyst with a high efficiency and a low contamination risk. Instead of the often used Au, it is also possible to work with In as a catalyst, which is enhancing the compatibility with a Si process line [29-30].

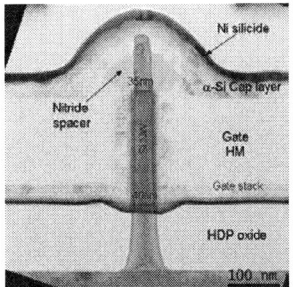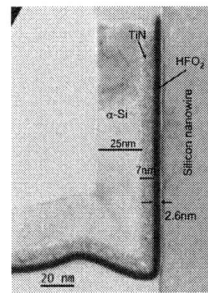

Figure 7. TEM cross-section of a 35 nm NW TFET (left) and details of the gate stack (right) [28].

Extensive research is on going to optimize the TFET electrical device characteristics. The tunnel efficiency depends on the fin width. Device optimization has to take into account both a large variety of technological parameters such as e.g. implantation profiles, anneal conditions (RTA, spike, laser, SPER etc), gate stack and spacer engineering, and design aspects [31].

The large and indirect Si band gap, however, leads to a low tunneling efficiency limiting the achievable on-current. Therefore, research is triggered towards hetero-structures consisting of a Si drain and intrinsic region, and a low band gap material source such as e.g. SiGe, Ge or even an III-V compound. High performing CMOS TFET process flows based on using Ge and GaAs for n- and pTFET, respectively, are being investigated. Recent reviews on this topic are found in [5, 32].

The main challenges for TFETs remain device performance with a high on-current, a low-off current and a very low subthreshold swing, while keeping the process complexity limited. Low substhreshold swings have been reported but mostly over a rather restricted bias range. There are strong indications that hetero-structure TFETs have a great potential for future low-power/high-performance applications. Therefore, TFETs are considered as the building blocks for coming to a green technology.

E. Carbon Based Devices

Single or multi-wall carbon nanotubes (CNT) technologies have made much progress and both doping types can be achieved. An early discussion of different type of CNT designs has been published by Appenzeller [26]. A good transistor performance has already been demonstrated several years ago. A very promising application of CNT is as interconnect material whereby it may even replace the Cu metallization. Much effort is going on to develop the appropriate catalysts to grow both lateral and vertical CNTs in a well-controlled manner. CNTs have a 15 times higher thermal conductivity than Cu (6000 W/m.K compared to 400 W/m.K for Cu), in addition to their 1000 times higher current capacity (10^9 A/cm^2 compared to 10^6 A/cm^2 for Cu). The concept of using CNTs for via filling is illustrated in Fig. 8. It is essential to have a good nano-particle growth efficiency, a good CNT quality and a low defect density.

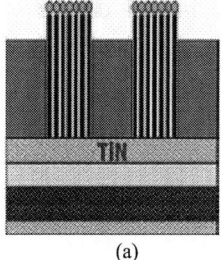 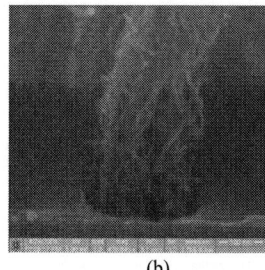

(a) (b)

Figure 8. CNTs used for via filling: concept (a) and a 200 nm via filled with about 10^{11} cm^{-2} CNTs (b).

Graphene, a two-dimensional carbon structure, has a very high intrinsic carrier mobility and is emerging as an interesting material for future electronics. Extensive research is done for ultra-high speed and low power devices. The pioneering work in the field was performed by Geim and Novosolev [33]. Top gated FETs with a mobility of 7600 cm^2/V and a dG/dV$_{TG}$ = 210 µS/V have been demonstrated at room temperature [34]. Figure 9 compares the cut-off frequency of experimental graphene MOSFETs (all with gapless large-area graphene channels) with the best f$_T$ values reported for other types of RF FETs: InP HEMTs, GaAs mHEMTs, SiMOSFETs, GaAs pHEMTs [35]. The highest reported cut-off frequency for an epitaxially grown graphene FET is around 100 GHz [36]. The gate length was 240 nm. Graphene can also be used for the fabrication of TFETs. Although some circuits have already

been demonstrated, there is a still a long way to go before achieving manufacturable graphene-based technologies.

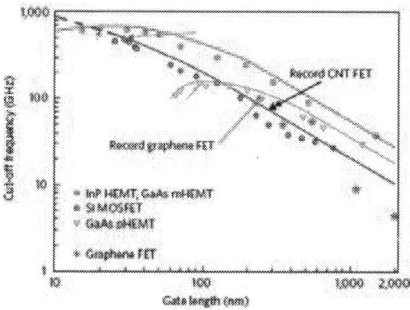

Figure 9. Cut-off frequency versus gate length for different RF FET types, indicating the excellent performance of the graphene FET [35].

III. MORE THAN MOORE

More than Moore is directly linked to the different applications driven by societal needs and isrelying on an increased functionality. Therefore, there is also a strong emphasis on packaging related issues. Another key challenge to face is the interface with the outside world, which is different for the different application fields. Two examples of extreme requirements are e.g. automotive and biomedical applications. As shown in Fig. 1 (diagonal), combined with scaling the way is paved towards System-in-Package (SiP) and System-on-Chip (SoC) applications.

For More than Moore (MtM) the driving force is not scaling but increasing the functionality of the circuits, which can be achieved by combining standard CMOS with other technologies such as photonics, chemical sensors, RF BiCMOS, MEMS, thermal sensors etc. This is schematically illustrated in Fig. 10. For the different combinations several demonstrators have already been successfully fabricated.

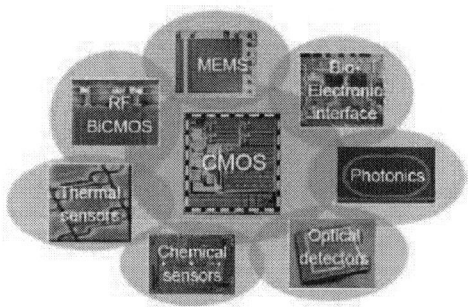

Figure 10. Combination of CMOS with other technologies to increase the overall functionality, enabling a myriad of new applications in the MtM field.

An important application field is automotive since the amount of electronics in cars is not only strongly increasing but also becoming an important cost element of the total cost of the present-day car. However, not only environmental concerns (e.g. CO$_2$ emission, fuel consumption) but also safety (ABS, EBS, airbag sensors, cruise control, etc) and comfort

(GPS, board computer, local networks, etc) lead to an increased need of electronic components. In this field MEMS devices are finding their way.

There is also strong interest in after standard CMOS processing to use the layer above for the integration of passive components, such as highly-linear capacitors and integrated inductors. An example of an application is a shunt inductor realized in a 90 nm above-IC technology, to give a better ESD projection as compared to a back-end inductor realization for a LNA and VCO [37]. Also other RF MEMS modules can be processed above-IC, as was e.g. demonstrated for a bulk acoustic wave resonator [38].

This approach, leading to a saving of silicon real estate, may replace some the presently System-in-Package (SiP) concepts used. The MEMS modules have to be tuned for the specific applications.

The More than Moore domain is surely requiring 3D packaging to achieve a higher packing density, to shorten the interconnects and to lower the interconnect density. Beside technological challenges an important aspect is the control of the thermal budget and to enable a cost effective processing. Nowadays, there is strong interest in 3D wafer level packaging (3D-WLP), whereby through silicon vias (TSV) are needed.

Semiconductor expertise can also be used to boost the sustainable energy generation. Photovoltaic research is covering a large variety of semiconductor materials such as Si, Ge, III-V, II-VI, etc. Depending on the application different requirements are put forward. The light-weight high efficient GaAs/Ge tandem solar cells used in space applications have different requirements than the technologies used for terrestrial applications, for which the main driving force is the cost/kw-peak. For the latter research is on going to develop thin layer high efficient and low-cost solar cells. Future approaches may be based on organic materials such as e.g. the low band gap polymers (thiphenes) [39-41].

Another field with a huge potential is related to healthcare. The population is becoming older thanks to the great progress in medical sciences. However, it is essential not only to increase the lifetime but to ensure the quality of life. This necessitates to diagnose and monitor and whenever possible to cure the deceases. Typical examples of important deceases are Alzheimer and Parkinson, which are frequently associated with the elderly society. The world of bio-electronics is making large progress and opens new application fields. The nanometer geometry of the state-of-the-art devices is very similar to the size of a virus or an anti-body. Using the so-called neurons on a chip or artificial synapse approach enables the bi-directional communication between neurons and an integrated circuit. The interaction can be stimulated either electrically or chemically. An illustration of this exciting new field is given in Fig. 11, showing the direct coupling of a neuron to the gate of a transistor.

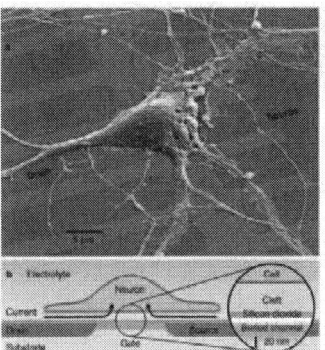

Figure 11. Processing of neurons on top of a Si chip: (a) Electron micrograph of a neuron on a p-type buried-channel transistor and (b) Schematic cross-section of the structure [42].

The bio-electronics applications require a strongly multi-disciplinary approach with a research team formed by experts in microelectronics, chemistry, biology and medicine in order to fabricate the different system components needed: sensors, tranducers, the on-chip cellular microenvironment, the in-vitro hybrid devices, the in-vivo neuroprobes, the modeling and signal processing, and the appropriate packaging and integration aspects.

IV. CONCLUSION

Only selected illustrations were given of the exciting research activities on going in the field of 'More Moore' and 'More than Moore'. Important topics such as e.g. quantum dots, spintronics, polymer electronics, GaN for power and lightning applications, molecular electronics, single electron devices, etc were not mentioned at all. Many disruptive technologies are appearing at the horizon. The future electronics systems needed to comply with the societal needs will be based on advanced nanoelectronics devices combined with an increased functional diversity of the systems.

ACKNOWLEDGEMENT

The author has based this manuscript mainly on research carried out at imec and would therefore like to thank K. Baert, E. Beynen, R. Cartuyvels, S. De Gendt, G. Groeseneken, M. Heyns, D. Leonelli, A. Mercha, R. Mertens, J. Poortmans, R. Rooyackers, P. Vereecken, A. Verhulst and C. Van Hoof for many discussions and the use of some of their material.

REFERENCES

[1] J. Bardeen, W.H. Brattain, "Physical principles involved in transistor action," Phys. Rev, vol. 75, pp. 1208-1225, 1949.
[2] J.S. Kilby, "Invention of the integrated circuit," IEEE Trans. Electron Dev., vol. 23, pp. 648-654, 1976.
[3] G.E. Moore, "Cramming more components onto integrated circuits," Electronics Mag., vol. 38, pp. 114-117, 1965.
[4] European Nanoelectronics Initiative Advisory Council (ENIAC), Strategic Research Agenda, November 2007, http://www.eniac.eu
[5] C. Claeys, D. Leonelli, R. Rooyackers, A. Vandooren, A. Verhulst, M.M. Heyns, G. Groeseneken and S. De Gendt, ECS Trans., vol. 35(5), pp. 15-26, 2011.
[6] A.M. Ionescu and H. Riel, Nature, vol. 479, pp. 329-337, 2011.
[7] J. Gea-Banacloche, "Quantum computers: A status update," Proc. IEEE, vol. 98, pp. 1983-1985, 2010.

[8] F. Schwierz, Nature Nanotechn., vol. 5, pp. 487-496, 2010.

[9] J.J.L. Morton, D.R. MacCamey, A. Eriksson and S.A. Lyon, Nature, vol. 479, pp. 345-353, 2011.

[10] S. Sugahara and J. Nitta, "Spin-transistor electronics: An overview and outlook,", Proc. IEEE, vol. 98, pp. 2124-2154, 2010.

[11] S. Deleonibus, "Electronics Device Architectures for the Nano-CMOS Era," Pan Standford Publ., 2009.

[12] C. Auth, A. Cappellani, J. S. Chun, A. Dalis, A. Davis, T. Ghani, G. Glass, T. Glassman, M. Harper, M. Hattendorf, P. Hentges, S. Jaloviar, S. Joshi, J. Klaus, K. Kuhn, D. Lavric, M. Lu, H. Mariappan, K. Mistry, B. Norris, N. Rahhal-orabi, P. Ranade, J. Sandford, L. Shifren, V. Souw, K. Tone, F. Tambwe, A. Thompson, D. Towner, T. Troeger, P. Vandervoom, C. Wallace, J. Wiedemer, and C. Wiegand, "45nm high-k + metal gate stain enhanced transistors," in VLSI Technology Symp. Tech Dig., pp. 128-129, 2008.

[13] P. Verheyen, G. Eneman, R. Rooyackers, R. Loo, L. Eeckhout, D. Rondas, F. Leys, J. Snow, D. Shamiryan, M. Demand, Th. Y. Hoffman, M. Goodwin, H. Fujimoto, C. Ravit, B-C. Lee, M. Caymax, K. De Meyer, P. Absil, M. Jurczak, and S. Biesemans, "Demonstration of recessed SiGe S/D and inserted metal gate on HfO_2 for high performance pFETs," Technical Digest of IEDM, pp. 886-889, 2008.

[14] Z. B. Ren, G. Pei, J. Li, B. Yang, R. Takalkar, K. Chan, G. Xia, Z. Zhu, A. Madan, T. Pinto, T. Adam, J. Miller, A. Dube, L. Black, J. W. Weijtmans, B. Yang, E. Harley, A. Chakravarti, T. Kanarsky, R. Pal, I. Lauer, D. G. Park, and D. Sadana, "On implementation of embedded phosphorous-doped SiC stressors in SOI nMOSFETs," VLSI Technology Symp. Tech. Dig., pp. 172-173, 2008.

[15] P. Verheyen, V. Machkaoutsan, M. Bauer, D. Weeks, C. Kerner, F. Clemente, H. Bender, D. Sharniryan, R. Loo, T. Hoffmann, P. Absil, S. Biesemans, and S. G. Thomas, "Strain enhanced nMOS using in situ doped embedded $Si_{1-x}C_x$ S/D stressors with up to 1.5% substitutional carbon content grown using a novel deposition process," IEEE Electron Dev. Lett., 29, pp. 1206-1208, 2008.

[16] S. Ito, H. Namba, K. Yamaguchi, T. Hirata, K. Ando, S. Koyama, S. Kuroki, N. Ikezawa, T. Suzuki, T. Saitoh, and T. Horiuchi, "Mechanical stress effect of etch-stop nitride and its impact on deep submicron transistor design," Techn. Dig. IEDM, pp. 247-250, 2000.

[17] C. Claeys, E. Simoen, S. Put, G. Giusi, and F. Crupi, "Impact of strain engineering on gate stack quality and reliability," Solid-State Electron., 52, pp. 1115-1126, 2008.

[18] C. Claeys, E. Simoen, "Germanium-Based Technologies" From Materials to Devices", Elsevier, 2007.

[19] J. Mitard, B. De Jaeger, F.E. Leys, G. Hellings, K. Martens, G. Eneman, D.P. Brunco, R. Loo, J.C. Lin, D. Shamiryan, T. Vanderweyer, G. Winderickx, E. Vrancken, C.H. Yu , K. De Meyer, M. Caymax, L. Pantisano, M. Meuris and M.M. Heyns, "Record Ion/Ioff performance for 65-nm pMOSFET and novel Si passivation scheme for improved EOT scalability," Technical Digest of IEDM, pp. 873-875, 2008.

[20] J. Mitard, B. De Jaeger, G. Eneman, A. Dobbie, M. Myronov, M. Kobayashi, J. Geypen, H. Bender, B. Vincent, R. Krom, J. Franco, G. Winderickx, E.Vrancken, W. Vanherle, W. Wang, J. Tseng, R. Loo, K. De Meyer, M. Caymax, L. Pantisano, D. R. Leadley, M. Meuris, P. Ansil, S. Biesemans and T. Hoffman, "High hole mobility in 65 nm Ge p-channel field effect transistors with HfO_2 gate dielectric," Jpn. J. Appl. Phys., vol. 50, pp. 04DC17, 2011.

[21] E. Simoen, J. Mitard, G. Hellings, G. Eneman, B. Vincent, R. Loo, A. Delabie, S. Sioncke, M. Caymax and C. Claeys, "Challenges and opportunities in advanced Ge pMOSFETs," Mat. Sci. Sem. Proc., unpublished.

[22] M. Heyns, A. Alian, G. Brammertz, M. Caymax, Y.C. Chang, L.K. Chu, B. De Jaeger, G. Eneman, F. Gencarelli, G. Groeseneken, G Hellings, A. Hikavyy, T.T. Hofmann, M. Houssa, C. Huyghebaert, D. Leonelli, D. Lin, R. Loo, W. Magnus, C. Mercking, M. Meuris, J. Mitard, L. Nyns, T. Orzali, R. Rooyackers, S. Sioncke, B. Soree, X. Sun, A. Vandooren, A.S. Verhulst, B. vincent, N. Waldron, G Wang, W.E. Wang and L. Witters, "Advancing CMOS beyond the Si roadmap with Ge and III/V devices," Technical Digest of IEDM, pp. 299-302, 2011.

[23] M. Yokoyama, S.H. Kim, R. Zhang, N. Taoka, Y. Urabe, T. Maeda, H. Takagi, T. Yasuda, J. Yamada, O. Ichikawa, N. Fukuhara, M. Hata, M. Sugiyama, Y. Nakano, M. Takenaka, and S. Takagi, "CMOS integration of InGaAs nMOSFETs and Ge pMOSFETs with self-aligned Ni-based metal S/D using direct wafer bonding," VLSI Technology Symp. Tech Dig., pp. 164-165, 2011.

[24] N. Collaert, M. Aoulaiche, M. Rakowski, A. Redeolfi, B. De wahter, J. van Houdt,a nd M. Jurczak, "Optimizing the readout bias for the capacitorless 1T FinFET RAM cell," IEEE Electron Dev. Lett., 30, pp. 1377-1379, 2009

[25] N. Collaert, M. Aoulaiche, B. De Wachter, M. Rakowski, A. Redolfi, S. Brus, A. De Keersgieter, N. Horiguchi, L. Altimime, and M. Jurczak "A low-voltage biasing scheme for agressively scaled bulk FinFET 1T-DRAM featuring 10s retention at 85C," VLSI Technology Symp. Tech Dig., pp. 161-162, 2010.

[26] J. Appenzeller, Y.M. Lin, J. Knoch, Z. Chen, and P. Avouris, "Comparing carbon nanotube transistors – The choice: A novel tunnel device devising", IEEE Trans. Electron Dev., vol. 52, pp. 2568-2576, 2005.

[27] W.Y. Choi, B.-G. Park, J.D. Lee and T.-J. K. Liu, "Tunneling field-effect transistors (TFETs) with subthreshold swing (SS) less than 60 mV/dec," IEEE Electron Dev. Lett., vol. 28, pp. 743-745, 2007.

[28] A. Vandooren, R. Rooyackers, D. Leonelli, F. Iacopi, E. Kunnen, D. Nguyen, M. Demand, P. Ong, L. Willie, J. Moonens, O. Richard, A.S. Verhulst, W.G. Vandenberghe, G. Groeseneken, S. De Gendt, M.M. Heyns, "A 35nm diameter vertical silicon nanowire short-gate tunnelFET, Proc. 2009 Silicon Nanoelectronics Workshop, p. 21, 2009.

[29] F. Iacopi, O. Richard, Y. Eichhammer, H. Bender, P.M. Vereecken, S. De Gendt, and M.M. Heyns, "Size-dependent characteristics of Indium-seeded Si nanowire growth," Electrochem. Soc. Lett., vol. 11, pp, K98-K100, 2008.

[30] F. Iacopi, P.M. Vereecken, M. Schaekers, M. Caymax, N. Moelans, B. Blanpain, O. Richard, C. Detavernier, and H. Griffiths, "Plasma-enhanced chemical vapour deposition growth of Si nanowires with low melting point metal catalysts: an effective alternative to Au-mediated growth," Nanotechnol., vol. 18, pp. 505307/1-7, 2008.

[31] D. Leonelli, A. Vandooren, R. Rooyackers, A.S. Verhulst, S. De Gendt, M.M. Heyns and G. Groeseneken, "Optimization of tunnel FETS: Impact of gate oxide thickness, implantation and annealing conditions," Proc. ESSDERC, pp. 170-173, 2010.

[32] A.C. Seabaugh and Q. Zhang, "Low voltage tunnel transistors for beyond CMOS logic," Proc. IEEE, 98, pp. 2095-2110, 2010.

[33] A.K. Geim and K.S. Novoselov, "The rise of graphene," Nature Materials, vol. 6, pp. 183-191, 2007.

[34] C.-Y. Sun, "Carbon based graphene nanoelectronics technologies," Proc. Int. Conf. on Solid-State and Integrated Circuit Techn., Eds. T.-A. Tang and Y.-L. Jiang , pp. 1192-1193, 2010.

[35] F. Schwierz, "Graphene transistors – A new contender for future electronics," Proc. Int. Conf. on Solid-State and Integrated Circuit Techn., Eds. T.-A. Tang and Y.-L. Jiang , pp. 1202-1205, 2010.

[36] Y.-M. Lin, C. Dimitrakopoulos, K.A. Jenkins, D.B. Farmer, H.-Y. Chiu, A. Grill, and Ph. Avouris, "100-GHz transistors from wafer-scale epitaxial graphene,",Science, vol. 5, pp. 5962, 2010.

[37] P. Wambacq, T. Nakaie, and S. Decoutere, "Low-power low-noise highly ESD robust LNA and VCO design using above IC inductors," Proc. Custom Integrated Circuits Conference, pp. 492-495, 2005.

[38] M.A. Dubois, J.F. Carpentier , P. Vincent, C. Billard, G. Parat, C. Muller, P. Ancey, and P. Conti, "Monolithic above-IC resonator technology for integrated architectures in mobile and wireless communication," IEEE J. Solid State Circuits, vol. 41, pp. 7-17, 2006.

[39] T. Aernouts, P. Vanlaeke, I. Haeldermans, J. D'Haen, P. Heremans, J. Poortmans and J. Manca, "P3HT:PCBM bulk heterojunction solar cells: morphological and electrical characterization and performance optimization," Organic and Nanoparticle Hybrid Photovoltaic Devices," pp. 1013-Z01-02, 2007.

[40] T. Aernouts, C. Girotto, H. Gommans, D. Cheyns, C. Rolin, J. Genoe, and J. Poortmans, "Processing technologies for organic solar cells," Proc. 22nd European Photovoltaic Solar Energy Conference and Exhibition, pp. 560-565 , 2007.

[41] D. Cheyns, H. Gommans, M. Odijk, J. Poortmans and P. Heremans, "Stacked organic solar cells based on pentacene and C60," Solar Energy Materials and Solar Cells, vol. 91, pp. 399-404, 2009.

[42] M. Spira, D. Kamber, A. Dormann, A. Cohen, C. Bartic, G. Borghs, J. Langendijk, S. Yitzschaik, K. Shabthai and J. Shappir, "Improved neuronal adhesion to the surface of electronic device by engulfment of protruding micro-nails fabricated on the chip surface," Transducers & Eurosensors '07, pp. 1247-1270, 2007.

Compact Modeling Support for Nanoscaled IC Technology and Design

44

Analog Circuits Sizing Using the Fixed Point Iteration Algorithm with Transistor Compact Models

Farakh Javid[1], Ramy Iskander[1], François Durbin[2] and Marie-Minerve Louërat[1]

[1] LIP6 Laboratory, University Pierre and Marie Curie, Paris, France

[2] CEA DAM/DIF, Bruyères-le-Châtel, France

Abstract—This paper presents an algorithm, based on the fixed point iteration, to solve for sizes and biases using transistor compact models such as BSIM3v3, BSIM4, PSP and EKV. The proposed algorithm simplifies the implementation of sizing and biasing operators. Sizing and biasing operators were originally proposed in the hierarchical sizing and biasing methodology [1]. They allow to compute transistors sizes and biases based on transistor compact models while respecting designer's hypotheses. Computed sizes and biases are accurate, and guarantee the correct electrical behavior as expected by the designer. Sizing and biasing operators interface with a Spice-like simulator, allowing possible use of all available compact models for circuit sizing and biasing over different technologies. To illustrate the effectiveness of the proposed algorithm, a folded cascode OTA was efficiently sized with a 130nm process, then was migrated to a 65nm technology. Both sizing and migration were performed in a few milliseconds.

Index Terms—Analog IP, analog sizing, design reuse, bipartite graphs, transistor compact models, technology migration.

I. INTRODUCTION

Over the past few decades, research in analog synthesis led to the emergence of two major schools : the first school pushing towards *Full Design Automation (FDA)* and the second school pushing towards *Full Design Handcrafting (FDH)*. Many academic and commercial tools have been introduced by the FDA school such as OASYS [2], IDAC [3], OPASYN [4], DELIGHT.SPICE [5], ASTRX/OBLX [6], AMGIE [7], MAELSTROM [8], ANACONDA [9]. Except for OASYS, IDAC and AMGIE which are knowledge-based, the tools were mainly simulation-based. On the other hand, few FDH school academic tools have been introduced, such as COMDIAC [10], PAD [11], [12] and [13], which provide analog designers with sufficient insight for full trade-offs optimization.

Nowadays, many EDA companies develop tools that help to explore design trade-offs. To assess this research direction, we developed the hierarchical sizing and biasing methodology [1] to hierarchically size and bias analog circuits. This methodology generates suitable sizing procedures that respect the circuit topology and designer's constraints. It elaborates an intermediate design representation, called *bipartite dependency graphs*. These are used to represent sizing procedures and ensure their consistency, reusability and technology independence.

The hierarchical sizing and biasing methodology consists of three main tasks: *transistor sizing and biasing*, *device sizing and biasing* and *circuit sizing and biasing*. Transistor sizing and biasing is performed using computational routines called

sizing and biasing operators. These operators are used to size and bias elementary transistors. Device sizing and biasing is performed by constructing dedicated device sizing procedures based on elementary transistor operators. Circuit sizing and biasing combines sizing procedures of children devices or lower level subcircuits in order to construct the sizing procedure of the whole circuit. The circuit sizing procedure is an enumerated sequence of sizing and biasing operators of all devices forming the circuit.

In this paper, we focus on the transistor sizing and biasing task. This task was originally performed using Newton-Raphson algorithm as suggested in [1], [14]. We propose a new formulation of the sizing and biasing operators using derivative-free fixed point iteration. We construct and evaluate a complete sizing procedure in the form of a bipartite graph for a folded cascode amplifier. We show that industrial transistor compact models such as BSIM3v3, BSIM4, PSP and EKV can be directly used with the fixed point algorithm for the bipartite graph evaluation.

The paper is organized as follows : section II recalls the sizing and biasing operators. Section III gives an overview of the fixed point iteration features. The methodology to implement the fixed point iteration within sizing operators is presented in section IV. Sizing operators with fixed point iteration are applied to the design of a folded cascode amplifier in section V. The paper is concluded in section VI.

II. SIZING AND BIASING OPERATORS

A. Principle

Sizing and biasing operators are based on the inversion of the transistor compact model equations. Each operator has a set of input parameters that are set by the designer and computes unknown widths and biases (Table I, where $V_{EG} = V_{GS} - V_{TH}$). An operator computes either :

$$W = f_W(Temp, I_D, L, V_{GS}, V_{DS}, V_{BS}) \tag{1}$$

either :

$$V_{GS} = f_{V_{GS}}(Temp, I_D, W, L, V_{DS}, V_{BS}) \tag{2}$$

f_W and $f_{V_{GS}}$ are two inverse functions of the transistor compact model given in equation (3) :

$$I_D = f_{MODEL}(Temp, W, L, V_{GS}, V_{DS}, V_{BS}) \tag{3}$$

where *MODEL* is a standard transistor model like BSIM3v3 [15], BSIM4 [15], PSP [16] and EKV [17]. f_W and $f_{V_{GS}}$

are monotonic functions, thus equations (1) and (2) are currently solved with the Newton-Raphson method. Convergence criteria for the Newton-Raphson method are the same as in commercial simulators.

TABLE I

CLASS DEFINITION OF SIZING & BIASING OPERATORS

Operator	Definition
$OPVS(V_{EG}, V_B)$	$(Temp, I_D, L, V_{EG}, V_D, V_G, V_B) \mapsto (V_S, W, V_{TH})$
...	...
$OPVG(V_{EG})$	$(Temp, I_D, L, V_{EG}, V_D, V_S) \mapsto (V_G, W, V_{TH}, V_B)$
...	...
$OPVGD(V_{EG})$	$(Temp, I_D, L, V_{EG}, V_S) \mapsto (V_G, V_D, W, V_{TH}, V_B)$
...	...
$OPW(V_G, V_S)$	$(Temp, I_D, L, V_D, V_G, V_S) \mapsto (W, V_{TH}, V_B)$
...	...
$OPID(V_G, V_S)$	$(Temp, W, L, V_D, V_G, V_S) \mapsto (I_D, V_{TH}, V_B)$
...	...

B. Simulator Encapsulation

As shown in Fig.1, an electrical simulator is encapsulated within the sizing and biasing operators [14]. Thus the sizing operators directly interface with industrial design kits to ensure the accuracy of the computed results. At the bottom in Fig.1, there is an electrical netlist that specifies the suitable design kit and contains only two transistors : one PMOS and one NMOS. Both transistors refer to a transistor compact model and are entirely sizable and biasable through simulator interactive commands. This two-transistor netlist is loaded by the electrical simulator launched in interactive mode, to perform sizing and biasing. Three types of interactive commands are evaluated : *set*, *get* and *run*. The *set* command allows to set transistor known parameters (sizes and biases) at simulator level. The *get* command enables to retrieve currents, voltages and small signal parameters computed by the simulator. After a *set* command, a simulation must be run using the *run* command, in order to compute the DC operating point of the transistor. An API (Application Programming Interface) is developed using *expect* library [18] to automate the *set*, *get* and *run* commands execution. The API is used within sizing and biasing operators, that implement the fixed point iteration described in section IV. Sizing and biasing operators are optimized to minimize the number of calls to the simulator, which can reach several hundreds during sizing.

III. FIXED POINT ITERATION ALGORITHM

The fixed point iteration [19] is an algorithm that allows to solve nonlinear equations that have the following form :

$$x = g(x), \quad g : [a,b] \to [a,b] \tag{4}$$

i.e. find the root $x^* \in [a,b]$ that verifies : $x^* = g(x^*)$. The fixed point iteration consists in choosing an initial estimate $x_0 \in [a,b]$, and generating the sequence $\{x_n\}$ recursively, using the relation below :

$$x_n = g(x_{n-1}) \tag{5}$$

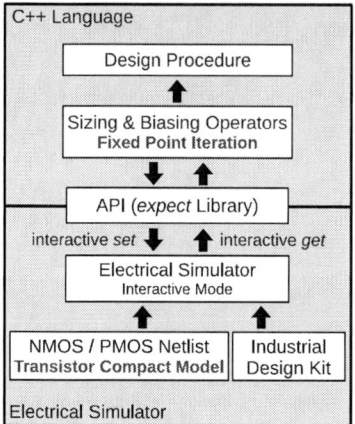

Fig. 1. Simulator encapsulation within sizing and biasing operators.

where n is the current iteration. The recurrent equation (5) is performed until it converges, under some condition, towards $x^* \in [a,b]$. Thus x^* is called a *fixed point* of the iteration defined by the function g. The convergence condition for the fixed point iteration is given by :

$$|g'(x)| < 1, \forall x \in [a,b] \tag{6}$$

The stopping criteria for the fixed point iteration is given as follows :

$$|x_n - x_{n-1}| < \epsilon \tag{7}$$

where ϵ is usually related to x_0.

Another root-finding approach for nonlinear equations is the Newton-Raphson method, used to solve equations in the form of $f(x) = 0$. The Newton-Raphson iteration is defined by :

$$x_n = x_{n-1} - \frac{f(x_{n-1})}{f'(x_{n-1})} \tag{8}$$

As shown in equation (8), the Newton-Raphson iteration requires to compute a derivative f', whereas the fixed point iteration (equation (5)) does not require any derivative computation. Thus, using the fixed point iteration highly simplifies the root-finding algorithm implementation in the sizing and biasing operators, which is an improvement over the Newton-Raphson method. Moreover, according to [20], it is possible to accelerate the fixed point convergence using the *Aitken method*, that consists in computing the $\{x_n, n \geq 2\}$ sequence using :

$$x_{n+1} = \frac{x_{n-2} \cdot x_n - (x_{n-1})^2}{x_n - 2x_{n-1} + x_{n-2}}, n \geq 2 \tag{9}$$

IV. FIXED POINT ITERATION IMPLEMENTATION WITHIN SIZING AND BIASING OPERATORS

In this section we describe the methodology to implement the fixed point iteration within sizing and biasing operators. To compute sizes, we only focus on solving the problem described in equation (1). We need to solve for the width W^* that satisfies a target current I_{D_T} chosen by the designer. W^* is found by solving the following equation :

$$I_D(W) - I_{D_T} = 0 \qquad (10)$$

where $I_D(W)$ is computed from a given transistor compact model. In a first approach, equation (10) was solved using the Newton-Raphson method. However, as stated in section III, the main drawback of this method is the compulsory computation of the derivative $I'_D(W)$. To avoid this, we first transform equation (10) into :

$$\frac{I_D(W)}{I_{D_T}} - 1 = 0 \qquad (11)$$

Then, using the fixed point iteration formulation (see equation (4)), we rewrite equation (11) as :

$$W = \frac{I_D(W)}{I_{D_T}} - 1 + W = g(W) \qquad (12)$$

The fixed point iteration corresponding to equation (12) is given below :

$$W_n = \frac{I_D(W_{n-1})}{I_{D_T}} - 1 + W_{n-1} \qquad (13)$$

We chose the initial value W_0 to be :

$$W_0 = 10 \cdot W_{min} \qquad (14)$$

where W_{min} is the minimum width for a given technology.

Considering the convergence condition stated in equation (6), the derivative of $g(W)$ in equation (12) is given by :

$$g'(W) = \frac{I'_D(W)}{I_{D_T}} + 1 \qquad (15)$$

Assuming that $I_D(W)$ is linear and observing that $I'_D(W) \gg I_{D_T}$, imply that $|g'(W)| > 1$ (see Fig. 2), *i.e.* the convergence condition in equation (6) is not respected. Thus the fixed point iteration written in equation (13) will not converge.

Now, using the method described in [20], it is possible to modify equation (4) to make the corresponding iteration converge. Adding αx on both sides of equation (4) gives :

$$x + \alpha x = g(x) + \alpha x \qquad (16)$$

Equation (16) leads to :

$$x = g_1(x) \qquad (17)$$

where $g_1(x)$ is expressed as follows :

$$g_1(x) = \frac{\alpha x + g(x)}{1 + \alpha}, \ \alpha \neq -1 \qquad (18)$$

Equation (17) becomes the new equation to be solved with the fixed point iteration. The derivative of $g_1(x)$ is given by :

$$g'_1(x) = \frac{\alpha + g'(x)}{1 + \alpha}, \ \alpha \neq -1 \qquad (19)$$

Equation (19) shows that α parameter can be tuned to make convergent the fixed point iteration with g_1 function. A suitable value for α parameter is :

$$\alpha = -g'(x) \qquad (20)$$

Indeed, using equation (20) in equation (19) gives : $g'_1(x) = 0$, what respects the convergence condition in equation (6).

Now, applying equation (16) to our case (equation (12)) gives :

$$W + \alpha W = \frac{I_D(W)}{I_{D_T}} - 1 + W + \alpha W \qquad (21)$$

Rewriting equation (21) leads to :

$$W = \frac{1}{1 + \alpha}\left(\frac{I_D(W)}{I_{D_T}} - 1\right) + W = g_1(W) \qquad (22)$$

The new fixed point iteration, corresponding to equation (22), is given below :

$$W_n = \frac{1}{1 + \alpha}\left(\frac{I_D(W_{n-1})}{I_{D_T}} - 1\right) + W_{n-1} \qquad (23)$$

This iteration is guaranteed to converge when using the suitable α parameter computed with equation (20). Knowing that $g(W)$ is approximately linear, we compute α as the slope of $g(W)$ using :

$$\alpha = -g'(W) \approx -\frac{g(10 \cdot W_{min}) - g(W_{min})}{10 \cdot W_{min} - W_{min}} \qquad (24)$$

Moreover, the Aitken acceleration method (equation (9)) is also implemented to decrease the computation time.

Examples of $g(W)$ (equation (12)) and $g_1(W)$ (equation (22)) are represented in Fig. 2, with the same I_{D_T} and transistor biasing. $f(x) = x$ is the function that intersects $g(W)$ and $g_1(W)$ when $W = W^*$. It can be seen on this figure that $|g'(W)|$ is very high (typical values of $|g'(W)|$ are around $1e5$), whereas $|g'_1(W)| < 1$.

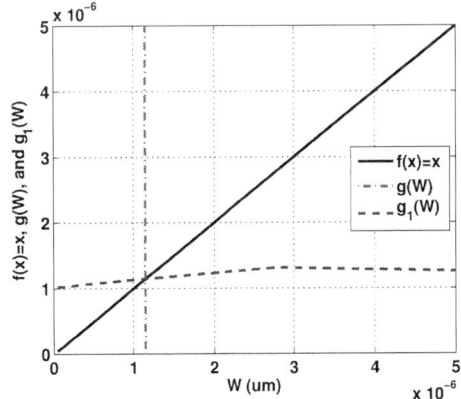

Fig. 2. Representation of $g(W)$ and $g_1(W)$. The solution W^* computed by the fixed point iteration is at the intersection of $f(x) = x$ and $g_1(W)$. Note that $|g'(W)|$ is very high, whereas $|g'_1(W)| < 1$.

V. CASE STUDY

Using the fixed point iteration within sizing and biasing operators, we design the folded cascode OTA represented in Fig. 3. The folded cascode OTA is composed of six devices : D1, D1c, D2, D2c, D3 and D4. C_L is the load capacitance.

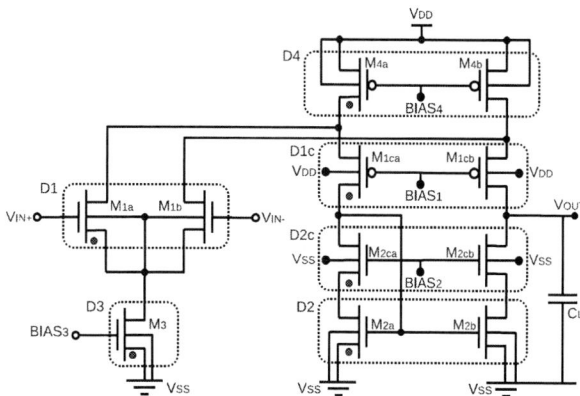

Fig. 3. Folded cascode OTA. Each dashed-box is a device, in which the reference transistor is marked with a dot.

A. Folded Cascode OTA Bipartite Graph

First, the bipartite graph associated to the folded cascode OTA is generated, using the methodology described in [1], [21]. The bipartite graph is shown in Fig. 4. It is the sizing procedure for the circuit. Thus, its evaluation from top to bottom provides transistors sizes and biases. The designer allocates the input parameters P_{in} (at the top of the graph in Fig. 4), then their value (see Table II) is spread in the graph and used by the sizing and biasing operators to compute unknown sizes and biases. Rectangle nodes named "eq" represent designer defined equations, they can be used to define the I_D current of a transistor from the reference biasing current I_{BIAS} of transistor M_3.

B. Folded Cascode OTA Sizing in a 130nm Technology

The folded cascode OTA is sized using a 130nm technology (BSIM3v3 model) and the input parameters P_{in} listed in Table II. The reference current I_{BIAS} is defined as the current of transistor M_3. The bipartite graph evaluation provides the widths that are listed in Table III. Using these computed widths, the folded cascode amplifier is then simulated. Simulation results are shown in Fig. 5 : the DC gain is equal to 60.8dB, phase margin is equal to 75.4° and the transition frequency is 94.4MHz. The load capacitance C_L is set to 0.5pF.

C. Folded Cascode OTA Migration to a 65nm Technology

The folded cascode OTA is then migrated to a 65nm process (BSIM4 model), using the same bipartite graph (i.e. the same design procedure, see Fig. 4). Our goal is to keep the same transition frequency and phase margin. Thus we tune the input parameters P_{in} (Table II) and evaluate again the bipartite graph. I_{BIAS} is increased and current ratios between transistors are modified to ensure a suitable DC gain compared to the 130nm technology. $V_{EG,D3}$ is decreased to ensure the operation of M_3 in saturation region. $V_{EG,D1}$ is set to 0V, thus M_{1a} and M_{1b} transistors are operating near subthreshold in saturation region. C_L is set to 0.6pF. Simulation results are

TABLE II
INPUT PARAMETERS P_{in} FOR THE FOLDED CASCODE OTA.

Parameter	130nm Sizing	65nm Migration
$Temp$(Kelvin)	300.15	300.15
V_{DD}(V)	1.2	1.2
V_{SS}(V)	0.0	0.0
I_{BIAS}(μA)	50	66.6
L_{ref}(μm)	0.5	0.5
$V_{EG,D4}$(V)	-0.12	-0.1
$V_{EG,D1c}$(V)	-0.12	-0.1
$V_{EG,D2c}$(V)	0.12	0.1
$V_{EG,D1}$(V)	0.12	0.0
$V_{EG,D3}$(V)	0.12	0.05
$V_{D,D4}$(V)	0.9	0.95
$V_{D,D2}$(V)	0.3	0.3
$V_{G,D1}$(V)	0.6	0.6
$V_{G,D2}$(V)	0.6	0.6
(eq1$^{(*)}$) $K\_I1 = I_{D\{D2,D2c\}}/I_{BIAS}$	0.5	0.25
(eq2$^{(*)}$) $K\_I2 = I_{D\{D1c\}}/I_{BIAS}$	-0.5	-0.25
(eq3$^{(*)}$) $K\_I3 = I_{D\{D1\}}/I_{BIAS}$	0.5	0.5
(eq4$^{(*)}$) $K\_I4 = I_{D\{D4\}}/I_{BIAS}$	-1	-0.75

$^{(*)}$ See Fig. 4.

TABLE III
COMPUTED WIDTHS FOR THE FOLDED CASCODE OTA.

Device	Computed Width (μm) 130nm Technology	Computed Width (μm) 65nm Technology
D1	4.9	11.6
D1c	17.3	7.4
D2	1.4	1.7
D2c	5.2	2.2
D3	11	17
D4	33.6	22.8

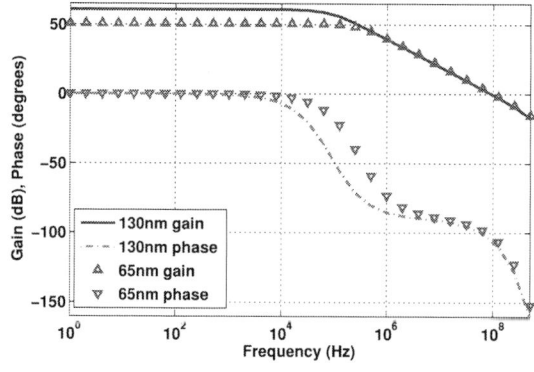

Fig. 5. Simulation results for the folded cascode OTA in 130nm and 65nm technologies.

shown in Fig. 5 : the DC gain is 50.5dB, the phase margin is 76.9°, and the transition frequency is 101MHz. Computed widths are listed in Table III. We succeed to maintain transition frequency and phase margin, but DC gain is lowered. Thus design trade-offs become very important when migrating. The very low evaluation time of the bipartite graph allows to explore design trade-offs in an interactive way.

48

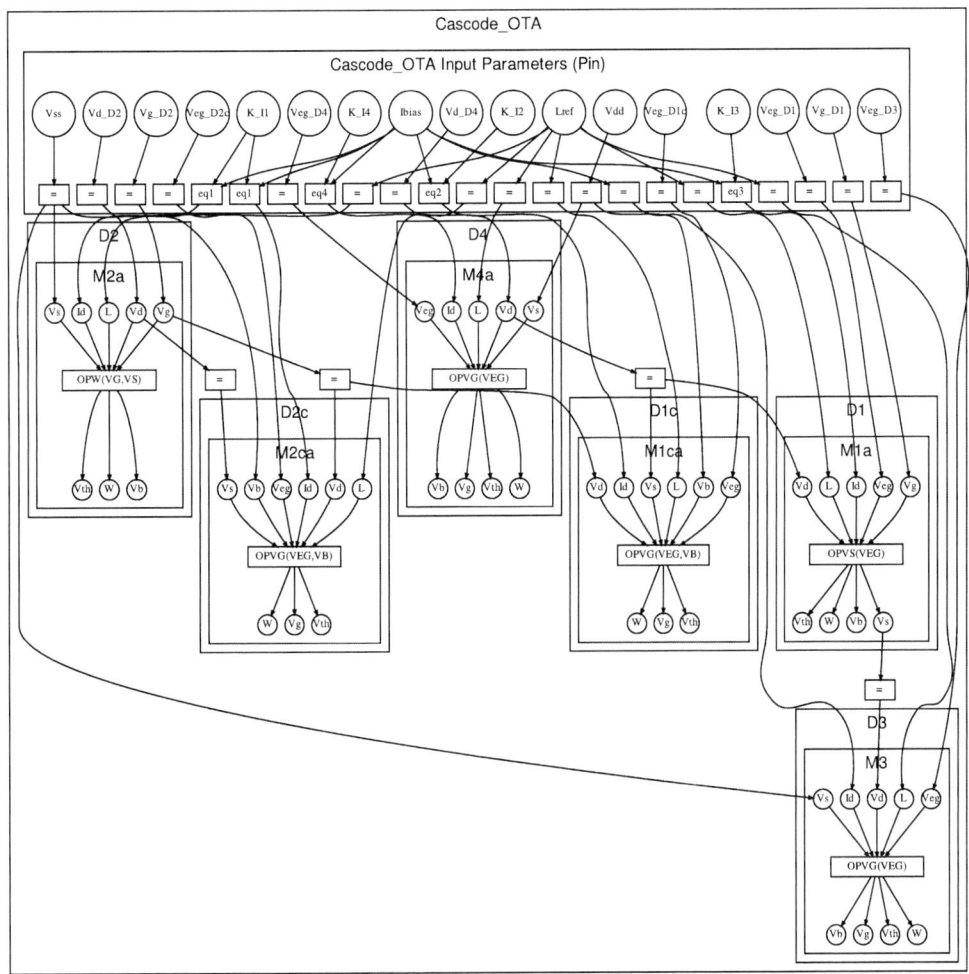

Fig. 4. The bipartite graph (*i.e.* the sizing procedure) associated to the folded cascode OTA. Sizing and biasing operators are part of the bipartite graph. Input parameters P_{in} (see Table II) are on the top of the graph.

D. Fixed Point Iteration Computational Efficiency

Table IV presents the computation time (*i.e.* the evaluation time of the bipartite graph in Fig. 4 from top to bottom) during the sizing and migration of the folded cascode OTA, for the fixed point iteration and the Newton-Raphson method. The fixed point algorithm takes a few more milliseconds compared to the Newton-Raphson method, this is due to the computation of α parameter for each transistor, that requires two more simulations at the fixed point algorithm initialization. The main advantage of the fixed point iteration algorithm is its efficient and simple implementation into the sizing and biasing operators, what enhances its stability. This is very promising for our future work with nanoscale CMOS technologies [22].

TABLE IV
COMPUTATION TIME FOR THE FOLDED CASCODE OTA SIZING AND
MIGRATION : NEWTON-RAPHSON METHOD VS FIXED POINT ITERATION

130nm Technology	
Algorithm	Computation Time (ms)
Newton-Raphson	16
Fixed Point	20
65nm Technology	
Algorithm	Computation Time (ms)
Newton-Raphson	16
Fixed Point	21

VI. CONCLUSION

We presented an efficient analog sizing algorithm that uses transistor compact models within fixed point iteration. The fixed point iteration does not require any derivative computation, thus highly simplifies the root solving procedure implemented into sizing and biasing operators. Moreover, it has been demonstrated that the fixed point iteration is guaranteed to converge, and that convergence can be highly accelerated. Using sizing and biasing operators implementing the fixed point iteration algorithm, a folded cascode OTA was successfully sized in a 130nm process (BSIM3v3 model), and migrated to a 65nm process (BSIM4 model). Sizing and migration were performed using the same bipartite graph (*i.e.* design procedure). Design trade-offs were rapidly explored thanks to the very low evaluation time of the bipartite graph.

REFERENCES

[1] Ramy Iskander, Marie-Minerve Louërat, and Andreas Kaiser. "Hierarchical sizing and biasing of analog firm intellectual properties". *Integration, the VLSI Journal*, 2012. in press, DOI 10.1016/j.vlsi.2012.01.001.

[2] R. Harjani, R. A. Rutenbar, and L. R. Carley. "OASYS: A Framework for Analog Circuit Synthesis". *IEEE Transactions on Computer-Aided Design of Integrated Circuits*, 8(12):1247–1266, December 1989.

[3] M. G. R. DeGrauwe, E. Dijkstra O. Nys, J. Rijmenants, S. Bitz, B. L. A. G. Goffart, E. A. Vittoz, S. Cserveny, C. Meixenberger, G. van der Stappen, and H. J. Oguey. "IDAC: An Interactive Design Tool for Analog CMOS Circuits". *IEEE J. of Solid-State Circuits*, SC-22(6):1106–1116, December 1987.

[4] Han Young Koh, Carlo H. Sequin, and Paul R. Gray. "OPASYN: A Compiler for CMOS Operational Amplifiers". *IEEE Transactions on Computer-Aided Design of Integrated Circuits*, 9(2):113–125, February 1990.

[5] W. Nye, D.C. Riley, A. Sangiovanni-Vincentelli, and A. L. Tits. "DE-LIGHT.SPICE: An Optimization-Based System for Design of Integrated Circuits". *IEEE Transactions on Computer-Aided Design of Integrated Circuits*, 7(4):501–518, April 1988.

[6] E. S. Ochotta, R. A. Rutenbar, and L. R. Carley. "Synthesis of High-Performance Analog Circuits in ASTRX/OBLX". *IEEE Transactions on Computer-Aided Design of Integrated Circuits*, 15(3):273–294, March 1996.

[7] G. Van der Plas, G. Debyser, F. Leyn, K. Lampaert, J. Vandenbussche, G. Gielen, W. Sansen, P. Veselinovic, and D. Leenaerts. "AMGIE-

A synthesis environment for CMOS analog integrated circuits". *IEEE Trans. on Circuits Syst.*, 20(9):1037–1058, September 2001.

[8] M. Krasnicki, R. Phelps, R. A. Rutenbar, and L. R. Carley. "MAEL-STROM: Efficient Simulation-based Synthesis for Custom Analog Cells". *Proc. Design Automation Conf.*, pages 945–950, June 1999.

[9] R. Phelps, M. Krasnicki, R. A. Rutenbar, L. R. Carley, and J. R. Hellums. "ANACONDA: Robust Synthesis of Analog Circuits Via Stochastic Pattern Search". *IEEE Transactions on Computer-Aided Design of Integrated Circuits*, pages 567–570, 2000.

[10] Jacky Porte. "*COMDIAC: Compilateur de Dispositifs Actifs*". Ecole Nationale Supérieure des Télécommunications, Paris, September 1997.

[11] D. Stefanovic, M. Kayal, and M. Pastre. "PAD: A new interactive Knowledge-Based Analog Design Approach". pages 291–299, March 2005.

[12] Danica Stefanovic and Maher Kayal. "*Structured Analog CMOS Design*". Kluwer Academic Publishers, 2009.

[13] D. M. Binkley, C. E. Hopper, S. D. Tucker, B. C. Moss, J. M. Rochelle, and D. P. Foty. "A CAD Methodology for Optimizing Transistor Current and Sizing in Analog CMOS Design". *IEEE Transactions on Computer-Aided Design of Integrated Circuits*, 22(2):225–237, February 2003.

[14] F. Javid, R. Iskander, and M-M. Louërat. "Simulation-Based Hierarchical Sizing and Biasing of Analog Firm IPs". *IEEE International Behavioral Modeling and Simulation Conference*, pages 43–48, September 2009.

[15] William Liu. "*MOSFET Models for SPICE Simulation Including BSIM3v3 and BSIM4*". Wiley-Interscience, 2001.

[16] NXP, MOS Model PSP level 103, 2011.

[17] Christian Enz, François Krummenacher, and Eric Vittoz. "An Analytical MOS Transistor Model Valid in All Regions of Operation and Dedicated to Low-Voltage and Low-Current Applications". *Analog Integrated Circuits and Signal Processing Journal*, Vol. 8(No. 1):83–114, July 1995.

[18] D. Libes. EXPECT. 2010.

[19] P. Henrici. "*Elements of Numerical Analysis*". Wiley, John Sons, 1964.

[20] J. Legras. "*Méthodes et techniques de l'analyse numérique*". Dunod, 1971.

[21] F. Javid, R. Iskander, M-M. Louërat, and D. Dupuis. "Analog Circuits Sizing Using Bipartite Graphs". *IEEE Midwest Symposium on Circuits and Systems*, August 2011.

[22] L. Lewyn, T. Ytterdal, C. Wulff, and K. Martin. "Analog Circuit Design in Nanoscale CMOS Technologies". *Proceedings of the IEEE*, Vol. 97(No. 10):1687–1714, October 2009.

Analytical Drain Current Core Model for Undoped Double Gate MOS Transistor

Paweł Sałek, Lidia Łukasiak, Andrzej Jakubowski

Institute of Microelectronics and Optoelectronics
Warsaw University of Technology
ul. Koszykowa 75, 00-662 Warszawa, Poland

Abstract—A new analytical drain current model is proposed for symmetrical undoped double gate MOS transistors. It contains no fitting parameters and is physically-based.

Index Terms—double gate MOS, current model, potential

I. INTRODUCTION

Modern electronic devices are going to be smaller to meet the increasing market requirements for performance and price of integrated circuits. With decreasing transistor dimensions statistical fluctuations of doping concentration become more significant [1]. Moreover, aggressive scaling affects the rise of second-order effects, such as short channel effects (SCE) or drain induced barrier leakage (DIBL). To make extremely small devices working it is necessary to use the double gate structure. The classical MOS model is not suitable for such devices. This justifies the need for a new core model for undoped double gate MOS transistors (UDGMOS). The new model should be continuous and should cover all the operating states of the transistor, since the concept of inversion loses its meaning in this case. The model should permit direct calculation of the drain current for any terminal voltages. In this paper we propose such core model. It is physically-based and fits numerical simulation with good accuracy.

II. MODEL

A. Defining the problem

The cross-section of a symmetrical DGMOS is shown schematically in Fig. 1. The approximate 1D Poisson equation neglecting the presence of holes in an n-channel device is described as [2]:

$$\frac{d^2\phi}{dx^2} = \frac{qn_i}{\varepsilon_s} \cdot \exp\left(\frac{\phi - V_b}{v_t}\right) \qquad (1)$$

Where ϕ is electrostatic potential, x - position in the cross-section of the channel, V_b is channel voltage and other symbols have their usual meaning.

Implicit solution to the Poisson equation was proposed by Taur [3]:

$$\phi(x) = \phi_0 - 2v_t \cdot \ln\left\{\cos\left[\exp\left(\frac{\phi_0 - V_b}{2v_t}\right) \cdot x \cdot \sqrt{\frac{q^2 n_i}{2kT\varepsilon_s}}\right]\right\} \qquad (2)$$

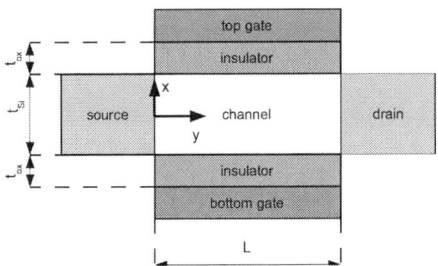

Figure 1.   Simplified cross-section of a symmetrical double-gate MOSFET

Equation (2) has to be solved iteratively together with the boundary condition [3]:

$$\varepsilon_s \cdot \frac{d\phi}{dx}\bigg|_{x=\frac{t_{Si}}{2}} = \varepsilon_{ox} \cdot \frac{V_G - \phi\left(\frac{t_{Si}}{2}\right)}{t_{ox}} \qquad (3)$$

Numerical solution of the Poisson equation yields better accuracy [4][5], however is time-consuming. Several models describing only one region of operation, i.e. sub-threshold or over-threshold region were proposed [3][6], but the final formula is more complicated due to the use of several functions. Models using smoothing functions [7][8] have a similar disadvantage.

B. Modeling the potential

It is difficult to propose any physically-based approximation of (2). For this reason we analyze the electric field across the channel at y=L/2, which is presented in Fig. 2.

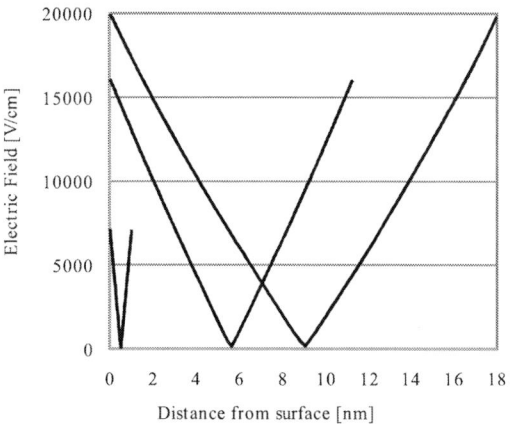

Figure 2. Electric field distribution across the channel for tSi=1nm, 11nm and 18nm

Since electric field distribution shown in Fig. 2 is almost linear across the channel, one can assume that the potential is a parabolic function of x. Combining this phenomenon with (2) and (1) solved in the middle of the channel (x=0) one obtains:

$$
\phi(x, V_G) =
$$

$$
v_t \cdot \left(\frac{x^2}{t_{Si}^2 + \frac{\varepsilon_s}{\varepsilon_{ox}} t_{ox} t_{Si}} - 1 \right) \cdot W \left[\frac{t_{Si}^2 + \frac{\varepsilon_s}{\varepsilon_{ox}} t_{ox} t_{Si}}{v_t} \cdot \frac{q n_i}{2 \varepsilon_s} \cdot \exp \left(\frac{V_G - V_b}{v_t} \right) \right] + V_G \quad (4)
$$

where W stands for the principal branch of the Lambert W-function [9]. The accuracy of this model is presented in Fig. 3. Apart from its analytical formula the new model shows good accuracy when compared with ATLAS [10] numerical simulations.

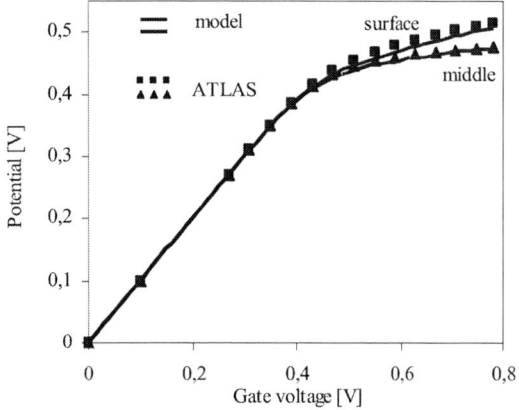

Figure 3. Comparison of potential model and ATLAS simulations for t$_{Si}$=11nm, t$_{ox}$=5nm

C. Drain current

Drain current dependence on the potential distribution in the channel has been already proposed and verified by several authors [11-16]:

$$
I_D =
$$

$$
\mu \frac{W}{L} \cdot \left\{ 2 C_{ox} \left[V_G (\phi_{sL} - \phi_{s0}) - \frac{1}{2} (\phi_{sL}^2 - \phi_{s0}^2) \right] + \right. \quad (5)
$$

$$
\left. + 4 v_t C_{ox} (\phi_{sL} - \phi_{s0}) - t_{Si} k T n_i \left[\exp \left(\frac{(\phi_{sL} - V_{DS})}{v_t} \right) - \exp \left(\frac{\phi_{s0}}{v_t} \right) \right] \right\}
$$

where ϕ_{SL}, ϕ_{S0}, ϕ_{0L} and ϕ_{00} are surface potentials calculated at (x,y) = (tSi/2, L), (tSi/2, 0), (0, L) and (0, 0), respectively. Other symbols have their usual meaning.

III. RESULTS

Equations (5) and (4) form an analytical drain current model. A comparison with the results of numerical simulations is shown in Fig. 4. The worst accuracy is observed around the threshold voltage, where the relative error may reach 150 percent.

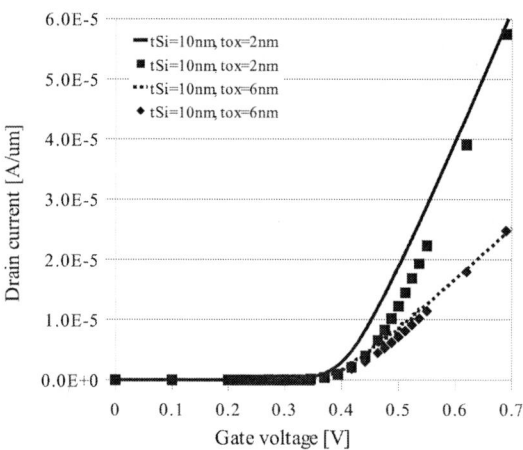

Figure 4. Comparison of the new drain current model (lines) and ATLAS simulations (symbols)

If better accuracy is required, our model may be improved with the third order correction method proposed by Yu et al. [17]. In our case the procedure consists of the following steps:

- Use (4) to calculate the potential ϕ_0 in the middle of the channel (x=0) as a function of gate voltage:

$$
\phi_0 = V_G - v_t \cdot W \left[\frac{t_{Si}^2 + \frac{\varepsilon_s}{\varepsilon_{ox}} t_{ox} t_{Si}}{v_t} \cdot \frac{q n_i}{2 \varepsilon_s} \cdot \exp \left(\frac{V_G - V_b}{v_t} \right) \right] \quad (6)
$$

- Using (6) and the definition of z variable [17]:

$$z = \tan\left[\frac{\exp\left(\dfrac{\phi_0 - V_b}{v_t}\right)}{L_D} \cdot \frac{t_{Si}}{2}\right] \quad (7)$$

calculate z_1 (the first approximation of z):

$$z_1 = $$
$$= \tan\left\{\frac{t_{Si}}{2L_D}\exp\left[\frac{1}{2v_t}\cdot\left(V_G - v_t \cdot W\left(\frac{t_{Si}^2 + \dfrac{\varepsilon_s}{\varepsilon_{ox}}t_{ox}t_{Si}}{v_t} \cdot \frac{qn_i}{2\varepsilon_s} \cdot \exp\left(\frac{V_G - V_b}{v_t}\right)\right) - V_b\right)\right]\right\} \quad (8)$$

where L_D is intrinsic Debye length.

- Calculate the correction terms [17] to obtain z_2 (the second approximation of z):

$$\gamma_2 = \arctan(z_1) \quad (9)$$

$$\delta_2 = \frac{1}{1 + z_1^2} \quad (10)$$

$$F = \frac{V_G - V_b}{2v_t} - \ln\left(\frac{2L_D}{t_{Si}}\right) \quad (11)$$

$$\eta_2 = \frac{1}{2}\cdot\ln\left(\frac{\gamma_2^2}{\delta_2}\right) + r\cdot z_1 \cdot \gamma_2 - F \quad (12)$$

$$\lambda_2 = \frac{\delta_2}{\gamma_2} + (1 + r)\cdot z_1 \cdot \delta_2 + r\cdot\gamma_2 \quad (13)$$

$$\vartheta_2 = -\frac{\delta_2^2}{\gamma_2^2} - \frac{2z_1\delta_2^2}{\gamma_2} + \left(1 + 2r - z_1^2\right)\delta_2^2 \quad (14)$$

$$\nu_2 = \frac{2\delta_2^2}{\gamma_2^3} + \frac{6z_1\delta_2^3}{\gamma_2^2} + \left(6z_1^2 - 2\right)\frac{\delta_2^3}{\gamma_2} + \left(2z_1^2 - 6 - 8r\right)z_1\delta_2^3 \quad (15)$$

$$z_2 = z_1 - \frac{\eta_0}{\eta_1}\left(1 + \frac{\eta_0\eta_2}{2\eta_1^2} + \frac{\eta_0^2\left(3\eta_2^2 - \eta_1\eta_3\right)}{6\eta_1^4}\right) \quad (16)$$

- In (9-16) replace z_1 with z_2, η with λ and δ with γ and calculate z_3 (the third approximation of z) [17]:

$$z_3 = z_2 - \frac{\lambda_0}{\lambda_1}\left(1 + \frac{\lambda_0\lambda_2}{2\lambda_1^2} + \frac{\lambda_0^2\left(3\lambda_2^2 - \lambda_1\lambda_3\right)}{6\lambda_1^4}\right) \quad (17)$$

- Calculate the corrected value of the potential [17]:

$$\beta = \arctan(z_3) \quad (18)$$

$$\phi(x) = V_b - \frac{2kT}{q}\ln\left(\frac{t_{Si}}{2L_D\beta}\cos\left(\frac{2\beta x}{t_{Si}}\right)\right) \quad (19)$$

The corrected model is compared with numerical simulations in Figs. 5 – 6 for different body and gate dielectric thicknesses. The relative error is reduced more than 3 times in comparison to the non-corrected model.

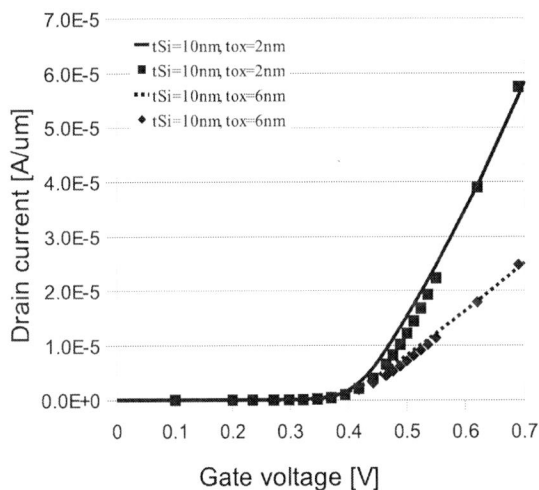

Figure 5. Comparison of the corrected current model (lines) and ATLAS simulations (symbols) for constant t_{Si}. L=100nm, V_{DS}=10mV.

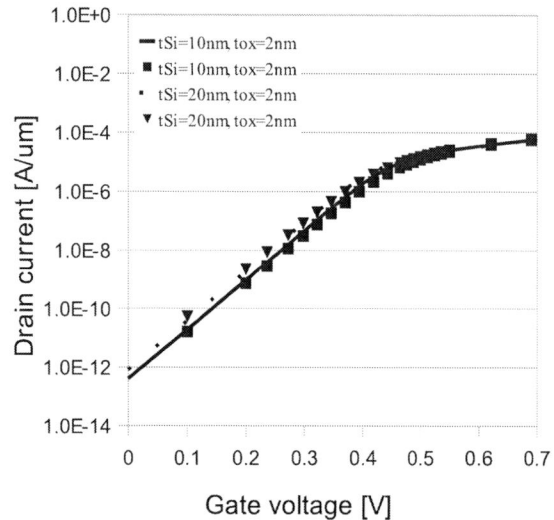

Figure 6. Comparison of the corrected current model (lines) and ATLAS simulations (symbols) for constant tox. L=100nm, V_{DS}=10mV.

IV. SUMMARY

A new analytical, physically based core model for undoped double gate MOS transistors was proposed. It has no fitting parameters and shows good agreement with the values extracted from the simulated characteristics in the range of device parameters (channel and gate-oxide thickness) interesting from the point of view of aggressive miniaturization.

REFERENCES

[1] F. J. Garcıa Sanchez , A. Ortiz-Conde, J. Muci, "Understanding threshold voltage in undoped-body MOSFETs: An appraisal of various criteria," Microelectronics Reliability, vol. 46, pp. 731–42, 2006

[2] Yuan Taur, "Analytic Solutions of Charge and Capacitance in Symmetric and Asymmetric Double-Gate MOSFETs", IEEE Transactions on Electron Device Letters, vol.48, pp. 245-247, 2001

[3] Y. Taur, "An Analytical Solution to a Double-Gate MOSFET with Undoped Body", IEEE Electron Dev. Lett., vol. 21, pp. 245-7, 2000

[4] C. Sheng, W. Gaofeng, H. Qijun, W. Hao, "A Novel Measure on Inversion Degree of Undoped Symmetric Double-Gate MOSFETs", Chinese Journal of Electronics, vol.19, 2010

[5] S. Malobabic, A. Ortiz-Conde, F. J. J. García Sánchez, "Modeling the Undoped-Body Symmetric Dual-Gate MOSFET", Proceedings of the Fifth IEEE International Caracas Conference on Devices, Circuits and Systems, vol.1, pp. 19-25, 2004

[6] V. Hariharan, R. Thakker, K. Singh, A. B. Sachid, M. B. Patil, "Drain current model for nanoscale double-gate MOSFETs", Solid-State Electronics, vol.53, pp. 1001–1008, 2009

[7] J. He, W. Bian, Y. Tao, F. Liu, K. Lu, W. Wu, T. Wang, M. Chan, "An explicit current–voltage model for undoped double-gate MOSFETs based on accurate yet analytic approximation to the carrier concentration", Solid-State Electronics, vol.51, pp. 179–185, 2007

[8] F. Lime, et al., "A physical compact DC drain current model for long-channel undoped ultra-thin body (UTB) SOI and asymmetric double-gate (DG) MOSFETs with independent gate operation", Solid-State Electronics, vol.57, pp. 61–66, 2011

[9] R. M. Corless, G. H. Gonnet, D. E. G. Hare, D. J. Jeffrey and D. E. Knuth, "On the Lambert W function", Advances in Computational Mathematics, vol.5, pp. 329-359, 1996

[10] ATLAS User's Manual, SIVACO International, 2005

[11] A. Ortiz-Conde, F. J. García Sánchez, "Unification of asymmetric DG, symmetric DG and bulk undoped-body MOSFET drain current", Solid State Electronics, pp. 1796-1800, 2006

[12] A. Ortiz-Conde, F. J. García Sánchez, S. Malobabic, J. Muci, and R. Salazar, "Drain Current and Transconductance Model for the Undoped Body Asymmetric Double-Gate MOSFET", ICSICT '06. 8th International Conference on Solid-State and Integrated Circuit Technology, pp. 1239 – 1242, 2006

[13] A. Ortiz-Conde, F. J. García-Sánchez, J. Muci, "Rigorous analytic solution for the drain current of undoped symmetric dual-gate MOSFETs", Solid-State Electronics, vol. 49, pp. 640–647, 2005

[14] A. Ortiz-Conde, F. J. García-Sánchez, J. Muci, "A Unified View of Drain Current Models for Undoped Double-Gate SOI MOSFETs", NSTI-Nanotech, vol.3, pp. 526-531, 2007

[15] A. Ortiz-Conde, F. J. García-Sánchez, J. Muci, S. Malobabic, J. J. Liou, , "A Review of Core Compact Models for Undoped Double-Gate SOI MOSFETs", IEEE Transactions on Electron Devices, vol.54, pp. 131-140, 2007

[16] R. Yadav, R. P. Prakash, S. C. Bose, "Modified Surface Potential Based Current Modeling of Thin Silicon Channel Double Gate SOI FinFETs", 2nd International Workshop on Electron Devices and Semiconductor Technology, 2009

[17] B. Yu, H. Lu, M. Liu, Y. Taur, "Explicit Continuous Models for Double-Gate and Surrounding-Gate MOSFETs", IEEE Transactions on Electron Devices, vol. 54, pp. 2715-2722, 2007

MIXDES 2012, 19th International Conference *"Mixed Design of Integrated Circuits and Systems"*, May 24-26, 2012, Warsaw, Poland

Analytical Model for Predicting Subthreshold Slope Improvement versus Negative Swing of S-shape Polarization in a Ferroelectric FET

Alexandru Rusu, Adrian M. Ionescu

Ecole Polytechnique Fédérale de Lausanne, Switzerland

{alexandru.rusu, adrian.ionescu} @ epfl.ch

Abstract—**In this paper we present and validate a model for predicting the inverse subthreshold slope, SS, improvement in a ferroelectric FET as a function of the negative swing (Landau's α parameter), α=dP/dE of the polarization under negative capacitance operation. The model is experimentally validated at room temperature based on Fe-FET using a P(VDF-TrFE) copolymer as dielectric and showing sub-thermal swings over more than two orders of magnitude.**

Index Terms—**Negative capacitance, Fe-FET, ferroelectric, surface potential amplification, PVDF**

I. INTRODUCTION

The ferroelectric transistor (Fe-FET) is under intense exploration today due to the negative capacitance effect [1] that was recently demonstrated experimentally [2]. Recently, subthreshold swings lower than 60mV/dec on several decades of current on relatively large transistors have been reported together with an S-shape of the polarization of the ferroelectric gate stack. However, there is no existing design rule or analytical prediction that related the value of the subthreshold slope to the negative value of the capacitance (or swing of the S-shape polarization). In this paper we report the first analytical model that can predict the subthreshold slope improvement of the Fe-FET compared to conventional CMOS transistor. P(VDF-TrFE) copolymer was used as a ferroelectric due to the low leakage current [3], good reliability and CMOS compatibility [4].

II. DESIGN AND FABRICATION OF FE-FET WITH NEGATIVE CAPACITANCE

The experimental device consists in a conventional p-MOS transistor which has integrated in the gate stack a ferroelectric ultra-thin film (60nm). The utilized ferroelectric was P(VDF-TrFE) (70%-30%) copolymer, that was spincoated and baked for 5 minutes at 137°C. A particular attention was paid to the isolation of the device by using an optimized n-well (N_D=3.3×10^{17}cm^{-3}) and STI isolation, which minimize all sources of leakage. Different size devices were fabricated with W and L varying from 3μm to 200μm.

For modeling and characterization purposes we have introduced an internal contact between the ferroelectric layer and the oxide[2], that can be probed (see Figure 1), in order to

measure the potential at that point and to compare the entire stack transistor with the conventional p-MOS transistor.

By integrating the internal metal, the static characteristics of the transistor will not change, considering the entire metal equipotential.

The ferroelectric capacitance can be defined as:

$$C_{Fe} = \frac{dP}{dV_{ferro}} = \frac{dP}{dV_{int}} \frac{G}{1 - G} \quad (1)$$

where P is the polarization, V_{ferro} is the voltage applied on the ferroelectric, V_{int} is the voltage between the PVDF and the SiO$_2$ and G is the gain of the V_{int} with respect to V_g ($G=dV_{int}/dV_g$). In case of negative capacitance, the gain is greater than one.

The cross-section of the test device and the equivalent capacitive divider schematics is presented in Figure 1.

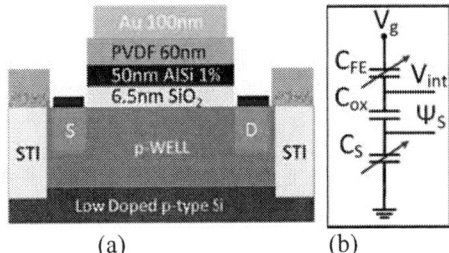

(a)   (b)

Figure 1. (a) Cross section of the Ferroelectric FET, (b) equivalent capacitive divider.

III. MEASUREMENT SETUP

Based on the internal potential we could make a comparison of the p-MOS transistor and the Fe-FET transistor presenting the same size and same parameters. Quasi static CV (QSCV) and IV measurements were performed on the device using both gates in order to compare the Fe-FET transistor and the equivalent MOS transistor. The p-MOS transistor was characterized by applying signal to the internal contact and leaving the top gate floating. The full stack transistor was first measured leaving the internal contact floating and performing CV measurements of the gate and afterwards the internal metal was contacted and a 0A current was injected by the SMU into it in order to measure the potential at that point while sweeping a voltage on the top gate [2]. The internal potential (V_{int}) is dependent on the polarization of the ferroelectric implying that it is history dependent.

55

IV. NEGATIVE CAPACITANCE AND S-SHAPE POLARIZATION

In case of negative capacitance effect, the variation of the V_{int} potential will be higher than the variation of the gate voltage V_g, determining the drain current of the Fe-FET to vary faster than the drain current of the equivalent MOS transistor. This type of variation was measured and is reported in Figure 2.

Due to the thin oxide, the voltage applied on the gate could not have such high values necessary to switch completely the ferroelectric consequently the hysteretic loops were not saturated.

The internal voltage V_{int} presents a hysteresis shape therefore the negative capacitance effect could be observed on both branches. By applying the conservation of the electric displacement [5] at the interface of the SiO_2 and the PVDF the polarization could be extracted and is reported in Figure 3.

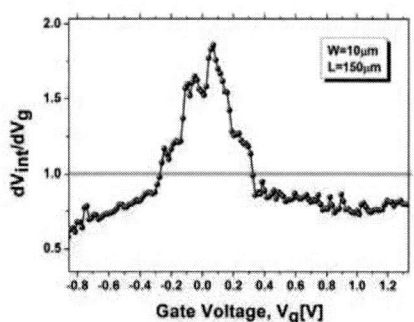

Figure 2. Experimental internal voltage gain, dV_{int}/dV_g versus gate voltage in our fabricated Fe-FET. From eq. (1), G>1 is obtained for a negative capacitance.

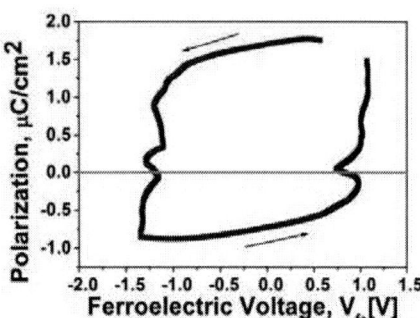

Figure 3. Extracted polarization based on measurements. It can be observed the presence of the negative capacitance in area close to 0 polarization.

The negative capacitance zones can be observed on both branches close to the zone with P=0, which is consistent with the theory.

In order to calculate the polarization, the surface potential of the silicon was calculated by fitting the internal gate capacitance using the Tsividis model [6]. In Figure 4 we

present the measured capacitance and the extracted surface potential.

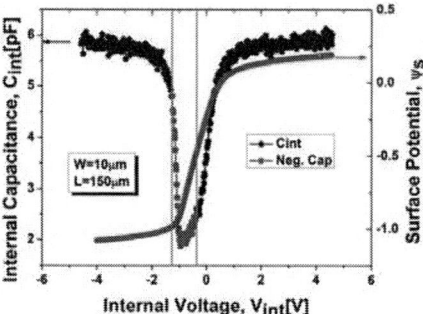

Figure 4. QSCV measurement of the internal gate capacitance with source and drain grounded. The highlighted zone corresponds to the negative capacitance region.

In the capacitance measurement a zone is highlighted that corresponds to the negative capacitance region. One can observe that this region corresponds to the weak-inversion operation of the MOS transistor. The presence of the voltage amplification in weak inversion presents a major advantage that off-on switching will be assisted by the ferroelectric voltage amplification. The drain current measurement is presented in Figure 5.

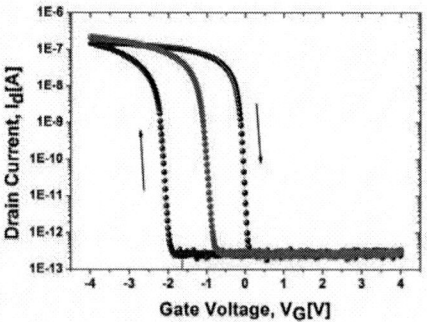

Figure 5. Drain current of the Fe-FET with the gate voltage applied on the internal contact subsequently to the top gate contact.

The blue curve represents the drain current of the reference MOS transistor that is obtained by applying the sweeping voltage on the internal contact. The black curve represents the drain current of the full stack ferroelectric transistor. This way we can make a direct comparison between the MOS transistor with and without the influence of the ferroelectric layer. The hysteresis can be observed also in the drain current. The polarization loop is relatively symmetric with respect to P=0, which leads to a symmetric drain current with respect to the reference MOS drain current. The threshold voltage V_T is changing according to the value of the polarization. In Figure 6 there are compared the drain current curves of the transistor with the signal applied on the internal contact and the signal applied on the top gate, taking in account the descending branch.

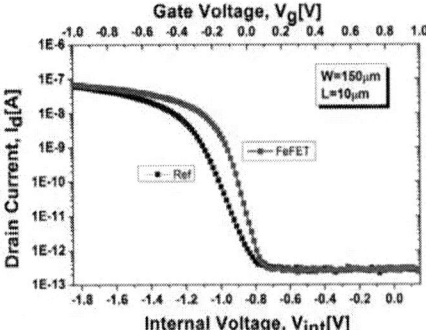

Figure 6. Comparison of Id-Vg of a Fe-FET and a reference MOSFET with W=150µm and L=10µm. Significant improvement of the subthreshold swing can be observed due to the negative capacitance effect.

The subthreshold swing is given by the variation of the surface potential with respect to the gate voltage, in this manner a variation higher than one will give a subthermal characteristic. The extracted surface potential [7] is presented in Figure 7.

One can clearly observe the improvement of the variation of the surface potential in the subthreshold region.

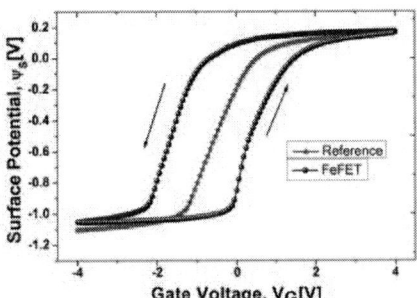

Figure 7. Surface potential extraction.

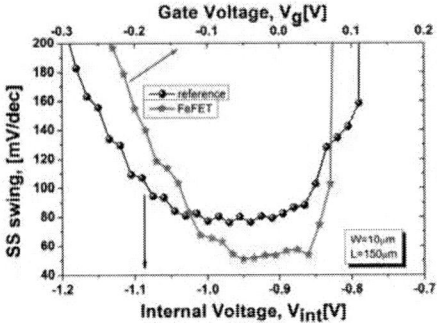

Figure 8. Subthreshold swings of the reference transistor and of the full stack Fe-FET.

The presence of the ferroelectric dielectric determines a more abrupt of the transistor, therefore the lowering of the subthreshold swing. In Figure 8 we report the comparison of

the swings of the Fe-FET and reference MOSFET on the descending branch of the hysteretic characteristics.

The minimum value of the subthreshold swing of the Fe-FET is 50.86mV/dec and the one of the reference transistor is about 76mV/dec. On the ascending branch of the I_d current of the Fe-FET the minimum SS swing has the value 63.66 mV/dec.

The distinct values of the SS swing of the two branches can be explained by the asymmetry of the polarization.

V. ANALYTICAL MODEL: CURRENT SWING VERSUS NEGATIVE POLARIZATION

Starting from the assumption that the negative capacitance effect is obtained in the weak inversion region of operation, we derive a simple analytical model that can predict the improvement of the SS swing. It is known that the negative slope of the polarization with respect to the field applied on the ferroelectric corresponds to the α parameter of the Landau theory for ferroelectrics [8]:

$$E = \alpha(T - T_C)P + \beta P^3 \qquad (2)$$

where E is the electric field, T_c is the Curie temperature, T is the temperature, α and β are material parameters.

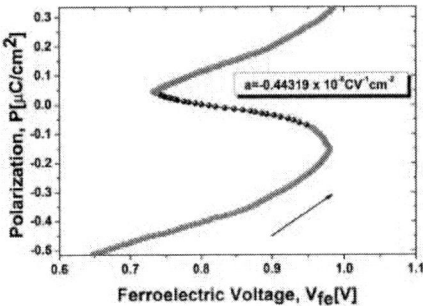

Figure 9. Data points considered for the calculation of the a-parameter correspond to the negative swing of the polarization, dP/dV_{Fe} (=CFe) <0.

We have considered the descending branch of the current, which corresponds to the ascending part of the polarization and we calculated the slope. The polarization in the part of interest is presented in Figure 9.

The Landau theory states that the polarization negative swing, dP/dV_{Fe}, is constant and the curve is passing thru 0; therefore we can approximate the curve in the negative region with a linear function [8]:

$$P = a * V_{Fe} = \alpha * \frac{V_{Fe}}{t_{Fe}} \qquad (3)$$

where α is the Landau parameter and t_{Fe} is the thickness of the oxide layer. For the simplicity of the calculation we will use in the following equations the variable a ($=\alpha/t_{Fe}$). From linear fitting on the experimental data, we extract the value $a=-0.43\times10^{-6}CV^{-1}cm^{-2}$ approximately.

Based on the measurements we observed that the negative capacitance region is present mainly in the subthreshold region. Therefore we will make the calculation of the surface potential only in that region and we can approximate it with a linear function[6]:

$$\psi_S = b * V_{int} + c \qquad (4)$$

57

where b and c are parameters that can be express in function of the device parameters. The b parameter represents the variation of the surface potential with the internal voltage and the reverse of b is called n_0 and can be calculated from the following equation[6]:

$$(n_0 - 1)C_{ox} = \frac{\sqrt{2q\varepsilon_S N_D}}{2\sqrt{2\phi_F}} \qquad (5)$$

where C_{ox} is the oxide capacitance, q is the elementary charge, ε_S is the permittivity of the silicon, N_D is the substrate doping concentration and ϕ_F is the Fermi potential. The c parameter is not of direct interest since it will be demonstrated that the SS improvement does not depend on it.

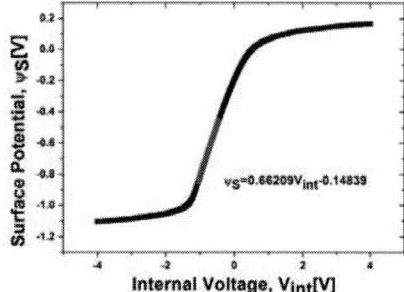

Figure 10. Data set of points from the surface potential where the negative capacitance was present.

In Figure 10 there is presented the surface potential with respect of the internal voltage and it is highlighted the area where the negative capacitance is present.

The b and c parameters were extracted directly from the measurements having the values: b=0.66 and c=-0.14.

Next we apply the conservation of the electric displacement at the junction between the ferroelectric and the oxide[5]:

$$D_{Fe} = D_{ox}$$
$$\varepsilon_0 E_{Fe} + P = \varepsilon_0 \varepsilon_{ox} E_{ox} \qquad (6)$$

Utilizing the two approximations we can extract the surface potential with respect to the gate voltage:

$$\psi_S = \frac{V_G\left(-a - \frac{\varepsilon_0}{t_{Fe}}\right) - \frac{ca}{b} - \frac{c \times \varepsilon_0 \varepsilon_r}{b \times t_{ox}} + \frac{\varepsilon_0 \times c}{b \times t_{Fe}}}{\frac{(1-b) \times \varepsilon_0 \varepsilon_r}{b \times t_{ox}} + \frac{a}{b} + \frac{\varepsilon_0}{b \times t_{Fe}}} \qquad (7)$$

We have obtained a linear equation therefore the derivative of the surface potential with respect to the gate voltage will have the following value:

$$\frac{d\psi_S}{dV_G} = \frac{\left(-a - \frac{\varepsilon_0}{t_{Fe}}\right)}{\frac{(1-b) \times \varepsilon_0 \varepsilon_r}{b \times t_{ox}} + \frac{a}{b} + \frac{\varepsilon_0}{b \times t_{Fe}}} \qquad (8)$$

The slope is dependent of the a(α) parameter, specific to the ferroelectric layer. If $1 > \frac{d\psi_s}{dVg} > b$ then the subthreshold slope of the transistor will be improved but it will not overcome the 60mV/dec limit. If $\frac{d\psi_s}{dVg} > 1$ then the subthreshold slope will have values lower than the sub thermal

barrier imposed by the silicon. From the slope of the surface potential we can extract the subthreshold swing:

$$SS = \frac{dV_{GB}}{dlog_{10}I_D} = ln10 \cdot \frac{d\psi_S}{dlnI_D} \cdot \frac{dV_{GB}}{d\psi_S} \qquad (9)$$
$$\cong ln10 \cdot \frac{kT}{q} \cdot \frac{dV_{GB}}{d\psi_S}$$

In our case, the calculated subthreshold slope is 50.5mV/dec and the measured one is 50.86mV/dec.

Figure 11 reports the predicted and measured inverse subthreshold slope for different measured transistors, showing a good agreement between the predicted values based on our model (simulated) and the measured ones. The a and b parameters were calculated for each transistor.

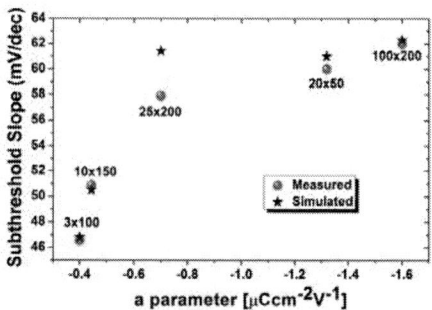

Figure 11. Comparison between model-based prediction and measured values of subthreshold swing for various Fe-FET transistors (the channel size W[µm]xL[µm] is illustrated next to data each point).

The prediction errors can be due to the linear approximation of the negative swing region of the $P(V_{Fe})$ and also of the assumption that a is independent of the transistor dimensions. Another assumption was that negative capacitance appears only in the weak inversion and a linear approximation of the surface potential is suitable.

In Figure 12 we present the analytical dependence of the SS on the a-parameter, as resulting from our model.

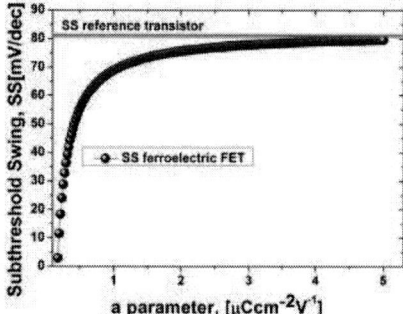

Figure 12. Dependence of the subthreshold swing, SS, of a Fe-FET on the a parameter.

A reference MOSFET with the SS of 82mV/dec, at room temperature, is considered and the a-parameter varies in the range of 0.15 to $5\mu CV^{-1}cm^{-2}$. To our best knowledge this is the first quantitative prediction related the subthreshold swing and the Landau's parameters (reflected in our a-parameter).

58

VI. CONCLUSIONS

A novel model that can predict the subthreshold swing of the ferroelectric transistor as a function of the negative polarization swing, dP/dV_{Fe}, was developed and experimentally validated. The model is based on the assumption that the negative capacitance effect is exploited in the weak-inversion regime and the polarization varies quasi-linearly with the voltage on the ferroelectrics.

REFERENCES

[1] M. Ershov et al, "Negative Capacitance Effect in Semiconductor Devices," IEEE TED, Vol. 45, No 10, pp. 2196-2206, Oct. 1998

[2] A. Rusu, G. A. Salvatore, D. Jiménez, and A. M. Ionescu, " Metal-Ferroelectric-Metal-Oxide-Semiconductor Field Effect Transistor with Sub-60mV/decade Subthreshold Swing and Internal Voltage Amplification," IEDM 2010, pp. 16.3.1-16.3.4, December 2010.

[3] S. Horiuchi, "Organic ferroelectrics", Nat. Mater., vol. 7, pp. 357–366, 2008.

[4] H.Ishiwara, "Current status of ferroelectric-gate Si transistors and challenge to ferroelectric-gate CNT transistors", Current Appl. Phys., Vol.9, pp.S2-S6, 2009,

[5] Al. Rusu et al, " A study of polarization effects in metal-ferroelectric-oxide-semiconductor capacitors",CAS 2009, pp. 517-520, 2009

[6] Y. Tsividis, Operation and Modeling of the MOS Transistor, 2nd Edition, McGraw Hill, New York, 1999.

[7] R. Langevelde et al, " An explicit surface-potential-based MOSFET model for circuit simulation," SSE, V. 44, pp. 409-418, 2000

[8] M. IWATA et al," Landau Theory of Phase Transition in Ferroelectric Vinylidene Fluoride/Trifluoroethylene Copolymer Single Crystals: Hysteresis Loop," JJAP, Vol. 44, No. 9A, p 6667-6669, 2005

MIXDES 2012, 19th International Conference *"Mixed Design of Integrated Circuits and Systems"*, May 24-26, 2012, Warsaw, Poland

Characterization of Parameter Variability and Correlations for FD SOI CMOS Technology

Daniel Tomaszewski, Grzegorz Głuszko, Jolanta Malesińska, Krzysztof Kucharski

Division of Silicon Microsystem and Nanostructure Technology
Institute of Electron Technology
Warsaw, Poland
e-mail: dtomasz@ite.waw.pl, ggluszko@ite.waw.pl, jmales@ite.waw.pl, kucharsk@ite.waw.pl

Abstract—**A fully-depleted SOI CMOS technology, being implemented in ITE, Warsaw, has been briefly described. Extensive wafer-scale electrical measurements have been done towards investigation of the technology stability. The measurements have been followed by I-V characteristics processing based on compact models of MOS transistors. The calculated electrophysical MOSFET parameters have been mapped and their distributions have been determined. The MOSFET parameter extraction has been supplied with additional measurements of the poly-Si resistors. Correlations between determined parameters have been also investigated.**

Index Terms—**CMOS, FD SOI, low-voltage, low-power, test structure, compact modeling, variability**

I. INTRODUCTION

A fully-depleted silicon-on-insulator (FD SOI) CMOS technology has been investigated towards fabrication of low-power, low-voltage CMOS integrated circuits (ICs) since 90s (e.g. [1], and its later editions). However after invention of multi-gate structures [2] offering an almost ideal gate control over channel conduction, increased transconductance and current efficiency, and minimized short-channel effects the multigate MOSFETs have become not only an object of extensive research (e.g. [3]) but their practical use in mass ULSI integrated circuit production at a so-called 22 nm node has been announced (e.g. [4]).

However recently, based on extensive measurements and calibrated numerical calculations it has been reported [5], that the FD SOI CMOS technology is still a promising solution for the 28 nm node and even beyond. Obviously, appropriate SOI substrates with very thin, and uniform device and buried oxide (BOX) layers, and perfect properties of the Si/BOX interface are required.

As compared to bulk CMOS the FD SOI CMOS devices offer much better performance and significantly decreased static and dynamic power consumption. This is due to correct operation at reduced supply bias, due to almost eliminated junction leakage and reduced parasitic capacitances. Moreover the bulk CMOS designs may be easily "translated" into FD SOI

CMOS. This aspect makes also the FD SOI a cost-effective solution. As compared to FinFETs, the planar FD-SOI devices also exhibit a number of advantages. Their design and fabrication techniques are well established, compatible with CMOS and do not contain steps for 3D structures formation. In some applications FD SOI MOSFETs are also more flexible than FinFETs via possibility of back-biasing, which may be considered as an additional freedom degree

Interest in SOI CMOS technology in ITE dates to the late 90s. In the presented paper we have described recent works on characterization of the FD SOI CMOS technology, which is under development in ITE-Warsaw.

II. FD SOI CMOS TECHNOLOGY IN ITE

Within a collaboration with Université catholique de Louvain (UCL), and supported by FP7 project [6], the FD SOI CMOS process [7] has been implemented in ITE laboratory. Processing has been done on two types of SOI substrates from SOITEC, namely on four-inch wafers with 340 nm-thick Si film and 400 nm-thick BOX, and on four inch wafers with 135nm-thick Si film and 1μm-thick BOX. The 135 nm/1 μm substrates have been cut out from 200 mm SOI wafers. The processing has been preceded by thinning of the Si-film down to ca 90 nm The main features of this process are as follows: supply voltage 3 V, nominal threshold voltage ±0.4..0.5 V, and min poly-silicon gate width 1.5 μm. In Fig.1. the FD SOI p- and n-channel MOSFETs manufactured using this process are

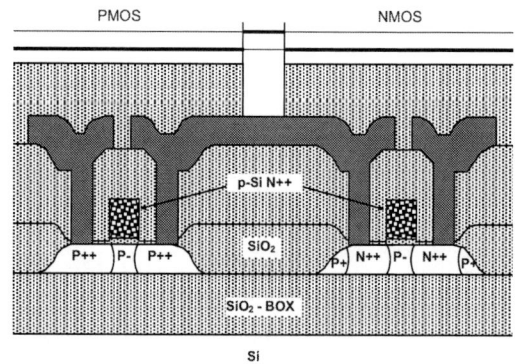

Figure 1. Cross-section of the structure of a pair of the p- and n-channel FD SOI MOSFETs.

The work was supported partly by EC within a Marie-Curie project PIAP-GA-2008-218255 ("COMON") and partly by EC and Polish Ministry of Science and Higher Education within a project ACP7-GA-2008-212859 ("TRIADE").

schematically shown. As may be seen the n-channel MOSFETs are inversion-mode devices, whereas the p-channel MOSFETs are accumulation-mode ones. Different aspects of the process implementation in ITE were reported e.g. in [8].

Based on this technology the CMOS test structures have been manufactured. Individual devices as well as digital blocks have been tested. In spite of the fact that correctly operating digital blocks i.e. 161-stage CMOS ring oscillators have been obtained, a stability of the technology remains a big issue. This topic is relevant while the process is dedicated towards manufacturing of low-power analog circuits for stand-alone as well as sensor applications. Works on statistical modeling of these devices have been initiated and reported in [9]. In the next paragraphs further wafer-scale measurements done in ITE followed by data processing towards process variability characterization are described.

III. MOSFET MEASUREMENTS AND PARAMETER EXTRACTION

Extensive measurements of the devices embedded in the FD SOI CMOS test structures have been focused on different transistor types and on poly-Si resistor measurements.

In the test structure four different types of n- and p-channel MOS transistors are embedded. They differ with channel doping concentrations, which are determined by the boron implantation doses. In the n- and pMOSFETs which are implanted with the same dose the channel doping concentrations are the same. This allows for investigation of doping concentration effect on electrical properties of the MOSFETs, for eventual tuning of the threshold voltages, and finally for design and use of multi-threshold analog blocks. In our measurements we have concentrated on non-implanted n- and pMOSFETs, called NIN, and PIP respectively, on pMOSFETs with boron implantation into channel with a dose=$4.5 \cdot 10^{11}$ cm^{-2}, and energy=20 keV denoted by P1, and on nMOSFETs with boron implantation of dose=$8 \cdot 10^{11}$ cm^{-2}, and energy=20 keV denoted by P2. This was imposed by the fact that P1 pMOSFETs and P2 nMOSFETs are expected to have symmetrical threshold voltages what is important particularly from digital IC design point of view. In Fig.2 the non-implanted nMOSFET module layout is illustrated. The layouts of the implanted test transistors are the same. Also the module with poly-Si and Al resistors is shown.

The MOSFET transfer I_D-V_G characteristics for small drain voltage equal to ± 0.05 V (depending on channel conduction type), and for substrate bias equal to 0 V (corresponding to full depletion operating conditions) have been measured on the whole wafers using the semi-automatic characterization set-up. Based on these characteristics the following parameters have been calculated: ON/OFF current, threshold voltage, transconductance factor, mobility degradation, subthreshold slope, channel shortening and source/drain series resistance. Threshold voltage has been determined using two methods, a standard one, where the threshold voltage is an intersection point of a line tangent to I_D-V_G curve at maximum transconductance, and a method based on linear fit of the $I_D/g_m^{0.5}$ characteristics in strong inversion [10]. In the second method series resistance effects are partially taken into account. Extraction of the channel shortening and source/drain series

resistance is based on processing of characteristics of devices with different lengths [11]. Such device series are available in the modules shown in Fig.2. In the extracted transconductance factors the evaluated channel shortenings have been taken into account.

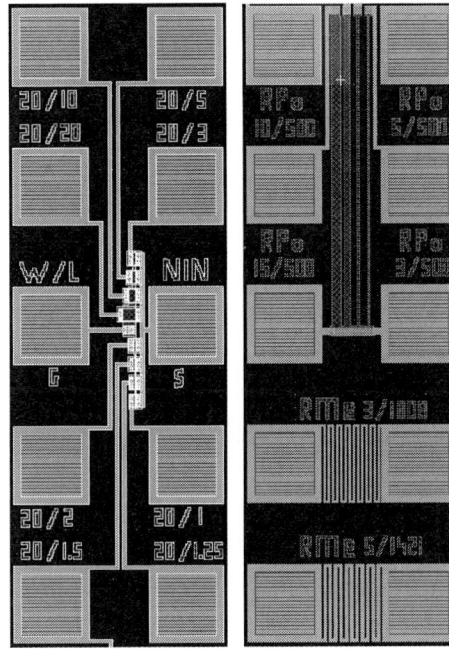

Figure 2. Layouts of two modules of the FD SOI CMOS test structure: n-channel transistors (left), poly-Si and metal resistors (right); prepared with LayoutEditor.

The applied procedure allows to determine basic parameters of the MOSFETs. The parameter values, which differ significantly from the projected ones and result from the process faults and non-mature status may be determined. Devices with unacceptable parameters may be localized. Next, neglecting them the typical devices may be indicated, and localized. The typical I-V curves and parameter values which characterize the given technology may be determined. As they are subject to statistical variations sets of typical parameters may be next used for statistical modeling purpose.

In Fig. 3, 4 sets of the measured I_D-V_G characteristics of n- and pMOSFETs are shown. In the charts worse characteristics corresponding to the edge structures have been neglected. In these devices channel width W is 20 μm, channel length L is 3 μm, BOX thickness t_{BOX} is 1 μm. A small asymmetry between the threshold voltages of n- and pMOSFETs may be stated. Moreover, contrary to the pMOSFETs with a reasonably good I_D-V_G repeatability the nMOSFETs exhibit a large spread, although a dominating set of curves may be distinguished. In the next chapter analysis of the data will be presented.

Similar sets of characteristics have been measured for non-implanted devices. In terms of characteristics spread conclusions are the same. Obviously, the transfer I_D-V_G characteristics of non-implanted and implanted devices are shifted along the gate voltage axis.

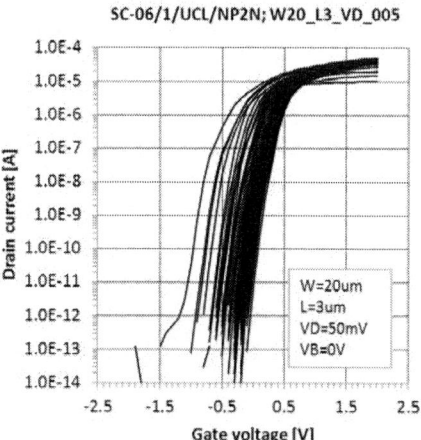

Figure 3. I_D-V_G characteristics of P2-implanted FD SOI nMOSFETs with channel width W=20 μm, length L=3 μm.

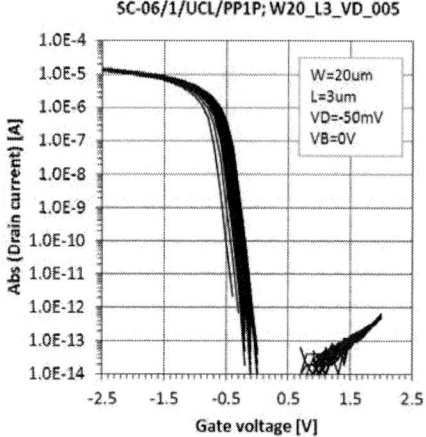

Figure 4. I_D-V_G characteristics of P1-implanted FD SOI pMOSFETs with channel width W=20 μm, length L=3 μm.

IV. MOSFET PARAMETER DISTRIBUTIONS

According to the characterization procedure mentioned earlier sets of device parameters have been extracted. Threshold voltages are of primary concern, In Fig. 5, 6 the area distributions of the threshold voltages of non-implanted pMOSFETs and nMOSFETs with channel width W=20 μm, length L=1.5 μm are shown. In accordance with the I-V characteristics (not shown here) the nMOSFETs suffer from a large spread of the threshold voltage. The incorrectly working devices come from the wafer edge, whereas in the middle of the wafers a much better device operation has been stated The p-type devices exhibit significantly better behavior. The processing conditions are mostly identical for both types of devices except of channel stopper formation (only for nMOSFETs), channel implantations (in fact almost identical, differing only with a dose), and source/drain formation (boron

implantation for pMOSFETs, phosphorus implantation for nMOSFETs). Hence it is expected that a structure channel-channel stopper-source/drain seems to be responsible for incorrect n-type device behavior. This is the subject of further investigation remaining beyond the scope of this work. In Fig. 7 a histogram of the threshold voltages of the considered MOSFETs is shown. In all cases dominating threshold voltages may be noticed. Their values for P1-implanted pMOSFETs and P2-implanted nMOSFETs are reasonably symmetrical, thus it may be stated that channel implantation doses have been set correctly. However a significantly larger spread of the threshold voltage of intrinsic and implanted nMOSFET than of their p-type counterparts is clearly visible. Additionally, the nMOSFETs with channel implantation exhibit significantly larger spread of the threshold voltage than the intrinsic devices.

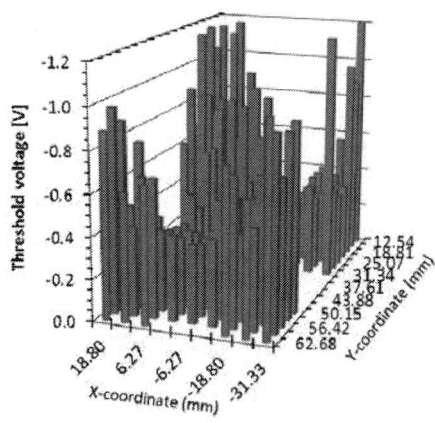

Figure 5. Threshold voltages of P2-implanted FD SOI nMOSFETs with channel width W=20 μm, length L=1.5 μm.

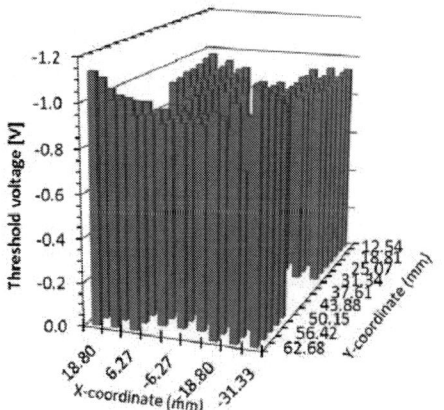

Figure 6. Threshold voltages of P1-implanted FD SOI pMOSFETs with channel width W=20 μm, length L=1.5 μm.

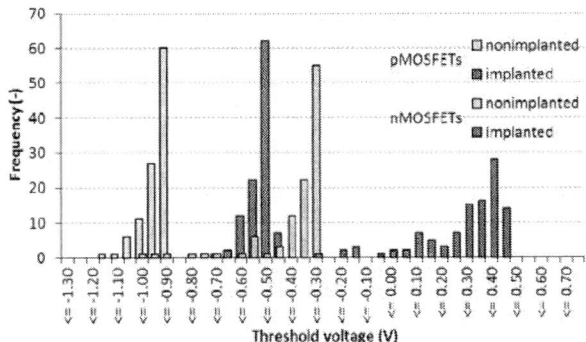

Figure 7. Histograms of threshold voltages of n- and p-channel non-implanted and implanted MOSFETs with channel width W=20 µm, length L=3 µm.

A very important parameter characterizing the process, particularly gate photolithography and source/drain formation is the gate length change with respect to the projected value. This parameter determines a so-called electrical channel length, influences short-channel effect, static and dynamic MOSFET operation [12]. The channel shortening for the n- and pMOSFETs has been evaluated together with the series resistances using the approach described in [11]. The results are arranged as histograms shown in Fig. 8, 9. For illustration purpose the area distributions of channel shortening (channel length change is negative) for non-implanted devices are shown in insets.

It may be stated that the pMOSFETs exhibit much better behavior with respect to this parameter too. Their distribution is much more narrow. On the other hand it is worthwhile to mention that the applied extraction procedure [11] is sensitive to the overall device characteristics quality. It is based on two-step linear regression procedure. The first linear fit is done for drain-source resistance vs channel nominal length data for a number of gate-source voltages. The second one is done for A vs B coefficients of fits from step 1 obtained for different gate voltages, and allows to calculate a unique value of channel length change. However, if certain devices used for the first fit exhibit improper behavior (e.g. non-monotonous dependence of S-D resistance on channel nominal length) then the final extraction results may be seriously disturbed.

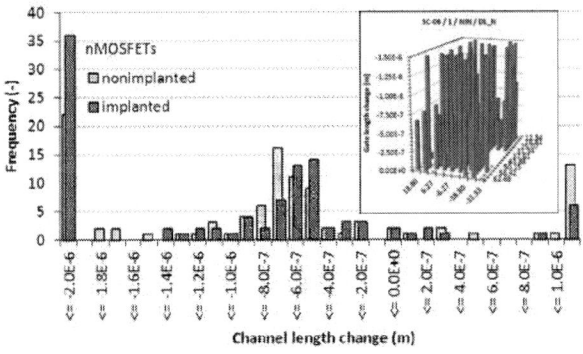

Figure 8. Histograms of channel length change of n-channel MOSFETs.

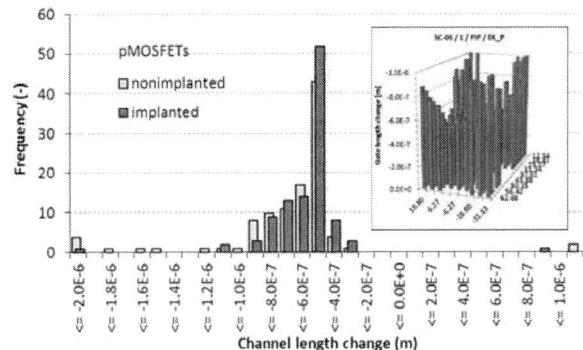

Figure 9. Histograms of channel length change of p-channel MOSFETs.

V. POLY-SI RESISTOR PARAMETER MEASUREMENTS AND DISTRIBUTIONS

The MOS transistors are complex devices, in which it is difficult to estimate an effect of different structural parameters on the device electrical behavior variability. Effective gate length is of primary concern for MOSFET operation. In order to estimate the poly-Si gate length spread we have used the poly-Si resistors placed on the same chips as the MOS transistors (Fig.2). Their I-V measurements have been done for series of resistor widths. In Fig. 10 a set of 3 µm wide and 500 µm long resistor I-V characteristics is shown. They exhibit a perfectly linear shape. In parallel they are subject to a noticeable spread. Characteristics of wider resistors, namely 5, 10, and 15 µm wide ones are ideally linear, and much more repeatable.

For extraction of ΔW_{poly}, i.e. poly-Si line width change with respect to the projected size a linear fit expressed by formula (1) to experimental data has been done. Resistances R_{poly} of poly-Si resistors with different widths W and constant length L have been used. The length of 500 µm ensures, that eventual resistor length variations may be neglected. This procedure

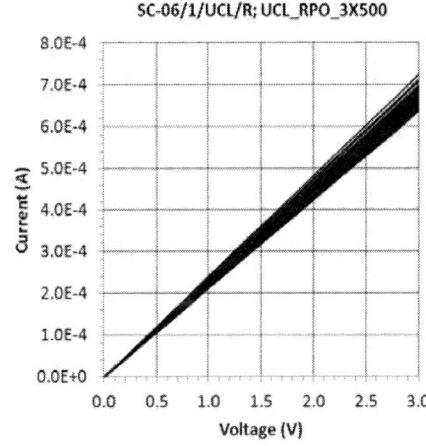

Figure 10. I-V curves of 3 µm wide and 500 µm long poly-Si resistors.

allows for direct estimation of the sheet resistance $R_{s,poly}$ of the poly-Si layer too. In the formula (1) negative value of ΔW_{poly} parameter means poly-Si line narrowing, whereas positive value means line widening. It is worthwhile to mention that the linear fit (1) to experimental data has been ideal ($R^2=1$).

$$L \big/ R_{poly} = \left(W + \Delta W_{poly}\right)\big/ R_{s,poly} \qquad (1)$$

In Fig. 11 on-wafer distributions of poly-Si line width change are shown. From the histogram it results that the poly-Si patterns are mostly subject to the narrowing by a small amount not larger than 0.1 μm. Taking into account a potential of the ITE photolithography equipment the result is satisfactory. In Fig. 12 distributions of the extracted sheet resistance are shown. A small radial-type variation of this parameter may be noticeable. It may be attributed to the method of poly-Si doping by phosphorus diffusion prior its photolithography step. Nevertheless the obtained poly-Si sheet resistance is almost constant across the wafer. From the point of view of the sheet resistance extraction the evaluation of the poly-Si width is relevant. Namely, if $R_{s,poly}$ parameter is calculated without account to the line narrowing, directly from the resistor resistance, the results are worse than those, calculated with account to the line narrowing. Both distributions are shown on histogram in Fig.12.

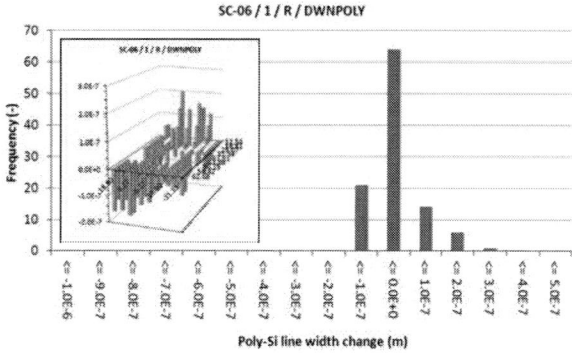

Figure 11. Distribution (area and histogram) of poly-Si line width change with respect to the designed one.

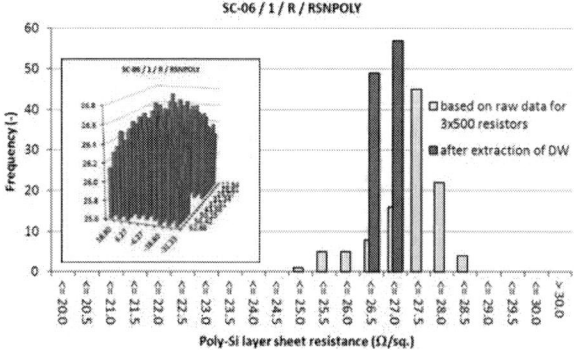

Figure 12. Distribution (area and histogram) of sheet resistance of polysilicon; for comparison there is a histogram of sheet resistance calculated directly based on I-V measurements of 3 μm ×500 μm resistors.

The results shown above indicate directly, that the MOSFET gate length shortening of the order 0.5-0.6 μm for n- and p-type devices, and discussed in chapter IV is dominated by S/D junction formation steps. It may be expected that lateral diffusion of dopants in these areas is of the order of 0.2 μm on each channel side.

VI. PARAMETER CORRELATIONS

For a detailed analysis of MOSFET parameters for process characterization and IC design an investigation of correlations between the parameters is required. Also, evaluation of the correlations between parameters of different devices is valuable, while it allows to select and determine uncorrelated process parameters which may be used for Monte-Carlo simulation of ICs and for yield maximization-oriented IC design methodology [13].

In Fig. 13 a scatterplot of the channel length changes determined for all the considered MOS transistor sets (see chapter IV) vs poly-Si line width change (see chapter V) is shown. It may be noticed that the gate length change remains uncorrelated with respect to ΔW_{poly} parameter. It has been stated earlier, that the absolute value of ΔW_{poly} is much smaller than gate length shortening determined mostly by S/D lateral diffusions. So the plot confirms, that poly-Si gate and S/D formation parameters remain uncorrelated.

In parallel it is expected that gate length shortenings for the devices of the same type are correlated. This is illustrated in Fig. 14. Unexpectedly, the correlation between these parameters for nMOSFETs is very week. On the other hand it should not be a surprise having in mind results shown in Fig. 8. The gate shortenings for p-type MOSFETs are strongly correlated. This confirms the earlier statement about much better reproducibility of p-channel non-implanted and implanted devices.

Finally, relations between extracted threshold voltages of the n- and p-type intrinsic and implanted MOSFETs have been analyzed. The results for 3 μm long devices are shown in Fig. 15. Correlations between corresponding groups of devices may

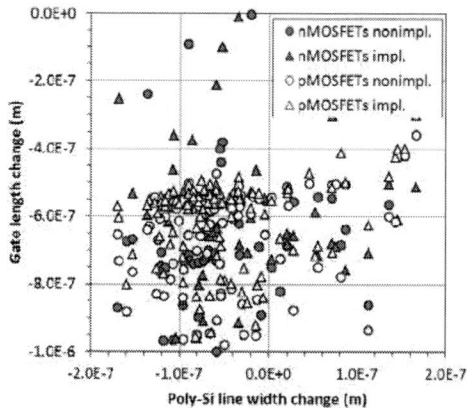

Figure 13. Scatterplot of the MOSFET channel length change vs poly-Si line width change deternmined based on different sets of devices.

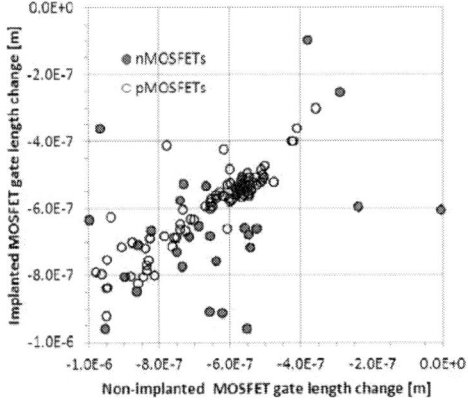

Figure 14. Correlations between gate length variations for n- and p-type MOSFETs

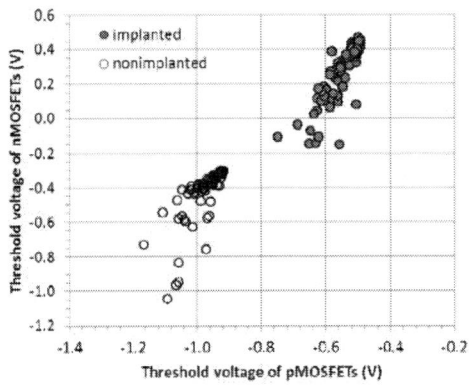

Figure 15. Correlations between threshold voltages for n- and p-type MOSFETs

be noticed. The data in Fig. 15 show, that, if evidently wrong intrinsic nMOSFETs are neglected, then channel implantation increases the spread of the nMOSFET threshold voltages. This effect seems to be weaker in pMOSFETs, which are accumulation-mode junctionless devices. Th

VII. CONCLUSIONS AND SUMMARY

Methodology of data acquisition and analysis has been presented and illustrated based on the FD SOI CMOS technology. The work has been focused on investigation of the process stability via the wafer-scale measurements of the MOSFET I-V characteristics followed by their post-processing and parameter extraction. The MOS transistor data analysis has been supplied with additional wafer-scale measurements of the poly-Si resistors, followed by poly-Si line parameter extraction.

The spread of the selected MOSFET parameters and the correlations between them have been briefly analyzed. These steps are particularly important from the point of view of both technology development and IC design.

It is worthwhile to mention that wafer-scale measurements and parameter extraction is time-consuming, but it seems to be the best way to evaluate not only the process variability. Also having a large number of characteristics and by making their statistical analysis it is possible to select data corresponding to the nominal/typical device performance. Both aspects are important from the point of view of the nominal IC design as well as for the IC design oriented towards yield optimization by taking the process variability into account (e.g. [13]).

REFERENCES

[1] J.-P. Colinge, "Silicon-on-Insulator technology: materials to VLSI", Springer, 1991

[2] J. P. Colinge, "FinFETs and other multi-gate transistors", Springer, 2008

[3] http://www.compactmodelling.net (COMON project web-page)

[4] http://newsroom.intel.com/docs/DOC-2032

[5] http://www.soiconsortium.org/

[6] http://triade.wscrp.com (TRIADE project web page)

[7] D. Flandre, et al., "Fully depleted SOI CMOS technology for heterogenous micropower, high-temperature or RF microsystems", Solid-State Electronics, 45 (2001), 541-549

[8] K. Kucharski, et al., "Implementation of FD SOI CMOS technology in ITE", Elektronika: konstrukcje, technologie, zastosowania, 2011, Vol. 52, no. 3, p. 13-15

[9] M. Yakupov, D. Tomaszewski, K. Kucharski, W. Grabinski, "BPV method as a tool for statistical compact modeling of SOI CMOS technology", MOS-AK/GSA ESSDERC/ESSCIRC Workshop, Seville, 17th Sept, 2010

[10] G. Ghibaudo, "New method for the extraction of MOSFET parameters", Electronics Letters, 28th April 1988, Vol.24, No.9

[11] D. K. Schroder, "Semiconductor material and device characterization", 3rd ed., Wiley Interscience, 2006

[12] N. Arora, "MOSFET modeling for VLSI simulation: theory and practice", World Scientific Publishing Company, 2007

[13] M. Yakupov, D. Tomaszewski, "Analysis of Selected Methods for CMOS Integrated Circuit Design for Yield Optimization", Proc. 18th Int. Conference Mixed Design of Integrated Circuits and Systems, Gliwice, 16-18 June 2011

Impact Ionization Effect in Deep Submicron MOSFET Features Simulation

Dmitry Speransky

Radiophysics and Computer Technologies Department
Belarussian State University
Minsk, Belarus
dmitry.speransky@gmail.com

Tran Tuan Trung

Microelectronics Department
Belarussian State University of
Informatics and Radioelectronics
Minsk, Belarus

Abstract—**Within the framework of Keldysh impact ionization model the calculation of effective threshold energy for silicon MOSFET with 100 nm channel length by means of ensemble Monte-Carlo simulation is performed. The possibility of impact ionization rate treatment with one-parameter Keldysh model in pre-breakdown and breakdown transistor operation mode using calculated effective threshold energy value is proposed.**

Index Terms—**MOSFET, Monte-Carlo, impact ionization, threshold energy**

I. INTRODUCTION

It is known that in numerical simulations of integrated circuit elements with the reduction of their dimensions, particularly MOSFETs, an account of impact ionization process is essential. The latter is caused by the fact that the rate of impact ionization in such elements can be comparable or even greater than the rates of other considered scattering processes as a result of the presence of high electric field strengths.

It is also known that one of the most powerful submicron MOSFET simulation methods with account of all dominant mechanisms of charge carrier scattering including impact ionization is ensemble Monte-Carlo procedure (see, for example, [1–3] and references therein).

The main purpose of this study is the estimation of impact ionization effective threshold energy in deep submicron silicon n-channel MOSFET with 100 nm channel length in the framework of Keldysh impact ionization model.

II. THRESHOLD ENERGY OF IMPACT IONIZATION

Impact ionization is a threshold process [4]. In the simple case, the value of threshold energy E_{th} is defined using energy and momentum conservation rules. Subject to this in [5] the method of threshold energy estimation in several semiconductors with different band structures is offered. It was taken into account that different values of the threshold energy are possible and the conclusion that effective (or some averaged) threshold energy for charge carriers may depend on electric field strength was drawn.

In Monte-Carlo simulations of semiconductor electrical properties the so-called Keldysh formula is widely used for calculation of impact ionization scattering rate $W_{II}(E)$ with given threshold energy E_{th} [6]:

$$W_{II}(E) = A W_{ph}(E_{th}) \left(\frac{E - E_{th}}{E_{th}} \right)^2, \qquad (1)$$

where A is a fitting parameter, $W_{ph}(E_{th})$ is the total electron-phonon scattering rate at the energy equal to E_{th}. Thus the model has two fitting parameters A and E_{th} with $E_{th}^{soft} = 1.2$ eV for "soft" and $E_{th}^{hard} = 1.8$ eV for "hard" thresholds [2]. There is no common point of view on what values of E_{th} and A should be chosen exactly. The value of A can vary, and we used $A = 0.38$ and $A = 100$ for soft and hard thresholds respectively [7].

III. MOSFET FEATURES SIMULATION

Monte-Carlo simulation of the MOSFET is performed to determine the influence of the drain voltage V_D on effective threshold energy value E_{theff} at two gate voltages $V_G = 1$ and 2 V. Considered MOSFET structure and some aspects of the simulation algorithm are described, in particular, in [3, 7–10]. Electron transport simulation in silicon conduction band includes X and L valleys with account of nonparabolicity. Electron scattering processes are intravalley acoustic and intervalley phonon scattering, ionized impurity scattering, optical phonon scattering in L valley, plasmon scattering and impact ionization process. Holes are treated in quasi-static approximation for simplicity. Source and drain are modeled as ideal ohmic contacts and metal gate is supposed to be aluminum. The simulation is done for the temperature $T = 300$ K and MOSFET dimensions denoted in Fig. 1 with the following parameters: gate oxide thickness is 10 nm, acceptor doping levels of the channel and substrate are equal to $5 \cdot 10^{23}$ m^{-3} and 10^{24} m^{-3}, donor doping level of the source and drain regions is equal to $8 \cdot 10^{24}$ m^{-3}.

In Fig. 1 the device cross-section along the channel is shown. In Fig. 2 impact ionization scattering rates $W_{II}(E)$ with electron distribution functions $f(E)$ near the drain end of the transistor channel, where impact ionization rate is the most sufficient, versus electron energy E at $T = 300$ K are presented.

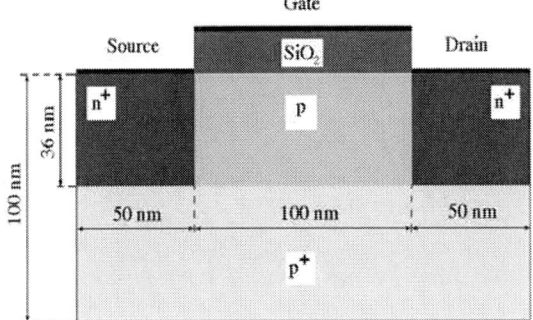

Figure 1.   Cross-section of simulated MOSFET

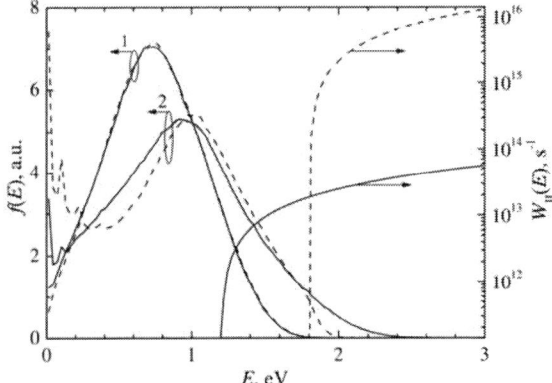

Figure 2.   Electron distribution functions $f(E)$ and impact ionization scattering rates $W_{II}(E)$ versus electron energy E in the transistor channel for soft threshold (solid curves) and hard threshold (dashed curves) models at V_G = 2 V. Curves 1 – V_D = 2 V, and curves 2 – V_D = 3 V

As appears in Fig. 2, monotonous growth of $W_{II}(E)$ with the increase of the primary carrier energy from 1.2 to 3 eV for soft threshold indicates that effective impact ionization threshold energy can be significantly higher than minimum threshold which is approximately equal to the energy gap of 1.1 eV in silicon at $T = 300$ K.

In [11] it is suggested that effective threshold energy value can be estimated utilizing the criterion of maximum product of impact ionization scattering rate $W_{II}(E)$ and electron distribution function $f(E)$ for energy values grater than minimum threshold. It should be noticed that in this case the value of impact ionization effective threshold energy E_{theff} can be treated as energy value for which the maximum number of scattering events in a unit of time for simulated ensemble of primary electrons occurs.

In Fig. 3 the dependence of E_{theff} versus the drain voltage V_D for the gate voltages V_G = 1 and 2 V is shown. As can be seen from the figure it is possible to mark out three regions in the dependences of E_{th} versus V_D:

I) V_D is in the interval $0.5 \div 1$ V. It is the region of constant values of $E_{th}^{soft} = E_{theff}^{soft} = 1.2$ eV and $E_{th}^{hard} = E_{theff}^{hard} = 1.8$ eV, and they do not depend on V_D.

II) 1 V $< V_D < 3$ V. It is the region of almost linear growth of effective threshold energy both for soft and hard thresholds.

III)   It is the region of $V_D > 3$ V and almost constant $E_{theff}^{hard} = 1.88$ eV, which is close to $E_{theff}^{soft} = 1.7$ eV and differs from the last not more than by approximately 0.2 eV.

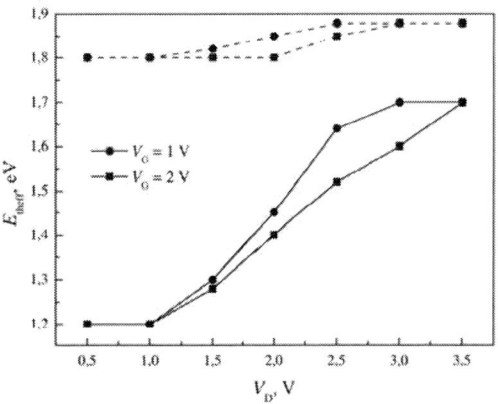

Figure 3.   Effective threshold energy E_{theff} versus drain voltage V_D dependence for soft (solid curves) and hard (dashed curves) thresholds

Current-voltage characteristics of the simulated MOSFET are presented in Fig. 4. Comparison of Fig. 3 and Fig. 4 allows one to make a conclusion that region I corresponds to the nearly linear transistor operation mode with saturation region passage. Region III corresponds to pre-breakdown mode when strong avalanche multiplication occurs. Region II corresponds to the transition range from a saturation to pre-breakdown mode.

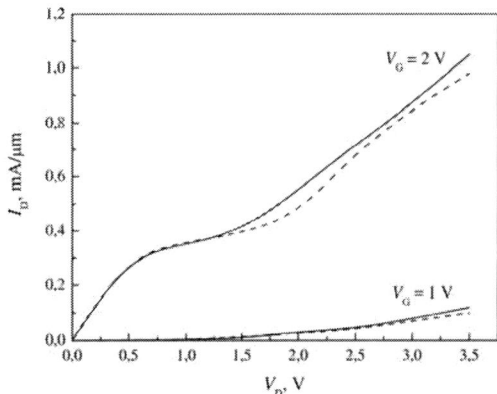

Figure 4.   Simulated current-voltage characteristics of the MOSFET with the use of soft threshold (solid curves) and hard threshold (dashed curves) models

Thus for a rather high drain bias when a strong electric field exists in the MOSFET channel the effective threshold energy of impact ionization process does not depend sufficiently on the choice of Keldysh model type. The latter may be explained by the fact that high electric field is localized near the drain end of the channel (in approximately 20 nm region) and electrons gain enough energy during the free flight time to make an impact ionization act according to both soft and hard threshold models. Obtained results agree to some extent with those of [12] where effective threshold energy in Si was calculated taking into account its dependence on the wave vector of primary electron. It was shown that in a uniform electric field of strength $\leq 5 \cdot 10^7$ V/m the effective threshold energy was approximately 1.7 eV. Also according to [13] for $A = 100$ and electric field strength of $4 \cdot 10^7$ V/m the deviation of the effective threshold energy from the hard threshold appears to be approximately 0.12 eV which is close to our value of approximately 0.08 eV.

IV. CONCLUSION

In conclusion it should be noticed that in current work the impact ionization effective threshold energy in deep submicron MOSFET with channel length of 100 nm was defined in the framework of Keldysh model. Calculations were done by means of ensemble Monte-Carlo simulation for drain biases in the interval from 0.5 to 3.5 V and gate biases $V_G = 1$ and 2 V. Corresponding transistor current-voltage characteristics were calculated. Analysis and comparison of calculated $E_{\text{theff}}(V_D)$ and $I_D(V_D)$ dependences for soft and hard threshold models let us see that in pre-breakdown operation mode the difference between $E_{\text{theff}}^{\text{hard}}$ and $E_{\text{theff}}^{\text{soft}}$ is approximately 0.2 eV. The latter allows us to conclude that in numerical simulations of the MOSFET characteristics in pre-breakdown and breakdown regimes it is possible to use Keldysh impact ionization model with only one fitting parameter A. Effective threshold energy in that case is *a priori* known and may be taken as average value of effective energies for soft and hard thresholds, which is approximately equal to 1.8 eV (see Fig. 3). Also the validity of application of one-parameter model for the simulation of very short channel MOSFETs in various regimes of operation, apparently, must be investigated.

ACKNOWLEDGMENT

The work was supported by the Ministry of Education of Belarus, project "Electronics 1.1.03". The authors would like to thank Prof. V.M.Borzdov with the Radiophysics and Computer Technologies Department of Belarussian State University and Prof. V.V.Nelayev with the Microelectronics Department of Belarussian State University of Informatics and Radioelectronics for valuable suggestions and support for this work.

REFERENCES

[1] M. V. Fischetti, S. E. Laux, Monte Carlo analysis of electron transport in small semiconductor devices including band structure and space-charge effects, *Phys. Rev. B*, 1988, Vol. 38, pp. 9721–9745J. Clerk Maxwell, A Treatise on Electricity and Magnetism, 3rd ed., vol. 2. Oxford: Clarendon, 1892, pp.68–73.

[2] C. Jacoboni, P. Lugly, *The Monte Carlo Method for Semiconductor Device Simulation*, Wien–New York: Springer-Verlag, 1989

[3] V.M. Borzdov, O.G. Zhevnyak, V.O. Galenchik, F.F. Komarov, *Monte-Carlo simulation of integral electronics device structures*, Minsk: Belarus State University Publishing, 2007 (in Russian)

[4] F. Capasso, Physics of avalanche photodiodes, in *Semiconductors and semimetals* (*Lightwave communication technology, vol 22, Part D. Photodetectors*), ed. R. K. Willardson and A. C. Beer, Orlando: Academic Press, 1985, pp. 2–168.

[5] C.L. Anderson, C.R. Crowell, Threshold energies for electron-hole pair production by impact ionization in semiconductors, *Phys. Rev. B*, 1972, Vol. 5, No. 6, pp. 2267–2272.

[6] L.V. Keldysh, Concerning the Theory of Impact Ionization in Semiconductors, *Soviet Physics JETP*, Vol. 21, No. 6, 1965, pp. 1135–1144.

[7] N.A. Bannov, O.I. Kaz'min, Simulation of highly nonequilibrium electronic phenomena in submicron silicon MOSFETs, *Mikroelektronika*, Vol. 18, No. 2, 1989, pp. 112–125 (in Russian).

[8] D. Vasileska, The role of quantization effects on the operation of 50 nm MOSFET and 250 nm FIBMOS devices, *Phys. Status Solidi b*, 2002, Vol. 223, No. 1, pp. 127–133.

[9] R. Rengel, J. Mateos, D. Pardo, T. Gonzalez, and M.J. Martin, Monte Carlo analysis of dynamic and noise performance of submicron MOSFETs at RF and microwave frequencies, *Semicond. Sci. Technol.*, 2001, Vol. 16, pp. 939–946.

[10] D.S. Speransky, A.V. Borzdov, V.M. Borzdov. Impact ionization effective threshold energy estimation in MOSFET with 50-nm channel length / Proc. 21st International Crimean Conference " Microwave and Telecommunication Technology". CriMiCo 2011, Sevastopol, Ukraine, 2011, pp. 810-811.

[11] E. O. Kane, Electron scattering by pair production in silicon, *Phys. Rev.*, 1967, Vol. 159, No. 3, pp. 624–631.

[12] N. Sano, M. Tomizawa, A. Yoshii, Monte Carlo analysis of ionization threshold in Si, *Appl. Phys. Lett.*, 1990, Vol. 56, No. 7, pp. 653–655.

[13] R. Chwang, C.R. Crowell, Effective threshold energy for pair production in nonpolar semiconductors, *Solid State Commun.*, 1976, Vol. 20, No. 3, pp.169–172.

KLU Sparse Direct Linear Solver Implementation into NGSPICE

Francesco Lannutti, Paolo Nenzi, Mauro Olivieri
DIET
Sapienza – University of Roma
Roma, Italy
Email: nicolati@fastwebnet.it, nenzi@diet.uniroma1.it, olivieri@diet.uniroma1.it

Abstract—**The simulation of large digital and mixed-signal integrated circuits is one of the challenges of the electronics design automation industry. In this work, a fast linear solver, KLU, is implemented into NGSPICE circuit simulator and its performances have been verified on standard netlists.**

Index Terms—**NGSPICE; KLU; circuit simulation; transient analysis.**

I. INTRODUCTION

The electrical-level simulation of large circuits, consisting millions of devices, is still an open issue that limits the possibility of full chip verification and one of the grand challenges in ITRS roadmap [1, 2]. The universal tool (still after 40 years from its conception) for time domain verification of ICs is SPICE (Simulation Program with Integrated Circuit Emphasis). In this context, SPICE refers to all the circuit simulators that are based on the same algorithms of the original UC Berkeley SPICE simulator.

The simulation time of transient analysis in SPICE grows super-linearly with the number of equations that describe the circuit. In [3], it has been reported that the run time per transient analysis iteration scales as $O(N^{1.2})$ (with "N", the number of equations), while the double precision FLOPS (Floating Point Operation per Second) of CPUs only as $O(N^{0.96})$ (with "N" the number of transistors in the CPU). This difference in the exponents implies that time spent for a single transient iteration will continuously increase with the technology scaling.

The efforts to reduce this widening gap can be classified in two categories: specialized algorithms and parallelization. The former approach gave birth to a class of simulators often called "fast-spice", presenting significant speedups against standard simulators for some class of circuits (e.g. RAM). The latter approach has been pursued since the early days of SPICE [4] and parallel implementations are still developed today, in particular for the newly available GPU architectures [5]. Speedups obtained by the parallelization of the transient analysis in SPICE are in the range 2x to 24x, lower than the ones of fast-spices, nevertheless the resulting simulators are not tailored to any specific circuit topology.

The parallelization of SPICE is not trivial because of the structure of the SPICE algorithm for transient analysis, shown in figure 1 (after [4]). The three highlighted operations in the inner loop are the most demanding in term of computing resources: linearization of non-linear devices around the trial operating point, conductances stamping into the modified nodal analysis (MNA) matrix and linear system solution.

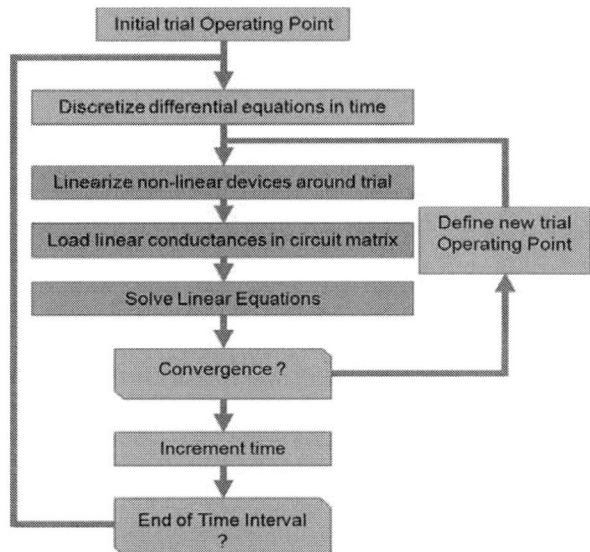

Figure 1. The SPICE algorithm for transient analysis, adapted from [4]. The three boxes in the inner loop are the most time consuming operations

The linearization of non-linear devices is the most easily parallelizable operation because it is intrinsically SIMD (Single Instruction Multiple Data). Furthermore, being the most time consuming operation, it's the most relevant contribution to speedup. Figure 2 shows the percent of the time spent in transient simulation for the netlists in the ISCAS 85 suite [6,7]. The *BSIM3load*, *CAPload* and *RESload* are the functions that compute the linear conductances and load them into the system matrix for BSIM3, capacitors and resistors devices, respectively. Their execution time sums up to 54% of the analysis time. Parallel implementations of linearization phase have been done both for CPUs using OpenMP [8] and GPUs [5]. The NGSPICE circuit simulator [9] implements, as option, the parallelization of device linearization as described in [8] for BSIM3 and BSIM4 devices (data in figure 2 have been computed with this option turned off). The speedup achieved on multicore CPUs is 2.

The next important contribution, from figure 2, is the linear system solution that accounts for 36% of total analysis time (the sum of *spFactor*, *spSolve* and *spClear*). The parallelization of a direct sparse solver (SPICE based circuit simulators use a direct solver) is not trivial because of the irregular structure of the sparse matrix and its sparsity. Most of the elements of a matrix resulting from MNA are zero and the non-zero pattern has no regularity. Thus the solution of a sparse linear system is an I/O intensive problem. Nevertheless the parallelization of sparse direct linear solvers has been attempted since the early days of SPICE [4] or, recently in [10].

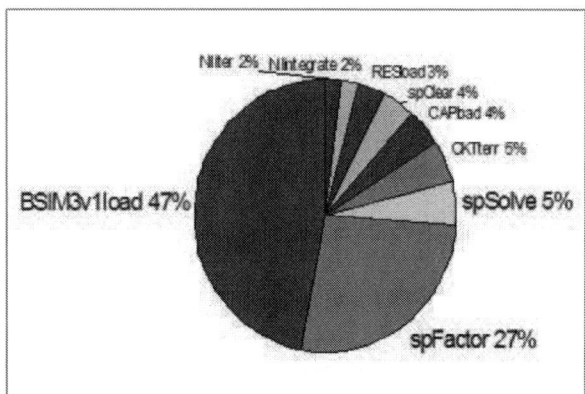

Figure 2. Time spent in different functions of the ngspice code during transient analysis on the ISCAS 85 [6,7] suite. The names correspond to the function names in the code and the time is computed as percent of the time spent for all the netlists in the suite

In order to complete the analysis of time spent in the transient analysis from figure 2, 5% of the time is spent in *CKTterr*, the truncation error calculation, 2% in *NIiter*, the code for Newton-Raphson iterations and 2% in *NIintegrate*, the integrator routine. The remaining 1% (not shown in the figure) is spent in the rest of the simulator.

In this work the implementation of a recently developed serial direct sparse linear solver, providing speedups comparable to parallel implementations is presented. The NGSPICE circuit simulator is used as driver for testing the implementation. In the modified version of NGSPICE, the original solver Sparse (version 1.3) [11] has been replaced with KLU [12] developed by Timothy Davis (University of Florida). KLU is direct solver optimized for the structure of linear systems resulting from the nodal analysis of electronic circuits that has been demonstrated [12] to be up 1.000 times faster than Sparse 1.3 in solving an MNA linear system.

II. ALGORITHMS ANALYSIS

The matrices generated by the modified nodal analysis of electronic circuits (MNA matrices) posses some properties that can be exploited to enhance the performances of direct solvers [12]. The main diagonal of an MNA matrix is almost free of zero-valued elements. The MNA matrix has un-symmetric values with a roughly symmetric non-zero pattern, it is very sparse and if ordered properly leads to sparse LU factors. Moreover, it can be permuted to a block triangular form (BTF) with a single diagonal block that contains most of the non-zero

elements, surrounded by other, much smaller, diagonal blocks. The smallest circuit in the ISCAS 85 suite is the c432 ("c" stands for "combinational", and 432 is the number of lines describing the circuit). The circuit is a 27-channel interrupt controller [6] with 36 input and 7 outputs, consisting of 160 logic gates. The transient analysis MNA matrix for c432 has 5007 rows and columns with 27521 non-zero elements (corresponding to 0.11%). The largest block of the matrix consists of 27445 non-zero elements spanning 4930 rows and columns. The remaining elements 76 are singletons (a singleton is a single non-zero element on the main diagonal).

A. Analysis of the Sparse linear solver

Sparse is the default solver in the NGSPICE circuit simulator, developed by Kenneth Kundert [11]. It solves the linear system

$$Ax = b \qquad (1)$$

with "*A*" sparse and "*b*" dense or sparse, and is optimized to solve repeatedly systems with the same non-zero pattern. This is the case of the transient analysis, where the circuit structure (that defines the non-zero of the matrix) does not change during the analysis, and only the numerical values of elements change. Sparse considers the sparse matrix as linked list of nodes each one pointing to the next element in column and to the next element in row. This structure allows the circuit simulator to build the overall system matrix starting from the stamps of the single instances in the circuit. It is not necessary to know a priori the final size of the matrix. The structure of the matrix in sparse is optimized for O(1) access operations, achieved by storing the pointer to each element in the calling routine. Sparse uses the Markowitz preordering to reduce the fill-ins [12] and factor the entire matrix in LU form using partial pivoting. In addition, Sparse can exploit the sparsity of the RHS vector. The solution of a linear system is divided into two steps: factorization and solve.

Sparse has two factorization functions, one for computing the LU factors, called only at the start of the analysis, and the other one for updating the factors with new numerical values, called at each iteration (the *spFactor* in figure 2).

The solve step (the *spSolve* function in figure 2), is the forward and backward substitution of the RHS in the lower and upper triangular matrices L and U to compute the "x" vector.

B. Analysis of the KLU algorithm

KLU is a direct sparse matrix solver, designed by Timothy Davis at the University of Florida [12] tailored to efficiently solve the linear systems resulting from the MNA analysis. In KLU the system matrix is stored in compressed storage column format (CSC). All the KLU operations are performed in a "workspace" memory area that is allocated with the matrix, to reduce the costly memory allocation and copy operations. The algorithm solves the linear system in five steps:

1. Maximum transversal,

2. Block triangular form permutation,

3. Column approximate minimum degree reordering,

4. LU factorization using Gilbert-Peierls algorithm,

5. Forward/backward substitution.

The Maximum Transversal [14] uses a DFS (Depth First Search) on the system matrix to find a permutation that removes all the zeroes on the main diagonal. If the matrix isn't singular, it's always possible to find an un-symmetric column permutation Q that removes all the zeroes on the main diagonal.

After this step the following system is solved:

$$AQy = b \qquad (2)$$

where $Qx = y$. In the next step, the BTF algorithm, based on the Tarjan algorithm [15] is used to find the strongly connected components of the graph representing the circuit. In this step, a symmetric row/column permutation P that converts the matrix into an upper block triangular form is found. This form is particularly advantageous because all the elements below the diagonal blocks are zero and do not contribute to fill-ins and the LU factorization is needed only for the diagonal blocks. The off-diagonal blocks are only linear combination of the unknowns of corresponding diagonal blocks.

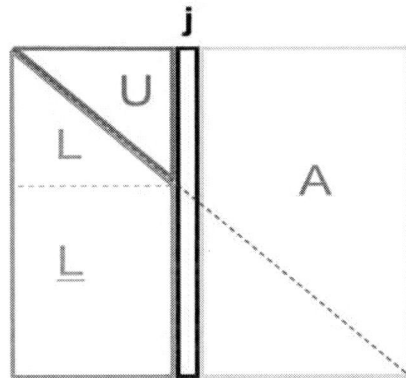

Figure 3.   Gilbert Peierls LU factorization process, the picture shows the jth step of the factorization. The LU factorization is built one column at time. The U and L are the LU factors at the j_{th} step, where $\underline{L}$ represents the part of the matrix not factored yet. The j vector corresponds to the jth column of the generic block A.

The linear system to be solved becomes:

$$PAQP^T y = Pb \qquad (3)$$

The COLAMD (COLumn Approximate Minimum Degree) algorithm [16] is then applied to each diagonal block to reduce the fill-ins in the following factorization step.

KLU uses the Gilbert-Peierls LU factorization (GPLU) [16], to compute the LU factors for each diagonal block. GPLU algorithm factors each block by repeated solution of linear systems. It produces a column of L and a column of U at every step, using the j_{th} column of the generic diagonal block A as right hand side vector (RHS) of the j_{th} triangular system, composed by the $(j-1)_{th}$ columns of L (see Figure 3). The solution vector, at each step, becomes the j_{th} column of L and U (the upper part pertains to U and lower parts to L). The pseudocode in figure 4 describes the algorithm.

The version of GPLU implemented in KLU exploits the sparsity of the block to reduce the number of operations required for factorization to O(flops(LU)) [17].

The system solve step compute the dense solution vector x, through a chain of forward and backward substitutions applied to each LU factored block and block-back substitutions in the off-diagonal blocks.

Gilbert Peierls LU factorization algorithm

```
L = I
for each column j
    b = A(:,j)
    x = L \ b          % see below
    (do partial pivoting)
    U(1:j,j) = x(1:j)
    L(j+1:n, j) = x(j+1:n) / U(j,j)
    end

% Solution of x = L \ b

x = b
for columns I in (1:j-1)
    x(i+1:n) = x(i+1:n) − L(i+1:n, i) * x(i)
end
```

Figure 4.   Pseudocode for the basic GP algorithm used by KLU to compute the LU factors of each diagonal block.

C. Comparison of the performances of the two linear solvers

The first analysis we did was to compare the performances of both solvers, Sparse and KLU outside of the circuit simulator, using the test tools included in each solver. We dumped the system matrix corresponding to the netlists in the ISCAS 85 suite, with the RHS vector. The netlists characteristics are reported in table 1. For each circuit, the number of equations for the non back-annotated version (std) and for the back-annotated one (ann) is reported.

TABLE I.   CHARACTERISTICS OF THE ISCAS85 NETLISTS

ISCAS 85	Netlists characteristics		
	Logic gates	*Functional Blocks*	*Equations (std/ann)*
C432	160	5	5007/9789
C499	202	2	10221/19243
C880	383	7	7998/14975
C1355	546	2	10509/19485
C1908	880	6	15252/23818
C2670	1193	7	23433/40452
C3540	1669	11	33229/49218
C5315	2307	10	50294/78941
C6288	2406	240	45571/67045
C7552	3512	8	67695/101564

Back-annotation has been done using the standard parameters described in [7] and only RC parasites have been considered. The results of this initial analysis are presented in figure 5. For every netlist, KLU executes faster than Sparse 1.3 and, the time difference increases with the dimension of the matrix, as shown in the figure 5. It is interesting to notice that both curves present the same features as the peaking corresponding to the c499_ann and the c5315_ann netlists, due to the fact that the annotated netlist has more elements than the next (not annotated) one. This behavior suggests that both solvers execute in a time proportional to the number of non-zero elements in the matrix, with KLU being more efficient and asymptotically more efficient (as the curves diverges). The obtained results are consistent with the ones reported by Davis [12].

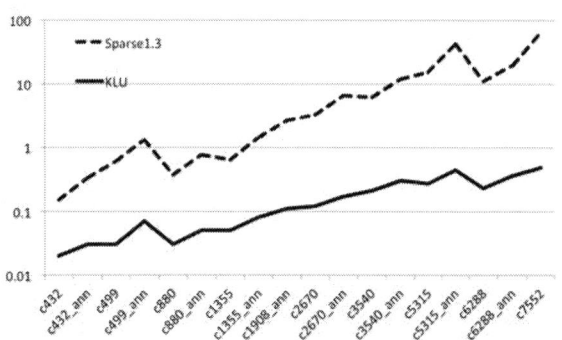

Figure 5. The execution times for the solution of linear systems obtained from the ISCAS85 suite. The "_ann" netlists contain RC back-annotation.

III. NGSPICE IMPLEMENTATION RESULTS

Afterwards KLU solver has been implemented into the NGSPICE circuit simulator to test its performances on the solution of non-linear systems in the transient analysis loop.

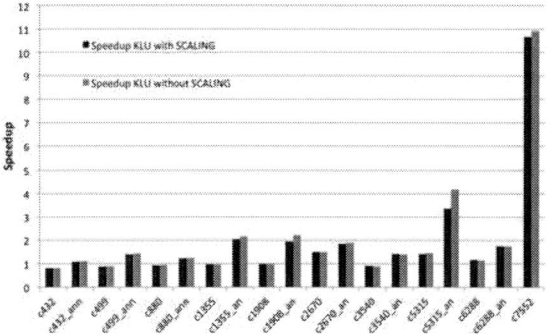

Figure 6. Speedup obtained by KLU (with and without scaling) over Sparse 1.3 solver during the transient analysis on the netlists of ISCAS85.

The results are shown in figure 6, where the speedup for the single transient analysis iteration is reported. KLU results slower than Sparse on the six smaller (of the ten) non-annotated netlists and always faster for the annotated ones. The maximum speedup obtained is 11 for the c7552 netlist with a mean value of 1.8. This behavior on small netlists is imputable to the higher cost of re-factorization in KLU. Sparse function *spFactor* is more performant than it's KLU counterpart *klu_refactor*.

The impact matrix scaling, to improve pivoting, has been verified. Matrix scaling solves numerical stability at the cost of additional computation at each re-factorization. This is clearly visible in figure 6 where, the execution times are higher in fifteen of the nineteen netlists and, almost equal in the remaining cases.

IV. CONCLUSIONS

The implementation of the KLU linear solver into NGSPICE has been presented as a possible solution to the ITRS requirements of reducing the simulation time of large digital and mixed-signal ICs. The KLU solver has been verified to be up to 100 times faster than Sparse in the solution of a linear system coming from MNA analysis, and only 11 times faster in the execution of a transient analysis, when implemented into the circuit simulator. This behavior has been attributed to the higher complexity of KLU refactoring function, compared to the Sparse one, making KLU an effective replacement for very large circuits only. In figure 6 it is evident that considerable speedup is obtained for the c7552 netlist only.

Additional speedup can be obtained by optimizing the implementation of KLU into NGSPICE. The actual research is focused on implementing other linear solvers (UMFPACK and SuperLU) to compare performances on serial and, where available, parallel implementations.

ACKNOWLEDGMENT

The authors would acknowledge Zia Abbas, Antonio Mastrandrea and Francesco Menichelli, for the fruitful discussions, their invaluable support in the implementation of KLU code in NGSPICE and in the automation of simulations and reports that greatly reduced the data analysis time.

REFERENCES

[1] ITRS, "International Technology Roadmap for Semiconductors (Design)," ITRS Design Roadmap, www.itrs.net, p. 19, 2009.

[2] ITRS, "International Technology Roadmap for Semiconductors (Modeling and Simulation)," ITRS Modeling and Simulation Roadmap, www.itrs.net, p. 19, 2009.

[3] N. Kapre, "SPICE² – A Spatial Parallel Architecture for Accelerating the SPICE Circuit Simulator", Ph.D. Thesis, California Institute of Technology, 2010.

[4] A. Vladimirescu, "LSI Circuit Simulation on Vector Computers", Memorandum No. UCB/ERL M82/75, 1982.

[5] R. Poore, "GPU-accelerated time-domain circuit simulation," *Custom Integrated Circuits Conference, 2009. CICC*, no. Cicc, pp. 629-632, 2009.

[6] M. C. Hansen, H. Yalcin, J. P. Hayes, "Unveiling the ISCAS-85 Benchmarks: A Case Study in Reverse Engineering,", *IEEE Design & Test*, 16(3), 1999.

[7] J. Xu, "SPICE Simulation of ISCAS85 Benchmark Circuit for research" available online at:http://www.ece.uic.edu/~masud/iscas2spice.htm.

[8] T. H. Weng, R.K. Perng, and B. Chapman, "OpenMP Implementation of SPICE3 Circuit Simulator," *International Journal of Parallel Programming*, vol. 35, no. 5, pp. 493-505, Jul. 2007.

[9] Ngspice Circuit Simulator, avalable online at: http://www.ngspice.org.

[10] S. Hutchinson, et al., "The Xyce Parallel Electronic Simulator – An Overview," *Proceedings of the International Conference ParCo2001*, Naples, Italy, September 2001.

[11] K. S. Kundert. Sparse matrix techniques. In *Circuit Analysis, Simulation and Design*, part 1, A. E. Ruehli (editor), North-Holland 1986.

[12] T. Davis and E. P. Natarajan, "Algorithm 907: KLU, a direct sparse solver for circuit simulation problems," *ACM Transactions on Mathematical Software*, vol. 37, no. 3, pp. 36:2 - 36:17, 2010.

[13] H. M. Markowitz, "The elimination form of the inverse and its application to linear programming", *Management Sci.*, 3 (1957), pp. 255–269.

[14] I. S. Duff, "On algorithms for obtaining a Maximum Transversal," *ACM Transaction,* vol. 7, no. 3, pp. 315-330, 1981.

[15] R. E. Tarjan, "Depth-first search and linear graph algorithms," *SIAM Journal,* 1972.

[16] T. A. Davis, J. R. Gilbert, S. Larimore and E. Ng, "A column approximate minimum degree ordering algorithm,", *ACM Trans. Math. Softw.,* 30, pp. 35-376, 2004.

[17] J. R. Gilbert, and T. Peierls, "Sparse partial pivoting in time proportional to arithmetic operations,", *SIAM J. Sci. Statist. Comput.* 9, pp. 862–874, 1988.

MIXDES 2012, 19th International Conference *"Mixed Design of Integrated Circuits and Systems"*, May 24-26, 2012, Warsaw, Poland

Simulation Study of Nanoscale Double-Gate CMOS Circuits Using Compact Advanced Transport Models

Muthupandian Cheralathan[1], Esteban Contreras[2], Joaquín Alvarado[3], Antonio Cerdeira[2] and Benjamin Iñiguez[1]

[1] DEEAE, Universitat Rovira i Virgili, 26 Av. Paisos Catalans, Tarragona 43007, Spain

[2]SEES, CINVESTAV, Av. IPN No.2508, Apto.Postal 14-740, 07360 DF, Mexico

[3]CIDS, Benemérita Universidad Autónoma de Puebla, 72570 Puebla, México

muthupandian.cheralathan@urv.cat

Abstract—In this paper we present the results of the implementation of a nanoscale double-gate (DG) MOSFET compact model, which includes hydrodynamic transport model, in Verilog-A in order to carry out circuit simulation. The model in Verilog-A is used with the SMASH circuit simulator for the analysis of the DC and transient behavior electrical CMOS circuits. A template device representative for a downscaled symmetric double-gate MOSFET was used to validate the model. A CMOS inverter has been analyzed. Comparison between the drift-diffusion (DD) and hydrodynamic transport model within the practical range of bias voltages has been highlighted.

Index Terms—**Double-gate MOSFET, Hydrodynamic, Verilog-A, SMASH.**

I. INTRODUCTION

The device-scaling concept has been the main guiding principle of the MOS-device engineering over the past few decades [1]. As the conventional bulk MOSFET technology is scaling down towards the practical limit, DG MOSFETs appeared as a promising technological alternative that has attracted substantial research interests due to superior short channel control, volume inversion, etc. There has been work dedicated to modeling and simulation of DG MOSFETs. Compact core models for undoped and doped symmetric DG MOSFETs have been presented [2-6]. The development of models to simulate circuits containing new devices is an important task to allow the introduction of these devices in practical applications. Accurate and time computationally efficient compact models of DG MOSFETs are required to predict or simulate circuit performance. For a circuit simulation the main task is to count with a precise transistor model to reproduce the transistor behavior, which is either already introduced or can be implemented in the commercial circuit simulators to be used. In recent years, some work has been done on the implementation of DG MOSFET compact models in circuit simulators [7-10].

In this work, we present the implementation of a DG MOSFET compact model including hydrodynamic transport model in Verilog-A for circuit simulators. The model is based on an analytical expression that models the variation of surface potential as well as the difference of potential at the surface and at the middle of the silicon layer [2]. This model for the potentials is used in an analytical compact model for the drain current of a double-gate MOSFETs derived from a core charge

control model which results from the solution of the 1D Poisson's equation [3]. Electrostatic short-channel effects (threshold voltage roll-off, DIBL, subthreshold swing degradation) were introduced in the core model using scalable and geometry dependent equations. This model is valid from lightly doped to highly doped devices. We extended this model to include hydrodynamic transport model [11]. Besides, charge and capacitance models consistent with the DC model were developed. The quantum effects are not considered in this model, since we have targeted silicon thicknesses larger than 10 nm. The entire model is implemented in Verilog-A, which allows the use of the model in commercial circuit simulators for circuit design of both digital and analog applications. We have shown an example of DG MOSFET based CMOS inverter circuit. Finally, we compare results between drift-diffusion and hydrodynamic transport models within the practical range of voltages.

II. SIMULATED DEVICE

The 22 nm template transistor we considered is shown in Figure 1. The 22 nm DG MOSFET with a gate length of 22 nm, a gate stack consisting of 2.4 nm of HfO_2 on top of 0.7 nm of SiO_2 (EOT = 1.1 nm). The silicon film thickness is 10 nm. The channel is undoped (10^{15} cm^{-3}). We have considered both n-channel and p-channel DG MOSFETs with these dimensions.

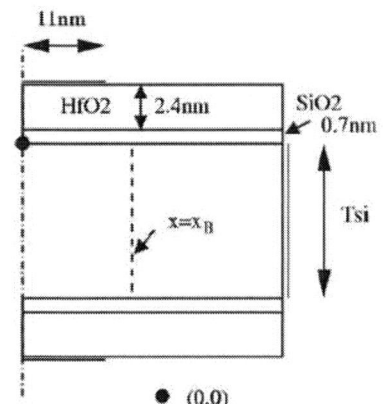

Figure 1 Structure of the 22 nm template DG MOSFET considered in the work. One half of the symmetric structure is shown. All dimensions are in nm.

III. DG MODEL

The potentials at the surface Φ_S and in the center Φ_O of the silicon layer can be calculated analytically as shown in [2]. The surface potential in the subthreshold Φ_{SBT} and in the above threshold Φ_{SAT} regimes are calculated analytically using the Lambert function. Finally, the overall surface potential in all regions can be calculated as [3]:

$$\phi_s = \phi_{sBT}\frac{1}{2}\left\{1 - \tanh\left[10\left(V_{gs} - V_T - V_{ch}\right)\right]\right\}$$
$$+ \phi_{sAT}\frac{1}{2}\left\{1 + \tanh\left[10\left(V_{gs} - V_T - V_{ch}\right)\right]\right\} \quad (1)$$

where V_{gs} is the applied gate voltage, V_T is the threshold voltage and V_{ch} is the channel voltage. The Lambert function is represented in Verilog-A language as a built-in function conserving the requirements of a compact model. This analytical surface potential calculation gives the possibility of writing charge carrier calculation at source, q_s, and at the drain q_d as explicit functions of the applied voltages.

The drain current in a DG MOSFET is calculated as a function of the mobile charge densities at the source q_s and at the drain q_d as [3], assuming a drift-diffusion transport:

$$I_{DS} = 2\frac{WC_{ox}\phi_t^2\mu_s}{L}\left[2\left(q_s - q_d\right) + \frac{q_s^2 - q_d^2}{2} + q_{dep}\ln\left[\frac{q_d + q_{dep}}{q_s + q_{dep}}\right]\right] \quad (2)$$

where W and L are width and length of the device respectively. C_{ox} is the gate capacitance; Φ_t is the thermal voltage, q_b is the normalized fixed charge concentration in the silicon layer, μ_s is the surface mobility. The expression for the drain current in eqn. (2) is used as the core model for DG MOSFET. In the complete model, the effects of velocity saturation, channel length modulation, threshold voltage roll-off, DIBL and subthreshold swing degradation are all included.

In extremely short channel devices, the transport regime is quasi-ballistic; thus, an important overshoot velocity is expected. The velocity overshoot is included in the model using a one dimensional energy-balance model [11]. The velocity overshoot is modeled assuming a hydrodynamic transport model which is included in the core drain current model. The final drain current expression accounting for the hydrodynamic transport model is given as:

$$I_{DS} = 2\frac{WC_{ox}\varphi_t^2\mu_s}{L(1+\gamma_n V_{dss})}\left[2\left(q_s - q_d\right) + \frac{q_s^2 - q_d^2}{2} + q_{dep}\ln\left[\frac{q_d + q_{dep}}{q_s + q_{dep}}\right]\right] \quad (3)$$

where, $\gamma_n = \frac{\mu_{eff}}{v_{sat}L}\left(\frac{1}{1 + 2\lambda_w/L}\right)$ takes into account both the velocity saturation effect and the hydrodynamic transport [11], through which the velocity overshoot is also modelled, $\lambda_w \approx 2v_{sat}\tau_w$ being the energy relaxation length, τ_w the energy relaxation time constant, v_{sat} the saturation velocity and V_{dss} is the effective drain-source voltage. The charge and capacitances expression are developed following the procedure presented in [6] for undoped devices, but considering the doping in the charge control model.

IV. CMOS INVERTER

We have studied the behavior of a CMOS inverter based on DG MOSFETs with the technological features given in Figure 2. The supply voltage is set to 1V. The load capacitance used for this inverter is 7fF.

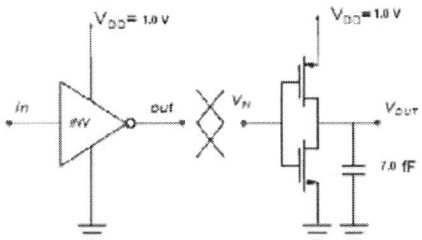

Figure 2 Simulated CMOS Inverter

V. RESULTS AND DISCUSSIONS

We have used SMASH circuit simulator [12] to carry out the simulations presented in this work. The model has been implemented in Verilog-A code for both n-channel and p-channel DG MOSFETs and compiled to include the DG devices as new active components of the circuit simulator.

In [11] it was shown that our hydrodynamic compact model agreed very well with Monte Carlo (MC) simulations of nanoscale devices, including the same template device we are considering in this paper.

Figure 3 shows the transfer characteristics of drift-diffusion and hydrodynamic transport model at low and high V_{DS}. From the curves it can seen that the hydrodynamic transport model included in the core model gives higher drain current than the DD model, due to the velocity overshoot. This can be clearly seen at higher drain bias.

Figure 4 shows the output characteristics of drift-diffusion and hydrodynamic transport model at high V_{GS}. As expected, (because of the velocity overshoot effect) it can be seen from the curves that the hydrodynamic transport model has higher drain current than the DD model in the saturation region.

Figure 5 shows the voltage transfer of a CMOS inverter obtained using our DG MOSFET model in SMASH by Verilog-A. The channel width of the p-channel device is twice the one of the n-channel device. The curves show the DD and hydrodynamic model. It can be seen that the switching voltage is higher using the hydrodynamic model than the DD model.

Figure 6 shows the transient response of a CMOS inverter with DD and hydrodynamic model. It can be seen that the rise time is much shorter in the hydrodynamic model than in the DD model. The hydrodynamic model gives a therefore a smaller delay than the DD model.

Figure 7 shows capacitance characteristics for the DG MOSFET as a function of gate voltage. We show that the approximation considered in the calculation of capacitances in the Verilog-A code using SMASH gives practically the same result as the values directly obtained from the capacitance model (Fig. 7a-b).

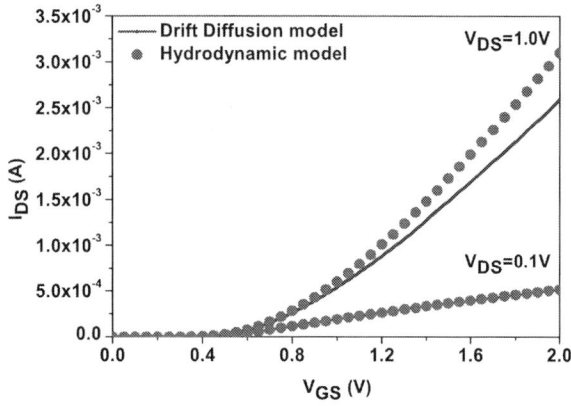

Figure 3 Transfer characteristics obtained for an n-channel DG MOSFET L_g=22 nm T_s=10 nm and EOT=1.1 nm.

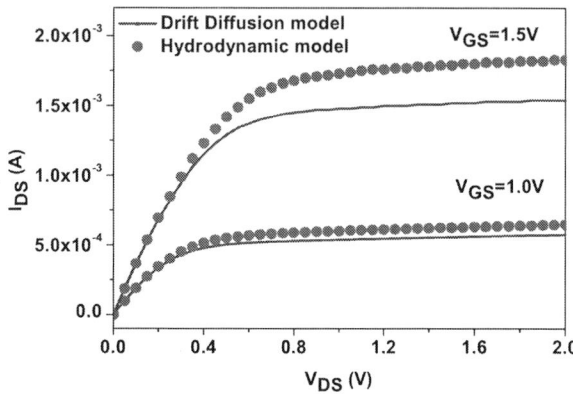

Figure 4 Output characteristics obtained for an n-channel DG MOSFET L_g=22 nm T_s=10 nm EOT=1.1 nm.

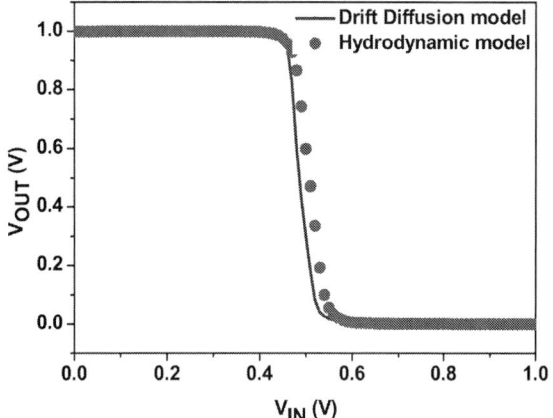

Figure 5 Voltage transfer characteristics of a CMOS inverter using the DG MOSFET model in Verilog-A with SMASH.

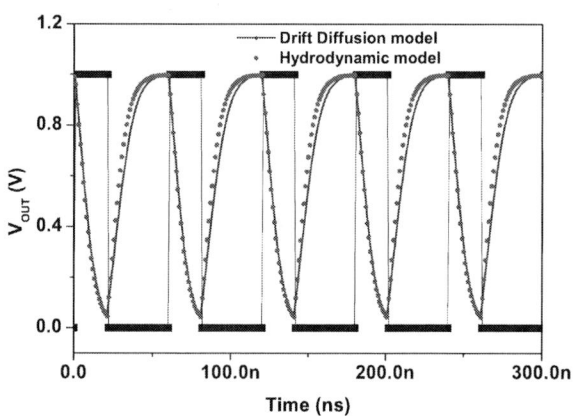

Figure 6 Transient response of CMOS inverter using our DG MOSFET model in SMASH by means of Verilog-A.

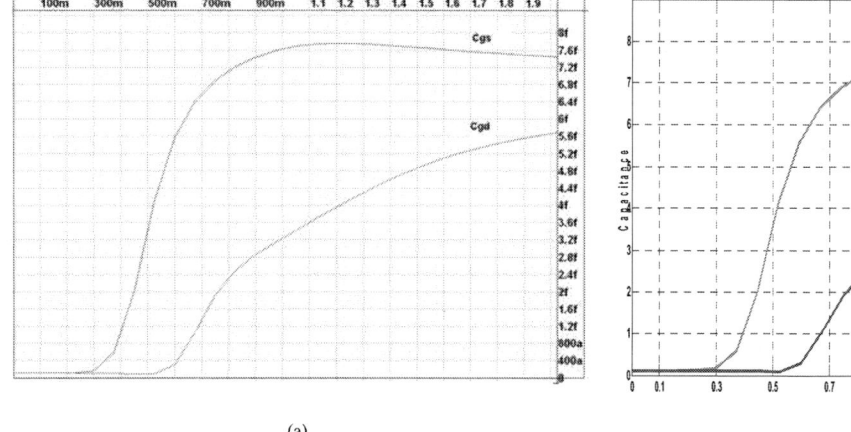

Figure 7 Gate-to-source capacitance (C_{gs}) and gate-to-drain capacitance (C_{gd}), for V_{DS}=0.5V, obtained using the method implemented in Verilog-A with SMASH (a) and the entire capacitance model developed (b).

To use the capacitance model in Verilog-A with SMASH, we have to calculate the charges called Q_{gs} and Q_{gd} from the numerical integration of the device capacitances C_{gs} and C_{gd}, obtained from the model, from $V_{gs}=0$ to V_{GS}, and from $V_{gd}=0$ to V_{GD}, respectively, where V_{GS} and V_{GD} are the applied gate-source and gate-drain voltages:

$$Q_{gs} = \left\{ \begin{array}{c} \sum_{i=0}^{n} C_{gs}(i) * (V_{gs}(i) - V_{gs}(i-1)) \\ where \; V_{gs}(i=0) = 0 \\ V_{gs}(i=n) = V_{gs} \end{array} \right\}$$

$$Q_{gd} = \left\{ \begin{array}{c} \sum_{i=0}^{n} C_{gd}(i) * (V_{gd}(i) - V_{gd}(i-1)) \\ where \; V_{gd}(i=0) = 0 \\ V_{gd}(i=n) = V_{gd} \end{array} \right\}$$

The simulator then obtains the charging/discharging currents by differentiating of charges with respect to the time. This modeling is used to obtain the capacitances C_{gs} and C_{gd} in Verilog-A with SMASH, shown in Fig. 7(a).

VI. CONCLUSIONS

In this paper, we present the implementation of a compact nanoscale DG MOSFET model, including an hydrodynamic transport model, in Verilog-A for circuit simulation. We carry out a comparison of the simulated performance of a CMOS inverter using both the hydrodynamic transport model and the drift diffusion model for nanoscale DG MOSFETs are presented. As expected, smaller delays are obtained with the hydrodynamic model.

ACKNOWLEDGMENT

This work was supported by Ministerio de Ciencia e Innovacion under project TEC2011-28357-C02-01, by the European Commission under contract FP7-PEOPLE-2007-3-1-IAPP No. 218255 "Compact Modelling Network (COMOn)", by the ICREA Academia Award and by the PGIR Grant from URV, the project Conacyt no. 5646 and by the project 2010 CONE2 00061 from Catalan Government. J. Alvarado thanks to CONACyT for its support by the program "Programa de Apoyo Complementario para la Consolidación Institucional (Fondo Institucional) Repatriación y Retención". Authors would like to thank, Gilles Depeyrot, Frederic Poullet and Cedric Valla of Dolphin Integration for their advice.

REFERENCES

[1] International Technology Roadmap for Semiconductor 2009 and the 2010 update, http://public.itrs.net

[2] Cerdeira A, Moldovan O, Iñiguez B, Estrada M. Modeling of potentials and threshold voltage for symmetric doped double-gate MOSFETs. Solid-State Electronics 52 (2008): 830-837.

[3] Cerdeira A, Iñiguez B, Estrada M. Compact model for short channel symmetric doped double-gate MOSFETs. Solid-State Electronics 2008; 52: 1064-1070.

[4] F. Lime, B. Iniguez and O. Moldovan , "A quasi-two-dimensional compact drain-current model for undoped symmetric double-gate MOSFETs including short-channel effects," IEEE Trans. Electron Devices, vol. 55, no. 6, pp. 1441-8, June 2008.

[5] O. Moldovan, F.A. Chaves, D. Jiménez, J. P. Raskin and B. Iñiguez, "Accurate prediction of the volume inversion impact on undoped double gate MOSFET capacitances," Int. J. Numer. Model., vol.23, pp. 447-457, Jan. 2010.

[6] O. Moldovan, D. Jiménez, J.R. Guitar, F.A. Chaves, and B. Iñiguez, "Explicit analytical charge and capacitance models of undoped double-gate MOSFETs," IEEE Trans. Electron Devices, vol.57, no.7, pp. 1718-1724, Jul. 2007.

[7] Xingye Zhou; Jian Zhang; Zhize Zhou; Lining Zhang; Chenyue Ma; Wen Wu; Wei Zhao; Xing Zhang; , "An improved computationally efficient drain current model for double-gate MOSFETs," Solid-State and Integrated Circuit Technology (ICSICT), 2010 10th IEEE International Conference on , vol., no., pp.1874-1876, 1-4 Nov. 2010.

[8] E. Contreras, A. Cerdeira, J. Alvarado and M.A. Pavanello, "Application of the symmetric doped double-gate model in circuit simulation containing double-gate graded-channel transistors," J. Inte. Circuits and Systems, vol.5, no. 2, pp 110-115, 2010.

[9] O. Cobianu, O. Soffke and M. Glesner, "A verilog-a model of an undoped symmetric dual-gate MOSFET," Adv. Radio Sci., vol. 4, pp 303-306, 2006.

[10] J. Alvarado, B. Iñiguez, M. Estrada, D. Flandre and A. Cerdeira, "Implementation of the symmetric doped double-gate MOSFET model in verilog-a for circuit simulation," Int. J. Numer. Model, vol.23, pp. 88-106, Jul 2009.

[11] M. Cheralathan, C. Sampedro, J. B. Roldan, F. Gamiz, G. Iannaccone, E. Sangiorgi and B. Iniguez, "Compact drain-current model for reproducing advanced transport models in nanoscale double-gate MOSFETs," Semicond. Sci. Technol., vol.26, 095015(7pp), Jul 2011.

[12] SMASH User Manual Version 5.18 release 2012.

MIXDES 2012, 19th International Conference *"Mixed Design of Integrated Circuits and Systems"*, May 24-26, 2012, Warsaw, Poland

Standardization of Multigate MOSFET Modeling

Nicolas Chevillon, Fabien Prégaldiny, Christophe Lallement
InESS / Université de Strasbourg
Parc d'innovation, BP 10413,
67412 Illkirch Cedex, France
Email: f.pregaldiny@unistra.fr

Jean-Michel Sallese
Ecole Polytechnique Fédérale de Lausanne (EPFL)
CH-1015 Lausanne, Switzerland
Email: jean-michel.sallese@epfl.ch

Abstract—**A fully explicit and physics-based model has been developed for arbitrary shapes of lightly doped long-channel non-planar multigate MOSFETs (MuGFETs). Through the definition of equivalent geometrical parameters, MuGFETs as quadruple-gate (QG), triple-gate (TG), triangular gate, cylindrical gate-all-around (GAA), and FinFETs, are mapped into the symmetric double-gate (DG) MOSFET topology without the need to introduce any unphysical parameter. Based on this modeling approach, any multigate architecture inherits of the fundamental relationships that have been developed for planar DG MOSFETs, including the normalization of all electrical quantities that considerably simplifies its analysis. We propose here to extend the validity of this MuGFETs standard model to short channels through a 2-D short-channel effect modeling of the DG FinFET. 3-D numerical simulations of small geometries of TG MOSFETs are compared with the model.**

Index Terms—**multigate MOSFET, short-channel, 3-D TCAD simulations.**

I. INTRODUCTION

The challenge of the CMOS scaling is mainly oriented to the multigate MOSFET devices, which propose better immunity against short-channel effects than bulk MOSFET devices. Triple-gate MOSFET are foreseen as next-generation devices for microprocessors [1]. Concerning their compact modeling, the simplest multigate MOSFET device to model is undoubtedly the fully depleted symmetric double-gate (DG) MOSFET. Different approaches have been proposed. Yu et al. initially developed an exact solution [2] that is still difficult to use in the context of compact modeling, where more simple relationships are preferred. Next, a more compact formulation was proposed by Sallese et al. [3], where an approximate solution was sought and led to a more compact but still accurate formulation. However, these models concern the symmetric DG MOSFET, which technology is quite challenging and not yet adopted by the industry.

Conversely, multigate MOSFETs such as FinFETs and Ω-FETs [4] have proven their compatibility with standard CMOS technology and lithography techniques, leading to basic circuits [5]. However, modeling multigate architectures is quite complex, the main reason being that the structure is 2-D, and hence, no simple solution exists for the Poisson equation. An empirical approach was proposed by Yu et al. [6], who introduced some arbitrary smoothing functions and fitting parameters in order to fit the charges and the current in nonplanar multigate topologies.

Next, an interesting and complete analytical modeling of multiple-gate MOSFETs in subthreshold was proposed by Ritzenthaler et al. [7]. The potential in the volume of the channel was obtained by solving the 3-D Laplace equation, where the electrostatic potential along the "most leaky path" from source to drain was used to get an analytical subthreshold current expression. As for [6], the drain current characteristics from weak to strong inversion were obtained by means of smoothing functions.

Recently, we have proposed a simple solution to model in a compact way multigate MOSFETs regardless of their architecture without the need to introduce any unphysical parameter [8]. Undoped QG, cylindrical GAA, TG MOSFETs can be modeled with a common formalism through a DG model. This standard model is currently valid only for long-channel devices. In this paper, we propose to include in the standard model, a modeling of short-channel effects initially developed for the DG FinFET [9]. We also evaluate the accuracy of the new compact model by comparisons with 3-D TCAD simulations of short-channel TG MOSFETs.

The paper is organized as follows. Section II describes the principle of the multigate standard model. Validations on typical surrounding-gate FETs show the performances of the model. Section III presents the compact modeling of non-surrounding-gate FETs such as TG MOSFETs and DG FinFETs. Finally, we present and discuss the relevance of the use of a 2-D short-channel modeling for devices with 3-D architectures.

II. STANDARD COMPACT MODEL FOR LONG-CHANNEL MuGFET

Unlike planar DG MOSFETs, strictly speaking, the Poisson equation has to be solved in two dimensions in more complex MuGFET architectures for long-channel devices. Unlike its 1-D counterpart, this nonlinear differential equation has no exact analytical solution. To overcome this problem, it is interesting to introduce some simplifications that are based on two simple assumptions. Likewise for the DG MOSFET, in strong inversion, mobile charges are located at the Si/SiO$_2$ interfaces, the volume of silicon having little impact on the global charge density. This means that the device reverts to a quasi 1-D system depending on the silicon channel perimeter of the cross section area, which suggests that the solution obtained for the long-channel DG MOSFET may still be

78

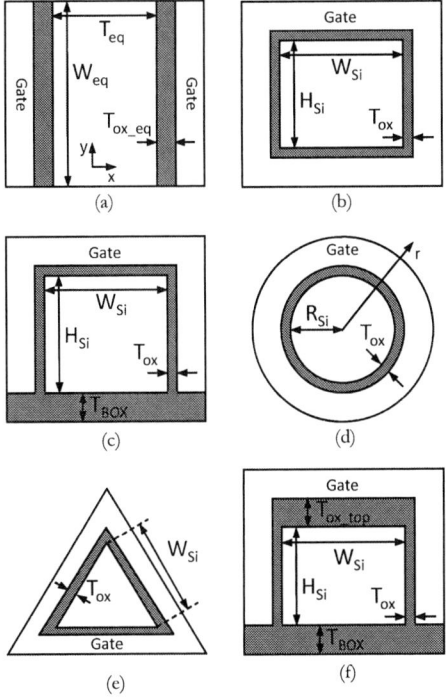

Fig. 1. Schematic of multigate MOSFET cross sections. (a) DG MOSFET. (b) QG MOSFET. (c) TG MOSFET. (d) GAA MOSFET. (e) Triangular MOSFET. (f) DG FinFET.

accurate enough to estimate the charge density in strong inversion for the MuGFETs.

Now, considering weak inversion (although subthreshold would be more adopted when considering undoped devices, we propose to still use "weak inversion" for simplicity), the charge density is negligeable, so the Poisson equation becomes a laplacian, and we note that setting the potential to a constant value is a viable solution. Therefore, the charge density is uniform in the silicon, and the current only depends on the area of the cross section of the body. The device is in volume inversion.

To some extent, these arguments justify that the multigate MOSFET can be "planarized" and considered as a DG MOS-FET having the same Si/SiO$_2$ interface perimeter, ensuring consistency with strong inversion, and also the same volume of silicon to match subthreshold characteristics. In this attempt, we simply ignore corner effects, which are known to be negligible in a lightly doped channel [10].

Based on these considerations, we propose to model the MuGFETs as a simple double-gate MOSFET, by using equivalent geometrical parameters [8]. Assuming a constant oxide capacitance along the silicon channel perimeter, we propose a compact modeling standardization of multigate MOSFETs of Surrounding-Gate (SG) geometries [i.e. Figs. 1(a,b,d,e)]. Starting from a device having a silicon channel with section S and perimeter P, we define an equivalent thickness T_{eq} for

the double-gate compact model [3] as:

$$T_{eq} = \frac{2 \cdot S}{P} \tag{1}$$

According to the compact model of the symmetric DG MOSFET [2], the normalized charge potential relationship in a MuGFET is given by:

$$v_g - \Delta\phi - v_{to} - v_{ch} = 4 \cdot q_g + \ln(q_g)$$
$$+ \ln\left(1 + q_g \cdot \frac{C_{ox}}{C_{Si\_eq}}\right) \tag{2}$$

where $C_{Si\_eq} = \varepsilon_{Si}/T_{eq}$ is the equivalent silicon capacitance and q_g is the gate charge density for one gate.

In the definition of the drain current, we use the same specific current I_{SP} as in [3] where the silicon width is half parameter of the silicon perimeter, $W_{eq} = P/2$, so the specific current is given by:

$$I_{SP} = 4 \cdot \mu \cdot C_{ox} \cdot U_T{}^2 \cdot \frac{W_{eq}}{L} \tag{3}$$

and the corresponding normalized current becomes :

$$i = -q_m^2 + 2 \cdot q_m + 2 \cdot \frac{C_{Si\_eq}}{C_{ox}} \cdot \ln\left(1 - q_m \cdot \frac{C_{ox}}{2 \cdot C_{Si\_eq}}\right)\Big|_{q_{ms}}^{q_{md}} \tag{4}$$

where $q_m = -2 \cdot q_g$ is the mobile charge density.

III. EXAMPLES OF MuGFET MODELING

In this section, we propose to assess the analytical model for Quadruple-Gate (QG) MOSFETs [see Fig. 1(b)] and Gate-All-Around (GAA) MOSFETs [see Fig. 1(d)] based upon the equivalent thickness and width versus full 3-D numerical simulations [11]. In order to minimize the number of parameters, a constant mobility of 1000 cm^2/V·s was used. Physical and equivalent parameters were obtained from the same set of geometrical and technological parameters. Simulations have been carried out in a QG MOSFET having a square cross section of 40 nm × 40 nm. The channel length was set to 1 μm. In this case, the equivalent silicon thickness (Eq. 5) and channel width (Eq. 6) are 20 and 80 nm, respectively.

$$T_{eq(QG)} = \frac{H_{Si} \cdot W_{Si}}{H_{Si} + W_{Si}} \tag{5}$$

$$W_{eq(QG)} = H_{Si} + W_{Si} \tag{6}$$

Figs. 2 and 3 show the current versus the gate voltage at low and high drain voltages, as well as the current versus the drain voltage for different gate voltages. For both characteristics, a comparison of the standard model with 3-D simulations confirms the validity of the modeling approach presented so far. With regard to the drain current dependence on the gate voltage (see Fig. 2), the accuracy is pretty good, both in linear and log scales, and without the need to introduce any empirical parameter. Similarly, the matching is very good for the output characteristics, as shown in Fig. 3. Therefore, the transformation of the QG MOSFET into a DG MOSFET based

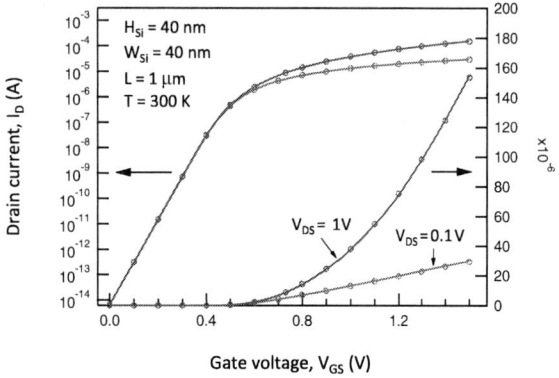

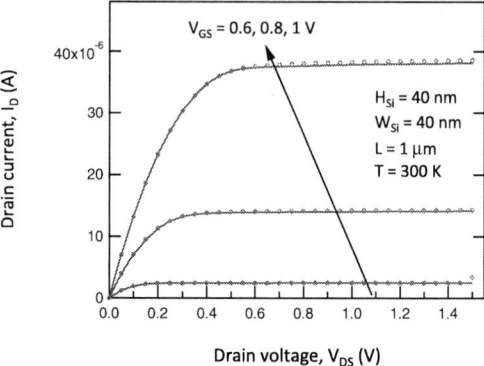

Fig. 2. Drain current of a QG MOSFET with a square cross section as a function of gate voltage in linear and saturation regimes. Symbols: 3-D simulations, lines: compact model.

Fig. 3. Drain current of a QG MOSFET with a square cross section as a function of drain voltage in linear and saturation regimes. Symbols: 3-D simulations, lines: compact model.

on the equivalent geometrical parameter definitions is totally justified.

About the GAA MOSFET, we ran TCAD simulations for different silicon radii of the current versus gate voltages. Fig. 4 shows that the standard model is very accurate. Indeed, the definition of the equivalent thickness is fully consistent with the exact solution of the Poisson-Boltzmann equation for the GAA MOSFET [8]. Considering the correct definition of the gate oxide capacitance, it comes out that the electrical characteristics of a GAA with radius R_{Si} can be predicted from a planar DG model using equivalent geometrical parameters $T_{eq(GAA)} = R_{Si}$ and $W_{eq(GAA)} = \pi \cdot R_{Si}$.

As far as the gate oxide thickness is uniform, the equivalent thickness has proven to be able to simulate nonplanar multigate geometries with great accuracy. The case of nonconstant oxide capacitance is however important and merits some attention. We address this nonideal case in the next section.

IV. CASE OF A PARTIALLY SURROUNDING-GATE FET: THE FinFET

When the gate is not uniform all around the channel, as for DG FinFETs [see Fig. 1(f)] or TG MOSFETs [see Fig. 1(c)], the gate capacitance is no longer constant and, *a priori*, the model can no longer be used. In the weak inversion regime, this topology is expected to give the same current as for the QG since it only depends upon the cross-section area of the silicon channel. However, in strong inversion, the gate perimeter facing the thin gate oxide T_{ox} is smaller than for the QG, and implies that the ON-state current will be decreased with respect to the QG. The drain current appears mostly at the lateral Si/SiO$_2$ interfaces of C_{ox} oxide thickness.

However, even for an oxide layer thicker than its nominal value, the related interface could still be a conducting channel when the gate bias exceeds threshold voltage at this interface. This is indeed supported by 3-D simulations. Therefore, additional currents located at thick oxide interfaces should also be taken into account when accurate modeling of non-SG architectures is carried out.

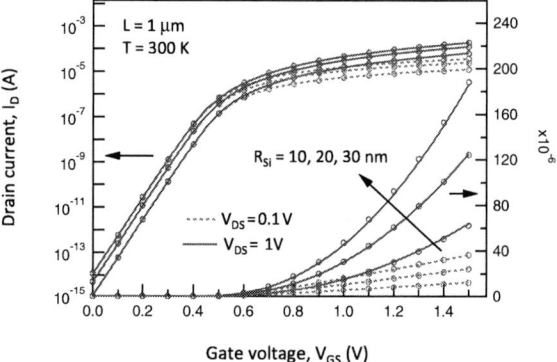

Fig. 4. Drain current of a GAA MOSFET as a function of gate voltage in linear and saturation regimes for different silicon radii. Symbols: 3-D simulations, lines: compact model.

Basically, if we consider that, in strong inversion, each channel is "independent," meaning that the normal electric field at the Si/SiO$_2$ interfaces dominates, the total current of the DG FinFET (see Fig. 5) can be considered as the sum of all Si/SiO$_2$ channel currents. In addition, if the oxide thickness is different at each interface, it will affect both the related threshold voltage and the normalized current [see (4)].

At this point, it is interesting to introduce a basic assumption that consists in considering that the threshold voltages for thin and thick gate oxide interfaces will be almost the same. This assumption means that the strong inversion current at each interface will appear for the same gate voltage.

Besides, in strong inversion, since the normalized mobile charge density q_m is now assumed not to depend on the gate oxide thickness, so will be the normalized current i_{SI}, which, in strong inversion, is approximated by the integral from source to drain of $-q_m^2 + 2 \cdot q_m$. This means that the total current in strong inversion $I_D{}^{SI}$ for the DG FinFET can be expressed via the sum of specific current I_{SP} for each

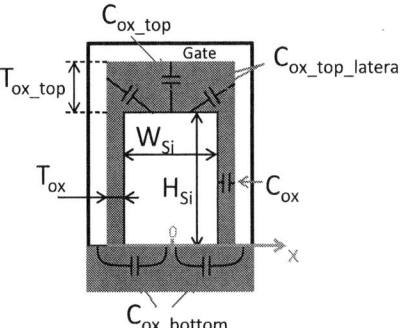

Fig. 5. Equivalent gate oxide capacitance network of the DG FinFET.

channel.

$$I_D{}^{SI} = i^{SI} \cdot (2 \cdot I_{SP_{\text{Lateral}}} + I_{SP_{\text{Top}}} + I_{SP_{\text{Bottom}}}) \qquad (7)$$

After transformations [8], the specific current of the DG FinFET is

$$I_{SP_{\text{FinFET}}} = 4 \cdot \mu \cdot C_{ox\_eq} \cdot U_T{}^2 \cdot \frac{H_{Si} + W_{Si}}{L} \qquad (8)$$

According to this expression of the specific current, we can consider the DG FinFET as a QG MOSFET [see (3)] with an oxide capacitance C_{ox_eq}, which is the equivalent oxide capacitance of the DG FinFET that reverts to the mean value of the capacitances *per unit surface* from each Si/SiO$_2$ interface [8]:

$$C_{ox\_eq} = \frac{C_{ox} \cdot H_{Si} + C_{ox\_top\_eq} \cdot W_{Si}/2 + C_{ox\_bottom} \cdot W_{Si}/2}{H_{Si} + W_{Si}}$$
$$(9)$$

where $C_{ox\_top\_eq}$ and C_{ox_bottom} are the equivalent gate oxide capacitance values *per unit surface* for both top and bottom interfaces, respectively [8]. Fig. 5 describes each elementary capacitance in the case of the DG FinFET.

The equivalent gate oxide capacitance has been derived under the strong inversion condition, but since the subthreshold charge density is almost independent of the capacitance, we can still adopt the equivalent capacitance regardless of the mode of operation.

The normalized current is still given by:

$$i = -q_m{}^2 + 2 \cdot q_m + 2 \cdot \frac{C_{Si\_eq}}{C_{ox\_eq}} \cdot \ln\left(1 - q_m \cdot \frac{C_{ox\_eq}}{2 \cdot C_{Si\_eq}}\right)\Big|_{q_{ms}}^{q_{md}}$$
$$(10)$$

with the equivalent silicon capacitance $C_{Si\_eq} = \varepsilon_{Si}/T_{eq(QG)}$.

V. STANDARD MODEL EXTENTED TO SHORT-CHANNEL TRIPLE-GATE MOSFETs

A. Short-channel modeling

Now that we have a standard compact model for lightly doped long-channel MuGFETs, one step further is to include the short-channel effect modeling to move towards device geometries needed in real applications. We propose to verify

the standard model validity on short-channel TG MOSFETs by including the short-channel effect modeling from the DG FinFET compact model [9] which is based on the same DG model core.

This short-channel effect modeling was developed for DG FinFET, so it takes into account the electrostatic effects only in 2-D. Nevertheless, given the rectangular form of the silicon channel of the TG, we assume that this modeling can give good results on short channels of this architecture. In a more accurate way, we should include the influence of the top gate, but the geometries considered in this paper have a silicon height H_{Si} higher than the silicon width W_{Si}.

Briefly, the short-channel modeling uses the variation of the minimum potential in the channel compared to the potential in a long channel. The short-channel effects are included as a correction of the gate voltage v_g with the minimim potential variation $\Delta\Psi_{Smin}$ according to an explicit calculation without the need of any empirical parameter [9]:

$$v_{gN} = v_g + \Delta\Psi_{Smin} \qquad (11)$$

This correction depends on the width W_{Si}, the channel length L and the bias voltages. The channel length modulation (CLM) is also based on the DG FinFET compact model [9], and thus, is a 2-D modeling. As for the mobility degradation model used here, we keep it as in [8].

B. Results and discusion

We have studied the behavior of undoped short-channel Triple-Gate MOSFETs for channel lengths from 40 nm to 1 μm for a silicon width of 15 nm and a silicon height of 20 nm and 100 nm. The oxide thickness T_{ox} is 1.5 nm. We have run 3-D TCAD simulations taking into account the mobility degradation by using the CVT mobility model within Altlas simulator [11].

Fig. 6 shows the threshold voltage roll-off varying with the channel length. The threshold voltage is extracted according to a constant current method, $V_{th} = \arg(I_D(V_{GS}) = I_{\text{threshold}})$. The threshold current depends on the device geometry according to the following expression:

$$I_{\text{threshold}} = \alpha \cdot H_{Si} \cdot W_{Si} \cdot (L_0/L) \qquad (12)$$

with $\alpha = 1.4 \cdot 10^7$A/m^2 and $L_0 = 1\mu$m. So, the roll-off represents exclusively the short-channel effects, and an accurate comparison between any geometry is possible. The degradation of the threshold voltage is compared between the 3-D simulations results for two different silicon heights, and the standard model results where the TG is modeled as an equivalent QG. We observe a negligeable influence of the silicon height reduction on the threshold voltage degradation, between the device geometries with $H_{Si} = 100$ nm where the 2-D short-channel modeling can be accurately used, and the devices with a small height of 20 nm. Then, the use of the 2-D short-channel modeling is still relevant when the silicon height is close to the silicon width, while higher than or equal to it.

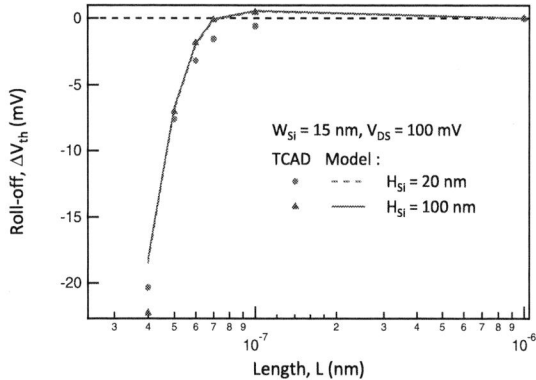

Fig. 6. Comparison of roll-off between TCAD simulations and model with threshold extraction depending on the gate length.

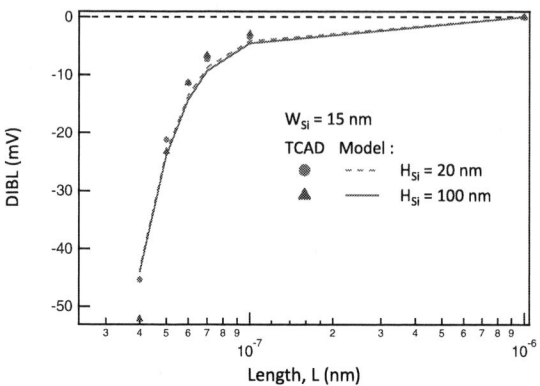

Fig. 8. Comparison of DIBL between TCAD simulations and model with DIBL depending on the gate length.

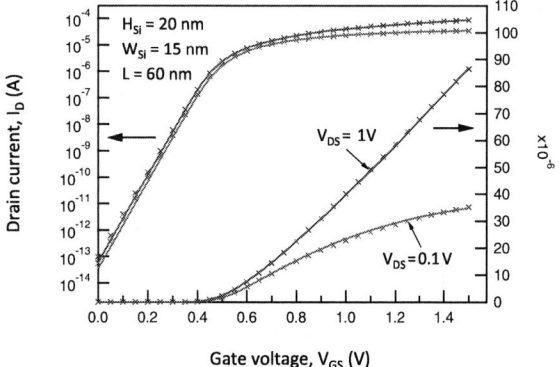

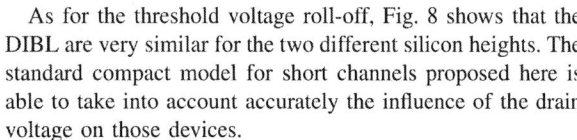

Fig. 7. Drain current of a TG MOSFET as a function of gate voltage in linear and saturation regimes. Symbols: 3-D simulations, lines: compact model.

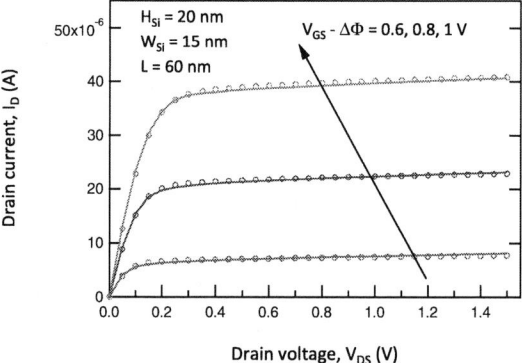

Fig. 9. Drain current of a TG MOSFET as a function of drain voltage in linear and saturation regimes. Symbols: 3-D simulations, lines: compact model.

As for the threshold voltage roll-off, Fig. 8 shows that the DIBL are very similar for the two different silicon heights. The standard compact model for short channels proposed here is able to take into account accurately the influence of the drain voltage on those devices.

Fig. 7 shows the drain current versus the gate voltage at low and high drain voltages for the width of 15 nm and the length of 60 nm. This length is the smallest one for a width of 15 nm that is in agreement with the range of validity of the short-channel modeling. The accuracy is very good, both in linear and log scales, and without the need to introduce any empirical parameter. The mobility degradation is in good agreement at both low and high drain voltages.

The drain current dependance on the drain voltage (Fig. 9) is accurate and the CLM is well taken into account.

VI. CONCLUSION

In this paper, we describe how a multigate FET topology having a common gate electrode could be merely considered as a symmetric DG MOSFET through the definition of an equivalent silicon thickness, an equivalent gate oxide capacitance,

and an equivalent channel width. All these transformations are explicit and only rely on physical and technological parameters, with no empirical relationships. We propose the use of a short-chanel effect modeling developed for the DG FinFET to model short-channel effects in TG MOSFETs. The reduction of the silicon height down to a value close to the silicon width has a negligible influence on the short-channel effects, and so the proposed model is still valid for TG with a square cross section. Likewise, the 2-D CLM modeling takes accurately into account the current slope in the saturation region of the output characteristics. In order to validate the standard model on smaller geometries, a study of the quantum mechanical effects needs to be done according to the both axis of the channel cross section, as well as a 3-D consideration of the electrostatic potential in the channel.

REFERENCES

[1] *Intel Announcement for Microprocessor Production With 22 nm FinFET Technology*, http://newsroom.intel.com/community/intel_newsroom/blog/2011/05/04/intel-reinvents-transistors-using-new-3-d-structure.

[2] B. Yu, M. Lu, Huaxin Liu, and Y. Taur, "Explicit continuous models for double-gate and surrounding-gate MOSFETs," *IEEE Trans. Electron Devices*, vol. 54, no. 10, pp. 2715–2722, Oct. 2007.

[3] J.-M. Sallese, F. Krummenacher, F. Prégaldiny, C. Lallement, A. Roy, and C. Enz, "A design oriented charge-based current model for symmetric DG MOSFET and its correlation with the EKV formalism," *Solid-State Electron.*, vol. 49, no. 3, pp. 485–489, 2005.

[4] J.-P. Colinge, "Multiple-gate SOI MOSFETs," *Solid-State Electron.*, vol. 48, no. 6, pp. 897–905, Jun. 2004.

[5] K. von Arnim *et al.*, "Low-power multi-gate FET CMOS technology with 13.9 ps inverter delay, large-scaled integrated high performance digital circuits and SRAM," in *Proc. IEEE Symposium on VLSI Technology*, Jun. 2007, pp. 106–107.

[6] B. Yu, J. Song, Y. Yuan, W.-Y. Lu, and Y. Taur, "A unified analytic drain current model for multiple-gate MOSFETs," *IEEE Trans. Electron Devices*, vol. 55, no. 8, pp. 2157–2163, Oct. 2008.

[7] R. Ritzenthaler, F. Lime, B. Iñiguez, E. Miranda, F. Martinez, F. Pascal, M. Valenza, O. Faynot, and S. Cristoloveanu, "Analytical modeling of multiple-gate MOSFETs," in *Proc. 8th Spanish Conference on Electron Devices (CDE)*, 2011, pp. 1–4.

[8] N. Chevillon, J.-M. Sallese, C. Lallement, F. Prégaldiny, M. Madec, J. Seldmeir, and J. Aghassi, "Generalization of the concept of equivalent thickness and capacitance to multigate MOSFETs modeling," *IEEE Trans. Electron Devices*, vol. 59, no. 1, pp. 60–71, Jan. 2012.

[9] A. Yesayan, F. Prégaldiny, N. Chevillon, C. Lallement, and J.-M. Sallese, "Physics-based compact model for ultra-scaled FinFETs," *Solid-State Electron.*, vol. 62, no. 1, pp. 165–173, Aug. 2011.

[10] W. Xiong, J. Park, and J. Colinge, "Corner effect in multiple-gate SOI MOSFETs," in *Proc. IEEE Int. SOI Conference*, sep. 2003, pp. 111–113.

[11] *Atlas User's Manual*, SILVACO International, 1997.

MIXDES 2012, 19th International Conference *"Mixed Design of Integrated Circuits and Systems"*, May 24-26, 2012, Warsaw, Poland

Time-domain Waveform Based Extraction of FinFET Nonlinear I-V Model

Dominique Schreurs and Gustavo Avolio
Division ESAT-TELEMIC
University of Leuven
Leuven, Belgium
Dominique.Schreurs@esat.kuleuven.be

Antonio Raffo and Giorgio Vannini
Department of Engineering
University of Ferrara
Ferrara, Italy

Giovanni Crupi and Alina Caddemi
Dipartimento di Fisica della Materia e Ingegneria Elettronica
University of Messina
Messina, Italy

Abstract—**This work presents a straightforward approach to model the dynamic I-V characteristics of microwave FET transistors. Since the main cause of the transistor nonlinearity can be attributed to the drain-source current generator, its correct modeling is fundamental for predicting accurately the device behaviour under realistic operating conditions, namely large-signal operation. The experimental data required for the proposed modelling strategy consist of a small set of low-frequency time-domain waveform measurements. Numerical optimization is adopted to identify the model parameters. The validity of the proposed method is verified by its application to a silicon FinFET. Very good agreement between model predictions and nonlinear measurements is demonstrated.**

Index Terms—**FinFET; microwave transistors; nonlinear time-domain measurements; nonlinear models.**

I. INTRODUCTION

In the past decade, particular attention has been paid to develop fully vector calibrated nonlinear microwave measurement systems [1]. This measurement capability has stimulated researchers to make use of such measurements not only for compact model validation [2], but also for model construction [1][3]. In particular, the direct extraction of the nonlinear model from large-signal measurements has triggered interest. The existing approaches make use of high-frequency measurements only [4][5]. This is restrictive as it requires particular measurement conditions, such as loadpull, to distinguish between the current and charge contributions. Also dispersion is not accounted for. Thanks to the recent development of the low-frequency (LF) vector calibrated nonlinear measurement system [6][7], we will demonstrate that it has become possible to extract the I-V current source straightforwardly. An important difference compared to using DC measurements is that potential dispersion effects [8] can be taken into account.

This work has been supported by FWO-Vlaanderen projects and by KU Leuven GOA project.

The modelling approach is explained in Section II, and experimental results on a FinFET are presented in Section III. Conclusions are drawn in Section IV.

II. MODEL EXTRACTION

Fig. 1 represents the nonlinear equivalent circuit of a field-effect transistor (FET) at microwave frequencies. The gate current source has been omitted, because its value is very small in case of MOS devices. Commonly, the transistor nonlinearities are represented by current (I-V) and charge (Q-V) functions [9], which describe the current conduction and the variation of electrical charge within the transistor. In case of FETs, the drain-source current generator I_{DS} represents the most critical nonlinear component as its characteristics define the device performance in terms of gain and deliverable output power, as well as it defines the reliability boundaries under dynamic operation. Moreover, charge trapping related low-frequency dispersion and thermal effects may directly impact the dynamic response of the current generator I_{DS} [8]. In this perspective, the extraction of the global I-V nonlinear model encompassing also dispersive phenomena may not constitute a trivial task.

When time-domain waveforms measured in the megahertz range are exploited, the network in Fig. 1 simplifies to only I_{DS} and the drain (R_D) and source (R_S) parasitic resistances. The gate resistance R_G is neglected as the conductive current at the gate is very small.

As regarding the charge sources and the capacitor C_{DS}, they can be neglected at low-frequency. Their values can be determined on the basis of multi-bias S-parameter measurements [8][10], by which it is generally assumed that the effect of dispersive phenomena is negligible on the response of the charge sources.

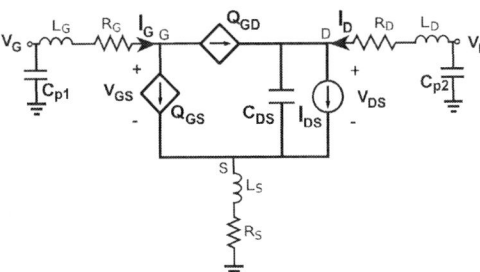

Figure 1. FET nonlinear microwave equivalent circuit. Under low-frequency operation the circuit simplifies to the drain-source current generator I_{DS} and the parasitic resistances R_S and R_D.

The model for the RF I_{DS} adopted in this work is based on the empirical Angelov I_{DS} model [11]. It has been shown in literature that this model is well applicable to describe both the DC and RF I_{DS} characteristics of FinFETs, e.g., [12]. The analytical expression is:

$$I_{DS}=IPk0(T)(1+\tanh(\psi))\tanh(\alpha V_{DS})(1+\lambda V_{DS}) \quad (1)$$

$$\psi=P1m(T)\left(V_{GS}-V_{pkm}\right)+P2\left(V_{GS}-V_{pkm}\right)^2+P3\left(V_{GS}-V_{pkm}\right)^3 \quad (2)$$

where the instantaneous I_{DS} is expressed as an algebraic function of the instantaneous intrinsic gate-source and drain-source voltages V_{GS} and V_{DS}. P1m, P2, and P3 are fitting coefficients, and IPk0 is the current corresponding to the maximum transconductance value. λ is the slope of the drain-source current in the saturation region, and α is a nonlinear function of the intrinsic voltages. The parameters IPk0 and P1m are dependent on the temperature (T).

The RF I_{DS} model parameters are extracted through numerical optimization and using a set of LF time-domain active load pull measurements. The optimization procedure consists of a random search method, followed by a fine-tuning step using a gradient based method. The loadpull measurements are obtained by injecting low-frequency CW signals at both the input and output ports of the DUT [6][7]. The configuration exploited for the measurements reported in this work is schematically illustrated in Fig. 2. The four ADC channels' receiver enables vector calibrated measurements over the 10 kHz-24 MHz bandwidth.

III. EXPERIMENTAL RESULTS

A. RF validation

The I-V nonlinear model is extracted for a FinFET (W = 45.6 μm, L = 60 nm). The bias points of the experiments are $V_{GS0} = 0.3$ V, $V_{DS0} = 0.9$ V, and $V_{GS0} = 0.6$ V, $V_{DS0} = 0.6$ V.

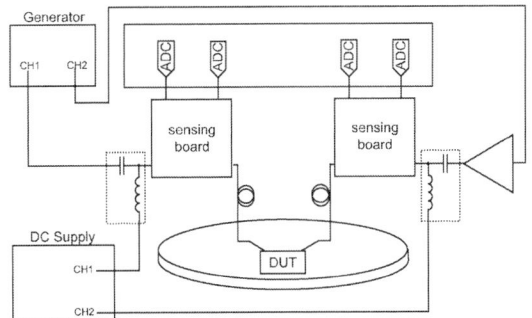

Figure 2. LF characterization system for time-domain active load pull measurements. The DC supply biases the gate and drain of the DUT.

The measured load lines are illustrated in Fig. 3. A fundamental frequency (f_0) of 2 MHz has been chosen, as this value is generally considered to be above the cut-off of the low-frequency dynamics of microwave devices [8]. In addition, it is assumed that microwave non-quasi-static (NQS) effects can be neglected at the microwave frequency adopted in the experimental validation (4 GHz).

On the same plot, the measurements are compared with the simulated I-V model predictions after the optimization procedure. It is noted that a very good agreement has been obtained. Moreover, the output static characteristic is also reported. This characteristic is measured for a DC gate-source voltage (V_{GS0}) equal to the maximum value of the gate-source voltage waveforms of the load lines.

As the maxima of the dynamic characteristics touch the static characteristic, we can conclude that the FinFET does not manifest significant low-frequency dispersion effects, as can be expected from literature [10].

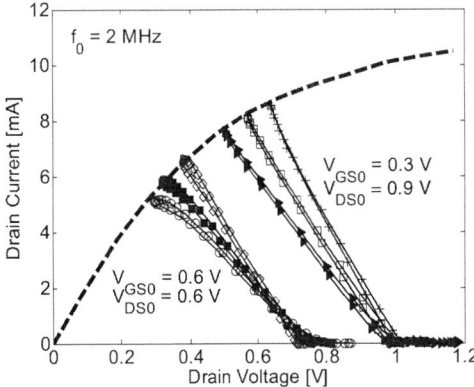

Figure 3. Measured load lines (symbols) and simulations (lines) for the considered FinFET at: $V_{GS0} = 0.6$ V, $V_{DS0} = 0.6$ V, and $f_0 = 2$ MHz; $V_{GS0} = 0.3$ V, $V_{DS0} = 0.9$ V, and $f_0 = 2$ MHz. Measured static characteristic (dashed line) at $V_{GS0} = 1.2$ V.

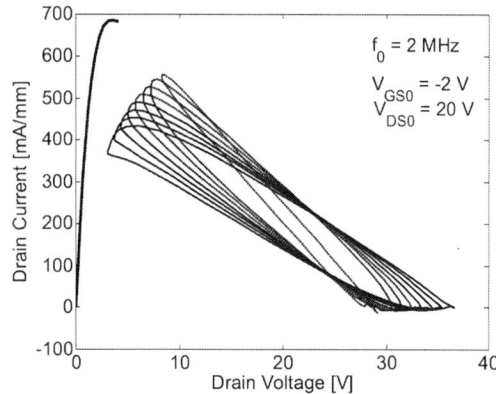

Figure 4. Measured load lines (dashed lines) of an AlGaN/GaN HEMT at: $V_{GS0} = -2$ V, $V_{DS0} = 20$ V, and $f_0 = 2$ MHz. Measured output static characteristic (continous line) at $V_{GS0} = 0$ V.

On the other hand, when dealing with transistors fabricated with III-V compounds, dispersive effects are more pronounced, as clearly illustrated in Fig. 4. The 2 MHz load lines significantly deviate from the static characteristic and such discrepancy is mainly due to low-frequency dispersion [6][8]. Therefore it can be concluded that the use of low-frequency time-domain waveform measurements is unavoidable for an accurate identification of the nonlinear I_{DS} model when dispersion is significant.

Figs. 5-7 show simulations results of the developed FinFET model versus measurements at 4 GHz. The bias point is set to $V_{GS0} = 0.3$ V and $V_{DS0} = 0.9$ V, and the load impedance is 50 Ω. In summary, the RF FinFET model is composed of the RF I_{DS} as extracted at low-frequency (2 MHz), complemented by the charge sources and C_{DS}, which are derived from S-parameter measurements.

Fig. 5 shows the very good agreement between the measured and the simulated output power at the first three harmonics versus input power. Higher harmonics are not shown due to the limited dynamic range of the RF large-signal measurement set-up. Fig. 6 shows the comparison between the simulated and measured average drain current. It can be observed that the dependency on input power can be predicted very well by the model. The average drain current increases with increasing input power due to the self-biasing effect. The latter arises from the bottom clipping of the drain current waveform in the pinch-off region, as clearly visible in Fig. 7 illustrating a comparison in time domain. It can be noticed that the simulated and measured RF time domain waveforms of the drain current correspond very well, which confirms the validity of the proposed modelling approach.

B. Nonlinear de-embedding

In this section the accuracy of the I-V model is verified by applying a de-embedding procedure to RF nonlinear measurements [13]. The de-embedding allows one to retrieve the actual current and voltage waveforms at the I_{DS} plane starting from RF nonlinear measurements which include the contribution of the transistor's nonlinear capacitances and the

parasitics due to the device layout and access structures. Evidently, both the transistor capacitances and the parasitics influence the RF I-V measurements and, consequently, they mask the I-V characteristics as defined at the I_{DS} plane.

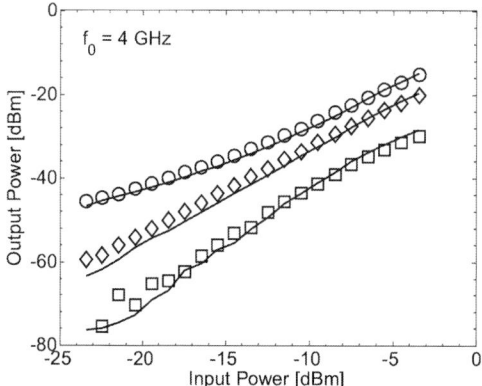

Figure 5. Measured output power and simulations (lines) at $V_{GS0} = 0.3$ V and $V_{DS0} = 0.9$ V, $f_0 = 4$ GHz; load impedance is 50 Ω. The first three harmonics are shown: f_0 (circles), $2f_0$ (diamonds), and $3f_0$ (squares).

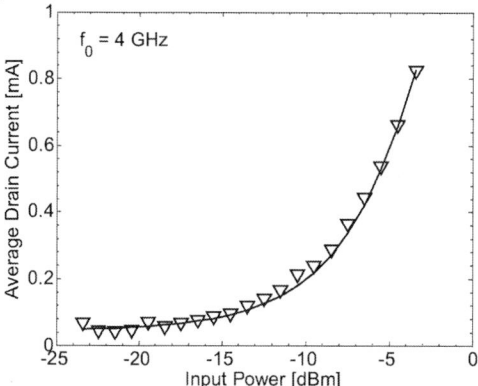

Figure 6. Measured average drain current (triangles) and simulations (line) at $V_{GS0} = 0.3$ V and $V_{DS0} = 0.9$ V, and $f_0 = 4$ GHz.

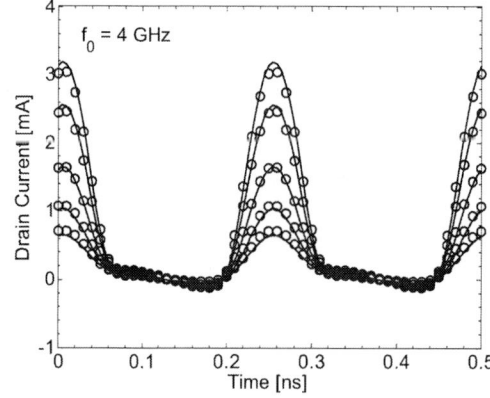

Figure 7. Measured drain current waveform (circles) and simulations (lines) at: $V_{GS0} = 0.3$ V and $V_{DS0} = 0.9$ V, and $f_0 = 4$ GHz. The input power is swept from -23.4 dBm to 3.4 dBm.

In Fig. 8 the measured RF transcharacteristic and the one obtained after de-embedding are illustrated. In Fig. 9, the de-embedded drain-current waveform is compared with the one obtained by simulating the I-V model extracted in Section II. Good agreement between the current waveforms is achieved thus confirming the accuracy of the model.

IV. CONCLUSIONS

In this work a straightforward and accurate approach oriented to the modeling of the dynamic I-V function of microwave transistors has been investigated. It relies on the combination of a small set of low-frequency nonlinear measurements and numerical optimization. The latter is used for the determination of the parameters of an empirical nonlinear model, which is based on Angelov's formulation in this work. The proposed approach has been validated on a silicon FinFET. Good agreement between the model predictions and the measurements under operating conditions different from the ones used within the optimization process has been obtained. It should be pointed out that the approach can be straightforwardly applied to other empirical and compact models.

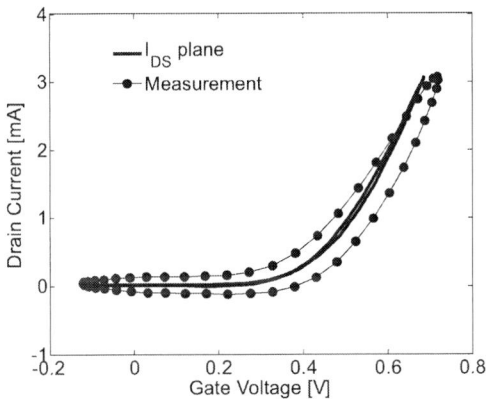

Figure 8. Measured transcharacteristic (circles) and de-embedded transcharacteristic (continuous line) at: V_{GS0} = 0.3 V and V_{DS0} = 0.9 V, and f_0 = 4 GHz. The input power is 2.5 dBm.

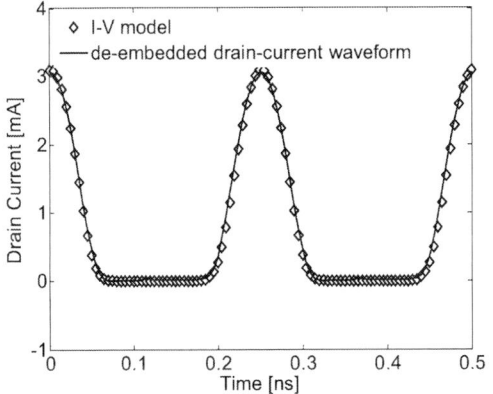

Figure 9. Simulated drain-current waveform (diamonds) and de-embedded drain-current waveform (continuous line) at: V_{GS0} = 0.3 V and V_{DS0} = 0.9 V, and f_0 = 4 GHz. The input power is 2.5 dBm.

ACKNOWLEDGMENT

The authors would like to thank Prof. I. Angelov for valuable discussions and suggestions. The authors thank IMEC for supplying the devices.

REFERENCES

[1] D. Schreurs, "Applications of vector non-linear microwave measurements," IET Microwaves, Antennas and Propagation, vol. 4, no. 4, pp. 421-425, April 2010.

[2] D. Schreurs, S. Vandenberghe, and E. Vandamme, "Thorough verification of large-signal RF MOSFET models by means of vectorial large-signal measurements," International Conference on Mixed Design of Integrated Circuits and Systems (MIXDES), pp. 41-44, 2001.

[3] E.P. Vandamme, W. Grabinski, and D. Schreurs, "Large-signal network analyzer measurements and their use in device modelling," International Conference on Mixed Design of Integrated Circuits and Systems (MIXDES), pp. 65-71, 2002.

[4] M.C. Curras-Francos, P.J. Tasker, M. Fernandez-Barciela, Y. Campos-Roca, and E. Sanchez, "Direct extraction of nonlinear FET Q-V functions from time domain large signal measurements," IEEE Microwave and Guided Wave Letters, vol. 10, no. 12, pp. 531-533, Dec. 2000.

[5] D. Schreurs, J. Verspecht, S. Vanderberghe, and E. Vandamme, "Straightforward and accurate nonlinear device model parameter-estimation method based on vectorial large-signal measurements," IEEE Transactions on Microwave Theory and Techniques, vol. 50, no. 10, pp. 2315-2319, Oct. 2002.

[6] A. Raffo, S. Di Falco, V. Vadalà, and G. Vannini, "Characterization of GaN HEMT low-frequency dispersion through a multiharmonic measurement system," IEEE Transactions on Microwave Theory and Techniques, vol. 58, no. 9, pp. 2490-2496, Sept. 2010.

[7] G. Pailloncy, G. Avolio, M. Myslinski, Y. Rolain, M. Vanden Bossche, and D. Schreurs, "Large-signal network analysis including the baseband," IEEE Microwave Magazine, vol. 12, no. 2, pp. 77-86, April 2011.

[8] A. Raffo, V. Vadalà, D. Schreurs, G. Crupi, G. Avolio, A. Caddemi, and G. Vannini, "Nonlinear dispersive modeling of electron devices oriented to GaN power amplifier design," IEEE Transactions on Microwave Theory and Techniques, vol. 58, no. 4, pp. 710-718, April 2010.

[9] D. Root and B. Hughes, "Principles of nonlinear active device modeling for circuit simulation," Automatic RF Techniques Group (ARFTG) Conference, pp. 1-24, 1998.

[10] G. Crupi, D. Schreurs, A. Caddemi, I. Angelov, M. Homayouni, A. Raffo, G. Vannini, and B. Parvais, "Purely analytical extraction of an improved nonlinear FinFET model including non-quasi-static effects," Microelectronic Engineering, vol. 86, no. 11, pp. 2283-2289, Nov. 2009.

[11] I. Angelov, N. Rorsman, J. Stenarson, M. Garcia, and H. Zirath, "An empirical table-based FET model," IEEE Transactions on Microwave Theory and Techniques, vol. 47, no. 12, pp. 2350-2357, Dec. 1999.

[12] G. Crupi, D. Schreurs, I. Angelov, A. Caddemi, and B. Parvais, "Non-linear FinFET modelling: lookup table and empirical approaches," International Journal of Microwave and Optical Technology, vol. 3, no. 3, pp. 157-164, July 2008.

[13] G. Avolio, D. Schreurs, A. Raffo, G. Crupi, G. Vannini, and B. Nauwelaers, "A de-embedding procedure oriented to the determination of FET I-V charcateristics from high-frequency large-signal measurements," Automatic RF Techniques Group (ARFTG) Conference, pp. 1-6, 2010.

Two-dimensional Physics-based Modeling of Dopant-segregated Schottky Barrier UTB MOSFETs

Mike Schwarz[1,2,✠], Thomas Holtij[1,2], Alexander Kloes[1], and Benjamín Iñíguez[2]

[1]Technische Hochschule Mittelhessen, Giessen, Germany
[2]Universitat Rovira i Virgili, Tarragona, Spain
✠mike.schwarz@ei.thm.de

Abstract—**In this paper, we present an analytical modeling approach to predict the tunneling and thermionic current of Schottky barrier UTB MOSFETs with dopant-segregated source and drain junctions. Essential 2D effects on the currents are included in the model which are combined with diffusion effects in the channel region. The model is compared with measurement data for a dopant segregated fully-depleted Schottky barrier MOSFET and is in a good agreement.**

Index Terms—**2D Poisson, conformal mapping, compact modeling, dopant segregation, Schottky barrier, ultra-thin body (UTB) MOSFET.**

I. INTRODUCTION

Nowadays, research focuses on these SOI-type structures to reduce short channel effects and improved the gate control [1], [2]. So called DG-MOSFETs, FinFETs and GAA (Gate-all-around) FETs - allow for low channel doping concentrations whilst still avoiding the subthreshold performance degradation.

Precise compact models describing the behavior and characteristics of these devices are required for implementation in circuit simulators and similar design tools to ensure the wide-spread uptake of these devices. In multi gate devices these models have to take into account two- or even three-dimensional effects [3].

Schottky barrier (SB) MOSFET structures are very attractive MOSFET devices for improved performance due to its high scalability even down to the sub-10nm region. The main benefits are low specific resistances [4], which might improve the drive current.

In order to further improve the Schottky barrier performance, techniques as Fermi level depinning with an insulator between the metal and semiconductor, low Schottky barrier silicides and dopant segregation (DS) are applied. This effect results in a thin highly doped layer which causes a strong band bending at the silicide-to-silicon interface and hence in an increased tunneling probability for the carriers, due to the effectively reduced Schottky barrier height [5].

This paper presents the model for an SB UTB MOSFET, a further promising candidate in the MOS technology to replace the bulk-MOSFET, which requires a two-dimensionally modeling approach too.

Based on the 2D conformal mapping technique [6] for the electrostatic potential [7] and the electric field [8], the ambipolar behavior is predicted by the superposition of the separately estimated electron and hole currents. The applied tunneling current model is solved with respect to the coordinates and taking into account for effects of a drift-diffusion, presented in [9] in contrast to [10], where the tunneling current is solved with respect to the energy.

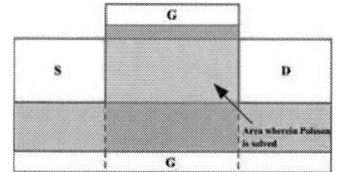

Fig. 1.   General geometry of a SB UTB MOSFET.

II. PRELIMINARIES

In this work we assume a p-type device, nevertheless all evolved equations can easily be converted for an n-type device.

In general the electrostatic problem for a SB UTB MOSFET can be described by Poisson's equation for the potential ϕ, where the Poisson equation in the channel region with space charge ρ is

$$\Delta\phi_{2D} = -\frac{\rho}{\epsilon} = -\frac{Q_{inv} + Q_{dep}}{\epsilon}. \tag{1}$$

Considering a non or lightly doped device results in a negligible depletion charge ($Q_{dep} \approx 0$). Furthermore, we calculate Poisson's equation within the subthreshold region of the device and therefore the inversion charge within the channel region can be neglected ($Q_{inv} \approx 0$). These assumptions lead to the simplification of Poisson's equation to a Laplacian equation

$$\Delta\phi_{2D} \approx 0. \tag{2}$$

In Fig. 1 the general structure of a SB UTB MOSFET is presented. By cutting out the orange shaded rectangle, the region of interest, the assumed model structure of the SB UTB MOSFET device is shown with the coordinates and conditions in Fig. 2. To keep mathematics simple, a two corner structure is used. The applied boundary conditions for the case of a SB-DG-MOSFET have been discussed in detail in [7].

Here, $V_{g,top}$ represents the top gate potential, and V_{fb} the flat band voltage, while V_s and V_d describe the source and drain potentials. The built-in potential at source and drain contacts is represented by ϕ_{bi}. The bottom gate potential $V_{g,bottom}$ is applied to the vertical boundaries starting at a depth corresponding to the position of the bottom gate electrode in the original structure, which is an approximation. In the oxides linear boundaries are applied. The origin of

TABLE I
SETTINGS OF THE SIMULATED TCAD SENTAURUS [11] DEVICE.

l_{ch}	80	[nm]	effect. channel length
t_{ch}	20	[nm]	channel thickness
$t_{ox,top}$	3.5	[nm]	top oxide thickness
$t_{ox,bottom}$	100	[nm]	bottom oxide thickness
ϕ_{Bn}	0.28	[eV]	SB electrons
N_B	10^{15}	[cm^{-3}]	substrate doping
$\epsilon_{ox,top}$	7	[As/Vcm]	permittivity top oxide
$\epsilon_{ox,bottom}$	3.9	[As/Vcm]	permittivity bottom oxide

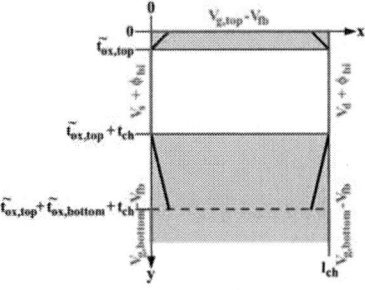

Fig. 2. Simplified geometry and boundaries of a SB UTB MOSFET.

the coordinates is set in the left top corner. In x direction we have the distance l_{ch} on the top right corner, while in y direction the oxide is represented by the grey region with the thickness $\tilde{t}_{ox,top}$. Then the channel begins with thickness t_{ch} and ends with the distance $\tilde{t}_{ox,top} + t_{ch}$. With the oxide on the bottom we receive the distance $\tilde{t}_{ox,top} + \tilde{t}_{ox,bottom} + t_{ch}$. The parameters $\tilde{t}_{ox,bottom}$ and $\tilde{t}_{ox,top}$ are transformed oxide thickness's, which are described in detail in [7]

$$\tilde{t}_{ox,bottom} = \frac{\epsilon_{Si}}{\epsilon_{ox,bottom}} \cdot t_{ox,bottom},$$
$$\tilde{t}_{ox,top} = \frac{\epsilon_{Si}}{\epsilon_{ox,top}} \cdot t_{ox,top}. \quad (3)$$

III. DEVICE PHYSICS

Understanding the device physics and the resulting behavior are very important elements during the evolving procedure of a compact model. Therefore, and to concentrate on the main contributions, the SB UTB MOSFET device is simulated in TCAD Sentaurus (Table I) and analyzed in detail. In Fig. 4 the electrostatic potential for three different gate potentials is given, which range from accumulation to inversion mode, with a gate potential at the bottom $V_{g,bottom}=0V$.

The behavior is as expected. While one increases $V_{g,top}$ from -0.5V to 1V underneath the top oxide the energy bands are bend downwards, see Fig. 3. The influence onto the bottom channel from the bottom gate is lower compared to the one of the top, which due to the thicker oxide. Nevertheless, the

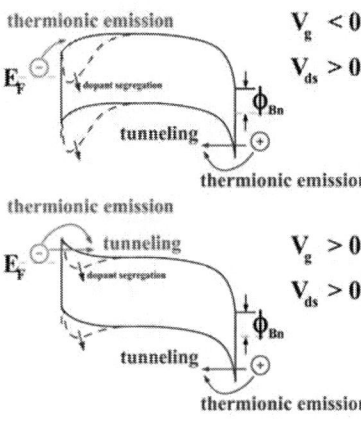

Fig. 3. Energy bands of a SB MOSFET device, showing the current contribution in accumulation and inversion and the influence of dopant segregation at the source electrode.

asymmetrical behavior forces that the main contribution to the tunneling current is generated underneath the top oxide. Similar is the influence onto the electric field as illustrated in Fig. 5. Due to the correspondence between the electric field and the electrostatic potential the electric field underneath the top oxide is higher compared to the bottom oxide, Figs. 5(a) to 5(c). Due to this circumstance the effect of tunneling at the bottom should be less than at the top. This effect may be interpreted based on the WKB approximation [12] used to estimate the tunneling probability. If the gradient of the electric field is higher, the probability for a charge carrier to tunnel through the channel region is higher too.

Of course more physics describe the generation of tunneling charge, as the carrier distributions which are also responsible for the description of the tunneling current, and linked to the electrostatic potential. However, in general the electrostatics are responsible for the tunneling and thermionic mechanisms in the device.

The above described effects result in the generation rate of the corresponding charge carriers electrons and hole in the channel, respectively. In Fig. 6 and in Fig. 7 the electron and hole tunneling generation rates are illustrated for the bias conditions introduced above. From the before made observations the behavior for both is as expected.

For electrons follows from the Figs. 6(a) to 6(c) a movement of the generation rate from the middle or center of the device towards the source junction. The bands are bend downwards by increasing the gate bias and the tunneling probability for the electrons at the source electrode rise due to the high electric field. Furthermore, it can be observed that the electron generation rate moves from the bottom oxide to the top oxide and concentrates at the top corner close to the source junction.

A similar behavior results for the hole generation rate. Due to the existing voltage drop between source and drain the situation differs slightly.

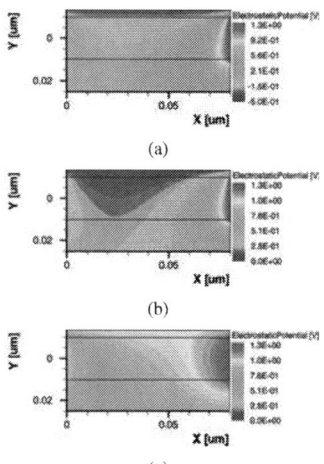

(a)

(b)

(c)

Fig. 4. Electrostatic potential: (a) $V_{g,top}$=-0.5V. (b) $V_{g,top}$=0V. (c) $V_{g,top}$=1V. Bias conditions: $V_{g,bottom}$=0V, V_{ds}=1V.

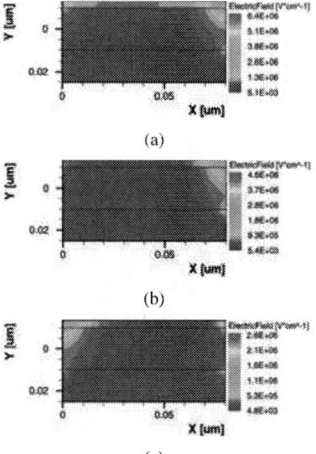

(a)

(b)

(c)

Fig. 5. Absolute value of electric field: (a) $V_{g,top}$=-0.5V. (b) $V_{g,top}$=0V. (c) $V_{g,top}$=1V. Bias conditions: $V_{g,bottom}$=0V, V_{ds}=1V.

The energy bands (Fig. 3) are bend upwards and at the drain junction already a modulated barrier exists, which is thin enough that holes are able to tunnel into the channel region. The main concentration exists in the corner of the device, here at the drain junction, Fig. 7(a). Increasing the gate bias results in a less bending of the energy bands. Due to this, the barrier becomes thicker at the drain junction and the tunneling probability is reduced. This results in a less generation of hole charges at the drain electrode, Figs. 7(a) to 7(c). Finally, the generation rate moves towards the bottom oxide, Fig. 7(c). This effect has a minor influence, due to the small amount of the generated charges.

On the other hand, the thermionic emission mechanism is strongly influenced by the maximum of the potential barrier in the channel region. Therefore, the potential is responsible for the thermionic current behavior.

Depending on the electrostatic potential and operation mode of the device, it is expected for accumulation, that the current flow due to the thermionic mechanism concentrates near the

bottom oxide and the channel middle of the device, where the influence from both gates is low. Here the potential barrier is lowered by the source an drain junctions. If the device operates in inversion, the thermionic current determining barrier is the Schottky barrier itself, neglecting the Schottky barrier lowering effect. Furthermore, the main contribution results at the source electrode, due to the high Schottky barrier height at the drain junction. To conclude the device behavior,

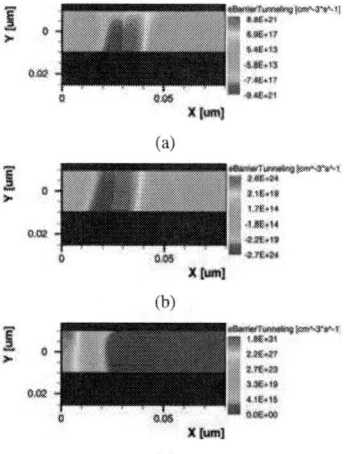

(a)

(b)

(c)

Fig. 6. Electron Tunneling Generation Rate: (a) $V_{g,top}$=-0.5V. (b) $V_{g,top}$=0V. (c) $V_{g,top}$=1V. Bias conditions: $V_{g,bottom}$=0V, V_{ds}=1V.

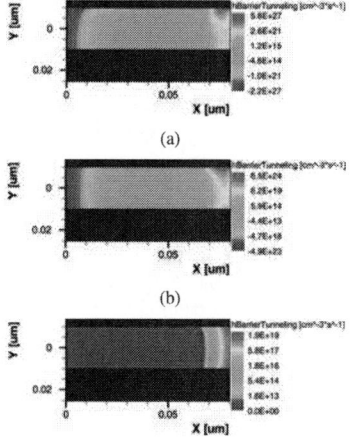

(a)

(b)

(c)

Fig. 7. Hole Tunneling Generation Rate: (a) $V_{g,top}$=-0.5V. (b) $V_{g,top}$=0V. (c) $V_{g,top}$=1V. Bias conditions: $V_{g,bottom}$=0V, V_{ds}=1V.

one can say that in accumulation, tunneling holes underneath the top oxide are the main contribution to the device current. In subthreshold range, thermionic emission of electrons from the source close to the bottom oxide describes the current. In inversion tunneling electrons concentrated underneath the top oxide are the responsible charges for the current.

For the case of a device with dopant segregation at the Schottky barrier a thin highly doped layer results a the junction, which causes a strong band bending, indicated in Fig. 3 at the silicide-to-silicon interface. As a result the tunneling probability for the carriers is increased.

90

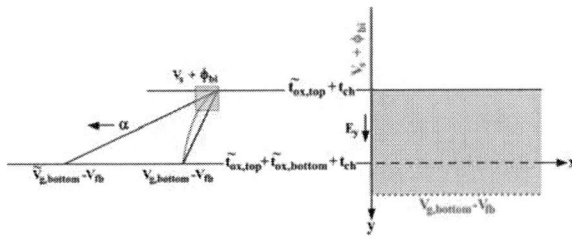

Fig. 8. Empirical fitting parameter α corrects the inaccuracy in the gradient and leads to a modified boundary $\tilde{V}_{g,bottom}$. The red curve illustrates the correct boundary as it results from TCAD simulations including fringing fields.

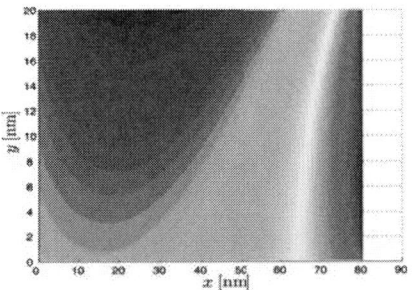

Fig. 9. Electrostatic potential of a SB UTB MOSFET. Bias conditions: $V_{g,top}$=-0.5V, $V_{g,bottom}$=0V, V_{ds}=1V. Model simulation.

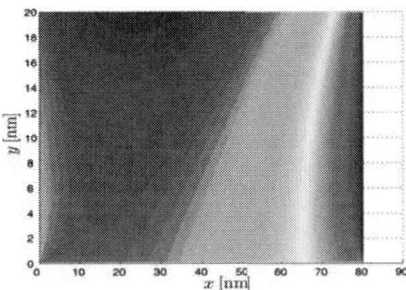

Fig. 10. Electrostatic potential of a SB UTB MOSFET with correction parameter α. Bias conditions: $V_{g,top}$=-0.5V, $V_{g,bottom}$=0V, V_{ds}=1V. Model simulation.

IV. MODELING APPROACH

In our model the effect of dopant segregation will be accounted for by an effective Schottky barrier height $\phi_{B,eff}$. This height has to be used for the calculation of the tunneling and thermionic current, while the electrostatics in the model are predicted with the non-effective Schottky barrier height ϕ_B, included in the built-in potential of the boundary [7]. The model presented here is compared with measurement data for a dopant segregated fully-depleted Schottky barrier MOSFET [5].

A. Electrostatics

The analytical framework presented in [13] was used to model the SB UTB MOSFET device. Here the conformal mapping technique, especially the Schwarz-Christoffel transformation [14], is applied to solve the 2D electrostatics in an analytical way. From the closed-form solutions the following results for the electrostatics in a UTB SB-MOSFET device have been calculated.

In Fig. 9 the electrostatic potential is illustrated. If one compares the potential from Fig. 9 with the corresponding simulated potential of TCAD from Fig. 4(a) a difference between both solutions can be observed. Especially at the source junction close to the bottom silicon-to-oxide interface a huge difference between the simulated electrostatic potential of TCAD Sentaurus and the estimated one of the model can be observed. This inaccuracy results from three approximations. The first is the neglecting of fringing fields, while the structure is simplified as shown in Fig. 2. Therefore, the gradient of the potential close to the bottom oxide is influenced and predicted not accurately enough. In general this influence is low compared to the following approximations.

The second reason of inaccuracy is the transformed oxide in equation (3) for the simplification of the conformal mapping technique. Applying the transformed oxide thickness results in an inaccuracy for thick oxides. The third and last reason of inaccuracy is an analytical approximation in the framework in [13] for the case of linear boundary conditions.

The influences resulting from all these approximations depend on the gradient of the potential at the bottom oxide. This can be improved by introducing an empirical fitting parameter α, which modifies the gradient of the potential along the boundary in the oxide and therefore the electrostatic potential, see Fig. 8.

It is very important to correct this influence, otherwise the thermionic emission current which is mainly concentrated close to the bottom oxide, saturates too early in the $I_d - V_g$ curve, due to the wrong potential. The effect of this behavior onto the tunneling currents is of minor interest, because they are mainly concentrated underneath the top oxide, where the prediction is accurate.

As shown in Figure 8, the shaded rectangular area shows the region of interest, where the most influence from the inaccuracy results. Here, the aim to correct with the empirical fitting parameter the boundary in the bottom oxide, which leads finally to a corrected gradient of the potential.

Therefore, follows for the modified boundary condition

$$\tilde{V}_{g,bottom} = V_s + \phi_{bi} - \tilde{t}_{ox,bottom} \cdot \tilde{E}_y, \qquad (4)$$

with

$$\tilde{E}_y = \alpha \cdot \frac{V_s + \phi_{bi} - (V_{g,bottom} - V_{fb})}{\tilde{t}_{ox,bottom}}. \qquad (5)$$

By consideration of the empirical fitting parameter α as in Fig. 8, the electrostatic potential results in Fig. 10. Comparing the new result with the electrostatic potential from the TCAD Sentaurus simulation in Fig. 4(a), one can observe an improvement in the prediction of the potential.

In Fig. 11 the corresponding electric field is given. If one compares the prediction with the electric field from TCAD Sentaurus in Fig. 5(a), it can be observed, that the prediction is in a good agreement with the TCAD simulation results. In contrast to the electrostatic potential, the solution of the electric field from the framework in [13] is solved by an approach without the need of an analytical approximation.

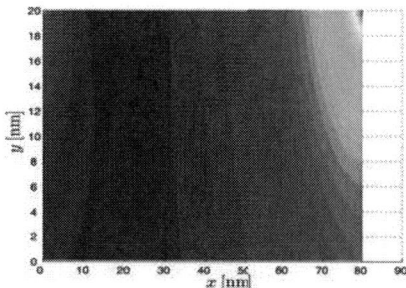

Fig. 11. Electric field of a UTB SB-MOSFET. Bias conditions: $V_{g,top}$=-0.5V, $V_{g,bottom}$=0V, V_{ds}=1V. Model simulation.

B. Current Calculation

The current is calculated with the above corrected solution for the electrostatics as explained in [15]. In the model the contributions to the current result from electrons at the source electrode and holes at the drain electrode depending on the bias condition. It follows

$$I_{TE} = I_{tun} + I_{therm}. \tag{6}$$

All other components are neglected. The contribution of the thermionic current is calculated as presented in [10]

$$J_{therm}(y) = A^{\star}T^2 \exp\left(-\frac{q\phi_B(y)}{kT}\right)\left[1 - \exp\left(\frac{-qV_{ds}}{kT}\right)\right]. \tag{7}$$

$A^{\star}$ represents the Richardson constant, $\phi_B(y)$ describes the barrier height or maximum barrier which the charge carriers have to surmount in the slice at position y. For the case of accumulation $\phi_B(y)$ is represented by the minimum of the electrostatic potential in the channel region. In inversion $\phi_B(y)$ is obtained by the effective Schottky barrier height $\phi_{B,eff}$. The thermionic current results from

$$I_{therm} = \int\limits_0^{w_{ch}} \int\limits_{t_{ox}}^{t_{ox}+t_{ch}} J_{therm}(y) \cdot dy \cdot dz \tag{8}$$

with channel thickness t_{ch} and width w_{ch}.

The tunneling probability is predicted with the WKB approximation [12]. The Schottky barrier lowering effect is neglected and it follows

$$T(E,x,y) = \exp\left(\frac{-4\sqrt{2m \cdot m_0}\left(|E(x,y)| \cdot x\right)^{3/2}}{3q\hbar|E(x,y)|}\right). \tag{9}$$

Here m is the effective mass, m_0 the electron rest mass, q the elementary charge, $\hbar$ the reduced Planck constant and E the

electric field at the point of interest which results from the 2D analytical solution via the conformal mapping technique presented in [13].

With the prediction of (9) and the Fermi-Dirac distributions of the charge carriers for $f_m(\xi(x))$ and $f_s(\xi(x))$ taking into account the effective Schottky barrier height $\phi_{B,eff}$, one is able to calculate the tunneling current density. In contrast to [10], instead of integrating with respect to the energy, here an integration with respect to coordinate x along the channel takes place. As explained in [15] the tunneling current density is rewritten to account for additional drift-diffusion effect. It follows the tunneling current density according to the diffusion theory at metal-semiconductor contacts [10] but with account for an individual electric field E(x, y)

$$J_{tun}(y) = \frac{q\mu_n N_C}{kT} \int_0^{l_{ch}} |E(x,y)| \cdot T(E,x,y) \cdot$$
$$f_m(\xi(x))[1 - f_s(\xi(x))]\frac{\partial \xi}{\partial x}dx, \tag{10}$$

whereby the total current is obtained by integration according to equation (8) and in the subsequent plots is indicated by I_{TE}.

The tunneling current results from

$$I_{tun} = \int\limits_0^{w_{ch}} \int\limits_{t_{ox}}^{t_{ox}+t_{ch}} J_{tun}(y) \cdot dy \cdot dz. \tag{11}$$

In (11) a drift-diffusion mechanism affects the tunneling current via electron effective μ_n. Furthermore, in the tunneling current density model (10) the mobility degradation is taken into account as in [16]. Thus the effective mobility may be expressed by formula (12)

$$\mu_n = \frac{\mu_{n_0}}{1 + \Theta(V_{g,top} - \phi(l_{ch}/2))}, \tag{12}$$

where Θ is the mobility degradation coefficient and treated as fitting parameter, $V_{g,top}$ the gate potential and $\phi(l_{ch}/2)$ the channel potential in the middle of the device.

The thermionic emission current equation (7) is controlled by Richardson's constant. Both parameters are treated as fitting parameters.

Finally a smoothing as presented in [3] from the linear to saturation region is applied.

V. RESULTS

In this section the model explained above is compared with the $I_d - V_g$ measurement data for a SB UTB device from Jülich [5] in log-scale and lin-scale, Fig. 12. The used model parameters are listed in Table II. As one can observe the model predicts the measured device characteristics quite well. The slope is reproduced accurately and for high V_{ds} a good agreement can be observed. Nevertheless, some inaccuracies exist at close to the V_g=0V range.

These inaccuracies result from the transition between the thermionic and tunneling current and requires further improvements in the smoothing of the model equations. However, the model is able to estimate the behavior of real SB

TABLE II
DEVICE MODEL PARAMETERS FOR A DRAWN CHANNEL LENGTH $l_{ch} = 80nm$.

Structural	l_{ch}	60	[nm]	effect. channel length
	t_{ch}	20	[nm]	channel thickness
	$t_{ox,top}$	3.5	[nm]	top oxide thickness
	$t_{ox,bottom}$	100	[nm]	bottom oxide thickness
Mobility	μ_{n0}	11.0	[cm^2/Vs]	elec. tunneling mobility
	μ_{p0}	1.0	[cm^2/Vs]	hole tunneling mobility
	Θ	0.8	[-]	transverse field dependency
Misc	N_B	10^{15}	[cm^{-3}]	substrate doping
	V_{fb}	-0.56	[V]	flatband voltage
	ϕ_{bi}	0.56	[V]	built-in potential s/d
	$\phi_{Bn,eff}$	0.15	[eV]	effect. SB electrons
	ϕ_{Bn}	0.6	[eV]	SB electrons
	A_n^*	62	[A/cm^{2}K^2]	elec. Richardson constant
	A_p^*	32	[A/cm^{2}K^2]	hole Richardson constant
	m_n	0.28	[-]	elec. effect. mass
	m_p	0.30	[-]	hole effect. mass
	α	1.8	[-]	gradient correction parameter

UTB MOSFET and is in a good agreement with measurement data over a large bias range.

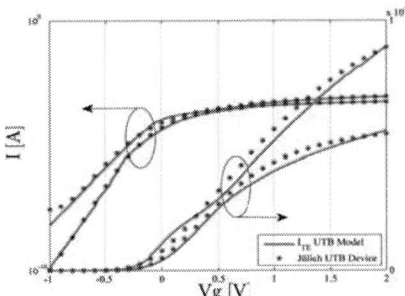

Fig. 12. $I_d - V_g$ measurement data of the Jülich SB UTB MOSFET vs. I_{TE} SB UTB model in log-scale and lin-scale. Bias conditions:$V_{g,top}$=-1V to 2V, $V_{g,bottom}$=0V, V_{ds}=0.5V and 1V.

VI. CONCLUSION

An analytical approach to calculate the tunneling and thermionic current of Schottky barrier UTB MOSFETs with dopant-segregated source and drain junctions was presented, which shows a good agreement with with measurement data for a SB UTB MOSFET device. As a result, the model is able to predict the current well. The approach in particular inherently includes 2D effects on the device current and takes into account the drift-diffusion effects.

ACKNOWLEDGMENT

This project was supported by the German Federal Ministry of Education and Research under contract No. 1779X09, by the German Research Foundation (DFG) under Grant KL 1042/3-1, and by the European Commission under IAPP-218255 ("COMON"), by the Spanish Ministerio de Ciencia y Tecnología under Projects TEC2011-28357-C02-01, and also by the PGIR/15 Grant from URV and by the ICREA Academia Prize. Particularly we would like to thank the IBN1-IT from the Forschungszentrum Jülich for providing the measurement data.

REFERENCES

[1] D. Jimenez et al., "Continuous analytic I-V model for surrounding-gate MOSFETs," IEEE Trans. Electron Devices, vol. 25, no. 8, pp. 571–573, 2004.

[2] N. Collaert et al., "Multi-gate devices for the 32nm technology node and beyond," in Proc. 37th European Solid State Device Research Conference ESSDERC 2007, 2007, pp. 143–146.

[3] A. Kloes et al., "MOS^3: A New Physics-Based Explicit Compact Model for Lightly Doped Short-Channel Triple-Gate SOI MOSFETs," IEEE Trans. Electron Devices, vol. 59, no. 2, pp. 349–358, 2012.

[4] J. M. Larson et al., "Overview and status of metal S/D Schottky-barrier MOSFET technology," IEEE Trans. Electron Devices, vol. 53, no. 5, pp. 1048–1058, 2006.

[5] C. Urban et al., "High-frequency performance of dopant-segregated NiSi S/D SOI SB-MOSFETs," in Proc. European Solid State Device Research Conf. ESSDERC '09, 2009, pp. 149–152.

[6] A. Kloes et al., "A new analytical method of solving 2D Poisson's equation in MOS devices applied to threshold voltage and subthreshold modeling," Solid-State Electronics, vol. 36, no. 12, pp. 1761–1775, 1996.

[7] M. Schwarz et al., "2D Analytical Calculation of the Electrostatic Potential in Lightly Doped Schottky barrier Double-Gate MOSFET," Solid-State Electronics, vol. 54, no. 11, pp. 1372–1380, 2010.

[8] ——, "2D Analytical Calculation of the Electric Field in Lightly Doped Schottky Barrier Double-Gate MOSFETs and Estimation of the Tunneling/Thermionic Current," Solid-State Electronics, vol. 63, no. 1, pp. 119–129, 2011.

[9] ——, "2D Analytical DC Model for Nanoscale Schottky Barrier DG-MOSFETs," in Proceedings of ISDRS 2011, Stamp Student Union, College Park, University of Maryland, USA, 2011.

[10] K. K. N. S. M. Sze, Physics of Semiconductor Devices. John Wiley & Sons, 2007.

[11] Synopsys, TCAD Sentaurus, c-2009.06 ed., Synopsys, Inc., 2009.

[12] G. Wentzel, "Eine Verallgemeinerung der Quantenbedingungen für Würfel Zwecke der Wellenmechanik," Z. Physik, vol. 38, pp. 518–529, 1926.

[13] M. Schwarz et al., "Analytical Compact Modeling Framework for the 2D Electrostatics in Lightly Doped Double-Gate MOSFETs," Solid-State Electronics, vol. 69, no. 1, pp. 72–84, 2012.

[14] E. Weber, Electromagnetic fields, Vol.I., Mapping of fields. John Wiley, New York, 1950.

[15] M. Schwarz et al., "I-V Model for Lightly Doped Schottky Barrier DG-MOSFETs Including 2D Effects," in ESSDERC Fringe, Hesinki, Finland, 2011.

[16] N. Arora, Mosfet Modeling for VLSI Simulation: Theory and Practice. World Scientific Pub Co, 2007.

MIXDES 2012, 19th International Conference *"Mixed Design of Integrated Circuits and Systems"*, May 24-26, 2012, Warsaw, Poland

Verilog-A Compact Semiconductor Device Modelling and Circuit Macromodelling with the QucsStudio-ADMS "Turn-Key" Modelling System

Mike E. Brinson
Centre for Communications Technology
London Metropolitan University
London N78DB, UK
mbrin72043@yahoo.co.uk

Michael Margraf
Qucs and QucsStudio Project Founder
Berlin, Germany
michael.margraf@alumni.tu-berlin.de

Abstract—The Verilog-A "Analogue Device Model Synthesizer" (ADMS) has in recent years become an established modelling tool for GNU General Public License circuit simulator development. Qucs and ngspice being two examples of open source circuit simulators that use ADMS. This paper presents a "turn-key" compact device modelling and circuit macromodelling system based on ADMS and implemented in the QucsStudio circuit design, simulation and manufacturing environment. A core feature of the new system is a modelling procedure which does not require users to manually patch the circuit simulator C++ code. At the start of a QucsStudio simulation the software automatically detects any changes in Verilog-A model code, re-compiling and dynamically linking the modified code to the body of the QucsStudio code. The inherent flexibility of the "turn-key" system encourages rapid experimentation with analogue and RF compact device models. In this paper QucsStudio "turn-key" modelling is illustrated by the design of a single stage RF amplifier circuit.

Index Terms—QucsStudio, ADMS, Verilog-A, compact device modelling, turn-key component modelling.

I. INTRODUCTION

Until the adoption of Verilog-A as the preferred analogue hardware description language for compact semiconductor device modelling by the Compact Model Council [1], C had been the standard modelling language. However, hand coding of compact device models in C was often found to be very tedious, time consuming and subject to error, particularly when determining the partial derivatives of the device currents and charges needed in DC and transient simulation of non-linear circuits. In contrast to C, the Verilog-AMS hardware description language provides built-in tools which automatically generate partial derivatives, making compact device modelling a much more straight forward process. Current trends suggest that there is growing acceptance by the compact modelling community of the Verilog-AMS subset Verilog-A as the preferred compact modelling language. The standardization of Verilog-AMS [2] and specifically the addition of a number of compact modelling enhancements to

its analog Verilog-A subset [3] have also greatly influenced Verilog-A usage. The release of the Verilog-A "Analogue Device Model Synthesizer" (ADMS) software [4] under the GNU General Public License has also accelerated the rate at which Verilog-A has been accepted and used by the modelling community as a viable replacement for C. Moreover, the growing number of commercial [5] [6] and open source circuit simulators [7] [8] which use Verilog-A for compact semiconductor device and circuit macromodelling is a testimony to the importance of Verilog-A in the development of circuit simulator technology. This paper outlines the structure and operation of a new "turn-key" Verilog-A compact model development system which automatically re-compiles and dynamically links modified Verilog-A model code to the C++ body of a circuit simulator prior to the start of a simulation sequence. The new compact modelling system has been implemented in a freely available circuit design, simulation and manufacturing environment called QucsStudio [9]. QucsStudio is released under the GNU General Public License for use with the Microsoft Windows® operating system and includes a second generation version of the popular Qucs circuit simulator plus other important circuit design, simulation and manufacturing features.

II. VERILOG-A COMPACT DEVICE MODELLING WITH QUCS-ADMS

Verilog-A based compact device modelling [10] was first implemented in Qucs version 0.0.11. In the original Qucs modelling technique the ADMS Verilog-A to C++ synthesizer was used to compile Verilog-A model code to C++ code manually. After conversion the C++ code also had to be manually merged with the main body of the Qucs circuit simulator [11]. Similarly, the Qucs graphical user interface code needed to be patched to add a new model symbol to the simulators library of built-in component symbols. Finally, due to the fact that the Qucs simulator uses C++ static model libraries the entire simulator C++ code had to be re-compiled and re-linked to generate a new extended simulator each time a compact device model was added to the software. In principle, it was possible to add compact device models to Qucs using the previously described procedure. In practice, the modeling process required users to have an advanced knowledge of C++ programming coupled with a good understanding of the Qucs model application interface, making the process of adding new compact models one which was more suited to Qucs developers rather than the wider Qucs user community. The original Qucs compact device modelling process was further complicated in that it was designed primarily to function with software development tools supplied with the Linux operating system rather than more universally available Microsoft Windows® operating system.

III. QucsStudio-ADMS "Turn-Key" Verilog-A Compact Device Modelling

A primary aim of the QucsStudio Verilog-A compact device modelling system is to provide the circuit simulation software with a simple modelling tool that does not require the main body of the circuit simulator C++ code to be patched by hand when adding new device models. In contrast to the original Qucs modelling scheme the QucStudio version is based on dynamic linked model libraries rather than static model libraries. This change has had a major effect. The implemented modelling system has been called a "turn-key" system to emphasize that it takes over responsibility for determining when a compact device model needs to be re-compiled and re-linked. Changes in Verilog-A model code act like a key turning on the compilation and linking of changed code. When changes take place, edited models are automatically updated at the start of the next user requested circuit simulation. Fig. 1 presents a simple flow chart, outlining each of the QucsStudio modelling stages. In this diagram the modelling process is shown starting from Verilog-A code entry at the top, using a built-in text editor, followed by attachment of a model Verilog-A code file (XXXX.va) to a QucsStudio "C++ compiled model " icon, through synthesis of the C++ model code, using the ADMS software, to C++ compilation and finally construction of a model subcircuit at the bottom. At the start of the modelling process equations representing the different physical aspects of device operation are entered into the QucsStudio software as Verilog-A "module" code. A convenient colour highlighted text editor is provided with QucsStudio for this task. Phase two involves the synthesis of the C++ code from the entered Verilog-A code. With the Verilog-A model code visible on the QucsStudio text editor display window, pressing key F2 causes the generation of the compact model C++ code to take place, followed by C++ compilation using the MinGW tools to form a dynamic linked library (XXXX.dll) for the model under construction. The last two stages only take place if the original Verilog-A module code is error free. On successful generation of the "C++ compiled model" it is linked to the main body of QucsStudio software and becomes a standalone simulation component. It can also be combined with other QucsStudio components by attaching it to a QucsStudio schematic diagram, and indeed to other "C++ compiled models", to form a subcircuit or macromodel. During the simulation of circuits which include C++ compiled models changes to their Verlog-A code will automatically trigger the "turn-key" modelling process

ensuring that the Verilog-A compact device models are kept up to date at all times.

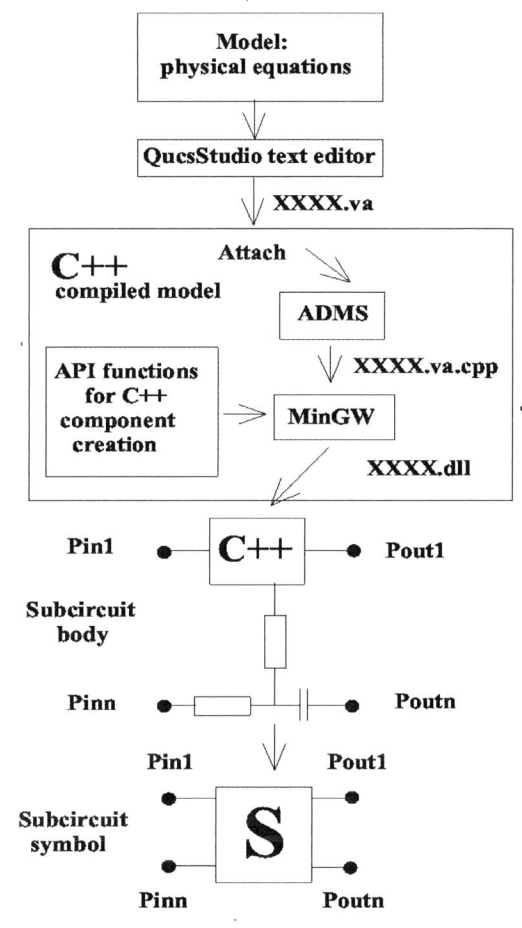

Figure 1. A flow chart outlining the QucsStudio "Turn-key" Verilog-A compact model development system

To demonstrate the QucsStudio Verilog-A "turn-key" modelling procedure the construction of a simple RF npn BJT compact device model is presented next. The model information given in Fig. 2 is based on a large signal Ebors-Moll bipolar transistor equivalent circuit, a simplified set of non-linear device equations, including second order high-level current injection effects and internal capacitance, plus a subcircuit schematic showing external inductance and capacitance. The Verilog-A code for the RF npn BJT is listed in Fig. 3. On attaching this code to a QucsStudio C++ compiled model icon the software tries to extract the external node names and parameters. If successful QucsStudio draws a group of named terminals attached to the C++ compiled model icon. These correspond in name and list order to the "inout" terminals given in the Verilog-A module statement. Shown in Fig.4 is a single stage class A RF npn BJT amplifier circuit, with collector feedback, configured as a small signal AC simulation over the frequency band 1MHz to 8GHz.

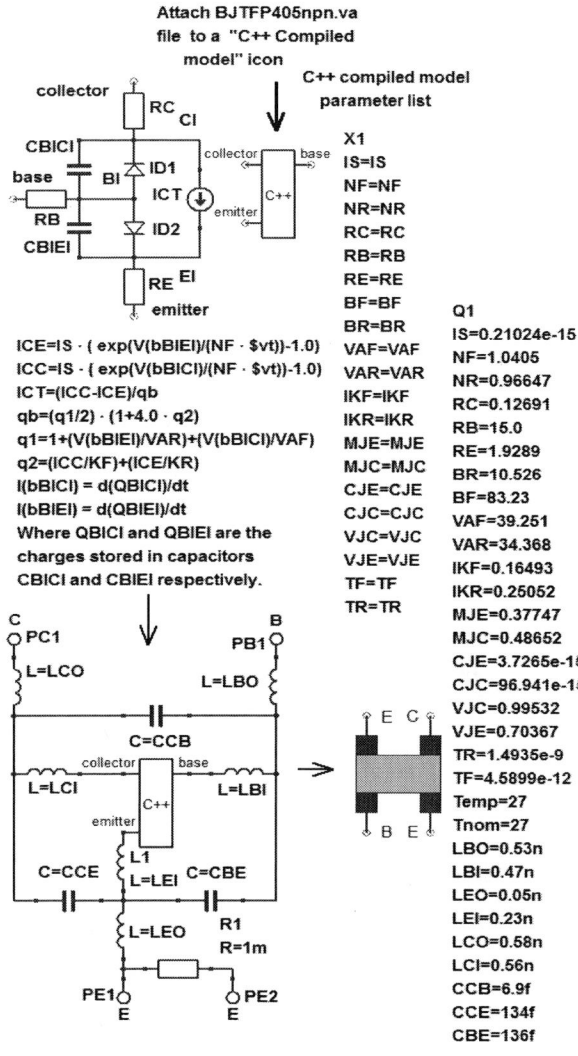

Figure 2. A QucsStudio "turn-key" RF npn BJT model outline showing: Ebors-Moll equations and equivalent circuit plus second order high-level current injection effects and internal capacitance equations; C++ compiled model icon with parameters (X1); subcircuit body and symbol plus parameters (Q1).

IV. QucsStudio C++ compiled Model Programming Interface

Central to the operation of the QucsStudio Verilog-A "turn-key" modelling system is a C++ compiled model component. An outline of the structure and content of a compiled model coponent is listed in Fig.5. In general QucsStudio built-in component models are defined by a C++ code template which lists model properties. The list contains amongst other things the number of external and internal nodes as well as pointers to the parameter list and to the schematic symbol. It also contains

function calls, like (tEvaluate)Matrix in Fig.5, that determine the physical operation of a component. Many of these are optional. The production version of the QucsStudio software is provided with a number of detailed examples of component model template entries. These are fully documented and provide a wealth of important model building data. The values for the variables listed in the component template are generated by ADMS during synthesis of the model C++ code.

```
// Verilog-A npn RF BJT model
`include "disciplines.vams"
`include "constants.vams"
module BJTFP405npn(collector,base,emitter);
inout collector,base,emitter ;  electrical collector,base,emitter;
electrical CI, BI, EI, nI1; // Internal nodes.
`define CTOK 273.15
`define GMIN 1e-12
`define TWOQ 2*`P_Q
parameter real IS = 0.21024e-15 from [1e-20 : inf);
parameter real NF = 1.0405 from [0.5 : inf) ;
parameter real NR = 0.96647 from [0.5 : inf) ;
parameter real RC = 0.12691 from [1e-20 : inf);
parameter real RB = 15.0 from [1e-20 : inf);
parameter real RE = 1.9289 from [1e-20 :inf;)
parameter real BF = 83.23 from [1e-20 : inf) ;
parameter real BR = 10.526 from [1e-20 : inf);
parameter real VAF = 39.251 from [1e-20 : inf);
parameter real VAR = 34.368 from [1e-20 : inf);
parameter real IKF = 0.16493 from [1e-20 : inf);
parameter real IKR = 0.25052 from [1e-20 : inf);
parameter real MJE = 0.37747 from [1e-20 : inf);
parameter real MJC = 0.48652 from [1e-20 : inf);
parameter real CJE = 3.7265e-15 from [1e-20 : inf);
parameter real CJC = 96.941e-15 from [1e-20 : inf);
parameter real VJC = 0.99532 from [1e-20 : inf);
parameter real VJE = 0.70367 from [1e-20 : inf);
parameter real TF = 4.5898  from [1e-20 : inf);
parameter real TR = 1.4935 from [1e-20 : inf);
parameter real Temp = 27 from [-273.15 : inf);
real con1, con2, con3, con4, con5, con6, con7;
real  x1, y1, x2, y2, z1, z2, con8, con9, QBICI, QBIEI, q1O2;
real IEC, ICC, q1, q2, T1, T2,  VJCO2, VJEO2;
// Model branches
branch (collector, CI)  bcollectorCI;  branch (base, BI) bbaseB1;
branch (EI, emitter) bElemitter;  branch (BI, CI) bBICI;
branch (BI, EI) bBIEI;  branch (CI, EI) bCIEI;
analog begin
  con1 = 1.0/(NF*$vt);  con2 = 1.0/(NR*$vt);
  VJCO2 = VJC/2;  VJEO2 = VJE/2;  con3 = 1.0-MJE;
  con4 = 1.0-MJC;  con5 = exp(MJE*ln(2));  con6 = exp(MJC*ln(2));
  // Current contributions
  I(bcollectorCI) <+ V(bcollectorCI)/RC;  I(bbaseBI) <+ V(bbaseBI)/RB;
  I(bElemitter)  <+ V(bElemitter)/RE;
  IEC = IS*(limexp(V(bBICI)*con2)-1.0);   ICC = IS*(limexp(V(bBIEI)*con1)-1.0);
  q1=1.0 + V(bBICI)/VAF + V(bBIEI)/VAR;   q2 = ICC/IKF + IEC/IKR;
  I(bBICI) <+ IEC/BR +  `GMIN*V(bBICI);   I(bBIEI) <+ ICC/BF +  `GMIN*V(bBIEI);
  q1O2 = q1/2;  I(bCIEI)  <+ (ICC-IEC)/(1e-20+q1O2*sqrt(1.0+4*q2));
  y1 = 1.0-V(bBICI)/VJ   y2 = 1.0-V(bBIEI)/VJE;
  z1 = exp(con4*ln(y1));   z2 = exp(con3*ln(y2));
  QBICI = (V(bBICI)>=VJCO2)
      ? TR*IEC+CJC*con6*(V(bBICI)*V(bBICI)+con4*V(bBICI))
      : TR*IEC+CJC*((VJC/con4)*(1.0-z1));
  QBIEI = (V(bBIEI)>=VJEO2)
      ? TF*ICC+CJE*con5*(V(bBIEI)*V(bBIEI)+con3*V(bBIEI))
      : TF*ICC+CJE*((VJE/con3)*(1.0-z2));
  I(bBICI) <+ ddt(QBICI);  I(bBIEI) <+ ddt(QBIEI);
end
endmodule
```

Figure 3. Verilog-A code for a simplified RF npn BJT model: the model parameters have the same meaning as those defined in the SPICE 3f5 BJT model [12].

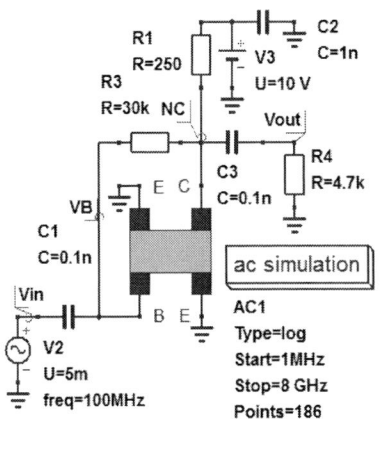

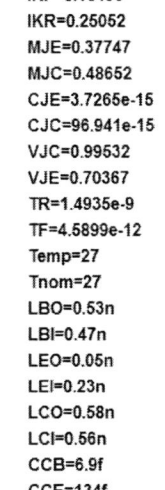

Q1
IS=0.21024e-15
NF=1.0405
NR=0.96647
RC=0.12691
RB=15.0
RE=1.9289
BR=10.526
BF=83.23
VAF=39.251
VAR=34.368
IKF=0.16493
IKR=0.25052
MJE=0.37747
MJC=0.48652
CJE=3.7265e-15
CJC=96.941e-15
VJC=0.99532
VJE=0.70367
TR=1.4935e-9
TF=4.5899e-12
Temp=27
Tnom=27
LBO=0.53n
LBI=0.47n
LEO=0.05n
LEI=0.23n
LCO=0.58n
LCI=0.56n
CCB=6.9f
CCE=134f
CBE=136f

```
// component definition
EXPORT tComponentInfo compInfo = {
  isNonLinear,        // component type ('isNonLinear' or 'isLinear')
  "BJTFP405npn",      // model identifier
  "VerilogAMS model of BJTFP405npn", // component description
  3,                  // number of external nodes
  3,                  // number of internal nodes
  0,                  // number of inputs (system simulations only)
  20,                 // number of parameters
  params,             // pointer to list of parameters
  0,                  // size of global variable buffer in bytes
  0,                  // pointer to component icon (0 = unused)
  0,                  // size of component icon
  -1,                 // index of parameter determining schematic symbol (-1 = unused)
  0,                  // pointer to list of schematic symbols
  (tEvaluate)fillMatrix, // function calculating analog model (0 = no model exists)
  0,                  // function calculating noise model (0 = noise free)
  0,                  // function calculating system model (0 = no model exists)
  0,                  // string with digital Verilog model (0 = no model exists)
  0                   // string with digital VHDL model (0 = no model exists)
};
```

Figure 5. QucsStudio component definition template.

```
1. setA(Node1, Node2, num);

      sets a single linear matrix element

2. setG(Node1, Node2, conductance);

      sets a linear conductance

3. setR(Node1, Node2, resistance);

      sets a linear resistance

4. setC(Node1, Node2, capacitance, initial voltage);

      sets a linear capacitance

5. setL(Node1, Node2, internal Node, inductance, initial current);

      sets a linear inductance

6. setM(Node1, Node2, mutual inductance, initial current);

      sets a linear mutual inductance

7. setI(Node1, Node2, current);

      sets a linear current

8. setDelayedI(Node1, Node2, line constant, delay, buffer pointer);

      sets a linear time-delayed current

9. setIQ(Node1, Node2, current, charge);

      sets a non-linear current and charge

10. setGC(Node1, Node2, Node3, Node4, current derivative, charge derivative);

      sets a non-linear conductance and capacitance

11. setNoiseA(Node1, Node2, noise current density);

      sets a single linear noise matrix element

12. setNoiseG(Node1, Node2, noise current density);

      sets a linear noise current

13. setNoiseNG(Node1, Node2, noise current density);

      sets a non-linear noise current
```

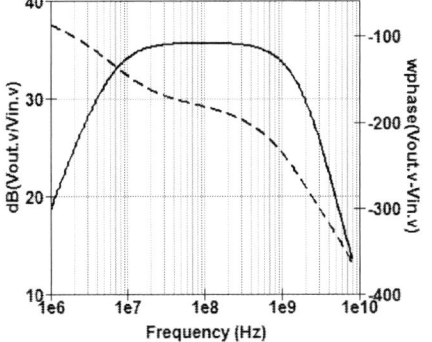

Figure 4. A class A npn BJT RF amplifier with collector feedback: small signal AC test circuit and example gain and phase response curves; legend: solid line = dB(Vout.v/Vin.v) the amplifier gain in dB and dotted line = wphase(Vout.v/Vin.v) the unwrapped phase of the amplifier gain in degrees.

In the case of the RF npn BJT the translated C++ model code is stored in file BJTFP405npn.va.cpp and compiled by the MinGW tools to produce a BJTFP405npn dynamically linkable library. In order to synthesize the C++ code needed for simulation of an analogue model the QucsStudio-ADMS-MinGW tools undertake the required operations in terms of a model application programming interface (API), specifically designed for QucsStudio C++ component model creation. The currently available API model functions are defined in Fig.6.

V. ADDING VERILOG-A NATURES TO QUCSSTUDIO

A high percentage of compact device models include a mixture of linear and non-linear R, L and C components. The 2.3.0 version of ADMS appears to treat all R and C components as non-linear elements. Also in this version of ADMS it is not permitted to express the connection of inductance L between circuit nodes p and n as V(p,n) <+ L*ddt(I(p,n)). However, this limitation can be overcome by combining a capacitor with a gyrator [13] to form a linear or non-linear inductance. In those compact models with a significant number of R, C and L elements simulation run

Figure 6. QucsStudio C++ model application programming interface functions: function parameters have their usual meaning as implied by their names.

times can be reduced, especially transient simulation, by ensuring that the QucsStudio "turn-key" modeling system uses linear R, C and L components whenever possible rather than non-linear devices. In the QucsStudio software the selection of linear or non-linear fundamental R, C or L components has been implemented by adding three new Verilog-A "natures" to the standard disciplines and natures file "disciplines.vams". These are listed in Table I. A parameter, called insideQucsStudio, is also defined by QucsStudio to allow automatic linear or non-linear component selection from within

TABLE I. QUCSSTUDIO NATURES PLUS MODIFIED ELECTRICAL DISCIPLINE FOR LINEAR R AND C COMPONENTS

nature Resistance
    units = "ohms";
    access = R;
endnature
	nature Conductance
	    units = "s";
	    access
	endnature
		nature Capacitance
		    units = "F";
		    access = C;
		endnature

	discipline electrical
	    potential Voltage;
	    flow Current;
	    flow Conductance;
	    flow Resistance;
	    flow Capacitance;
	end discipline	

Verilog-A model code, for example in the resistive case:

```
`ifdef insideQucsStudio
    R(b1)  <+ Rvalue;
`else
    I(b1)  <+ V(b1)/Rvalue;
`endif;
```

VI. TRANSIENT SIMULATION OF A CLASS A RF NPN BJT AMPLIFIER

The circuit diagram drawn in Fig. 7 shows a single stage class A npn BJT RF amplifier with signal outputs taken directly from the BJT base and collector terminals respectively. Fig. 7 also illustrates a typical set of base and collector transient simulation waveforms for a 50mV peak, 10MHz sinusoidal signal applied to the amplifier input. Although the QucsStudio "turn-key" modelling system is primarily designed to be a fast simple to use development tool for constructing equation-defined compact device models it also works along-side an advanced post-simulation data processing and visualization package incorporating the GNU GPL Octave numerical analysis software [14]. The current version of QucsStudio allows simulation output data to be numerically processed by the Octave package, following completion of a circuit simulation task. For example, the voltage amplitude spectra shown in Fig.8 were computed using the Octave "m" script listed in Fig. 9. This illustrates the use of Octave functions and statements for converting QucsStudio simulation data into the Octave data format, the use of the Octave function fft to perform a fast Fourier transform of output data NB.Vt and NC.Vt and finally how Octave visualization statements can generate the output data plots shown in Fig.8.

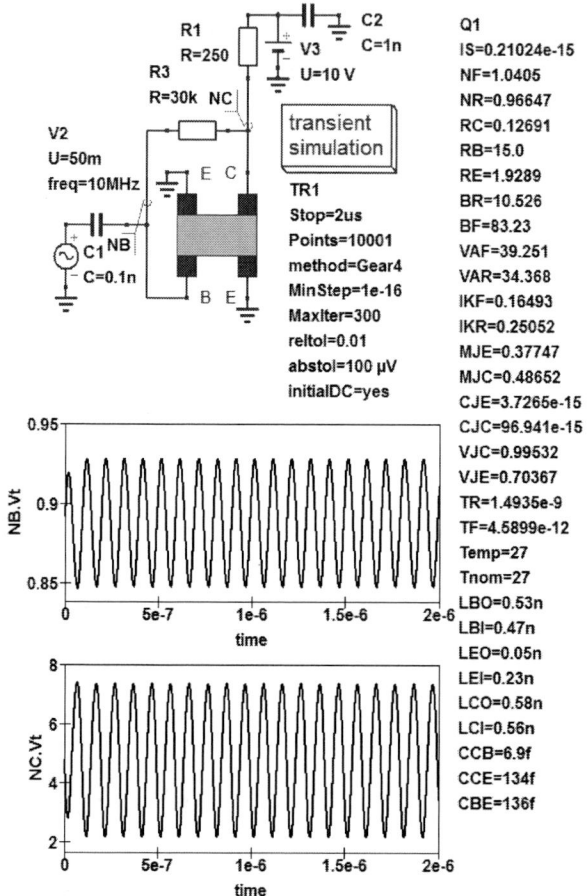

Figure 7. A class A npn BJT RF amplifier with collector feedback: transient simulation test circuit with directly coupled voltage test probes NB and NC and time response output waveforms for a sinusoidal input signal of 50mV peak, 10MHz frequency and zero input phase.

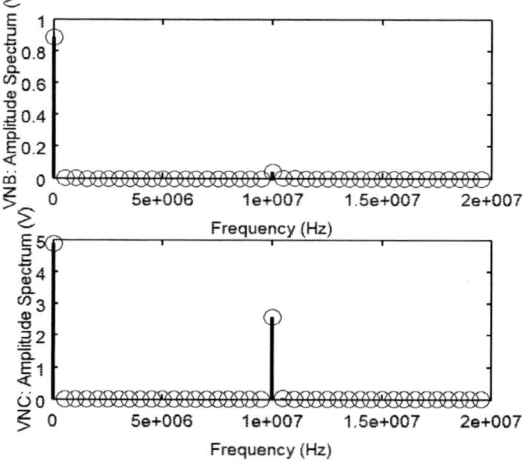

Figure 8. Voltage amplitude spectra plotted against frequency for amplifier probe signals NB and NC.

```
% File testfeedbacknpnTRAN.m file
% Control file called on completion of transient simulation.
% Calls functions loadQucsDataset,loadQucsVariable
% and stemfft.
function stemfft(data, points, FinTim)
  yfft = abs(fft(data/points));
  no2  = length(yfft)/2;
  yvec(1) = yfft(1);
  yvec([2:no2]) = 2*yfft([2:no2]);
  fc   = linspace(0, no2, no2)/FinTim;
  stem(fc, yvec, "linewidth", 4 , "color", "black");
endfunction


qucsFilename = 'TestfeedbacknpnTRAN.dat';
loadQucsDataset;
whos
clf()
newplot()
[VNB,Dep]=loadQucsVariable("TestfeedbacknpnTRAN.dat","NB.Vt");
[VNC,Dep1]=loadQucsVariable("TestfeedbacknpnTRAN.dat","NC.Vt");
[Time,Dep3]=loadQucsVariable("TestfeedbacknpnTRAN.dat","time");


subplot(2,1,1)
stemfft(VNB, 2048, 2e-6);
set(gca, "linewidth", 4, "fontsize", 20, "fontname", "TimesRoman",
  "fontweight", "bold", "xlim", [0,20e6], "xlabel", text("string",
  "Frequency (Hz)", "fontsize", 20, "fontname", "TimesRoman",
  "fontweight", "bold"), "ylabel",  text("string",
  "VNB: Amplitude Spectrum (V)", "fontsize", 20,
  "fontname", "TimesRoman", "fontweight", "bold", "rotation", 90));


subplot(2,1,2);
stemfft(VNC, 2048, 2e-6);
set(gca, "linewidth", 4, "fontsize", 20, "fontname", "TimesRoman",
  "fontweight", "bold", "xlim", [0,20e6], "xlabel", text("string",
  "Frequency (Hz)", "fontsize", 20, "fontname", "TimesRoman",
  "fontweight","bold"), "ylabel",  text("string",
  "VNC: Amplitude Spectrum (V)", "fontsize", 20,
  "fontname", "Arial", "fontweight", "bold", "rotation", 90});
print("TestfeedbacknpnTRAN.png","-dpng");
```

Figure 9. Octave "m" script for post-simulation amplifier data processing: Functions "loadQucsDataset" and "loadQucsVariable" are provided with the QucsStudio software for conversion of simulation output data to Octave internal format.

VII. Conclusions

Compact device modelling with Verilog-A has become standard practice among commercial and GNU General Public License circuit simulators, with many packages adopting the ADMS Verilog-A to C++ model synthesizer as the central core in their device modelling strategy. Initial open source implementations of the ADMS model synthesizer often depended on model developers patching simulator C++ code when constructing new models. This approach not only requires developers to have a good understanding of a particular circuit simulators model application interface but is likely to be error prone. The introduction of the QucsStudio "turn-key" approach to compact device modeling provides for the first time, as far as the authors are aware, a freely available, fast and simple to use GNU General Public License modelling tool which does not require users to manually patch circuit simulator C++code.

References

[1] Compact Model Council, TechAmerica, Arlington, VA. http://www.gela.org/About-TechAmerica, 2009. [accessed January 2012].

[2] Accellera, "Verilog-AMS Language Reference Manual, version 2.2", 2004, http://www.accellera.org, 2010. [accesssed January 2012].

[3] L. Lemaitre, G. Coram, C. McAndrew and K. Kundert, "Extensions to Verilog-A to support compact device modeling", Proceedings of the IEEE International Workshop on Behavioural Modeling and Simulation, BMAS, 7-8 Oct. 2003, pp, 134-138.

[4] L. Lemaitre. ADMS, http://adms.noovela.com:8001/, 2007. [accessed January 2012].

[5] Smash mixed signal simulator, Version 5.18.0, Dolphin Integration, France, http://www.dolphin.fr/medal/smash/smash_overview.php, 2011. [accessed January 2012].

[6] Symica Custom IC Design Toolkit, Symica LLC, 2009-2012, http://www.symica.com/products/symica-de, 2012. [accessed January 2012].

[7] P Nenzi, ngspice release 23, http://ngspice.sourceforge.net/index.html, 2011. [accessed January 2012].

[8] A. Davis, Gnucap, Version 0.35, http://www.gnu.org/software/gnucap/, 2008. [accessed January 2012].

[9] M. Margraf, QucsStudio, Production Version 1.3.0, http://www.mydarc.de/DD6UM/QucsStudio/qucsstudio.html, 2012. [accessed 2012].

[10] M. Margraf, S.Jahn, J. Flucke, R. Jacob, V. Habchi, T. Ishikawa, A. Gopala Krishna, M. Brinson, H. Parruitte, B.Roucaries and G. Kraut, Qucs (Quite universal circuit simulator), Version 0.0.16, 2011, http://qucs.sourceforge.net/index.html, [accessed January 2012].

[11] M. Brinson and S. Jahn, Building device models and circuit macromodels with the Qucs GPL circuit simulator, COMON project meeting, IHP, Frankfurt (Oder), Germany, 2009, http://www.mos-ak.org/frankfurt_o/papers/M_Brinson_Qucs_COMON_April_2_2009_final.pdf. [accessed anuary 2012].

[12] P. Antognetti and G. Massobrio (Editors), "Semiconductor device modeling with SPICE", McGraw-Hill Book Company, New York, 1988.

[13] S. Jahn and M.E. Brinson, "Interactive compact device modelling using Qucs equation-defined devices", International journal of Numerical Modelling: Electronic Networks, Devices and Fields, 2008, 21:335-349.

[14] J.W. Eaton et al., GNU Octave high level language for numerical computations, http://www.gnu.org/software/octave/about.html. [accessed January 2012].

Especial TRAMS (Terascale Reliable Adaptive Memory Systems) Project

MIXDES 2012, 19th International Conference *"Mixed Design of Integrated Circuits and Systems"*, May 24-26, 2012, Warsaw, Poland

Enhancing 6T SRAM Cell Stability by Back Gate Biasing Techniques for10nm SOI FinFETs under Process and Environmental Variations

Zoran Jakšić, Ramon Canal
Department of Computer Architecture
Universitat Politecnica de Catalunya
Barcelona, Spain
zjaksic@ac.upc.edu, rcanal@ac.upc.edu

Abstract—Process variations have shown to be critical for future Si-based bulk technologies. FinFETs have shown to be an alternative. In this work, we characterize the performance of the 6T SRAM cell by estimating read static noise margin and world line write margin for future FinFET-based 10nm technology. For simulation, we used HSPICE tool and SOI BSIM-CMG model card of 10nm FinFET previously developed by the University of Glasgow, Semiconductor Device Modeling Group. Process variations are based on ITRS predictions and they are modeled at the device level. On top of that, we also add environmental variations such as temperature and voltage. We also propose an effective procedure to apply the back-gate biasing technique for independent gate FinFET to the BSIM-CMG model card. As a specific example, we show how the RSNM of 6T SRAM cell can be improved by using back gate biasing techniques for independent gate FinFETs. First we show how WLMN is increased by reducing strength of pull up transistor with reverse back gate biasing of PU transistors. Then, we show how the RSNM can be increased by reducing the strength of access transistor by reverse back gate biasing of PG transistors. When combining these two techniques RSNM can be improved up to 25% without compromising cell write ability for any sample.

Index Terms—FinFETs, 6T SRAM cell, back gate biasing, static noise margin, process variation

I. INTRODUCTION

Scaling technology beyond 32nm impose the problem of variations. They manifest as fluctuations in power consumption, stability and maximal speed degradation. [1]–[3]. Variations can be divided in two groups - process and environmental. Process variations are consequence of imperfections in development process that are translated to a non-deterministic behavior of the transistors. These imperfections are not scaled with technology so their effect on circuit functionality increases as technology shrinks. Temporal variations of temperature and voltage as well as aging (NBTI, PBTI, and HCI) can be put under the category of environmental variations [4].

On the other hand, novel multi-core processor architectures demand more on-chip caches for effectively sharing information across parallel processing units. These memories are mainly designed with SRAMs and since they occupy the greatest part of the chip area it is of crucial importance that SRAM cell, as their major unit, is scaled to the minimal

dimensions. Parameter fluctuations have greatest influence on circuits that rely on perfectly matched transistor pairs [2] and if the small dimensions of SRAM cells are added to previous statement it is clear why process variations have the greatest influence on them.

FinFETs are the most promising candidate for replacement of classical bulk technology. Better channel control is achieved because of their 3D structure, and higher I_{on}/I_{off} can be reached with lightly doped channel. Lightly doped channel reduce the effect of random dopant fluctuations (RDF) which is the greatest source of variability in classical bulk technology. However, some sources of variability caused by LER, FER and GER still remain [1], [2].

In this paper, we investigate the characteristics of 6T SRAM cell when they are designed in 10nm SOI FinFETs technology. Assuming that the cell is scaled to the minimal dimensions, we simulated its stability (Read Static Noise Margin and World Line Write Margin) when it is exposed to the different sources of variability (process variability caused by LER, FER, GER, random discrete dopants and environmental variation that is regarded to the variation of chip temperature).

We analyzed a technique for World Line Write Margin enhancement by lowering the strength of the pull up transistors. This can be achieved using a reverse back gate biasing technique when independent gate FinFETs are placed as pull up transistors. Also the Read Static Noise Margin can be increased by lowering strength of the access transistors which can be done by using the reverse back gate biasing technique for PG FinFET or by simple weak-writing (accessing the cell with lower word line voltage).

Using the back gate biasing technique is very attractive method to fight variability since it can be applied dynamically after chip fabrication.

There are two main contributions with this paper. The first contribution is the method that we proposed how the HSPICE BSIM-CMG model card can be used for simulation of independent gate FinFETs when back-gate biasing technique is applied. The second contribution is the analysis which shows how the RSNM of 6T SRAM cell is improved when back gate biasing technique is applied in cell design.

The rest of the paper is organized as follows. In section II we present basic information regarding FinFET technology and HSPICE model card that we used in our work. In section III we explain procedure of variability calibration of FinFET model in order that it can be incorporated in HSPICE for speed up of cell evaluation. Classical 6T SRAM cell is presented in section IV. In section V, we present the related work regarding the characterization of FinFET SRAM cells and methods to fight process variations. In section VI, we present the simulation results of 10nm FinFET SRAM cells and the evaluation of our proposals. Finally, we draw the main conclusions.

II. FinFET Technology

The physical and electrical characteristics of FinFETs that we used in our study are presented in this section. Figure 1 shows the 3D structure of a FinFET. The major characteristic of this device type is that gate wraps channel from 3 sides. This structure gives better control of short channel effects and high dopant concentration is not necessary anymore as it was case with traditional bulk MOSFETs [5], [6]. This is a huge advantage because variability caused by random discrete dopants becomes insignificant [7]. Also the leakage current is lowered comparing to the classical bulk technology. Knowing that leakage current is the major source of static power consumption SOI FinFETs pose great advance in solution of this problem.

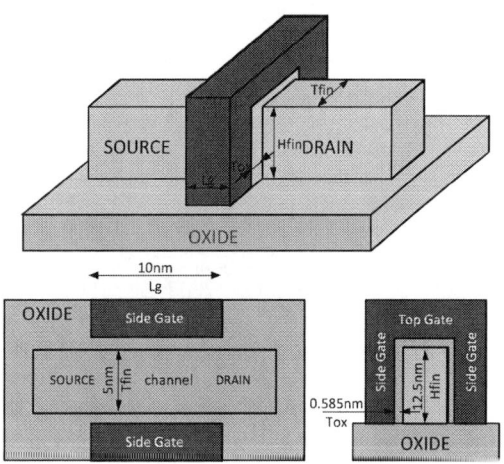

Fig. 1.   3D structure of one fin tri-gate FinFET

However, some sources of variability still remain. Variation in circuit parameters caused by the effects of Line Edge Roughness (LER), Fin Edge Roughness (FER) and Gate Edge Roughness (GER) still pose significant challenge for these devices usage [1], [7].

Dimensions of FinFETs used in this work can be seen from the Figure 1. Our model is designed for 10nm channel length (L_g). Fin height (H_{fin}) and fin thickness (T_{fin}) are 12.5nm and 5nm respectively. Effective oxide thickness is 0.585nm [7].

For each transistor, length is fixed to (L_g) and width can be computed as:

$$W = N_{fin}(2H_{fin} + T_{fin}) \qquad (1)$$

Where N_{fin} is the number of fins of the transistor.

The top gate of the channel can be etched and by doing so two independent gates can be formed [6]. This gives different design opportunities. The most common application of this structure is called back gate biasing. This technique involves the following:

One gate is used as a control node as in classical devices while the other is biased to the constant voltage. For example, applying constant negative voltage on one N-type FinFET I_d can be reduced. This technique is similar to the technique of body biasing in classical bulk technologies. [8], [9]

For the simulation of this technique on single gate FinFETs, we used the following approach. Every back gate biased transistor is transformed into the transistor as in Figure 2. VG source is back gate bias and in general it is dependent of the voltage V_{GS} that the best I_D curve fitting can be achieved.

For simplicity, we assume that V_G is constant (independent of V_{GS}). This should not induce big error in our simulation results since we are focused only on one part of I_D curve (I_{on} current is main reason of cell stability). For better understanding of this technique it is useful to note that I_{on} current is reduced by 10% per 0.05V of applied bias (Figure 3) for N and P type FinFET.

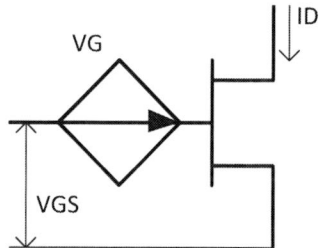

Fig. 2.   Equivalent scheme of back gate biased FinFET

III. FinFET Variability

Numerical simulations based on finite-elements methods or TCAD (Technology CAD) tools are useful for technology evaluation. TCAD simulations can be combined with Monte Carlo simulations to predict the impact of device variation on circuit performance. However these simulations are very time-consuming and applying them for circuit with large number of devices (transistors) is practically impossible [5], [10]. Efficient compact model (like SPICE BSIM-CMG model) is more suitable [10], [11].

Using the model-based approach, we can drastically reduce the simulation time. For instance, the simulation of one $V_{gs}I_d$ curve with 15 points takes approximately 6 hours on one processor core when TCAD simulation is run. For achieving same curve in HSPICE when BSIM-CMG model card is used, it takes just a few seconds.

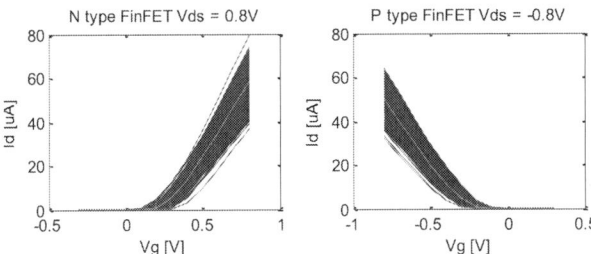

Fig. 3. FinFET transfer characteristics

In this section, we propose a procedure to obtain variability characteristics from TCAD simulations in order that they can be incorporated in HSPICE simulations.

Figure 3 presents transfer characteristics of N and P type FinFETs. Plots have been obtained by TCAD simulations of 10nm FinFET when variability of LER, FER, GER and random discrete dopants are incorporated. For both FinFET types, 1000 instances have been analyzed. Nominal supply voltage is $V_{DS} = 0.8V$.

The most common procedure for variability simulation using the HSPICE tool is to do Monte Carlo simulation when some of the model parameters are Gaussian random variables with mean equal to the nominal value [3], [12]. Finding a standard deviation of specific parameter is more complicated. The procedure for determination parameters standard deviation is illustrated on Figure 4.

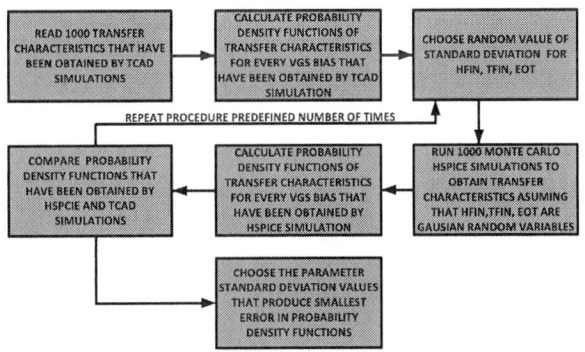

Fig. 4. FinFET variability calibration procedure

The first step is to load 1000 TCAD transfer plots and calculate I_d probability density functions for every value of gate-source bias. Next, we pick up a random value of standard deviation for different parameters in HSPICE model card (HFIN, TFIN, and EOT), perform DC simulation with 1000 Monte Carlo samples and compare results with the ones that have been obtained by TCAD simulations. The term "compare results" refers to comparing probability density function of I_d for every value of gate-source bias. This procedure is repeated iteratively until the difference (or similarity) between the outputs is below a certain threshold. For the measure of similarity (i.e. error) between distribution functions, we choose

number of values that fall in 3σ and 5σ range of the I_d distribution function that is obtained by TCAD simulation.

After the iterative process, this procedure achieves an error (as defined in the previous paragraph) of less than 5% for 3σ and less than 2% for 5σ, when TFIN and HFIN parameters are considered as Gaussian random values with relative standard deviation of 16% and 12% respectively.

In case of independent back-gate biased FinFET we assumed same variability metrics same as for Tri-Gate FinFET which is realistic considering that same device dimensions are used.

IV. FINFET SRAM CELLS

Over the years 6T cell (Figure 5) was the defacto choice for SRAM design. As the need for bigger (and denser) memories is always present, it is of crucial importance that the cell is scaled to minimum dimensions. However, some constraints have to be satisfied. A detailed explanation of functionality of classical 6T SRAM cell can be found in [13]. For proper dimensioning of the cell, two basic constraints have to be satisfied. PU transistors have to be weaker than PG transistor (this constraint is called write ability) and PD transistor have to be stronger than PG (readability). According to this, two design metrics were defined.

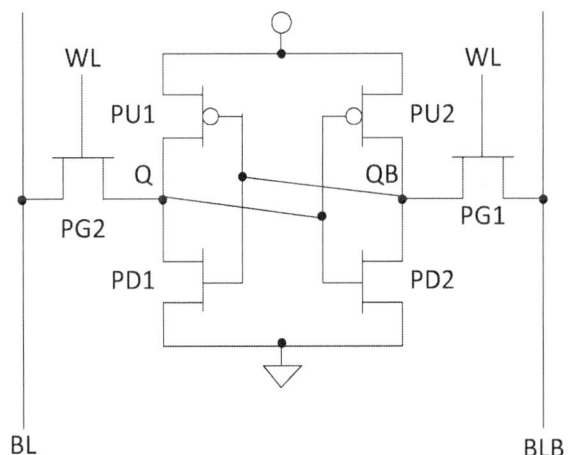

Fig. 5. Classical FinFET 6T SRAM cell

A *read static noise margin* (RSNM) characterizes the read stability of the SRAM cell. It can be defined as the minimal noise voltage that can flip the cell state during the read process. RSNM is calculated as the maximum square that can fit in the butterfly curve. The butterfly curve is formed by plotting the voltage transfer characteristics of the two inverters in an SRAM cell when both bit lines (BL and BLB) and word line (WL) are biased at V_{DD}. If the two squares inside the butterfly curve don't have equal side lengths, RSNM is the side length of the smaller one [10].

Word line write margin (WLWM) characterizes the cell's write ability. A write operation in SRAM is typically carried out at a WL voltage of V_{DD}. In some cases WL voltage can be lowered during the cell access in order to improve cell Read Stability (RSNM) or reduce dynamic power consumption. WLWM is the maximal value by which the WL voltage can be lowered than V_{DD} that still allows a successful write [10].

The operation of a 6T FinFET SRAM cell is same as conventional 6T single-gate SRAM circuit and same constraints have to be satisfied. However, previously defined dimension for single-gate bulk devices can't be directly applied on FinFET technology due to the quantization of the FinFET width [6], [14].

In general, as it can be seen from the Figure 3, P-type FinFET is weaker than N-type so by setting 1 fin transistors at PU and PG the write-ability constraint is satisfied.

The memory cell is most sensitive to the reading process. In order to improve the read static noise margin, the number of fins in the PD transistors should be increased but this increase area penalty. For the minimum satisfying cell performance regarding read stability, 2 fins are needed for PD transistor [6], [14].

In this paper, we show how the RSNM can be improved further when the strength of PU and PG transistor is reduced without threatening the cell write ability. Using the back gate biasing technique, the strength of the PU transistor can be reduced by applying the positive voltage on back gate. This leads to an increase of the WLWM.

The strength of the access transistor can be reduced to increase RSNM. There are two possibilities for reducing access transistor current. One possibility is to use negative back gate bias on PG transistor when independent gate FinFET is used. The other is to simply lower the word line voltage when cell is accessed. These two techniques are the same regarding to the back gate biasing procedure simulation that we proposed.

V. RELATED WORK

Body biasing techniques have been successfully demonstrated for reducing V_{th} variation. In general, this is the best method to fight variability since it gives possibilities for dynamic modification of V_{th} after chip fabrication that is achieved by adjusting the body-to-source voltage. Forward body biasing (FBB) decreases the threshold voltage V_{th} of transistors, increasing maximum frequency and leakage, while reverse body biasing (RBB) has the opposite effect. [9].

FinFETs have become especially interesting in last years because classical bulk technology has reached its limits due to high leakage caused by short channel effects. The design of 3D transistors gives the possibility of independent gate FinFETs implementations where techniques similar to classical body biasing can be applied. Idea of applying different voltage on one gate gives grater possibility of current manipulation since it can be applied to the transistor level (instead to the group of transistors as it is case of classical body biasing). However routing of these devices is more complicate since independent gate FinFETs have one more pin.

Current research regarding implementation of independent gate FinFETs show that reducing the channel length and

supply voltage according to the ITRS prediction keep RSNM and WLWM of 6T SRAM cell in range of 150-250mV.

Some of those works can be found in [6], [14], [15] where authors reached 240mV of RSNM for 32nm FinFETs. Similar results can be found in [16], [17] where RSNM of SRAM cells for 25nm SOI FinFETs was evaluated.

As far as we know the smallest designed FinFETs are published in the work of Gupta et al. [18]. They present 6T SRAM cells designed with FinFETs of 10.8nm channel length and shown the results of RSNM up to 250mV and WLMN up to 300mV for 0.7V power supply.

In our work, we reduced device dimensions further, and we show a complete framework for variability simulation of Tri-Gate FinFETs using HSPICE tool. Also we show an original method for simulating back-gate biasing technique for independent gate FinFETs using Tri-Gate model card that can give very accurate results when it is applied in simulation of SRAM cell stability.

VI. SIMULATION RESULTS

In this section, we present the results of FinFET 6T SRAM cell simulations. The cell is designed for 0.8V power supply. Plots are made for different variations of PU and PG back gate biased voltage and different temperatures. Cell is scaled to the minimal dimensions. PU and PG transistor have one fin and PD transistor have 2 fins as this has been shown as optimal [6].

Since the worst case analysis tends to show too pessimistic results we present distribution plots of the RSNM and WLMN when back gate biasing is applied on PU and PG transistors.

Figure 6a show how the Word Line Write Margin (WLMN) is increased when the strength of the PU transistor is reduced and Figure 6b shows how the RSNM is reduced for the same.

Reverse back gate biasing can be applied on PG transistor, too. In order to reduce it's strength we should apply negative voltage on its back gate. Figure 7a shows how the WLMN is decreased when the strength of the access transistor is reduced and Figure 7b shows how the RSNM is increased for the same. It is important to notice that for bias of 100mV write ability is compromised for some samples.

By combining previous two techniques we can increase RSNM more. It can bee seen that for specific case of PU bias=100mV (about 20% of its I_{on} current reduction) and PG bias=-150mV(about 30% of its I_{on} current reduction) median value of RSNM can be increased up to 195mV which is more than 25% higher than RSNM of the classical 6T SRAM cell.

Increasing back gate voltage (reducing its I_{on}) of PU transistor doesn't offer further improvement of RSNM without compromising write ability which is our fundamental demand. Considering that, previous results can be considered optimal regarding RSNM.

For the best illustration of RSNM improvement on the Figure 8 we show butterfly curves of the classical 6T SRAM cell and the cell where back gate biasing is applied . Red lines show the ideal butterfly curves (cell without process variations and applied back gate bias).

106

Figure 10 illustrates RSNM temperature dependency of proposed 6T SRAM cells.

VII. CONCLUSION

In this paper we presented possibilities of using 10nm FinFETs in SRAM cell design. Our work has shown that these

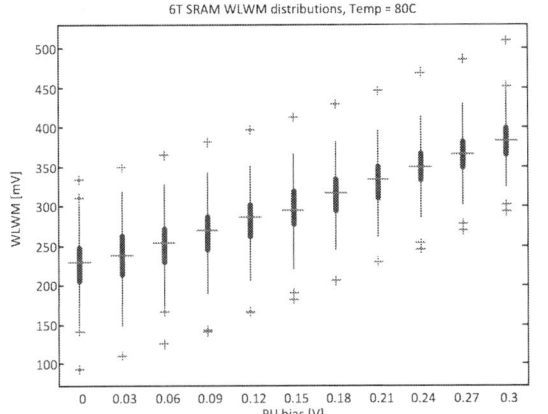 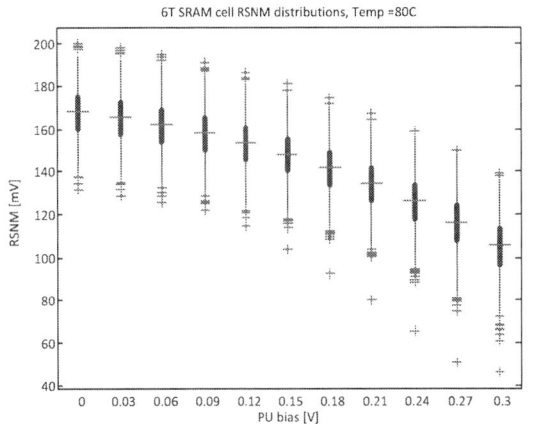

Fig. 6. 6T SRAM cell WLMN and RSNM probability distributions when PU back gate biasing is applied

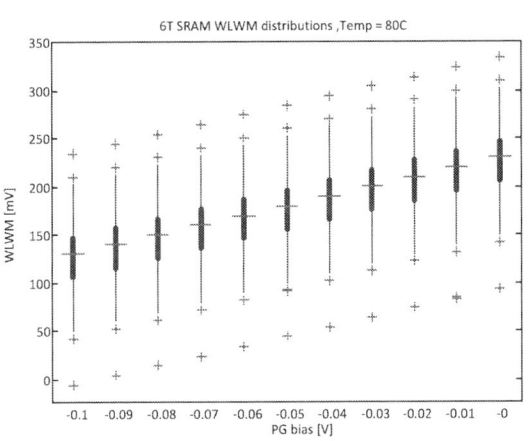 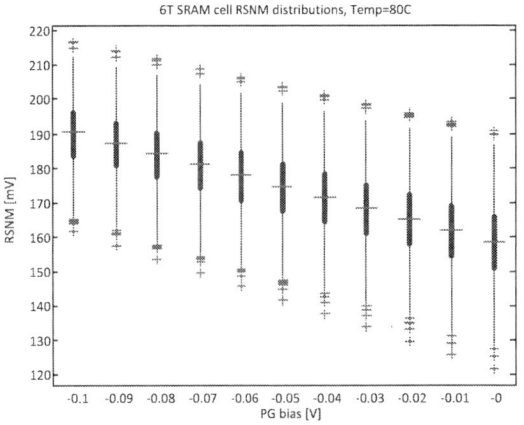

Fig. 7. 6T SRAM cell WLMN and RSNM probability distribution functions: when PG back gate biasing is applied

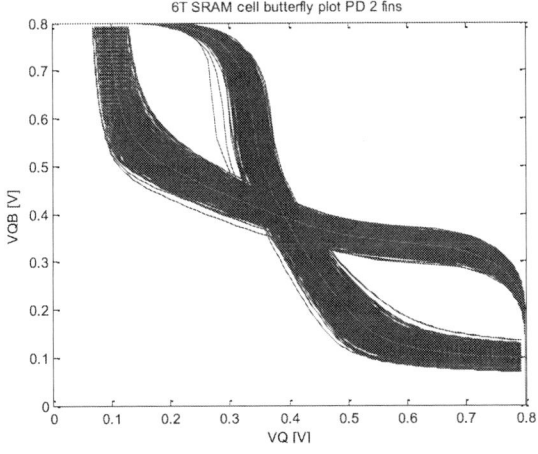 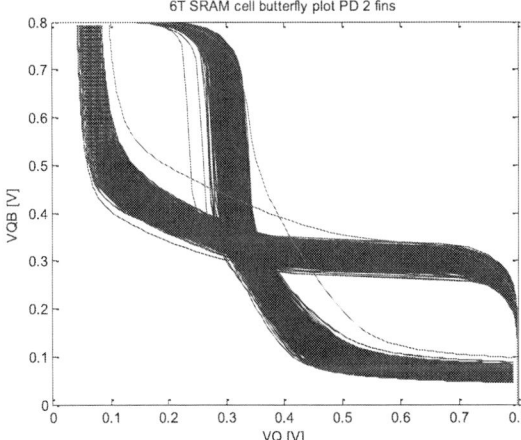

Fig. 8. 6T SRAM butterfly curves for: a) classical 6T SRAM cell, b) 6T cell with PU Vbias = 0.1V and PG Vbias=0.15V

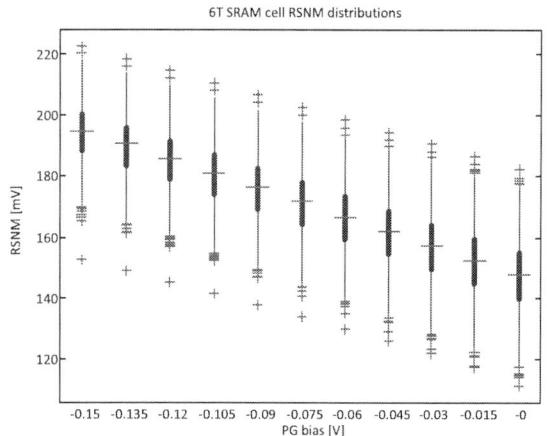

Fig. 9. 6T SRAM cell RSNM probability distribution functions: when PU and PG biasing is applied, PU bias=0.1V

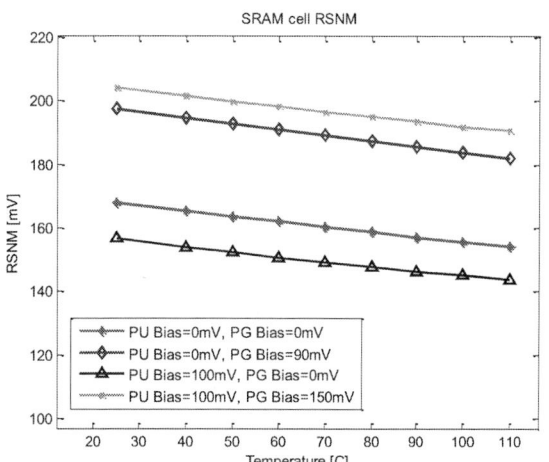

Fig. 10. 6T SRAM cell RSNM Temperature dependency

circuits demonstrate great stability when they are exposed to the effects of process and temperature variations. Our major goal was to show how the Read Static Noise Margin can be improved using independent gate FinFETs when back gate biasing technique is applied. Results have shown that RSNM can be improved up to 25% when PU and PG transistors' currents are reduced by back gate biasing and (or) weak writing without compromising write ability, not even when cell is exposed to the effects of process and environmental variation. The good side of this technique is that can be applied dynamically for every cell but the greatest disadvantage is that it demands additional power supplies and complicated routing.

ACKNOWLEDGMENT

This work has been partially supported by the Spanish Ministry of Education and Science under grant TIN2010-18368, the TRAMS project of the FP7 program of the Eu-

ropean Commission under agreement 248789, the Generalitat of Catalunya under grant 2009SGR1250 and Intel Corporation.

REFERENCES

[1] A. Asenov, A. Brown, J. Davies, S. Kaya, and G. Slavcheva, "Simulation of intrinsic parameter fluctuations in decananometer and nanometer-scale mosfets," *IEEE TRANSACTIONS ON ELECTRON DEVICES*, vol. 50, no. 9, pp. 1837–1852, September 2003.

[2] E. Baravelli, M. Jurczak, N. Speciale, K. Meyer, and A. Dixit, "Impact of 1er and random dopant fluctuations on finfet matching performance," *IEEE TRANSACTIONS ON NANOTECHNOLOGY*, vol. 7, no. 3, pp. 291–298, March 2008.

[3] S. Ganapathy, R. Canal, A. Gonzalez, and A. Rubio, "Circuit propagation delay estimation through multivariate regression-based modeling under spatio-temporal variability," in *Design, Automation, and Test in Europe (DATE)*.

[4] S. Ghosh and K. Roy, "Parameter variation tolerance and error resiliency: New design paradigm for the nanoscale era," *Proceedings of IEEE*, vol. 98, no. 10, pp. 1718–1751, October 2010.

[5] J. P. Colinge, *FinFETs and Other Multigate Transistors*. Springer, 2008.

[6] S. Tawfik, Z. Liu, and V. Kursun, "Independent-gate and tied-gate finfet sram circuits:design guidelines for reduced area and enhanced stability," in *Internatonal Conference on Microelectronic (ICM)*.

[7] X. Wang, A. Brown, B. Cheng, and A. Asenov, "Statistical variability and reliability in nanoscale finfets," in *IEDM*.

[8] Y. Yasuda, Y. Akiyama, Y. Yamagata, Y. Goto, and K. Imai, "Design methodology of body-biasing scheme for low power system lsi with multi- vth transistors," *IEEE TRANSACTIONS ON ELECTRON DEVICES*, vol. 54, no. 11, pp. 2946–2952, November 2007.

[9] A. Bonnoit, , and L. Pileggi, "Reducing variability in chip-multiprocessors with adaptive body biasing," in *Low-Power Electronics and Design (ISLPED)*.

[10] D. Lu, A. Niknejad, C. Hu, and C. Lin, "Compact modeling in variations of finfet sram cells," *Design and Test of Computers*, vol. 27, no. 2, pp. 45–50, February 2010.

[11] "Hspice reference manual," September 2011.

[12] A. Agawal, D. Blauw, and V. Zolotov, "Statistical timing analysis for intra-die process variations," in *International Conference on Computer Aided Design (ICCAD)*.

[13] N. Weste and D. Haris, *CMOS VLSI Design*. Pearson Addison Welsy, 2005.

[14] Z. Liu, S. Tawfik, and V. Kursun, "Statistical data stability and leakage evaluation of finfet sram cells with dynamic threshold voltage tuning under process parameter fluctuations," in *International Symposium on Quality Electronic Design (ISQED)*.

[15] S. Tawfik and V. Kursun, "Work-function engineering for reduced power and higher integration density an alternative to sizing for stability in finfet memory circuits," in *IEEE International Symposium on Circuits and Systems (ISCAS)*.

[16] M. Fan, Y. Wu, V. Hu, P. Su, and C. Chuang, "Investigation of cell stability and write ability of finfet subthreshold sram using analytical snm model," *IEEE TRANSACTIONS ON ELECTRON DEVICES*, vol. 57, no. 6, pp. 1357–1381, June 2010.

[17] ——, "Investigation of stability and ac performance of sub-threshold finfet sram," in *International Symposium on VLSI Technology Systems and Applications (VLSI-TSA)*.

[18] S. Gupta, S. Park, and K. Roy, "Tri mode independent-gate finfets for dynamic voltage/frequency scalable 6t srams," *IEEE TRANSACTIONS ON ELECTRON DEVICES*, vol. 58, no. 11, pp. 3837–3846, November 2011.

MIXDES 2012, 19ᵗʰ International Conference *"Mixed Design of Integrated Circuits and Systems"*, May 24-26, 2012, Warsaw, Poland

Mitigating Lower Layer Failures
with Adaptive System Reconfiguration

Tanausú Ramírez, Enric Herrero, Nicholas Axelos, Javier Carretero,
Nikos Foutris, Daniel Sanchez and Xavier Vera
Intel Barcelona Research Center (IBRC), Intel Labs
Barcelona, Spain
{tanausu.ramirez, enric.herrero, nicholasx.axelos, javier.carretero.casado,
nikos.foutris, danielx.sanchez, xavier.vera}@intel.com

Abstract—**Future terascale systems based on sub-22nm technologies will show significant variability and reliability challenges from the transistor to the circuit level. On this upcoming scenario, a reliable system must be built on top of unreliable components, which will degrade and even fail during the normal lifetime of the chip. To achieve this target, we present a high-level reconfiguration approach for future heterogeneous systems that mitigates the possible lower layer shortcomings and adapts the processor to the user's requirements.**

Index Terms—**heterogenous multicore; variability; resiliency; system reconfiguration**

I. INTRODUCTION

Technology scaling has driven the evolution of computer systems over past decades resulting in faster, cheaper and denser processor chips. However, there is a growing concern about reliability as we move into the late CMOS technologies. The sensitivity to errors and variation issues on future systems will rise exponentially as the feature sizes shrink and voltage drops to reduce power consumption. On this upcoming scenario, systems would have very low yield and a poor operational performance and reliability.

The great challenge for future technologies is building reliable systems on top of unreliable components, which will degrade and even fail during the lifetime of the chip. The need to find cost-effective reliability mechanisms has driven a broad field of solutions to mitigate the effects of errors. In this sense, the level of unreliability expected for sub-22nm technologies calls for cross-layer approaches that make the best use of solutions coming from the different levels of the system stack.

Traditionally, manufacturers enforce reliability with overly conservative methodologies during device, circuit and chip design. However, microarchitectural and system intervention offers a novel way to manage reliability and testing at high-level without significantly sacrificing cost and performance. Using the microarchitecture knowledge of hardware and application runtime behavior, we can qualify the lifetime reliability based on the target application and user requirements. Likewise, this offers an opportunity to decrease cost and increase performance of the system.

In this work we present a high-level approach to deal with the variability and reliability issues from the lower layers

following the aforementioned insights. Here, we describe a proposal for a microarchitectural and system level solution based on a dynamic reconfiguration mechanism for future heterogeneous multicore processors.

The proposed reconfiguration mechanism tries to grant three main requirements: performance, reliability and energy efficiency. Unlike previous works, we take into account user requirements as a novel way to manage the traditional axis of these three factors. Thus, the level of these requirements is established according to the application and user needs, which define the reconfiguration target for the system.

In heterogeneous multicore systems, each core may behave differently based on its features and usage, and workload expectations can shift dynamically at runtime. In order to assess dynamically the state of the system, we take information from the hardware behavior and application characteristics. These inputs are used to evaluate the system status with regard to the desired target and act accordingly. For that, we also provide different countermeasure techniques for applying dynamic reconfiguration, such as application to core allocation or voltage and frequency scaling.

The combination of all these procedures in the proposed adaptive system reconfiguration mechanism allows the multicore processor to adapt its components as needed, enabling reliable and cost-efficient system.

II. THE RECONFIGURATION SYSTEM

The reconfiguration engine is the global controller responsible for dynamically adapting the multicore processor to different runtime requirements. The purpose of our reconfiguration mechanism is to assign every application to a proper core in such a way that resource allocation is optimized, the defined requirements are enforced, and we provide a minimum quality of service (QoS). The reconfiguration mechanism must also adapt to different application execution phases by dynamically reallocating resources in order to maximize throughput.

A. Defining User Requirements

In order to take the user requirements into account, we propose to lead the reconfigurations based on what the user expects from the platform. For that, we have identified three axes in which the user may be interested: performance, power and reliability.

Most of the existing reconfiguration mechanisms focus only on maximizing performance such as instruction throughput (i.e., performance in instructions per second -IPS) or energy efficiency without taking into consideration the user needs. For instance, a high instruction throughput does not always grant the best user experience. Consider the case of a video decoding application combined with an antivirus program with very high throughput. It is clear that what matters to the user is ensuring an adequate frame rate in the video decoder and not the overall processor throughput. We also want to note that usually the user is not aware about which are the actual performance requirements of the applications to be run. Therefore, it is important to define an interface that is meaningful to the user and provides a successful user experience in the system.

We want to manage the user requirements in a way that it is easy to understand for both the user and the hardware. Next, we explain the three factors used to express the user requirements; performance, power and reliability.

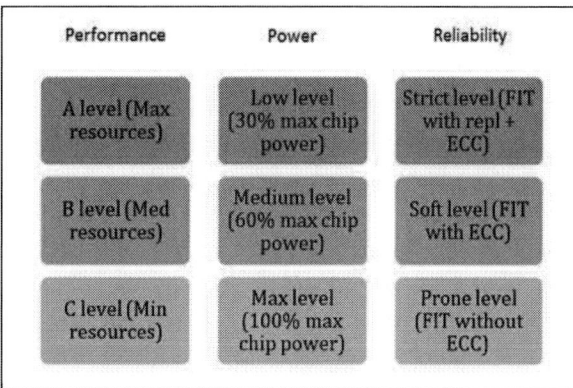

Figure 1.   User Options examples.

a) **Performance.** We propose a new way to allow the user to select different performance levels based on priorities. The main purpose of these priorities is twofold: (i) to lead the application allocation to get the proper performance and (ii) to ensure fairness among the applications utilizing the resources based on the user demands (this can be seen as a kind of QoS). This division in user priorities takes into account the user requirements and is suited for all market segments. Another benefit of this prioritization is that it allows to easily taking advantage of the increasingly heterogeneous CMPs without directly exposing this heterogeneity to the user.

b) **Power.** We offer to the user three different levels of power. Each level defines the maximum amount of power that a core can consume in each moment based on the total chip power budget. Notice that numbers shown in Figure 1 are only reference numbers.

c) **Reliability.** Likewise, three different levels of reliability are offered to the user. Each reliability level allows the user to set the proper fault tolerance rate (FITs) according to the corresponding protection mechanisms. An example on how these levels would be implemented is shown in Figure 1.

Therefore, we propose a definition of user requirements based on levels rather than actual numbers. Overall, the benefits are twofold: (i) it is easy to understand for the user, and (ii) it offers enough flexibility to the reconfiguration engine for guaranteeing the different required levels.

B. Defining the Reconfiguration Engine

The reconfiguration mechanism should dynamically control the multicore processor and reconfigure it to obtain the best configuration in terms of performance, power and reliability based on the defined user requirements.

An overview of the reconfiguration mechanism can be seen in Figure 2, where we combine solutions at a coarse grain and at a fine grain. Our proposal is based on a 2-level engine in which the reconfiguration mechanism is divided into:

Chip-level mechanism. The chip-level mechanism is in charge of assigning applications to cores (only when new applications start or end, including context switches) and distributing resources.

Core-level mechanism. The core-level mechanism is responsible for reconfiguring each individual core for maximum throughput (IPS) with the available resources according to the executed application and core characteristics.

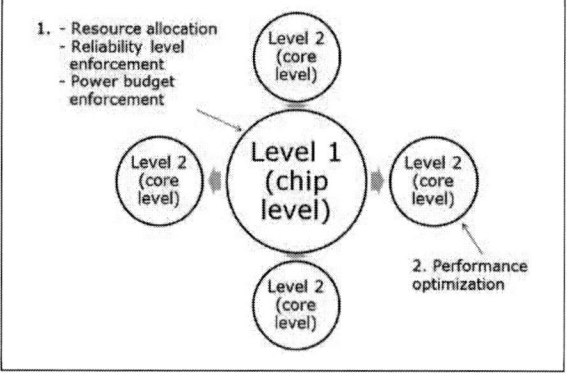

Figure 2.   Reconfiguration Mechanism.

Having a two-level reconfiguration mechanism allows to take into account the global constraints in the resource allocation problem and also apply fine grain adaptability at the core level. Thus, for a given set of applications to be executed, the mechanism selects the configuration that gives maximum performance and meets the user and application requirements. In the following section we describe each of the proposed mechanisms.

1) Chip-level Engine

The chip level reconfiguration mechanism is responsible for three duties, (i) performing the application-core mapping, (ii) setting the chip voltages and (iii) allocating chip resources to the different cores. Figure 3 depicts an overview of the algorithm followed by the reconfiguration engine that shows each of the steps of this chip-level mechanism.

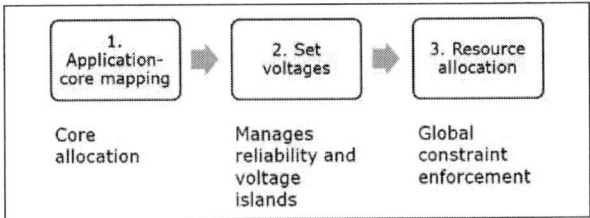

Figure 3.   Chip-level algorithm overview.

a)   Application-to-Core Mapping

The first step is to allocate the applications to the cores. In order to perform the core mapping at the chip level, we use cost functions. These cost functions will depend on the final target. For example, we can have a cost function defined to allocate the more demanding applications to the best cores or define another function to minimize the overall chip power consumption. Next, we discuss two different cost functions that we consider: performance oriented and energy oriented.

Performance-oriented cost function. Cores can perform differently because variations and degradation cause chip multiprocessor to be heterogeneous. This cost function focuses on the maximum frequency that yields the best application performance under heterogeneous conditions.

For this case, we define the performance cost function as the time needed to execute N instructions. Implementing this cost function is not easy because it requires knowing how applications benefit from different core characteristics. Therefore, this approach requires profiling the applications being executed for different configurations. For instance, if we consider the heterogeneity of the cores in terms of maximum operating frequency we would require profiling applications for the different available maximum frequencies.

Energy-oriented cost function. This function focuses on a different type of heterogeneity that can appear in chip multiprocessors, the minimum operating voltage (V_{min}). In this case, the cost function is defined to save energy. One easy option is to operate cores at minimum voltages, which would maximize the power savings. However, this comes at a cost in reliability since lower voltages are more prone to errors.

To balance these two factors, we define the energy cost as the absolute difference between the minimum operating voltage of the core and the minimum voltage associated to the application reliability requirements. As this cost function is designed to minimize power consumption, the mapping does not take into account application performance priorities.

As mapping decisions are made based on *cost* functions, we propose to use a cost matrix (W) to implement the application-to-core allocation, where rows are the applications, and columns are the cores. Each cell W_{ij} in the matrix represents the cost (calculated with previous cost functions) of running application A_i in core C_j. Based on the target, the proper cost function is selected to create the matrix. We assume that for each cost function, 0 is the optimal match and it is unbounded on the upper limit.

Once the cost matrix is defined the problem of application-to-core assignment can be solved as a typical resource allocation problem. A common way of solving it would be through the Hungarian Algorithm [1]. The result of this operation is the assignment of one application per core, optimizing the resource allocation according to the application requirements and core features represented by the cost function.

b)   Voltage Setting

This part of the mechanism sets the voltage of all cores, taking into account the reliability requirements of each application. The reconfiguration system controls the reliability in terms of FITs (Failure in time) based on a model that measures dynamically the Architectural Vulnerability Factor (AVF) [2] of cores. Notice that the granularity of the voltage settings will be done based on physical limitations according to the chip layout. In some implementations, we may be able to set voltage per-core basis, and for others, we will be using voltage islands that may include multiple cores. In case of having independent voltage regulators per core the AVF adaptation could be done at a fine grain level.

The voltage setting procedure has two phases:

Initial voltage setting. First, voltage for each island is determined by the maximum voltage requested by a core to guarantee its reliability requirements. This means that if one application in a voltage domain requires the maximum reliability (and therefore a higher voltage) and the rest a lower reliability, this domain will be set at the highest required voltage for providing that maximum reliability.

Dynamic Adaptation. At runtime, the mechanism takes into account the dynamic reliability margin calculated for each application estimated through performance counters. The overall reliability estimation is given by $FIT_{obtained} = AVF*FIT_{to\_enforce}$, which means that depending on the AVF value at a given point in time, the reliability constraints imposed to the core can be relaxed. The required voltage levels are calculated for each application and then, the most restrictive for that voltage domain is selected. This dynamic adaptation allows saving power during the execution phases with less vulnerability without compromising the requested reliability levels.

c)   Resource Allocation

The last task of this chip-level mechanism consists in a resource allocation algorithm that decides the amount of resources each core can use. This resource assignment depends on application priorities and user needs.

This part distributes the available resources among the cores which are limited by the performance, power and reliability constraints imposed by the user requirements. These resources can be shared hardware structures or shared constraints, like the power budget.

2) Core-level engine

In addition to the chip-level mechanism, the reconfiguration engine has a core-level mechanism as well. This *second level* of reconfiguration works at a fine grain by detecting phase changes of applications and using only the assigned amount of shared resources by the first level.

Once an application is allocated to a core, a core-level reconfiguration is applied. This reconfiguration has the purpose of maximizing the application throughput (IPS) with the available resources in order to exploit program phase characteristics. Figure 4 shows the basic structure of such core-level mechanism. At this level, every given number of cycles (e.g, quantum) a reconfiguration is triggered. The core-level reconfiguration takes as inputs the reliability level and power budget provided by the first level reconfiguration engine and the core performance counters[1].

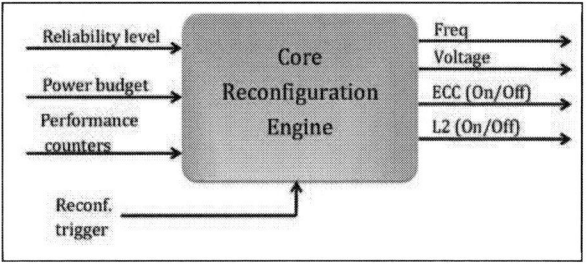

Figure 4. Core-level reconfiguration mechanism.

The decisions that the engine takes will depend on the core-level reconfigurations that are available. In the example of Figure 4, cores can change the frequency, voltage, enable and disable ECC protection in some structures, and completely disable the L2 cache for power savings.

In order to take the right decisions, the engine requires some kind of information describing the impact of the possible reconfigurations in terms of performance, power and reliability.

a) Maximizing performance

The core level mechanism must ensure that the power budget allocated to the core is not exceeded while applies the available configurations to improve the performance. The reconfiguration mechanism decides which structures to activate through a multithreshold decision mechanism. This mechanism is based on the power budget per core and a set of performance counters that decides which structures must be activated/deactivated. For every reconfiguration, an activation

---

[1] We assume that events monitored to lead the reconfigurations will be available as performance counters. This is not restricted to the existing ones.

cost (in power) is defined and some performance counters and thresholds are associated.

The mechanism that decides which structures need to be activated works iteratively. First, it checks the first level thresholds. First level thresholds are the most difficult ones to reach, but the ones yielding the highest benefit in case they are reached. Once all the first level thresholds have been verified, if there is additional power budget, the second level thresholds are checked. Every time that the engine discovers that a threshold has been reached it checks if there is enough available power budget. If that is the case, it activates the corresponding structure.

The engine can use as many reconfigurations and input parameters as desired and has the main advantages of not prioritizing some mechanisms over others. It also allows the addition of extra reconfigurations without requiring a modification of the whole reconfiguration mechanism.

b) Guaranteeing power and reliability at core-level

It is important to mention that the core-level engine is also the one in charge of guaranteeing that the core satisfies the constraints in terms of power and reliability. Next, we discuss these two issues.

Power. The core reconfiguration mechanism is the responsible for enforcing the power budget that is assigned to a given core. This can be done at a fine granularity if the mechanism is able to measure the power consumption for each core separately. Otherwise, the mechanism reconfigures the system based on a table of maximum power consumptions for all configurations.

Reliability. Reliability at core level is obtained through Q-AVF [3]. This Q-AVF is calculated at runtime, based on the use of different hardware counters and linear regression models. Based on the Q-AVF and the raw error rate models, we obtain the actual FIT rate of the core. This value is used to adjust the voltage and enable/disable the reliability techniques so the required reliability level is guaranteed. Notice that if the FIT rate is below the required one, the reconfiguration engine will free resources that could be used by other cores. For example, the mechanism can detect and exploit this excess of reliability margin and extract additional performance.

III. SELECTION OF RECONFIGURATIONS

We evaluated which are the most interesting reconfigurations that should fit the proposed reconfiguration system. We have considered reconfigurations to improve performance, reduce power consumption or increase reliability. The ones that have more potential in this proposal are:

- *Dynamic Voltage—Frequency Scaling* (DVFS) [4]: This technique is currently implemented in most of the commercial processors and allows adapting the voltage and frequency to the application requirements, saving power when the processor is idle.

- *Power-gating*: A straightforward solution for saving power in current processors is disabling idle parts. In our case we apply this technique to cores, caches and error correction mechanisms.

- *Adaptive caches*: This technique can be used to save power or increase reliability of caches regarding overall system performance. Adaptive caches are caches that can change their size depending on application requirements. Many of them use the column caching technique [5] like the Utility-based Cache Partitioning [6] and Elastic Cooperative Caching [7].

- *Dynamic protection*: With this technique the level of protection can be changed dynamically. This is based on the fact that not all data and applications require the same amount of protection and reliability requirements change for different data and execution phases. For instance, some existing techniques modify the level of protection at the register level [8] or at the cache level [9].

- *Redundant Execution*: Finally, redundant execution such as RMT [10] is another common dynamic technique for increasing processor reliability through redundant application executions. This technique allows the processor to increase its reliability by comparing the results obtained in the redundant executions in order to detect possible errors.

IV. RELATED WORK

In the recent years many techniques have appeared that deal with the increasing heterogeneity and reconfiguration options of current and future chip multiprocessors. Reconfiguration mechanisms have different optimization goals that can be the overall system throughput, energy-efficiency, thermal distribution or reliability.

A. Throughput oriented.

Among the techniques that optimize throughput with a given power budget Isci et al. [11] present a technique for managing a multicore voltage and frequency. They propose different policies like the usage of application priorities or maximization of the overall throughput. Adaptation is achieved through sampling of suitable configurations and assuming that performance has a linear relation with frequency and power a cubic relation with V/F scaling. Choi et al. [12] also do sampling for estimating the optimal solution but through the hill climbing technique. In this case the technique is oriented to SMT processors and uses an algorithm which samples different resource partitions similar to the current one and selects the best performing. This solution is limited by the learning time to find the best solution and that it may be limited by local maxima. LinOpt [13], on the other hand, solves the reconfiguration problem as a linear optimization problem, maximizing the average throughput. This solution also imposes a power budget constraint but assumes that power depends linearly on supply voltage. Other proposals use machine learning techniques in order to reconfigure the system like Bitirgen et al. [14]. This technique proposes to manage resizable caches, DVFS and memory schedulers based on the output of an artificial neural network. This technique is focuses in optimizing the memory throughput through stochastic hill climbing.

B. Energy oriented.

Another metric to consider when reconfiguring systems is energy. Some works have focused in minimizing the amount of energy required to run all tasks in a system. Many solutions have appeared taking this approach [15] [16] [17] [18]. Dubach et al. [15] propose a predictive model which adapts to application phases at runtime. This technique maximizes the overall energy-efficiency and detects program phases through program counters. Chen et al. [17] also propose a power oriented reconfiguration mechanism although their objective function to minimize is the dollar cost of operation. They propose three strategies based on steady state queuing analysis, feedback control theory and a hybrid mechanism. This solution calculates the desired mapping of applications and their frequency settings. Ioannou et al. [18], on the other hand, detect MPI application phases online through the message passing interface and apply DVFS adjustments in order to minimize energy consumption within a performance window.

Several other papers [19] [20] [21] [22] focus more on controlling where energy is spent since this plays an important role in hotspot generation and in the required thermal solutions. Donald et al. [19] propose the usage of a PI controller based on thermal inputs to maximize performance without exceeding temperature constraints. This technique controls core temperatures through DVFS and application migration. Wang et al. [20], on the other hand, also take into account the overall consumed power and make use of model predictive control theory to adjust the system both for an overall power budget and core-level temperature constraints. Finally, Power [21] proposes a technique that also combines adaptation for energy and temperature constraints. This technique is based on a distributed mechanism which performs reconfiguration at different levels with a small amount of information exchange.

C. Dealing with heterogeneity and errors.

Other techniques [23] [24] [25] [26] use reconfiguration to hide processor heterogeneity and failures from the end user. Throughput-Driven Fairness (TDF) scheduling [24] for example, reallocates applications among a set of heterogeneous cores in order to increase the apparent mean frequency of the cores. This technique detects program phases in order to execute the frequency sensitive on high-frequency cores and evaluating the expected performance of each application at the desired average frequency in order to provide enough resources to all applications. Architectural Core Salvaging [23], on the other hand, takes into account the fact that cores have an increasing number of hard faults that appear during its lifetime that can leave the core unusable. In order to solve this problem they propose to deactivate the regions of the cores that do not operate properly and schedule applications to these cores that don't require the usage of these parts of the core.

V. CONCLUSIONS

This works presents a high-level reconfiguration approach for heterogeneous multicore systems that helps mitigating variability and improving reliability costs while guaranteeing performance and power requirements. The 2-level reconfiguration mechanism tries to maximize the throughput of all applications taking into consideration not only the underlying state of the system but also the user requirements and application characteristics. Working at two levels allows the combination of solutions at a coarse grain and at a fine grain to obtain the best configuration in terms of performance, power and reliability.

ACKNOWLEDGMENT

This work has been partially supported by the EC FET TRAMS (Terascale Reliable Adaptive Memory Systems) project, 248789.

REFERENCES

[1] H. Kuhn, "The Hungarian method for the assignment problem," vol. 2, no. 1-2, 1955.

[2] S. S. Mukherjee, C. Weaver, J. Emer, S. K. Reinhardt and T. Austin, "A Systematic Methodology to Compute the Architectural Vulnerability Factors for a High-Performance Microprocessor," in Proceedings of the 36th annual IEEE/ACM International Symposium on Microarchitecture, 2003.

[3] A. Biswas, N. Soundararajan, S. Mukherjee and S. Gurumurthi, "Quantized AVF: A Means of Capturing Vulnerability Variations over Small Windows of Time," in Proceedings of Workshop on Silicon Errors in Logic -System Effects (SELSE), 2009.

[4] P. Macken, M. Degrauwe, M. Van Paemel and H. Oguey, "A voltage reduction technique for digital systems," 1990.

[5] D. T. Chiou, Extending the Reach of Microprocessors: Column and Curious Caching, Ph.D. Dissertation. Massachusetts Institute of Technology., 1999.

[6] M. K. Qureshi and Y. N. Patt, "Utility-Based Cache Partitioning: A Low-Overhead, High-Performance, Runtime Mechanism to Partition Shared Caches".

[7] E. Herrero, J. González and R. Canal, "Elastic cooperative caching: an autonomous dynamically adaptive memory hierarchy for chip multiprocessors," in ISCA '10 Proceedings of the 37th annual international symposium on Computer architecture, 2010.

[8] P. Montesinos, W. Liu and J. Torrellas, "Using Register Lifetime Predictions to Protect Register Files against Soft Errors," 2007.

[9] S. Wang, J. Hu and S. Ziavras, "Self-Adaptive Data Caches for Soft-Error Reliability," vol. 27, no. 8, 2008.

[10] S. Mukherjee, M. Kontz and S. Reinhardt, "Detailed design and evaluation of redundant multi-threading alternatives," 2002.

[11] C. Isci, A. Buyuktosunoglu, C.-Y. Cher, P. Bose and M. Martonosi, "An Analysis of Efficient Multi-Core Global Power Management Policies: Maximizing Performance for a Given Power Budget".

[12] S. Choi and D. Yeung, "Learning-Based SMT Processor Resource Distribution via Hill-Climbing," 2006.

[13] R. Teodorescu and J. Torrellas, "Variation-Aware Application Scheduling and Power Management for Chip Multiprocessors," 2008.

[14] R. Bitirgen, E. Ipek and J. F. Martinez, "Coordinated management of multiple interacting resources in chip multiprocessors: A machine learning approach," 2008.

[15] C. Dubach, T. M. Jones, E. V. Bonilla and M. F. P. O'Boyle, "A Predictive Model for Dynamic Microarchitectural Adaptivity Control," 2010.

[16] Q. Wu, P. Juang, M. Martonosi and D. W. Clark, "Formal online methods for voltage/frequency control in multiple clock domain microprocessors," 2004.

[17] Y. Chen, A. Das, W. Qin, A. Sivasubramaniam, Q. Wang and N. Gautam, "Managing server energy and operational costs in hosting centers," 2005.

[18] N. Ioannou, M. Gries, M. Kauschke and M. Cintra, "Phase-Based Application-Driven Power Management on the Single-chip Cloud Computer," 2011.

[19] J. Donald and M. Martonosi, "Techniques for Multicore Thermal Management: Classification and New Exploration," 2006.

[20] Y. Wang, K. Ma and X. Wang, "Temperature-constrained power control for chip multiprocessors with online model estimation," 2009.

[21] R. Raghavendra, P. Ranganathan, V. Talwar, Z. Wang and X. Zhu, "No "power" struggles: coordinated multi-level power management for the data center," New York, NY, USA, 2008.

[22] D. Brooks and M. Martonosi, "Dynamic Thermal Management for High-Performance Microprocessors," 2001.

[23] M. D. Powell, A. Biswas, S. Gupta and S. S. Mukherjee, "Architectural core salvaging in a multi-core processor for hard-error tolerance".

[24] K. Rangan, M. Powell, G.-Y. Wei and D. Brooks, "Achieving uniform performance and maximizing throughput in the presence of heterogeneity," 2011.

[25] A. Tiwari and J. Torrellas, "Facelift: Hiding and slowing down aging in multicores," 2008.

[26] S. Herbert and D. Marculescu, "Variation-aware dynamic voltage/frequency scaling," 2009.

MIXDES 2012, 19th International Conference *"Mixed Design of Integrated Circuits and Systems"*, May 24-26, 2012, Warsaw, Poland

SRAM Lifetime Improvement by Using Adaptive Proactive Reconfiguration

Peyman Pouyan, Esteve Amat, Antonio Rubio
Department of Electronic Engineering
UPC Barcelona Tech, Barcelona Spain
peyman.pouyan, esteve.amat, antonio.rubio@upc.edu

Abstract—**Modern generations of CMOS technology nodes are facing critical causes of hardware reliability failures, which were not significant in the past. Such vulnerabilities make it essential to investigate new robust design strategies at the Nano-scale circuit system level. In this paper we have introduced an adaptive proactive reconfiguration technique that considers the inherent process variability (variability-aware) and BTI aging, and effectively enlarges the SRAM lifetime.**

Index Terms—**reconfiguration; SRAM; process variability; multi-spare proactive reconfiguration; BTI aging**

I. INTRODUCTION

When the device dimensions have been scaled down to the nanoscale regime, the device variability has taken a crucial relevance into the devices behavior, and consequently on integrated circuits reliability and performance. These variations such as process fluctuations and aging mechanisms could become relevant in Embedded Static Random-Access Memories (SRAM) [1-3], since they are usually designed in technology node sizes and occupy a significant area in integrated circuits [4].

Conventionally, to improve the manufacturing yield in memories the reconfiguration has been stated as one of the most common techniques [5, 6], using redundancy elements to replace the faulty ones. But with the increase of unreliable devices new strategies have been developed, recently, a new reconfiguration technique named proactive reconfiguration has been presented [7], which enlarges the lifetime of the memories. In this reconfiguration scenario all available resources, the operational and spare parts take part in the system operation. Nevertheless, in this paper we present a novel technique that improves the existing techniques, since our adaptive proactive reconfiguration technique takes into account the process variability fluctuation.

Hence, this work has been divided as follows: Section II presents the importance of the variability in the current and future technologies and the main aging mechanism will be introduced. Section III reviews the concept of the conventional reconfiguration and the proactive reconfiguration technique and introduces our methodology to improve the proactive mechanism. Finally, the results of the impact of our proposal

on the lifetime are exposed in Section IV and general conclusions of the work are stated in Section V.

II. VARIATIONS IN NANO-SCALE CIRCUIT DESIGN

One of the major sources of unreliability in nanoscale circuit design is existence of different fluctuation sources [8], which affect a wide set of the transistor model parameters. In this paper we will only consider the variability phenomenon on the threshold voltage (V_T) of the transistors. On the first section we explain the static fluctuations and the dynamic variations are introduced in the second section.

A. Static variations

Static variations fall into two main category of fluctuations [8], the interdie variation which results from different runs in manufacturing exist between wafers, and the intradie variations which is the variation in transistor parameters exist inside the same die. The interdie variations are because of process fluctuations in length, width, oxide thickness and etc., while line edge roughness or random dopant fluctuations cause intradie variations. These variations pose a significant attention and design considerations when the devices scale down to nanoscale sizes of sub-22nm [9] as the levels of variability in the threshold voltage may arise to levels of the standard deviation of the distribution around a 35% of the average value [9], causing an important yield drop.

B. Dynamic variations

Dynamic variability can be categorized into two parts: environmental and aging. Environmental variations include fluctuations of voltage and temperature [10]. Meanwhile several aging mechanisms are well reported, i.e. Bias Temperature Instability (BTI) aging, hot-carrier injection. But in this work, we only consider BTI aging, since it is the main one that leads to a relevant V_T-shift that reduces the lifetime and finally could result in a SRAM failure [11]. There are various models to describe it [12, 13] and we have assumed stress and recovery properties for it. In this model the device under stress experiences some V_T-shift and part of this could be recovered when the device is not stressed. To model such a behavior, firstly we linearize the stress phase to different slopes in life periods, as our work is an analysis of circuit behavior during years. Secondly, we assume that the minimum time for an almost complete BTI degradation recovery is about 10^4 seconds; both linear slopes and minimum recovery time are obtained from experimental measurements [11, 12]. Additionally, we introduce a parameter named Recovery factor (*Rf,* expressed in percentage) that is related to the potential V_T-recovery after a long time stress phase and its corresponding degradation. The recovery factor means that when a device experiences a long enough recovery phase [14-17] some part of the V_T-shift, caused by stress phase, could be reversed.

III. RECONFIGURATION TECHNIQUES

The reconfiguration technique is an important and effective way for increasing lifetime and improving reliability in memories [5, 6, 7]. There could be different scenarios for using the redundancy in the system and here we revisit two of those and later we introduce our methodology.

A. Reactive reconfiguration

Reactive reconfiguration is an effective method to increase the yield in memories [5, 6] employed along years. The system is usually divided in working elements and spare ones, which could be columns or rows. The first ones are always working until they fail and then are replaced by the spare ones, improving then the system lifetime and performance. There exist different algorithms to implement such a reconfiguration scenario [18]. Though the reactive reconfiguration is very effective and important, but with scaling down the technology and with emerge of new reliability phenomenon in nanoscale circuits such as BTI aging and strong process variations, new reconfiguration scenarios need to be developed.

B. Proactive reconfiguration

Instead of reserving the spare elements in the system up-to the time that a failure happens, in proactive reconfiguration the spare elements take part in normal operation of the system. This scenario of reconfiguration was firstly introduced in [7] and extended in [19] to improve the memory lifetime. In this technique the existence of redundant resources in the operation of system provides the possibility for the elements of the system to have time periods for being released from being under stress and get benefit from recovery properties. This technique can be implemented in the system [7] allowing the elements having two modes of active and deactivated on a rotating basis.

In addition to lifetime enhancement, the proactive reconfiguration also cause a balanced distribution of wear-out among the elements in the system and this characteristic makes it interesting to use in systems which are affected by process variation. In this work, we introduce a novel extension to proactive reconfiguration by adapting the proactive reconfiguration in such a scenario that elements of the system recover with various time periods according to their initial value affected by process variation and their BTI aging behavior.

C. Adaptive Proactive Reconfiguration

The adaptive proactive reconfiguration is a method, which improves the proactive reconfiguration technique, since it considers the process variability along with the BTI aging of the system. The system behavior continuously adapts the recovery time of the elements with respect to their process variability at the beginning of unit life and their own variations due to BTI aging during their lifetime. In this work such a system and units are a memory and its columns [20, 21].

To implement such a methodology the first step is a monitoring test in order to be aware of V_T status of each column in the fresh memory (not stressed). Therefore, a test is applied to all SRAM cells in all columns and each column is recognized by the V_T value of its cell, determining which presents the highest value (worst cell) [20, 22]. Note that, this operation does not necessarily causes an idle time in the memory system because the measurement can be performed while the corresponding column is in the recovery phase. The next step is to apply appropriate recovery time to each column in the memory. This is done in a round robin method in which all columns (spare and operational columns) participate in the memory operation and have recovery ratios that considers their V_T value, affected by process variation and BTI degradation. In a set of columns, which only have one spare column among them, a loop allows the columns to enter in recovery mode, one at each time, but when having more than one spare, the loop is between different classes of columns and depending to the number of available spares a number of columns can go on recovery mode. The round robin loop is called a reconfiguration cycle and each step of it is named a reconfiguration step. The reconfiguration cycle, could present a magnitude of several days (as the minimum recovery time is almost 3 hours). In a single spare mode of column reconfiguration, at each step the spare column will replace the working column that goes to recovery and the column's data is copied in the spare column. This copied data will be written back to the column before the next column reconfiguration step. Figure 1 shows an example for an adaptive proactive reconfiguration scenario with 4 simultaneous operational columns (C_1, C_2, C_3, C_4) and one spare (*SP*).

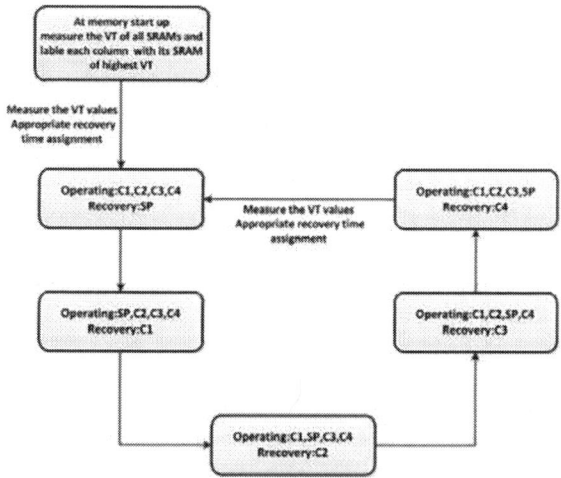

Figure 1. Adaptive proactive reconfiguration methodology having one spare

Figure 2 shows an example of a reconfiguration scheme for a sub-memory block of 1kB, with 128 columns of 64 rows.

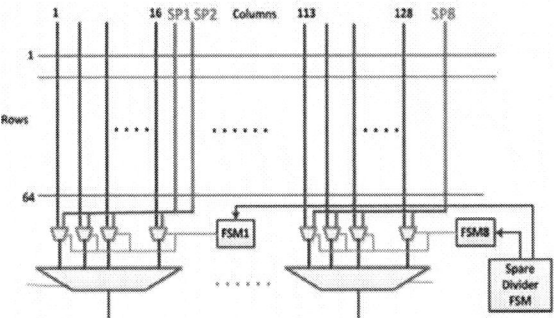

Figure 2. An architectural implementation example of adaptive proactive reconfiguration in 1kB memory block, each set of columns can have different number of spares according to their process variability

In this case if there are some remaining spare columns after the reactive reconfiguration, each set of 16 columns will have one or multiple number of spare columns allocated in respect to process variability of each set, in such a way that the set with highest variability will have higher number of proactive spare columns.

IV. RESULTS AND COMPARISON

In order to show the benefits of our methodology we compare our results of adaptive proactive reconfiguration (case C) with the basic proactive reconfiguration (case B) and non-proactive reconfiguration (case A). Firstly, lifetime enhancements of our methodology is presented in an graphical example, in which we show the aging behavior of memory columns using our method, then the lifetime improvements of single spare adaptive proactive reconfiguration are demonstrated by using Monte-Carlo analysis. This analysis is used in order to demonstrate the enhancements of our implementation in various process variability scenarios and compares the results in different reconfiguration scenarios. Finally the last section describes the modified adaptive proactive reconfiguration when having more than one proactive spare in the system and presents the results for multi-spare adaptive proactive reconfiguration. All the simulations and analysis are done using Matlab. To determine and measure the system lifetime, we randomly generate fresh V_T of SRAM cells for a set of memory columns under normal distribution with a given mean and standard deviation (*300 mV* and *30 mV*, respectively). The maximum V_T value acceptable (H) to predict the system lifetime after an aging process to a non-failed SRAM is imposed at 400mV. These numbers are taken from literature and used as an example to show the benefits of our methodology and can be adapted to any technology node.

In one example, the behavior of a sample system is analyzed using our methodology. The sample system contains 4 operational columns and one spare, and the wear-out recovery (Rf) is considered at 30%. We assume one set of arbitrary fresh threshold voltage values: $V_{T1}=300mV$,

$V_{T2}=280mV$, $V_{T3}=290mV$, $V_{T4}=310mV$ and $V_{T5}=320mV$. Figure 3 presents the system behavior expressed in lifetime (months) for the two cases of non-proactive reconfiguration (case A) and our adaptive proactive reconfiguration (case C). Our method presents, in this example, an improvement of the system lifetime about 2.7X, showing the benefits to use the adaptive recovery phase ratios during the columns lifetime. The reason for such a significant enhancement in our method is that the lifetime in the non-proactive methodology depends on the maximum threshold voltage (worst) column value of the SRAM (here V_{T5}), and a parallel behavior evolution is observed between them and the worst column is which arrives first at the system failure criterion. Meanwhile, our method presents a non-parallel evolution of the V_T-shift as the aging ratio of the different columns with different values adapts in such a way that all V_T values reach the point of failure together and at the same time, getting a longer lifetime.

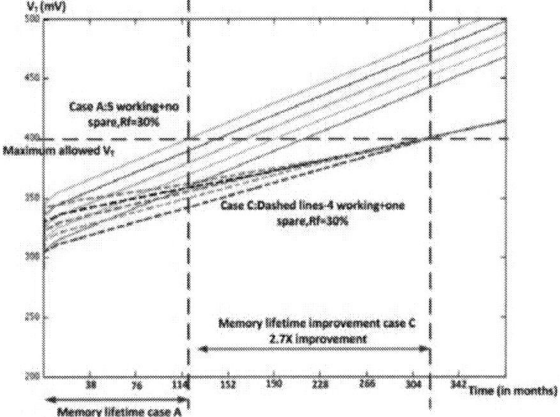

Figure 3. Comparison of lifetime plots of a memory system based on 5 columns between a non-proactive (case A, line) and an adaptive proactive (case C, dashed lines) methods. A relevant lifetime improvement for case C has been obtained.

A. Lifetime improvments in single spare proactive reconfiguration

Figures 4 and 5 present the values of Monte Carlo simulations that are performed to evaluate the average lifetime of a memory columns system for the three configuration cases (A, B and C). In this figures the lifetime improvements are shown by being normalized to the lifetime of the system with no proactive reconfiguration. In this sense, Figure 4 shows the lifetime enhancements of a memory system implemented by 5 columns (4 working plus 1 spare) under all configuration cases and for different recovery factors. It shows that our methodology (case C) is the best scenario in extending the columns lifetime. It can enhance the lifetime higher in comparison with basic proactive technique (case B, with equal recovery times) in all different recovery factors. The lifetime enhancement becomes higher with increase of the recovery factor in the system, for instance our technique shows 5X improvement over the no proactive case and about 40% improvement in compare with basic proactive reconfiguration (case B).

117

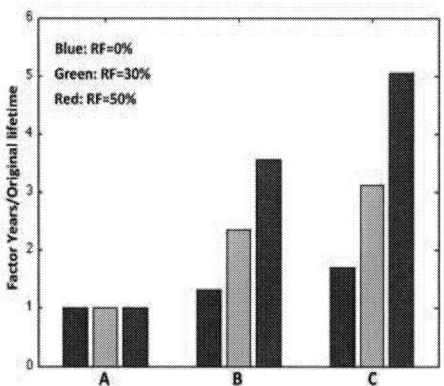

Figure 4.Lifetime behavior in a memory system with 5 operational columns and having one spare, with different recovery factors at the 3 cases (A, B, C)

Then, Figure 5 presents the results for a system based on 9 columns, 8 working and 1 spare. The analysis shows that both proactive strategies present a lifetime enhancement as the recovery factor increases, but our adaptive proposal (case C), presents the largest improvement on system lifetime.

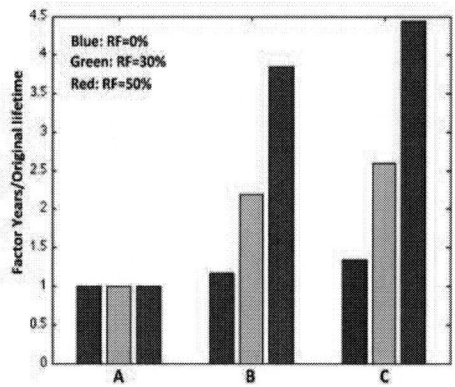

Figure 5. Lifetime behavior in a memory system with 9 operational columns and single spare, with different recovery factors at the 3 cases (A, B, C)

Note that, if both figures are compared, we observe that the system lifetime improvement decreases when the number of working units increases without a corresponding increase of proactive spare units. For instance in this system and when the recovery factor is 50% lifetime improvement of 4.5X times is achieved in comparison with non-proactive (case A) and an improvement of 20 % in compare with basic proactive (case B). This behavior requires a study to evaluate the lifetime improvement in highly process variability environments when multiple spare units are used proactively.

B. Multi-spare parts proactive reconfiguration

There could be different principles to get advantage of multi spare proactive columns. One of the methods that shows significant lifetime enhancements is based on differentiating

the columns at the beginning of memory operation in accordance with their V_T status, which are the voltage threshold values affected by process variation. In this technique the columns that have V_T values close to each other are recognized by the first test and organized in different classes of columns. Therefore during the memory lifetime the columns in each one of classes go to recovery mode together, which means all the columns in one class will experience the same recovery period. The recovery time of each class is adapted with process variability and degradation status of the column in the class with the highest V_T value. Figure 6 presents the applied methodology for a case with 4 proactive spare columns and 16 operational columns as example.

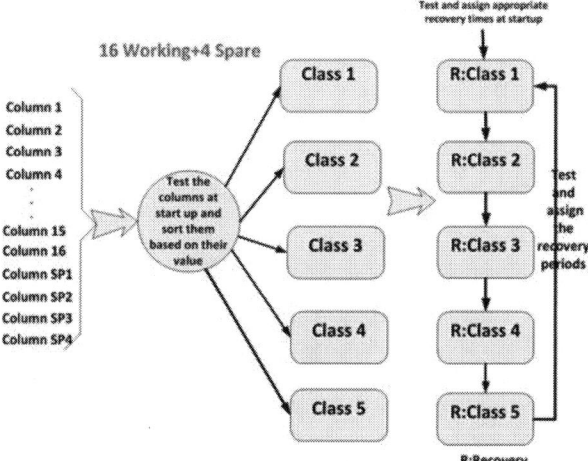

Figure 6. Multi-spare process aware proactive reconfiguration

This scenario can be implemented in the memory system by assigning a dynamic number of spares to each set of columns at the beginning of memory lifetime for instance in the architectural example depicted in Figure 2. Table I presents the results of the Monte Carlo analysis showing the average lifetime without using our technique and the lifetime improvements of memory columns in multi spare adaptive technique when having a set of 16 operational columns.

TABLE I. THE LIFETIME AVERAGES, IN YEARS ARE PRESENTED FOR A MEMORY SYSTEM BASED ON 16 COLUMNS AND WITH DIFFERENT NUMBER OF PROACTIVE SPARE PARTS IN ADAPTIVE PROACTIVE TECHNIQUE.

Working Units	Spare Units	Rf=0%	Rf=30%	Rf=50%
16	0	3.5	3.5	3.5
	1	4.7	9.6	16.6
	2	5.5	11.5	19.3
	4	7	14.6	25

It is observed that as the number of operational columns increases the lifetime could be enhanced by using a higher number of proactive spare columns. For instance when the recovery factor is 50% using one spare column can increase the lifetime about 4.5X times while using 2 spare columns with the same recovery factory can enhance the lifetime about 5.5X and finally having 4 spare columns can improve the lifetime of the column set around 7.2X times. Therefore it can be concluded that an efficient lifetime improvement is achieved by using a higher number of available proactive columns in our adaptive technique.

V. CONCLUSIONS

In this work an adaptive proactive reconfiguration methodology has been presented, using single and multiple numbers of proactive spares. This method enhances the SRAM cells lifetime when they are subjected to the presence of process variability and BTI aging. This technique is based on distributing the active operation of all columns of SRAM cells in respect to their status degradation due to process variation and BTI aging. The lifetime improvement is obtained when the recovery phase ratios are adapted and optimized to the cells variability parameters that are caused by process fluctuations. The results for single spare adaptive proactive technique show a memory lifetime enlargement with a maximum around 5X over no proactive reconfiguration case and a maximum of 40% respect to previous proactive techniques, not taking into account process variability. Finally our adaptive technique can be extended to have multi number of proactive spare columns and it shows around 4.5X to 7X of lifetime improvement over the no proactive method.

ACKNOWLEDGMENT

This work was supported by the European TRAMS project (FP7 248789) and the Spanish MICINN (JCI-2010-07083 and TEC2008-01856).

REFERENCES

[1] A. Bansal; et al. "Impact of NBTI and PBTI in SRAM bit-cells: Relative sensitivities and guidelines for application-specific target stability/performance ," *IEEE IRPS*, pp. 745-749;2009

[2] S. Mukhopadyay, Chen Qikai, K.Roy; "Memories in Scaled technologies: A Review of Process Induced Failures, Test methodologies, and Fault Tolerance ," *IEEE DDECS*, pp. 1-6;2007

[3] K. Kang, H.Kufluoglu, K.Roy, M.Ashraful Alam; "Impact of negative-bias temperature instability in nanoscale SRAM array: Modeling and analysis," *IEEE TCAD*, vol. 26, pp. 1770-1781, 2007.

[4] Y. Nagagome, M.Horiguchi, T.Kawahara, K.Itoh ; "Review and future prospects of low-voltage RAM circuits," IBM J. Res. & Dev., vol. 47, pp. 525-552,2003.

[5] K.N. Ganapathy, A.D.Singh, D.K.Pradhan ; "Yield optimization in large RAM's with hierarchical redundancy," *IEEE Journal of Solid-State Circuit*, vol. 26, pp. 1259-1264, 1991.

[6] A. Chen; "Redundancy in LSI memory array," *IEEE Journal of Solid-State Circuit*," vol. 4, pp. 291-293, 1969.

[7] J. Shin,V.Zyuban, P.Bose, T.M.Pinkston; "A Proactive Wearout Recovery Approach for Exploiting Microarchitectural Redundancy to Extend Cache SRAM Lifetime," *Proc of 35th ISCA*, pp. 353-362, 2008.

[8] Sapatnekar, S.S.; "Overcoming Variations in Nanometer-Scale Technologies," *IEEE Journol of Emerging Topics in Circuits*, vol.1 (1), pp. 5–18, 2011.

[9] W.Xingsheng; et al. "Statistical threshold-voltage variability in scaled decanometer bulk-HKMG MOSFETs: a full-scale 3D simulation scaling study", IEEE TED, pp. 1-9, 2011.

[10] B.Lasbouygues, R.Wilson, N.Azemard, P.Maurine; "Temperature and Voltage-Aware Timing Analysis ability," *IEEE Transaction on CAD*, vol .26(4), pp. 801–815, 2007.

[11] A. Bansal; et al. "Impacts of NBTI and PBTI on SRAM static/dynamic noise margins and cell failure probability," *Microelectronics Reliability*, vol.49(6), pp. 642–649, 2009.

[12] S. Zafar; et al. "A comparative study of NBTI and PBTI (Charge trapping) in SiO_2/HfO_2 stacks with FUSI, TiN, re gates," *VLSI Tech., Digest of Technical Papers*, pp. 23–25, 2006.

[13] M. Cho; et al. "Positive and negative bias temperature instability on sub-nanometer eot high-K MOSFETs," *IEEE IPRS*, pp. 1095-1098, 2010.

[14] P. Hehenberger, H.Reisinger, T.Grassser; "Recovery of negative and positive bias temperature stress in pMOSFETs," *IEEE IRW*, pp. 8-11, 2010.

[15] T. Grasser; et al. "Simultaneous Extraction of Recoverable and Permanent Components Contributing to Bias-Temperature Instability," IEEE IEDM, 2007.

[16] K. Zhao, J.H.Stathis, A.Kerber, E.Cartier; "PBTI relaxation dynamics after AC vs. DC stress in high-k/metal gate stacks," *IEEE IRPS*, pp. 50-54, 2010.

[17] T. Grasser, W.Gos, V.Sverdlov, B.Kaczer; "The universality of NBTI relaxation and its implications for modeling and characterization", IEEE IRPS, pp. 268-280, 2007.

[18] Sy-Yen Kuo, K.W.Fuchs; "Efficient Spare Allocation for Reconfigurable Arrays," *IEEE Design and Test of Computers*, pp. 24-31, 1987.

[19] L. Lin, Z.Youtao, Y.Jun; "Proactive Recovery for BTI in High-*K* SRAM Cells," *IEEE DATE*, pp. 1-6, 2011.

[20] P.Pouyan, E.Amat, A.Rubio; "Process-Variability Aware Proactive Reconfiguration Technique for Mitigating Aging Effects in Nano-scale SRAM Lifetime" *Proc VTS*, 2012.

[21] I. Kim; et al. "Built in Self Repair for Embedded High Density SRAM," *Proc Test*, pp. 1112-1119, 1998.

[22] F. Ahmed, L.Milor; "Reliable cache design with on-chip monitoring of NBTI degradation in SRAM cells using BIST," *Proc VTS*, pp. 63-68, 2010.

Strain Relevance on the Improvement of the 3T1D Cell Performance

Esteve Amat, Carmen Garcia Almudéver, Nivard Aymerich, Ramon Canal and Antonio Rubio

Electronic Department
Universitat Politècnica de Catalunya (UPC)
Barcelona, Spain
esteve.amat@upc.edu

Abstract—**3T1D-DRAM cell has been stated as a valid alternative to be implemented as a L1 memory cache and substitute 6T-SRAM, highly affected by variability. While scaling down capacitor-less DRAM cells is a challenging trend, in this paper we show how it can be compensated the scaling drawbacks through the channel strain of the cell devices and the proposal of new cell configurations to further enhance the circuit behavior.**

Index Terms—**strained channels, DRAM and temperature**

I. INTRODUCTION

The 3T1D memory cell [1] has become a valid alternative to the classical 6T-SRAM one, as the device variability takes larger relevance due to the device dimensions shrink to the nanometer range. The 3T1D-DRAM comparable speed with 6T-SRAM, non-destructive read, larger device variability tolerance and higher compact cell structure are presented as its main benefits [2] in front of the well established memory cell structures. Moreover, to extend the 3T1D performance beyond 65nm node the substitution of SiO_2 by high-k dielectrics as a gate oxide material is required [1], to reduce the intolerable gate dielectric leakage, and improving then the retention time of the dynamic memory cell.

In this sense, the high-k dielectric introduction has involved a relevant reduction of the device mobility, due to the charge trapping [3] appeared on the device. This added to the mobility degradation caused by the large vertical electric fields on the nanoscale devices has represented a serious inconvenient for the devices and circuits performance. To overcome it, the strained channel technology, where III-V materials are introduced on the device producing a band gap narrowing and causing a mean free path increase, have improved the device behavior [4] and it has been adopted as a reliable solution. Note that, the strain influence on a memory cell has been studied before for the 6T-SRAM cell giving a significant cell behavior improvement [5], but has not been yet studied for 3T1D-DRAM cell.

For this, in this work, a study about the 3T1D memory cell performance is done based on the influence of an increase of the strain (mobility) into the 3T1D cell devices. We have focused on several aspects: device configuration, strain relevance, and also the environment temperature have been also analyzed. Finally, some strategies to improve the cell results have been proposed.

II. EXPERIMENTAL

The conventional 3T1D-DRAM cell, presented in Fig. 1, is simulated using the 22nm HP (High Performance) Predictive Technology Models (PTM) [6], as a reference. The optimal dimension values of the cell devices have been extracted from [7] and a supply voltage (V_{DD}) of 1V has been applied. Some reliability scenarios have been analyzed: influence of the gate dielectric material (SiO_2 vs. high-k), channel configuration (unstrained vs. strained) and different V_{DD} values. To do this, some PTM versions (1.0, 2.0 and 2.1) have been used. To complete the study sub-22nm technology nodes have been employed, i.e. 16nm HP PTM and 13nm TRAMS project models [8].

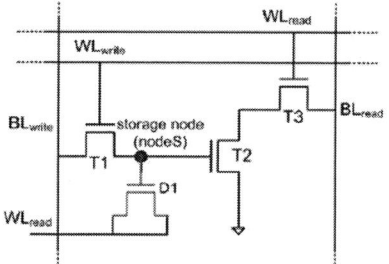

Figure 1.   Schematic structure for a 3T1D-DRAM memory cell.
WL: wordline   BL: bitline

In order to study the 3T1D-DRAM performance all the studies have been focused on the following cell parameters:

a) <u>Write Access Time</u> (WAT) defined between $V(WL_w) = 0.5*V_{DD}$ and $V_s=0.9*(V_{DD}-V_T)$.

b) <u>Read Access Time</u> (RAT) defined between $V(WL_w) = 0.5*V_{DD}$ and $V(BL_r)=0.9*V_{DD}$.

c) <u>Dynamic Power consumption</u> (PW) obtained by the average value along one cycle.

d) <u>Retention Time</u> (RT) that it is the time required for the storage node voltage (V_s) in the cell to decay to V_{Smin} [9].

In particular, to observe the strain influence on cell, the silicon straining has been simulated by multiplying the nominal mobility parameter (U0) in SPICE model by a new parameter k_n, ranged from 1 to 5, that will reproduce the strain (mobility) increase. Finally, the influence of the environment temperature on the 3T1D memory cell behavior has been analyzed as well. For this, the temperature is raised up from room temperature to 125°C.

III. RESULTS

A. Selection of the study conditions

For a better study determine the adequate cell conditions and device model is necessary. For this, first, the influence of the gate dielectric material on the 3T1D-DRAM cells has been studied using the available 22nm HP PTM versions (1.0, 2.0 and 2.1). Then, Fig. 2 compares unstrained cells (empty symbols) based on SiO_2 or high-k gate dielectrics (squares and circles, respectively). The results show that the SiO_2 samples present worst performance for all the cell parameters in front of high-k ones, mainly, related to the lower leakage of the latter. In this context, when high-k samples (circles) based on unstrained or strained channel (empty and full symbols, respectively) have been also compared, and better behavior for the last ones has been obtained (lower WAT and higher RT), due to its better reliability. In this sense, the strained channel model based also on high-k dielectrics will be selected as our reference model for the rest of this work.

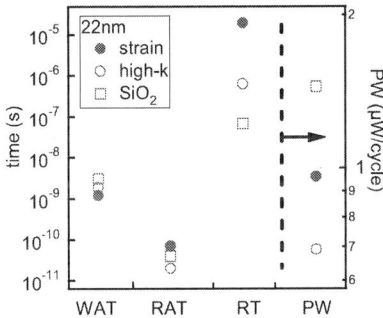

Figure 2. Study of the 3T1D cell performance, when differents PTM device models are employed (1.0, 2.0 and 2.1). The high-k samples based on strained channels present the best behavior.

The supply voltage is another relevant factor on the cell performance. For this, Fig. 3 analyzes its influence for all the technology nodes (22, 16 and 13nm). The lower V_{DD}, the worst performance for all the analyzed strained values, since higher access times (WAT and RAT) smaller retention times are observed, whereas a reduction of the PW is obtained. For this, a supply voltage of 1V has been selected to perform all our studies. Moreover, we have observed that when the technology node is reduced, the 3T1D-DRAM cells present improvements on access times (lower WAT and RAT) and PW (lower consumption), as it is expected, but worst (lower) retention times are obtained. This last result (lower RT) has to take it into account when sub-22nm nodes are simulated, since the retention time is the key parameter to introduce the 3T1D-DRAM cell as a L1 memory cache component [2].

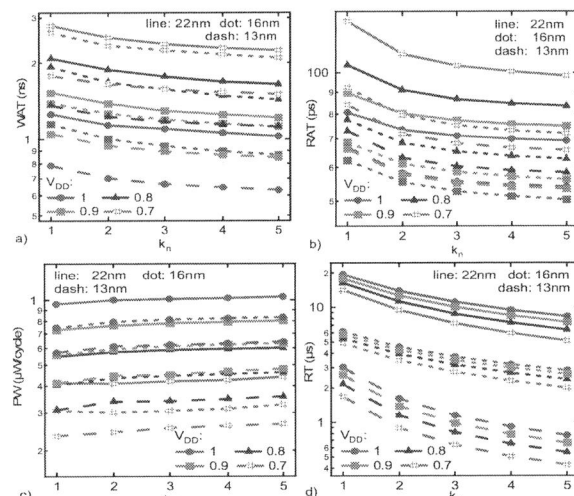

Figure 3. Study of the V_{DD} influence for the different technology nodes (22, 16 and 13nm) and strained channel values on 3T1D memory cell. The smaller nodes present faster cells but also lower retention times.

B. Strain relevance on cell performance

The strained channels have been presented as a valid option to improve the devices mobility, and as a consequence the memory cell performance. Then, analyze the influence of the mobility increase in the cell performance is required. For this, Fig. 4 presents the study about how the strain improves the performance of 3T1D-DRAM cells based on 22nm HP PTM devices. Two scenarios have been developed: 1) when the strain of all the cell devices has been modified at the same time (all HP), and 2) when only one device strain has been altered keeping the original value for the rest. Note that, larger strain value also will lead to a worst aging performance [10], so a carefully decision about the correct strain value must be taken.

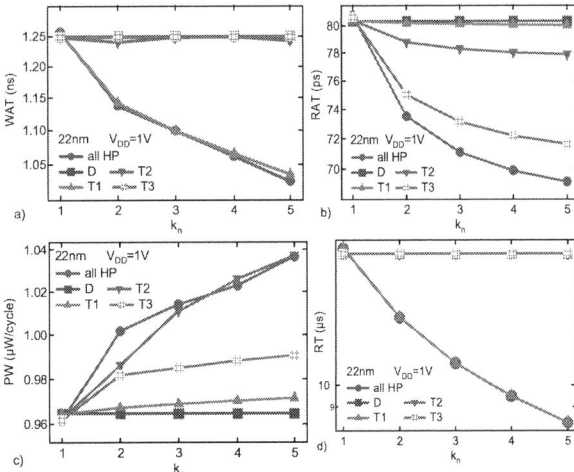

Figure 4. Influence of the strain on the 22nm cell performance when the strain magnitude is modified on one device at a time. T1 device shows the highest impact on 3T1D-DRAM performance.

121

Thereby, when the global cell strain (mobility) is increased equally for all the cell devices, then, faster cells have been obtained. Since lower WAT and RAT values are observed, in contrast to the higher PW and lower RT values. So then, a tradeoff it seems to be required about the cell requirements, faster or less power consumption. Then, to study more in depth, we have analyzed the influence of each cell device in the global cell performance, when the strained channel value (mobility) is modified only at one device at a time. And all the results show that, as the strain increases ($k_n > 1$), T3 has a significant impact on RAT performance, T2 on PW, and finally T1 on WAT and RT. Meanwhile, the strain relevance is much lower for the gate diode (D1). The significant influence of T1 (write access transistor) in the retention time will be taking it into consideration for the 3T1D development, since this is the key parameter of the cell.

The strain increase (higher mobility) has been also applied to all the technology nodes used in this work, from 22 to 13nm. Fig. 5 compares its performance, and similar tendency for every cell parameter has been observed. When the strain of all devices is modified an improvement on WAT and RAT (lower values) but a worsening on PW (higher) and RT (lower) have been presented. The represented small error bars point out the difference between the obtained cell parameters for all the technology nodes, showing similar impact of the strain for every technology node.

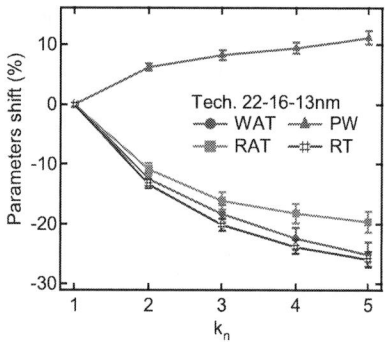

Figure 5. Study of the strain influence for all the technology nodes, 22, 16 and 13nm. Improvement on WAT and RAT (high values), unlike a worst PW (higher) and RT (lower) perfomnce has been obtained. The low relevance of the technology node has been observed, pointed out by the small error bars.

C. Temperature influence on 3T1D cell perfromance

To complete the study about the strain influence on 22nm HP devices the environment temperature relevance has been analyzed, since it is an important parameter that could alters the cell performance. Fig. 6 shows that as the temperature increases, from 25 to 125ºC, worst results as a whole have been obtained, since both access times increase (slower cell) and retention time shrinks. This could be caused by the V_{Smin} dependence on working temperature that directly affects the leakage current [9]. On the other hand, the increase of the device strain (higher mobility) could be presented as a mitigation factor of the worsening performance on some cell parameters, since better access times (lower WAT and RAT) has been obtained, whereas worst retention time (lower

values) have been observed for higher values of channel strain.

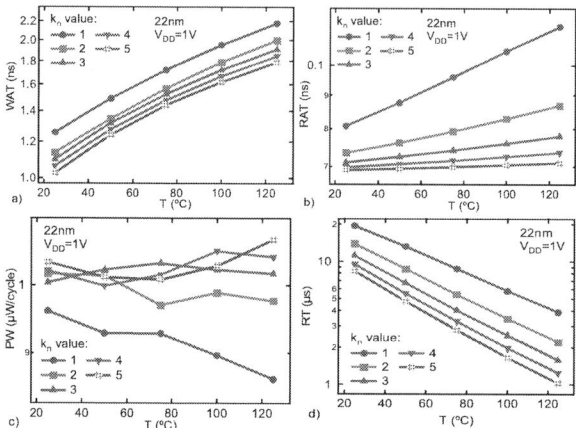

Figure 6. 3T1D behavior when strain (k_n) and temperature are raised up. Worsening performance is observed

D. Cell strategies to improve the cell bahavior

Finally, some strategies to improve 3T1D-DRAM cell behavior could be taken into account. The first option could be mix in a same 3T1D cell the PTM HP and Low Power (LP) [11] models (HPLP), like a dual V_T strategy [12]. Since the HP models provides fast devices, ideal for the access transistors (T1-T3), where high speed is required, whereas the LP ones involves more robust devices against aging and lower power consumption, ideal for the devices connected to the storage node (T2-D1, Fig. 1), providing a more robust S node to mitigate the cell aging. Note that, the use of HP and LP models in a same memory cell could involve an enlargement of the cell area, and consequently a higher fabrication cost but an adequate study is required. The second option has been developed solely modifying the strain of T1 and T3 (HPT13) devices, since these are the ones that show the more relevant

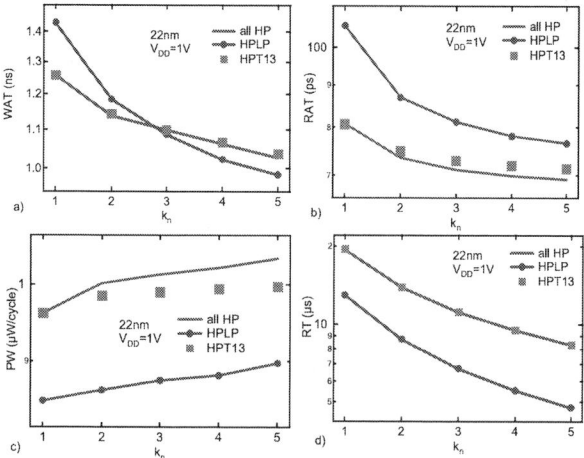

Figure 7. Impact of the strain increse on the new proposed (HPLP, HPT13) 3T1D cell structures. To select someone a trade-off seams to be required, i.e. HPT13 shows some improvements (better retention time and access times), but also other drawbacks (higher PW).

influence on circuit performance when the channel strain has been modified. In this sense, Fig. 7 shows the 3T1D cell behaviors based on these two proposed combinations, and they have been also compared with the reference scenario where all the cell devices have the same HP configuration and the strain value is modified at a time (all HP). Thereby, both new proposals present positive aspects, but finally a tradeoff will be required between cell speed and retention time requirements. For example, the HPT13 option shows an enough high RT and a promising WAT-RAT, which added to the easy implementation, makes this option a feasible solution, but higher dynamic power consumption has been observed. Meanwhile, the HPLP shows faster access time and worst retention time, but could be interesting to get higher aging tolerance cells and less power consumption close to a 12%.

IV. CONCLUSIONS

The 3T1D-DRAM cell performance has been analyzed in front of different device configuration models. Promising results about the introduction of high-k materials as a gate dielectrics and strained channels technology have been observed. The increase of the channel strain amount of the cell devices on 3T1D memories provides faster circuits and increases the cell performance. Moreover, the environment temperature has been presented as a cell worsening factor when is raised up, so, a control and reduction of the temperature of the cell has to take into account to enlarge the cell behavior. Finally, feasible and promising new cell configurations to improve the 3T1D-DRAM performance beyond 22nm have been proposed. After a careful study of the influence of each transistor in the memory cell behavior, two different circuit designs to enhance the overall cell speed (HPT13) or aging tolerance of memory circuits (HPLP) have been presented.

ACKNOWLEDGMENT

This work was supported by the European TRAMS project (FP7 248789) and the Spanish MICINN (JCI-2010-07083 and TEC2008-01856).

REFERENCES

[1] W. K. Luk, J. Cai, R. H. Dennard, M. J. Immediato, and S. Kosonocky, "A 3-Transistor DRAM Cell with Gated Diode for Enhanced Speed and Retention Time," *Symposium on VLSI Circuits*, pp. 184-185, 2006.

[2] X. Liang, R. Canal, G.-Y. Wei, and D. Brooks, "Replacing 6T SRAMs with 3T1D DRAMs in the L1 Data Cache to Combat Process Variability," *IEEE Micro*, vol. 28, no. 1, pp. 60-68, 2008.

[3] D. L. Kwong, "CMOS integration issues with high-k gate stack," *International Symposium on the Physical and Failure Analysis of Integrated Circuits*, pp. 17-20, 2004.

[4] E. Amat et al., "Processing dependences of channel hot-carrier degradation on strained-Si p-channel metal-oxide semiconductor field-effect transistors," *Journal of Vacuum Science & Technology B: Microelectronics and Nanometer Structures*, vol. 29, no. 1, pp. 01AB07-01AB07-4, 2011.

[5] R. Kuchipudi and H. Mahmoodi, "Strain Silicon Optimization for Memory and Logic in Nano-Scale CMOS," *International Symposium on Quality Electronic Design*, pp. 27-32, 2007.

[6] W. Zhao and Y. Cao, "New generation of predictive technology model for sub-45nm design exploration," *International Symposium on Quality Electronic Design*, pp. 590-595, 2006.

[7] K. Lovin, B. C. Lee, X. Liang, D. Brooks, and G.-Y. Wei, "Empirical performance models for 3T1D memories," *IEEE International Conference on Computer Design*, pp. 398-403, 2009.

[8] X. Wang, A. R. Brown, N. Idris, S. Markov, G. Roy, and A. Asenov, "Statistical Threshold-Voltage Variability in Scaled Decananometer Bulk HKMG MOSFETs: A Full-Scale 3-D Simulation Scaling Study," *IEEE Transactions on Electron Devices*, vol. 58, no. 8, pp. 2293-2301, 2011.

[9] N. Aymerich, S. Ganapathy, A. Rubio, R. Canal, and A. González, "Impact of positive bias temperature instability (PBTI) on 3T1D-DRAM cells," *Integration, the VLSI Journal*, no. 0, 2011.

[10] E. Amat et al., "Channel hot-carrier degradation on strained MOSFETs with embedded SiGe or SiC Source/Drain," *IEEE International Conference on Solid-State and Integrated Circuit Technology*, pp. 1648-1650, 2010.

[11] "Predictive Technology Models (PTM) [On line]," *Http://ptm.asu.edu*.

[12] Y. Nakagome, M. Horiguchi, T. Kawahara, and K. Itoh, "Review and future prospects of low-voltage RAM circuits," *IBM Journal of Research and Development*, vol. 47, no. 5-6, pp. 525-552, 2003.

MIXDES 2012, 19ᵗʰ International Conference *"Mixed Design of Integrated Circuits and Systems"*, May 24-26, 2012, Warsaw, Poland

Variability and Reliability Analysis of CNFET in the Presence of Carbon Nanotube Density Fluctuations

Carmen G. Almudéver and Antonio Rubio
Electronic Engineering Department
UPC, BarcelonaTECH
Barcelona, Spain
Email: carmen.garcia.almudever@upc.edu

Abstract—Current carbon nanotube (CNT) synthesis processes are not perfect. One of the most critical issue is the presence of density variations in CNT growth. These variations are due to the lack of precise control of CNT location during the synthesis and the presence of metallic CNTs (m-CNTs). In this work we analyze the impact of CNT density fluctuations on carbon nanotube field effect transistor (CNFET) performance. A CNFET reliability analysis is also presented because of CNT density variations can cause a complete failure of CNFET.

Index Terms—CNT, CNT density variability, CNFET.

I. INTRODUCTION

Carbon Nanotube Field Effect Transistors (CNFETs) are promising candidates as a potential extension to silicon transistors. With extraordinary electrical properties, such as quasi-ballistic transport or higher carrier mobility [1][2], CNFETs exhibit characteristics rivaling those of state-of-the-art Si-based MOSFETs.

An ideal CNFET technology (meaning all carbon nanotubes are semiconducting, have same diameter and are perfectly aligned and well-positioned) is predicted to be 5x faster than silicon CMOS, while consuming the same power [3]. However, several significant challenges must be overcome before such benefits can be realized. There are some imperfections inherent in carbon nanotube (CNT) growth process that results in CNT diameter and density variations, a mixture of metallic and semiconducting CNTs and mis-positioned and mis-aligned CNTs that affect CNFET performance and must be solved.

This work is focused on density variations that is responsible for the number of CNTs that forms the transistor and then can comprise reliability of CNFETs. These variations are due to the non-uniform inter-CNT spacing and the presence of metallic CNTs that must be removed.

In this paper a CNFET variability and reliability analysis in the presence on CNT density fluctuations is presented. It is organized as follows. In section II the main CNT synthesis process challenges are introduced. In section III the impact of CNT density variations on current characteristics and different CNFET parameters is presented. Finally, a CNFET reliability analysis is shown in section IV.

II. CNT SYNTHESIS PROCESS VARIATIONS

Exceptional I-V characteristics have been demonstrated in ideal CNFETs. However, actually, there are several limitations in the CNT production methods that affect its performance and could eclipse the expectations. Current CNT synthesis processes are far from being perfect.

One challenge in CNT growth process is the alignment and the positioning of CNTs. Although CNT growth on quartz yields a large fraction of aligned CNTs ($> 99\%$) [4], there exits a non-negligible fraction of mis-positioned CNTs that may cause incorrect logic functionality. However, CNFETs circuits immune to such mis-positioned CNTs have been developed [5].

Great improvements have done in the field of CNT synthesis process, however the accurate control of chirality (angle of the atom arrangement along the tube) is an unresolved task. Chirality is responsible for the random distribution of CNT diameters. Depending of the CNT growth method different mean diameters and diameters distributions can be observed [6]. These diameter variations strongly affects the bandgap of a CNT (energy gap is inversely proportional to CNT diameter) and cause fluctuations in the threshold voltage of a CNFET.

Another challenge is the presence of CNT density variations [7]. They are due to the non-uniform spacing between CNTs (non-uniform pitch) and cause not only large variations in CNFET performance, but also have a significant probability of complete failure in cases where there is no CNT present in the CNFET (*open*). Furthermore, a CNT density increase is needed. The average CNT density obtained today is between 10-50 CNTs/μm but for logic circuits at least 250 CNTs/μm are required.

The presence of metallic CNTs (m-CNTs) among semiconducting CNTs (s-CNTs) is also a handicap. Metallic CNTs should not be used to make CNFETs because they have a high conductivity that makes that its current cannot be controlled by the gate, causing source-drain shorts-like in the transistor. Therefore, the presence of m-CNTs increase the probability of CNFET failure (*short*) as we will see in section IV. With a typical CNT synthesis process 1/3 of CNTs are metallic and

2/3 are semiconducting. In order to reduce the proportion of m-CNTs different processing options can be used.

One option is to grow predominantly s-CNTs. With enhanced CNT growth methods a percentage between 96% and 90% of s-CNTs is achieved [8][9]. Another alternative is to separate m-CNTs from s-CNTs after CNT growth to obtain mostly s-CNTs. A considerable reduction in percentage of m-CNTs (1%-5% m-CNTs) has been achieved by CNT self-sorting techniques [10][11]. However this improvement in percentage of m-CNT is not enough for VLSI-scale digital circuits. For VLSI CNFET circuits to meet leakage, noise margins, and delay variation targets, the percentage of m-CNTs must be reduced to less than 0.01% [3]. Then a third processing option is to remove m-CNTs after CNT growth.

Existing techniques for m-CNT removal include Single-Device electrical Breakdown (SDB) [12], gas phase and chemical reaction-based removal techniques [13][14] or VLSI-compatible Metallic-CNT Removal (VMR) [15]. In SDB, s-CNTs are switched off using gate voltage so that a high current only flows through m-CNTs causing them to break down. With this technique $\sim 100\%$ of m-CNTs are removed but it is not VLSI compatible. Gas phase chemical reaction-based removal techniques are highly compatible with VLSI semiconductor processing but the m-CNT removal is dependent on CNT diameters (narrow CNT diameter distributions are required). VMR is a VLSI-compatible technique but can impose significant area penalties up to 200%. Summarizing, existing techniques for m-CNT removal are either not VLSI compatible, or they are VLSI compatible but paying a high cost of area. Hence, co-optimization of m-CNT removal and circuit design is necessary for enabling robust digital logic using CNFETs.

It is worth noting that m-CNTs removal results in a decrease of the number of CNTs present in the transistor, or in other words, they causes CNT density variations as well. Furthermore, these m-CNTs removal techniques are no perfect. They also inadvertently remove some s-CNTs further reducing the density of CNTs and increasing the probability of failure (*open*) as we will see in section IV.

As we mentioned, in this work we focus on CNT density variations characteristic of the CNT synthesis process because of non-uniform spacing between CNTs and the presence of m-CNT that must be removed.

III. CNT DENSITY VARIABILITY ANALYSIS

A. CNFET structure

Figure 1 shows the CNFET structure used in this work. It is a MOSFET-like CNFET based on the Stanford University CNFET nominal compact model [16] [17]. Observe that it is a top gate structure and uses source and drain ohmic metal contacts. It is composed by 8 perfectly aligned and positioned semiconducting CNTs whose section under the gate is intrinsic and source and drain extension regions are n or p doped (p-type or n-type transistor). The choice of 8 CNTs is based on [3]. In [3] they conclude that with 8% of CNTs grown as metallic-type, at least eight CNTs per device are required to

bound the failure probability (caused in case all CNTs are metallic) below 2×10^{-9}. The chirality of the CNTs is (19,0), that is, a semiconducting CNT with a diameter of $\sim 1.5nm$. The inter-CNT spacing or pitch is 3nm (inter-CNT electrostatic charge screening effect is taken into account [18]). The length of the gate, source and drain (L_{ch}, L_{ss}, L_{dd}) is 16nm and the width of the gate is (W_{gate}) is 32nm.

Figure 2 shows the output and transfer characteristics of a n-type CNFET. These characteristics have been obtained using the CNFET HSPICE simulation model developed by Stanford University. Some key transistor parameters are in Table I. They have been calculated as follows:

- ON current (I_{ON}): it is measured when $V_{DS} = V_{GS} = 0.9V$
- OFF current (I_{OFF}): it is measured when $V_{DS} = 0.9V$ and $V_{GS} = 0V$
- ON-OFF currents ratio (I_{ON}/I_{OFF})
- Threshold voltage (V_{TH}):): in a MOSFET-like CNFET the threshold voltage can be obtained through this expression

$$V_{TH} = \frac{E_g}{2q} = \frac{\sqrt{3}}{3} \frac{aV_\pi}{eD_{CNT}} \qquad (1)$$

where $a = 2.49\text{Å}$ is the carbon to carbon atom distance, $V_\pi = 3.033eV$ is the carbon $\pi - \pi$ bond energy in the tight bonding model, e is the unit electron charge, and D_{CNT} is the CNT diameter.

- Transconductance factor (K):): it is evaluated from the V_{DD} saturated current level using the expression of the Sah [19] model for a equivalent Si-MOS transistor neglecting channel modulation and carriers saturation effects (as correspond to a CNFET).

$$K = \frac{2I_{DS}}{(V_{GS} - V_{TH})^2} \qquad (2)$$

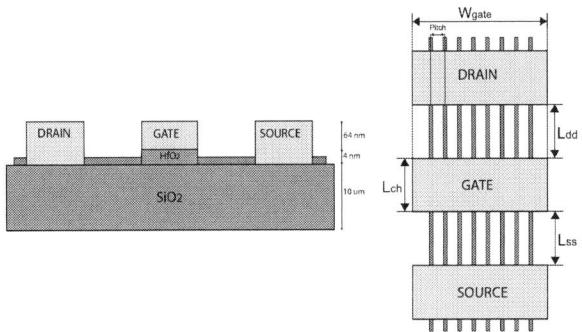

Fig. 1. CNFET structure showing (left) a front-view and (right) a top-view sections.

B. CNT density variability procedure

In order to analyze the impact of density variations on CNFET performance we have developed a MATLAB script.

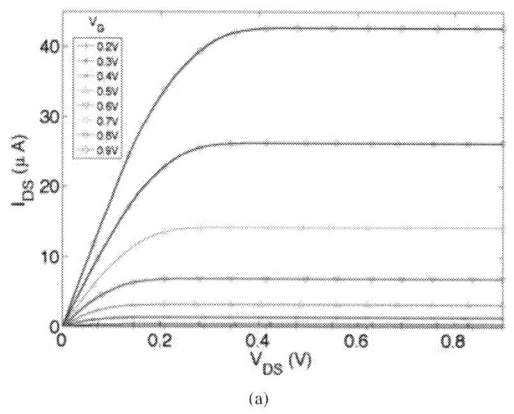
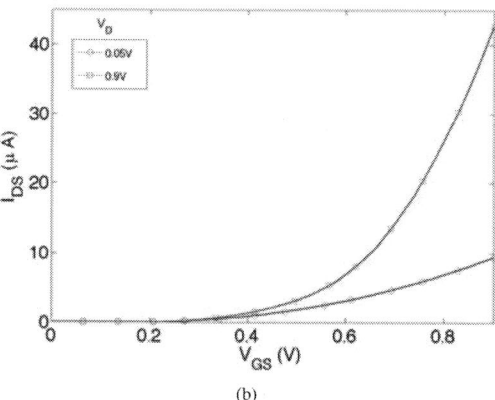

(a)   (b)

Fig. 2.   Simulation results. (a) Output and (b) transfer characteristics of a n-type CNFET.

TABLE I
KEY TRANSISTOR PARAMETERS

I_{ON}	I_{OFF}	I_{ON}/I_{OFF}	V_{TH}	K
42.70 μA	65.60 pA	6.51×10^5	0.29 V	0.23 mA/V^2

We have simulated 1000 different n-type CNFETs (with different density of tubes) obtaining the $I_{DS} - V_{DS}$ and $I_{DS} - V_{GS}$ characteristics and I_{ON}, I_{OFF}, I_{ON}/I_{OFF}, V_{TH} and K distributions.

The procedure used to analyze the effect of CNT density variations in CNFET is displayed in a simplified way in Fig. 3.

In the CNFET sample extraction stage, given a width of the gate (W_{gate}) and a pitch distribution (P), the density of CNTs (n) in the transistor is obtained. It is worth noting that we consider a Chi square pitch distribution which mean (μ) and standard deviation (σ) are 2.69nm and 0.13nm respectively. Subsequently, given a probability of CNT be metallic (T_M),

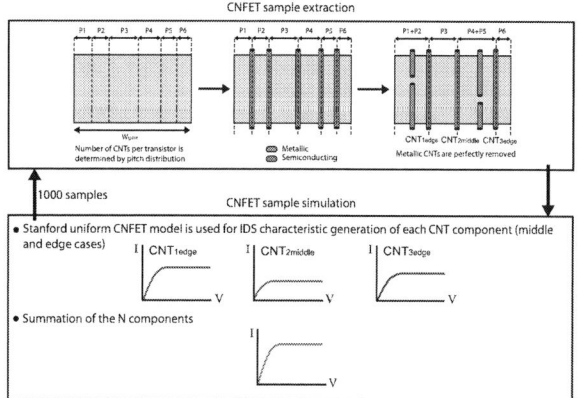

Fig. 3.   Monte Carlo process simulation using our MATLAB script.

the proportion of m-CNTs and s-CNTs is established. In step 3 we assume that all m-CNTs are removed whereas all s-CNTs keep intact; that is, we assume an ideal m-CNT removal process (optimistic assumption). Then the pitch of the remaining s-CNTs is recalculated. Hence the result of the CNFET sample extraction stage is a sample of n-tubes CNFET, with m no active m-CNTs and s active s-CNTs with a variable distance between them.

In the CNFET sample simulation phase, the I-V characteristic of the n-tubes CNFET sample is obtained. This is done through the summation of the $(n-m)$ I_{DS} current components (each s-CNT that forms the CNFET) as it is shown in Figure 3. Observe that the metallic ones do not contribute ($I_{DS} = 0$). S-CNTs are evaluated using the Stanford CNFET compact model taking into account the charge screening effects and the position of the tube in the transistor (edge or middle).

C. Simulation results of CNT density variations

A 1000 samples Monte Carlo analysis for different widths of gate ($W_{gate} = 32nm$ and $W_{gate} = 16nm$) and several m-CNT probabilities ($T_M = 5\%$ and $T_M = 33\%$) has been done using the previously procedure. Figure 4 shows the $I_{DS} - V_{DS}$ and $I_{DS} - V_{GS}$ characteristic distributions for a 32nm and 16nm n-type CNFET and a probability of a CNT be metallic of 33%. Black lines correspond to 1000 CNFET devices under density variability and red curve is the variability average. Note the green lines that appear in 16nm CNFET (Figure 4c and 4d). These are $I_{DS} \approx 0A$, that is, transistors without any CNT (4 CNFETs in this case).

In Table II the variability ($100 \times 3\sigma/\mu$) of key transistor parameters due to density variations is shown. In general it increases as the m-CNT probability increases and as the width of the gate decreases. Currents ratio (I_{ON}/I_{OFF}) has the highest variability followed by I_{OFF}. Note that I_{ON} and K show similar variability, whereas V_{TH} it is not affected by CNT density fluctuations.

So the density of CNTs, given by pitch distribution and the proportion of m-CNTs that are subsequently removed and also

126

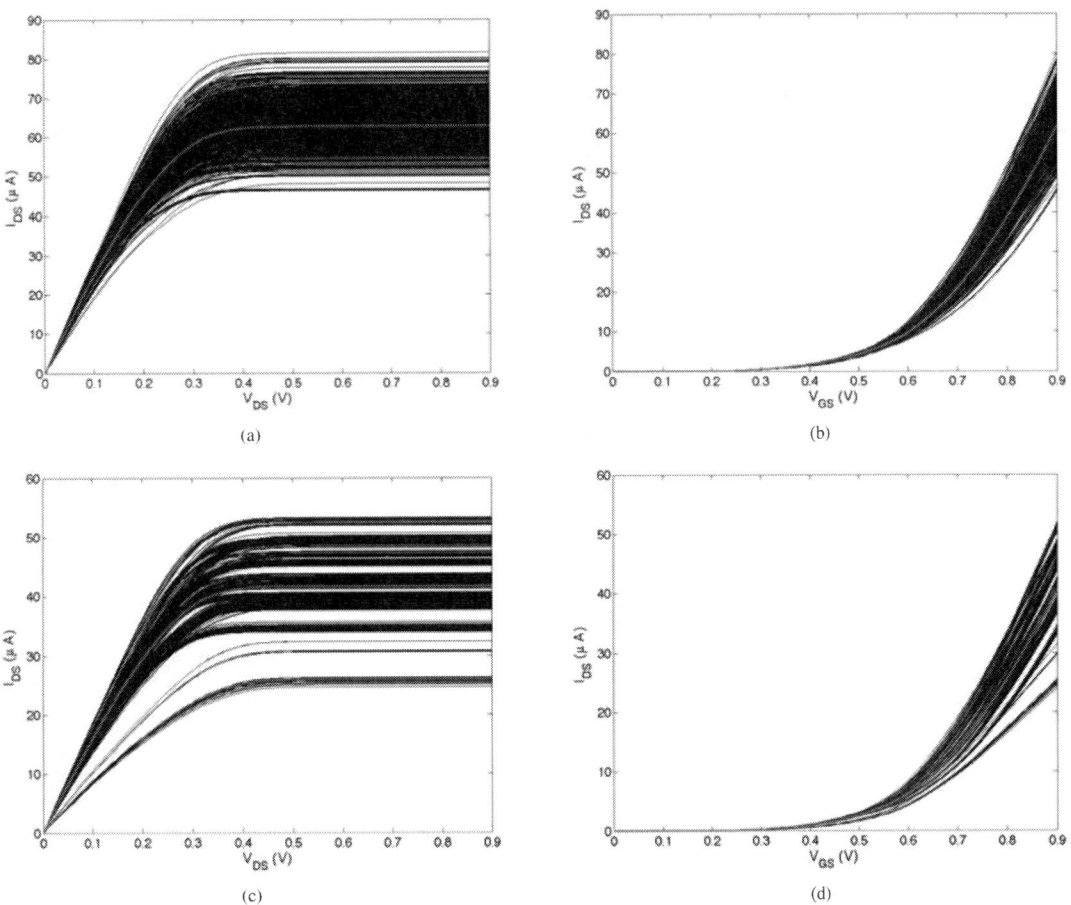

Fig. 4. (a) $I_{DS} - V_{DS}$ and (b)$I_{DS} - V_{GS}$ characteristic at $V_D = 0.9V$ for a n-type 32nm CNFET and $T_M = 33\%$. (a) $I_{DS} - V_{DS}$ and (b)$I_{DS} - V_{GS}$ characteristic at $V_D = 0.9V$ for a n-type 16nm CNFET and $T_M = 33\%$. The curves for the 1000 simulated devices (black curves) are shown, along with average (red curve). Green lines in (c) and (d) are 16nm transistors without any CNT (open).

TABLE II
TRANSISTOR PARAMETERS VARIABILITY

	$W_{gate} = 32nm$		$W_{gate} = 16nm$	
T_M	**5%**	**33%**	**5%**	**33%**
I_{ON}	22.85 %	31.79 %	22.61 %	38.67 %
I_{OFF}	27.45 %	83.29 %	60.64 %	139.87 %
I_{ON}/I_{OFF}	52.34 %	159.42 %	117.76 %	156.06 %
V_{TH}	0 %	0 %	0 %	0 %
K	22.60 %	31.68 %	22.41 %	38.55 %

dependent on the width of the gate, plays an important role not only in CNFET performance but also in CNFET reliability as we have demonstrated in this section and as we will see in section IV.

IV. STATISTICAL RELIABILITY ANALYSIS

In the previous section we have presented a methodology to study the impact of density variations, inherent in CNT synthesis, on CNFET technology. As we mentioned, in this

analysis we consider an ideal m-CNT removal process. That is all m-CNTs present in the transistor are removed whereas all s-CNTs keep intact. But this is an optimistic assumption because in reality removal techniques are non-ideal; some m-CNTs still survive after m-CNT removal and some s-CNTs can be removed accidentally during the process. Although close to 100% m-CNT removal is realizable, a non-negligible fraction (typically, 10%-40%) of the s-CNTs is removed as well. As a result, density variations increase and most important the CNFET may fail. This fact has a key impact on the reliability of the CNT technology.

In this section a CNFET reliability analysis is presented. With this purpose we consider two types of failures in our CNFETs (Figure 5):

1) Catastrophic failures: CNFET fails to have a transistor-like behavior (*open* or *short*) or shows a degenerate behavior.

2) Parametric failures: the CNFET behaves as a transistor but it does not satisfy some boundary conditions required

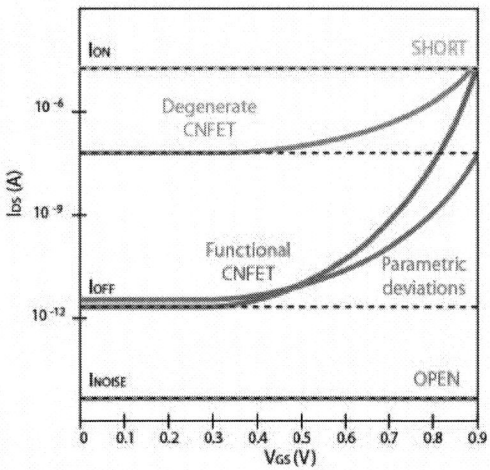

Fig. 5. Catastrophic failures, parametric failures and functional CNFET characteristics.

by the designer.

A. Catastrophic failures

As we showed in section II different m-CNT removal techniques are used in CNFET technology in order to have only s-CNTs in the transistor. However, these techniques are no perfect. Some m-CNTs keep intact after m-CNT removal process and some s-CNTs are inadvertently removed. So realistic removal process can produce different catastrophic failures in CNFET. An *open* failure occurs when there is no CNT bridging the source and drain contacts, that is when all CNTs are removed. On the other hand, a predominant presence of m-CNTs in the transistor causes a *short* because of its high conductivity makes that its current cannot be controlled by the gate. Finally the presence of m-CNTs among s-CNTs may cause that n-tubes CNFET shows a degenerate transistor-like behavior. In short, we can classify the catastrophic failures as: *open*, *short* and *degenerate* transistor-like behavior (Figure 5).

Lets consider that a CNT has a probability of being metallic (p_m) and of being semiconducting ($p_s = 1 - p_m$). Furthermore we define the probability of a CNT to be removed during the process being s-CNT (m-CNT) as p_{sR} (p_{mR}).

A 1-tube CNFET is non-functional (fails) if it is an m-CNT (*short*), or if it is an s-CNT or m-CNT and it is removed (*open*), that is, there is no any contacting CNT between source and drain. So we can define the failure probability (p_f) for the case of 1-tube CNFET as:

$$p_{f(1tube)} = p_m p_{mR} + p_s p_{sR} \qquad (3)$$

For a CNFET with n independent CNTs (n-tube CNFET), the CNFET fails if all contacting CNTs are predominantly metallic (*short*), if there is no any CNT that bridge S/D contacts (*open*), or if there are m-CNTs between s-CNTs

(*degenerate transistor*). So the overall probability failure probability with n CNTs is (considering uniform CNT density):

$$p_{f(ntubes)} = p_{f(1tube)}^n = (p_m p_{mR} + p_s p_{sR})^n \qquad (4)$$

As we mentioned with typical CNT growth methods 1/3 of CNTs are metallic and 2/3 are semiconducting. In order to reduce the proportion of m-CNTs, enhanced CNT synthesis methods and CNT sorting and m-CNT remove techniques are used. With enhanced CNT growth methods a percentage between 4% and 10% of m-CNTs is achieved. Reduction in percentage of m-CNTs has also been demonstrated by CNT self-sorting techniques (1%-5% m-CNTs). High m-CNT removal rates ($\sim 100\%$) may be obtained by m-CNT removal techniques, however these techniques also inadvertently remove some s-CNTs, typically 10%-40%. Taking into account these percentages and considering a $p_{mR} \sim 100\%$, Figure 6 shows the failure probability for different scenarios (SC) and for different number of CNTs per transistor (n). Observe that the failure probability decreases as the number of CNTs per transistor increases and as p_m decreases. Obviously SC7 ($p_m = 5\%$ and $p_{sR} = 0\%$), where an ideal m-CNT removal process is considered, presents the minimum p_f and SC3 ($p_m = 33\%$ and $p_{sR} = 40\%$) shows the maximum p_f (worst case). Horizontal black line is the minimum reliability requirement (2×10^{-9}) mentioned in [10]. See that in order to obtain a $p_f \leq 2 \times 10^{-9}$ at least 7 CNTs are required for the most favorable scenario (SC7) and 27 CNTs are necessary for scenario 6 (dashed lines). Note that for the worst scenario (SC3) more than 30 CNTs ($n > 30$) are needed to bound this failure probability.

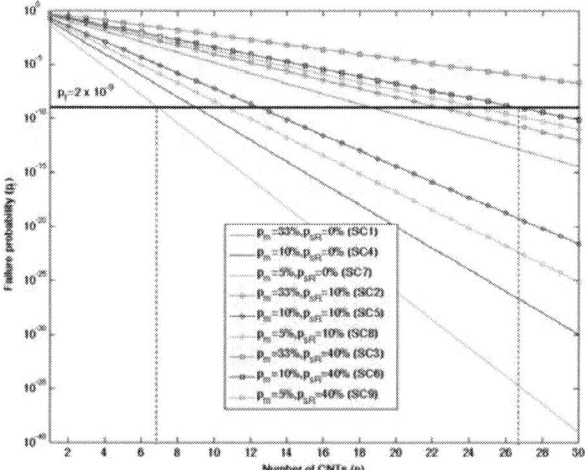

Fig. 6. Failure probability for different number of CNTs per transistor and for different m-CNT removal scenarios.

B. Parametric failures

As we demonstrated in section III, CNT density variability causes changes in CNFET characteristics and parameters but it still behaves as a transistor. However, some of them may not satisfy some boundary conditions defined by the designer. That is, they may suffer parametric deviations that are cause of failure. In order to determine which proportion of the 1000 CNFETs samples studied in our density variability analysis is functional we have established some boundary conditions based on the variability results shown Table II. We consider that a CNFET is functional if it accomplishes that:

- $I_{ON} > \mu - 2\sigma$
- $I_{OFF} < \mu + 2\sigma$

Given these conditions, we calculate the percentage of simulated CNFETs that verify both conditions (i.e. they are functional transistors) and how many CNFETs violates one or both of them. Table III shows the percentage of functional and non-functional (catastrophic+parametric failures) CNFETs. For a low probability of m-CNTs ($T_M = 5\%$) all transistors are functional whereas for a high m-CNT probability ($T_M = 33\%$) 3.2% and 6.6% of transistors are non-functional for a 32nm and 16nm CNFETs respectively.

TABLE III
PERCENTAGE OF FUNCTIONAL AND NON-FUNCTIONAL CNFETs

	$W_{gate} = 32nm$		$W_{gate} = 16nm$	
T_M	5%	33%	5%	33%
Catastrophic failure	0%	0%	0%	0.4% (open)
Parametric failure	0%	3.2%	0%	6.2%
Functional CNFETs	100%	96.8%	100%	93.4%

V. CONCLUSION

A study of the impact of CNT density fluctuations in CNFET performance and reliability have been presented.

CNT density variations, one of the main sources of CNFET fluctuation and inherent in CNT growth process, are due to non-uniform inter-CNT spacing and the presence of metallic CNTs that must be subsequently removed. They cause large variations in CNFET behavior and increase the probability of CNFET failure, because there is no CNT bridging source and drain contacts (*open*) or all contacting CNTs are predominantly metallic (*short*).

We have analyzed the impact of these variations in several CNFET parameters (I_{ON}, I_{OFF}, I_{ON}/I_{OFF}, V_{TH} and K). As we expected, the variability increases as the proportion of m-CNTs increases and as the width of the gate decreases. Furthermore, a statistical reliability analysis have been done. The CNFET failure probability for different m-CNT removal and m-CNT probability scenarios has been calculated, showing that the better the m-CNT removal process and the lower the m-CNT probability, the lower the failure probability.

Hence, CNT density variations play a key role not only in CNFET performance but also in CNFET reliability.

ACKNOWLEDGMENT

This research work has been partially supported by the Spanish Ministry of Science and Innovation (MICINN) through the project TEC2008-01856 with the additional participation of FEDER founds and by the EU project TRAMS(248789). The group of research is considered a consolidated group by the MICINN.

REFERENCES

[1] T. Durkop, S.A. Getty, E. Cobas, and M.S, Fuhrer, "Extraordinary mobility in semiconducting carbon nanotubes", *Nano Letters*, vol. 4, pp. 35-39, 2004.

[2] Y.-M. Lin, J. Appenzeller, J. Knoch, and P. Avouris, "High-performance carbon nanotube field-effect transistor with tunable polarities", *IEEE Trans. on Nanotechnology*, vol. 4, pp. 481-489, 2005.

[3] N. Patil, J. Deng, S. Mitra, and H.-S.P. Wong, "Circuit-Level Performance Benchmarking and Scalability Analysis of Carbon Nanotube Transistor Circuits", *IEEE Trans. on Nanotechnology*, vol. 8, no.1, pp. 37-45, 2009.

[4] S. J. Kang et al., "High-Performance Electronics Using Dense, Perfectly Aligned Arrays of Single-Walled Carbon Nanotubes", *Nature Nanotechnology*, vol. 2, pp. 230-236, 2007.

[5] N. Patil et al., "Integrated Wafer-scale Growth and Transfer of Directional Carbon Nanotubes and Misaligned- Carbon-Nanotube-Immune Logic Structures", *Proc. Symp. VLSI Technology*, pp. 205-206, 2008.

[6] X. Liu et al., "Detailed analysis of the mean diameter and diameter distribution of single-wall carbon nanotubes from their optical response", *Phys. Rev. B*, vol. 66, no. 4, 2002.

[7] J. Zhang, N. Patil, A. Hazeghi, and S. Mitra, "Carbon Nanotube Circuits in the Presence of Carbon Nanotube Density Variations", *in DAC*, 2009.

[8] Y. Li et al., "Preferential Growth of Semiconducting Single-Walled Carbon Nanotubes by a Plasma Enhanced CVD Method", *Nano Letters*, vol. 4, pp. 317-321, 2004.

[9] L. Qu, D. Feng, and L. Dai, "Preferential Syntheses of Semiconducting Vertically Aligned Single-Walled Carbon Nanotubes for Direct Use in FETs", *Nano Letters*, vol. 8, no. 9, pp. 2682-2687, 2008.

[10] M. S. Harold, A. A. Green, J. F. Hulvat, S. I. Stupp, and M. C. Hersam, "Sorting carbon nanotubes by electronic structure using density differentiation", *Nature Nanotechnology*, vol. 1, no. 1, pp. 60-65, 2006.

[11] R. Krupke, F. Hennrich, H. V. Lhneysen, "Separation of metallic from semiconducting single-wall carbon nanotubes", *Science*, vol. 301, no. 5631, pp. 344-347, 2003.

[12] P. Collins, S. Arnold, and P. Avouris, "Engineering carbon nanotubes and nanotube circuits using electrical breakdown", *Science*, vol. 292, pp. 706-709, 2001.

[13] A. Hassanien et al., "Selective etching of metallic single-wall carbon nanotubes with hydrogen plasma", *Nanotechnology*, vol. 16, pp. 278-281, 2005.

[14] C. Yang et al.., "Preferential etching of metallic single- walled carbon nanotubes with small diameter by fluorine gas", *Phys. Rev. B*, vol. 73, p. 75419, 2006.

[15] N. Patil et al., "VMR: VLSI-compatible metallic carbon nanotube removal for imperfection-immune cascaded multi-stage digital logic circuits using Carbon Nanotube FETs", *IEDM*, pp. 1-4, 2009.

[16] J. Deng and H.-S. Wong, "A Compact SPICE Model for Carbon-Nanotube Field-Effect Transistors Including Nonidealities and Its ApplicationPart I: Model of the Intrinsic Channel Region", *IEEE Transactions on Electron Devices*, vol. 54, no. 12, pp. 3186-3194, 2007.

[17] J. Deng and H.-S. Wong, "A compact spice model for carbon-nanotube field-effect transistors including nonidealities and its application part II: Full device model and circuit performance benchmarking", *IEEE Transactions on Electron Devices*, vol. 54, no. 12, pp. 3195-3205, 2007.

[18] J. Deng and H.-S. Wong, "Modeling and analysis of planar gate capacitance for 1-D FET with multiple cylindrical conducting channels", *IEEE Transactions on Electron Devices*, vol. 54, no. 9, pp. 2377-2385, 2007.

[19] C. T. Sah, "Characteristics of the metal-oxide-semiconductor transistors", *IEEE Transactions on Electron Devices*, vol. 11, no. 7, pp. 324345, 1964.

Technologies towards Cognitive Transceivers - Par4CR Project

MIXDES 2012, 19ᵗʰ International Conference *"Mixed Design of Integrated Circuits and Systems"*, May 24-26, 2012, Warsaw, Poland

Digital Hardware Resources for Steering a Nonlinear Interference Suppressor

Erwin J.G. Janssen, Dusan Milosevic,
Peter G.M. Baltus and Arthur H.M. van Roermund
Eindhoven University of Technology
Department of Electrical Engineering
Mixed-signal Microelectronics
Email: E.J.G.Janssen@tue.nl

Hooman Habibi
Eindhoven University of Technology
Department of Electrical Engineering
Signal Processing Systems
Email: H.Habibi@tue.nl

Abstract—**In this paper the requirements and resulting costs for the digital hardware are discussed to steer a nonlinear interference suppression circuit (NIS). This NIS circuit suppresses a strong unwanted RF blocker by exploiting a nonlinear transfer function in a radio receiver. Nonlinear transfer functions enable frequency-independent amplitude discrimination because they do not obey to the principle of superposition. Such an approach is considered because it can save considerable power and cost in the analog front-end of multi-radio devices. But, to this end an investment is required in the digital part of the system. The analysis in this paper leads to an estimated area consumption of 125k gates and 12k SRAM cells resulting in ≈0.25 mm² and ≈100 μW/MHz if implemented in CMOS 65nm. Also the future development of the cost figures is analyzed.**

Index Terms—**Nonlinear filters, multi-radio coexistence, interference suppression, digital signal processing.**

I. INTRODUCTION

In the recent years the number of different wireless standards supported by handheld devices have increased steadily. Therefore, the coexistence of multiple standards becomes an increasingly important issue [1], [2]. Conventional filtering in the frequency domain is often inadequate and also not cost effective because of bulky off-chip components, so time-sharing concepts are often used to enable the simultaneous activity of different standards. This time-shared approach requires a challenging synchronization between the data packets of different standards, and therefore leads to a reduction of the data throughput. Therefore, there is interest in finding different solutions to the coexistence problem.

To enable true simultaneous activity of closely separated transceivers, some means of self-interference suppression must be turned to. The implementation of such a system must be small, and low-power. One solution to this end is to make use of nonlinear interference suppression (NIS) in an early stage of the receiver. This principle enables filtering in the amplitude domain, thereby suppressing strong undesired signals and amplifying weak desired signals, independent of their frequency. Successful implementation of this system requires an analog circuit with an adaptive nonlinear transfer [3], and

The research leading to these results has received funding from the [European Union] Seventh Framework Programme ([FP7/2007-2013] under grant agreement no. 230688)

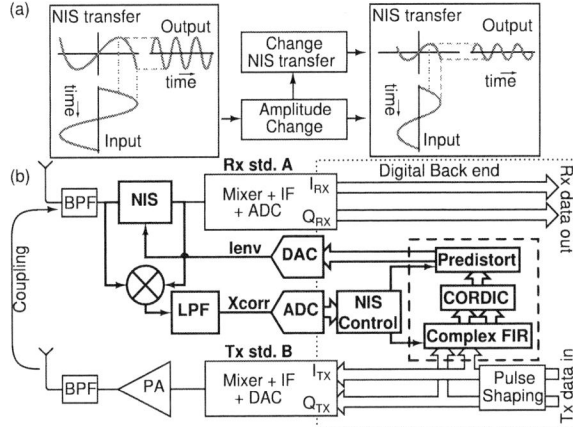

Fig. 1. Operating principle (a) and system diagram (b) showing the application of the NIS principle in a multi-standard transceiver. The circuitry inside the dashed box is discussed in more detail in this paper.

some digital circuitry to steer the analog circuit properly. In this paper an investigation is done to estimate the cost of the digital circuitry used in an algorithm to steer an NIS circuit, in terms of chip area and power consumption. Also an analysis is done that shows the future trend of the findings.

The paper is organized in the following manner. In section II the NIS application area is described and the analysis of the required accuracy is discussed. Section III presents the block level analysis, and section IV discusses the various elementary operations required for the sub-blocks. Then, in section V, the cost aspects of the digital functions and the total NIS algorithm are combined leading to an estimate for the cost in terms of chip area and power consumption. Finally, the conclusions are drawn in section VI.

II. SYSTEM LEVEL CONSIDERATIONS NONLINEAR INTERFERENCE SUPPRESSION

As is shown in Fig 1(a), the transfer function of the NIS is changed according to the actual amplitude level. Therefore, the application of the NIS requires the knowledge of the amplitude of the large signal to be suppressed. In handheld devices such as tablets and smartphones, this situation is encountered if two transceivers of different standards are active simultaneously [1], [2]. Problem scenarios are e.g. the coexistence of IMT-2000 [4] and WiMAX [5], DVB-H/T and GSM [6] as well as the simultaneous operation of Bluetooth with WLAN [7]. In addition to this multi-standard operation, also frequency division duplex scenarios suffer from simultaneous transmit/receive activities [8], [9].

As can be seen in Fig. 1(b), the large signal's amplitude at the input of the receiver of standard A due to the transmitter of standard B is determined in the digital back-end. The digital sub-blocks that must be present to steer the NIS circuit are indicated by the dashed box. The baseband signals are tapped from the transmitter after pulse shaping, and then passed through a complex filter. This complex filter models the (linear) behavior of the baseband electronics, transmitter mixer, PA, the bandpass filter (BPF), the antenna coupling and the BPF at the input of the receiver. Next, the magnitude of the resulting signal is determined using a "coordinate rotation digital computer" (CORDIC) algorithm [10]. As a final step before the DAC is addressed, predistortion takes place on the signal. This is done because the NIS control signal I_{env} is not entirely proportional to the magnitude of the interference seen at its input, because of non-idealities in the NIS circuit.

The digital sub-blocks are controlled by another sub-block "NIS control". This sub-block receives information about the cross-correlation of the output and the input of the NIS circuit. By analyzing this cross-correlation, the actual settings of the complex filter are updated to track the changes in the coupling characteristics. The output of the ADC digitizing the cross-correlation signal has a low bandwidth, so the investment in this part can be low. In this paper the feed-forward path indicated by the dashed box is discussed in more detail, because here the main amount of digital hardware is expected.

A NIS circuit enabling the desired functionality has been designed using a 140nm CMOS technology [3]. Measurements have shown the strongest signal it is able to suppress is about 11dBm, and the weakest signal that can sufficiently be suppressed is in the order of 0dBm. Therefore a dynamic range of strong signal suppression of approximately 11dB is observed. For lower strong signal levels, the suppression becomes less, and eventually the circuit converges to linear behavior.

A. Required resolution and speed

The accuracy of I_{env} must be such that sufficient blocker suppression can be guaranteed. Simulations and measurements on the hardware reported in [3] show that an error of 0.5% in I_{env} corresponds approximately to 40dB suppression. To guarantee that the suppression is not limited by the error of the control signal, this signal must have an accuracy less than 0.5%. Therefore the accuracy that is aimed for is chosen to be 0.25%. Choosing the requirement of the dynamic range equal to 15 dB to leave some margin, the signal-to-noise ratio of the control signal must be at least:

TABLE I
HARDWARE RESOURCES OF THE COMPLEX FIR.

Type of Operation	Number of Operations
16-bit Shift Register, M taps	4
16-bit Multiplications	$3 \cdot M$
16-bit Additions, 2 operands	$3 \cdot M$
32-bit Additions, 2 operands	$2 \cdot M$
32-bit Additions, M operands	2

$$ \text{SNR} = 15 - 20\log_{10}(0.25\%) \approx 67\text{dB} \tag{1} $$

The SNR specification translates into a combination of the amount of bits n_{DAC} with the over-sampling rate (OSR) to be used in the DAC:

$$ n_{\text{DAC}} \geq \frac{67 - 10\log_{10}(\text{OSR})}{6.02} \tag{2} $$

Although increasing the OSR reduces the number of bits, the benefit thanks to increasing the sample rate is not that high because of the logarithm in (2). The number of bits the signal must be represented by is therefore >11. To guarantee the required performance, choosing a 14-bit DAC such as e.g. [11] is sufficient. Furthermore, because rounding errors will be introduced in the various digital sub-blocks, two additional bits are added leading to a total of 16 bits in the digital circuitry.

The clock speed required in the DAC is mandated by the bandwidth of the aggressor's signal and the OSR. The clock speed of the digital circuitry must be such it can properly interface the DAC, and can be traded against the area consumption by applying pipelining. Concerning power consumption, no such scaling is possible. Furthermore this choice strongly depends on the characteristics of the transmitter standard. To provide a uniform analysis, the choice is made to minimize the amount of clock cycles required, which results in a worst case area estimate. The results can always be further adapted in case the area versus speed trade-off is chosen differently.

III. BLOCK LEVEL ANALYSIS

In this section various sub-blocks in the NIS algorithm are discussed and analyzed, resulting in an estimate for the number of elementary functions required.

A. Complex FIR filter

As mentioned before, to accommodate for the (linear) distortion incurred to the transmitted signal, a complex FIR filter has to be included. This should be a complex filter because it models an RF bandpass characteristic on the baseband signal. The implementation of such a filter requires several complex multiply-accumulate operations. By using Gauss's complex multiplication algorithm [12], a complex multiplication requires three multiply operations and five additions, of which three n-bit additions and two $2n$-bit additions. The results of these operations must be summed. Furthermore two n-bit shift registers must be used for both the I and Q signals, and two n-bit shift registers for the real and imaginary parts of

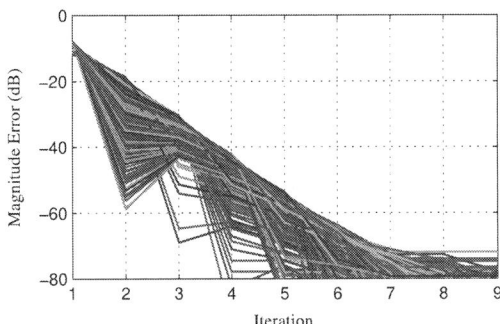

Fig. 2. Magnitude error evolution using the CORDIC algorithm on 500 randomly generated vectors having a peak-to-minimum ratio of 10dB, represented by 16 bits.

TABLE II
HARDWARE RESOURCES OF THE CORDIC.

Type of Operation	Number of Operations
16-bit Additions, 2 operands	2 per cycle-1
16-bit Multiplication	1

the filter characteristics. An overview of the operations that the complex FIR filter using Gauss's complex multiplication algorithm requires is listed in Table I.

B. CORDIC algorithm

The CORDIC algorithm is an algorithm to determine trigonometric and hyperbolic functions using simple arithmetic procedures such as addition, bit shift and table look-up [10]. Because of these basic functions, it is very well suited to be implemented in digital circuitry. The described procedure is an iterative process, and the magnitude of the error decreases with the number of iterations.

1) Convergence speed: For the application in mind (steering the NIS circuit) it is needed to convert the I and Q signals to magnitude. Therefore, the cartesian-to-polar conversion is investigated in more detail here. As can be understood from the previous sections, the amount of iterations used in the CORDIC algorithm is a choice. The plot shown in Fig. 2 shows the result of 500 randomly generated vectors represented by 16 bits, and the evolution of the resulting magnitude error after each step. The vectors have a peak to minimum ratio of 10dB. After the first iteration, the maximum error is in the order of -10dB. Then, the error decreases with 12dB (two bit) per iteration. After eight iterations the maximum error in the magnitude estimation saturates and is below -72dB, which corresponds to 12 bits. So, eight iterations are required in the CORDIC algorithm to guarantee sufficient precision. Table II lists all the different pieces of hardware required for the CORDIC operation.

C. Steering the DAC; Predistortion

To implement the predistortion, various approaches can be considered such as a mathematical (e.g. polynomial) fit, a complete input-output mapping using a look-up table or a

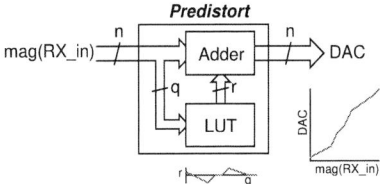

Fig. 3. Implementation of the DAC predistortion.

TABLE III
HARDWARE RESOURCES OF THE PREDISTORTION.

Type of Operation	Number of Operations
16-bit Additions, 2 operands	1
Read from 2^q-by-r look-up table	1

look-up table modeling the deviation from a nominal curve. The latter is chosen here and shown in Fig. 3. The reason for this choice is the fact this enables relatively high flexibility for limited cost. A complete input-output mapping requires a big look-up table with lots of redundancy, and a mathematical description results in limited flexibility.

As can be understood by looking at Fig. 3, this approach requires one adder and one look-up table. The adder should be able to add two binary representations of which one having n-bits and the other having r-bits, which requires an n-bit addition. The look-up table requires 2^q entries that represent r bits. A listing of the required operations in the predistortion can be found in Table III.

IV. ELEMENTARY OPERATIONS

In this section, various elementary operations to be used in the NIS algorithm are being discussed and analyzed, with the emphasis on their complexity in terms of number of equivalent gates. In section IV-A additions are discussed, which is on itself an elementary function to be used in section IV-B where multipliers are being discussed. Section IV-C treats the shift register, and finally the look-up table is discussed in section IV-D.

A. Addition

The "add" operation is crucial in digital circuits, because it finds its use in many operations. For example a multiplication uses addition to give its end result. In this subsection a small overview of implementations of this mathematical operation is given. Further information can be found in [13].

1) Full/half adders: Basic building blocks used in adder structures are the half and full adder [13]. The area consumption of a half adder is equal to three logic gates and the area consumption of a full adder is equal to seven logic gates. Using these building blocks, n-bit adders can be implemented.

2) Parallel (ripple-carry) adder: The most easy and straightforward way of implementing an n-bit addition is using a parallel adder. The area consumption of this structure is $7n$ gates.

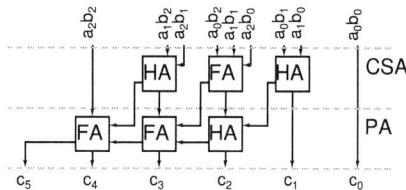

Fig. 4. Implementation of the addition of the partial products using $(n-2)$ CSA adders and one PA adder (array configuration).

3) Adding multiple n-bit numbers: To add m numbers each represented by n-bits, several fast approaches exist making use of the "carry-save" adders. The carry-save structure also consumes an area of $7n$ gates, but is faster than parallel adders. The area of a such a structure is equal to $(m-2)\text{Area}_{CSA} + \text{Area}_{PA}$ gates.

B. Multiplications

When considering the cost of digital signal processing algorithms, the number of multiplications are of importance. This is because a multiplication uses a series of two basic binary functions, namely AND and ADDITION. A simple implementation of a multiplication in digital is done using the procedure described below. Here the case of multiplying two three bit ($n = 3$) numbers a and b is considered.

$$a = \begin{bmatrix} a_2 & a_1 & a_0 \end{bmatrix} \tag{3}$$
$$b = \begin{bmatrix} b_2 & b_1 & b_0 \end{bmatrix}$$
$$c = a \cdot b$$

$$
c = \begin{array}{cccccc}
 & & & a_0b_2 & a_0b_1 & a_0b_0 \\
 & & a_1b_2 & a_1b_1 & a_1b_0 & 0 \\
 & a_2b_2 & a_2b_1 & a_2b_0 & 0 & 0 \quad + \\
\hline
[c_5 & c_4 & c_3 & c_2 & c_1 & c_0]
\end{array}
$$

In the example the binary values of a_i are multiplied with b_i, and can be implemented using a_i AND b_i. The results of these operations are then summed, leading to c. In general it can be concluded from this simple example that the number of AND operations are equal to n^2, so this first step can be done with n^2 logic gates. The main burden is in the implementation of the addition of the partial products. This can be performed using an array addition as described in subsection IV-A3.

In Fig. 4 the implementation of the addition of the partial products using CSA and PA adders is shown. The total number of gates of the multiplier including the summation of the partial products then becomes:

$$\#\text{Gates}_{\text{MULT}} = 9n^2 - 11n \tag{4}$$

To verify the amount of gates used in the multiplier, several multipliers have been implemented in VHDL, and compared to the model. This comparison in shown in Fig. 5.

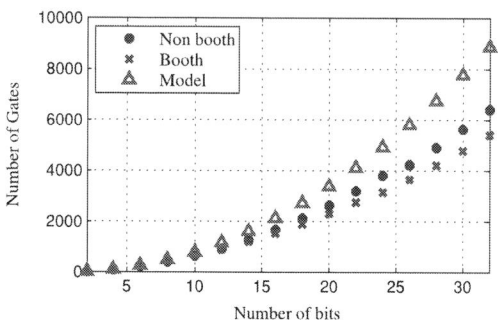

Fig. 5. Number of gates present in a multiplier for different sizes and two different implementation choices in VHDL (Booth/Non booth) compared to the modeling discussed in this document. The analysis presented here can be seen as a worst case estimate.

C. Shift register

A shift register can be implemented using D flip-flops. It is seen the amount of D flip-flops used in the shift register is equal to $n \cdot (K - 1)$ (first tap $\rightarrow x_0$). The implementation of a D flip-flop can be done in several fashions, with a different amount of hardware usage. The number of gates used in a single (conventional master-slave) D flip-flop is 8, apart from an inverter used in the clock branch which can be shared among all the flip-flops. This leads to a total number of gates of the n-bits shift register using K taps equal to:

$$\#\text{Gates}_{\text{SR}} = 8n \cdot (K - 1) \tag{5}$$

D. Look-up Table (LUT)

The implementation of a look-up table requires a memory block. The implementation of memory in an ASIC can be performed in different ways, such as e.g. flip-flops, SRAM, DRAM or Flash. The SRAM option seems the best choice because for the memory size required in our application it provides a cheap, fast and relatively small memory that can be changed (written) online. After power shutdown the values can be stored in Flash, which can be available at another location, but connected to a common controlling unit. In general an SRAM bit is smaller than a single logic gate (in Fig. 6 different memory types are compared concerning chip area for the coming years [14]). Using these single bit memory elements, a multi-bit multi-address memory can be composed. Besides the memory elements, also an address decoder is required. In general the size of the address decoder is much smaller than the size of the SRAM memory cells, and is therefore neglected in the analysis. The total amount of hardware usage in the LUT then becomes:

$$\#\text{SRAM Cells}_{\text{LUT}} = \left\lceil 2^n \cdot \text{bits/word} \right\rceil \tag{6}$$

V. TECHNOLOGY ASPECTS AND TOTAL COST EVALUATION

The area and power consumption of digital circuits is governed by the product of the number of gates and respectively

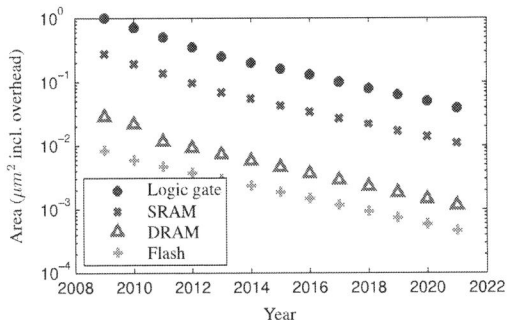

Fig. 6. Prediction of the effective area consumption of digital functions versus year, including an estimate for overhead (wiring etc.).

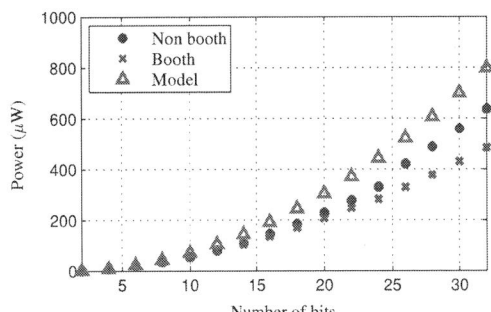

Fig. 7. Power consumption of a multiplier for different sizes and two different implementation choices in VHDL (Booth/Non booth) compared to the modeling discussed in this document. In this example a toggle rate of 0.02 per nanosecond is assumed, along with a power supply of 1.08 V using CMOS 65nm.

TABLE IV
HARDWARE RESOURCES REQUIRED FOR THE NIS ALGORITHM WITH n THE NUMBER OF BITS.

Sub-block	# Gates	# SRAM cells
Complex FIR	$M(27n^2 + 76n) - 60n$	-
CORDIC	$14Cn + 9n^2 - 18n$	-
Predistortion	$7n$	$2^q \cdot r$

the area and power consumption per gate including some overhead due to e.g. wiring. As discussed in the previous sections, the required functionality in combination with the required precision translates into the number of gates that must be used. Therefore, to minimize the cost of digital circuitry it is of great concern that the cost of an individual gate is minimized, which therefore has been the important goal in semiconductor industry since several decades [15].

A. Area

Fig. 6 shows the future development of the amount of area consumed by various digital elements including an estimate for the overhead [14]. The miniaturization as observed in the past continues for some more years, according to this prediction. To derive the area consumption of the functionality discussed in this paper the number of elements (gates/SRAM cells) can be multiplied by the area per element.

B. Power consumption

The total DC power consumption of digital circuits can be modeled by:

$$P_{DC} = \frac{1}{2} C \cdot V_{dd}^2 \cdot f_{clk} \cdot \alpha \qquad (7)$$

where C denotes the total capacitance of the gates and wiring of the circuit, V_{dd} is the power supply voltage and f_{clk} is the clock frequency. The remaining variable α denotes the toggle probability, and models the ratio of the amount of capacitance switched during a single clock period over the total capacitance. The amount of capacitance scales with the die size, so for fixed toggle probability the power consumption increases proportional to the amount of gates.

In Fig. 7 the power consumption is shown of the multipliers which were already discussed in section IV-B. It is indeed seen that the power consumption is strongly related to the amount of hardware, as expected. The characteristics shown here are valid for a power supply voltage of 1.08 V (1.2 - 10%) and a toggle rate of 0.02 per nanosecond. The power consumption per gate in this technology (65nm CMOS) in a 16-bit multiplier is then derived to be ≈ 5.37 fJ/gate/toggle. Using (7) this can be translated into nominal behavior (V_{dd} = 1.2 V), leading to a nominal power consumption of ≈ 6.63

fJ/gate/toggle. In general the nominal power consumption is equal to:

$$P_{DC} = \alpha \cdot 6.63 \text{ nW/gate/MHz} \qquad (8)$$

C. Cost of the digital circuitry for steering the NIS circuit.

By combining the results of the previous sections, the amount of hardware required to implement the NIS steering circuitry can be calculated. The equations listed in Table IV show the total number of gates in respectively:

- The complex FIR with M the number of taps.
- The CORDIC algorithm with C iterations.
- The predistortion with q and r the size of the LUT.

For this analysis is chosen: $M = 15$, $C = 8$, $q = 10$, $r = 12$ and $n = 16$. The anticipated cost based on the analysis presented in this paper then results in 124880 gates and 12288 SRAM cells resulting in ≈ 0.25 mm^2 and ≈ 100 μW/MHz of the clock frequency if implemented in CMOS 65nm. Assumed is a toggle probability of 12% for the gates, and power supply of 1.2 Volt. Typically the toggle rate ranges from 6% - 12 % for data-path logic [16]. The power consumption of the SRAM cells is neglected because of its low expected toggle rate. Fig. 8 shows the prediction of the amount of hardware usage of the NIS algorithm using future technologies. Finally, Fig. 9 shows the resources breakdown among the various sub-blocks of the algorithm. The cost is dominated by the complex

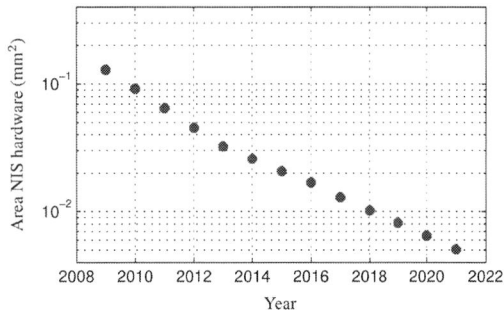

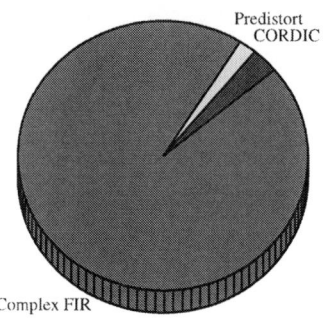

Fig. 8. Prediction of the amount of hardware required to implement the NIS algorithm using past and future technologies (the technology belonging to 2009 is CMOS 45 nm, the successor of CMOS 65nm).

Fig. 9. Division of area usage between the different functions in the NIS algorithm.

FIR structure, whereas both the CORDIC and the predistortion have a relative small contribution to the total.

VI. CONCLUSION

In this paper an analysis is presented concerning the required cost in terms of chip area and power consumption of the digital hardware to be used in an algorithm to steer the NIS circuit. The required functionality is translated into several sub-blocks that are analyzed individually. Combined with the accuracy demands in the NIS algorithm, the results of the analysis lead to an estimate of the cost figures. The hardware usage is derived to be 125k gates and 12k SRAM cells that, if implemented in CMOS 65nm, result in ≈ 0.25 mm^2 and ≈ 100 μW/MHz. The main consumer of digital hardware is the complex FIR, whereas the CORDIC and the predistortion are relatively small. The cost of the NIS IC is $P_{DC} = 3$-35mW and area consumption of 0.18mm^2 in CMOS 140nm [3]. The DAC presented in [11] consumes 16.7mW and uses 0.1mm^2 chip area in CMOS 130nm. The future developments show a trend of decreasing chip area and power consumption for the digital hardware because of decreased device sizes, but for the analog part this trend is not expected because of physical limitations. So, especially when considering the expected future development of the cost of the digital circuitry with respect to the cost of the analog circuitry, we conclude that the overhead introduced by the digital circuitry is minor.

ACKNOWLEDGMENT

The authors would like to thank the Dutch technology foundation STW and the Par4CR project for financial support and David Westberg, Max Bjurling, Henk ten Pierick and Hendrik van der Ploeg of Catena for their fruitful discussions.

REFERENCES

[1] S. Sheng, "RF coexistence - challenges and opportunities," in *Radio Frequency Integrated Circuits Symposium*, Jun. 2011.

[2] J. Zhu, A. Waltho, X. Yang, and X. Guo, "Multi-radio coexistence: Challenges and opportunities," in *IEEE Proc. of the 16th International Conference on Communications and Networks*. IEEE, Aug. 2007, pp. 358–364.

[3] E. Janssen, D.Milosevic, and P. Baltus, "An RF amplifier with frequency-independent self-interference blocker suppression," in *Radio Frequency Integrated Circuits Symposium, RFIC*. Montreal: IEEE, Jun. 2012.

[4] M. Konrad and W. Koch, "Blanking gaps in uplink cellular UMTS in the IMT-2000 extension band to solve the bluetooth coexistence problem," in *International Symposium on Wireless Communication Systems*, Sep. 2006, pp. 333–337.

[5] S. Suansook, S. Aramvith, and P. Prapinmongkolkarn, "Comparative performance analysis of WiMAX and WLAN with WPAN coexistence in UL band," in *International Symposium on Intelligent Signal Processing and Communication Systems*, Nov. 2007, pp. 526–529.

[6] D. L. H. Tong, R. Lababidi, F. Baron, A. Louzir, and B. Jarry, "Low-pass active filter enabling DVB-H/T and GSM standard coexistence," in *International Microwave Symposium*, Jun. 2007, pp. 1047–1050.

[7] A. Palin and M. Honkanen, "VoIP call over WLAN with Bluetooth head-set multiradio interoperability solutions," in *Proc. Int. Symp. Personal, Indoor Mobile Radio Commun.*, vol. 3, Sep 2005, pp. 1560–1564.

[8] E. Keehr and A. Hajimiri, "Equalization of third-order intermodulation products in wideband direct conversion receivers," *Solid-State Circuits, IEEE Journal of*, vol. 43, no. 12, pp. 2853 –2867, dec. 2008.

[9] A. Mirzaei, X. Chen, A. Yazdi, J. Chiu, J. Leete, and H. Darabi, "A frequency translation technique for SAW-less 3G receivers," in *VLSI Circuits, 2009 Symposium on*, june 2009, pp. 280 –281.

[10] J. Volder, "The CORDIC trigonometric computing technique," *Electronic Computers, IRE Transactions on*, no. 3, pp. 330–334, 1959.

[11] Y. Cong and R. Geiger, "A 1.5-V 14-bit 100-MS/s self-calibrated DAC," *Solid-State Circuits, IEEE Journal of*, vol. 38, no. 12, pp. 2051 – 2060, dec. 2003.

[12] (2011) Complex multiplication. [Online]. Available: http://mathworld.wolfram.com/ComplexMultiplication.html

[13] R. Zimmerman, "Binary adder architectures for cell-based VLSI and their synthesis," Ph.D. dissertation, Swiss Federal Institute of Technology, Zurich, 1997.

[14] (2010) International technology roadmap for semiconductors. [Online]. Available: http://www.ITRS.net

[15] G. E. Moore, "Cramming more components onto integrated circuits," *Electronics*, pp. 114 – 117, April 1965.

[16] (2008) Xilinx website. [Online]. Available: http://www.xilinx.com/ise/power_tools/wpt_help/app_docs/calculating_toggle_rates.htm

MIXDES 2012, 19ᵗʰ International Conference *"Mixed Design of Integrated Circuits and Systems"*, May 24-26, 2012, Warsaw, Poland

LINC Architecture: A Discussion

Ernst Habekotté and Floris van der Wilt
Catena Microelectronics B.V.
Delft, The Netherlands

Abstract—**A discussion about the LINC architecture with alternative views on the power combiner merged with the antenna, reduction of gain mismatch with less impact of PA output impedance and broadband behavior.**

Index Terms—**LINC, Chireix power combiner, constant envelope PA, peak amplitude detection, integrated antenna.**

I. INTRODUCTION

When looking for highly compact and lightweight radios that comply well with new generations of mobile equipment and that offer at the same time long operating times by efficiently using the battery power, it is the transmitter that requires special attention because it is the most power hungry part of mobile equipment.

At the same time there is interest in realizing most of the transmitter on one chip for
- high reliability,
- small volume fitting in mobile phone sized housing and
- low costs preferably in silicon based IC-processes.

In multiband applications a broadband transmitter would be the ultimate goal.

The so-called LINC architecture (LInear with Non-linear Components) (Figure 1) is one out of several suitable architectures improving the power efficiency, while being able to handle complex modulated RF signals. A signal separator at its input splits the original input signal into two constant envelope out-phased signals in such a way that the linear combination of these two signals forms the original AM and PM modulated input signal ((1) and (2)).

$$S_{in}(t) = V(t)\, e^{j(\omega t + \varphi(t))} = S_1(t) - S_2(t) \qquad (1)$$

$$V(t) = V_{max} \sin(\theta(t)) \qquad (2)$$

$$S_1(t) = \frac{V_{max}}{2}\, e^{j(\omega t + \varphi(t) - \theta(t))} \qquad (3)$$

$$S_2(t) = \frac{V_{max}}{2}\, e^{j(\omega t + \varphi(t) + \theta(t))} \qquad (4)$$

Part of the activities is financed by the European Commission within the FP7 Marie Curie IAPP project Par4CR (Grant Agreement Number 230688)

Two highly non-linear amplifiers (PA) for high power efficiency increase the signal power of these two constant envelope out-phased signals (3) and (4). For their combination at the output of the two PAs usually a non-isolating power combiner is used in order to sustain the full benefits of the power efficiency realized by the two non-linear PAs and offer the complex modulated RF signal in amplified form to the antenna (5) with G as the gain.

$$S_{out}(t) = G\, S_{in}(t) \qquad (5)$$

The essential functions in the LINC transmitter architecture are as follows:

1. Signal separator
2. Constant envelope power amplifiers
3. Power combiner
4. Antenna

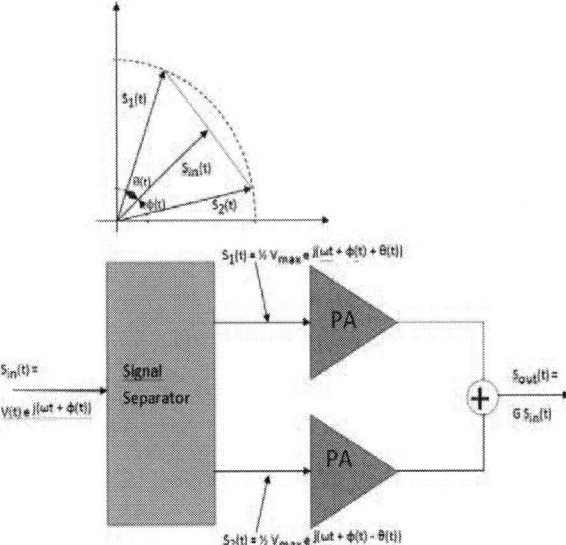

Figure 1. LINC architecture.

All these functions have particular design aspects to consider:

- The signal separator may be realized in an analogue manner and in a digital one. Each one has its own advantages and disadvantages [1], [2]. The digital implementation offers to adjust the out-phased, constant envelope output signals rather easily according to measured differences between the actually transmitted signal and the intended one (pre-distortion [3]). This topic is here assumed to be sufficiently covered in the literature.

- The finite output impedances of the PAs affect the output power, which affects the power efficiency and the non-linearity of the complete architecture [4]. A simple model shown later on allows direct insight in how this works and suggests a rather straightforward solution in the form of a PA that in itself behaves more or less as a voltage source.

- The power combiner is usually based on quarter wave length (¼ λ) transmission lines allowing for practical implementations in which the antenna is connected to a common electrode ("ground"). Due to the non-isolating character of such a power combiner and the phase component with opposite polarities in the PA output signals each PA has a reactive component in its load, which can be compensated for in order to extend the power efficiency over a relatively wide band of amplitude values [5], [6] (i.e. "Chireix" power combiner). Below 5 - 10 GHz these ¼ λ transmission lines become uncomfortably large in size given the limited physical room within mobile equipment and especially when considering integration on a chip [7].

- In relation to the previous item it is to be noted that the transmitting antenna must be physically large as well because for efficient radiation it should have dimensions that are in the order of the RF carrier wavelength. A good compromise is ¼ λ, which however at low frequencies stays a formidable size of in mm if not cm range. In that sense the antenna forms a similar "problem" as the ¼ λ transmission lines in the power combiner. The unavoidably bulky antenna invites to seriously think about merging all bulky components.

II. DISCUSSION

It is important to get a full picture of the above mentioned topics, i.e. on the preferable type of PA, constraints due to the power combiner on volume and broadband behavior and what role the antenna may play solving some of these issues.

A. Constant envelope PA

Due to the fact that the two PAs influence each other when using the Chireix combiner, it is important to select the type of PA for the LINC architecture that copes in the best manner with this. This is often done by means of simulations with which, among others, the best possible power efficiency is investigated for different common types of PAs [8], such as:

- Current-Mode Class D
- Voltage-Mode Class D
- Class F
- Class F^{-1}

For looking into alternative types of PAs it is more convenient to use an analytical approach because simulations only cover whatever type of PA has been selected for the

verification, whereas an analytical approach allows insight in what mechanism plays an important role.

Let's therefore follow the analysis presented in [4] using the simplified model for the LINC architecture as shown in Figure 2. V_1 and V_2 can have arbitrary phase with respect to each other and have non-zero source impedances Z_{g1} and Z_{g2} respectively. Superposition allows analyzing first what happens if one of the two voltage sources is put to zero.

The ¼ λ transmission line acts as impedance transformer relating the characteristic impedance $Z_0{}^2$ to the product of the terminating impedance on one end and what appears at the other end. As soon as the PAs introduce non-zero output impedance this is also reflected at the output of the transmission lines, as shown in (6).

$$Z_0{}^2 = Z_{e1} Z_{g1} \text{ and } Z_0{}^2 = Z_{e2} Z_{g2} \tag{6}$$

Or alternatively:

$$Z_{e1} = \frac{Z_0{}^2}{Z_{g1}} \text{ and } Z_{e2} = \frac{Z_0{}^2}{Z_{g2}} \tag{7}$$

This implies also that at point B Z_2 must be the converted parallel combination of the load Z_L and Z_{e1} as shown in (8).

$$Z_2 = Z_0{}^2 / (\frac{Z_0{}^2}{Z_{g1}} // Z_L) = Z_0{}^2/Z_L + Z_{g1} \tag{8}$$

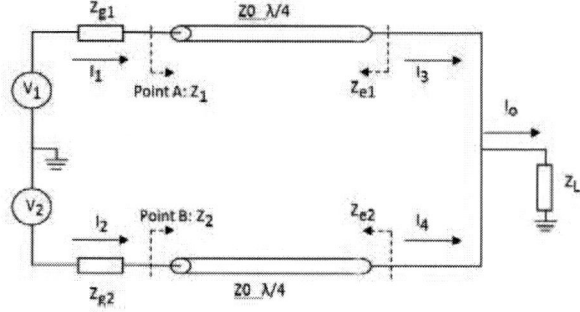

Figure 2. Simplification of the LINC architecture [4].

With $V_1 = 0$, I_2 can be found with V_2 divided by the total load consisting of the series connection of Z_{g2} and the just found Z_2, as follows:

$$I_2(V_1 = 0, V_2) = \frac{V_2}{\left(\frac{Z_0{}^2}{Z_L} + Z_{g2} + Z_{g1}\right)} \tag{9}$$

The same is found as well for I_1 when $V_2 = 0$:

$$I_1(V_2 = 0, V_1) = \frac{V_1}{\left(\frac{Z_0^2}{Z_L} + Z_{g2} + Z_{g1}\right)}$$

(10)

This leads to the conclusion that if both voltage sources V_1 and V_2 generate a current to flow, this current must be the sum of the individual currents due the each individual voltage source and that the current will be shared by both PAs (Figure 3). This implies that the output impedance of the PAs affects the magnitude and the phase of the output voltage with impact on the actual output power and possibly also on Error Vector Magnitude, Adjacent Channel Leakage ratio and AM/AM input/output relation.

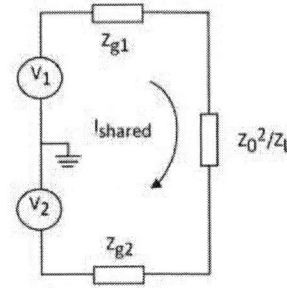

Figure 3. LINC model found by analyzing the voltage and current relations.

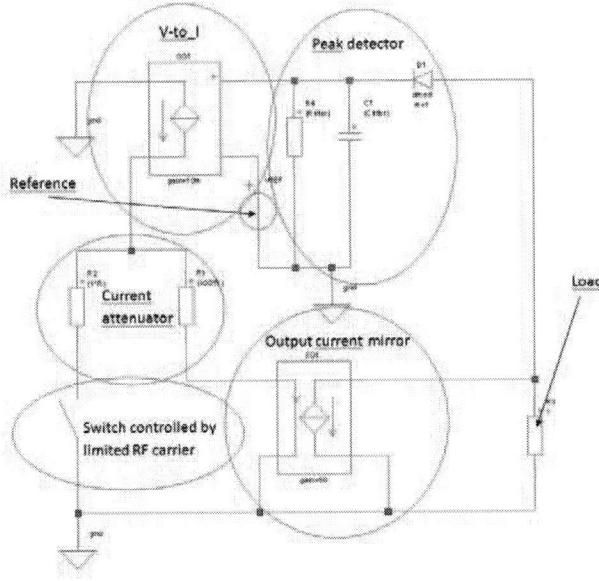

Figure 4. Broadband PA with envelope detection.

The largest impact occurs with modulation because the source impedances will differently respond to the actual modulation frequency. This takes place at a much lower rate than the RF carrier though. A peak detection feedback loop can make the PA behave very much like a voltage source and allows at the same time broadband behavior. On top of that such a loop makes the gain adjustable and reduces the gain mismatch between the two PAs.

Note that in the commonly used non-linear PAs low output impedance comes mainly from the required filtering at the output of the internally switched current source or current switch. This makes these PAs only usable at the particular RF carrier frequency for which they have been designed.

B. Bulky power combiner and antenna

Below 5 GHz RF carrier frequency a Chireix power combiner would consume uncomfortably large silicon real estate so that it can only be considered to be realized off-chip on a printed-circuit-board (PCB). It should be noted though that some length reduction of ¼ λ transmission lines is possible with the so-called slow wave technique, but not more than 25 % [9]. Whereas the bulky antenna is unavoidable given its required capability to radiate with sufficient overall efficiency, one may consider using on-chip transformers or coupled inductors instead of ¼ λ transmission lines in order to reduce the size of the power combiner [10], [11]. Although feasible, the losses are usually higher due to constraints on the number of metal levels and thicknesses in combination with the required coupling. Moreover, the RF carrier frequency would thus be fixed on-chip, which does not suit very well a broadband application. When considering the RF frequency defining filter components and the power combiner, the transmitter antenna directly draws the attention as well because of its required resonance at RF frequency and its physical size related to the RF frequency wave-length.

C. Power combiner incorporated in the antenna(s)

It would be much more convenient if the bulky and RF frequency defining power combiner could be merged with the antenna, which has to be bulky anyway in order to be able to radiate efficiently. This still leaves the problem of how to handle broadband applications unsolved, but the frequency selectivity is then moved to the location where it should be solved anyway.

A merged power combiner antenna using one or multiple antennas connected to single-ended or balanced PA outputs may give quite some advantages:

- Less losses by simply reducing what's between the PA outputs and the actually radiating antenna(s)
- Less matching issues because matching is reduced to what's needed to transfer the PA output powers to the antenna(s)
- Less interconnects with less parasitic components between PA outputs and the antenna(s)

- More freedom with antenna impedance levels offering a chance for a better compromise between supply voltage and power efficiency than with fixed 50 Ω
- Frequency band definition at the location where it should be anyway, i.e. with the antenna(s) and related interconnect, so that the PAs themselves do not necessarily have to contain any frequency selectivity, which is perfect for integration on chip
- Smaller overall physical dimensions by merging the active part in the form of an IC (flip-chip) with the unavoidable physically large antennas on the same PCB, which may also support "booster" PAs if higher output radiation levels are needed
- Improved isolation between PA outputs as claimed in different articles about integrated antennas

In [12] a solution is presented. Two types of antennas are of particular interest, as these were specifically developed for LINC architecture (Figure 5). The claims in [12] are:

- Improvement of efficiency by removal of the conventional power combiner
- More compact in size
- Higher levels of circuit-antenna integration, thus better in reliability and performance

The proposed antennas have a dual-feed microstrip patch antenna with a resistor r-loaded microstrip line connecting the two feeding lines to the antenna replacing the power combiner and the antenna at the same time.

Both antennas are four-port structures with two ports P_1 and P_2 as input ports feeding the antenna, a third port P_3 as "bleeder" and a fourth port as invisible radiation output port. A further improvement in terms of efficiency should be possible by making the bleeder function to return this power to the battery instead of dissipating it only.

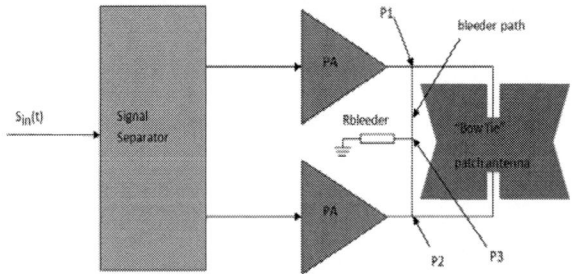

Figure 5. Integrated antenna merged with Wilkinson power combiner [12].

The design is based on "even-odd" mode analysis in which the odd-mode excitation is used to effectively radiate and even-mode excitation is eliminated and dissipated in the bleeder resistor.

This is realized by making the patch antenna a half-wavelength resonator so that it complies with odd-mode excitation.

At the same time the signal path to the bleeder resistor has a length of 3/2 λ at the design frequency (Figure 6) so that it presents an open at the feeding points during odd-mode excitation.

During even-mode excitation the patch antenna forms an open as both feeding points move in terms of driving voltages in the same way. Changing the shape of the antenna or a multi-layered implementation allow altering certain features of the antenna, such as radiating in-phase signals instead of out-phase signals, suppressing harmonics and offer broad-band features [13]. In [14] it is shown that a patch antenna can be made switchable for different frequency band by adjusting the inductively coupled feeder to the antenna using pin-diodes.

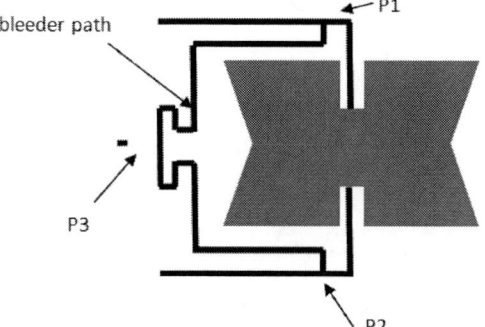

Figure 6. Bow Tie patch antenna [12] with bleeder path.

For broadband and multiband applications the fractal antenna offers here excellent alternatives because of their compact design compared to their conventional counterparts and their robustness due to using simple repetitive, scaled geometries. An example is shown in Figure 7.

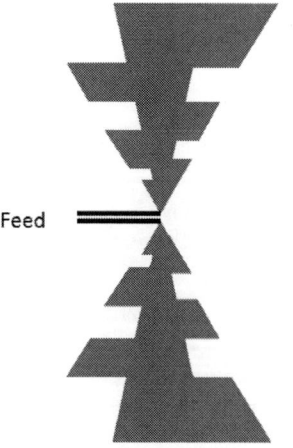

Figure 7. Example of the fractal antenna concept.

It is the self-similarity of repeated, scaled geometries and the origin symmetry that makes fractal antennas broadband, whereas the complementary aspect of the layout smoothes out any impedance variation over frequency [15].

The latter is clear from Babinet's principle saying that the product of the complex input impedances related to antenna structures based on an opening in a metal screen and its complement in metal must be equal to the square of the intrinsic impedance of the medium, in which the antennas are immersed, divided by 4 [16].

D. Power combiner in the ether

One step further is to move the combination of out-phased constant envelope PA output signals in the ether [17]. Two or more antennas connected individually to the PAs in the LINC architecture radiate the two out-phased constant envelope signals, which combine in the ether. The antennas in close vicinity of each other will pick-up radiated energy from each other so that in this sense again a non-isolated power combiner is realized. How antennas interact can be taken into account in the form of an antenna model with lumped components [18] so that this can be accounted for during the design of the LINC architecture. Furthermore, the impact of the distance between the antennas on the near and distant field needs to be investigated.

III. CONCLUSION

From the previous discussion it may be clear that only recently the potential advantages are recognized of incorporating the antenna design more directly into the design of the LINX architecture. This observation should be stated more generally, as follows:

Bringing antenna design into the PA design flow (or vice versa) allows for much better trade-offs towards optimum broadband transmitters instead of connecting an in itself optimized PA to an again in itself optimized antenna by means of a transmission line or an impedance match.

Merging the power combiner into the antenna or beyond into the ether requires insight in antenna design as to optimize the combination of antenna and PAs.

As already mentioned above, antennas may be conveniently modeled with lumped components not only reflecting how the RF power is fed into the radiating part of the antenna but also accounting for its radiation aspects and interaction with other antennas in close vicinity so that including antenna design in a PA design flow should be feasible.

ACKNOWLEDGMENT

Discussions with O. Venard, C. Berland and R. Montesinos from ESIEE are very much appreciated. Input from many colleagues and in particular from F. Sessink on the modeling of

the PA with amplitude peak detection are greatly acknowledged.

REFERENCES

[1] B. Shi and L. Sundström, "A 200-MHz IF BiCMOS Signall Component Separator for Linear LINC Transmitters", IEEE Journal of Solid-State Circuits, Vol. 35, No. 7, July 2000.

[2] W. Gerhard and R. H. Knoechel, "LINC Digital Component Separator for Single and Multicarrier W-CDMA Signals, IEEE Transactions on Microwave Theory and Techniques, Vol. 53, No. 1, January 2005.

[3] Mazen Abi-Hussein, Vivek Ashok Bohara, Olivier Venard and Corinne Berland, "Two-Dimensional Memory Selective Polynomial Model for Digital Predistortion", submitted to NEWCAS2012.

[4] J. Yao and S. I. Long, "Power Amplifier Selection for LINC Applications", IEEE Transactions on Circuits and Systems-II: Express Briefs, Vol. 53, No. 8, August 2006.

[5] Jijun Bi, "Chireix's/LINC Power Amplifier for Base Station Applications Using GaN Devices with Load Compensation", Thesis Delft University of Technology, September 2008.

[6] I. Hakala, D. K. Choi, L. Gharavi, N. Kajakine, J. Koskela and R. Kaunisto, "A 2.14-GHz Chireix Outphasing Transmitter", IEEE Transactions on Microwave Theory and Techniques, Vol. 53, No. 6 June 2005.

[7] A. Pham and C. G. Sodini, "A 5.8GHz, 47% Efficiency, Linear Outphase Power Amplifier with Fully Integrated Power Combiner", Radio Frequency Integrated Circuits (RFIC) Symposium, 2006.

[8] R. Montesinos, C. Berland, O. Venard et P. Descamps , "Analyse des Amplificateurs de Puissance RF dans les Systèmes LINC", 17èmes Journées Nationales Microondes, Brest, 18-19-20 Mai 2011.

[9] D. Huang, W. Hant, N.-Y. Wang, T. W. Ku, Q. Gu, R. Wong and M.-C. F. Chang, "A 60GHz CMOS VCO Using On-Chip Resonator with Embedded Artificial Dielectric for Size, Loss and Noise Reduction", ISSCC 2006.

[10] A. D. Pye and M. M. Hella, "Analysis and Optimization of Transformer-Based Series Power Combining for Reconfigurable Power Amplifiers", IEEE Transactions on Circuits and Systems-I: Regular Papers, May 2010.

[11] M. Tarsia, J. Khoury and V. Boccuzzi , "A low Stress 20 dBm Power Amplifier for LINC Transmission with 50 % PAE in 0.25 µm CMOS", ESSCIRC 2000.

[12] S. Gao and P. Gardner , "Integrated Antenna/Power Combiner for LINC Radio Transmitters", IEEE Transactions on Microwave Theory and Techniques, Vol. 53, No. 3, March 2005.

[13] S. C. Gao and P. Gardner , "Compact Harmonics-Suppressed Integrated Antenna for LINC Transmitters", IEEE 2002.

[14] G. Mansour, P. S. Hall, P. Gardner and M. K. A. Rahim, "Switchable Multi-Band Coplanar Antenna", Loughborough Antennas & Propagation Conference, 2011.

[15] R. G. Hohlfeld and N. Cohen, "Self-similarity and the Geometric Requirements for Frequenc Independence in Antennae", Fractals, Vol. 7, No. 1, 1999.

[16] C. A. Balanis, "Antenna Theory: analysis and design", 2nd edition, John Wiley & Sons, INC.ISBN 0-471-59268-4.

[17] F. Benahmed Daho, G. Neveux, M. Mouhamadou, P. Vaudon, C. Decroze and D. Carsenat, "An Operational Modified-LINC Demonstrator for Wireless Communication", Proceedings of the 5th European Conference on Antennas and Propagation (EUCAP), 2010.

[18] A. Hafiane, H. Aïssat and O. Picon, "Simple electrical model to calculate patch array antenna S-parameters", Electronics Letters, Vol. 39, No. 14, 2003.

xTCA for Instrumentation

MIXDES 2012, 19th International Conference *"Mixed Design of Integrated Circuits and Systems"*, May 24-26, 2012, Warsaw, Poland

Development of uTCA Hardware for BAM system at FLASH and XFEL

Samer Bou Habib, Dominik Sikora
Insitute of Electronic Systems
Warsaw University of Technology
Warsaw, Poland

Jaroslaw Szewinski, Stefan Korolczuk
Department of Detectors and Nuclear Electronics
National Center for Nuclear Research
Otwock-Swierk, Poland

Abstract—This paper describes the design of a uTCA modular card system suited for conversion, sampling and processing of optical pulses. The system consists of a uTCA carrier card along with a double width FPGA Mezzanine Card (FMC) with a changeable optical frontend. The cards were designed for the needs of the BAM system of the FLASH and XFEL accelerators at the DESY facility in Hamburg. The carrier board contains a very powerful FPGA, all required uTCA circuits along with digital interfaces. The FMC card mainly contains four 16-bit fast Analog-to-digital converters (up to 250 MSPS), ADC clock generation and distribution modules, two SFP connectors and a specialized dual RS-485 connection. This paper describes such issues as system organization into universal digital circuits and specialized analog and clock circuits to allow high speed real-time analysis of the properties of thehigh-bandwidth optical signals of the BAM system and better control of the accelerator beam.

Index Terms—BAM, Fast ADC, uTCA, FMC, FLASH, XFEL.

I. INTRODUCTION

The modern superconducting linear accelerator facilities such as the FLASH and XFEL require a very accurate RF field amplitude and phase stability[1]. The field stabilization is assured by the LLRF system and is based on the precision of the field measurement at the frequency of 1.3 GHz [2]. Yet the high requirements are set assuming that the more stable the RF field then the same applies to the accelerator beam. This is not always true, as shown in Figure 1, the beam fluctuations increase with the higher loop gain values due to the amplification of noises. Beam feedback system improves the LLRF control by providing information about the actual beam parameters in real time. With this information, the feedback loop can be closed to maximize the beam stability in addition to the RF field.

To estimate the energy of the electron bunch, the signal from the BAM detector is used to modulate the laser pulses by electro-optical modulator (EOM). These laser pulses are generated with the frequency of 216 MHz. During regular operation in FLASH, bunches are generated with frequency of 1 MHz, which means that every 216-th laser pulse will be modulated, and these modulated pulsed are in the center of out interest (Figure 2).

The height of the modulated laser pulse is proportional to the arrival time. The arrival time is used here as an estimate of the electron bunch energy. It is possible, because depending

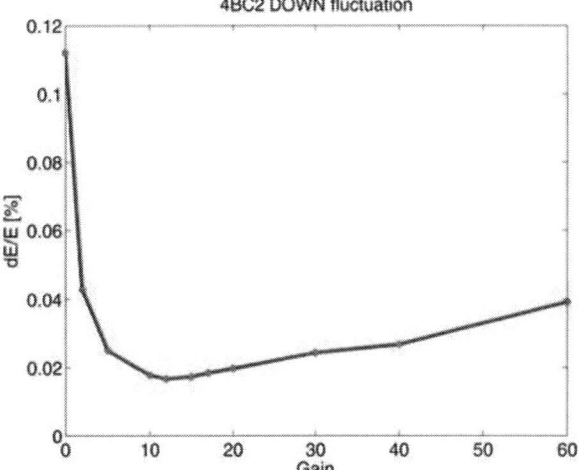

Figure 1. Beam Stability measurement at FLASH

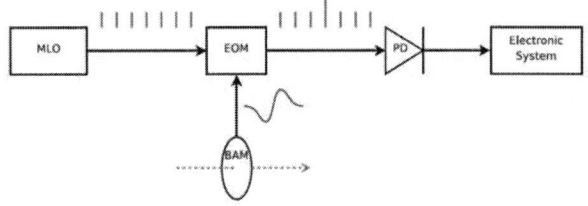

Figure 2. The general concept of the laser pulses modulation

on the energy, electrons travel on different-length ways in the bunch compressor - electrons with higher energy travels using the shorter path, and will arrive earlier, and the electrons with the lower energy, will travel over the longer way, which will result in later arrival. More information about the BAM operation can be found in [3] .

To regulate and minimize the beam fluctuations, proper correction of the RF field amplitude has to be estimated, and applied to the LLRF system. The absolute height of the modulated pulse is not accurate enough, because of the drifts of the signal. Better results are obtained from the relative height of the pulse. To have the reasonable correction value, relative height of the modulated pulse must me compared with the unmodulated one. The the method of the amplitude correction estimation is show in Figure 3 and equation 1.

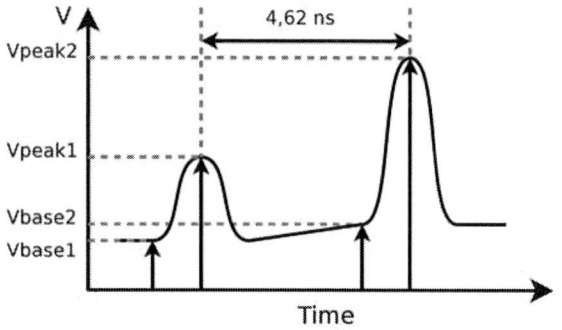

Figure 3.   Laser pulses sampling for relative height estimation

$$A_{corr} = \frac{V_{peak2} - V_{base2}}{V_{peak1} - V_{base1}} \qquad (1)$$

The VME based system was able to sample laser pulses with maximum frequency of 125MHz, so the only available scheme for sampling was sampling with 108 MHz which means every second pulse. The disadvantage of this method was that incorrect synchronization caused sampling wrong samples, and the modulated samples were not seen. New design based on uTCA with the usage of modern fast FPGAs and ADCs opens the door for introducing much improvement to the system, solving the problem with higher sampling rates without loosing any of the high precision, i.e. by increasing the sampling frequency to the 216 MHz, and then every laser pulse will be sampled.

II. SYSTEM CONCEPTION

The main objective of the project was the design of a two channel optical-signal receiver allowing to sample pulse amplitude. The goal of the system is to obtain from the received optical pulses the information needed for the BAM system. The designed receiver consists of two main modules: a double width uTCA FPGA carrier board and a double width FMC ADC board. An option was foreseen that the receiver can be even used as a stand-alone system with no need for the uTCA crate. The main system conception is illustrated on Figure 4.

Three optical pulsed signals with a repetition rate of 216 MHz and a bandwidth of ~800MHz are fed to the board frontend where they are converted to RF electrical signals. The RF pulses produced must be impedance matched and amplified to reach maximum ADC performance. The front end should also insure very low amplitude and phase noise conversion. One of the resulting signals is used as a reference clock for synchronization of the ADCs. The other two are split to feed four ADCs, i.e. 2 ADCs per channel, to sample the peaks

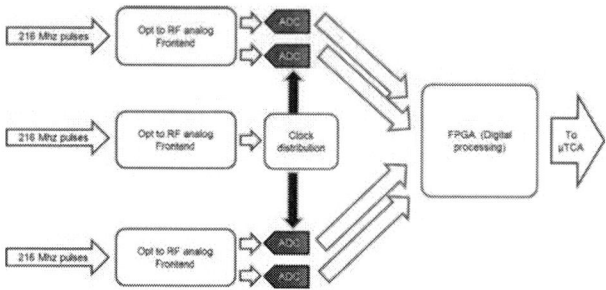

Figure 4.   System general conception

and base lines. The data is then sent to an FPGA for signal processing and the calculated information is then passed on to the uTCA synchronization system.

III. SYSTEM DESIGN AND ARCHITECTURE

In order to achieve proper functioning and desired performance the FMC standard was used and very careful and precise designing techniques were applied to the delicate optical, analog, clocking and even digital circuits of the project.

As stated earlier, the system mainly consists of two boards:

- the uTCA carrier board – containing mainly the digital parts of the system needed for proper acquisition of the data and calculation of the desired correction,
- the FMC ADC board – containing the optical-to-RF conversion, clock distribution, ADCs and specialized communication and control circuits for the BAM system.

The System architecture and layout are shown on Figure 5.

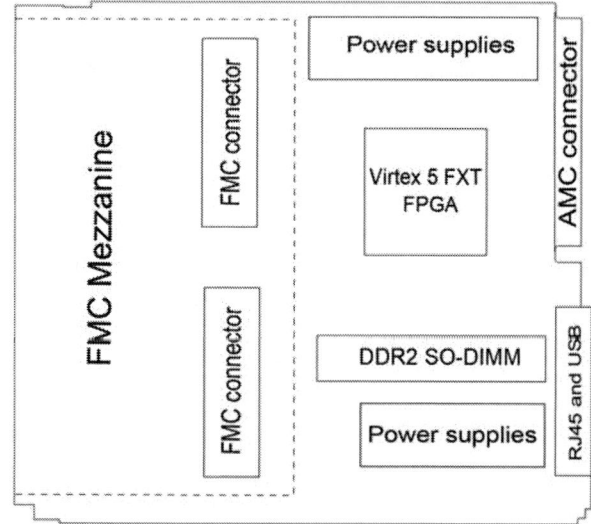

Figure 5.   System Architecture and layout

IV. UTCA CARRIER BOARD

The main aim of building this board was to provide a platform, which would be able to host 4 (or in minimal approach 2) high-speed ADCs up to 1 GSPS, with reserved 16 lines for each device. Additionally, the design fulfills the following requirements:

- Support for a double width FMC;
- AMC.4 compatibility with IPMI support [4];
- High performance FPGA on board;
- Possible operation as a standalone device (full-featured uTCA crate costs 10k Euro).

A. Block Diagram

The block diagram of the carrier board is shown on Figure 6.

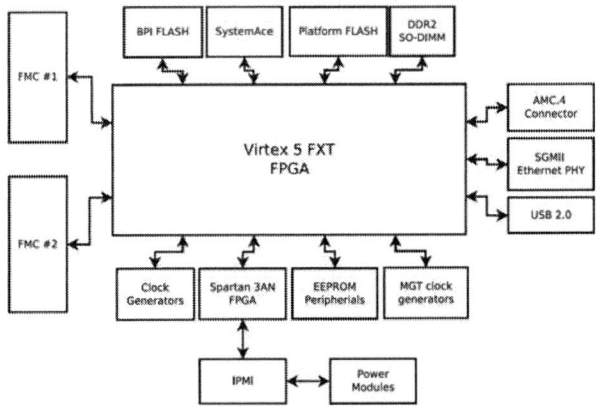

Figure 6.   Block diagram of carrier board

B. Main Features

The main features of the carrier board are:

- Two single or one double size FMC mezzanine support;
- Over 70 differential pairs or 140 single ended signals available for each FMC slot;
- Four gigabit serial channels per FMC slot;
- Virtex5 FXT70 FPGA with embedded PowerPC440 processor;
- DDR2 SO-DIMM connector;
- IPMI unit;
- Advanced configuration modes with fail-safe configuration memory;
- CF card as a configuration memory, and non-volatile storage for embedded systems;
- Serial port and JTAG over USB;
- Gigabit Ethernet PHY and RJ45 socket for stand-alone operation;
- Low noise power supply modules;
- Watchdog and configuration supervisor implemented in separate small FPGA;
- AMC.4 communication lanes:
  - Gigabit Ethernet;
  - PCI Express;
  - Two point-2-point links;
  - Two M-LVDS links;

- CLKA and CLKB routed to FPGA;
- JTAG.

The assembled board is shown on Figure 7.

V. FMC ADC BOARD

The ADC board is a double width FPGA Mezzanine Card (FMC)[5] which is specialized for the FLASH and XFEL BAM system. It consists of the optical-to-RF conversion circuits, clock generation circuits, ADCs, a specialized dual RS-485 connection, and communication ports.

A. Block Diagram

The block diagram of the FMC ADC board is shown on Figure 8.

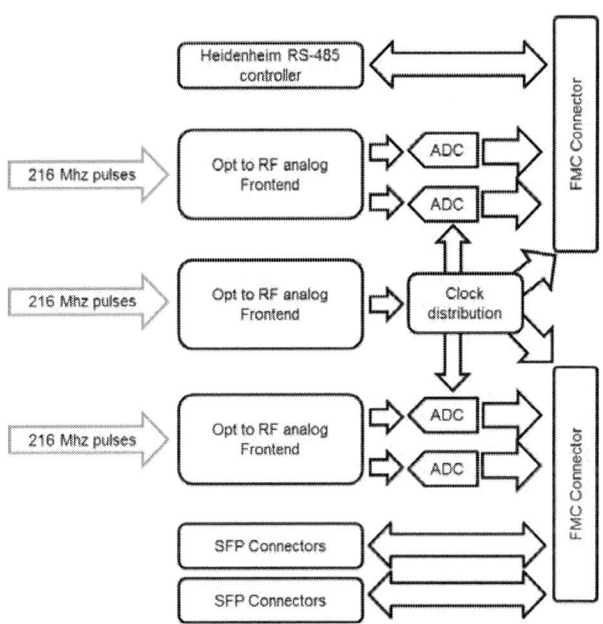

Figure 8.   Block diagram of FMC card

B. Main Features

The circuits of the FMC board can be divided into three main groups:

1) The pulse sampling circuits: The pulse sampling circuits consist of an optical-to-RF conversion scheme using a photodiode and a transimpedance amplifier. This part is designed as a piggyback allowing the switching between different photodiodes and transimpedance amplifiers. This procedure increases the modularity of the system and allows choosing the best parts according to bandwidth, lowest noise, distortion and matching.[1] The obtained RF signals are then split into two channels and afterwards amplified, differentiated and matched to the input of the ADCs using a very low noise and distortion

---

[1]The piggyback has an additional feature to use the designed hardware for purposes different than the BAM system where high precision fast sampling of analog signals is needed.

149

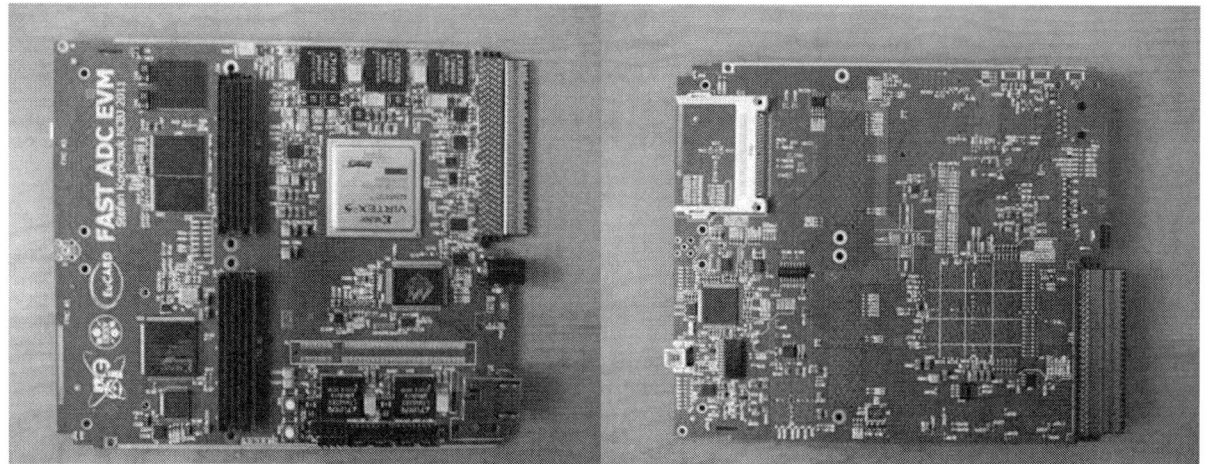

Figure 7. Assembled uTCA carrier board

amplifier with very low phase and amplitude imbalance to maximize ADC performance. The differential signals are then sampled with 16-bit fast analog-to-digital converters at a sampling rate of 216 MSPS (same as incoming pulses). Each channel consists of two ADCs for sampling the peak and baseline of the incoming pulse.

The same optical-to-RF conversion is used for the reference pulses that are used to synchronize and clock the ADCs. The two split channels are fed into a low jitter PLL chip that uses one input for clock distribution and the other as reference for the built-in PLL. The choice of operation is set by the FPGA on the carrier board. There are two clocks from the chip connected to each ADC. One is a phase-tunable low noise LVDS clock and the other an even lower noise LVPECL clock which additionally has the possibility to install fixed delays on the lines. The generated clock signals are also sent to the carrier board to insure synchronized data acquisition and processing. The programmable and fixed delays are inevitable to set the sampling points to exact time values, i.e. to sample the peak and baseline of the incoming pulses.

2) The Heidenheim controller circuits: The controller is a specialized communication port used in the BAM system. It consists of two Half-duplex RS-485 transceivers. One is used for sending the clock to the Heidenheim controller, while the other for sending and receiving data form it.

3) SFP communication ports: The board has two SFP connector added to allow fast two-way optical communication with the rest of the uTCA systems.

C. PCB Design

The design of the FMC board had to be strictly prepared according to the rules of signal integrity and RF analog design to insure low noise, low jitter, and low distortion high precision ADC, clocking, and digital circuits [6]. The designed FMC board is shown in 9.

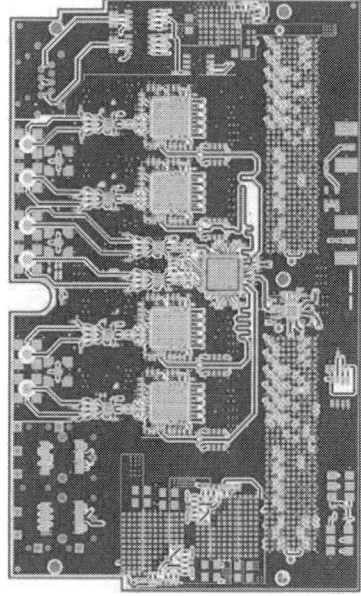

Figure 9. FMC PCB design

VI. MEASUREMENTS

To insure the feasibility of the designed modules, some tests and measurements were prepared and made in the specialized laboratory.

A. Front-end Measurements

The front-end consisting of the photodiode, optical-to RF conversion and power splitting and amplification was measured using laser pulses identical to the ones in the FLASH accelerator (and XFEL in the future). The pulses were analyzed with a very fast oscilloscope and a signal source analyzer to insure proper functioning and low amplitude and phase noise.

150

Example differential pulses from the measurement are shown in Figure 10.

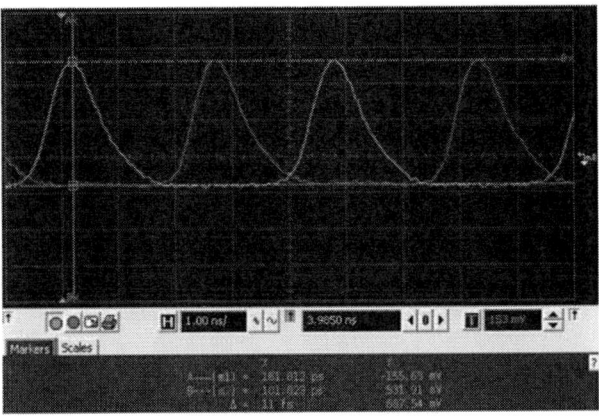

Figure 10.   Front-end pulses after conversion and amplification

The measured phase noise of the pulses after conversion was around 170 fs integrated over a bandwidth from 10Hz to 1MHz.

B. ADC Measurements

The designed front-end was connected to the used ADC evaluation board to measure precision and feasibility of the design. Two signals were used: one as an input and the other for clocking the ADC through a chip for tuning the phase. 16K samples were made at every time point and the averages were use to prepare the plot shown on Figure 11.

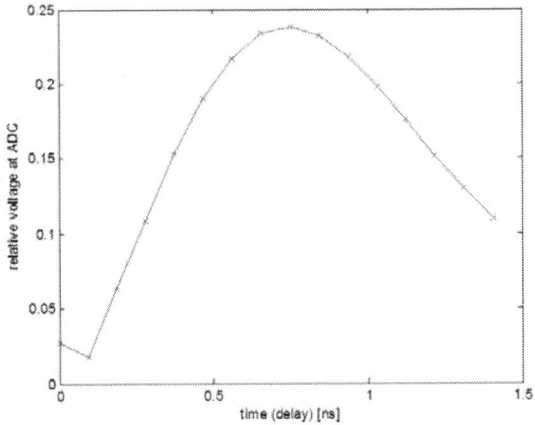

Figure 11.   Sampled converted optical pulse

The measurements showed a sampling instability of below 0.19%. To insure even better precision in the designed system a different differential amplifier was used with much better characteristics in terms of noise and distortion. The clocking circuits on the FMC board are also prepared to be of lower jitter. Moreover, the clock and analog circuits are better matched at the given frequencies to obtain the best possible performance.

VII. Summary

A set of boards was prepared for the BAM system for the FLASH and XFEL accelerators. It consists of a universal uTCA carrier board containing a powerful FPGA and digital circuits along with a specialized FMC board with high precision fast ADCs that converts and samples the optical signals used in the BAM system. The designed circuits were tested for feasibility with signals identical to the ones in the accelerator. Moreover the designed modules can be used as a stand-alone system to allow fast sampling, processing and analysis with very high precision of various optical and RF signals.

Acknowledgment

The research leading to these results has received funding from the European Commission under the EuCARD FP7 Research Infrastructures grant agreement no.227579 and support of Polish National Science Council Grant 1288/7.PR UE/2010/7

References

[1]   V. Ayvazyan et al. Requirements for rf control of ttf2 fel user facility. PAC03, page 2342, 2003.

[2]   S. Simrock et al. Digital low-level rf controls for future superconducting linear colliders. PAC05, pages 515–519.

[3]   F. Lohl, et. al, Electron bunch timing with femtosecond precision in a superconducting free-electron laser, Physical Review Letters 2010.

[4]   AMC, uRTM and uTCA Shelf for Physics, PICMG Specification MTCA.4, PCI Industrial Computer Manufacturers Group, http://www.picmg.com.

[5]   American National Standard for FPGA Mezzanine Card (FMC) Standard, ANSI/VITA 57.1-2008, VMEbus International Trade Association, http://www.vita.com.

[6]   Samer Bou Habib et all, Design of eight-channel ADC card for GHz signal conversion, Mixdes 2010.

MIXDES 2012, 19th International Conference *"Mixed Design of Integrated Circuits and Systems"*, May 24-26, 2012, Warsaw, Poland

FMC-based Neutron and Gamma Radiation Monitoring Module for xTCA Applications

Tomasz Kozak, Dariusz Makowski, Andrzej Napieralski
Department of Microelectronics and Computer Science
Technical University of Lodz
Lodz, Poland
email: tkozak@dmcs.pl

Abstract—**The machines used in High Energy Physics (HEP) experiments, such as accelerators or tokamaks, are sources of gamma and neutron radiation fields. The radiation has a negative influence on electronics and can lead to the incorrect functioning of complex control and diagnostic system designed for HEP machines. Therefore, in most cases the electronic equipments is installed in radiation-safe areas, but in some cases this rule is omitted to decrease costs of the project. The European X-ray Free Electron Laser (E-XFEL), being under construction at DESY research center, is a good example. The E-XFEL uses single tunnel and part of the electronic system will be installed next to main beam pipe and exposed to radiation. The modern Advanced/Micro Telecommunications Computing Architecture (ATCA/µTCA) standards are foreseen as a base for control and diagnostic system for this new project. These flexible standard provides high reliability, availability and usability for the system which can be decreased by negative influence of parasitic radiation field. The additional shielding will be introduced to protect racks with electronics, but during commissioning and, in case of control systems errors, the assumed radiation levels can be exceeded. Therefore, it is highly recommended to monitor doses absorbed by electronics. Moreover, it could be helpful for estimating system lifetime, scheduling maintenance periods and protecting machine from unexpended failures. The paper describes a Radiation Monitoring Module (RMM) based on FPGA mezzanine card standard capable of monitoring gamma radiation and neutron fluence in real-time.**

Index Terms—**Micro Telecommunications Computing Architecture, FPGA Mezzanine Card, RF Control System, gamma radiation dosimetry, neutron radiation dosimetry, linear accelerator, X-ray Free Electron Laser**

I. INTRODUCTION

The High Energy Physics experiments use sophisticated machines like linear and circular colliders, linear accelerator or tokamaks. The electronic systems needed to control them must fulfil very strict requirements concerning computation power, flexibility, reliability and availability. Therefore, old system architectures such as VME are slowly substituted by modern ones. The Advanced Telecommunications Computing Architecture (ATCA) and Micro Telecommunications Computing Architecture (MTCA) - xTCA family - gain popularity in many leading research centres. The Deutsches Elektronen-Synchrotron (DESY), Joint European Torus (JET), CERN or SLAC consider one of the xTCA standards as a base architecture for their control systems [1].

The HEP machines are sources of gamma and neutron parasitic fields. In most cases control electronic is highly isolated from them to avoid destructive influence by placing it in radiation-safe areas. The European X-ray Free Electron Laser (E-XFEL) is a good example of the exception. In order to decrease the cost of the project, it will use a single tunnel and most of the control system electronics will be placed in the same tunnel as the main beam pipe. Therefore, the control system will be exposed to gamma and neutron fields generated by the machine [2]. It could have a negative influence on electronics reliability and therefore machine availability. The negative influence of radiation on electronics has been proven by many research [3][4]. The electronic equipment will be installed in shielded racks. The FLASH accelerator was used to measure doses of radiation and to design the radiation shielding for electronics [5].

A radiation monitoring system capable of measuring doses absorbed by electronics inside the rack could be helpful during accelerator studies and commissioning. It can also help to schedule the maintenance periods in more effective way which could decrease costs of accelerator operation.

II. E-XFEL AS AN EXAMPLE FOR REQUIREMENTS FOR RADIATION MONITORING SYSTEM

The E-XFEL accelerator will consist of 116 accelerating modules and other special components such as bunch compressors or arrays of magnets. Each single RF station will be driving four accelerating modules [2]. The electronics required for the Low Level RF (LLRF) system responsible for controlling the parameters of the accelerating field and for other diagnostic systems will be placed in the shielded racks distributed along the tunnel. Therefore, a distributed architecture of radiation monitoring system is recommended. The on-line accessibility of measurement data is required, because it is impossible to enter the linac tunnel during normal operation of the machine. The measured data should be collected in an external database for further analysis. The dosimeters for the system should fulfill the requirements presented in Table I [4]. The developed radiation monitoring system will have a distributed architecture and need to be integrated with control system of the E-XFEL which will be based on one of the xTCA architecture. The xTCA family has been designed for telecommunication industry.

TABLE I. REQUIREMENTS FOR GAMMA RADIATION AND NEUTRON FLUENCE DOSIMETERS

Detection ability	Gamma and neutron fluence
Fluence range	$10^6 - 10^{10}$ neutrons·cm^{-2}
The lowest fluence	$10^4 - 10^5$ neutrons·cm^{-2}
Dose range	$10^2 - 10^3$ Gy
The lowest dose	$10^{-3} - 10^{-2}$ Gy
Energy range	up to 20 MeV

Therefore, nowadays an additional effort is made to fit xTCA into specific requirements of HEP experiments. As a result of this work the PICMG 'xTCA for Physics' standard extension has been introduced. To provide bigger flexibility and reusability of the electronics an additional standard was developed. The FGPA Mezzanine Card (FMC) specification (ANSI/VITA 57.1-2008) has been proposed. The standard defines form factor of PCB and pinout of the FMC connector [6]. As a main idea the FMC modules carry the unique electronic components which provide front-end functionality in the system, but not computation power and communication interfaces which are supplied by a Carrier Board (CB). The CB could have different form factors such as standalone boards or Advanced Mezzanine Card (AMC) which fits the FMC standard into xTCA. The FMC carrier module is already a part of the LLRF control system. Therefore, the solution where the radiation dosimeters and readout sub-circuits will be designed as a FMC module has been chosen. It can be easily integrated with control system electronics and readout infrastructure.

III. RADIATION MONITORING FMC MODULE

A. Dosimeters for FMC Radiation Monitoring Module

Various dosimetry methods allows to detect and measure radiation doses. They are based on a wide spectrum of electrical, thermal, luminescent and chemical effects triggered by radiation in different kind of materials. They include gaseous detectors, scintillators, bubble dosimeters and thermoluminescent dosimeters (TLD) [4]. Detectors chosen for the designed module should provide a good selectivity, dynamic ranges and cover dose ranges specified in Tab. 1. Moreover, selected dosimeters need to have small dimensions to decrease occupied space volume and their readout circuit should not utilize high voltages, which are not available in the xTCA standards. The semiconductor-based dosimeters seems to be natural candidates which fulfill two latter requirements. Therefore, the Radiation sensitive Field Effect Transistor (RadFET) was chosen as gamma radiation dosimeter.

The principle of operation of the RadFET dosimeter is based on the electron-hole generation process present in the transistor oxide layer exposed to gamma radiation and, in minority, by neutrons through secondary effects. It leads to charge build-up which influences the electric parameters of the device such as threshold voltage [3]. The shift of threshold voltage is a function of absorbed dose. The process occurs faster in P-type devices, thus RadFETs are P-type transistors. The sensitivity of the device to radiation shows dependence on the oxide layer thickness and can be adjusted due to application requirements. The RadFET operates in two modes – irradiation and reader. In the former one the source and drain of

the transistor are shorted. The voltage on the gate is called the bias voltage (V_{bias}) and has significant influence on the device sensitivity. This feature allows to adjust the sensitivity and dose ranges on-the-fly. For threshold voltage readout the transistor needs to be switched to the reader mode. The typical readout circuit is presented in Figure 1. In this mode the transistor source is connected to a constant current source and the constant current flows through the device. The measured voltage shift allows to calculate the absorbed gamma radiation dose.

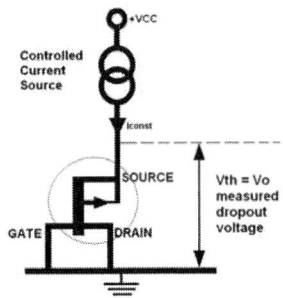

Figure 1. RadFET dosimeter readout circuit.

The main advantages of RadFET are small size of the detector, simple readout circuit and low unit cost. The biggest disadvantage is temperature dependence which can be mitigated by setting current to Minimum-Temperature-Coefficient (MTC) value. Several RadFETs available on the market, produced by different companies, have been analyzed to determine the gamma detector most suitable for the FMC module. The example of RadFETs has been presented in Figure 2. The RFT-300-CC10G1 produced by REM Oxford Ltd. has been finally selected. The device has 0.3 µm oxide. The sensitivity parameters for the REM RadFETs family is presented in Table II [7].

TABLE II. RESPOSIVITY FOR DIFFERENT EXPOSURE MODES IN FUNCTION OF OXIDE THICKNESS

t_{ox} [µm]	Sensitivity [mV/cGy]		
	$V_{bias} = 9$ V	$V_{bias} = 18$ V	$V_{bias} = 0$ V
0.2	0.65	0.85	0.12
0.25	0.95	1.20	0.16
0.3	1.25	1.75	0.20

Figure 2. RadFET dosimeters from REM Ltd. (two from left) and from Tyndall National Institule (two from right).

The gamma detector can be also used to detect neutrons, but special converters are essential for this process which makes the sensor much more complex [8]. This is not suitable for measuring neutrons in the fields with strong gamma background. To avoid this the neutron fluence measurement method will be based on counting the number of generated SEU events in specially selected SRAM memories. The previous researches proved that neutrons can trigger the SEUs in the digital memories. Moreover, the number of generated errors can be recalculated to the value of fluence intensity [9]. The SEU can be easily detected by consecutive memory scanning when the reference pattern stored in the memory is known. The sensitivity of the SRAM depends on many parameters such as manufactured technology, size, supply voltage and vendor. Several memories have been tested to determine the most suitable chip. The research proved that the newer memories are more immune to radiation than the older ones. Moreover, sensitivity can be increased by decreasing the supply voltage below the value specified in the datasheet [4]. Advantages of SRAM memories include accessibility, digital interface and low cost of the unit. Unfortunately, each chip should be calibrated with a reference neutron source which is the main disadvantage of chosen method. For needs of the project the 512 kB Samsung K6T4008C1B SRAM has been chosen.

The selected dosimeters can be easily integrated with digital readout subsystems, represent low unit cost and high selectivity, which is an important factor in radiation mixed environment characteristic for linear accelerators.

B. Radiation Monitoring Module hardware

The designed FMC Radiation Monitoring Module has a single width form factor with Low Pin Count (LPC) connector. The chosen factor limits available area on the PCB, but makes the module more compact and compatible with all solutions which support the FMC standard.

The module is divided into two sections – analog and digital. The first one holds two RadFETs used as gamma radiation dosimeters. Each detector has a separate readout circuit. Individual adjustable current sources feed each of the transistors with a constant current of 490 µA. It is the MTC value for the selected type of RadFETs [7]. The current sources are supplied from +11.5 V power supply which sets the upper threshold voltage (V_{th}) level which can be achieved on the irradiated RadFETs. The maximum V_{th} value together with sensitivity value determine the maximum doses which

can be detected by the designed detector. Both of the detectors has a adjustable bias voltage (V_{bias}) used in irradiation mode. The +10 V or 0 V can be chosen and for one of the FETs this value can be changed on-the-fly. It provides a possibility to configure the detector with different sensitivity levels. Utilization of different V_{bias} values allow to receive coarse and fine type of measurement. The device with positive bias will provide more precise data due to larger sensitivity, but lifetime of the detector will be shorter than with $V_{bias} = 0$ V. The voltage signal from RadFET in reader mode (see Fig. 1) is passed, through an operation amplifier configured as a voltage follower, into an analog to digital converter (ADC) which close the analogue path of the signal. Selected ADC has a 16-bit resolution and provides data from two independent channels by digital serial interface in TTL standard.

The digital section includes eight Samsung SRAM K6T4008C1B memories, with a total capacity of 4 MB. The memory is used as neutron fluence detector. The normal operation supply voltage is decreased from +5 V to increase sensitivity of detector. The values from +1.8 V to +3.3 V can be applied depending on the used batch of memory. Additional integrated circuits mounted on the FMC are serial PROM memory required by FMC standard and simple USART-USB converted which provides a flexible communication channel between control logic and the external world. Moreover a digital temperature sensor has been installed. The electronics on the FMC use TTL logic standard for ADC and thermometer chips and voltages from 1.8 V to 3.3 V (depends on configuration) for SRAM digital interface. The set of buffers has been introduced to provide voltage level conversion between FMC logic standards and logic standard used on the carrier board. After conversion all signals are passed directly to the FMC connector.

The designed FMC module, which PCB layout is presented in Figure 4, does not provide any computation power or supply voltages, therefore it needs to be connected to an additional board called carrier board. The FMC standard assumes that the computation power, communication interfaces and supply power will be provided by FPGA-based CB.

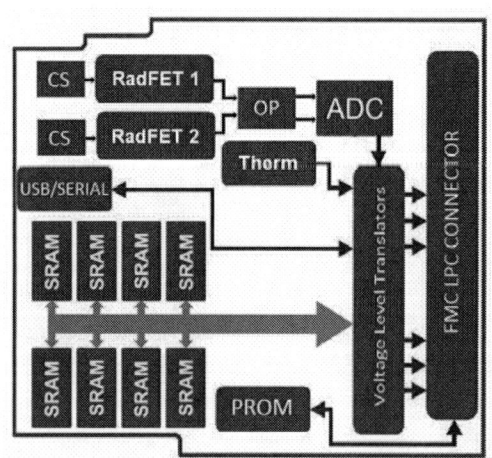

Figure 3. FMC Radiation Monitoring Module block diagram (CS - current source, OP - operational amplifier, Therm – digital temperature sensor).

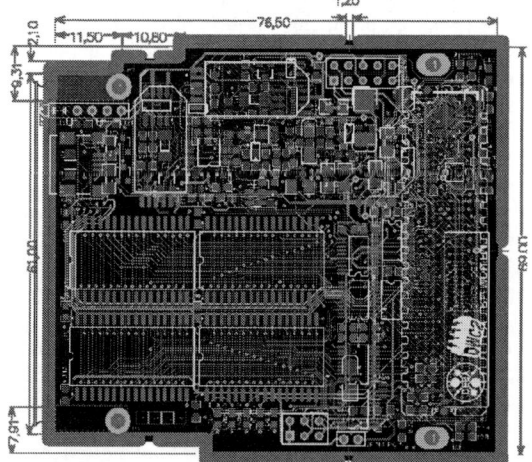

Figure 4. FMC Radiation Monitoring Module PCB layout.

154

Nevertheless, each board equipped with an FMC connector and compliant with FMC specification as far as the signal layout on the FMC connector is concerned, can be used as a carrier. The versatile AMC carrier DAMC2, designed at DESY, will be used as the carrier board for the described FMC Radiation Monitoring Module. The DAMC2 is general purpose carrier proposed for different applications in the E-XFEL Machine Protection Systems [10]. The FMC module is also compliant with some Xilinx evaluation kits such as SP605 or ML605 which provides a suitable development platform for the firmware and software needed for the Radiation Monitoring System.

C. FMC Radiation Monitoring Module firmware structure

The designed module was in fabrication stage during preparation of this paper. Nevertheless, the firmware architecture for FPGA-based CB was designed. The block diagram of VHDL modules implemented in the FPGA is presented in Figure 5.

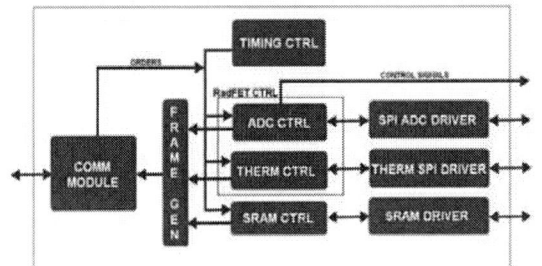

Figure 5. Block diagram of VHDL modules for Radiation Monitoring FMC.

The SPI_ADC_DRIVER and THERM_SPI_DRIVER are dedicated blocks which perform a role of driver for the ADC and thermometer used in the project. The driver will provide a possibility to write configuration register in each device and read measurement data. The third driver needed is SRAM_DRIVER which will be able to write to and read data from K6T4008C1B SRAM memory. The ADC_CTRL and THERM_CTRL, which together can be considered as the RadFET controller, are responsible for providing higher level logic.

They control, initialize and gather data from ADCs and thermometers. The data from ADCs needs to be gathered every several minutes. The TIMING_CTRL module is designed to measure the time periods between consecutive measurements and generate trigger signals for others VHDL modules. The RadFET controller will change the mode of the transistors from irradiation to reader mode and perform the voltage and temperature readouts. Afterwards, the gathered raw data should be transferred to the frame generator module (FRAME_GEN), where communication frames are created according to the custom communication protocol. Each frame will carry information about module ID number, type of frame, raw data and calculated checksum. Then, the frames will be transferred to the communication module (COMM_MODULE) responsible for building frames in communication standard used in the system such as PCIe bus. The SRAM_CTRL block will constantly scan connected SRAM memories in order to find SEUs generated by neutrons. The number of detected SEUs can be accessed via a register. The modular construction of the code allows to make fast changes in the project such as changing communication interface.

IV. CONCLUSIONS AND PLANS

Presented FMC Radiation Monitoring Module provides two types of dosimeters suited for gamma and neutron field measurement in an xTCA-based system which can be exposed to these kinds of radiation. It may concern control systems in High Energy Physics facilities such as linear accelerators, colliders, tokamaks, where control electronics can be exposed to radiation. The prepared module can be easily integrated with system based on xTCA and FMC standards. It can be also used in an independent system with suitable standalone carrier board. It makes it highly flexible and reusable which decreases costs and is the main advantage of the proposed solution. Also, low cost of single unit, simple digital interface, good selectivity and ranges of proposed dosimeters speak in favor of this design. The module requires a calibration routine and has limited lifetime which can be consider as the main drawbacks.

In the near future the module should be assembled, tested and ready for calibration. Tests will be performed at DESY research center.

ACKNOWLEDGMENT

The research leading to these results has received funding from the European Commission under the EuCARD FP7 Research Infrastructures grant agreement no. 227579.

REFERENCES

[1] B. Goncalves et al., "ATCA Advanced Control and Data acquisition systems for fusion experiments", 16th IEEE-NPSS Real Time Conference, 2009
[2] M. Altarelli et al., "The European X-Ray Free-Electron Laser Technical design report", 2006
[3] G. Barbottin, A. Vapaille, "Instabilities in Silicon Devices, New Insulators, devices and Radiation Effects", Elsevier, 1999
[4] D. Makowski, "The impact of Radiation on Electronic Devices with the special consideration of Neutron and Gamma Radiation Monitoring", PhD Thesis 2007.
[5] F. Schmidt-Foehre et al., „A new embedded Radiation Monitor System for Dosimetry at the European XFEL", Proceedings of IPAC2011, 2011
[6] ANSI/VITA 57.1-2008, "American National Standard for FPGA Mezzanine Card (FMC) Standard", July 2008
[7] RFT-300-CC10G1 datasheet, REM Oxford Ltd., September 2009
[8] T.E. Manson., "Neutron detectors for materials research. Technical report", OAK Ridge National Laboratory, 2000.
[9] D. Makowski et al., "The Application of SRAM Chip as a Novel Neutron Detector", NSTI-Nanotech 2005 Proceeding, 2005
[10] P. Vetrov, "Versatile AMC for XFEL Machine Control", ATCA Workshop presentation, RT2010, 2010

MIXDES 2012, 19ᵗʰ International Conference *"Mixed Design of Integrated Circuits and Systems"*, May 24-26, 2012, Warsaw, Poland

Image Acquisition Module for uTCA Systems

Aleksander Mielczarek, Dariusz Makowski,
Grzegorz Jabłoński, Piotr Perek, Andrzej Napieralski
Department of Microelectronics and Computer Science
Technical University of Lodz
Lodz, Poland
[amielczarek, dmakow, gwj, pperek, napier]@dmcs.p.lodz.pl

Abstract—The paper presents a prototype of Image Acquisition Module (IAM) dedicated for plasma monitoring. The complete frame grabber together with camera controller, timing and communication interfaces is fitted in one Advanced Mezzanine Card (AMC) to maintain compatibility with the Micro Telecommunications Computing Architecture (µTCA) standard.

Index Terms—image acquisition, frame grabber, xTCA, µTCA, AMC

I. INTRODUCTION

The Micro-Telecommunication Computing Equipment (µTCA) and Advanced Telecommunication Computing Architecture (ATCA) standards (referred to as xTCA) are gradually gaining popularity in the industrial control systems. The number of compatible COTS Advanced Mezzanine Card (AMC) modules increases every year but there is still an important gap on the market. Currently there are no high-speed video acquisition cards available. The machine vision plays an important role in many industrial processes, hence the effort was taken to develop an AMC module offering possibility to interface high-speed high-resolution cameras.

Most Image Acquisition Systems (IAS) have to be scalable by design. Consider, for instance, the ITER thermonuclear reactor Instrumentation and Control (I&C) system. It is supposed to include about 20 video cameras with frame sampling frequencies of 1 kHz and approximately 400 high-resolution imaging devices providing data with throughput reaching almost 7 Gb/s [1]. Extending system by one camera shall require as small changes to the existing infrastructure as it is only possible.

II. GENERAL SYSTEM ARCHITECTURE

A. High-throughput Image Acquisition System

To fulfill the scalability requirement, the system shall have modular construction. This paper focuses on using the AMC standard as the base for Image Acquisition Module (IAM). The compliance with AMC standard helps to ensure high Reliability, Accessibility and Serviceability (RAS) of the final device. Such a module may interface with virtually any µTCA crate or ATCA carrier blade. The communication with the host system relies on the high-speed serial connectivity. The example AMC module is shown in the Figure 1.

The IAM has to interface one or more video cameras. Considering the multi-gigabit video streams and the connectivity available for AMC module it is justified to acquire data from a single camera at a time. The Camera Link has been selected as the video transmission interface. It has become an industrial standard for the low-distance high-throughput raw video transmission. It is relatively easy to be implemented in hardware, an off-the-shelf chipset is commercially available. The extended link provides throughputs up to 6.8 Gb/s [3] and allows for cable length up to 10 meters.

For a reliable transmission over larger distances the video data has to be captured and encapsulated in frames of higher-level data transmission protocol like Gigabit Ethernet or PCI Express. This defines the main task of the described IAM.

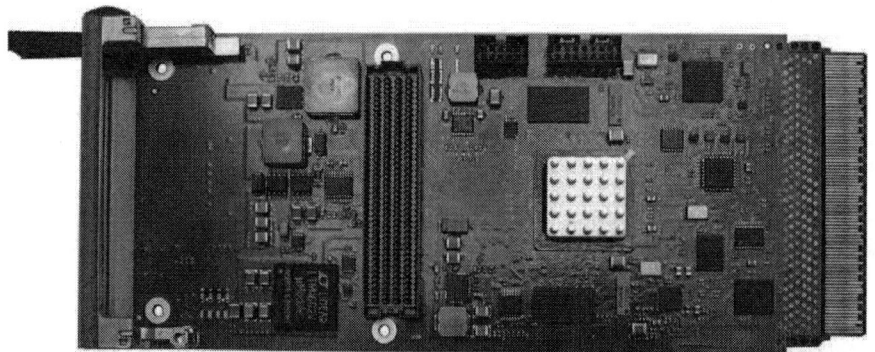

Figure 1. Commercial FMC carrier module compliant with AMC standard.

156

B. Camera and Its Interface

The two types of fast cameras are available on the market. First type has an internal storage and offers high speed recording to its own RAM memory. These devices are usually capable of storing up to 32 GB of video stream and then make it available through some general purpose communication interface. Assuming resolution of 3 Mpx at the frame rate of 1000 FPS the memory can store only 11 seconds of recording.

The second type of camera provides whole video stream with latency kept as small as possible. The real-time operation is a fundamental requirement for plasma monitoring hence only this type of camera will be further considered. The available sensor resolutions are currently of the order of 1 to 5 megapixels with 8 to 10 bits per pixel. The sampling speed varies largely, but in most cases it is naturally limited by the throughput of the communication interface. The image sensor is usually sensitive to light in the spectral range of visible light or infrared. Several companies specialize also in the ultraviolet and X-ray wavelengths, but these devices are mainly low speed specialized sensors.

The following communication standards are often used in high-throughput machine vision systems:

- GigE Vision (up to 1 Gb/s)
- IEEE 1394 (up to 3.4 Gb/s)
- USB 3.0 (up to 5 Gb/s)
- Camera Link (6.8 Gb/s)

The Camera Link data transmission standard is hence the only choice for the top-notch video cameras.

C. AMC Framegrabber Card

There are no COTS video acquisition modules with Camera Link dedicated for the xTCA available. Nevertheless, there is an option to construct such a solution without having to design it from scratch. The AMC card can be an universal carrier board for other extension modules. The FPGA Mezzanine Card (FMC) standard addresses this need. The AMC / FMC modules provide board with an FPGA circuit together with many supporting blocks, like memory banks or clocking circuits.

The commercially available carrier modules offer Xilinx or Altera FPGA circuit. Considering Xilinx devices, the Virtex-5, Spartan-6 and recently Virtex-6 are mainly used. The boards offer a variety of communication standards. The most common options are Gigabit Ethernet, PCI Express and SATA. The actual number of links of each type is dependent on particular board design and availability of high-speed communication interfaces in particular FPGA device.

The boards usually provide SDRAM memory with DDR2 or DDR3 interface, well suited for supporting embedded processor designs and for use as general purpose data buffers. Several boards offer memory expansion by use of the SODIMM memory modules. The QDR SRAM memory is less common due to its high price per megabyte and low capacity. Nevertheless it does not require refresh cycles and supports simultaneous read and write operations. Higher bandwidth and low deterministic latency render it a convenient solution for implementing fast FIFO queues [2].

The plain FMC carriers are generally not able to acquire any data from the outside world. This functionality has to be provided by the small overlay modules. The FMC specification defines two types of connectors. A High-Pin Count (HPC) 400-pin connector is capable of passing up to 160 differential user-defined signals, several clocks and reference voltages. Optionally a Low-Pin Count (LPC) connector may be used. It has the same dimensions as HPC connector but only four out of ten contact rows are populated [5].

Integre Technologies developed an FMC module with Camera Link interface. The FMC-200 board, shown in Figure 2, was selected for the prototype system.

Figure 2. Commercial Camera Link FMC module.

III. IMAGE ACQUISITION MODULE

A. System Usage

The IAS system described in the paper is dedicated for observing the plasma in the toroidal thermonuclear reactor. Its task is to detect visible plasma instabilities and provide a detailed visual log of any unexpected event in the reactor chamber. The module will be further used for evaluating implementations of the lossless video compression and feature extraction algorithms.

B. High-Speed Camera

The developed IAS system uses Mikrotron MC3010 (grayscale) or MC3011 (color) camera shown in Figure 3. The device is capable of capturing frames of maximum resolution of 1696 x 1710 at the rate of 285 frames per second [3]. The user can reduce its region of interest increasing the frame rate up to 180 000 frames per second. The camera is capable of generating data streams reaching 6.6 Gb/s. The data is serialized over 15 pairs of Camera Link interface with the ratio of 7:1 and sent to host together with 3 independent clocks and control signals.

Figure 3. Mikrotron MC3010/MC3011 camera.

C. Camera-Link Interface Module

The FMC-200 from Integre Technologies is the only FMC module with Camera Link interface that is currently available commercially. The module first captures high-speed video data from three links composed of four serial data lines and clock signal. Then it provides a bit permutation to get more intuitive data representation and finally serializes the data again using clock frequency of about 86 MHz. The video stream is transmitted farther using 11 links with serialization ratio of 8:1 with bit clock of 668 MHz. The module provides a set of configuration registers through legacy parallel interface (address bus, data bus, strobes).

The module simplifies the data capture process, by synchronizing data arriving from all the three channels. In case of cooperation with MC3010 camera this interface works properly only up to the frequency of 70 MHz. The tests performed in the 75 MHz to 85 MHz frequency range indicated presence of errors on one or more links to the carrier board. There are two evidences for this problem to be related to the link between camera and FMC-200.

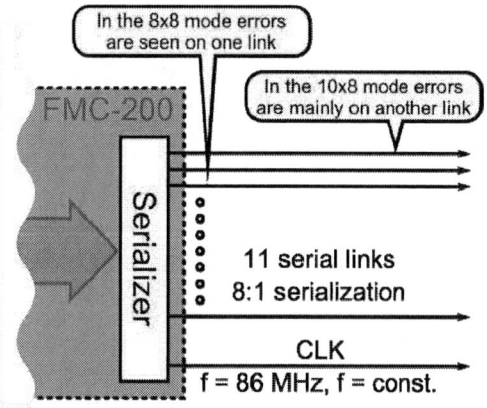

Figure 4. Errors seen at the FMC module output.

The errors concentrate on different FMC-200 data output channels depending on the serialization configuration selected – see Figure 4 for reference. During the test the configuration is

updated in the camera and in the FMC-200 module, but the deserializers implemented in TAMC631 remain unchanged. The FMC-AMC link parameters do not change during the switch hence the error probably appears earlier in the data path. If it were caused by wrong routing or errors in TAMC-631 firmware it should be always observed at the same channels.

Second evidence is that the errors presence is dependent on the camera pixel clock. The FMC to carrier board interface operates at constant bit rate. Its timing is independent of the camera to FMC link frequency, therefore this link should operate identically despite the change.

The FMC-200 also provides basic UART functionality for controlling the camera. This interface is only accessible by the module internal registers. The implemented controller cannot be bypassed, so the host system is forced to access this unit, even if it contains native UART interfaces – which is the case with MicroBlaze-based system.

D. Data Acquisition Module

The TEWS Technologies TAMC631-12R board was selected as a FMC carrier module. It offers a modern Spartan-6 FPGA circuit capable of capturing a high-speed serial transmission from FMC-200. The board is equipped with two banks of DDR3 memory (128 MB each). There is no QDR memory available. The majority of local clocks is provided by the universal configurable integrated clock generator. The module block diagram is shown in Figure 5.

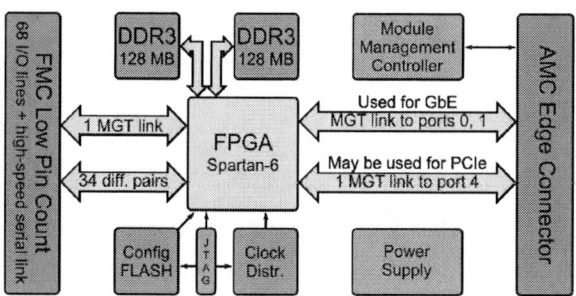

Figure 5. TAMC631 block diagram.

The chosen FPGA circuit offers only four multi-gigabit links. One of them is connected to the FMC slot, others to the AMC connector. Two links are provided on the AMC ports 0 and 1 and may be used for Gigabit Ethernet connectivity. For backward compatibility reasons only one of them is used in the prototype system. The board offers one link on AMC port 4, which may be used to establish a PCI Express x1 connection.

The board is supervised by the Module Management Controller (MMC), according to the requirements of the AMC standard. The MMC circuit monitors the voltages, currents, temperatures and provides the hot-swap functionality. It is also responsible for identifying the FMC module and controlling the voltages provided to the overlay. Although the MMC seemed operating properly, it was unable to initialize the FMC module. The solution was to override the power requirement settings manually through the Intelligent Platform Management Bus.

IV. Module Firmware

A. Overview

The FPGA firmware prepared for IAM is divided into four major components: data write path, memory controller, data read path and embedded microcontroller unit. Every component is composed of smaller functional modules, shown in the block diagram in the Figure 6.

Every of the major blocks is constrained to be placed in a distinct FPGA circuit area to avoid timing problems. The clock domains are not crossing the block boundary, until it is not absolutely necessary. The hardware debugging is possible with use of the integrated logic analyzers of the ChipScope Pro tool.

B. Write Path

The data arrives from the FMC-200 module in the form of 11 channels of 8:1 serialized data together with the reference clock signal. The data is first deserialized, and then the start of frame and data valid markers are extracted and provided to the write sequencer. Status bits are removed and the remaining bytes are aligned into the 128-bit long words. The aligner may be sourced from the actual camera data stream or use the built-in test data generator with six different patterns.

C. Video Frame Format

Every video frame is prepended with a header containing a number of 128-bit long rows (see Figure 7). The first row contains the frame signature, length of the payload data, count of provided sub-headers and header check sum. In the following row the video frame is characterized by its resolution, number of frames per second and a 48-bit long sequence number. Finally a time stamp row is appended. Currently it contains value of a free-running counter incremented by one every 50 ns. This value will be replaced by timestamp provided by dedicated synchronization IP-core.

The video data is transmitted in grayscale with 8 bits per pixel in the row by row manner. At the end, the data is padded with zeros to the next 128-bit word boundary.

Finally a 128-bit footer with 112 bits for the checksum is provided. The video stream with headers and footer is transferred to the memory controller module.

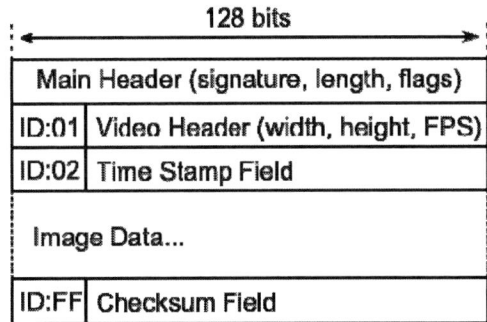

Figure 7. Video frame format.

D. Memory Controller

The DDR3 memory interface is too slow to handle simultaneous data read and write operations at the full required throughput. For this reason, at a time only one memory may be written and only one memory may be read. The memory usage is governed by the memory arbiter. To utilize the available bandwidth effciently all operations are performed in the burst mode, therefore a dedicated read and write controllers are required.

The write path sequencer is instructed by the embedded mirocontroller how many frames it is permitted to store in the memory before it have to realease it. After that, the memory is marked as full and is made available for the reader module.

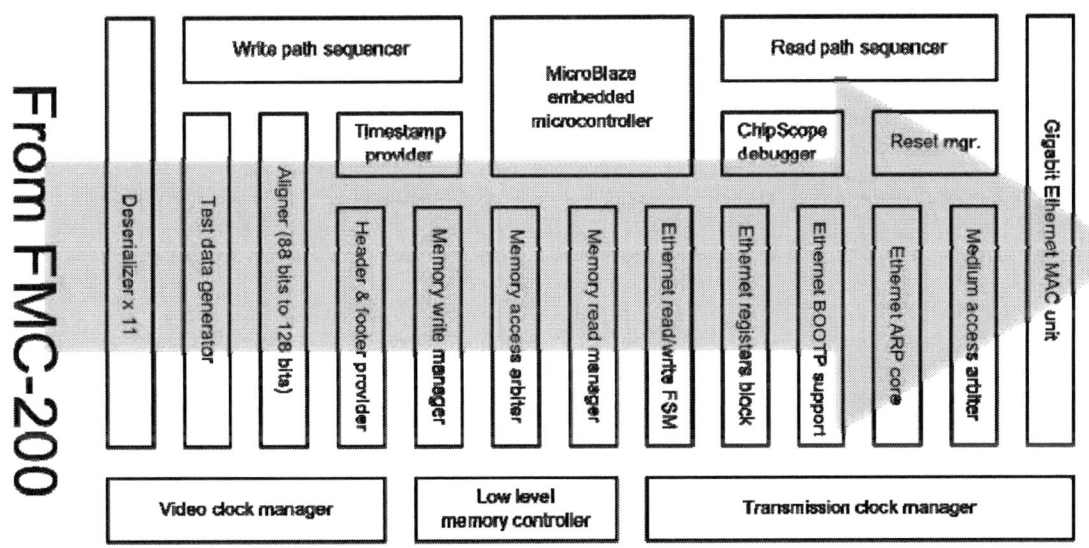

Figure 6. Image Acquisition System FPGA firmware block diagram.

159

E. Read Path

The Ethernet module transmits the data using custom UDP-based protocol. The reliability and efficiency is ensured by data completeness test and retransmission mechanism [7]. The module is configured through a set of Ethernet-accessible registers. The firmware implements the BOOTP protocol client to automatically obtain IP addres. The ARP protocol is used to discover the MAC address of data destination host. The module also provides responses to the external ARP requests. The code employs the TriMode Ehernet MAC hardware block available in the utilized FPGA [8].

F. Integrated Microcontroller

The embedded microcontroller system uses the MicroBlaze soft-processor and provides multiple functionalities. The processor and its peripheral are shown in the Figure 8. The CPU governs the module initialization and reads the registers of Ethernet module. When a request for new configuration is detected it adjusts the FMC-200 module and the video camera to provide the selected data format. Alternatively it may switch the writer module to provide one of the test images. Finally it sets the number of frames to store in each memory and fills the dynamic fields of the data headers. When the system is ready it starts the acquisition with the next start-of-frame marker.

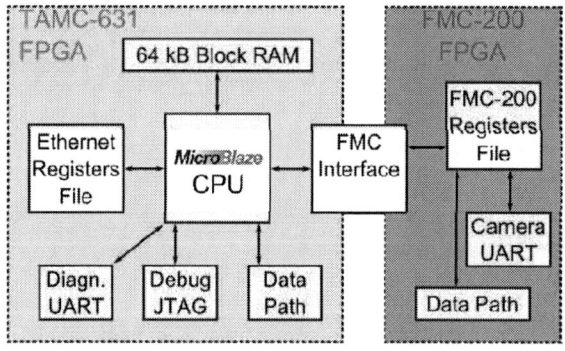

Figure 8. MicroBlaze-based control system

The system contains number of diagnostic features. The program execution can be monitored using the JTAG interface. The diagnostic messages and several test modes can be accessed via the traditional UART interface. The most timing critical signals are also monitored with the ChipScope Pro tool.

V. RESULTS AND FUTURE PLANS

A. System Evaluation

The IAM was tested in the laboratory and was proven to operate properly. In the test setup the module was inserted into the μTCA shelf. The data was transmitted over the backplane to the CPU module located in the same shelf, that also provided the BOOTP service. The demonstration shown that it is possible to acquire data stream from a top-notch high-speed camera using an AMC module operating in μTCA crate, using only off-the-shelf components. The full data stream (up to 7 Gb/s) was not transmitted to the CPU module due to limitation of using only single Gigabit Ethernet link. For the demonstration the frame rate was reduced to provide about 90% link saturation.

B. Conclusion

To achieve the full video stream throughput the external communication interface has to be upgraded. The PCIe x4 interface, with data transmission speeds of about 8 Gb/s, will be used in the next revision.

The system latency can be significantly reduced by replacing the DDR3 SDRAM memory with the QDR SRAM. The QDR memories allow for simultaneous read and write operations, therefore there is no need of filling the memory with preprogrammed number of frames. The data may be stored and retrieved in parallel with almost no delays.

The FMC-200 module errors seem impossible to be compensated. The new FMC module with Camera Link interface is under development.

ACKNOWLEDGMENT

The research leading to these results has received funding from the European Commission under the FP7 Research Infrastructures project EuCARD, grant agreement no. 227579.

REFERENCES

[1] D. Makowski, S. Simrock, "xTCA Initiatives for ITER", 5th Workshop on ATCA and MicroTCA for Physics, Oct. 2011

[2] T. Granberg, "Handbook of digital techniques for high-speed design"

[3] Mikrotron GmbH, "Eosens 3cl camera manual", 2010

[4] PULNiX America, Inc., "Specifications of the Camera Link Interface Standard for Digital Cameras and Frame Grabbers", Oct. 2000

[5] PICMG, "Advanced Mezzanine Card Base Specification", Nov. 2006

[6] ANSI/VITA 57.1-2008, "American National Standard for FPGA Mezzanine Card (FMC) Standard", July 2008

[7] A. Piotrowski, M. Orlikowski, T. Kozak, P. Predki, G. Jablonski, D. Makowski, A. Napieralski, "Performance Optimisation in Software for Data Acquisition Systems", MIXDES 2011

[8] Xilinx, "Virtex-5 Family Overview", DS100, Feb. 2009

MIXDES 2012, 19th International Conference *"Mixed Design of Integrated Circuits and Systems"*, May 24-26, 2012, Warsaw, Poland

Image Visualisation and Processing in DOOCS and EPICS

Piotr Perek, Jan Wychowaniak, Dariusz Makowski, Mariusz Orlikowski, Andrzej Napieralski

Technical University of Łódź

Department of Microelectronics and Computer Science

Łódź, Poland

Abstract—The High Energy Physics (HEP) experiments, due to their large scale, required performance and precision, have to be controlled by complex, distributed control systems. The systems are responsible for processing thousands of signals from various sensors of different types. Very often, one of the data sources applied in such systems are visible light/infrared cameras or other imaging sensors. They can provide additional information about studied phenomena, which is not available on the basis of analysis data from other sensors. However, they require dedicated mechanisms for data collecting and processing. Moreover, often the images from cameras should be available to system operator. It needs the support from both operator panels interface and control application which should provide data in the dedicated format.

The paper presents two different approaches to image distribution, processing and visualisation applied in distributed control systems. Discussed is the issue of support for cameras and image data implemented in the Distributed Object Oriented Control System (DOOCS) and an example control system designed to the needs of image acquisition system on the base of the Experimental Physics and Industrial Control System (EPICS) environment.

Index Terms—DOOCS, EPICS, HEP, MicroTCA, distributed control system, image acquisition system, visualisation, cameras, data processing

I. Introduction

In today's industrial and scientific environments, processes and investigations require integrated control and supervision solutions. Such means would be responsible for ensuring reliability of the processes, maintaining efficiency, increasing productivity or results, which would also have a reducing impact on costs. Distributed Control Systems (DCS), which answer to this need, provide production plants and research machinery with a net of controlling units of a particular kind distributed throughout. This puts various parts of a supervised system under control of such units, which are connected in a way enabling communication.

A DCS performs control over a system e.g. by accessing sensors, raising alarms on occurrence of abnormal conditions, triggering actions or altering properties of the supervised subsystem devices etc. In this context the advantage of DCS comes from the peer-to-peer-like communication between the controlling units and also the workstations or operator terminals.

Among various sensors, which the DCS controlling unit can be connected to, dedicated video cameras can be distinguished. This is particularly important in the High Energy Physics

(HEP) science context. Data from cameras can be useful in determining parameters of investigated phenomena (e.g. electron beams) or supporting early abnormalities detection.

The paper discusses the experience of the authors with regard to diagnostic image acquisition in HEP experiments at Deutsches Elektronen Synchrotron (DESY) and ITER. The DESY is a German research center for particle physics, with main purposes focused on fundamental research in particle physics and synchrotron radiation. The ITER is international organization appointed to construct an experimental fusion reactor based on the "tokamak" concept. The ITER tokamak will be the world's largest and the most technologically advanced reactor. It was designed in order to produce more power than it consumes what has never been achieved. It is assumed that the ITER will be able to produce 500 MW of energy for 50 MW of input power [1].

Cameras used for the diagnostic image acquisition purpose work under control of the Distributed Object Oriented Control System (DOOCS) and Experimental Physics and Industrial Control System (EPICS) in the above mentioned centers, respectively.

DOOCS is a distributed control system developed by DESY to satisfy the needs of operating the HERA and FLASH experiments hardware. This object-oriented entity is a complex solution for all along the way from the physical hardware up to operator terminals. The system follows the server-client architecture model [2].

The EPICS environment is a set of software tools dedicated for creation of distributed soft-real time control systems for large-scale scientific experiments. It is open source, well-proved and reliable solution, developed by the most significant organizations involved with HEP experiments on the world e.g. ANL, FNAL, KEK, DESY, SLAC etc. [3] For this reason, the EPICS was chosen by ITER organization as the baseline for development of the ITER tokamak control system [4].

A comparison of approaches to cameras-based diagnostics in both control systems is also drawn.

II. Key Elements of DOOCS Architecture

The structure of DOOCS is coined around the notion of reflecting and representing the physical hardware devices in a form of software objects. This results in a model, in which such object - a device server - associated with a given device,

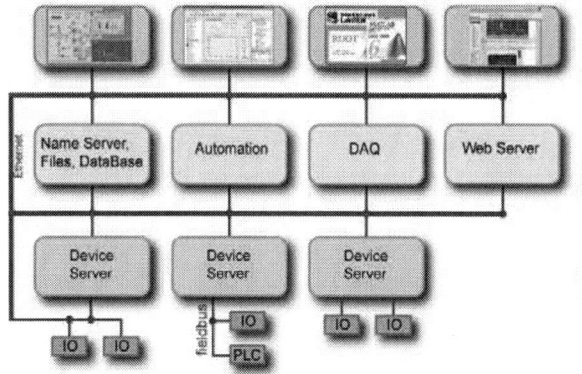

Fig. 1. The DOOCS architecture overview

controls its operation and exposes its properties to other parts of the system [5].

The system is written in the C++ programming language. The aim in the design process was to provide support for creating a device server as an autonomous software entity that would perform complete control over a group of devices of a given type and expose certain sets of its properties to the network. A server, upon start-up, creates an instance of the device representation for each supervised device. Then the properties of each device are made visible in the network [5], [6].

A collection of independent device servers, instead of a central software entity or database, is what conveys the notion of a distributed system. The system is composed of three main layers (Fig. 1) [7]:

- The layer of front-end servers (also referred to as device servers). These servers are what provides the software representation of the devices and makes their properties visible to the network.
- The middle-layer servers. This layer is responsible for facilitating resource-demanding computations, hosting the data processing and acquisition (DAQ) mechanism, databases and other services.
- The client applications layer.

A. Front-end Servers

The front-end servers, composing the lowest layer in the control system, perform direct communication with the hardware. For this purpose, different buses may be used, which include VME, CAN, PROFIBus, GPIB, or Ethernet-based solutions [5].

A device server in DOOCS operates as a UNIX process and represents and communicates with one or more hardware devices of a particular type. All the server classes in the server source code are a specialization of a base class. This base class provides a definition of a standard set of functions and properties that are later a common part of all the servers. It also implements the routines for communication with the network. Furthermore, maintaining the local configuration

of a server in a form of a configuration file, as well as data archiving, is also implemented. Access to the local configuration makes it convenient for the server to smoothly recover from possible erroneous operation situations (like crashes etc.). The information exchange between the network and a server properties is provided with dedicated data objects.

B. Middle-layer Servers

The middle-layer, responsible for data processing, but also referred to as the service tier, is where the global services are based. These include the Equipment Name Server (ENS), the Data AcQuisition system (DAQ), the web services and the databases. This layer aims to implement all the complex functionality required in the system. The purpose of this approach is to possibly relieve the client layer applications from stepping beyond the responsibility to present data and interact with operators [5].

The mentioned ENS is responsible for responding to name queries by resolving protocols and host names. The responses from the ENS after requests initiating communication within the system help routing the subsequent calls to the proper destinations and with using proper data exchange protocols (RPC, shared memory, TINE etc.).

The DAQ is an approach for data integration derived from solutions in accelerator controls of experiments from the HEP domain. The intentions towards, and gains from, employing such structure include:

- providing correlation and synchronization of readings from the supervised hardware
- aggregating data obtained from the hardware in a single repository for the use of feedback and further processing
- preserving data for further investigation performed offline
- archiving

The DAQ provides a common communication interface shared by all the services and servers in the middle layer. Device servers are provided with the capability to route the collected data to the DAQ unit. There the data is later available in the central shared memory. Processes like feedback can then take advantage of the availability of this data and operate basing on using the stored values. This way the DAQ may be seen as an efficient data access path for the services and middle layer processes to the device servers. Synchronized data flow coming from the instruments via the device servers and data availability can be taken advantage of by advanced measurement or further processing activities.

C. Client Applications

The client applications layer is what is responsible for representation of the data passing through the control system to the operators. Client programs do not require any particularly sophisticated logic, as the middle layer relieves them from data processing. Thanks to this, it is possible to create the user interfaces with help of graphical editors. Such a tool is provided with the control system. It is a lightweight DOOCS Data Display (DDD) editor. It facilitates creating graphical

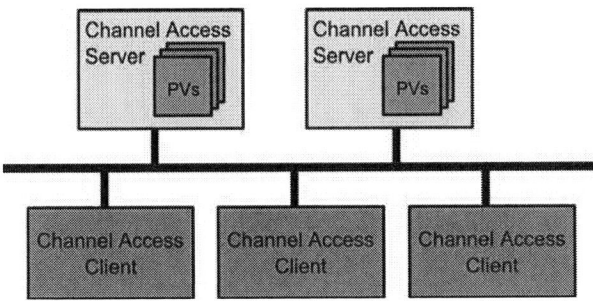

Fig. 2.   EPICS architecture

panel, via which an operator is able to access data, read devices properties and issue commands.

The control system libraries for the client applications also provide integration with engineering and scientific software like LabVIEW or Matlab. What is more, an application programming interface (API) is provided to make possible creating separate applications for cooperation with the control system, in a selection of programming language [5].

III. EPICS-BASED CONTROL SYSTEM INFRASTRUCTURE

The EPICS architecture is based on the server-client model (see Fig. 2). The model allows for existence of several clients and servers in a single control system. Data between servers and clients are exchanged using network-based protocol called Channel Access (CA) Protocol. The fundamental piece of data in EPICS is called the Process Variable (PV). And thus, the servers are programs that define and provide access to the PVs. On the other hand, the clients are programs that require access to the PVs. The entire set of Process Variables provided by all servers in the system are treated as distributed real-time database of information and control parameters.

The special type of EPICS server is Input/Output Controller (IOC). It implements real-time control algorithms and ensures interface to hardware [3]. The IOC provides a set of specific structures that store hardware status and control parameters. These data structures, referred to as records, have particular functionality associated with them (inputs, outputs, calculations, alarm detection, etc.). Different records types have different functions and applications. The data within a records is accessible via PVs. The records are frequently associated with I/O equipment which require unique 'device support'. The EPICS defines dozens of records which allow to exchange data of various types. However, in case of hardware which is not supported it is necessary to reimplement a set of records which allow for communication with the hardware.

The EPICS can also be viewed as a set of tools and applications. Each program which meets the requirements of CA Protocol can be described as compliant with EPICS. Therefore, many applications for various purposes are developed by EPICS collaborators. Each user can select the tools appropriate to their needs or develop their own.

The data that is made available by an EPICS application by the means of records has to be presented to system operator in a graphical form. Furthermore, operator should have possibility to control the system using friendly graphical interface. For this purpose, a variety of display managers for EPICS (e.g. MEDM, EDM, EDD/DM, dm2k) have been developed. One of the most recent graphical user interfaces is BOY (Best OPI, Yet). The BOY is Operator Interface (OPI) environment, as EPICS, recommended by ITER and supported by the Control System Studio (CSS) – tool for control applications development [8], [9].

The BOY environment is a modern user interface which enables fast development of complex user interfaces. Operator panels prepared using BOY connect to an EPICS application and allow for displaying collected data in user-friendly way. They also provide interface for data input.

IV. CAMERAS SUPPORT AND IMAGE DATA PROCESSING IN DOOCS

The selection of hardware devices that the DOOCS control system communicates with may include cameras of diverse flavours. Their different types and hardware connection standards are also handled by the system with the object-oriented approach. The system provides a base class that implements all the functions that are common to various camera types. This class also contains implementation of parameters that are common to all camera types (like exposure, brightness, etc). From this base class derive specialized classes, that are dedicated and specific to particular cameras. They define interfaces for communication with the hardware via USB, Ethernet, FireWire etc., as well as other camera-specific features (functions, parameters) that do not exist in the base class [10].

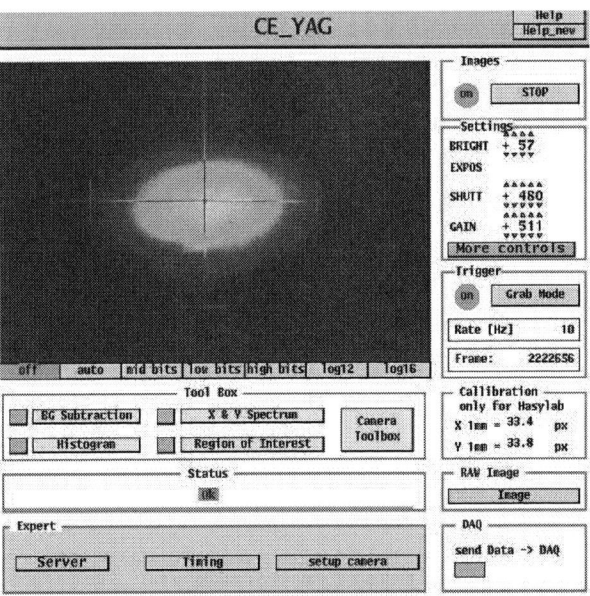

Fig. 3.   An exemplary DOOCS camera GUI panel

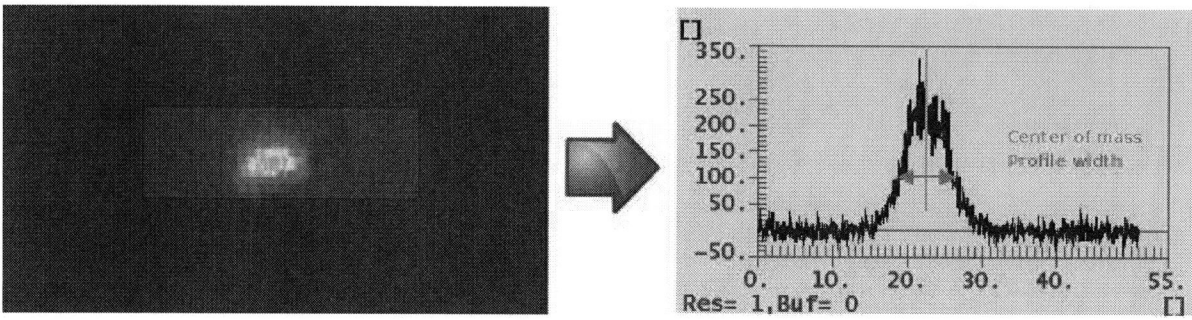

Fig. 4.   Image data at the front-end server (left) and operations performed by the middle layer servers (right)

The server classes provide complete control over the camera properties and settings. The acquired images can be read directly by the client programs, preserved inn files or stored in the DAQ. There is also support for selected simple operations on images, like background subtraction, histogram calculation, etc. More advanced calculations/transformations can be implemented in middle layer servers.

The camera servers have the possibility to be connected to timing systems, which enables the support for synchronization of the image taking process with the events in the observed phenomenon. Therefore if a camera supports external triggering, images can be taken at moments specified by the timing pulses. The system can be configured in a way that timing-triggered images can contain additional data, like timestamps or event numbers. This enables for correlation of such images with other sorts of diagnostic data [11].

DOOCS provides the IMAGE data type, dedicated to use with cameras. The property of a camera server of the type IMAGE allows to access the most recently taken image. The IMAGE type is composed of a header and a byte array storing the image itself. The header contains image metadata, including resolution, image format, rotation and other. It can also contain a timestamp and event number. The IMAGE property of the camera servers can be directly used by the client applications, but also more advanced processing (like offline investigation or extracting information for feedback actions) can take place. For the latter purpose the DAQ system is a suitable environment.

As mentioned earlier, front end servers for the cameras are sometimes capable of performing simple initial processing of the acquired data. For more advanced or specialised tasks dedicated middle layer servers can be used. The operation of such servers may yield results including image prepared in a specific way for display with using the DOOCS Data Display (ddd) panels. Another output may be specific values, or other information, extracted from the image data.

An exemplary middle layer server operating by the means of utilising image data acquired by a camera concerns processing a longitudinal profile of an electron beam in the DESY FLASH facility. The server pulls data from a front end server communicating with an ICCD sensor-based camera

observing diagnostic laser emitted in pulses synchronised with the electron beam. The longitudinal charge distribution of the electron beam is reflected by the laser [12]. The longitudinal profile of the beam is produced by the front end server and made available for further use in the DOOCS system in a form of a data array representing the beam charge against time range. The discussed middle layer server acquires the data and performs processing aimed at removing the background noise from the profile, computing its width or centre of mass (Fig. 4). The obtained values are stored in order to track changes of these beam parameters in time.

V. EPICS-RELATED APPROACH TO IMAGE ACQUISITION

The following section of the paper presents a control application for image acquisition system, which was prepared with using the EPICS framework. Due to the fact, that EPICS does not provide any native mechanisms for image processing, the authors have developed a custom solution for the needs of the described system.

The image acquisition system is a prototype solution developed as part of cooperation with the ITER organization. It was designed as a first step to built complex diagnostic system based on visible light/infrared cameras for ITER tokamak. About 20 cameras comprising the system will observe plasma and internal surfaces of the reactor [13], [14]. Images collected from the cameras will be useful for recognition and classification of thermal events occurring inside the tokamak [15].

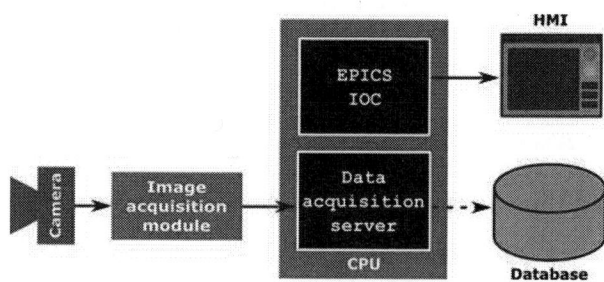

Fig. 5.   Structure of image acquisition system

The system presented in this paper section was designed in order to investigate the performance of the Micro Telecommunications Computing Architecture (MicroTCA) system in applications of high-speed image acquisition and processing. It consists of a central processing unit, an image acquisition module and a high speed camera. The system structure is presented in Fig. 5.

The high speed, high resolution camera has been applied in the system. It is distinguished by following features:

- 3 Megapixel high speed CMOS sensor
- Resolution: 1696x1710 pixel
- 8 bit resolution
- Up to 285 frames/s at full resolution
- Up to 180 000 frames/s with reduced resolution
- CameraLink interface

Data from the camera is collected by Advanced Mezzanine Card (AMC) equipped with frame grabber made in FMC format. The FPGA device on AMC module receives data from the camera and transmits them to the CPU via Gigabit Ethernet (GbE) interface. According to the ITER requirements, images collected from the camera are to be sent to the database as well as presented on operator control panel. For this purpose, a data server application for CPU has been prepared. It is responsible for receiving data from the acquisition module and preparing separate streams for a database and the control application.

The control application for the system has been prepared using the EPICS framework. The application is an IOC which gives the possibility to acquire data from the camera. Communication with the image acquisition module takes place through the agency of the data server. The EPICS IOC connects to the data server via local Ethernet connection. Information between them is exchanged using higher-layer protocol based on self-defined commands. The commands sent from the IOC to the server allow to control data acquisition parameters e.g. video source, resolution, refresh rate etc. On the other hand, the commands sent from the server inform about the current status of the acquisition.

As it was mentioned, to support new hardware it is necessary to develop 'device support'. This includes records which provide input/output operations. For the described acquisition system, a set of dedicated records was prepared. It includes Binary Input (BI), Binary Output (BO), Long Input (longin), Multi-bit Binary Output (MBBO), String Input (stringin) and Waveform record. The input records allow to read data from the server. On the other hand, the data can be written to the server, and then to the module, using the output records.

Besides the EPICS application, also a graphical interface for the system operator has been prepared. The operator panel prepared for the image acquisition system using the BOY environment is shown in Fig. 6.

The panel presents the image which was collected from the camera and information about its resolution. Below the image the timestamp is shown. The timestamp is generated by the data acquisition module and written into the image header. The EPICS application extracts this information and makes it available in the form of input record. Furthermore, on the panel there are two status LED indicators. The first one informs whether the data acquisition module is connected to the data server, while the other one indicates the current status of the acquisition process (running/stopped). The remaining components on the panel are responsible for the system control. First of all, the operator has the possibility to start/stop data acquisition. Moreover, on the panel there are also three drop down lists which allow the operator to modify following properties:

- video source (camera images, camera test images, test images generated by acquisition module),
- video profile (image resolution and camera frame rate),
- refresh rate (frequency of sending images from the data server to the EPICS application).

The raw image data in the EPICS application is presented in the form of waveform record. The waveform record can store data of any type supported by EPICS. The data is stored in an array. In case of images required from the camera the waveform record contains unsigned 8-bit values. Each cell of the array represents particular pixels of the image. In OPI environment the images are represented using the Intensity Graph component. It displays the numeric array in a form of 2D plot using colors defined in color map. For images presented in the described system the values from the waveform record are mapped to grayscale. Height and width of the intensity graph is adjusted to the actual image size on the basis of dedicated records ensured by EPICS application.

Due to the fact, that the camera that was used in the system has bayer pattern filter on the sensor glass lid, it is possible to get color information from the acquired data. Values of red, green and blue components of the pixel color are calculated on the base of neighbouring pixels. For this reason, there has been also prepared an EPICS application that converts data collected from the camera from grayscale to RGB format. In this case, the visualization also differs. The RGB data is provided by

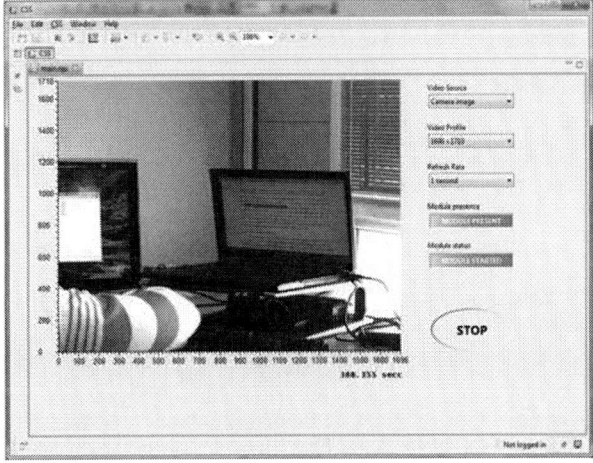

Fig. 6. Operator panel for image acquisition system

the EPICS application using waveform which components are 32-bit wide. The Intensity Graph component on the operator panel presents the image using self-defined color map. It is noteworthy, that this solution of color image presentation is not efficient, because 8 bits from every 32-bits value are lost. Therefore, the authors intend to examine another components both for EPICS application and BOY operator panels, which will be more accurate for the presented system.

VI. CONCLUSIONS

The above discussion upon two approaches to the question of taking advantage of the distributed control systems draws significant differences between them. Both the solutions have been present for more than 17 years now, therefore they managed to establish solid foundations for the operation principles they implement. They are, however, not the same in terms of the adopted model. DOOCS can be seen as a concrete example of how a DCS may operate within a complex scientific environment. This is due to the fact that it was developed for the needs of particular facilities, therefore it can directly consider their requirements. EPICS, in contrast, focuses more on providing a generic, not location-bounded framework, enabling to create control systems that would then adhere to specific environment conditions. One more consequence of this is that DOOCS was able to develop and incorporate direct support for image data processing and visualisation, from the hardware level up to software for operator terminals. At the same time a new DCS created from EPICS needs customisation that comprises providing hardware support, appropriate data records or components for operator panels. However, such approach has also advantages, because the users can exchange their solutions with others. Tools developed by one user can be easily applied by all members of community involved with EPICS development.

These essential differences between the discussed solutions affects also the topic of using cameras in diagnostic image acquisition. The DOOCS DCS, developed from the very start with the intention of adhering to particular hardware base, was able to incorporate a generic form of direct support for cameras. Appropriate software structures (e.g data types) are present, ready to be taken advantage of. This yields a basic unified and succinctly defined application interface for communicating with selection of camera types directly at the control system level, ready for its user to utilise.

On the other hand EPICS, as a framework ensuring the generic components for DCS development, does not provide support for camera handling and image processing. For this reason, the user is forced to implement their own 'record support' or to use existing records for image representation, as it has been done in the described image acquisition system. Another approach is application of modules for EPICS developed by other users and dedicated for operation on images. An example of such module is the areaDetector, which supports image acquisition and processing and ensures interface for cameras control.

Using such modules provides a partial remedy to the camera support lack in EPICS-based DCS'es, and it can substantially limit the user experience differences in both systems, when it comes to diagnostic image processing. Also, despite the generic interface being available in DOOCS, the implementation of hardware support, encompassing issues like communication (buses, data transfer protocols) is a requirement in both systems.

ACKNOWLEDGMENTS

The research leading to these results has received funding from the European Commission under the FP7 Research Infrastructures project EuCARD, grant agreement no. 227579.

REFERENCES

[1] Y. Shimomura and W. Spears, "Review of the ITER Project," *IEEE Transactions on Applied Superconductivity*, vol. 14, no. 2, June 2004.
[2] TESLA Report 2008-03, "LLRF System Components Development (I)," Editor: R.Romaniuk, ISE, WUT.
[3] "EPICS Main Page." [Online]. Available: http://www.aps.anl.gov/epics/
[4] A. Wallander, F. D. Maio, J.-Y. Journeaux, W.-D. Klotz, P. Makijarvi, and I. Yonekawa, "Baseline Architecture of ITER Control System," *IEEE Transactions on Nuclear Science*, vol. 58, no. 4, August 2011.
[5] K. Rehlich, "Status Of The Ttf VUV-FEL Control System," *PCaPAC 2005 conference talk*.
[6] G. Grygiel, O. Hensler, and K. Rehlich, "DOOCS: A Distributed Object-Oriented Control System on PC's and Workstations," *PCaPAC conference*, 1996.
[7] S. Goloboroko, G. Grygiel, O. Hensler, V. Kocharyan, K. Rehlich, and P. Shevtsov, "DOOCS: an Object-Oriented Control System as the Integrating Part for the TTF Linac," *ICALEPCS 97*, Beijing.
[8] "BOY - Control System Studio." [Online]. Available: http://sourceforge.net/apps/trac/cs-studio/wiki/BOY
[9] X. Chen and K. Kasemir, "Boy, a modern graphical operator interface editor and runtime," *Particle Accelerator Conference*, 2011.
[10] K. Rehlich, "Status of the FLASH Free Electron Laser Control System," *Proceedings of ICALEPCS'07*, 2007, Knoxville, Tennessee, USA.
[11] G. Grygiel and V. Rybnikov, "DOOCS Camera System," *Proceedings of ICALEPCS'07*, 2007, Knoxville, Tennessee, USA.
[12] B. Steffen, et al., "A Compact Single Shot Electro-Optical Bunch Length Monitor for the SwissFEL," *Proceedings DIPAC09*, 2009, Basel.
[13] I. Yonekawa, "Data acquisition and management requirement for ITER," *Fusion Engineering and Design*, vol. 43, no. 3-4, pp. 321–325, January 1999.
[14] M. Walsh, et al., "ITER Diagnostic Challenges," *IEEEINPSS 24th Symposium on Fusion Engineering*, 2011.
[15] V. Martin, J.-M. Travere, V. Moncada, and F. Brémond, "Towards Intelligent Video Understanding Applied to Plasma Facing Component Monitoring," *Contrib. Plasma Phys. 51*, no. 2-3, pp. 152–255, 2011.

Design of Integrated Circuits and Microsystems

Gap in pagination due to withheld paper.

Pages 169-177

MIXDES 2012, 19th International Conference *"Mixed Design of Integrated Circuits and Systems"*, May 24-26, 2012, Warsaw, Poland

A Balun Transimpedance Amplifier with Adjustable Gain for Integrated SP0$_2$ Optic Sensors

José Carvalho, Luis B. Oliveira, João P. Oliveira, João Goes
CTS-UNINOVA and Faculty of Sciences and Technology (FCT)
Universidade Nova de Lisboa (UNL)
Campus de Caparica, Portugal
E-mail: jpao@fct.unl.pt

Manuel M. Silva
INESC-ID
Technical University of Lisbon
Lisbon, Portugal
E-mail: manuel.silva@inesc-id.pt

Abstract—**The oxygen level in blood, usually referred as SPO$_2$ (Saturation of hemoglobin with oxygen as measured by pulseoximetry) is an essential medical information. Measuring the oxygen level of the human blood using non-intrusive techniques is a vital achievement in modern medicine. This can be performed by processing the infrared and red light transmitted by the patient's finger and received by a photoreceptor. Before being applied to an analog-to-digital converter (ADC), the incoming light has to be converted to a voltage and the range should be dynamically adjusted in order to use always the full input range of the ADC. Since the photoreceptor generates an output current, a transimpedance amplifier (TIA) with gain control is required. The two-stage TIA proposed in this paper, uses a regulated common-gate in first stage employing noise cancellation and balun operation using an additional CS stage, while the adjustable gain is implemented in the second-stage, which is based on an intrinsically noiseless MOS parametric amplifier (MPA). This MPA operates in the discrete-time domain, thus eliminating the need of an input sample-and-hold (S/H) block in the ADC. The proposed circuit has been designed in a 130 nm digital 1.2 V CMOS technology. The electrical simulations show that the overall power consumption is lower than 250 μ W and input referred noise power density is extremely low.**

Index Terms—**Transimpedance amplifier (TIA), oximeter, noise cancelation.**

I. INTRODUCTION

Nowadays, there is a need for ubiquitous healthcare remote monitoring. Economic and reliable devices are required, able to continuously and remotely observe the health of patients. The blood oxygen level is one of the most important vital signs, alongside with blood pressure, heart beat rate, breathing rate, and body temperature, and it's constant monitoring can give early warning of problems in the circulatory and respiratory systems.

Typical applications are home care monitoring of the elderly or chronically ill, wireless medical sensing, remote monitoring of the health status of military personnel in the battlefield and fire-fighters engaged in fire control and rescue missions. One key target feature of the equipment it that it must be non-invasive, eliminating the need of surgical intervention to implant the sensors.

Pulse oximetry is a non-invasive method to measure the percentage of oxygenated hemoglobin in a patient blood. Pulse oximeters are widely used in intensive care, operating rooms, emergency care, birth, neonatal and pediatric care, sleep studies, and also in veterinary care. A transducer is usually clipped or taped to a translucent area of the patient, such as an earlobe or a finger. The transducer generates an electrical current signal, which is converted to a voltage signal by a transimpedance amplifier (TIA), as shown in Fig. 1.

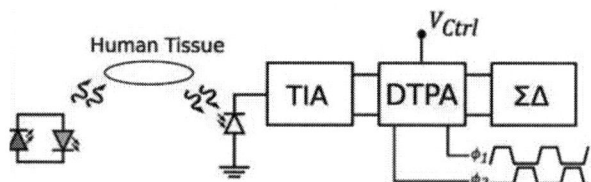

Figure 1. Pulse oximeter microsystem architecture.

In this work, a CMOS differential output TIA with adjustable gain is proposed. The proposed pulse oximeter architecture includes, besides the TIA, a photo-detector (PD) and an ADC, as shown in Fig. 1. The TIA must have low power dissipation and gain control. Adjustable gain is necessary to accommodate variations due to different light sources and intensities and transducer positioning, thus adapting the signal swing to the ADC's input dynamic range (DR).

This work was supported in part by the Portuguese Foundation for Science and Technology (FCT) under projects IMPACT (PTDC/EEA-ELC/101421/2008) and OBiS FRET (PTDC/CTM/099511/2008).

The paper structure is as follows. In Section II, the architecture of a pulse oximeter is reviewed. In Section III, the continuous-time first stage of the TIA is described. Section IV describes the variable gain discrete-time second stage of the TIA, using MOS parametric amplification. In Section V simulation results obtained in a 130 CMOS technology are given and, finally, the conclusions are drawn in Section VI.

II. REVIEW OF PULSE OXIMETRY

In the oximeter measuring system shown in Fig. 1, two light-emitting diodes (LEDs), for red and infrared range light are pulsed at a low frequency with a low duty cycle pulse signal, in order to reduce the pink-noise and the power consumption. The light transmitted through the patient skin and tissues, generates a current in the silicon photo-diode. This current is then converted into a voltage by the first stage of the TIA and in the second stage it is amplified using a Discrete Time MOS Parametric Amplifier (DT MPA) block, which holds the signal that is then quantized by the ADC.

The percentage of oxygen in blood (SPO2), is expressed as the ratio between oxygenated hemoglobin and the total hemoglobin (HbO_2) according to,

$$SP02 = \frac{HbO_2}{Hb + HbO_2} \qquad (1)$$

where Hb is the amount of hemoglobin with reduced oxygen. Oxygenated hemoglobin is bright red, so it absorbs more infra-red (IR) light, letting more red photons pass through in contrast to the dark red hemoglobin with reduced oxygen. It is this different absorption of red light (660 nm) and IR light (940 nm) that makes it possible to measure the blood oxygenation by measuring the ratio of absorption of red and IR light.

The light sources, usually LEDs, emit red and IR light through a translucent part of the patient's body, such as an earlobe, a finger or a toe. The light passes through the patient's skin and tissue, and is then measured by the photo-diode. Light has to travel through different tissues, such as skin, bones and muscle. Blood vessels expand and contract with the heartbeat, so the oximeter signal appears modulated, making it possible to effectively separate the blood transmitted signal, an AC signal, from the signal from other tissues in the background, a DC signal [1].

The physics behind the working principle of the pulse oximeter is the Beer-Lambert law. This law relates the transmitted and incident light through a medium that contains an absorbing substance of concentration (C), length(L), and a wavelength dependent absorption coefficient $\varepsilon(\lambda)$ according to

$$I_{TRANS} = I_{INC}e^{-\varepsilon(\lambda)CL}. \qquad (2)$$

where I_{TRANS} and I_{INC} are the intensity of the transmitted and incident light, respectively.

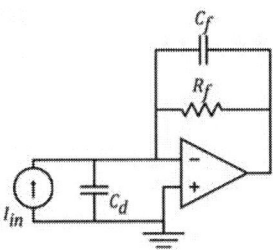

Figure 2. Shunt feedback transimpedance amplifier and photodiode equivalent circuit.

One can define the transmittance T as I_{TRANS}/I_{INC} and the un-scattered absorbance A as $-ln(T)$, resulting in the total light absorbance of blood, A_{bld}, given by

$$A_{bld} = [\varepsilon_{HbO2}(\lambda_R)C_{HbO2} + \varepsilon_{Hb}(\lambda_R)C_{Hb}]L. \qquad (3)$$

when the blood vessels expand and contract, their thickness varies, and so does the absorbance of the blood in them with L in (3). If R is the ratio of relative absorbance at the red and IR wavelengths, we get

$$R = \frac{\ln(I_{H,Red}/I_{L,Red})}{\ln(I_{H,IR}/I_{L,IR})} \approx \frac{I_{ac}^{Red}/I_{DC}^{Red}}{I_{ac}^{IR}/I_{DC}^{IR}} \qquad (4)$$

which leads to

$$R = \frac{\varepsilon_{HbO2}(\lambda_{Red})C_{HbO2} + \varepsilon_{Hb}(\lambda_{Red})C_{Hb}}{\varepsilon_{HbO2}(\lambda_{IR})C_{HbO2} + \varepsilon_{Hb}(\lambda_{IR})C_{Hb}}. \qquad (5)$$

It's known that the approximation in equation (4) is good, as long as the AC signals are small when compared to the DC component [1]. We can combine (1) and (5) to obtain

$$SpO_2 = \frac{\varepsilon_{Hb}(\lambda_R) - \varepsilon_{Hb}(\lambda_{IR})R}{\varepsilon_{Hb}(\lambda_R) - \varepsilon_{HbO2}(\lambda_R) + [\varepsilon_{HbO_2}\lambda(_{IR}) - \varepsilon_{Hb}(\lambda_{IR})]R} \times 100\% \qquad (6)$$

The absorbance for Hb and HbO_2 are known for the red and IR wavelengths, so it's now clear that the pulse oximeter measures the ratio R, which is related to SP02 by means of (4, 5).

III. TIA FIRST STAGE: I/V CONVERTER

To convert the light signal at reduced cost an integrated CMOS photodiode is used as the transducer. A photodiode can be modeled as a current source in parallel with a capacitor [2], as shown in Fig. 2. The capacitance value is usually high, and is bias dependent: typical values are of the order of the pico-farad (pF) for a device of $50\mu m \times 50\mu m$ [3]. This capacitance is dependent on the technology, device size and biasing [3], and the value of 1 pF is assumed for the design and simulation here.

The current of an integrated photodiode is highly dependent on its junction area [3]. In this work, it is assumed that the current is between 0.1 – 10 µA. The sizing of the photodiode is outside the scope of this work, and it should be done according to this current specification.

In this application the TIA should satisfy several requirements, high gain-bandwidth product (GBW), low noise and low power consumption.

The choice of the TIA is considered derivation of the proposed TIA circuit is supported in the following subsections.

A. Feedback I/V converter

The feedback I/V converter is commonly used in optoelectronic integrated circuits [4]. It consists of an operational amplifier (OA) with feedback, as illustrated in Fig. 2. Making the same assumptions as in [5] the following transimpedance function is

$$\frac{V_o}{I_{in}} = \frac{R_f}{1 + sR_f(C_f + \frac{C_d}{A_o}) + s^2 R_f C_d B^{-1}}, \qquad (7)$$

where B is the gain bandwidth product of the OA (assuming it has a dominant Pole) and A_o is the low frequency gain. Although the feedback amplifier has a good noise performance [5], the power consumption would be high to have a high gain OA, so alternative topologies are considered.

B. Common-Gate I/V Converter

Another basic I/V converter is the common-gate (CG) stage represented in Fig. 3(a), where I_{B1} and V_{Bias} are a bias current and voltage, respectively. This topology has the advantage of low input impedance, broadband, and well behaved time response [4].

The transimpedance is R_x, and is required to have a value of tens of kOhm, thus creating a high DC voltage drop. To keep the power low and headroom for a voltage swing, a low current has to be used to bias the input transistor. This leads to a low g_m, which increases the input impedance

$$Z_{inCG} \approx \frac{1}{g_{m1}}. \qquad (8)$$

In addition, a low g_m compromises the noise and stability performance of the amplifier [6].

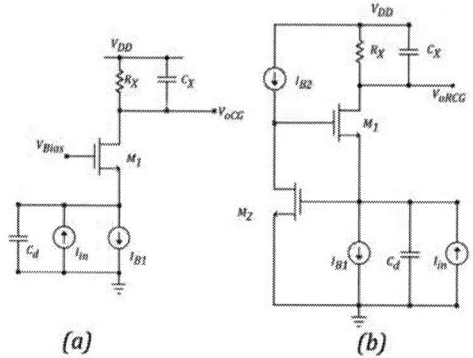

Figure 3. CG TIA (a) and RCG TIA (b).

C. Regulated Common-Gate I/V converter

To lower the input impedance without increasing g_m, a regulated common-gate (RCG) circuit can be used. This is presented in Fig. 3(b). This can be viewed as a CG stage to which an amplifier loop is added, which has the effect of boosting g_m [5], which is multiplied by the added amplifier gain. The input impedance is approximately

$$Z_{in} \approx \frac{1}{Ag_{m1}} \qquad (9)$$

A is the gain of the common-source transistor M_2 (with active load I_{B2}) in Fig. 3 (b)

$$A = g_{m2}(r_{o2} // R_{oB2}) \qquad (10)$$

where R_{oB2} is the incremental resistance of current source I_{B2}.

D. Noise Canceling

It is convenient that the TIA, with a single-ended input, has a differential output (balun operation). By implementing the I/V conversion and balun together, noise cancelling of the thermal noise of the input transistor can be achieved. The thermal noise of the CG transistor (M_1) is represented by the current source I_{n1} in Fig. 4, which generates a noise voltages $V_{n,in}$ at the input and $V_{n,out1}$ at the output, which are in anti-phase, while the signal at the input and output is in-phase. Since the gain of the CS is negative, the output signals V_{out1} and V_{out2} are in anti-phase, effectively doubling the gain, while $V_{n,out1}$ and $V_{n,out2}$ are cancelled. The gain matching between the two stages is crucial to obtain the noise cancelling [7].

E. Proposed I/V Converter

The TIA proposed here, comprises a RCG with noise cancelling, shown in Fig. 5.

The current sources are implemented by current mirrors sized so that $I_{B2} = 50$ µA, $I_{B1} = 10$ µA and $I_{B3} = 200$ µA. To balance the gain and DC offset the load resistors of both the

180

RCG and CS stages are equal. The CS transconductance was chosen to achieve output balancing and noise cancelling. The high transimpedance gain required is obtained by having $R_{RCG} = 20 k\Omega$.

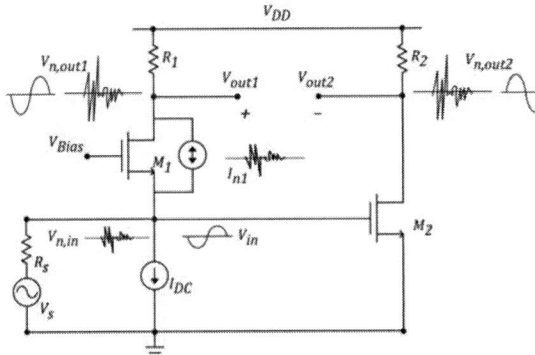

Figure 4. Noise cancellation technique, [6].

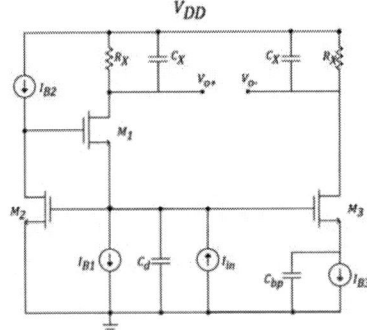

Figure 5. Proposed RCG TIA with noise cancellation.

IV. SECOND STAGE: DISCRETE-TIME PARAMETRIC AMPLIFIER WITH VARIABLE GAIN

From the capacitor equation $Q = CV$, if the capacitance of a varactor is changed from an initial value C_1 to a lower value C_2, while conserving the charge, a voltage gain is achieved, given by,

$$A_v = \frac{V_o}{V_i} = \frac{C_1}{C_2}. \quad (11)$$

This principle can be implemented in MOS technology by controlling the gate-to-bulk capacitance of an MOS device. This control is achieved through a voltage that is applied simultaneously at the drain and source [8, 9] terminals. As explained in [9], considering a discrete-time signal configuration, an input signal sampled at one phase can be amplified and held in a second phase if one can reduce the

gate-to-ground capacitance, while maintaining the total gate charge.

The principle of this amplifier is illustrated in Fig. 6. In the sampling phase Φ_1, the input signal is sampled by the PMOS varactor in strong inversion and it is held at the beginning of Φ_2, by maintaining the gate terminal floating. Amplification is then obtained, during Φ_2, when the capacitance value is reduced, since the device is forced to enter into depletion state. A first order analysis indicates a maximum amplification gain of $v_o / v_i = C_{ox} / C_{gb}$, where C_{ox} is the total oxide capacitance and C_{gb} is the gate-bulk capacitance during the amplification (boost) phase [6].

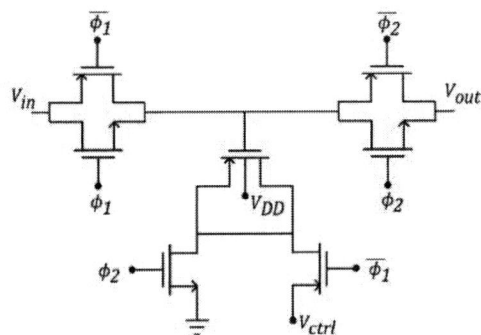

Figure 6. The DT MOS parametric amplifier based on a PMOS varactor.

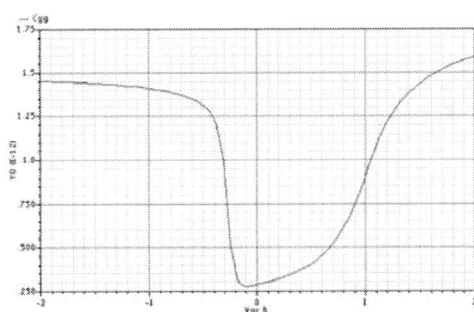

Figure 7. Total gate capacitance of PMOS varactor device.

The total gate capacitance of a PMOS device as a function of the applied common-mode gate voltage is shown in Fig. 7. One can conclude that biasing the device in moderate inversion during the sampling phase, a lower value (lower than C_{ox}) gate capacitance is obtained. Therefore, by controlling this capacitance, the gain of the circuit can be adjusted. This gain control voltage has been included in the proposed circuit, shown in Fig. 6. By varying the voltage applied to the drain/source terminal during the sampling phase (V_{ctrl}), a variation of the device sampling capacitance is possible, thus enabling a voltage gain control.

The maximum gate capacitance when in strong inversion was chosen to be 1 pF (in order to minimize load and parasitic capacitance effects on the gain) and the MOS device was sized accordingly using a PMOS structure. The non-overlapping phases are assumed to be provided by the existing clock driving the ADC. In our simulations a 100 kHz sampling frequency was used. A simple resistive ladder is used, together with an analog multiplexer, to provide V_{ctrl} : this circuit is basically a 2-bit voltage-mode digital-to-analog converter, DAC. The gain is adjustable in 4 steps, by controlling the two input bits of the DAC.

V. SIMULATION RESULTS

The regulated common-gate topology with noise cancellation (Fig. 5) was used as the first stage balun I/V converter. The second stage is a fully-differential DT MPA (two circuits similar to that in Fig. 6. The circuit is designed in a standard (purely digital) 130 μm CMOS process.

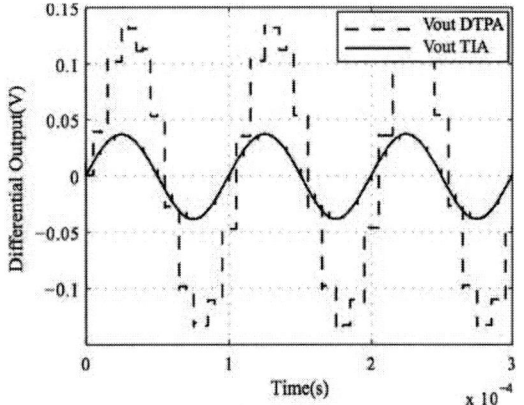

Figure 8. Differential output signal.

In Fig. 8 the simulated differential output signals of the I/V converter and of the DTPA can be observed, when an input 10 kHz sine wave with 1 μA amplitude is assumed as the photodiode input.

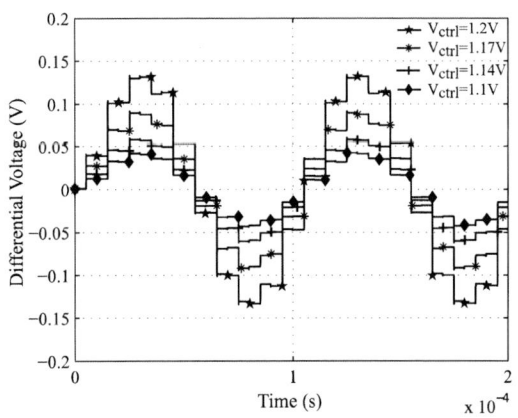

Figure 9. Simulated output of the DT MPA circuit for different gains.

In Fig. 9, different gains of the DT MPA can be clearly observed for the same input signal. By varying the PMOS varactor drain voltage between 1.2 V to 1.1 V, the different gains are obtained. Since low noise is a requirement for the TIA the input referred noise is also simulated and is shown in Fig. 10.

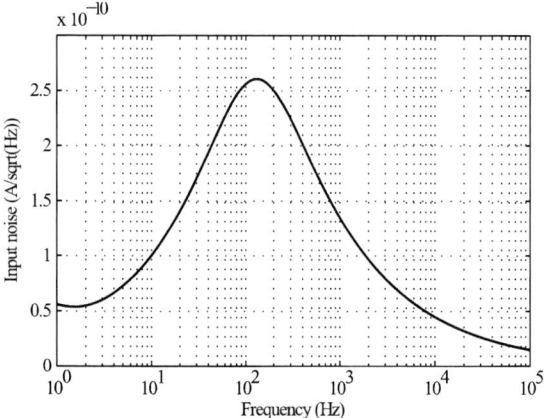

Figure 10. Input referred noise power density.

VI. CONCLUSIONS

This paper presents a two-stage TIA with gain control. The proposed amplifier uses a RCG employing a balanced noise cancellation technique in the first stage. For the second state a DT variable gain MPA circuit was presented. Simulations results obtained with Cadence tools using BSIM V3.3 models were given with a supply voltage of 1.2 V for the circuit designed in UMC 130 nm digital CMOS technology. These results demonstrate the feasibility of the proposed circuit with a power consumption lower than 250 μW.

REFERENCES

[1] R. Sarpeshkar, *Ultra Low Power Bioelectroncis*, Cambridge, 2010.

[2] J. Savoj and B. Razavi, *High-Speed CMOS Circuits for Optical Receivers*, Kluwer Academic Publishers, 2001.

[3] S. Radovanovic, *High-speed photodiodes in standard CMOS technology*, Phd Thesis, Print Partners Ipskamp, 2004.

[4] J. Gao, *Optoelectronic Integrated Circuit Design and Device Modeling*, Wiley, 2011.

[5] L. B. Oliveira, C. M. Leitão and M. M. Silva, "Noise Performance of a Regulated Cascode Transimpedance Amplifier for Radiation Detectors," to appear in *IEEE Trans. Circuits and Systems-I*, 2012.

[6] S. M. Park and H. J. Yoo, "1.25-Gb/s regulated Cascode CMOS Transimpedance Amplifier for Gigabit Ethernet Applications," *IEEE J. Solid-State Circuits*, vol. 39, no. 1, pp. 112-121, Jan. 2004.

[7] F. Bruccoleri, E. A. M. Klumperink and B. Nauta, "Wide-Band CMOS Low-Noise Amplifier Exploiting Thermal Noise Canceling," *IEEE J. Solid-State Circuits*, vol. 39, no. 2, pp. 275-282, Feb. 2004

[8] Y. Tsividis and K. Suyama, "Strange ways to use the MOSFET," Proc. *IEEE Int. Symp. Circuits and Systems*, pp. 449–452, June 1997.

[9] S. Ranganathan and Y. Tsividis, "Discrete-time parametric amplification based on a three-terminal MOS varactor: Analysis and experimental results," *IEEE J. Solid-State Circuits* vol. 38, no. 12, pp. 2087–2093, Dec. 2003.

MIXDES 2012, 19th International Conference *"Mixed Design of Integrated Circuits and Systems"*, May 24-26, 2012, Warsaw, Poland

A Calibratable Capacitance Array Based Approach for High Resolution CR SAR ADCs

Jan Henning Mueller, Sebastian Strache, Laurens Busch, Ralf Wunderlich and Stefan Heinen
Chair of Integrated Analog Circuits and RF Systems
RWTH Aachen University, Aachen, Germany
Email: jmueller@ias.rwth-aachen.de

Abstract—This paper describes widely used capacitor structures for charge-redistribution (CR) successive approximation register (SAR) based analog-to-digital converters (ADCs) and analyzes their linearity limitations due to kT/C noise, mismatch and parasitics. Results of mathematical considerations and statistical simulations are presented which show that most widespread dimensioning rules are overcritical. For high-resolution CR SAR ADCs in current CMOS technologies, matching of the capacitors, influenced by local mismatch and parasitics, is a limiting factor. For high-resolution medium-speed CR SAR ADCs, a novel capacitance array based approach using in-field calibration is proposed. This architecture promises a high resolution with small unit capacitances and without expensive factory calibration as laser trimming.

Index Terms—Analog-digital conversion, Analog-digital integrated circuits, Calibration, CMOS integrated circuits, Mathematical model, MATLAB, Mixed analog digital integrated circuits, Noise, Numerical simulation, Prediction methods

I. INTRODUCTION

A. CR SAR ADC Principle

The SAR ADC basic structure is shown in Fig. 1. A digital-to-analog converter (DAC) generated signal is compared to the analog input voltage. Using binary search algorithm, the generated signal is changed until the difference at the comparator inputs is below a certain threshold and the digital output word is determined. As the input voltage should be constant for the conversion time, a sample and hold (S&H) element is usually placed previous to the analog input.

The CR SAR ADC (CSA) principle has been published first in 1975 [1]. Fig. 2 shows the basic single-ended structure. Here, the capacitors represent both the S&H element and the DAC. During the sampling phase, the upper plates of the capacitors are connected to GND by S_A and the input voltage is sampled via connection to the bottom plates. Then, S_A is opened and S_B is connected to V_{ref}. $S_{0...N}$ are connected to GND which forces the comparator negative input voltage to $-V_{in}$. By switching $S_N...S_1$, the DAC operation is performed.

B. Background and Recent Trends

CSAs are widely used for medium-resolution and medium-speed applications because of their simple structure and their low power consumption [2]. In the last years, an increasing number of papers about this topic has been published [3], proposing several approaches to reduce the power consumption of the capacitor recharging by use of other reference

voltages, a fully- or semi-differential structure or an advanced switching scheme [4]–[6]. The split-capacitor named fragmentation of the most-significant bit (MSB) capacitor [7] significantly reduces the power dissipation for down transitions. Furthermore, a small overall capacitance decreases area and power consumption and increases speed. To reduce the overall capacitance of the ADC without reduction of the unity capacitance, one or more attenuation capacitors can be added in serial [8], but in [2] it was shown that this approach requires a larger unit capacitance resulting in an even higher overall area consumption. This assumption was confirmed by the simulations in [9]. Therefore, the most-promising approach seems to be a differential, conventional split-capacitor structure with a small unit capacitance, e.g. 0.5 pF as described in [10]. However, a small capacitance limits the linearity and the resolution of the ADC due to kT/C noise and mismatch, whereas mismatch means not only mismatch due to production accuracy but also due to parasitics.

C. Paper Structure

The paper is structured as follows. In Section II, the unit capacitor sizing limits are considered. The principle of a calibratable structure is introduced and its advantages are discussed. Finally, the mismatch and linearity simulations of the different concepts are explained and the simulation results are presented. In Section III, a hardware realization of a calibratable CSA and its calibration concept are proposed. Section IV gives an overview over the next development steps; the paper is concluded in Section V.

II. LIMITATIONS TO UNIT CAPACITOR SIZING

The smaller the capacitance, the lower are area, power consumption and conversion time. But for higher resolutions, the voltage accuracy gets more critical, which also depends on the capacitance.

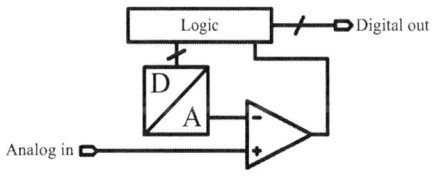

Fig. 1. Principle of SAR ADC.

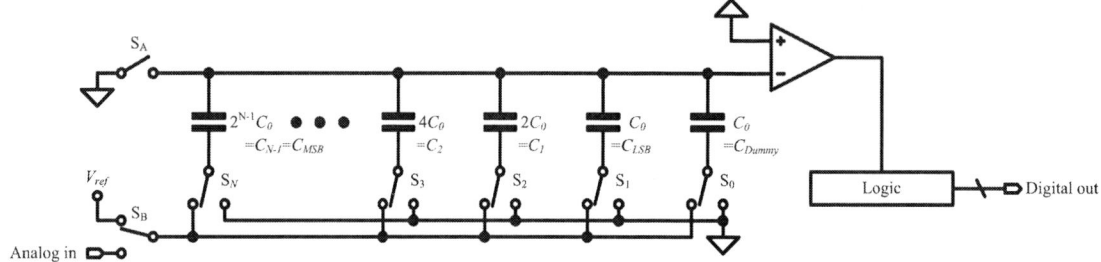

Fig. 2.   CR SAR ADC.

A. kT/C Noise

kT/C noise is a special case of the thermal noise of a resistor given by

$$\overline{v^2} = 4 \cdot k \cdot T \cdot R \cdot \Delta f, \qquad (1)$$

resulting from the capacitor and the parasitic resistances building a low pass filter. The resulting formula

$$\overline{v^2} = \frac{k \cdot T}{C} \qquad (2)$$

is independend of the resistance. The kT/C noise is higher for smaller capacitances. This value has to be considered during the S&H phase. The corresponding capacitance is the overall capacitance

$$C = \sum_i C_i = 2^N \cdot C_0, \qquad (3)$$

whereas the last term is only valid for a conventional structure without attenuation capacitor. The kT/C noise should be much smaller than the least-significant bit (LSB) voltage. Often, it is assumed that it should be at most half the value of V_{LSB}:

$$V_{LSB} = \frac{V_{ref}}{2^N} >> \sqrt{\overline{v^2}} = \frac{1}{2} \cdot \frac{V_{ref}}{2^N}, \qquad (4)$$

which allows to give the achievable resolution N as a function of the unit capacitance C_0:

$$N(C_0) = 2 \cdot \mathrm{ld}\left(\frac{V_{ref}}{2} \cdot \sqrt{\frac{C_0}{k \cdot T}}\right). \qquad (5)$$

This relation is shown in Fig. 3 which exposes that kT/C noise is not a problem for a conventional CSA architecture.

B. Matching

Mismatch of the unity cells is caused by production process non-idealities such as varying capacitor size and dielectric thickness. The parasitics of the capacitors themselves and their wiring can also affect the capacitance of each unit capacitor. If a normal distribution is assumed, the capacitance standard variation σ_{C0} can be used to describe all these effects. It has to be considered that with increasing capacitance, usually this standard variation in relation to the capacitance decreases due to the law of area:

$$\frac{\sigma_C}{C} = \frac{A_C}{\sqrt{W \cdot L}}, \qquad (6)$$

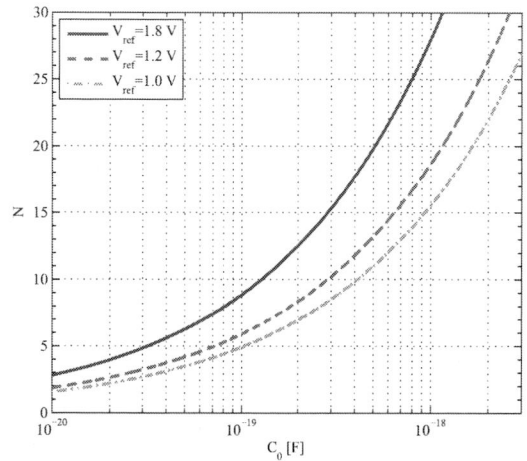

Fig. 3.   Achievable resolution for conventional CSA architectures due to kT/C noise, $T = 300\,\mathrm{K}$.

with A_C as technology specific constant. If parallel unit cells are used instead of larger capacitors, this equation also can be used because of the root-square addition of random variables. This is important for proper mismatch estimation of the larger capacitors in the CSA structure.

1) Worst-Case and Normal-Distribution Estimation: To get a relation between the standard variation σ_{C0} and the resulting nonlinearities, the most critical transistion point at MSB/2 has to be considered [11]. The relative capacitance mismatch error is

$$|\Delta C_x| = |C_{x,target} - C_{x,error}|. \qquad (7)$$

For worst-case analysis [11] it is assumed, that the MSB capacitor is at maximum size while all other capacitors are at minimum size:

$$C_{N,error} = 2^{N-1}(C_0 + |\Delta C_0|) \quad \text{and} \qquad (8)$$
$$C_{0...N-1,error} = 2^{N-1}(C_0 - |\Delta C_0|). \qquad (9)$$

Demanding the differential non-linearity (DNL) to be $< \frac{1}{2}$ LSB, this equations lead to

$$|\Delta C_0| < \frac{C_0}{2^{N+1} - 2} \qquad (10)$$

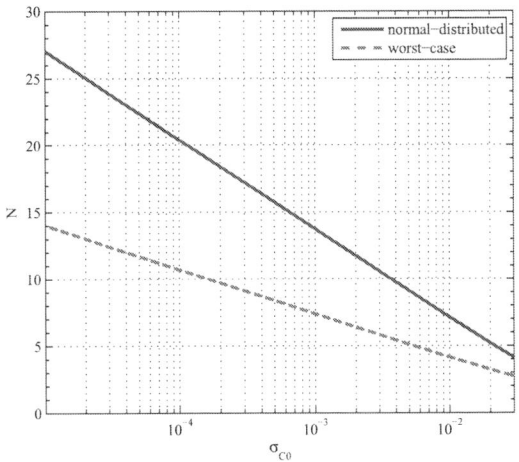

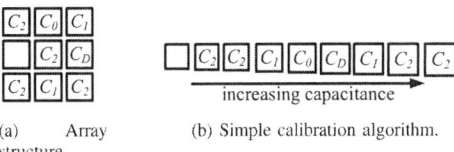

(a) Array structure.

(b) Simple calibration algorithm.

Fig. 5. Individually usable unit capacitor cells for a 3-bit CSA with elements C_{Dummy} and $C_0...C_2$ composed of unit cells. The blank cells are unused.

Fig. 4. Achievable resolution for conventional CSA architectures due to mismatch with $|\Delta C_0| = 3\sigma_{C0} \cdot C_0$.

with e. g. the assumption $|\Delta C_0| = 3\sigma_{C0} \cdot C_0$. Simulations show, that this worst-case estimation is overcritical because it neglects the law of area (or the square-root addition of random variables, respectively). Respecting Eq. (6), Eq. (10) turns into

$$|\Delta C_0| < \frac{\sqrt{2^{N-1}}}{2^{N+1} - 2} \cdot C_0. \qquad (11)$$

The simulations in [9] are also based on this approach. Fig. 4 shows the achievable resolution as a function of σ_{C0} following Eq. (10) and (11), which are slightly more severe in contrast to [9] because of the DNL approach.

Compared to current technology capabilities, for CSA designs with small unit capacitances, obviously matching becomes a problem even for medium resolutions.

2) Calibratable Architectures: To face this problem, laser trimmig can be used to adjust the capacitors. However, this post-production process is expensive and not suitable for co-integrated CSAs. Furthermore, variations during the product lifetime cannot be compensated. Therefore, this paper proposes a CSA structure which can be calibrated in field. This section only describes the model.

The proposed structure consists of an array of unit capacitor cells which can be individually connected during an initial calibration as shown in Fig. 5(a). A cell can be used as a single capacitor, or bigger capacitances can be created connecting 2 or more cells in parallel. This allows to compose binary weighted capacitances (as in a conventional design). While C_0 and C_{Dummy} are represented by only 1 of the unit cells, $C_{1...N-1}$ are represented by $2^{1...N-1}$ unit cells. Overall, 2^N unit cells are needed for this purpose, even if there may be more available for calibration redundancy. Assuming that the capacitances of the unit cells are normally distributed, it will be shown that it is possible to compose nearly perfect conventional capacitances. Energy-minimizing techniques as split capacitor are also possible because of the flexible unit

cell usage.

3) Simulation: A MATLAB® simulation has been set up to test different calibration algorithms for the capacitor array design mentioned before and to compare the achievable performance with conventional designs. The simulation creates an array of normal distributed unit capacitances with the mean value C_0 and the standard deviation σ_{C0}. For simulations of a conventional design with the capacitors $C_{0...N-1}$, the capacitance of C_n is calculated by adding randomly 2^n of these unity capacitances. For the calibratable structure, the capacitances are created using different calibration algorithms which try to select the optimum unit cells. A so-called *simple algorithm* sorts the unit capacitances in increasing order and combines the smaller capacitances with the larger to generate the capacitances consisting of more then 1 C_0. C_0 and C_{Dummy} are selected from the middle, where the capacitors with the smallest deviations are supposed to be located. The algorithm selection for a 3-bit CSA is illustrated in Fig. 5(b). An *advanced algorithm* uses a recursive sorting function to allow an improved selection. Both algorithms are only used to identify the limits of calibratable structures in general and not considered to be realized in hardware. A suitable hardware realization approach, which works as described in Section III-B, has also been implemented.

With the composed capacitances created in these manners, analog-to-digital conversions are simulated using the conventional algorithm as described in Section I-A. The resulting performance values integral non-linearity (INL), DNL and signal-to-noise-ratio (SNR)/effective number of bits ($ENOB$) are obtained using a ramp as input voltage over time. Unfortunately, with this approach a very slow ramp (and so a lot of conversions) are needed to get precise results. To minimize simulation time, hence a combination of this numerical and an analytical approach has been used with only a medium number of comparisions. After every conversion, the closest comparision and thus the most critical conversion phase is determined, allowing to analytically find the critical voltage when the output word toggles. A comparision between known cases [11] shows that this algorithm works very accurately ($ENOB$, INL and DNL divergence $< 10^{-6}$). The Monte Carlo worst case and typical case simulation results are shown in Tab. I and II for 10-bit ADCs with $\sigma_{C0} = 2\%$. The calibration algorithms use a unit cell overhead of about 20 %.

Results show, that even in typical case the performance of CSAs can be significantly improved with calibratable capacitances. The worst case results for the conventional architecture

TABLE I

SIMULATION RESULTS FOR $\sigma_{C0} = 2\%$ AND $N = 10$, WORST CASE.

Architecture	SNR [dB]	$ENOB$ [bit]	INL_{max} [LSB]	DNL_{max} [LSB]
Conventional	48.88	7.83	1.85	1.66
Attenuation Capacitor	36.31	5.74	6.23	11.10
Simple Calibration	60.36	9.73	0.35	0.35
Advanced Calibration	61.97	9.99	< 0.01	< 0.01
Hardware Calibration	61.87	9.98	0.08	0.11

TABLE II

SIMULATION RESULTS FOR $\sigma_{C0} = 2\%$ AND $N = 10$, TYPICAL CASE.

Architecture	SNR [dB]	$ENOB$ [bit]	INL_{max} [LSB]	DNL_{max} [LSB]
Conventional	58.17	9.37	0.71	0.62
Attenuation Capacitor	47.71	7.63	2.53	2.71
Simple Calibration	61.79	9.97	0.09	0.09
Advanced Calibration	61.97	9.99	< 0.01	< 0.01
Hardware Calibration	61.95	9.99	0.04	0.04

comply very well to the results presented in [9]. Comparative simulations between conventional and attenuation capacitor structures with an equal overall capacitance (instead of equal unit capacitance) substantiate the conclusion of [2], that attenuation capacitor designs do not provide an area or power advantage over conventional designs.

The simulation does not only provide an estimation of the performance for a given architecture, unit capacitance and technology, but also allows suggestions which standard deviations and thereby unit capacitances are required for a desired resolution as Fig. 4. If e.g. INL and DNL are claimed to be < 0.5 LSB, simulations show that σ_{C0} has to be < 0.48 % for the conventional structure, which complies very well to the analytical approach plotted in the mentioned graph. With the advanced calibration algorithm, nearly all mismatch errors can be removed. Nevertheless, also the hardware calibration can handle a σ_{C0} of up to 5.21 %, that makes calibratable structures highly promising for CSA designs which use low-area unit capacitors.

III. REALIZATION OF CALIBRATABLE CR SAR ADC

In Section II-B.2, a calibratable CSA model has been described. In this Section, a feasible hardware realization is proposed.

A. ADC Structure

The fully-differential calibratable CSA is depicted in Fig. 6 and consists of 2 arrays with each $35 \times 35 = 1225$ capacitor unit cells (for a 10-bit CSA at least $2^{10} = 1024$ are required) which are described in Section III-A.1. Further, there are a variable offset comparator (see Section III-A.3) and a digital

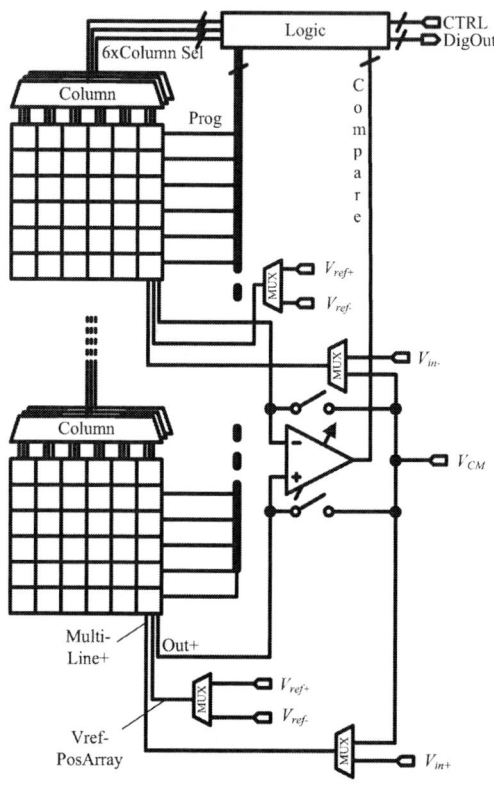

Fig. 6. Block diagram of a fully-differential calibratable array based CSA.

control logic. The arrays are controlled via vertical column lines (3 per column) and horizontal programming lines (1 per row). The column lines are controlled by column decoders as in random access memory (RAM) arrays. The top plates of the capacitor cells are connected to the comparator input ($Out+/-$), while the bottom plates can be connected to V_{CM}/V_{in} (common mode, CM) and to $V_{ref+,-}$ (to handle negative input voltages) or can be shorted to the top plates.

1) Capacitor Unit Cell: Fig. 7 shows the implementation of a unit cell. The unit capacitor C_0 is shown in the bottom left corner. The lower port of the capacitor, which is actually the top plate in layout, is connected to the comparator as in the conventional design. This net is named *Out*. The upper port of the capacitor, actually the bottom plate in layout, can be connected to the connection lines shown on the right via the pass transistors M_{Px}, whereby the *Outline* shortes the capacitor (it is externally connected to *Out*) and the reference line V_{ref} and the *Multiline* (connected to V_{in} or V_{CM}) act as the suitable connections for the capacitors as the switches S_n in the conventional design (see Fig. 2). *Out* and the connection lines are unique for the whole array.

On the left side, the column lines are shown, whereas the programming line *Prog* is shown on top. The column lines are used to select the whole column for programming the unit cell with the *Prog* line. Therefore, one of the column lines is set

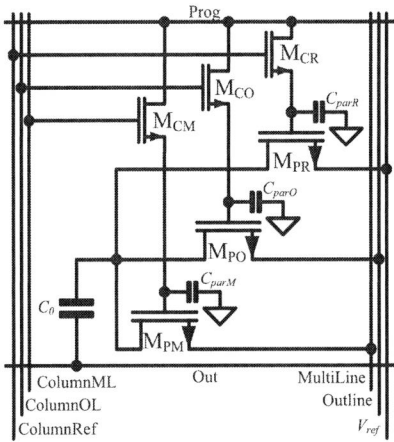

Fig. 7. Unit cell of the calibratable array based CSA.

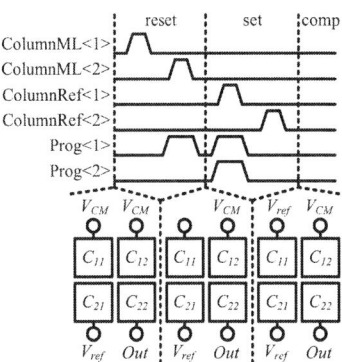

Fig. 8. Example programming of a 2×2 array for an up transistion: the C_{11} connection is changed from V_{CM} to V_{ref}, C_{12} stays connected to V_{CM} and C_{21} to V_{ref}, C_{22} is unused.

to HIGH potential opening one of the column transistors M_{Cx}, allowing the *Prog* line to charge or discharge the parasitic capacitance C_{parx} which anon establishes or removes the connection of the attached pass transistor M_{Px}.

With the grid of horizontal *Prog* lines and vertical column lines, the cells can be individually connected to V_{in}/V_{CM}, to V_{ref} or shorted. Although it is possible to open more then one pass transistor at the same time, this does not make sense for the normal operation.

All transistors can be regarded as transfer gates, which do not forward the full voltage from drain to source or vice versa, because the voltage is reduced by the threshold voltage. To overcome this voltage loss, the *Prog* lines and the column lines have to be driven with higher voltages than V_{ref} (known as bootstrapping). It is also important that the *Prog* high period overlaps the column line high period to avoid problems with a finite edge slope.

In the unit cell layout, the transistors and the lines can be placed below a metal-insulator-metal (MIM) capacitor, so no area overhead is added.

2) Capacitor Array Operation: For the conversion operation, each used unit cell belongs to a capacitance $C_{Dummy,0...N-1}$ as in a conventional approach, whereby e.g. C_{Dummy} and C_0 are represented by each only one unit cell and C_{N-1} by 2^{N-1} unit cells used in parallel. The unused unit cells are shorted. The employed unit cells can be connected either to $V_{in+,-}/V_{CM}$ (*Multiline*) or to $V_{ref+,-}$ as described before. The information which cells belong to which capacitance is saved in the digital control logic during calibration. This allows to run the ADC with the capacitor unit cells as a conventional CSA after an initial calibration.

In each conversion step (S&H, one step for sign detection and one step per bit form a analog-to-digital conversion) old connections (from the last step) are removed, and new connections are established. Therefore, each conversion step consists of 3 phases: *reset* phase, *set* phase and *comparator* phase. In the *reset* phase, now unused connections are

removed, such as the V_{CM} connection of C_n for an up transistion. In the *set* phase, new connections are set, such as the required V_{ref} connections for the same up transistion. In the *comparator* phase, the comparator determines the current bit. In the *reset* and the *set* phase, the set of column lines of the connection, which has to be removed or established, strobe all corresponding columns from one side to the other, while the *Prog* lines reset or set the corresponding connections of the unit cells in the current column. The order of the *reset* and the *set* phase is important to avoid shorts, i. e. in the mentioned example in the *reset* phase the V_{CM} connections of the C_n cells are removed (the capacitors are floating for a moment), and the V_{ref} connections are established in the *set* phase, hence in the *reset* phase the *ColumnML* lines strobe, while the *ColumnRef* lines strobe in the *set* phase. This example is illustrated in Fig. 8 for a 2×2 array. The *ColumnOL* lines are not used during this conversion step, but for shortening the unused cells during S&H. For a down conversion, the order of the column line sets must be the other way round. Because there is no way of preserving the old connection, during the *reset* and *set* phases all connections are refreshed.

Because the strobe clock needs to be faster than the conversion step clock, this architecture is only suitable for low and medium speed CSAs.

3) Variable Offset Comparator: The variable offset comparator acts as a normal comparator (indicate which input is at higher voltage), when the variable offset is set to 0. During ADC operation, the comparator always works as a normal comparator. The comparator inputs can also be connected to V_{CM}.

For calibration, it is possible to vary the comparator input offset a little bit (about a few mV). This can be done by connecting capacitors to one output node as an asymmetric load. If two voltages are compared, the comparator outputs HIGH potential if the positive input voltage is higher than the negative input voltage. If the difference is large enough, this output is always the same, independent of a small comparator input offset. But if the input voltages are close to each other,

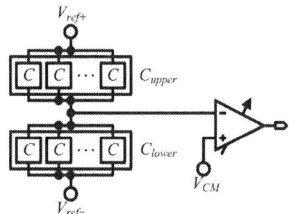

Fig. 9. Capacitive divider for calibration.

e. g. if the positive voltage is only a few mV higher than the negative input voltage, the comparator output will toggle if a negative offset is applied and vice versa. In this way it is not only possible to say which voltage is higher, but also if the input voltages are close to each other. For calibration, at first the input is compared with a positive offset and then with a negative offset. If the comparator toggles, the voltage difference is in a small window.

B. Calibration

The calibration of CSAs is nontrivial because of the limited accuracy of the comparator. If a conventional CSA is calibrated, the comparator needs to detect voltage differences which are much smaller than in normal operation, since the mismatch of the small capacitors leads only to small voltage changes at the middle of the capacitive divider which is connected to the comparator input. In this design, this problem can be avoided by shortening the other capacitors, hence only the current calibrated capacitors form the capacitive divider.

The calibration can be executed directly before conversion operation. Therefore, well-matching binary capacitances as in a conventional CSA have to be determined. This is done using a capacitive voltage divider between V_{ref+} and V_{ref-} with 2 groups of 1 or more capacitor cells, C_{upper} and C_{lower}, as shown in Fig. 9. The divided voltage is compared to V_{CM} using the window technique described in Section III-A.3. If the voltage is within the window, the upper and lower capacitances match each other well.

For calibration, at first C_{Dummy} is selected randomly and used as C_{upper}. Then, the logic seeks a cell with a good matching by selecting different cells for C_{lower}, one after the other using trial and error. If a good matching cell is found, it is assigned to C_0. Then, it is connected in parallel to the C_{Dummy} cell and the logic seeks for two cells which match well to $C_{upper} = C_{Dummy} + C_0$ and are then assigned to C_1. This algorithm continues until all capacitances up to C_{N-1} have been composed. For a capacitor splitting approach, instead of one C_{N-1} group, binary single-controllable capacitances can be determined building the MSB capacitance. The calibration technique has also been implemented in the simulation described in Section II-B.3.

IV. OUTLOOK

A proof-of-concept prototype of the proposed structure with 100 kS/s has been designed and is currently in fabrication. The

prototype uses a common-centroid non-calibrated capacitance composition to test ADC operation only, and can also be calibrated as described in Section III-B. This will allow to evaluate the benefits of the calibration. The chip is expected for III/2012.

For an efficient and area- and power-competitive hardware realization, it may be required to reduce the number of free configurable unit cells in order to reduce circuit complexity and the required clock speed, thus some capacitors must be hard-wired. The simulation has to be adapted finding a trade-off between accuracy and efficiency. It may also be possible to use a modified unit cell design, e. g. with additional control lines which allow to select a whole column.

V. CONCLUSION

It has been shown that kT/C noise does not constitute a problem to conventional CSAs, but matching is a resolution limiting factor for a small C_0. An in-field calibration approach has been proposed which relaxes the matching reqirements by more then a factor of 10. Thereby, either the required area for the ADC can be minimized or the achievable resolution can be increased. The possible hardware realization has been proposed. Due to the higher CMOS integration in the future the design will even become more interesting.

REFERENCES

[1] McCreary, J. L. and Gray, P. R.: *All-MOS Charge Redistribution Analog-to-Digital Conversion Techniques – Part I.* Solid-State Circuits, IEEE Journal of, vol. 10, no. 6, pp. 371-379, Dec. 1975.

[2] Saberi, M., Lotfi, R., Mafinezhad, K. and Serdijn, W. A.: *Analysis of Power Consumption and Linearity in Capacitive Digital-to-Analog Converters Used in Successive Approximation ADCs.* Circuits and Systems I: Regular Papers, IEEE Transactions on, vol. 58, no. 8, pp. 1736-1748, Aug. 2011.

[3] Cho, S.-H., Lee, C.-K., Kwon, J.-K. and Ryu, S.-T.: *A 550-uW 10-b 40-MS/s SAR ADC With Multistep Addition-Only Digital Error Correction.* Solid-State Circuits, IEEE Journal of, vol. 46, no. 8, pp. 1881-1892, Aug. 2011.

[4] Chang, Y., Wang, C. and Wang, C.: *A 8-bit 500 kS/s Low Power SAR ADC for Bio-Medical Applications.* IEEE Asian Solid-State Circuits Conference, pp. 228-231, Nov. 2007.

[5] Liu, C., Chang, S., Huang, G. and Lin, Y.: *A 0.92 mW 10-bit 50-MS/s SAR ADC in 0.13 µm CMOS Process.* VLSI Circuits Digest of Technical Papers, Symposium on, pp. 236-237, Jun. 2009.

[6] Hariprasath, V., Guerber, J., Lee, S.-H. and Moon, U.-K.: *Merged capacitor switching based SAR ADC with highest switching energy-efficiency.* Electronic Letters, vol. 46, no. 9, pp. 620-621, Apr. 2010.

[7] Ginsburg, B. P. and Chandrakasan, A. P.: *An Energy-Efficient Charge Recycling Approach for a SAR Converter with Capacitive DAC.* Circuits and Systems 2005 (ISCAS 2005), IEEE International Symposium on, pp. 184-187, May 2005.

[8] Kim, H., Min, Y.J., Kim, Y. and Kim, S.: *A Low Power Consumption 10-bit Rail-to-Rail SAR ADC Using a C-2C Capacitor Array.* Electron Devices and Solid-State Circuits 2008 (EDSSC 2008), IEEE International Conference on, pp. 1-4, Dec. 2008.

[9] Haenzsche, S., Henker, S. and Schuffny, R.: *Modelling of Capacitor Mismatch and Non-Linearity Effects in Charge Redistribution SAR ADCs.* Mixed Design of Integrated Circuits and Systems (MIXDES), 2010 Proceedings of the 17th International Conference, pp. 300-305, Jun. 2010.

[10] Harpe, P., Zhou, C., Wang, X, Dolmans, G. and de Groot, H.: *A 30 fJ/Conversion-Step 8b 0-to-10MS/s Asynchronous SAR ADC in 90nm CMOS.* Solid-State Circuits Conference Digest of Technical Papers (ISSCC), 2010 IEEE International, pp. 388-389, Feb. 2010.

[11] Baker, J. R.: *CMOS Circuit Design, Layout and Simulation.* Second Edition. Wiley-IEEE Press, 2004.

MIXDES 2012, 19th International Conference *"Mixed Design of Integrated Circuits and Systems"*, May 24-26, 2012, Warsaw, Poland

A Charge Pump with Power-on Reset Circuit

Adam Gołda, Andrzej Kos

AGH University of Science and Technology
Department of Electronics
Krakow, Poland
golda@agh.edu.pl, kos@agh.edu.pl

Abstract—**A charge pump with power-on reset (POR) circuit is presented. The proposed circuit generates reset signal during supply voltage rising. It uses common capacitor, voltage follower, and current source to make POR signal and later on as usual charge pump. So, it saves the chip area and ensures initial charging of capacitor. Furthermore, the proposed circuit guarantees that reset signal appears when supply voltage achieves strictly specified value. As a result all elements are set to their initial states correctly.**

Index Terms—**Power-on reset; power-up control; charge punp; CMOS technology**

I. INTRODUCTION

Both the charge pump (CHP), that constitutes phase-locked loop (PLL) or delay-locked loop (DLL), and power-on reset (POR) circuits are the common parts of the integrated systems. There are several kinds of POR circuits. Their principle of operation is based on capacitor charging in RC circuit and inverter switching after its input voltage exceeds gate threshold voltage [1], [2]. However, such implementations do not work properly for long rising times of supply voltage. Better results are obtained in circuit with charge clamp [1], [3]. Such circuit is based on transistors in diode configuration and current mirrors instead of the resistors [1], [4]. One could also find implementation that generates power-on reset signal using voltage divider and inverters [5]. But it can be used in circuits with low voltage supply in which threshold voltage of MOS transistor has relatively high value in comparison to supply voltage V_{DD}. Implementation with voltage divider and comparator [2] has better properties and it works properly even for high supply voltage. There are also circuits based on current mirrors [6], [7]. The work of current mirror is controlled using threshold voltages of other MOS transistors and its current pulls down or up the voltage that switches inverter. Such circuits work properly even for infinity range of supply rise times [6]. The common bandgap cells are used to generate reset signal, too [8]. However in such implementations threshold voltage, which decides about POR signal generation, depends on velocity of supply voltage rising and can take insufficient values. There are also implementations dedicated to circuits with multiple supply voltage. The application proposed in [9] is based on NAND structure.

The presented charge pump with power-on reset circuit is characterized by twofold utilization of common elements to reset signals generation and as the charge pump. As a result unnecessary growth of chip area is averted.

II. IDEA OF CHARGE PUMP WITH POWER-ON RESET CIRCUIT

The presented charge pump (CHP) with POR circuit (Fig. 1) utilizes common current source (CS_1), voltage follower (VF), and capacitor (C) to generate power-on reset signals (impulse POR, and continuous POR-C) during rising of supply voltage and to work as usual charge pump afterwards. In the first phase, POR-C takes the logic value 1 which almost equals rising supply voltage. In the same time, capacitor C is charged. After capacitor voltage exceeds V_0 value, what is recognized using comparator (CMP), POR-C takes logic value 0 and control unit (CNTR) reconfigures circuit to usual charge pump. Simultaneously, the impulse POR signal is generated by pulse generator (PG).

During common work, the charge pump is controlled by signals taken from phase-frequency detector (PFD).

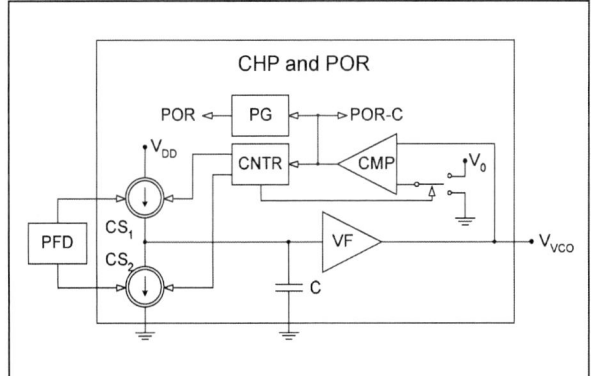

Figure 1.   Simplified diagram of proposed charge pump with power-on reset circuit

The big advantage of proposed circuit is twofold utilization of common charge pump elements, i.e. capacitor C, voltage follower VF, and current source CS_1. As a consequence the chip area is saved, because there is no need to use another capacitor. Moreover, the charge pump capacitor is pre-charged to value close to V_0. Thus it takes less time to achieve required frequency and probable discharging of uncharged capacitor is avoided (such situation is possible as it depends on output signals of phase and frequency detector). Moreover, proposed approach guarantees that power-on reset signal is generated for voltages close to supply voltage V_{DD}. As a result, the sequential logic circuits are properly initialized.

III. IMPLEMENTATION

The general diagram of proposed charge pump with power-on reset circuit is presented in Fig. 2.

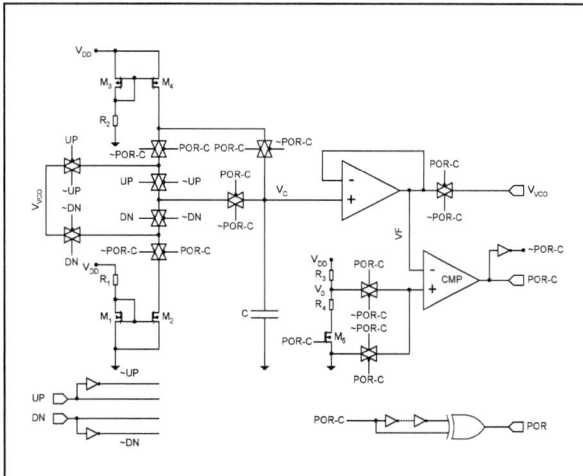

Figure 2. Example of implementation of proposed charge pump with power-on reset circuit

It uses an effect of voltage follower saturation to generate power-on reset signals. The V_0 value is set by means of R_3, R_4, and M_1 voltage divider and has to take into account aforementioned phenomenon of VF saturation. The task of the M_1 transistor is to disconnect resistor ladder after reconfiguration of the circuit into common charge pump. It saves the energy as the static current does not flow from supply voltage through resistors R_3 and R_4 to the ground. As a result, the additional transmission gates have to reconnect input of comparator from V_0 (which is flowing after turning of the M_1 transistor) to the ground.

The example implementation have been performed in 0.35 μm technology with 3.3 supply voltage. The voltage divider is set in such way that V_0 takes the value of about $0.77 \cdot V_{DD}$ (≈2.55 V).

The principles of operation of proposed approach depend on velocity of supply voltage rising. At first let us consider the case for very short rising times of V_{DD}. Figures 3 and 4 present waveforms of the most important signals for 2 μs rising time of supply voltage.

The supply and V_0 voltages very quickly attain their final values. The capacitor and VF output voltages are smaller than V_0, so POR-C equals 1 and circuit is configured to charge C from current source SC_1. When output voltage of VF exceeds V_0, POR-C changes to 0, POR impulse is generated, and circuit is reconfigured to usual charge pump.

The use of voltage follower saturation phenomenon means that voltage V_C across capacitor can attain higher values even equaled to V_{DD}. Therefore, the proper selection of V_0 value has to meet a problem of this saturation effect and proper selection of input voltage of VCO.

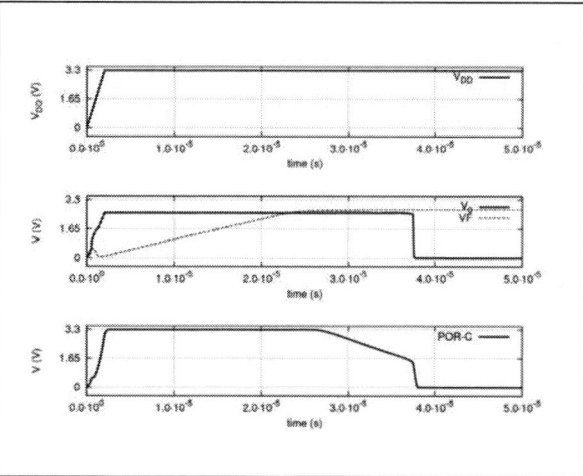

Figure 3. Simulation results of power-on reset signals generation for short (2 μs) rising time of V_{DD}; waveforms: V_{DD}, V_0, VF, and POR-C

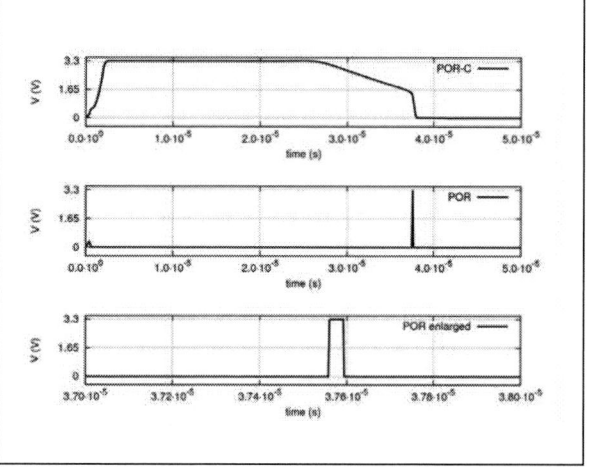

Figure 4. Simulation results of power-on reset signals generation for short (2 μs) rising time of V_{DD}; waveforms: POR-C, POR, and enlarged POR

The charging process of capacitor is slightly different for longer rising times of supply voltage, Fig. 5 and 6. Supply voltage rises so slowly that SC_1 charges capacitor to voltage that is almost equal to V_{DD}. Therefore, this current source in not properly biased until POR signal appears and the circuit is reconfigured. The aforementioned effect of voltage follower saturation causes that VF output voltage is smaller than V_{DD}. The V_{DD} and output voltage of VF have the same slope, whereas V_0 has smaller gradient (defined by R_3, R_4, and M_1 voltage divider), Fig. 5. So, generation of POR impulse signal starts when the V_0 exceeds V_F output voltage. Admittedly, in this case the instantaneous value of supply voltage is smaller than the nominal one and equals 2.98 V but this difference, which is less than 10%, is relatively small.

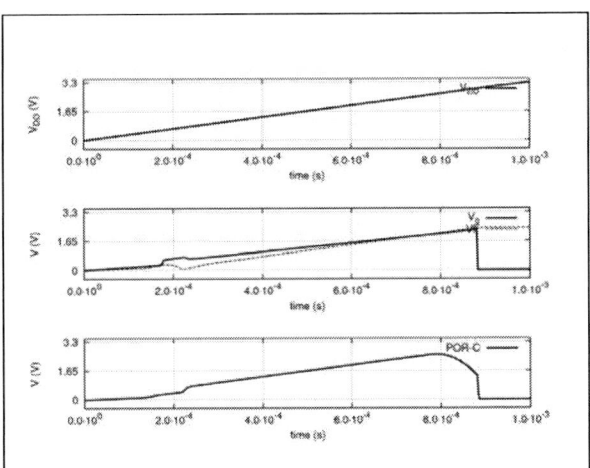

Figure 5. Simulation results of power-on reset signals generation for long (1 ms) rising time of VDD; waveforms: VDD, V0, VF, and POR-C

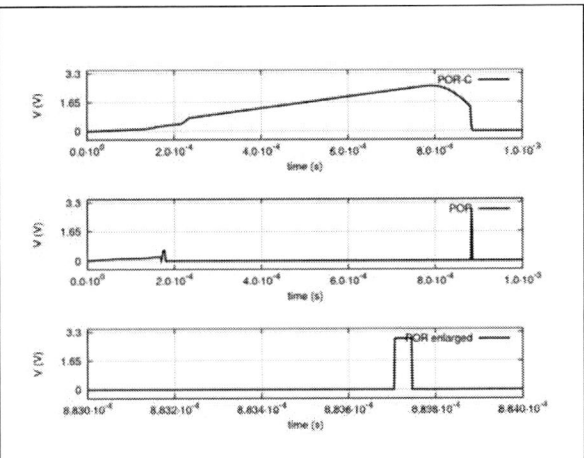

Figure 6. Simulation results of power-on reset signals generation for long (1 ms) rising time of VDD; waveforms: POR-C, POR, and enlarged POR

Thus, it guarantees that all digital devices which require initiation (e.g. register and other sequential units) are properly set.

IV. CONCLUSIONS

Presented charge pump with power-on reset circuit assures that POR and POR-C signals are generated for the threshold voltage that is defined by voltage divider, and it can be set to value close to V_{DD} what guarantees correct initialization of all elements in chip. Utilization of common elements into two stages of circuit operation allows reduction of chip area. Moreover, the charge pump capacitor is pre-charged during the first phase, what assures faster setting of desired frequency, and avoids improper charging of capacitor during the first steps of charge pump work.

REFERENCES

[1] T. Yasuda, M. Yamamoto, T. Nishi, "A Power-on Reset Pulse Generator for Low Voltage Applications," Proc. of the IEEE Int. Symp. on Circuits and Systems ISCAS 2001, Sydney, Australia, 6-9 May 2001, Vol. 4, pp. 598-601

[2] L. Xinquan, Y. Weixue, Ligang, C. You, "A Low Quiescent Current and Reset Time Adjustable Power-on Reset Circuit," Proc. of the Int. Conf. on ASIC ASICON 2005, Shanghai, China, 24-27 October 2005, pp. 568-571

[3] K.-H. Chen, Y.-L. Lo, "A Fast-Lock DLL with Power-on Reset Circuit," Proc. of the IEEE Int. Symp. on Circuits and Systems ISCAS 2004, Vancouver, Canada, 23-26 May 2004, Vol. 4, pp. 357-360

[4] M.-L. Hsia, Y.-S. Tsai, O. T.-C. Chen, "An UHF Passive RFID Transponder Using A Low-Power Clock Generator without Passive Components," Proc. of the 49th IEEE Int. Midwest Symp. on Circuits and Systems, San Juan, Puerto Rico, 6-9 August 2006, Vol. 2, pp. 11-15

[5] S. K. Wadhwa, G. Siddhartha, K., A. Gaurav, "Zero Steady State Current Power-on-Reset Circuit with Brown-Out Detector," Proc. of the Int. Conf. on VLSI Design VLSID'06, Hyderebad, India, 3-7 January 2006

[6] A. Katyal, N. Bansal, "A Self-biased Current Source Based Power-on Reset Circuit for On-chip Applications," Proc. of the Int. Symp. on VLSI Design, Automation & Test, Hsinchu, Taiwan, 26-28 April 2006

[7] A. Dieguez et al., "A Wake-Up Circuit With Temperature Compensated Clock In 1.2V-0.13um CMOS Technology," Proc. of the IEEE Int. Conf. on Electronics, Circuits and Systems ICECS 2007, Marrakech, Morocco, 11-14 December 2007, pp. 367-370

[8] T. Tanzawa, "A Process- and Temperature-Tolerant Power-on Reset Circuit with a Flexible Detection Level Higher than the Bandgap Voltage," Proc. of the IEEE Int. Symp. on Circuits and Systems ISCAS 2008, Seattle, USA, 18-21 May 2008, pp. 2302-2305

[9] Q. A. Khan, G. K. Siddhartha, "A Sequence Independent Power-on-Reset Circuit for Multi-Voltage Systems," Proc. of the IEEE Int. Symp. on Circuits and Systems ISCAS 2006, Island of Kos, Greece, 21-24 May 2006, pp. 1271-1274

MIXDES 2012, 19ᵗʰ International Conference *"Mixed Design of Integrated Circuits and Systems"*, May 24-26, 2012, Warsaw, Poland

A High Speed A/D Converter for Using in Low Power CMOS Image Sensors

Masood Teymouri
Innovation Research Center
Urmia University of Technology
Urmia, Iran

Banafsheh Alizadeh
Islamic Azad University
Miandoab Branch
Miandoab, Iran

Ali Dadashi
Urmia University
Urmia, Iran

Azad Mahmoudi
Islamic azad university
Sardasht branch
Sardasht, Iran

Abstract—**This paper presents a high speed, high resolution column-parallel ADC with global digital error correction. Proposed A/D converter is suitable for using in high-frame-rate CMOS image sensors. This new method has more valuable than conventional ramp ADC from viewpoint of speed and resolution. A prototype 11-bit ADC is implemented in 0.25μm CMOS technology. Moreover, an overall SNR of 63.8dB can be achieved at 0.5Msample/s. The power dissipation of all 320 column-parallel ADCs with the peripheral circuits consume 76mW at 2.5V supplies.**

Index Terms—**Single slope ADC, Column parallel ADC, Comparator, Digital correction, High frame rate image sensor, Two Step ADC**

I. INTRODUCTION

Application of high speed image sensor extends over various fields such as scientific imagers and industrial high-speed machine vision sensors. Many applications require high-frame-rate imagers that this motivates the development of high-speed image sensors with on-chip analog-to-digital converters (ADCs).

Digital output by an on-chip ADC has become the most popular scheme for CMOS image sensor because it can make an imager design extremely simple by eliminating the analog interface from the image sensor [1]. Early commercial CMOS imager modules were usually equipped with a single chip-level ADC, i.e. all pixel outputs are input to a single ADC [2], [3]. Chip level ADC will become increasingly difficult to implement when the number of pixels is high. Another method is pixel-level ADC. Here, an ADC is dedicated for each pixel or each group of neighboring pixels. Therefore, very low-speed converters can be used. This method is especially advantageous for the high-speed operation required for machine vision applications but this method causes the pixel structure to be complicated [4]. An alternative approach is a column-level ADC that each column of pixel array has an ADC.

II. COLUMN-PARALLEL ADC ARCHITECTURE

When multiple ADCs are implemented in the columns, this structure is called column-parallel ADCs. As shown in Fig. 1 every column of pixel array has an A/D converter and every column pixel signal (S_1, S_2, .., S_n) in a row are digitized by each ADC channel so in this method ADCs can be operated at a lower speed than chip level ADC for same frame rate.

Whereas, the required sampling rate in the column-parallel ADCs can be slower than chip level ADC (commonly is as slow as a few hundred Ksamples/s). Therefore, high speed ADC schemes [5] and slow ADC schemes such as successive approximation and single slope ADC can be employed [6], [7].

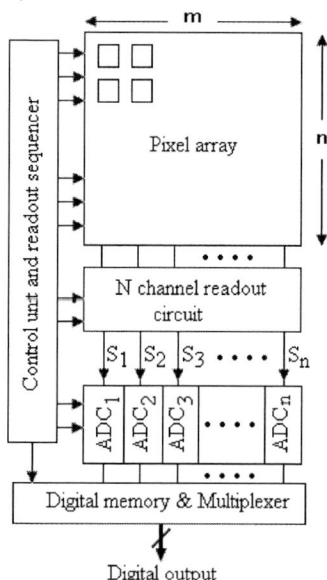

Figure 1. Column parallel CMOS imager block diagram.

III. SINGLE SLOPE A/D CONVERTER SYSTEM

One of the simplest ADC configurations is the single slope or ramp ADC. The circuit consists of a reference generator or ramp generator, which generates the accurate reference ramp signal, a comparator, and a counter. As shown in Fig. 2 the output of a binary counter is connected to the input of the reference generator and the ramp generator outputs a slightly higher voltage by increasing counter. For normal operation, the input signal is applied to one input of the comparator, the counter is set at zero and the ramp generator block is reset. Then counter counts up with each clock pulse and ramp voltage is compared against the input voltage by a comparator. If the input voltage is greater than the ramp output, the counter will continue counting normally. However, eventually the ramp voltage in the negative input of the comparator will exceed the input voltage, the counter stops. The analog input signal is converted into a time that is measured by counting clock pulses until the counter is stopped.

Furthermore, the counter counts from zero for each analog input signal. Therefore sampling rate of the analog signal is very slow for high resolution conversion that is given by

$$F_s = \frac{F_{clock}}{2^N} \qquad (1)$$

Where F_s, F_{clock}, and N are sampling rate, input clock pulse and number of bits respectively. Since one of the main advantages of the single slope converter is very low power consumption in compared with other A/D converter structures. Therefore this is most popular A/D converter for using in column-level ADC for using in CMOS imagers [7].

IV. TWO-STEP SINGLE SLOPE A/D CONVERTER

To avoid some of the problems encountered with the single slope converter, the new two-step single slope A/D converter is proposed. A block diagram of the ADC is shown in Fig. 3. This converter consist of a only one comparator, which has two input differential pairs, a binary counter, two memory blocks which have five and six bits memory cell, a ramp generator which outputs two ramp signals, and a digital correction block.

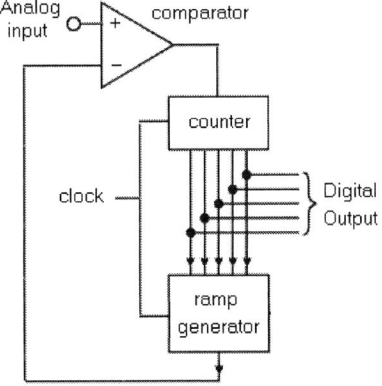

Figure 2. Single slope A/D converter.

A. Proposed A/D Converter Description

As shown in Fig. 3, the proposed A/D converter is able to digitize analog input signal into 11-bit digital output. Digitizing process is done sequentially at four phases which are reset, φ_r, course φ_c, fine, φ_f, and digital correction φ_e. As shown in timing diagram of the A/D converter, see Fig. 4, for each conversion process, reset signal is firstly applied to reset and clear all of the A/D converter blocks. In the second phase, by activating φ_c, the analog input signal is applied to negative input of the comparator, i.e. (node A-), and ramp generator block executes first ramp function, at this time second ramp value is constant and equal to $V_{cm}=1V$. Since analog input signal, V_{input}, is greater than initial value of the first ramp hence negative output of the comparator is high. In this case φ_{L1} is high and switch, S_{sh} becomes on and 6-bit memory cell is in the transparent mode. Thus first ramp is applied to the positive input of the comparator, i.e. (node A+). By applying every clock pulse, counter counts and the first ramp voltage increases and compares against the input signal until the ramp voltage exceeds the input signal. In this time, t_1 the negative output of the comparator and φ_{L1} become low. Therefore the counter output data and the value of the first ramp at t_1 ($Vref_i$) is stored in 6 bit memory cell and capacitor respectively. Since the sampling scheme with S_{ch} and C_{sh} is boot strapped switching, so proper value of V_{refi} is stored in the C_{sh} [8]. After course phase, counter should be set to zero to be ready for next process. In the second or fine phase, only second ramp function is performed and the first ramp generator becomes idle. Note that, in this phase comparator compares two differential input voltages ($V_{second\_ramp} = V_{second\_ramp^+} - V_{second\_ramp^-}$) and ($Vref_i$ - V_{input}). Counter counts and differential second ramp is generated until differential input voltage of B, V_{second_ramp} become greater than differential input voltage of A, ($Vref_i$ - V_{input}), in this time φ_{L2} goes high and 5 bit memory cell latches the binary input data. Therefore $V_{residue}$= $Vref_i$ - V_{input} which is called residue voltage is digitized at t_2.

B. Preamplifier circuit implementation

Fig. 5 shows the circuit diagram of a comparator-gain stage with two differential pairs which are named A and B. Such a stage is something that would be used in schemes of Fig. 3. By considering the circuit, differential output voltage of the gain stage is given by

$$V_{output} = V_{out+} - V_{out-} = r_{out} \cdot (gm_A \cdot V_{in1} - gm_B \cdot V_{in2}) =$$
$$- r_{out} \cdot ((i_{11} - i_{12}) - (i_{21} - i_{22})) \qquad (2)$$

Regarding to the above equation, comparison between two input differential voltages, Vin1 and Vin2, can be performed in current mode. [9][10]. We know that input signal is converted into six most significant bits, (H_6, H_5, H_4, H_3, H_2, H_1) by using first ramp and the other five least significant bits, (L_5, L_4, L_3, L_2, L_1) is achieved by applying second ramp. Therefore step voltage of the first ramp, (V_{LSBH}) and second ramp, (V_{LSBL}) are equal to $V_{LSBH}=(V_{in-pp})/64$ and $V_{LSBL}=V_{LSBH}/32$ respectively.

Where V_{in-pp} is peak to peak voltage of the analog input signal which is 1V. For normal comparison between V_{in1} and V_{in2}, relation between gm_A and gm_B must be selected from (2).

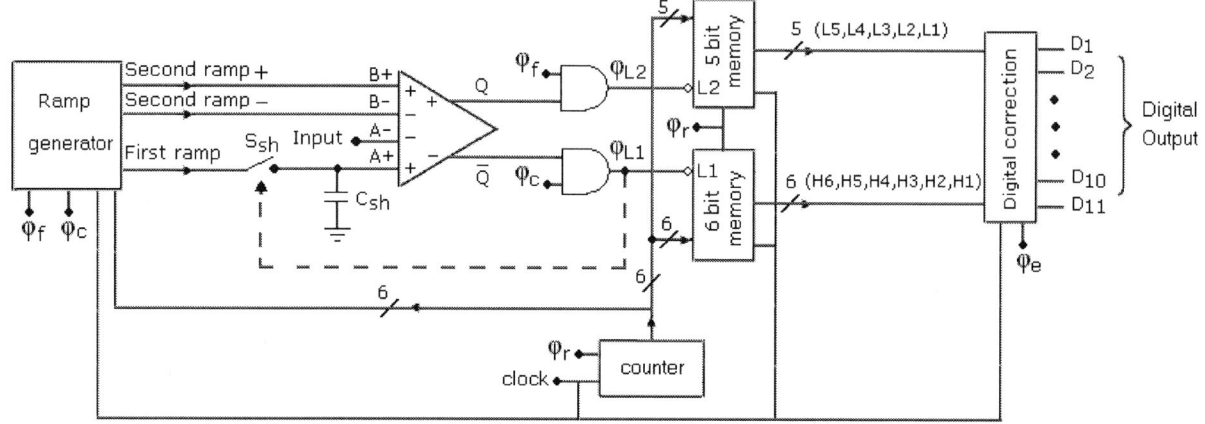

Figure 3. Proposed two step single slope A/D converter block diagram.

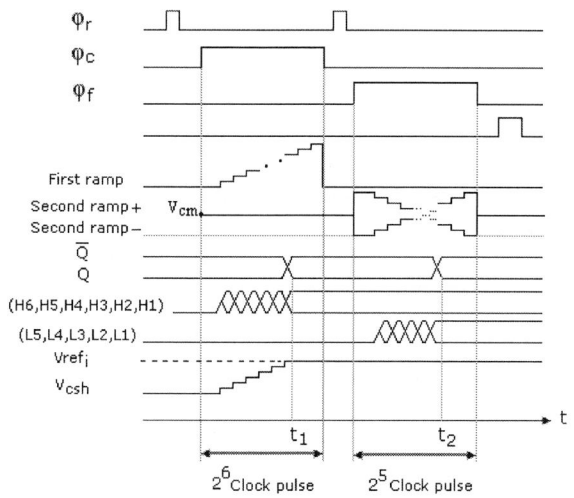

Figure 4. Timing diagram for one conversion.

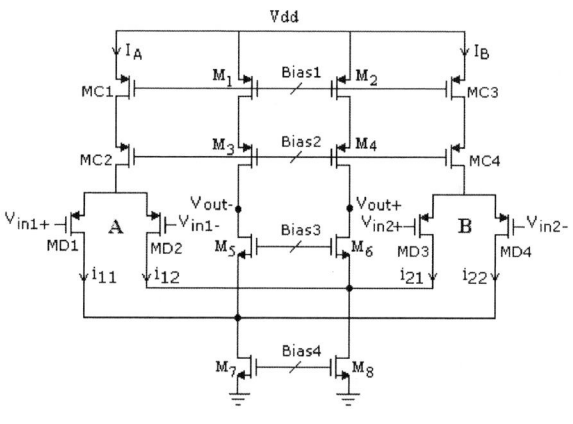

Figure 5. Comparator gain stage.

If the gm_A selects equal to the gm_B, so the transistors size of the differential pair is identical with the differential pair B. Therefore the current of MC_1, is equal to the current of MC_2, Iwhich are $I_A = I_B = 15\mu$. To guarantee this relation between I_A and I_B, Fig. 5 shows, cascode current source that is formed by transistors MC_1 series with the MC_2 and MC_3 series with the MC_4 is best choice.

V. DIGITAL CORRECTION BLOCK

A powerful technique to cancel error among stages of the multistage A/D converters is to provide overlap between the quantization range of successive stages and digitally correct the binary data produced by each stage [11]. Consider the two-step 11 bit ADC depicted in Fig. 3, at first process resolves 6 bits and in the next process resolves 5 bits of the output data. The reference voltages for first ramp are precisely 64 LSBH (least significant bit of the coarse process) apart, and the quantization range of the second ramp is 32 LSBL (least significant bit of the fine process) wide. Fig. 6 shows that if the input voltage, V_{input} lies between $Vref_{i-1}$ and $Vref_i$, at the coarse phase by opening of the switch, S_{sh}, the upper reference, $Vref_i$, will be stored in the capacitor, C_{sh} and residue voltage which is $V_{residue} = (Vref_i - V_{input})$ is resolved in the fine process.

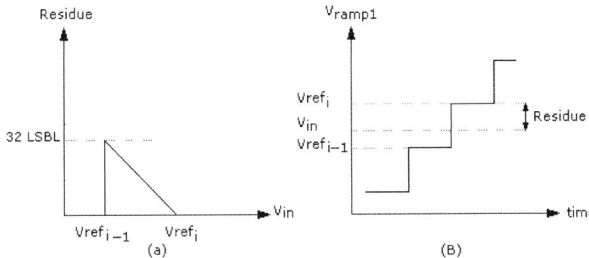

Figure 6. Residue plot for ADC. (a) Residue and input voltages of two successive steps. (b) Residue for sample V_{input}.

To produce the final binary output, the data generated by the two processes must be digitally corrected.

If the digital output of the first process is represented by a binary number $H=H_6,H_5,H_4,H_3,H_2,H_1$ then its value is normalized to the 11 bits A/D converter which is $H=H_6 2^{10}+H_5 2^9+H_4 2^8+H_3 2^7+H_2 2^6+H_1 2^5$. Similarly, the digital output of the second process, $L=L_5,L_4,L_3,L_2,L_1$, can be expressed as $L=L_5 2^4+L_4 2^3+L_3 2^2+L_2 2^1+L_1 2^0$.

Note that real residue voltage must be equal to $V_{input} - V_{ref_{i-1}}$ but $V_{ref_i} - V_{input}$ is resolved in the fine process. Hence the final 11 bit digital output, D must be calculated by equation below,

$$D = H - L$$

| H_6 | H_5 | H_4 | H_3 | H_2 | H_1 | 0 | 0 | 0 | 0 | 0 |

—

| 0 | 0 | 0 | 0 | 0 | 0 | L_5 | L_4 | L_3 | L_2 | L_1 |

=

| D_{10} | D_9 | D_8 | D_7 | D_6 | D_5 | D_4 | D_3 | D_2 | D_1 | D_0 |

Where output data, D is represented by binary number $D=D_{10},D_9,D_8,D_8,D_7,D_6,D_5,D_4,D_3,D_2,D_1,D_0$.

VI. COLUMN PARALLEL ADC

One of the most common techniques to increase the frame rate of the CMOS imager is to process in parallel as many pixels in the array as possible. The single slope ADC is used as a most popularity ADC in column parallel CMOS imagers because of two reasons which are low power and small layout size [1]. But the single slope ADC has the disadvantage of having a slow conversion speed, so it is not good candidate for high speed and high resolution ADCs. This paper introduced a new two-step single slope ADC that offers a significantly higher conversion speed with maintaining the advantages of a single slope ADC.

To implement column parallel ADCs, only one comparator and two memory cells are placed in each column. Binary counter, ramp generator block and the other common blocks are shared to all column ADCs. The main focus during the column ADCs design was to reduce its power consumption and layout size as much as possible. A best way to escape large layout size and huge power consumption of the ADCs, it is useful that common blocks among ADC columns become

shared to all of the comparators. Such as bias circuit, binary counter, ramp generator, digital correction circuit and control unit.

The pixel array is placed in standard row-column structure so that all pixel signals (S_1, S_2, ... S_{320}) in a row are read out into the column comparators in parallel. Therefore all pixel signals in a row are converted to digital simultaneously. For digitizing all of the pixel signals, at first, pixel output is sampled, and then the ramp generator outputs set to the lowest value. In the first stage, the binary 6-bit counter performs the A/D conversion by counting digital clock cycles from 0 to 63. At the end of the first stage, every pixel signal is digitized into 6 bit and captured in column-six-bit memory cell. Note, every column-reference holder capacitor, C_{sh1}, C_{sh2}, . . C_{sh320} stores a voltage which is equal to nearest and bigger reference than its pixel signal. In the second stage counter counts clock pulses again from 0 to 31. This cause second ramp applies to the all comparators and difference between each column-signal and holder capacitor voltage resolves into 5-bit binary data. The 320×11-bit digitized image data for one row are multiplexed and read out using 11-bit digital output ports into the error correction block.

VII. SIMULATION RESULT OF THE A/D CONVERTER

The A/D converter circuit is a part of the 320- column -parallel ADC which is designed based on 0.25 μm CMOS, 1P3M standard technology. Each column comparator circuit consumes 170uW power provided from the 2.5V power supply. Other blocks consume less than 22mW so all of the comparators with peripheral circuits consume about 76mW. The circuit of the ADC is simulated by HSPICE with performing a post-layout simulation on the extracted circuit net-list, Fig. 7 and Fig. 8 show chip layout and the FFT of the ADC output at 0.5MSPs with a full scale sine-wave input. The electrical performance of the A/D converter is summarized in table I.

TABLE I. SPECIFICATIONS OF THE PROPOSED ADC

Power supply	2.5V
ENOB	10.3 Bit
Sampling rate	0.5Msample/s
Analog input	1V
SNR@ 100KHz	63.87dB
SFDR@100KHz	79.2dB
THD@100KHz	-78.7dB
SNR @Nyquist	62.6dB
THD@Nyquist	-75dB
ADC power consumption for all columns	76mW

VIII. COLUMN PARALLEL ADC

The main difference among this work and published papers [12][13] is method of comparing input voltage and reference voltages. In the comparator of the reference [12] the dynamic reference generator and extra circuitry attached to single slope structure cause to consume huge power and layout size and increase conversion time. Because the dynamic reference generator outputs M different ramp voltages in compared to the conventional structure that generates one ramp voltage. M row line with many switches is needed to apply multi ramp voltages to all of the comparators. Numerous row line and switches increases ramp generator output parasitic capacitance and consumes much power to handle proper ramp voltages. In the reference of the [13] another two step structure is introduced. By focusing on the two-step SS ADC with single ramp (see Fig. 3a from [13]), to compare input voltage with the two references in coarse and fine phases, single ended voltage mode structure is used. But in this work differential current mode structure is introduced. In this new method simplicity, low layout size, low power consumption and high speed conversion rate is main advantages in compared to two reported papers. Table II shown the performance summary of the proposed A/D converter and recently reported result [12][13].

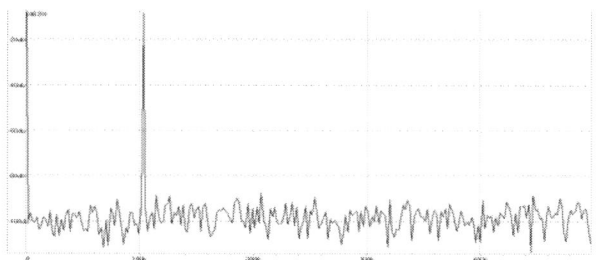

Figure 8. Output spectrum for an input signal with 1Vpp signal swing and 100-kHz frequency.

CONCLUSION

One of the bottlenecks of high frame rate CMOS image sensors is posed by ADCs. Therefore, for high speed applications, column-parallel ADCs appear as a very reasonable option. In this paper a column parallel two step ADC with digital correction is presented. The resulting structure is simple and compact, and the two step conversion introduces an alternative approach to improve power consumption with high speed and high resolution. The overall ADC achieves 10.3 bit of ENOB at 0.5 MS/s with 76mW of power dissipation for consuming of 320 column-parallel ADCs and peripheral circuits. Fig. 9 gives a comprehensive schematic diagram of a 320 column parallel ADCs.

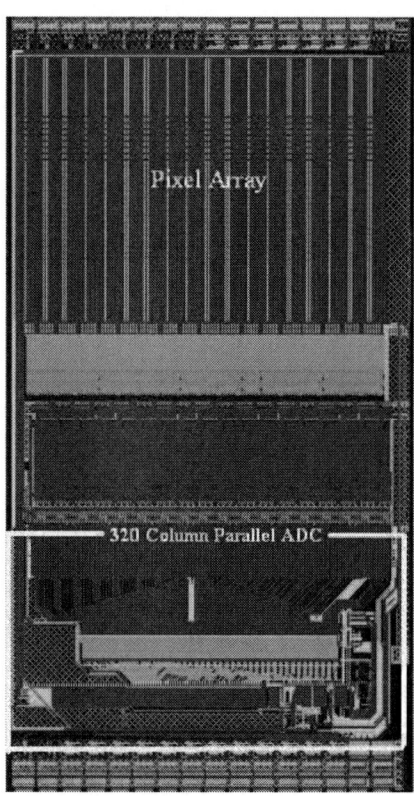

Figure 7. Chip layout

TABLE II. PERFORMANCE SUMMARY

	[12]	[13]	This Work
Technology	0.25 μm CMOS	0.35 μm CMOS	0.25 μm CMOS
ADC Column Number	330	240	320
Power Supply	2.5V	2.8V	2.5V
A/D Consumption for one column	157 μW	150μW	170μW
A/D Conversion time	16us	4us	2us
A/D Digitizing phases	3b-coarse/8b-fine	5b-coarse/6b-fine	6b-coarse/5b-fine
A/D Resolution	10b	10b	11b
A/D Resolution	10b	10b	11b

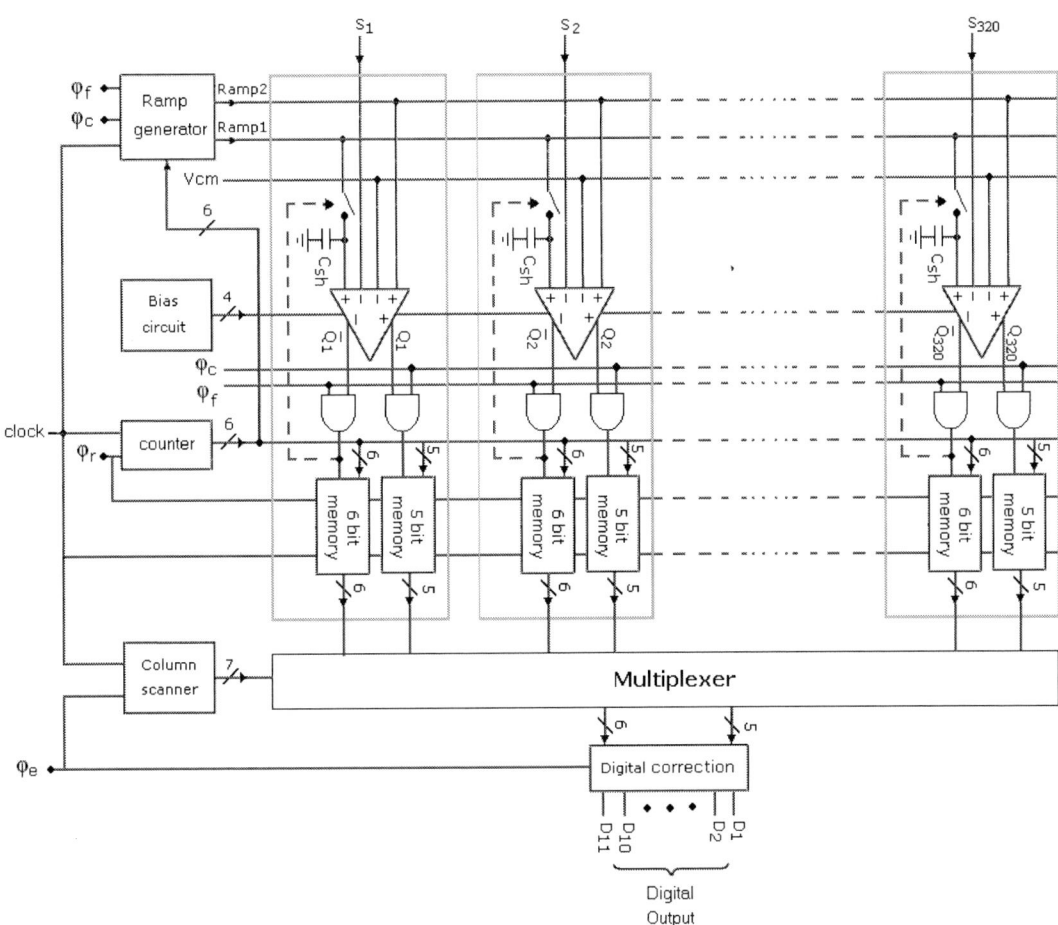

Figure 9. 320-column-parallel ADC architecture.

REFERENCES

[1] J. Nakamura, "IMAGE SENSORS and SIGNAL PROCESSING for DIGITAL STILL CAMERAS", CRC Press, 2006.

[2] Smith, S. Hurwitz, J. Torrie, M. Baxter, D. Holmes, A. Panaghiston, M. Henderson, R. Murray, A. Anderson, S. Denyer, "A single-chip 306×244-pixel CMOS NTSC video camera", IEEE ISSCC Dig. Tech. Papers, Feb. 1998.

[3] Loinaz, M. Singh, K. Blanksby, A. Inglis, D. Azadet, K. Ackland, B. Lucent Technol, "A 200 mW 3.3 V CMOS color camera IC producing 352×288 24 b video at 30 frames/s", IEEE ISSCC Dig. Tech. Papers, Feb. 1998.

[4] Stuart Kleinfelder, SukHwan Lim, Xinqiao Liu, and Abbas El Gamal, "A 10000 Frames/s CMOS Digital Pixel Sensor", IEEE JOURNAL OF SOLID-STATE CIRCUITS, VOL. 33, NO. 12, DECEMBER 1998

[5] Masanori Furuta, Yukinari Nishikawa, Toru Inoue, and Shoji Kawahito, "A High-Speed, High-Sensitivity Digital CMOS Image Sensor With a Global Shutter and 12-bit Column-Parallel Cyclic A/D Converters," IEEE Journal of Solid State Circuits, Vol. 42, No. 4, April 2007.

[6] Z. Zhou, B. Pain, "CMOS active pixel sensor with on-chip successive approximation analog-to-digital converter", IEEE Trans. Electron Devices, 1997.

[7] F. Snoeij, A. Theuwissen, Kofi A. Makinwa, and J. Huijsing, "A CMOS Imager With Column-Level ADC Using Dynamic Column Fixed-Pattern Noise Reduction," IEEE Journal of Solid State Circuit, vol. 41, NO. 12, December 2006.

[8] Christian Jesus B.Fayomi "Low-Voltage Analog Switch in Deep Submicron CMOS: Design Technique and Experimental Measurements" IEICE TRANS.Fundamentals,VOL.E89-A,2006

[9] J. Sobhi, KH. Hadidi, A. Khoei, " A New Method for Offset Cancellation in High-Resolution High- Spedd Comparators" IEICE Trans. June 2005.

[10] Kh. Hadidi and G.C. Temes, " A high resolution, low offset,high speed comparator," Proc. IEEE Custom Imtegrated Circuits Conference, May 1992.

[11] B.Razavi, Principlees of Data Convertion System Design, IEEE Press.1995.

[12] Martijn F. Snoeij,"Multiple-Ramp Column-Parallel ADC Architectures for CMOS Image Sensors" IEEE JOURNAL OF SOLID-STATE CIRCUITS, VOL. 42, NO. 12, DECEMBER 2007.

[13] Seunghyun Lim,Jeonghwan Lee "A High-Speed CMOS Image Sensor With Column-Parallel Two-Step Single-Slope ADCs"IEEE TRANSACTIONS ON ELECTRON DEVICES, VOL. 56, NO. 3, MARCH 2009.

MIXDES 2012, 19th International Conference *"Mixed Design of Integrated Circuits and Systems"*, May 24-26, 2012, Warsaw, Poland

A High Speed Two-Stage Dual-Path Operational Amplifier in 40nm Digital CMOS

Hong Chen, Vladimir Milovanovic, Horst Zimmermann

Institute of Electrodynamics, Microwave and Circuit Engineering

Vienna University of Technology

Vienna, Austria

{hong.chen, vladimir.milovanovic, horst.zimmermann}@tuwien.ac.at

Abstract—This paper presents a design of a high speed operational amplifier using 40nm digital CMOS technology. The proposed two-stage dual-path fully differential topology is based on the Improved Recycling Folded Cascode (IRFC) topology. The IRFC first stage provides a moderate DC gain, and the high efficient dual-path push-pull output stage provides ultra-high unity-gain frequency bandwidth (UGBW) with relatively low power consumption. It could find wide application in high bandwidth high resolution analog-to-digital converters (ADCs). Under 1.1V supply voltage, the simulation results show that the proposed operational amplifier topology could achieve 56.3dB DC gain, 3GHz UGBW, 24.8µV RMS noise integrated from DC to 50MHz and 2.9ns settling time with 1V peak-to-peak differential input signal.

Index Terms—operational amplifier; high speed; dual-path; recycling folded cascode.

I. INTRODUCTION

As the CMOS technology is continuously scaling down, the design of ultra-high speed wired or wireless communication system is becoming possible. However, the advanced digital CMOS technology brings increasingly challenge for analog designers when designing mixed-signal systems [1]. Operational amplifier, which is one of the key analog modules, could achieve larger bandwidth due to the scaled transistor model, but at the cost of dramatically decreased gain, limited signal swing, poorer matching and so on. To efficiently increase operational amplifier's gain and output swing, multi-stage fully-differential operational amplifier topology is appreciated. The operational amplifier with three or even more stages equipped with the Nested-Miller compensation or the Reversed Nested-Miller [2][3] compensation shows high efficiency in the gain enhancement, while they require additional large compensation capacitors compared to the traditional two-stage operational amplifier, which will lead to a larger die area and the limited slew rate. Besides, additional common mode feedback (CMFB) circuits would consume additional power. The traditional two-stage telescopic or folded cascode operational amplifiers could hardly provide adequate gain in Nanometer CMOS technology [4]. In [5], using 65nm technology, 56.1dB DC gain is obtained in "opamp2b", which is at the cost of low DC current, therefore leaving very small headroom of making tradeoff between the unity-gain frequency bandwidth (UGBW) and the DC gain by adjusting the current. A novel topology called "Recycling Folded Cascode" (RFC) operational amplifier is proved to be able to double the gain, the bandwidth and the slew rate without increasing the power dissipation [6].

Furthermore, the so-called "Improved Recycling Folded Cascode" (IRFC) operational amplifier provides a larger gain and bandwidth enhancement [7].

For the high performance analog-to-digital converter (ADC) in the Nanometer digital CMOS technology with low supply voltage, the requirements of operational amplifier in the sample/hold (S/H) circuit, the multiplying digital-to-analog convertor (MDAC) and the integrator are very critical. Both the gain and the bandwidth should be large to meet the system's speed and accuracy demands. Targeting at the application in high performance ADCs, a two-stage fully-differential operational amplifier using IRFC topology as the input stage is presented in this paper. The proposed dual-path driving topology enables this operational amplifier to achieve a large UGBW, high slew rate and keep the power consumption low. Besides, the IRFC input topology enables this operational amplifier to provide moderate gain in 40nm CMOS technology.

The paper is organized as follows. The Section II reviews the principle of the IRFC topology briefly and then discusses the principle operation of the proposed topology. The Section III shows the key characteristics analysis of the proposed topology. The Section IV shows the simulation results. The Section V concludes the paper.

II. PRINCIPLE OPERATION OF THE PROPOSED TOPOLOGY

The Improved Recycling Folded Cascode amplifier (IRFC) topology is shown in Fig.1. Compared to the original Recycling Folded Cascode operational amplifier (RFC), the IRFC topology uses two additional DC paths (M3c M3e and M4c M4e) to increase the width ratio between M3a and M3b, as well as M4a and M4b, thus increase the driving ability of these two current mirrors. No additional current for upper cascode branch is needed to satisfy Kirchhoff's law at the nodes X and Y. By adjusting the width ratio M3a:M3b:M3c and M4a:M4b:M4c (3: α: β in Fig.1, and α+β=1), different value of input transconductance enhancement could be chosen.

Since two additional input transistors are needed, the RFC/IRFC topology provides two signal paths at each single half by its nature, which could be used to construct a two-stage push-pull operational amplifier. The first signal path starts from the input node to the conventional output node. The second signal path starts from the input node to the node P or Q. A "hidden" operational amplifier builds this second signal path, which is extracted from the Fig.1 and shown in Fig.2.

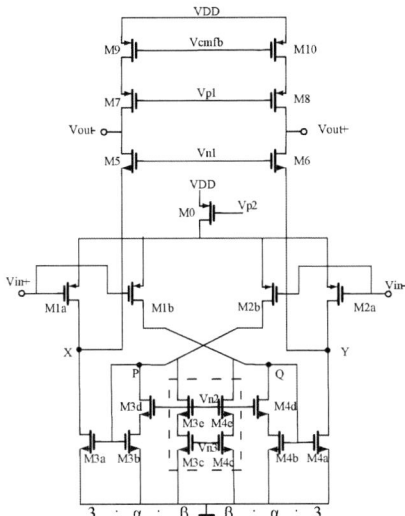

Figure 1. The IRFC amplifier topology

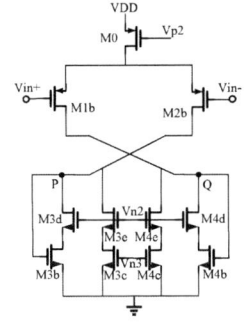

Figure 2. The hidden amplifier of the IRFC topology

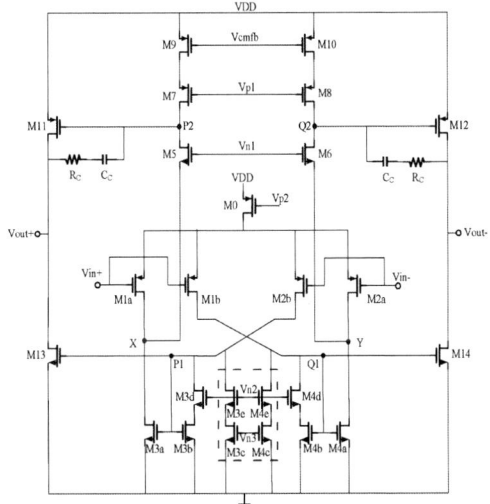

Figure 3. The proposed two-stage dual-path operational amplifier topology

The proposed two-stage dual-path operational amplifier is shown in Fig.3. The NMOS and PMOS transistor of the output stage are driven by the first and the second signal path of the first stage respectively. The second signal path would enlarge the equivalent transconductance of the output stage so that the pole at the output node would be pushed to higher frequency. As a result, larger UGBW could be achieved. The second signal path could not contribute the same gain as the first signal path does since the impedance at the nodes P1 and Q1 is much lower than their counterparts at the nodes P2 and Q2. So the output stage would not show an ideal push-pull characteristic as the inverter output stage does [8]. However, in the proposed topology, the DC voltage value of P1 (Q1) and P2 (Q2) could be adjusted according to transistor's threshold voltage, so the output stage here could achieve much better efficiency than the inverter output stage does. For the inverter output stage, the efficiency of the output stage depends on the supply voltage and the technology. For example, if the threshold voltage of NMOS and PMOS is 0.35V and 0.4V respectively, and the DC input voltage of the inverter output stage is set to be 0.55V (half of the 1.1V supply voltage), then there would be a considerate amount of quiescent current dissipation, making this Class AB topology less attractive.

A single common mode feedback (CMFB) loop from the second stage's output to the first stage's current source is used for the proposed operational amplifier topology. For the frequency compensation, the Miller capacitor with the nulling resistor topology is used.

III. CHARACTERISTICS OF THE PROPOSED TOPOLOGY

A. Low Frequency Gain

The low frequency gain of the proposed operational amplifier is

$$A_V(0) = \left[g_{m1a}(1+3/\alpha)g_{m11}R_{11} + g_{m1b}g_{m13}R_{12} \right]R_2 \quad (1)$$

in which $R_{11} = g_{m5}r_{o5}(r_{o1a} \| r_{o3a}) \| g_{m7}r_{o7}r_{o9}$, representing the resistance at the node P2; $R_{12} = g_{m4b}^{-1} \| r_{o1b} \approx g_{m4b}^{-1}$, representing the resistance at the node P1; and $R_2 = r_{o11} \| r_{o13}$, representing the resistance at the output node.

The parameter $3/\alpha$ represents the width ratio M3a:M3b and M4a:M4b, the g_{mi} represents the transconductance and the r_{oi} represents the output resistance. The "$g_{m1a}(1+3/\alpha)g_{m11}R_{11}$" is the gain contribution from the first signal path and the "$g_{m1b}g_{m13}R_{12}$" is from the second signal path. As discussed in the Section II, the R_{12} is much smaller than R_{11}. Besides, indicated from (1), the boosting factor of $1+3/\alpha$ in the first signal path enlarges this gain contribution gap further. The calculation and simulation results both prove that about 2.0~2.5 dB gain enhancement could be achieved through the second signal path. This small gain enhancement is anyway appreciated in the operational amplifier design with Nanometer CMOS technology.

B. Slew Rate

The slew rate is an important indicator of operational amplifier's large signal settling performance. The positive and the negative slew rate of the fully differential RFC topology is not absolutely symmetric, as briefly mentioned in [6]. Here detailed discussion regarding the proposed topology is given. Suppose Vin+ goes high, then M1a M1b will be turned off and M2a will be driven into deep linear region, so most of the current $2I_b$ in the tail current source M0 will be injected into M2b. Since the DC current in M3c and M3e branch is fixed to be $I_b\beta/2$, the current flow through M3d and M3b branch will be $(2-\beta/2)I_b$. Amplified by a factor of $3/\alpha$, the current flow through M5 will be $3I_b(2-\beta/2)/\alpha$. Since the current in M7 and M9 is I_b (suppose the CMFB loop's response does not alter the V_{cmfb} at this moment), the current that drives the Miller capacitance C_C at the node P2 is $3I_b(2-\beta/2)/\alpha-I_b$, leading to a slew rate of

$$SR^+=\frac{3I_b\left(2\text{-}\beta/2\right)/\alpha\text{-}I_b}{C_C} \quad (2)$$

at the positive output node. Meanwhile, since M6 has been turned off, the current that drives the Miller capacitance C_C at the node Q2 is I_b, leading to a slew rate of

$$SR^-=-\frac{I_b}{C_C} \quad (3)$$

at the negative output node. The differential slew rate at this moment is

$$SR=\frac{3I_b\left(2\text{-}\beta/2\right)/\alpha C_C}{C_C} \quad (4)$$

From (2) and (3) it is clear that the slew rate at the two output nodes is not symmetric. After the CMFB loop starting to adjust the V_{cmfb}, the slew rate difference between the positive and the negative output node will be lowered until reaching zero, which indicates that the slew rate becomes symmetric. To shorten this feedback propagation time, the CMFB circuit should have large enough loop bandwidth. The differential slew rate in (4), however, stays unchanged all the time, so the differential output signal's total rising and falling settling time of the proposed operational amplifier should be same.

C. Phase Margin

The phase margin analysis will be based on the operational amplifier's pole-zero position discussion. The dominant pole ω_{p1} of the proposed operational amplifier is contributed by the nodes P2 and Q2

$$\omega_{p1}=-\frac{1}{R_{11}(g_{m11}R_2C_C+C_{p1})} \quad (5)$$

in which $R_{11}=g_{m5}r_{o5}(r_{o1a}\|r_{o3a})\|g_{m7}r_{o7}r_{o9}$, $R_2=r_{o11}\|r_{o13}$, C_{p1} represents the total capacitance at the node P2 (or Q2). There are three non-dominant poles and two left half plane zeros in the signal path, which are

$$\omega_{p2}=-\frac{g_{m5}r_{o5}}{g_{m7}r_{o7}r_{o9}C_{p2}} \quad (6)$$

$$\omega_{p3}=-\frac{g_{m11}+g_{m13}}{C_L} \quad (7)$$

$$\omega_{p4}=-\frac{g_{m3b}}{C_{p3}} \quad (8)$$

$$\omega_{z1}=(\frac{3}{\alpha}+1)\omega_{p4} \quad (9)$$

$$\omega_{z2}=-\frac{1}{C_C(R_C\text{-}g_{m11}^{-1})} \quad (10)$$

The C_{P2} and the C_{P3} represent the total capacitance at the node X (or Y) and the node P1 (or Q1) respectively, and the C_L represents the load capacitance. The relative position between poles ω_{p2} and ω_{p3} could be interchangeable due to different application requirements. For the operational amplifier design that aims at high bandwidth applications (this paper), the equivalent transconductance of the second stage ($g_{m11}+g_{m13}$) would be very large, thus pushing the ω_{p3} to high frequency. Meanwhile, due to large transistor sizes, the parasitic capacitance C_{P2} at the nodes P2 and Q2 would be pretty large, so the ω_{p2} would be pushed towards the origin point. As a result, the ω_{p2} would be the first non-dominant pole and the ω_{p3} would be the second non-dominant pole. On the other hand, for the operational amplifiers that operate in relatively low-speed systems or with large load capacitances, the smaller transconductance of the second stage, smaller C_{p2} and larger C_L might make the ω_{p3} the first non-dominant pole and then the ω_{p2} becomes the second non-dominant pole. The ω_{p4} and the ω_{z1} are the pole-zero pair contributed by the current mirror M3a:M3b and M4a:M4b. This pole-zero pair is located at very high frequency so their effect would be negligible. The LHP zero ω_{z2} is associated with the compensation resistor R_C, and it would be used to cancel out the first non-dominant pole ω_{p2}. Then the phase margin would mainly be determined by the ω_{p3}. To achieve 60 degrees phase margin, the ω_{p3} should be located at 1.73 times greater than the UGBW. If the UGBW requirement is 3.0GHz, then $\omega p3$ should be put at 5.2GHz. Then for a Miller compensation capacitance of 1.5pF, the second stage's transconductance should be about 49mS. It is very difficult to achieve this large transconductance with low current consumption if using traditional Class A two-stage operational amplifier topology, especially when supply voltage is as low as 1.1V. The proposed dual-path topology enhances the second stage's transconductance by a 100% if $g_{m11}=g_{m13}$, which makes the frequency compensation a much easier task. Fig.4 shows pole-zero locations where ω_u refers to the unity-gain frequency point.

At the first glance on (1) and (4), the value of the parameter α should be as small as possible to make the ratio M3a:M3b and M4a:M4b large, so the gain and the slew rate could be maximized.

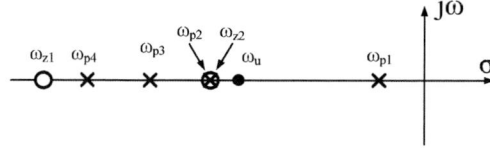

Figure 4. The pole-zero location of the proposed topology in the s-domain

However, with the unchanged power consumption, the smaller the α is, the smaller the g_{mb3} becomes, which implies a smaller ω_{p4} and consequently worse phase margin. In this paper, α is designed to be 0.5, so β is 0.5 and the ratio M3a:M3b (M4a:M4b) is 6:1.

D. Noise

Noise is one of the key specifications when designing ADC. Since the operational amplifier is one of the main noise contributors, the low noise consideration is vital when designing operational amplifier for ADCs. The noise analysis regarding the RFC topology could be found in [6], where the influence of the transistors M3b and M4b in Fig.3 has not been included. From the Fig.2 it is clear that the transistors M3b M4b are also two noise sources. Besides, M3c and M4c on the extra DC paths would also contribute noise. The input-referred noise brought by the hidden operational amplifier is

$$\overline{V_{n,in}^2} = \frac{16}{3} kT \left(\frac{1}{g_{m1b}} + \frac{g_{m3b}}{g_{m1b}^2} + \frac{g_{m3c}}{g_{m1b}^2} \right) +$$

$$\frac{2}{C_{OX}} \left[\frac{k_P}{(WL)_{1b}} + \frac{k_N g_{m3b}^2}{(WL)_{3b} g_{m1b}^2} + \frac{k_N g_{m3c}^2}{(WL)_{3c} g_{m1b}^2} \right] \frac{1}{f} \quad (11)$$

The transconductance of M3b and M3c should be small to decrease both the thermal and the flicker noise. Since $g_m = 2I_D/(V_{gs}-V_{th})$ (long channel model equation is used for analysis here because of the long channel length of these current mirror transistors), either decrease the drain current or increase the overdrive voltage could decrease the transconductance of M3b and M3c. Once the tail current in the M0 is fixed for the high slew rate requirement, the total current in the M3b and M3c would be fixed as well, leaving the gate voltage of the M3b and M3c the only variable to be adjusted. However, as discussed in Section II, the voltage at the nodes P and Q should be set near the threshold voltage of the NMOS transistor so that the quiescent current in the output stage can be kept low. To further decrease the noise, the M3c M3e and the M3e M4e could be replaced by two resistor branches, as shown in Fig. 5. The relatively low accuracy of poly and diffusion resistor would cause some biasing offset, which may cause inaccuracy in the frequency compensation so enough extra phase margin should be left in that case.

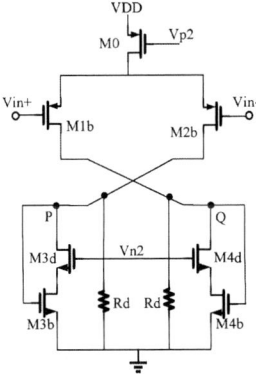

Figure 5. Using resistors to build the DC path in the hidden amplifier

IV. SIMULATION RESULTS

The schematic design is based on the TSMC's 40nm LP digital CMOS technology, and the simulation is done with the Synopsys's HSPICE simulator. The supply voltage is 1.1V. Two 2pF capacitors are used as the load elements at each differential output node. The input and the output common mode voltage are set to be half of the supply voltage for the convenience of cascading and the maximization of the output swing. The following simulation results are based on the transistor DC path version shown in the Fig.2.

From Fig.6 it could be found out that the DC gain is 56.3dB. The UGBW is 3.0GHz, and the phase margin is 61 degrees. Fig.7 shows the DC characteristic. The linear range of the output from 0.13V to 0.96V is obtained for input voltage ranging from 549mV to 551mV. Fig.8 depicts the step response characteristic of the closed loop unity gain operational amplifier configuration. The input signal is a 1 Vpp pulse signal at 50MHz. The positive and the negative slew rates are as high as 1.28kV/μs and 1.2kV/μs, respectively. The rising and the falling settling time are 2.90ns and 2.94ns, respectively within 0.1% error range. Fig.9 depicts that the input referred noise spectral power density at 5MHz is 2.92nV, and the input RMS noise integrated from 1Hz to 50MHz is 24.8μV. The total DC bias current of the core operational amplifier is 6.5mA. Table I shows the performance comparison of the recent published high speed operational amplifiers.

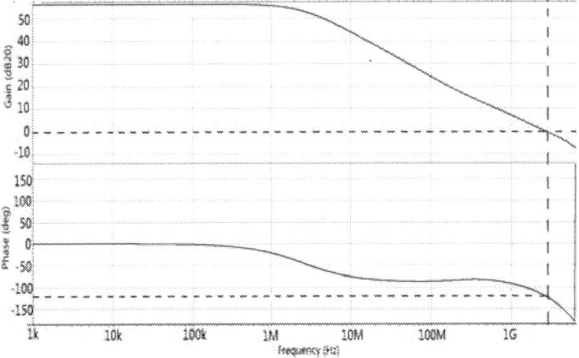

Figure 6. Open loop frequency response

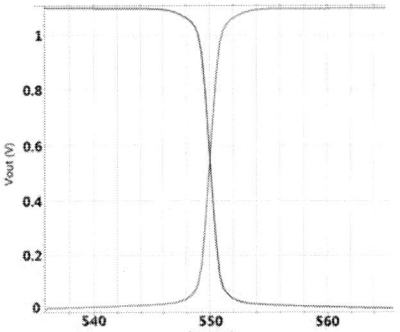

Figure 7. DC characteristics

TABLE I. PERFORMANCE COMPARISON

Parameter	[5] (opamp2b)	[8]	[9]	[10]	This work
Technology	65nm CMOS	0.18μm CMOS	0.18μm CMOS	65nm CMOS	40nm CMOS
Supply (V)	1	1.8	1.8	1.2	1.1
Load	2pF	2pF	300fF//1kΩ	5pF//10 kΩ	2pF
DC gain (dB)	56.1	65	50	58	56.3
UGBW (GHz)	0.45	2.3	2.6	1	3.0
PM (degs)	77	58	35	62	61
Settling time (ns)	10/6 @1%	7.2 @0.01%	-	-	2.90/2.94 @0.1%
Slew Rate (V/μs)	-	>450	-	-	1290/1200
Power (mW)	1.6	25	7.2	11.4	7.2
FoM (MHz*pF/mA)	2250	331	390	1052	923

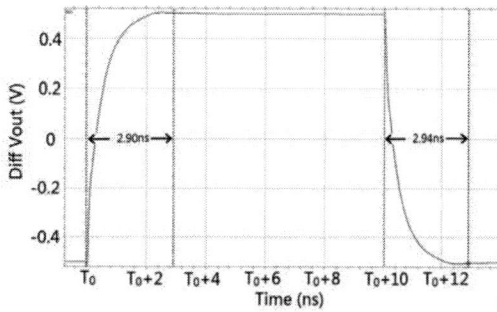

Figure 8.   Large signal step response

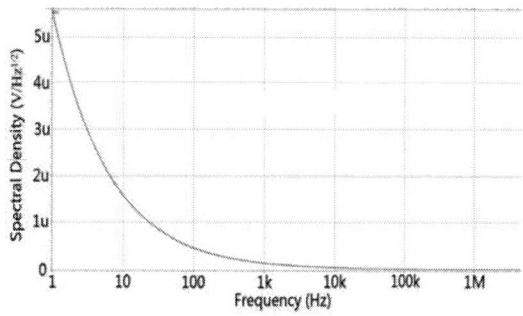

Figure 9.   Input referred noise spectral power density

V. CONCLUSION

This paper presents a high speed operational amplifier designed with 40nm digital CMOS technology for 1.1V supply voltage. The proposed dual-path two-stage topology is proven to be capable of enhancing both the DC gain and the UGBW with relatively low power consumption. The 56.3dB DC gain, 3GHz UGBW, 1290/1200V/μs rise/fall slew rate, fast step response and low input referred RMS noise make the proposed operational amplifier suitable for the application like video rate ADC, active filter and other high bandwidth high resolution fields.

ACKNOWLEDGMENT

The authors would like to thank Lantiq A GmbH and Austrian BM VIT for their support of the FIT-IT xPLC project via FFG.

REFERENCES

[1] A.-J. Annema, B. Nauta, R. van Langevelde and H. Tuinhout, "Analog circuits in ultra-deep-submicron CMOS," IEEE J. Solid-State Circuits, vol. 40, pp. 132 - 143, Jan. 2005.

[2] Ka Nang Leung and Philip K. T. Mok, "Nested Miller Compensation in Low-Power CMOS Design," IEEE Trans. Circuits and Systems, II, vol. 48, pp. 388 - 394, Apr. 2001.

[3] Alfio Dario Grasso, Gaetano Palumbo and Salvatore Pennisi, "Advances in Reversed Nested Miller Compensation," IEEE Trans. Circuits and Systems, I, vol. 54, pp. 1459 - 1470, Jul. 2007.

[4] R. Jacob Baker, "CMOS Circuit Design, Layout and Simulation," John Wiley & Sons, Inc., 2008.

[5] Mohammad Taherzadeh-Sani and Anas A. Hamoui, "A 1-V Process-Insensitive Current-Scalable Two-Stage Opamp With Enhanced DC Gain and Settling Behavior in 65-nm Digital CMOS," IEEE J. Solid-State Circuits, vol.46, pp. 660 - 668, Mar. 2011.

[6] Rida S. Assaad and Jose Silva-Martinez, "The Recycling Folded Cascode: A General Enhancement of the Folded Cascode Amplifier," IEEE J. Solid-State Circuits, vol.44, pp. 2535 - 2542, Sep. 2009.

[7] Y.L. Li, K.F. Han, X. Tan, N. Yan and H. Min, "Transconductance Enhancement Method for Operational Transconductance Amplifiers," Electronics Letters, vol. 46, pp. 1321 - 1323, Sep. 2010.

[8] Daibashish Gangopadhyay and T.K. Bhattacharyya, "A 2.3 GHz g_m-boosted High Swing Class-AB Ultra-Wide Bandwidth Operational Amplifier in 0.18μm CMOS," IEEE International Midwest Symposium on Circuits and Systems, pp. 713 - 716, 2010.

[9] J.Harrison and N. Weste, "350MHz Opamp-RC Filter in 0.18μm CMOS," Electronics Letters, vol. 38, pp. 259 - 260, May. 2002.

[10] Heimo Uhrmann, Franz Schlögl, Kurt Schweiger and Horst Zimmermann, "A 1GHz-GBW Operational Amplifier for DVB-H Receivers in 65 nm CMOS," International Symposium on Design and Diagnostics of Electronic Circuits & Systems, pp. 182 - 185, 2009.

MIXDES 2012, 19th International Conference *"Mixed Design of Integrated Circuits and Systems"*, May 24-26, 2012, Warsaw, Poland

A Low Noise Low Offset Current Mode Instrumentation Amplifier

Andriana Voulkidou, Stylianos Siskos, Theodoros Laopoulos
Electronics Laboratory of Physics Department
Aristotle University of Thessaloniki
Thessaloniki, Greece
avoulk@physics.auth.gr, siskos@physics.auth.gr, laopoulos@physics.auth.gr

Abstract—This paper presents a low noise current mode instrumentation amplifier for low frequency temperature sensing applications. The circuit utilizes a positive second generation current conveyor (CCII+) as a basic building block. Nested chopping technique is used to totally cancel output offset voltage and eliminate 1/f noise. The amplifier is designed in 0.18μm CMOS process. Post-layout simulations on the extracted circuit net-list provide an equivalent input referred noise *of 20nV/√Hz, a 0.582μV offset and a CMRR of 110dB. The power consumption is 420μW at ±1.65V.*

Index Terms—**Current mode instrumentation amplifier, nested chopping, current conveyor.**

I. INTRODUCTION

Instrumentation amplifiers are widely used in a large number of signal processing ranging from healthcare to environmental monitoring. Given the strict cost limitations imposed by the sensor market, the integration of actuation and sensing parts, as well as the read out circuit in a single chip, also needs a highly sensitive instrumentation amplifier to suppress any unwanted signal and amplify the extremely small signal to appropriate level where it can be further processed.

An infrared (IR) temperature sensor can be constructed as shown in the schematic of Fig.1. It consists of a dielectric membrane with a thermopile on it. Several thermocouples are connecting together and forming a thermopile. The basic heat sensing element consists of a thermopile which detects the temperature difference between its hot and cold junction. Our application targets to be part of the total sensor-system with *42μV/W* sensitivity and thermopile resistance of *60kΩ*. These specifications impose a challenge on the design of an accurate and reliable instrumentation amplifier.

Conventional instrumentation amplifiers based on voltage operational amplifier topology exhibit bandwidth dependent gains and common mode rejection ratio (CMRR), due to the fixed gain bandwidth product of the operation amplifier. Moreover, these topologies require precise resistor matching to achieve high CMRR [2],[3]. Better performance with respect to frequency range of operation, CMRR and matching issues provide the current mode instrumentation amplifiers [4]-[6].

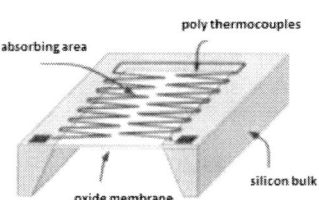

Figure 1. Scematic structure of the IR sensor [1].

Current mode circuits are excellent alternatives to voltage mode circuits. If the information is conveyed as a current, the square root in MOS transistor circuits is proportional to the square root of the signal, if saturation region operation is assumed for the devices. Furthermore, all nodes are near virtual ground resulting to low impedance at each node and then very small time constants and high values of parasitic poles. For these reasons, the design in a current mode leads to simpler circuit realization and lower power consumption.

This paper describes an improved current mode instrumentation amplifier and is organized as follows: In Section II, offset and noise reduction technique is briefly reviewed. The circuit implementation and post-layout results are presented in sections III and IV, respectively.

II. THEORY OF OPERATION

A. Nested Chopping Technique

In the conventional CMOS amplifier the input-referred noise becomes Gaussian at high frequencies (thermal noise) whereas at low frequencies the noise power increases almost linearly with decreasing frequency and is called *1/f* noise or flicker noise. The frequency at which the flicker noise becomes dominant over thermal noise is called the *1/f* noise corner frequency.

One way to reduce low-frequency noise and residual offset is the nested chopping technique. In this chopping technique, the input signal is modulated firstly by a low frequency square-ware signal and then by a high frequency signal. The modulated signal is then amplified and demodulated to the baseband with reduced offset and 1/f noise. The demodulation is realized by two pairs of choppers with the respective frequencies to the input choppers. During the signal's demodulation, the noise and offset are also demodulated to the chopping frequencies and its odd harmonics.

The baseband noise of chopper amplifiers is almost equal to the wideband white noise [7]. This condition occurs only when the chopping frequency is greater than the noise corner frequency and greater than twice the bandwidth of the input signal. The noise power spectral density (PSD) after chopping is given by [8],

$$S(f) = A_o^2 \left(\frac{2}{\pi}\right)^2 \sum_{-\infty}^{+\infty} \frac{1}{n^2} S_n(f - n f_{chop}) \quad (1)$$

where A_o is the amplifier gain, and f_{chop} is the chopping frequency.

B. Modulators and Residual Offset

The choppers are implemented by a set of switches surrounded by half-sized dummy transistors as shown in Fig 2. The residual offset of a chopper amplifier originates mainly from the non-idealities of the input modulators [9]. The clock feedthrough introduced by the switches in the input modulators is the dominant source of offset. At each switching instant a certain amount of charge ΔQ is injected on the input capacitance, causing a spike equal to $V_{inj} = (\Delta Q)/(C_{in})$. The amount of the injected charge is calculated by,

$$\Delta Q = \frac{WLC_{ox}}{2}(V_{ck} - V_{in} - v_{th}) \quad (2)$$

where W, L, C_{ox} v_{th}, are transistor parameters, V_{in} and V_{clk} are the signal and clock amplitudes respectively.

This parasitic signal, spike, is periodic with $T=1/f_{chop}$, as well as the demodulation signal. Therefore, a substantial part of this spike energy will be amplified and then translated back to dc resulting to a residual offset. This offset is given by,

$$V_{of} = \frac{2\tau}{T} V_{sp} \quad (3)$$

where τ is the time constant of the spikes and V_{sp} is their amplitude. Consequently, the residual offset is determined by the number of the spikes and theirs energy content.

In the nested chopping technique, since the one pair of the input modulator operates at high frequency and suppresses the 1/f noise, the other pair can operate at lower frequency. Therefore, the residual offset caused by these choppers is much lower. The corresponding signals of the nested-chopper amplifier are shown in Fig.3. The spikes of the high chopping frequency are modulated by the low chopping frequency diminishing the total offset.

III. CIRCUIT IMPLEMENTATION

The proposed instrumentation amplifier shown in Fig.4 targets to perform the critical element in the interface circuitry

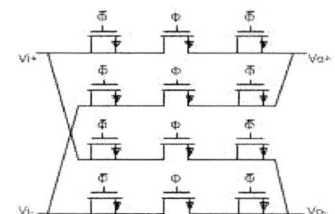

Figure 2.   One pair of modulator schematic [8].

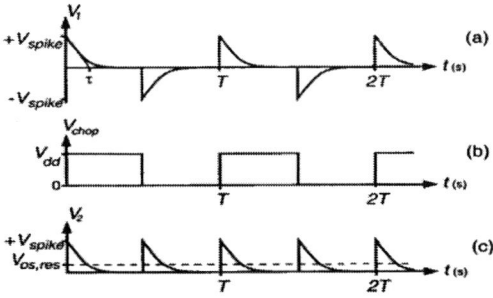

Figure 3.   Residual Offset caused by spike a.A spike signal b. Demodulation signal c.Demodulate signal [7].

to an infrared temperature sensor. The small SNR of the minimum detectable signal imposes severe noise conditions and the amplification of microvolt signals dictate the highest offset that can be tolerated.

To meet the desired design specifications the topology of the amplifier is based on current conveyors CCIIs and utilizes the nested chopping technique. Specifically, the instrumentation amplifier transfers its input voltages Vy from the high impedance terminals to terminals X developing a current across the resistor R_i. This current is conveyed at the output terminals Z and is converted into voltage across the resistor R_o. The gain is determined by the ratio of the resistors an is expressed as

$$A_D = \frac{(V_{in+}) - (V_{in-})}{(V_{out+}) - (V_{out-})} = \frac{R_o}{R_i} \quad (4)$$

The inner choppers (Φ_{ihigh}) operate at high frequency of *100kHz* in order to remove *1/f* noise and the outer choppers (Φ_{ilow}) operate at a much lower frequency of *40Hz*. The low chopping frequency f_{low} imposes limitations to the input signal frequency. The maximum input signal frequency is reduced to the half f_{low}. However, a bandwidth of a few tens of Hertz is sufficient for our application.

Using CCIIs as basic building block for the instrumentation amplifier contribute to the circuit's stability as the whole topology is designed without any external feedback loop. The current conveyor's architecture is depicted at Fig.5 and is based on [10]. The input transistors M1 and M2 are implemented with PMOS transistors and large gm1 and gm2 for better flicker-noise performance. Increasing the input transistors size reduces the flicker-noise corner frequency, however, the input

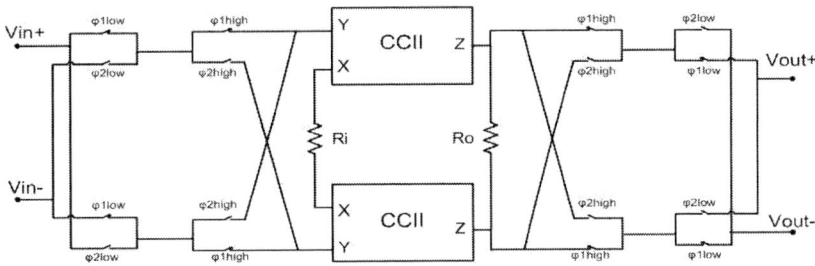

Figure 4. Chopping Instrumentation Amplifier.

capacitance is increased by the same amount leading to a higher offset. The detailed transistors dimensions are shown in Table I. Simple current mirrors retain the power consumption low. By connecting the negative output to node X the circuit converted into a non-inverting CCII.

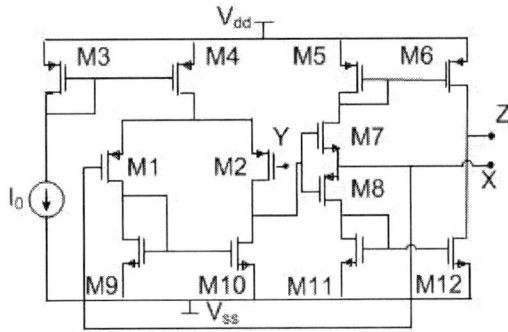

Figure 5. Second generation current conveyor.

TABLE I. TRANSISTORS DIMENSIONS

Mosfet Dimensions of the CCII			
M1, M2	120μm/7μm	M7	150μm/0.4μm
M3	65μm/5μm	M8	50μm/0.4μm
M4	115μm/5μm	M9, M10	50μm/2μm
M5,M6	150μm/2μm	M11,M12	4μm/16μm

IV. SIMULATION RESULTS

The chopped instrumentation amplifier was simulated on layout level in the 0.18um CMOS process. The power supplies were $V_{DD}=-V_{SS}=1.65V$. To demonstrate the features of the proposed circuit design all the simulations were performed using SPECTRE simulator.

The periodic steady-state (PSS) and periodic noise analysis (PNOISE) capabilities are used to simulate the noise spectre with or without chopping. Choosing a chopping frequency much higher than the flicker noise-corner frequency would ameliorate the noise performance. On the other hand, it would degrade the residual offset performance. A good trade-off between the noise and offset performance is to choose the high chopping frequency equal to the corner frequency.

The amplifier input referred noise response is initially extracted without chopping and illustrated at Fig. 6. The noise corner frequency is at *100kHz* and the white noise is $22nV/\sqrt{Hz}$. Without chopping, the *1/f* noise is dominant at frequencies below *100KHz*. Fig. 7 shows the noise spectre when the choppers are enable. The high chopping frequency selected to be at *100kHz* and the low chopping frequency at *40Hz*. The input signal at the differential input of the instrumentation amplifier during the noise performance had a frequency at *20Hz*. The suppress of the flicker noise is clearly visible and the white noise floor is dominant both to the low and high frequencies.

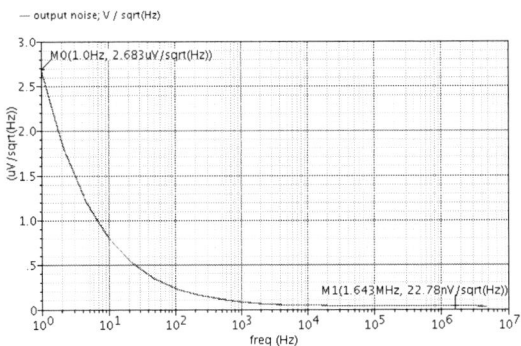

Figure 6. Noise spectral density without chopping.

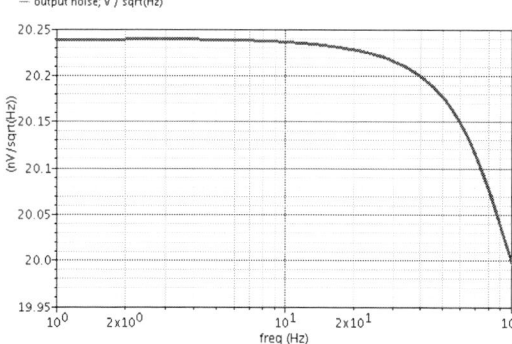

Figure 7. Noise spectral density with chopping.

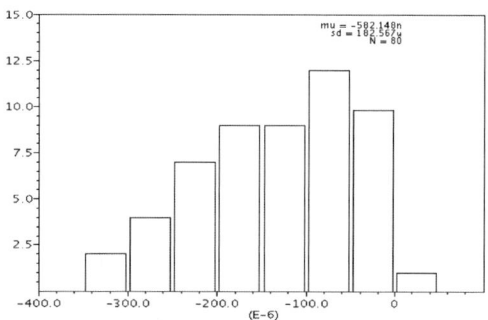

Figure 8. Offset distribution of the designed IA.

The residual offset of the instrumentation amplifier is measured by Monte Carlo analysis and is shown in Fig. 8. The amplifier has an average offset of 582nV. The simulations results are summarized in Table II.

TABLE II.   SIMULATION PERFORMANCE OF THE IA

Supply Voltage	±1.65V
Input Noise Density	$22nV\sqrt{Hz}$
Input Offset Voltage	582 nV
Noise Corner Frequency	100kHz
CMRR	110dB
Power Consumption	420µW
Input Common Mode Voltage	-0.9 – 1 V

The proposed design achieves an excellent noise performance and a significant low offset. Since the CMRR is independent of the resistors matching, the good CMRR performance of the current mode instrumentation amplifiers is

also verified in this design. Moreover, it should be highlighted that the full instrumentation amplifier consumes only 420µW.

The instrumentation amplifier layout is shown in fig. 9. The layout was designed as symmetric as possible to eliminate mismatches between the differential parts of the device. Common centroid techniques were implemented for better performance.

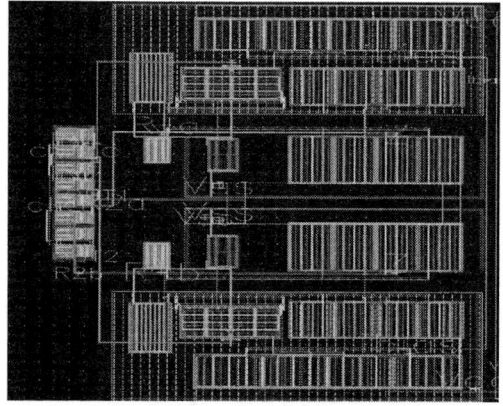

Figure 9. Layout of the Instrumentation Amplifier.

Table III shows a performance comparison between the proposed and currently used instrumentation amplifiers [11],[12]. It can be seen that this work acheives the lowest input offset. Also, we can observe that the proposed circuit has a lower input noise voltage than the other topologies [11],[12]. Thus, it has a wider dynamic range and higher signal-to-noise ratio when compared with [11],[12]. Notwithstanding that this work provides the lowest CMRR, its CMRR performance is considered satisfactory for many applications.

TABLE III.   COMPARISON BETWEEN OTHERS IA CIRCUITS

	Instrumentation Amplifiers		
	This work	**[11]**	**[12]**
Input offset voltage	582nV	1µV	3µV
Input noise $nV\sqrt{Hz}$	22	60	27
CMRR	110dB	134dB	140dB

V. CONCLUSIONS

In this paper, a current mode instrumentation amplifier is implemented with good performance related to the noise, and offset. The non-existence of the ground connection contributes to high CMRR. However, many improvements could be investigated in relation to the output low impedance using an output buffer and power consumption as it is one of the critical characteristic for sensors application.

ACKNOWLEDGMENT

The research activities that led to these results were co-financed by Hellenic Funds and by the European Regional Development Fund (ERDF) under the Hellenic National Strategic Reference Framework (NSRF) 2007-2013, according to Contract no. MICRO2-49 of the Project "Development of Innovative sensor systems offering distributed intelligence – MEMSENSE" within the Programme "Hellenic Technology Clusters in Microelectronics – Phase-2 Aid Measure".

REFERENCE

[1] LC. Menofi and Q. Huang, "A low CMOS instrumentation amplifier for thermoelectric infrared detectors," IEEE J. of Solid State Circuits, vol.32, no.7,July 1997.

[2] J. Szynowski, "CMRR anlysis in instrumentation amplifier",Electron. Lett., vol.19, no.14, pp. 547-549, 1983.

[3] R.P. Areny and J.G.Webster, "Common-mode rejection ratio in differential amplifier stages", IEEE Trans. Instrum. Meas., vol. 40, no. 4, pp. 669-676, Aug.1991.

[4] S.J. Azhari and H. Fazalipoor, "A novel current-mode instrumentation amplifier (CMIA) topology", IEEE Trans. Instrum. Meas., vol. 49, no. 6, pp. 1272-1277, Dec. 2000.

[5] S. J. G. Gift, "An enchanced current-mode instrumentation amplifier", IEEE Trans. Instrum. Meas., vol. 50, no. 1, pp. 85-88, Jan. 2001.

[6] K. Koli and K. A. I. Halonen, "CMRR enchanced techniques for current-mode instrumentation amplifiers", IEEE Trans. Circuits Syst., vol. 47, no. 5, pp. 622-632, 2000.

[7] A. Baker , K. Thiele and J. H. Hustijsing, "A CMOS nested chopper instrumentation amplifier with 100nV offset", IEEE J. of Solid State Circuits, vol.35, pp. 1877-1883, 2000.

[8] J. H. Nielsen and E. Bruun, "A CMOS low-noise instrumentation amplifier using chopper modulation", Analog Integr. Circuits and Signal Proc., vol 42, pp. 65-76, 2005.

[9] C. C. Enz and G. C. Temes, "Circuit techniques for reducing the effects of opamp imperfections: autozeroing, correlated double sampling, and and chopper stabilization", Proc.IEEE, vol. 84, pp.1584-1614, Nov. 1996.

[10] Wilson,"Recent developments in current conveyors and current mode circuits", Proc.IEEE, vol. 135, no.2, pp.63-77, April 1990.

[11] F. Qinwen, F. Sebastiano, H. Huijsing, K. Makinwa, "A 1.8µW 1µV-offset capacitively-coupled chopper instrumentation amplifier in 65nm CMOS," Proc. of ESSCIRC 2010, pp.170-173, 14-16 Sept. 2010.

[12] M.A.P. Pertijs and W.J.Kindt, "A 140 dB-CMRR Current-Feedback Instrumentation Amplifier Employing Ping-Pong Auto-Zeroing and Chopping," IEEE J. of Solid-State Circuits, vol.45, no.10, pp.2044-2056, Oct. 2010.

MIXDES 2012, 19ᵗʰ International Conference *"Mixed Design of Integrated Circuits and Systems"*, May 24-26, 2012, Warsaw, Poland

A Novel Charge Recycling Approach to Low-Power Circuit Design

Chandradevi Ulaganathan, Charles L. Britton, Jr., Jeremy Holleman, and Benjamin J. Blalock

Department of Electrical Engineering and Computer Science
The University of Tennessee
Knoxville, Tennessee, USA

Abstract—**A novel charge-recycling scheme has been designed and implemented to demonstrate the feasibility of operating digital circuits using the charge scavenged from the leakage and dynamic load currents inherent to digital logic. The proposed scheme uses capacitors to efficiently recover the ground-bound charge and to subsequently boost the capacitor voltage to power up the source circuit. This recycling methodology has been implemented on a 12-bit Gray-code counter within a 12-bit multi-channel Wilkinson ADC. The circuit has been designed in 0.5μm BiCMOS and in 90nm CMOS processes. SPICE simulation results reveal a 46–53% average reduction in the energy consumption of the counter. The total energy savings including the control generation aggregates to an average of 26–34%.**

Index Terms—**Charge recycling, dynamic power supply, low-power, virtual power supply.**

I. INTRODUCTION

Reducing power consumption has been an important design paradigm that facilitates highly integrated, energy efficient, portable systems. Battery-operated sensor-based mixed-signal systems necessitate low-cost, low-power operation to extend the battery lifetime. Often, the generation of multiple supply voltages or the use of sophisticated power management ICs presents a large overhead on these systems. The aim of this work is to explore a charge-recycling approach to design energy-efficient digital cells in mixed-signal systems.

The power dissipation in a conventional CMOS digital circuit is given by [1]

$$ P = \left(I_{lkg} V_{DD} \right)_{static} + \left(C_L V_{DD}^2 f_{clk} \alpha + I_{sc} V_{DD} \right)_{dynamic} \quad (1) $$

where the first term represents the static power dissipation due to leakage current, while the second term represents the switching and short-circuit dynamic power dissipation. The switching power is required to charge and discharge the output node and is determined by the load capacitance C_L, supply voltage V_{DD}, frequency of operation f_{clk}, and the activity factor α, which is the probability of a transition per clock. The short-circuit power consumption is due to the flow of current from V_{DD} to ground that is caused by a slow rising or falling input signal.

As achieved by CMOS scaling, power supply voltage (V_{DD}) reduction is one of the most effective methods to reduce power consumption in digital logic. However, aggressive technology scaling has increased the speed and area efficiency of ICs, but has also resulted in larger leakage currents that offset the reduction in switching energy by increasing the static power consumption. Hence, the challenge is to design low-power energy-efficient circuits that meet the required performance.

Several power reduction techniques employing dynamic V_{DD} and threshold voltage (V_{TH}) scaling, frequency scaling and adiabatic computation techniques have been successfully implemented in various digital designs [2-9]. Dynamic scaling techniques vary the power supply voltage or threshold voltage to improve the performance or reduce power consumption in digital blocks depending on the circuit requirements. Adiabatic operation minimizes the dynamic switching energy by reducing the charge or discharge voltage transition and through very slow operation. In another adiabatic approach, the energy expended for charging capacitive nodes may be recycled during discharge and stored for reuse [1, 7]. This paper focuses on reducing the power consumption of digital circuits by charge-recycling as well as by dynamic V_{DD} reduction.

Charge-recycling based vertically stacked computation logics have been reported in [2, 3]. Implicit charge-recycling is accomplished by vertically stacking similar logic units such that the ground-bound charge from upper domain logic cells can be reused in the lower stack (domain). Hence, current balancing between the logic stacks is critical to ensure that the charge from higher stacks is efficiently recycled in the lower domain stacks. Additionally, silicon-on-insulator (SOI) or triple-well processes are required to prevent excessive body effect. Thus, this scheme is effective for large computation systems that are made of several identical logic unit blocks with similar energy consumption, performance and concurrent operation.

A charge-pump based recycling scheme using a virtual V_{DD} has been published in [4-6]. This method employs an adiabatic charge pump to boost the voltage associated with the collected charge to V_{DD}. The delay involved with the charge-pump action is overlapped with the computational delays within the digital system. Hence, the application of this method is constrained to digital systems with considerable computation time and pipelined operation.

In this paper, we explore charge-recycling means to lower digital power consumption within mixed-signal systems. The idea borrows, in part, from the use of recycled charge to supply power to digital cells in [4-6]. The proposed scheme employs a virtual ground to capture the dissipated switching charges and leakage currents that would typically be lost to the ground node. The collected charge is then boosted to the supply voltage and is used as a virtual V_{DD} to power the source circuit. This work is novel by virtue of the absence of any delay in the charge-recycling path.

As with any charge-recycling based system, this scheme is effective when the source digital circuits are operated for at least the minimum time required for charging the virtual ground capacitors to a usable voltage. The use of virtual supply nodes also provides the choice of different voltage-islands based digital operation without the need for on-chip V_{DD} generation or external V_{DD} pins. However, power reduction due to V_{DD} scaling also results in degradation of circuit performance [10]. Hence, this scheme is suitable to be employed within medium-speed, low propagation-delay circuits and also in complex pipeline stages of a computational system. Another advantage of implementing virtual supply and ground is the reduction of leakage currents [4].

Low-power sensor-based systems such as multi-channel ADCs that operate under tight power constraints are ideal applications for this methodology. Furthermore, the absence of delay in the charge-boosting technique renders its integration within low-power portable systems that have long idle states and intermittent time periods of operation. Additionally, this scheme could also be used as an energy harvester that supplies power to a suitable target or perhaps to the chip during standby.

Section II presents an overview of the proposed charge-recycling architecture. Section III discusses in detail the design and implementation of the individual components, and an analysis of the energy efficiency is included. Section IV presents different topologies and the implementation of the proposed charge-recycling scheme. Section V presents the simulation results and performance analysis of the system. Finally, section VI concludes this work.

II. PROPOSED ARCHITECTURE

An important figure-of-merit in energy efficient systems is the minimization of the energy per cycle of operation, E_{TOT}. E_{TOT} is given by the total power-delay product, $P_{TOT}{\cdot}T_{clk}$, which is the power consumed to perform all required computations within one clock cycle. The minimization of E_{TOT} maximizes the energy efficiency [1]. Fig. 1 shows a block level schematic of the proposed charge-recycling scheme.

The source logic block and the target logic block could be different components of a digital system or the same logic circuit. Identification of potential blocks in a system is essential to maximize charge-recycling and power reduction. The choice of the source block is governed by the rate of switching activity in the logic and also by the ease with which the system can be separated into functional blocks. Also, it is critical to ensure that the required circuit performance is achieved with the reduced supply voltage resulting from charge-recycling.

During the charge-accumulation phase the source logic is powered up from V_{DD} to a virtual ground (V_{VGND}) to enable the collection of ground-bound charge by the charge-recycling (CR) capacitor bank at the V_{VGND} node. The V_{VGND} node is monitored to ensure the functionality of the circuit is not degraded by the continuous reduction in the available power

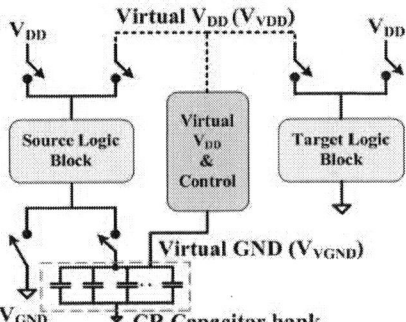

Fig. 1. Block level schematic of the charge-recycling scheme.

supply voltage. Once the V_{VGND} node reaches a predetermined voltage, V_{RG}, control signals are generated to switch to the charge-recycling phase.

The charge-recycling phase begins with the boosting of the capacitor voltage to a virtual V_{DD} (V_{VDD}). This is achieved by stacking the charged V_{VGND} node capacitors to generate the required V_{VDD} value. The target logic block is then powered up from V_{VDD} to ground, V_{GND}. Similar to the charge-accumulation phase, the V_{VDD} node is monitored to prevent the supply voltage from discharging below a voltage level, V_{RD}, which might impede the circuit's functionality.

The proposed architecture offers the advantage of power reduction without any design change to preexisting digital circuits. Simulation results on the architecture verify the benefits of the proposed power reduction methodology. An added advantage of power supply scaling by the use of virtual supply nodes is a reduction in the leakage currents. This reduction is the result of increased threshold voltages due to body biasing, and also lower drain-to-source voltages that cause a reduction in the Drain-Induced Barrier Lowering (DIBL) component of leakage current [10].

III. DESIGN AND ENERGY ANALYSIS

The design and operation of the system, along with the generation of the control signals, are discussed in this section.

A. Charge-Recycling Process – Design and Control

The schematic during the charge-accumulation phase is shown in Fig. 2. A low-power comparator generates the control signals to enable the charging of the capacitors until the V_{VGND} node reaches the maximum allowed voltage, V_{RG}. V_{RG} is determined such that the circuit's performance requirements are met at the reduced virtual supply voltage, even in the presence of comparator offset. This voltage in turn sets the number of CR capacitors (N) required to obtain V_{VDD}. For this application, a moderate value of $V_{DD}/4$ is chosen such that four equal-sized CR capacitors (N=4) are placed in series to provide the V_{VDD}.

When V_{VGND} reaches the fixed voltage V_{RG}, the comparator output, ϕ goes high and the switches (S2-S8) connecting the CR capacitors to V_{VGND} are opened and the circuit is directly connected to V_{GND} via S1. This initiates the charge-recycling phase. Since the switches connect from ground to a sufficiently low voltage, NMOS-only switches have been employed for

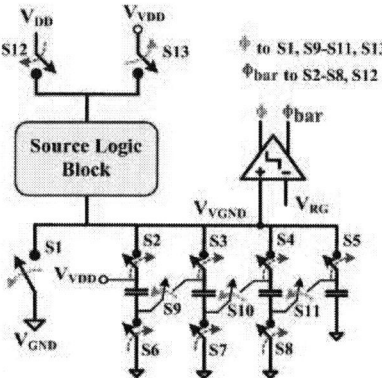

Fig. 2. Schematic illustrating operation during charge-accumulation phase.

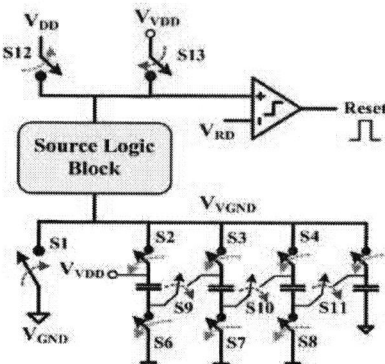

Fig. 3. Schematic illustrating operation during charge-recycling phase.

switches S1 to S8.

Fig. 3 illustrates the switch settings of the capacitive stack necessary to establish the V_{VDD} voltage. Switches S9 to S11 connect the top and bottom plates of adjacent capacitors. Transmission gate switches are used here to provide low resistance as the voltages change due to the capacitors discharging during the circuit's operation. The V_{DD} switches, S12 and S13, are PMOS-only versions. Once the comparator detects the V_{VDD} voltage to be below V_{RD}, a reset pulse is generated and the circuit reverts to the V_{DD} power rail. Also, the CR capacitors are returned to the accumulation phase configuration, as in Fig. 2. The end of the charge-recycling phase initiates the charge-accumulation phase and the cycle repeats. The value of V_{RD} is conveniently set at $3V_{DD}/4$.

The first cycle of the charge-accumulation phase consumes more time when compared to subsequent cycles. This is due to the initial need to charge from ground to V_{RG} while the following cycles only need an incremental charge from V_{RD}/N to V_{RG}. The system repeats the charge-accumulation and charge-recycling phases as long as the clock to the source block is enabled, i.e. until the source block goes to idle mode or is powered OFF.

B. Estimation of the Charge-Recycling Capacitor Size

The charge-recycling capacitors need to be large enough to supply the energy required by the target logic blocks and maintain a virtual supply voltage of more than V_{RD} to deliver power to the target block. Conventional CMOS logic cells consume energy of $C_L V_{DD}^2$ to charge a load capacitor C_L to V_{DD}. A worst case estimate of the required charge can be given by $C_L V_{DD}^2$ times the number of PMOS transistors in the target design [4, 11]. So the CR capacitors that form the virtual V_{DD} need to be large enough to sustain the worst case energy requirements of the target logic.

Another design parameter that needs to be studied for the capacitor sizing is the frequency of charge-accumulation and charge-recycling phases versus the power-reduction efficiency of the proposed scheme. The frequency of the recycling phases depends on the energy available for reuse in the CR capacitors as well as the activity factor and the power consumption of the source and target blocks. Since the control logic would also dissipate energy for generating the control signals, operating at

a high cycle rate would mean more energy dissipation and lower energy efficiency. However, increasing the accumulation-recycle time implies large CR capacitors and longer charging times. Hence, an optimum value of CR capacitors and accumulation-recycle time is chosen to permit maximum energy efficiency.

A reasonable estimate of the average power consumed by the digital cell is used to determine the effective size of the CR capacitors. Since the topology employs stacking of CR capacitors, the effective capacitance decreases with the number of stacks, N. Hence, the individual capacitors are sized by N times the required value. The price paid for eliminating charge-pump induced delay (as in [4]) between the accumulation and recycle phase is an increase in chip area needed for the CR capacitors.

C. Low-Power Comparator

The low-power dynamic comparator used in this design is a regenerative latch that is commonly used as a sense amplifier in SRAM cells [11, 12]. The schematic of the high-speed, low-power comparator is presented in Fig. 4. The positive regenerative action of the cross-coupled inverter pairs provides the high-gain necessary for the accurate comparison.

The control signals of $\phi 1$ and $\phi 1b$ are derived from the input clock of the source logic block, thus enabling the comparator to run at the same frequency as the source digital block. During the sample/reset phase, $\phi 1$ is low, transistors MBN, MBP are OFF, thus disabling the latch. In this phase

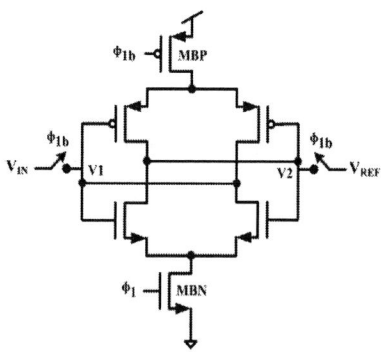

Fig. 4. Schematic of the low-power comparator.

210

input switches S1 and S2 are closed and the inputs (V_{VGND} and V_{RG} for virtual GND comparison, V_{VDD} and V_{RD} for monitoring virtual V_{DD}) are sampled at the high impedance nodes V_1 and V_2. Next, in the regeneration/active phase, $\phi1$ goes high to activate the regenerative action and to open the input switches to isolate the nodes V_1 and V_2 from the input. The voltage difference between the sampled input signals is amplified and the output is available at V_1 and V_2 nodes.

In order to ensure low power dissipation in the nano-watt range, the transistors MBN and MBP are sized to limit the maximum current available for regeneration. Two comparators are employed in the system to monitor the virtual ground and virtual V_{DD} voltages. The outputs from the comparators are used to generate the control signals to switch between the charge-accumulation and the charge-recycle phases.

D. Energy Consumption in the Charge-Recycling Scheme

To evaluate the efficiency of the system, it is pertinent to analyze the energy consumption and energy savings in the charge-recycling scheme. The total energy E_{IN} stored by the CR capacitors at the end of the accumulation phase is given by

$$E_{IN} = \left(\frac{1}{2} N \cdot C_{CR} V_{RG}^{2} \right) N \qquad (2)$$

where $N \cdot C_{CR}$ is the individual charge-recycling capacitance, V_{RG} is the maximum voltage at the virtual ground and N is the number of CR capacitors in parallel ($V_{DD} = N \cdot V_{RG}$). Without the CR scheme, this energy E_{IN} would have been lost as discharge to ground.

Let $E_{T,AVG}$ be the average energy consumed per computation by the target block. During the charge-accumulation phase, $E_{T,AVG}$ is provided by V_{DD}, whereas, during the charge-recycling phase, the CR capacitors furnish the required $E_{T,AVG}$. As long as the available energy, E_{OUT}, is greater than the required, $E_{T,AVG}$, the recycling system provides energy to the target block provided V_{VDD} is above V_{RD}. Thus only a portion of the total recovered charge is used in each accumulation-recycle phase while the rest of the charge resides on the CR capacitors. Note that from the second cycle onwards, the CR capacitors need to recover only the difference in charge required to reach to V_{RG} and so energy recovered can be written as

$$E_{IN}' = \frac{1}{2} C_{CR} \left(V_{DD}^{2} - V_{RD}^{2} \right) \qquad (3)$$

Let the energy consumed by the CR scheme be given by E_{DISS} which encompasses the energy dissipated by the clock to drive the switches, energy consumed by the comparators, control signal generation and also the resistive loss within the switches. The process-related leakage currents and parasitic coupling to the substrate also add to E_{DISS}. Note that $E_{T,AVG}$ determines the size of CR capacitors and switches. Now, the actual recycled energy E_{RCYC} is given by

$$E_{RCYC} = E_{T,AVG} \cdot N_{T,RCYC} = E_{IN}' - E_{DISS} \qquad (4)$$

where $N_{T,RCYC}$ is the number of clock cycles in the target logic during the charge-recycling phase. Thus, the system's energy efficiency is maximized by reducing E_{DISS}. The percentage energy saved using the charge-recycling scheme is approximately given by [4]

$$E_{SAVED} = \frac{E_{RCYC}}{E_{IN}'} \cdot 100\% \qquad (5)$$

IV. IMPLEMENTATION

Different implementations of the proposed scheme are possible depending on the application. The charge-recycling approach can be employed at the circuit-level or at the system-level. At the system level, a time multiplexing approach, wherein multiple CR capacitor banks can be used to provide continuous virtual V_{DD} to target blocks. Also, based on the energy requirement, one virtual V_{DD} can be used to supply power to different target blocks. At the circuit-level, since the recovered charge is readily available as virtual V_{DD}, it is possible to conceive a partially self-powered circuit.

A 12-bit Gray-code counter within a 12-bit, 8-channel low-power Wilkinson ADC [13] designed to operate at 10Ksps, has been used to illustrate the effectiveness of the recycling scheme in a 0.5μm process. Since the counter consumes approximately 30% of the ADC's total power [13], power reduction at the counter would be highly beneficial. Additionally, the CR scheme has been implemented in a 90nm process where leakage effects are more pronounced.

Fig. 5 and Fig. 6 present the layouts of the CR based Gray-code counter in the 90nm and 0.5μm processes, respectively. Analog buffers are included to monitor the virtual supply and ground nodes for testing purposes. The CR capacitors dominate the additional area required for the charge-recycling system. High-density, low-leakage metal-insulator-metal (MIM) on-chip capacitors are used for the CR capacitors. In the 90 nm implementation, the area of the CR system (counter, CR capacitors and control logic) is 13,500 μm² while the counter alone occupies an area of 7,000 μm². This increase is mainly due to the capacitors and they occupy an area of 6,300 μm² for total of 15 pF. In the 0.5μm design, the counter alone occupies

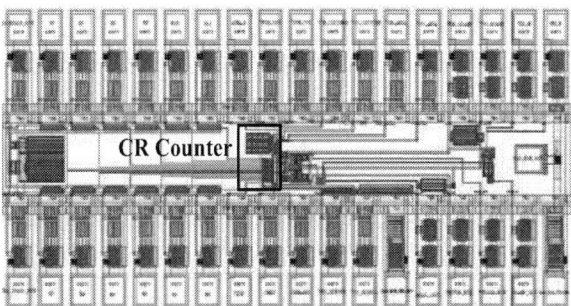

Fig. 5. Layout of Gray-code counter with CR scheme in 90nm process.

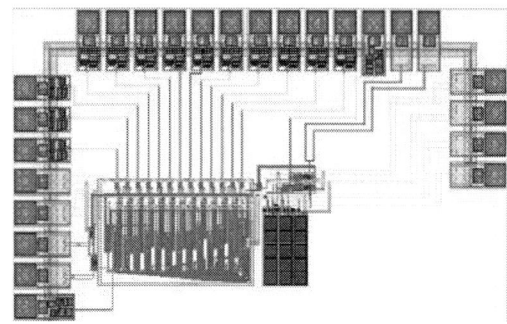

Fig. 6. Layout of Gray-code counter with CR scheme in 0.5μm process.

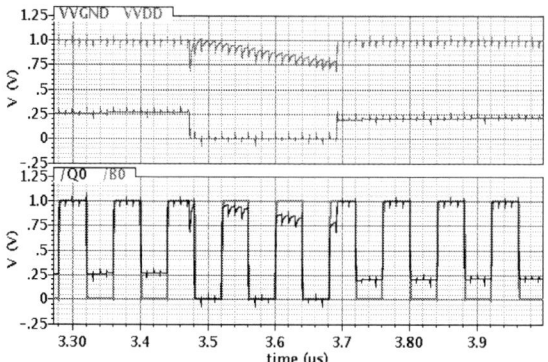

Fig. 7. Transient simulation results of CR counter in 90 nm process.

an area of 0.471 mm², the 75pF total capacitance and logic take 0.077 mm² which is a 16% increase in the total area. Note that the source block used in this work is a simple 12-bit counter and the increase in overall area would be less substantial with a more complex source block or a system such as the ADC [13] where the area increase is only 1.3%. The power savings realized using the charge-recycling scheme justifies the increase in area.

V. SIMULATION RESULTS AND PERFORMANCE ANALYSIS

SPICE simulations were performed on the system to evaluate the power reduction and energy efficiency of the proposed scheme. For the 0.5μm design, the power supply was set at 2.5V, with the minimum virtual V_{DD} of 1.75V. The 90nm implementation had V_{DD} of 1V with minimum V_{VDD} of 0.7V. At the maximum ADC conversion rate, the Gray-code counter runs at 44 MHz. Hence the CR Gray-code counter was characterized at 50 MHz. Further, the effectiveness of the recycling-scheme was also assessed using different target logic block such as a 10-bit Binary counter and operating at a higher frequency of 100 MHz. Fig. 7 presents the virtual V_{DD}, ground and counter output bit (before and after level-shifting) from simulations at 50 MHz. As seen in Fig. 7, once the V_{VGND} reaches 0.25V, V_{VDD} is used to power up the circuit and V_{VGND} is connected to ground. The V_{VDD} node discharges to about 0.7V before the control switches over to charge-accumulation phase where V_{VGND} collects charge.

A. Energy Saving

The energy reduction of partially self-powered counters was estimated by simulating at a fixed performance (frequency of 50 MHz) with and without charge-recycling. The circuit's energy consumption and the percentage energy reduction are presented in Table I. The results show that the counter's energy consumption per cycle has been reduced by 50% with charge-recycling. The energy saved including the energy dissipated by the control circuitry is more than 25% in the CR Gray-code counter. Since 90nm process has more leakage current contribution, more charge is recycled and the dynamic V_{DD} reduction also reduces leakage currents thus contributing to increased energy savings. Further, the comparable energy savings between the two designs verifies that the CR scheme can be employed to efficiently recycle switching-dominated charge as well as charge from leakage currents. The percentage savings can be further improved by minimizing the power dissipation in the comparator and control signal generation.

Also, the energy efficiency using different source and target blocks was investigated with the Gray-code counter configured as the source and a different logic (10-bit Binary counter) as the target. Table II presents a comparison of the percentage energy saved by recycling charge from the Gray-code counter to power-up the Binary counter and vice versa. The system was simulated for the time required to complete one full cycle of counting at the target block at a frequency of 100 MHz. Since both the counters have similar energy requirements and some of the recycled charge is dissipated at the switches, the charge-accumulation phase lasts longer than the charge-recycle phase. So, the energy reduction at the source block is more than that at the target block. Improving the efficiency of virtual V_{DD} generation would certainly increase the total energy saved at the target counter.

TABLE I. PARTIALLY SELF-POWERED CIRCUIT'S PERFORMANCE WITH AND WITHOUT CHARGE-RECYCLING AT 50 MHZ.

Gray-Code Counter Design	Without CR (nJ)	With CR (nJ)	%Energy Reduction (Counter)	%Energy Reduction (incl. Control)
0.5μm	15.29	8.19	− 46%	− 26%
90nm	0.241	0.114	− 53%	− 34%

TABLE II. ENERGY REDUCTION OF SYSTEM WITH AND WITHOUT CHARGE-RECYCLING AT 100 MHZ (90NM DESIGN).

System Configuration	Parameter	Without CR (pJ)	With CR (pJ)	%Reduction
Source: Gray-code Counter Target: Binary Counter	Energy (source)	58.85	35.15	− 40.3%
	Energy (target)	60.84	45.47	− 25%
	Energy (system)	119.69	96.77	− 19%
Source: Binary Counter Target: Gray-code Counter	Energy (source)	243.3	131.4	− 46%
	Energy (target)	235.5	190.6	− 19%
	Energy (system)	478.8	384.2	− 20%

212

B. Effect on Circuit's Speed and Delay

Powering digital blocks from virtual V_{DD} and virtual ground introduces variation in the available supply voltage. In this CR scheme, the reduction in the available power supply and the use of body effect to increase V_{TH} as a means to reduce leakage currents would result in increased circuit delay [11]. The propagation delay (t_d) of the counters operating at 50 MHz is presented in Table III. The change in delay is expected due to the variation in power supply during both the charge-accumulation and the charge-recycling phases. It should be emphasized that the increase in delay does not degrade the performance of the counters and there are no missing counts at the output. Since this scheme is targeted for medium-speed low-power circuits, the small increase in delay does not affect the system's operation.

TABLE III.   PROPAGATION DELAY WITH AND WITHOUT CHARGE-RECYCLING.

Gray-Code Counter Design	Without CR (ns)	With CR (ns)	Delay increase (ns)
0.5μm	2.00	2.66 – 3.13 (Avg: 2.9)	0.90
90nm	1.11	1.22 – 1.86 (Avg: 1.54)	0.43

C. Leakage Current Reduction

Another advantage of implementing a virtual supply and ground is the reduction in leakage currents [4]. The leakage current is dominated by sub-threshold leakage and DIBL currents and is modeled as [10]

$$I_{OFF} = Ae^{\frac{V_{GS}-V_{TH0}-\gamma V_{SB}+\eta V_{DS}}{nv_T}} \cdot \left(1 - e^{\frac{-V_{DS}}{v_T}} \right) \quad (6)$$

where $A = \mu_0 C_{ox} \dfrac{W}{L_{eff}} v_T^2 e^{1.8}$, μ_0 is the zero-bias carrier mobility, C_{ox} is the gate-oxide capacitance, L_{eff} is the transistor effective channel length, W is the transistor width, η is the DIBL coefficient, γ is the linearized body-effect coefficient, n is the transistor sub-threshold swing coefficient and v_T is the thermal voltage (kT/q) [10].

The supply voltage reduction lowers the drain-source voltage and thus reduces the DIBL current. The leakage is also suppressed by the reverse body bias voltage (V_{SB}). The smaller feature sizes in the 90nm process node have more leakage compared to the 0.5μm process. Therefore, the two CR-based designs present a good opportunity to study the leakage current reduction with virtual supplies in these processes.

VI.   CONCLUSION

This paper has demonstrated the feasibility of operating digital circuits using the charge scavenged from the leakage and dynamic load currents inherent to digital design. A novel charge recycling scheme has been designed and implemented for a 12-bit Gray-code counter in 0.5μm and 90nm processes.

Simulation results demonstrate an average energy reduction of 34% with the charge-recycling scheme. The average energy of the counter alone was decreased by 53% by recycling its charge. These figures are better than the 18% to 24% energy savings previously reported in CR-based designs [4-6]. The fabricated charge-recycling counter designs are currently being characterized for reduction in power and leakage currents. In the future, optimizing the energy efficiency of the recycling scheme would boost the energy saved and further enhance the application of charge-recycling approach. Another direction of interest is to develop an energy-efficient recycling scheme that dynamically adapts to changes in temperature and process variations.

ACKNOWLEDGEMENT

The authors wish to thank Robert L. Greenwell for his valuable comments on the manuscript.

REFERENCES

[1] Rahul Sarpeshkar, *Ultra Low Power Bioelectronics: Fundamentals, Biomedical Applications and Bio-inspired Systems*, Cambridge University Press, New York, 2010.

[2] S. Rajapandian, Z. Xu, and K. L. Shepard, "Energy-Efficient Low-Voltage Operation of Digital CMOS Circuits Through Charge-Recycling," *Symposium on VLSI Circuits Digest of Technical Papers*, pp. 330–333, 2004.

[3] J. Gu and C. H. Kim, "Multi-Story Power Delivery for Supply Noise Reduction and Low Voltage Operation," *ISLPED*, pp. 192–197, 2005.

[4] K. Keung, V. Manne, and A. Tyagi, "A Novel Charge Recycling Design Scheme Based on Adiabatic Charge Pump," *IEEE Trans. Very Large Scale Integr. (VLSI) Syst.*, vol. 15, no. 7, pp.733–745, Jul. 2007.

[5] K. Keung and A. Tyagi, "SRAM CP: A Charge Recycling Design Schema for SRAM," *PATMOS 2006, Lecture Notes in Computer Science*, vol. 4148, pp.95–106, 2006.

[6] V. Manne and A. Tyagi, "An Adiabatic Charge Pump Based Recycling Design Style," *PATMOS 2003, Lecture Notes in Computer Science*, pp.299–308, 2003.

[7] A. G. Dickinson and J. S. Denker, "Adiabatic Dynamic Logic," *IEEE Journal of Solid-State Circuits*, vol. 30, no. 3, pp. 311–315, Mar. 1995.

[8] W. C. Athas, L. J. Svensson, J. G. Koller, N. Tzartzahis, and E. Chou,"Low-power digital systems based on adiabatic-switching principles," *IEEE Trans. Very Large Scale Integr. (VLSI) Syst.*, vol. 2, no. 4, pp.398–407, Dec. 1994.

[9] L. J. Svensson and J. G. Koller, "Driving a capacitive load without dissipating fCV²," in *Proc. IEEE Symp. Low-Power Electronics*, pp. 100–101, 1994.

[10] K. Roy and S. C. Prasad, *Low-Power CMOS VLSI Circuit Design*, Wiley, New York, 2000.

[11] J. M. Rabaey, A. Chandrakasan, and B. Nikolic, *Digital Integrated Circuits*, 2nd ed., Pearson Education, Delhi, 2004.

[12] B. Razavi and B. A. Wooley, "Design Techniques for High-Speed, High-Resolution Comparators," *IEEE Journal of Solid-State Circuits*, vol. 27, no. 12, pp. 1916–1926, Dec. 1992.

[13] N.Nambiar *et al.*, "SiGe BiCMOS 12-bit 8-channel low power Wilkinson ADC," *Midwest Symposium on Circuits and Systems*, pp. 650-653, Aug. 2008.

MIXDES 2012, 19th International Conference *"Mixed Design of Integrated Circuits and Systems"*, May 24-26, 2012, Warsaw, Poland

A Switched-Capacitor Implementation of Short-Term Synaptic Dynamics

Marko Noack, Christian Mayr, Johannes Partzsch, Michael Schultz and René Schüffny
Endowed Chair for Highly Parallel VLSI Systems and Neuromorphic Circuits
Institute of Circuits and Systems, Technische Universität Dresden, Germany
Email: marko.noack@tu-dresden.de

Abstract—In this paper we present a novel switched-capacitor implementation of short-term synaptic dynamics with simultaneous depression and facilitation. The developed circuit model is a modified version of a model of neurotransmitter release derived from biological measurements. Despite the simplicity of the circuit the rich dynamics of the original model can be delivered. By completely relying on SC techniques for all calculations, our circuit is significantly less sensitive to process variations and easier to calibrate than commonly employed subthreshold circuits. The circuit makes use of a technique for minimizing leakage effects allowing for real-time operation with time constants up to several seconds. Functionality and robustness of the circuit are verified by simulations and comparisons to the original model.

Index Terms—Switched-capacitor, short-term synaptic dynamics, depression, facilitation, silicon synapse.

I. INTRODUCTION

Biological synapses employ a range of short-time adaptation mechanisms in their stimulus transmission. This so-called short-term plasticity has been identified as a crucial constituent of neural information processing, allowing for temporal filtering [1] and pattern classification in attractor networks [2]. The quantal model of neurotransmitter release introduced in [3] is a well-established model of these synaptic dynamics that has been directly derived from biological evidence. Furthermore, it has been thoroughly analyzed in terms of information processing [4], [5].

Despite the functional importance of biologically realistic short-term dynamics, there are only few neuromorphic circuits implementing simultaneously acting short-term depressing and facilitating mechanisms as described in [3]. The circuits presented in [6] produce either depression or facilitation, while the more complex circuit presented in [7] is at least capable of switching between the two. A combination of facilitation and depression mechanisms is shown in [8], but this implementation is rather aimed at hardware efficiency than biological relevance. In [9], we prepared the model in [3] for circuit design, where the depression part is implemented with OTA-C circuits.

When it comes to neuromorphic implementations of neural dynamics, the main challenge is replication of large time constants in the order of milliseconds up to several seconds. Most circuits make use of transistors working in the subthreshold region [6], [8], exploiting the low drain currents in this regime to discharge capacitors over a long period of time. A serious disadvantage resulting from this approach is the sensitivity

to threshold voltage mismatch, resulting in large variations on the fabricated chip [10]. As an alternative, the time base of the circuits may be accelerated for increased current amplitudes [7], but this prevents interfacing to real-time operating sensors and biological substrates.

In this paper, we aim at tackling both the modeling and circuit implementation problems introduced above. We take the full quantal model of [3] and adapt it for an optimized circuit implementation, deriving equations for mapping parameters to the adapted model. By this modification, we retain the full power of the original model while greatly reducing circuit effort. In contrast to the OTA-C based circuit presented in [9], we employ switched-capacitor (SC) circuit techniques for realizing large time constants, whose behaviour is mainly determined by capacitance ratios instead of absolute transistor parameters, which greatly decreases their mismatch sensitivity. This approach has already been successfully applied to neuromorphic neuron implementations [11], [12]. We carried out a sample design in 180 nm CMOS, relying on SC techniques for all calculations required by the model to maximally reduce reliance on analog performance. The main limit on the achievable time constants are the leakage currents through the employed switches, which can be minimized with simple circuit extensions to reach values in the order of seconds. With all these characteristics, our proposed circuit can be easily ported to sub-100 nm technologies, taking full advantage of their higher integration density, while still being easily controllable by digital logic.

In section II we present our modified model of short-term synaptic dynamics, obviating the need for expensive multiplication circuits. The implemented design is presented in section III followed by simulation results in section IV.

II. MODEL OF SHORT-TERM SYNAPTIC DYNAMICS

A. Original Quantal Model

The quantal model of neurotransmitter release introduced in [3] describes the amplitude of excitatory postsynaptic currents (PSC) as a function of the timing of presynaptic spikes and their history. Equations (1) – (3) show the iterative description of the model, where n is the number of the spike and Δt_n is the time between n-th and $(n+1)$-th spike.

$$u_{n+1} = u_n \cdot (1 - U) \cdot e^{-\frac{\Delta t_n}{\tau_{facil}}} + U \qquad (1)$$

$$R_{n+1} = R_n \cdot (1 - u_n) \cdot e^{-\frac{\Delta t_n}{\tau_{rec}}} + 1 - e^{-\frac{\Delta t_n}{\tau_{rec}}} \qquad (2)$$

$$PSC_n = A \cdot R_n \cdot u_n \qquad (3)$$

Two mechanisms acting simultaneously are modulating the PSC amplitude. An amplifying mechanism called facilitation is modeled by variable u, which increases by a certain amount at every presynaptic spike and decays back with time constant τ_{facil} to its resting state U. At high spiking frequencies u_n saturates at 1. R, which describes a depression mechanism is decreased at presynaptic spikes by a certain amount, which is influenced by u. The depression recovers to its resting state 1 with time constant τ_{rec}. At high frequencies R_n saturates at 0. The amplitude of the n-th PSC is calculated by a multiplication of the two variables u_n and R_n and a scaling factor A, which represents the absolute synaptic efficacy.

B. Proposed Model

In order to develop a neuromorphic circuit which is capable of reproducing the quantal model [3] the approach presented in [9] could be taken. Therefore, exponentially decaying voltage curves have to be generated which are updated and triggered at the occurrence of presynaptic spikes. The major drawback of this approach concerning a switched-capacitor implementation is the need for a wide-range voltage multiplier for calculating the product of u_n and R_n. Existing multipliers are rather complex, very area consuming [13] or need large operational amplifiers driving resistive loads [14]. In contrast, our proposed model is capable of approximately reproducing the original quantal model without any multiplier circuit and with a minimum effort on analog circuitry in general.

The iterative description of the proposed model is shown in eqs. (4) − (6).

$$\tilde{u}_{n+1} = \tilde{u}_n \cdot (1 - \tilde{U}) \cdot e^{-\frac{\Delta t_n}{\tilde{\tau}_{facil}}} + \tilde{U} \qquad (4)$$

$$\tilde{R}_{n+1} = ((1 - \alpha) \cdot \tilde{R}_n + \alpha \cdot \tilde{u}_n) \cdot e^{-\frac{\Delta t_n}{\tilde{\tau}_{rec}}} \qquad (5)$$

$$P\tilde{S}C_n = \tilde{A} \cdot (\tilde{u}_n - \tilde{R}_n) \qquad (6)$$

From Eq. (1) it is obvious that $\tilde{u}$ is equivalent to u. Only a parameter mapping has to be performed from U to $\tilde{U}$. In contrast to the original model, where R_{n+1} depends on the product of u_n and R_n, $\tilde{R}_{n+1}$ only depends on a weighted sum of $\tilde{u}$ and $\tilde{R}$, which simplifies the circuit enormously. Factor α determines the strength of the depression. Here $\tilde{R}$ is inverted to R meaning the 'recovered' state is at $\tilde{R} = 0$. This implies that large values of $\tilde{u}$ lead to a stronger increase of $\tilde{R}$ which has the same effect as in Eq. (2) where large values of u lead to a stronger decrease of R. Also it can be seen that $\tilde{R}$ never exceeds $\tilde{u}$ so that the difference $(\tilde{u} - \tilde{R})$ is always positive. The amplitude of the postsynaptic current $P\tilde{S}C$ is determined by this difference scaled by $\tilde{A}$. The time course of the PSC is modeled by an exponential decay with time constant $\tilde{\tau}_{PSC}$.

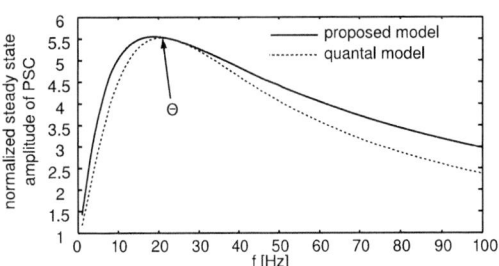

Fig. 1. Normalized steady-state amplitude of PSC at constant presynaptic pulse rates for the quantal model and the proposed model. The peak frequency is marked with Θ. Used parameters are $U = 0.03$, $\tau_{rec} = 130$ ms, $\tau_{facil} = 530$ ms, $\tilde{U} = 0.055$ $\alpha = 0.44$. $\tilde{\tau}_{rec} = 87$ ms, $\tilde{\tau}_{facil} = 864$ ms.

C. Parameter Mapping

A characteristic property, which allows for comparing both models is the steady-state behavior at constant presynaptic firing rates. If both mechanisms, facilitation and depression, are acting simultaneously the steady-state amplitude PSC_{st} of the PSC shows a bell-shaped curve (see Fig. 1). In order to fit our model against the quantal model, parameters can be adjusted, so that the peaks of the curves are matching. The peak frequency Θ of the quantal model is determined by

$$\Theta \approx \frac{1}{\sqrt{U \cdot \tau_{facil} \cdot \tau_{rec}}}. \qquad (7)$$

In the proposed model the peak frequency $\tilde{\Theta}$ is

$$\tilde{\Theta} \approx \frac{1}{\sqrt{\tilde{U} \cdot \alpha \cdot \tilde{\tau}_{facil} \cdot \tilde{\tau}_{rec}}}. \qquad (8)$$

The peak amplitude, normalized to the first PSC amplitude is

$$\frac{PSC_{st}(\Theta)}{PSC_1} \approx \sqrt{\frac{1}{4U} \cdot \frac{\tau_{facil}}{\tau_{rec}}} \qquad (9)$$

for the quantal model and

$$\frac{P\tilde{S}C_{st}(\tilde{\Theta})}{P\tilde{S}C_1} \approx \frac{1}{\sqrt{4 \cdot \tilde{U} \cdot \alpha \cdot \frac{\tilde{\tau}_{rec}}{\tilde{\tau}_{facil}} + \alpha \frac{\tilde{\tau}_{rec}}{\tilde{\tau}_{facil}}}} \qquad (10)$$

for the proposed one. With equations (7) − (10) the time constants of the proposed model can be derived to

$$\tilde{\tau}_{facil} = \tau_{facil} \cdot \left(1 + \frac{\alpha}{2} \sqrt{\frac{\tau_{rec}}{U \cdot \tau_{facil}}}\right) \qquad (11)$$

$$\tilde{\tau}_{rec} = \tau_{rec} \cdot \frac{1}{\frac{\alpha^2 \cdot \tilde{U}}{2U} \cdot \sqrt{\frac{\tau_{rec}}{U \cdot \tau_{facil}}} + \alpha \cdot \frac{\tilde{U}}{U}} \qquad (12)$$

The remaining parameters $\tilde{U}$ and α, whose ranges are between 0 and 1, are used for better curve fitting. The absolute PSC amplitude has to be scaled with $\tilde{A} = A \cdot \frac{U}{\tilde{U}}$.

III. SWITCHED-CAPACITOR IMPLEMENTATION

The switched-capacitor circuit implementing the proposed model is depicted in Fig. 2(a). It consists of three analog blocks for facilitation, depression and PSC generation and a state machine (FSM) for controlling the switches. The values $\tilde{u}$ and $\tilde{R}$ are stored on capacitors C_u and C_R respectively. C_{PSC} stores the trace of the PSC.

A. Circuit Operation

As intended by equations (4) and (5) an exponential decay of $\tilde{u}$ and $\tilde{R}$ as well as of the PSC curve has to be performed. This is implemented by generic switched capacitor resistor emulations with switches S_{u1} and S_{u2} and capacitor C_{Ru} for the facilitation block, S_{R1}, S_{R2} and C_{RR} for depression and S_1, S_2 and C_{RPSC} for the PSC generation circuit, which discharge C_u, C_R and C_{PSC} towards ground. The corresponding time constants of decay $\tilde{\tau}_{facil}$, $\tilde{\tau}_{rec}$ and $\tilde{\tau}_{PSC}$ are determined by the switching frequency and can be adjusted externally. Therefore signals DECAY_u, DECAY_R and DECAY_PSC are provided, which send pulses to the state machine in certain intervals (see Fig. 2(b)). If the FSM receives one of these signals two non-overlapping switch phases are triggered. In the case of DECAY_u, first S_{u1} is switched on, which completely discharges C_{Ru}. After S_{u1} is switched off, S_{u2} is switched on leading to a charge equalization on C_{Ru} and C_u. The voltage V_u over C_u drops from an initial voltage V_{u0} to $V_{u0} \cdot \frac{C_u}{C_u+C_{Ru}}$. After n switching events a value of $V_{u0} \cdot (\frac{C_u}{C_u+C_{Ru}})^n$ is reached. With $(\frac{C_u}{C_u+C_{Ru}})^n = e^{-1}$ the number of switching events needed for one $\tilde{\tau}_{facil}$ period can be derived, so the pulse frequency of DECAY_u is determined by $f_{DECAY\_u} = -(\tilde{\tau}_{facil} \cdot \ln \frac{C_u}{C_u+C_{Ru}})^{-1}$. Same applies for depression with and PSC generation circuits. In our design we chose $C_u/C_{Ru} = 1.12\,\text{pF}/32\,\text{fF} = 35/1$ and $C_R/C_{RR} = C_{PSC}/C_{RPSC} = 480\,\text{fF}/32\,\text{fF} = 15/1$. Thus, for $\tilde{\tau}_{facil} = 500$ ms signal DECAY_u is sent with a frequency of 71 Hz.

If a presynaptic spike occurs (signal PRESPK is sent to the FSM), the values $\tilde{u}$, $\tilde{R}$ and $\tilde{PSC}$ have to be updated. By definition the current value of $\tilde{u}$ is represented by V_u after the update and the current value of $\tilde{R}$ is represented by V_R before the update. This ensures that $\tilde{u}$ decays towards $\tilde{U}$ and $\tilde{R}$ towards 0. Thus, V_u is updated first, then the PSC amplitude is calculated and finally V_R is updated.

In order to calculate Eq. (4) the FSM alternately switches S_{u3} and S_{u2} (see Fig. 2(b)), which increases V_u exponentially towards V_A and thereby implements a constant update factor $(1 - \tilde{U})$. The number of switching events is provided by the 6 bit wide signal UTIL. The relationship between $\tilde{U}$ and UTIL depends on the capacitance ratio and can be expressed as

$$\text{UTIL} = \left\lfloor \frac{\log(1 - \tilde{U})}{\log(\frac{C_u}{C_u+C_{Ru}})} \right\rfloor . \qquad (13)$$

After charging of C_u, switches S_4 and S_5 are closed and the voltage difference $(V_u - V_R)$ is stored on C_{PSC}, which corresponds to Eq. (6). Finally V_R is charged up towards V_u. As can be seen in Eq. (5) the strength of the charging is determined by α which is performed by alternately switching S_{R2} and S_{R3} similar to the facilitation mechanism. The

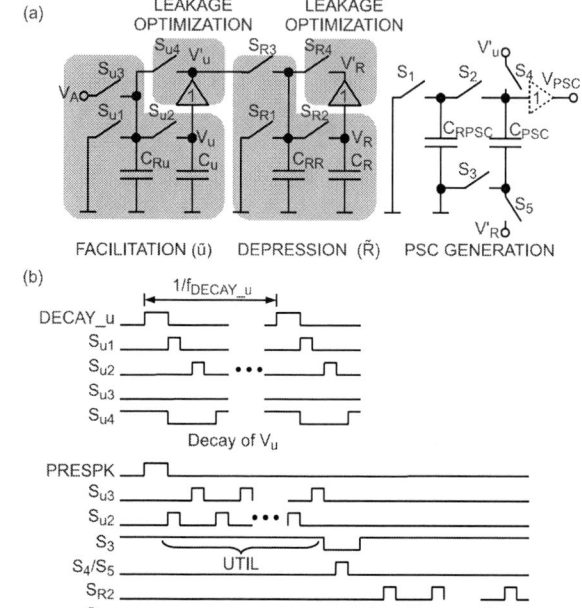

Fig. 2. (a) SC implementation of the proposed model of synaptic dynamics. Capacitors C_u and C_R store the values $\tilde{u}$ and $\tilde{R}$ respectively. C_{PSC} stores the trace of the PSC. (b) Diagram of switch signals for decay of V_u and an update at a presynaptic spike. Decay of V_R and V_{PSC} follow the same principle as for V_u.

number of switching events can be calculated by

$$\text{ALPHA} = \left\lfloor \frac{\log(1 - \alpha)}{\log(\frac{C_R}{C_R+C_{RR}})} \right\rfloor . \qquad (14)$$

Except for the state where S_4 and S_5 are closed, S_3 is always closed to get a defined voltage V_{PSC} by grounding capacitor C_{PSC}.

The voltage V_A controls the absolute synaptic efficacy $\tilde{A}$ and represents the upper limit for V_u and V_R. If V_A stays one threshold voltage below supply voltage, C_u, C_{Ru}, C_R and C_{RR} can be implemented as PMOS capacitors which then always operate in the strong inversion region. Also all switches can be implemented with single NMOS transistors and the buffer amplifiers do not need to provide rail-to-rail operation. As can be seen in Fig. 3 the amplifiers are implemented in a single ended single stage folded cascode topology to allow low supply voltage and a sufficiently high gain to reduce the output voltage offset. Since the state machine operates at a low clock frequency of about 160 kHz slew rate is not a major issue. This allows a compact design with a low power consumption.

B. Leakage Optimization

Due to the large time constants in our implementation, leakage effects at high impedance nodes have to be considered. The main source for leakage effects at C_u and C_R are the

216

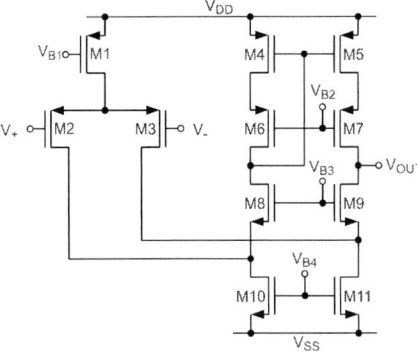

Fig. 3. Amplifier in single ended single stage folded cascode topology used as buffer in Fig. 2.

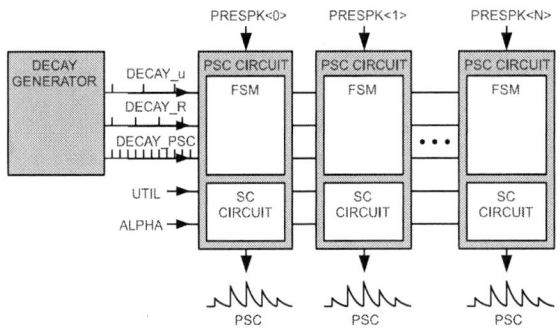

Fig. 4. System of several PSC circuits receiving pulse signals from the decay generation circuit. The PSC circuits consist of the SC circuit shown in Fig. 2(a) and a finite state machine for controlling the switches. PSCs are triggered by the input signal PRESPK.

switching transistors S_{R2} and S_{u2} in off-state. As shown in [15] the main leakage effects that appear at a turned off transistor are junction leakage, channel leakage and subthreshold leakage. They scale with the potentials between the transistor's terminals. Therefore, in the interval when neither a decay nor an update is performed, the voltages across C_u and C_R are fed back via switches S_{u4} and S_{R4} to C_{Ru} and C_{RR} respectively. So V_{DS} of the transistors implementing S_{u2} and S_{R2} is defined by gain and offset voltage of the buffers, which is close to 0. Consequently, subthreshold leakage of S_{u2} and S_{R2} is strongly reduced. As bulk and gate of the transistors are at 0 V while they are turned off, the leakage currents caused by reverse biased PN junctions at times of high V_{GB} and V_{DB} cannot be influenced. However, these are much smaller than the subthreshold leakage in the employed technology [15].

C. Scalability

The circuit is designed for a special architecture of neuromorphic synapse arrays presented in [16], where the circuits generating the PSCs are separated from the individual synapses. Thus, several synapses receive the same presynaptic input, which is usually the case in neural networks. This leads to a smaller number of PSC circuits and smaller synapses, significantly improving the scalability of the overall system. As depicted in Fig. 4 the PSC circuits with the same parameterization can share one decay generation circuit to keep them as small as possible.

IV. RESULTS

Figure 5 shows the simulation results of PSC amplitudes of the original quantal model [3], the proposed model and a circuit simulation for a sequence of constant pulse rates of 15 Hz, 30 Hz, 80 Hz and 15 Hz with instantaneous transitions. All curves are normalized to the first PSC amplitude. As can be seen the proposed model well approximates the quantal model in terms of steady state amplitudes and the behavior at spike frequency transitions. At frequencies up to 30 Hz the circuit simulation results match the model results. At higher frequencies the PSC amplitudes underlie deterministic

jumps which are caused by time discretization of the decay mechanism of V_u and V_R, since at these presynaptic spike frequencies the pulse frequency of the signal DECAY_u is in the same order of magnitude. In order to suppress the jumps and to get a finer discretization the ratio of capacitances C_{Ru}/C_u can be made smaller which would raise two issues: first, the counter in the FSM, which controls the update of V_u has to be larger since more switching phases are needed to charge the capacitor C_u and second, the matching between these capacitors gets worse. But since future technologies tend to smaller transistor sizes, the digital building blocks carry less weight and could be more complex.

An example spike train, which shows both facilitation and depression, is depicted in Fig. 6. The voltage V_{PSC} has been sampled on valid data only, meaning the state where S_4 and S_5 are closed is excluded because C_{PSC} is not grounded in this moment. In order to verify the robustness of the circuit concerning device mismatch and process variations a Monte Carlo simulation with 100 runs has been performed. The amplitude deviations, which are similar to those in subthreshold circuits [17], are mainly caused by the amplifiers whereas the time constants can be robustly reproduced due to the well matched capacitors and the externally provided switching frequency. Note that due to the simplification of the model by replacing the multiplication in Eq. 2 with the difference in Eq. 5 some cases can be covered less accurately. Thus, the response to a pulse train when $\tau_{facil} = \tau_{rec}$ is slightly smoother than shown in [18]. However, most cases can be replicated well.

For the design in a 180 nm CMOS technology the required chip area is approximately 2400 μm^2 including the FSM, which consumes about 75 % of the entire area. When moving towards smaller technologies shrinking of analog components like capacitors and switches is limited, whereas the digital parts scale fully. Thus, in a 65 nm technology, the area could be shrunk to 850 μm^2, which is slightly smaller than the depressing synapse presented in [17].

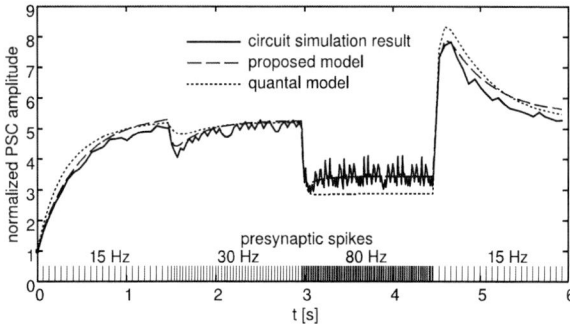

Fig. 5. Simulation results of the quantal model [3], the proposed model and the switched-capacitor circuit with a sequence of instantaneous spike frequency transitions. Parameters are the same as in Fig. 1.

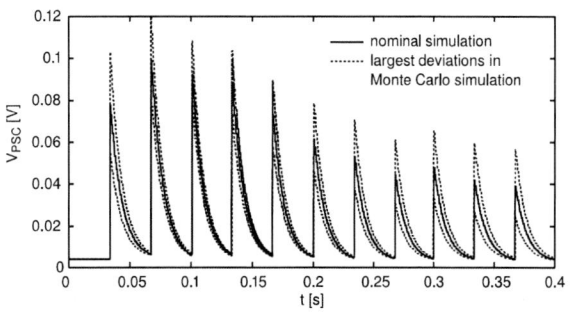

Fig. 6. Simulation results of V_{PSC} for an example spike train including results of a Monte Carlo simulation with 100 runs. The grey lines represent the largest deviations from the nominal simulation. Parameters are $\tilde{U} = 0.081$ $\alpha = 0.36$, $\tilde{\tau}_{rec} = 500$ ms, $\tilde{\tau}_{facil} = 100$ ms.

V. CONCLUSION

In this paper we presented an SC implementation of synaptic dynamics derived from a biologically realistic model. Our model adaptation reduces the implementation complexity to simple subtractions, while keeping the computational richness of the original quantal model. The circuit allows for real-time operation and is robust in terms of process variation on the chip. Especially the time constants, which govern dynamic behavior of a neural network, exhibit very little spread. Since digital building blocks dominate the overall circuit area, our design will scale well with advanced CMOS technologies.

ACKNOWLEDGMENT

This research has received funding from the European Union Seventh Framework Programme (FP7/2007- 2013) under grant agreement no. 269459 (CORONET)

REFERENCES

[1] L. Grande and W. Spain, "Synaptic depression as a timing device," *Physiol*, vol. 20, pp. 201–210, 2005.

[2] J. Mejias and J. Torres, "Maximum memory capacity on neural networks with short-term synaptic depression and facilitation," *Neur Comput*, vol. 21, no. 3, pp. 851–871, 2009.

[3] H. Markram, Y. Wang, and M. Tsodyks, "Differential signaling via the same axon of neocortical pyramidal neurons," *PNAS*, vol. 95, no. 9, pp. 5323–5328, 1998.

[4] L. Abbott, G. Wade, and W. Regehr, "Synaptic computation," *Nature*, vol. 431, no. 14, pp. 796–803, 2004.

[5] V. Matveev and X.-J. Wang, "Differential short-term synaptic plasticity and transmission of complex spike trains: to depress or to facilitate?" *Cereb Cortex*, vol. 10, pp. 1143–1153, 2000.

[6] S.-C. Liu, "Analog VLSI circuits for short-term dynamic synapses," *EURASIP J. Appl. Signal Process.*, vol. 2003, pp. 620–628, 2003.

[7] J. Schemmel, D. Brüderle, K. Meier, and B. Ostendorf, "Modeling synaptic plasticity within networks of highly accelerated IF neurons," in *ISCAS*, may 2007, pp. 3367 –3370.

[8] E. Chicca, G. Indiveri, and R. Douglas, "An adaptive silicon synapse," in *Circuits and Systems, 2003. ISCAS '03. Proceedings of the 2003 International Symposium on*, vol. 1, may 2003, pp. I–81 – I–84 vol.1.

[9] M. Noack, C. Mayr, J. Partzsch, and R. Schüffny, "Synapse dynamics in CMOS derived from a model of neurotransmitter release," in *20th ECCTD 2011*, 2011, pp. 197–200.

[10] J. Pineda de Gyvez and H. Tuinhout, "Threshold voltage mismatch and intra-die leakage current in digital CMOS circuits," *Solid-State Circuits, IEEE Journal of*, vol. 39, no. 1, pp. 157 – 168, jan. 2004.

[11] J. Elias and D. Northmore, "Switched-capacitor neuromorphs with wide-range variable dynamics," *Neural Networks, IEEE Transactions on*, vol. 6, no. 6, pp. 1542 –1548, nov 1995.

[12] F. Folowosele, R. Etienne-Cummings, and T. Hamilton, "A CMOS switched capacitor implementation of the mihalas-niebur neuron," in *BioCAS 2009*, nov. 2009, pp. 105 –108.

[13] Z. Hong and H. Melchior, "Four-quadrant CMOS analogue multiplier," *Electronics Letters*, vol. 20, no. 24, pp. 1015 –1016, 22 1984.

[14] N. Khachab and M. Ismail, "A nonlinear CMOS analog cell for VLSI signal and information processing," *Solid-State Circuits, IEEE Journal of*, vol. 26, no. 11, pp. 1689 –1699, nov 1991.

[15] K. Roy, S. Mukhopadhyay, and H. Mahmoodi-Meimand, "Leakage current mechanisms and leakage reduction techniques in deep-submicrometer CMOS circuits," *IEEE Transactions on Electron Devices*, pp. 1393–1400, 2000.

[16] M. Noack, J. Partzsch, C. Mayr, S. Henker, and R. Schuffny, "Biology-derived synaptic dynamics and optimized system architecture for neuromorphic hardware," in *MIXDES*, 2010, pp. 219–224.

[17] C. Bartolozzi and G. Indiveri, "Synaptic dynamics in analog VLSI," *Neural Computation*, vol. 19, no. 10, pp. 2581–2603, 2007.

[18] Y. Wang, H. Markram, P. H. Goodman, T. K. Berger, J. Ma, and P. S. Goldman-Rakic, "Heterogeneity in the pyramidal network of the medial prefrontal cortex," *Nature Neuroscience*, vol. 9, no. 4, pp. 534–542, Mar. 2006.

Area Efficient Front-end Readout Electronics for Pixel Detector Based on Inverter Amplifier

Rafal Kleczek, Piotr Otfinowski, Pawel Grybos

Department of Measurement and Instrumentation
AGH University of Science and Technology
30-059 Cracow, Poland
rafal.kleczek@agh.edu.pl, piotr.otfinowski@agh.edu.pl, pgrybos@agh.edu.pl

Abstract—**Modern X-ray imaging applications require low noise and power, high rate readout front-end electronics. A widely used, dedicated for semiconductor detectors analog part of readout front-end architecture consists with charge sensitive amplifier and pulse shaping amplifier. To meet the requirements of pixel applications the simple architectures of front-end electronics and used amplifiers, low power dissipation and occupied chip area are demanded. We present the design of readout front-end electronics dedicated for pixel detectors, based on an inverter amplifier. It is characterized by very low power dissipation level P = 13µW, low noise performance ENC = 59e⁻ rms and small occupied chip area A = 850µm².**

Index Terms—**low noise electronics, CMOS front-end readout**

I. INTRODUCTION

Nowadays an X-ray radiation is widely used in medicine, industrial and scientific applications. Starting with a single semiconductor detector and a single readout channel system, the extensively development of VLSI technology has led to the multichannel architecture of the modern systems, where each detector electrode is read by an independent electronic channel. The development of the technology allows also to build higher rate, lower both noise and power dissipation level readout front-end electronics, which is the crucial part in any systems for an X-ray imaging.

The way to extend a spatial resolution of an X-ray imaging system is to concentrate more pixels in a given detector volume. The higher number of pixels implies the lower power dissipation and smaller occupied area by pixel, which impacts on both noise and matching from channel to channel of an overall system. The trend (see Tab. 1) is a minimization of the pixel dimensions while keeping the power consumption and noise at required level at the same time.

TABLE I. COMPARISON OF PIXEL CHIPS IN SUBMICRON TECHNOLOGY

Chip	Medipix II [1]	Pilatus 2 [2]	XPAD3S [3]	EIGER [4]	FPDR90 [5]
Technology	250 nm	250 nm	250 nm	250 nm	90 nm
Pixel size [µm²]	55×55	172×172	130×130	75×75	100×100
Power per pixel [µW]	8	10	40	8,8	42
Noise [e-rms]	141	123	127	180	106

This work was supported by Ministry of Science and Higher Education of Poland with project no. N515 243 037 in the years 2009-2011.

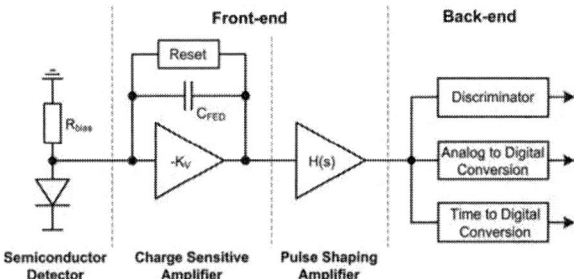

Figure 1. The generic architecture of readout electronics for semiconductor detectors: a front-end and a back-end parts.

The widely used readout front-end systems are designed according to the generic architecture, which is consisted of two main blocks: a front-end part and back-end one (see Fig. 1). The front-end part of the system is responsible for acquiring signal generated by the detector and analog processing of it. The incoming quasi Dirac current pulse is integrated by the charge sensitive amplifier (CSA) creating a voltage step at its output. The reset circuit discharges the feedback capacitor C_{FED} to prevent saturation of the input stage. The voltage step is applied to the pulse shaping amplifier (shaper) which amplifies, improves signal-to-noise (SNR) ratio and shortens the output pulse width. The back-end part changes the analog preprocessed signal to a digital domain. The way of conversion (binary readout with discriminators, analog-to-digital conversion, time-to-digital conversion) depends on a desired application.

This paper presents the design of readout front-end electronics for semiconductor pixel detectors in CMOS 130 nm technology. It is built of: a CSA, a simple CR-RC shaper and a discriminator based on simple inverter amplifiers. The designed readout electronics occupies very small silicon area and characterizes both low power dissipation and noise performance at the same time.

II. THE DESIGNED FRONT-END ELECTRONICS

The measurement resolution of any X-ray imaging system depends on properties of an analog front-end part, which determines time and noise parameters of an overall system. From this point of view the used voltage amplifier should have an infinite voltage gain and an unlimited bandwidth, what in

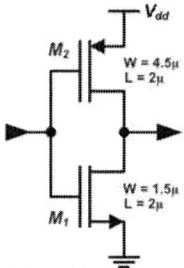

Figure 2. The core of the CSA and the shaper stage – an inverter amplifier.

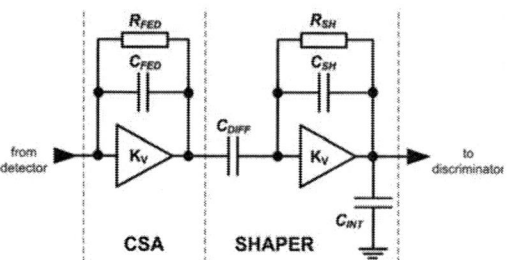

Figure. 3. The architecture of the analog part of the designed readout front-end electronics.

practice cannot be implemented. To ensure that nearly all of generated by the detector electron-hole pairs flows into the input stage of front-end circuit, the input impedance of the electronics must be much lower than the detector impedance. It implies that the detector capacitance C_{DET} has to be much lower than the input capacitance C_{in}, which as a result of Miller multiplication is equal to $C_{in} = K_V \cdot C_{FED}$, where: K_V – amplifier voltage gain; C_{FED} – CSA feedback capacitance. From AC behavior point of view the amplifier has to have sufficient large bandwidth to process of very fast incoming input current pulse. To conclude, a high value of amplifier gain bandwidth product GBW is required, but usually high GBW means high power dissipation, more complicated architecture of an amplifier and larger chip occupied area, which cannot be applied in applications dedicated for silicon pixel detectors.

A. Inverter Amplifier

Having in mind the limitation of area in a single pixel we decided to choose an inverter amplifier as a core of the CSA and the shaper (see in Fig. 2). The idea of using an inverter stage as an analog amplifier in front-end electronics was implemented before by other groups [4, 8, 9]. The low area of silicon required by an inverter architecture is a strong advantages for pixel application. From power dissipation point of view the longer transistor length L and smaller transistor width W, the lower level of power is consumed. To obtain the circuit which is quite insensitive to the imperfections of fabrication process the minimum dimensions available by technology should be avoided. As a result of these constraints the presented transistors' dimensions were chosen. Table 2 presents performance of the inverter amplifier as a function of power supply voltage V_{dd}. The threshold voltage for used NMOS and PMOS transistors are respectively 373mV and -380mV.

TABLE II. PERFORMANCE OF THE INVERTER AMPLIFIER AS A FUNCTION OF POWER SUPPLY VOLTAGE V_{DD}

Power supply [V_{dd}]	K_V [V/V]	GBW [GHz]	P_{DISS} [μW]
0.8	88.4	0.28	0.29
0.9	93.9	0.59	0.77
1.0	94	0.99	1.79
1.1	90.4	1.37	3.52
1.2	85.6	1.74	5.94
1.3	80.9	2.09	9.29
1.4	77.1	2.43	13.6
1.5	73.9	2.75	18.8

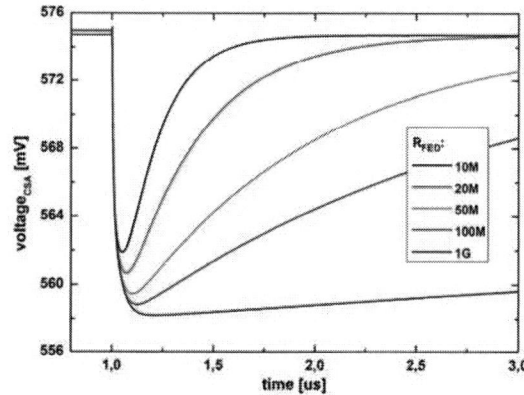

Figure 4. The CSA output for input charge q_{in} = 1/3fC as a function of feedback resistance R_{FED}.

To obtain a higher value of voltage gain K_V the power supply voltage V_{dd} should be decreased, but the GBW decreases at the same time, which limits the speed of processing of an input signals. Additionally, the lower supply voltage limits the power consumption. As a result of these limitations the voltage supply was selected as V_{dd} = 1.2V.

B. Charge Sensitive Amplifier and Pulse Shaping Amplifier

The architecture of the analog part of designed readout electronics is shown in Fig. 3. It consists with the CSA (amplifier, R_{FED}, C_{FED}) and the shaper (amplifier, C_{DIFF}, C_{SH}, C_{INT} and R_{SH}).

The CSA uses the described inverter as a core with capacitance of C_{FED} and resistance R_{FED}. As a resistance R_{FED} we use a MOS transistor working in linear region, which effective resistance can be achieved in the range from 10MΩ to 1GΩ. The influence of the R_{FED} resistance on the CSA output for an average input charge equals to q_{in} = 1/3fC is shown in Fig. 4.

The charge amplification of a CSA is inversely proportional to the effective feedback capacitance C'_{FED}. To have this gain enough high, C'_{FED} should be small. However, the gate-drain capacitance C_{GD1} of M1 transistor and C_{GD2} of M2 transistor should also be taken into account, because they contribute to the effective capacitance visible in the CSA feedback $C'_{FED} = C_{FED} + C_{GD1} + C_{GD2}$. The sum of gate-drain capacitances obtained in simulations equals to $C_{GD1} + C_{GD2} \approx 4fF$. To make the CSA amplification more

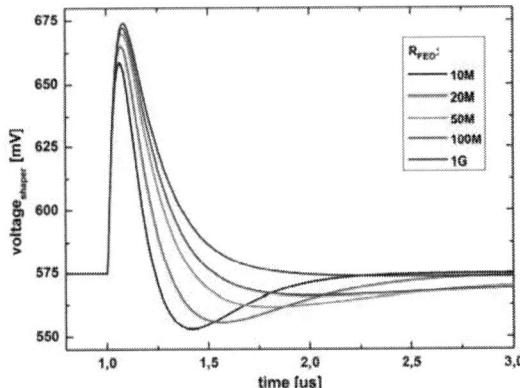

Figure 5. The shaper output for input charge q_{in} = 1/3fC as a function of the CSA feedback resistance R_{FED}.

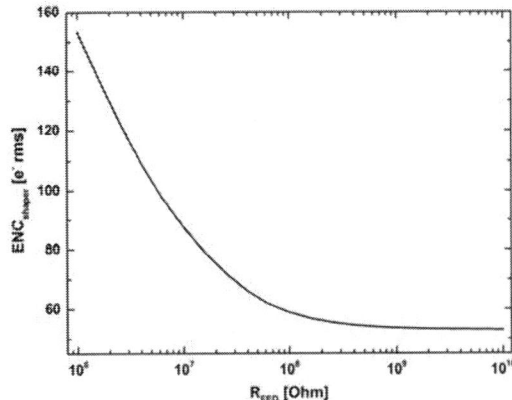

Figure 6. The ENC value at the shaper output dependence on the CSA feedback resistance R_{FED} value.

insensitive to the mismatch of the transistor dimensions the selected capacitance C_{FED} is equal to 15fF.

The second part of the described analog front-end electronics is the shaper, which in our project is a CR-RC pass-band filter. The shaper not only improves the SNR by filtering the signal noise out of an unwanted bandwidth, but also shorten the output signal to meet the timing requirements, which is necessary to operate with high rate of input pulses. Additionally, it provides voltage gain to make the output voltage signal sufficient enough to give required information about X-rays detected by a semiconductor sensor. The presented in Fig. 2 inverter amplifier is also the core of the shaper. The resistance R_{SH} = 10MΩ (a MOS transistor working in linear region) and the capacitor C_{SH} = 15fF set the time constant of the filter. The amplification of the CSA output voltage step is set by the ratio values of the capacitor C_{SH} = 15fF and the capacitor C_{DIFF} = 165fF and it is equal to 11. It was set to obtain an amplitude of about 100mV for an input charge q_{in} = 1/3fC. The shaper is loaded by the capacitance C_{INT} = 100fF to decrease the noise level of the analog front-end electronics. The influence of the R_{FED} resistance on the shaper output for an average input charge equals to q_{in} = 1/3fC is shown in Fig. 5.

Noise performance of a front-end electronics is usually defined by an Equivalent Noise Charge (ENC) parameter. The ENC is defined as an amount of charge that should be delivered to the input of the circuit to obtain SNR at the shaper output equals to unity. The ENC limits the measurement resolution of the overall system. The total equivalent noise charge can be expressed using three components:

$$ENC^2 = ENC_i^2 + ENC_v^2 + ENC_f^2 \quad (1)$$

where ENC_i^2 formula for the CR-RC filter can be expressed as:

$$ENC_i^2 = 0.92 \cdot t_p \cdot \left(2qI_{det} + \frac{4kT}{R_{FED}} \right)$$ - current parallel noise

component: t_p – peaking time, q – elementary charge, I_{det} – detector leakage current, k – the Boltzman constant, T – temperature, R_{FED} – the CSA feedback resistance.

From both time and noise point of view the crucial part of the presented circuit is the resistance R_{FED}. Fig. 6 shows ENC

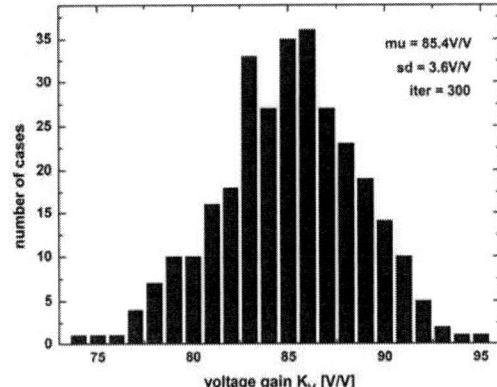

Figure 7. The inverter amplifier voltage gain spread – MC analysis.

parameter as a function of the R_{FED} resistance. Operating with lower rate of input pulses allows to achieve the lower ENC value, because R_{FED} resistor can have larger value. So large R_{FED} resistor is desirable for lower noise performance, however it makes pulse width long. In case of high rate of input pulses and keeping the ENC at the low level the reset circuit has to be applied.

The measured channel parameters deviate from the nominal values as a result of statistical and deterministic effects, which are an inherent feature of the VLSI technology. The designed circuit should be enough immune to mismatch to allow the proper operation of all channels. The result of statistical analysis – Monte Carlo (MC) spread of the inverter amplifier voltage gain is presented in Fig. 7. Even for the lowest value of voltage gain K_V = 74V/V the CSA input impedance is high enough to collect nearly all electrons/holes generated by a semiconductor detector.

The shaper output dc voltage spread is shown in Fig. 8. We decided to use a coupling capacitor C_{coup} between the shaper and the following discriminator to minimize the influence of dc voltage shift.

The designed analog part of readout front-end electronics parameters are: ENC = 59e⁻ rms, power dissipation

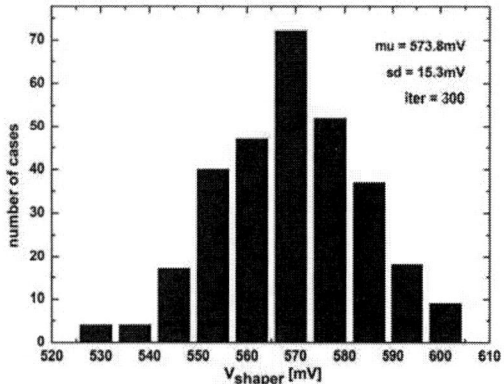

Figure 8. The shaper output dc voltage spread – MC analysis.

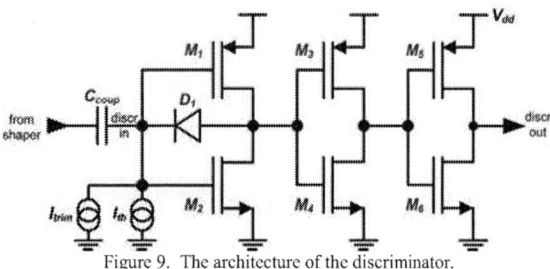

Figure 9. The architecture of the discriminator.

$P = 11.9\mu W$, occupied silicon area $= 630\mu m^2$, peaking time $t_p = 90ns$.

C. Discriminator

Further signal processing at the shaper outputs in the multichannel chip depends on the specific requirements of the foreseen applications. In various imaging techniques employing X-rays it is sufficient to measure spatial distributions of the X-rays of energies above a given threshold or within a given energy window. For such applications one can use the binary readout architecture. In this architecture each pixel of the front-end electronics is equipped with an amplitude discriminator.

The discriminator can also be based on inverters [6] as it is shown in Fig. 9. The discriminator is AC coupled to the shaper output and its generates 1-bit information in response to each signal above a given threshold - see Fig. 10. The threshold is set in the first inverter by the voltage drop on the diode D1. This voltage drop is controlled by global current I_{th}. Because of the possible threshold spread from pixel to pixel an additional current source I_{trim} is added in each pixel separately.

Thanks to using the diode the generated threshold has a smaller variation from pixel to pixel, however the possible threshold setting is limited [7].

The designed discriminator parameters are: power dissipation $P = 1.1\mu W$, occupied silicon area $= 220\mu m^2$.

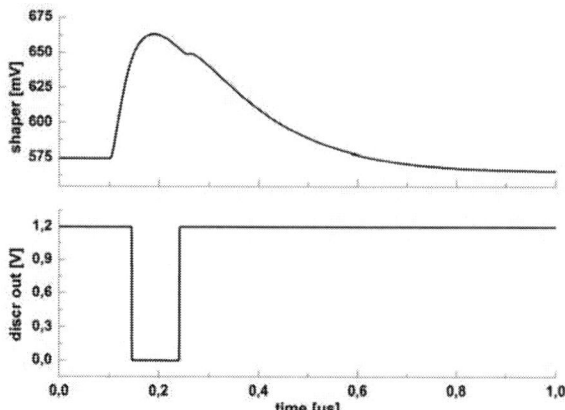

Figure 10. The waveforms at the shaper output, the discriminator input and output for input charge $q_{in} = 1/3fC$.

III. CONCLUSIONS

This paper presents the description and the design of a readout front-end electronics for a semiconductor pixel detectors, implemented in CMOS 130 nm technology. The designed front-end electronics is built with: charge sensitive amplifier with continuous resistive feedback, pulse shaping amplifier and discriminator. The low area of silicon required by an inverter amplifier, which is used as a core of the system components, is a strong advantages for pixel application. To minimize power consumption and noise performance the appropriate transistor dimensions and supply voltage V_{dd} are recommended.

REFERENCES

[1] X. Llopart, M. Campbell, R. Dinapoli, D. San Segundo, E. Pernigotti, "Medipix2: a 64-k Pixel Readout Chip With 55-μm Square Elements Working in Single Photon Counting Mode,", *IEEE Trans. Nucl. Sci.*, vol. 49, no. 5, 2002, pp. 2279 - 2283.

[2] P. Kraft, *et al.*: "Characterisation and calibration of Pilatus detectors." *IEEE Trans. Nucl. Sci.*, vol. 56, no. 3, 2009, p. 758764.

[3] P. Pangaud, *et al.*, "First Results of XPAD3, a New Photon Counting Chip for X-Ray CT-Scanner with Energy Discrimination," *IEEE NSS-MIC 2007 Conference Record*, vol. 1, p. 14-18.

[4] R. Dinapoli, et al.: "A new family of pixel detectors for high frame rate X-ray applications" *Nucl. Instr. and Meth.* A617, 2010, p. 384-386.

[5] P. Maj, P. Grybos, R. Szczygiel, "Development of a Fast Readout Chip in Deep Submicron Technology for Pixel Hybrid Detectors", *Proceedings ot the 20th European Conference on Circuit Theory and Design (ECCTD 2011)*, 29-31 August 2011, Linkoping, Sweden, p. 409-412.

[6] Ch. Bronnimann, et al.: "A pixel read-out chip for PILATUS project", *Nucl. Instr. and Meth. A*, vol. 465, 2001, pp. 235-239.

[7] L. Rossi, P. Fischer, T. Rohe, N. Wermes, "Pixel Detectors. From Fundamentals to Applications", Springer-Verlag, Berlin Heilderberg, 2006.

[8] R. Horisberger, D. Pitzl, "A novel readout chip for silicon strip detectors with analog pipeline and digitally controlled analog signal processing", *Nucl. Instr. and Meth. A*, vol. 326, 1993, p. 92-99.

[9] P. Kraft, "Characterization of the readout chip for the Pilatus 6M Detector", Diploma thesis, ETHZ-IPP Internal Report 03, 2005, Switzerland.

MIXDES 2012, 19th International Conference *"Mixed Design of Integrated Circuits and Systems"*, May 24-26, 2012, Warsaw, Poland

Area Efficient Low Power Neural Amplifiers Using MOS and MIM Capacitors in Submicron Technologies for Ultra Low Corner Frequencies

Piotr Kmon, Pawel Grybos, Robert Szczygiel, Miroslaw Zoladz

AGH University of Science and Technology

Krakow, Poland

pgrybos@agh.edu.pl, kmon@agh.edu.pl, robert.szczygiel@agh.edu.pl, zoladz@agh.edu.pl

Abstract—**This paper presents the design methods that allow to drastically limit an area of a recording channel in multichannel integrated circuits dedicated to neurobiology experiments. The techniques that are presented in this paper allow to apply them in a 3D pixel multichannel integrated systems where area limitations are very strict. Furthermore, these allow one to minimize main problems e\ xisting in modern submicron processes i.e. leakage currents, an ability of obtaining very large MOS based resistances or a uniformity of main parameters of recording channels. For further improvement of the recording channels we designed and processed in a 180nm CMOS technology two recording channels that differ from one other in channel feedback components. The measurement results show that our methods allow to tune the lower cut-off frequency in a very large range, i.e. 10 mHz – 300 Hz. The upper cut-off frequency can be changed into two different modes, i.e. neural spike recording mode where it is equal to 9 kHz or slow biomedical signals recording mode where it can be changed in the 10 Hz - 280 Hz range. The voltage gain of the recording channels can be switched either to 260 V/V or to 1000 V/V. The input referred noise of the recording channel is equal to 5 µV while its power consumption is equal to only 11 µW. The single recording channel occupies only 0.06 mm² of the chip area and together with its large functionality allows one to adapt it into modern 3D pixel multichannel neurobiology applications.**

Index Terms—**multichannel recording, ASIC, neurobiology experiments, 3D pixel systems**

I. INTRODUCTION

Access to modern technologies made it possible to build minimally invasive monitoring systems for observing electrical activity of specific brain areas. Such systems have already been presented as promising tools in diagnosing, detecting and identifying neural patterns that are specific to behavioral phenomenon [1, 2]. In particular, there is a need to develop brain machine interfaces that could help people suffering from neurological disorders such as blindness, spinal cord injuries, dystonia, tremor, Parkinson's disease and some of such systems have already been reported [3, 4]. One of the main requirements of such a system is that it should be chronically implantable, occupy small area and have wireless capabilities.

The Fig. 1 shows a conceptual scheme of such a system. It is built of a multielectrode array (MEA) that is bump bonded with an electronic chip. The most popular are MEAs proposed

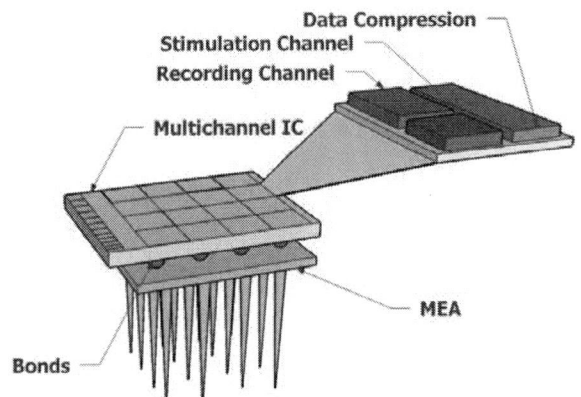

Fig. 1. Conceptual scheme of the 3D multichannel system dedicated to the neurobiological experiments.

by University of Utah [5] where interelectrode pitch is 400 µm. To satisfy functionality requirements such systems should comprise many operating blocks like recording and stimulating units, data compression, wireless power and data transmission or neural pattern classification. A crucial part of such systems is a neural recording amplifier that have to fulfill strict requirements regarding to the low noise and low power operation. Furthermore, taking into account large functionality of the final system, the recording channel should occupy small silicon area. Nevertheless, because recording channel should have ability to form a frequency bandwidth in a very broad range (i.e. lower cut-off frequency should be tuned from mHz up to the hundreds of Hz) methods to minimize area occupied by the filters should be introduced.

Based on our experience in designing multichannel systems for neurobiology applications [6 - 8] and in utilizing modern submicron processes [9], we propose to use a submicron process to address all the requirements mentioned above. The paper is organized as followed. Section II shows description of the common recording channel structure and introduces main obstacles in decreasing the area of the recording site. Section III consists of the description of the two recording approaches compared, while the section IV consists of the measurement results. The conclusions are drawn in section V.

II. Architecture of the Recording Channel

Most of neurobiology experiments need to record different types of neural signals at the same time. Specifically, these are the neural spikes and the LFP signals that often have to be recorded on different recording sites. The lower cut-off frequencies of the recording channels for these type of signals have to be set to 500 Hz and much below 1 Hz respectively. Thus, the recording channel shall has a possibility to shift its frequency bandwidth from one bandwidth setting to another.

The most common design approach of the recording channel that is used in discussed multichannel systems, is an operational amplifier with a capacitive feedback, schematically shown in the Fig. 2. The lower cut-off frequency of this circuit is inversely proportional to the RC_1 constant while the voltage gain is given by C_0/C_1 ratio. A very popular solution for controlling the upper cut-off frequency is to vary the current sourcing the operational amplifier.

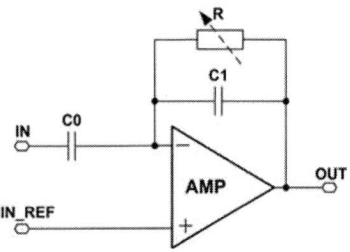

Fig. 2. Schematic idea of the front end amplifier commonly used in a neural recording experiments.

It can be seen that both the voltage gain and the lower cut-off frequency have one common element, namely this is the capacitor C_1. This clarifies the problems of obtaining extremely small lower cut-off frequencies in the recording channels of the systems reported in [10 - 12]. Let's now consider two solutions of building the recording channel with the voltage gain of 100 V/V and the lower cut-off frequency set to the 1 Hz. If one chooses the C_1 equal to 500 fF the C_0 should be equal to the 50 pF what is a reasonable value of the on chip capacitance. On the other hand, this influences the feedback R resistance that shall be on the order of 320 GΩ what in the multichannel chip fabricated in modern submicron processes may be hard to obtain. Next, if one would like to limit the necessary feedback resistance to 500 MΩ then the C_1 should be equal to 320 pF resulting in unacceptable area consumption of the C_0 capacitor with its capacitance equal to 32 nF. To avoid such a big area consumption the MOS based capacitors working in the strong inversion region are often used (in the 180nm CMOS process it allows to save almost six time smaller area comparing to the MIM capacitors). However, as we reported in [6] it introduces problems with the MOS transistors leakage currents that can drastically deteriorate performance of the recording channel.

The common technique that is utilized to form band pass filters is to use passive elements, i.e. resistors R and capacitors C. Therefore, to build filters with an ability to change its band pass to record either LFP or neural signals R and C elements shall be controllable in an appropriate range. For instance, in

order to set filter bandwidth for LFP signals recording (lower cut-off frequency set to 100 mHz) or neural spikes (lower cut-off frequency set to 500 Hz) the CR constant has to be set respectively to 1.6 s or to 0.3 ms what gives almost 5000 difference in the extreme elements values.

Having looked for passive elements that are produced in integrated electronic processes (only MIM – Metal Insulator Metal capacitors and high poly resistances which have the highest area to value ratio are considered) one can see that area to its value ratio of these have not been improved so effectively as the dimensions of the MOS transistors (see Table I). Thus, having in mind the necessity of building the controllable band pass filters with very large RC constants, using passive components influences the area consumption of the final filter.

TABLE I. Capacitances and Resistances to Area Ratios for Different Integrated Technologies.

Technology [μm]	MIM capacitor capacity/area ratio [fF/μm²]	High poly resistor resistance/area ratio [kΩ/sq]
0.7	0.75	2
0.5	0.9	1
0.35	1	1
0.25	1	0.75
0.18	1	1.03
0.13	1	0.98
0.09	1.5	1.01
0.065	2	1.05

Having looked for the projects reported to date the common technique to form a high pass filter is to use a bulk diode of the PMOS transistor [13]. Unfortunately, this method does not allow to control the resistance of this element and introduces large spread of the lower cut-off frequency in the multichannel architecture. There are also solutions that employ MOS transistor which channel effective resistance is controlled by one controlling input [14] but these do not show ability of recording both LFP signals and neural spikes. Furthermore, these designs complain that there is a large spread of the lower cut-off frequency while setting it to single Hz level [10 - 12]. Finally, taking into account above obstacles a new method of obtaining very large RC constants based on the components occupying very small silicon and very precisely controlled is necessary.

III. Comparison of the Two Recording Channels

As it can be seen the main elements that influence the area consumption of the recording channel are the passive components. Thus, obtaining very large resistance/area ratio together with its high uniformity will give one ability to employ smaller values of the feedback capacitance C_1 finally reducing the area occupied by the recording channel. Thus, in order to obtain large tuning range of the lower cut-off frequency, low spread of its value in the wide tuning range and to minimize area occupation of the single recording site we proposed to use MOS transistor working in the linear regime which effective channel resistance would be controlled by the on chip circuits independently in each channel [7]. Additionally, in order to comply large tuning range and high uniformity of the effective channel resistance of the MOS transistor in the whole tuning range, we proposed to employ correction *DAC* that controls the gate source drop out voltage of the MOS transistor. In our former design [6] as an input capacitors we were using MOS transistors working in the strong inversion region what comparing to the MIM capacitor allowed to save 83% of the area. Nevertheless, if such a solution is adopted, one has to consider the transistors leakage currents that dramatically increase with modern submicron processes. That currents generate the drop out voltage on the feedback resistance R (see Fig. 2) and shall be compensated by additional blocks [7].

In order to improve this approach we decided to compare two recording channels that differ from one another in feedback capacitors (see Fig. 3). One of the feedback of these amplifiers is built of the capacitors based on the MOS transistors working in the strong inversion region while the second amplifier's feedback is built of the MIM capacitors. We assumed that both amplifiers should have comparable voltage gain and should occupy the same silicon area. The *LF_DAC* is a correction DAC that is controlled by the on chip digital register. It is responsible for controlling the resistance of the M_{RF} PMOS transistor which dimensions are equal to $W/L=$ 0.4 µm/50 µm and which works in the linear region. Operational amplifier *AMP* is based on the folded cascode amplifier that consumes 3.46 µA from ±0.9 V supply voltage.

IV. MEASUREMENT RESULTS

We designed two test channels that are processed in the CMOS 180nm technology. Each of the recording channels consist of the two amplifying stages where the first stages are depicted in the Fig. 3. The second stage is based on the operational amplifier working with a resistive feedback. It adds additional voltage gain that can be programmable by the user and can be either 12.5 V/V or 50 V/V. Outputs of both first stages are provided to the test points as it is shown in the microphotograph Fig. 4. All of the measurements were performed by the use of the NI-6351 card and the ALESSI REL-6100 probe station.

We wanted to check if the input transistors leakage currents influence the output DC voltage of the first stage. Such currents may flow through the large feedback resistance R (see Fig. 2) and generate harming drop out voltage. Thus, we measured how the DC voltage at the outputs of these amplifiers depends on the effective channel resistance of the M_{RF} PMOS transistor. In order to perform this measurements we were writing to the 8-bit correction *LF_DAC* different register values and the results of these measurements are presented in Fig. 5. It can be seen that the DC voltage at the output of the amplifier with MIM based feedback does not depend on the *LF_DAC* setting while in the MOS based capacitor amplifier version DC voltage changes over 150 mV. Such a large DC voltage variation may easily saturate the following stages of the recording channel and should be compensated. In our previous design we showed the methods that allow to minimize this effect [7]. Nevertheless, as it is depicted in the Fig. 5 MIM based feedback allows to avoid using such a technique and to further minimize the area occupied by such an additional block.

We also checked whether it is possible to obtain extremely small lower cu-off frequencies in the channel version where MIM based capacitors are used. As it can be seen the tuning range of the lower cut-off frequency in the MIM channel version is much wider than in the version where input capacitor is based on the MOS transistor. Additionally, it can be noticed that starting from about LF_DAC=60 the lower cut-off frequency correction circuit in the capacitor the MOS based version begins to saturate what is a result of the output DC voltage changings. Each of the tested recording channels

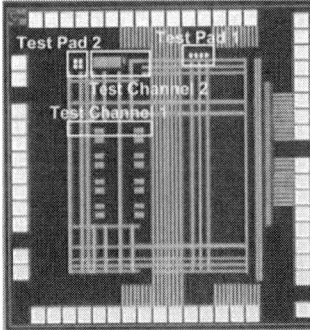

Fig. 4. Microphotograph of the chip with test channels marked.

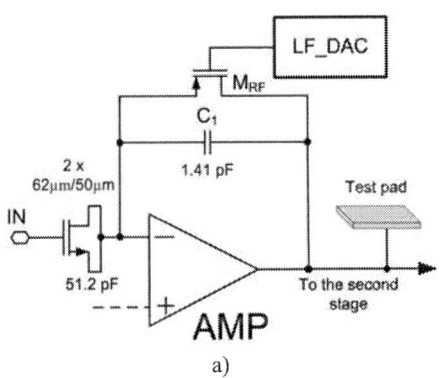

a)

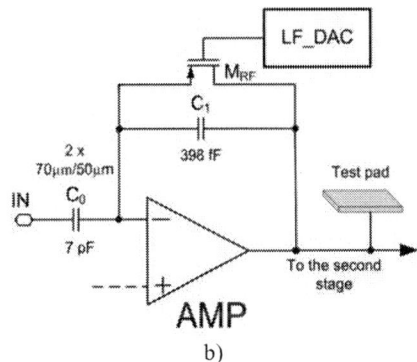

b)

Fig. 3. Schematic idea of the two test preamplifiers: a) version with the MOS based input capacitor, b) version with the MIM based input capacitor.

consumes 11 μW of the power from ±0.9 V supply voltage and occupies 0.06 mm² of the silicon area. Voltage input referred noise are equal to 5 μV while the bandwidth is set to 1 Hz – 9 kHz. The upper cut-off frequency can be changed by controlling the current sourcing the second stage of the amplifier and it can be set to either 280 Hz or 9 kHz.

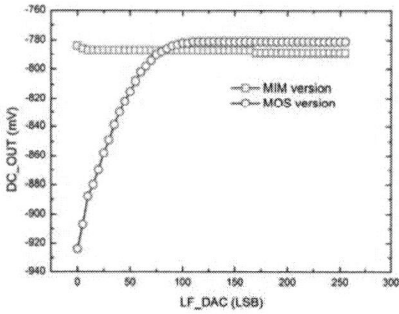

Fig. 5. Measurements results of the DC output voltage of the test channels vs. LF_DAC correction DAC setting.

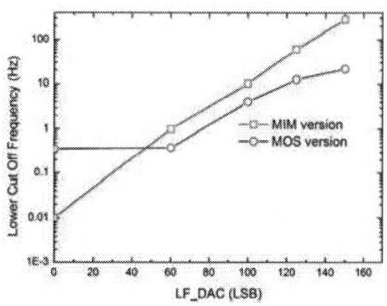

Fig. 6. Measurement results of the lower cut-off frequency of the two test channels vs. LF_DC correction DAC setting.

TABLE II. SUMMARIZED PARAMETERS OF THE MIM BASED RECORDING CHANNEL.

Employed technology	CMOS 180nm
Voltage Gain [V/V]	260 / 1000
Lower cut-off frequency [Hz]	0.01 ÷ 300
Upper cut-off frequency [kHz]	0.01-0.28 / 9
Input referred noise [μV] (1 Hz ÷ 9 kHz)	5.0
Area of the recording channel [mm²]	0.06
Power consumption [μW/channel]	11

V. CONCLUSIONS

We have compared two different approaches of designing the recording channels dedicated to the neurobiology experiments. Measurement results show that employing the MIM-based capacitors as an elements of the controllable analogue filters is a very attractive solution. Despite of its main disadvantage regarding to the poor capacitance/area ratio it has many benefits. Firstly, the MIM capacitors in controllable filters do not introduce transistor leakage currents what may be a limiting factor in obtaining very small lower cut-off frequencies in the recording channel. Furthermore, supplementary circuits do not have to be utilized to minimize

the effects of these currents thus allowing to save additional silicon area. Moreover, the CMOS 180nm process that we used permits to put MIM capacitors above the active elements what allows to save considerable silicon area. Summary of the measurements of the recording channel with the MIM capacitors used in the feedback are presented in the table Table II.

ACKNOWLEDGMENT

This research and development project was supported by Polish Ministry of Science and Higher Education in the years 2011-2013 (2011/01/N/ST7/01256).

REFERENCES

[1] A. V. Nurmikko, J. P. Donoghue, L. R. Hochberg, W. R. Patterson, Y. – K. Song; C. W. Bull, D. A. Borton, F. Laiwalla, S. Park, Y. Ming, J. Aceros, "Listening to Brain Microcircuits for Interfacing With External World—Progress in Wireless Implantable Microelectronic Neuroengineering Devices", Proceedings of the IEEE, 2010, Vol. 98, No. 3, pp. 375 – 388

[2] M. A. Lebedev and M. A. Nicolelis, "Brain-machine interfaces: Past, present and future," Trends in Neurosciences, vol. 29, no. 9, pp. 536–546, Sep. 2006

[3] A. Litke, N. Bezayiff, E. Chichilinsky, W. Cunnigham, W. Dabrowski, A. Grillo, M. Grivich, P. Grybos, P. Hottowy, S. Kachiguine, R. Kalmar, K. Mathieson, D. Petrusca, M. Rahman, A. Sher, "What does the eye tell the brain?: Development of system for the large-scale recording of retinal output activity." IEEE Transactions on Nuclear Scienc, vol. 51, 2004, pp. 1434-1440

[4] T. W. Berger et al., "Restoring lost cognitive function: Hippocampalcortical neural prostheses," IEEE Eng. Med. Biol. Mag., vol. 24, no. 5, pp. 30–44, Sep./Oct. 2005

[5] C. T. Nordhausen, E. M. Maynard, and R. A. Normann, "Single unit recording capabilities of a 100-microelectrode array," Brain Res., Vol. 726, pp. 129–140, 1996

[6] P. Grybos, P. Kmon, M. Zoladz, R. Szczygiel, M. Kachel, M. Lewandowski, T. Blasiak, "64 Channel Neural Recording Amplifier with Tunable Bandwidth in 180 nm CMOS Technology", Metrol. Meas. Syst., Vol. XVIII (2011), No. 4

[7] M. Zoladz, P. Kmon, P. Grybos, R. Szczygiel, R. Kleczek, P. Otfinowski, "A Bidirectional 64-channel Neurochip for Recording and Stimulation Neural Network Activity", IEEE EMBS Neural Engineering Conference, 2011, Cancun, Mexico

[8] P. Grybos, "Low Noise Multichannel Integrated Circuits in CMOS Technology for Physics and Biology Applications". Monography 117, AGH Uczelniane Wydawnictwa Naukowo-Dydaktyczne, Kraków 2002, available at: www.kmet.agh.edu.pl/www/asics

[9] R. Szczygiel, P. Grybos, P. Maj, "A Prototype Pixel Readout IC for High Count Rate X-Ray Imaging Systems in 90 nm CMOS Technology", IEEE Transactions on Nuclear Science, 57(3), 1664–1674, 2010

[10] Brown, E.A., Ross, J. D., Blum, R. A., Nam, Y., Wheeler, B. C., DeWeerth, S. P. : 'Stimulus-artifact elimination in a multi-electrode system', IEEE Transactions on Biomedical Circuits and Systems, 2008, Vol. 2, No. 1

[11] Perelman, Y., Ginosar, R., : 'Analog frontend for multichannel neuronal recording system with spike and LFP separation', Journal of Neuroscience Methods 153, 2006, pp. 21–26

[12] Gosselin, B., Sawan, M., Chapman, C. A.: 'A Low-Power Integrated Bioamplifier With Active Low-Frequency Suppression', IEEE Transactions on Biomedical Circuits and Systems, 2007, Vol. 1, No. 3R.

[13] R. Harrison and C. Charles, "A low-power low-noise CMOS amplifier for neural recording applications," IEEE J. Solid-State Circuits, vol. 38, no. 6, pp. 958–965, Jun. 2003

[14] G.E. Perlin, K. D. Wise, "An Ultra Compact Integrated Front End for Wireless Neural Recording Microsystems", Journal of Microelectromechanical Systems, Vol. 19, No. 6, December 2010

MIXDES 2012, 19th International Conference *"Mixed Design of Integrated Circuits and Systems"*, May 24-26, 2012, Warsaw, Poland

Convex Combination Initialization Method for Kohonen Neural Network Implemented in the CMOS Technology

Rafał Długosz[*,†], Tomasz Talaśka[†], Pierre-André Farine[*] and Witold Pedrycz[‡]

* Swiss Federal Institute of Technology (EPFL)
Institute of Microtechnology
Rue A.-L. Breguet 2, CH-2000, Neuchâtel, Switzerland
Email: rafal.dlugosz@epfl.ch and pierre-andre.farine@epfl.ch

† University of Technology and Life Sciences
Faculty of Telecommunication and Electrical Engineering
ul. Kaliskiego 7, 85-796, Bydgoszcz, Poland
E-mail: tomasz.talaska@gmail.com

‡ University of Alberta
Department of Electrical and Computer Engineering
Edmonton, AB T6G 2V4, Canada
E-mail: wpedrycz@ualberta.ca

Abstract—The paper presents a new CMOS implementation of the initialization mechanism for Kohonen self-organizing neural networks. A proper selection of initial values of the weights of the neurons exhibits a significant impact on the quality of the learning process. A straightforward realization of the initialization block in software is simple, but in hardware it requires providing the programming signal to all weights of each neuron. This makes the layout of the chip very complex, especially in case of large networks. This paper presents a new approach, in which to program particular neuron weights we use the same lines that are used by the adaptation block. This proposal is the first known transistor level implementation of the Convex Combination Method (CCM) that so far was implemented only in software.

Index Terms—Kohonen neural network, initialization mechanism, CMOS implementation, ASIC

I. INTRODUCTION

Efficiency of training of Artificial Neural Networks (ANN) depends on many parameters. One of them is a proper polarization of neuron weights before starting the learning phase. The neuron weights can be viewed as coordinates that determine the location of particular neurons in an n-dimensional input data space, where n is the number of the inputs of the neural network (NN). A proper distribution of neurons over this space before proceeding with the learning phase has a direct influence on the convergence speed of the learning process, as well as on the final results of learning [1], [2], [3], [11]. In this work we focus on the NNs trained without the supervision realized in the CMOS technology. In such networks the properly selected initial values of the weights have a direct influence on the number of, so called, dead neurons. Such neurons take part in the competition during the learning phase, but never win and therefore do not

become representatives of any data class. This increases the quantization error of the network.

In the literature one can find various initialization methods, as no universal method suitable for all learning algorithms does exist. Proper initial values of the weights strongly depend upon the training data set and those data are not known in advance. For this reason, the values of the weights usually are determined either empirically or randomly. There exist also more sophisticated methods, but they are usually complex.

The problem of the initialization of the weights can be considered from two different points of view. As ANNs are usually implemented in software, the problem of the computational complexity of a given initialization method is of less relevance in this case. Taking into account the computational capabilities of computers today, the most important criterion is the efficiency of a given algorithm. The problem is quite different in case of the ANNs realized at the transistor level as application specific integrated circuits (ASICs). In this case various hardware limitations have to be additionally taken into account. In ASICs such parameters as the power dissipation and the chip area are usually of particular importance. For this reason, the initialization mechanism has to be realized in a relatively simply way. This requires additional investigations whose purpose is to simplify the algorithm is such a way that will not affect the overall initialization process. This makes the design process of such NNs much more challenging.

Many initialization methods have been proposed over the past twenty years [3], [4], [5], [6], [13], [14]. In a very common and simple approach a random and uniform distribution of the weights over the input data space is being applied [4], [5], [6]. This approach is very fast, which is one of its main advantages. However it is not always effective, as it does not

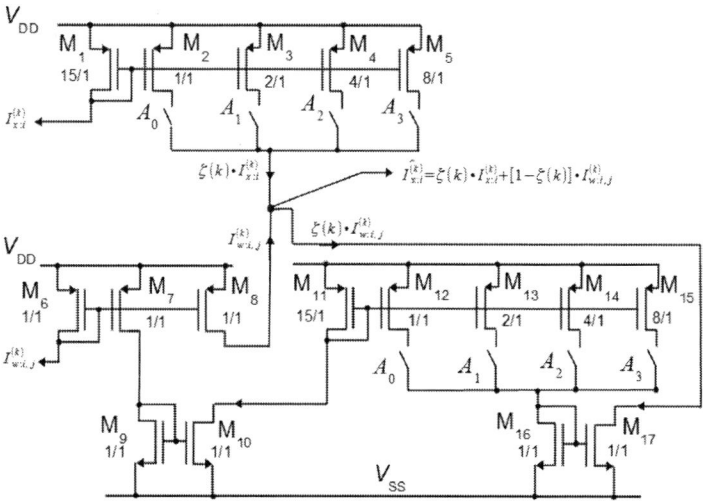

Fig. 1. The proposed circuit used in the Convex Combination Initialization Method

reflect a distribution of data over the input data space. In case of the ANNs realized in hardware, especially when the neurons operate in parallel that requires a separate circuits for each neuron, this method gives rise to two additional problems. One of them is especially visible in case of large networks, in which the necessity of providing the programming lines to each weight makes the layout very complex. Due to the limited number of pins on the chip, the weights are programmed sequentially that requires an additional circuitry responsible for addressing particular memory cells.

We have applied this method in our previous prototype chip realized in the CMOS 0.18 μm technology with the Winner Takes All (WTA) NN, but as it was a small network designed to verify the concepts of other blocks, the described problem was insignificant in that case [9].

The second problem is visible in analog networks, in which the weights are stored in analog memory cells. Such cells are programmed through switches that on their second sides are connected to a common programming line connected to an external pin. If particular weights have substantially different values, the leakage problem occurs that reduces the storage time of the memory cells [10].

Another initialization method can be referred to as linear approach. In this case a signal linearly increasing over time is provided sequentially to particular weights [6]. Looking from the hardware complexity point of view, this approach is similar to the one in which random values are being used. The only difference relies on the type of the external programming signal provided to the chip. This method does not take into account the distribution of data in the learning set.

More sophisticated initialization algorithms take into account a distribution of data in the input data space. One of them relies on locating the neurons in the place of the first m learning patterns X drawn from a given training set, where m is the number of neurons of the NN. As patterns X in a typical set are placed in a random order, the distribution of the neurons after the initialization phase is more representative for the overall data set than in the methods described above. At the system level, this approach is more efficient than the previous two but in the hardware realization the problems with the necessity of providing additional programming lines to particular neurons still exists. For this reason this approach is also not suitable for large NNs.

The particular methods described above can be combined together in a single NN, creating various hybrid approaches, but this makes sense only in software systems.

In this paper, we focus on another approach called Convex Combination Method (CCM). This method, although proposed many years ago [7], was never implemented at the transistor level. In the literature one can find only software implementations [15]. The concept is attractive in hardware implementation, as it does not require additional programming lines. Additionally it takes into account the input data set. In this approach to initialize the weights it is possible to use the same adaptation mechanism, after some modifications, that is next used during the learning phase.

In the WTA NNs it is reasonable to use the so called, conscience mechanism. Our previous investigations show that this mechanism when used together with the initialization block is able to eliminate the dead neurons, even when the initialization is not optimal [9], [11].

II. FUNDAMENTALS OF THE CONVEX COMBINATION INITIALIZATION METHOD

In the learning algorithm used in the WTA NN, as well as in the Kohonen self-organizing map (SOM) the weights vectors $W_j(l)$ of a particular neuron in an lth cycle are adapted toward a given input learning pattern $X(l)$ as follows:

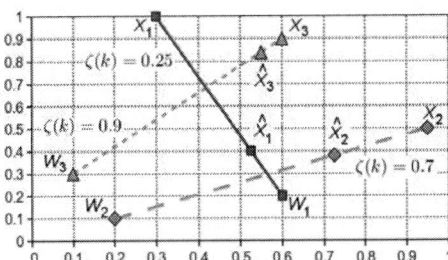

Fig. 2. Location of selected weights vectors W and corresponding input patterns X and $\hat{X}$ in an example case of two NN inputs.

$$W_j(l+1) = W_j(l) + \eta G(R, d(j,p))(X(l) - W_j(l)) \quad (1)$$

Here η is the learning rate, R is the neighborhood range, while $G()$ is the neighborhood function. The difference between the WTA NN and the SOM concerns the usage of different parameters in the function $G()$. In the WTA NN only the winning neuron is allowed to adapt the weights. In this approach the neighborhood range $R = 0$. As a result, $G() = 1$ only for the winning neuron, while for all other neurons $G() = 0$. In the SOM, on the other hand, R is usually greater than 0, which means that in addition neighboring neurons, j, at a distance $d(j,p)$ to the winning neuron, p, are adapted with a strength that depends upon the value of the function $G()$.

In the Convex Combination Method (CCM), to modify the neuron weights during the initialization phase we use the same adaptation block that is used to modify the weights in the subsequent learning phase. The difference relies on substituting signals x_i that are provided to particular inputs i of the NN with new signals $\hat{x}_{i,j}$ that are modified copies of x_i. The $\hat{x}_{i,j}$ signals are calculated as follows:

$$\hat{x}_{i,j}(k) = \zeta(k) \cdot x_i(k) + [1 - \zeta(k)] \cdot w_{i,j}(k) \quad (2)$$

In case of the current-mode hardware implementation of the CCM, which is the subject of this paper, particular signals in (2) are represented by currents, as shown in Fig. 1. For this reason, (2) can be rewritten as follows:

$$\hat{I}_{x:i,j}(k) = \zeta(k) \cdot I_{x:i}(k) + [1 - \zeta(k)] \cdot I_{w:i,j}(k) \quad (3)$$

In (2) and (3) $\zeta(k)$ is a function whose value is increasing from the value that at the beginning of the initialization phase is close to zero to one. As can be noticed, for $\zeta(k) = 0$ particular modified patterns $\hat{X}_j(k)$ would cover corresponding vectors $W_j(k)$ of particular neurons:

$$\hat{x}_i(k) = w_{i,j}(k) \rightarrow \hat{I}_{x:i}(k) = I_{w:i,j}(k) \quad (4)$$

This situation is not allowed, as in case of using CCM the adaptation in the initialization phase is performed toward the $\hat{X}_j(k)$ signals, instead of $X(k)$. If these signals cover the corresponding weights vectors, the initialization is stopped.

For this reason the initial values of $\zeta(0)$ are some small numbers greater than zero. For $\zeta(k) = 1$ the modified patterns $\hat{X}(k)$ become equal to real patterns $X(k)$:

$$\hat{x}_i(k) = x_i(k) \rightarrow \hat{I}_{x:i}(k) = I_{x:i}(k) \quad (5)$$

Once this happens the function $\zeta(k)$ is no longer modified, and the network starts the adaptation phase. For $\zeta(k) = 1$ the neurons theoretically should be located in the areas close to particular learning patterns X that represent particular data classes.

Each neuron in the NN has its own modified copy of a given pattern $X(k)$. These signals in particular neurons are "bend" toward local vectors $W_j(k)$, that at the beginning of the initialization phase are either zero or have some small values. Particular $\hat{X}_j(k)$ signals form a kind of attractors that, as the $\zeta(k)$ function increases, due to adaptation mentioned above attract particular neurons during the initialization phase. Note that modified patterns $\hat{X}_j(k)$ are always located on the line connecting given vectors $X(k)$ and $W_j(k)$, as shown in Fig. 2 for selected values of the $\zeta(k)$ function. For this reason, particular patterns $\hat{X}_j(k)$ always pull the corresponding vectors $W_j(k)$ toward particular input data.

In each of the k cycles of the initialization phase, and next in each of the l cycles of the learning phase, the neurons calculate distances between the vectors $\hat{X}$ and own vectors W. In hardware it is realized by a special distance calculation circuit (DCC). In the WTA and the SOM learning algorithms the next operation after calculating these distances is a detection of the winning neuron i.e. the one with the smallest distance. At the beginning of the initialization phase, as all vectors W_j are equal, all resultant distances are also equal. For this reason, the "winning" neuron has to be selected arbitrary. In fact, particular neurons are in this phase chosen in a certain order that is repeated until $\zeta(k) = 1$. If this order is known, it enables "co-joining" given neurons with given data.

III. HARDWARE IMPLEMENTATION OF THE CONVEX COMBINATION METHOD

The CCM circuit proposed in this paper, shown in Fig. 1, can be viewed as an extension of the WTA NN realized earlier by us in the CMOS 0.18 μm technology. Th details of this implementation are presented in [9], [10]. The CCM blocks are located at the inputs of each neuron, as each neuron requires its own modified pattern $\hat{X}_j(k)$. The circuit operates in the current-mode. The value of $\zeta(k)$, equal for all neurons, is controlled by logic signals $A0-A3$ that control two multi-output current mirrors composed of transistors M_1-M_5 and M_{11}-M_{15}, respectively. In this configuration, for given signals $I_{x:i}(k)$ and $I_{w:i,j}(k)$ we obtain 16 different levels of the resulting signals $\hat{I}_{x:i,j}(k)$, given by (3).

In the CCM, at the beginning of the initialization phase, we should theoretically assume some small values of all neuron weights, as mentioned earlier. In hardware it is not necessary to program the weights, as some small signals are usually stored in the analog memory cells, for example, due to presence of noise or other inaccuracies.

As described in Section II, in the initialization phase particular neurons have to be selected arbitrarily to modify the weights. In the hardware implementation, on the other hand, this operation is performed by a Winning Selection Circuit (WSC). If all distances are equal, the winning neuron for a given $X(k)$ is selected randomly by the WSC. In the WTA NN realized by us as an ASIC we have used a binary-tree WSC that always only one neuron [12]. In the chip realization the fixed order can be realized very easy. The neurons on the chip are located in a given order. The "winning" neuron can send a 1-bit signal to the closest neighbor in the layout that will be activated in the next, $k + 1$ cycle.

A. transistor level verification of the proposed circuit

Transistor level verification of the proposed CCM circuit is shown in Fig. 3. The $\zeta(k)$ function increases from 0 to 1 with a step of 1/15. The output signal, $\hat{I}_{x:i,j}(k)$ for selected values of the input current $I_{x:i}(k)$, and the current $I_{w:i,j}(k)$ that represents a corresponding weight, is shown in diagram (a). The circuit operates properly independently on which signal (x or w) is greater.

As shown in diagram (b) the power dissipation depends on the values of particular signals. The stronger impact on this parameter exhibits the $I_{w:i,j}(k)$ signal. An average power for the x and w signals varying in-between 1 and 7 μA equals 30 μW. The smallest power dissipation is always at the beginning of the initialization phase for $\zeta(k) \approx 0$ and then increases for $\zeta(k) \to 1$. To minimize the energy consumption of the overall NN, which is important for example in portable devices, the CCM circuit should be turned off immediately after completing the initialization phase. To make it possible additional switches have to be used to enable the signals x bypassing the CCM circuit. Turning off this circuit is realized simply by setting the x and w signals to 0. This situation is shown in the period in-between 160–240 μs.

Diagram (c) illustrates a difference between the output signal $\hat{I}_{x:i,j}(k)$ and the theoretical signal calculated on the basis of (3). The error usually does not exceed 1%, with maximum values of 2.5%, which is acceptable in this case.

IV. CONCLUSIONS

A new circuit that enables hardware implementation of the Convex Combination Initialization Method has been proposed. The circuit is a simple solution which does not substantially increase the complexity of the overall NN, while it improves the quality of the learning process of the NN. In the CMOS 0.18 μm technology a single CCM circuit occupies 300 μm^2.

REFERENCES

[1] D. Nguyen, B. Widrow, "Improving the learning speed of 2-layer neural networks by choosing initial values of the adaptive weights", *Int. Joint Conference on Neural Networks*, San Diego, USA, pp.21-26 (III), 1990

[2] K. Kenni, K. Nakayama, H. Shimodaira, "Estimation of initial weight and hidden units for fast learning of multi-layer neural network for pattern classification", *Int. Joint Conference on Neural Networks* (IJCNN), Washington, USA, vol.3, pp.1652 - 1656, 1999

[3] Y.K. Kim, J.B. Ra. "Weight Value Initialization for Improving Training Speed in the Backpropagation Network", *Int. Joint Conference on Neural Networks* (IJCNN), Seattle, USA, Vol. 3, pp. 2396 - 2401, 1991

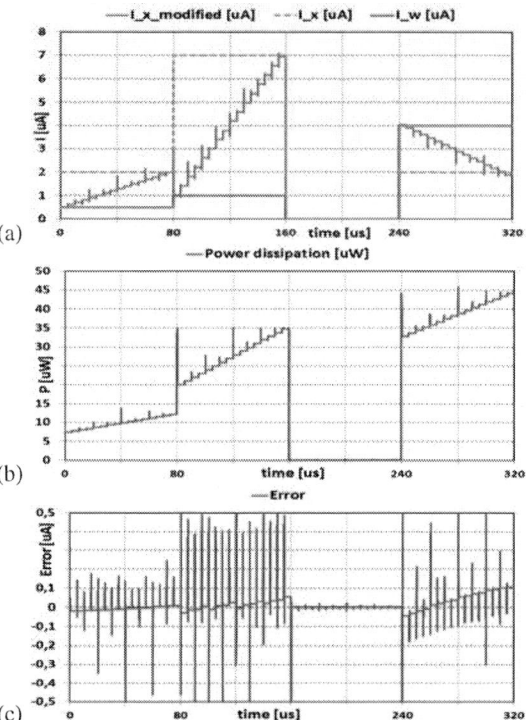

Fig. 3. Transistor level verification of the proposed CCM circuit: (a) selected values of the $\zeta(k)$ function, (b) resultant $\hat{x}_{i,j}$ signals for given $w_{i,j}$ andd x_i signals, (c) power dissipation of the circuit.

[4] G. Thimm, E. Fiesler, "Neural network initialization In "From neural to artificial neural computation" ", Editors: J. Mira, F. Sandoval *Int. Workshop on Artificial Neural Networks*, pp. 535.542, Malaga, 1995

[5] Y. Chen, F. Bastani, "ANN with two-dendrite neurons and its weight initialization", *Int. Joint Conference on Neural Networks* (IJCNN), Baltimore, USA, Vo. 3, pp. 139 - 146 ,1992

[6] T. Kohonen, *Self-Organizing Maps*, Springer Verlag, Berlin, 2001

[7] H. Nielsen, "Counterpropagation networks", *Applied Optics*, Vol. 26, pp. 4979- 4984, 1987

[8] D.E. Rumelhart, D. Zipser, "Feature discovery by competitive learning", *Cognitive Sciences*, Vol. 9, 1985

[9] R. Długosz, T. Talaśka, W. Pedrycz, R. Wojtyna "Realization of the Conscience Mechanism in CMOS Implementation of Winner-Takes-All Self-Organizing Neural Networks", *IEEE Transactions on Neural Networks*, Vol. 21, Iss.6, pp.961–971, June 2010

[10] R. Długosz, T. Talaśka, W. Pedrycz, "Current-Mode Analog Adaptive Mechanism for Ultra-Low Power Neural Networks", *IEEE Transactions on Circuits and Systems–II: Express Briefs*, Vol. 58, Iss. 1, pp. 31–35, January 2011

[11] T. Talaśka, R. Długosz, "Initialization mechanism in Kohonen neural network implemented in CMOS technology", *11th European Symposium on Artificial Neural Networks* (ESANN), Bruges, Belgium, pp. 337–342, 2008

[12] R. Długosz, T. Talaśka, "Low Power Current-Mode Binary-Tree Asynchronous Min/Max Circuit", *Microelectronics Journal*, Elsevier, Vol.41, No.1, pp. 64–73, January 2010

[13] D. Graupe "Principles of artificial neural networks", *World Scientific*, Advanced Series on Circuits and Systems, vol. 6, 2007

[14] N. Vora, S. S. Tambe, B. D. Kulkarni "Counterpropagation neural networks for fault detection and diagnosis", *Computers & Chemical Engineering* vol. 21, No. 2, pp.177-185, 1997

[15] S. N. Sivanandam, S. Sumathi, S. N. Deepa "Introduction to neural networks using MATLAB 6.0", *Tata McGraw-Hill* Computer engineering Series, 2006

Crystal Oscillator with Dual Amplitude Stabilization Feedback Loop

Krzysztof Siwiec

Institute of Microelectronics & Optoelectronics
Warsaw University of Technology
ul. Koszykowa 75, 00-662 Warszawa, POLAND
e-mail: K.Siwiec@imio.pw.edu.pl

Abstract—In this paper crystal oscillator realized in 90 nm Low Leakage UMC CMOS technology is presented. It is based on the Pierce architecture and uses dual feedback loop to limit the amplitude and generate bias point. Dual feedback architecture allows to achieve almost rail to rail signal with low harmonic levels for wide range of supply voltage. Both of those features have positive impact on phase noise level. The oscillator was designed to work with 16 MHz crystal. It can work with supply voltages from 0.6 to 1.8 V and consumes 78.4μA for 1.2 V supply. In typical case phase noise for 1.2 V supply voltage is -160 dBc/Hz at 10 kHz offset.

Index Terms—Crystal oscillator, quartz resonator, amplitude stabilization

I. INTRODUCTION

Crystal oscillators are one of the most common circuits used in electronic devices. They can be found in digital and analogue circuits. Over the years many architectures were developed, but the most common are those based on the Pierce oscillator (see Fig. 1). In almost all practical implementations addition circuitry for amplitude limitation are added. There are many reasons to do so. As it was shown in [1] amplitude limitation is needed to minimize crystal nonlinearity impact on frequency stability. Limited amplitude has also positive impact on power consumption and phase noise performance.

In this work a new dual loop amplitude limitation is presented. Its main advantages are simple design, automatic bias generation, wide supply voltage operation and low phase noise performance.

This paper is organized as follows. Proposed architecture is presented in Section II. Simulation results are shown in Section III and conclusion in Section IV.

II. CIRCUIT ARCHITECTURE

Presented architecture is based on the Pierce oscillator. Its architecture and small signal equivalent are shown in Fig. 1. To find oscillation start-up condition small signal analysis has to be performed. Based on negative resistance oscillator condition, it can be found that minimum transconductance, which ensures sustainable oscillations, is equal to [1]:

$$g_{m\min} = \frac{\omega}{Q_s C_s} \frac{\left(C_L^2 + 2C_L C_0\right)}{C_L^2}, \qquad (1)$$

This work was supported in part by the Polish National Centre for Research and Development under project No. NR02-0096-10.

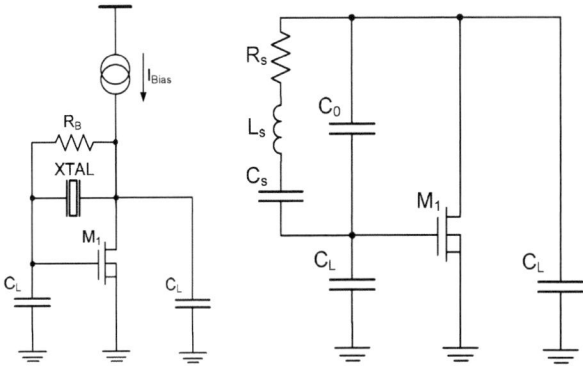

Fig. 1. Pierce crystal oscillator and its small signal equivalent.

where ω is oscillation pulsation and Q_s is crystal quality factor:

$$Q_s = \frac{1}{\omega_s R_s C_s}. \qquad (2)$$

It is usually assumed that to ensure proper startup of oscillator the transconductance value should be three to four times bigger than $g_{m\min}$. Moreover, if one can take into account process, temperature and supply voltage variations, one will be able to find out that it should be much more. Such a big transconductance will lead to high overdrive of a crystal oscillator circuit and a crystal itself. High overdrive leads to higher nonlinearity effects and harmonic distortion, which has negative impact on frequency stability [1]. Moreover bigger bias current is needed to achieve big transconductance, so the power consumption is increased. What is more as it can be found from general theory of phase noise in oscillators [2], situation where drain to source voltage of active circuit transistor drops below saturation threshold, has negative impact on phase noise. From those reasons amplitude limitation is necessary. It is usually done by using an amplitude detector and comparing its output with a reference voltage in a feedback loop [3], [4]. This solution allows to regulate output amplitude, but needs much additional circuitry. In this work different solution was employed. The idea is to limit the amplitude by a diode (or diode connected transistor) connected between drain and gate of M1 [5], [6]. At the begging architecture presented in Fig. 2 was tested.

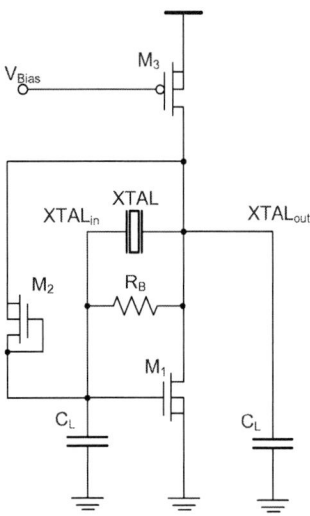

Fig. 2. Schematic of the Pierce oscillator with amplitude limitation.

The principle of operation is simple. When the drain voltage of M1 drops below the value, for which diode connected transistor M_2 starts to conduct, capacitor C_L is being discharged and the voltage on M1 gate is dropping. This results in reduction in M1 drain current thus limiting the amplitude. The main advantage of such solution is simplicity.

In Fig. 3 transient characteristics (upper and lower envelope of the $XTAL_{out}$ signal and the dc voltages of the $XTAL_{in}$ and $XTAL_{out}$ nodes) of the oscillator start-up are presented. It can be seen that when the lower envelope is dropping, the dc voltage at the gate of M1 is being reduced. This limits the lower envelope value to 0.22 V. But at the same time the dc voltage at the output of the oscillator ($XTAL_{out}$) is rising. This is caused by the fact, that while the drain current of M_1 is reduced, the drain current of M_3 remains constant. When the upper envelope of the $XTAL_{out}$ value crosses the vdd voltage the M_3 transistor is being cut off for some part of a period, thus its mean drain current value is reduced and matched to the drain current value of M_1. This cut off mechanism introduces high nonlinearities to the circuit, which as it was mentioned before, has negative impact on frequency stability and phase noise. Voltage spectrum of the output signal is presented in Fig. 4. Two measures of circuit nonlinearity were used. A ratio of second to first harmonic (V_{21}) and a THD (Total Harmonic Distortion) defined as:

$$THD = \frac{\sum_i V_i^2 - V_1^2}{V_1^2}, \quad (3)$$

where V_i is the i-th harmonic. The V_{21} ratio and THD are -20.73 dB and 8.5 ‰ respectively.

To reduce the harmonic level in the circuit, second feedback was added to control the V_{Bias} signal. Schematic of proposed solution is presented in Fig. 5. In this case M_3 PMOS transistor is used as a transconductor. M_1, M_5 and M_6 form

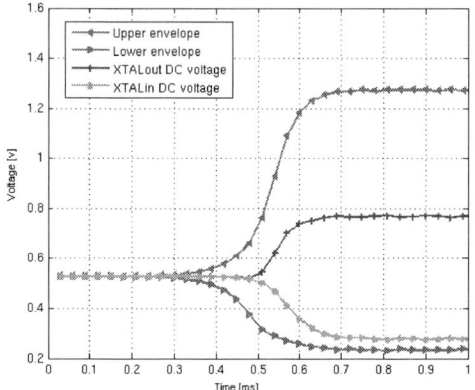

Fig. 3. Transient characteristics of the oscillator startup.

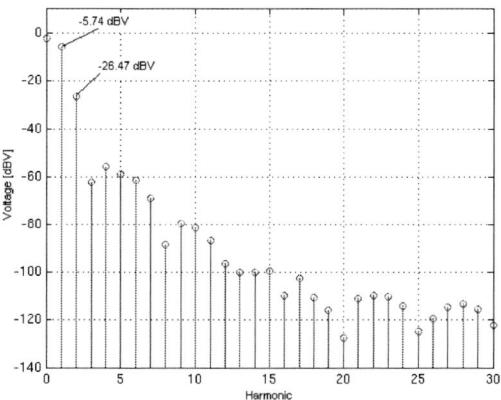

Fig. 4. Voltage spectrum of the XTALout signal.

biasing circuit. M_2 and M_4 create two feedback loops. Because in this case both M_1 and M_3 gate voltages are regulated, the bias circuit can be design so that the startup g_m is much higher than the minimum value. This first of all simplifies the design, because the bias current does not have to be chosen so carefully. The M3 transistor width should be chosen so that the startup g_m is easily achieved with the current consumption below the specification. The length of M1 and M3 transistors should be much above minimum technological value to ensure high output resistance. Moreover high immunity to process, temperature and supply voltage variations can be achieved with this architecture. If the bias circuit will be designed so that the g_m of M3 transistor in worst case corner will be high enough to start the oscillator, the circuit will work properly in all corners. Thanks to the feedback loops those features are achieved, with low power consumption.

Fig. 6 presents transient characteristic (upper and lower envelope of the $XTAL_{out}$ signal, the dc voltages of the $XTAL_{in}$, $XTAL_{out}$ and V_{Bias} nodes) of designed oscillator. It can be seen that when the amplitude is raising, both v_{gs} voltages of M_1 and M_3 are reduces (see V_{Bias} and $XTAL_{in}$ signals). The mechanism is similar as in the previous circuit.

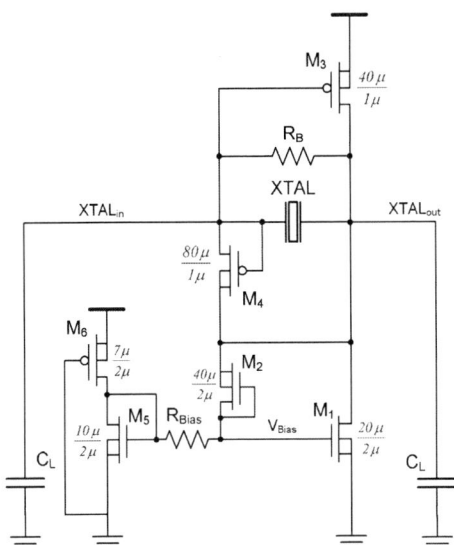

Fig. 5. Schematic of the proposed crystal oscillator.

The difference is that when the upper envelope is rising, second diode connected transistor (M4) also starts to conduct reducing bias voltage of the M3, thus limiting upper envelope. Thanks to this both upper and lower envelope is limited to 1 V and 0.04 V respectively. Moreover the dc voltage of XTAL$_{out}$ is stabilized at a value of 0.52 V, which is close to the half of the supply voltage. This allows to achieve high amplitude compared with the supply voltage value, which has positive impact on phase noise level. Moreover harmonic levels are reduced, which is shown in Fig. 7. The ratio of second to first harmonic is -36.24 dB, which is substantially lower value than -20.73 dB achieved in the one loop oscillator architecture. Furthermore THD was reduced to 0.25 ‰, which is more than ten times less than before.

III. SIMULATION RESULTS

In this section simulation results of the designed circuit are presented. All results were achieved from post-layout simulations. Circuit was simulated with 16 MHz, 26000 quality factor crystal, modeled as presented in Fig. 1. Corner analysis was performed for five process corners: TT, SS, FF, SF, FS (where the first and second letter describes the corner of NMOS and PMOS transistors respectively, and T stands for Typical, S stands for Slow, F stands for Fast process corner); three temperatures: -40 °C, 27 °C, 125 °C; and three supply voltage values: 1.08 V, 1.2 V, 1.32 V. Together it gives 45 simulations. Presented circuit was designed in Low Leakage UMC CMOS 90 nm technology.

One of the goals during design was to ensure proper work of the oscillator in wide supply voltage range. To prove that it was achieved, oscillator was simulated for 0.6 to 1.8 supply voltage range. In all cases the oscillator started properly. Fig. 8 presents peak-peak output amplitude and total harmonic distortion as a function of supply voltage. As it can be seen the

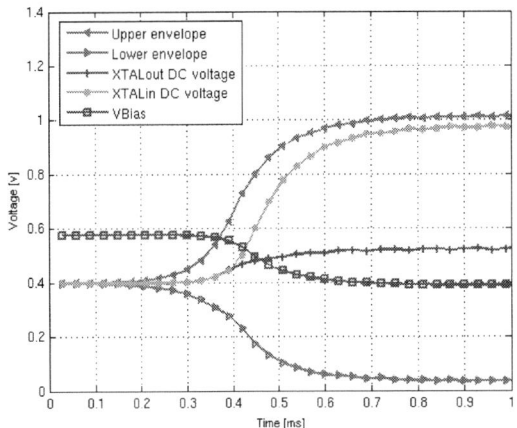

Fig. 6. Transient characteristics of the oscillator startup.

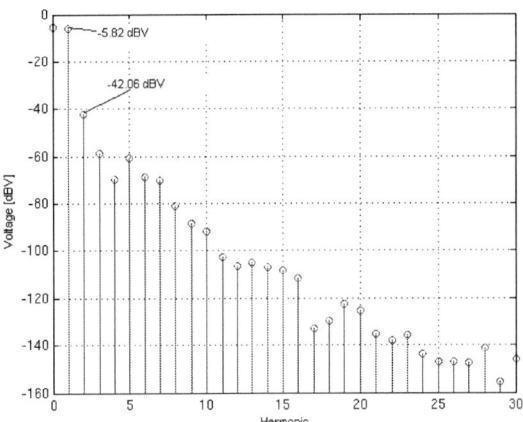

Fig. 7. Voltage spectrum of the XTAL$_{out}$ signal.

output amplitude is proportional to a supply voltage. It is not surprising, because the PMOS feedback loop (M$_3$ and M$_4$) is starting to work, when the output voltage upper envelope approaches supply voltage value. Moreover THD is also proportional to supply voltage. It is caused by the fact, that the bias generated by the M$_5$ and M$_6$ transistors is proportional to supply voltage. For higher bias, transistors M$_2$ and M$_4$ have to be open for longer time, in order to compensate higher current flowing thru R$_B$ and R$_{Bias}$ resistors. To reduce THD for higher supply voltages M$_6$ transistor width should be reduced. However, it may lead to startup problems for low supply voltages. Another possible solution is to increase resistance values of R$_B$ and R$_{Bias}$. Although it may increase the phase noise level in the circuit.

As it was mentioned before, presented circuit should have high immunity to process, supply voltage and temperature variations. To prove that, corner simulations were performed. Fig. 9 shows the output amplitude and the THD values for 45 corners. Simulation results were sorted by supply voltage value. As it can be seen the output amplitude does not change

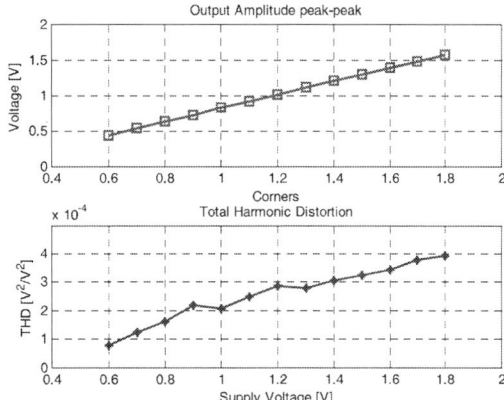

Fig. 8. Output peak to peak amplitude and total harmonic distortion as a function of supply voltage for designed oscillator.

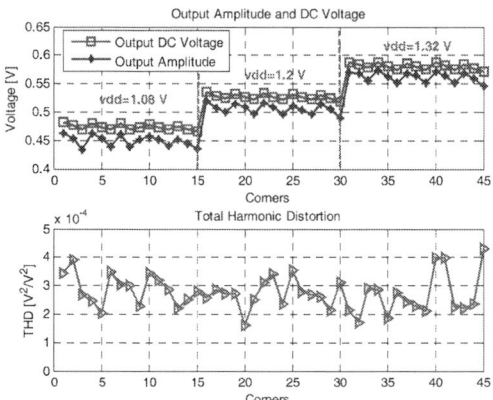

Fig. 9. Output amplitude and total harmonic distortion for 45 corners.

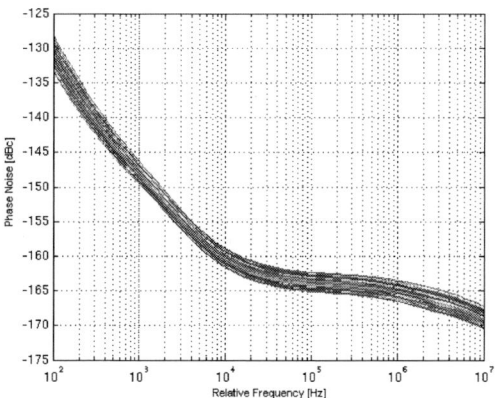

Fig. 10. Phase noise of the crystal oscillator for 45 corners.

much with process and temperature. It depends only on supply voltage. The output amplitude variations caused by process and temperature are within ±3.5 % from the mean value. Furthermore output DC voltage is kept close to the half of supply voltage, ensuring optimum operation. The THD value in the worst case is around 0.45 ‰, which is still more then ten times less than in the first circuit.

Fig. 10 shows phase noise characteristics of designed crystal oscillator for 45 corners. In all cases noise floor level is below -162 dBc and phase noise at 1 kHz offset is lower than -146 dBc. In typical case noise floor and phase noise at 1 kHz offset are -163.8 dBc and -148 dBc respectively, while consuming 78.4 μA.

IV. CONCLUSION

In this paper new dual feedback loop CMOS crystal oscillator has been presented. The oscillator was designed in 90 nm Low Leakage UMC CMOS technology. To verify circuit performance post-layout corner simulations in Cadence Spectre simulator were carried out.

The main goal, which was reduction of harmonic distortion without increasing circuit complexity, has been achieved. Moreover thanks to the self bias feature, the initial biasing does not have to be chosen so carefully, which simplifies the design.

Presented oscillator can work with a wide range of supply voltage (from 0.6 V to 1.8 V). This is especially important in modern technologies where low voltage operation is often necessary. Thanks to amplitude stabilization mechanism presented circuit is immune to process and temperature variations.

Designed oscillator achieves low phase noise performance. The circuit noise floor is below -162 dBc and the. phase noise level at 1 kHz offset is lower than -146 dBc for all process corners.

ACKNOWLEDGEMENT

The author would like to thank Prof. Witold Pleskacz from Warsaw University of Technology for his help and valuable feedback.

REFERENCES

[1] E. Vittoz, M. Degrauwe and S. Bitz, "High-performance crystal oscillator circuits: Theory and application", *IEEE J. solid-state circuits,* vol. 23, pp. 774-783, June 1988.

[2] A. Hajimiri, T. H. Lee, "A General Theory of Phase Noise in Electrical Oscillators", *IEEE J. solid-state circuits,* vol. 33, pp. 179-194, February 1998.

[3] T. Wey, "On Amplitude and Operating Point Control of a Voltage-Controlled Crystal Oscillator", *IEEE MWSCAS 50th Midwest Symposium on Circuits and Systems,* pp. 221-224, August 2007.

[4] R. A. Bianchi, J. M. Karam, and B. Courtois, "Analog ALC Crystal Oscillators for High-Temperature Applications", *IEEE trans. Solid-State Circuits,* vol. 35, no. 1, pp. 2-14, January 2000.

[5] S. Yao, H, Zhu, X, Wu, "Design of Low Power CMOS Crystal Oscillator with Tuning Capacitors", *IAEND Engineering Letters,* 12 February 2007.

[6] J. A. T. M van den Homberg, "A Universal 0.03-mm^2 One-Pin Crystal Oscillator in CMOS", *IEEE J. solid-state circuits,* vol.. 34, no. 7, pp. 956-961, July 1999.

MIXDES 2012, 19th International Conference *"Mixed Design of Integrated Circuits and Systems"*, May 24-26, 2012, Warsaw, Poland

Dc-Gain Enhanced Fast-Settling Triple Folded Cascode Op-Amp

Azad Mahmoudi
Islamic Azad University
Sardasht Branch
Sardasht, Iran
azadmahmoudei@gmail.com

Ali Dadashi
Urmia University
Urmia, West Azerbaijan, Iran
urrmia@gmail.com

Saeid Masoumi
Islamic Azad University
Tasouj Branch
Tasouj, Iran
s.masoumi.ee@gmail.com

M Khaleghi Kouzehkanan
Islamic Azad University
Khameneh Branch
Khameneh, Iran
m.kouzehkanan@gmail.com

Abstract—A new operational amplifier based on the conventional TFC op-amp structure is presented. A novel method is used to increase the dc-gain of the triple folded cascode op-amp. This method does not limit the bandwidth, output voltage swing range and the phase margin of the triple folded cascode op-amp.

Index Terms—positive feedback; linearity; triple folded cascode; dc-gain

I. INTRODUCTION

Operational amplifier is one of the basic and important circuits which has a wide range of applications in several analog circuits, such as filters and data converters, etc. The rapid demand for high speed and high resolution applications such as ADCs and DACs results in an increasing demand for high speed and high gain amplifiers. For high-accuracy circuits, op amps with very high open loop gain and high unity gain frequency are required in order to meet both accuracy and fast settling requirements. Satisfying both of these requirements is difficult with short-channel CMOS processes, since the intrinsic gain of the devices is limited. Depending on the required specifications, several op-amp structures have been designed. The new structure proposed in this paper is based on the conventional triple folded cascode (TFC) amplifier. To increase the dc-gain of the op-amp a new method is introduced that uses positive feedback concept. The true performance of the op-amp (accuracy) in the higher output voltage swings and improved speed, stability and linearity are the other advantages of the proposed structure in the comparison with conventional designs (e.g. [1]).

II. PROPOPSED STRUCTURE

To achieve higher DC-gain in the TFC Op-Amp, positive feedback method can be used [1][2][3][4]. Fig. 1 shows the complete structure of the proposed Op-Amp without Common Mode Feedback Block (CMFB) and bias circuits.

A. Differential DC-Gain

As we can see, the total current injected to the output node is approximately as:

$$i_{out} = \alpha_f \times (g_{mM1,2} \times V_{in} + \beta \times g_{mMf1,f2} \times V_{out}) \quad (1)$$

where β is the signal attenuation coefficient from output nodes to O nodes ($\frac{V_o}{V_{out}}$). In fact $1/\beta$ is equal to intrinsic gain of a PMOS device ($g_m \times r_o$), and also α_f is:

$$\alpha_f = \frac{(r_{oMf1,f2} \| r_{oM_{1,2}} \| r_{oM3,4} \| \dfrac{1}{g_{mMf1,f2}})}{(\dfrac{1}{g_{mM5,6}} \| r_{oM5,6}) + (r_{oMf1,f2} \| r_{oM_{1,2}} \| r_{oM3,4} \| \dfrac{1}{g_{mMf1,f2}})} \quad (2)$$

and V_{out} can be calculated as:

$$V_{out} = R_{out} \times i_{out} \quad (3)$$

$$V_{out} = R_{out} \times \alpha_f \times (g_{mM1,2} \times V_{in} + \beta \times g_{mMf1,f2} \times V_{out}) \quad (4)$$

The differential DC-gain of this Op-Amp can be written as:

$$A_{vf0} = \frac{V_{out}}{V_{in}} \approx \frac{\alpha_f \times g_{mM1,2} \times R_{out}}{k} \quad (5)$$

$$k = 1 - \alpha_f \times \beta \times g_{mMf1,f2} \times R_{out} \quad (6)$$

In Eq. 5, $\alpha_f \times g_{mM1,2} \times R_{out}$ is DC-gain of the simple TFC Op-Amp, and k can be controlled by choosing appropriate sizes for *Mf1*, *Mf2*. To increase DC-gain, k must be less than one. Also to have a stable structure, k cannot take a negative value. So k must be *0 < k < 1*. It is clear that by decreasing k the total of A_{vf0} will increase. But it cannot take a value very close to zero, because it might take a negative value due to process variations. Furthermore, before fabrication, the proposed Op-Amp must be tested in the corners of fabrication process. As mentioned before k must be *0 < k < 1*. So from Eq. 6 transconductance of the feedback devices (*gmMf1,2*) should be less than $\dfrac{1}{\alpha_f \times \beta \times R_{out}}$. This small transconductance achieves by using small devices, biased with small currents. So in comparison with simple TFC Op-Amp, power consumption increase is negligible.

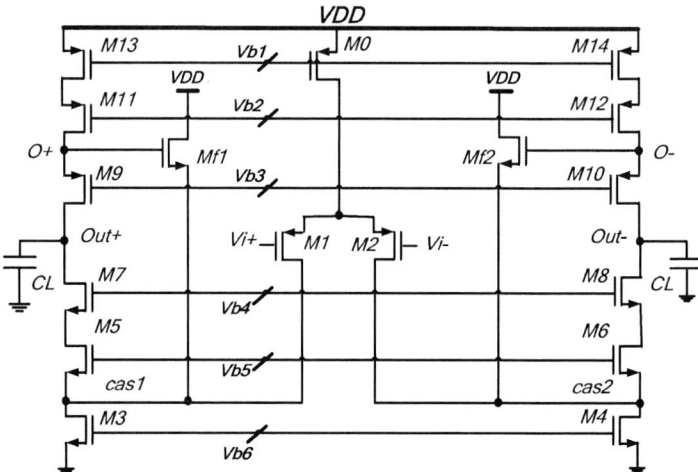

Figure 1. Proposed op-amp structure.

B. Frequency Response

The proposed op-amp has many poles and zeros, but two first poles or zeros are important in the frequency response of the op-amp. The total transfer function of the proposed op-amp can be estimated as:

$$Av_f(s) = \frac{A_{vf0}}{(1+\frac{S}{P_{f1}})(1+\frac{S}{P_{f2}})} \qquad (7)$$

P_{f1} is the first pole which occurs in the output node of the Op-Amp and P_{f2} is the second pole which occurs in the cascode node.

To calculate the frequency response we can write:

$$V_{out} = \frac{\alpha_f \times (g_{mM1,2} \times V_{in} + \beta \times g_{mMf1,f2} \times V_{out})}{(1+\frac{s}{p_2'})} \times \frac{R_{out}}{(1+\frac{s}{p_1'})} \qquad (8)$$

P'_1 is the first pole of the TFC Op-Amp which occurs in the output node and P'_2 is the second pole of the TFC Op-Amp which occurs in the cascode node. P'_1 and P'_2 are:

$$P_1' = -\frac{1}{R_{out} \times C_{out}}, \quad P_2' = -\frac{1}{R_{cas} \times C_{cas}} \qquad (9)$$

$$R_{out} = R_{out1} \| R_{out2} \qquad (10)$$

$$R_{out1} \approx g_{mM9,10} \times r_{oM9,10} \times g_{mM11,12} \times r_{oM11,12} \times r_{oM13,14} \qquad (11)$$

$$R_{out2} \approx g_{mM7,8} \times r_{oM7,8} \times g_{mM5,6} \times r_{oM5,6}$$
$$\times (r_{oM3,4} \| r_{oM1,2} \| r_{oMf1,f2} \| \frac{1}{g_{mMf1,f2}}) \qquad (12)$$

$$C_{out} = C_L + C_{dM9,10} + C_{dM7,8} \qquad (13)$$

$$R_{cas} \approx r_{oM1,2} \| r_{oM3,4} \| r_{oM5,6} \| \frac{1}{g_{mM5,6}} \| r_{oMf1,f2} \| \frac{1}{g_{mMf1,f2}} \qquad (14)$$

$$C_{cas} = C_{dM1,2} + C_{dM3,4} + C_{sM5,6} + C_{sMf1,f2} \qquad (15)$$

So considering Eq. 8 the transfer function of the Op-Amp calculated as:

$$A_{vf}(s) = \frac{V_{out}(s)}{V_{in}(s)} = \frac{(\alpha_f \times g_{mM1,2} \times R_{out})}{(1+\frac{s}{p_1'}) \times (1+\frac{s}{p_2'}) - (\beta \times \alpha_f \times g_{mMf1,2} . R_{out})}$$

$$\approx \frac{\dfrac{\alpha_f \times g_{mM1,2} \times R_{out}}{(1-(\beta \times \alpha_f \times g_{mMf1,2} \times R_{out}))}}{(1+\dfrac{s}{(1-(\beta \times \alpha_f \times g_{mMf1,2} \times R_{out})).p_1'}) \times (1+\dfrac{s}{p_2'})}$$

$$\approx \frac{\dfrac{\alpha_f \times g_{mM1,2} \times R_{out}}{k}}{(1+\dfrac{s}{k.p_1'}) \times (1+\dfrac{s}{p_2'})} \qquad (16)$$

Hence the final transfer function of the proposed Op-Amp is:

$$A_{vf}(s) \approx \frac{\dfrac{\alpha_f \times g_{mM1,2} \times R_{out}}{k}}{(1+\dfrac{s}{k.p_1'}) \times (1+\dfrac{s}{p_2'})} \qquad (17)$$

DC-gain of the Op-Amp is:

$$A_{vf0} = \frac{\alpha_f \times g_{mM1,2} \times R_{out}}{k} \qquad (18)$$

and considering Eq. 17 and Eq. 27, P_{f1} and P_{f2} are:

$$P_{f1} = k \times P_1' = -\frac{k}{R_{out} \times C_{out}} \qquad (19)$$

$$P_{f2} = P_2' = -\frac{1}{R_{cas} \times C_{cas}} \qquad (20)$$

It is clear that the second pole is not affected by feedback devices ($Mf1$, $Mf2$) to a great extent and first pole is reduced to P_{f1}.

UGBW of the simple TFC Op-Amp (ω_{u1}) is:

$$\omega_{u1} = P_1 \times A_{v0} \tag{21}$$

where A_{v0} is $\alpha \times g_{mM_{1,2}} \times R_{out}$, and α is:

$$\alpha = \frac{(r_{oM_{1,2}} \| r_{oM9,10})}{(\dfrac{1}{g_{mM7,8}} \| r_{oM7,8}) + (r_{oM_{1,2}} \| r_{oM9,10})} \tag{22}$$

It is considerable that P_1 is approximately equal to P_1'. *UGBW* of the proposed Op-Amp (ω_{u2}) is:

$$\omega_{u2} = P_{f1} \times A_{vf0} = k \times P_1' \cdot \frac{\alpha_f \times g_{mM_{1,2}} \cdot \times R_{out}}{k} = \frac{\alpha_f}{\alpha} \times \omega_{u1} \tag{23}$$

As mentioned before, transconductance of $Mf1,f2$ device is very small. Therefore, according to Eq. 2 and Eq. 22, α_f is very close to α. So the *UGBW* of the proposed Op-Amp (ω_{u2}) is equal with the *UGBW* of the TFC Op-Amp (ω_{u1}). Hence the speed of proposed Op-Amp in the close loop configuration does not reduce in comparison with the TFC Op-Amp.

C. Speed Consideration

In the conventional designs (e.g. [1], [2] and [3]), one extra node is added to the topology of folded cascode op-amp by PFB. Also it is clear that introducing an additional node to op-amp structure introduces an extra pole to the transfer function of the op-amp which increases the settling time of the op-amp in the close loop configuration. Also in [2] and [3] a pole-zero doublets introduced with PFB causes a reduction in the speed of the op-amp in the closed loop configuration. In the proposed op-amp in this paper positive feedback devices do not add any additional node to the TFC topology and pole to the transfer function of the TFC op-amp. Furthermore in this design there is not any pole–zero doublet, so opposite to [1], [2] and [3] the settling time of the proposed op-amp is not increased in the comparison with TFC op-amp.

D. Linearity Consideration

The main drawback of the structures which use positive feedback is distortion introduced by *PFB*. In Op-Amp presented in [1] a differential pair with a cascode current source connected to the output nodes is used as *PFB* which has poor linearity for large voltage swings in the output nodes of the Op-Amp. The nonlinearity of transconductance of *PFB* caused by large voltage swing in the output nodes reduces the performance of the Op-Amp in the rails. Moreover single ended structure of the Op-Amp degrades the linearity drastically. In the proposed Op-Amp, positive feedback devices form source followers which are more linear than a differential pair [5]. In the proposed op-amp opposite to [1][4], only a portion of the output voltage is applied to the positive feedback devices, so the linearity of the feedback devices and also the linearity of op-amp is improved significantly.

E. Output Voltage Swing Range

The main drawback of the technique used in [4] is the reduction of the output voltage swing range of the op-amp by positive feedback devices. In the proposed op-amp, positive feedback devices never restrict the range of output voltage swing.

III. SIMULATION RESULTS

In this section, the post-layout simulation results of the proposed op-amp are shown and the proposed op-amp is compared with the conventional TFC and op-amp presented in [4]. The op-amps, have been designed in a typical 0.35μm CMOS process (*Vdd=3.3*) with the same capacitor load and the same power consumption and are simulated by HSPICE software using *level* 49 parameters (BSIM3v3). The HSPICE AC simulation result of the proposed and TFC op-amp is shown in Fig. 2. Proposed op-amp has a UGBW of 850 MHz with 65° phase margin. As demonstrated in Fig. 2, the proposed op-amp achieves a dc-gain of more than 87dB which is 20 dB greater than the dc-gain of the conventional TFC op-amp in the same power consumption. The closed loop configuration shown in Fig. 3 is used to study the linearity and step response of the op-amps. Step response simulations of the designed op-amps demonstrates that the proposed op-amp has the accuracy of more than 14 bit for up to 3.5 Vp-p output

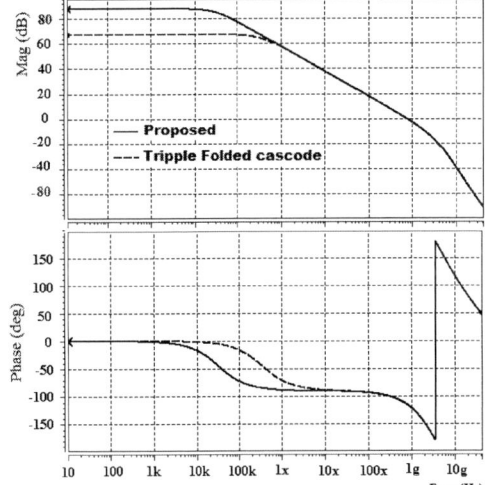

Figure 2.   Frequency response of the Op-Amps.

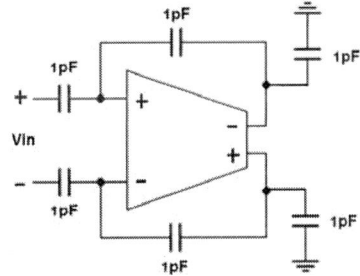

Figure 3.   Closed loop configuration.

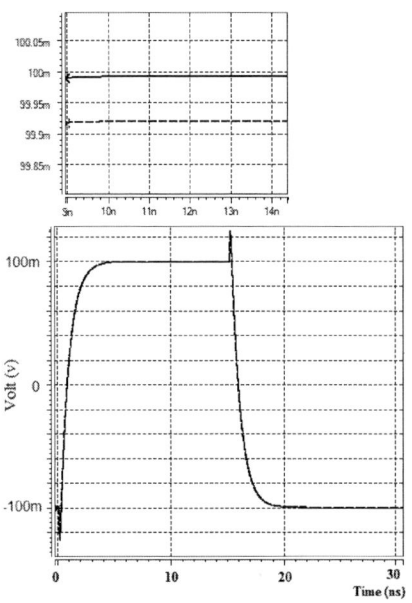

Figure 4. Step response comparison.

TABLE I. SPECIFICATIONS COMPARISON

Parameters	TFC	[4]	Proposed
DC-gain (dB)	67	67	87
THD(f=5MHz, Vop-p=2.6V)	-71 dB	-66 dB	-68 dB
Output Voltage Swing Range (V)	$V_{DD}-6.V_{ds,sat}$	$V_{DD}-4.V_{ds,sat}-\lvert V_{thn}\rvert$	$V_{DD}-6.V_{ds,sat}$
Phase Margin (°)	65	65	65
Power Consumption (mW)	9.5	9.5	9.5
UGBW (CL=0.4PF) (GHz)	0.85	1.25	0.85

voltage swings; but the conventional TFC has the accuracy of only 11bit in the same output voltage swing and in the approximately same power consumption. Monte-Carlo simulation shows a 3 dB reduction in the dc-gain for a *20mV* variation in the threshold voltage of all of the devices used in the proposed amplifier. Frequency response of the proposed Op-Amp in the different power supply voltages is studied. For a *100mV* variation in the supply voltage, dc-gain reduction is 3dB and UGBW and phase margin remains constant to a great extent. Simulation in the process corners shows that, in *FF* corner the bandwidth of the Finally, proposed amplifier is increased by *9%* and phase margin improves *2°* and dc-gain decreases about *6dB*. In *SS* corner the bandwidth of the proposed amplifier is decreased by *10%*, phase margin degrades approximately *2°* and dc-gain increases about *5 dB*. The accuracy of the proposed amplifier in the mentioned close loop configuration degrades just 1 bit only in *FF* corner and settling time of the proposed Op-Amp degrades *10%* in *SS* corner. Finally, the characteristics of the designed op-amps are summarized in Table 1 for comparison.

IV. CONCLUSION

In this paper, a new op-amp is presented which uses positive feedback concepts in the conventional TFC op-amp to increase the dc- gain of the op- amp. The proposed method does not limit the bandwidth, output voltage swing range and the phase margin of the triple folded cascode op-amp. HSPICE simulations confirm the theoretical estimated improvements.

REFERENCES

[1] Carlos A. Laber and Paul R. Gray, "A Positive-Feedback Transconductance Amplifier with Applications to High-Frequency, High- Q CMOS Switched- Capacitor Filters", IEEE J. Solid-State Circuits, VOL. 23. NO. 6, pp. 1370-1378, DECEMBER 1988.

[2] M. Asloni, Kh. Hadidi And A. Khoei, "Design of a New Folded Cascode Op-Amp Using Positive Feedback and Bulk Amplification" IEICE TRANSACTIONS on Electronics Vol.E90-C No.6 pp.1253-1257, June 2007.

[3] A. Dadashi, Sh. Sadrafshari, Kh. Hadidi and A. Khoei "An enhanced folded cascode Op-Amp using positive feedback and bulk amplification in 0.35 μm CMOS process", Analog Integrated Circuits and Signal Processing,, Vol. 67, pp. 213–222, May 2011.

[4] A. Dadashi, Sh. Sadrafshari, Kh. Hadidi and A. Khoei "Fast- settling CMOS Op-Amp with improved DC-gain", Analog Integrated Circuits and Signal Processing, Vol. 70, No. 3, pp. 283-292, February 2012.

[5] B. Razavi, Design of Analog CMOS Integrated Circuits, McGraw- Hill, New York, 2001.

Gap in pagination due to withheld paper.

Pages 239-242

MIXDES 2012, 19th International Conference *"Mixed Design of Integrated Circuits and Systems"*, May 24-26, 2012, Warsaw, Poland

Design of a Fully Programmable Analog Interval Type-2 Triangular/Trapezoidal Fuzzifier

Hossein Yazdanjouei, Hossein Feizy, Abdollah Khoei, and Khayrollah Hadidi

Microelectronics Research Laboratory

Urmia University

Urmia, Iran

st_h.yazdanjouei@urmia.ac.ir, st_h.feizy@urmia.ac.ir, a.khoei@urmia.ac.ir, kh.hadidi@urmia.ac.ir

Abstract—In this paper, analog implementation of a fully programmable Interval Type-2 (IT2) fuzzifier is presented. This fuzzifier operates in current mode and provides triangular, trapezoidal, S and Z-shape Interval Type-2 membership functions precisely by means of just one Type-1 fuzzifier and a Type-2 Generator circuit. In this circuit, shapes of upper and lower membership functions could be adjusted separately and they may have different shapes. The amount of Footprint of Uncertainty (FOU), slope, height and width of generated Interval Type-2 membership functions are fully programmable and each of these properties has the ability to be tuned easily and independently. Hence this circuit is suitable for use as fuzzifier block in Interval Type-2 fuzzy logic controllers (IT2FLC). Simulation results in HSPICE using 4M2P TSMC 0.35μm CMOS technology shows, the power consumption of proposed circuit with supply voltage of 3.3V is 657μW and it occupies an area about 0.01mm².

Index Terms—Interval Type-2 fuzzy; fuzzifier; CMOS; membership function; Triangular/Trapezoidal.

I. INTRODUCTION

Since the introduction of Type-1 fuzzy set theory by Prof Lotfi Zadeh [1] in 1965 many applications have been reported in literature concerning modeling and control [2], [3], [4], data mining [5], [6], [7], time-series prediction [8], [9], [10], etc. But in some cases, research has shown that Type-1 fuzzy sets do not have the ability to handle uncertainties existing in systems appropriately since they have crisp and predefined membership functions [11], [12], [13].

Therefore almost ten years later again Prof Lotfi Zadeh introduced a more powerful type of fuzzy sets called Type-2 fuzzy sets [14]. The prominence of Type-2 fuzzy sets owes the fuzziness in their membership function.

They provide additional design degrees of freedom and can handle the uncertainties in a more convenient way. Although Type-2 fuzzy logic systems (T2FLS) outperform Type-1 fuzzy logic systems (T1FLS) in many aspects but general Type-2 fuzzy systems are computationally complicated and they cannot be easily implemented in software and hardware forms. Because of that, there has not been lots of enthusiasm and effort to use them in practical applications. Hence a special case of T2FLSs was born, called Interval Type-2 fuzzy logic systems (IT2FLS) that is a simplified form of T2FLSs.

An example of an Interval Type-2 fuzzy set, Ã, is depicted in Fig. 1(a). As can be seen, an IT2FS is bounded from the above and below by two Type-1 fuzzy sets, Ā and A̲, which are called Upper membership function (UMF) and Lower membership function (LMF) respectively. The area between UMF and LMF is called footprint of uncertainty (FOU). As shown in Fig. 1(b), UMF and LMF of Interval Type-2 fuzzy sets could have arbitrary shapes.

Fig. 2 shows a general Interval Type-2 fuzzy logic system. An IT2FLS consists of 4 major building blocks as shown in Fig. 2: fuzzifier, fuzzy inference engine and rule base, type reducer and defuzzifier. IT2FLSs are similar to their Type-1 counterparts except that they use Interval Type-2 fuzzy sets, therefore the output of inference engine is an Interval Type-2 set and the system needs a type reducer block to convert the output of inference engine in to a Type-1 fuzzy set before defuzzification stage.

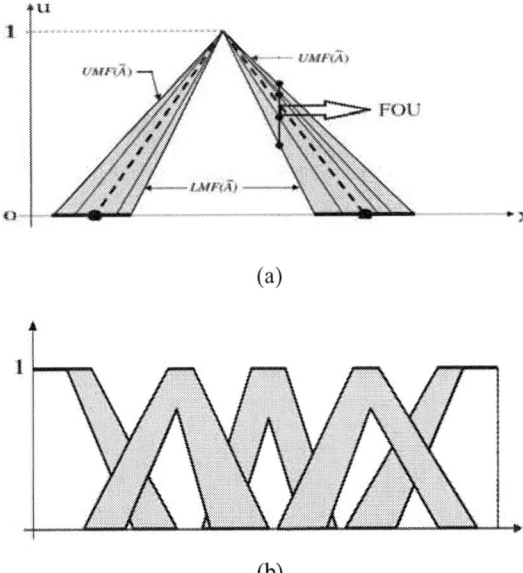

(a)

(b)

Figure 1.   (a) Interval Type-2 fuzzy sets. (b) Upper membership function and lower membership function of Type-2 fuzzy sets could be different.

243

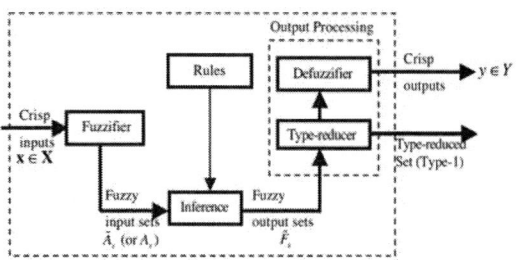

Figure 2.   Interval Type-2 fuzzy logic system

IT2FLSs are currently the most widely used for their reduced computational cost but their hardware implementation is still a nascent research area.[15] Unlike T1FLSs that are often implemented in hardware as digital, analog and mixed-signal circuits, most of the practical applications of IT2FLSs are based on software implementation of them and a few hardware implementations are mostly based on FPGAs, microcontrollers or digital VLSI chips [16]- [19].

In this paper we proposed an accurate and fully programmable Interval Type-2 fuzzifier circuit. This circuit produces triangular, trapezoidal, S and Z shape IT2 sets by means of just one T1 fuzzifier and a Type-2 Generator circuit. The shapes of upper and lower membership functions could be adjusted separately, resulting in a flexible Type-2 fuzzifier. The amount of Footprint of Uncertainty (FOU), slope, height and width of generated IT2 membership functions are fully programmable and each of them has the ability to be tuned easily and independently.

The remainder of this paper is arranged as follows: Section 2 focuses on the implementation of proposed Type-2 fuzzifier, Section 3 shows the simulation results of proposed circuits and eventually section 4 concludes the paper and compares it with few available similar works.

II. CIRCUIT DESCRIPTION

Fig. 3 shows the basic structure of proposed fuzzifier. It consists of two major blocks: a Type-1 fuzzifier and a Type-2 Generator circuit. Type-1 fuzzifier generates the basic membership functions and Type-2 Generator circuit converts them to Type-2 fuzzy set.

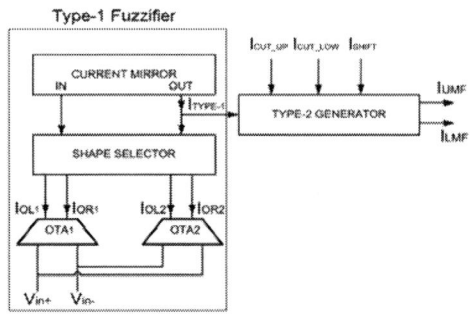

Figure 3.   Basic structure of proposed fuzzifier

A. Type-1 Fuzzifier

Type-1 fuzzifier consists of two operational transconductance amplifiers (OTA), a shape selector circuit and a current mirror. The CMOS implementation of OTAs is depicted in Fig. 4(a). It is based on flipped voltage follower (FVF) circuit introduced in [20] and [21].

In this circuit the nodes A and B follow the input differential voltage through transistors MnL1 and MnR1. Thus a current equal to $I = V_d / 2R_i$ passes through the resistors selected by $V_{cntrl}(i)$. This current subtracts (adds) to the bias current (I_b) in transistors MnL2 (MnR2) and then will be mirrored to output of OTA through transistors MnL3, MnL4 and MnL5 (MnR3, MnR4 and MnR5) leading to $I_{OL} = I_b - I$ and $I_{OR} = I_b + I$. The transistors MnL6 and MnR6 are used as a level shifter to increase the input range of OTA.

The DC transfer characteristics of OTA without any current source form node A to ground is shown in Fig. 4(b). By placing I_b at node A, output currents of OTA will be shifted as shown in Fig. 4(c). The slope of I_{OL} and I_{OR} could be adjusted by resistor values selected by $V_{cntrl}(i)$ and the height of I_{OL} and I_{OR} could be adjusted by amount of I_b.

To generate a Type_1 fuzzy set, the inputs of two OTAs must be connected together as shown in Fig. 3. The output currents of OTAs are shown in Fig. 6(a). These currents pass through a shape selector circuit and this circuit decides on the OTA output currents that must be added together at the input and output nodes of current mirror. Therefore shape selector circuit determines the shape of Type-1 membership function to be triangular, S-shape or Z-shape. Fig. 5 depicts shape selector circuit.

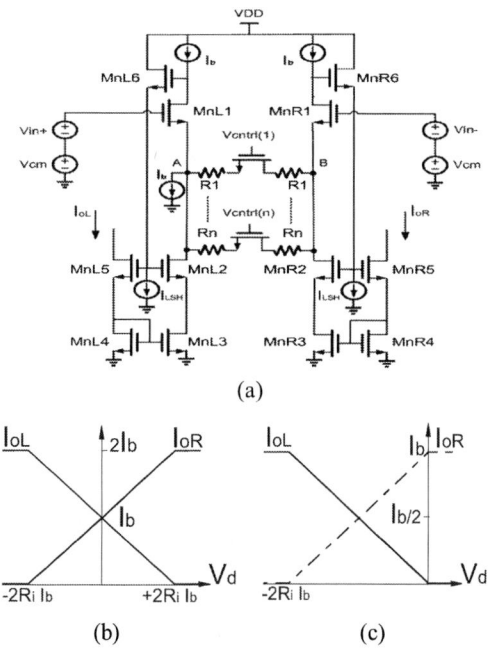

(a)

(b)   (c)

Figure 4.   (a) Highly linear OTA. (b) Output currents of OTA. (c) Shifted output currents of OTA by placing I_b from node A to ground.

244

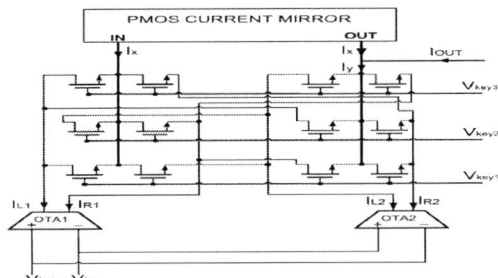

Figure 5.   Shape selector circuit

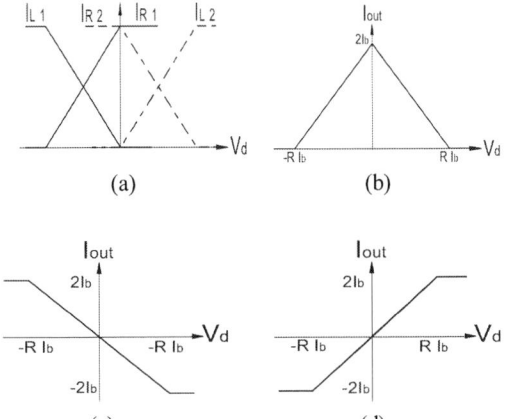

Figure 6.   (a) Outputs of OTA1 and OTA2. (b) Triangular output
of Type-1 fuzzifier. (c) Z shape output of Type-1 fuzzifier. (d)
S-shape output of Type-1 fuzzifier.

When V_{key1} of shape selector circuit is active, the currents I_{L1} and I_{L2} will be added together at the input node and currents I_{R1} and I_{R2} will be added together at the output node of PMOS current mirror. Therefore $I_{out} = I_y - I_x = (I_{R1} + I_{R2}) - (I_{L1} + I_{L2})$ and it will have triangular shape as shown in Fig. 6(b). When V_{key2} is active, $I_{out} = I_y - I_x = (I_{L1} + I_{R2}) - (I_{R1} + I_{L2})$ and it has a Z-shape output as shown in Fig. 6(c) and eventually when V_{key3} is active, $I_{out} = I_y - I_x = (I_{R1} + I_{L2}) - (I_{L1} + I_{R2})$ and the output has an S-shape membership function as shown in Fig. 6(d).

B. Type-2 generator circuit

Type-2 generator circuit is shown in Fig.7. It consists of a current mirror (Mm1-Mm6) that distributes the output current of Type-1 fuzzifier (I_{Type1}) in the circuit, a shift circuit (Ms1-Ms4) that generates desired amount of FOU and two cut circuits (Mcu1-Mcu8 and Mcl1-Mcl8) that cut the UMF or LMF currents in the desired values determined by I_{cut_up} and I_{cut_low}. The cut circuits are based on min circuit introduced in [22] and are used to convert triangular membership function to trapezoidal membership function whenever needed and also to eliminate negative currents in S and Z-shape functions.

To produce upper membership function (I_{UMF}), I_{Type1} will be compared with I_{cut_up}. If I_{Type1} is smaller than I_{cut_up}, I_{UMF} will be

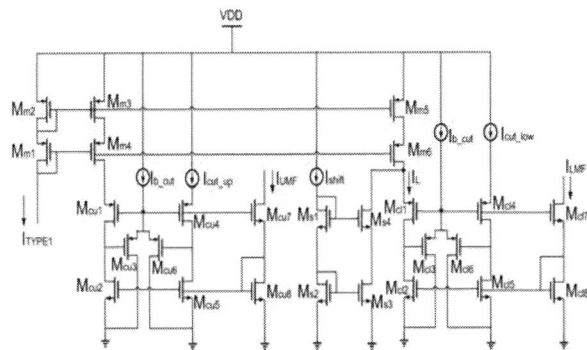

Figure 7.   Type-2 generator circuit

equal to I_{Type1}, else it will be fixed at I_{cut_up}. To produce lower membership function (I_{LMF}), at first I_{shift} will be subtracted from I_{Type1} through transistors Ms1-Ms4 to create desired amount of FOU between I_{UMF} and I_{LMF}, then the resultant current (I_L) will be transferred to second cut circuit and will be compared with I_{cut_low}. If I_L is smaller than I_{cut_low}, I_{LMF} will be equal to I_L, else it will be fixed at I_{cut_low}.

By using this scheme, both of I_{UMF} and I_{LMF} has the ability to be converted into trapezoidal membership function with desired height whenever needed. This property makes UMF and LMF flexible and allows them to have arbitrary shapes.

III. SIMULATION RESULTS

In order to test the performance of proposed Type-2 fuzzifier, prelayout netlists of above circuits have been simulated in HSPICE using 4M2P TSMC 0.35μm CMOS technology. Table I lists the parameters used for simulation.

As mentioned above, the proposed fuzzifier has the ability to assign different shapes to UMF and LMF. Fig. 8 depicts variant forms of Type-2 fuzzy sets the fuzzifier can produce. Fig. 8(a) shows the output currents when both of UMF and LMF are triangular. Fig. 8(b) shows the outputs when UMF and LMF have been cut at 25μA and 15μA respectively thus both of them are trapezoidal. In Fig. 8(c) and Fig. 8(d) UMF is triangular and LMF is trapezoidal and vice versa. Finally V_{key2} and V_{key3} have been activated respectively to generate Z-shape and S-shape functions as shown in Fig. 8(e) and Fig. 8(f). Table II lists the parameters used to generate membership functions shown in Fig. 8.

TABLE I.   PARAMETERS USED FOR SIMULATION

Parameter name	Value
I_b	20μA
I_{LSH}	10μA
I_{b_cut}	5μA
VDD	3.3V
V_{cm}	1.65V

245

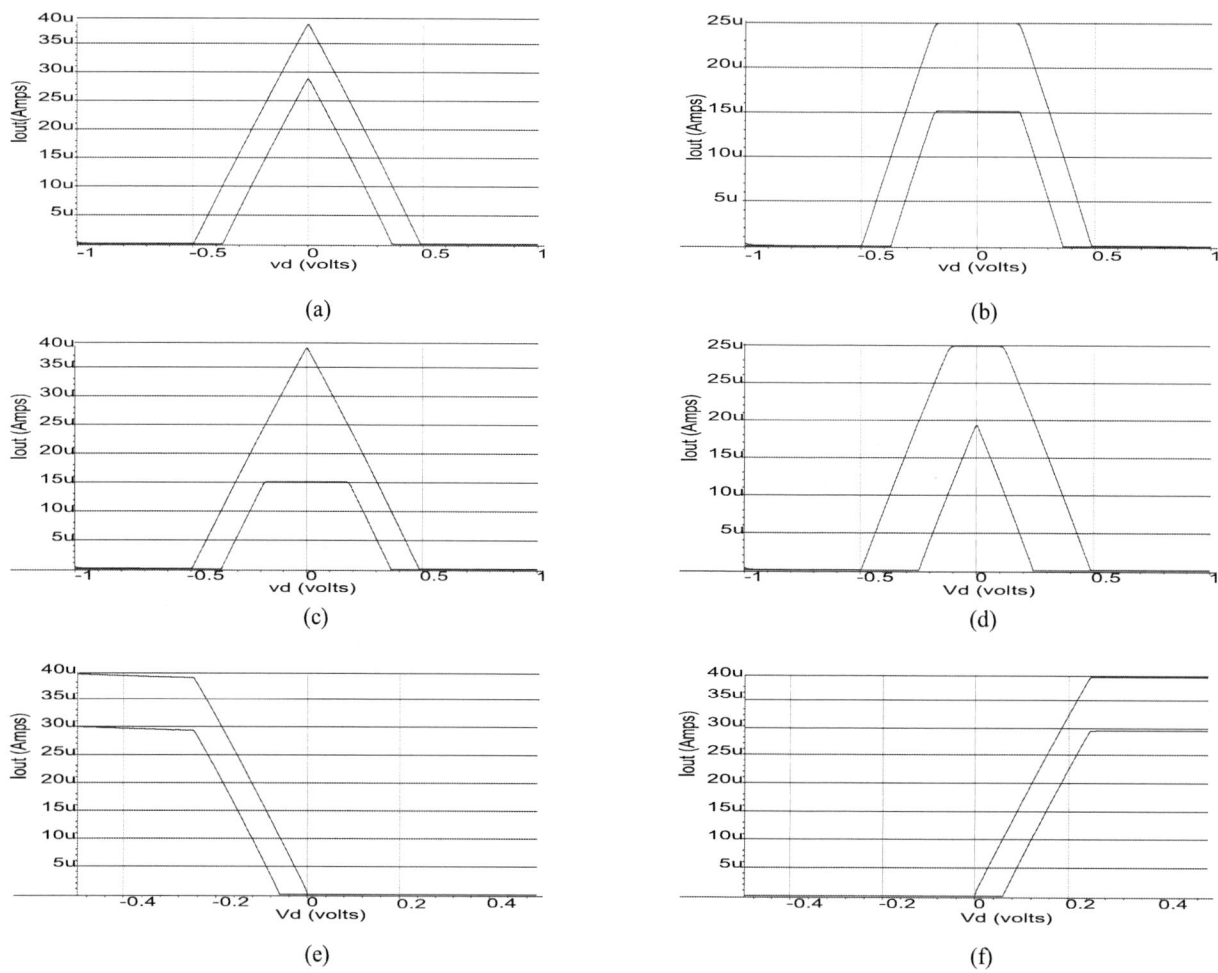

Figure 8.   Variant output forms of proposed Type-2 fuzzifier. (a) Both of UMF and LMF are triangular. (b) Both of UMF and LMF are trapezoidal. (c) UMF is triangular and LMF has been cut at 15μA thus is trapezoidal. (d) UMF has been cut at 25μA and is trapezoidal but LMF is triangular. (e) V_{key2} has been turned on thus UMF and LMF are Z-shape. (f) V_{key3} has been turned on thus UMF and LMF are S-shape.

Fig. 9 illustrates programmability of width, slope, height and horizontal displacement of membership functions produced. To adjust the slopes of membership functions, desired resistor value must be selected by $V_{cntrl}(i)$ between node A and B in OTA1 and OTA2. Fig. 9(a) shows the output of fuzzifier when the resistor values in OTA1 and OTA2 is equal and they have been changed simultaneously from 10kΩ to 40kΩ in 5kΩ steps.

In Fig. 9(b) just the value of resistor in OTA1 has been stepped from 5kΩ to 20kΩ in 5kΩ steps while the resistor in OTA2 is fixed at 10kΩ, thus just left slope of membership functions have changed.

In Fig. 9(c) resistor values are same as Fig. 9(b) but at this time the resistor value in OTA1 is fixed and resistor values in OTA2 have been stepped, thus right slope of membership function changes while left slope remains constant.

In Fig. 9(d) height of membership functions have been tuned varying I_b from 10μA to 30μA in 10μA steps.

Horizontal displacement of Type-2 membership functions has been achieved by fixing the negative input of fuzzifier (Vin-) in desired values and applying the input signal to positive input of fuzzifier (Vin+).

In Fig.9 (e) Vin- has been stepped from -0.5V to 0.5V in 0.1V steps.

TABLE II.   PARAMETERS USED TO GENERATE FIG.8 FMFS

Fig	R1	V_{key1}	V_{key2}	V_{key3}	I_{cut_up}	I_{cut_low}	I_{shift}
8(a)	10k	3.3	0	0	>40μ	>30μ	10μ
8(b)	10k	3.3	0	0	25μ	15μ	10μ
8(c)	10k	3.3	0	0	>40μ	15μ	10μ
8(d)	10k	3.3	0	0	25μ	>20μ	20μ
8(e)	5k	0	3.3	0	>40μ	>30μ	10μ
8(f)	5k	0	0	3.3	>40μ	>30μ	10μ

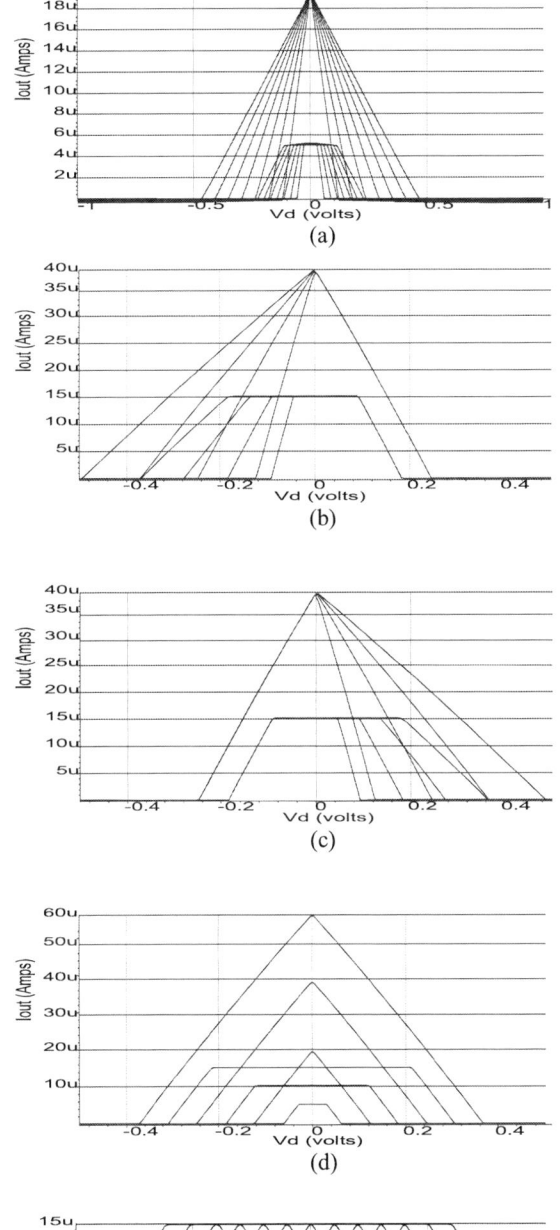

Figure 9. (a) width sweep. (b) left slope sweep. (c) right slope sweep. (d) height sweep. (e) horizontal displacement.

Fig. 10 exhibits simulation results of proposed fuzzifier in different process corners (TT, FF, FS, SF and SS) and Fig. 11 shows the proposed layout.

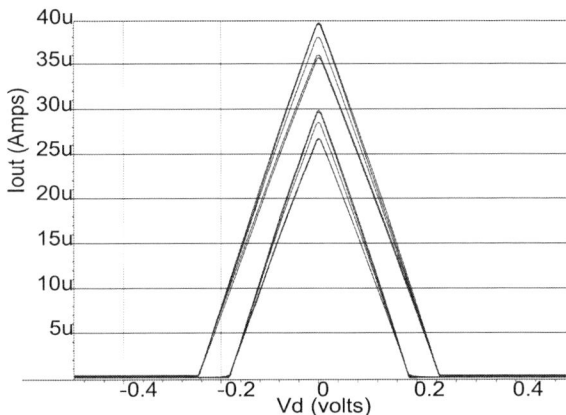

Figure 10. corner simulation results of proposed Type-2 fuzzifier

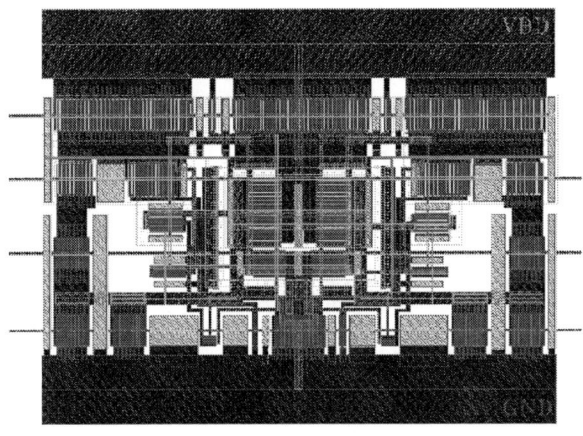

Figure 11. layout of proposed fuzzifier

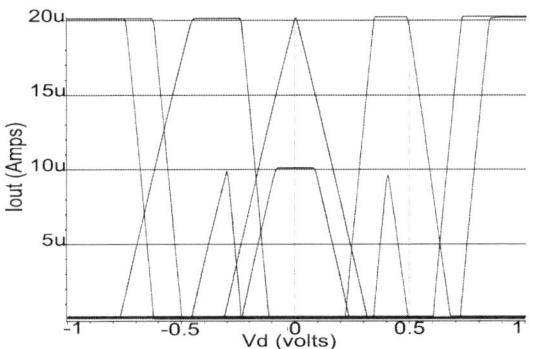

Figure 12. outputs of proposed fuzzifier to demonstrate flexibility of it to generate different shapes for UMF and LMF

247

Eventually Fig. 12 depicts several Type-2 membership functions produced by proposed circuits beside each other to demonstrate flexibility of introduced Type-2 fuzzifier.

IV. CONCLUSIONS

In this paper a fully programmable and accurate analog Type-2 fuzzifier has been introduced. The proposed fuzzifier has the ability to produce triangular, trapezoidal, Z-shape and S-shape membership functions as UMF and LMF of Type-2 fuzzy sets. The shapes of UMF and LMF could be determined independently. This property makes the proposed circuit so flexible. The amount of Footprint of Uncertainty (FOU), slope, height and width of generated Interval Type-2 membership functions are also fully programmable and each of them has the ability to be tuned easily and independently. This circuit uses just one Type-1 fuzzifier to generate a Type-2 membership function; hence its power consumption is low compared with Type-2 fuzzifiers that use two individual Type-1 fuzzifiers.

Prelayout simulation results in HSPICE using 4M2P TSMC 0.35μm CMOS technology shows, power consumption of proposed fuzzifier, with supply voltage of 3.3V is 657μW and it occupies an area about 0.01mm2.

Table III compares this work with two other similar circuits.

TABLE III.   COMPARISON WITH PREVIOUS WORKS

Ref	[23]	[24]	This work
Tech	0.35μm	0.18μm	0.35μm
Supply	3.3V	3.3V	3.3V
Power	1.29mW	2.64mW	657μW
Area	$0.02mm^2$	$0.006mm^2$	$0.01mm^2$
MF types	Tri, Trap, Z, S	Tri, Trap, Z, S	Tri, Trap, Z, S
Diff shapes for UMF and LMF	no	yes	yes
Num of Type-1 fuzzifiers	one	two	one
Input type	current	voltage	voltage

REFERENCES

[1] L. A. Zadeh, "Fuzzy sets", Information and Control, vol. 8, pp. 338–353, 1965.

[2] J. Bezdek, "Fuzzy models–what are they, and why?", IEEE Trans. on Fuzzy Systems, vol. 1, no. 1, pp. 1–5, 1993.

[3] J. Buckley and H. Ying, "Expert fuzzy controller", Fuzzy Sets and Systems, vol. 43, pp. 127–137, 1991.

[4] T. A. Johansen, "Fuzzy model based control: Stability, robustness, and performance issues", IEEE Trans. on Fuzzy Systems, vol. 2, no. 3, pp. 221–234, 1994.

[5] J. F. Baldwin, "Knowledge from data using fuzzy methods", Pattern Recongnition Letter, vol. 17, pp. 593–600, 1996.

[6] K. Hirota and W. Pedrycz, "Fuzzy computing for data mining", Proc. IEEE, vol. 87, no. 9, pp. 1575–1600, 1999.

[7] W. Pedrycz, "Fuzzy set technology in knowledge discovery", Fuzzy Sets and Systems, vol. 98, pp. 279–290, 1998.

[8] N. K. Kasabov and Q. Song, "DENFIS: Dynamic evolving neural-fuzzy inference system and its application for time-series prediction", IEEE Trans. on Fuzzy Systems, vol. 10, no. 2, pp. 144–154, 2002.

[9] S. S. Liao, T. H. Tang, and W.-Y. Liu, "Finding relevant sequences in time series containing crisp, interval, and fuzzy interval data", IEEE Trans. on Systems, Man, and Cybernetics–B, vol. 34, no. 5, pp. 2071–2079, 2004.

[10] M. Versaci and F. C. Morabito, "Fuzzy time series approach for disruption prediction in tokamak reactors", IEEE Trans. on Magnetics, vol. 39, no. 3, pp. 1503–1506, 2003.

[11] J. M. Mendel, "Uncertain Rule-Based Fuzzy Logic Systems: Introduction and New Directions.", Upper Saddle River, NJ: Prentice-Hall, 2001

[12] Wu D. and W. W. Tan, "Genetic Learning and Performance Evaluation of Type-2 Fuzzy Logic Controllers", IEEE Trans. on Systems, Man, and Cybernetics—Part B: Cybernetics, 2005.

[13] Jerry M. Mendel, "Type-2 fuzzy sets and Systems: anOverview", IEEE Computational Intelligence Magazine, pp. 20-29, February 2007, doi: 10.1109/MCI.2007.380672R.

[14] L. A. Zadeh, "The concept of a linguistic variable and its application to approximate reasoning ", Inform. Sci., vol. 8, pp.199-249, 1975.

[15] M . Melgarejo and C. A. Pena-Reyes, "Implementing Interval Type-2 Fuzzy Processors [Developmental Tools]", Computational Intelligence Magazine, IEEE , vol.2, no.1, pp.63-71, Feb. 2007

[16] J. Bulla, G. Sierra and M. Melgarejo "Implementing a Simple Microcontroller-Based Interval Type-2 Fuzzy Processor", Proceedings of 51st Midwest Symposium on Circuits and Systems (MWSCAS), Knoxville (TN), 2008, pp. 69-72.

[17] M. A. Melgarejo and C. A. Pena-Reyes, "Hardware architecture and FPGA implementation of a type-2 fuzzy system", Proc. ACM GLSVLSI, Boston, MA, 2004, pp. 458–461.

[18] M. A. Melgarejo, R. A. Garcia and C. A. Pena-Reyes, "Pro-two: A hardware based platform for real time type-2 fuzzy inference", Proc. IEEE Int. Conf. Fuzzy Syst., vol. 2, 2004, pp.977–982.

[19] Shih-Hsu Huang, Yi-Rung Chen, "VLSI implementation of type-2 fuzzy inference processor", Circuits and Systems, 2005. ISCAS 2005. IEEE International Symposium on , vol., no., pp. 3307- 3310 Vol. 4,23-26 May 2005.

[20] R.G. Carvajal, J. Ramirez-Angulo, A.J. Lopez-Martin, A. Torralba, J.A.G. Galan, A. Carlosena, F.M. Chavero, "The flipped voltage follower: a useful cell for low-voltage low-power circuit design", Circuits and Systems I: Regular Papers, IEEE Transactions on , vol.52, no.7, pp. 1276- 1291, July 2005.

[21] M. Kachare, J. Ramirez-Angulo, R. G. Carvajal and A. J. Lopez Martin, "New low-voltage fully programmable CMOS triangular/trapezoidal function generator circuit", Circuits and Systems I: Regular Papers, IEEE Transactions on, vol.52, no.10, pp. 2033–2042, October 2005.

[22] T. Temel, "High-performance current-mode multi-input loser-take-all minimum circuit", Electronics Letters , vol.44, no.12, pp.718-719, June 5 2008.

[23] P.M.S. Rocha Rizol, L. Mesquita, O. Saotome, G. Botura, "Hardware implementation of type-2 programmable fuzzifier", Circuits and Systems (LASCAS), 2011 IEEE Second Latin American Symposium on , vol., no., pp.1-4, 23-25 Feb. 2011.

[24] M. Khosla, R.K. Sarin, M. Uddin, A. Sharma, "Analog realization of fuzzifier for IT2 fuzzy processor", Electronics Computer Technology (ICECT), 2011 3rd International Conference on , vol.1, no., pp.239-245, 8-10 April 2011.

MIXDES 2012, 19th International Conference *"Mixed Design of Integrated Circuits and Systems"*, May 24-26, 2012, Warsaw, Poland

Design of an Analog Output Buffer for Active Matrix Displays Using Low-Temperature Polycrystalline Silicon Thin-Film Transistors

Ilias Pappas, Stylianos Siskos
Electronics Lab, Physics Dept.
Aristotle University of Thessaloniki
54124 Thessaloniki, Greece
ilpap@auth.gr

Alkis A.Hatzopoulos
Dept. of Electrical and Computer Eng.
Aristotle University of Thessaloniki
54124 Thessaloniki, Greece
alkis@eng.auth.gr

Abstract—A new source follower type analog buffer used as an output buffer for the data / column driver of an Active Matrix display is presented in this paper. The proposed buffer is implemented with low temperature polycrystalline silicon thin film transistor (LT poly-Si TFTs) and the main advantage of the buffer is its high immunity to the threshold voltage variation of the LT poly-Si TFTs. The functionality of the buffer is verified through simulation with HSpice, using for the simulations parameters extracted from fabricated LT poly-Si TFTs in order to obtain realistic simulation results.

Index Terms--analog buffer; data / column driver; flat panel displays; LT poly-Si TFTs.

I. INTRODUCTION

Low-temperature polysilicon thin-film transistors (LT poly-Si TFTs) have been widely investigated due to their potential applications in large area and flexible electronics [1], such as displays, memories, x-ray detectors and scanners [2], [3]. Despite the similarities between polysilicon TFTs and the commonly used MOSFETs, a number of key differences exist. The main reason causing these differences is that, instead of a single-crystal silicon wafer, a typical heat-sensitive material, like glass or quartz, is usually used. Due to such a substrate, TFTs are three terminal devices, with the substrate contact being absent. The presence of an insulating substrate provides an ideal isolation of each device and negligible parasitic capacitance [4].

At the beginning of the fabrication process, amorphous silicon is being deposit and then it is crystallized by either Solid-Phase Crystallization (SPC) or laser annealing procedure. The crystallize nature of the channel material results in the better electrical performance of the poly-Si TFTs compared to the amorphous silicon TFTs. Therefore, the higher carrier mobility of polysilicon TFTs, compared to the mobility of amorphous silicon TFTs [5], leads to integration of the driving and the peripheral circuits of a display device by using only the polysilicon TFT technology, resulting in reduction of the fabrication cost.

The major disadvantages of polysilicon TFTs are the large variations of the threshold voltage and carrier mobility from transistor to transistor [6]. These variations are caused because of the random distribution of the grain boundaries within the channel of the transistor [7], during the crystallization of the amorphous silicon. Across a 2.7-inch diagonal display [8], the threshold voltage variation is about 300 mV and it can be up to ±1 V in some displays with large substrate area [5]. These disadvantages make the design of analog drivers difficult and leads to non-uniformity of the brightness and poor gray-scale accuracy.

Last decades, Flat Panel Displays (FPD) have been emerged on almost all types of display applications; from watches to computers and projection TVs. Display addressing (driving) techniques have a major influence on the display image quality. The addressing techniques can be classified in three essential types, namely, direct (static) addressing, passive matrix (PM) addressing and active matrix (AM) addressing. The AM addressing technique is used in high-information contents displays [9] and is the most commonly used addressing technique in LCDs.

Since active matrix displays (AMDs) pixels use voltage as driving quantity, the non-uniformity of the polysilicon TFT threshold voltage is critical. Many compensation methods for the variation of the polysilicon TFTs characteristics have been proposed [10]-[13]. These methods can be separated into voltage driving, current driving and digital driving. However, these methods either require complex compensation circuitry or they are not directly applicable to AMDs [8].

In this paper, an improved source follower type analog buffer for AMD applications is proposed. The new buffer presents high immunity to threshold voltage and mobility variations. Also, it is capable to drive large load capacitance in large area panels. It consists of five n-type TFTs and two additional control signal. The proposed buffer does not require storage capacitor for the threshold voltage variation cancellation, leading to high speed operation. It can also be used to as the output buffer of a data / column voltage driver in an AMD, in order to apply the accurate data voltage to each sub-pixel of the display array, despite the variations of the threshold voltages of the poly-Si TFTs. In this way, the brightness and the color-scale of the sub-pixel will be ensured.

II. DATA / COLUMN DRIVER ARCHITECTURE

The conventional column driver for a voltage driven AM FPD consists of a shift register, a data latch, a shift register, a level shifter, a digital to analog converter (DAC) and an output analog buffer, as shown in fig.1. The functionality of the data / column driver is based on converting the digital word of a specific color to a corresponding analog voltage which will be delivered to the appropriate pixel. The output analog buffer is used to transfer the data voltage to the pixels array. The color-scale that will be produced from the pixel depends on the voltage level of the data voltage which will be applied to the pixel. If the analog buffer is sensitive to the threshold voltage variation, the data voltage level will be inaccurate. In this way, the produced color-scale from the pixel will not be well-controlled, since the threshold voltage variation has a large and random distribution along the pixel panel. Thus, the pixel panel will suffer from non - uniformity and poor color-scale. The analog buffer has to be designed in such way, so that the threshold voltage variations of the TFTs will not affect its functionality and performance.

Many compensation methods for the variation of the polysilicon TFTs characteristics have been proposed [10]-[15]. These methods can be separated into voltage driving, current driving [16], [17] and digital driving. However, these methods either require complex compensation circuitry or they are not directly applicable to AM FPDs. The rest of the digital modules can be regularly designed because the poly-Si TFTs are used

as switches and their threshold voltage variation has negligible effect on their performance.

III. PROPOSED ANALOG OUTPUT BUFFER

The proposed buffer consists of five n-type poly-Si TFTs, one multiplexer 2-to-1 and two control signals, as shown in fig. 2. The source follower type buffer presents high immunity to threshold voltage variation and, also, it is capable to drive large load capacitance in large area panels. The proposed buffer does not require storage capacitor for the threshold voltage variation cancellation, leading to high speed operation and to the reduction of the silicon real estate. Furthermore, it does not require multi-level data voltage during its operation time or complex bias voltages such as [11], which will cause the change of the column driver architecture.

The operation of the buffer is based on applying the data voltage V_{data} to the load capacitance C_L. Therefore, the gate voltage of the driver transistor M4 needs to be: $V_{G4} = V_{data} + V_{TH}$, where V_{TH} is the threshold voltage of M4. For the right operation of the buffer, we assume that M1 and M4 share the same electrical characteristics, such as threshold voltage. In practical applications, this can be achieved by placing the transistors close enough on the same wafer, fabricated under the same polysilicon growth and process conditions [12]. The detailed operation of the buffer is divided in three phases.

Phase 1: (Capacitance Discharge and Node A charge)

During this phase, only Reset signal is "high" causing

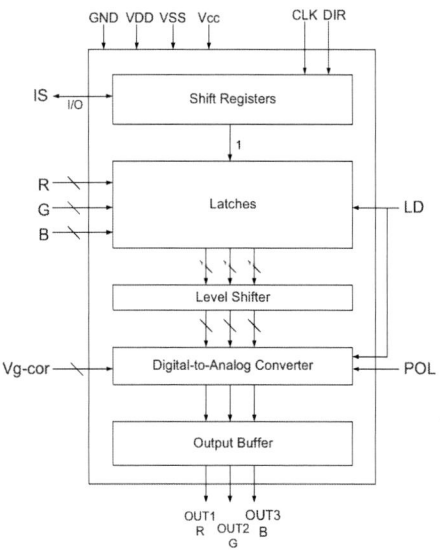

Figure 1.   Conventional data / column driver of an Active Matrix Flat Panel Display.

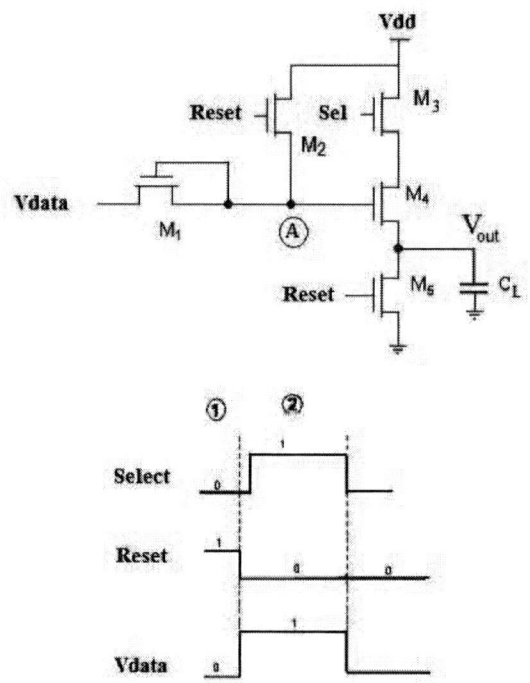

Figure 2.   Proposed analog buffer and its timing diagram.

250

transistors M2 and M5 to turn ON. Therefore, the load capacitor CL is discharged through M5 and node A will be charge to a voltage almost equal to the supply voltage Vdd.

Phase 2: (Capacitance Charge)

At the beginning of the second phase, the Select signal is turned to "low" level and the data voltage V_{data} is applied to the input of the buffer. Since the voltage at node A is higher than the data voltage, node A will be discharge through M1. Transistor M1 is diode-connected causing node A to be discharged until its voltage will be equal to: $V_A = V_{data} + V_{TH}$. When the voltage at node A is stable at this value, the Select signal is turn to "high" level and M4 is connected in a source follower topology. Therefore, the load capacitance will be charge until its voltage will be equal to V_{data}.

IV. SIMULATIONS SET UP

A. Transistor Parameters

In order to ensure that the simulation results are realistic, the values of the RPI model parameters were extracted from fabricated TFT devices using the Silvaco tools (ATLAS). The used polysilicon TFTs were fabricated on fused quartz glass substrates, covered by 200 nm thick SiO2 which was deposited by electron cyclotron resonance – plasma enhanced chemical vapor deposition (ECR-PECVD). First, amorphous Si films (about 50 nm thick) were deposited by low pressure chemical vapor deposition at the temperature of 425 °C and pressure of 1.1 Torr using Si2H6 as reactant gas. Then, the amorphous Si films were transformed into polycrystalline phase by furnace annealing at 600 °C for 24 hours in nitrogen ambient (Solid Phase Crystallization, SPC). Such polysilicon films have a high density of intra-grain defects that cause decrease in the field effect mobility and increase in the threshold voltage of the TFTs [17]. However, application of excimer laser annealing to the SPC polysilicon films was formed to effectively reduce the intra-grain defect density, while the mean grain size remains unchanged [17]. For this reason, the SPC films were irradiated by XeCl excimer laser (λ=308 nm, 23 shots) with energy density 435 mJ/cm2. A standard self-aligned NMOS process was used to fabricate devices with a gate width W varying from 10 to 100 μm and gate length varying from 4 to 20 μm. A SiO2 layer of thickness 60 nm deposited by ECR-PECVD at 100 °C was used as a gate insulator.

From a large number of fabricated TFTs, we have selected transistors closely located on the same wafer with channel dimensions W/L = 10 μm / 10 μm and 100 μm / 10 μm, which were used for the design of the proposed buffer. The output and transfer characteristics of these transistors were measured at room temperature using a computer-controlled system including a Keithley 617 electrometer and two Keithley 230 voltage sources.

In the next step, using the ATLAS program of the Silvaco tools [18], the cross-sections of the transistors were described with exactly the same type of materials for each region of the transistors. Then, the measured characteristics of the transistors were imported into the ATLAS program. Optimization of the parameters was performed in order to achieve good correlation between measured characteristics and characteristics reproduced with ATLAS. The threshold voltage (parameter VTO in HSpice) of the transistor with size W/L = 10 μm / 10 μm was found to be 1.29 V and its high field mobility (parameter MUO in HSpice) was found to be 90.7 cm^2 / Vs. For the transistor with size W/L = 100 μm / 10 μm, the same parameters were found to be 1.31 V and 85.6 cm^2 / Vs, respectively. The extracted values of the RPI model parameters were inserted into HSpice and the characteristics of each transistor were reproduced again. Figs. 3 (a) and (b) show the measured and simulated output characteristics of two TFTs and Fig.4 the corresponding transfer characteristic. Both ATLAS and HSpice simulated input and output characteristics are represented in Figs. 3 and 4, respectively. The measured and simulated characteristics are in good agreement, with a deviation between them by about 2%. The good correlation between measured and simulated characteristics confirms that HSpice simulations are realistic and, thus, the functionality of the proposed analogue buffer can be ensured.

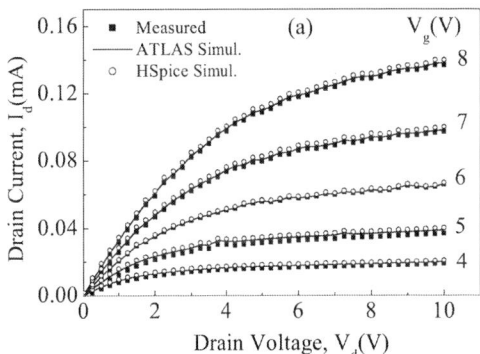

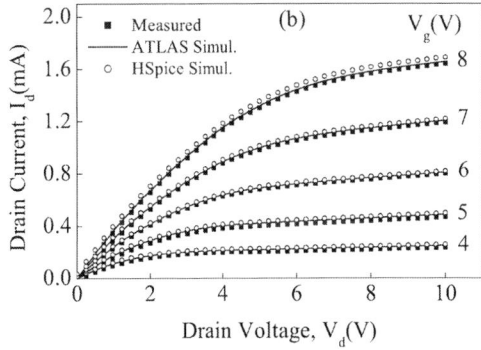

Figure 3. Measured and simulated output characteristics of polysilicon TFTs with channel dimensions: (a) W/L = 10 μm/10 μm, (b) W/L = 100 μm / 10 μm

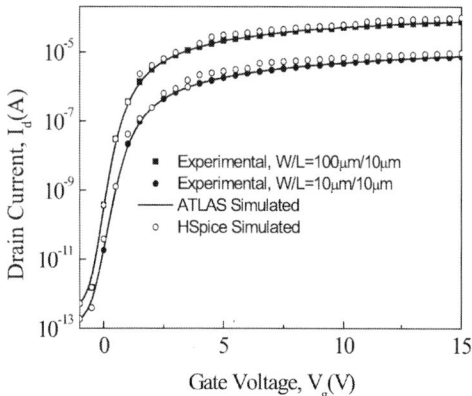

Figure 4. Measured and simulated transfer characteristics of polysilicon TFTs at $V_d = 0.1$ V.

B. Circuit Parameters

The channel length of all transistors was equal to 10 μm. The channel width of the driving TFT (M4) is equal to 100 μm in order to be able to source the necessary current to load the capacitance. Furthermore, transistor M1 has the same channel width as M4 because the two transistors have to share the same electrical characteristics. All the other transistors (M2, M3, M5) act as switches, therefore, their channel length is equal to 10 μm. The supply voltage (Vdd) is equal to 15 V while the data voltage range starts from 2 V up to 10 V. The output capacitance is 20 pF which corresponds to the data line of a 4-inch QVGA display. The control signals have pulse waveform with 15 V magnitudes and 5 μs pulse duration.

V. SIMULATION RESULTS

Fig. 5 (a) shows the simulation results (transient analysis) of the proposed buffer, with the input voltage varying from 2 to 10 V. From fig. 5 (a), it is clear that the functionality of the buffer is verified since the output voltage follows the input levels. Figure 5 (b) shows the offset voltage ($V_{out} - V_{data}$) versus the input data voltage. In the worst case, the error between the input data voltage and the output voltage is 46 mV, when the input data voltage is equal to 5 V. This means that for an 8 V input data voltage range and 46 mV error voltage, the output voltage delivered to the pixels can have 174 voltage levels resulting in 174 different color-scales (more than 7-bit color scale). Therefore, for a QVGA using this output data buffer and having three sub-pixels per pixel, the total color palette is 5.5M colors.

The inaccuracy between the input and output voltage is caused because of the non-ideally value of the node A voltage. The theoretical value of the node A voltage is $V_{data} + V_{TH}$, however, the simulation results have shown that there is a

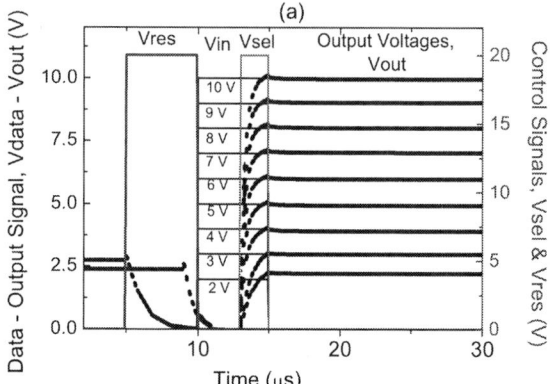

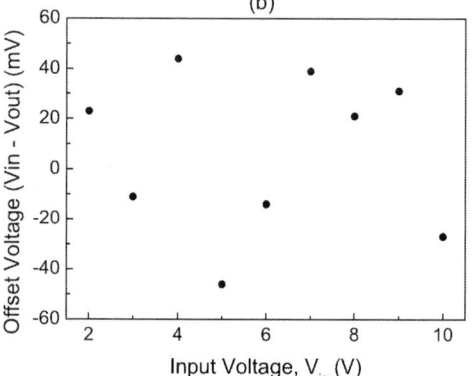

Figure 5. (a) Simulation results of the proposed analog buffer when the input levels vary fom 2 up to 10 V. (b) Offset voltage ($V_{out} - V_{in}$) versus the input voltage V_{in}.

small difference at this value because the drain-to-source voltage of diode-connected M1 is not exactly V_{TH}.

The response time of the buffer, i.e. the time required for the output voltage to reach the 90 % of its nominal value, is 3 μs and the time required for the two phases of operation and the charge of the capacitance is less than 10 μs. A benefit of the proposed buffer is its fast response due to the absence of a storage capacitance for the threshold voltage variation cancellation.

Fig. 6 shows the transient analysis of the proposed buffer, for the two different input voltage levels of 4 V and 8 V and for three different values of the transistors' threshold voltage. As it can be seen, the functionality of the proposed buffer is not affected from the different threshold voltages.

During the theoretical approach and the simulations, tit was assumed that the threshold voltage of all the buffer's transistors is the same. However, due to the poly-Si TFT nature, this assumption is not always valid. This is the reason for performing Monte Carlo simulations.

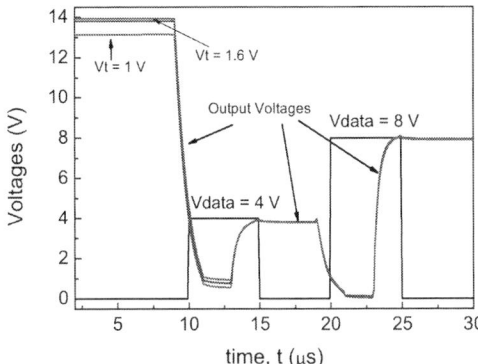

Figure 6. Transient response of the proposed buffer for two different data voltage (V_{data} = 4 and 8 V) and for three different threshold voltage values (V_{TH} = 1, 1.3 and 1.6 V)

The Monte Carlo analysis was used in order to determine the effect of the threshold variation of the transistors M1 and M4, on the performance of the buffer, for input voltage level of 8.5 V. For Monte Carlo analysis, we assumed Gaussian distribution of the threshold voltage of the transistors M1 and M4, with standard deviation of the threshold voltages ± 20% from their nominal values of 1.29 V and 1.31 V, respectively. Such a distribution for the threshold voltage was found from the statistical analysis of the device parameters, obtained from measurements in a large number of TFTs of the same technology [22]. The variation of the threshold voltages of these two transistors was considered, because identical value for the threshold voltages is crucial for the functionality of the proposed buffer. This is the worst case scenario, since the threshold voltage of both transistors is varied up to 40% ($V_{TH,M1} \pm 20\%$ and $V_{TH,M4} \pm 20\%$).

Fig. 7 shows the output voltage error range of 1000 Monte Carlo simulations. As it can be seen from fig 7, the output voltage error is less than ±65 mV from its nominal value indicating an error of 7.6 %. This result shows that even for the worst case scenario, the performance of the proposed buffer remains stable.

The simulations have shown the good performance of the proposed analog buffer. Compared to previous similar works [12] - [14] the main advantage of the proposed analog buffer is that operation phase for sampling and storing the threshold voltage of the drive TFT is not required. In this way, no capacitance for storing the threshold voltage is necessary, resulting in less silicon real estate, smaller response time and higher operation frequency. Furthermore, less control signals [12] - [14] and no bias voltages [11], [13] are needed. Therefore, no changes have to be made to the architecture of the conventional row and column drivers of the AM FPD, since the control signals can be easily produced from the conventional row driver. Finally, the maximum offset voltage of the proposed buffer is $V_{offset,MAX}$ = 46 mV which is much smaller than the offset voltage of about 0.4 V for the same

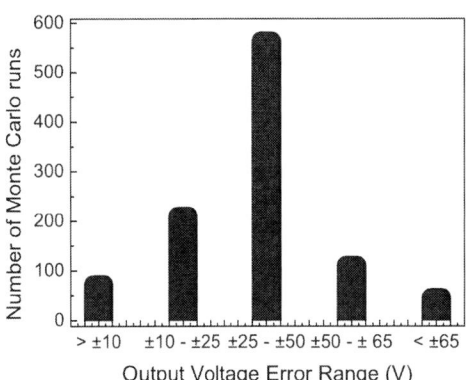

Figure 7. Number of Monte Carlo simulations of the output voltage error range of total 1000 Monte Carlo simulations for the threshold voltage variations.

data voltage: V_{DATA} = 10 V, presented in [13] which has a smaller supply voltage (Vdd = 10 V).

VI. CONCLUSION

A new analogue buffer for AMLCDs applications was proposed in this paper. The proposed buffer consists of five n-type poly-Si TFTs and two control signals. The functionality of the buffer was verified from simulations with HSpice. In order to obtain realistic simulations, parameter's extraction of fabricated poly-Si TFTs, with Silvaco's tools, was made. The simulations have shown that the buffer has high immunity to the threshold voltage variations of its transistors.

REFERENCES

[1] W. G. Hawkins, "Polycrystalline-silicon device technology for large area electronics," IEEE Trans. Electron Devices, vol. 33, pp. 447-450, 1986.

[2] H. Hayashi, M. Negishi, and T. Matsushita, "A thermal printer head with CMOS thin-film transistors and heating elements integrated on a chip," Tech. Dig. IEEE Int. Solid-State Circuits Conf., pp.266-270, 1988.

[3] N. D. Young, G. Harkin, R. M. Bunn, C. McCulloch, and I. D. French, "The fabrication and characterization of EPROM array on glass using low-temperature poly-Si TFT process," IEEE Trans. Electron Devices, vol. 43, pp.1930-1936, 1996.

[4] S.W. Lee, H. J. Han, and J. W. Lee, "High performance low-power integrated 8-bit digital data for poly-Si TFTLCDs," Society for Information Display'99 Digest, pp. 76-79, San Jose, California.

[5] M. D. Jacunski, M. S. Shur, and M. Hack, "Threshold voltage, field effect mobility and gate-to-channel capacitance in polysilicon TFT's," IEEE Trans. on Electron Devices, vol. 43, pp. 1433-1440, 1996.

[6] S. Jagar, C. F. Cheng, S. Zhang, H. Wang, and M. C. Poon,, "A SPICE model for thin-film transistors fabricated on grain-enhanced polysilicon film," IEEE Trans. on Electron Devices, vol. 50, pp. 1103-1108, 2003.

[7] C.T. Angelis, C. A. Dimitriaids, M. Miyasaka, F. V. Farmakis, G. Kamarinos, J. Brini and J. Stoemenos, "Effect of excimer laser annealing on the structural and electrical properties of polycrystalline

silicon thin-film transistors", Journal of Applied Physics, vol. 86, No. 8, p.p. 4600-4606, October 1999.

[8] M. Miyasaka and J. Stoemenos, "Excimer laser annealing of amorphous and solid-phase- crystallized silicon films", Journal of Applied Physics, vol. 86, No. 0, p.p. 5556-5565, November 1999.

[9] S. D. Zhang, K. O. Sin, J .N. Li, and P. T. K. Mok, "Ultra-thin elevated channel poly-Si TFT technology for fully-integrated AMLCD system on glass," IEEE Trans. on Electron Device, vol. 47, pp. 569-575, 2000.

[10] R. Dawson, Z. Shen, D. A. Furst, S. Connor, J. Hsu, M. G. Kane, R. G. Stewart, A. Ipri, C. N. King, P. J. Green, R. T. Flegal, S. Rearson, W. A. Barrow, E. Dickey, K. Ping, C. W. Tang, S. Van Slyke, F. Chen, J. Shi, J. C. Sturm, and M. H. Lu, "Design of an improved pixel for a polysilicon active matrix organic light emitting diode display," in SID Int. Symp. Digest of Technical Papers, vol. 29, pp. 11-14, 1998.

[11] I. Pappas. S. Siskos and C. A. Dimitriadis, "A new analog buffer using low-temperature polysilicon thin-film transistors for Active Matrix Displays", IEEE Transaction on Electron Devices, vol. 54, No. 2, February 2007.

[12] L. Y. Chun and P. K. T. Mok, "Process-independent analog data driver for polysilicon TFT AMLCD," Int. J. Electronics, vol. 91, pp. 199-210, 2004.

[13] Ya-Hsing Tai, C. C. Pai, Bo-Ting Chen, and Huang-Chung Cheng, "A Source-Follower Type Analog Buffer Using Poly-Si TFTs With Large Design Windows", IEEE Electron Devices Letters, vol. 26, No. 11, November 2005.

[14] Hoon-Ju Chung, S.-W. Lee and C.-H. Han, "Poly-Si TFT push pull analogue buffer for integrated data drivers of poly-Si TFT-LCDs", Electronics Letters, vol. 37, No. 17, p.p. 1093- 1095, August 2001

[15] S. H. Jung, W. J. Nam, and M. K. Han, "A new voltage-modulates AMOLED pixel design compensation for threshold voltage variation in poly-Si TFTs," IEEE Electron Device Letters, vol. 25, pp. 690-692, 2004.

[16] Y. Si, L. Lang, Yi Zhao, X. Chen and S. Liu, "Improvement of pixel electrode circuit for Active-Matrix OLED by application of reversed-biased voltage", IEEE Transactions on Circuits and Systems II: Express Briefs, vol. 52, No. 12, December 2005.

[17] Sanjiv Sanbandan and Arokia Nathan, "Stable Organic LED displays using RMS estimation of threshold voltage dispersion", IEEE Transactions on Circuits And Systems II: Express Briefs, vol. 53, No. 9, September 2006.

[18] N. H. Weste and K. Esthraghian, "Principles of CMOS VLSI design", Addison-Wesley publishing company, second edition, 1988.

[19] Synopsys HSPICE version U-2003.03-SPI manual.

[20] E. G. Colgan, et al., "A 10.5-in.-diagonal SXGA active-matrix display," IBM Journal Res. Develop., vol 42, pp 427-444, 1998.

[21] M. Valdinoci, L. Colalongo, G. Baccarani, G. Fortunato, A. Pecora, and I. Policicchio, "Floating body effects in polysilicon Thin-Film Transistors", IEEE Transactions on Electron Devices, vol. 44, No. 12, p.p. 2234-2241, December 1997.

[22] I. Pappas, A. T. Hatzopoulos, D. H. Tassis, N. Arpatzanis, S. Siskos, C. A. Dimitriaidis and G. Kamarinos, "A simple and continuous polycrystalline silicon thin-film transistor model for SPICE implementation", Journal of Applied Physics, vol. 100, 064506, 2006.

MIXDES 2012, 19th International Conference *"Mixed Design of Integrated Circuits and Systems"*, May 24-26, 2012, Warsaw, Poland

Fast-Settling Gain Stage Using Replica Amplification for High Performance Pipeline ADCs

M Khaleghi Kouzehkanan
Islamic Azad University
Khameneh Branch
Khameneh, Iran
m.kouzehkanan@gmail.com

Ali Dadashi
Urmia University
Urmia, West Azerbaijan, Iran
urrmia@gmail.com

Masood Teymouri
Innovation Research Center
Urmia University of Technology
Urmia, Iran

Saeid Masoumi
Islamic Azad University
Tasouj Branch
Tasouj, Iran
s.masoumi.ee@gmail.com

Abstract—**This paper presents a new gain stage based on the Replica gain enhancement method. The proposed gain stage operates 2.35 times faster than a similar size two-stage gain stage in the same precision, power consumption, and the same load capacitor. Proposed structure has been simulated by HSPICE software using level 49 parameters (BSIM3v3) in a typical 0.18µm CMOS technology. HSPICE simulation confirms the theoretical estimated improvements.**

Index Terms—**gain Stage; replica amplification; positive feedback block; linearity**

I. INTRODUCTION

Finite Op-Amp Gain Bandwidth product and output swing are the limitations for precision analog circuits. These limitations are especially serious at lower supply voltages where limited headroom prevents the use of cascode devices to improve gain. Less swing means that more power must be spent reducing noise by using larger capacitors and higher current in the active components. In lower supply voltages using simple single-stage amplifier is an attractive choice because of its high output swing. Also, the simple single pole architecture makes this amplifier the most attractive for high speed, low accuracy applications. The single pole amplifier is also of a great deal of interest because of its inherent stability. But, for high accuracy applications amplifiers with higher DC gain is required. Cascading of individual gain stages gives a high gain amplifier, but each stage introduces a low frequency pole, which produces a negative phase shift and degrades the phase margin. Many phase compensation schemes for multi-stage amplifiers have been reported. All the reported schemes are a variation of the basic Miller compensation scheme for a two-stage amplifier. The dominant pole is pushed to lower frequencies due to Miller effect, resulting in lower bandwidth structures. In this paper replica amplification method is used to implement a high speed gain stage. The proposed gain stage operates 2.35 times faster than a similar size two-stage gain stage in the same precision, power consumption, and the same load capacitor.

This paper is organized as follows: in Sect. 2, proposed gain stage structure is presented; in Sect.3, simulation results for the proposed gain stage are given, also the proposed gain stage is compared with the conventional two-stage gain stage, finally Sect. 4 concludes the paper.

II. PROPOSED DESIGN

A. Replica Amplification

Fig. 1 shows the gain enhancement technique using a replica amplifier RA. This method is analyzed in detail in [1]. For simplicity, we first assume the replica amplifier (RA) is identical to the main amplifier (MA) and has the same feedback network as MA. The input voltage is applied to both MA and RA in parallel. Since RA is used in the same inverting configuration as MA, the output voltage of the replica amplifier Vor is already very close to the ideal output voltage. Since the coupling transconductance amplifier (CA) is connected in parallel with the inputs of the replica amplifier, CA produces a current *ix,* which is the same as the current *ix* produced by RA. This current *ix* from CA is injected into the output resistance (***rom***) of the main amplifier, producing a voltage that is already close to the ideal voltage. Therefore, MA provides only a small amount of error current Δ*ix* to bring the output voltage even closer to the ideal voltage. Since the input voltage needs to change by only a small amount to produce Δ*ix*, the effective open-loop gain is increased. As mentioned in [1], the output voltage of the main amplifier is:

$$Vo_m \approx -\beta \cdot \left(\frac{1}{1 + (\frac{1+\beta}{a_o})^2} \right) . V_{in} \qquad (1)$$

corresponding to an error given by:

$$e \cong \left(\frac{1+\beta}{a_o} \right)^2 \qquad (2)$$

where β is $\frac{Z_f}{Z_i}$ and a_o is $R_{om,r} \cdot g_m$. Considering Eq. 2, we can

see that the error term has been reduced by the factor of $((1 + \beta) / a_0)$, and the effective open loop gain is increased by the same factor. When the closed-loop gain is small and a_o is large, the enhancement factor is potentially very large. In such cases, the actual improvement is limited by the matching between the main and the replica amps. As it is illustrated in [1] and [2], only a negligible increase in settling time of the main amplifier occurs when the replica amplifier utilizes to increase the accuracy of the main amplifier. Measurement and

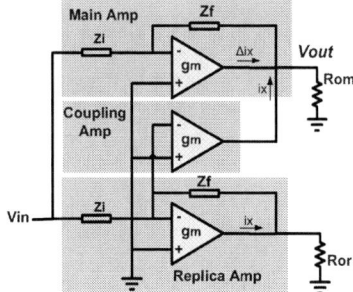

Figure 1. Replica gain enhancement method.

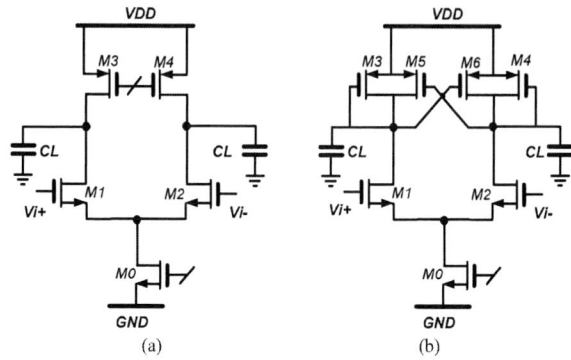

Figure 2. Single-stage amplifier.

simulation results in [1] and [2] show that when the main and the replica amps have the same CL, there is a 30% increase in the settling time, decreasing to a zero increase when CL, decreases to zero. Contrary to other gain enhancement methods such as cascading and gain boosting, replica gain enhancement method improve the accuracy of amplifiers without any speed reduction.

B. Single-stage Amplifier

Fig. 2(a) shows simple single-stage amplifier with active load. As CMOS feature sizes continue to scale, intrinsic device gain of the newer processes has decreased, so dc-gain of the amplifier is intrinsically low. To increase the dc-gain, positive feedback can be used. Fig. 2(b) shows single-stage amplifier using positive feedback. Dc gain of this amplifier is:

$$A_{v0} = -g_{mM_{1,2}} \times R_{out} \qquad (3)$$

where:

$$R_{out} \approx (r_{oM_{1,2}} \| r_{oM_{3,4}} \| r_{oM_{5,6}} \| \frac{1}{g_{mM_{3,4}}}) \| (\frac{-1}{g_{mM_{5,6}}})$$
$$\approx \frac{1}{g_{mM_{3,4}} - g_{mM_{5,6}}} \qquad (4)$$

Dominant pole and Unity Gain Bandwidth (UGBW) of the amplifier are respectively:

$$P = -\frac{1}{R_{out} \times C_{out}} \approx -\frac{1}{R_{out} \times C_L} \qquad (5)$$

$$\omega_u \approx \frac{g_{mM_{1,2}}}{C_L} \qquad (6)$$

Phase margin of this amplifier is 90° because it has only one pole. Another advantage of this structure is the better performance of amplifier in the rails, because transconductance variation of active load devices (M3-M6) is similar in the rails. So the Output resistance remains constant to some extent in the rails. This variation in output resistance causes a gain reduction in the rails in the simple single-stage with simple active load (Fig. 2(a)), because output resistance of active load devices decreases in the rails [3][4].

C. Two-stage Amplifier

The Fig. 3 shows designed two-stage amplifier. DC gain of the both stages of the two-stage amplifier is:

$$A_{v0} = -g_{mi} \times R_{out} \qquad (7)$$

where g_{mi} is the transconductance of the differential pair devices and R_{out} is the output resistance of the stages which is calculated in Eq. 4. DC gain of this amplifier is A_{v0}^2. Frequency response of the amplifier has been analyzed in detail in many references [4] and is:

$$Av_f(s) = \frac{A_{vf0}}{(1+\frac{S}{P_1})(1+\frac{S}{P_2})} \qquad (8)$$

$$P_1 = -\frac{1}{R_{out1} \times C_{out1}} \qquad (9)$$

$$C_{out1} \approx C_{dM_{11,12}} + C_{dM_{13,14}} + C_{dM_{15,16}} + C_{gM_{13,14}}$$
$$+ C_{gM_{15,16}} + C_{gM_{21,22}} + A_0 \times C_{c1,2} \approx A_0 \times C_{c1,2} \qquad (10)$$

where R_{out1} is the output resistance of the first stage and is calculated in Eq. 4, and p_2 is:

$$P_2 = -\frac{1}{R_{out2} \times C_{out2}} \qquad (11)$$

$$C_{out2} = C_{dM_{21,22}} + C_{dM_{23,24}} + C_{dM_{25,26}} + C_{gM_{21,22}}$$
$$+ C_{gM_{23,24}} + C_{gM_{25,26}} + C_L \approx C_L \qquad (12)$$

Considering Miller effect and bypassing R_{out1} by C_{out1} in frequencies upper than P_1 (calculated in [4]), R_{out2} is:

$$R_{out2} \approx \frac{1}{g_{mM_{21,22}}} \qquad (13)$$

Hence P_2 is:

$$P_2 \sim -\frac{g_{mM21,22}}{C_{out2}} \qquad (14)$$

and UGBW of the two-stage amplifier is:

$$\omega_{u2} = P_1 \times Av_0 = P_2 \times \cot(\varphi_m) \qquad (15)$$

Considering Eq. 6 and Eq. 15 ω_{u2} is:

$$\omega_{u2} = \frac{g_{mM21,22}}{C_{out2}} \times \cot(\varphi_m) = \omega_{u1} \times \cot(\varphi_m) \qquad (16)$$

where ω_{u1} is the UGBW of the single-stage amplifier and calculated in Eq. 6. In the optimum phase margin ($\varphi_m = 68°$), ω_{u2} is:

$$\omega_{u2} = 0.42 \times \omega_{u1} \qquad (17)$$

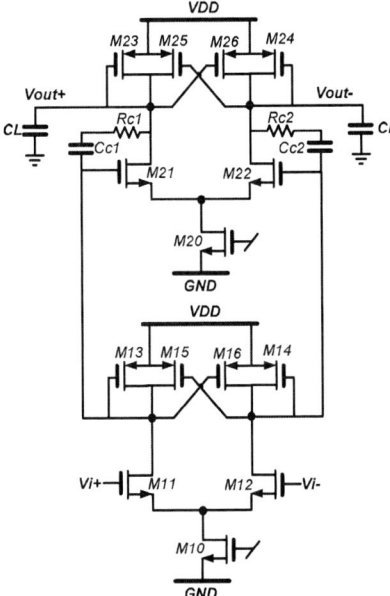

Figure 3.   Two-stage amplifier.

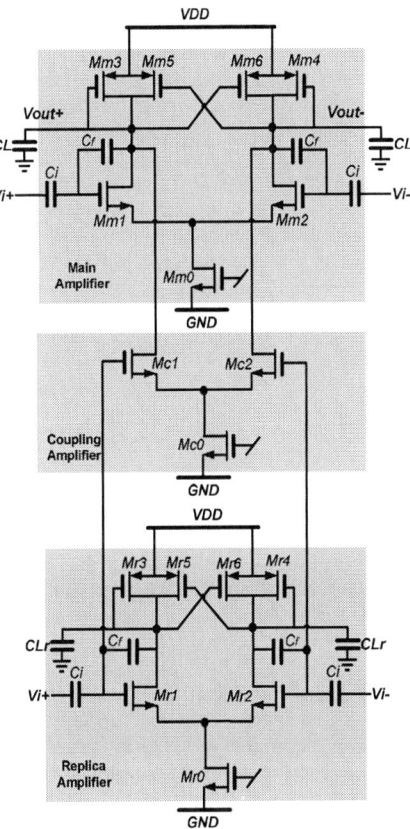

Figure 4.   Designed replica gain stage.

As shown in Eq. 17, for the same CL, UGBW of two-stage amplifier is less than UGBW of single-stage amplifier shown in Fig. 2(b). While the power consumption of the two-stage amplifier is two times of the single-stage one.

D. Designed Replica Gain stage

Fig. 4 shows designed amplifier using replica gain enhancement method. The structure is consists of a main amplifier and replica amplifier. Both main and replica amplifiers implemented by using single-stage amplifier shown in Fig 2(b). Output voltage and error of replica amplifier can be estimated as:

$$V_{Or} = \frac{A_{Vr}.V_{in}}{(1 + \frac{A_{Vr}}{\beta})} \qquad (18)$$

$$e_r = \frac{V_{in}}{(1 + \frac{A_{Vr}}{\beta})} \qquad (19)$$

This error voltage (e_r) appears between the gates of the differential pair devices of replica amplifier (Mr1, Mr2). As shown in fig. 4 the error voltage applied to the differential pair of the coupling amplifier. Assuming the same output resistance for both main and replica amplifiers the coupling differential pair reaches output voltage of the main amplifier to V_{Or}. So using this method, output voltage of main amplifier can reach to V_{Or} only by replica amplifier and a coupling differential pair (Mc1, Mc2), without any effect of main amplifier. The main amplifier only corrects the remained error and decreases it to final error (e). The total error of the main amplifier (e) is equal to:

$$e \cong \left(\frac{1+\beta}{a_o} \right)^2 \qquad (20)$$

In this structure feedback network is formed by feedback capacitors (C_i, C_f) shown in Fig. 4. The value of β is $\frac{C_i}{C_f}$ and a_o is the dc gain of the single-stage amplifiers (main and replica) used in the proposed structure which is calculated in Eq. 3 and Eq. 4. It is clear that the main and replica amplifiers settle independently, to a great extent. As it is shown in Fig. 4 both main and replica amplifiers has simple single pole and has 90° phase margin, so total structure performs such a single pole single-stage amplifier with 90° phase margin. UGBW of both replica and main amplifiers is:

$$\omega_u = \frac{g_{mM_{1,2}}}{C_{out}} \qquad (21)$$

where, C_{out} is the total output capacitance of amplifiers. C_{out} of main amplifier consists of CL and output capacitance of the main amplifier and loading effect of feedback capacitors in the output nodes. As it is clear, CL is connected to the main amplifier and has not any effect on the speed of replica amplifier. Therefore, C_{out} of replica amplifier consist of the output capacitance of the replica amplifier and loading effect of feedback capacitors in the output nodes. So C_{out} of replica amplifier is very smaller than C_{out} of main amplifier, and in the same power consumption replica amplifier operates faster than main amplifier. As it is mentioned in [1] and [2], when the main and the replica amplifiers have the same CL, there is a 30% increase in the settling time, decreasing to a zero increase when CL, decreases to zero. Hence UGBW of designed replica amplifier is equal to UGBW of a single-stage amplifier for the same CL, to some extent. Therefore, considering Eq. 17, UGBW of designed replica amplifier is 2.35 times greater than UGBW of a two-stage amplifier for the same CL and same power consumption. So in comparison with two-stage amplifier in the same power consumption and CL, designed replica amplifier achieves same accuracy while it operates 2.35 times faster than the two-stage amplifier.

III. SIMULATION RESULTS

In this section, simulation results of the designed circuits are presented and designed replica gain stage is compared with designed two-stage gain stage. All circuits have been designed in a typical 0.18μm CMOS process (Vdd=1.8V) and simulated by HSPICE software using *level* 49 parameters (BSIM3v3). Fig. 5 shows AC response of the single-stage gain stage with positive feedback (Fig. 2(b)) in comparison with a simple single-stage (Fig. 2(a)) and a two-stage gain stage (Fig. 3). Single-stage amplifiers (Fig. 2) have approximately the same UGBW with 90° phase margin. As demonstrated in Fig. 5, the single-stage gain stages achieve unity gain bandwidth about 2.35 times greater than UGBW of the two-stage gain stage. Closed loop configuration shown in Fig. 6 is used to study the settling time and accuracy of the single-stage, and two-stage gain stage. Closed loop gain of the gain stages is about 6 dB. Fig. 7 shows the step response of the gain stages. Fig. 5 and Fig. 7 demonstrate that the settling time of the designed replica gain stage is lower than the two-stage gain stage in the same accuracy and even in the lower power consumption, and is approximately equal to the settling time of the single-stage gain stages. Finally, the characteristics of the designed gain stages are summarized in Table 1 for comparison.

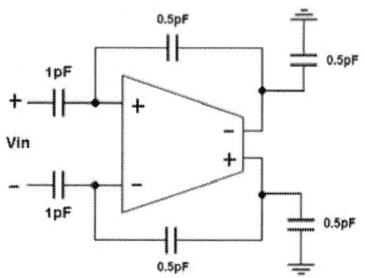

Figure 6. Closed loop configuration

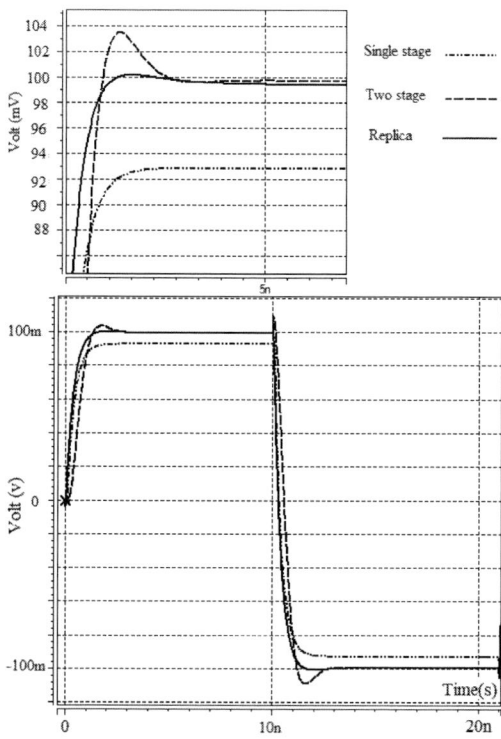

Figure 7. Step response comparison

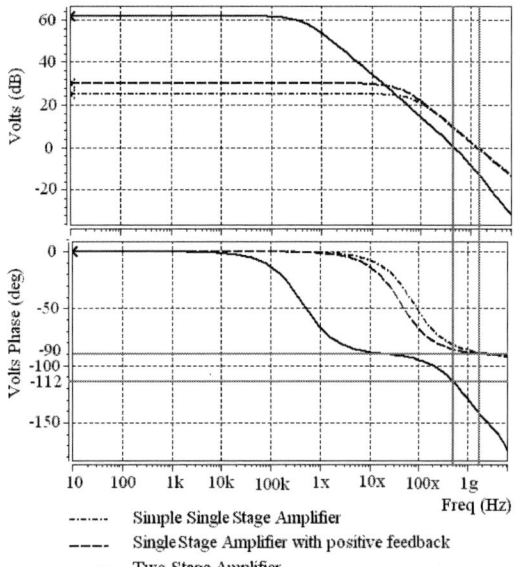

Figure 5. Frequency response comparison.

TABLE I.   SPECIFICATIONS OF THE DESIGNED GAIN STAGES

Parameters	Single Stage	Two Stage	Replica
Open Loop DC-gain (dB)	30	60	60
Gain Error (Vop-p = 1V) (%)	1	0.1	0.1
Phase Margin (°)	90	68	90
Power Consumption (mW)	2.5	5	5
UGBW (CL=1 PF) (GHz)	1.2	0.89	1.2

IV. CONCLUSION

A new gain stage based on the Replica gain enhancement method is presented. The proposed gain stage operates 2.35 times faster than a similar size two-stage gain stage in the same precision, power consumption, and the same load capacitor. HSPICE simulation confirms the theoretical estimated improvements.

REFERENCES

[1] P. C. Yu and H. Lee, "A high-swing 2-volt CMOS operational amplifier with replica-amp gain enhancement," IEEE J. Solid-state Circuits, vol. 28, pp. 1265-1272, Dec 1993.

[2] Paul C. Yu, Hae-Seung Lee "Settling Time Analysis of a Replica-Amp Gain Enhanced Operational Amplifier" IEEE Transaction on Circuits and Systems-11: Analog and Digital Signal Processing, vol. 42, NO. 3, March 1995

[3] R. Gregorian, Introduction to CMOS Op-Amps and Comparators. New York: Wiley, 1999.

[4] B. Razavi, Design of Analog CMOS Integrated Circuits, McGraw-Hill, 2001.

MIXDES 2012, 19th International Conference *"Mixed Design of Integrated Circuits and Systems"*, May 24-26, 2012, Warsaw, Poland

FPGA Implementation of Chaotic Pseudo-Random Bit Generators

Pawel Dabal
Faculty of Electronics
Military University of Technology
Warsaw, Poland
pdabal@wat.edu.pl

Ryszard Pelka
Faculty of Electronics
Military University of Technology
Warsaw, Poland
rpelka@wat.edu.pl

Abstract—**We present recent results of our studies on the FPGA implementation of pseudo-random bit generators (PRBGs) based on a chaotic behavior of nonlinear systems. A number of different PRBG architectures have been considered, including logistic mapping, Hénon mapping, and frequency dependent negative resistor (FDNR). All versions of PRBGs have been implemented in five FPGA families of devices from *Xilinx* (Spartan 3 and 6, Virtex 4, 5, and 6). We present detailed comparison of FPGA resources required for PRBG implementation and evaluation of maximum operating frequencies. The pseudo-random bit generators presented in this paper can be used for key generation in stream ciphers in secure, real-time transmission of digital signals.**

Index Terms—**pseudo-random bit generator; chaotic systems; logistic map, Hénon map**

I. INTRODUCTION

Modern communication systems (including mobile systems) require the use of advanced methods of information protection against unauthorized access. Therefore, one of the essential problems of modern cryptography is the generation of keys having relevant statistical properties. In recent years, the cryptographers pay an increasing attention to digital systems based on chaos theory. The use of chaotic signals to carry information was first proposed in 1993 by Hayes et al. [1] and, since then, chaotic systems have been given much attention and become an important topic in both nonlinear science and engineering.

Nonlinear chaotic systems are described by many authors, e.g. in [2, 3, 4], and have some specific properties, such as a very high sensitivity to relatively small changes of the parameters, non-periodicity, unpredictability, and ability to reciprocal synchronization [5]. The deterministic nature of the chaotic number series can be used for generation of keys for digital ciphering algorithms. However, in practice these solutions have also some important drawbacks and limitations, because the chaotic nature of generated series is non-ideal, due to the limited precision of arithmetic operations and quantization. As a result, instead of random number sequences we get pseudo-random or periodic series.

Despite these difficulties, an idea of using a nonlinear

This work has been supported by the Military University of Technology, Warsaw, Poland, as a part of the project PBS 936.

chaotic dynamic system for design of cryptographic secure pseudo-random number or bit generator (PRNG or PRBG) seems to be interesting from a practical reasons. For this purpose we should formulate two basic requirements for a cryptographic secure PRNG (PRBG): (1) backward and forward prediction cannot be possible, and (2) the system should be resistant to the cryptographic attacks. Of course, the PRNGs are commonly used in many other applications, e.g. in numerical analysis, integrated circuit testing, computer games and scientific simulations. Some examples of the chaotic-based PRNGs have been described in [6, 7, 8]. The fact, that the orbit of chaotic system has the property of being irregular, aperiodic, unpredictable and having sensitive dependence on initial conditions is useful in pseudo-random number generation to generate a binary key stream that is used for data encryption. In this paper we present implementation of three different PRBG architectures based on: (1) logistic mapping, (2) Hénon mapping, and (3) frequency dependent negative resistor (FDNR). These architectures has been tested using 16-, 32-, and 64-bit precision of arithmetic and implemented in popular FPGA devices from *Xilinx*.

The paper is organized as follows. In sec. II we present three architectures of generators and discuss some problems concerning their operation principle. Then, in sec. III we describe some implementation issues related to the *Xilinx* FPGA platform. Experimental results of implementation and throughput measurements are given in sec. IV. Finally, sec. V contains a brief summary and conclusions.

II. CHAOS AS A SOURCE OF RANDOMNESS

Deterministic chaos is a property of equations or systems of equations, involving the high sensitivity of solutions for arbitrarily small changes of parameters. This usually applies to nonlinear differential and difference equations that describe dynamic systems. We used three different types of equations (mappings): logistic [2], Hénon [3], and oscillator whit FDNR [9].

A. Logistic mapping

One of the earliest known and most studied functions having chaotic nature is logistic mapping, the simplest and described by many authors nonlinear system, defined by a polynomial of second degree. Logistic mapping was originally proposed by P.F. Verhulst in 1845, but has become widely known through the work by R. May, who proposed the use of its properties to generate a chaotic sequence of numbers according to the recursive equation [2]:

$$x_{n+1} = rx_n(1 - x_n) \qquad (1)$$

where $0 \le x_n \le 1$, $0 < r \le 1$, $n = 0, 1, 2, \ldots$.

Depending on the value of the parameter r, the dynamics of the related sequence may change dramatically. When r is in the range $3.569945672 < r \le 4$, the numbers generated in successive iterations of the mapping become chaotic, and there is no constant pattern in the derived series. Of course, the generated sequence is also affected by the choice of the initial value x_0.

Fig. 1 shows the basic concept of the generation a pseudo-random binary sequence. As a result of successive iterations of equation (1) we get a sequence of binary words of a specified length, depending on the arithmetic precision. As it has been shown in [10], for 48-bit precision, we can say that the generated sequence can be considered as random, and it can be accepted by an appropriate statistical tests.

B. Hénon mapping

The Hénon mapping is a discrete-time dynamical system. It is one of the most studied examples of dynamical systems that exhibit chaotic behavior. The Hénon map takes a point (x_n, y_n) within the plane and maps it into a new point (x_{n+1}, y_{n+1}), according to the equations:

$$
\begin{aligned}
x_{n+1} &= y_n + 1 - ax_n^2 \\
y_{n+1} &= bx_n
\end{aligned}
\qquad (2)
$$

The map depends on two parameters, a and b, which for the canonical Hénon map have values of $a = 1.4$ and $b = 0.3$. For the canonical values the Hénon map is chaotic. For other values of a and b the map may be chaotic, intermittent, or converge to a periodic orbit. An overview of the type of behavior of the map at different parameter values may be obtained from its orbit diagram. As a dynamical system, the canonical Hénon map is interesting because, unlike the logistic map, its orbits defy a simple description.

C. Oscilator with FDNR

The oscillator is based on the chaos system described in [9], which is defined as,

$$-\ddot{X} = \dddot{X} + B\dot{X} + X \qquad (3)$$

where the nonlinear element is defined as,

$$B(\dot{X}) = \begin{cases} \alpha, & \dot{X} \ge 1 \\ 0, & \dot{X} < 1 \end{cases} \qquad (4)$$

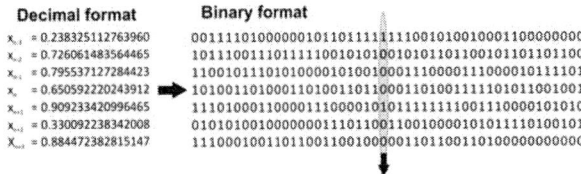

Figure 1. The basic idea of random bit sequence generation.

Chaotic generator based on the differential equations can be digitally implemented by realizing the numerical solution of its differential equations. The system given by equation (3) was solved using Euler technique. Let $Y = \dot{X}$ and $Z = \ddot{X}$, then the numerical solution of equation (3) is evaluated as,

$$
\begin{aligned}
X_{t+h} &= X_t + hY_t \\
Y_{t+h} &= Y_t + hZ_t \\
Z_{t+h} &= Z_t - h(Z_t + Y_t B(Y_t) + X_t)
\end{aligned}
\qquad (5)
$$

where t denotes the time and h is the time step. These equations can be realized using a simple register transfer module (combinational logic unit), where the state variables X, Y and Z are implemented as registers.

III. IMPLEMENTATION

The function generators were designed using MATLAB/Simulink with System Generator tool which offers ready to use library of fixed-point arithmetic blocks, that can be directly implemented into the FPGA device. The VHDL code produced by this tool has been exported to the Xilinx ISE environment as VHDL described IP-core blocks. Using this software we connected designed generator to I/O ports.

To check the performance of implemented generators we selected five of FPGA device families from *Xilinx*. They are mounted on the development boards and differ in terms of the basic design of configurable logic blocks (CLB), manufacturing process technology (from 90nm for Spartan 3 to 40nm for Virtex 6), and the number of available resources: flip-flops, LUTs, BlockRAMs or DSP blocks. We used the following devices: XC3SD1800A (Spartan 3 XtremeDSP™ Starter Platform), XC6SLX45T (Spartan 6 FPGA SP605 Evaluation Kit), XC4VFX12 (Virtex 4 V4FX12 Evaluation Board), XC5VFX30T (Virtex 5 FXT Evaluation Kit) and XC6VLX240T (Virtex 6 FPGA ML605 Evaluation Kit). The last updated version of *Xilinx* ISE Design Suite (13.3) was used to generate programming files for devices.

A. Design of logisitic PRBG

In order to simplify the design, we set the parameter $r = 4$, and the multiplication by factor r is calculated by simple 2-bit shift of an argument to the left. It should be noted that cycle states occur at $x_0 = 0$, 0.25, 0.75 and 1.0. The proper choice of r guarantees the existence of a chaotic orbit that can be described by only one map. Fig. 2 shows block diagram of this generator. The 32- and 64-bit fixed-point representations are used to describe the numbers (all bits are used by the fractional part).

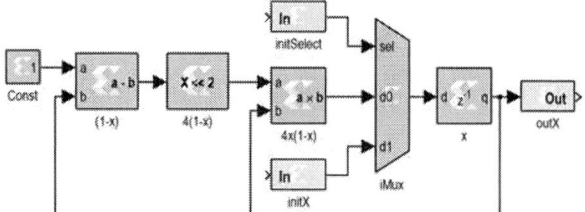

Figure 2. Flowchart of the logistic mapping generator.

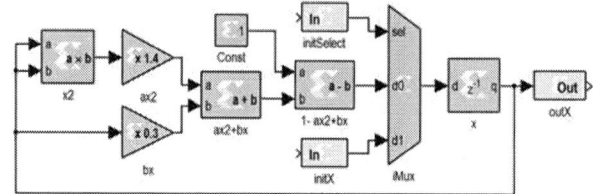

Figure 3. Flowchart of the Hénon mapping generator.

The generated sequence depends on the one parameter only, i.e. an initial value x_0, which should be loaded before the start of the generator.

B. Design of Hénon PRBG

Fig. 3 shows block diagram of the generator. The 32- and 64-bit fixed-point representation are used to describe the numbers. Unlike in logistic generator, we can not replace the multiplication by simple shift operation. The most significant bit is used for the sign, the following three bits for the integer part, and the rest of the bits for the fractional part. The generated sequence depends on the one parameter, i.e. an initial value x_0, which should be loaded before the start of the generator.

C. Design of PRBG based on oscillator with FDNR

Equations in the native form (5), require four multiplication operations. Elimination of these multiplications can significantly reduce the system's complexity. Since the system is chaotic for intervals of h and α, they are selected as $h = 2^{-a}$ and $\alpha = 2^b$, where a and b are positive integers. This transforms multiplications into simple shifts. The outputs of the chaotic generator are within intervals of bounded maximums and minimums. Therefore, fixed-point numbers representation is an excellent solution for the system realization. The 16-, 32-, and 64-bit fixed-point representations are used to describe the numbers. The most significant bit is used for the sign, the following three bits for the integer part, and the rest of the bits for the fractional part. The nonlinear element is simply realized by a comparison block and a multiplexer as shown in Fig. 4. The *sel* signal will be set in a case of $Y < 1$, and passes zeros to a shifter block. Otherwise, the *sel* will be reset, which passes Y.

The generated sequence depends on three parameters, i.e. an initial values x_0, y_0, z_0 which should be loaded before the start of the generator.

D. The interface of chaos generator

We designed a separate IP-core block for aggregation of generated numbers (bits) using VHDL hardware description language. This block communicates with the environment via the Processor Local Bus (PLB), and takes two parameters: the length of the words, which is associated with the precision of the generator, and number of ports. It is also capable to set the initial value for generator.

E. Embedded system for testing PRBG

To retrieve data from generators, and send them to PC in order to calculate the statistical parameters, a dedicated microprocessor system has been designed using Embedded Development Kit (EDK) software. The main part of the system is 32-bit soft-processor MicroBlaze. It is configured to provide minimum resource requirements. The processor is connected to the small internal RAM (8kB) and external DDR RAM (64MB or more). The use of external memory results from the use of TCP/IP protocol, which requires an area exceeding the size of the available memory in the programmable device. TCP/IP needs also a timer to generate time base. In addition to the memory controllers there are also two interfaces: serial RS232 and Ethernet connected to the controller and CPU via interrupt lines. The last element is the internal interface connecting generators. Internal communication is performed via Processor Local Bus (PLB). A separate project has been created for each development board. Block diagram of the test platform is shown in Fig. 5.

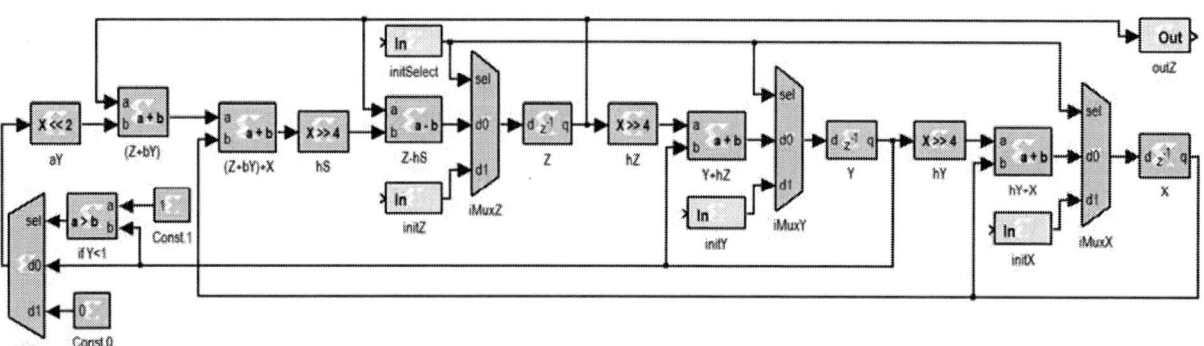

Figure 4. Flowchart of the oscillator with FDNR generator.

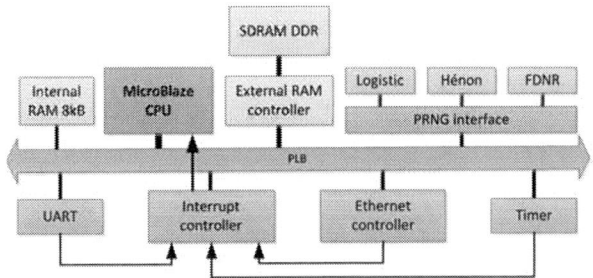

Figure 5. Block diagram of the test platform.

The software running on MicroBlaze processor has been written using the C programming language and TCP/IP library.

IV. EXPERIMENTAL RESULTS

All three versions of generators (logistic, Hénon, and FDNR) have been implemented in each of the five selected FPGA devices. The logistic mapping and Hénon mapping have been tested at 32- and 64-bit precision, while the FDNR generator has been verified at 16-, 32-, and 64-bit arithmetic.

In our experimental tests we focused on determination of two key parameters: (1) amount of FPGA resources required by each individual version of chaotic generator, and (2) maximum operating frequency. These two parameters are essential for the designer who wants to use the generators in a specific application. It should be noted, that in this paper we do not discuss the results of tests related to the statistical properties of generators. These tests are commonly performed using the standard NIST SP 800-22 statistical test suite [11], and can be easily performed by computer simulation of generators (e.g. using MATLAB), even without the need of hardware implementation. Detailed discussion of NIST statistical tests performed for chaotic generators based on the logistic mapping has been published in our earlier work [10].

The amount of FPGA resources required by the generators and test system are listed in Table I. It should be noted, that the number of required resources depends on the assumed precision. The number of used flip-flops depends on the word length and the number of variables in a selected generator. The multiplier unit was built using DSP blocks that can perform 18x25 (Virtex) or 18x18 (Spartan) multiplication. In the case of Virtex 4, the available number of DSP blocks was insufficient to implement all proposed generators in one project. Therefore, the Hénon generator @64b has been tested in a separate project. The additions and subtractions are calculated using LUTs. The most complex and demanding a relatively large number of resources generator is the generator based on Hénon mapping. This follows from the fact that this generator uses three multiplication operation, which can not be replaced by a simple bit shift. This is especially troublesome for 64-bit precision. It should be noted, that doubling the length of the word causes tripling of the number of required LUT blocks. Replacing the four-input LUT block architecture in Spartan 3 and Virtex 4 devices by six-inputs LUT architecture (Virtex 5 and 6 and Spartan 6) causes a significant reduction of the required LUT resources (about 25%).

Table II shows the maximum operating frequency f_{max} of the generators, obtained from the logical synthesis report. It can be seen that newer FPGA technologies are significantly faster and provide higher operating frequencies of generators.

V. CONCLUSION

In this paper we described implementation of three versions of chaotic pseudo-random bit generators in five selected FPGA devices offered by *Xilinx*. A comparative study of required number of FPGA resources and maximum operating frequencies has been presented. The smallest amount of required resources is reported for PRBG based on the logistic mapping while the largest hardware requirements has PRBG using the Hénon mapping. We also examined the impact of arithmetic precision on the number of FPGA resources required

TABLE I. FPGA RESOURCES REQUIRED BY THE SYSTEM AND GENERATORS

Device	xc6vlx240t (Virtex 6)			xc5vfx30t (Virtex 5)			xc4vfx12 (Virtex 4)			xc6slx45t (Spartan 6)			xc3sd1800a (Spartan 3)		
Resource type	*Flip-flop*	*LUT*	*DSP*	*Flip-flop*	*LUT*	*DSP*	*Flip-flop*	*LUT*	*DSP*	*Flip-flop*	*LUT*	*DSP*	*Flip-flop*	*LUT*	*DSP*
Aviable	301440	150720	768	20480	20480	64	10944	10944	32	54576	27288	58	33280	33280	84
Used (total)	6369	9589	38	5718	8427	38	5288[b]	8230	24	4156	7714	40	5054	9634	40
System[a]	5837	6271	0	5188	4893	0	4817	6250	0	3628	4493	0	4523	5406	0
Logistic @ 32b	32	48	4	32	64	4	32	64	4	32	48	4	32	64	4
Logistic @ 64b	64	372	14	64	368	14	64	288	16	64	289	16	64	288	16
Hénon @ 32b	32	539	4	32	556	4	32	717	4	32	538	4	32	717	4
Hénon @ 64b	64	1600	16	64	1636	16	64	2259	16	64	1592	16	64	2259	16
FDNR @ 16b	48	115	0	48	139	0	48	129	0	48	115	0	48	129	0
FDNR @ 32b	96	208	0	96	256	0	96	257	0	96	208	0	96	257	0
FDNR @ 64b	192	427	0	192	512	0	192	514	0	192	431	0	192	514	0

a. test platform without generators

b. without Hénon @64b generator

TABLE II.   MAXIMUM OPERATING FREQUENCIES [MHZ]

Device	Virtex-6	Virtex-5	Virtex-4	Spartan-6	Spartan-3
System	150,0	125,0	100,0	75,0	62,5
Logistic @ 32b	151,1	78,7	67,1	55,6	41,6
Logistic @ 64b	100,6	59,1	45,8	44,5	28,3
Hénon @ 32b	58,2	49,2	41,8	37,0	24,6
Hénon @ 64b	25,7	21,6	18,3	16,2	11,5
FDNR @ 16b	189,5	148,8	117,7	143,3	68,7
FDNR @ 32b	183,0	142,2	107,6	134,7	63,0
FDNR @ 64b	135,3	103,8	101,1	110,8	57,9

by each PRBG version. Furthermore, we analyzed the maximum operating frequencies of each individual PRBG architecture implemented on five FPGA selected devices. The fastest PRBG based on the FDNR architecture can operate at approximately 190 MHz (FDNR @16b, Virtex 6), while the slowest PRBG achieves the maximum operating frequency of 11,5 MHz (Hénon @64b, Spartan 3).

We have shown that the proposed PRBG architectures can be relatively easy implemented in FPGA devices an used in embedded SoC systems. It is also possible to implement dedicated interfaces for different bus standards.

Our further work will be focused on in-depth analysis and optimization of PRBG's statistical properties using not only the standard NIST test suite (as we already reported in [10]) but also some other tools and methods [12, 13].

REFERENCES

[1] S. Hayes, C. Grebogi, and E. Ott, "Communicating with chaos", Phys. Rev. Lett. 70, pp 3031–3034, 1993.

[2] R. M. May, "Simple mathematical models with very complicated dynamics", Nature, vol. 261, pp. 459, June 1976.

[3] M. Hénon, "A two-dimensional mapping with a strange attractor", Communications of Mathematical Physics, vol. 50, no. 1, pp. 69-77, 1976.

[4] O.E. Rössler, "An Equation for Continuous Chaos", Physics Letters A, vol. 57, no. 5, pp. 397-398, July 1976.

[5] L. M. Pecora and T. L. Carroll, "Synchronization in chaotic systems", Phys. Rev. Lett., vol. 64, no. 8, pp. 821-824, 1990.

[6] T. Addabbo, M. Alioto, A. Fort, A. Pasini, S. Rocchi, and V. Vignoli, "A class of maximum-period nonlinear congruential generators derived from the Rényi chaotic map", IEEE Trans. Circuits Syst. I, Regular Papers, vol. 54, no. 4, pp. 816–828, 2007.

[7] C. Y. Li, J. S. Chen, and T. Y. Chang, "A chaos-based pseudo random number generator using timing-based reseeding method", in IEEE Proc. ISCAS, pp. 21–24, 2006.

[8] X. Wang, J. Zhang, and W. Zhang, "Chaotic keystream generator using coupled NDFs with parameter perturbing", in Proc. CANS, vol. 4301, pp. 270–285, 2006.

[9] A. S. Elwakil and M. P. Kennedy, "Chaotic oscillator configuration using a frequency dependent negative resistor," Int. J. Circuit Theory Applicat., vol. 28, pp. 69–76, 2000.

[10] P. Dabal, R. Pelka, „A chaos-based pseudo-random bit generator implemented in FPGA device", [Online]: IEEE Xplore, April 2011.

[11] National Institute of Standards and Technology, "A statistical test suite for random and pseudorandom number generators for cryptographic applications", Special publication 800-22, Revision 1a, August 2010.

[12] G. Marsaglia, "The Marsaglia random number CD-ROM including the DieHard battery of test of randomness", [Online]: http://stat.fsu.edu/pub/diehard/.

[13] P. L'Ecuyer and R. Simard, "TestU01: A C library for empirical testing of random number generators", AMC Trans. Math. Softw., vol. 33, no. 4, Art. No. 22, 2007.

MIXDES 2012, 19th International Conference *"Mixed Design of Integrated Circuits and Systems"*, May 24-26, 2012, Warsaw, Poland

Fully Integratable 4-Phase Charge Pump Architecture for High Voltage Applications

Lufei Shen
Integrated Electronic Systems Lab
TU Darmstadt
Darmstadt, Germany
Lufei.Shen@ies.tu-darmstadt.de

Klaus Hofmann
Integrated Electronic Systems Lab
TU Darmstadt
Darmstadt, Germany
Klaus.Hofmann@ies.tu-darmstadt.de

Abstract—**This paper presents a 4-phase charge pump circuit architecture, which is based on the high voltage CMOS technology with isolated transistor modules from Austriamicrosystems, and can be applied in the applications up to 120V. Due to the introduction of 4-phase clock scheme with dead time techniques, the drawbacks of reverse current at the Pelliconi charge pump are significantly reduced. Correspondingly, the voltage gain and efficiency at each stage are improved. Different configurations of the high voltage sandwich capacitors in this new charge pump architecture are discussed. The feasibility of this proposed charge pump structure is proved by the Post-Layout simulation result.**

Index Terms—**charge pump; Pelliconi; high voltage generation; CMOS; body effect; sandwich capacitor; DC-DC conversion**

I. Introduction

DC-DC step-up conversion for high voltage generation is becoming a very important part of power management systems. Especially, in portable devices, some integrated circuits can merely operate under high voltage DC supply, while only low voltage batteries are available.

The traditional DC-DC Boost converter with relatively simple structures is widely adopted in the form of discrete circuits. One of the main obstacles in achieving the integration of the DC-DC Boost converter is its commonly large-sized inductors, which cannot be easily integrated on the chip. For the current CMOS technologies, the fabricated on-chip inductors show also low inductance and quality factors.

Charge pump circuits are composed of switches and capacitors, which are both easy to be integrated on a single chip. Therefore, it exhibits more promising perspective in the design of fully integrated DC-DC conversion circuits. Thanks to the recent studies on charge pump circuits, a large variety of circuit structures have already been proposed in order to improve some certain performances of charge pumps, for example, power efficiency, stability, single stage voltage gain...etc., but the study on high voltage generation by using charge pumps was still limited by the CMOS technologies in the past. The newly developed high voltage CMOS technologies (in this paper, the technology H35 of Austriamicrosystems) with the maximum operating condition

above 100V have made the research on integrated high voltage charge pump circuits possible.

II. Charge Pump Circuits and High Voltage CMOS Technology

A. Dickson Charge Pump

Among different charge pump structures, the Dickson charge pump [1] is usually considered as the classical solution to design the low voltage charge pump circuits. Since the integrated diodes provided in most CMOS technologies are typically parasitic diodes, which occupy large chip area, and exhibit low saturation current density but high leakage current, diode-connected MOSFETs are used as the replacement of the diodes in Dickson charge pump based circuits.

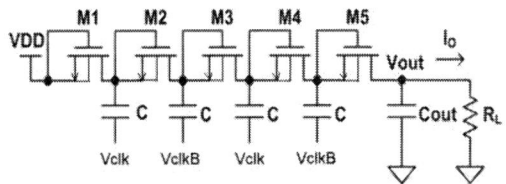

Figure 1. A 4-stage Dickson charge pump using diode-connected NMOSs

$$V_{out} = V_{DD} + N \bullet \left[\left(\frac{C}{C + C_S} \right) \bullet V_{CLK} - V_{th} \right] - V_{th} - \frac{N \bullet I_{out}}{(C + C_S) \bullet f} \quad (1)$$

The main difficulty to generate high output voltage by using Dickson charge pump is the body effect that the threshold voltage of MOSFETs increases with the value of the voltage difference **Vsb** between bulk and source terminals of the transistor. In the example circuit in Fig. 1, the bulk terminal of NMOS can be connected to the source terminal to maintain **Vsb** = 0 and the threshold voltage, however, as a consequence, the body diode of the transistor will conduct during the operation, and it usually cannot stand high current.

If the bulk terminal is connected to some certain constant potential, e.g. the ground, then **Vsb** changes with the source potential **Vs**. In high voltage charge pump circuits, **Vs** and also

265

Vsb of transistors at higher stages will be very high, so that the threshold voltage **Vth** becomes too high to transfer the charges to later stages, and the whole circuit saturated at some certain output voltage level. Even if the body diode of MOSFETs in SOI CMOS processes can be adopted to replace diode-connected MOSFETs, however, its large chip area, high technology cost and low saturation current density make it incompetent in the market.

B. Pelliconi Charge Pump

Pelliconi charge pumps [2] are another type of charge pumps, which use control signals to switch on and off the MOS-switches in the circuit, while the MOS-diodes in Dickson charge pump based circuits are passively turned on and off by the voltage difference between drain and source terminals. For circuit like the Pelliconi charge pump, the most important thing is to design the control signal of every MOS-switch with always correct timing and amplitude.

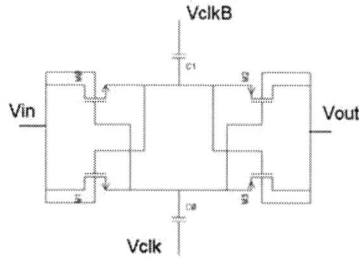

Figure 2.   Single stage Pelliconi charge pump circuit

$$V_{out} = V_{DD} + N \bullet \left[V_{CLK} - \frac{I_{out}}{(C + C_S) \bullet 2f} \right] \qquad (2)$$

The two-phase non-overlapping clocks **Vclk** and **VclkB** operate also mainly as the charge supply, and transfer charge from stage to stage. At the same time, potentials at the both pumping capacitors' terminals, which are connected to the MOS-switches and charged up to certain voltage level, serve as the control signals for the corresponding NMOS and PMOS switches. Because the voltage drop over each switch is not the threshold voltage **Vth,** compared with Dickson charge pump based circuits, but the conduction voltage drop of the MOS-switches, which is normally very small, therefore, the stage voltage gain is considerably improved. Furthermore, since the two pumping capacitors transfer the charge to the load capacitor at each half period of the clock, the output voltage ripple is much lower than that of the Dickson charge pump, in which the charges are only transferred to the load capacitor during each second half clock cycle. Unfortunately, the body effect problem remains unsolved, the non-ideal control signals of both NMOS and PMOS switches result in large reverse current through the both conducting NMOS and PMOS switches at each clock switching time, and consequently reduced voltage gain and efficiency.

C. High Voltage CMOS Technology

Most charge pump architectures are developed and analyzed using the low voltage CMOS technologies. In contrast, the components such as transistors and capacitors in high voltage CMOS technologies due to their complex structures and thicker gate oxide show worse performance than the low voltage ones, e.g. higher channel resistance, larger parasitics, much larger area... etc. Thus, lots of design concepts become unpractical in the high voltage applications.

With the development of high voltage CMOS technologies, components with operation conditions above 100V are now available. One of them is the 0.35 μm high voltage CMOS technology H35 from Austriamicrosystems. Since the high voltage transistors with **Vds_max** =120V in H35 have tied source and bulk terminals (**Vsb** = 0) to avoid latch-up effect and to reduce body effect, their body diodes will conduct in the application of MOSFET switches in charge pumps. Therefore, the low voltage isolated transistors with triple-well structures are more suitable choices in the high voltage charge pump design. Although these isolated transistors have **Vds_max** = **Vgs_max** = 5.5V, they can provide **Vd_psub_max** = **Vs_psub_max** = 120V due to the deep NWELL under the transistors. For charge pump circuits with the stage voltage gain below 5V, these transistors work properly as MOS-switches.

There are also two types of high voltage sandwich capacitors CWPM (**Vterm1_term2_max** = 70V) and CPM (**Vterm1_term2_max** = 120V). Both of them can also be put on the deep NWELL, which ensures a maximal 120V voltage difference between single terminal and p-substrate. These high voltage sandwich capacitors have comparably large parasitics and need large chip area, even so they are the only choice for integrated high voltage capacitors above 50V at the moment for this technology.

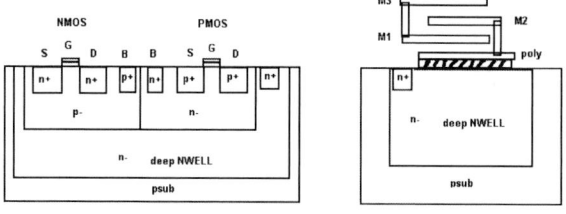

Figure 3.   Cross-sections of low voltage isolated NMOS and PMOS transistor and high voltage sandwich capacitor

III. PROPOSED 4-PHASE CHARGE PUMP FOR HIGH VOLTAGE GENERATION

To utilize the Pelliconi charge pump structure further in the high voltage charge pump design, the dynamic bulk-biasing technique [3] is adopted to reduce the output voltage saturation caused by body effect of MOS-switches and to avoid damages of the device from the large current through the body diode of the MOS-switches. The bulk terminals of MOS-switches are biased to the corresponding lowest or highest potential in the same charge pump stage. For NMOS transistors, this technique is only possible in the triple-well process, because in a CMOS bulk process, NMOS transistors share usually the same P-substrate as bulk, whose potential should normally always be the lowest potential in the whole circuit. A 4-phase charge pump structure is also added to the modified Pelliconi charge pump in order to decrease the reverse current.

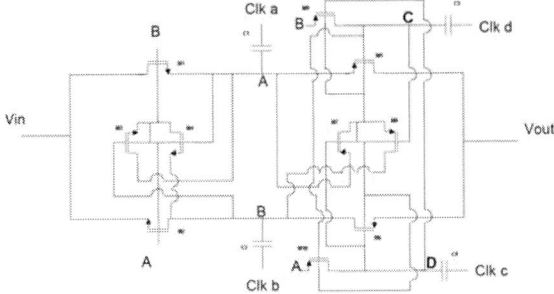

Figure 4. Proposed 4-phase charge pump block diagram

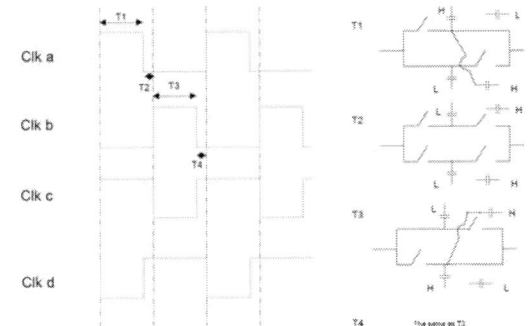

Figure 5. Proposed 4-phase clock scheme

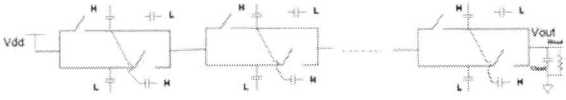

Figure 6. Charge transfer in cascaded proposed 4-phase charge pump

In Fig. 4, the stage block diagram of the proposed 4-phase charge pump is shown. For all the NMOS-Switches in the same stage, the bulk terminals are dynamically biased by the two small extra NMOS transistors M3, M4 to the lower potential between node A and B, which are changeably to be the lowest potential in the actual stage at each half period. The same case is for all the PMOS transistors. The control signals of the both large NMOS switches M1, M2 are the potential of node A and B at the large pumping capacitors C1 and C2, respectively, similar to that in the Pelliconi charge pump. For the large PMOS switches M5, M6, the control signals come from node C and D, whose potentials are charged up to the certain control level by the small control capacitor C3 and C4, respectively.

In Fig.5, one can find the proposed 4-phase clock scheme. Between charge transfer phase (T1,T3), all the switches will be shortly open, so that the reverse current in the direction from **Vout** to **Vin** at the switching time of the control signals can be significantly reduced due to the introduction of the short dead time(T2,T4). The small control capacitors C3, C4 are charged up by the large pumping capacitors C1, C2 through small PMOS-switches M9 and M10 during charge transfer phase (T1, T3). The control signals at node A, B, C, D can then always remain at the same potential level. Fig.6 demonstrates the charge transfer in the proposed 4-phase charge pump.

Other 4-phase charge pump structures have been previously described in literatures. For example, in [4], a 4-phase charge pump structure is proposed for low voltage applications. However, its small control capacitors are connected to the corresponding large pumping capacitors during the dead times, when the terminals of the large pumping capacitors are at lower potentials, but the terminals of the small control capacitors are at higher potentials. In this paper, the proposed 4-phase charge pump has better circuit configuration to ensure that the small control capacitors are charged up correctly, due to the fact that all capacitors are disconnected with each other during the dead times. It is also evident that the low voltage isolated transistors work properly at the condition that the stage voltage gain remains under their maximal **Vds** ratings. Furthermore, this structure can be applied up to 120V in the technology H35, namely, the maximal voltage difference between deep NWELL and p-substrate.

A. Simulation Results Comparision

In the high voltage circuit design, the complete understanding of the technology and high voltage components is essential. The high voltage sandwich capacitor CWPM in H35 is used in the default configuration with the terminal of metal 1-metal 3 connected to the deep NWELL underneath the device. The simulation results of different capacitor terminal connections are thus not identical. The similar situation exists for the sandwich capacitor CPM. To compare the proposed 4-phase charge pump and the Pelliconi charge pump, the assumption here is that the clock buffers are connected with the terminals of the capacitors, which are not connected to the deep NWELL at the same time.

Schematic simulations including clock buffers are under the following conditions: stage number **N** = 36, **Vdd** = **Vclk** =3.7V, **f** =10MHz, **C_pumping** = 10pF, **C_load** = 30pF, **R_load** = 1.05M Ω . For the proposed charge pump additionally: **C_control** = 78fF, dead time **T2** = **T4** = 8ns, charge transfer time T1 = T3 = 42ns. The voltage drop at the dead time $\Delta V_{out-dead\_time} = I_{out} \bullet T_{dead\_time} / C_{load} \approx 0.025\text{mV}$. (This is only the output voltage drop during the dead time, not the overall output voltage ripple.) Shorter dead time ensure a more stable output voltage, however, too short dead time will be a huge challenge for the clock distribution design. Fig.7 shows the improved performance (solid line) comparing the Pelliconi charge pump (dashed line) at the first stage of the 36 stage charge pump. Due to the large pumping capacitors and high stage voltage, the changes of the control voltages at higher stages become slow, thus, the reverse current becomes higher at later stages.

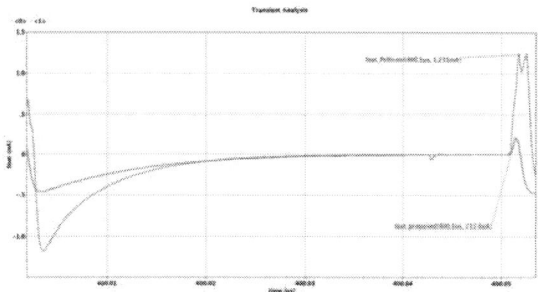

Figure 7.   Reverse current comparison

For the two configurations of the high voltage sandwich capacitors, there are different schematic simulation results. (**config.1**: deep NWELL connected with one terminal of the capacitor and clock buffers. **config.2**: deep NWELL connected with one terminal of the capacitor, but not with clock buffers.)

TABLE I.   SIMULATION RESULTS OF VARIOUS 36 STAGE CHARGE PUMPS

36 Stage Charge Pump Circuits	Vout (V)	Efficiency
Pelliconi Config.1	90	13 %
Pelliconi Config.2	84	26 %
Proposed charge pump Config.1	116	21 %
Proposed charge pump Config.2	105	39 %

The difference between the simulation results of these two configurations can be explained in the following way: For **config.1**, the large deep NWELLs are charged and discharged alternately by the clock buffers, which lowered down the power efficiency. However, the other terminal of the capacitor with only two layers of metals has less parasitic capacitance, which guarantees sufficient stage voltage gain. For **config.2**, the parasitic capacitors of the large deep NWELLs connected directly to the circuits, have significant effect on the stage voltage gain, the effective pumping capacitor value is decreased by these parasitic capacitors.

B. Layout of the Proposed 4-Phase Charge Pump using H35

The layout of high voltage circuits is more complicated than that of the low voltage applications, because the required guard rings to avoid latch-up and leakages, and the wide metal lines to stand high peak current make the components in high voltage technologies larger and difficult to be connected. In Fig 8, the large sandwich capacitors are implemented by drawing capacitor arrays of small unit capacitors (No large area capacitors are allowed). The 4 small control capacitors (4 X 78 fF = 312fF) are used to improve the control signals, due to the large voltage drop of one single small control capacitor (78fF) during dead time at the Post-layout simulation. Again, the Post-layout simulation result proves the explanation for the difference between **config.1** and **config.2** of the sandwich capacitors. The final Post-layout simulation result based on

H35 shows that the cascaded 36 stage proposed charge pump with **config.1** can reach the output voltage of approx. 116V.

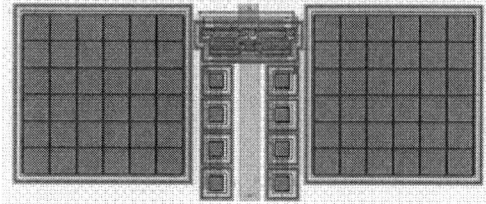

Figure 8.   Layout of one stage proposed 4-phase charge pump using H35

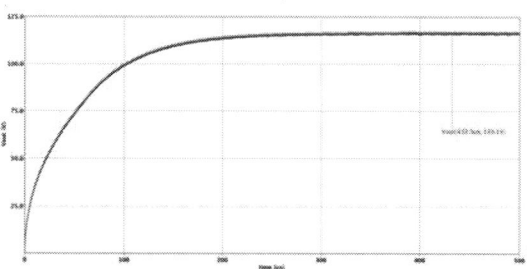

Figure 9.   Post-layout simulation result of the proposed 36 stage charge pump

IV. CONCLUSION

This paper describes a new 4-phase charge pump structure, which can be adopted in high voltage applications. By employing appropriate dead time techniques, isolated MOS-switches and suitable configuration of the high voltage sandwich capacitors, this architecture can reduce the reverse current problem in conventional Pelliconi charge pumps, and its output voltage can reach approx. the maximal operation voltage offered by the high voltage CMOS technologies.

ACKNOWLEDGMENT

The authors thank the **LOEWE** research initiative of the state of Hesse/Germany for supporting this research and development within the **LOEWE** Priority Program Cocoon.

REFERENCES

[1] J. F. Dickson, "On-chip high-voltage generation in NMOS integrated circuits using an improved voltage multiplier techniques," IEEE J. Solid-state Circuits, vol. 11, pp. 374–378, June 1976.

[2] R.Pelliconi, I. David, B. Andrea, P. Marco, and L.R. Pier, "Power Efficient Charge Pump in Deep Submicron Standard CMOS Technology," Solid-State Circuits Conference, 2001. ESSCIRC 2001. Proceedings of the 27th European.

[3] S. Jongshin, C. In-Young, Y. June Park, and H. Shick Min, "A New Charge Pump Without Degradation in Threshold Voltage Due to Body Effect," IEEE Journal of Solid-State Circuits, vol. 35, No. 8, August 2000.

[4] W. Yi-Hsin, T. Hui-Wen and K. Ming-Dou, "Design of Charge Pump Circuit in Low-Voltage CMOS Process with Suppressed Return-Back Leakage Current," Institute of Electronics, National Chiao-Tung University, Hsinchu, Taiwan, Department of Electronic Engineering, I-Shou University, Kaohsiung, Taiwan, 2010.

High-resolution Hold-off Time Control Circuit for Geiger-mode Avalanche Photodiodes

Shijie Deng
Department of Electrical and Electronic Engineering and
Tyndall National Institute
University College Cork, Ireland.
Cork, Ireland
shijie.deng@tyndall.ie

Alan P. Morrison
Department of Electrical and Electronic Engineering and
Tyndall National Institute
University College Cork, Ireland.
Cork, Ireland
a.morrison@ucc.ie

Abstract—**A high-resolution hold-off time control circuit for Geiger-mode avalanche photodiodes (GM-APDs) that enables linear changes to the hold-off time from several nanoseconds to microseconds is presented. The resolution of the hold-off time can be varied from nanoseconds to tens of nanoseconds with a range up to 1.2 µs to cater for a variety of GM-APDs. This circuit allows setting of the optimal 'afterpulse-free' hold-off time for any GM-APD through digital inputs or additional signal processing circuitry. The layout area is 95 µm × 55 µm which makes it suitable for use with APD arrays. The APD is automatically reset following the end of the hold-off period.**

Index Terms—**Geiger-mode avalanche photodiodes, Hold-off time, High-resolution, Afterpulsing**

I. INTRODUCTION

Geiger-mode avalanche photodiodes (GM-APD) are commonly used where high sensitivity low-light intensity detection is required. Typical applications include DNA sequencing, quantum key distribution, LIDAR and medical imaging. In the Geiger-mode, the APD is biased above its breakdown voltage. When a photon is absorbed by the APD an avalanche event is triggered and the event is counted. After every avalanche event, some residual charge is stored in traps in the APD. The release of this stored charge when the GM-APD is active often leads to an avalanche current correlated to a previous avalanche event, but not related to a new photon arrival. This is an unwanted source of noise and is typically termed "afterpulsing". The afterpulsing is reduced by waiting for all the trapped charge to dissipate before resetting the GM-APD. This is generally achieved using an appropriate control circuit that quenches the APD avalanche current by lowering it's bias voltage below the breakdown voltage. The APD is kept in the OFF state for fixed period of time, called the hold-off time, before resetting the device to its original bias voltage to await the next avalanche event. If the hold-off time is less than the mean trap lifetime then afterpulsing will significantly affect the photon counting statistics. If the hold-off time is much greater than the mean trap lifetime then the counting rate will be limited and the validity of the counting statistics will also be affected. The hold-off time must be set according to the

nature and density of the traps present in a particular device to minimize the significance of afterpulsing.

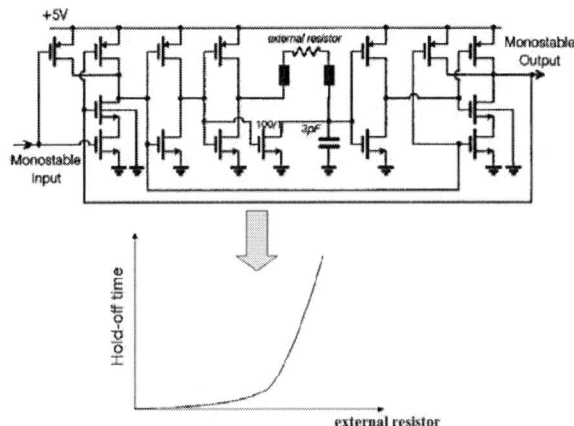

Fig. 1. Schematic and the performance of the monostable

The most popular method used for setting hold-off time is to use monostables [1], [2], [3], [4]. The monostable can offer a wide range of hold-off times (from nanoseconds to microseconds), however adjustment of the hold-off time is non-linear and difficult to control, as illustrated in Fig. 1. In [5], the delay line technique is used to set the hold-off time, as shown in Fig. 2. This circuit uses separate ramp voltage generators that create pre-defined pulse-widths for setting the hold-off time. This technique makes it easier to select an appropriate hold-off time, but this design is costly with increased layout area and complexity. In addition, the layout area required with this approach increases with each discrete hold-off time added. An additional disadvantage to all the traditional techniques for setting the hold-off time is the need for an additional monostable or pulse generator to reset the APD, which adds to the complexity of the control circuit.

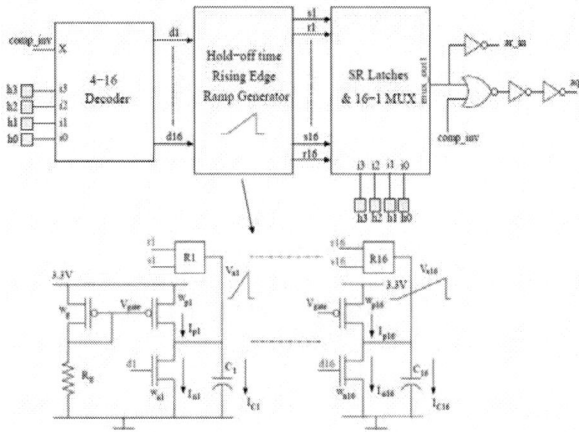

Fig. 2. Diagram of delay line technique from [5]

This paper describes a high-resolution hold-off time control circuit that allows linear setting of the hold-off time from several nanoseconds to microseconds. Its small size, reduced complexity and automatic reset makes it attractive for use with a wide variety of GM-APD architectures. The step-size of the hold-off time can be altered from several nanoseconds to dozens of nanoseconds by varying the period of the external clock. This clock controls a counter through which the hold-off time can be varied linearly. The circuit layout, implemented using L-Foundry 0.15 μm CMOS process, has dimensions of only 95 μm × 55 μm. This relatively small size facilitates its integration with arrays of GM-APDs. This circuit, shown in Fig. 3, will reset the APD automatically at the end of the hold-off period without the requirement for an additional monostable or delay line circuit.

II. CIRCUIT DESCRIPTION

Fig. 3 shows the block diagram of the high-resolution hold-off time control circuit. The non-inverting input of the comparator is connected to the anode of the APD, which is biased at voltage between the avalanche breakdown voltage, V_{break}, to ($V_{break} + V_{dd}$). The comparator is used to sense the avalanche current at the anode of the APD which also has an inverse output, $\overline{compo}$, that is connected to an external bond pad for readout. One PMOS and one NMOS transistor are used as the switches for quenching or resetting the APD. An external clock signal provides the counting clock during the hold-off period and a counter is used to control the hold-off time.

Initially, when there is no avalanche current, $compo$ is low, the external clock is blocked and the 6-bit counter is reset to 0 ("000000"), both PMOS and NMOS transistors are turned off. When an avalanche event happens in the APD, current flows through the load resistor, R_L, and the voltage increases at the anode of APD. The comparator senses the voltage rise and $compo$ goes from low to high. Qp goes low to turn on the PMOS transistor and the anode of the APD is connected to V_{dd} for quenching. Meanwhile, the counter is receiving clocks from the external clock (Clk_in).

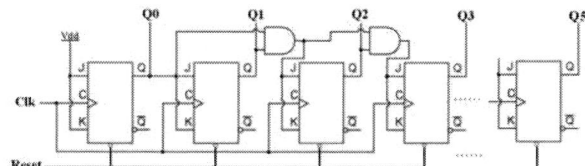

Fig. 4. Schematic of the 6-bit synchronous binary counter used

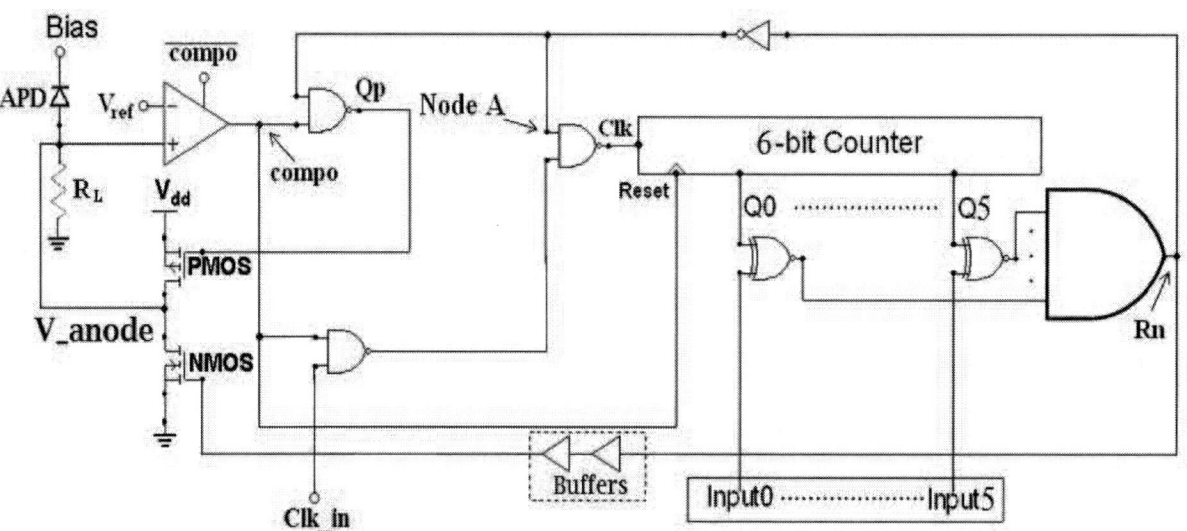

Fig. 3. Block diagram of the high-resolution hold-off time control circuit

270

The counter used here is a 6-bit synchronous binary counter which consists of 6 J-K flip-flops with the clock signal connected to the clock input of every flip-flop and the J and K inputs are tied together, see Fig. 4. The J and K inputs of the first flip-flop are connected to V_{dd}; the J and K inputs of the other flip-flops are connected to the output Q of each front end. When the reset signal *compo* is high, the counter receives clocks from *Clk_in* and counts upwards from 0 ("000000") to 63 ("111111"). Each output of the counter is connected to one input of an XNOR gate. The other input of the XNOR is connected to an external input (controlled by end user). When the output of the counter is equal to the external inputs, all the outputs of the XNOR gates go to logic "1" (high). Then *Rn* goes high which makes *Qp* go high to stop the hold-off process and turn on the NMOS transistor to reset the APD (two buffers are used here to make sure the reset process starts after the hold-off process is finished). At this time, the '*Node A*' goes low to stop the clock to the counter and the counter is stopped. This then makes *Rn* remain high for resetting. When the anode of the APD is reset back to ground, *compo* is low, the *Clk_in* is blocked again and the counter is reset to 0 ("000000"). Now the outputs of the counter do not match the external inputs, *Rn* goes low and the NMOS transistor is turned off to complete the reset process. The APD is then ready to detect the next photon. By setting the external inputs, the counting number can be determined and the hold-off time can be altered. The step resolution is decided by the counting speed, which depends on the period of the external clock *Clk_in*.

III. LAYOUT AND SIMULATIONS

The layout of the proposed circuit was completed using L-Foundry 0.15 μm CMOS process and is illustrated in Fig.5. The overall chip dimension is about 700 μm × 700 μm which mostly occupied by the bond pads. The dimensions of IC core (without bond pads) are 95 μm × 55 μm. All the simulations reported are post-layout simulations.

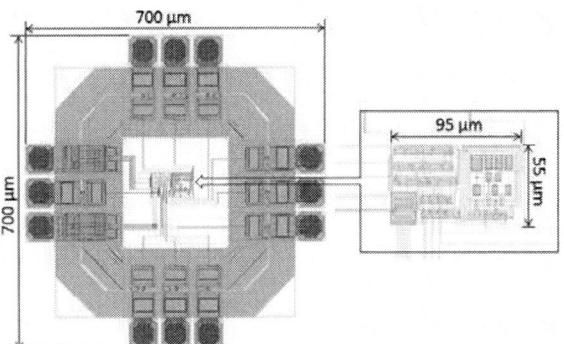

Fig. 5. Layout of the proposed circuit

For circuit simulations, a linear model of the GM-APD is used, as illustrated in Fig. 6 [6]. V_b is a voltage source that represents the breakdown voltage, which is set at 27 V. The bias voltage is set to 30 V. R_d is the internal resistance, which is set to 250 Ω. Cd, is the junction capacitance, which is set to 2 pF. The simulations were run in the Cadence design environment with V_{dd} = 3.3 V.

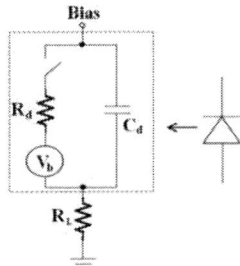

Fig. 6. Simulation model of the GM-APD

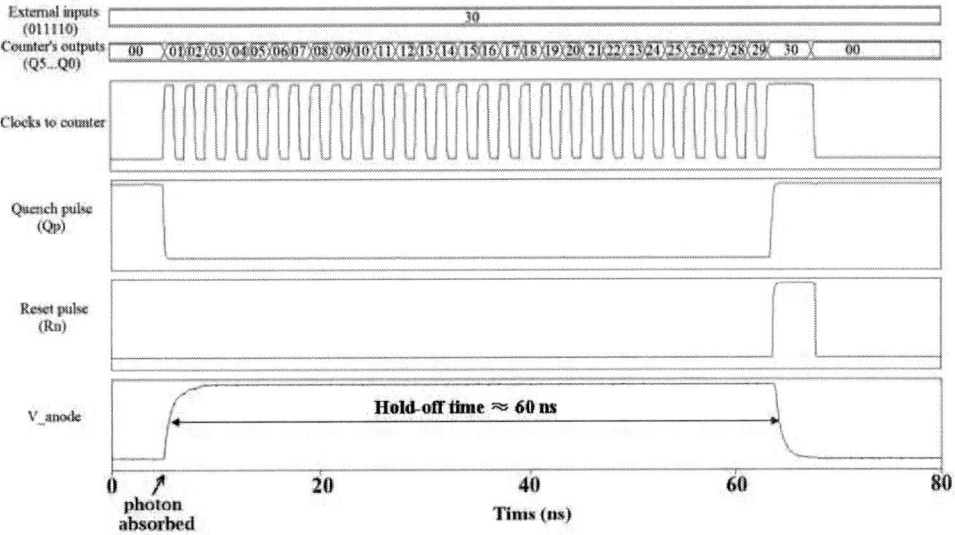

Fig. 7. Example of the circuit operation when the external inputs are set to 30 ("011110")

271

Fig. 7 shows an example of the circuit operation when the external inputs are set to 30 ("011110"). The period of external clocks is set to 2 ns. As can be seen from the figure, at 5 ns when a photon absorbed, the comparator senses the voltage change from the anode of the APD and Qp goes low for quenching. Meanwhile, the external clock is not blocked and providing the clocks to the counter. The 6-bit counter counts upwards from 0 ("000000") at a rate set by the external clocks (here is set to 2 ns). When the outputs of the counter match the external inputs, which in this case are set to 30 ("011110"), Rn goes high which makes Qp go high to stop the hold-off process and turn on the NMOS transistor to reset the APD. At this time, the clocks to the counter are blocked and the counter is stopped thereby making Rn high for resetting the APD. When the anode of the APD is set back to ground, the counter is reset to 0 ("000000") and Rn goes low to stop the reset process. In this way, with the external inputs of 30 ("011110"), the hold-off time is set to around 60 ns.

Fig. 8 shows the simulation results of varying the external input codes versus the resultant hold-off time. It shows when the input code increases from 1 ("000001") to 63 ("111111") the hold-off time linearly increases to more than 120 ns with a step resolution of about 2 ns.

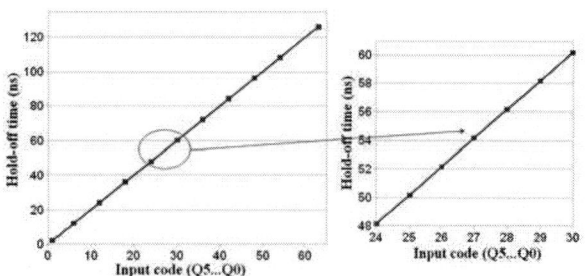

Fig. 8. External input codes versus resultant hold-off time when the step resolution is set to 2 ns

Fig. 9 shows the setting range of the hold-off time for different step resolutions. As can be seen from the figure, when the step resolution is varied from 2 ns to 20 ns, the range of the hold-off can be altered by more than a microsecond.

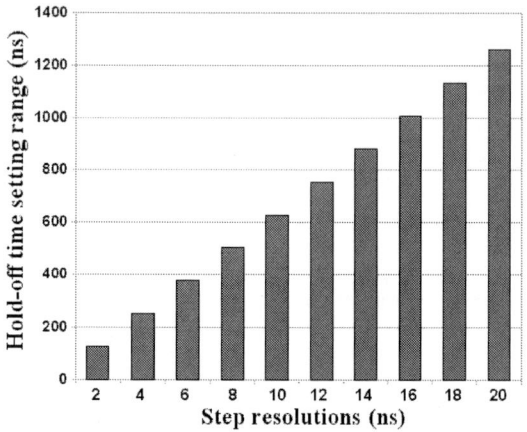

Fig. 9. Setting range of the hold-off time for different step resolutions

IV. CONCLUSION

A high-resolution hold-off time control circuit for Geiger-mode avalanche photodiodes is described in this paper. With this circuit, the hold-off time can be linearly varied from several nanoseconds to microseconds with a user-set resolution. The optimal 'afterpulse-free' hold-off time for any GM-APD can be easily set through the circuit's digital inputs or via an additional signal processing circuit. A layout of this circuit was completed using a conventional CMOS process, resulting in a small layout area that makes it suitable for integration with arrays of GM-APDs. The circuit also incorporates a facility designed to reset the APD automatically at the end of the hold-off time that further simplifies the control for the end-user.

ACKNOWLEDGMENT

This work was supported by Science Foundation Ireland under grant 07/SRC/I1173.

REFERENCES

[1] F. Zappa, M. Ghioni, S. Cova, C. Samori and A.C. Giudice, "An integrated active-quenching circuit for single-photon avalanche diodes," IEEE Transactions on Instrumentation and Measurement, vol.49, no.6, pp.1167-1175, Mar.2000.

[2] F. Zappa, A. Lotito, A.C. Giudice, S. Cova and M. Ghioni, "Monolithic active-quenching and active-reset circuit for single-photon avalanche detectors," IEEE Journal of Solid-State Circuits, vol.38, no.7, pp.1298-1301, Jul.2003.

[3] S. Tisa, F. Guerrieri and F. Zappa, "Variable-load quenching circuit for single-photon avalanche diodes," Optics Express, vol.16, no.3, pp. 2232-2244, Feb.2008.

[4] Rosario Mita and Gaetano Palumbo, "High-Speed and Compact Quenching Circuit for Single-Photon Avalanche Diodes," IEEE Transactions on Instrumentation and Measurement, vol. 57, no. 3, pp.543-547, 2008.

[5] D. Cronin and A. P. Morrison, "Intelligent System for Optimal Hold-Off Time Selection in an Active Quench and Reset IC," IEEE Journal of Selected Topics in Quantum Electronics, vol.13, no.4, pp.911-918, Jul./Aug. 2007.

[6] .S. Cova, M. Ghioni, A. Lacaita, C. Samori, and F. Zappa, "Avalanche photodiodes and quenching circuits for single-photon detection," Applied Optics, vol.35, no.12, pp.1956-1976, 1996.

Integrated Circuit for Wireless Inductive Powering Implemented in 180nm CMOS Process

Mirosław Żołądź, Piotr Kmon, Piotr Otfinowski, Jacek Rauza, Paweł Gryboś

AGH University of Science and Technology,
Krakow, Poland
zoladz@agh.edu.pl, pgrybos@agh.edu.pl

Abstract—**Our report is on the simulation results and the design of the integrated circuit dedicated for inductive powering of a neural recording system on a chip. Four different voltage rectifiers were considered. designed to compare their power efficiency, drop out voltages, and ability to work in a megahertz range. Each of the full bridge rectifiers is equipped with voltage clamps to prevent it from the high coil voltages. Presented circuits are dedicated for carrier signals in the range of a few MHz and load current of 10mA. Performance of the circuits is discussed based on the post-layout simulations. The circuit was designed in 180nm CMOS process and sent to fabrication.**

Index Terms—**wireless energy transmission, ASIC, neurobiology experiments, voltage recitifier, voltage regulator**

I. INTRODUCTION

System on a Chip (SoC) implementation is currently a hot topic in the field of biomedical devices [1, 2]. It is due to the benefits can be obtained mainly when device implantation for chronic recording or stimulation of a neural system is necessary. Wireless inductive powering is often the only alternative, because of a limited life span of a battery or infections caused by the wires necessary for data and power transmission. Many of wireless integrated circuits reported in the literature have similar constructions [3 - 5] (Fig. 1). Namely, external inductive – capacitance tank circuit supplies AC voltage to rectifier inputs that are protected by voltage limiters. Rectified voltage is filtered with integrated or an external capacitor and subsequently stabilized by voltage regulator that is supported by either external voltage reference or on chip band-gap voltage reference. In some solutions clock recovery with data demodulation circuit is also integrated [5].

The circuit presented in this paper is intended to be a part of a SoC for wireless multichannel recording and stimulating neural systems [6, 7]. Such a system is built of an analog front end, A/D converter, digital control circuits, and RF transmitter and requires diverse supply voltages. Amplifiers utilized in analog front end require 1.8 V supply voltage and an additional 0.9 V potential for reference voltage of a front end preamplifiers [7]. Furthermore, the stimulation circuits require 3.3 V supply voltage to guarantee proper voltage swing on the output of current sources. We assumed that the output voltage swing at the output of the stimulation circuit to be at least +/-

1 V for typical stimulating current and electrode reference. We also calculated that the 90 % of the whole power consumed by the SoC will be delivered to the blocks that are supplied with the 1.8 V. The clock recovery system is required for wireless chip control.

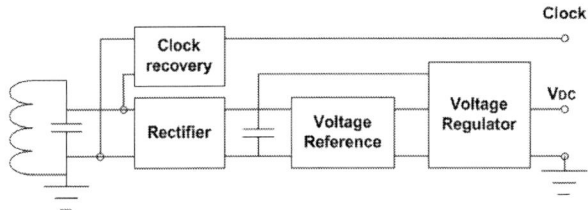

Fig. 1. General block diagram of inductive supply circuit [3-5].

The paper is organized as follows. In Section II four rectifiers with different constructions are presented. In Section III the voltage limiter is described while in Section IV simulation results of these circuits are discussed. Section V contains conclusions.

II. VOLTAGE RECTIFIERS

The simplest way to rectify carrier signal provided from the coil is to use a half-wave rectifier which introduces small drop out voltage between its input and output terminals and is easy to implement in a chip. Nevertheless, the need of implementing large functionality of the biomedical Systems on Chip and strict power limitations involve using modern submicron technologies. However, one has to take into account that in such technologies the oxide and junction breakdown voltages are limited to only few volts. Moreover, because the distance between the transmitting and responding coils may vary, the rectified voltage may change in a large range. This may result in exceeding the voltage limits before mentioned. Therefore, the better solution is to use full wave-rectifiers that have following advantages over half-wave rectifiers fabricated in the same process:

- increased allowed input carrier voltage,
- increased output current efficiency.

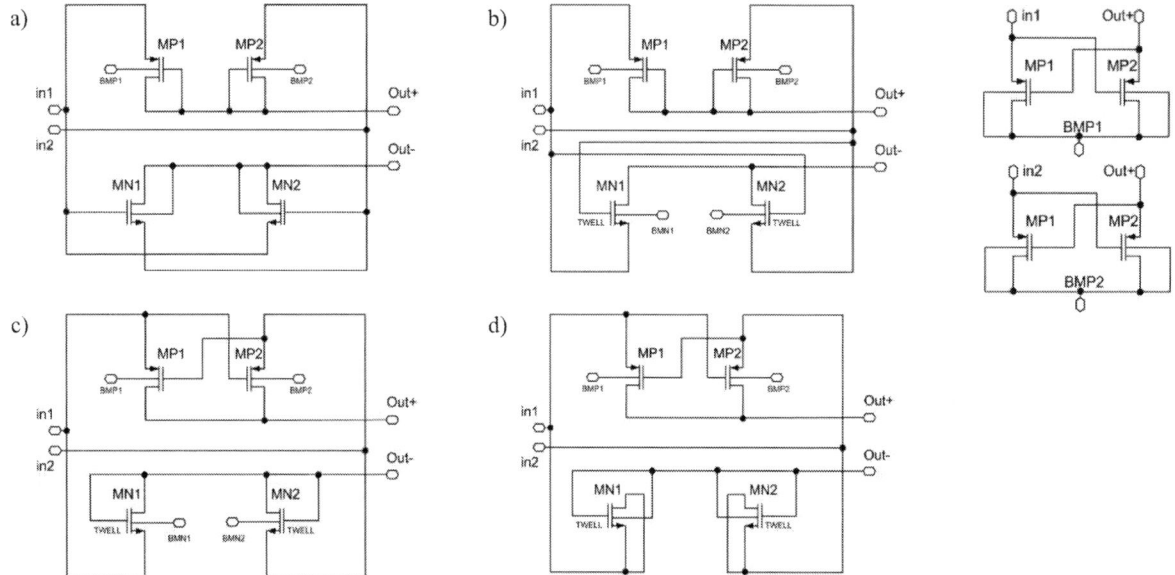

Fig. 2. Schematic idea of the voltage rectifiers: a) Ver1, b) Ver2, c) Ver3, d) Ver4.

In our project we would like to compare four different constructions of the voltage rectifiers. During the design we assumed that all versions of rectifiers shall occupy the same silicon area and shall have similar current efficiency. To compare different rectifiers we took into account its ability to work in a megahertz range, power dissipation through the substrate, dropout voltage, and its resistance to the latch-up effect.

All circuits versions are full-wave rectifiers and are presented in the Fig. 2. The first three approaches (called Ver1, Ver2, and Ver3) are combinations of two blocks. These are MOS transistors working as MOS diodes or as switches. The Ver1 and Ver2 are rectifiers with PMOS transistors working as a MOS diodes and NMOS transistors working as switches. The Ver2 differs from the Ver1 with the NMOS transistors which are placed in an additional well. The Ver3 is based on the PMOS transistors working as a switches while the NMOS transistors work as a diodes. The Ver1, Ver2, and Ver3 are based on the MOS transistors with low threshold voltage to decrease the rectifier dropout voltage. The Ver4 is based on the PMOS switches and a junction bulk substrate diodes of the NMOS transistors placed in an additional well. The 180nm CMOS process that was utilized in this project allow to employ either 1.8V or 3.3V transistors thus the latter one used to increase maximum acceptable coil voltage.

One of the very important aspects in such circuits is a need to limit the substrate leakage currents that can induce latch up effect. Because direction of the receiving coil voltage changes very quickly we employed dynamic control of the bulks nodes to minimize substrate leakage. MOS transistors bulks nodes voltages are therefore controlled by the additional circuits that are presented only in the Fig. 2 a). The similar structures are also used in the all voltage rectifiers versions.

During the design of a rectifier one has to take into account aspects influencing the dropout voltage. It is a very important parameter because it impacts the power dissipated in the rectifier and the maximum acceptable distance between the transmitting and receiving coils. Thus, one has to consider the resistance r_{ON} of the MOS switches and the gate source dropout voltage V_{GS} of the diode connected MOS transistors. These are given respectively:

$$r_{ON} = \frac{1}{\mu C_{OX} \dfrac{W}{L} (V_{GS} - V_{TH})} \qquad (1)$$

$$|V_{GS}| = |V_{TH}| + \sqrt{\frac{2|I_D|}{\mu C_{OX} \dfrac{W}{L}}} \qquad (2)$$

where μ is the carrier mobility, C_{OX} is a oxide capacitance per gate area, W and L are transistor channel dimensions, V_{TH} is a threshold voltage of the transistor, and V_{GS} is a gate source dropout voltage.

Keeping in mind also the parasitic capacitances that are introduced to the overall rectifier circuit and which may seriously decrease the rectifiers efficiency we decided to choose following dimensions of the transistors (see table Tab. I).

TABLE I. Summarized Dimensions of the MOS Transistors employed in the Presented Voltage Rectifiers.

Version of the voltage rectifer	Effective W/L of the PMOS transistors [µm/ µm]	Effective W/L of the NMOS transistors [µm/ µm]
Ver1	4000/0.5	1000/0.5
Ver2	4000/0.5	1000/0.5
Ver3	2000/0.5	3000/0.5
Ver4	3000/0.5	2000/0.5

III. Voltage Limiter

Because the distance between transmitting and receiving coils may vary the voltage across the receiving coil may exceed the maximum breakdown voltage for a given technology. Thus, in order to prevent this effect we employed voltage limiter at each input of the voltage rectifiers. Its schematic idea is presented in the Fig. 3. When input voltage is higher than the sum of the gate source dropout voltages of the diode connected PMOS transistors MD1-MD8 that $W/L = 50$ µm/0.5 µm, current starts to flow what results in generation of the voltage dropout on the resistance R1 built of the high poly resistor with its resistance equal to 325 kΩ. This voltage causes that transistor MP1 ($W/L = 50$ µm/1 µm) starts to conduct current what finally turns on the transistor MN1 ($W/L = 1000$ µm/1 µm) and decreases the voltage across the receiving coil. By setting numbers of the diode-connected transistors and values of the resistances R1, R2, one is able to control the voltage at which the voltage limiter starts to work. During the design of this stage we gave special attention to the uniformity of the crossing voltage from one voltage limiter to the other and on satisfying the current requirements that can flow through the transistor MN1 while the voltage limiter is working.

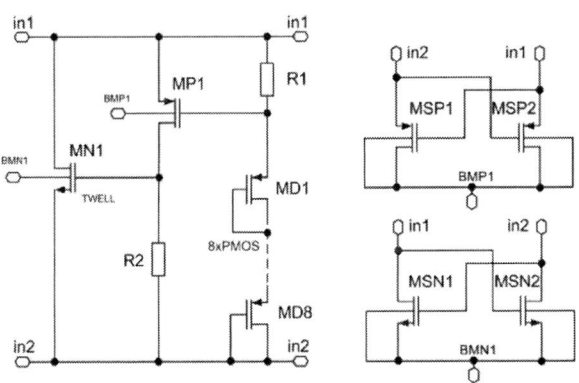

Fig. 3. Schematic idea of the voltage limiter.

IV. Post layout Simulation

The post layout simulations were performed to check performance of the voltage rectifiers that were designed. Thus, in the final simulation results parasitic capacitances are taken into account. We assumed that the loading capacitor of the voltage rectifiers is equal to 1 nF and the current provided by the rectifiers to the load is equal to 10 mA what is large enough taking into account our previous projects of the front-end electronics [6]. Rectified voltages for 4 MHz carrier frequency for each of the presented approaches are depicted in Fig. 4. It can be seen that the biggest RMS voltage rectified is for Ver4 while the smallest one is for Ver3. Very similar are rectified voltages at the outputs of the Ver1 and Ver2 because these rectifiers differ from each other only with the NMOS transistors that for Ver2 are placed in the additional well. Although the rectified voltages are very similar for these versions, benefits of using the additional well are the smallest substrate noise in a future system.

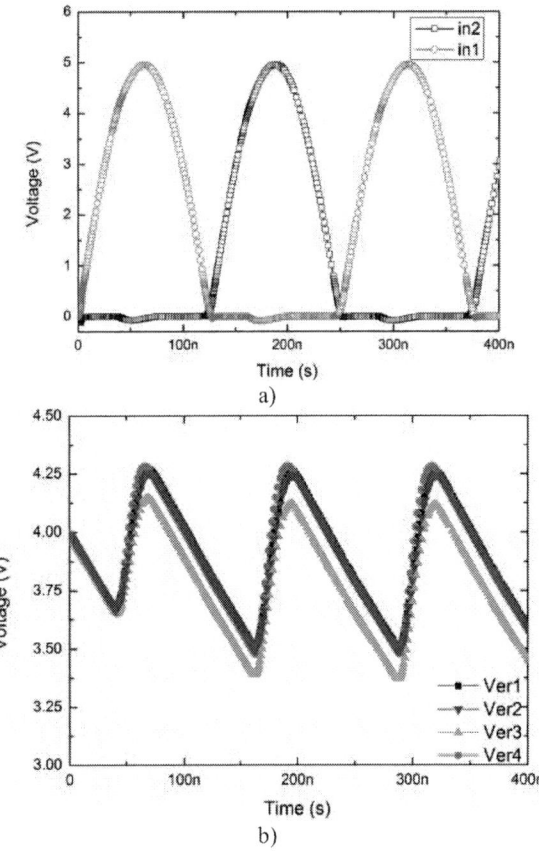

Fig. 4. Post layout simulation results of the rectified voltages for different rectifiers approaches.

We also checked how the carrier frequency influences the rectifiers output RMS voltage. Results of these simulations are presented in the Fig. 5. It can be seen that for a given loading current the optimal carrier frequency may be found. This remark is very important because increasing the carrier frequency does not introduce better efficiency of the voltage rectifier. Moreover, the higher carrier frequencies are more supressed by the patients skin and tissue which results in

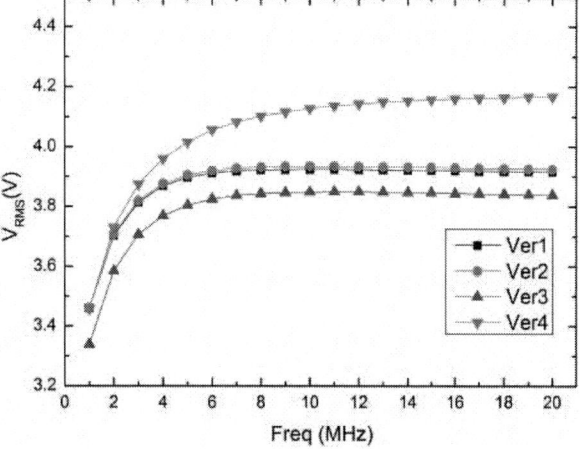

Fig. 5. Carrier frequency influence on rectifiers output voltage RMS.

decreasing the rectifiers efficiency.

The voltage limiter functionality was also checked. To find out how it is susceptible on process variations we performed corner analysis of the voltage at which voltage limiter starts to shunt the receiving coil current. Results of these simulations are presented in the Fig. 6.

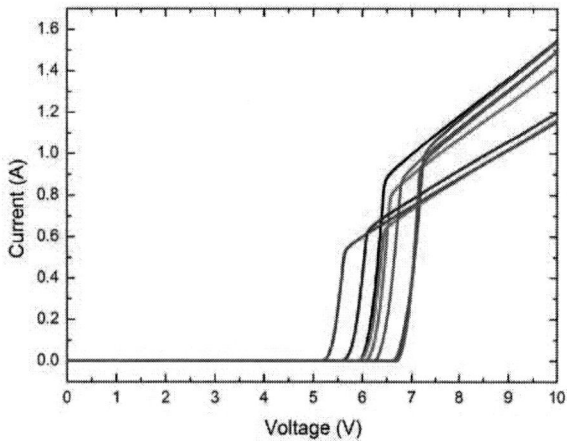

Fig. 6. Corner simulation results of the voltage limiter.

It can be seen that crossover voltage does not exceed 7 V what taking into account corresponding breakdown voltages of the process used is a satisfactory value.

V. CONCLUSIONS

We showed the design and simulation results of the integrated circuit for wireless inductive powering. Described

voltage rectifiers and voltage limiter were designed in the 180nm CMOS process and the chip was sent for fabrication. Layout view of the chip is presented in the Fig. 7. To compare crucial parameters of the voltage rectifiers we decided to design four different approaches. Simulation results of the described blocks were also presented showing good agreement with requirements for such blocks.

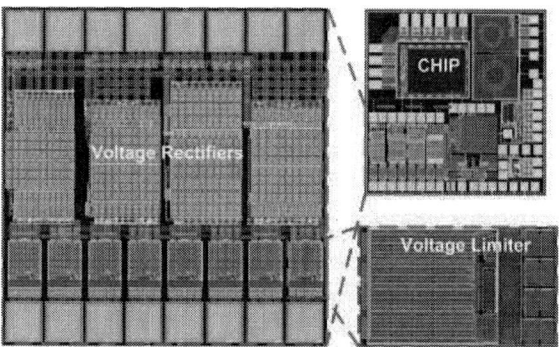

Fig. 7. Layout view with voltage rectifiers and voltage limiter marked.

ACKNOWLEDGEMENTS

This research and development project was supported by Polish Ministry of Science and Higher Education in the years 2010-2011

REFERENCES

[1] M. A. Lebedev and M. A. Nicolelis, "Brain-machine interfaces: Past, present and future," Trends in Neurosciences, vol. 29, no. 9, pp. 536–546, Sep. 2006

[2] A. V. Nurmikko, J. P. Donoghue, L. R. Hochberg, W. R. Patterson, Y. – K. Song; C. W. Bull, D. A. Borton, F. Laiwalla, S. Park, Y. Ming, J. Aceros, "Listening to Brain Microcircuits for Interfacing With External World—Progress in Wireless Implantable Microelectronic Neuroengineering Devices", Proceedings of the IEEE, 2010, Vol. 98, No. 3, pp. 375 – 388

[3] M. Ghovanloo, K. Najafi, "Fully Integrated Wideband High-Current Rectifiers for Inductively Powered Devices", IEEE Journal of Solid-State Circuits, Vol. 39, No. 11, November 2004

[4] M. Mahdi Ahmadi, G. A. Jullien, "A Wireless-Implantable Microsystem for Continuous Blood Glucose Monitoring", IEEE Transactions on Biomedical Circuits and Systems, Vol. 3, No. 3, June 2009

[5] Harrison, R.R.; Kier, R.J.; Chestek, C.A.; Gilja, V.; Nuyujukian, P.; Ryu, S.; Greger, B.; Solzbacher, F.; Shenoy, K.V., "Wireless Neural Recording With Single Low Power Integrated Circuit", IEEE Transactions on Neural Systems and Rehabilitation Engineering, Volume 17, 2009, pp 322 – 329

[6] P. Grybos, P. Kmon, M. Zoladz, R. Szczygiel, M. Kachel, M. Lewandowski, T. Blasiak, "64 Channel Neural Recording Amplifier with Tunable Bandwidth in 180 nm CMOS Technology", Metrol. Meas. Syst., Vol. XVIII (2011), No. 4

[7] M. Zoladz, P. Kmon, P. Grybos, R. Szczygiel, R. Kleczek, P. Otfinowski, "A Bidirectional 64-channel Neurochip for Recording and Stimulation Neural Network Activity", IEEE EMBS Neural Engineering Conference, 2011, Cancun, Mexico

MIXDES 2012, 19th International Conference *"Mixed Design of Integrated Circuits and Systems"*, May 24-26, 2012, Warsaw, Poland

RF Varactor Design Based on Evolutionary Algorithms

Pedro Pereira, Helena Fino, M. Ventim-Neves

CTS, Uninova, Departamento de Engenharia Electrotécnica
Faculdade de Ciências e Tecnologia, FCT, Universidade Nova de Lisboa
2829-516 Caparica, Portugal
{pmrp, hfino}@ieee.org, ventim@uninova.pt

Abstract—This paper introduces an optimization methodology for the design of RF varactors. The characterization of the varactor behaviour is supported by a set of equations based on technological parameters, granting the accuracy of the results, as well as the adaptability of the model to any technology. The varactor design is achieved through the implementation of a Genetic Algorithms (GA) optimization methodology, which is able to deal with continuous and/or discrete variables, making possible to suit both technological and layout constraints. A set of working examples for UMC130 technology are addressed. The results presented, spotlight the potential of varactor analytical model, combined with a GA optimization procedure, when integrated in optimization design tools. The accuracy of the results is checked against HSPICE simulator.

Index Terms—Technology-aware Varactor Design; Discrete-Variable Optimization; Evolutionary Algorithms

I. INTRODUCTION

The study of VCOs applications has attracted designers' attention for the use of integrated capacitors, aiming at a fully on-chip circuit. Designing integrated capacitors is a well known technique, and can be implemented in any IC process. For the implementation of monolithic capacitors to be used in *LC*-VCOs, CMOS technology is used rather than BJT or BiCMOS technology. Generally, CMOS technology offers lower manufacturing costs, higher packing density, better performance with digital circuits, and less power dissipation when compared with BJT technology. Regarding BiCMOS technology, it has been shown that in the lower gigahertz frequencies, it exhibits poor Q [1].

The most elementar capacitor is the linear capacitor, which offers good linearity in a form of a *poly-to-poly* (PIP) capacitor or as a *metal-insulator-metal* capacitor (MIM), when implemented in CMOS-processes. While the PIP capacitor shows lower sensitivity to the parasitic capacitances, MIM capacitors take advantage of capacitances that appear between two metal layers. Most technologies offer MIM capacitors that reach high density (1-2 fF/μm^2), high Q (great than 100 @1 GHz), and low parasitic bottom plate capacitance [2]. A third topology, a metal-to-metal capacitor, where metal fingers are combined in order to achieve the desired capacitance, can also be implemented in CMOS process [3]. In [4] a performance comparison of MIM and metal finger capacitors is presented, showing that metal finger capacitors can have better Q, despite the lower process tolerance.

For implementing an *LC*-VCO, by the way of a tunable *LC* tank, two approaches can be done; either using a variable inductance or a variable capacitor. The variable capacitor (varactor) is usually the chosen element, since it is more easily controlled than the inductor. Varactors can be implemented either as a junction diode or as a MOS transistor. The junction diode operates in its reverse-biased region, taking advantage of the parasitic capacitances between the diffusion layer and the substrate wells. Since them show a small tunable capacitance range, which goes down as the supply voltage scales down, junction diode varactors are adequate for applications with limited tuning needs [2]. The Q of a junction diode varactor is usually quite good, reaching values higher than 50 @ 1 GHz. If a reasonable large tuning range is needed, the MOS transistor should be used. Even so, some literature states that in practice, this range is no more than five times higher than that achieved with the junction diode [5]. Additionally, Q remains fairly good across the full tuning range. For a varactor implemented with a CMOS transistor, the inversion and accumulation modes are the most common configurations.

In the literature several varactor models have been proposed aiming its integration in the tank characterisation of RF LC-VCOs. In [6] an analysis of the impact of the VCO signal swing in the varactor capacitance over time is presented. However, results of the analysis have not been used as feedback inputs to the design process. A varactor model, based on the transistor equivalent circuit utilizing BSIM3v3 model, with the addition of an overlap capacitance, is proposed in [7]. As a drawback, this model has to be supported either by adjustments inside the SPICE model or adding a negative power supply to the model as a way to guarantee results accuracy. A physically based MOS varactor model, where charge modelling, physical geometry and parasitics are taking into account, is offered in [8] and [9]. More recently, the design of a varactor by the means of evolutionary algorithms was proposed in [10] and [11]. However, this process is a time consuming task, taking several hours, since it relies on an exhaustive enumeration of designs where each combination of design parameters is validated through simulation.

Genetic Algorithms (GA) have been widely used in general circuit design [12] and [13]. This paper introduces a GA optimization based tool for the design of CMOS varactors. The efficiency of the tool, as well as the accuracy of the results, is guaranteed by the use of a scalable varactor analytical model. The proposed tool offers the possibility for obtaining the varactor layout parameters for a given capacitance value. The solution is obtained considering constraints in the design variables which are defined by the designer. Further constraints imposed by the technology used may also be considered. The designer may also choose which performance parameter is to be optimized, such as maximizing the quality factor, Q, at a predefined operation frequency, or the capacitance tuning range, C_{TR}. The tool was developed in Matlab and the GA optimization toolbox was used. The validity of the solution obtained is checked against results from simulation with HSPICE simulator.

Besides the Introduction, this paper comprises five additional sections. Section II presents the varactor analytical model used in the proposed tool. Then, in section III, a description of the algorithm used in the optimization procedure is presented. A set of varactor design examples is presented in section IV, where the solutions obtained with the proposed tool are compared against HSPICE simulation results. Finally, conclusions are offered.

II. CMOS VARACTOR ANALYTICAL MODEL

This section will be dedicated to present a mathematical model for the CMOS varactor characterization. This model has two main advantages. Firstly, the analytical model for the varactor capacitances are based on process and technological parameters, avoiding the undesired empirical/fitting factors. Secondly, the analytical transistor model, which is needed to determine the transistor current, is based on the well know EKV MOS [14], guaranteeing the accuracy of the results for low-voltage circuit design.

In the EKV model the drain current is given by a single expression for linear and saturation regimes of operation and valid from weak to strong inversion. The drain current is decomposed in a forward and a reverse current I_F and I_R that can be obtained with

$$I_{F(R)} = [\ln (1 + \exp[(V_P - V_{S(D)}) / (2U_T)])] , \quad (1)$$

where V_p is the *Pinch-off voltage*.

For the automatic generation of the EKV model parameters a Matlab script was developed. This script accepts as input the sizes of the transistor, and the specification of the technology to be used. Simulation files of the DC characterization of the transistor are automatically generated, and then the methodology presented in [15] and [16] was adopted.

The EKV transistor model is suitable to perform the CV-characterisation of varactors, when integrated in a computer design process, due to the continuity of the model and the reduced number of model parameters. According to [14], the intrinsic capacitances of a varactor, are obtained through the relative variation of the nodes charge to the node voltage

$$C_{xy} = \pm \partial Q_x / \partial V_y , \quad \text{with } x,y = G,D,S,B . \quad (2)$$

In a varactor, the total capacitance is usually referred as the gate capacitance, since the drain, source and bulk are connected to a fixed voltage, which allows neglecting the drain/source – bulk capacitance. Additionally, overlap and fringing

capacitances – extrinsic capacitances – must be accounted for. The varactor total capacitance can be obtained through

$$C_{total} = C_{GB} + C_{GD} + C_{GS} + C_{S(D)B} + C_{extrinsic} . \quad (3)$$

In [14] and [17] simplified expressions to determine each of the intrinsic capacitances are proposed. The varactor intrinsic capacitances are obtained through the set of equations

$$C_{GS,s} = 2/3 \cdot C_{ox} \cdot [1 - (I_{rev}^2 + I_{rev} + 0.5 \, I_{for}) / (I_{rev} + I_{for})^2] , \quad (4)$$

$$C_{GD,s} = 2/3 \cdot C_{ox} \cdot [1 - (I_{for}^2 + I_{for} + 0.5 \, I_{rev}) / (I_{rev} + I_{for})^2] , \quad (5)$$

$$C_{GB,s} = C_{ox} \cdot [(n_q - 1) / n_q] \cdot [1 - C_{GS,s} / C_{ox} - C_{GD,s} / C_{ox}] , \quad (6)$$

$$C_{SB,s} = (n_q - 1) \cdot C_{GS,s} , \quad (7)$$

$$C_{DB,s} = (n_q - 1) \cdot C_{GD,s} , \quad (8)$$

where I_{rev} and I_{for} are the normalised reverse and forward current, respectively; n_q is the slope factor, Γ is the body effect parameter, and V_P is the pinch-off voltage, obtained by

$$I_{rev} = \sqrt{0.25 + I_R} , \quad (9)$$

$$I_{for} = \sqrt{0.25 + I_F} , \quad (10)$$

$$n_q = 1 + \frac{\Gamma}{2 \cdot \sqrt{V_P + \Phi + 10^{-6}}} . \quad (11)$$

In deep submicron CMOS technologies, besides the intrinsic capacitances, the extrinsic (parasitics) capacitances can be a major player in the varactor total capacitance. In the extrinsic region the extrinsic capacitance is bias dependent, and thus essentially influenced by the gate voltage [18]. The extrinsic capacitance can be obtained through

$$C_{extrinsic} \approx C_{ext}(V_g) , \quad (12.a)$$

$$C_{ext}(V_g) = 2 \cdot (C_{ov}(V_g) + C_{if}(V_g) + C_{of}) , \quad (12.b)$$

In (12.b) $C_{ov}(V_g)$ is the parallel plate capacitance associated with the electric field in the gate-to-drain/source overlap region; $C_{if}(V_g)$ is the inner fringing capacitance associated with the inner electric field emerging from metallurgical junction source/drain to the underside of the poly-gate. C_{of} is the outer fringing capacitance, independent of the gate voltage, related to the electric field emerging from the sidewall of the poly-gate, ending at the source/drain region [18]. The gate overlap capacitance is here defined as

$$C_{ov}(V_g) = C_{ox} \cdot L_{ov}(V_g) , \quad (13)$$

where L_{ov} is the effective diffusion length, given by

$$L_{ov}(V_g) = A(V_g) \cdot L_d , \quad (14)$$

with

$$\begin{cases} A(V_g) = 1 & V_g \geq 0 \text{ V} \\ A(V_g) = \frac{1}{1-\lambda \widetilde{V_g}} & V_g < 0 \text{ V} \end{cases}, \quad (15)$$

where λ takes the value of 0.75. Moreover, a smoothing function should be introduced to make the bias dependent overlap capacitance converging to its maximum value

$$\widetilde{V_g} = V_g - 0.5 \cdot V_g + \sqrt{V_g^2 + 0.05} . \quad (16)$$

Finally, the overlap capacitance can be obtained by

$$C_{ov}(V_g) = W_g \cdot C_{ox} \cdot A(V_g) \cdot L_d . \quad (17)$$

Concerning the fringing capacitances, the model proposed in [19] is adopted. For the inner fringing capacitance the equation proposed for submicron technologies, suffers from not being bias dependent, since it is strongly influenced by the gate voltage. However, and based on the mentioned model, a slight different expression in order to predict more accurately the inner fringing capacitance is presented in [18],

$$C_{if} = C_{if,max} \cdot \exp\left[- \left(\frac{V_g - V_{fb} - \emptyset_f/2}{3\emptyset_f/2} \right)^2 \right]. \quad (18)$$

The outer fringing capacitance is bias voltage independent, and can be obtained by

$$C_{of} = \frac{2 \cdot \varepsilon_{ox}}{\pi} \cdot \ln\left(1 + \frac{T_{poly}}{T_{ox}} \right). \quad (19)$$

Finally, and gathering the previous equations, the total extrinsic capacitance can be written as

$$C_{ext}(V_g) = 2W_g \cdot \left[\begin{array}{c} C_{ox} \cdot A(V_g) \cdot L_d + \\ \frac{2 \cdot \varepsilon_{ox}}{\pi} \cdot \ln\left(1 + \frac{T_{poly}}{T_{ox}}\right) + \\ C_{if,max} \cdot \exp\left[- \left(\frac{V_g - V_{fb} - \emptyset_f/2}{3\emptyset_f/2} \right)^2 \right] \end{array} \right]. \quad (20)$$

Fig. 1 shows the characteristic of an inversion MOS varactor obtained with the proposed varactor analytical model, for different tuning voltages, namely 0.2V (A), 0.4V (B) and 0.6V (C), against simulations obtain with HSPICE software. The relative error for the mean capacitance was less than 4% in all cases. For those examples, the UMC130 CMOS technology, a transistor width of 10 µm and length of 0.8 µm, with a supply voltage of 1.2 V, was considered. The results in Fig. 1, show the accuracy of the varactor model thus guaranteeing its adequacy to be integrated in an optimization based design tool.

III. OPTIMIZATION METHODOLOGY

Genetic Algorithms are a stochastic search method that mimics the natural biological evolution, operating on a population of potential solutions, applying the principle of

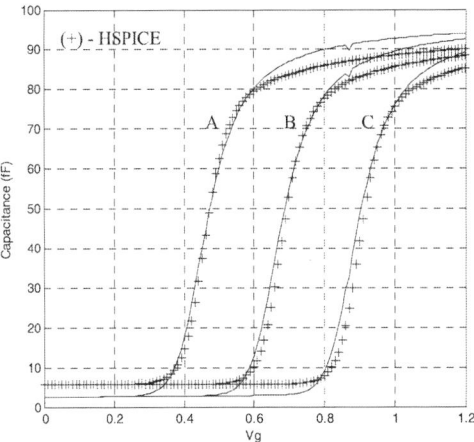

Figure 1. Tuning characteristics for an I-MOS varactor.

survival of the fittest to produce better and better approximations to a solution. As result, GAs have several fields of applications that go through non-linear problems, defined on discrete, continuous or mixed search spaces, constrained or unconstrained optimization problems. The main advantages of using GAs in optimization problems instead of the more traditional search and optimization methods reside in the fact that they search a population of points in parallel instead of a single point, thus making results less sensitive to the point chosen. Furthermore, GAs do not require derivative information or previous knowledge, they handle noisy functions well, and are resistant to becoming trapped in local optima [20].

In this paper a varactor design methodology, which is supported by a GA optimization procedure, is proposed. The optimization-based design flow for the proposed methodology is represented in Fig. 2. The design process starts with the creation of the population related to the variables that are going to be optimized, which can be discrete or continuous. The following step makes use of the EKV model parameters, previously obtained, in order to evaluate the varactor characteristics. Each combination of individuals is classified into a ranking, and a set of the best combinations are chosen for reproduction and mutation. This step is performed in loop until the optimization algorithm conditions are satisfied. Finally, the optimization procedure will test if specifications, as well as circuit feasibility are met.

Depending on the application several characteristics may be used for evaluating the quality of the varactor design, such as the quality factor (Q) or the tuning range (C_{TR}). These characteristics are influenced by varactor geometrical parameters as well as the technological parameters and constraints (if any). For the evaluation of the quality factor we consider

$$Q = \frac{stored\ energy\ per\ cycle\ (cap.)}{dissipated\ energy\ per\ cycle\ (res.)} \approx \frac{1}{2\pi f R C} , \quad (21)$$

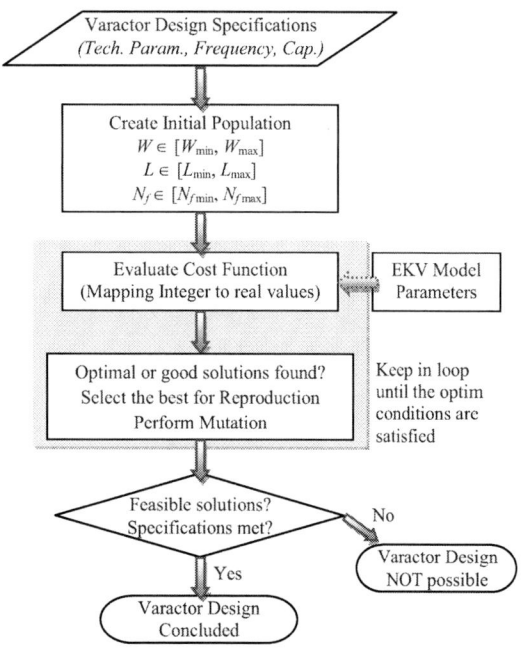

Figure 2. Varactor optimization-based design flowchart.

where

$$R = R_{poly/square} \cdot \frac{1}{N_f^2} \cdot \frac{W}{L} \; , \qquad (22)$$

and the tuning range is deemed as the variation range around the average capacitance in a oscillation period, and obtained by

$$C_{TR} = \pm \frac{C_{max} - C_{min}}{C_{max} + C_{min}} \; . \qquad (23)$$

The optimization design tool was implemented with the Matlab GA toolbox. Since this toolbox shows several limitations for discrete optimization problems, where each variable has an imposed step-size, additional functions that deal with discrete variables were added [21]. If we define a generic cost function $Cost(W, L, N_f)$ as the cost function of the varactor design, the optimization problem can be symbolically represented by (24), with W as the transistor width, L the transistor length and N_f the number of fingers

$$Maximize \quad Q(W, L, N_f) \text{ or } C_{TR}(W, L, N_f) \; , \qquad (24.a)$$

$$Subject \text{ to } (1-\delta) \cdot C_{expect} \leq C(W, L, N_f) \leq (1+\delta) \cdot C_{expect} \; . \qquad (24.b)$$

Additionally, and regarding the optimization constraints, limits to the number of fingers were imposed. The finger length is fixed, as well as the space between two consecutive fingers, to two times de minimal length. Finally, the maximum number of fingers is limited by the transistor width (and maximized by the user, see Table I), as shown in (25).

$$N_{f\,max} = W / (4 \cdot L_{min}) \, , \qquad (25.a)$$

$$N_f = 2 \cdot L_{min} \, . \qquad (25.b)$$

IV. VARACTOR DESIGN

In an LC-VCO, the tank circuit is of major importance since it is responsible for producing the required oscillatory signal. Moreover, the varactor is the element which gives to the oscillator the capability of being tunable. This section aims to emphasize the skill of the proposed tool, gathering the varactor model presented and a genetic algorithm optimization procedure, as the path to reach optimal design of a varactor.

In this section the design of a varactor aiming its integration in a LC tank circuit, with an inductor of 7 nH. Two examples considering the LC tank working at central operating frequencies of 1.0 and 2.5 GHz, for UMC130 technology are addressed. The design concerns the evaluation of three independent parameters, namely the transistor width (W), the transistor length (L) and the number of gate fingers (N_f). The GA optimization algorithm will deal with both continuous and discrete variables. The transistor width and length are considered as continuous variables, but the number of gate's fingers is an integer value. Technological and physical parameters, as well as optimization constraints, are shown in Table I. For each of the two mentioned cases, two different objective functions are addressed. In example A, the varactor layout parameters maximizing the quality factor, Q, are computed; example B deals with designing a varactor with maximum tuning range. In both cases, to determine the required varactor capacitance, (26) is used. The capacitance tolerance, δ, is an additional constraint, with maximum value of 2.5%. The results obtained are presented in Table II. For the varactor design solution obtained when maximizing the quality factor, the corresponding capacitance over all range of the gate voltage is depicted in Fig. 3.

$$f = \frac{1}{2 \cdot \pi \cdot \sqrt{L \cdot C}} \; . \qquad (26)$$

TABLE I.   TECHNOLOGICAL-PHYSICAL PARAMETERS AND OPTIMIZATION CONSTRAINTS

Parameters	Value
Supply voltage (Vdd)	1.2 V
Tuning voltage (V_{tune})	0.4 V
$R_{poly/square}$	7.60 Ohm/square
Minimum Length (L_{min})	0.39 μm
Transistor Width (W)	$W_{min} = 2 \cdot L_{min}$ and $W_{max} = 300$ μm
Transistor Length (L)	$L_{min} = L_{min}$ and $L_{max} = 4 \cdot L_{min}$
Number of fingers (N_f)	$N_{f\,min} = 1$ and $N_{f\,max} = 40$
GA inital population	150 individuals
GA Max. Generations	1000

280

TABLE II.   OPTIMIZATION RESULTS AND ITS COMPARISON WITH HSPICE SIMULATIONS

Design Parameters [a]	Capacitance @ 1.0 GHz 3.62 pF		Capacitance @ 2.5 GHz 0.58 pF	
	A	B	A	B
W (µm)	233.68	299.53	64.64	133.93
L (µm)	1.44	1.14	0.84	0.44
N_f	39	26	38	10
C_{av_model} (pF)	1.78	1.76	0.283	0.294
C_{av_HSPICE} (pF)	1.67	1.70	0.274	0.312
Error (%)	7.13	3.51	2.90	5.86
Q	110.0	30.0	556.0	9.5
C_{TR} (%)	92.71	93.22	94.31	98.26

a. Considering a varactor pair in back-to-back configuration

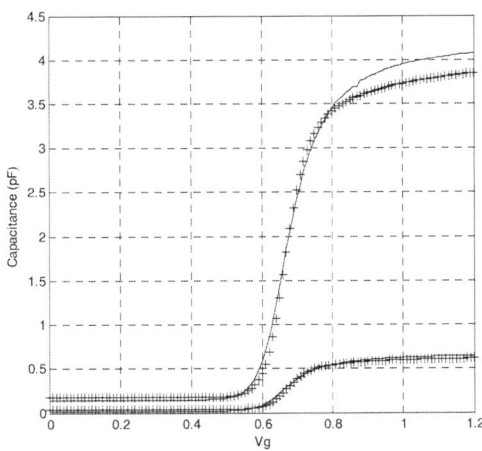

Figure 3.   Capacitance behaviour for the designed varactor.

The results presented in this section highlight the capacity of the proposed design methodology to achieve optimal design of varactors. It is our conviction, that the results obtained are quite acceptable for an optimization methodology, without the needs of regular use of simulators during the optimization process, aiming its integration in a full LC-VCO optimization design procedure.

V.   CONCLUSIONS

In the present work the varactor design based on evolutionary algorithms is proposed. The varactor analytical model is supported by a set of equations, mostly based on technological parameters, instead of fitting parameters obtained through exhaustive measures and/or simulations. The optimization tool presented in this work has three main advantages: i) as the varactor model is based in technological parameters, it may be straightforwardly adapted to new technologies; ii) reduced computation time, if compared with electromagnetic simulators; and iii) it can be easily integrated into an LC-VCO optimization based design tool.

In this work the design of a varactor is supported by a GA optimization methodology, which is able to deal with continuous and/or discrete variables, making possible to satisfy both technological and layout constraints. A set of working examples showing the design of four varactors for four different cases is shown. The results presented, even if in some cases with errors above 5%, point out the potential of varactor analytical model, combined with a GA optimization procedure, when integrated in optimization design tools. The accuracy of the results was checked against HSPICE simulator.

ACKNOWLEDGMENT

This work is co-financed by the Portuguese National Funding through the FCT (Science and Technology Foundation) PEst-OE/EEI/UI0066/2011 project.

FCT Fundação para a Ciência e a Tecnologia

MINISTÉRIO DA EDUCAÇÃO E CIÊNCIA

REFERENCES

[1] P. Andreani and S. Mattisson, "On the use of MOS varactors in RF VCOs," IEEE Journal of Solid-State Circuits, vol. 35, pp. 905-910, 2000.

[2] A. D. Berny, R. G. Meyer and A. Niknejad, "Analysis and Design of Wideband LC VCOs," Technical Report No. UCB/EECS-2006-50, EECS Department, University of California, Berkeley, 2006.

[3] M. Tiebout, Low Power VCO Design in CMOS (Springer Series in Advanced Microelectronics), Springer-Verlag New York, Inc., 2005.

[4] Q. S. Lim, A. V. Kordesch and R. A. Keating, "Performance comparison of MIM capacitors and metal finger capacitors for analog and RF applications," Proc. of RF and Microwave Conference - RFM, pp. 85-89, 2004.

[5] A. S. Porret, T. Melly, C. C. Enz and E. A. Vittoz, "Design of high-Q varactors for low-power wireless applications using a standard CMOS process," IEEE Journal of Solid-State Circuits, vol. 35, pp. 337-345, 2000.

[6] R. L. Bunch and S. Raman, "Large-signal analysis of MOS varactors in CMOS-Gm LC VCOs", IEEE Journal of Solid-State Circuits, vol. 38, pp. 1325-1332, 2003.

[7] P. Sameni, et al., "Modeling and characterization of VCOs with MOS varactors for RF transceivers" EURASIP Journal on Wireless Communications and Networking, pp. 1-12, 2006.

[8] J. Victory, Z. Yan, G. Gildenblat, C. McAndrew and J. Zheng, "A physically based, scalable MOS varactor model and extraction methodology for RF applications" IEEE Transactions on Electron Devices, vol. 52, pp. 1343-1353, 2005.

[9] J. Victory, et al.,"PSP-Based Scalable MOS Varactor Model", Proc. of IEEE Custom Integrated Circuits Conference - CICC, pp. 495-502, 2007.

[10] L. Mendes, et al., "Design Optimization of Radio Frequency Discrete Tuning Varactors," in Applications of Evolutionary Computing, Lecture Notes in Computer Science, vol. 5484, Springer Berlin/Heidelberg, pp. 343-352, 2009.

[11] E. J. Pires, et al., "Design of Radio-Frequency Integrated CMOS Discrete Tuning Varactors Using the Particle Swarm Optimization Algorithm," Proc. of the 10th International Work-Conference on Artificial Neural Networks - IWANN, pp. 1231-1239, 2009.

[12] M. A. Stelmack, N. Nakashima and S. M. Batill, "Genetic Algortihms For Mixed Discrete/Continuous Optimization In Multidisciplinary Design". Proc. of the 7th AIAA/USAF/NASA/ISSMO Symposium on Multidisciplinary Analysis and Optimization, pp., 499-509, 1998.

[13] M. Wankhedeand and A. Deshmukh, "Optimization of Cell-based VLSI Circuit Design using a Genetic Algorithm: Design Approach" Proc. of the International MultiConference of Engineers and Computer Scientists - IMECS, vol. 2, pp. 854-860, 2009.

[14] M. Bucher, C. Lallement, C. Enz, F. Théodoloz and F. Krummenacher, "The EPFL-EKV MOSFET model equations for simulation, Version 2.6," 1998.

[15] S. G. A. Machado, C. C. Enz and M. Bucher, "Estimating key parameters in the EKV MOST model for analogue design and simulation," Proc. of IEEE International Symposium on Circuits and Systems - ISCAS, vol. 3, pp. 1588-1591, 1995.

[16] P. Pereira and M. H. Fino, "CMOS Delay and Power Estimation for Deep Submicrometer Technologies Using EKV Model", Proc. of Xth International Workshop on Symbolic and Numerical Methods, Modeling and Applications to Circuit Design - SMACD, pp. 253-257, 2008.

[17] J. Bremer, T. Peikert and W. Mathis, "Analytical inversion-mode varactor modeling based on the EKV model and its application to RF VCO design," Proc. of the 17th International Conference on Mixed Design of Integrated Circuits and Systems - MIXDES, pp. 64-69, 2010.

[18] F. Prégaldiny, C. Lallement and D. Mathiot, "A simple efficient model of parasitic capacitances of deep-submicron LDD MOSFETs," journal of Solid-State Electronics, vol. 46, no. 12, pp. 2191-2198, 2002.

[19] R. Shrivastava and K. Fitzpatrick, "A simple model for the overlap capacitance of a VLSI MOS device," IEEE Transactions on Electron Devices, vol. 29, no. 12, pp. 1870-1875, 1982.

[20] S. Sivanandam and S. Deepa, "Introduction to Genetic Algorithms". Springer, 2008.

[21] P. Pereira, M. Fino, F. Coito and M. Ventim-Neves, "GADISI – Genetic Algorithms Applied to the Automatic Design of Integrated Spiral Inductors," in Emerging Trends in Technological Innovation, vol. 314, Springer Boston, pp. 515-522, 2010.

Temperature and Supply Voltage Compensated Biasing for Digitally Controlled Oscillators

Sebastian Höppner, Stefan Haenzsche, Stephan Hartmann, Stefan Schiefer, René Schüffny

Faculty of Electrical Engineering and Information Technology
Technische Universität Dresden, Germany
Emails: {hoeppner, haenzsch, schueffn}@iee.et.tu-dresden.de

Abstract—**This paper presents a current bias circuit for digitally controlled oscillators (DCOs). It consists of three individual current components with different supply voltage and temperature sensitivities and thereby allows to effectively compensate period variations with respect to supply voltage and temperature. This reduces the required fine tuning range for ADPLL application of the DCO. The concept is proven by a DCO circuit implementation in 65nm CMOS technology and verified by simulations and testchip measurements.**

Index Terms—**current bias, temperature compensation, supply voltage compensation, DCO, ADPLL**

I. INTRODUCTION

Digitally-controlled oscillators are an important part of all-digital phase-locked loop (ADPLL) clock generators which are widely used in modern systems-on-chip. Their oscillation frequency is strongly effected by process, voltage and temperature (PVT) variations. Often different tuning mechanisms are employed within the same circuit, where coarse tune elements are used for compensation of process variations and fine tune elements track the phase and frequency during system operation for noise, supply voltage variations and temperature drifts. Therefore the trade-off between a small fine tuning step for reduced jitter in the ADPLL output and a wide fine tuning range for compensation of supply voltage and temperature variations must be considered.

Several approaches for compensation of supply voltage and temperature influences on the oscillation period of ring oscillators have been reported previously. [1] and [2] compensate the ring oscillator frequency for temperature and process by adaptive biasing using threshold voltage sensing circuits. [3] presents a compensation technique using an addition based current source. However this approach requires a reference gate-source voltage which can not be used for tuning because the temperature compensation is optimized for a fixed gate-source voltage. In [4] an all-digital low frequency reference oscillator with PVT compensation is presented, which is based on on-chip evaluation of the relative delay of different logic gate types. This topology is not suited for high-speed ring-oscillations with current-starved tuning mechanism. [5] presents a special DCO ring topology to reduce the supply voltage influences on the oscillation period.

The drawback of the previous work is that sensitivity versus supply voltage changes are not compensated separately from the process and temperature related effects, which is especially critical for circuits in small CMOS technologies, where short channel effects increase significantly. This work presents selective compensation for supply and temperature related effects. Thereby the fine tune range of DCO can be reduced significantly or the fine tune step size can be decreased with the same number of control bits. This additionally can reduce the DCO gain and therefore the output jitter of the ADPLL clock generator. Compensation for process variations is achieved by ADPLL closed loop regulation. The presented technique can be used for high-speed multi phase DCOs.

II. DIGITALLY CONTROLLED OSCILLATOR

A. DCO circuit

We consider a multi-stage ring oscillator as shown in Fig. 1, which can be used for multi-phase clock generation for frequency synthesis [6], [7]. Its stages are built up using current-starved inverters as shown in Fig. 1(b). The tuning voltages (tp/tn) are generated by a current-based digital-to-analog converter (IDAC) and are applied to all stages, which ensures symmetry of the phases.

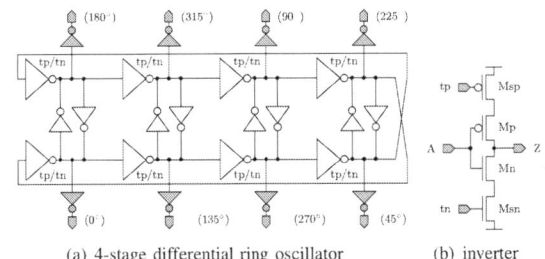

(a) 4-stage differential ring oscillator   (b) inverter

Fig. 1. DCO schematic

Although the period of oscillation is defined by the current-sources in the inverter-cells, it depends on the supply voltage V_{DD} and the temperature θ by

$$T_{\mathrm{DCO}} = F(\theta, V_{\mathrm{DD}}, I_{\mathrm{ref}}) \quad (1)$$

where F denotes a nonlinear function. I_{ref} is the reference current of the IDAC used for tuning the DCO. Fig. 2 shows an example simulation result of the temperature and supply voltage dependency of the DCO core for a constant reference current. With increased supply voltage the period decreases because the current source devices provide increased output current, due to their finite output resistance. With increasing temperature the threshold voltage of the switching devices decreases, leading to reduced oscillation period as well.

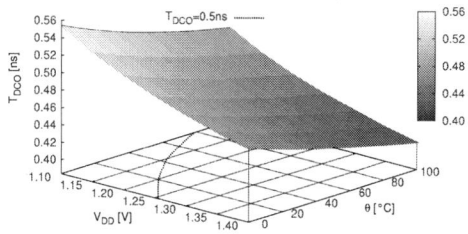

Fig. 2. DCO period for constant $I_{ref} = 30\mu A$

B. IDAC

Tuning of the DCO is performed by a current-based analog-to-digital converter (IDAC) shown in Fig. 3. It consists of binary weighted switched current sources for coarse tuning c_{coarse} and minimum sized thermometer coded switched current sources for fine tuning c_{fine}. The small fine tune stages effectively reduces ripple in the control voltage nodes (tp/tn) of the DCO when changing the fine tune code (c_{fine}) during operation (e.g. by ADPLL), because only little parasitic charge is injected. Thereby control signal related jitter (i.e. reference feedthrough) is minimized. Additionally the fine tuning characteristic is monotonic. The gain of the fine tune stage is adjustable by a 2-bit control signal ftgain.

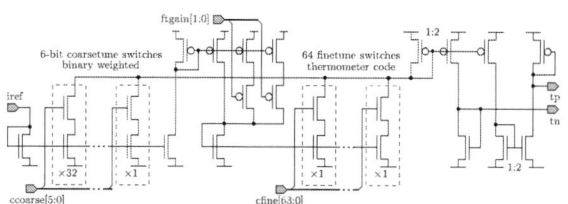

Fig. 3. IDAC schematic

C. ADPLL Application Scenario

We consider the DCO to be applied in an ADPLL based clock generator in a system-on-chip (SoC) which provides a clock signal with period T_0. As illustrated in Fig. 4 in an initial coarse lock-in phase the coarse tune value is determined such that $T_{DCO} \approx T_0$, e.g. by binary search [8]. In that phase the fine tune signal is at middle position and the DCO frequency changes in a wider range. Therefore the output clock of the ADPLL is gated, such that the SoC components are not clocked during coarse lock-in. After that fine lock-in is performed by closed-loop ADPLL operation, which adjusts c_{fine} such that $T_{DCO} = T_0$, thereby compensating the remaining coarse tune error. When the ADPLL is locked the fine tune mechanism tracks supply voltage and temperature variations

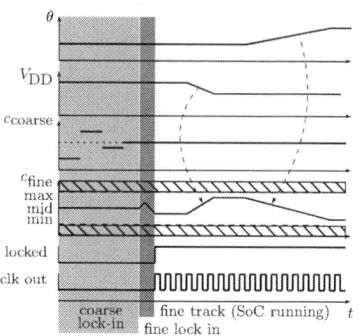

Fig. 4. ADPLL operation phases

during system operation. The output clock has a defined frequency and is used to clock the SoC components.

The SoC has specified operating parameters θ and V_{DD}, which change during system operation within specified min/max regions, e.g. due to environmental temperature changes, system heating or IR drop in the supply networks. To maintain phase and frequency lock, the fine tune signal must stay within its min/max region during system operation. The operating condition of the SoC at initial coarse lock-in is not known. Therefore the fine tuning mechanism must be capable to compensate for all variations within the specified operating parameter range.

As illustrated in Fig. 5, the DCO period fine tune characteristics $T_{DCO} = f(c_{fine})$ can change with respect to (V_{DD}, θ) variations in the grey regions, where the upper and lower boundaries denote best-case and worst-case (V_{DD}, θ) corners with respect to the DCO oscillation period T_{DCO}. The criterion for safe system operation within the specified operating parameter ranges are:

- coarse lock-in at $(V_{DD}, \theta)_{best}$ and occurrence of $(V_{DD}, \theta)_{worst}$ during operation, $\rightarrow c_{fine} \leq c_{fine,max}$
- coarse lock-in at $(V_{DD}, \theta)_{worst}$ and occurrence of $(V_{DD}, \theta)_{best}$ during operation, $\rightarrow c_{fine} \geq c_{fine,min}$

These conditions can be met either by providing a wide fine tuning range at the cost of chip area or larger jitter due to increased fine tuning steps, or by minimizing the DCO period variations with respect to supply voltage and temperature.

III. BIAS COMPENSATION CIRCUIT

A. Overview

When the oscillator is locked to a specified period the V_{DD} θ variations are compensated by the fine tuning mechanism during closed-loop PLL operation. As an example Fig. 6 shows the tuning current for a constant period of $T_0 = 0.5$ns versus temperature and supply voltage variations. The maximum required current tuning range ($\Delta I_{ref} = I_{ref,max} - I_{ref,min}$) must be covered by the fine tuning stage of the IDAC. If ΔI_{ref} is large, a large number of fine tune switches is required, which leads to larger chip area, or the fine tune step size must be increased, which leads to larger jitter. To circumvent this trade-off, it is proposed to provide a reference current

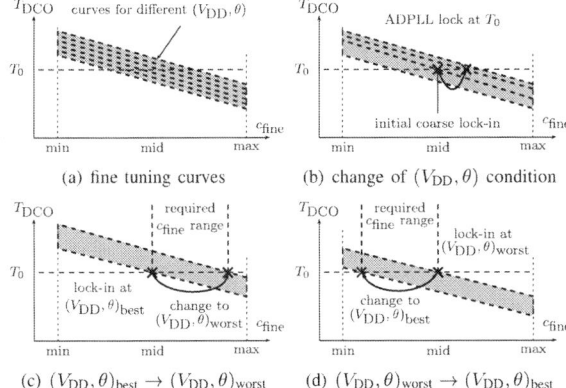

(a) fine tuning curves (b) change of (V_{DD}, θ) condition

(c) $(V_{DD}, \theta)_{best} \rightarrow (V_{DD}, \theta)_{worst}$ (d) $(V_{DD}, \theta)_{worst} \rightarrow (V_{DD}, \theta)_{best}$

Fig. 5. Illustration of DCO coarse lock-in at different (V_{DD}, θ) conditions and fine tune variations during system operation

I_{ref}, which is compensated for temperature and supply voltage variations, thereby decreasing the required IDAC fine tuning range. To compensate the supply voltage and temperature dependency $T_{DCO} = F(\theta, V_{DD}, I_{ref})$ of the oscillator core, the current bias source must have the *inverse* characteristics $I_{ref} = |G(\theta, V_{DD})|_{T_0}$.

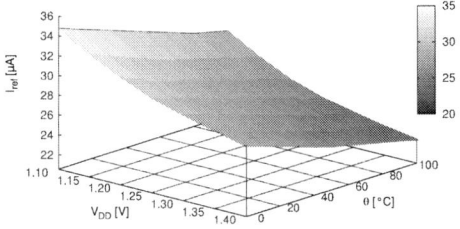

Fig. 6. I_{ref} for $T_{DCO} = 500$ps, from reverse interpolation of simulation data

Obviously the function $I_{ref} = |G(\theta, V_{DD})|_{T_0}$ is nonlinear as well. It is proposed to employ a current source which has linearized characteristics of G versus θ and V_{DD} and cancels out first-order supply voltage and temperature variation effects. The higher order error remains and is compensated by the fine tuning mechanism of the DCO. The linearized compensated reference current has *three* degrees of freedom a_{comp}, b_{comp} and c_{comp} with

$$I_{ref,comp}(V_{DD}, \theta) = \underbrace{(a_{comp}, b_{comp}, c_{comp})}_{\mathbf{a}} \cdot (V_{DD}, \theta, 1)^T. \quad (2)$$

Therefore, the architecture of this reference current source as shown in Fig. 7 consists of *three* independent bias currents with different supply voltage and temperature characteristics. The first component $I_{ref,0}$ is independent from θ and V_{DD}, whereas the second ($I_{ref,1,ptk}$) and third ($I_{ref,2,pvk}$) components show strong dependency on θ and V_{DD} respectively. The reference current for the DCO IDAC is generated by summing

up these three components

$$I_{ref}(V_{DD}, \theta) = \underbrace{(k_0, k_1, k_2)}_{\mathbf{k}} \cdot (I_{ref,0}, I_{ref,1,ptk}, I_{ref,2,pvk})^T. \quad (3)$$

The weighting factors $\mathbf{k} = (k_0, k_1, k_2)$ are adjustable by a programmable current bank based on switchable current sources. The output current of the three individual bias current sources are considered in a first order (linear) approximation as

$$\begin{pmatrix} I_{ref,0} \\ I_{ref,1,ptk} \\ I_{ref,2,pvk} \end{pmatrix} = \underbrace{\begin{pmatrix} a_0 & b_0 & c_0 \\ a_1 & b_1 & c_1 \\ a_2 & b_2 & c_2 \end{pmatrix}}_{\mathbf{A}} \cdot \begin{pmatrix} V_{DD} \\ \theta \\ 1 \end{pmatrix} \quad (4)$$

Therefore the total reference current is

$$I_{ref}(V_{DD}, \theta) = \mathbf{k} \cdot \mathbf{A} \cdot (V_{DD}, \theta, 1)^T \quad (5)$$

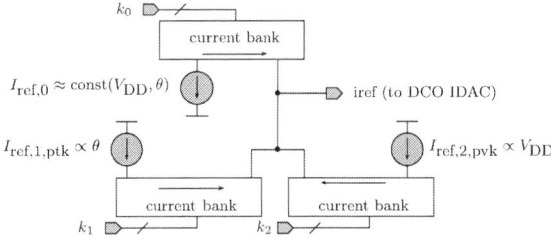

Fig. 7. Bias current source with adjustable temperature and supply voltage dependency

B. Current bias component 0

The first bias current component $I_{ref,0}$ has a low sensitivity with respect to θ and V_{DD}. A beta-multiplier based current reference [9] as shown in Fig. 8 is used here. The current I_0 is determined by M1, M2 and the resistor R. The amplifier circuit ensures that

$$V_{GS,1} = V_{GS,2} + R \cdot I_0. \quad (6)$$

and the equally sized devices M3 and M4 ensure that the M1 and M2 have the same current I_0. The width ratio of M2 and M1 is $W_2 = K \cdot W_1$. Assuming M1 and M2 to operate in saturation region it is

$$I_0 = \frac{\beta}{2} \cdot (V_{GS,1} - V_{th})^2 = K \cdot \frac{\beta}{2} \cdot (V_{GS,2} - V_{th})^2 \quad (7)$$

with $\beta = KP \cdot W_1/L$. Thereby I_0 can be expressed as

$$I_0 = \frac{2}{R^2 \cdot \beta} \cdot \left(1 - \frac{1}{\sqrt{K}}\right)^2. \quad (8)$$

The reference current does not depend on the supply voltage V_{DD} in a first-order approximation. The temperature dependency of I_0 can be written as

$$\frac{1}{I_0} \cdot \frac{\delta I_0}{\delta \theta} = -\frac{2}{R} \cdot \frac{\delta R}{\delta \theta} - \frac{1}{\beta} \cdot \frac{\delta \beta}{\delta \theta}. \quad (9)$$

285

It is $\delta\beta/\delta\theta < 0$ because the charge mobility is decreasing with increasing temperature. In order to achieve $\delta I_0/\delta\theta = 0$, the resistor must exhibit a postitive temperature dependency $\delta R/\delta\theta > 0$ according to Eq. 9. An n-well resistor is used for this purpose. The reference current $I_{\text{ref},0}$ is generated from I_0 by a switchable PMOS current bank.

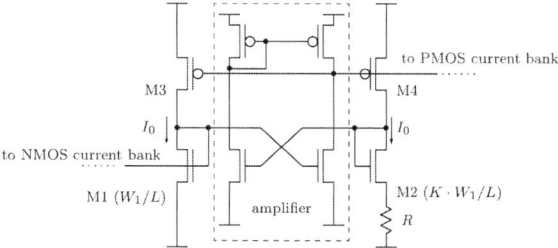

Fig. 8. Beta-multiplier current reference for $I_{\text{ref},0}$ and $I_{\text{ref},1}$, power-down switches and start-up circuit not shown

C. Current bias component 1

The second bias component $I_{\text{ref},1}$ provides a significant temperature dependency but a low supply voltage dependency. The same beta-multiplier circuit as for $I_{\text{ref},0}$ is employed, except that a poly resistor with negative temperature dependency $\delta R/\delta\theta < 0$ is used. Thus the temperature sensitivity is $\delta I_1/\delta\theta > 0$ according to Eq. 9. The reference current $I_{\text{ref},1}$ is generated from I_1 by a switchable NMOS current bank.

D. Current bias component 2

The third bias component $I_{\text{ref},2}$ shows a significant dependency on V_{DD}. Fig. 9 shows its schematic realization. The resistive divider R1,R2 defines a reference voltage V_{ref} which linearly depends on V_{DD}. The transistor M2 sources a current I_2 through R3, with $R_3 = R_1$. An error amplifier senses the voltage difference over R1 and R2 and adjusts the $V_{\text{GS},2}$ of M2 until $V_{\text{R3}} = V_{\text{R1}}$ and therefore the currents through R1 and R2 are equal (I_2). It is

$$I_2 = \frac{1}{R_3} \cdot \frac{R_1}{R_1 + R_2} \cdot V_{\text{DD}} = \frac{1}{R_1 + R_2} \cdot V_{\text{DD}} \quad (10)$$

The sensitivities with respect to V_{DD} and θ read

$$\frac{\delta I_2}{\delta V_{\text{DD}}} = \frac{1}{R_1 + R_2} \quad (11)$$

$$\frac{1}{I_2} \cdot \frac{\delta I_2}{\delta\theta} = -\frac{1}{R_1 + R_2} \cdot \frac{\delta(R_1 + R_2)}{\delta\theta} \quad (12)$$

Poly resistors are employed here, because they show a low absolute temperature dependency $\delta R/\delta\theta$, such that I_2 is mainly sensitive to V_{DD}. The reference current $I_{\text{ref},2}$ is generated from I_1 by a switchable NMOS current bank.

E. Parameter Extraction

For a given circuit realization the reference current weighting factors $\mathbf{k} = (k_0, k_1, k_2)^T$ must be determined. This can be done either by circuit simulation or lab characterization of samples of the manufactured chips.

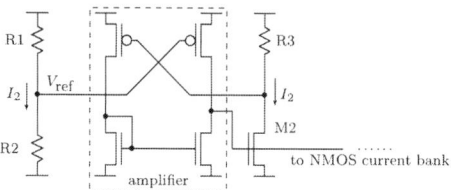

Fig. 9. Current reference for $I_{\text{ref},2}$, power-down switches not shown

First, the (V_{DD}, θ) characteristics of the DCO for constant reference current $T_{\text{DCO}} = F(V_{\text{DD}}, \theta)$ are determined (see Fig. 2). Second, the inverse characteristics $I_{\text{ref}} = |G(\theta, V_{\text{DD}})|_{T_0}$ is determined numerically by inverse interpolation. A two dimensional plane is fitted to G by least square method which results in the linear approximation of the targeted bias current characteristics in Eq. 2 (determine $\mathbf{a}$). This linear characteristics must be reproduced by the bias circuit $I_{\text{ref}}(V_{\text{DD}}, \theta) = I_{\text{ref,target}}(V_{\text{DD}}, \theta)$. Therefore the three individual bias componentes are characterized for their linear (V_{DD}, θ) characteristics, thereby determining the matrix $\mathbf{A}$. Combining Eq. 5 with Eq. 2 and solving for $\mathbf{k}$ leads to

$$\mathbf{k} = \mathbf{a} \cdot \mathbf{A}^{-1}. \quad (13)$$

By determination of $\mathbf{k}$, the bias configuration signals cizero, ciptk and cipvk can be determined.

IV. RESULTS

A. Testchip Implementation

The proposed DCO has been implemented in 65nm LP CMOS technology. Fig. 10(a) and Fig. 10(b) show the layouts of the current bias circuit and the IDAC and oscillator core respectively. Fig. 11 shows a chip photo of the MPSoC testchip including several ADPLL clock generators with the proposed biasing technique. The DCO clock is fed to a frequency divider of configurable ratio. The divided output signal is connected to an on-chip LVDS output pad and measured by a LeCroy WavePro 7300a oscilloscope. Fig. 12 shows the measured DCO output signal. An additional current output pad is available for direct measurement of the IDAC output current.

Fig. 13 shows the measured DCO fine tuning curves for different c_{coarse} values and ftgain settings.

B. Bias circuit parameters

Tab. I shows the Monte Carlo simulation results for the current bias block including global (process) and local (mismatch) variations (696 samples). The relative variability of the main sensitivity parameters (in bold font) of the three components are suitably low.

Calibration of the bias circuit has been performed based on simulation data at the typical process corner (TT) using the method presented in Sec. III-E. The resulting optimum settings for the bias block are $k_1 = 1.2$ and $k_2 = 0.9$ (normalized with respect to $k_0 = 1.0$). Note that these settings are adjustable with 4-bit accuracy only. Thus there is a remaining weighting

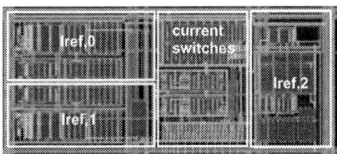

(a) current bias circuit (24μm $\times$ 54μm)

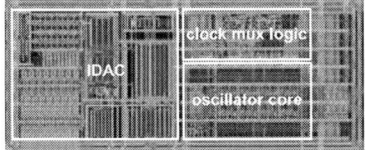

(b) IDAC and oscillator core (24μm $\times$ 58μm)

Fig. 10.   DCO layouts

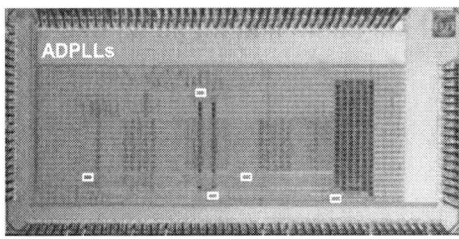

Fig. 11.   Chip photo of MPSoC testchip containing 5 ADPLLs

factor error $\mathbf{k}/\mathbf{k}_{\text{ideal}} = (0.9848, -0.9045, -1.0395)$, i.e. the calibration of the linearized current bias is accurate within approximately $\pm 4\%$.

C. Compensation Results

Fig. 14 and Fig. 15 show the simulated and measured DCO periods respectively at fixed tuning values with and without bias current compensation. Tab.II summarizes the results for period compensation of a DCO with fixed c_{coarse} and c_{fine} for $T_{\text{DCO}} \approx 0.5$ns.

Fig. 16(a) and Fig. 16(b) show the simulated fine tuning curves for lock-in at $(V_{\text{DD}}, \theta)_{\text{best}}$ and $(V_{\text{DD}}, \theta)_{\text{worst}}$ respectively

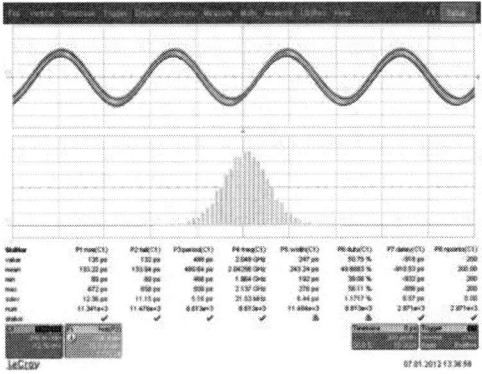

Fig. 12.   Measured DCO clock signal waveform

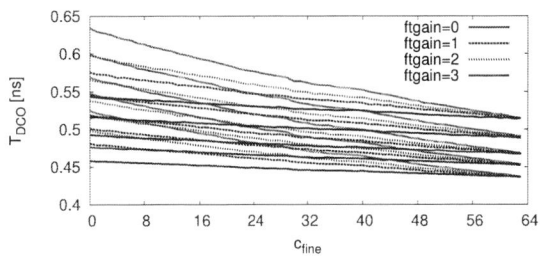

Fig. 13.   Measured DCO tuning curves at $V_{\text{DD}} = 1.2$V, $\theta = 25^{\circ}C$, $c_{\text{coarse}} = [20, 25, 30, 35, 40]$

TABLE I
CURRENT SOURCE MONTE-CARLO SIMULATION RESULTS

	mean	std	std/mean
a_0 [μA/V]	-2.918	0.6985	-0.239
b_0 [μA/K]	0.05654	0.01129	0.200
$\mathbf{c_0}$ [μA]	100.49	13.00	0.131
a_1 [μA/V]	-0.5084	0.1407	-0.277
$\mathbf{b_1}$ [μA/K]	0.1149	0.01066	0.093
c_1 [μA]	41.97	4.069	0.097
$\mathbf{a_2}$ [μA/V]	16.33	1.116	0.068
b_2 [μA/K]	0.003123	0.002018	0.646
c_2 [μA]	2.97	0.207	0.070

according to the illustration in Fig. 5. Measurement of the required fine lock ranges within the (V_{DD}, θ) range $V_{\text{DD,min}} \leq V_{\text{DD}} \leq V_{\text{DD,max}}$ and $\theta_{\text{min}} \leq \theta \leq \theta_{\text{max}}$ is performed by the following procedure:

1) initially lock ADPLL at one (V_{DD}, θ) corner
2) apply resulting c_{coarse} to DCO in open loop mode
3) measure $T_{\text{DCO}}(c_{\text{fine}})$ for three remaining (V_{DD}, θ) corners
4) repeat 1) to 3) for lock-in at all 4 (V_{DD}, θ) corners

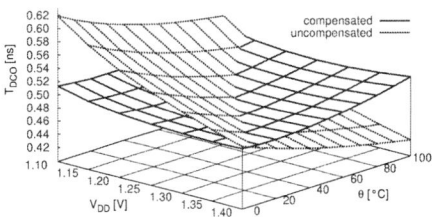

(a) 3D visualization

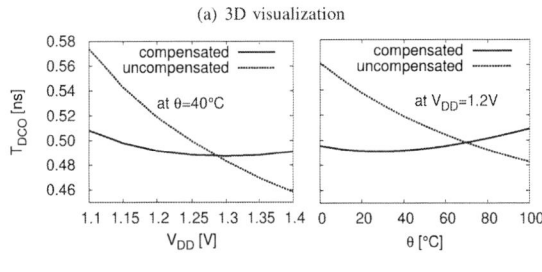

(b) 2D visualization

Fig. 14.   DCO period simulation results with ($k_1 = 1.2$ and $k_2 = 0.9$) and without ($k_1 = 0$ and $k_2 = 0$) compensated biasing

287

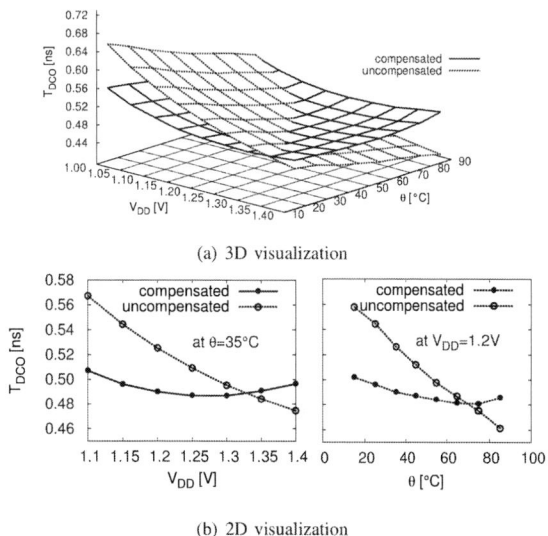

(a) 3D visualization

(b) 2D visualization

Fig. 15. DCO period measurement results with ($k_1 = 1.2$ and $k_2 = 1.1$) and without ($k_1 = 0$ and $k_2 = 0$) compensated biasing

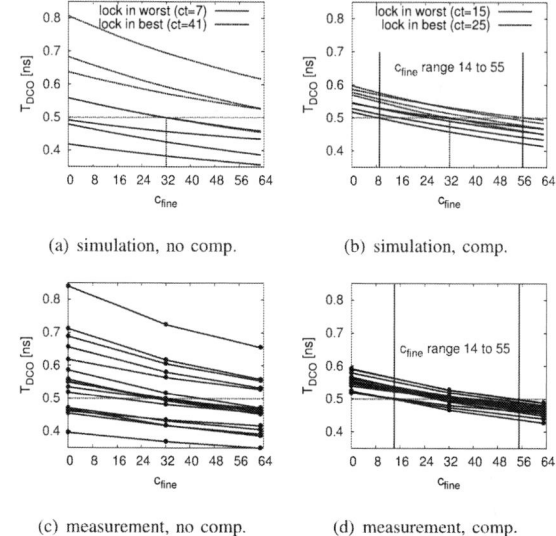

(a) simulation, no comp.

(b) simulation, comp.

(c) measurement, no comp.

(d) measurement, comp.

Fig. 16. Fine tune lock-range results at $1.08V \leq V_{DD} \leq 1.32V$ and $0°C \leq \theta \leq 85°C$ ($15°C \leq \theta \leq 85°C$ for measurement)

The resulting 16 tuning curves are plotted in Fig. 16(c) and Fig. 16(d) without and with bias compensation respectively. Measurement and simulation results are consistent.

Without compensation for supply voltage and temperature variations it is not possible to maintain fine-lock for all possible changes of V_{DD} and θ during SoC operation. The bias compensation circuit allows to keep fine-lock during circuit operation with a maximum required c_{fine} range from 9 to 56 (simulation) and 14 to 55 (measurement), thereby enabling safe SoC operation under all specified V_{DD} and θ conditions.

TABLE II
DCO PERIOD COMPENSATION RESULTS

(1.08V; 1.32V) (15°C; 85°C)	$\overline{T}_{DCO}$ [ps]	ΔT_{DCO} [ps]	$\overline{T}_{DCO}/\Delta T_{DCO}$
sim. uncomp.	503	196	39.0 %
meas. uncomp.	513	156	30.4 %
sim. comp.	499	36	7.4 %
meas. comp.	497	39	7.8 %

V. CONCLUSION

In this work a configurable current bias circuit with adjustable supply voltage and temperature dependency for the use in digitally controlled oscillators has been presented. The circuit allows effective compensation of period variations. It enables the realization of a DCO circuit, which fine tuning stage is capable to compensate all supply voltage and temperature variations in a wide operation parameter range to ensure save SoC clock generator operation. This eases ADPLL design because the required fine tuning range can be reduced significantly. The circuit has been integrated in an ADPLL clock generator in 65nm CMOS technology. The proposed

methodology is verified successfully by (statistical) circuit simulations and measurements.

ACKNOWLEDGMENT

This work is supported by the German Ministry of Education and Research BMBF under grant number 13N10788 (CoolBaseStations). The authors are responsible for the content of this publication.

REFERENCES

[1] K. Sundaresan, P. Allen, and F. Ayazi, "Process and temperature compensation in a 7-MHz CMOS clock oscillator," Solid-State Circuits, IEEE Journal of, vol. 41, no. 2, pp. 433 – 442, 2006.

[2] C.-F. Tsai, W.-J. Li, P.-Y. Chen, Y.-Z. Lin, and S.-J. Chang, "On-chip reference oscillators with process, supply voltage and temperature compensation," in Next-Generation Electronics (ISNE), 2010 International Symposium on, 2010, pp. 108 –111.

[3] X. Zhang and A. Apsel, "A low-power, process-and- temperature- compensated ring oscillator with addition-based current source," Circuits and Systems I: Regular Papers, IEEE Transactions on, vol. 58, no. 5, pp. 868 –878, may 2011.

[4] C.-Y. Yu, J.-Y. Yu, and C.-Y. Lee, "A low voltage all-digital on-chip oscillator using relative reference modeling," Very Large Scale Integration (VLSI) Systems, IEEE Transactions on, vol. PP, no. 99, pp. 1 –6, 2011.

[5] S.-Y. Seo, J.-H. Chun, Y.-H. Jun, S. Kim, and K.-W. Kwon, "A digitally controlled oscillator with wide frequency range and low supply sensitivity," Circuits and Systems II: Express Briefs, IEEE Transactions on, vol. 58, no. 10, pp. 632 –636, oct. 2011.

[6] S. Höppner, R. Schüffny, and M. Nemes, "A low-power, robust multi-modulus frequency divider for automotive radio applications," in Mixed Design of Integrated Circuits Systems, 2009. MIXDES '09. MIXDES-16th International Conference, 2009, pp. 205 –209.

[7] S. Höppner, S. Henker, H. Eisenreich, and R. Schüffny, "An open-loop clock generator for fast frequency scaling in 65nm CMOS technology," in Mixed Design of Integrated Circuits Systems, 2011. MIXDES '11. MIXDES-18th International Conference, Jun. 2011.

[8] H. Eisenreich, C. Mayr, S. Henker, M. Wickert, and R. Schüffny, "A novel ADPLL design using successive approximation frequency control," Microelectron. J., vol. 40, pp. 1613–1622, Nov. 2009.

[9] R. J. Baker, CMOS Circuit Design, Layout and Simulation, Second Edition. IEEE, WILEY-INTERSCIENCE, 2005.

Wideband Low-Noise RF Front-End for CNT-NEMS Sensors

Christian Kauth, Marc Pastre and Maher Kayal
STI-IEL-Electronics Lab
Ecole Polytechnique Fédérale de Lausanne (EPFL)
CH-1015 Lausanne, Switzerland
christian.kauth@epfl.ch

Abstract—**Hybrid NEMS interfaces are the key to systems combining the benefits of highly sensitive miniaturized mechanical sensors with the vast functionalities available in electronics. In this context, we analyze diverse RF front-ends meant for interfacing high-impedance carbon nanotube based NEMS. Given the feeble signals from the NEMS, their high output impedance and non-negligible interconnect parasitics, front-end design must imperatively focus on minimal noise figure. Limits on minimal detectable signal are extracted via design, simulation and characterization of a 3-stage common-emitter front-end.**

Index Terms—**Signal to noise ratio, Signal detection, Noise figure, Low-noise amplifiers, Radiofrequency amplifiers, Wideband, Nanoelectromechanical systems.**

I. INTRODUCTION

Carbon nanotube nanoelectromechanical systems (CNT-NEMS) start finding their way to sensor applications. Their small dimensions and important resonance frequencies make them candidates for high resolution sensing of physical and chemical phenomena [1]. Electronic detection of the mechanical resonance plays a crucial role in the sensing process. So far, most experiments relied on expensive and cumbersome RF equipment like spectrum and network analyzers, to read the fragile signal out of the NEMS. The small signal strength, high resonance frequencies and large parasitic capacitances of the measurement setup generally do not allow for a direct read-out. This justifies the widespread mixer setup used to detect resonance [2]. Although mixing allows to detect resonance, it suffers from phase information loss, which limits the use of the NEMS sensor to open-loop configurations. With the goal of closing the loop and avoiding the use of rich RF lab equipment, we assess in this paper the potential of simple RF front-ends to sense the signal from the NEMS via a direct measurement. The equivalent hybrid electromechanical interface is presented in section II, while section III provides a prediction of the achievable noise figure. System bias is the topic of section IV and the predictions are validated via measurements in section V. Alternative front-end implementations are finally highlighted in section VI.

II. HYBRID ELECTROMECHANICAL INTERFACE

In the realm of electronics, carbon nanotubes have successfully been used in transistor configurations, operating the CNT as the channel material [3]. Defining the tube terminations as source and drain, in analogy to the MOS transistor, a nearby gate may control the current flow through the device. Clamping drain and source, while suspending the tube over a trench, adds mechanical degrees of freedom to the system. Driving the gate with a control voltage, induces an electrostatic force that sets the CNT into motion. This mechanical motion translates then back into an electrical signal via field- and piezoresistive-effects, which the front-end is to read in order to decode the motional information [4]. Although parasitic coupling from the gate to the drain is inevitable, this problem is to be solved either on the NEMS design side or at the filter stage, following the front-end. The major concern in this paper is to design a low-noise front-end, providing the subsequent circuit stages with the best possible signal to noise ratio (SNR).

As indicated in Fig.1, the CNT, biased to saturation, presents an output resistance, which comes with a noise current spectral density

$$S_{i^2,CNT} = \frac{4kT}{R_{CNT}} \qquad (1)$$

and a parasitical source-drain capacitance, which circuit designers cannot impact. Follows in series a relatively large, noisy contact resistance $R_{contact}$, caused by mode reduction at the 3-dimensional metal to unidimensional CNT interface. For the following discussion, we suppose a decent NEMS design with a device that can be operated as a transistor and has measurable saturation, meaning that the contact resistance $R_{contact}$ is inferior to the CNT's output resistance R_{CNT}. This output resistance R_{CNT} is in the order of MΩ [5] and the parasitical capacitance C_{CNT} will typically not exceed hundreds of aF, guaranteeing an output impedance above MΩ up to GHz frequencies. Pads and wirebonding to take the signal from the chip out of the package onto the printed circuit board (PCB), hosting the electronic front-end, will add in each case 1nH of inductance and a 200fF capacitance. On the PCB, the cumulative parasitical pad and track capacitance easily reaches C=2pF. Although this capacitance can be cancelled out via a parallel resonant circuit in narrow-band applications, we here target a wide-band front-end, that can adapt to the CNT's changing resonance frequency (over 2 decades) upon sensing. With the aim of minimizing the front-end's noise figure (NF) and hence maximizing the system's overall SNR, we look next at an implementation of the generic front-end of Fig.1, namely the 3-stage common-emitter (CE) amplifier presented in Fig.2. Following a preliminary study of the noise-critical first stage

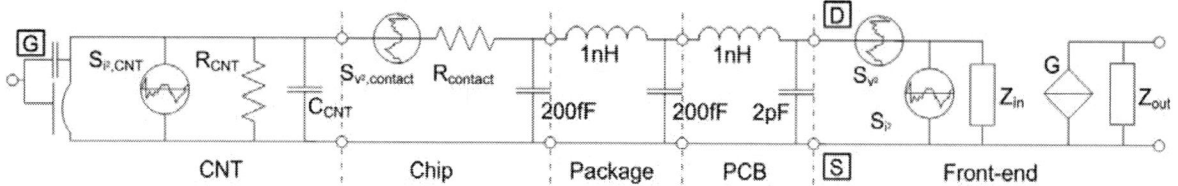

Fig. 1. Small signal equivalent circuit from NEMS to front-end

in section III, the full 3-stage amplifier is treated in sections IV & V.

III. NOISE FIGURE PREDICTION

Motivated by Friis' formula for the noise factor of a cascade of stages,

$$F_{total} = F_1 + \sum_{n=2}^{N} \frac{F_n - 1}{\prod_{i=1}^{n-1} G_i}, \qquad (2)$$

where F_n is the noise factor of stage n and G_n the power gain of that same stage, we limit our analytical study to single-stage front-ends, loaded by a noisy resistance R_L. At sufficient gain, this first stage virtually defines the whole circuit's noise factor and its design merits special care. The considered topology uses a bipolar junction transistor (BJT) in a common emitter configuration. Its small signal schematic, including all white noise sources, is shown by the inset of Fig.3. CNT-NEMS operating generally in the MHz to GHz frequency range, hypothesizing operation above the device 1/f-corner frequency [6] is reasonable and one may thus safely neglect Flicker noise.

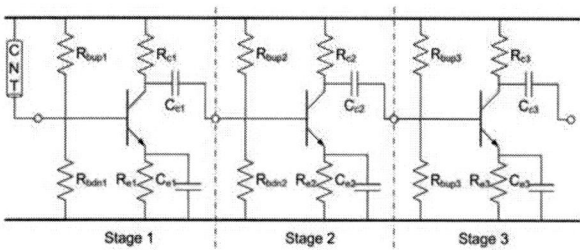

Fig. 2. 3-stage biased CE low-noise amplifier

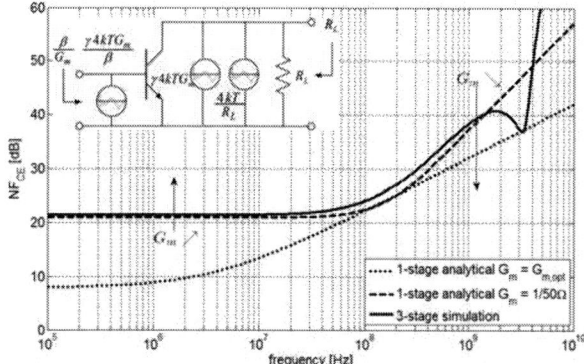

Fig. 3. NF of CE front-end with C=2pF, R_{CNT}=1MΩ, R_L=1kΩ

The circuit's noise contributions come from the BJT's shot noise at base and collector, completed by the load's thermal noise. Connecting this front-end to the CNT as suggested in Fig.1, leads to the following noise factor expression

$$
\begin{aligned}
F_{CE} = \quad & 1 & \text{by CNT} \quad (3)\\
+ \quad & \frac{\gamma}{\beta^2} R_{CNT} G_m \left[1 + \left(\omega \frac{\beta C}{G_m}\right)^2\right] & \text{by collector}\\
+ \quad & \frac{\gamma}{\beta} R_{CNT} G_m & \text{by base}\\
+ \quad & \frac{R_{CNT}}{\beta^2 R_L} \left[1 + \left(\omega \frac{\beta C}{G_m}\right)^2\right] & \text{by load,}
\end{aligned}
$$

where G_m is the transconductance, γ the noise excess factor, β the current gain, and C is the parasitic interconnect capacitance, overshadowing any base-emitter coupling. The noise figure

$$NF = 10 \log F, \qquad (4)$$

is plotted against frequency on Fig.3, indicating the existence of an optimal bias-point according to the targeted operation frequency. The important NF values result from the combination of the CNT's large output impedance and the pF interconnect capacitance. Given that CE topologies possess enough power gain to subvert the load's noise contribution, the NF is limited by the base noise at low frequencies and by the collector noise at high frequencies. The continuous line corresponds to the simulated NF of the 3-stage CE front-end, designed to operate at 100MHz and presented subsequently. One can observe that the 3-stage simulation follows rather well the 1-stage analytical prediction, portending that the noise contribution of stages 2 and 3 remains negligible with respect to the first stage's noise.

IV. SYSTEM BIAS

The 3-stage CE front-end henceforth considered is depicted in Fig.2. Insensitive to the transistors' current gain β, the bias networks of each stage are decoupled via capacitors C_{ci}. The base potential is a free design variable and hence presents a degree of freedom that, in the case of the first stage, can be exploited to adjust the base voltage as to properly set the CNT's DC bias to roughly 1 to 10 μA [7]. The emitter capacitance C_{ei} grounds the bias resistor R_{ei} over the widest feasible bandwidth to avoid noise contributions and negative feedback. Choosing R_{up1} and R_{dn1} larger than the BJT's input impedance $\frac{\beta}{G_m}$, renders their contribution to the noise figure insignificant. To keep the biasing independent from the current gain β, current through the base biasing resistors shall render

290

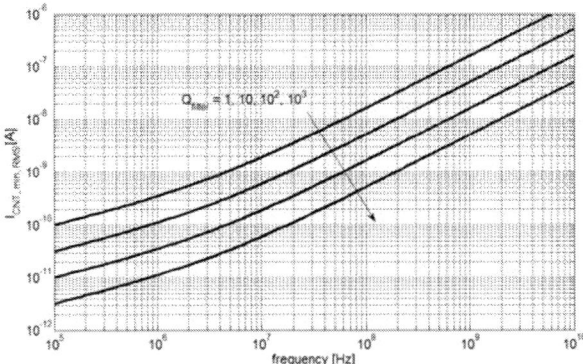

Fig. 4. Minimal detectable current from CNT with R_{CNT}=1MΩ through 2pF interconnects with a CE (β=100) front-end at ambient temperature for SNR$_{out}$=1.

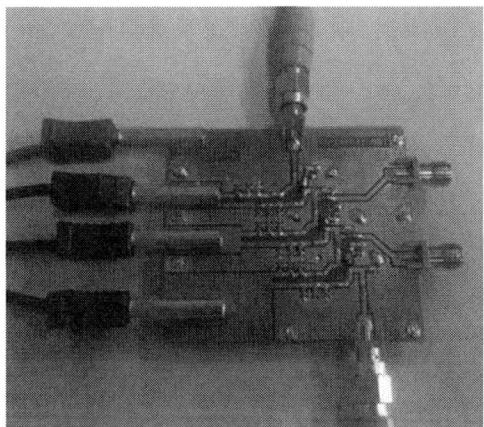

Fig. 5. 3-stage discrete component CE front-end

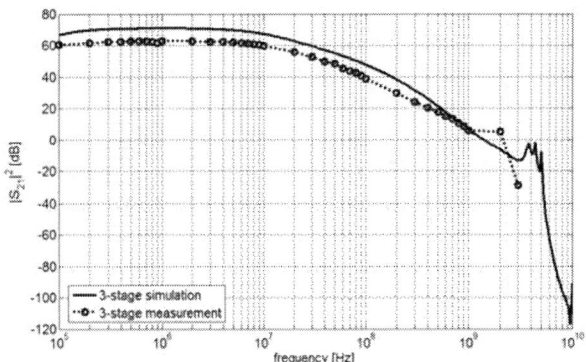

Fig. 6. 3-stage CE forward power gain in 50Ω framework

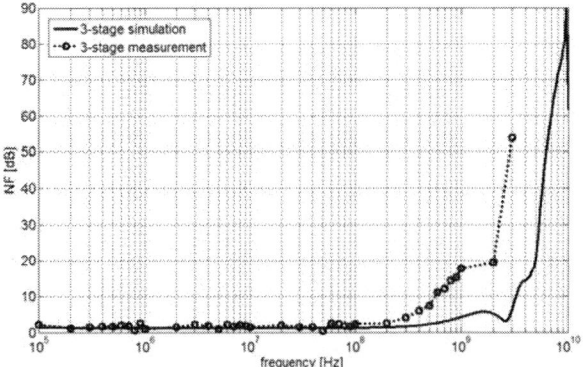

Fig. 7. 3-stage CE noise figure in 50Ω framework

the BJT base current negligible. The last two constraints can be achieved simultaneously by providing sufficient voltage supply. The optimal stage-1 collector current follows from derivation of expression (3) with respect to G_m, yielding

$$I_{C0,opt} = \sqrt{2\beta}\frac{kT}{q}\omega C, \qquad (5)$$

with q the elementary positive charge. The bias of the subsequent stages may be chosen as to provide maximum stable gain (large I_{C0}), as their impact on the NF is less significant. For such an optimal bias, the minimal detectable current variation out of the CNT is given by

$$I_{CNT,RMS} \geq \sqrt{\frac{4kTB}{R_{CNT}} \cdot F_{opt} \cdot \text{SNR}_{out}}, \qquad (6)$$

where B is the bandwidth of the subsequent filter and SNR$_{out}$ the desired signal to noise ratio at the output of the filter. Fig.4 suggests that the optimally biased CE front-end is reliant on the use of filters to detect CNT currents of nA amplitude at resonance frequencies above tens of MHz.

V. MEASUREMENTS

With the objective of validating the previous conclusions drawn for the CE front-end interfacing a CNT-NEMS, exclusively based on analytical analyses and simulations, this section confronts the simulations with measurement results for the 3-stage CE front-end in the common 50Ω RF framework. Although the absolute values of gain and NF happen to be very different from what they would be in the CNT MΩ framework, the mere fact that the measurements match the simulation (see Fig.6 & 7), provides the necessary provisional faith into the previously drawn conclusions on noise figure (see Fig.3) and minimal detectable signal from the CNT-NEMS (see Fig.4). A next step consists in verifying those predictions via measurement on a real CNT-NEMS.

The PCB in question is depicted on Fig.5. 50Ω impedance matching was omitted, as the front-end is primarily meant for MΩ NEMS interfaces. The measured and simulated forward power gain is plotted on Fig.6. The low-pass cut-off is due to the pole defined by $\omega_{p1} = \frac{G_m}{C_e}$. Although it is possible to extend the passband to lower frequencies, a second pole limits this endeavour at $\omega_{p2} = \frac{1}{(R_{in}+R_c)C_c}$, R_{in} being the input impedance of the next stage, speak $\frac{\beta}{G_m}$. While the measured pass-band frequencies were accurately predicted by the simulation, its gain presents an 8dB discrepancy. The

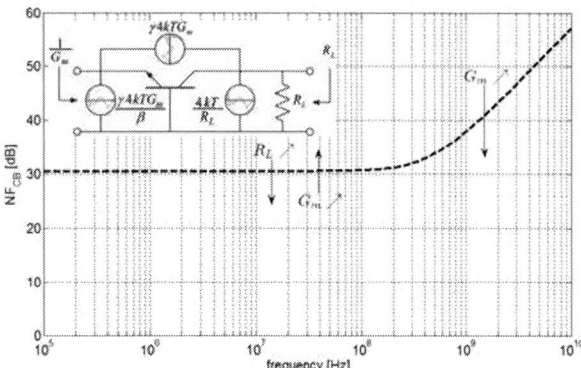

Fig. 8. NF of CB front-end with C=2pF, R_{CNT}=1MΩ, R_L=1kΩ, $1/G_m$=50Ω

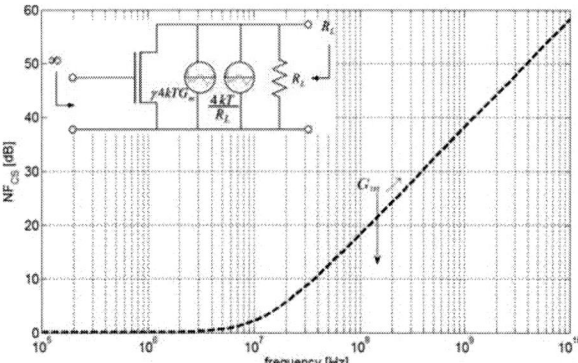

Fig. 9. NF of CS front-end with C=2pF, R_{CNT}=1MΩ, R_L=1kΩ, $1/G_m$=50Ω

power gain in the passband writing

$$|S_{21}|^2 = 20 \log \frac{2\beta G_m R_c}{\beta + G_m R_c}, \quad (7)$$

the difference between measured and simulated current gains β accounts for twice 1.9dB and the cabling losses for 0.4dB, leaving 3.8dB of yet unexplained discrepancy. More importantly does the NF, shown in Fig.7, confirm that the simulation predicts the measurement trend up to the front-end's gain decay. By this validation principle, we can prognosticate for the CE front-end an NF similar to the simulated one of Fig.3 when interfacing the CNT-NEMS, implying an upper bound on signal sensitivity provided by Fig.4.

VI. ALTERNATIVE TOPOLOGIES

To conclude this round-trip on RF front-ends for CNT-NEMS sensors, we briefly highlight 3 alternate topologies:

A. Common base

The common base (CB) front-end, represented in the inset of Fig.8, also relies on bipolar technology. Compared to the CE, it presents a larger pass-band at the cost of a unity current gain. This implies a stronger noise contribution from the load,

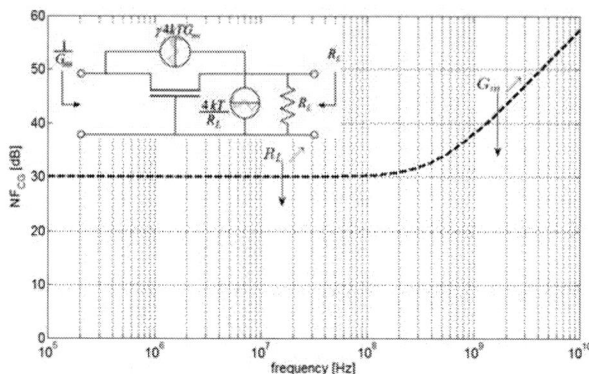

Fig. 10. NF of CG front-end with C=2pF, R_{CNT}=1MΩ, R_L=1kΩ, $1/G_{ms}$=50Ω

as emanates from

$$
\begin{aligned}
F_{CB} = \quad & 1 & \text{by CNT} \quad (8) \\
+ \quad & \frac{\gamma}{\beta^2} R_{CNT} G_m \left[1 + \left(\omega \frac{\beta C}{G_m} \right)^2 \right] & \text{by collector} \\
+ \quad & \frac{\gamma}{\beta} R_{CNT} G_m & \text{by base} \\
+ \quad & \frac{R_{CNT}}{R_L} \left[1 + \left(\omega \frac{C}{G_m} \right)^2 \right] & \text{by load.}
\end{aligned}
$$

At low frequencies, the NF is limited by load and base shot noise, while the collector shot noise dominates at high frequencies. Although a large load enhances gain and noise performance at low frequencies, the performance of a CB front-end remains inferior to the CE in terms of noise.

B. Common source

The noise of metal-oxide-semiconductor (MOS) front-ends can be summarized to the channel thermal noise, with negligible gate-induced noise in the considered frequency range. Its immense input impedance makes the common source (CS) an ideal candidate for low NF at relatively low frequencies, as arises from Fig.9. The NF, given by

$$
\begin{aligned}
F_{CS} = \quad & 1 & \text{by CNT} \quad (9) \\
+ \quad & \frac{\gamma}{R_{CNT} G_m} \left[1 + (\omega R_{CNT} C)^2 \right] & \text{by channel} \\
+ \quad & \frac{1}{G_m^2 R_{CNT} R_L} \left[1 + (\omega R_{CNT} C)^2 \right] & \text{by load,}
\end{aligned}
$$

can be enhanced at any frequency at the cost of a larger biasing current. At low frequencies, the SNR is limited by the CNT noise, while the channel noise is responsible for the increasing NF at high frequencies.

C. Common gate

A large load and hence large voltage gain is beneficial to a common gate (CG) topology at low frequencies. As for the CS front-end, at high frequencies, increased bias current enhances the NF

$$F_{CG} = \quad 1 \qquad \text{by CNT} \quad (10)$$
$$+ \quad \frac{\gamma}{R_{CNT}G_{ms}}\left[1 + (\omega R_{CNT}C)^2\right] \quad \text{by channel}$$
$$+ \quad \frac{R_{CNT}}{R_L}\left[1 + \left(\omega\frac{C}{G_{ms}}\right)^2\right] \quad \text{by load.}$$

Its noise performance is comparable to the one of CB.

VII. CONCLUSION

We designed, analyzed and characterized a 3-stage common emitter front-end and found it to present large noise figures in the MHz to GHz frequency range, when interfacing a carbon-nanotube NEMS. This significant noise contribution of the circuit is inherently ascribed to the NEMS's high output impedance in combination with a pico-farad interconnect capacitance.

The minimal detectable signal out of a CNT-NEMS resonating at 100MHz was evidenced to be in the order of tens of nano-amps for a wide-band common emitter front-end, optimally biased for minimum noise. This limit can be bypassed in narrow-band applications either by resonating out the interconnection parasitical capacitance with a parallel LC circuit or by using a subsequent band-pass filter stage.

Finally the common source front-end was shown to be a promising candidate for low frequency sensing (up to tens of MHz), due to its insignificant contribution to the system's signal to noise ratio in this frequency range.

ACKNOWLEDGMENT

This research has been funded by Nano-Tera.ch, a program of the Swiss Confederation, evaluated by SNSF.

REFERENCES

[1] H. B. Peng, C. W. Chang, S. Aloni, T. D. Yuzvinsky, and A. Zettl. "Ultrahigh frequency nanotube resonators," *Phys. Rev. Lett.*, vol. 97, no. 8. p. 087203, Aug 2006.

[2] V. Sazonova, Y. Yaish, H. Ustunel, D. Roundy, T. A. Arias, and P. L. McEuen, "A tunable carbon nanotube electromechanical oscillator," *Nature*. vol. 431, no. 7006, pp. 284–287, Sep. 2004.

[3] J. Chaste, L. Lechner, P. Morfin, G. Feve, T. Kontos, J.-M. Berroir. D. C. Glattli, H. Happy, P. Hakonen, and B. Placais, "Single carbon nanotube transistor at ghz frequency," *Nano Letters*, vol. 8. no. 2, pp. 525–528. 2008, pMID: 18229967. [Online]. Available: http://pubs.acs.org/doi/abs/10.1021/nl0727361

[4] G. Y. Guo, L. Liu, K. C. Chu, C. S. Jayanthi, and S. Y. Wu, "Electromechanical responses of single-walled carbon nanotubes: Interplay between the strain-induced energy-gap opening and the pinning of the fermi level," vol. 98, no. 4, p. 044311, 2005. [Online]. Available: http://dx.doi.org/doi/10.1063/1.2011781

[5] *Carbon Nanotubes: Properties and Applications.* CRC, 2006, p. 93.

[6] P. G. Collins, M. S. Fuhrer, and A. Zettl, "1/f noise in carbon nanotubes," *Applied Physics Letters*, vol. 76, no. 7, pp. 894–896, 2000. [Online]. Available: http://link.aip.org/link/?APL/76/894/1

[7] A. Javey, J. Guo, D. B. Farmer, Q. Wang, E. Yenilmez, R. G. Gordon. M. Lundstrom, and H. Dai, "Self-aligned ballistic molecular transistors and electrically parallel nanotube arrays," *Nano Letters*. vol. 4, no. 7, pp. 1319–1322, 2004. [Online]. Available: http://pubs.acs.org/doi/abs/10.1021/nl049222b

Thermal Issues
in Microelectronics

MIXDES 2012, 19th International Conference *"Mixed Design of Integrated Circuits and Systems"*, May 24-26, 2012, Warsaw, Poland

1/f Noise Temperature Behaviour of Poly Resistors

Walter C. Pflanzl, Ehrenfried Seebacher

Austriamicrosystems/TAOS

Unterpremstaetten, AUSTRIA

walter.pflanzl@austriamicrosystems.com

Abstract— **This paper presents the 1/f noise behaviour over a temperature range from -50°C to +200°C for poly resistor devices with sheet resistance 50 Ω/□ and 1.2 kΩ/□. Based on statistical measurement data a classical approach of 2th order is used to model the flicker noise characterization data as function of temperature with sufficient accuracy.**

Index Terms—1/f noise, temperature modeling, poly resistor

I. INTRODUCTION

Modern silicon processes dramatically improve the switching speed [1], the power consumption, driver capability and RF noise. Beside these parameters others like flicker noise are of interest especially in analog applications optimized for low power, weak signal sensor interfacing or even audio circuits with emphasis for low noise requirements.

MOS devices in general can be easily optimized by length/width variation; higher the value, lower the flicker noise e.g. the flicker noise formula in [2]. Beside geometry also electrical parameters determine the noise performance as well as layout and design of the circuit. Proper optimization drastically reduces the noise thus another group of devices will play a key role: poly resistors.

Resistors, especially poly resistors, do not show thermal noise alone, they also contribute significantly with flicker noise to modern silicon circuits [3]. Accurate modeling of flicker noise of poly resistors becomes a heavy demand for high performance analog design kits. This paper tries to go a step further by temperature behaviour modeling of resistor flicker noise.

II. THEORETICAL ASPECTS OF TEMPERATURE BEHAVIOUR OF FLICKER NOISE OF RESISTORS

A. Thermal noise

Every resistor generates unavoidable thermal noise, the spectral noise current density S_{ir} writes as

$$ S_{ir} = \frac{4kT}{R(V,T)} \quad \left[\frac{A}{Hz^2}\right] \tag{1} $$

with k the Boltzmann constant, T the ambient temperature in Kelvin and the resistance R as a function of voltage (self-heating) and temperature. The S_{ir} increases proportionally with temperature rise.

B. CMC flicker noise formula for resistors

There are several flicker noise formulations available, but the compact modeling council (CMC) approach [4] seems to be accurate enough at nominal temperature for the spectral noise current density S_{ir}:

$$ S_{ir} = KF(T)\left(\frac{i}{W}\right)^{AF} \frac{W_{eff}}{L_{eff}} \frac{1}{f^{BF}} \quad \left[\frac{A}{Hz^2}\right] \tag{2} $$

KF(T), AF, BF ... noise parameters with AF fixed to 2 and BF fixed to 1, i ... resistor current, W_{eff}, L_{eff} ... effective width and length of the resistor, f ... frequency

The temperature model is a linear approach, but can be easy extended to higher orders, if necessary. We propose a second order formulation as sufficient:

$$ KF(T) = KF_{T_0}\left(1 + TC1KF(T - T_0) + TC2KF(T - T_0)^2\right) \tag{3} $$

KF(T) ... noise parameter as function of temperature, T_0 ... nominal temperature (27°C), KF_{T0} ... KF for T_0, TC1KF, TC2KF ... linear and quadratic temperature coefficient, T ... temperature in Kelvin.

In [5] we can find hints about the expectable temperature evolution for flicker noise: Higher the temperature than nominal will show lower LF noise whereas lower the temperature than nominal will show higher LF noise.

III. CHARACTERISATION

A. Measurement setup

Good quality models can only be generated with good quality characterization data. For this paper a new type of probe station is used from CASCADE™ with emphasis of proper external noise suppression. From the same supplier there is a measurement module (EDGE™ system [6]) available with proper filtering circuits, LNA and spectrum analyzer for flicker noise measurements adopted specially for this very probe station.

For comparison purpose a setup for resistor bridge flicker noise measurement (similar to [7]) was used. With this setup only packaged devices could be measured without temperature control. Also the measurement capacity was very limited compared to an automated on wafer solution. On the other

297

hand the flicker noise data measured on wafers are more disturbed by interferences, nevertheless in good agreement with the resistor bridge measurements.

B. Devices used

Two types of resistors have been taken into account: one with low ohmic sheet resistance (LOSR) of 50 $\Omega/\square$ and the second one with a high ohmic sheet resistance (HOSR) type with 1.2 kΩ /$\square$.

C. Biasing & measurement strategy

A fixed voltage across the resistor has been used to bias the DUT. The resistor current has been measured in order to determine the resistance and the corresponding temperature has been recorded too. The characterization is divided into two parts:

1) Noise parameter determination: Three different geometries have been measured at different bias (figure 1 and 3).

2) Noise parameter temperature coefficient determination: The temperatures used for this analysis were -50°C, 27°C (nominal temperature), 100°C (only for LOSR type) and 200°C. The minimum width resistor structure was selected for measurements, because it delivered best "clean" measurement noise data. At least 10 devices have been measured in order to get an impression of possible device stray (figure 2 and 4).

IV. MODELING APPROACH

The modeling methodology is separated into two parts: the DC modeling and the flicker noise modeling.

A. DC modeling methodology

In formula (2) it may be seen that the noise is directly proportional to the current flowing through it. Therefore a precise DC model is necessary. This was done with the help of a special test chip to cover a wide range of widths and lengths for geometry scaling as well as to investigate the stationary self-heating and the back bias effect. More details are given in [8].

B. Flicker noise modeling methodology

As it has been mentioned before, the flicker noise parameter determination is divided into two steps: KF extraction at nominal temperature and temperature coefficient TCxKF determination. Due the simplicity of the noise parameter KF(T) formula (2), and by keeping the values EF=1 and AF=2 constant, the extraction of KF and of both temperature coefficients was straightforward.

Results of the model parameter extraction for two types of resistors are presented in Table I:

TABLE I. MODEL PARAMETRS

	KF	*TC1KF*	*TC2KF*
LOSR	2.45E-22	-8.90E-03	2.80E-05
HOSR	7.00E-22	-4.80E-03	3.20E-06

C. Model versus measurement

All extractions have been done for the spectral noise voltage density Svr, because the measurement setup directly measures it as function of frequency (figures 1-4).

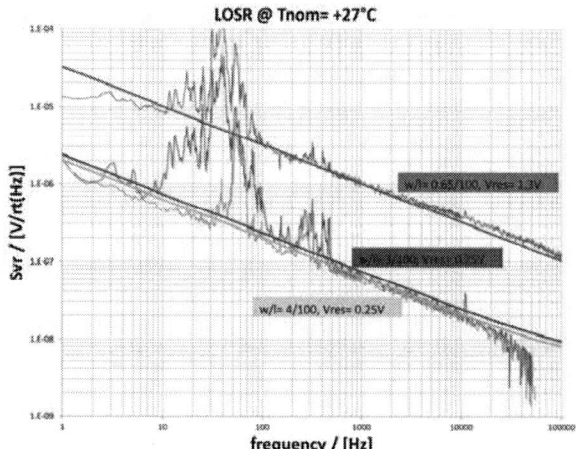

Fig. 1.   Voltage noise spectral density Svr over frequency for low ohmic sheet resistors (LOSR) for three different geometries (solid line= model).

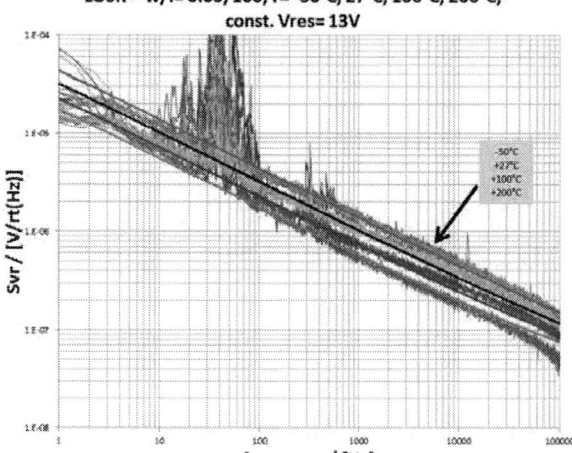

Fig. 2.   Voltage noise spectral density Svr over frequency for LOSR @ Vres= 13V of one geometry at four different tempertures. For each temperature up to 15 devices have been measured.

V. NOISE CONTRIBUTORS - MODEL IMPROVEMENT?

As already demonstrated, formula (2) describes in quite good agreement the evolution of flicker noise over temperature. Further investigations showed that the flicker noise behaviour cannot be explained by the temperature slope of the resistor and thus moving the bias point for const. Vres. The addition for KF with formula (3) is a must which is displayed in fig. 5 and 6.

The model improvement by use of the temperature model (3) for KF is up to $\pm 40\%$ (fig. 5) for the LOSR type resistor and for the HOSR it varies from $+10$ to -50% (fig. 6).

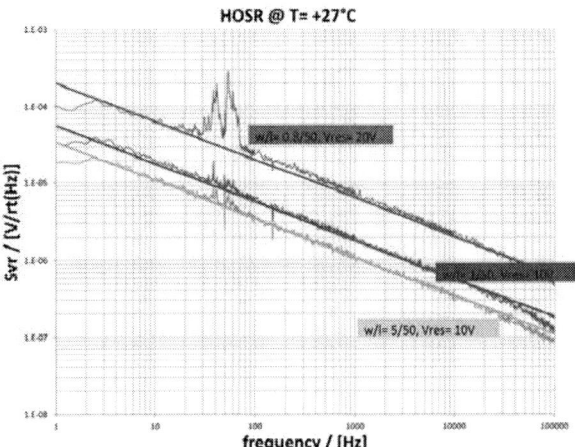

Fig. 3. Voltage noise spectral density Svr versus frequency for high ohmic sheet resistors (HOSR) for three different geometries (solid line= model).

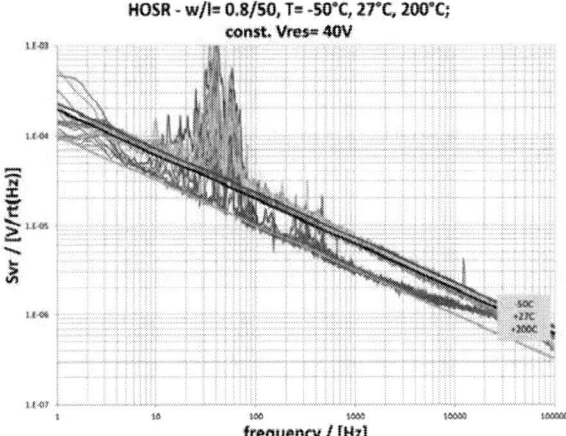

Fig. 4. Voltage noise spectral density Svr over frequency for HOSR @ Vres= 40V of one geometry at four different tempertures. For each temperature up to 15 devices have been measured.

VI. SUMMARY

This paper demonstrates the feasibility to measure and model the flicker noise of poly resistors over the temperature range from -50°C to $+200^\circ$C. The change of resistance over temperature is not capable to explain the flicker noise temperature behaviour, thus the temperature model for KF is required. The temperature coefficients for KF have been extracted for two types of sheet resistances: 50 $\Omega/\square$ and 1.2 $k\Omega/\square$ resistors. The temperature model fits reasonable good with the measurement data. The model improvement is in the range of up to 50%.

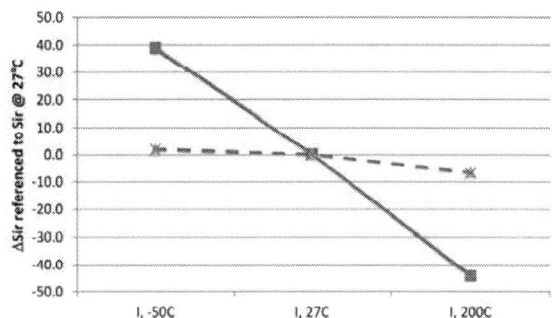

Fig. 5. The dashed blue line corresponds only the resistance change (LOSR) over temperature in relation to the value at T_0 in percentage thus Sir is caluclated with fixed KF. The solid red line includes the KF(T) model and demonstrates the model improvement.

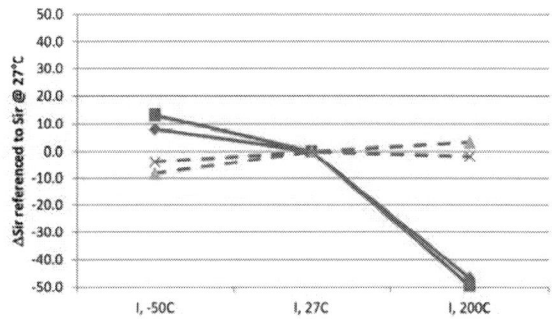

Fig. 6. The dashed lines correspond only the resistance change (HOSR) over temperature in relation to the value at T_0 in percentage thus Sir is caluclated with fixed KF. The solid lines include the KF(T) model and demonstrate the model improvement (blue corresponds Vres= 4V and red Vres= 40V).

REFERENCES

[1] http://en.wikipedia.org/wiki/IBM_z196_(microprocessor)

[2] http://www-device.eecs.berkeley.edu/~bsim3/BSIM4/BSIM470/BSIM470_Manual.pdf

[3] X.Y. Chen, J.A. Johansen, C. Salm, A.D. van Rheenen, "On low-frequency noise of polycrystalline GexSi1-x for sub-micron CMOS technologies", Solid-State Electronics 45 (2001) 1967-1971

[4] Eldo Device Equations Manual, Release AMS 2010.2, p. 123

[5] Nitin K. Rajan, David A. Rotenberg, Jin Chen and Mark A. Reed, „Temperature depenendece of 1/f noise mechanisms in silicon nanowire bichemical field effect transistors" 14. December 2010, Applied Physics Letters 97, 243501 (2010)

[6] http://www.cmicro.com/products/probe-systems/dedicated-systems/flicker-noise

[7] F. Seifert, "Resistor current noise measurements", April 14, 2009 https://dcc.ligo.org/public/0002/T0900200/001/current_noise.pdf

[8] W.C. Pflanzl and E. Seebacher, "Poly resistor modeling over a wide range of geometries and their different temperature and voltage behaviour for a HV CMOS process", MIXDES 2008, pp. 421–424

MIXDES 2012, 19ᵗʰ International Conference *"Mixed Design of Integrated Circuits and Systems"*, May 24-26, 2012, Warsaw, Poland

Compensation of the Temperature Fluctuations in the Silicon Photomultiplier Measurement System

Mateusz Baszczyk, Piotr Dorosz, Sebastian Głąb, Wojciech Kucewicz, Łukasz Mik, Maria Sapor

AGH University of Science and Technology
Faculty of Electrical Engineering, Automatics, Computer Science and Electronics
Department of Electronics
al. A. Mickiewicza 30
30-059 Krakow, Poland
e-mail kucewicz@agh.edu.pl

Abstract—**This paper presents the method for the compensation of the temperature fluctuations in the measurement system based on the Silicon Photomultipliers. The system was created by the authors of the paper. Temperature has very strong influence on the gain of the detectors. Because single photons are detected and the system is very sensitive, its parameters has to be steady during the measurements. Temperature is being stabilized by changeable bias voltage of the detector. This paper describes how the algorithm of compensation has been determined during series of measurements and calculations.**

Index Terms—**Silicon Photomultiplier; SiPM; tempeature influence; photon detector; photon counting**

I. INFLUENCE OF THE TEMPERATURE ON THE P-N JUNCTION

Independently of the direction the p-n junction is polarized in, the currents in it are strongly dependent on temperature. The total current of the p-n junction can be described as a sum of the diffusion current in the neutral areas and the carrier generation-recombination current in the depletion region [3].

$$J \cong q \sqrt{\frac{D_p}{\tau_p}} \frac{n_i^2}{N_D} + \frac{q n_i W_D}{\tau_g} \qquad (1)$$

D_p is the diffusion coefficient of the holes, W_D is the width of depletion region, k_B is the Boltzman constant, τ_p is the lifetime of the excessive minority carriers and τ_g is the lifetime of electron-hole pairs. D, n_i and τ are strongly dependent on the temperature T. But changes of the diffusion coefficient and the lifetime of the carriers can be acknowledged as negligibly small in comparison to the strong dependence of n_i and the temperature, expressed by [3].

$$n_i \propto \left(T^{3/2} \cdot exp\left(\frac{-E_g}{2k_B T}\right) \right) \qquad (2)$$

It has been estimated that each increase of the temperature by 10°C causes the reverse current to grow twice. In the state of the avalanche breakdown, the increase of the temperature contributes to the more vivid vibrations of the particles of the crystal lattice. The vibrating atom occupies more space and the probability of the collision with an accelerated carrier increases. The collisions occur earlier so the free path is shorter. It means that the carriers are accelerated on shorter path and have smaller kinetic energies. Insufficient energy results in a reduction of the probability of knocking out carriers pairs. Avalanche multiplication becomes weaker and the avalanche current decreases. With steady voltage avalanche current decreases and with steady current avalanche breakdown voltage increases.

If the temperature of the Silicon Photomultiplier is higher than 0°K, inside the detector, due to vibrations of the lattice, pairs of the electron-hole carriers are created. It is called the thermal generation of the carriers. The probability of detecting the photon (detecting the absorption of the photon resulting in the generation of avalanche current) is directly proportional to the value of bias voltage of the detector. The more this voltage exceeds the breakdown voltage of the photodiode, the higher is the chance the avalanche of the charged carriers will appear. During the absence of light (lack of photons), high bias voltage enables single, charged carrier coming from thermal generation in the depletion region to trigger the ionization process resulting in the creation of an avalanche.

II. MEASUREMENT SYSTEM

The measurements were carried out with the use of acquisition system presented in Fig. 1. The signals from Silicon Photomultipliers are processed through the ASIC designed for this system, then converted in Analog to Digital Converters and sent to FPGA module [4]. The system is able to detect the number of photons ranging from a single one to over hundred of them. Used detectors have been tested and the light has been measured in the variety of temperatures: from -150 °C to 30 °C. This enabled us to proceed to the stage where the compensation algorithm could be determined.

Figure 1. System of data acquisition.

III. EXPECTED DEPENDENCIES

All of the measurements are performed for small intensity of light. Single photons are being registered and their statistical distribution is calculated and presented in the form of a histogram (ADC - Analog to Digital Converter units).

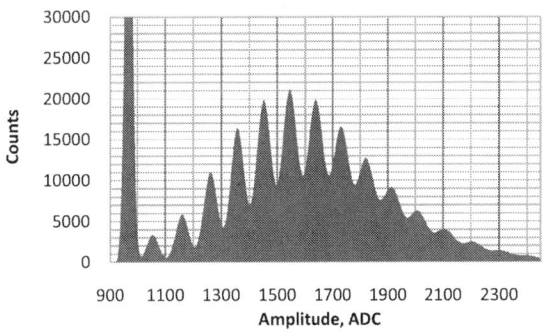

Figure 2. Measurement data presented in a form of the histogram.

The histogram from Fig. 2 consists of Gaussian curves. Each curve corresponds to the number of photons registered by the detector simultaneously. The first peak (for the lowest ADC number) represents the number of times when only one photon appeared in the Silicon Photomultiplier. The second peak shows how often two photons were present, the third - three photons, etc.

In the stable temperature and bias voltage, the gain of the charge created as a result of the absorption of a photon is always constant for the specific detector. It means that the voltage corresponding to the generation of a single avalanche (single photon detected) is stable. This value of voltage is referred as the gain of a single photon. Its evaluation is based on the distances between following local maxima (peaks) on the histogram. Gain per photon is the arithmetic mean of these distances expressed in mV.

Because it is known that gain depends on bias voltage and temperature it can be characterised as a function G(V,T), where V is the bias voltage and T is the temperature of the detector. From the former experiments [2] it is known that the gain is lineary dependent on both voltage and temperature.

$$G(V,T) = \frac{dG}{dV}\left(V - V_{BD}(T)\right) \quad (3)$$

$$G(V,T) = \frac{dG}{dT}T + G(T_0,V) \quad (4)$$

V_{BD} is the breakdown voltage and T_0 is a reference temperature. Even a very small change of either temperature or bias voltage results in different gain per photon value in the measurement. Because of the fact that the measurement system is extremely sensitive, its working conditions should be as stable as possible. It is not possible to create an ideal measurement station with unchangeable temperature. The most practical solution is to use the value of the bias voltage of the detector to correct temperature influence.

Gain per photon is directly proportional to the bias voltage and it linearly decreases with the increase of the temperature. It is possible to create an algorithm which would compensate the changes in gain caused by the fluctuations of the temperature by adjusting bias voltage.

IV. MEASUREMENTS IN VARIOUS TEMPERATURES

All measurements were taken with the use of Silicon Photomultiplier 4020 produced by FBK Foundation (Italy).

Firstly, the dependencies showing that the gain is a linear function of voltage and temperature had to be confirmed.

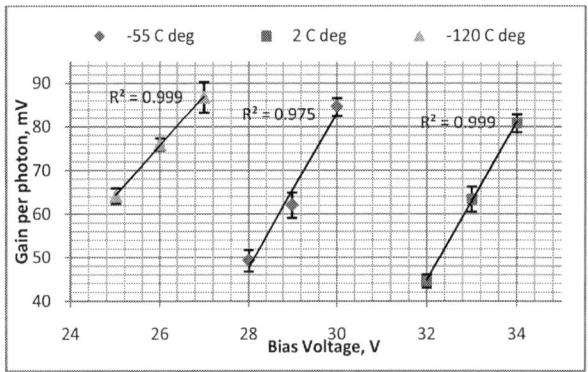

Figure 3. Gain per photon as a function of bias voltage (the temperature is stable).

Fig. 3 shows that gain increases with bias voltage. The increase stays linear for various temperatures of measurements environment.

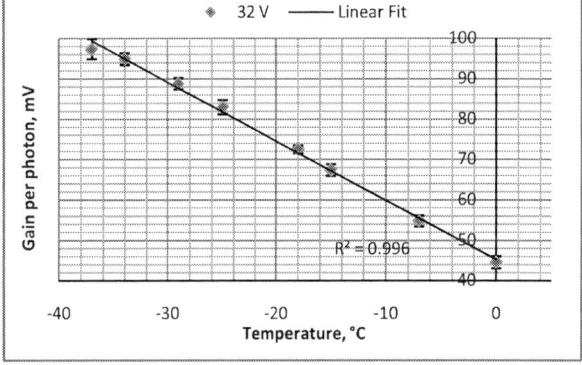

Figure 4. Gain per photon as a function of the temperature (bias voltage is stable).

Higher temperature causes a decrease of the gain of the detector (Fig. 4). This happens due to the diminishment of the avalanche current at steady voltage. The decrease of the gain is strongly linear as well. Differences in histograms from Fig. 5 are direct results of the temperature value thus change of gain per photon (Fig. 4).

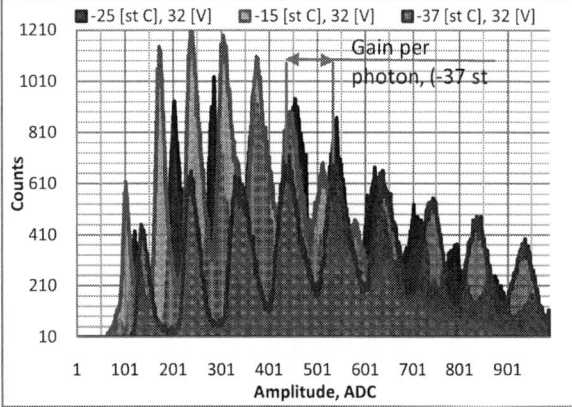

Figure 5. Influence of the temperature on the shape of the histogram. Histograms calculated for the measurements in different temperatures (but identical bias voltage) have various gains (distances between peaks).

The plots in Fig. 5 present three measurements taken in temperatures: -37 °C, -25 °C, and -15 °C. The red (or salmon) plot represents measurement that took place in the lowest temperature (gain per photon was the highest - refer to Fig. 4). The distance between the local maxima (peaks) is larger than in case of histograms for higher temperatures. It can be easily confirmed by counting how many maxima appear on the histogram in a fixed range of amplitude. Also, the highest peak shifts to the lower amplitudes with the increase of temperature.

Any fluctuations of the temperature at the time of conducting the measurements have an impact on the results. In order to describe how temperature and bias voltage affects the acquisition system the function that contains these dependencies has to be determined. Because both bias voltage and temperature bring linear influence on the gain of the detector we can describe the function that binds these parameters:

$$G(V,T) = aV + bT + c \qquad (5)$$

$$a = \frac{dG}{dV} \qquad (6)$$

$$b = \frac{dG}{dT} \qquad (7)$$

The values of the parameters a,b and c from (5) need to be calculated.

V. CALCULATION METHODS AND RESULTS

A. Determinant method

First method of determining the parameters a, b and c was a very simple one. Measurement series was performed for 36

different sets of temperature-bias values. This creates a possibility of writing 7140 independent simultaneous equations. The values of G, V and T variables have been substituted and determinants have been calculated. Value of a, b, c have been averaged.

B. Fit Function

The second method of finding a, b and c values is based on Levenberg-Marquardt best fit function (5). The parameters of this function are determined so as to minimize the weight mean square error (8) between the measured data y_i and the best fit function $G(V_i, T_i, a, b, c)$:

$$\sum_{i=0}^{N}(y_i - G(V_i, T_i, a, b, c))^2 \qquad (8)$$

Where N is the number of measured data points.

Although fit function method is more sophisticated it gives results very similar to the first one that uses determinants (Table 1). Both methods could be applied interchangeably.

TABLE I.   TABLE OF EQUATION'S PARAMETERS

Methods	Parameters of equation (5)		
	a	b	c
Determinant method	-1,424	19,176	-567,4
Fit function	-1,404	19,179	-567,087

The 3D model has been created (Fig. 6, 7). It presents the spatial arrangement of measurement points. The function from (5) has been also included in the form of a plane. 3D graph shows how well measured data cover calculated fit model (5).

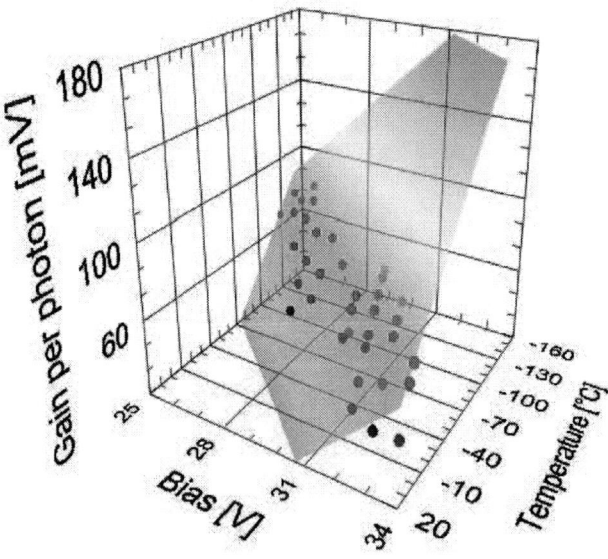

Figure 6. 3D graph. Solid spheres represent measured data, the surface represents the best fit function form equation (5). View presents described linear relation between gain per photon, bias and temperature.

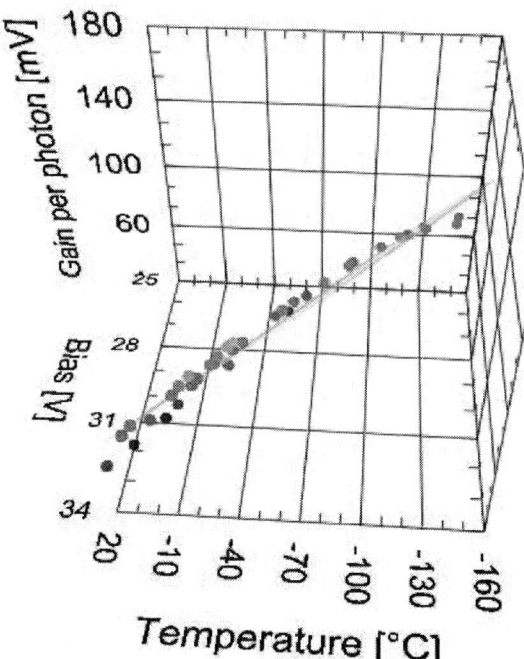

Figure 7. The same 3D graph as in Fig. 6 but shown at different angle.

It is difficult to explicitly determine which method is better (Fig. 8, 9). Relative errors of gain are very similar for both of them.

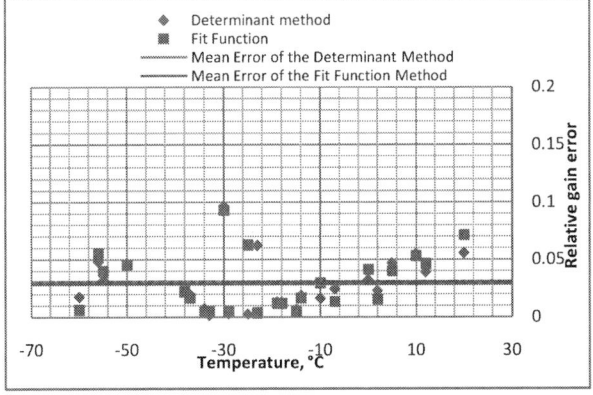

Figure 8. Relative gain error as a function of temperature.

Measurements with Silicon Photomultiplier are done in room temperature. That is why we are mostly interested in the range above 0°C. Calculated error is smaller than 8%, what makes parameters a, b and c credible enough.

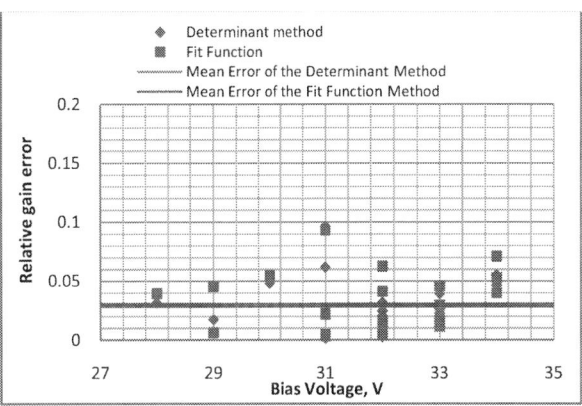

Figure 9. Relative gain error as a function of bias voltage.

VI. CONCLUSIONS

The measurements proved the possibility of the compensation of temperature fluctuations by bias voltage.

Parameters of the function that describes the relation between these physical quantities and gain of the detector have been calculated. Both methods of determining a, b and c have given similar results.

An automatic temperature influence compensation algorithm could be written in LabView software. The program would measure the temperature and set bias voltage so as to intended gain level of the detector would remain the same. The value of proper bias voltage would be calculated form equation (5). The performance of this compensation tool would be checked also on different models of the detectors and on Silicon Photomultipliers from various producers. That way we could evaluate what are the differences between parameters a, b and c for different photo detectors. This values are not included in any datasheets and it is not known what are the actual dependencies between temperature, bias voltage and gain of Silicon Photomultipliers. This knowledge could be very useful in the process of choosing the device optimal in given conditions.

ACKNOWLEDGMENT

This work has been partially supported by MNS-DIAG project (POIG 01.03.01.-00-014/08-00).

REFERENCES

[1] V.Golovin, V.Saveliev, "Novel type of avalanche photodetector with Geiger mode operation", Nuclear Instruments and Methods in Physics Research A 518 (2004) 560–564.

[2] M. Ramilli, "Characterization of SiPM: Temperature dependencies.", Nuclear Science Symposium Conference Record, 2008. NSS '08. IEEE.

[3] D. A. Neamen, "Semiconductor Physics and Devices. Basic Principles", McGraw-Hill Higher Education, 3rd Edition, New York, 2003.

[4] J. Barszcz et al., "Four Channels Data Acquisition System for Silicon Photomultipliers", Mixdes 2011.

MIXDES 2012, 19th International Conference *"Mixed Design of Integrated Circuits and Systems"*, May 24-26, 2012, Warsaw, Poland

DC Measurements Method of the Thermal Resistance of Power MOSFETs

Krzysztof Górecki, Janusz Zarębski
Department of Marine Electronics
Gdynia Maritime University
Gdynia, Poland
gorecki@am.gdynia.pl, zarebski@am.gdynia.pl

Abstract—**In this paper a new direct-current measuring method of thermal resistance of power MOS transistors is proposed. The conception of this method and the way of its realization are presented. The discussion on the influence of selected factors on the accuracy of the elaborated method is included in this paper. The correctness of the method is verified by comparing the results of measurements obtained with the use of the new method with the results obtained with the infrared method.**

Index Terms—**thermal resistance, power MOSFETs, measurements, selfheating**

I. INTRODUCTION

Temperature affects electrical properties (characteristics) of semiconductor devices very strongly [1 - 6]. Therefore, the estimation of the device thermal resistance describing its ability of heat dissipation at the steady state is of a great importance [7 -9].

According to the definition [7]: the thermal resistance R_{th} of any semiconductor device is expressed as the quotient of an increase of the device inner (junction) temperature T_j over the ambient one T_a and the device dissipated power P_{th}, which caused this temperature to increase

$$R_{th} = \frac{T_j - T_a}{P_{th}} \qquad (1)$$

The values of the ambient temperature and the dissipated power can be measured in a simple manner. The value of the temperature T_j must be measured at the steady state.

With regard to the way the value of the temperature T_j is obtained, one can distinguish two kinds of methods measuring the device thermal resistance. The first are optical (infrared) methods [1, 2, 7], in which the value of the device inner temperature is obtained by measuring infrared radiation energy emitted by the investigated device [10]. The other are electrical methods, in which the information about the device inner temperature results from the measurements of the device temperature-sensitive parameter of the known temperature dependence.

The advantage of the optical methods is a possibility of estimating the temperature distribution on the device surface, but then a free access to the device surface is indispensable.

Apart from this, the problem is to estimate precisely surface emissivity of the investigated device. The incorrectly estimated value of this parameter can result in the incorrect estimation of the excess of the device inner temperature over the ambient one, which differs from the actual value by even more than 50% [10, 11]. In turn, the advantage of the electrical methods is a possibility of estimating the thermal resistance of both capsulated and uncapsulated devices.

As results from [11], the optical methods are dedicated mainly to the uncapsulated devices, yet they can be also used to measure the thermal resistance of the capsulated device, provided that the investigated device operates without any heat sink. In such operation conditions of the device, the difference between the case and the inner device temperature is negligibly small compared with the difference between device inner temperature and the ambient temperature [12 - 15].

Two groups of electrical methods of measuring the device thermal resistance are known and described in numerous papers, e.g. [1, 2, 7, 12, 13 14, 16]. In pulse methods the investigated device is excited by the power rectangular pulses train and the values of the thermo-sensitive parameter are measured during the break of the pulse excitation [1, 7, 14, 16]. In turn, direct-current methods need to measure the coordinates of a few device-operating points and later the value of the device thermal resistance is calculated from any analytical dependence [1, 2, 12, 13]. The advantage of the direct-current method is the lack of interferences connected with switching the investigated device and the simple manner of the measuring procedure.

In the paper a new dc method of measuring the thermal resistance of power MOS transistors, which is the enhanced version of the method described in [17], is proposed. In the next sections the description of the proposed method and the results of the verification of this method are presented.

II. DESCRIPTION OF THE METHOD

In the new method the thermal resistance R_{th} of the MOS transistor is measured as the quotient of the increase of the internal temperature of the investigated transistor and the increase of the transistor dissipated power, which caused the increase of this temperature. As the temperature-sensitive parameter the voltage u_{GS} between the gate and the source of the transistor at the determined value of the drain current i_D is made use of.

Therefore, the thermal resistance R_{th} of the considered transistor can be expressed by the following dependence [17]

$$R_{th} = \frac{u_{GS1} - u_{GS2}}{i_D \cdot (u_{DS1} - u_{DS2}) \cdot \alpha_{GS}} \qquad (2)$$

where u_{GS1}, u_{GS2}, u_{DS1}, u_{DS2}, i_D are the coordinates of two selected points $A(u_{GS1}, u_{DS1}, i_D)$ and $B(u_{GS2}, u_{DS2}, i_D)$ of the considered transistor operation lying in the saturation region, whereas α_{GS} designates the slope of the thermometric characteristics $u_{GS}(T)$.

In the method the following fact was made use of: at the fixed value of the drain current i_D of the transistor operating in the saturation region the dependence of the gate-source voltage u_{GS} is a linear function of the temperature, which is shown in Fig.1.

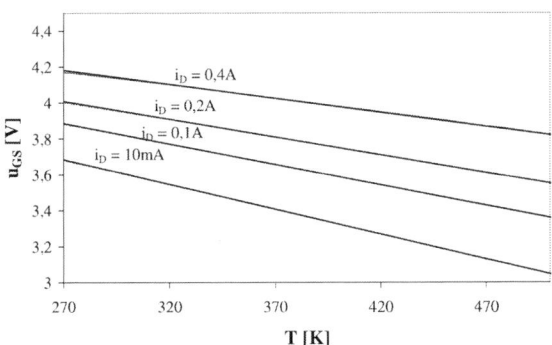

Fig. 1. The dependence of the MOSFET gate-source voltage on the transistor inner temperature

As seen from Fig.1, the slope α_{GS} of the characteristic $u_{GS}(T)$ is a decreasing linear function of the drain current. Using the Shichman-Hodges model [18] of MOSFET for the saturation region the u_{GS} voltage can be expressed in the form

$$u_{GS} = \sqrt{\frac{i_D}{KP \cdot \left(1 + \frac{u_{DS}}{U_E}\right)}} \cdot \left(\frac{T_j}{T_0}\right)^{-\kappa/2} + U_{PO} + \alpha_U \cdot (T_j - T_0) \qquad (3)$$

then, the parameter α_{GS} can be expressed in the form

$$\alpha_{GS} = \frac{\partial u_{GS}}{\partial T_j} = -\frac{u_{GS} - U_P(T_j)}{T_j} \cdot \frac{\kappa}{2} + \alpha_U \qquad (4)$$

where $U_P(T_j)$ – the threshold voltage of the device at the temperature T_j, u_{GS} – the gate-source voltage of the transistor at the temperature T_j, α_U – the temperature coefficient of the changes of the threshold voltage and κ – the exponent existing in the temperature dependence of the MOSFET transconductance parameter.

As it results from the authors' analysis with the use of the electrothermal model of the MOS transistor [3, 19], the linear dependence between the gate-source and drain-source voltages

at the fixed value of the drain current is observed (Fig.2). Therefore, the value of the voltage u_{GS} in Eq. (4) can be obtained from Fig.2 as the value of the voltage $u_{GS}(u_{DS} = 0)$ corresponding to the ambient temperature T_a

$$\alpha_{GS}(u_{DS} = 0) = u_{GS1} - \frac{u_{GS1} - u_{GS2}}{u_{DS1} - u_{DS2}} \cdot u_{DS1} \qquad (5)$$

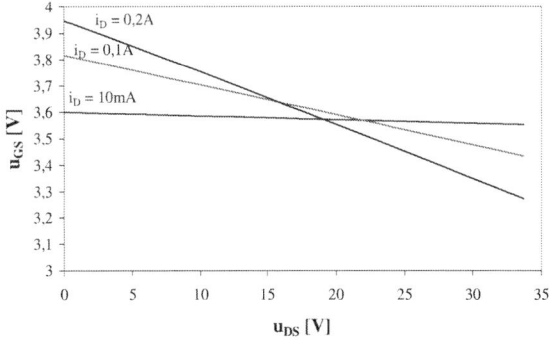

Fig. 2. The dependence of the voltage u_{GS} on the voltage u_{DS} of the transistor

Thus

$$\alpha_{GS} = -\frac{u_{GS1} - U_{P0} - u_{DS1} \cdot \dfrac{u_{GS1} - u_{GS2}}{u_{DS1} - u_{DS2}}}{T_a} \cdot \frac{\kappa}{2} + \alpha_U \qquad (6)$$

where T_a – the ambient temperature, U_{P0} – the device threshold voltage corresponding to T_a.

As results from the authors' investigations described e.g. in [3], a good agreement between d.c. characteristics of the various kinds of MOSFETs, obtained from simulations and measurements is achieved for the stated value of κ equal to - 1.5. Therefore, in further considerations such a value of κ is taken into account.

In order to measure the values of the parameters U_{P0} and α_U appearing in the formula (6), it is necessary to perform extra measurements in the points $C(u_{GS3}, u_{DS3}, i_{D3})$ and $D(u_{GS4}, u_{DS4}, i_{D4})$. The coordinates of the points A, B, C and D must meet the following conditions

$$\begin{cases} u_{DS3} \cdot i_{D3} = u_{DS1} \cdot i_D \\ u_{DS4} \cdot i_{D4} = u_{DS2} \cdot i_D \end{cases} \qquad (7)$$

Then, transforming the formula describing the characteristics of the MOS transistor in the saturation region for particular points, we obtain

$$U_{P0} = \frac{u_{GS3} \cdot \sqrt{i_D} - u_{GS1} \cdot \sqrt{i_{D3}}}{\sqrt{i_D} - \sqrt{i_{D3}}} + \\ + \frac{\alpha_U \cdot R_{th} \cdot \left(u_{DS1} \cdot i_D \cdot \sqrt{i_{D3}} - u_{DS3} \cdot i_{D3} \cdot \sqrt{i_D}\right)}{\sqrt{i_D} - \sqrt{i_{D3}}} \qquad (8)$$

$$\alpha_U = \frac{i_D \cdot (u_{GS3} - u_{GS4}) + \sqrt{i_D \cdot i_{D4}} \cdot (u_{GS2} - u_{GS3}) + \sqrt{i_D \cdot i_{D3}} \cdot (u_{GS4} - u_{GS1}) + \sqrt{i_{D3} \cdot i_{D4}} \cdot (u_{GS1} - u_{GS2})}{R_{th} \cdot i_D \cdot [u_{DS2} \cdot (\sqrt{i_{D4}} \cdot (\sqrt{i_{D1}} - \sqrt{i_{D3}}) + \sqrt{i_D} \cdot (\sqrt{i_{D3}} - \sqrt{i_D})) + u_{DS1} \cdot (\sqrt{i_{D3}} \cdot (\sqrt{i_{D4}} - \sqrt{i_D}) + \sqrt{i_D} \cdot (\sqrt{i_D} - \sqrt{i_{D4}}))]} \tag{9}$$

As seen in the formulas (8-9), we obtain the thermal resistance R_{th}, the value of which is looked for. In order to measure the value of this parameter, it is necessary to solve the set of equations, which consists of the dependences (3, 6, 8, 9) using, for instance the method of simple iteration. Practically, the result is obtained after performing 4 –5 iterations.

III. LIMITATIONS OF THE METHOD

Due to some simplifications assumed in the method described in Section II, this method can be used only in some restricted range of MOSFET operation. These restrictions are discussed below.

As results from [3], the sub-threshold current is a substantial component of the power MOSFET drain current. The value of this component can exceed even 100 mA. Therefore, the Shichman-Hodges model of MOSFET used in the process of formulating Eqs (2, 6, 8, 9) describes correctly the device characteristics at the drain current range much higher than the value of the sub-threshold drain current. Consequently, the error of the presented method can be of a high value at small values of the drain current.

In Eqs. (2, 6, 8, 9) the voltage drop on the device series drain and source resistances are not taken into account either. The values of these resistances are increasing functions of temperature, causing the error of the measuring method – the measured value of the device thermal resistance is understated in comparison to its true value. The importance of this error increases with an increase of the series resistance of the transistor source. So, the measuring error discussed here can be of a great impedance, especially for high-voltage MOSFETs having a high value of resistivity of the quasi-neutral regions [20].

The last phenomenon omitted in the method equations is the voltage modulation of the channel length causing an increase of the transistor drain current with an increase of the drain-source voltage. As a result of this phenomenon the measured values of the device thermal resistance are smaller than in the reality.

IV. VERIFICATION OF THE METHOD

The verification of the method was carried out in two stages. First, the correctness of the formulas was verified with the use of computer simulation; then, the measurements predicted in the proposed method were carried out and the obtained values of the thermal resistance were compared to the values measured with the pirometric method.

In order to verify the correctness of the formulas applied in the new method the coordinates of the points A, B, C, D were determined with computer simulations. These simulations were carried out with the use of the model of the power MOS transistor described by the formula (2) and taking into account the phenomenon of selfheating with the following formula

$$T_j = T_a + R_{th} \cdot i_D \cdot u_{DS} \tag{10}$$

The following arbitrary selected values of the transistor model parameters were used: $KP = 1$ A/V^2, $U_E = 50$ kV, $T_0 = 300$ K, $U_{PO} = 3.5$ V, $\kappa = 1.5$, $\alpha_U = -3$ mV/K, $T_a = 300$ K, $R_{th} = 50$ K/W. The obtained coordinates of the points A, B, C and D are given in Table I.

TABLE I. COORDINATES OF THE POINTS A, B, C, D USED IN CALCULATIONS

point	A	B	C	D
u_{GS} [V]	3.7050	3.5924	3.8520	3.7549
u_{DS} [V]	10	20	5	10
i_D [A]	0.1	0.1	0.2	0.2

Due to the use of the formulas (1, 5, 7, 8) the value $R_{th} = 51$ K/W is obtained, which is different from the standard value by merely 2%. The calculated values of the parameters of the transistor model are respectively $U_{PO} = 3.50013$ V, $\alpha_U = -3.0004$ mV/K and they are not very different from the nominal values by more than 0.03%. The agreement between the parameters values used in the calculations and the values obtained with the formulated in this work formulas confirms the correctness of these formulas.

The rightness of the proposed method in the range of medium drain current values was also checked by means of the following numerical experiment. Using the nonlinear electrothermal model (ETM) of the power MOSFET described in [3], the nonisothermal dc characteristics of the transistor MTD20N06V were simulated. Next, the coordinates of four points A, B, C and D lying on these characteristics were used in the method. In the electrothermal model used in the simulations apart from the fundamental phenomena existing in the device, there are also selfheating and device series resistances, voltage modulation of the channel length and sub-threshold and avalanche phenomena. The correctness of the model is verified experimentally showing in [3] a very good agreement between the simulated and measured characteristics of various kinds of the considered class of transistor in the wide range of changes of temperature and the device terminal voltages and currents. One of the important parameters of the model is the device thermal resistance, the value of which is treated as the standard value of this parameter in our further considerations.

Using ETM along with values of its parameters [3], the values of the coordinates of the characteristics points A, B, C and D of the transistor MTD20N06V situated on the large heat-sink were calculated. For these points the drain current and the drain-source voltage are as follows: 0.4 A / 10 V, 0.4 A / 20 V, 0.8 A / 5 V and 0.8 A / 10 V, respectively.

According to the proposed measuring method the calculated device thermal resistance is equal to 5.49 K/W, whereas the standard value of this parameter is equal to 5.5 K/W. So, very good accuracy of estimating the device thermal resistance is achieved, which confirms the correctness of the presented method.

Such an excellent agreement between the values of the thermal resistances obtained both from the method and the model, respectively, were achieved at the drain current values from 300 mA to 800 mA only. For currents of higher values than 800 mA, overestimated values of the thermal resistance were obtained due to the fact that the transistor source resistance is omitted in the method. In turn, at the drain current the values smaller than 300 mA downside values of the thermal resistance were received because of drain sub-threshold current is omitted in the method.

The method was used to measure the thermal resistance of a few samples of power MOS transistors. The obtained results are very close to the results obtained by the classical pulse method of measuring the thermal resistance of semiconductor devices [16] and the infrared method [10], which confirms the correctness of the proposed method.

The experimental verification of the new method was carried out for the power MOS transistor of the IRFR024N type [21], placed in the case D-PAK and operating without a radiator. Table II compares the measured coordinates of the points A, B, C and D used in measuring the value of the thermal resistance of this transistor.

TABLE II. COORDINATES OF THE MEASURING POINTS

point	A	B	C	D
u_{GS} [V]	3.343	3.156	3.462	3.347
u_{DS} [V]	8.21	15.96	3.96	8.36
i_D [mA]	48.5	50.1	103.6	102.6

As a result of the realization of the new measuring method the value of the investigated transistor R_{th} = 75 K/W was obtained. Moreover, the thermal resistance of this transistor was measured with the pirometric method and the value R_{th} = 72 K/W was obtained.

Moreover, the measurements of the thermal resistance of the transistor IXFH12N100Q situated on a large heat-sink were performed. At the drain current value equal to 1A the value of the thermal resistance equal to 3.03 K/W was obtained. The value of the thermal resistance of the considered transistor measured by the infrared camera was equal to 3.1 K/W.

Thus, the proposed method ensures obtaining the value of the thermal resistance of the MOS transistor with an error not exceeding a few percentage.

V. CONCLUSION

In the paper the new method of measuring the value of the thermal resistance of power MOS transistors is proposed. The realization of the method is simple and involves only performing the measurements of the coordinates of the four points of the investigated transistor operation, which lie in the saturation region and carrying out calculations with the formulas given in section 2.

The correctness of the method was proved by computer simulations and influence of the chosen phenomena on the method measurements accuracy was discussed. It results from the simulations that the proposed method assures proper

values of the device thermal resistance at the medium values of the drain current. The current values have to be higher than the sub-threshold drain current component and on the other hand, so small that the voltage drop on the device source series resistance is negligible in comparison to the changes of the gate-source voltage value caused by selfheating.

It was shown experimentally that the measurements results obtained with the presented direct-current method are convergent with the results obtained with the infrared method with the differences between them not exceeding a few percentage.

Another advantage of the use of the new method is measuring the value of the threshold voltage of the investigated transistor U_{PO} and the coefficient of temperature changes in this voltage α_U.

Currently, the authors are investigating the influence of the selected factors on the measurement error of the new method. The obtained results will enable optimization of the choice of location of the points on the characteristics of the investigated transistor ensuring at the same time the minimal measurement error.

ACKNOWLEDGEMENTS

This project is financed from the funds of National Science Centre which were awarded on the basis of the decision number DEC-2011/01/B/ST7/06740.

REFERENCES

[1] J. Zarębski: Modelowanie, symulacja i pomiary przebiegów elektrotermicznych w elementach półprzewodnikowych i układach elektronicznych. Prace Naukowe Wyższej Szkoły Morskiej w Gdyni, Gdynia, 1996.

[2] W. Janke: Zjawiska termiczne w elementach i układach półprzewodnikowych. WNT, Warszawa, 1992.

[3] J. Zarębski, K. Górecki: The electrothermal large-signal model of power MOS transistors for SPICE. IEEE Transaction on Power Electronics, Vol. 25 , No. 5-6, 2010, pp. 1265 – 1274.

[4] K. Górecki, W.J. Stepowicz: Electrothermal model of optocoupler for SPICE. International Journal of Numerical Modelling Electronic Networks, Devices and Fields, Vol. 22, No.4, 2009, pp. 321-333.

[5] J. Zarębski, K. Górecki: SPICE-aided modelling of dc characteristics of power bipolar transistors with selfheating taken into account. International Journal of Numerical Modelling Electronic Networks, Devices and Fields, Vol. 22, No. 6, 2009, pp. 422-433.

[6] M. Rashid, "Power Electronics Handbook", Elsevier, 2007.

[7] S. Rubin and F. Oettinger, Thermal Resistance Measurement on Power Transistors. National Bureau of Standards, NBS 400-14, U.S. Dept. of Commerce, 1979.

[8] F.N. Masana: Die Attach Thermal Monitoring of IGBT Devices. Mixed Design of Integrated Circuits and System, MIXDES 2006, Gdynia 2006, pp. 421 – 424.

[9] M. Rencz, V. Szekely, Z. Kohari, B. Courtois: Thermal evaluation and modelling of the SIP9 and SP10 MEMS packages. The Seventh Intersociety Conference on Thermal and Thermomechanical Phenomena in Electronic Systems ITHERM 2000, Vol. 1, p. 126.

[10] K. Górecki, J. Zarębski: Pomiary rezystancji termicznej tranzystorów mocy z wykorzystaniem metod pirometrycznych. Pomiary, Automatyka, Kontrola PAK, Nr 1, 2003, ss. 41-44.

[11] K. Górecki, J. Zarębski: Porównanie elektrycznych i pirometrycznych metod pomiaru parametrów termicznych elementów półprzewodnikowych. Elektronika, Nr 11, 2005, ss. 55-57.

[12] J. Zarębski, K. Górecki: A New Measuring Method of the Thermal Resistance of Silicon P-N Diodes. IEEE Transaction on Instrumentation and Measurement, Vol. 56, No. 6, 2007, pp. 2788-2794.

[13] J. Zarębski, K. Górecki: A Method of the Thermal Resistance Measurements of Semiconductor Devices with P-N Junction. Measurement, Vol.41, No.3, 2008, pp. 259-265.

[14] J. Zarębski, K. Górecki: A New Method for the Measurement of the Thermal Resistance of the Monolithic Switched Regulator LT1073. IEEE Transaction on Instrumentation and Measurement, Vol. 56, No. 5, 2007, pp. 2101-2104.

[15] J. Zarębski, K. Górecki: A Method of Measuring the Transient Thermal Impedance of Monolithic Bipolar Switched Regulators. IEEE Transactions on Components and Packaging Technologies, Volume 30, No. 4, 2007 pp:627 – 631.

[16] F. F. Oettinger, D. L. Blackburn: Semiconductor Measurement Technology: Thermal Resistance Measurements, U. S. Department of Commerce, NIST/SP-400/86, 1990.

[17] K. Górecki, J. Zarębski: A New DC Measuring Method of the Thermal Resistance of Power MOS Transistors. VIII International Conference on Microtechnology and Thermal Problems in Electronics Microtherm 2009, Łódź, 2009, pp. 165-170.

[18] B.M. Wilamowski, R.C. Jaeger: Computerized circuit Analysis Using SPICE Programs. McGraw-Hill, New York, 1997.

[19] J. Zarębski, K. Górecki, D. Bisewski: A New Electrothermal Model of the Power MOSFET for SPICE. 11-th International Conference Mixed Design of Integrated Circuits and Systems MIXDES 2004, Szczecin, 2004, pp. 89-93.

[20] J. Zarębski: ON-Resistance of Power MOSFETs. Informacije MIDEM, Vol. 35, No. 1, 2007, ss. 1-4.

[21] IRFR/U024N. Catalogue data, International Rectifier, www.irf.com

Overheat Security System
for High Speed Embedded Systems

Maciej Frankiewicz, Adam Gołda, Andrzej Kos
AGH University of Science and Technology
Department of Electronics
Kraków, Poland
frankiew@agh.edu.pl, golda@agh.edu.pl, kos@agh.edu.pl

Abstract—The paper describes design and structure of the Temperature-Controlled Oscillator for high frequency processors which work is based on information of present chip temperature. The task of the circuit is to stabilise required value of temperature of the silicon die and ensure safe work of the processor. The circuit consists of: temperature sensor, ring oscillator and some additional blocks to control the circuit behaviour. The prototype chip was designed and fabricated in CMOS LF 0.15 um (1.8 V) technology and is cooperating with an 8-bit OctaLynx microcontroller.

Index Terms—CMOS integrated circuits, Mixed analog digital integrated circuits, Power dissipaion, Ring Oscillators, Temperature sensors

I. INTRODUCTION

Considering the aspiration to make modern digital circuits faster and scaling down the technology researchers must take into account thermal effects. The goal is to find the optimal balance state between keeping the throughput of the system on the possible maximum level and decreasing power consumption of the device. Assuming that circuit cannot exceed acceptable maximum temperature for proper work it should be controlled in that way to keep its temperature as close to the permissible limit as possible with minimal fluctuations. To achieve that some dynamic control systems have been developed including: dynamic voltage scaling (DVS), dynamic clock throttling (DCT) and dynamic frequency scaling (DFS) [1]. Despite former achievements on the field some further investigations must be done to improve thermal behaviour of the circuit.

The paper describes a structure of the wide-range oscillator which generates the frequency dependent on the present temperature of the silicon die. The structure of the circuit designed and fabricated in CMOS LF 150 nm (1.8 V) technology will be presented in the next sections. The temperature-controlled oscillator (TCO) works with the OctaLynx 8-bit microcontroller which is a prototype created for test purposes [2].

II. IDEA OF DYNAMIC CLOCK CONTROL

The concept of temperature-controlled oscillator [3] origins from the fact that circuit working with higher frequency consumes more power. That means that if the silicon die reaches high temperature range it is necessary to lower the clock rate and as a consequence cool the chip down. The problem is how to manage the frequency of the clock to minimize temperature fluctuations and let the circuit work possibly fast. Simpler dynamic systems sensed the temperature and when it crossed fixed reference level just divided the frequency of clock signal. This action can be not efficient enough because as it is commonly known for digital circuits (e.g. processors) distribution of power consumption in time is not uniform. What is more, such discreet frequency division with one or more reference levels can result in quite big changes of the frequency and - as a consequence – the temperature. As been previously mentioned, fluctuations of the temperature inside the circuit should be as small as possible. That means that rapid steps of power dissipation resulting from discreet division should be avoided and more fluent frequency control must be used. Structure presented in the paper is an implementionof combination of both described clock control methods. Designed oscillator is continosly tuned based on information from the temperature sensor and additionally frequency is divided with some discreet reference levels for more efficient action.

III. CHIP STRUCTURE

In order to verify the theoretical investigation a prototype circuit has been designed and fabricated. The TCO chip consists of three main blocks: temperature sensor, voltage-controlled oscillator (VCO) and a control block. Block diagram of the system is presented in Fig. 1. The sensor delivers information of present chip temperature which tunes the oscillator. The control block consists of three sub-blocks: temperature range recognition block, decoder block and generator driver. All of them will be described in next paragraphs.

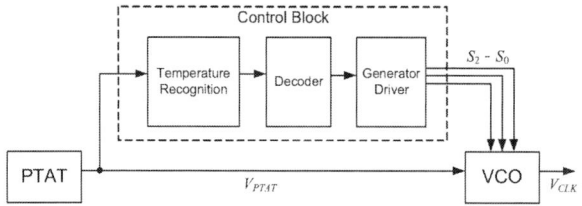

Figure 1. Block diagram of the designed TCO.

In the first approach the temperature-dependent signal from the sensor continuously tunes the oscillator frequency. If the temperature is rising the frequency of the clock should be lower and as a consequence the chip temperature will decrease. As an effect of this thermal-frequency feedback for small temperature changes the circuit should obtain steady state which is a compromise between high frequency and low temperature. This state can be different depending on how much power-consuming tasks have to be executed by the circuit. If this method is not efficient enough the second way of frequency control is provided. The temperature recognition block compares present temperature with several reference levels and produces digital information in which temperature range the circuit currently is. Based on this information proper frequency division factor can be set which is done by the decoder block. Generator driver is a circuit which ensures proper control of the frequency division in the oscillator circuit and will be described more precisely in next sections.

A. PTAT sensor

The PTAT sensor task is to give the voltage proportional to the temperature of the chip. Its work is based on the phenomenon that difference between voltages of two diodes or bipolar transistors, which have different areas and conduct the same current, is proportional to the absolute temperature. Implemented structure, presented in Fig. 2, has been chosen because the accuracy is sufficient for this application and presented circuit is easy to implement in CMOS process with usage of planar transistors [4]. Presented PTAT sensor consists of 5 MOS transisors, 2 bipolar transistors and 2 resistors.

The sensor produces voltage given by (1)

$$V_{PTAT} = \frac{W_5}{W_2} \frac{kT}{q} \ln(n) \frac{R_2}{R_1} \qquad (1)$$

where W_2 and W_5 are widths of the gates of the M_2 and M_5 transistors, k is the Boltzmann's constant, T is temperature, q is the charge and n is the ratio between emitter areas of Q_1 and Q_2 planar transistors. Current conducted by the circuit is limited to prevent self-heating of the sensor. In presented case the temperature sensor was designed using two PNP planar transistors with the emitters area multiplication factor $n = 2$. The material of the resistors has significant influence to the parameters of this sensor [5]. In this work resistors were built of highly resistve polisilicon layer. The V_{PTAT} signal is increasing with slope of about 2,9 mV/°C and ranges from 212,10 to 473,92 mV at expected temperature range of 10 to

100°C. Temperature charcteristic of the PTAT sensor is shown in Fig. 3. Very good linearity of the temperature sensor response is clearly visible.

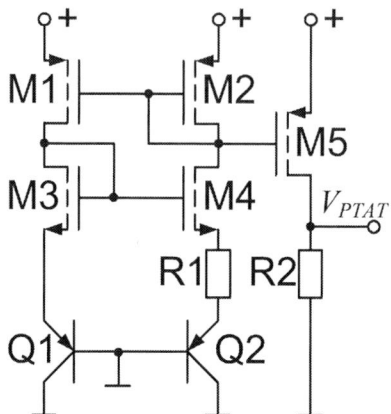

Figure 2. Schematic diagram of the PTAT sensor.

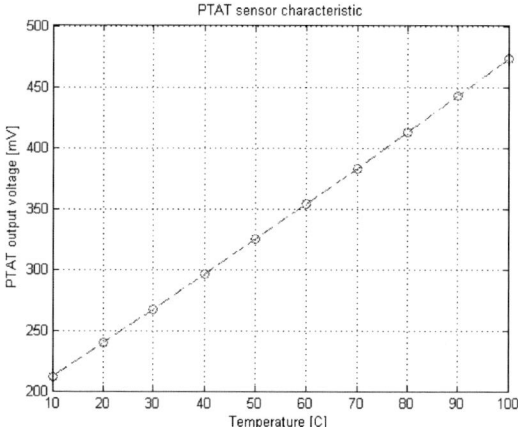

Figure 3. Measured thermal characteristic of PTAT sensor.

B. Control block

The path of processing the temperature-dependent signal consists of several sub-blocks which will be described separately. First of them is the temperature range recognition block. Its structure is similar to flash Analog-to-Digital Converter (ADC), as presented in Fig. 4 [6]. Present signal V_{PTAT} is compared with several reference levels and as an effect a digital signal in linear code is produced. In order to minimize the power consumption of the circuit the comparators were replaced by CMOS inverters with proper treshold voltages which fulfill similar conditions. This block produces information in which of several temperature ranges the silicon die of the chip is currently. Knowledge of temperature range allows the decoder block the real-time control of the frequency division factor. The output word was formed in 3-bit vector S_2-S_0 to make easier control of the generator and is presented in Table 1. Presented temperature reference levels are result of authors previous experiments. The temperature of 60°C is a

310

typical value at which digital circuits work and there should no problems with their behaviour. The maximum allowed temperature was set to 90°C because some of the tested circuits did not work properly at higher temperatures. As a result it has been decided to shut off the clock at this temperature to cool dow the chip (similarly to Dynamic Clock Throttling method). Third ference level of 75°C is a mean value of two previously described reference levels.

The role of generator driver will be explained in one of next sections.

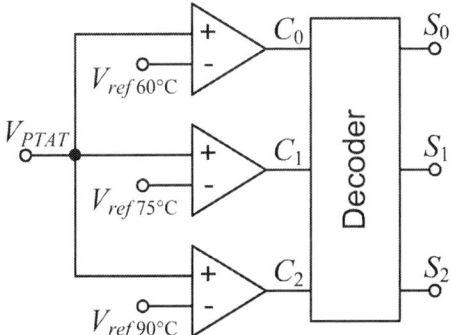

Figure 4. Schematic diagram of the temperature recognition block.

TABLE I. OUTPUT VECTOR OF THE DECODER BLOCK.

Temperature range		Output vector		
min	*max*	S_2	S_1	S_0
-	60	1	0	0
60	75	0	1	0
75	90	0	0	1
90	-	0	0	0

C. Ring oscillator

The heart of the system is its generator. VCO is realized as a ring oscillator and the goal is to slowdown the generator when chip tempereture gets high and let the silicon die cool down itself. Choice of the ring oscillator structure was a result of the necessity of low power consumption and ability to produce high frequeny signal. Frequency generated by the implemented circuit can be tuned in two different ways. First of them is continuous change of the supply voltage done by the signal coming directly from the PTAT sensor. Lowering the supply causes extending of the inverters propagation time. This way of control was realized as a nonlinear process: when the temperature is quite low changes of frequency are smaller than in case of relatively high temperatures. This comes out of the fact that there is no need for intervention when the chip teperature do not endanger its work and as an effect frequency control can be more fluent. Second method consists in changing the length of the ring: with greater number of inverters in the signal path the frequency gets lower according to (2)

$$f_0 = \frac{1}{2nt_p} \qquad (2)$$

where n is a number of inverters in the ring and t_p is a propagation time of single inverter. That way digital control of the frequency is enabled. Controlling signal S comes from the control block mentioned before. Moreover, it has been decided to turn off the generator when the silicon die reaches temperature range which can be dangerous for the circuit work. Turning off was realized by opening the oscillator ring.The output inverter is a buffer with constant 1.8 V supply voltage which task is to provide the clock signal with stable amplitude. Structure of described ring oscillator is presented in Fig. 5 while Fig. 6 shows tuning characteristic of the VCO in dependence of chip temperature for each of the ring lengths (the generator is controlled by voltage from the temperature sensor).

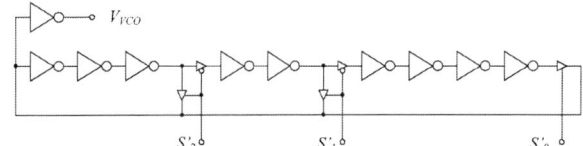

Figure 5. Schematic diagram of the ring oscillator.

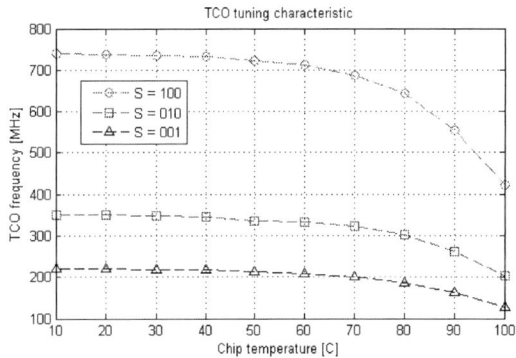

Figure 6. Measured frequency generated by the oscillator in dependence of temperature for each length of the ring.

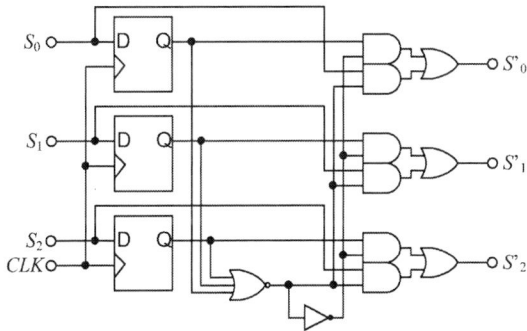

Figure 7. Schematic diagram of the generator driver.

D. Generator driver

Change of the oscillator ring legth during work of the processor could result in some unexpected problems and errors in execution of the tasks. To avoid that the change must be done only in a strictly specified moment. The generator driver is a structure which converts asynchronous signal from temperature recognition and decoder blocks into synchronous signal tuning generator only at falling edge of the clock. Additional task of this circuit is to avoid problems in the start of the TCO work. It always takes a while for the generator to start with the oscillations or it could happen that in the beginning ring would be opened and with no oscillatins signal from the decoder could not start tuning the generator. Presened generator driver ensures proper work of the TCO in described situations. Structure of the generator driver is shown in Fig. 7.

E. Chip structure summary

Layout of the designed overheat protection system is presented in Fig. 8 and covers area of about $100 \times 100 \ \mu m^2$. Main blocks of the circuit are marked in the figure below as: PTAT – temperature sensor, VCO – ring oscillator, CB – control block and TD – test driver which is additional circuit for verification purposes.

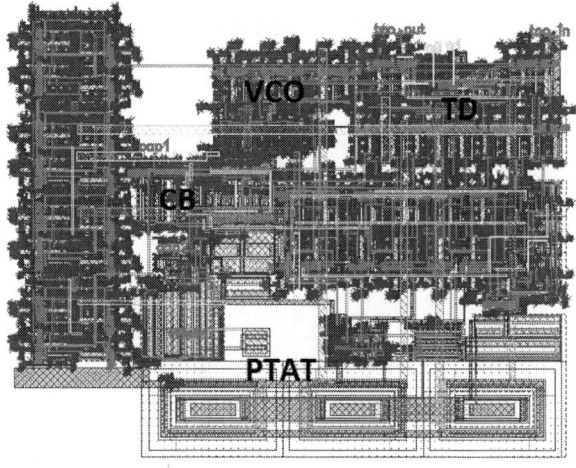

Figure 8. Layout of the presented overheat protection circuit.

Expected temperature range at which the circuit would work is from 10 to 100°C and behaviour of the circuit will vary for different temperatures of the chip. At this temperature range the system generates maximm frequency of about 740 MHz at low temperature and is shut off to prevent chip destruction above 90°C. Maximum power consumption also differs for each temperature range. That is mainly result of power consumed by temperature sensor. All main parameter of the system are gathered in Tab. 2. Presented power consumption is

the maximm value. It is important that this value is consumed only at vaery short moment in time of circuit work.

TABLE II.   CLOCK CONTROL SYSTEM PARAMETERS.

Parameter	Value	Unit
Layout area	100 x 100	μm^2
Expected temperature range	10 - 100	°C
Generated frequency (at expected temperature range)	740 - 0	MHz
Maximum power consumption (at expected temperature range)	1,96 – 3,26	mW

IV. CONCLUSIONS

In the paper method of frequency control in dependence of chip temperature was described and the prototype circuit structure was presented. The chip surface dimensions are about 100 um x 100 um and consumes about 1,5 mA at 1.8 V. Abovementioned parameters are addidtional cost which has to be paid if frequency control is needed. This extra cost can be considered as an disadvantage of described solution but ensures stablilisation of the chip temperature and safe work of the circuit. Presented approach can significantly improve thermal behaviour of the chip.

THE AUTHORS

MSc Maciej Frankiewicz, PhD Adam Gołda and Prof. Andrzej Kos are with the Department of Electronics, AGH University of Science and Technology, Kraków, Al. Mickiewicza 30, Poland

e-mail: frankiew@agh.edu.pl, kos@agh.edu.pl.

REFERENCES

[1] A. Gołda and A. Kos, "Predictive frequency control for low power digital systems", in *Proc. 13th Int. Conf. Mixed Design of Integrated Circuits and Systems MIXDES*, Gdynia, 2006, pp. 441-445

[2] R. Gał, A. Gołda, M. Frankiewicz, and A. Kos, „FGPA implementation of 8-bit RISC microcontroller for embedded systems", in *Proc. 18th Int. Conf. Mixed Design of Integrated Circuits and Systems MIXDES*, Gliwice, 2011, to be published

[3] K. Chan-Kyung, K. Bai-Sun, L. Chil-Gee, and J. Young-Hyun, "CMOS temperature sensor with ring oscillator for mobile DRAM self-refresh control", in *Proc. IEEE Int. Symp. on Circuits and Systems ISCAS*, Seattle, 2008, pp. 3094-3097

[4] A. Gołda and A. Kos, „Parameters Identification of embedded PTAT temperature sensors for CMOS circuits", in *Proc. 14th Int. Conf. Mixed Design of Integrated Circuits and Systems MIXDES*, Ciechocinek, 2007, pp. 392-395

[5] A. Gołda and A. Kos, „Analysis and design of PTAT temperature sensor in dgital CMOS VLSI circuits", in *Proc. 13th Int. Conf. Mixed Design of Integrated Circuits and Systems MIXDES*, Gdynia, 2006, pp. 415-420

[6] M. Frankiewicz, P. Mroszczyk, A. Gołda, and A. Kos, „Asynchronous 4-bit CMOS flash analog-to-digital converter with over- and underflow detection system", in *Proc. 16th Int. Conf. Mixed Design of Integrated Circuits and Systems MIXDES*, Łódź, 2009, pp. 234-237

MIXDES 2012, 19th International Conference *"Mixed Design of Integrated Circuits and Systems"*, May 24-26, 2012, Warsaw, Poland

Paths of the Heat Flow from Semiconductor Devices to the Surrounding

Krzysztof Górecki, Janusz Zarębski

Department of Marine Electronics
Gdynia Maritime University, Gdynia, Poland
gorecki@am.gdynia.pl, zarebski@am.gdynia.pl

Abstract—**In the paper paths of the heat flow generated inside a semiconductor structure to the surrounding are considered. On the basis of the literature information, typical construction of semiconductor devices cooling systems are analysed. Apart from this, the results of measurements of the thermal resistance and transient thermal impedance of such devices illustrating the influence of some factors on the value of thermal parameters are discussed. The general form of the device thermal model including the multipath of the heat flow and thermal properties of the elements of the heat flow path is proposed.**

Index Terms—**thermal resistance, semiconductor devices, thermal models, selfheating**

I. INTRODUCTION

One of the main problems restricting the development of the microelectronics is efficient abstraction of the heat generated in the semiconductor structure to the environment [1 - 3]. The limited efficiency of practical cooling systems causes that the internal temperature of semiconductor devices increases, attaining often the values considerably differing from the ambient temperature. The device internal temperature rise is a basic factor worsening the reliability of electronic elements and circuits comprising these elements [4 - 8]. Therefore, it is so important to develop efficient methods of cooling devices and electronic circuits.

Producers of semiconductor devices aim at reducing the thermal resistance between the semiconductor structure and the device case [9]. Therefore, new constructions of cases of semiconductor devices are proposed by producers. For example, in the paper [10] one paid attention to the fact that the admissible value of the temperature of the interior of the low-voltage semiconductor device is limited by the materials applied to the construction of the device case, particularly leadrich solder alloys. The main task of the classical device package is to protect it from corrosion and mechanical hazards and it has to guarantee the possible low value of the thermal resistance between the semiconductor chip and the case surface of the device, typically having an element which makes it possible to join external elements of the heat removing path. Usually such an element is of the form of a copper plate, to which the heat-sink can be added [9].

Cases of devices have different constructions depending, among others, on the semiconductor chip size, the manner of the device setting-up and the power dissipated into the device. For example, one observes essential differences between the appearance of the case SOT-23 for low-power transistors and the case TO-220 of power transistors, as well as the case of electro-insulated modules, e.g. QM50DY-H used in power electronics.

The construction of the case of semiconductor devices is very important, but seldom it has a decisive meaning in the global thermal resistance between the chip structure and the environment. Depending on the applied system of the device cooling it is necessary to take into account thermal proprieties of the other elements of the heat abstraction path [1, 11 - 13].

In the literature comparatively a lot of places are dedicated to the description of thermal phenomena occurring in semiconductor structures, eg. [14 - 18], on the other hand thermal proprieties of printed circuit boards (PCB), heat-sinks or systems of the fluidic cooling are considered only in a small number of publications, eg. [1, 19 - 22]. The offered thermal models of semiconductor devices typically refer to the structures or the semiconductor device together with its case [15 - 17]. This encouraged the authors to investigate the influence of the elements of the heat flow path being found outside the case of the device on the efficiency of abstraction of the heat generated in this device. The aim of this research is the elaboration of the compact thermal model of the thermal semiconductor device taking into account, except the property of the structure along with the case, also the manner of the montage, the size of paths on the printed circuit board, the size of the heat-sink, systems of the affected cooling and the thermal property of the case of the whole electronic equipment. In the present paper the initial effects of the research are presented.

II. ELEMENTS OF HEAT FLOW PATH

The heat generated in the semiconductor chip is dissipated to the device case due to the thermal conduction phenomenon. The construction of the device case is determined by the device producer, whereas further elements of the heat flow path depend on the constructor-engineer of the electronic equipment.

Generally, the heat can be transported from the case to the surrounding by conduction, convection and radiation [23, 24]. Conduction is realized by the device metal terminals, next by the solder areas and conductive paths on PCB. The second possibility of the heat transport by conduction is realized from the device case surface through the insulating washer to the heat-sink.

Convection exists on each surface of the contact between the solid-state situated on the heat flow path and the surrounding fluid. The simplest case of convection is the natural one occurring on the device case surface devoid of any contact with the heat-sink, on the surface of the PCB or on the heat-sink. On some surfaces convection appears also on the surface of pipes with the cooling fluid, existing inside the heat-sink [25] or inside the device case [19].

In turn, radiation comes from each surface of the elements

existing in the heat flow path, beginning from the device case, through the PCB and the heat-sink to the case of whole the electronic equipment.

The heat transport from the device case to the surrounding can be also assisted by the thermo-electric phenomenon [25].

Usually, the device is the component of the electric equipment situated inside its case (equipment case – EC). This EC protect the considered equipment from mechanical shocks, but on the other hand, it makes the generated heat abstraction difficult. Depending on the size, construction and the kind of material from which the device case is made, convection and radiation have the dominant role in the heat transport from the device case to the surrounding.

As results from the literature [26, 27], such parameters as the heat conductivity, thermal emissivity, the heat transfer coefficient characterizing the heat flow path have various values depending on the kind of material and the geometrical sizes of these elements. Additionally, the temperature of the cooling surface or the difference between the temperatures of the cooled surface and the cooling fluid influence the values of the parameters mentioned above. Therefore, the description of the device heat properties are a very important and complex task.

III. THERMAL MODELS OF SEMICONDUCTOR DEVICES

To estimate the values of the devices internal temperatures, at known courses of the power generated inside them, the device/circuit thermal models, describing the heat transport from the chip to the device case or to the surrounding are used.

The effectiveness of the heat transport in the semiconductor structure to the surrounding can be described using the heat conduction equation with appropriate boundary and initial conditions [4, 16, 24], or by lumped thermal models, based on the concept of transient thermal impedance $Z_{th}(t)$ [27 – 31]. Knowing the transient thermal impedance of the semiconductor device, it is possible to determine the temporary course of temperature in the device interior at any course of power excitation p(t) (thermal analysis), using the thermal model of the device.

Currently, much attention is paid to the investigations of thermal phenomena existing inside the semiconductor devices [4, 16, 18]. Meanwhile, in typical situations the semiconductor assembly greatest influence on the temperature inside the instrument has a thermal efficiency of dissipating heat from the device case to the surrounding [32, 33]. The phenomena that determine the effectiveness of this process are not easy to model, and in many works are oversimplified, which leads to the incorrect determination of the inner device temperature [27, 28, 32]. A typical simplification is to assume linearity of the thermal model, i.e. assuming that the cooling efficiency does not depend on the power dissipated in the device or the ambient temperature. Actually, due to the dependence of the efficiency of heat dissipation mechanisms to the ambient temperature on the device temperature, the thermal model of such a semiconductor device is nonlinear [34, 35], which can be seen, among others, in the dependence of $Z_{th}(t)$ on the power dissipated in the device and the ambient temperature [32, 36].

Many works [18, 29, 36, 37, 38] are devoted to the formulation of an analytical description of lumped linear thermal models of semiconductor devices and integrated circuits that can be used in a thermal analysis. This analysis can determine the values of the temperature inside the device at a known course of power dissipated in it. In the thermal analysis of the

semiconductor device a form of the electrical analog [27, 28, 32, 34, 39] is commonly used, because it can be directly implemented into computer programs to analyze electronic circuits, such as SPICE. The electrical analog is of the form of the RC network: Cauer network or Foster network.

Both the networks are considered fully equivalent from the viewpoint of the terminal T_j (representing the inner device temperature) [16, 40], although the Foster network has not direct physical interpretation, while the Cauer network structure results directly from the discretized one-dimensional heat conduction equation, and voltage on particular nodes of this network correspond to the temperatures of structural components of the considered semiconductor device, such as assembly bases or housing. In the literature there are examples of the use of both the Foster network [16, 41] and the Cauer network [16, 42] in thermal models of semiconductor devices.

In the paper [34] the non-linear semiconductor device thermal model was presented by the authors. This model takes the form of electrical analog in the form of the Cauer network, as shown in Fig. 1. The circuit presented in the figure consists of controlled voltage sources e_{Ri} and controlled current sources i_{Ci}, representing non-linear resistances and thermal capacities of individual components of the semiconductor device and modeling phenomena, which are responsible for the heat transport from the case to the surrounding. This model was used successfully to describe thermal properties of semiconductors made of both silicon [34], and of silicon carbide [35]. The disadvantage of this model is a way of determining the values of its parameters, which requires a series of results of the measurements of the thermal impedance of the modeled device. Thus, this model can be used only to describe the existing cooling systems, but is not suitable to design such systems.

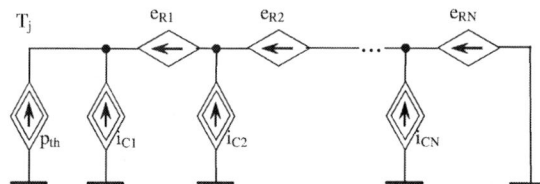

Fig. 1. The network representation of the non-linear semiconductor device thermal model

Often, semiconductor device thermal models characterizing the heat flow path from the semiconductor structure to the device case are provided by the manufacturers of these devices, e.g. [43]. For example, for power devices there is a thermal model in the form of thermal impedance in relation to the point on the housing. Yet, in the case of SMD devices – in relation to the soldering point. However, such models do not take into account a component which includes and describes the heat transfer between the device housing and the surrounding. The main reason for this is the dependence of the mentioned component on the method of assembling the device and its applied cooling system. As a result, these models do not allow including in the design the impact of the PCB (printed circuit board), heat sink, pins, the case of electronic equipment comprising the considered device, on the course of the transient thermal impedance of the device under consideration and the significant role of these factors in the process of heat dissipation [32, 33]. On the other hand, some manufacturers of semiconductor devices, such as power LEDs and power integrated circuits provide the data sheets proposals of shapes of the PCB mosaic dedicated to the device under consideration and suggestions concerning the nature of the PCB surface laminate, on which investigated device is embedded.

Heat flow paths from the device case to the surrounding may be diverse, and for each of these routes a different mechanism of heat dissipation (conduction, convection, radiation) is dominant. From the perspective of a designer of electronic equipment is important to determine the total thermal transient impedance from the structure of the semiconductor device to the surrounding, taking into account all the mechanisms of heat dissipation and all the ways of its movement.

Often, the accurate thermal analysis of the electronic equipment shows that the laminate with a connections mosaic may act as a heat sink, and the use of an additional component of the cooling system may be just an unnecessary occupation of the space and increasing the dimensions of the equipment. The effectiveness of heat transfer from the semiconductor device case to the surrounding is affected by many factors, the inclusion of which is not trivial. Such factor, among others, may be: temperature, power dissipated in the investigation device, thermal coupling between devices, the size of the heat sink and other elements making up the path of the heat flow and its spatial orientation, the coolant flow rate in the cases of liquid cooling components, the length of leads, solder surface fields, or properties of the case of electronic equipment containing the investigated semiconductor device [30, 32, 43]. Taking into account the impact of these phenomena in cooling semiconductor devices in the form of analytical formulas allows easy optimization of the cooling systems of these devices at the stage of their designing.

There are not currently known or presented in the available literature thermal lumped models that would describe correctly the heat transfer between the case of the semiconductor device and its surrounding and take into account the impact of the mentioned above factors on the efficiency of this transport, for example in the context of the electrical device structure and the possibility of using PCBs as scattering and heat dissipating elements. Of course, for each semiconductor device, which is working in specific cooling conditions, the thermal parameters of the lumped model, describing heat transfer between the semiconductor structure and the surrounding, or between this structure and the device case, can be measured. However, this requires the prior construction of an appropriate cooling system of the device, and the experimental determination of the thermal parameters of the model can only ex-post confirm (or not) the correctness of the choice of the investigated device cooling system.

The general form of the device thermal model is shown in Fig.2. As seen, the multipath flow of the heat generated in the device is taken into account in this model. The structure of each component of the considered model (blocks A – H in Fig.2) is of the form of the nonlinear networks presented in Fig.1. The influence of the external equipment case on the device thermal properties are modeled by changing the temperature of the air existing inside the equipment case depending on the air temperature outside the whole equipment and the power dissipated in the device.

At present, the detailed description of the thermal component models representing the respective parts of the heat flow path is investigated by the authors. To formulate the proper dependencies describing the model, some measurements of the thermal resistances and the transient thermal impedances of devices operating at various cooling conditions, dissipated powers,

ambient temperatures, constructions of the natural and forced cooling systems, are indispensable.

In the next Chapter, the results of measurements of the thermal parameters of the selected devices illustrating the influence of some factors on the values of these parameters are presented and discussed. These dependences will be taken into account in the device thermal model being presented in the elaboration.

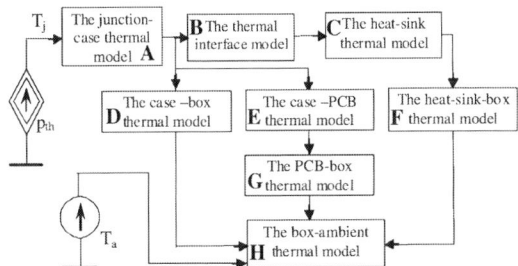

Fig. 2. The general form of the device thermal model with heat flow multipaths taken into account

IV. RESULTS OF INVESTIGATIONS

In the Chapter, the results of the authors' investigations, showing the influence of the chosen factors on the device thermal resistance and the transient thermal impedance are presented.

The thermal resistance of low-power devices operating without any heat-sink depends, among others, on the device leads length [32]. The measured dependence of the thermal resistance R_{th} of the low-power silicon diode BAVP17 with the glass case CE02 on its leads length l at two values of the solder area S situated on the PCB is shown. As seen, increasing the leads length causes also an increase of the diode thermal resistance value, but if l > 50 mm, the change of this dependence is observed. This means that the thermal resistance slightly decrease with an increase of l. This is due to an increase of the area of convection and extension of the heat flow path to the solder areas.

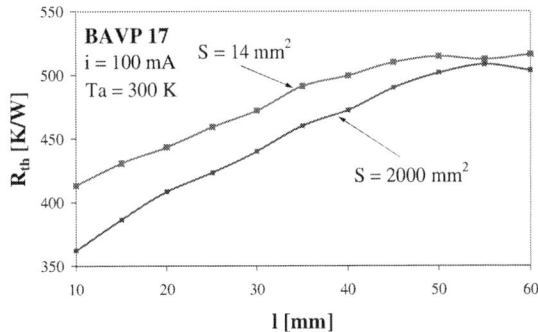

Fig. 3. The dependence of the diode thermal resistance on its metal leads length

In turn, Fig.4 presents the results of measurements of the dependence of the thermal resistance of the diode ZPY56 with the glass case DO-41 on the diode current. The measurements were carried out with the nominal length (l = 25 mm) of the metal leads (dashed lines) and with the shortened ones when l = 5 mm (solid lines), at various sizes of the solder areas: A – the copper leaf of the area S = 38x15 mm, B – the cooper leaf of the area S = 15x3 mm and C – the monolithic cooper of the dimensions 3x27x75 mm. As seen, increasing the solder areas causes a decrease of the device thermal resistance, similarly to an increase of the diode current or shortening of its metal leads.

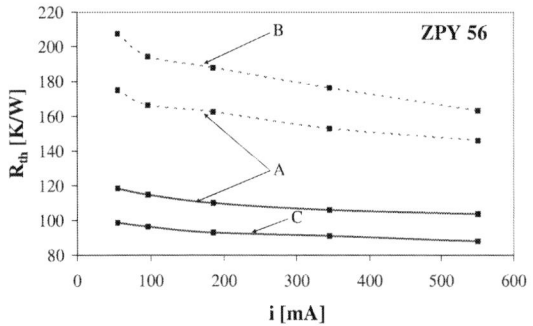

Fig. 4. The dependence of the thermal resistance of the diode ZPY56 on its current for various cooling conditions

It is worth mentioning that in the considered situation, after increasing the diode metal leads 15 times for the device having long leads, the thermal resistance value was reduced over a dozen or so percent, whereas shortening these metal leads from 30 to 5 mm at the same value of the solder areas, caused a decrease of the considered thermal parameter to about 30%.

In Fig.5 the results of measurements of the thermal resistance of the bipolar transistor BC109 as a function of the device dissipated power are presented. The investigated transistor in the case CE22 operated on the universal PCB of the dimensions 25x45 mm. The measurements were performed for the transistor: without any heat-sink and operating on two heat-sinks of various sizes: the small heat-sink (the aluminum rectangular of the dimensions 10x10x3 mm) and the large one (the black aluminum rectangular of the dimensions 100x90x10 mm).

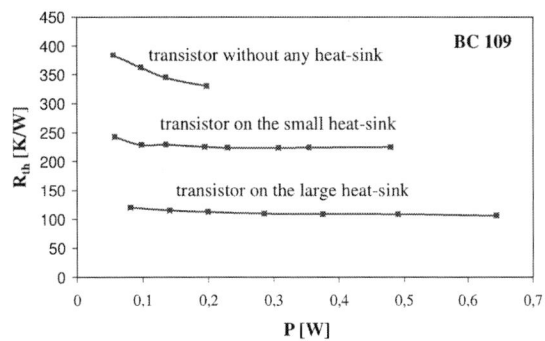

Fig. 5. The dependence of the thermal resistance of the bipolar transistor BC109 at various conditions of its cooling

As seen, the device thermal resistance distinctly decreases with an increase of the device power. For the transistor without any heat-sink the thermal resistance changes even more than 20%, while for the transistor operating on the large heat-sink these changes are about a few percentage only. The observed differences result from the changes of effectiveness of the convection phenomenon on the device case surface. The share of the convection in the process of the heat abstraction increases with an increase of the difference of the device case and surrounding temperatures [33].

The device thermal transients are well described by the course of its transient thermal impedance $Z_{th}(t)$, which allows estimating the time indispensible to get the thermal steady-state in the device. Fig.6 shows the courses of the transient thermal impedance normalized with respect to the thermal resistance for the transistor BC109 at three kinds of the device cooling.

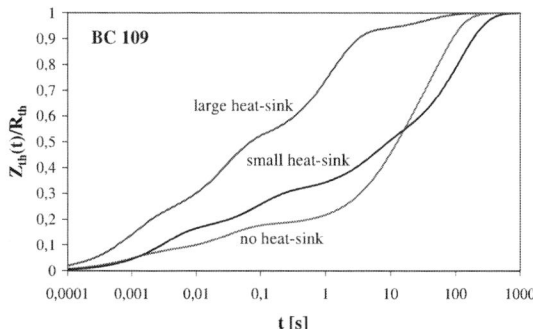

Fig. 6. Courses of the transient thermal impedance normalized with respect to the thermal resistance for the transistor BC109 at different cooling conditions

As seen, the set-up time t_{th} of the thermal steady-state (when $Z_{th}(t)/R_{th} \approx 1$) depends on the device thermal conditions. For the transistor BC109 without any heat-sink, with the small and large heat-sinks the set-up time is equal to 200 s, 600 s and 80 s, respectively. It is worth mentioning that the large heat-sink assures practically ideal cooling of the device case, therefore in such conditions the set-up time has the lowest value.

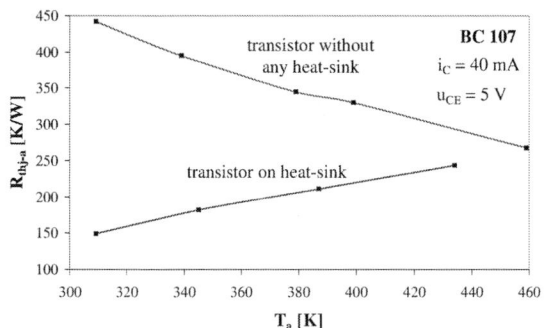

Fig. 7. The dependence of the thermal resistance of the low-power transistor on the ambient temperature T_a

In turn, in Fig.7 the influence of the ambient temperature on the value of the thermal resistance of the transistor BC107 with the case CE22 is shown. The investigations were performed for the transistor situated on the alumina heat-sink of the dimensions 100x100x10 mm and the same transistor operating without any heat-sink. As seen, for the transistor without any heat-sink its thermal resistance decreases with an increase of the temperature, whereas for the transistor situated on the heat-sink, this parameter value increases. In the first case, the increase of the effectiveness of convection with an increase of temperature is the main reason for the observed phenomenon, whereas in the other case – the decrease of the thermal conductivity with an increase of temperature is responsible for the observed results.

It is commonly known that dimensions of the heat-sink affect the device thermal resistance. Fig.8 illustrates the influence of the space orientation of the heat-sink on the power MOS transistor situated on the heat-sink made with the cast profile A-4240 of the length 60 mm. In this figure the courses a, b and c represent the heat-sink placed horizontally with the upwards and downwards cooling fins as well as the heat-sink situated vertically.

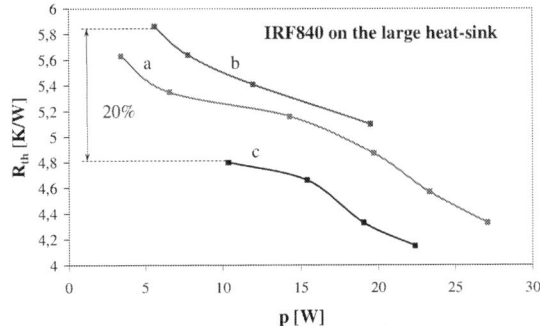

Fig. 8. The dependence of the thermal resistance of the transistor IRF840 situated on the large heat-sink on the dissipated power

As seen, the most efficient heat abstraction from the device on the heat-sink assures its vertical position, whereas the worse case of the heat abstraction is when the heat-sink is situated horizontally with downwards cooling fins. For these two cases the differences of the device thermal resistance are equal to about 20%, which results from various efficiencies of the convection phenomenon related to the heat-sink positions.

In some instances, when the electric insulation between the device and the heat-sink is needed, the insulating washers are used. At present, various kinds of polymer elastic tapes with various ceramic extendeus (e.g. alumina oxide, boron nitride) of very well electrical and thermal properties can be successfully employed.

Fig.9 shows the time courses of the transient thermal impedance of the power transistor BD285 (the case TO-220) situated on the heat-sink of the 60 mm length cast profile A-4240. The transistor is mounted: directly on the heat-sink (curve a), with the use of the silicon grease (curve b), with the mica plate (curve c) and with two various ceramic plates (curve d and e).

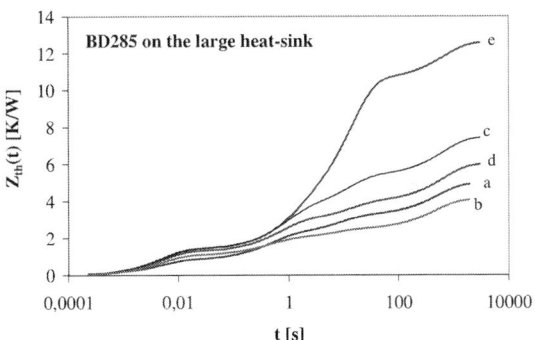

Fig. 9. The dependence of the transient thermal impedance of the power transistor BD285 mounted on the heat-sink with different insulating materials

As seen, the least value of the thermal resistance characterizes the transistor mounted directly to the heat-sink and the transistor operating with silicon grease, whereas using the silicon plates cause an increase of the device thermal resistance by even more than three times. The mica plate assures almost as good device cooling as the best ceramic plate.

A great influence on the device thermal parameter value has its case [44]. In Fig.10 the results of measurements of the dependence of the thermal resistance of the monolithic voltage regulator LT1073 on its input voltage V_{SUP} at different cooling conditions are presented. The investigated IC was situated on the PCB of the dimensions 110x105 mm. For the PCB situated horizontally (curve a) the thermal resistance of the device is by about 5% higher than in the vertical position of the PCB (curve b). On the contrary, situating the PCB inside the perpendicular metal box of the dimensions 83x148x150 mm (curve c) causes an increase of the device thermal resistance value by another 5%. Using the external heat-sink (curve d) causes a decrease of the thermal resistance value of the LT1073 by even more than 20%.

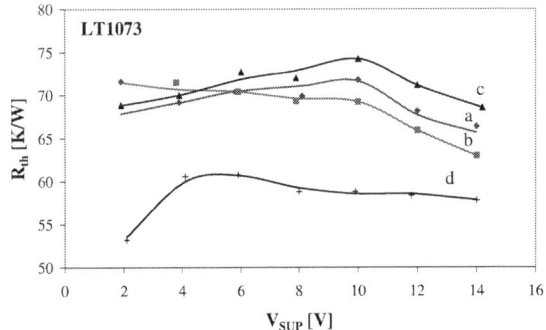

Fig. 10. The dependence of the LT1073 thermal resistance on the regulator input voltage for its various cooling conditions

V. CONCLUSIONS

From the presented results taken from the authors' investigations it is seen that the multipath flow of the heat dissipated in the device causes essential changes of its thermal parameters values. The complexity of the description of transportation of the heat dissipated in the device and removed to the surrounding often causes that the projects of the cooling systems are made by the method of "trial and error". Therefore, the sense of purpose of the investigations leading to formulate the device thermal model including the heat flow multipath is fully fulfilled. Such a model allows optimizing the device cooling system and reducing the costs of such system, assuring the operational reliability of the electronic equipment, assumed by the design engineer.

Formulating the multipath thermal model of the device is the main aim of the research project realized currently by the authors. This task demands among others performs a lot of measurements of the device thermal parameters with the use of various device cooling systems and formulating the analytical dependencies describing the influence of the cooling system technical parameters on the device thermal parameters. The results of the preliminary investigations show that the thermal model under test should be the nonlinear one, taking into account a lot of factors, as: the ambient temperature, the kind of the device case, the solder areas, the heat-sink dimensions and its space orientation and the device dissipated power. This thermal model will be useful for design engineers of electronic equipment.

ACKNOWLEDGEMENTS

This project is financed from the funds of National Science Centre which were awarded on the basis of the decision number DEC-2011/01/B/ST7/06740.

REFERENCES

[1] C. Sarno, G. Moulin: Thermal management of highly integrated electronic packages in avionics applications. Electronics Cooling, Vol. 7, No. 4, 2001, pp. 12-20

[2] A. Amerasekera, M-C. Chang, J.A. Seitchik, A. Chatterjee, K. Mayaram, J-H. Chern: Self-heating effects in basic semiconductor structures. IEEE Transactions on Electron Devices, Vol. 40, 1993, No. 10, pp. 1836-1844.

[3] A. Lidow, D. Knzer, G. Sheridan, D. Tam: The Semiconductor Roadmap for Power Managment in the New Millennium. Proceedings of the IEEE, Vol. 89, 2001, No. 6, pp. 803-812.

[4] A. Castellazzi, Y.C. Gerstenmaier, R. Kraus, G.K.M. Wachutka: Reliability analysis and modeling of power MOSFETs in the 42-V-PowerNet, IEEE Transactions on Power Electronics, Vol. 21, 2006, No. 3, pp.603-612,

[5] J. Parry, J. Rantala, C. Lasance: Temperature and reliability in electronics systems – the missing link. Electronics Cooling, Vol. 7, No. 4, 2001, pp. 30 – 36

[6] M.Ciappa, F.Carbognami,P. Cora, W. Fichtner: A novel thermomechanics-based lifetime prediction model for cycle fatigue failure mechnisms in power semiconductors. Microelectronics Reliability, Vol. 42, 2002, pp.1653-1658

[7] A. Castellazzi, R. Kraus, N. Seliger,D. Schmitt-Landsiedel: Reliability analysis of power MOSFET's with the help of compact models and circuit simulation. Microelectronics Reliability, Vol. 42, 2002, pp.1605-1610

[8] G. Coquery, S. Carubelli, J.P. Ousten, R. Lallemand: Power module lifetime estimation from chip temperature direct measurement in an automotive traction inverter. Microelectronics Reliability, Vol. 41, 2001, pp.1695-1700

[9] C.A. Happer: Electronic packaging and interconnection handbook McGraw-Hill Handbooks

[10] C. Buttay, D. Plason, B. Allard, D. Bergogne, P. Bevilacqua, Ch. Joubert, M. Lazar, Ch. Martin, H. Morel, D. Tournier, Ch. Raynaud: State of the art of high temperature power electronics. Materials Science and Engineering B, Vol. 176, No. 4, 2011, pp. 283-288.

[11] B.M. Guenin: The many flavors of ball grid array packages. Electronics Cooling, Vol. 8, No. 1, 2002, pp. 32-40

[12] K. Goodson, J. Santiago, T. Kenny, L.Jiang, S. Zeng, J-M. Koo, L. Zhang, S. Yao, E. Wang: Electroosmotic microchannel cooling system for microprocessors. Electronics Cooling, Vol. 8, No. 2, 2002, pp. 46-47

[13] R.E. Simons: Estimating temperatures in a water – to – air hybrid cooling system. Electronics Cooling, Vol. 8, No. 2, 2002, pp. 8-9

[14] M-C. Cheng, K. Zhang: An effective thermal circuit model for electro-thermal simulation of SOI analog circuits. Solid-State Electronics, Vol.62, 2011, pp.48-61

[15] D'Alessandro, V., Rinaldi, N. A critical review of thermal models for electro-thermal simulation. Solid-State Electronics, Vol. 46, 2002, No. 4, pp. 487-496.

[16] Gerstenmaier, Y.C. Castellazzi, A. Wachutka, G.K.M.: Electrothermal simulation of multichip-modules with novel transient thermal model and time-dependent boundary conditions. IEEE Transactions on Power Electronics, Vol. 21, No. 1, 2006, pp.45-55.

[17] A. Ammous, S. Ghedira, B. Allard, H. Morel, D. Renault: Choosing a thermal model for electrothermal simulation of power semiconductor devices. IEEE Trans. on Power Electronics, Vol.14, No.2, 1999, p.300-307

[18] P.E. Bagnoli, C. Casarosa, M. Ciampi, E. Dallago: Thermal resistance analysis by induced transient (TRAIT) method for power electronic devices thermal characterization. IEEE Trans. on Power Electronics, I. Fundamentals and Theory, Vol. 13, No. 6, 1998; s. 1208-19.

[19] E. Raj, Z. Lisik, W. Fiks: Influence of the manufacturing technology on microchannel structure efficiency. Materials Science and Engineering B, Vol. 176, no. 4, 2011, pp. 311-315.

[20] B.M. Guenin: Simplified transient model for IC packages. Electronics Cooling, Vol. 8, No. 3, 2002, pp. 13-15

[21] S.V. Garimella, V. Singhal, D. Liu: On-chip thermal management with microchannel heat sinks and integrated micropumps. Proceedings of the IEEE, Vol. 94, 2006, No. 8, pp. 1534-1548.

[22] H.Y. Zhang, D. Pinjala, T.N. Wong, K.C. Toh, Y.K. Joshi: Single-phase liquid cooled microchannel heat sink for electronic packages. Applied Thermal Engineering, Vol. 25, 2005, No. 10, pp. 1472-1487.

[23] Y. Yener, S. Kakac: Heat Conduction.Taylor &Francis, 2008.

[24] Z. Lisik: Zjawiska w strukturach półprzewodnikowych – metody ich modelowania. Wydawnictwo Politechniki Łódzkiej, Łódź 2005

[25] C.A. Gould, N.Y.A. Shammas, S. Grainger, I. Taylor: Thermoelectric cooling in microelectronic circuits and waste heat electrical power generation in a desktop personal computer. Materials Science and Engineering B, Vol. 176, No. 4, 2011, pp. 316-325.

[26] R. Prasher: Thermal Interface Materials: Historical Perspective, status and Future Directions. Proceedings of the IEEE, Vol. 94, No. 8, 2006, pp. 1571-1586.

[27] V. Szekely: A New Evaluation Method of Thermal Transient Measurement Results. Microelectronic Journal, Vol. 28, No. 3, 1997, pp. 277-292.

[28] P.A. Mawby, P.M. Igic, M.S. Towers: Physically based compact device models for circuit modelling applications. Microelectronics Journal, Vol. 32, 2001, pp. 433-447.

[29] M. Rencz, V. Szekely: Dynamic thermal multiport modeling of IC packages. IEEE Transactions on Components and Packaging Technologies, Vol. 24, No. 4, 2001, pp. 596-604.

[30] J. Zarębski, K. Górecki: A Method of Measuring the Transient Thermal Impedance of Monolithic Bipolar Switched Regulators. IEEE Transactions on Components and Packaging Technologies, Vol. 30, No. 4, 2007 pp. 627 – 631.

[31] J. Zarębski, K. Górecki: The electrothermal large-signal model of power MOS transistors for SPICE. IEEE Transaction on Power Electronics, Vol. 25 , No. 5-6, 2010, pp. 1265 – 1274.

[32] K. Górecki, J. Zarębski, W.J. Stepowicz: Wpływ wybranych czynników na parametry termiczne przyrządów półprzewodnikowych. Elektronika, No. 10, 2005, pp. 18-20.

[33] K. Górecki, J. Zarębski: Badanie wpływu wybranych czynników na parametry cieplne tranzystorów mocy MOS. Przegląd Elektrotechniczny, Vol. 85, No. 4, 2009, pp. 159-164.

[34] K. Górecki, J. Zarębski: Nonlinear compact thermal model of power semiconductor devices. IEEE Transactions on Components and Packaging Technologies, Vol. 33, No. 3, 2010, pp. 643-647.

[35] K. Górecki, J. Zarębski, D. Bisewski, J. Dąbrowski: Nonlinear compact thermal model of SiC power semiconductor devices. 17th International Conference Mixed Design of Integrated Circuits and Systems MIXDES 2010, Wrocław, 2010, pp. 365-370.

[36] K. Górecki, J. Zarębski: Estymacja parametrów modelu termicznego elementów półprzewodnikowych. Kwartalnik Elektroniki i Telekomunikacji, No. 3, 2006, pp. 347-360.

[37] S. Di Pascoli, P.E. Bagnoli, C. Casarosa: Thermal analysis of insulated metal substrates for automotive electronic assemblies. Microelectronics Journal, Vol. 30, No. 11, 1999, pp. 1129-1135.

[38] M. Rencz, V. Szekely, A. Poppe: Integration of a network solver and a field solver for the mixed level thermal simulation of MEMS problems. Proceedings of the SPIE, No. 4755, 2002, pp. 36-43.

[39] J. Zarębski, K. Górecki: SPICE-Aided Modelling of the UC3842 Current Mode PWM Controller with Selfheating Taken into Account. Microelectronics Reliability, Vol. 47, No. 7, 2007, pp. 1145-1152.

[40] J. Zarębski, K. Górecki: Modelling CoolMOS Transistors in SPICE. IEE Proc. on Cicuits, Devices and Systems, Vol. 153, No. 1, 2006, pp. 46-52.

[41] V. Szekely: A New Evaluation Method of Thermal Transient Measurement Results. Microelectronic J., Vol. 28, No. 3, 1997, pp. 277-292.

[42] F.N. Masana: Pseudo-3D Dynamic Thermal Model for Semiconductor Packages. 7th International Conference Mixed Design of Integrated Circuits and Systems MIXDES 2000, Gdynia 2000, pp. 357-360.

[43] Infineon Technologies web-site http://www.infineon.com.

[44] J. Zarębski, K. Górecki: A New Method for the Measurement of the Thermal Resistance of the Monolithic Switched Regulator LT1073. IEEE Transactions on Instrumentation and Measurement, Vol. 56, No. 5, 2007, pp. 2101-2104.

MIXDES 2012, 19th International Conference *"Mixed Design of Integrated Circuits and Systems"*, May 24-26, 2012, Warsaw, Poland

Technology Migration and Thermal Coupling

Marcin Janicki, Piotr Zajac, Michal Szermer, Andrzej Napieralski

Department of Microelectronics and Computer Science
Technical University of Lodz
Lodz, Poland
janicki@dmcs.p.lodz.pl

Abstract—**This paper presents the problem of ever increasing thermal coupling between integrated circuit components induced by the technology migration to newer technology nodes. All the discussions here are based on the results of static and dynamic thermal simulations of a test ASIC which contains a large matrix of heat sources.**

Index Terms—**Many-core architectures, technology migration, thermal coupling.**

I. INTRODUCTION

The ever growing density of dissipated power in integrated circuits causes serious thermal problems in newer technologies. The introduction of new gate materials having high dielectric constant values alleviated the problem with excessive static power dissipation due to the gate leakage. However, as it will be demonstrated in this paper, the constant shrinking of circuit dimensions soon is expected to pose new challenges, this time related to the dynamic power dissipation.

Currently, one of the most common ways of reducing local density of dissipated power is the use of multicore processor architectures. Unfortunately, almost all the publications studied so far the thermal implications of multi-core architectures only in the currently existing technologies, mostly using replications of an Alpha-like architecture as a standard benchmark [1]-[4].

The authors also used this benchmark in [5], however they tried to project the results onto future technologies. Quite the opposite, in this paper a novel approach is used where future architectures and technologies can be explored using the same test ASIC. The following section describes in detail the ASIC design. Then the results of its static and dynamic simulations are presented and discussed.

II. BENCHMARK GEOMETRY

The test ASIC which will be analyzed in this paper is not a replication of any real architecture, but it contains a 16 x 24 matrix of heat sources which are supposed to mimic different functional blocks in many-core architectures realized in various technologies. The main idea behind this approach is illustrated in Fig. 1. The size of individual transistor heat cells is chosen expressly to correspond to some integer or floating point units in a processor manufactured in the 16 nm technology. Then, the same functional block made in the 32 nm technology would correspond to 4 heat cells indicated by black squares.

Consequently, the entire chip containing originally 4 cores would consist of 16 cores after the migration through 2 future technology nodes, as shown in the figure. Taking into account that in the proposed ASIC all the heat cells can be individually switched on and off at any desired time instant and at different power levels, the same circuit can practically mimic numerous many core architectures realized in various technologies. Since the circuit itself is designed in an older 0.35 µm high voltage technology, thermal benchmarking is rendered much cheaper using the proposed approach. The detailed description of the design can be found in [6].

The thermal simulation results presented in the following sections were obtained using the analytical Green's function method [7], which is quite flexible and for simple geometries, e.g. for multilayered slabs, allows the computation of thermal influence coefficients linking power dissipation to temperature rise only for selected time instants and locations in a structure.

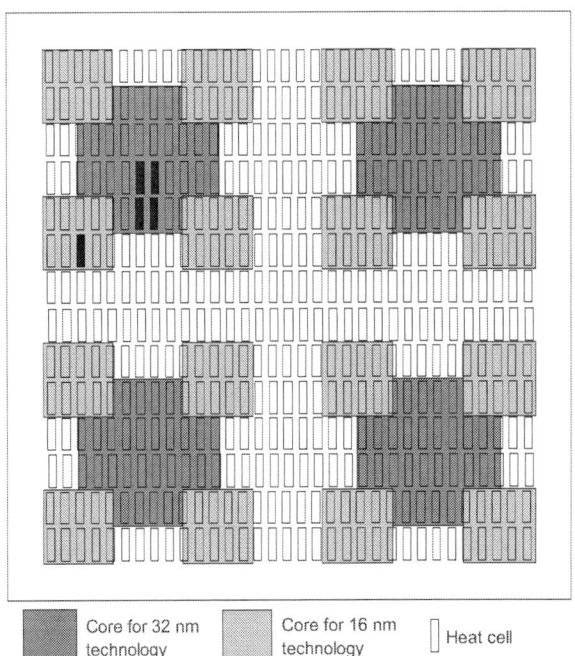

Figure 1.   Schematic layout of the test ASIC.

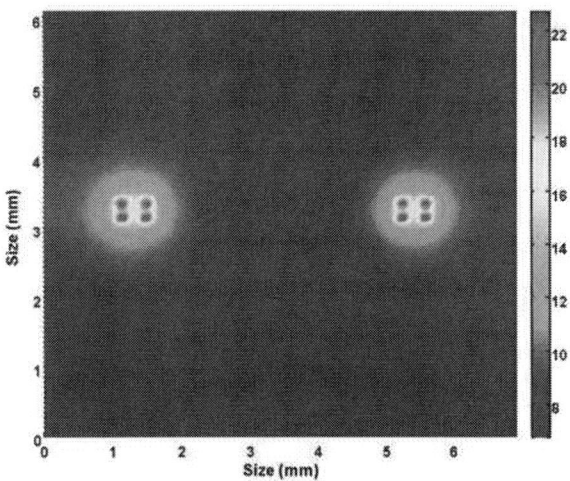

Figure 2.   Temperature rise map for 2 integer units placed apart in the 32 nm technology.

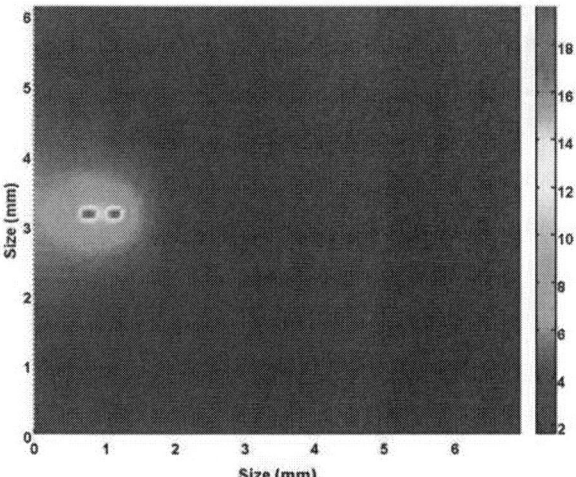

Figure 3.   Temperature rise map for 2 integer units placed close to each other in the 16 nm technology.

III. STEADY STATE THERMAL COUPLING

First, steady state simulation of the test ASIC were carried out. The goal of the simulation was to investigate the actual impact of technology migration on thermal coupling between particular components of a chip. For this purpose, based on the extracted parameter values from SPICE simulations with the Predictive Technology Models [8]-[9], the power in an integer arithmetic unit was estimated for the 32 nm technology and for the 16 nm technology yielding 1.14 W and 0.4 W respectively. Consequently, the first value, divided by 4 was assigned during the simulations of the 32 nm technology to 4 neighbouring heat cells and the latter one to a single heat cell for the simulations of the other one.

The thermal model assumed that the chip is mainly cooled from the back side and conduction cooled on the other one, which is typical for flip-chip processor assemblies. The heat transfer rate at the back surface was adjusted to model the presence of a heat sink with forced water flow, similar to the one described in [10]. Owing to the use of the Green's function solution method, it was possible to compute the temperature map in 10,000 locations in less than 1 minute.

For both technology nodes considered here, the steady state temperature maps were computed for 2 cases: when 2 similar integer units are located as close possible to each other and when they are shifted apart to opposite sides of the chip. The minimal, average and maximal surface temperature values for all the considered cases are given in Table 1. Moreover, two selected temperature mas are presented in Figs. 2-3. From the table it could be inferred that the change in the spacing between the arithmetic units does not influence the average temperature rise, but when the units are placed close to each other the minimal temperature decreases and the maximal one increases. Comparing the technologies, for the 16 nm technology node the chip remains cooler, however the hot spot temperature rise values are comparable.

Analysing now the thermal coupling between the arithmetic units, the simulations showed that for the 32 nm technology the contribution of the neighbouring unit to the total temperature rise is just 13 % when the units are placed apart and 23 % when they are located close to each other. For the 16 nm technology the respective numbers become 5 % and 21 %. Although these results might seem better, one should remember that each unit in the newer technology node will have more neighbours. Then the last number would increase to over 44 %, which means that almost half of the temperature rise will be due to the heating by neighbouring units what proves that in future technology nodes the thermal coupling among individual chip components will increase substantially.

IV. DYNAMIC THERMAL COUPLING

However, from the many-core architecture perspective, far more important is the analysis of the dynamic thermal coupling since the main purpose of using such architectures is to spread the dynamic power dissipation on the entire surface of a chip, thus avoiding the creation of hot spots. However, the benefits coming from multitasking or multithreading techniques quite soon might be limited by the increased thermal coupling, since when all the components heat also their neighbours it will not make any more difference where the power is dissipated.

TABLE I.   STEADY STATE TEMPERATURE DATA

Case	Temperature rise (K)		
	minimal	*average*	*maximal*
32 nm cores apart	6.7	8.2	22.7
32 nm cores close	5.3	8.2	24.8
16 nm cores apart	2.3	2.9	16.8
16 nm cores close	1.6	2.9	19.5

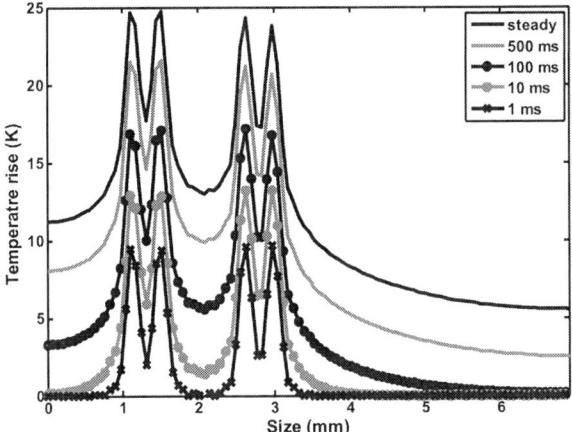

Figure 4. Evolution of temperature response in 32 nm technology.

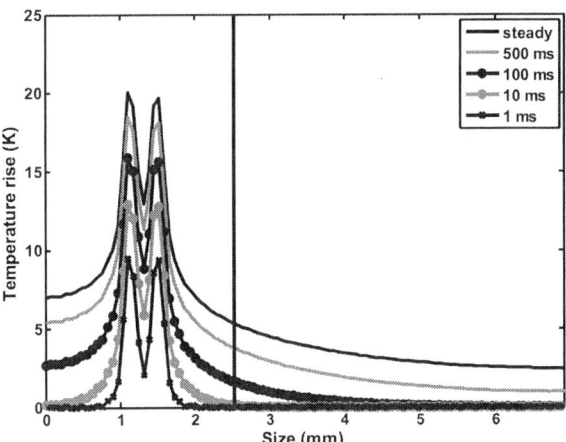

Figure 6. Single source temperature response in 32 nm technology.

Thus, in order to investigate this problem in more detail, the evolution of the temperature response in space and time was calculated along the lines joining the centres of arithmetical units for both of the considered technologies and with different spacing between the units. For the simulations it was assumed that the units start to operate at the same time at full load.

Indeed, the results shown in Figs.4-5 prove that the thermal coupling increases in new technologies. Looking in the figures at the midpoint between the heat sources, one can observe that the power dissipation in a source has some effect at this point only after a few milliseconds in the 32 nm technology, whereas for the 16 nm technology the thermal response in this location is visible already after much less than a millisecond.

This observation is confirmed also in the results presented in Figs. 6-7 showing the temperature response evolution from a single heat source. In order to facilitate the result analysis, the physical location of the edge of the neighbouring arithmetic unit is indicated in the figures by the vertical line. This time again one can observe that the temperature response reaches the neighbouring unit much faster in the 16 nm technology.

Namely, the heat diffusion time decrease from some 10 ms to around 1 ms, though the distance between the heat cells was reduced only by a half. This means that an arithmetic unit has 10 time less time to execute a particular task without affecting the temperature of its neighbour.

V. CONCLUSIONS

The simulations presented in this paper confirmed that the migration to future technology nodes might cause significant problems due to increased thermal coupling. This phenomenon will have in future serious implications, especially for many-core processor architectures. Moreover, as demonstrated in [5], the advances in cooling techniques may decrease the average temperature, but probably they will not be sufficient to reduce the dynamic thermal coupling between the cores.

The presented thermal simulations indicated also that when a single unit is supposed to consist of multiple heat cells the temperature in the unit is not uniform and the differences might reach 30 %. Thus, the authors decide to redesign the heat cells so that the heat source occupies more of the cell layout.

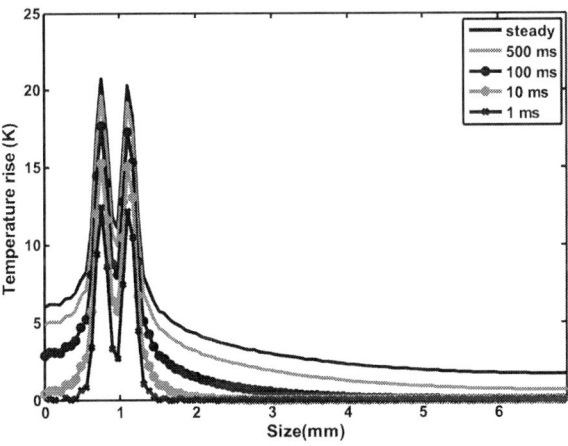

Figure 5. Evolution of temperature response in 16 nm technology.

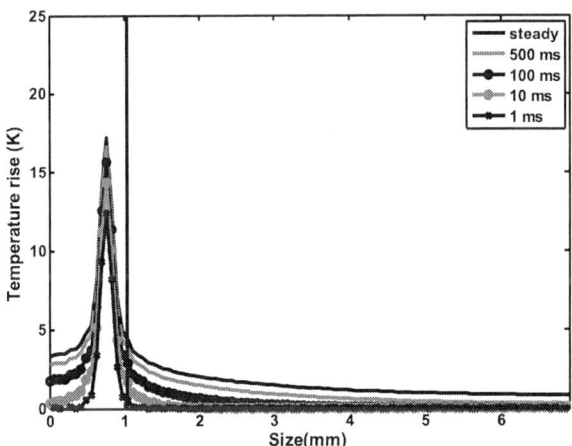

Figure 7. Single source temperature response in 16 nm technology.

321

ACKNOWLEDGMENT

This research was supported by the grant of Polish National Center of Science No. N515 5091 40.

REFERENCES

[1] K. Skadron, M. Stan, K. Sankaranarayanan, Huang, Velusamy, Tarjan, Temperature-aware micro-architecture: modelling and implementation, ACM Trans. on Arch. and Code Optim., vol. 1, pp. 94-125, 2004.

[2] W. Liao, L. He, and K. Lepak, "Temperature and supply voltage aware performance and power modeling at micro-architecture level," IEEE Trans. CAD Integrated Circuits and Systems, vol. 24, pp. 1042-1053, 2005.

[3] M. Monchiero, R. Canal, Gonzalez, Power/performance/thermal design space exploration for multicore architectures, IEEE Trans. Parallel and Distributed Systems, vol. 19, No. 5, pp. 666-681, 2008.

[4] J. Li, and J.F. Martínez, "Power-performance considerations of parallel computing on chip multiprocessors," ACM Trans. Arch. Code Optim., vol. 2 , pp. 397-422, 2005.

[5] M. Janicki, J. Collet, Louri, A. Napieralski, Hot spots and core-to-core thermal coupling in future multi-core architecture, 26th Semiconductor Thermal Measurement and Management Symposium SEMI-THERM, 2010, pp. 205-209.

[6] M. Szermer, C. Maj, P. Pietrzak, M. Janicki, P. Zajac, A. Napieralski, Test ASIC for the investigation of thermal coupling in many-core architectures, 28th Semiconductor Thermal Measurement, Modeling and Management Symposium SEMI-THERM, March 18-22, 2012, San Jose, USA.

[7] Janicki, G. De Mey, and A. Napieralski, Thermal analysis of layered electronic circuits with Green's functions, Microelectronics Journal, Vol. 38, pp. 177-184, 2007.

[8] W. Zhao, and Y. Cao, "New generation of Predictive Technology Model for sub-45nm early design exploration," IEEE T. on Electron Devices, vol. 53, no. 11, pp. 2816-2823, November 2006.

[9] P. Zajac, M. Janicki, M. Szermer, C. Maj, P. Pietrzak, A. Napieralski, Cache Leakage Power Estimation Using Architectural Model for 32 nm and 16 nm Technology Nodes, accepted for 28th IEEE Semiconductor Thermal Measurement, Modeling and Management Symposium SEMI-THERM, March 18-22, 2012, San Jose, USA.

[10] M. Janicki, Z. Kulesza, A. Napieralski, "Distributed Network of Remote Sensors for Real Time Prediction of Hot Spot Temperature Values", Proc. of 9th Conference IEEE Sensors 2010, November 1-4, 2010, pp. 656-659.

Thermal Models for Dynamic Clock Control

Maciej Frankiewicz, Andrzej Kos
AGH University of Science and Technology
Department of Electronics
Krakow, Poland
frankiew@agh.edu.pl, kos@agh.edu.pl

Abstract—The paper deals with the problem of dynamic control of clock signal freqeuncy in dependence of chip temperature. Some behavioural thermal models dedicated for high frequency circuits were created in MATLAB environment and tested. Additionally idea of prediction of temperature changes and its usage in frequency control was introduced and motivated.

Index Terms—temperature, frequency control, DCT, DFS, PFS

I. INTRODUCTION

Considering the aspiration to make modern digital circuits faster and scaling down the technology researchers must take into account thermal effects. The goal is to find the optimal balance state between keeping the throughput of the system on the possible maximum level and decreasing the power consumption of the device. Assuming that circuit cannot exceed acceptable maximum temperature for proper work it should be controlled in that way to keep its temperature as close to the permissible limit as possible with minimal fluctuations. To achieve that some dynamic control systems have been developed including: Dynamic Voltage Scaling (DVS) – scaling the supply voltage, Dynamic Clock Throttling (DCT) – shutting off the clock when overheating and Dynamic Frequency Scaling (DFS) – scaling clock frequency, which are often combined [1].

Idea of frequency control origins from the fact that circuit working with higher frequency consumes more power. In order to cool it down it has to be slowed down. When the chip returns to safe temperature range it can work faster again. This frequency-temperature feedback can be used to stabilize the temperature and ensure safe work of the circuit. Despite former achievements on the field some further investigations must be done to improve thermal behaviour of the circuit and find the optimal control system [2].

Popular electronic simulators are not suitable for thermal simulations. That causes the need to create an environment in which temperature of the circuit can be examined. Presented MATLAB models have been created to verify several different implementations of the dynamic frequency control systems before fabrication. In addition Predictive Frequency Scaling (PFS) method which is based on DFS and DCT but predicts changes in temperature in nearest future will be introduced and its advantages will be motivated.

II. THERMAL MODELS

All models were created to illustrate thermal processes inside the integrated circuits. For better comparison they were based on common fundamentals and were tested with the same input vector. The test vector represents power dissipated in the circuit during each task that have to be executed and can be divided into four periods: initialization, long period of an idle state with low power consumption to stabilize the temperature, awaking event with high power consumption which can cause overheating and finally period of tasks with medium power consumption. Important assumption is that power consumed by the circuit does not change during each task execution. It is also assumed that not working circuit dissipates only static power losses P_{stat} but while it is working the dynamic power losses P_{dyn} were limited to ones related to charging/discharging load capacitance which are the most important and depend on the frequency (1).

$$P_{dyn} = C_L f V_{DD}^2 \qquad (1)$$

In every case the temperature T_{stat} in static state (represented in figures by dashed line) can be calculated based on present power dissipated in the circuit P_{diss} (represented in Fig. 1 by signs 'x') corresponding to (2) where T_{amb} is ambient temperature and R_{th} is thermal resistance of the circuit and its package. When the circuit is working present chip temperature T_{chip} (represented by continuous line) goes to static temperature as an exponential function (3) where τ_{th} is thermal time constant, T_0 temperature at the beginning of the process and ΔT_{chip} is difference between T_0 and T_{stat} [3].

$$T_{stat} = T_{amb} + R_{th} P_{diss} \qquad (2)$$

$$T_{chip} = T_0 + \Delta T_{chip}\left(1 - \exp\left(-\frac{t}{\tau_{th}}\right)\right) \qquad (3)$$

Described behaviour of the integrated circuit is explained by the example presented in Fig. 1. Power consumend by the chip ('x' signs) causes that the circuit should have static temperature T_{stat} (dashed line). Because of the thermal time constant τ this temperature is not reached immediately but T_{chip} (continuous line) starts from the ambient temperature T_{amb} (22°C in this case) and exponentialy goes to specified static temperature.

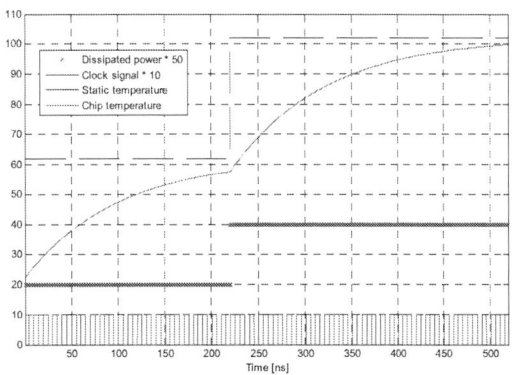

Figure 1.   Example of temperature change inside the integrated circuit package as a consequence of change of power dissipation.

Some temperature reference levels were used in the models and were took as a result of previous tests: maximum allowed temperature (represented by solid dashed line) was set at 90°C because some of tested circuits did not work correctly at this temperature while at 60°C all of them should work with no problems. Third reference level was set in the middle of them (75°C). Additionally on all figures the clock signal is presented (scaled 10 times) for better visualization of the processes.

Values of the physical parameters do not correspond to any real circuit but were chosen for good presentation of the results. Some quality indexes were introduced to compare results from different models. First of them is the gain in time of execution of all tasks. It is obvious that circuit should work as fast as it is possible. Another one is the maximum temperature of the circuit which should never exceed maximum allowed level. Interpretation of this parameter is not as simple as previous one. Too low temperature can lead to situation when circuit works too slow. As a result temperature should have a value that guarantees both safe and fast work.

A. Static model

The first described model is a static one. In this case frequency of the clock signal is independent on temperature. The model is necessary to compare its results with another ones. In Fig. 2 uniform clock frequency in the time of the execution is clearly visible. When the model works with its maximum frequency which is set to 1 GHz (fast mode) it takes 188 ns to execute all the tasks and the circuit temperature reaches 91.87°C which means overheating. If the frequency is set to the value at which temperature does not exceed maximum allowed level (safe mode) the circuit works for 192 ns. It has to be mentioned that the power needed to execute

the tasks will be different for every case and this maximum frequency would have to be calculated for each case separately which is impossible. That is the reason why usually the frequency is scaled down in regard to maximum value for example to 80%. In this case (slow mode) execution takes 236 ns. This time will be a reference value to calculate execution time gain in next models. Maximum temperature gain will be calculated in regard to 90°C level.

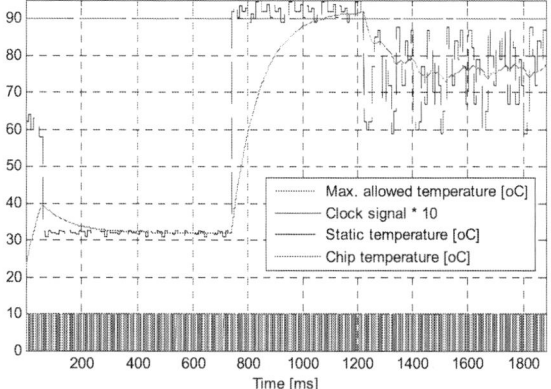

Figure 2.   Transient temperature response for static model at fast mode.

B. Discreet model

There are many possible physical realizations of the frequency control. First of them is division by several discreet values. Its application needs sensor which is source of information of present temperature [4], control block which process this information and controls the divider, clock signal comes from the generator. The block diagram of the discreet model is presented in Fig. 4 while in Fig. 3 modelled characteristic of the generator is presented. It has been decided to turn off the generator when the temperature reaches the maximum allowed value.

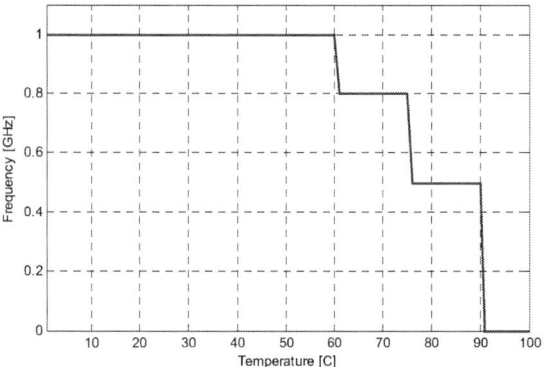

Figure 3.   Tuning characteristic of the generator with discreet frequency division.

324

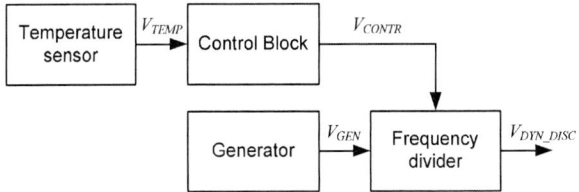

Figure 4.   Block diagram of the model with discreet frequency division.

The simulations showed that this method shortens the execution time to 219.62 ns and reduces the maximum temperature to 75.64°C, Fig. 5 That means that overheating was eliminated in this case. The greatest advantage of this method is simplicity of its implementation while biggest disadvantage are big fluctuations of the temperature when the frequency is changing.

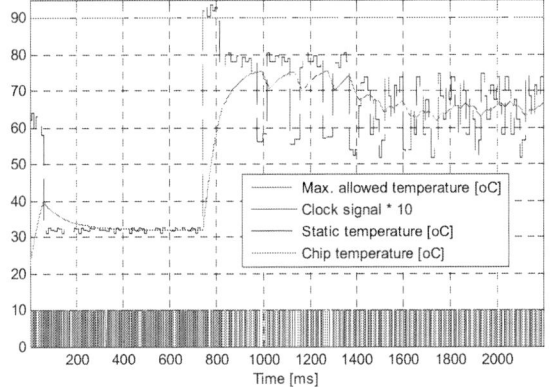

Figure 5.   Transient temperature response for dynamic model with discreet frequency division.

C. Continuous model

Another method of the frequency control is tuning it with continuous function. To realize that method the signal from the temperature sensor directly drives the VCO (Voltage-Controlled Oscillator). Block diagram of this structure is shown in Fig. 6.

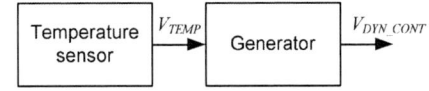

Figure 6.   Block diagram of the model with continuous frequency scaling.

The shape of the tuning function is an important problem to solve. If the temperature of the circuit is relatively low there is no need to rapidly change the temperature so it has been decided to tune the oscillator with nonlinear function. Shutting off the generator after crossing temperature limit would require steep fall of the function in high temperature range which is highly inconvenient and it has been resigned from this option. Implemented tuning characteristic of the generator is presented in Fig. 7.

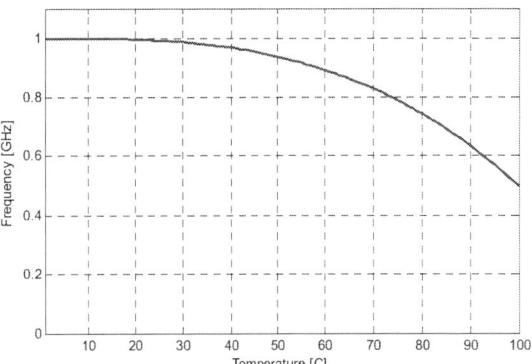

Figure 7.   Tuning characteristic of the generator with continuous frequency scaling.

The tests of this model (Fig. 8) showed that the execution time can be limited to 213.19 ns and the maximum temperature is reduced to 76.74°C. Usage of the continuous model leads to elimination of big temperature fluctuations by more fluent tuning while the quality parameters are similar to the discreet model.

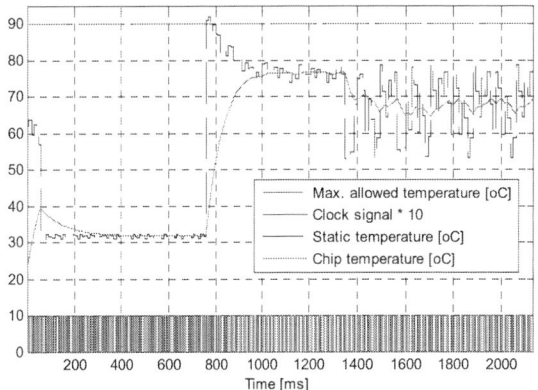

Figure 8.   Transient temperature response for dynamic model with continuous frequency scaling.

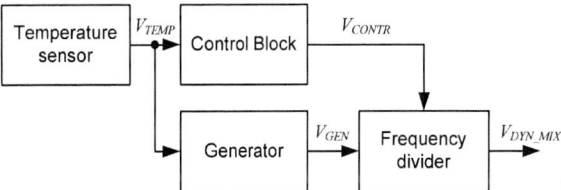

Figure 9.   Block diagram of the model with mixed frequency control.

D. Mixed model

It is possible to combine both previous methods in the mixed discreet-continuous model. In this case the temperature sensor tunes the oscillator and controls the frequency divider as it is shown in Fig. 9. Fig. 10 presents characteristic of such generator.

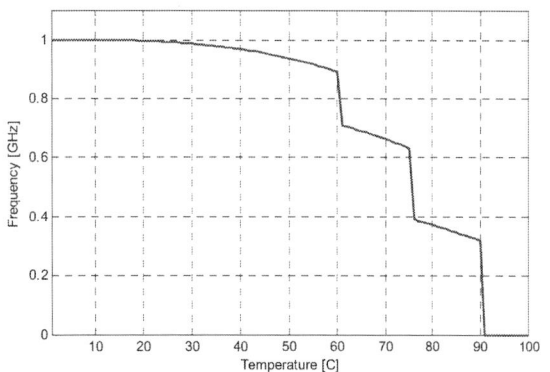

Figure 10. Tuning characteristic of the generator with mixed frequency control.

Described actions causes that 234.73 ns are needed to execute all tasks. So long time is a result of double frequency limitation. On the other hand maximum temperature is reduced to 69.21°C. The mixed model combines advantages of both dynamic methods (limited temperature fluctuations and possibility to switch off the generator) but the cost is quite low frequency of the clock signal. All of described effects are visible in Fig. 11.

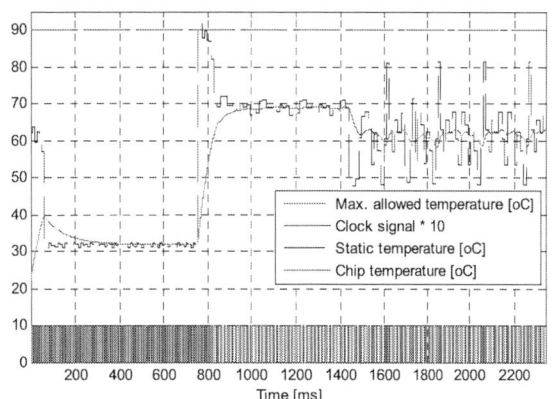

Figure 11. Transient temperature response for dynamic model with mixed frequency control.

E. Predictive models

The frequency control can be more efficient when changes of temperature in the nearest future are predicted. Even if the temperature is relatively high but the circuit is slowly cooling down there is no need to lower the clock rate. It can also happen that after a long time of sleep an awaking event happens and cooled circuit starts to work very fast. High temperature range will be reached in a short time and standard control methods could be not efficient enough. Prediction can be done by means of the operating system observing tasks which are waiting in the queue to be executed. The same goal can be achieved by the hardware means. A differentiator circuit

is quite easy to implement. The second solution will be explained. Information of changes of any function are carried by its differential. Differential of (3) is equal (4). If value of (4) is significantly greater then zero it means that the temperature value is rising fast, if lower than the temperature is falling. This function can be implemented and used to control the frequency divider as it is presented in Fig. 12.

$$T_{pred} = \frac{dT_{chip}}{dt} = \frac{\Delta T_{chip}}{\tau_{th}} \exp\left(-\frac{t}{\tau_{th}}\right) \qquad (4)$$

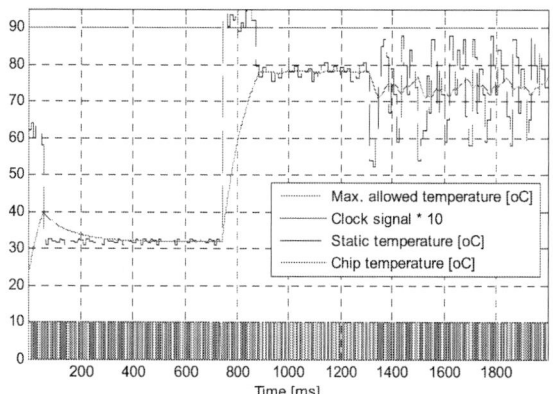

Figure 12. Block diagram of the model with discreet frequency division and temperature prediction.

Figure 13. Transient temperature response for predictive dynamic model with discreet frequency division.

The simulation results, presented in Fig. 13, showed that the time of execution is equal 199.63 ns and the maximum temperature reaches 78.41°C. The predictive model is faster than the discreet model without prediction and, moreover, has lower temperature fluctuations with no risk of overheating.

The predictive method can be also used in the mixed discreet-continuous model. Such situation is presented in Fig. 14. This model is faster than the one with no prediction (needs 215.43 ns to execute all tasks). The maximum temperature of 76.13°C is also higher than in non-predictive case but is kept in safe range.

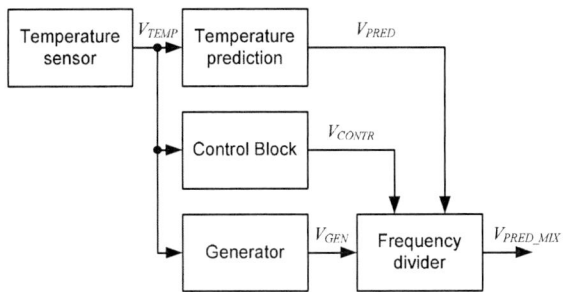

Figure 14. Block diagram of the model with mixed frequency control and temperature prediction.

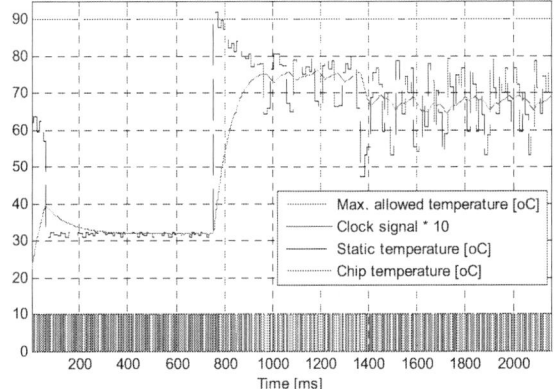

Figure 15. Transient temperature response for predictive dynamic model with mixed frequency control.

TABLE I. PARAMETERS OF TESTED THERMAL MODELS.

Model		Execution time [ns]	Execution time gain [%]	Max. temp. [°C]	Temp. gain [%]
Static	fast mode	188.00	20.00	91.87	-2.08
	safe mode	192.28	18.34	90.00	0.00
	slow mode	235.47	0.00	73.50	20.00
Discreet		219.62	6.73	75.64	15.96
Continuous		213.19	9.46	76.74	14.73
Mixed		234.73	0.31	69.21	23.10
Predictive – discreet		199.63	15.22	78.41	12.88
Predictive - mixed		215.43	8.54	76.13	15.41

III. SUMMARY

In the paper tests of models for the dynamic control of the clock frequency were presented. All models proved that usage of the dynamic methods can reduce time of execution of a program and improve thermal behaviour of integrated circuit. Main parameters of tested models are gathered in Tab. 1.

Additionally an idea of hardware implementation of the temperature prediction has been introduced and motivated.

The tests showed that the maximum reduction of execution time can be achieved by the predictive method with discreet frequency division (over 15% gain) but in regard to other significant advantages such as more fluent temperature changes predictive method with mixed frequency control is also an interesting case. All models ensured safe work without overheat and are suitable for hardware implementation as ASICs (Application Specified Integrated Circuits).

Additional motivation for using the predictive structures is power consumption. The tests proved that energy, defined by (5) and mean power, calculated from (6), consumed by circuits are lower when the temperature prediction is used than in case of non-predictive models.

$$E = \int_{o}^{T} P(t)dt \tag{5}$$

$$P_m = \frac{1}{T} \int_{o}^{T} P(t)dt \tag{6}$$

In simulated situation gain of energy consumption compared to non-predictive cases was respectively 15,10% for the discreet model and 14,71 for the mixed model. The mean power measured during the tests was 0,71% lower for the predictive discreet model and 3,91% lower for the predictive mixed model than in cases of non-predictive structures.

Some further work to improve presented models is planned in the nearest future. Chosen models will be fabricated to verify obtained results.

REFERENCES

[1] A. Gołda, A. Kos "Effective supervisors for predictive methods of dynamic power management", Proc. 14th Int. Conf. Mixed Desigh of Integrated Circuits and Systems MIXDES'2007, Ciechocinek, Poland, 21-23 June 2007, pp. 381-386

[2] M. Frankiewicz, A. Kos "Overheat Protection Circuit for High Frequency Processors", Materiały X Krajowej Konferencji Elektroniki, Darłówko, Poland, 4-9 June 2011, pp. 950-955

[3] A. Gołda, A. Kos "Predictive Frequency Control for Low Powerr Digital Systems", Proc. 13th Int. Conf. Mixed Desigh of Integrated Circuits and Systems MIXDES'2006, Gdynia, Poland, 22-24 June 2006, pp. 441-445

[4] A. Gołda, A. Kos "Analysis and Design of PTAT Temperature Sensor in Digital CMOS VLSI Circuits", Proc. 13th Int. Conf. Mixed Desigh of Integrated Circuits and Systems MIXDES'2006, Gdynia, Poland, 22-24 June 2006, pp. 415-420

MIXDES 2012, 19th International Conference *"Mixed Design of Integrated Circuits and Systems"*, May 24-26, 2012, Warsaw, Poland

Thermographic Measurements of Planar Inductors

Ioannis Papagiannopoulos,
Vasilis Chatziathanasiou
Department of Electrical and
Computer Engineers
Aristotle University of Thessaloniki
Thessaloniki, Greece
ipapagia@auth.gr

Gilbert De Mey
Department of Electronics and
Information Systems
Ghent University
Ghent, Belgium

Boguslaw Wiecek, Marcin Kaluza
Institute of Electronics
Technical University of Lodz
Lodz, Poland

Abstract—**Heating of circuits in electronics is an important design and operational aspect. Furthermore, there is an increasing need to check and evaluate temperature measurements acquired with the use of thermographic equipment. In other works, surface temperature distributions were captured and presented in an attempt to visualize the way heat is dissipated in the case of an integrated inductor. The aim of this paper is to interpret the thermograms captured with an infrared cooled image system during temperature measurements of Printed Circuit Board inductors.**

Index Terms—**thermal resistance; infrared imaging; planar inductors; temperature distribution**

I. INTRODUCTION

The use of infrared cameras is becoming more and more widespread as the technological advances allow higher frequency rates, spatial resolutions, and better signal to noise ratios. As presented in other works [1], it is possible to measure micrometer scale spatial temperature differences and high speed thermal transient phenomena with the use of thermographic equipment.

In this present work, an infrared cooled camera is used to measure the temperature distribution over a spiral planar inductor printed on a circuit board. Copper printed inductors are widely used as elements of power amplifiers, oscillators, switches etc. Various techniques are used to calculate the inductance of the inductors, such as empirical and mathematical formulations [2, 3]. On the other hand, calculating the current density distribution on the inductor turns is a different task. There are numerical solutions to the problem of current density distribution over rectangular cross section interconnects but few are mentioned about the current density over a spiral inductor [4, 5, 6]. This current density

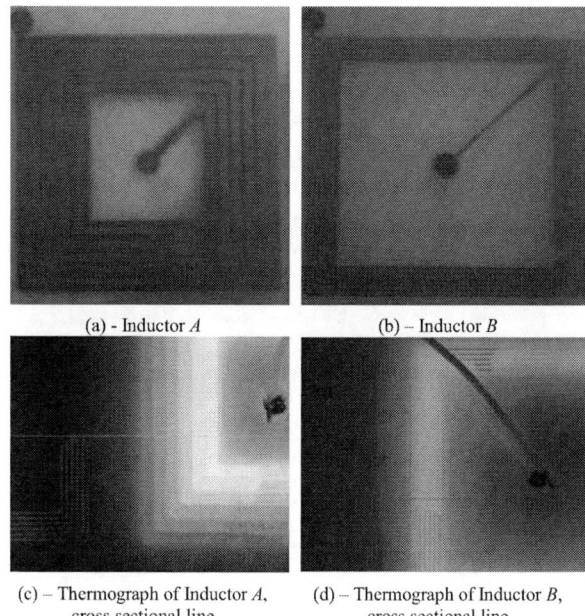

(a) - Inductor *A*   (b) – Inductor *B*

(c) – Thermograph of Inductor *A*,   (d) – Thermograph of Inductor *B*,
cross sectional line   cross sectional line

Figure 1.   Layout of the coils – example thermograms

defines the joule losses introduced on copper inductor – epoxy board thermal circuit.

In this current paper, the efforts are focused on the interpretation of thermograms of two 6-turn planar inductors of the same outer diameter but different inner diameter, operating at kHz frequency.

II. DESCRIPTION OF THE DEVICES UNDER TEST (DUT)

The inductors used during this work are two copper six turn inductors printed on FR4 boards. The specifications of these inductors are presented in Table 1. The geometry of their design is depicted in figure 1.   The excitation signals introduced in order to carry out the temperature measurements were alternating ones with frequency of 1300 kHz and 1280 kHz for inductor A and B respectively. The RMS value of the signal was of the magnitude of 0.5 A.

TABLE I.   INDUCTORS DESIGN PARAMETERS

	Number of turns	Width	Track width	Track spacing	Copper thickness
Inductor *A*	6	5.2 cm	2 mm	0,5 mm	30 um
Inductor *B*	6	5.05 cm	0,5 mm	0,5 mm	30 um

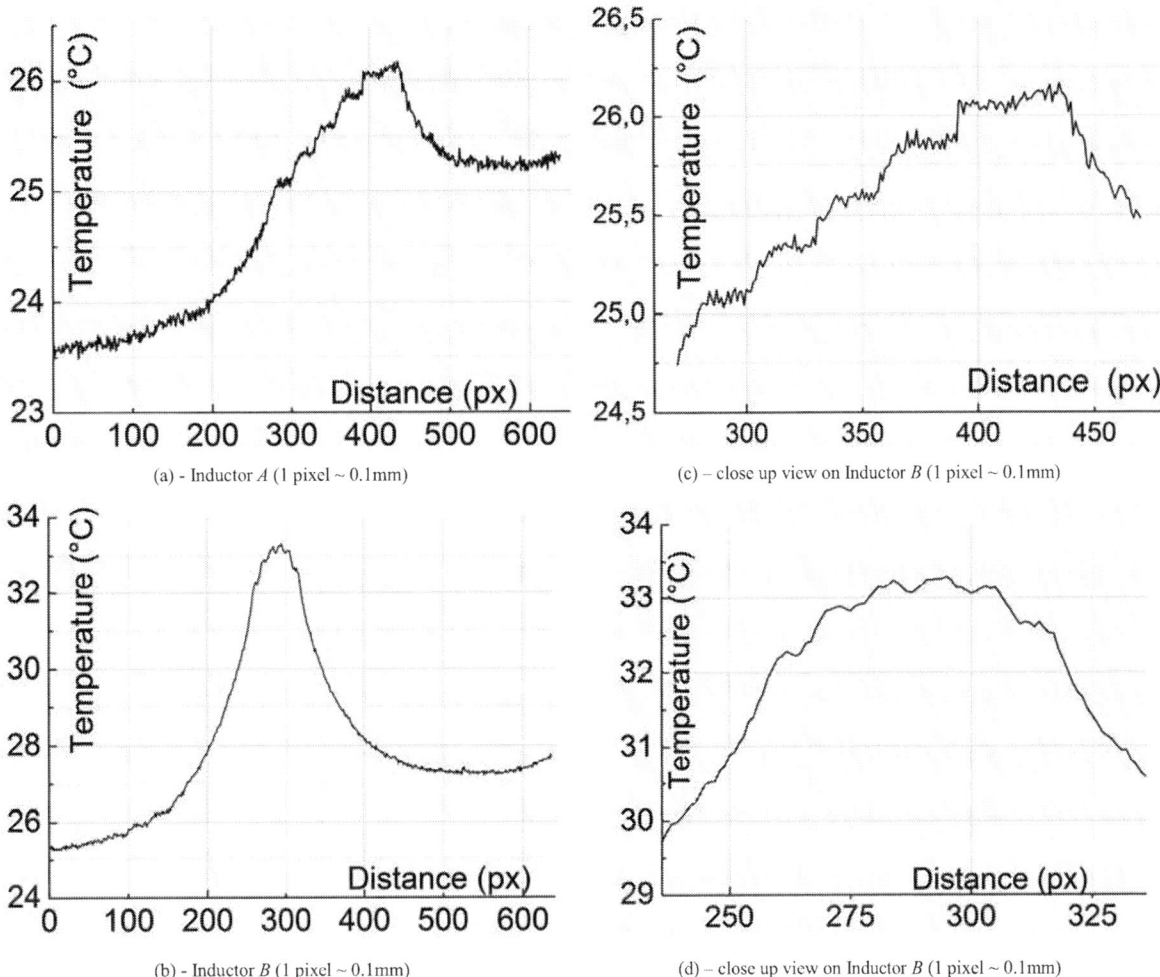

(a) - Inductor *A* (1 pixel ~ 0.1mm)

(c) – close up view on Inductor *B* (1 pixel ~ 0.1mm)

(b) - Inductor *B* (1 pixel ~ 0.1mm)

(d) – close up view on Inductor *B* (1 pixel ~ 0.1mm)

Figure 2.   Temperature distribution over cross section of Inductor A and (number of pixels in the horizontal axis). Each pixel corresponds to a distance of around 0.1 mm

In order to have an homogeneous emissivity on the surface of the specimens their surfaces were covered with black mat paint. The emissivity of the copper is comparatively low, therefore in the infrared range it acts as a mirror. In order to avoid this behavior and measure more precisely the surface temperature of the DUT, black mat paint was used to treat the surface and thus homogenize and improve the emissivity coefficient of the surface. This way it is possible to approximate the behavior of a black body.

From a thermal point of view, copper has a much higher thermal conductivity (k = 401 W/(m.K)) in comparison with the FR4 board (k = 0.25 W/(m.K)) that constitutes the substrate of the inductor. In addition, the free surface of the circuit board is much larger than the surface occupied by the inductor. Furthermore, the paint layer covering the surface of the inductor turns is characterized by a comparatively low thickness. Therefore, its contribution the thermal network is neglected and not considered as a diffusor of heat.

III.   TEMPERATURE MEASUREMENTS

The technique used to measure the temperature over the circuit surface is infrared thermography. A MWIR camera with a cooled InSb 640x512 pixel detector matrix was used. Throughout the experiment various setups were tested. Rings were also used in order to improve the focal distance and the consequently the resolution of the thermographic images captured.

In Figure 2 the temperature distribution over a cross section is presented. The temperature plots are derived from the data acquired from the cross sections shown in Figure 1 (c) and (d) for inductors *A* and *B* respectively. As we can see from the two plots, there is an obvious similarity between these two diagrams and the one from the integrated on silicon inductor [1]. As regards the thermal network formulation, the heat source is introduced as Joule losses on the turns of the inductor. The high thermal conductivity of copper leads to an almost uniform temperature distribution over each turn.

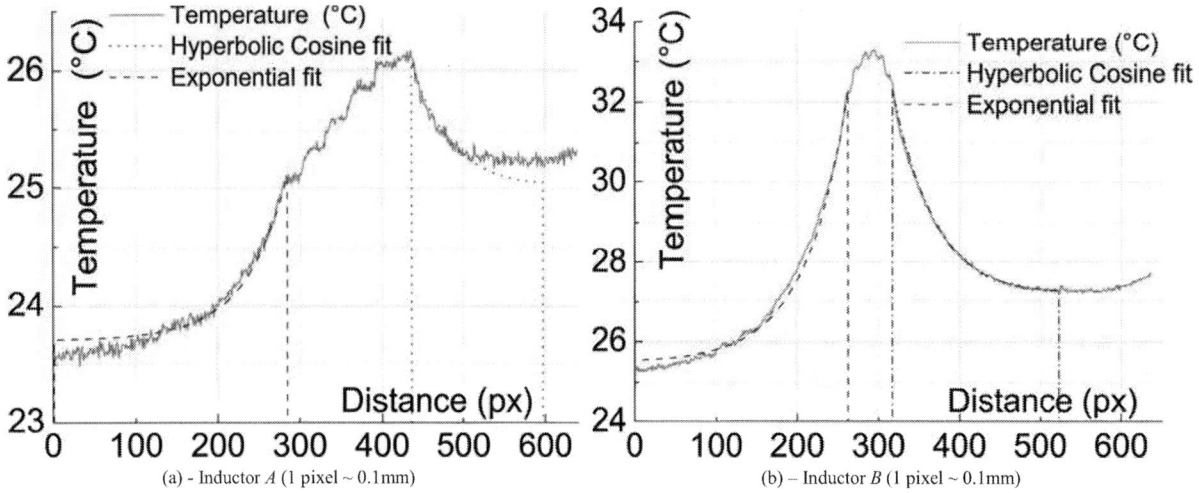

(a) - Inductor A (1 pixel ~ 0.1mm)   (b) – Inductor B (1 pixel ~ 0.1mm)

Figure 3. Temperature distribution fitting with exponential and over fitting

IV. RESULTS EVALUATION

A. Temperature gradient

The temperature distribution over the surface of the inductor can be approached by an exponential decay function with a certain characteristic length.

1) Characteristic Length

The characteristic length of the decay can be found using the following formula [7]:

$$L = \sqrt{\frac{kt_s}{2h}}, \qquad (1)$$

where k is the thermal conductivity of the FR4 board, t_s the thickness of the board, and h the heat transfer coefficient. If we assume that $h = 10$ W/(m²K), $k = 0.25$ W/(m.K), and $t_s = 1.6$mm then the characteristic length is 4.5mm. The characteristic length for the case of an FR4 board is around 5 mm, a value which agrees well with the decay we observe in figure 2.

2) Exponential fit

The temperature in the vicinity of the inductor turns is possible to be described by an exponential decay function:

$$T(x) = T_1 + \Delta T e^{-x/L}, \qquad (2)$$

where L is the characteristic length of the decay. T_1 is the temperature at infinite distance from the heat source and ΔT is the temperature difference between the heat source and the ambient (T_1).

In figure 3 we can see the fitting between the exponential function and the temperature measurements for both inductors tested.

3) Hyperbolic cosine fit

On the other hand, the temperature over the inner void of the inductor can be described by an hyperbolic cosine function:

$$T(x) = T_2 + A\cosh\left(\frac{x}{L}\right) = T_2 + A\frac{\left(e^{x/L} + e^{-x/L}\right)}{2}, \qquad (3)$$

where L denotes again the characteristic length of the heat dissipation, T_2 the temperature at the middle of the void surface and A is a fitting parameter defined by trial-error approach.

The fitting functions are the following:

$$T_A(x) = 0.1\cosh\left(\frac{x}{L}\right) + 24.95, \qquad (4)$$

and

$$T_B(x) = 0.2\cosh\left(\frac{x}{L}\right) + 27.1, \qquad (5)$$

for inductor A and B respectively.

In figure 3 the fitting function is plotted over the temperature distribution extracted from the thermographic measurements. For both inductors, it is obvious that it is possible to approximate the temperature distribution behavior using the exponential and the hyperbolic cosine fit.

TABLE II.   THERMAL RESISTANCES OF THE INDUCTORS

	R_{eff} [Ohm]	I [Amp]	ΔT_{mean} [K]	R_{TH} [K/W]
Inductor A	4.054	0.545	2.25	1.867
Inductor B	1.46	0.55	8.2	18.56

The minimum of this temperature distribution can be found at the middle of the inner void were the distance from the inner coil turn is maximized. Therefore, the distance x is zero at the same point.

B. Temperature Maxima

The temperature distribution over the turns of the inductor can be described as discrete steps over each turn. The temperature of on each turn surface can be considered homogenous as the thermal conductivity is comparatively high. The temperature differences between the different steps can be observed even if they are of the magnitude of 0.1 C°. In addition, it is possible to observe that the temperature maxima in all cases are above the turns of the inductor.

While examining inductor A, it is easily understood that the relaxation distance (characteristic length – L) of the heat conduction in the FR4 board is comparable to the dimensions of the inner void surface of the coil. Therefore, it is easy to explain the reason why the temperature difference to ambient over this void surface is higher in the case of inductor A than in the case of inductor B. As a direct consequence, the maximum temperature is observed over the inner turn in the case of inductor A and over the middle turn for inductor B.

C. Thermal Resistance

The thermal resistance of the DUT can be calculated by the following formula:

$$R_{th} = \frac{\Delta T}{P}, \qquad (6)$$

and can offer us information about the behavior of the thermal network after it reaches its steady state. In order to calculate the thermal resistance we need to evaluate the power dissipated in both inductors. Apparently, the power dissipated in these inductors is different.

Using the formula for effective resistance described in [8] :

$$R_{eff} = R_{DC}(1 + 0.1\left(\frac{\omega}{\omega_{crit}}\right)^2), \qquad (7)$$

we can approximate the electrical resistance of the inductors at the frequency of the excitation signal. R_{DC} denotes the DC electrical resistance of the inductor and ω_{crit} the critical frequency as described in [7]. Using this values and the formula $P = I^2 R_{eff}$ we can calculate the dissipated power on the inductor turns.

In Table 2, the results of these calculations are presented. We can observe that the thermal resistance of the two coils is different, because of the difference in the geometry of the heat sources.

V.   CONCLUSION

The use of infrared thermographic equipment can provide valuable information about the thermal network behavior of an inductor. It was possible to calculate the thermal resistance of the board, the characteristic thermal length of the printed circuit board. Furthermore, the accuracy of the temperature measurements follows the increase of the resolution of the thermographic equipment. Therefore, using temperature measurements one can pinpoint the maximum temperatures and its positions.

Measuring the temperature distribution of lower scale size equipment is nowadays possible. The study of the thermal behavior of a PCB inductor can give an insight to the thermal behavior of micro-scale inductors.

REFERENCES

[1] M. Kaluza, B. Wiecek, A. Hatzopoulos, V. Chatziathanasiou, I. Papagiannopoulos, G. De Mey, "Thermal Measurements of Silicon Integrated Spiral Inductors," PAK, vol. 54, no. 6, 2009.

[2] J-T. Kuo, K-Y. Su, T-Y. Liu, H-H. Chen, S-J. Chung, "Analytical Calculation for DC Inductances of Rectangular Spiral Inductors With finite Metal Thickness in the PEEC Formulation," IEEE Micr. and Comp. Lett., vol 16, no. 2, pp. 69-71, February 2006.

[3] S. Mohan, M. del Mar Hershenson, S. Boyd, T. Lee, "Simple Accurate Expression for Planar Spiral Inductances," IEEE Journ. of Solid-State Circ., vol. 34, no. 10, pp. 1419-1422, October 1999.

[4] G. Antonini, A. Orlandi, C. Paul, "Internal Impedance of Conductors of Rectangular Cross Section", IEEE Trans. on Microwave Theory and Techniques, vol. 47, no. 7, pp. 979-985, July 1999.

[5] C. Baojun, T. Zhen'an, Y. Tiejun, "Novel Closed-Form Resistance Formulae for Rectangular Interconnects," Chin. Inst. of Electr., Journ. of Semiconductors, vol. 32, no. 5, pp. 054008-1 – 6, May 2011.

[6] A. Barr, "Calculation Frequency-Dependent Impedance for Conductors of Rectangular Cross Section," AMP Jour. of Techn., vol. 1, pp. 91 – 100, 1991

[7] A. Bejan, D. Kraus, Heat Transfer Handbook, Hoboken, NJ: John Wiley & Sons, 2003, pp. 204.

[8] W. Kuhn, N. Ibrahim, "Analysis of Current Crowding Effects in Multiturn Spiral Inductors," IEEE Trans. Micr. Theor. And Tech., vol. 49, no. 1, pp. 31-38, January 2001.

332

Analysis and Modelling of ICs and Microsystems

MIXDES 2012, 19th International Conference *"Mixed Design of Integrated Circuits and Systems"*, May 24-26, 2012, Warsaw, Poland

A Specific Parameters Analysis of CMOS Hall Effect Sensors with Various Geometries

Maria-Alexandra Paun*, Jean-Michel Sallese, Maher Kayal

STI-IEL-Electronics Lab

Ecole Polytechnique Fédérale de Lausanne (EPFL)

CH-1015 Lausanne, Switzerland

* Corresponding author's e-mail: maria-alexandra.paun@epfl.ch

Abstract—In this paper, the performance of CMOS Hall Effect Sensors with four different geometries has been experimentally studied. Using a characteristic measurement system, the cells residual offset and its temperature behavior were determined. The offset, offset drift and sensitivity are quantities that were computed to determine the sensors performance. The temperature coefficient of specific parameters such as individual, residual offset and resistance has been also investigated. Therefore the optimum cell to fit the best in the performance specifications was identified. The variety of tested shapes ensures a good analysis on how the sensors performance changes with geometry.

Index Terms—Hall effect sensor; individual and residual offset drift; temperature coefficient

I. INTRODUCTION

One of the most widely used sensing technologies today consists of Hall effect sensors, based on magnetic phenomena. Many low-power applications like current sensing, position detection and contactless switching within automotive and industrial electronics use this kind of sensors [1, 2].

The sensitivity, offset and its temperature drift are important figures of merit in Hall sensors performance evaluation. There is a strong connection between the geometry and the Hall effect sensors performance as was studied by the authors in [3]. In order to predict and evaluate the sensors performance, three-dimensional physical simulations have been realized in order to facilitate the design process [4] and automated measurements systems have been developed [5].

Several of Hall effect sensors were simulated and evaluated for numerical offset, drift, Hall voltage and sensitivity using three-dimensional physical simulations in a recent paper by the authors [4].

The present paper analyzes the influence of the shape, dimensions, on the Hall effect sensors performances, including sensitivity, offset and drift for optimal design of Hall effect sensors. In this sense, different Hall effect sensors were integrated in a 0.35 μm CMOS technology. The diversity of

the shapes and dimensions analyzed allows us to have a wide range of Hall sensors amongst which to choose the one displaying the best performance for the project requirements. This is also a good opportunity to characterize the effect of temperature on individual, residual offset and resistance.

Section II presents the qualitative reasoning behind choosing the specific Hall effect sensors geometries and the design parameters for all proposed cells. The results corresponding to the residual offset and its behavior with the temperature are presented in Section III, with a comparative analysis on different Hall cell types.

The temperature coefficient of the individual offset, residual offset (using 2-phase and 4-phase current spinning technique) and resistance has been computed for several sensors. This section concludes which of the integrated shapes exhibited the best performance. The various shapes analyzed offer an overview on how the performance is related to the geometry.

II. METHODOLOGY

A. Hall cells integration

Different Hall effect sensors were integrated in a 0.35 μm CMOS technology. All these sensors are symmetric and orthogonal structures due to the fact that any geometrical mismatch could significantly increase the offset.

B1. The chosen structures

The classical Greek cross used for this type of sensors with progressive increase in cell dimensions (basic, L and XL cells) and borderless cell (with small contacts situated far away from the p-n junction) were integrated and analyzed.

B2. The qualitative reasoning behind the shapes choice

The Hall sensors are manufactured in a CMOS process and they basically have a p-substrate with a n-doped active region. On top they have a p+ diffusion layer to reduce the noise and prevent the current to flow under the contacts.

There is reasoning behind the choice of the nine different Hall sensors. The basic cell is taken as a reference, but it might suffer from a difference in the piezo-resistance due to the orientation of the axes. For the L and XL cells, the errors on the contour are less due to an averaging on a bigger size. The borderless shape might minimize the influence of any errors that might appear on the borders but the sensitivity is as well affected. The four tested Hall geometries (basic, L, XL and borderless cells) are presented in Fig. 1.

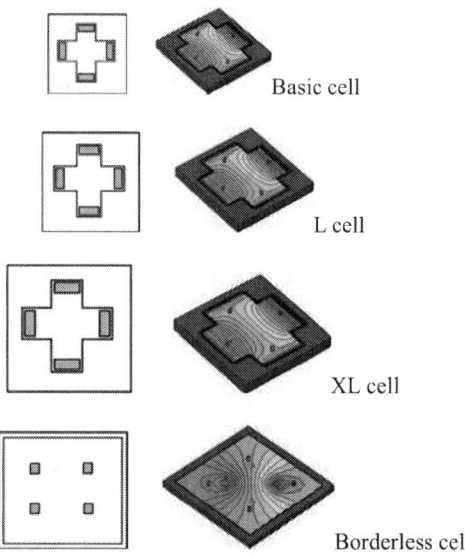

Figure 1.   The four analyzed Hall cells (2D and 3D simulated structures)

The design details are presented in Table I, together with measurements of the resistance, absolute sensitivity, offset drift (with four-phase current spinning technique). For each Hall cell structure, the geometrical correction factor G was computed according to the formula in [6].

TABLE I.   DESIGN PARAMETRS OF THE HALL EFFECT SENSORS

Hall Cell	Basic	L	XL	Borderless
R_0 (kΩ) @T=300 K, B=0 T	2.3	2.2	2.2	1.3
S_A (V/T) @ I_{bias}=1 mA	0.0807	0.0804	0.0806	0.0325
Offset drift (µT/°C)	0.409	0.264	0.039	0.526
L (µm)	21.6	32.4	43.2	50
W (µm)	11.8	17.8	22.6	50
L/W	1.83	1.82	1.91	1
s (µm)	11	16	20.7	2.3
G	0.913	0.912	0.924	0.76

L and W represent the cell length and width, respectively, of the active N-well region while s stands for contact length. The width of the contacts is in general imposed by the technology used in the Hall effect sensors fabrication process. The position of contacts with respect to borders is important in the offset analysis as contour errors might increase it.

Dimensions, via the geometrical correction factor as analyzed by authors in a recent paper [3], and distance

between the contacts and the active region borders are important in the evaluation of the cells offset and sensitivity.

Each structure is equipped with four contacts (denoted by a, b, c, d), among which two are for biasing the device and the other two opposite ones for measurement purposes, by collecting the voltage drop, as seen in the following figure.

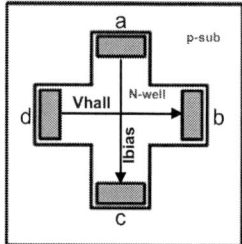

Phases	I_{bias}	V_{hall}
Phase 1	a to c	b to d
Phase 2	d to b	a to c
Phase 3	c to a	d to b
Phase 4	b to d	c to a

Figure 2.   Polarization of a Hall cell and the four phases of the current spinning technique

The sensitivity of Hall sensor is given by the ratio of the Hall voltage (V_{HALL}) to the magnetic field induction (B), as in the relation

$$S_A = \frac{V_{HALL}}{B} = \frac{G r_H}{n q t} I_{bias} \qquad (1)$$

where G is the geometrical correction factor, I_{bias} is the biasing current, r_H is the scattering factor of Silicon, usually 1.15, n is the carrier density and t is the thickness of the active region [6].

III. RESULTS AND DISCUSSION

All the sensors were designed for targeting specific objectives, namely the offset at T=300 K less than ±30 µT and the offset drift less than ±0.3 µT/°C. Our aim was to see which of the shapes fits the best within the specification interval and also has a behavior consistency. To this purpose, the measurements were performed on 11 samples. The offset drift value presented in Table I is therefore an average on 11 samples.

Previous tests were performed on all the Hall effect sensors, using an automated measurement system, presented in details by the authors in [5]. The advantage of the fully automatic system is the reliability, and the possibility to test all the cells, at the same time, under the same conditions. In this case the information on the residual offset was directly obtained.

In the present work, the structures were surrounded by the specific electronics and subsequently electrically tested but the phases were manually switched. In this way, we had information on each individual phase offset. 2-phase and 4-phase spinning current could be analyzed, but some accuracy can be lost. In this way, information on their circuit behavior was obtained. A current biasing (0 - 1 mA) was used for the Hall effect sensors polarization. Measurements in the absence and presence of the magnetic field were performed.

The offset measurements were performed in the absence of magnetic field while for Hall voltage and sensitivity estimations, the magnetic induction was B=0.5 T. The Hall cells offset was evaluated at room temperature and for certain temperatures in the interval -40° - 125°C. By accessing the latter information we extracted the offset drift, which is incorporated in Table I, after a 4-phase spinning of the sensors.

A. Residual offset temperature behaviour

The Hall voltage is affected by the offset by the relation

$$V_{out} = V_{HALL}(B) + V_{offset} \qquad (2)$$

Even though the shapes are symmetric we obtain a non-zero offset. In general, to reduce the Hall sensors offset, the current spinning technique is used, consisting in periodic commutation of current and voltage terminals [7].

The residual offset is an average of the four phases, as follows:

$$Offset_{residual\ (4\,phase)} = \frac{V_{P1} - V_{P2} + V_{P3} - V_{P4}}{4} \qquad (3)$$

The information on individual phase offset was obtained for each sensor on sample X12. Using relation (3) and dividing by the sensitivity, the magnetic field equivalent residual offset was obtained. The residual offset of the Hall sensors with the errors bars is presented for two currents, in Fig. 3 – Fig. 6. We can observe that the XL cell displayed the best behavior.

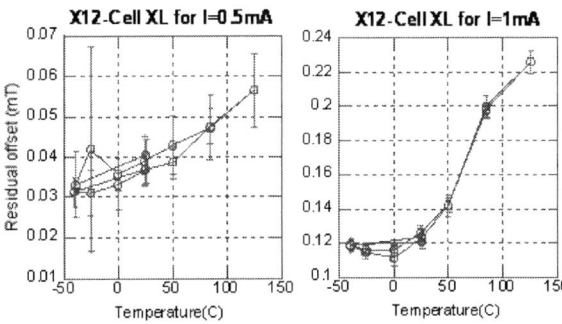

Figure 3.   Residual offset for XL cell

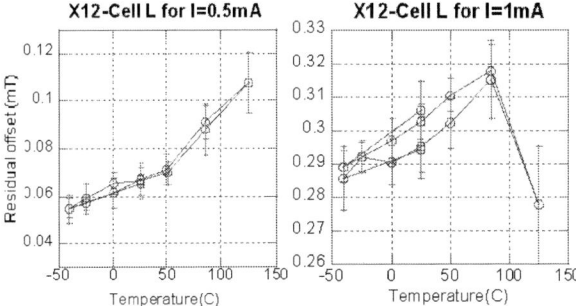

Figure 4.   Residual offset for L cell

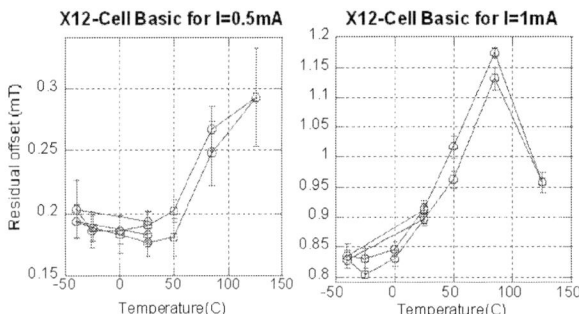

Figure 5.   Residual offset for basic cell

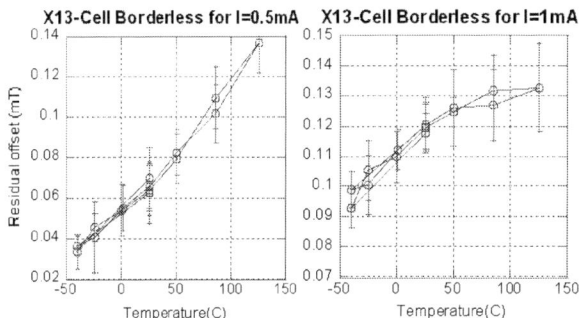

Figure 6.   Residual offset for borderless cell

We had several temperature cycles. We started from room temperature, decreased the temperature to 0° and then increased it to 125°, then decreased it again to -40° and then finally reverted to the room temperature. This is why for the same temperature point we have several measurements.

B. Temperature coefficient of specific quantities

We were interested to see if the initial offsets, residual offset (both 2 and 4-phase) and resistance have the same temperature drift coefficient, information useful for developing future temperature correction blocks.

From measurements, the coefficient of variation with temperature was computed for the above-mentioned parameters and the data for three cells and two biasing currents is summarized in the Tables II and III.

From the tables below, we can observe that for the resistance, the experimental results are in good agreement with the value for α provided by the technology used.

At low biasing current, the junction field effect is less, so we could expect a linear region in which the analyzed quantities to have a closer connection. At biasing current of 1.25 mA, indeed the residual offset for the 2 phases (for basic and XL cells) has a higher variation with temperature, like the other analyses for 2-phase vs. 4-phase current spinning technique have shown before.

TABLE II. Temperature coefficient for I_{BIAS} =1.25 mA

The temperature variation coefficient	$\alpha(T^{-1})$ X12-Basic R_0=2.37 kΩ	$\alpha(T^{-1})$ X12-XL R_0=2.34 kΩ	$\alpha(T^{-1})$ X12-Borderless R_0=1.36 kΩ
Resistance (kΩ)	0.004	0.004	0.0036
Residual offset (mV) 4 phases	0.002	0.001	0.090
Residual offset (mV) 2 phases	-0.011	0.02	0.016
Individual offset (mV) of phase P_1	-0.03	-0.03	0.016
Individual offset (mV) of phase P_2	-0.007	-0.01	0.013
Individual offset (mV) of phase P_3	-0.007	-0.01	0.0125
Individual offset (mV) of phase P_4	-0.027	-0.029	0.014

TABLE III. Temperature coefficient for I_{BIAS}=0.25 mA

The temperature variation coefficient	$\alpha(T^{-1})$ X12-Basic R_0=2.37 kΩ	$\alpha(T^{-1})$ X12-Basic R_0=2.34 kΩ	$\alpha(T^{-1})$ X12-Borderless R_0=1.36 kΩ
Resistance (kΩ)	0.004	0.004	0.004
Residual offset (mV) 4 phases	0.025	0.006	0.031
Residual offset (mV) 2 phases	0.011	0.011	-0.028
Individual offset (mV) of phase P_1	-0.012	-0.0009	0.031
Individual offset (mV) of phase P_2	-0.008	0.002	0.025
Individual offset (mV) of phase P_3	-0.008	0.002	0.013
Individual offset (mV) of phase P_4	-0.0123	0.001	0.015

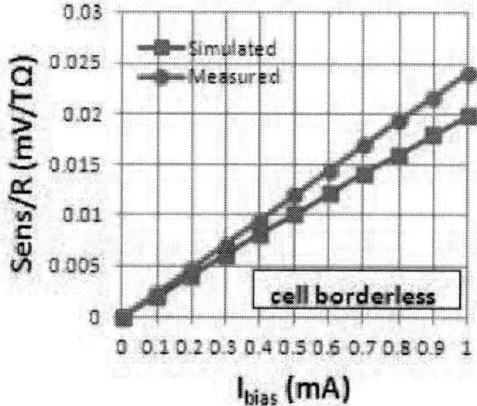

Figure 7. Ratio of absolute sensitivity to resistance (simulated and measurement results) vs. biasing current, for L cell

Figure 8. Ratio of absolute sensitivity to resistance (simulated and measurement results) vs. biasing current, for borderless cell

Among the four integrated and subsequently tested cells, XL cell displayed the minimum offset at room temperature, the lowest residual offset drift and the best sensitivity. So we can observe that this particular cell is the optimum one amongst the analyzed Hall effect sensors.

C. 3D physical simulations of the sensors

The TCAD Synopsys 3D physical simulator [8], based on numerical solutions to the carrier transport in semiconductors [9-11], proved to be a reliable tool to predict the Hall effect sensors performance, by investigating, amongst other parameters, their sensitivity.

In Figs. 7 and 8, the ratio of absolute sensitivity to resistance (simulated values versus measurement results) plotted against biasing current, for two different Hall cells, is presented. A good coherence between measurements and simulation results has been obtained, as you can see in the following graphs, for two of the analyzed sensors.

IV. Conclusion

Different Hall effect sensors were integrated in a CMOS technology and their performance evaluated. Four geometries (basic, L, XL and borderless cells) have been chosen and tested. The offset, its drift and the sensitivity are quantities that were measured to determine the sensors performance.

The results corresponding to the residual offset and its behavior with the temperature were presented, with a comparative analysis on different Hall cell types. The temperature coefficient of the individual offset, residual offset (using 2-phase and 4-phase current spinning technique) and resistance has been computed for several sensors. This information might be useful in designing future temperature correction blocks.

It was shown that there is coherence between the measurement and the simulation results. We finally concluded which of the integrated shapes exhibited the best performance.

The various shapes analyzed offer an overview on how the performance is related to the geometry.

As future work, we aim at investigating the residual offset prediction and spinning current technique efficiency by using the 3D physical simulator.

ACKNOWLEDGMENT

This work has been supported by Swiss Innovation Promotion Agency CTI (Project 9591.1) and the company LEM SA – Geneva, Switzerland.

REFERENCES

[1] E. Ramsden, "Hall-Effect Sensors – Theory and Applications", (2nd Edition), Elsevier, 2006

[2] S.-Y. Kim, C. Choi, K. Lee and W. Lee, "An Improved Rotor Position Estimation With Vector-Tracking Observer in PMSM Drives With Low-Resolution Hall-Effect Sensors", Industrial Electronics, IEEE Transactions on, Vol. 58 , Iss. 9 , 2011 , p. 4078 - 4086

[3] M.A. Paun, J.M. Sallese, and M. Kayal, "Geometry influence on Hall effect devices performance", U.P.B. Sci. Bull., Series A, Vol. 72, Iss. 4, 2010, p. 257-271

[4] M.A. Paun, J.M. Sallese, and M. Kayal, "Hall effect sensors performance investigation using three-dimensional simulations", Mixed Design of Integrated Circuits and Systems (MIXDES), 2011 Proceedings of the 18th International Conference , 2011 , p. 450 – 455

[5] M.A. Paun, J.M. Sallese, and M. Kayal, "Geometrical parameters influence on the Hall effect sensors offset and drift", Ph.D. Research in Microelectronics and Electronics (PRIME), 2011 7th Conference on 2011, p. 145 - 148

[6] R. S. Popovic, Hall Effect Devices, Second Edition, Institute of Physics Publising, 2004

[7] Y.Hu and W.-R.Yang, "CMOS hall sensor using dynamic quadrature offset cancellation", Solid-State and Integrated Circuit Technology, 2006, ICSICT '06, 8th International Conference on, 2006, p. 284 - 286

[8] Synopsys TCAD tools: http://www.synopsys.com/Tools/TCAD

[9] S. M. Sze, K. K. Ng, Physics of semiconductor devices, Third Edition, John Wiley and Sons, 2007

[10] I. S. Selberherr, "Analysis and Simulation of Semiconductor Device", Vienna, Austria, Springer-Verlag, 1984

[11] W. Allegretto, A. Nathan, and H. Baltes, "Numerical Analysis of Magnetic-Field-Sensitive Bipolar Devices", IEEE Transactions On Computer-Aided Design, Vol. 10, No. 4, 1991

MIXDES 2012, 19th International Conference *"Mixed Design of Integrated Circuits and Systems"*, May 24-26, 2012, Warsaw, Poland

Combined Hardware and Software Tracing of Real and Virtual Embedded System Parts

Christian Koehler[*], Albrecht Mayer[†]
and Maximilian Wurm[*]
[*]University of Siegen
Hölderlinstraße 3
57076 Siegen, Germany
[†]Infineon Technologies AG
Am Campeon 1-12
85579 Neubiberg, Germany

Abstract—Advanced debug support and trace capabilities are getting ever more important for overcoming the challenges of developing complex real-time embedded systems containing complex Systems-on-Chip (SoCs). Gathering and handling of very large software and system traces becomes a major challenge. We present an approach of recording and fusing software and system traces in combined simulated and real system development environment. This provides a complete, time aligned view of the whole system behavior. The development environment is based on the Chip-Hardware-In-The-Loop Simulation (CHILS) approach [1], which embeds a real microcontroller (MC) into different simulation environments.

Index Terms—Simulation, Modelling, Hardware Tracing, Software Tracing.

I. INTRODUCTION

System modeling of embedded systems becomes essential due to their rising complexity. Model-based development has become popular for that. It offers possibilities like early exploration of system design, rapid prototyping, verification, optimization and test. It allows the engineers to develop software and hardware in parallel to reduce the cycle times. Furthermore, these systems are often a recombination or extension of existing parts to reduce the development effort. The CHILS approach offers the possibility to embed a real MC into a system simulation of a larger technical context. This is an advantage especially for early software development, e.g. the MC software can be developed and tested within the system context before hardware prototypes are available.

Besides classical debugging, tracing becomes even more important for analyzing the systems to be developed. Software execution traces are useful instruments for debugging and optimization purposes. They are used for analysis and validation of real-time conditions of embedded systems [2], design coverage issues [3], software optimization, or for understanding the behavior of complex software systems [4]. One challenge is to not only trace single parts of a system, like the software executed on the MC, but to trace the entire system. The other is to get continuous fine grained traces without distorting the systems behavior. This is needed to understand the complete systems behavior and to analyze misbehavior which is caused by the combination of system parts.

The presented *Trace Fusion* approach allows recording and analyzing a system trace from a very heterogeneous system development environment, which consists of real and simulated system parts. The *Trace Fusion* approach allows continuous and fine-grain tracing from real and simulated trace sources. Its additional value in the design of embedded systems is underlined by a design example of an electric-motor-control application. Especially the visual exploration of the fused system traces adds new possibilities for debugging and understanding the system in development.

The paper is organized as follows: The next section presents related works on tracing solutions, especially on the problem of trace bandwidths. Section III introduces the hardware extended embedded systems simulation, which is realized by the CHILS approach. In addition, an overview of the new approach is presented. Sections IV to VI are dealing with the three trace sources, the real MC hardware, the interface between MC hardware and simulation, and the simulation itself. Section VII presents the trace fusion and analysis concept, while section VIII demonstrates the value of the approach by an application example.

II. RELATED WORKS

Tracing is an automated way of monitoring the system behavior by recording information about the current system state at regular intervals. Trace solutions for embedded system can be divided into two categories, software-based tracing and hardware-supported tracing.

A. Software-based Tracing and Hardware-supported Tracing

The classical approach to monitor the execution of software is to add printout functions to the source code. In purely software-based trace approaches the source code is extended by additional calls to some kind of trace libraries. If an operation-system-like execution base is used, a tracing approach could be implemented by an instrumented kernel as it is done for RTOS from QNX [5]. It is obvious that every extension of original source code by trace commands will change the execution of that code on the hardware. In real-time embedded systems this can be critical, since the timing behavior is changed. Nevertheless, software-based tracing solutions are very useful as long as the user is aware of the influence that the tracing extensions have.

The main advantage of hardware-supported tracing approaches is that the tracing system is mostly designed to not influence the system behavior, except the power consumption.

Advanced on-chip tracing support becomes even more important to overcome the challenges of developing real-time embedded systems driven by complex SoCs. The NEXUS 5001 standard [6] includes specifications for run-time control (debugging), code execution trace capture and data access trace capture, and calibration [7]. The standard defines special trace ports with a high bandwidth to an external storage unit.

The limitation of on-chip trace capabilities is the limited bandwidth of the interface to the tool environment. MC processor cores and bus systems can generate tens of Gbits of on-chip data, which is even higher than every external interface. Current on-chip trace mechanisms overcome this by adding additional on-chip trace memory, especially for buffering short bursts of trace data. Furthermore powerful on-chip filter capabilities, trigger logic and trace data compression mechanisms are implemented.

The Infineon Technologies AG (www.infineon.com) developed the Multi Core Debug Solution (MCDS) which is used in so called Emulation (ED) Devices. ED devices did not contain special high bandwidth trace ports, but low bandwidth standard debugging ports like JTAG or DAP. The MCDS approach scales with higher integration density and rising clock frequencies [8] unlike solutions with off-chip memory.

B. Simulation Tracing

In contrast to software-based tracing techniques on real hardware, simulation tracing will not change the system behavior. Trace hooks can be added as passive elements to the simulation system, without influencing the simulation results. The popular C-library SystemC [9] provides an easy to handle trace extension for tracing signals and variables of the simulation. Trace hooks can be added to every point of the simulation. Simulation environments like Matlab/Simulink can be easily extended by costumer build blocks, or build-in printout capabilities can be used to monitor the simulation results.

C. Trace Formats

Each kind of trace source provides its own trace format. Hardware-supported tracing often use proprietary high compressed binary trace formats to store the trace data. Other trace formats are ASCII-based files, like the SystemC trace output.

The amount of different solutions makes it difficult to combine traces from different sources to complete system traces. Standardized formats are needed to do so. One option is the Open Trace Format (OTF) ([10],[11]), which is a highly scalable trace format for massive parallel system. OTF is used as the common trace format in the HPC field. The presented approach uses OTF as exchange format to combine the traces from different trace sources.

III. HARDWARE EXTENDED EMBEDDED SYSTEM SIMULATION

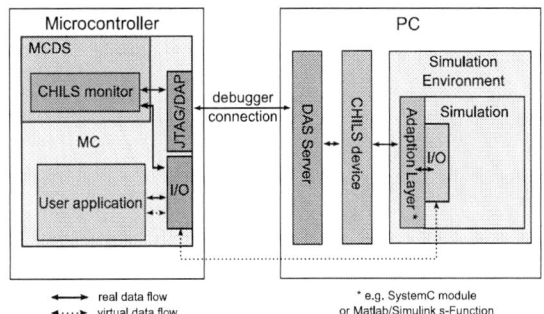

real data flow
virtual data flow

* e.g. SystemC module
or Matlab/Simulink s-Function

Fig. 1.   CHILS Coupling Setup

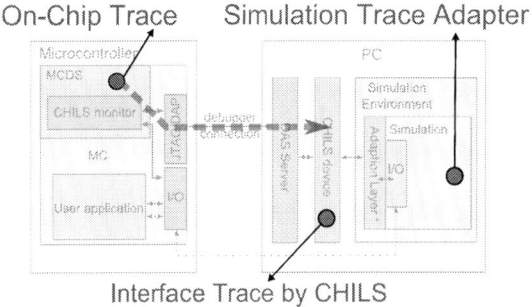

Fig. 2.   CHILS Setup with Trace Extensions

The CHILS approach embeds the real MC hardware into a system simulation to drop one complex part from the system model. The model will be smaller and faster, or a higher accuracy can be achieved. CHILS focuses on complex hardware/software system designed with existing MCs [1]. The system software can be developed from the beginning of the design phase of the entire system. Usually simulation models on a high level of abstraction, like Matlab/Simulink models, will be used for this purpose.

Figure 1 presents the general CHILS setup. The MC is connected to the PC via the standard debug interface. CHILS enables the data exchange between the MC user application and the simulation. The trace extension covers three different trace sources, the MC, the interface between MC and simulation, and the simulation (figure 2). Each trace source will provide OTF trace data streams which finally can be merged (see VII). The different sources require very specific trace implementations. The TC1767ED MC has on-chip fine grain trace functionalities, but the bandwidth to the simulation computer is very limited and the MC trace needs to be reconstructed from a highly compressed data format (see next section). Matlab/Simulink, as a popular simulation system representative, lacks the support for tracing. Especially constructed trace hooks are needed to derive traces from the simulation. The extension of the CHILS interface has to deal with the different MC interfaces and data types. It is not possible to determine the used data type without additional system knowledge, since each MC interface can transport different data types.

IV. TRACING THE REAL HARDWARE

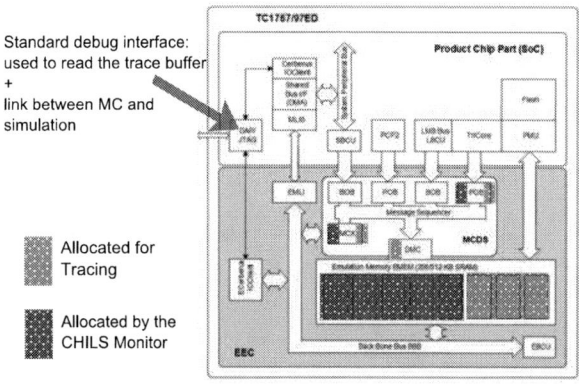

Fig. 3. TC1767ED Partitioning between CHILS and Tracing

The software execution trace of user applications is realized by the on-chip tracing capabilities of the Infineon TC1767ED device with MCDS. The MCDS provides diverse functionalities for instruction pointer (IP) traces and data traces of the on-chip units [8].

CHILS itself is using the MCDS resources, especially IP range triggers and the additional on-chip trace memory, in order to enable the nearly transparent data exchange between MC hardware and simulation [1]. The trace implementation shares the MCDS resources with CHILS (see figure 3. The on-chip trace memory is separated into the CHILS section, the so called CHILS monitor is located there, and the trace section. The trace section is divided into different tiles which can be configured separately for an external access (e.g. a PC) or for internal tracing. A continuous trace can be achieved by configuring the tiles as circular buffer[1].

CHILS uses a conservative synchronization principle for the connection where both sides, simulation and MC, are forced to execute identical time steps. During the data exchange, the user application on the MC as well as the simulation are suspended. So the time that is needed to synchronize the simulation and the MC user application only influences the "'simulation'" performance but not the accuracy of the system. The step size is chosen depending on the rate and the type of events that have to be exchanged between simulation and MC, and the needed simulation accuracy and performance. Larger step size will raise the overall performance, but they can reduce the accuracy [12]. An additional aspect occurs for the trace setup. Since the on-chip trace buffer is limited and the goal is to capture a continuous trace, the rate of reading the buffer from the PC has to be equal or higher than the production of on-chip trace data.

[1]The setup uses three $32Kbyte$ tiles as trace buffer.

The MC-trace needs to be reconstructed from a highly compressed and device specific data format. A nearly cycle accurate instruction pointer trace of all jump/call instructions can be derived by combining the trace data and the disassembled binary program code (see trace snippet below).

```
0:j d4004514 <_start+0x14>
0:ld.w  d1,f00005f0 <$wdtcon0>
7:jl d4004688 <asm_clear_endinit>
7:ld.w  d0,f00005f0 <$wdtcon0>
32:ji  all
33:movh.a  sp,53248
...
```

Afterwards the reconstructed program flow is converted into an OTF trace.

V. SIMULATION TO REAL HARDWARE INTERFACE TRACING

The CHILS approach defines four levels of interface abstraction for the MC interfaces: Message-Level, Byte-Level, Digital-Level and Analog Level [12]. For reducing the amount of exchanged data between MC hardware and simulation, the highest suitable level of abstraction for an interface is chosen. In addition, only the data from interfaces which are used by the simulation is transmitted. The trace configuration is automatically derived from the interface configuration, so only data from configured interfaces is traced. The data type of the traced interface values is has to be configured as well for each interface type[2]. For example, an 8 bit value could be interpreted as signed or unsigned value by the trace analyze environment. The interface trace generation is directly done by the so called CHILS device (see figure 2). The CHILS device is the synchronization component of CHILS on the PC. It uses the OTF library to generate the OTF traces without conversion from an other format.

VI. SIMULATION TRACING

As an example for simulation tracing, the Matlab/Simulink environment is chosen. The Simulink model is extended by special trace adapters which are acting as a kind of trace probe. Each adapter is implemented as a Simulink s-Function. The trace adapters are streaming out the traced values to a trace server, which is running on the same PC as the simulation. The trace adapters will be configured automatically depending on the output data type of the simulation part. In addition each adapter can be labeled and groups of adapters can be formed. The trace server collects the streamed trace data and generates OTF traces.

VII. TRACE FUSION AND ANALYSIS

Trace fusion enables the engineer to analyze traces from different systems, independent of whether they are real or virtual. We propose the *Trace Adapter* approach to realize a view on the complete system. Figure 5 presents the main

[2]The current implementation derives this configuration from an external configuration file.

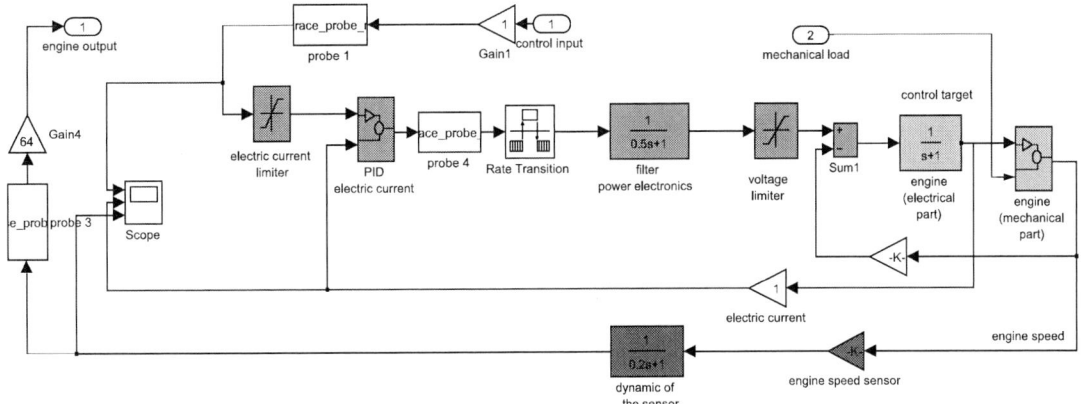

Fig. 4. Electric Motor Model in Simulink

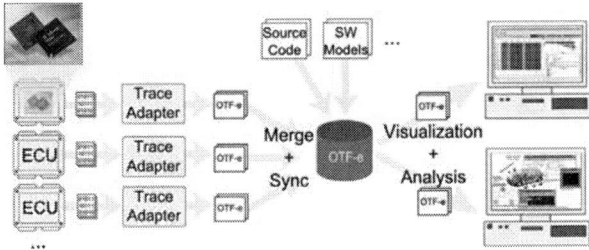

Fig. 5. Trace Fusion Concept

idea of the approach. Each kind of trace source has a specific trace adapter which generates OTF traces. This can be done by converting specific trace formats, as it is implemented in section IV, by the direct extension of the system with the OTF library (see section V), or by external trace generators, like in section VI. The trace fusion is realized by merging the different OTF files in combination with adapting the annotations of time. The OTF merging can be scripted easily. The first step is to synchronize the different trace files by defining a specific synchronization offset for each trace source. Furthermore the least common multiple (LCM) of the different trace source frequencies have to be found. Afterwards the timestamps of each trace file can be adapted using the LCM and the offset.

Since the OTF file format is used for tracing, the available tools from the HPC area can be used. Figure 6 shows a screen shot of the Vampir [13] environment. Vampir is a tool for visualization and analysis of parallel applications. It provides diverse timeline views and summary views like runtime statistics. The major drawback in using Vampir for the analysis is its focus on analysis of parallel processes in HPC environments. However, most of the visualization concepts can be reused easily for embedded systems, e.g. by adapting the naming of the views etc.

In contrast to other setups, the CHILS based approach allows a continuous and fine-grained trace of software execution

and hardware simulation. As mentioned before, the limit of on-chip tracing solutions is the bandwidth to the external trace storage system, whether it is a specialized hardware or a normal PC. Since CHILS is controlling the data exchange itself and the synchronization step size, it is possible to take the time for reading all produced trace data which is needed for a complete and continuous trace.

VIII. APPLICATION EXAMPLES

We evaluated the following three application examples.

The *PID-Model* is an electric motor control application. The system was initially modeled as a Matlab/Simulink model. In the next step, the proportional–integral–derivative (PID) controller block for engine speed control, was replaced by the TC1767 MC Simulink block. The Simulink electric motor model[3] provides the control input signal to the MC. The real MC processes the signal and generates the control output for the motor model. The motor speed is measured by a simulated speed sensor which provides the signal for the feedback loop to the MC. This is a typical in-the-loop setup which can be used to develop and to test the control software for the final system. Everything but the MC is simulated.

The *Dual-PID-Model* extends the *PID-Model* and represents a reduced simulation of an electrical vehicle. It includes two electric motor models (figure 4), one electric motor for the left front wheel drive, the other one for the right front wheel drive. In addition a simplified environmental model exits which generates the simulated mechanical load. The TC1767 MC Simulink block controls both electric motors depending on the control input. The simulation loop is completed by a control block which provides the stimulation for the simulated vehicle (figure 7).

The *PWM-Model* is especially designed to test the PWM (pulse-width modulation) signal generation. The PWM signal will be the control input for an electric motor in a final system. The direct usage of the PWM signal as part of the other scenarios would decrease the simulation performance since

---

[3]The original motor model is taken from [14].

343

TABLE I
SYSTEM PERFORMANCE

Scenario	Setup	Simulated Time (sec)	Simulation Time (sec)	Simulation Step Size (msec)	System Frequency (Mhz)	Performance (Mhz)	Multiple of Real Time
Dual-PID-Model	Tracing + decoding	50	20348	1.00	66.67	0.16	406.96
Dual-PID-Model	Tracing	50	17700	1.00	66.67	0.19	354
Dual-PID-Model	no tracing	50	233	1.00	66.67	14.31	4.66
PWM-Model	Tracing + decoding	4.4	13440	0.01	66.67	0.02	3054.55
PWM-Model	Tracing	4.4	8198	0.0072	66.67	0.04	1863.18
PWM-Model	no tracing	4.4	2904	0.0072	66.67	0.1	660
PID-Model	Tracing + decoding	20	7958	1.00	66.67	0.17	397.9
PID-Model	Tracing	20	6561	1.00	66.67	0.2	328.05
PID-Model	no tracing	20	92	1.00	66.67	14.49	4.6

TABLE II
TRACE PERFORMANCE

Scenario	Setup	Simulated Time (sec)	Raw MC Trace Size (MB)	Raw MC Trace Generation (MB/s)	Av. 32KB Block Transfer (sec)	Av. 32KB Block Decode (sec)	Av. No. Of 32KB Blocks transfered per Exchange
Dual-PID-Model	Tracing + decoding	50	1180.06	23.6	0.42	0.16	0.7552
Dual-PID-Model	Tracing	50	1180.06	23.6	0.42	-	0.7552
PWM-Model	Tracing + decoding	4.4	65.16	14.81	0.42	2.3	0.0034
PWM-Model	Tracing	4.4	65.16	14.81	0.42	-	0.0034
PID-Model	Tracing + decoding	20	483	24.15	0.42	0.11	0.7728
PID-Model	Tracing	20	483	24.15	0.42	-	0.7728

a relatively high exchange frequency is needed to capture the signal. The PWM base frequency is $666.67Hz$ while the pulse width is modulated in up to 100 steps. So to capture the signal the scan frequency has to be $133333.33Hz$ our higher. The *PWM-Model* uses a step size of $0.0072ms$ while a step size of $1ms$ is still suitable for the simulation of vehicles [15].

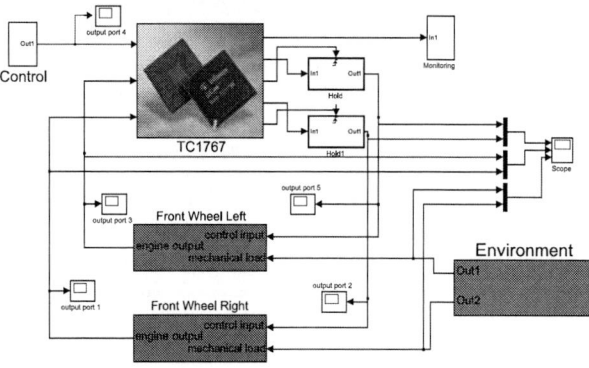

Fig. 7. Simulink Electrical Vehicle - Dual-PID-Model

A. Performance

Table I presents the performance of the combined trace approach. Each scenario is measured with three different setups: *tracing + decoding* (the MC trace is instantly decoded and converted into an OTF trace), *tracing* (the MC trace is just recorded) and *no tracing*. The simulation runtime of the *PID-Model* and the *Dual-PID-Model* is 71 to 87 times higher than the CHILS simulation without tracing. The performance difference is much smaller for the *PWM-Model* (factor 2.8 to 4.6). These findings can be explained by consulting table II. Primarily the reduced performance is caused by the time which is necessary to transfer raw trace from the MC trace buffer to the PC. At nearly each exchange between simulation and PC, a $32Kbyte$ block needs to be transferred in case of the *PID-Model* and the *Dual-PID-Model* scenario. A $32Kbyte$ block transfer takes approx. $0.42s$. The *PWM-Model* has a much smaller simulation step size ($0.0072ms$ vs. $1ms$) so less $32Kbyte$ blocks have to be transferred. As a result the performance difference to the *no tracing* setup is smaller. The limit is the relatively low debug-interface-bandwidth of approx. $76Kbyte/s$ of the low cost evaluation board. High end debug solution for Infineon TriCore MCs can reach up to $3.5MBytes/s$[4]. In that way, the approach can be speed up by a magnitude.

The results are acceptable comparing the demonstrated performance with complex simulations of MCs. The typical simulation performance of a cycle accurate model of a complex 32Bit MC is approx. $100KHz$, while an even more accurate register transfer level (RTL) model runs at a speed of less than $1KHz$ ([12] p.152). Since CHILS uses the real MC as replacement of a model, the accuracy is easy comparable

[4]E.g. the UAD2+ communication hardware from PLS Programmierbare Logik & Systeme GmbH.

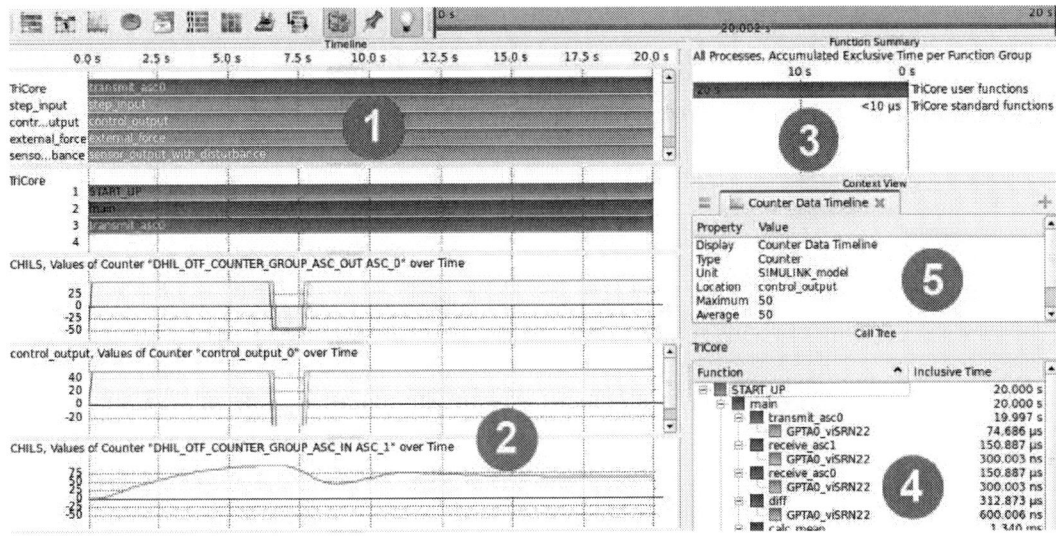

Fig. 6.   Vampir Screenshot of the Combined Trace

with highly accurate MC models.

The measurements underline that a continuous real-time IP-trace is not possible. The measured average MC IP-trace-bandwidth is between $15 MBytes/s$ and $24 MBytes/s$, while the measured debug-interface-bandwidth [5] is just $76 Kbyte/s$. Even a high end debug solution would not be sufficient.

B. Evaluating the Combined Trace

A combined $20s$ trace from the simulation, the Infineon MC and the hardware-to-simulation interface is shown in figure 6 In part 2 of figure 6 the traced signals can be seen. Part 4 of figure 6 shows the call tree of the MC application, while part 3 prints the accumulated run-time of the functions. More detailed views can be easily derived.

IX. CONCLUSION

The combined hardware and software tracing of real and virtual embedded system parts opens new possibilities for analysis complex embedded system to be developed. In combination with the CHILS approach, a continuous trace of the whole system is possible. The system behavior, especially the MC software, can be examined in detail by using these fine-granular combined traces. The open source format OTF allows flexible tracing and especially trace fusion and trace exchange between different systems. Available mighty tools from the HPC area can be used in order to visualize and analyse these traces. Nevertheless, adaptations of these tools are needed to provide full support for embedded systems analysis.

---

[5]System setup: Windows XP in an Oracle VirtualBox based virtual machine with 2GB RAM, the host machine is an Intel Core i7 920 with Windows 7 X64.

REFERENCES

[1] C. Koehler, A. Herkersdorf, and A. Mayer, "Chip hardware-in-the-loop simulation (chils) - embedding microcontroller hardware in simulation," in *Proceedings of the 19th IASTED International Conference on Modeling and Simulation*, R. Wamkeue. Ed.   IASTED, May 2008, pp. 297–302.

[2] J. Hill. "Context-based analysis of system execution traces for validating distributed real-time and embedded system quality-of-service properties," in *IEEE 16th International Conference on Embedded and Real-Time Computing Systems and Applications (RTCSA)*, 2010, pp. 92 –101.

[3] R. Lencevicius, E. Metz, and A. Ran, "Tracing execution of software for design coverage," in *Automated Software Engineering, 2001. (ASE 2001). Proceedings. 16th Annual International Conference on*, nov. 2001, pp. 328 – 332.

[4] L. Silva, K. Paixaando, S. de Amo, and M. de Almeida Maia, "Software evolution aided by execution trace alignment," in *Software Engineering (SBES), 2010 Brazilian Symposium on*, 27 2010-oct. 1 2010, pp. 158 –167.

[5] T. Fletcher, "Using system tracing tools to optimize software quality and behavior," QNX Software Systems, Tech. Rep., 2005. [Online]. Available: http://www.qnx.com/download/feature.html?programid=8092

[6] "Nexus5001," 1999. [Online]. Available: http://www.nexus5001.org

[7] H. O'Keeffe, "Ieee-isto 5001™-1999, the nexus 5001 forum™ standard providing the gateway to the embedded systems of the future." in *Proceedings of the Embedded Intelligence 2000 Conference*, 2000.

[8] A. Mayer, H. Siebert, and K. McDonald-Maier. "Boosting debugging support for complex systems on chip," *Computer*, vol. 40, no. 4, pp. 76 –81, april 2007.

[9] "Systemc." [Online]. Available: http://www.systemc.org

[10] "Open trace format (otf)." [Online]. Available: www.tu-dresden.de/zih/otf/

[11] A. Knüpfer, "Advanced memory data structures for scalable event trace analysis," Ph.D. dissertation, Technical University of Dresden, 2008.

[12] C. Koehler, *Enhancing Embedded Systems Simulation - A Chip-Hardware-in-the-Loop Simulation Framework*.   Vieweg+Teubner Verlag, 2011. [Online]. Available: http://www.viewegteubner.de/Buch/978-3-8348-1475-3/Enhancing-Embedded-Systems-Simulation.html

[13] "Vampir." [Online]. Available: http://www.vampir.eu/

[14] J. Hoffmann, *MATLAB und SIMULINK*.   Addison Wesley Longman Verlag GmbH, 1998.

[15] P. Waeltermann, T. Michalsky, and J. Held, "Hardware-in-the-loop testing in racing applications," *SAE Motor Sports Engineering Conference & Exhibition*, January 2004.

MIXDES 2012, 19th International Conference "*Mixed Design of Integrated Circuits and Systems*", May 24-26, 2012, Warsaw, Poland

Conducted Emissions Susceptibility Study for an High Precision LDO

Andreea Creosteanu
Technical Military Academy
Bucharest, Romania
andreea.caian@gmail.com

Laurentiu Creosteanu
Politehnica University
Bucharest, Romania

Abstract—**This paper presents a study on the susceptibility of an LDO towards wire conducted emissions on the battery line. Simulations have been performed to emulate the behavior of the LDO at different parasitical frequencies, at different loads on output and also with temperature.**

Index Terms— **Susceptibility; EMC; LDO; simulation.**

I. INTRODUCTION

The low drop out regulator is playing a very important role in nowadays circuitry. LDOs are used to maintain a certain supply voltage for different circuits in order to function properly.

Electromagnetic compatibility (EMC) has become very important in the last years due to the more and more electronics that can be inserted in a certain area [1-3]. In order to reduce the cost (by minimizing the number of redesigns due to EMC) we need to have a first time right design. To do this, we need to know from the design phase what will be the behavior of the chip at certain levels of emissions on the supply line.

This paper is focused on studying by simulation the behavior of an LDO when different emissions are present on the supply line.

The paper consists of six chapters. After a short introduction regarding the organization of the paper, chapter II presents a short description of the Electromagnetic compatibility and its influence on today's integrated circuits. In the next chapter, chapter III, we will focus on the integrated circuit which will be studied in the paper –an LDO, also making a brief description of its functional scheme, for better understanding the influence of the wire perturbations on its functioning. The fourth chapter discusses the simulation setups used for obtaining the results. In chapter V we will present the simulation results obtained. At the end of the paper we will make some conclusions of the study performed, in the sixth chapter of conclusions.

II. ELECTROMAGNETIC SUSCEPTIBILITY OF INTEGRATED CIRCUITS

Electromagnetic compatibility (EMC) can be defined as "the ability of an electronic system to function properly in its intended electromagnetic environment, and to not contribute with interference to other systems in the environment."[1-2].

Electromagnetic compatibility is concerned with the generation, transmission, and reception of electromagnetic energy.

The source (emitter) produces the emission, which is transferred through a coupling path to a receptor, where the energy from this emission can change the behavior of the receptor (named victim in this case) in a desired or undesired way, but most importantly, in an unpredictable way.

The interfering signal propagates mainly by two different ways: radiation and conduction (please see Figure 1 bellow).

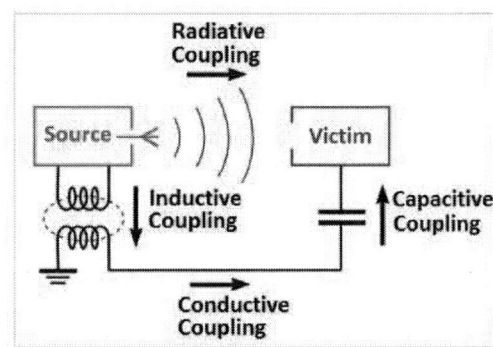

Figure 1. Coupling paths between the source and the victim

Nowadays, due to small chip dimensions and relatively low frequencies at which most circuits are operating, conduction seems to be the most relevant way of propagation.

Electromagnetic interference (EMI) has become a major problem for circuit designers, and it is likely to become even more severe in the future, if we take into account the following matters.

Diverse circuits operate in close proximity to each other, and can affect each other adversely.

Circuitry has become smaller and more sophisticated, so more circuits are being crowded into less space, increasing the probability of interference.

Clock frequencies have increased dramatically over the years.

Electronic circuits have more and more applications, from communication and computation to automation.

The continuous and constant increase of the number of electronic devices on vehicles has brought automotive products to fulfill very harsh EMC requirements, both for electromagnetic immunity and electromagnetic emission. This is even more severe on power train and safety products.

III. VOLTAGE REGULATORS AND APPLICATIONS

There are two types of linear regulators: standard linear regulators and low dropout linear regulators (LDOs). The difference between the two categories stands in the pass element and the amount of headroom (or dropout voltage) required to maintain a regulated output voltage. The dropout voltage is the minimum voltage required across the regulator to maintain regulation.

Figure 2 presents the basic form of an LDO block diagram. The input voltage is applied to a pass element, which is typically an N-channel or P-channel FET, but can also be an NPN or PNP transistor. The pass element operates in the linear region to drop the input voltage down to the desired output voltage. The resulting output voltage is sensed by the error amplifier and compared to a reference voltage. The error amplifier drives the pass element's gate to the appropriate operating point to ensure that the output is at the correct voltage. As the operating current or input voltage changes, the error amplifier modulates the pass element to maintain a constant output voltage. Under steady state operating conditions, an LDO behaves like a simple resistor. In a practical application, however, operating conditions are never static; therefore, feedback is necessary to change the LDO's effective resistance to maintain a regulated output voltage.

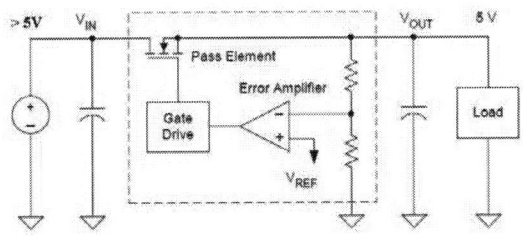

Figure 2. Basic LDO block diagram

IV. SUSCEPTIBILITY SIMULATIONS

Simulations were made to study the susceptibility of an LDO towards variations on its supply, corresponding to the most frequent perturbations of integrated circuits, the conductive ones.

In the study we used a basic LDO scheme, with the output voltage regulated at around 4.85V when having more than 5V on input, as marked on Figure 3 bellow.

On the output the load current was set either to 0.5mA or 5mA.

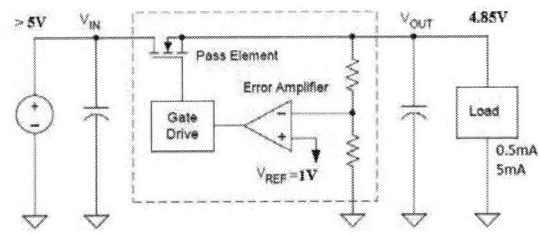

Figure 3. Particular LDO setup configuration used

In Figure 4 we present the concept schematic on which we ran the simulations in the proposed study.

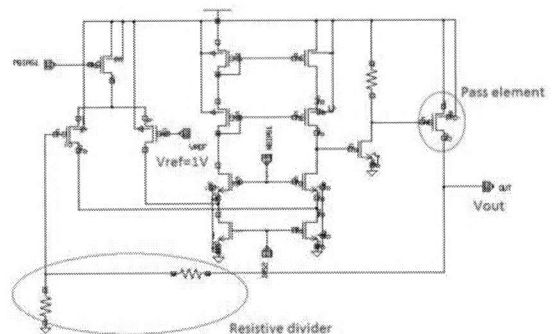

Figure 4. Concept LDO schematic used in simulations

The input voltage is compared by a resistive divider to a voltage reference of 1V.

For the operational amplifier we used a two-stage architecture, having a folded-cascode first stage and a common-source second stage.

The pass element in this case is a PMOS transistor.

The scheme was designed in a standard BiCMOS 0.18um technology.

On the supply voltage we applied a 0.5V variation, equivalent to the variations from wire conducted emissions affecting the supply of the circuit, and with a frequency starting from 150 kHz up to 10MHz, focusing mainly on frequencies bellow 2MHz, due to the difficulty in filtering these frequencies by external means.

V.   SIMULATION RESULTS

In each of the following figures describing the simulation results that will be presented, in plot A we have the input variations applied, and in B, C and D the effect on the output at -40C degrees, 25C degrees and 85C degrees respectively, unless otherwise specified.

On each plot, A, B, C and D it is also overlapped the simulated value of the parameter in normal functioning, without any supply variation. This value, since it is obtained with no oscillations on supply, is a relatively constant one (in the simulations performed it varies only with units of mV) and therefore appears as a line on each plot.

In Figure 5 bellow are plotted the simulation results with no load on output, with a frequency of the voltage supply variation of 1MHz, and at the three temperatures discussed: --40C degrees, 25C degrees and 85C degrees.

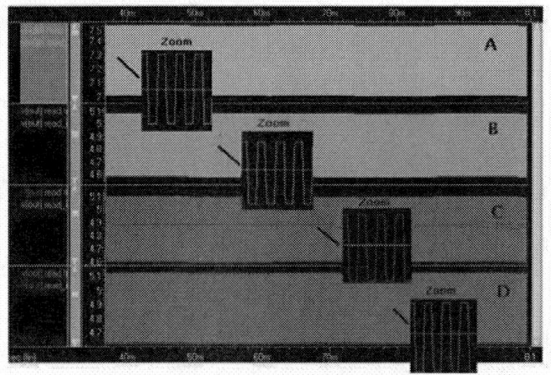

Figure 5.   Simulation results with no load on LDO output

As we can see on the figure above, the variation on the output voltage, when on the input we have a 0.5V variation (between 7V and 7.5V) is of about 300mV at 85C degrees and of about 380mV at -40Cdegrees and 25C degrees.

Also, in Figure 6 bellow are the results obtained when simulating the case when from the output we pull a current of 0.5mA, at the same 1MHz frequency and also in temperature corners.

On this figure it is visible that at low load no significant difference appears on the output voltage compared to the previous no-load simulated situation with or without variation on input.

Also there is no significant change in the range of the output voltage variation when variation on input is applied.

For the case when frequency on the input variation was increased to 10MHz, and keeping the same variation limits between 7V and 7.5V, results are presented in Figure 7.

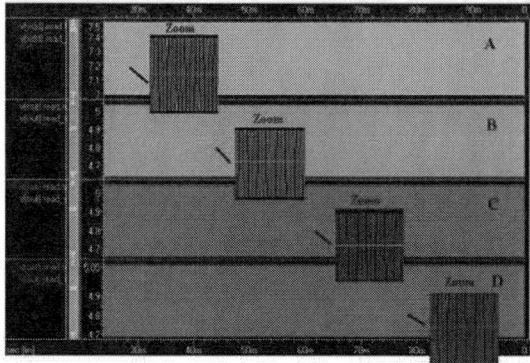

Figure 6.   Simulation results with 0.5mA current pulled from LDO

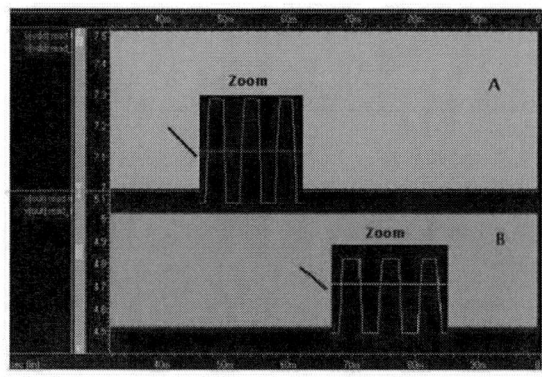

Figure 7.   Simulation results with increased input variation frequency

Because of the long simulation time required for this simulation due to the high frequency, the circuit was simulated only at one temperature -40C degrees since it seemed to represent the worst case for this study. Therefore, in this figure, plots C and D are not represented.

In this simulation, we can be observe a larger spread of the output voltage, of almost 500mV, more than 100mV higher than in the case of 1MHz frequency.

Next, we tried pulling a higher current of 5mA from the output, when varying the input voltage from 7V to 7.5V at the frequency of 1MHz. Results are presented in the figure bellow, Figure 8.

In this case, we obtain a larger output variation, comparable as spread with the one at 10MHz frequency. At -40C degrees on the output voltage we have almost 500mV variation, and at 25C degrees and 85C degrees 400mV variation.

In the figure below, Figure 9, are presented the results of the last simulation performed, with a 0.5mA current pulled from the output and with a frequency of the oscillation of 150 KHz, varying the input voltage between 7V and 7.5V.

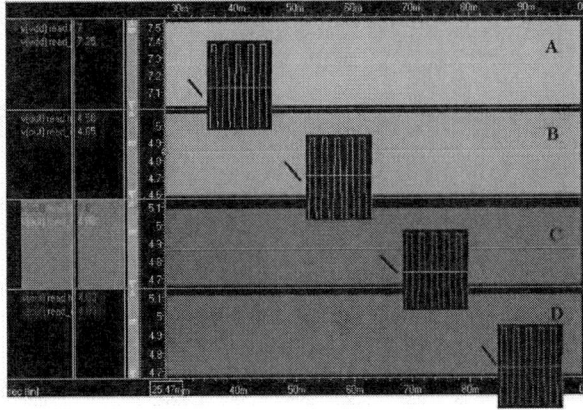

Figure 8. Simulation results with 5mA current pulled from LDO

Figure 9. Simulation results with decreased input variation frequency

Form this figure, the circuit seems to be more susceptible at this low frequency, if we look at the zoomed traces of the output voltage at different temperatures.

Also the variation of the output voltage is higher, of around 450mV at all temperatures.

VI. CONCLUSION

In this paper we presented the behavior at conducted emissions on the battery line of a 0.18um BCD technology LDO.

We simulated the behavior of the LDO circuit at different load currents on output, at different frequencies for the voltage supply variations and also at three temperatures that cover all the usual operating range for an LDO.

Significant output variations were observed at reduced temperatures, high frequencies, and also at low frequencies when pulling a higher current from the output.

Considering the high and still growing importance of undesired emissions in today's electronic circuits, this kind of studies can improve finding weak points in the susceptibility of electronic circuits even from the design phase, and also finding solutions to fix these problems.

REFERENCES

[1] Robert Warren Erickson, "Fundamentals of Power Electronics", Springer, New York, 2001

[2] H. W. Ott, "Electromagnetic Compatibility Engineering", Wiley Publication, 2009.

[3] M. Day, "Understanding Low Drop Out (LDO) Regulators", Texas Instruments.
Available at: http://focus.ti.com/download/trng/docs/seminar/Topic%

[4] O. Jovic, "Susceptibility of ICs to Conducted Electromagnetic Interference".

[5] D. L. Sengupta, V. V. Liepa, "Applied Electromagnetics and Electromagnetic Compatibility", Wiley Publication, 2006.

[6] B. Razavi, "Design of Analogue CMOS Integrated Circuits", McGraw-Hill, New York, 2001.

MIXDES 2012, 19ᵗʰ International Conference *"Mixed Design of Integrated Circuits and Systems"*, May 24-26, 2012, Warsaw, Poland

Device-Circuit Models for Extreme Environment Space Electronics

Marek Turowski and Ashok Raman
CFD Research Corporation (CFDRC)
Huntsville, Alabama, USA
Email: mt@cfdrc.com

Abstract—To design wide-temperature radiation-tolerant electronic systems for space missions and predict their characteristics and reliability in space, advanced models and simulation tools are required at multiple levels. Detailed, physics-based three-dimensional (3D) technology computer-aided-design (TCAD) device models, coupled in mixed-mode with external circuit models, enable accurate simulation and prediction of space electronics performance in extreme environments. We show the importance of correct device physics models for accurate computation of both steady-state and transient characteristics. Mixed-mode coupling of a realistic load circuit with the 3D TCAD device model is critical to be able to compute single-event transient waveforms and circuit characteristics that reflect well experimental results.

Index Terms—Radiation effects, extreme low temperature models, single-event effects (SEE), 3D, TCAD, mixed-mode simulation

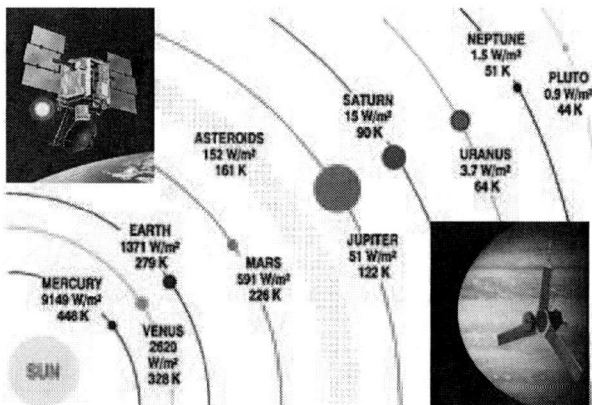

Figure 1. Solar intensity and calculated space probe temperature in the Solar System (after [1]).

I. INTRODUCTION

Space missions require avionic systems, components, and controllers that are capable of operating in the extreme temperature and radiation environments of deep space. Inside the Solar System, considering only the influence of the Sun on a black body in space, calculated spacecraft temperatures [1] range from a high of about +175°C (448 K) at the orbit of Mercury (nearest the Sun) to a low of about −230°C (44 K) at the orbit of Pluto (furthest from the Sun), as illustrated in Fig. 1. A thriving application at low temperatures is for scientific and military spacecraft. Many sensors, such as infrared or X-ray detectors, used for astronomical observations or surveillance must operate at very low temperatures [2].

Extreme environments for electronics primarily refer to operating conditions that span a wide temperature range (typically −230°C to +130°C) and a spectrum of high-energy incident radiation (including heavy ions, electrons, protons, and electromagnetic radiation). These conditions are well outside the current commercial and even military specifications, thereby necessitating focused research, design, and validation of devices, circuits, and systems for applications [1]-[3]. In addition, the escalating device count and circuit complexity in modern integrated circuit (IC) technologies, coupled with the inefficiency of fabrication and experimental testing of each variant design, mandate extensive use of hierarchical simplification and accurate technology computer aided design

(TCAD) tools in the design and analysis processes. This paper addresses both aspects mentioned above by describing the important features, applications, and limitations of TCAD tools, specifically, at the device physics level and mixed-mode (coupled device TCAD + circuit simulator), for modeling extreme environment electronics.

II. MODELING APPROACHES

In order to design cryogenic or wide-temperature, radiation-hardened (rad-hard) electronics, and to predict their characteristics and reliability in space, advanced models and simulation tools are required at multiple levels. Modeling the electrical behavior of devices and sub-circuits in extreme environment applications, just like in regular applications, is usually accomplished by one of the following three approaches:

a) Device level: Physics-based TCAD, or

b) Circuit level: Compact model based solution of electronic circuit/network, or

c) Mixed-mode (or 'mixed-level'): Coupled TCAD and compact model solution.

The following sections of this paper highlight the benefits and applications of each approach, while Fig. 2 presents a graphical illustration of each.

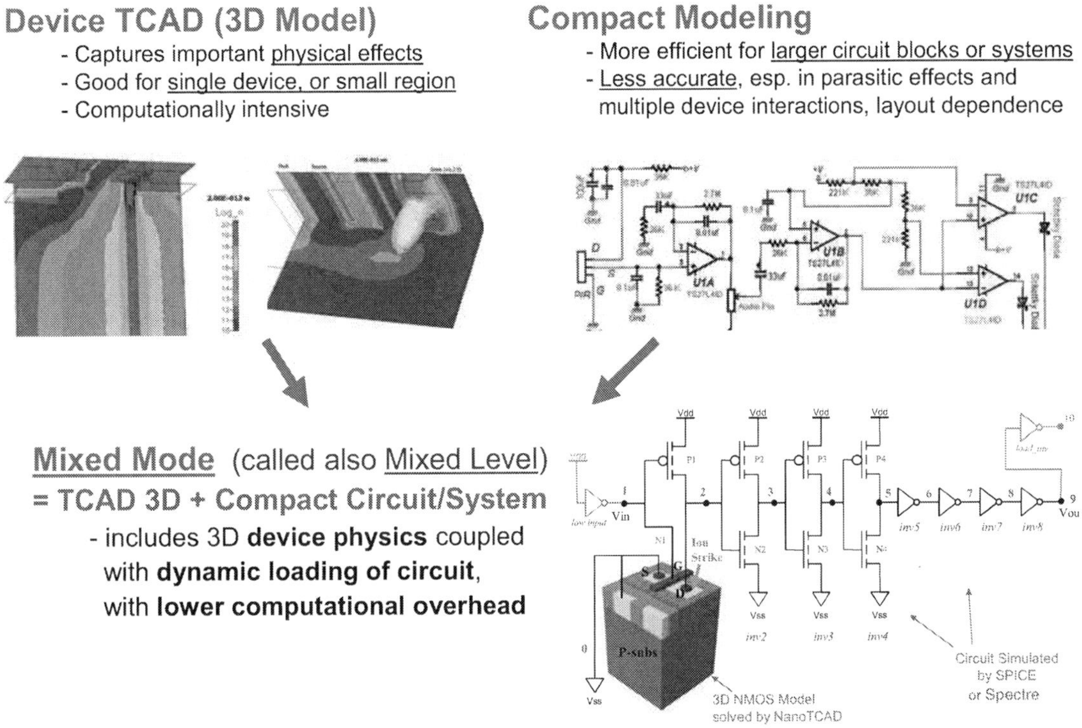

Device TCAD (3D Model)
- Captures important <u>physical effects</u>
- Good for <u>single device, or small region</u>
- Computationally intensive

Compact Modeling
- More efficient for <u>larger circuit blocks or systems</u>
- <u>Less accurate</u>, esp. in parasitic effects and multiple device interactions, layout dependence

<u>Mixed Mode</u> (called also <u>Mixed Level</u>) = TCAD 3D + Compact Circuit/System
- includes 3D **device physics** coupled with **dynamic loading of circuit**, with **lower computational overhead**

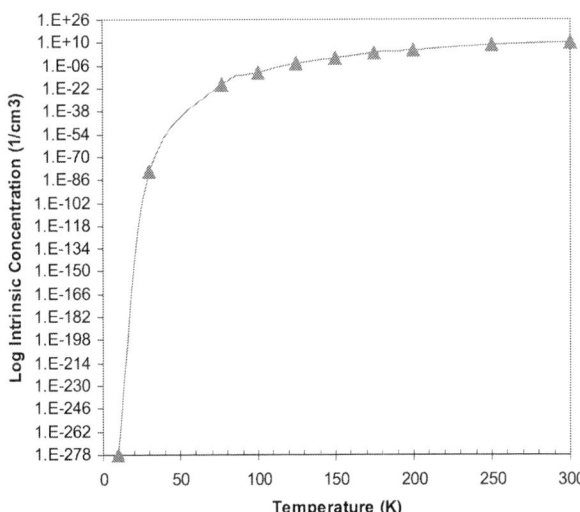

Figure 2. Various modeling approaches used for analyzing the electrical performance of devices and circuits.

A. *Physics-Based Device-Level Modeling (TCAD)*

The physics based TCAD semiconductor device simulation approach involves: (i) creation of a two-dimensional (2D) or three-dimensional (3D) geometrical model of the semiconductor structure to be analyzed, followed by generation of a discretized computational mesh (typically non-uniform and adapted to device features) to enable subsequent numerical solution,
(ii) specification of appropriate boundary conditions, material properties and related physics models (e.g., for electron/hole mobilities, lifetimes), and source terms (e.g., photogeneration, ionizing radiation-induced carrier generation, carrier traps, etc.), and (iii) setting numerical solution parameters and solving the underlying semiconductor device equations. The basic semiconductor device equations include the Poisson equation (to calculate electric potential distribution) and carrier transport equations (to calculate electron and hole concentrations), as described in Section III. The primary advantages of device-level TCAD modeling are the increased accuracy and detail of results due to inclusion of more physical phenomena and spatial effects and interactions, while a limitation is the computational resources required.

Such detailed modeling of semiconductor devices is highly reliant on the accuracy of the physics models used (e.g., for carrier mobilities, energy bandgap, carrier freeze-out at low temperatures, and so on). Although some physics models, applicable over wide temperatures, have been proposed in the past (*e.g.*, carrier mobilities in [4]-[7]), their applicability may be limited either by the temperature ranges encountered in

extreme environments or the technology generation. Another important factor for TCAD simulations at extreme low temperatures is the stability of the numerical solver, because certain parameter values such as intrinsic carrier densities (n_i) drop off exponentially and become extremely small in that regime (Fig. 3, Fig. 4) [8]. For example, at T = 40K, $n_i \approx 10^{-56}$ cm^{-3} and at T = 10 K, $n_i \approx 10^{-278}$ cm^{-3}, as illustrated in Fig. 3.

Figure 3. Variation of the intrinsic carrier concentration (n_i) in silicon with temperature. The very low values pose challenge to solvers.

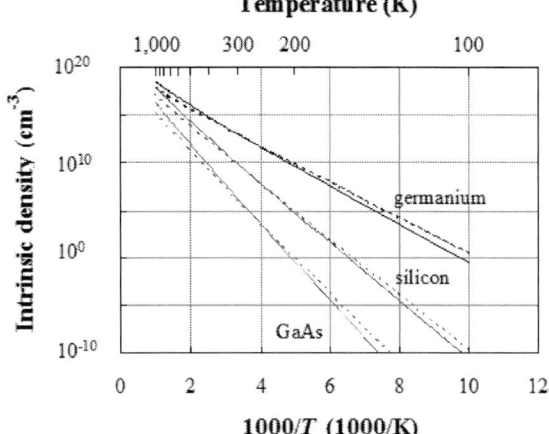

Figure 4. Intrinsic carrier density versus temperature in gallium arsenide (GaAs), silicon, and germanium. Compared is the calculated density with (solid lines) and without (dotted lines) the temperature dependence of the energy bandgap [9].

Considering that n_i^2 is often used in the drift-diffusion equations, typical numerical operations such as exponentiation and division performed on such extreme floating-point numbers can lead to arithmetic underflows and subsequent solver divergence. At the same time, the presence of very localized, extremely high charge densities due to radiation events such as heavy ion strikes and ionizing dose rate, may further destabilize the numerical solution mechanism. Therefore, even well established TCAD methods and solvers face serious challenges during simulations of semiconductor device behavior in extreme environments.

B. Compact Model Based Circuit-Level Modeling

This approach involves generating a description of the target circuit using appropriate compact models (essentially zero-dimensional, or 0D). These compact models are typically based on an empirical, but well-calibrated, set of analytical expressions that describe the electrical behavior of the device they represent. The classical, compact model-based circuit simulator that has been widely used in the electronics community over the past several decades is SPICE [10]. Other popular commercial circuit simulators include Spectre (from Cadence [11]), Eldo (from Mentor Graphics [12]), HSPICE (from Synopsys [13]), and so on. These simulators are periodically upgraded with compact models for the latest IC technologies – for instance, BSIM3v3, BSIM4, PSP (for CMOS), BSIMSOI (for SOI MOS), VBIC, Mextram, and HiCUM (for bipolar devices). The primary advantage of a compact model based approach is that larger circuits containing hundreds or thousands of devices can be analyzed using limited computational resources. The downside is that detailed physical or spatial effects (e.g., parameter distributions within the device geometry or in the spacing between devices) or inter-device interactions (e.g., due to substrate coupling effects or charge sharing) cannot be captured well by compact models (Fig. 2).

An additional limitation of most circuit simulators at present is that the compact models are not applicable at cryogenic temperatures encountered in extreme space environments [14]. Also, the details of radiation events (e.g., heavy ion strikes) cannot be accurately modeled using compact models alone, and commonly used circuit-level approximations (e.g., double-exponential current sources representing single events) often fail under various operating conditions [15]-[17].

C. Mixed-Mode (or Mixed-Level) Modeling

This hybrid approach implements a tight coupling between a physics-based TCAD solver and a compact-model-based circuit simulator. It refers to analyzing a specific device or sub-circuit, for which detailed parameter distributions and additional physical insight are sought, using 2D/3D models, while the remainder of the circuit is analyzed using a circuit simulator and compact models (Fig. 2). This approach is extremely important for analyzing fast radiation effects, such as Single Event Transients (SETs) in high-speed circuits, where capturing the interaction between the rapid external circuit response and the spatially-dependent physics in the devices of interest is critical [15]-[19].

An important requirement in a mixed-mode simulator is that the circuit solver should use the best available compact models with well-calibrated parameters, preferably from the PDK (process design kit) provided by the technology foundry, in order to ensure accuracy of the circuit solution, especially for very fast transient waveforms. Additionally, the numerical solution scheme must be robust enough to handle difficult operating conditions, such as very high speeds, steep rise/fall slopes of signals, and bi-stability.

In the following sections, we provide a detailed description of our model development work performed in the TCAD and mixed-mode regimes for the analysis of extreme environment electronics.

III. TCAD DEVICE EQUATIONS AND PHYSICS MODELS

TCAD device simulators typically offer two broad frameworks - drift diffusion (DD) and hydrodynamic (HD) - to analyze the transport of carriers (electrons and holes) as a function of the various operating conditions (e.g., applied biases and electric fields, external source terms such as photogeneration, radiation effects, etc.), and calculate the resulting device performance. The basic equations solved in either framework are: the Poisson equation and the carrier continuity equations for electrons and holes, respectively [20]. The basic unknown quantities in this equation system are the electrostatic potential (ψ) and the electron and hole concentrations (n, p). In the DD model carrier temperatures are assumed to be equal to the lattice temperature T_L. In the HD model additional energy balance equations are solved for electron and hole energies and temperatures (T_n, T_p), along with the lattice energy equation [20]. The lattice temperature, T_L, is solved by the heat conduction equation, where the heat source includes Joule heating and carrier recombination. Such a model accounts for self-heating effects in semi-conductor devices by solving the lattice heat flow equation self-consistently with the DD or HD transport equations. The self-heating effect may be very important for device behavior at extreme (low) temperatures [21].

352

Boundary Conditions: A TCAD device simulator typically provides a range of boundary conditions (BCs) to model various semiconductor devices and complex geometries. At each model boundary, BCs are needed for all dependent variables: ψ, n, p, T_n, T_p, and T_L. This applies to both external boundaries as well as internal boundaries (interfaces), e.g., hetero-interfaces between different materials. Some common BCs include semiconductor-metal contacts (Ohmic and Schottky), insulating walls, metal-oxide-semiconductor (MOS) gate contact, semiconductor-insulator interfaces, and so on. Further details about these boundary conditions, and particularly device contacts and how parameters are set at each of them, are provided in [20]. For *mixed-mode* simulations, a special type of contact is applied at the respective TCAD model boundaries, and it acts as an interface with the corresponding nodes of the external circuit [18], [22].

A. Physical Models for Extreme/Low Temperatures

In the semiconductor "transport equations" (DD or HD), several physical parameters are involved in the calculation of device electrical performance. These include charge carrier (electron and hole) mobilities, energy bands (band gaps), carrier effective masses, carrier lifetimes, thermal conductivities, etc. It is critical that appropriate models for each of these parameters are selected based on the semiconductor material composition, operating conditions, temperature range, etc. For a detailed discussion of how these parameters influence device operation and how they vary as a function of other parameters, the user is referred to classical textbooks, such as [20], [23], [24]. Here we will address a few models that are important for extreme environment, especially very low temperature, analyses.

Intrinsic Carrier Density: Defined as the concentration of thermally generated electron-hole pairs (EHPs) in an otherwise perfect semiconductor lattice with no impurities (*i.e.*, dopants) or lattice defects, the intrinsic carrier density (n_i) is an important factor in calculating the concentrations of charge carriers [23]. It is usually modeled as:

$$n_i = \sqrt{N_C \cdot N_V} \cdot \exp\left(-\frac{E_g}{2kT}\right) \qquad (1)$$

where N_C and N_V are the densities of states in the conduction and valence bands, respectively, E_g is the energy bandgap, k is the Boltzmann constant, and T is the absolute temperature in Kelvin. N_C and N_V are calculated as shown below, where $m_{n/p}^*$ are the electron/hole effective masses, and h is the Planck constant.

$$N_C = 2 \cdot \left(\frac{2\pi \cdot k \cdot T \cdot m_n^*}{h^2}\right)^{3/2}; \quad N_V = 2 \cdot \left(\frac{2\pi \cdot k \cdot T \cdot m_p^*}{h^2}\right)^{3/2} \qquad (2)$$

Note that the energy bandgap and the carrier effective masses themselves are highly dependent on temperature, necessitating accurate models for them so that the final intrinsic carrier density can be correctly determined [25]. Based on comparisons with experimental data, it is widely accepted that the intrinsic carrier density in silicon can vary by many orders of magnitude, e.g., from $\sim 10^{10}$ cm^{-3} at room temperature (300 K) down to $\sim 10^{-80}$ cm^{-3} at T = 30 K, or even less for lower temperatures (Fig. 3). Such dramatic variations create

significant numerical problems and convergence hurdles for TCAD programs [25], requiring innovative solution schemes and specialized algorithms [22].

Incomplete Ionization: At room temperature, it is reasonable to assume that all dopant atoms are fully ionized, implying that the concentration of free carriers (electrons or holes) is equal to the concentration of dopant atoms (donors or acceptors, respectively). However, this assumption is not valid at very low temperatures (T < 100 K) since electrons in dopant atoms do not possess sufficient thermal energy to get excited to the conduction band. The resulting effect is called incomplete ionization (or carrier "freeze-out") and can significantly impact device characteristics at low temperatures (Fig. 5).

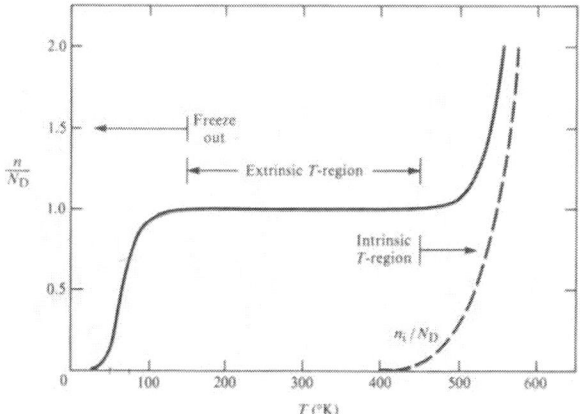

Figure 5. Ratio of ionized carriers (n) to total dopant density (N_D) as a function of temperature [26].

Another important factor that influences ionization is the doping levels in the semiconductor. For low doping values, incomplete ionization is a strong function of temperature. However, at very high doping, all the dopant atoms are always ionized, i.e., there is no freeze-out even at low temperatures. Over the years, several theories have been proposed to explain incomplete ionization [26], [27]. Recently, the significant effect of incomplete ionization even at room temperature has been demonstrated in transistors and solar cells [28], [29]. The model proposed there is based on detailed physics-based derivations and extensive validation against experimental results. Fig. 6 illustrates the trends of incomplete ionization (free carriers) with varying temperature and doping level.

Carrier Mobilities: Over the last several decades, a wide range of electron and hole mobility models for silicon and other materials have been published. These mobility models span a range of dependencies including doping densities, temperature, carrier-carrier scattering, oxide interface scattering, and lateral and transverse electric fields. Examples include the Lombardi [7] and Arora [6] models for MOSFETs, and the *Philips Unified* model [3], [5] for bipolar devices. In [3] the Philips mobility model is shown to yield good results in bipolar devices down to a low temperature of 40 K (-233°C). However, it was determined in subsequent tests (including some by the authors of this paper) that this match does not

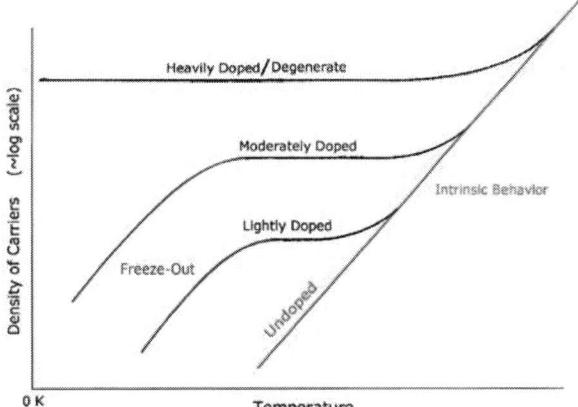

Figure 6. The ratio of ionized free carriers to total dopant density as a function of temperature and doping level in semiconductors [2].

extend to modern technologies such as the 180-nm SiGe BiCMOS [8]. Therefore, it is important to either derive new mobility models or update classical models using new device data to assure their continued applicability to TCAD analyses of new technologies and an extended range of temperatures.

Moen and Cressler [30] describe their detailed calibration of the Philips mobility model and Altermatt incomplete ionization models using extensive experimental resistivity measurements on the IBM 0.5μm SiGe BiCMOS technology to develop an improved set of parameters that are applicable to this modern technology. It appears that the lattice scattering mobility dominates the temperature dependence of carrier mobilities at lower doping concentrations and higher temperatures, while the impurity scattering and carrier-carrier scattering terms dominate carrier mobilities at higher doping concentrations and lower temperatures.

Other Physical Models: As mentioned earlier, several other parameters such as energy bandgaps and carrier effective masses exhibit temperature dependence that must be accurately captured. It is important to ensure that subsequent derived parameters such as densities of states and intrinsic carrier concentrations are correctly calculated. The papers [25] and [31] and references therein are a good starting point for these models.

IV. MIXED-MODE SIMULATIONS

The coupling of a device physics-oriented 2D/3D TCAD solver with a compact model-based fast circuit solver requires efficient exchange of numerical information between the two solvers at specific intervals (iterations, time steps) and at specific locations (interface boundary conditions). As an example, in the *MixCad* and *MixSpice* simulation tools from CFDRC, the coupling between the NanoTCAD 3D device simulator and Spectre (or Spice) circuit solver is implemented using a two-level Newton algorithm [18], [22], [32]. All TCAD models become a new type of element in the modified circuit netlist. Linearized descriptions of the TCAD models, required for integration into the Spice or Spectre circuit solution

algorithm, are computed at each step based on the 3D TCAD solution.

In several recent publications we have shown the important role of a robust mixed-mode simulator in the analysis and prediction of radiation effects in modern semiconductor devices and IC technologies ([15]-[19]). In particular, mixed-mode coupling of a realistic load circuit, including experimental parasitics, with the 3D TCAD device model, is critical to be able to compute single-event transient waveforms and circuit characteristics that reflect well experimental results [17], [19].

V. EXAMPLES OF LOW TEMPERATURE SIMULATIONS

The specific extreme-temperature models described above have been implemented in CFDRC NanoTCAD simulator and enabled modeling and simulation of radiation effects in extreme environments, relevant to space electronics. Sample transient results for heavy ion strikes into a 180-nm Si NMOS transistor at extreme low temperatures are presented in Fig. 7.

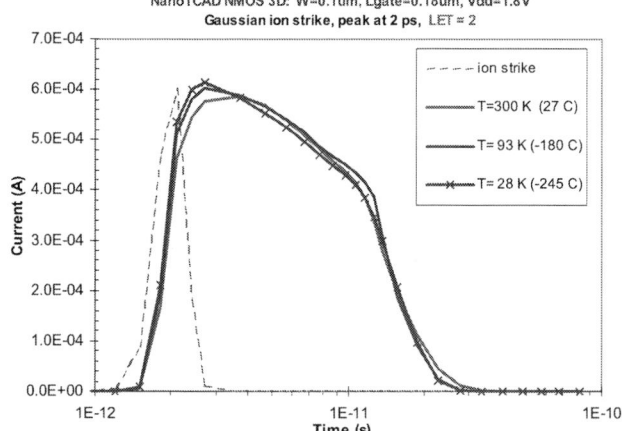

Figure 7. Sample NanoTCAD transient simulation of single-event currents in an NMOS FET at extremely low temperatures (down to -245°C).

Mixed-mode simulations using the CFDRC MixCad tool were successfully used to investigate the temperature sensitivity of the single event upset (SEU) threshold for a static random access memory (SRAM) cell designed and fabricated in the CMOS portion of the IBM 5AM BiCMOS process [33]. Newly calibrated cold temperature mobility models in CFDRC NanoTCAD, based on the data presented in [30], and enhanced, temperature-dependent BSIM3 compact models extracted explicitly for the cold temperature design, were used to simulate the temperature dependence of radiation effects. Detailed mixed-mode calculations in [33] indicated a 33% reduction in single-event upset threshold LET for the range of temperatures simulated, *i.e.* between 100 K (-173°C) and room temperature of 300 K.

VI. CONCLUSIONS

Advanced mixed-mode tools enable physics-based analyses of transient radiation effects in modern, high-speed, mixed-signal circuits (including the latest CMOS and SiGe BiCMOS

354

technologies) while using compact models directly from the IC manufacturer's PDK, and circuit designs directly from advanced EDA tools.

Special, enhanced models are required for extreme temperatures, as well as robust numerical solvers are needed for reliable computation of both steady-state and transient characteristics. Mixed-mode coupling of a realistic load circuit with a 3D TCAD device model is critical to be able to compute single-event transient waveforms and circuit characteristics that reflect well experimental results and real circuit behavior.

ACKNOWLEDGMENT

This work was supported by NASA Small Business Innovative Research (SBIR) programs.

REFERENCES

[1] NASA Glenn Research Center online article: "Cryogenic Electronics in Support of Deep-Space Missions", May 1997, http://www.grc.nasa.gov/WWW/RT1996/5000/5480di.htm.

[2] http://www.extremetemperatureelectronics.com

[3] J. D. Cressler, "On the Potential of SiGe HBTs for Extreme Environment Electronics," *Proceedings of the IEEE*, vol. 93, no. 9, pp. 1559-1582, September 2005.

[4] D. B. M. Klaassen, "A Unified Mobility Model for Device Simulation – I. Model Equations and Concentration Dependence", *Solid-State Electronics*, Vol. 35, No. 7, pp. 953-959, 1992.

[5] D. B. M. Klaassen, "A Unified Mobility Model for Device Simulation – II," *Solid-St. Electron.*, vol. 35, pp. 961-967, 1992.

[6] N. D. Arora, G. Sh. Gildenblat, "A Semi-Empirical Model of the MOSFET Inversion Layer Mobility for Low-Temperature Operation," *IEEE Trans. Electron Devices*, vol. 34, pp. 89-93, 1987.

[7] C. Lombardi, S. Manzini, A. Saporito, M. Vanzi, "A Physically Based Mobility Model for Numerical Simulation of Nonplanar Devices," *IEEE Trans. CAD*, vol. 7, pp. 1164-1171, 1988.

[8] A. Raman, M. Turowski, A. Fedoseyev, and J. D. Cressler, "Addressing Challenges in Device-Circuit Modeling for Extreme Environments of Space", *2007 International Semiconductor Device Research Symposium (ISDRS)*, College Park, MD, 2007.

[9] B. V. Zeghbroeck, *Principles of Semiconductor Devices*, online: http://ecee.colorado.edu/~bart/book/book/chapter2/ch2_6.htm

[10] http://bwrc.eecs.berkeley.edu/classes/icbook/spice/

[11] http://www.cadence.com/products/cic/spectre_circuit/

[12] http://www.mentor.com/products/ic_nanometer_design/analog-mixed-signal-verification/eldo/

[13] http://www.synopsys.com/Tools/Verification/AMSVerification/CircuitSimulation/HSPICE

[14] A. S. Kashyap, H. A. Mantooth, T. Vo, and M. M. Mojarradi, "Compact modeling of LDMOS transistors for extreme environment analog circuit design," *IEEE Trans. Electron Devices*, vol. 57, no. 6, pp. 1431-1439, June 2010.

[15] M. Turowski, D. Mavis, A. Raman, and P. Eaton, "Digital Single Event Transient Pulse Generation and Propagation in Fast Bulk CMOS ICs," *2006 IEEE Nuclear and Space Radiation Effects Conference (NSREC)*, July 2006, paper C-2.

[16] M. Turowski, A. Raman, and A. Fedoseyev, "Mixed-Mode Simulation of Single Event Upsets in Modern SiGe BiCMOS Mixed-Signal Circuits," *2009 Int'l Conf. on Mixed Design of Integrated Circuits and Systems- MIXDES 2009*, pp. 462-467.

[17] K. Moen, L. Najafizadeh, A. Raman, M. Turowski, and J. Cressler, "Accurate Modeling of Single-Event Transients in a SiGe Voltage Reference Circuit," *IEEE Trans. Nucl. Sci.*, vol. 58, no.3, pp. 877-884, June 2011.

[18] M. Turowski, A. Raman, and G. Jablonski, "Mixed-Mode Simulation and Analysis of Digital Single Event Transients in Fast CMOS ICs," *2007 Int. Conf. on Mixed Design of Integrated Circuits and Systems - MIXDES 2007*, pp. 433-438.

[19] M. Turowski, J. Pellish, K. Moen, A. Raman, J. Cressler, R. Reed, G. Niu, "Reconciling 3-D Mixed-Mode Simulations and Measured Single-Event Transients in SiGe HBTs," *IEEE Trans. on Nuclear Science*, vol. 57, no. 6, pp. 3342-3348, Dec. 2010.

[20] S. Selberherr, *Analysis and Simulation of Semiconductor Devices*, Vienna, Austria: Springer-Verlag, 1984.

[21] F. J. De la Hidalga, M. Jamal Deen, E. A. Gutiérrez, "Theoretical and Experimental Characterization of Self-Heating in Silicon Integrated Devices Operating at Low Temperatures" *IEEE Trans. Electron Devices*, Vol. 47, No. 5, pp.1098-1106, May 2000.

[22] M. Turowski, A. Raman, A. Fedoseyev, "Enabling Mixed-Mode Analyses of Nano-Scale SiGe BiCMOS Technologies in Extreme Environments", *10th Int. Conf. on Radiation and Its Effects on Components and Systems - RADECS 2009*, Paper C-8.

[23] B. G. Streetman and S. Banerjee, *Solid State Electronic Devices*, Upper Saddle River, NJ: Prentice Hall, 2000.

[24] S. M. Sze and K. K. Ng, *Physics of Semiconductor Devices – 3rd ed.*, Hoboken, NJ: Wiley – Interscience, 2006.

[25] S. Selberherr, "MOS Device Modeling at 77 K", *IEEE Trans. Elec. Dev.*, Vol. 36, No. 8, pp. 1464-1474, August 1989.

[26] R. F. Pierret and G. W. Neudeck, eds., *Advanced Semiconductor Fundamentals, Modular Series on Solid State Devices – Vol. VI*, Reading, MA: Addison Wesley Publishing Company, 1989.

[27] A. Akturk, J. Allnutt, Z. Dilli, N. Goldsman, and M. Peckerar, "Device Modeling at Cryogenic Temperatures: Effects of Incomplete Ionization", *IEEE Trans. Electron Devices*, Vol. 54, No. 11, pp. 2984-2990, Nov. 2007.

[28] P. P. Altermatt, A. Schenk, and G. Heiser, "A Simulation Model for the Density of States and for Incomplete Ionization in Crystalline Silicon. I. Establishing the Model in Si:P", *Journal of Applied Physics*, Vol. 100, 2006, pp. 113714-1-10.

[29] P. P. Altermatt, A. Schenk, B. Schmithusen, and G. Heiser, "A Simulation Model for the Density of States and for Incomplete Ionization in Crystalline Silicon. II. Investigation of Si:As and Si:B and Usage in Device Simulation", *Journal of Applied Physics*, Vol. 100, 2006, pp. 113715-1-7.

[30] K. Moen and J. D. Cressler, "Measurement and Modeling of Carrier Transport Parameters Applicable to SiGe BiCMOS Technology Operating in Extreme Environments", *IEEE Trans. Elec. Dev.*, Vol. 57, No. 3, pp. 551-561, March 2010.

[31] M. A. Green, "Intrinsic Concentration, Effective Densities of States, and Effective Mass in Silicon", *Journal of Applied Physics*, Vol. 67, No. 6, pp. 2944-2954, March 1990.

[32] *NanoTCAD* Software, Version 2011, CFD Research Corp., Huntsville, AL, (www.cfdrc.com)

[33] S. Sanathanamurthy, V. Ramachandran, M. L. Alles, R. A. Reed, L. W. Massengill, A. Raman, M. Turowski, A. Mantooth, B. Woods, M. Barlow, and J. Cressler "Simulation of SRAM SEU Sensitivity vs. Reduced Operating Temperature," *34th Annual GOMACTech Conf.*, March 2009, Orlando, Florida, Paper 32.9.

MIXDES 2012, 19th International Conference *"Mixed Design of Integrated Circuits and Systems"*, May 24-26, 2012, Warsaw, Poland

Estimating the Impact of Complete Analog Channel Selection on Zero-IF Multi-Standard Radio Receivers Power Consumption

Silvian Spiridon[1,2], Dan Claudius[1], Mircea Bodea[1]
[1]"POLITEHNICA" University of Bucharest, Bucharest, Romania
[2]Now with Broadcom, Bunnik, The Netherlands

Abstract—**This paper analyses the key trade-off that shapes the design of direct conversion radio receivers embedding analog signal conditioning: the trade-off between the receiver area, determined by its anti-alias Low Pass Filter (LPF) order, and its power consumption, constrained by the ADC specifications of resolution and speed. The paper's main goal is to determine the receiver's LPF order that enables the complete analog channel selection in the context of a multi-standard receiver implementation. Based on the multi-standard receiver generic blocker diagram analysis, a first order, system level analysis is used to determine the LPF order. The analysis is constructed from the circuit / transistor level designer perspective and is also used for estimating the impact of the complete analog channel selection on the receiver RF front-end power consumption and area. Thus, by using this analysis methodology the designer is enabled to handle efficiently the large amount of information required for designing multi-standard radio systems.**

Index Terms—**multi-standard receiver; filtering strategy**

I. INTRODUCTION

One of the major problems in today's mobile communication systems is every standard requires a different set of hardware equipment to allow a compatible dialogue. Hence, there is a need of reconfigurable radio front-ends, able to ensure compatibility with the envisaged standards.

The homodyne quadrature down-converter architecture provides the optimum solution for the implementation of multi-standard radio receivers, given its key advantages over the other receiver architectures, [1]:

- the useful signal is its own image, and, thus, the receiver image rejection requirements are much lower than for other architectures (e. g., superheterodyne, low-IF)
- all baseband processing, like analogue filtering, baseband amplification, analog-to-digital conversion and the digital demodulation, take place at the lowest possible frequency; thus minimizing the receiver power consumption.

Fig. 1 presents the multi-standard receiver implementing analog baseband signal processing principle block schematic, redrawn from [2]. This paper targets the major wireless standards with a frequency plan below 3 GHz (i. e., GSM, DECT, Bluetooth and W-LAN-802.11b,g), as described by Table I from [2].

The RF signal is amplified by the receiver Low Noise Amplifier and then down converted by the mixer directly to

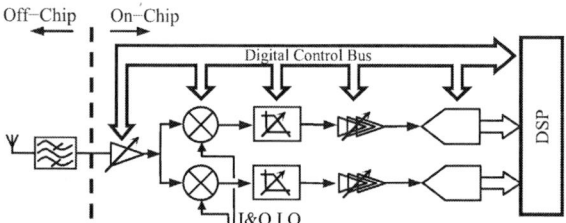

Figure 1. Principle block schematic of a re-configurable quadrature homodyne receiver embedding analog baseband signal conditioning, [2].

baseband, by mixing it with a complex Local Oscillator signal (I & Q LO).

Subsequently, the signal is conditioned by the LPF, which removes the un-wanted blockers and interferers. Finally, the baseband analog signal conditioning is completed by the Variable Gain Amplifier which boosts the wanted signal level providing the optimal Analog to Digital Converter (ADC) loading.

The front-end operation is managed by the Digital Signal Processor (DSP).

The paper's main goal is to use the standard independent analysis methodology developed in [3] to determine the LPF order that enables the complete analog channel selection for the envisaged multi-standard environment; and, subsequently, to determine its impact on the multi-standard receiver RF front-end power consumption.

Based on the analysis of the multi-standard receiver generic blockers diagram defined and introduced in [4], Section II analyses the multi-standard environment constraints on the channel selection.

Further on, in Section III, the LPF order required for the complete analog channel selection is calculated; while Section IV assesses the complete analog channel selection impact on the receiver power consumption. Finally, Section V concludes the paper.

II. MULTI-STANDARD CONSTRAINTS ON THE CHANNEL SELECTION

Besides the useful signal, other interferers and blockers can be present at the antenna input. The interferers and blockers particular characteristics (e. g., relative frequency position and level) are different for each particular wireless standard.

356

In order to facilitate the multi-standard implementation, a *receiver generic blockers diagram* was constructed in [4]. The diagram is derived based on the envisaged radio standards analysis, and consists of all blockers and interferers present at receiver's antenna input, under which influence the receiver must still be able to successfully demodulate the wanted signal.

Since the wanted signal can be received simultaneously with the unwanted blockers and interferers, the biggest threat is the receiver output clipping due (a) to the large difference in power levels between the blockers and the RF useful signal and (b) to the large receiver gain, [2].

In order to alleviate the issue, selectivity is enforced both at RF and in baseband.

On the high frequency side, the selectivity is fairly limited: the interferers and near-by blockers get transferred un-attenuated to the LNA input, as typically the antenna filter has a large enough bandpass to allow all the channels to be received properly; for the far-away blockers the attenuation corresponds to a first order low pass filter, depending on their relative frequency distance to the antenna filter bandpass edge.

Hence, the receiver baseband part will bear the biggest portion of the channel selection bourdon. The in-band blockers and interferers filtering is not easy, as these unwanted signals are quite close in frequency to the RF signal and the mixer will down-convert them un-attenuated to baseband.

Usually, a first-order low pass filter is implemented in the mixer output stage. The relative interferer power at the downconverter's output, $P_{blk,BB}$, is given by, [4]:

$$P_{blk,BB} = P_{blk} - 10\log\left[1 + \left(\Delta f_{blk} / BW_{MIX}\right)^2\right], \qquad (1)$$

where P_{blk}, respectively Δf_{blk}, are the relative interferer/blocker power at the receiver LNA input, respectively the relative interferer/blocker frequency, and BW_{MIX} is the mixer output stage LPF -3 dB frequency.

Given the inherent process variations of the filter's capacitors and resistors, in the worst case scenario, $f_{-3dB,MIX}$ can be even 30 % lower than its nominal value.

Thus, a bandwidth of 500 kHz is considered to allow the 100 kHz GSM signal to pass, while a bandwidth of 50 MHz is used for the 10 MHz W–LAN signal.

Other intermediate values for $f_{-3dB,MIX}$ can be considered. For instance, 2.5 MHz and 10 MHz will divide the mixer LPF bandwidth in equal slices on a logarithmic scale.

Still, nonetheless the first order LPF embedded in the mixer output stage will not provide sufficient filtering, given the large difference in power levels between the blockers and the RF useful signal (i. e., ≥ 70 dB, [4]).

Fig. 2.a depicts the blocker/interferer levels at the mixer output, respectively at the LPF input. In Fig. 2 B is the Adjacent Channel Interferer (ACI), C and D signals are the first, second, and so on, alternate adjacent channel interferers (AACI); the signals from E on represent the blockers.

Thus, an analog anti-alias low pass filter is required for the baseband signal conditioning completion.

Given the multi-standard application, a Butterworth LPF implementing a programmable cut-off frequency is chosen (e. g., [5]).

III.   SETTING THE RECEIVER LPF ORDER FOR THE COMPLETE ANALOG CHANNEL SELECTION

This research is focused to determine the LPF order that ensures the complete analog channel selection. Later on the impact of the complete analog channel selection on the receiver power consumption and area is estimated.

This corresponds to finding the LPF order that knocks down all blockers and interferers below the useful signal level with enough margin such as the signal can still be demodulated within the specified Bit Error Rate.

The required margin is at least equal to the minimum signal to noise ratio, SNR_0, required for the baseband processor to demodulate correctly the wanted signal.

The analysis presented in [6], computes the SNR_0 for different modulation techniques. The typical SNR_0 values for the envisaged standards are: 9 dB for the GSMK of GSM, 16 dB for the GFSK of Bluetooth, 13 dB for the GFSK of DECT and more than 20 dB for the 64QAM of W-LAN, [6].

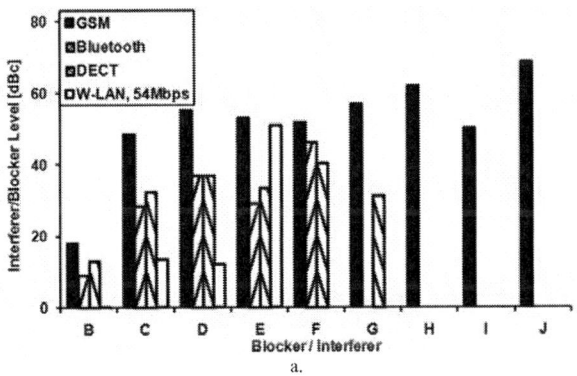
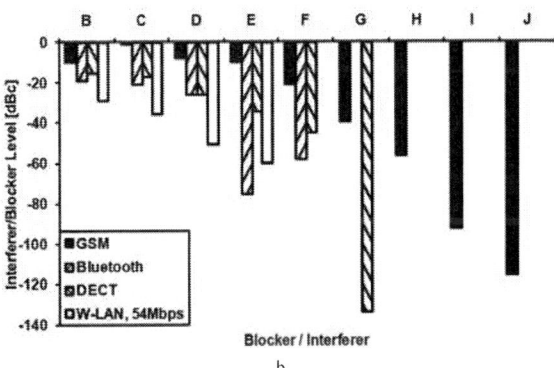

Figure 2. The relative interferers/blockers levels a. before the LPF (after the mixer); b. after a 4th order LPF.

TABLE I
LPF ORDER REQUIRED FOR COMPLETE ANALOG CHANNEL SELECTION

Standard	BW_{LPF} [MHz]	LPF Order								
		ACI	AACI		Blockers					
		B	C	D	E	F	G	H	I	J
GSM	0.1	4.5	4.8	4.1	4.0	3.4	2.7	2.4	1.7	1.7
Bluetooth	0.5	3.3	3.3	3.1	1.5	2.2	–	–	–	
DECT	0.87	4.3	3.8	3.2	2.8	2.5	1.1	–	–	–
W-LAN 802.11b,g (DSSS)	5.5	–	–	2.3	2.1	–	–	–	–	–
W-LAN 802.11g (OFDM)	10	3.3	3.0	2.3	0.4	–	–	–	–	–

The relative interferer/blocker level after the LPF, $P_{blk,LPF}$, can be calculated as, [4]:

$$P_{blk,LPF} = P_{blk,BB} - 10\log\left[1 + (\Delta f_{blk} / BW_{LPF})^{2n_{LPF}}\right], \qquad (2)$$

where BW_{LPF} is the LPF bandwidth and is set to half the baseband channel bandwidth, as for every zero-IF receiver and n_{LPF} represents the LPF order.

Scaling the LPF bandwidth according to the targeted standards baseband channel bandwidths enables a multi-standard implementation (e. g., [5]).

As mentioned, for the complete analog channel selection, n_{LPF}, is chosen such as the interferer or blocker level after filtering is below the useful signal with SNR_0. This translates to the condition:

$$P_{blk,LPF} \leq -SNR_0 \qquad (3)$$

Since the blocker frequency is always larger than the filter bandwidth, the n_{LPF} expressed in (2) can be approximated to:

$$n_{LPF} \geq \frac{P_{blk,BB} + SNR_0}{20\log(\Delta f_{blk} / BW_{LPF})} \qquad (4)$$

Eq. (4) has been used to calculate the LPF order for the envisaged major commercial wireless standard. Table I comprises the required n_{LPF} to filter-out the blockers and interferers of the envisaged wireless standards.

Based on the calculated LPF order from Table I, it results the envisaged standards interferers and blockers, except GSM standard's B, C and D interferers and DECT ACI are completely filtered out by a 4th order LPF. Figure 2.b depicts the interferers/blockers levels after a 4th order LPF.

Hence, in order to get a complete analog channel in the targeted multi-standard environment a 5th order LPF is required.

IV. ASSESSING THE COMPLETE ANALOG CHANNEL SELECTION IMPACT ON THE RECEIVER POWER CONSUMPTION

The key trade-off that shapes the receiver baseband part design is the trade-off between the receiver's LF part area and its power consumption, [3, 4].

First, the low noise requirements of wireless applications translate to small values for the LPF resistors and thus large values for its capacitors.

Hence, the receiver LF part area, A_{LPF}, is mainly determined by the LPF capacitors. Basically, the larger n_{LPF}, the larger is the receiver area.

Of course, this will also increase the LPF power consumption. In [3] several multi-standard LPFs were compared (e. g., [5], [8]).

The low pass filters are implemented as a cascade of second order cells or bi-quads. The reported power consumption per bi-quad was in the range of 5 mW when the filters are set to operate in W-LAN conditions (i. e., the standard with the largest signal bandwidth). Table II lists the estimated I and Q channels LPFs power consumption. For odd LPF orders, the presented analysis does not account any power consumption increase with respect to the preceding even order LPF, as the extra pole is passive.

Second, the receiver power consumption is constraint by the ADC specifications of resolution and speed. Basically, for larger n_{LPF}, the blockers and interferers are knocked down more.

So, the ADC requirements are relaxed. First, the ADC resolution is going to be smaller as less of the ADC dynamic range is occupied by the residual un-filtered blockers and interferers. Second, a smaller f_S can be chosen, as blockers located far away from the wanted signal will be completely filtered out. But, the larger n_{LPF}, the larger is the receiver area.

This key trade-off between the receiver's LF part area, A_{LPF} determined by the LPF order, n_{LPF}, and its power consumption, constraint by the ADC number of bits, n, and sampling frequency, f_S, has been in-depth analysed in [4].

The ADC power consumption is estimated by, [10]:

$$P_{ADC} = FOM_{ADC} \cdot f_S \cdot 2^n, \qquad (5)$$

where FOM_{ADC} is the ADC energy consumption per conversion cycle and represents the figure of merit in evaluating the ADC power consumption.

TABLE II. ESTIMATED LPF AND ADC POWER CONSUMPTION

LPF order	LPF power consumption for WLAN [mW]	ADC power consumption [mW]			
		GSM	Bluetooth	DECT	W-LAN
0	0	1120	115	260	15000
1	0	40	7	5	550
2	10	2	2	2	65
3	10	0.1	0.3	0.3	6
4	20	0.03	0.07	0.08	5
5	20	0.006	0.07	0.08	5

Based on the plot from [10], it is fair to estimate that bulk of today's ADCs will exhibit a FOM_{ADC} closed to 1 pJ per conversion cycle.

Both n and f_S depend on n_{LPF}, as calculated in [4]. Table II notes also the estimated ADC power consumption versus the LPF order.

In the case of the complete analog channel selection, that corresponds to $n_{LPF} = 5$, the ADC has to handle only the wanted signal since all the blockers and interferers have been filtered-out by the LPF. Hence, the ADC dynamic range does not need to account any additional headroom for residual interferers/blockers after the LPF. Also, the ADC sampling rate is chosen twice the effective signal bandwidth, rather than larger than twice the frequency of the largest blocker to avoid destructive folding. Basically, the ADC specifications of resolution and speed are at the minimal level and thus, also its power consumption.

But, the LPF area is directly linked to the amount of integrated capacitance and, thus, to n_{LPF}. Given the low bandwidth required for GSM, the amount of integrated capacitance represents a large portion of the overall receiver area, [3, 4]. So, considering the low frequencies of the GSM and DECT interferers that are not completely filtered out by a 4th order LPF (see Fig. 2.b), the ADC resolution and speed may be increased accordingly. This is preferred to implementing a 5th order LPF at the expense of about 20% more area.

Nonetheless, given the very good power consumption of today's ADC (see Table II), the need for an analog LPF is almost relinquished.

V. CONCLUSION

The paper analyzed the impact of the analog channel selection on the receiver power consumption and area from the circuit / transistor level designer perspective. Based on the multi-standard receiver generic blockers diagram analysis, the complete analog channel selection strategy is defined. It foresees a first order LPF incorporated in the down-converter and a separated 5th order analog LPF following the mixer.

By incorporating a first order LPF in the down-converter mixer output stage, the out-of band interferers/blockers levels are reduced before the actual channel selection is performed in the analog LPF. Based on the presented analysis it resulted a 5th order LPF enables the complete analog channel selection for the targeted multi-standard environment. But, by trading off the ADC power consumption, the overall LPF area can be reduced by about 20% if a 4th order LPF is implemented.

Moreover, as can be noticed from the values listed in Table II, the estimated power consumption of today's ADC is very low. Thus, the need of a dedicated active analog LPF in the RF front-end is almost relinquished.

Finally, given the very low power consumption of today's ADCs and the high rate at which their power consuption is improved, [11], an implicit conclusion of this study is that on short or medium term ADC-based Software Defined Radio Receivers will become reality.

ACKNOWLEDGMENT

The authors would like to express their acknowledgment to Dr. Frank Op't Eynde for the fruitful discussions on the topic.

REFERENCES

[1] T. H. Lee, *The Design of CMOS Radio-Frequency Integrated Circuits*, Cambridge University Press, 2nd Ed., 2004, pp. 710-713.

[2] S. Spiridon et. al, "Smart gain partitioning for noise – linearity trade-off optimization in multi-standard radio receivers," *Proceedings of the 18th International Conference Mixed Design of Integrated Circuits and Systems*, MIXDES 2011, June 2011, pp. 466-469.

[3] S. Spiridon, *Analysis and Design of Monolithic CMOS Software Defined Radio Receivers*, PhD Thesis, Ed. Tehnică, 2011.

[4] S. Spiridon, C. Dan, M. Bodea, "Filter partitioning optimum strategy in homodyne multi-standard radio receivers," *Proceedings of the 7th Conference on Ph.D. Research in Microelectronics and Electronics*, PRIME 2011, July 2011, pp.9-13.

[5] S. Spiridon, F. Op't Eynde, "Low power CMOS fully differential programmable low pass filter," *Proceedings of the 10th International Conference on Optimization Of Electrical And Electronic Equipment*, OPTIM 2006, May 2006, pp. 21-25.

[6] A. Tarniceriu, B. Iordache, and S. Spiridon, "An Analysis on Digital Modulation Techniques for Software Defined Radio Applications," *Proceedings of the 30th Annual International Semiconductor Conference*, CAS 2007, October 2007, vol. 2, pp. 451-454.

[7] S. Spiridon et. al, "Deriving the Key Electrical Specifications for a Multi-standard Radio Receiver," *Proceedings Of the First International Conference on Advances in Cognitive Radio*, COCORA 2011, April 2011, pp. 60-63.

[8] S. D'Amico, V. Giannini, A. Baschirotto, "A 4th-Order Active-Gm-RC Reconfigurable (UMTS/WLAN) Filter," *Journal of Solid State Circuits*, vol. 41, no. 7, July 2006, pp. 1630-1637.

[9] K. Bult, "Embedded analog-to-digital converters," *Proceedings of ESSCIRC 2009*, pp. 68-73, September 2009.

[10] B. Murmann. (2012, Jan 15th). "ADC Performance Survey 1997–2011," [Online]. Available:

http://www.stanford.edu/~murmann/adcsurvey.html.

[11] B. Murmann, "A/D Converter Trends: Power dissipation, scaling and digitally assisted architectures," *Proceedings of the Custom Integrated Circuit Conference*, CICC 2008, September 2008, pp. 105-112.

MIXDES 2012, 19th International Conference *"Mixed Design of Integrated Circuits and Systems"*, May 24-26, 2012, Warsaw, Poland

Extracting the Parameters of an EEHEMT Nonlinear Model for InP HEMT Operating at G-band Frequency

Sina Eskanadri, Farzad Tavakkol Hamedani

Communication Department
Faculty of Electrical and Computer Engineering, Semnan University
Semnan, Iran
S_Eskandari@sun.semnan.ac.ir

Abstract—This paper describes and demonstrates procedure of extraction parameters for a 2-finger with 30μm total gate periphery InP HEMT device by utilizing EEHEMT nonlinear model which has been defined in Agilent-Advanced Design System (ADS) software. We have analyzed complicated equations according to measured DC characteristics of typical 0.07 μm InP HEMT devices and also obtained initial values for AC parameters; then we optimized and tuned these parameters in order to achieve desired S-parameters and noise performance to operate at G-band frequency. Accuracy of extracted model and its simulated results have been investigated by comparing with measured and theoretical characteristics in order to use for design and development of LNA MMICs and other front-end Millimeter wave applications.

Index Terms—InP HEMT; EEHEMT nonlinear model; parameters extraction; Monolithic Millimeter wave Integrated Circuits (MMICs)

I. INTRODUCTION

Today's Indium Phosphide high electron mobility transistors (InP HEMTs) have been applied widely in monolithic millimeter wave integrated circuits (MMICs). It is because of their high electron mobility and transconductance with desired noise performance. In attention to recent advances in InP technology to decrease gate length and then increase electron mobility and transconductance, sensitivity and resolution of radiometer front-ends and astronomy applications have been improved [1].

In order to design and develop the MMICs, an accurate model should be provided which truly shows the HEMT devices performance at different bias and temperature conditions. Hence in this study, Agilent empirical nonlinear EEHEMT model has been analyzed and its parameters were extracted based on measured DC, AC characteristics and S-parameters of several 0.07 μm InP HEMT devices. It is notable that any expensive extraction software as IC-CAP (Integrated Circuit Characterization and Analysis Program) have not been utilized to extract these parameters. The simulation results have been verified by comparing with measured data and also in the cases which there were no experimental evidences, we indicated data to be followed by theoretical rules.

II. INP HEMT TECHNOLOGY

InP-based pseudomorphic HEMTs have demonstrated excellent high-frequency and low-noise performance as 0.07μm T-gate InP HEMT process which provides high gain and cut-off frequency required for the G-band MMICs. The InP HEMT epitaxial layer structures were grown by Molecular Beam Epitaxy (MBE) at Northrop Grumman Space Technology on 3-inch semi-insulating InP substrates. In conjunction with 0.07 μm gate length, a 75% indium percentage channel is used to provide enhanced mobility to 12000 cm^2/Vs and channel electron carrier density to 3.5×10^{12} cm^{-2} at room temperature as compared with lower indium percentage InGaAs channels. The analysis indicates that increasing electron mobility improves minimum noise figure (F_{min}) because of increasing f_t and also reduces source resistance.

The DC characteristics for 22 InP HEMT devices measured on wafer at 1V drain bias show the peak G_m of about 1500 mS/mm and the cut-off frequency (f_t) more than 250 GHz. The device breakdown voltage typically is above 2.5 V with a maximum drain source voltage of 2 V, likewise high maximum drain currents greater than 600 mA/mm have been achieved with careful epitaxial design to ensure good gate control and pinch-off through proper scaling with 0.07 μm gate [1–3].

III. INP HEMT NONLINEAR MODEL

Agilent EEHEMT is an empirical and nonlinear model for general MMIC HEMT applications, including a set of equivalent device elements when they are evaluated, predicts device performance. The model equations were developed concurrently with parameter extraction techniques to ensure the model would contain only parameters that were extractable from measured data. Although the model is amenable to automated parameter extraction techniques, it was designed to consist of parameters that are easily estimated (visually) from measured data such as G_m-V_{gs} plots [4]. The EEHEMT model parameters could be divided into two categories: the first category comprises DC parameters, namely drain source current parameters, G_m compression parameters, and gate forward conduction and breakdown parameters; the second category comprises AC parameters, namely dispersion current parameters and charge parameters.

The crucial parameters could be extracted for a specified HEMT from DC characteristic of transconductance (G_m) compression versus gate to source voltage (V_{gs}). It has been defined for Agilent EEHEMT nonlinear model (Fig. 1) in Advanced Design system (ADS) simulation software. To obtain the parameters which are illustrated at Fig. 1 for a 2-finger with 30 μm total gate periphery of InP HEMT, we compared it with G_m-V_{gs} characteristics which have been measured for 22 InP HEMT devices [1,2].

As it can be seen clearly, the measured transconductance curves start rising from about V_{gs}= -0.3 V and get to their maximum approximately 1500 mS/mm at V_{gs}= 0.2 V and then after around 0.05 V, they start to fall; so for achieving good fitting curves, we set V_{to} equal -0.32 V, V_{go}= 0.23 V, V_{co}= 0.27 V and GMMAX= 41 mS. However the G_m vs V_{GS} characteristics reported in [2] do not give information about Gm behavior above Vgs = 0.4V. So we have decided to estimate the remaining DC parameters based on equations and curves given for HEMTs elsewhere [5]. The following parameters have been extracted in this way: Alpha= 0.4, DELTGM= 0.18, V_{ba}= 0.82 V and V_{bc}= 0.37 V.

The isothermal output conductance is controlled by parameters such as Gamma and Kapa of the EEHEMT model. Since the value of Gamma controls threshold parameters as a function of V_{ds}, it plays an important role in determining maximum of transconductance compression and also its fall off for $V_{gs} > V_{go}$. G_m was swept versus V_{ds} according to equation 1, for different values of Gamma from 0 to 3V. As it shown in Fig. 2, by estimating Gamma= 0.075, the simulated transconductance and drain current truly would be fitted with measured data.

$$g_m = GMMAX\{1 + GAMMA(V_{dso} - V_{ds})\} \qquad (1)$$

The G_m characteristic and drain current versus gate to source voltage for extracted DC parameters are shown in Fig. 3. Matching with the measurement results for several typical 0.07 μm InP HEMT devices [1,2] is good. Moreover the general Gm curve shape (Fig.1) is followed. Similarly, it shows a peak Gm of about 1500 mS/mm and 360 mA/mm drain current at 0.2 V_{gs}. The important points are indicated by some arrows in Fig. 3.

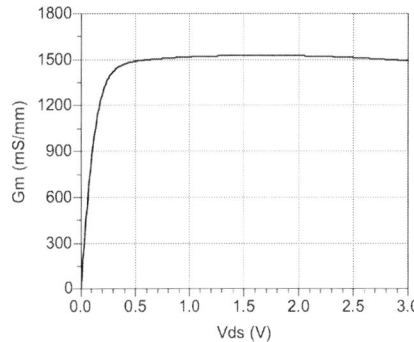

Figure 2. Trace of transconductance (G_m) versus drain voltage (V_{ds}) for Gamma= 0.075 at V_{gs}= 0.23 V to achieve the best fitting for maximum of transconductance compression.

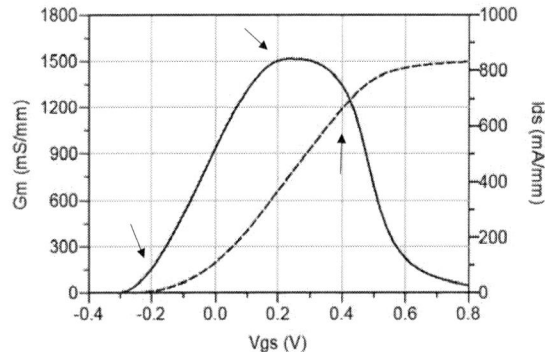

Figure 3. Drain current (I_{ds}, dash line) and transconductance (G_m, solid line) vs. gate voltage (V_{gs}) for 2f30 μm EEHEMT extracted model at 1V drain bias.

The DC modeling section of the EEHEMT model includes four different equations in four different regions of V_{gs} to map drain current of a transistor [4], and provides an excellent match to the characteristics of the device under test and also simulation results; therefore DC I-V curves are shown in Fig. 4, truly indicate drain to source currents versus drain to source voltages (V_{ds}) at various gate to source voltages, likewise low self-heating effects are appreciable in slope of curves as it was expected from typical HEMT devices [1,6].

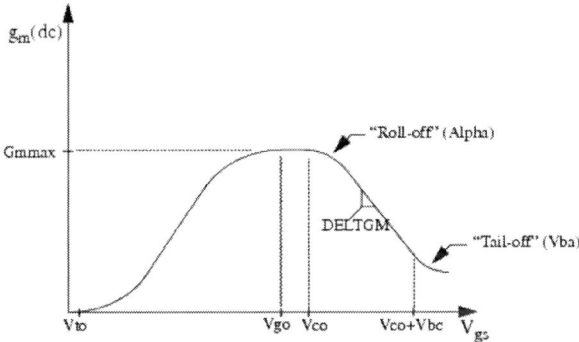

Figure 1. Transconductance compression (G_m) of EEHEMT model versus gate to source voltage (V_{gs}) at V_{ds}= V_{dso} [4].

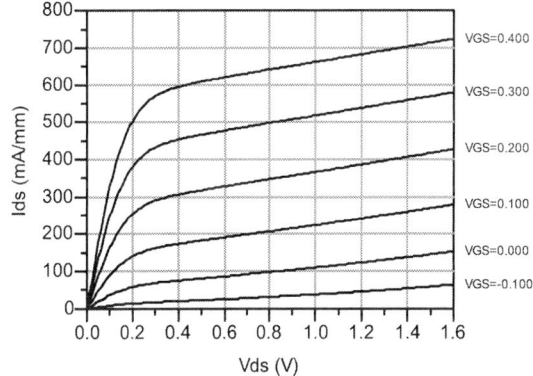

Figure 4. The simulated drain current (I_{ds}) versus drain voltage (V_{ds}) at various gate voltages (V_{gs}) for 2f30 μm EEHEMT nonlinear model.

The first step after DC characteristics of the InP HEMT device were carried out, is investigating dispersion effects based on EEHEMT equivalent circuit used to model dispersion consists of resistor R_{db}, capacitance C_{bs}, and nonlinear current source I_{db}. Dispersion effects are defined for DC property changes of HEMT devices at high frequency stimulation and can be attributed to the thermal and electron trapping. Under rapid signal variations, the traps cannot follow the applied signal and dispersion effect happens. We have selected default values of ADS-EEHEMT model for R_{db} and C_{bs} because they could make time constant ($\tau_{disp} = R_{db} \times C_{bs}$) large enough as it need for operating at G-band frequency range and also reduce contribution of R_{db} to the output conductance. Since G_m characteristic is fitted accurately with pervious section, the parameters G_{dbm}, K_{db} and V_{dsm} can be tuned to optimize the g_{ds} fit and then I_{db} would be obtainable by analyzing of (2)–(4) [4,7].

$$I_{ds} = I_{ds}(DC) + I_{db} \tag{2}$$

$$I_{db} = I_{ds}(AC) - I_{ds}(DC) + I_{dbp} \tag{3}$$

$$I_{dbp} = \sqrt{\frac{G_{dbm}}{K_{db}}} \tan^{-1}((V_{ds} - V_{dsm})\sqrt{K_{db}G_{dbm}})$$
$$+ G_{dbm}V_{dsm} \tag{4}$$

The charge model has been defined based on gate and drain of the device for EEHEMT model and contains a closed form expression which fits bias with its derivatives separated through node charges. The main gate charge parameters such as C_{11o}, C_{11th}, V_{infl}, and D_{eltgs} were extracted by tuning and optimization of S-parameters at V_{dso} drain bias because When $V_{ds} = V_{dso}$ and $V_{dso} \gg D_{eltds}$, the gate capacitance C_{11} reduces to a single voltage dependency in V_{gs} [4].

The parameters that have been described in procedure of extraction so far, play important role in characteristics and operation of InP HEMT devices. By adding other parameters of EEHEMT model which some of them set are equal with their typical values (as breakdown parameters [2]) and some of them are obtained by tuning and optimizing based on their initial values from different references, we have depicted fundamental curves and validated them with comparing measured data in next part.

IV. PARAMETER EXTRACTION VERIFICATION

The extracted nonlinear EEHEMT model was investigated by implementation of different simulation tests in ADS. Now some of fundamental curves would be illustrated to verify simulation results. It is noticeable that the results obtained for a 2-finger with 30 μm total gate periphery InP HEMT modeled, with 1 V drain bias and 360 mA/mm drain current at 0.2 V for gate voltage and did not use any bias and feedback networks.

Fig. 5 shows S-parameters of the EEHEMT from 1GHz to 220 GHz. Simulation results in Fig. 5a indicate a very good fitting for both magnitude and phase of S_{21} and S_{12} at the Whole frequency range with measured and simulated values have been done so far [6,8].

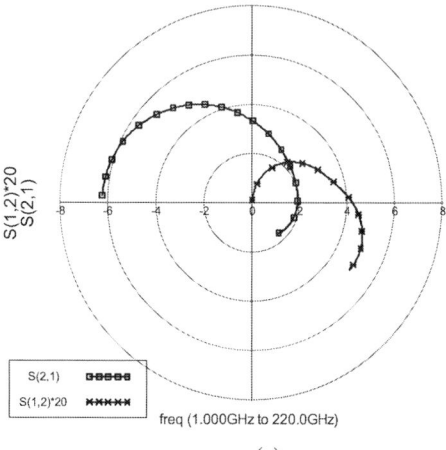

(a)

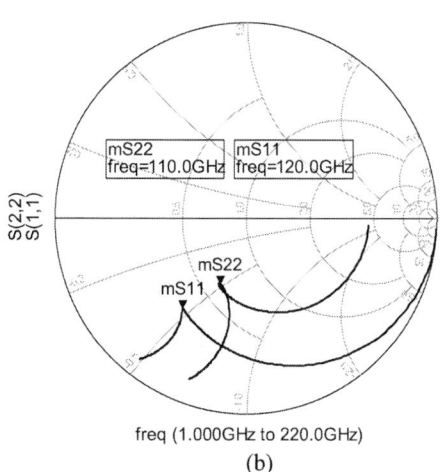

freq (1.000GHz to 220.0GHz)

(b)

Figure 5. The simulated S-parameters characteristics at V_{ds} = 1V and V_{gs} = 0.2V. (a) Variations of S_{21} and S_{12}, (b) Variations of S_{11} and S_{22}

However the changes for phase and magnitude of S_{11} and S_{22} (Fig. 5b) are according to the measured data at 1−110 GHz, but as it is seen clearly there is a different of phase changes at about 120 GHz and 110 GHz for S_{11} and S_{22} respectively. It is because of values which are selected for C_{12sat} and C_{gsat} to achieve desired operation for modeled HEMT at G-band frequency range.

H_{21} and maximum available gain (MAG) of this EEHEMT model are relatively smooth at 1−300 GHz and follow the theoretical slope of -20 dB/decade and -10 dB/decade, respectively. The simulated cut-off frequency (f_t) based on H_{21} for extracted model with 2-finger 30 μm total gate width shows greater than 310 GHz and also maximum oscillation frequency (f_{max}) is above 460 GHz as it is shown in Fig. 6 [9]. Here we can see about 2 dB difference between MAG of this model and the measured one in [9]; that is because of difference in transconductance of these two HEMTs. Likewise these similar results confirm the behavior of S_{21} and S_{12} at upper frequency range (110−220 GHz).

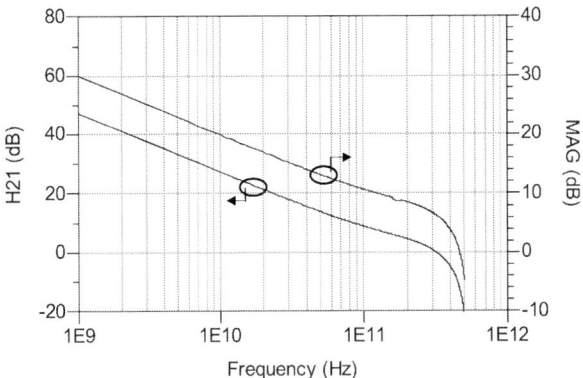

Figure 6. H21 and MAG values from simulated S-parameters for a 2f30 μm EEHEMT at a drain bias of 1 V and drain current of 360 mA/mm.

Noise performance of the extracted model is illustrated in Fig. 7, showing at low noise range as expected from HEMTs. Minimum noise figure has been obtained in simulation temperature of 290 K, according to IEEE standard and increases linearly with frequency; likewise it fulfills Pospieszalski's limitation given by the inequality :

$$1 \leq \frac{4R_{opt}}{(F_{min}-1)R_n} < 2 \qquad (5)$$

Where R_n is unnormalized noise resistance and R_{opt} is the real part of the Z_{opt} (optimum source impedance which corresponds to F_{min}) [10]. Since Pospieszalski's limitation is a necessary condition to predict practicality of noise simulation results based on the physics of the device, we expect the desired results would be accessible on wafer measurements.

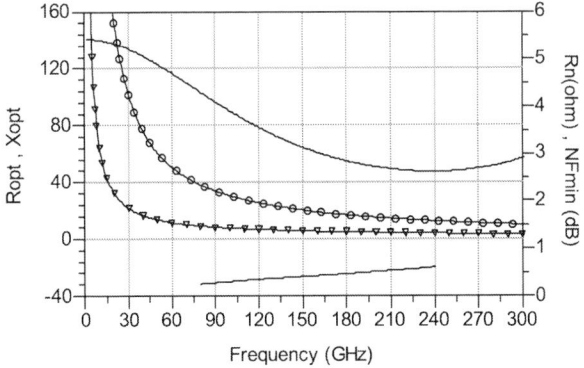

Figure 7. Noise performance of extracted model, minimum noise figure (solid line), noise resistance (dot line), real part of Z_{opt} (triangle line), image part of Z_{opt} (circle line).

V. CONCLUSIONS

In this paper, the parameters of Agilent EEHEMT model have been investigated and extracted for a 2-finger with 30 μm total gate periphery InP HEMT with 0.07 μm gate length. Since performance accuracy of DC and AC extracted parameters verified by comparing with measured curves and simulated data and also the model fulfilled theoretical rules and conditions like Pospieszalski's limitation for noise accessible values base on physics of devices, it shows good fitting behavior of the 0.07 μm InP HEMT devices and we predict that it would be a reliable model for using to design and improve radiometer front-end MMICs and astronomy applications.

REFERENCES

[1] R. Grundbacher, R. Raja, R. Lai, Y.C. Chou, M. Nishimoto, T. Gaier, D. Dawson, P.H. Liu, M. Barsky, and A. Oki, "A 150−215 GHz InP HEMT Low Noise Amplifier with 12 dB Gain," 2005 Indium Phosphide and Related Materials, pp. 613−616, 2005.

[2] P. Huang, R. Lai, R. Grundbacher, and B. Gorospe, "A 20 mW G-band monolithic driver amplifier using 0.07-μm InP HEMT," in IEEE MTT-S Int. Dig., San Francisco, CA, pp. 806–809, June 2006.

[3] Y. Ando, A. Cappy, K. Marubashi, K. Onda, H. Miyamoto, and M. Kuzuhara, "Noise parameter modeling for InP-based pseudomorphic HEMTs," IEEE Trans. Electron Devices, vol. 44, pp. 1367–1374, September 1997.

[4] Agilent- ADS EEHEMT Model Menu.

[5] Y.H. Chang, J. J. Chang, "Analysis of an EEHEMT Model for InP pHEMTs," IEEE Electron Devices Meeting, Tainan, pp. 237 – 240, December 2007.

[6] I. Angelov, H. Zirath, N. Rorsman, "A New Empirical Nonlinear Model for HEMT and MESFET Devices," IEEE Trans. Microwave Theory Tech., vol. 40, pp. 2258-2266, December 1992.

[7] M. Golio, M. Miller, G. Maracus, D. Johnson, "Frequency Dependent Electrical Characterisitics of GaAs MESFETs," IEEE Trans. Elec. Dev., vol. ED-37, pp. 1217-1227, May 1990.

[8] A. Orzati, D. Schreurs, L. Pergola, H. Benedickter, F Robin, O. J. Homan, and W. Bachtold, "A 110-GHz Large-Signal Lookup-Table Model for InP HEMTs Including Impact Ionization Effects," IEEE Trans. Microw. Theory Tech., vol. 51, NO. 2, pp. 468–474, February 2003.

[9] P. Kangaslahti, D. Pukala, T. Gaier, W. Deal, X. Mei, and R. Lai, "Low noise amplifier for 180 GHz frequency band," IEEE Int. Microwave Symp. Dig., pp. 451−454, June 2008.

[10] M. W. Pospiezalski, "Modeling of noise parameters of MESFET's and MODFET's and their frequency and temperature dependence," IEEE Trans. Microw. Theory Tech. , vol. 37, no. 9, pp. 1340–1350, September 1989.

MIXDES 2012, 19th International Conference *"Mixed Design of Integrated Circuits and Systems"*, May 24-26, 2012, Warsaw, Poland

Joint Simulation of Mixed-Signal Integrated Circuits and Printed Circuit Boards

Love Cederström, Achim Graupner
Zentrum Mikroelektronik Dresden AG
Dresden, Germany
{love.cederstroem, achim.graupner}@zmdi.com

Abstract—As system complexities and the market share of mixed signal products increase, design and test of integrated circuits (ICs) will become more difficult. This indicates that design of printed circuit boards (PCBs) for test must be integrated into the IC development flows. We propose software supported model translation to enable a method of joint simulation of a PCB and an IC in an integrated environment. The method employed can be shown to increase model coverage to more than 90 % for analog time domain models from some of the largest component suppliers. The benefits of this are demonstrated through analog simulations that reuse existing testbenches for an IC, uncovering potentially unstable transient effects arising from the combination of the IC and PCB.

Index Terms—joint simulation, circuit simulation, circuit boards, mixed analog digital integrated circuits, circuit testing, design methodology

I. INTRODUCTION

An issue of great importance for almost any company is how to efficiently test functionality and performance of a manufactured product. For the semiconductor industry, this is particularly true since a large portion of the end price can be directly traced back to how long a component must spend in testing [1]. Therefore, one important aspect of launching a new product is to successfully implement and efficiently utilize testing equipment.

The interface for the manufactured integrated circuit (IC) to the testing equipment is a load board, which is a printed circuit board (PCB). The load board can have a socket for packaged ICs or a test head for providing stimuli and doing measurements directly on the wafer before sawing and packaging. As complexity increases, and with an increasing percentage of complementary metal-oxide semiconductor designs with analog components, tool chains that support functional and performance verification are needed [2]; this also inflicts increasing demands on test boards. One piece of the puzzle would be to jointly simulate ICs with PCB circuitry, as they are being developed. To date, doing this in an integrated fashion is difficult.

With this paper, we will show how joint simulation of a PCB and a mixed-signal IC can be accomplished, starting with an overview of the issues for joint simulation in section II. Section III delves deeper into a practical approach. Section IV demonstrates the proposed method on a design example and section V summarizes with concluding remarks.

II. BACKGROUND

Load boards are often designed for multi-site testing where several mixed-signal ICs are tested in parallel. This leads to complex PCBs with as many as 20 metal layers and additional modules (illustrated in Fig. 1). The first step towards constructing a working PCB for test is knowledge and information transfer. The IC design engineers must transfer knowledge about how the IC works to the PCB design engineers. In many cases, multiple sets of schematics evolve for simulation and layout connectivity checks in different software environments. When using existing well-documented ICs and discrete devices as the active ingredients, this is a plausible approach. However, until silicon is available, the IC cannot be characterized with scrutiny, and the real behavior can be uncertain. It is under these conditions that we recognize the importance of well-integrated software environments, where the IC and PCB can be simulated together.

The widespread simulation of ICs began with the "Simulation Program with Integrated Circuit Emphasis" (SPICE) from UC Berkeley in the early 70's [3]. Since then a multitude of simulators and SPICE-dialects have seen the light. To bring the PCB and the IC worlds together and make joint simulation feasible in a seamless way, there are several plausible approaches. Our main thesis is that any simulation incorporating an IC (prototype under development) needs tools suited for simulating the IC. The circuits on the test PCB are in general less complex (i.e. defined by less detailed macro models) and have lower device count compared to ICs. As such, a PCB tool is not very likely to support advanced mixed-mode large-

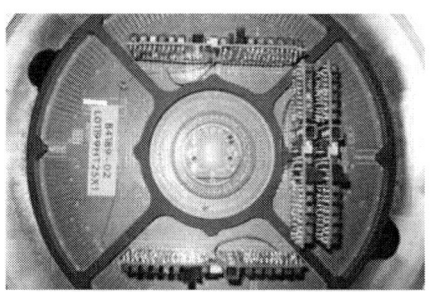

Fig. 1. Load board with VI probe and attached modules containing external analog circuitry for adapting tester signals.

scale simulations (i.e., mixing Verilog with a SPICE dialect) to the same extent as an IC design tool, so it is appropriate to perform simulations in an environment native to the IC design.

We have investigated how many models are available from several IC and discrete device manufacturers [4]–[16], as well as semiconductor foundries [17]–[22]. Our survey (Fig. 2) shows that the format availability differs between manufacturers and foundries. For semiconductor process design kits (PDKs), the model formats usually are available for IC-oriented simulators like Spectre, HSPICE or Eldo with roughly an equal distribution (Fig. 2a). Fig. 2b shows the number of models that can be directly acquired for commercially available ICs and discrete devices and how they are distributed over the different model formats. Within the total number of models, one can distinguish between digital-type models that describe input/output performance and analog-type models that describe the behavior of the device or circuit. Among digital-type models, the input/output buffer information specification (IBIS) and HSPICE are predominant. With the focus on PCBs for testing of mixed-signal ICs, analog-type models are of particular interest. Here regular SPICE and the proprietary versions PSpice and LTspice are most common (Fig. 2c). LTspice which is syntax-compatible with PSpice, is provided by Linear Technologies. A closer inspection of International Rectifier's Saber (7 %) [8] and Vishay's HSPICE (6 %) [15] shows that these models have counterparts in SPICE or PSpice format. This means that handling regular SPICE and PSpice would give more than 90 % model coverage from some of the largest suppliers of analog components, making the two the *de-facto* standard.

The diversity of simulators and models is the largest obstacle to achieving joint simulation. In this context one must bear in mind that there is a difference between a model and its format and the simulator, which possibly can handle model formats that are not of its native dialect. None of the three simulators with the best PDK model availability (Fig. 2a) support PSpice, as seen in TABLE I. There are two viable solutions to the model discrepancies: translation or using a simulator that handles models of all kinds. Translation can be done if care is taken, though such an approach is time consuming. To our knowledge, there are two applicable software tools available for handling both common PDK models and SPICE/PSpice: Agilent Advance Design System (ADS) [23] and Silvaco SmartSpice [24]. ADS is oriented towards radio frequency (RF) and brings its own user environment, while SmartSpice

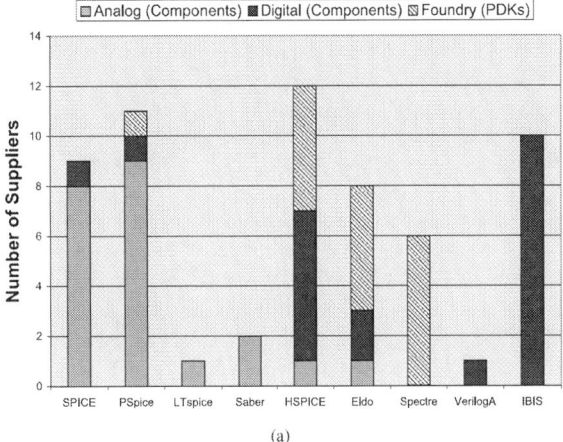

(a)

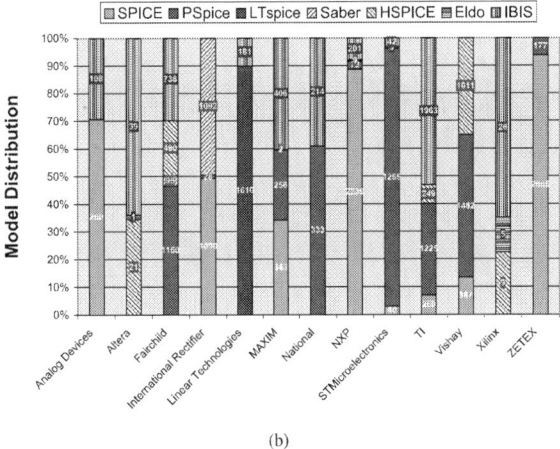

(b)

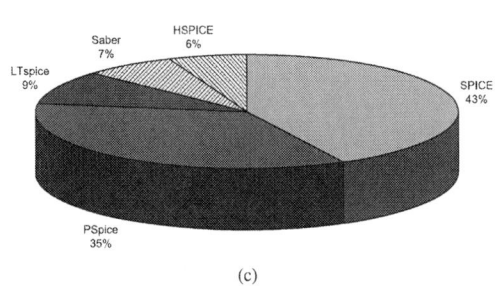

(c)

Fig. 2. The availability of models differs significantly depending on whether foundry PDKs or device manufacturers are considered (a), and there is also a large diversity for models of ICs and discrete devices (b). For analog ICs and discrete devices, regular SPICE and PSpice are dominant (c).

is a simulator aimed at large analog designs that can be run with command line entries. Given that integration is our goal, another user interface would be counterproductive, making SmartSpice more attractive. The problems with SmartSpice are that it does not have mixed-mode functionality, which would be preferable, and since it is a third party simulator, software

TABLE I
IC SIMULATOR MODEL SUPPORT

model:	simulator: HSPICE	Eldo	Spectre	SmartSpice
HSPICE	✓	✓	✓	✓
Eldo		✓		✓
Spectre		✓	✓	✓
SPICE	✓	✓	✓	✓
PSpice				✓

license costs would increase. This leaves the option of doing a model conversion from PSpice to one of the simulators that are strongest in IC design. As Spectre is part of one of the most prominent IC design suites for full custom analog-mixed signal design, has strong support of PDKs among foundries, and has support for both HSPICE and regular SPICE syntax, this is the recommended simulator.

III. Model Translation

The Spectre simulator from Cadence Design Systems (CDS) is aimed at analog IC simulations and can handle several SPICE dialects, but there is no compatibility between Spectre and PSpice [25], even though both Spectre and PSpice have evolved from SPICE. The first obvious incompatibility is the difference in syntax, and the second concern is model parameter discrepancies. One example where completely different syntax is used is the behavioral sources. In PSpice, E and G sources followed by the key word VALUE are used, while Spectre has arbitrary instance names and uses a bsource, as shown in TABLE II.

Next to the syntax differences, which are relatively straight forward to handle, the second, and more troublesome differ-

TABLE II
NONLINEAR BEHAVIORAL SOURCES

```
PSpice:
E<name> <out_p> <out_n> VALUE <expression>
G<name> <out_p> <out_n> VALUE <expression>
```

```
Spectre:
<name> (<out_p> <out_n>) bsource v=<expression>
<name> (<out_p> <out_n>) bsource i=<expression>
```

TABLE III
MOSFET MODEL PARAMETERS

Parameter		Default Value	
PSpice	Spectre	PSpice	Spectre
DELTA	delta	0	0
ETA	eta	0	0
GAMMA	gamma	*derived*	*0*
KP	kp	$2.0 \cdot 10^{-5}$	$2.0718 \cdot 10^{-5}$
KAPPA	kappa	0.2	0.2
LAMBDA	lambda	0.0	0.0
LD	ld	0.0	0.0
NEFF	neff	1.0	1.0
NFS	nfs	0.0	0.0
NSS	nss	*none*	*0.0*
NSUB	nsub	*none*	$1.13 \cdot 10^{16}$
PHI	phi	*0.6*	*0.7*
THETA	theta	0.0	0.0
TOX	tox	$1.0 \cdot 10^{-7}$	$1.0 \cdot 10^{-7}$
TPG	tpg	+1	+1
UCRIT	ucrit	*$1.0 \cdot 10^{4}$*	*0.0*
UEXP	uexp	0.0	0.0
UTRA	utra	0.0	0.0
UO	uo	600	600
VMAX	vmax	*0*	*infinity*
VTO	vto	0.0	0.0
WD	wd	0.0	0.0
XJ	xj	0	0
XQC	xqc	*1.0*	*0.0*

ence, is the use of model parameters. Even though both Spectre and PSpice implement models like MOS level 3 [26] and BSIM3v3 [27], characteristics like parameters' default values differ. One example is the MOS models where the default values of more than ten parameters differ; an excerpt of the parameters can be seen in TABLE III [28], [29]. The issue is that models might use parameters implicitly. That is, the model card may not express that a parameter should have a certain value, but relies on the default value; this means that a translation explicitly needs to state any discrepant parameters that are not present in the PSpice model.

To handle the translation from PSpice to Spectre, we have initiated an effort to develop a tool that reads PSpice netlists and models and gives Spectre-compatible syntax as an output. This has been done with Python, centered around the use of regular expressions. The tool reads the PSpice netlist at hand and performs the translation based on the input of rule files. The rule files contain the regular expressions to match; e.g., E and G sources followed by VALUE. The lines following the regular expression contain that which should be written to the new Spectre compatible netlist:

```
(^\s*E\w*)\s+(\w+\s+\w+)\s+VALUE\s?=\s*\{(.*)\}
+strbeg
\1 (\2) bsource v=\3
+strend
```

```
(^\s*G\w*)\s+(\w+\s+\w+)\s+VALUE\s?=\s*\{(.*)\}
+strbeg
\1 (\2) bsource i=\3
+strend
```

Another example of a rule file is for the model parameters, where columns separated with space describe parameters:

```
^\s*model\s+\w+\s+(mos|bsim|ekv)
+parbeg
L      l       100E-6    3E-6   #Default value
MJ     mj      0.5       1/2
MJSW   mjsw    0.33      1/3
N      n       1         1
RB     -       0         -      #Unsupported
RD     rd      0         0
RDS    -       infinite  -      #Unsupported
RG     -       0         -      #Unsupported
RS     rs      0         0
RSH    rsh     0         0
TT     -       0         -      #Unsupported
W      w       100E-6    3E-6   #Default value
+parend
```

The first line is a regular expression matching some MOSFET models. Following this are the parameters, where the first two columns would be the parameter names and the second two are the corresponding default values. The fifth column can contain a comment that will be used for reporting a command line warning if the corresponding parameter has been used.

This allows the tool to support model and netlist translation, making the overhead of bringing the whole PCB circuit schematic from the PSpice netlist into a Spectre simulation smaller. In this way, the PCB schematic can be entered into the IC design world and simulated.

IV. DESIGN EXAMPLE

The full-custom mixed-signal IC design suite provided by CDS is called Virtuoso. The Virtuoso schematics editor (VSE) is powerful and has extensive support for simulations through the Analog Design Environment (ADE). With an extension for RF system in package design (SiP), connectivity information transfer is possible from the VSE to the Allegro PCB design suite [30]. This methodology is state-of-the-art next to other tool chains for IC-package-PCB development such as Agilent's ADS. Because of CDS Virtuoso's strong support for mixed-signal design and because the PCB design software can be integrated using the methodology for SiPs, it is suitable as the front-end in our simulation flow.

To demonstrate our method, an example design with selected parts of the load board together with the relevant parts of an IC has been simulated. The application is a sensor interface IC with sensitive analog parts that must endure both high and low operating temperatures, electrostatic interference and a wide range of supply voltages (see Fig. 3). As with many sensor applications, one of the most critical parts is amplifying and converting a weak signal. In this specific case a configurable chain of variable gain amplifiers (VGAs) is responsible for this amplification. When testing the IC, the problem arises that the tester cannot generate low enough voltage levels for the measurements to be valid and the signal needs to be preconditioned. Fig. 4 shows how this can be done with an external operational amplifier, together with a few resistors, constituting an attenuator. The simulation was set up in Virtuoso using translated models (PSpice to Spectre), and the PCB circuitry could be simulated with available device models.

In this design example, simulation results showed that parasitic capacitors in sensitive nodes of the attenuator must be considered during PCB layout. Interestingly enough, large parasitics induced ringing at bidirectional analog pins when switching on the high-gain VGA chain. The manually inserted capacitors encircled in Fig. 4 represent these parasitics. For

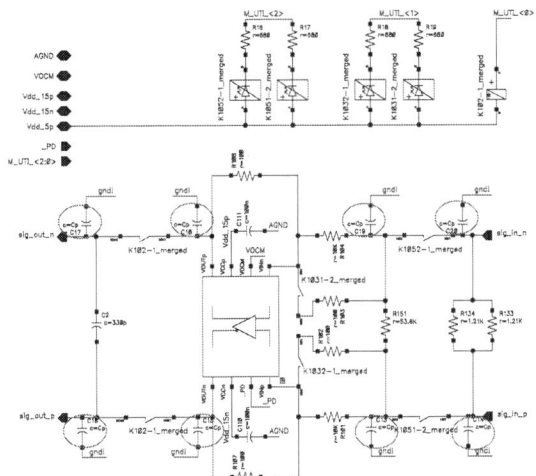

Fig. 4. Circuit schematic of the attenuation circuit on the PCB with inserted capacitors representing parasitics encircled. Using the VSE proved useful as the PCB could be modified and parameter sweeps could be performed within the same environment for both IC and PCB.

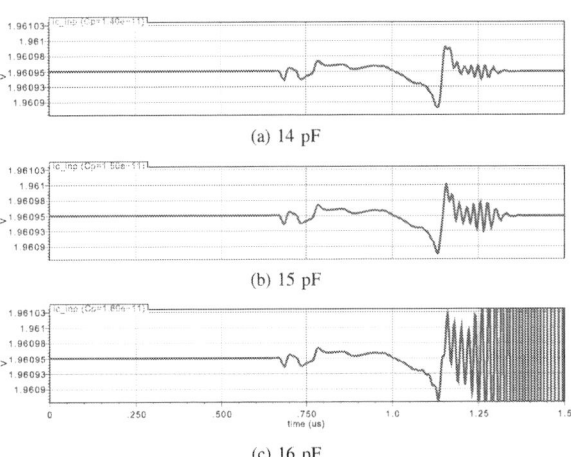

(a) 14 pF

(b) 15 pF

(c) 16 pF

Fig. 5. As capacitive values of certain nodes in the PCB were gradually increased, a tendency for ringing could be seen in the interface between the PCB and IC when the VGA chain was enabled. At lower values ($\lesssim 15$ pF) these died out, but at higher values ($\gtrsim 15$ pF), oscillation was maintained.

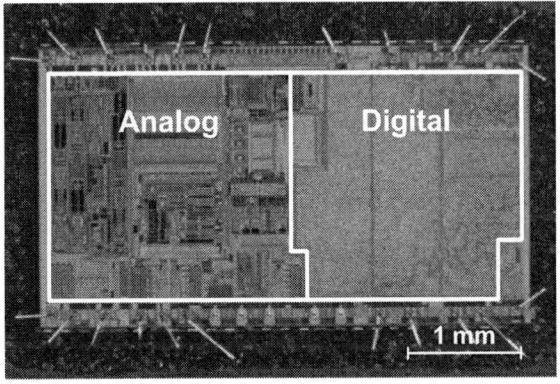

Fig. 3. Micrograph of the example mixed-signal sensor interface IC, the large feature size deals with process variation and operation under harsh conditions; e.g., at high temperatures.

lower values, a normal power on transient behavior was exhibited, but for higher values, the circuit started oscillating, as seen in Fig. 5. The potential oscillation was strong enough to manifest itself as a modulation on the IC output pins during simulation and could cause problems during test.

By using the same schematic entry tool for the PCB circuitry, we could reuse and extend testbenches developed during the IC design phase, as shown in Fig. 6. Furthermore, it allowed for an interactive circuit tweaking and re-simulation work flow; e.g., supported by the parametric sweep functionality of the ADE. With the support of the RF SiP methodology, the entered schematic could in the future not only be used for simulation but also as design entry for board layout.

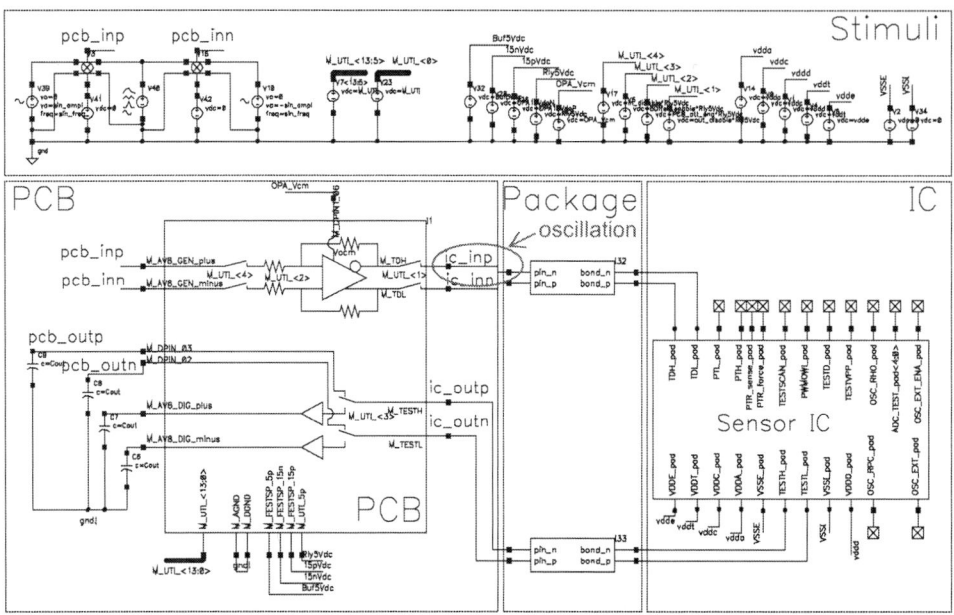

Fig. 6.   Testbench in VSE for both the PCB and the IC, which was based on reuse of a testbench for only the IC. With this setup, the full signal path can be simulated and the PCB symbol can point to either a schematic or a translated PSpice netlist. The oscillations observed during power-on (Fig. 5) occurred at the marked interface.

V. CONCLUSION

We have shown that a design environment enabling simulations of how a test PCB and IC affect each other can be beneficial. Our method requires custom model translation or use of a third-party analog simulator, where custom translation can be time consuming, but less costly in terms of licensing. The greatest advantage of model translation would be that existing simulation flows can be used; i.e., mixed-mode simulations, which no other integrated solution could manage. Therefore, we have implemented an automated support tool for translating models.

With the focus on an IC under development, an existing design flow could be expanded with PSpice translation capabilities, giving the Spectre simulator a largely increased coverage of manufacturer-supplied analog time domain models. To demonstrate and test the method, we set up a simulation environment, with which we successfully simulated the analog test signal path of a mixed-signal IC. With the enhanced simulation capabilities, the implemented flow enabled a higher level of software integration. This resulted in better communication between different stages of design, giving test engineers and PCB designers a better understanding of the IC.

ACKNOWLEDGMENT

The research leading to these results has received funding from the European Community's Seventh Framework Programme under grant agreement no. 237955 (FACETS-ITN). In addition, the authors would like to acknowledge test engineer Karlheinz Hübner, PCB layout engineer Ullrich Seltmann, and the IC design team under Dr. Ayman Ghazi, of ZMD AG.

REFERENCES

[1] J. Hu et al., "An industry-driven laboratory development for mixed-signal IC test education," in *Circuits and Systems (ISCAS), Proceedings of 2010 IEEE International Symposium on*, May 30 - June 2 2010, pp. 85–88.

[2] G. Gielen, "Design methodologies and tools for circuit design in CMOS nanometer technologies," in *Proceedings of the 32nd European Solid State Circuit Conference*. IEEE, September 2006, pp. 21–32.

[3] L. W. Nagel and D. Pederson, "SPICE (Simulation Program with Integrated Circuit Emphasis)," EECS Department, University of California, Berkeley, Tech. Rep. UCB/ERL M382, Apr 1973. [Online]. Available: http://www.eecs.berkeley.edu/Pubs/TechRpts/1973/22871.html

[4] Analog Devices Inc., model library as of Jan. 2012. [Online]. Available: http://www.analog.com/en/tools-software-simulation-models/resources/index.html

[5] Altera Corporation, model library as of Jan. 2012. [Online]. Available: http://www.altera.com/download/board-layout-test/hspice/hsp-index.html

[6] Diodes Inc., model library as of Jan. 2012. [Online]. Available: http://www.diodes.com/products/spicemodels/index.php#search

[7] Fairchild Semiconductor Corporation, model library as of Jan. 2012. [Online]. Available: http://www.fairchildsemi.com/models/

[8] International Rectifier, model library as of Jan. 2012. [Online]. Available: http://www.irf.com/product-info/models/

[9] Linear Technologies Corporation, model library as of Jan. 2012. [Online]. Available: http://www.linear.com/designtools/software/

[10] Maxim Integrated Products Inc., model library as of Jan. 2012. [Online]. Available: http://www.maxim-ic.com/tools/spice/

[11] National Semiconductor Corporation, model library as of Jan. 2012. [Online]. Available: http://www.national.com/en/software/

[12] NXP Semiconductors, model library as of Jan. 2012. [Online]. Available: http://www.nxp.com/models.html

[13] ST Microelectronics, model library as of Jan. 2012. [Online]. Available: http://www.st.com/internet/com/software/cae_models_and_symbols.jsp

[14] Texas Instruments Inc., model library as of Jan. 2012. [Online]. Available: http://focus.ti.com/adc/docs/midlevel.tsp?contentId=31690

[15] Vishay Intertechnology Inc., model library as of Jan. 2012. [Online]. Available: http://www.vishay.com/how/design-support-tools/

[16] Xilinx Inc., model library as of Jan. 2012. [Online]. Available: http://www.xilinx.com/support/download/index.htm

[17] "Mixed Signal/RF Technology," Taiwan Semiconductor Manufacturing Company Ltd., June 2010. [Online]. Available: http://www.tsmc.com/english/newsEvents/dc_brochure.htm

[18] "Tech files," Globalfoundries, January 2012. [Online]. Available: http://www.globalfoundries.com/design/tech_files.aspx

[19] "Advanced Technology," United Microelectronics Corporation, January 2012. [Online]. Available: http://www.umc.com/English/process/

[20] "Design kits," Samsung Electronics Co., Ltd, January 2011. [Online]. Available: http://www.samsung.com/global/business/semiconductor/products/strategicfoundry/Products_ProcessDesignKit.html

[21] "IBM Specialty Foundry Selection Guide," International Business Machines Corporation, January 2011. [Online]. Available: http://public.dhe.ibm.com/common/ssi/ecm/en/tgb03009usen/TGB03009USEN.PDF

[22] "0.18 um CMOS Process Family," X-FAB Semiconductor Foundries AG, July 2011. [Online]. Available: http://www.xfab.com/en/technology/cmos.html

[23] *Netlist Translator for SPICE and Spectre*, 2009th ed., Agilent Technologies, 2009.

[24] *SmartSpice User's Manual*, Silvaco Inc., February 2011.

[25] *Virtuoso® Spectre® Circuit Simulator User Guide*, 7th ed., Cadence Design Systems, December 2009.

[26] A. Vladimirescu and S. Liu, "The Simulation of MOS Integrated Circuits Using SPICE2," EECS Department, University of California, Berkeley, Tech. Rep. UCB/ERL M80/7, 1980. [Online]. Available: http://www.eecs.berkeley.edu/Pubs/TechRpts/1980/9610.html

[27] W. Liu, X. Jin, J. Chen, M.-C. Jeng, Z. Liu, Y. Cheng, K. Chen, M. Chan, K. Hui, J. Huang, R. Tu, P. Ko, and C. Hu, "BSIM 3v3.2 MOSFET Model Users' Manual," EECS Department, University of California, Berkeley, Tech. Rep. UCB/ERL M98/51, 1998. [Online]. Available: http://www.eecs.berkeley.edu/Pubs/TechRpts/1998/3486.html

[28] *PSpice A/D Reference Guide*, 16th ed., Cadence Design Systems, December 2009.

[29] *Virtuoso® Simulator Circuit Components and Device Models Manual*, 7th ed., Cadence Design Systems, May 2010.

[30] *Cadence® RF SiP Methodology Kit User Guide*, 7th ed., Cadence Design Systems, March 2008.

© Cadence, Virtuoso, Spectre, PSpice, LTspice, HSPICE, Saber, Eldo and SmartSpice are all properties of their respective holders.

MIXDES 2012, 19th International Conference *"Mixed Design of Integrated Circuits and Systems"*, May 24-26, 2012, Warsaw, Poland

Methodology for Development of LVS and LPE Rule Sets Adapted for MEMS Processes

Angel Pashev
Microelectronic Technologies Department
Smartcom Bulgaria AD
Sofia, Bulgaria
angel_pashev@smartcom.bg

Ivan Uzunov
Dept. of Telecommunication Networks
Technical University of Sofia
Sofia, Bulgaria
iuzunov@tu-sofia.bg

Dobromir Gaydazhiev
Microelectronic Technologies Department
Smartcom Bulgaria AD
Sofia, Bulgaria
dobromir_gaydajiev@smartcom.bg

Diana Pukneva
Microelectronic Technologies Department
Smartcom Bulgaria AD
Sofia, Bulgaria
diana_pukneva@smartcom.bg

Emil Manolov
Faculty of Electronic Engineering and Technologies
Technical University of Sofia
Sofia, Bulgaria
edm@tu-sofia.bg

Abstract—**The paper describes a methodology for development of LVS and LPE rule sets adapted for microelectromechanical technologies. The method is applied on an existing multipurpose MEMS process. To evaluate the accuracy and performance of the proposed method, parasitic extraction is performed on several different devices designed with the chosen MEMS technology. The LPE results are compared to reference results obtained through finite element analyses (FEA).**

Index Terms—**Electronic Design Automation (EDA), Microelectromechanical Systems (MEMS), simulation, modeling, Layout versus Schematic (LVS), Layout Parasitic Extraction (LPE)**

I. INTRODUCTION

The existence of accurate device models and appropriate automated technology rule checking is crucial in the design of microelectromechanical systems (MEMS) and for their microelectronic counterparts. Their use in the design phase for predicting the system behavior increases the chances of the designed system to work properly after the first design cycle, sparing in this way the time, engineering efforts and financial resources that would be spend in consecutive design cycles. In the MEMS design process, often finite element modeling (FEM) tools are used to model the behavior of the mechanical and electro-mechanical devices [1],[2]. Those tools can be very accurate, but also require a vast amount of computational power. Especially in large systems with many degrees of freedom, the run times can reach tens of hours or even days. This renders iterative, parametric or statistical analyses

practically impossible. Another approach is to create behavioral models of the microelectromechanical modules based on FEM analyses or purely theoretical formulas, and use them to simulate the behavior of the system. Those models require much less computational power, but do not model all aspects of the devices' behavior and are usually accurate only within certain part of the operating region [3].

In the existing electronic design automation (EDA) systems, layout parasitic extraction (LPE) tools are commonly used to extract the parasitic passive components introduced in the circuit during the layout design stage. A LPE tool is used in conjunction with a layout versus schematic (LVS) tool to extract from the layout of the designed circuit a netlist containing not only the components intentionally placed by the designer, but also the parasitic resistances, capacitances and inductances formed in the place and route phase. In this article a methodology for developing a LVS/LPE rule set adapted for use with MEMS processes is presented. The approach can be used for evaluation of the parasitic components introduced in the layout design phase or for evaluation of the static capacitance and resistance of different MEMS sensors and actuators based on its layout. Section II of the article introduces the approach to resistance and capacitance extraction, with their specifics. Section III describes the structures that are used to validate the methodology. Section IV discusses the results from the tests. In Section V the results are summarized and some conclusions are inferred.

II. ADAPTATION OF LVS/LPE RULE SET FOR MEMS TECHNOLOGIES

The prospect of using LVS/LPE tools to evaluate the parasitic components and static behavior of

---

This work is sponsored by the National Science Fund of the Bulgarian Ministry of Education, Youth and Science; Contract DDVU 02/6 from 17.12.2010.

microelectromechanical systems is investigated in this article. Those tools are commonly used in the EDA systems for different microelectronic processes (CMOS, BiCMOS etc.). They have undergone years of development and are standard part of the modeling and verification of integrated circuits showing good accuracy and fast runtimes. However they are not commonly applied for MEMS systems mainly because those processes introduce certain problems that have to be solved in order to develop a LVS/LPE rule set. The microelectromechanical processes often incorporate conformal conductors, layer anchoring and sacrificial dielectrics which usually do not exist in a standard CMOS process and are not commonly handled by the LPE tools. In the proposed methodology different measures in the LVS and LPE part of the rule sets are employed to work around those limitations.

The flow of operation is as follows:

- The layout of the design under test is provided in one of the industry standard formats (GDS or Oasis) as input information.

- LVS extraction is performed on the layout. Connectivity information is extracted and all necessary derived layers needed for predicting oxide removal, profile identification and net splitting are generated. The result from this step is output in a tool specific database.

- Layout parasitic extraction is performed using the database created by the LVS tool. The LPE tool extracts all passive components using a rule set that contains information for each conductive and dielectric layer for every profile of the technology. The output of this step is a standard Spectre or Hspice netlist containing the connectivity and components extracted during the LVS and LPE steps.

- The resulting netlist can directly be used in circuit simulation to investigate the behavior of the MEMS device or a bigger system in which the MEMS device is a building block.

This approach and its advantages over the currently used methods will be demonstrated using Mentor Graphics Calibre LVS and xRC [4],[5] rule sets developed for a general purpose MEMS process PolyMUMPs [6]. The chosen MEMS process is a mature technology allowing formation of wide variety of devices that also has all the "problematic" features. PolyMUMPs is a three-layer polysilicon surface and bulk micromachining process, with 2 sacrificial layers and one metal layer (Figure 1). The minimum feature size is 2µm. Devices that can be formed in PolyMUMPs include: Acoustics (microphones), Sensors, Accelerometers, Micro-fluidics and Display Technologies. Polysilicon (P0, P1 and P2) is used as structural material, deposited oxide (OX1 and OX2) is used to form the sacrificial layers, and silicon nitride is used as electrical isolation between the MEMS devices and the substrate. The process is designed to be as general as possible, and to be capable of supporting many different designs on a single silicon wafer [7].

PolyMUMPs features sacrificial layers, conformal conductors and anchoring of the structural layers, which

cannot be handled directly by the LPE tools. To work around those issues the option to define different profiles in Calibre xRC is used. An example of three profiles that are sufficient to describe a horizontal thermal actuator is shown in Figure 2. Profile 4 has the structural P1 layer anchored to P0 pads by the A1 anchor layer. This is modeling the area where the device is fixed to the substrate. The second profile represents an area where the arms do not have another conductor beneath them and the third profile represents the area where the arms have P0 plate beneath them that forms a sensing capacitor. The fact that there is an oxide removal step additionally increases the number of required profiles since the dielectric layers can be made of SiO_2 or air in different areas of the design.

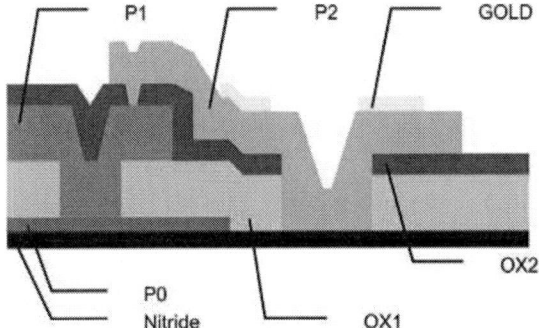

Figure 1. Cross-section of the PolyMUMPs layers: Nitride is an isolation layer separating the MEMS devices from the substrate. P0, P1 and P2 are polysilicon layers. OX1 and OX2 are sacrificial oxide layers [7].

Based on the allowed layer combinations and sacrificial oxide etching, all possible layer combinations (profiles) are defined in the LPE rule set. In the LVS part of the rule set, unique profile identification layers are generated to switch between the different profiles. Other important functions implemented in the LVS code include net naming, net splitting, virtual via generation and sacrificial dielectric removal prediction.

A corresponding net naming layer is defined for each conducting layer. Each net can be named by placing a label on the respective net naming layer over its conductive counterpart at the location where the net should originate.

Connectivity breaking layers are also defined for each conducting layer. They are used to break one net in two by placing a rectangle on the corresponding connectivity breaking layer that completely covers the width of a wire where the net has to be split. This rectangle is extracted as an ideal resistor with minimum resistance, thus breaking the network in two. An example list of P1 identification layers and their definitions for each profile are shown in Table I. Similar approach is followed for the other design layers.

Another model that is implemented in the rule set is the sacrificial oxide removal estimation. It is used to determine the type of dielectric that separates the different conductors. The type of dielectric depends on whether the sacrificial oxide layer has been removed, leaving air with relative dielectric permittivity (ε_r) of 1.0 or not - leaving SiO_2 with $\varepsilon_r=3.9$. There can be a structural polysilicon layer above each of the

371

sacrificial layers. In order to etch away the bottom sacrificial oxide the etching agent should first pass through the top polysilicon layer (P2) and the top sacrificial oxide (OX2). The estimation of the oxide removal is done by generating service layers (SIO and SIO2) for each sacrificial layer in the LVS rule set that mark the un-etched regions. Those layers are generated by inverting the expanded openings in the layout - etching holes and structure edges. The expansion is determined by the etching speed and the duration of the etching process. Both are defined as variables in the LVS rule set and can be changed independently for each run. The service layer SIO is used in profiles 3 and 5 to instruct the LPE tool that the first oxide has ε_r=3.9. In profiles 1 and 2 this layer is not present and the OX1 has the dielectric properties of air.

TABLE I.   P1 LAYER DEFINITIONS FOR DIFFERENT PROFILES.

Profile	LVS Layer	LVS Definition
1	P1_P0	((P1_met NOT ANCHOR1_DG_lvs) AND P0_met) NOT SIO
2	P1	((P1_met NOT P0_met) NOT ANCHOR1_DG_lvs) NOT SIO
3	P1_P0_SIO	((P1_met NOT ANCHOR1_DG_lvs) AND P0_met) AND SIO
4	P1_A1_P0	COPY P0_A1_P1
5	P1_SIO	((P1_met NOT P0_met) NOT ANCHOR1_DG_lvs) AND SIO

In the LPE tools, dielectric layers have to be defined between two neighboring conductors. In the fourth profile this rule is broken since P0 and P1 should be glued together. To avoid this violation a virtual VIA with minimal thickness and resistance is defined between the two layers in the LPE rule set. This virtual VIA layer is generated by the LVS rule set when P0, A1 and P1 layers overlap each other. This is actually part of the condition defining Profile 4.

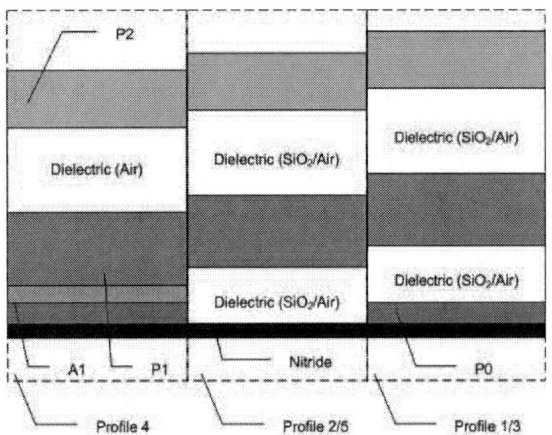

Figure 2.   Sample profile definitions in the LPE rule set.

The approach of using different profiles has a limitation that has to be taken into account –capacitances and resistances are not extracted between any two profiles. For the extraction of resistances this limitation is resolved by using virtual connecting of the respective conducting layers in the different profiles and net splitting. No solution is found for the capacitance extraction i.e. if there is P1 shape that does not have P0 below it (Profile 2) and it runs next to another P1 shape that is over P0 (Profile 1) – no capacitance will be extracted between those two shapes. Fortunately P1 has to be enclosed by P0 by not less than 4µm which leads to large separation between the shapes and decreased capacitance. Similar requirements are valid for the other conducting layers. They are driven by mask misalignment tolerances and are applicable for most processes, not only for PolyMUMPs. Based on those conclusions it is expected that the inaccuracy introduced by the limitation in capacitance extraction is not significant for most structures. To evaluate the accuracy of the proposed approach and especially the inter-profile capacitance significance, different test structures are used. The reference results are obtained by theoretical calculations or FEM simulations. The test structures and accuracy results are described in the next two sections.

III.   DESCRIPTION OF THE TEST STRUCTURES

Different types of plate capacitors, wiring stacks and a thermal actuator are used as testing vehicles to demonstrate the validity of the proposed approach and to illustrate the issues that MEMS processes introduce.

A. Resistance

The resistance extraction is examined using wire shapes on single or combination of conducting layers as the one shown in Figure 3. They represent the commonly used wiring configurations.

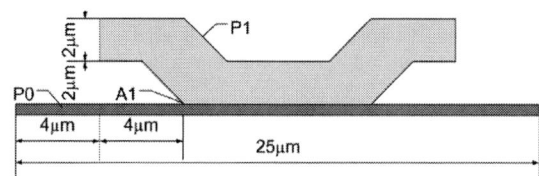

Figure 3.   Cross section of a P0-P1 wire. The A1 anchor layer is 9 µm wide, enclosed by 17 µm wide P1 wire. The P1 shape is enclosed by 25 µm wide P0 shape.

B. Capacitance

The sacrificial oxide removal model is tested with P0-P1 square plate capacitor that has a single etch hole in the middle (Figure 4). In PolyMUMPs this is a wet etch process. The etching agent dissolves the sacrificial oxide layers - OX1 and OX2, but does not dissolve any of the polysilicon layers. Etch holes should intentionally be placed on the overlying polysilicon layers to effectively remove OX2 or OX1 from certain areas of the design. The etching speed is about 30 µm/min. The usual duration of the process is between 0.5 and 2 min, which translates to oxide being removed within 15 to 60 µm radius from any Poly1 or Poly2 opening [8].

In the test structure the P1 layer is anchored to P0 along the perimeter of the plates so that no etching agent can enter from the sides of the device. The only opening is the hole in the middle from where the sacrificial oxide will be etched away and leave air filled cavity. The projected shape of that region is shown in Figure 4. Its area depends on the etching speed and the duration of the etching process. Often the designed structures require removal of the oxide from the entire device. To achieve this, enough etching holes have to be placed to cover the entire area of the device. If the polysilicon layer is not anchored to the layer below, etching agent can enter also from the sides of the device.

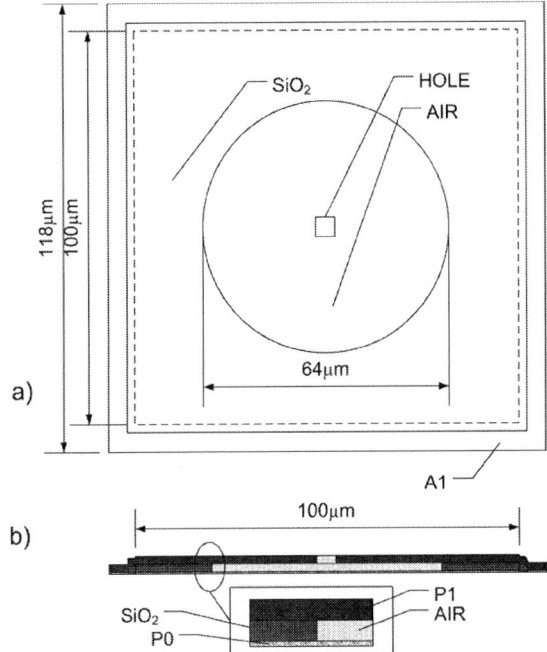

Figure 4. P0-P1 square capacitor with single etching hole in the middle: a) top view and b) cross-section. The bottom plate (P0) is a square shape with 100 μm long side.

Correct prediction of the dielectric etching is needed to ensure correct operation of the moving parts and to accurately estimate the capacitances in the circuit.

The structure from Figure 4 is used to test not only the oxide removal estimation but also capacitance extraction across different profiles. It uses profiles 1, 3, 4 and 5. To test the accuracy when only one profile is present similar square plate capacitors are used. They do not have etching holes and anchors and are 10μm by 10μm in size. Those structures are suitable to calculate the accuracy of a single profile structure and when compared to the structure from Figure 4 to estimate the magnitude of the error introduced by the missing inter-profile capacitances. Since the structures are relatively simple, both theoretical calculations, in which the fringe capacitance is neglected, and FEM tools accounting for the entire capacitance are used as reference results.

Lastly a thermal actuator design is used to verify both the resistance and capacitance extraction. Thermal actuators convert electrical potential difference into mechanical displacement. Potential difference applied on pads A and B generates electrical current which makes the material to heat and expand thermally, causing the tip of the actuator to deflect

[9]. A layout of a thermal actuator fabricated with the PolyMUMPs process and its cross section are shown in Figure 5. This actuator has an underlying plate on layer P0 that is aligned with the edge of the cold arm. It forms a capacitor with the thermal actuator. As the actuator deflects, the overlapping area and the capacitance decrease. In this way the amplitude of the deflection can be measured by measuring the capacitance between the thermal actuator and this plate. This device is a good example of the conformity of the P1 layer. It is anchored to P0 in the area of the pads and is spaced by 2 μm from P0 in the area of the cold and hot arms. There is a gap between the P0 shapes of the pads (A, B) and the plate (C). In that gap the P1 arms experience a drop of 0.5 μm. Between each of the regions described above there is a transition region whose width and cross-section depend on the material properties and layer thickness.

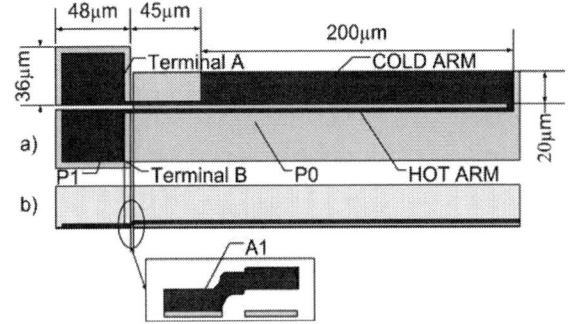

Figure 5. Vertical thermal actuator with underlying P0 plate. a) top view and b) cross-section.

IV. ACCURACY TEST RESULTS

A. Resistance testing

The resistance extraction is tested using four different conductor configurations – P0, P1, P2, P0-A1-P1. The conductor configurations for P0, P1 and P2 have simple geometry and the reference results are obtained with theoretical calculations using the well known formula:

$$R = \rho . \frac{1}{w.h}, \qquad (1)$$

where ρ is the material resistivity. The dimensions l, w and h are respectively the length, width and thickness of the resistor. The material sheet resistances (ρ/h) of the three polysilicon layers are given in Table II [6].

TABLE II.   SHEET RESISTANCE VALUES OF THE POLYSILICON LAYERS.

Layer	Resistance (Ω/sq)		
	Typical	*Min*	*Max*
P0	30	15	45
P1	10	1	20
P2	20	10	30

Nominal values as defined in the PolyMUMPs design manual are used in all resistance and capacitance calculations.

A future enhancement will include the development of corner rule sets using worst case values, thus allowing estimation of the device parameter deviation due to process variation.

The reference results for the P0-P1 conductor and thermal actuator are obtained using FEM analysis performed with ANSYS. A comparison of the results obtained with LVS/LPE extraction and the reference results is provided in Table III.

TABLE III.   RESISTANCE TEST RESULTS.

Tested Structures	Reference value, Ω	Extracted value, Ω	Relative error, %
P0 wire, L=100μm, W=2 μm	300	300	0
P1 wire, L=100μm, W=2 μm	100	100	0
P2 wire, L=100μm, W=2 μm	200	200	0
P0-A1-P1 wire, L=75μm, W=25 μm, (Figure 3)	29.605	29.607	0.01
Thermal actuator, (Figure 5)	1578.13	1578.85	0.04

Several profiles are used to define the P0-A1-P1 wire and the thermal actuator. Since resistance between the profiles is not directly extracted, net splitting layers are used to ensure that each of the profiles is separated. This is needed for the LPE tool to work properly and can be done manually or automatically in the LVS rule set. The extracted resistance values for the single layer wires exactly match the theoretical calculations. The comparison of the LPE extracted resistances and the results obtained with ANSYS also show good agreement both for the P0-A1-P1 wire and thermal actuator.

B. Capacitance testing

In Calibre xRC the capacitance is broken down in several types: near body, plate and fringe (Figure 6) [10]. The extraction of each of those types can be blocked. This option is very useful for debugging purposes and allows to easily evaluate the contribution of each capacitance type to the total capacitance. To verify that the layer stack definitions are correct, the extraction of the near body and fringe capacitances was blocked and the plate capacitance extracted by the LPE tool was compared to the reference values calculated by the formula:

$$C = \varepsilon_r . \varepsilon_r . \frac{w.l}{h}, \qquad (2)$$

where ε_r is relative permittivity of the dielectric material between the capacitor plates, $\varepsilon_o = 8.85419 . 10^{-12}$ $F \cdot m^{-1}$ is the vacuum permittivity, w, l and h are respectively the width, length and separation of the plates.

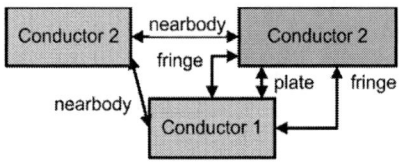

Figure 6.   Capacitance types [10].

Several single-profile plate capacitors are tested. The results are summarized in Table IV. It is obvious that the plate capacitance is extracted with errors only fractions of per mil. They are due only to the numerical accuracy of the software tools. From those results it can be concluded that larger error in the total capacitance comes from the missing inter-profile and other fringe or near-body capacitances.

TABLE IV.   CAPACITANCE TEST RESULTS - PLATE CAPACITANCE ONLY.

Tested Plate Capacitors	Plate Capacitance		Relative error, %
	Reference value, fF	Extracted Value, fF	
P0-P1, Air, w=100μm, l=100μm	44.271	44.271	0
P0-P1, SiO₂, w=100μm, l=100μm	172.657	172.66	-0.002
P1-P2, Air, w=10μm, l=10μm	1.18056	1.18056	0
P1-P2, SiO₂, w=25μm, l=50μm	57.551	57.5525	-0.003

To test the influence of the inter-profile capacitances on the accuracy, two more complex structures are used – the capacitor from Figure 4 and a horizontal thermal actuator. The capacitor has a perimeter of about 400 μm. Along the perimeter the LPE tool has to switch between three different profiles (Figure 2) and there is a significant P0 to P1 nearbody capacitance. Three different configurations of this capacitor are tested. The first one simulates a standard oxide removal process with duration 1 minute and etching speed of 30 μm/min. In this case there is a circular region in the middle of the device in which the OX1 dielectric is air. The OX1 material in the other parts of the device is SiO₂. The radius of the circle is 32μm. The second configuration considers the case when the entire OX1 is SiO₂. The third configuration uses air as a dielectric layer for the whole device.

To test the extraction of the fringe capacitances the part of the thermal actuator where there is P0 plate below it is tested. It has high perimeter to area ratio which results in high fringe capacitance value relative to the total capacitance. In this part of the actuator only Profile 3 is used thus no additional error from inter-profile capacitances is introduced. The reference results are calculated with ANSYS. A summary of the results is provided in Table V.

TABLE V.   CAPACITANCE TEST RESULTS - TOTAL CAPACITANCE.

Tested Plate Capacitors	Plate Capacitance		Relative error, %
	Reference value, fF	Extracted Value, fF	
P0-P1 (SiO₂/Air)	142,96	135.00	-5.57
P0-P1 (SiO₂)	184,157	176.16	-4,34
P0-P1 (Air)	47,2197	44,256	-6,26
Thermal Actuator	29.0805	29.4212	-1,17

The capacitance extraction error for the thermal actuator is 1.17%. It was already shown that the plate capacitance extraction is very accurate and since the investigated structure does not have nearbody capacitances, this error is entirely due to the fringe capacitances. The P0-P1 capacitor errors are combination of fringe capacitance errors and the missing inter-profile capacitances. The total relative errors are in the 5-6% range. For most applications an accuracy of 5 to 10% should be within the tolerable limits. The investigated structures use the minimum allowed spacing between P0 and P1 shapes, which maximizes the nearbody and fringe capacitances. In most structures minimum spacings and widths are avoided and even smaller errors are expected. The relative error can become higher if the investigated structure uses minimum spacing, has a very high perimeter to area ratio and there is a profile change along the perimeter.

V. CONCLUSION

The methodology discussed in this paper shows that industry standard LVS and LPE tools, widely used in the IC design systems, can be applied to MEMS processes if certain measures are taken to overcome issues posed by differences between the CMOS and MEMS processes.

The LVS/LPE rule sets can be used for two main purposes:
- to model the static behavior of MEMS devices with capacitive or resistive interfaces;
- to extract parasitic passive components from the layout of the designed circuit.

The first purpose serves as an additional tool in the MEMS designers' toolset, requiring less computational resources than FEM analyzers and being more accurate than behavioral models. The second, as in regular CMOS circuits, enhances the accuracy of the system model and increases the chances of a "first time right" design.

The output model is standard netlist compatible with most circuit simulators. In this way the MEMS models can easily be integrated with models of electronic systems, which usually are used to control the MEMS devices. The LVS/LPE tools are generally much faster and less resource intensive than any FEM analyzer, but do not have the ability to investigate the dynamic behavior of the mechanical systems.

A main limitation in the described methodology is that capacitance cannot be extracted on the boundary between two profiles. This limitation is not impeding the accuracy significantly, as capacitance extraction errors of less than 6% are easily achieved.

MEMS structures designed for the PolyMUMPs technology are used to demonstrate the proposed approach, illustrate the issues and evaluate the accuracy. However the same methodology can be applied to any MEMS process.

REFERENCES

[1] V. Kaajakari, Practical MEMS, Small Gear Publishing, 2009

[2] D.F. Ostergaard and M. Gyimesi, "Finite Element Based Reduced Order Modeling of Micro Electro Mechanical Systems (MEMS)", International Conference on Modeling and Simulation proceedings, pp. 684 – 687, 2000

[3] S.Beeby, G.Ensell, M.Kraft, N.White, MEMS Mechanical Sensors, Artech House, 2004

[4] "Calibre nmLVS datasheet", Mentor Graphics Corp., www.mentor.com, 2011

[5] "Calibre XRC datasheet", Mentor Graphics Corp., www.mentor.com, 2011

[6] J. Carter et al., PolyMUMPs Design Handbook Rev. 11.0, MEMSCAP Inc.

[7] S. Wilcenski, "Introduction to Prototyping Using PolyMUMPs", MEMSCAP Inc.

[8] B. Hardy, "PolyMUMPs General Release Instructions", MEMSCAP Inc.

[9] J. H. Comtois, V. M. Bright, and M. W. Phipps, Thermal microactuators for surface-micromachining process, Proc. SPIE, vol. 2642, pp. 10-21, 1995",

[10] "Calibre 2011.1 Standard Verification Rule Format (SVRF) Manual", Mentor Graphics Corp., 2011

MIXDES 2012, 19th International Conference *"Mixed Design of Integrated Circuits and Systems"*, May 24-26, 2012, Warsaw, Poland

Modeling and Optimization of a Ker Charge Pump Loaded by a Resistive Circuit

Jérôme Heitz, Norbert Dumas, Vincent Frick, Christophe Lallement and Luc Hébrard

Institut d'Électronique du Solide et des Systèmes (InESS), Université de Strasbourg - CNRS - UMR7163,

BP20 - 23, rue du Loess, 67037 Strasbourg Cedex - France

Email: n.dumas@unistra.fr

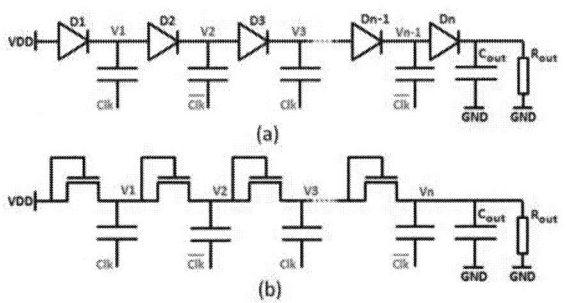

Fig. 1.   (a) Dickson charge pump with actual diodes; - (b) CMOS version of the Dickson charge pump

Abstract—**Many successive improvements of the charge pump principle introduced by Dickson have paved the way to efficient integrated DC-DC converters such as the one introduced by Ker. However the rapid optimization and prototyping of such circuits have not yet been proposed. This paper introduces a model of the Ker charge pump when loaded by a resistor and a two fold optimization method: area or efficiency/area. Simulations show the validity of the approach and the good accuracy of both the model and the method.**

Index Terms—**DC-DC converter, charge pump, high-voltage integrated circuit, CMOS HV.**

I. INTRODUCTION

Many IC modules operate with a control voltage higher than the power supply voltage (memories, sensors, MEMS ...). The efficient Buck-Booster solution requires at least a non-integratable external inductance which is prohibited in many applications because of integration issues. This has given rise to the use of switched capacitor converters, commonly known as charge pumps, which can be fully integrated on silicon. The first integrated voltage converter was based on the design proposed by Dickson in 1976 [1], i.e. a serial architecture, widely studied later by Cataldo [2]-[3] and Tanzawa [4]. There are also stacking architectures such as Favrat [5] with the clock booster of Nakagome [6]. Other topologies of charge pumps exist, but their principles are more complex and expensive [7], [8].

The principle of Dickson charge pumps (Fig. 1 (a)) is to place diodes in series in order to transfer electrical charges

only in one direction, toward the output. The negative electrodes of the load capacitors are connected alternatively to one of the clock signal and therefore their voltages are pushed from 0 to V_{DD}. Consequently the positive electrode voltage is also increased of V_{DD} and can charge the next stage. Yet, diode integration is not common in a standard CMOS process, therefore it has been replaced by MOS transistors connected in diode configuration (Fig. 1 (b)). Provided that the capacitors withstand the generated high voltage, this technique is easy to implement and was a success. Neglecting the parasitic effects, its output voltage is expressed by:

$$V_{out} = (N+1)V_{DD} - \sum_{k=1}^{N} V_{t(k)} \qquad (1)$$

where $V_{t(k)}$ is the threshold voltage of transistors at the stage k and N the number of stages. Generally, the bulk of the NMOS transistors is connected to ground. The increasing voltage from one stage to another causes an increase in the threshold voltage $V_{t(k)}$ due to the body effect. This limits the performance of the structure in terms of efficiency but also in terms of the maximum output voltage. To avoid this problem, Shin [9] proposes a pseudo-diode structure to maintain the threshold voltage constant. Wu [10] has it further improved by replacing the diode connected transistors by gate-voltage controlled transistors in order to completely cancel the effect of V_t. It has been achieved by controlling the gates with the output voltage from two stages further. This solution uses the facts that even or odd stages work synchronously and that the output voltages of the following stages are high enough to cause proper switching. However at initialization, when all capacitors are discharged, the transistors cannot switch and therefore a starting circuit must be implemented. To circumvent this drawback, Zhang [11] and Ker [12] propose a symmetrical architecture, which also has the advantage to provide the load with charges on each clock phase rather than every clock period. The Ker structure is more robust to clock overlapping. Furthermore, in addition of being a real improvement to the Dickson charge pump, it is quite adapted for being a synthetisable component which can be optimally sized with respect to the desired output voltage and current. However a model of this charge pump delivering an output current has never been proposed. It is the aim of this paper along with optimization techniques useful for the designer.

376

In the first section, this paper presents the operating principle of a Ker charge pump with resistive load. An accurate modeling will then be provided. This model allows rapid provisioning of a Ker charge pump that fulfills the specifications imposed. Next, thanks to this model, we propose two optimization methods: one to minimize the silicon area, the other to maximize the efficiency/area ratio. In the last section, simulation results demonstrate the accuracy of this approach before conclusion.

II. Ker Charge Pump

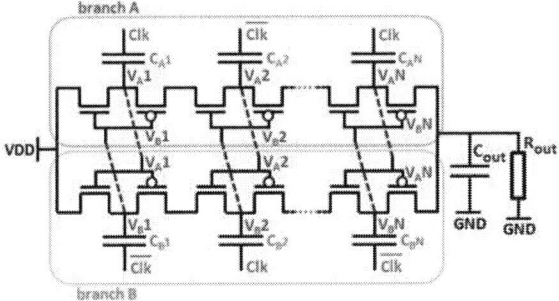

Fig. 2. Circuit of the Ker charge pump with N stages on resistive load

The charge pump introduced by Ker in 2006 [12] is based on the principle of the Dickson charge pump. The loading capacitors are configured in parallel contrary to the stacking architectures such as in Favrat [5]. The main innovation of this system is to cancel the threshold voltage of the Dickson diodes causing efficiency and voltage losses. Ker uses gate controlled MOS transistors, which command the charging of each capacitor. We proposed a model of the Ker charge pump, with resistive load, allowing an accurate estimate of the output characteristics (potential V_{out}, current I_{out}, efficiency, ...).

A. Operation Principle

The Ker structure has two perfectly symmetrical branches A and B, which work in phase opposition. Therefore at each clock phase the input supply voltage charges the first capacitor of one branch and one capacitor delivers its charges to the load. As shown in figure 2, one branch of the Ker structure is formed by an NMOS transistor, followed by a load capacitor in parallel and a PMOS transistor in series. These three elements form the basic stage of the structure. The charge pump is simply a series of elementary stages. The NMOS and PMOS transistors of a same stage are controlled by the same gate voltage. This gate voltage is the intermediary voltage $V_{A/B(k)}$ of the complementary stage of the opposite branch. Each stage of one branch commands or is controlled alternately by the complementary stage of the opposite branch.

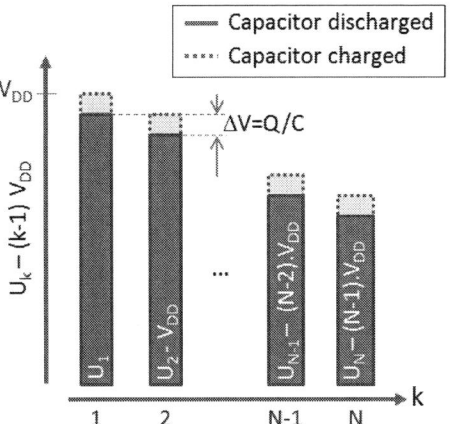

Fig. 3. Voltage drop across the capacitors where k is the stage number.

B. Steady-state output voltage without resistive load

In order to transform the charge pump into a Thévenin equivalent circuit, the output voltage without load has first to be evaluated. Considering the switching transistors as ideal, i.e. no ON resistance, one can express the output voltage $V_{out}(k)$ of stage k according to the input voltage $V_{in}(k)$ of this stage and the voltage applied to the electrode of the load capacitor connected to the clock:

$$V_{out}(k) = V_{in}(k) + V_{Clk} = V_{out}(k-1) + V_{DD} \qquad (2)$$

with $V_{out}(0) = V_{DD}$. Thus, the output voltage of the charge pump, $V_{out}(N)$, can be expressed as follows:

$$V_{out}(N) = \sum_{k=0}^{N} V_{out}(k) = (N+1) \times (V_{DD}) \qquad (3)$$

With the increasing number of stages, there is a voltage increase across the capacitors U_k. As a consequence, to make the charge pump technologically feasible, it is necessary to use high voltage capacitors, which withstand such a voltage.

C. Output Voltage with an Output Load at Steady State

Since there is an equivalence between a resistor and a switched capacitor, it is necessary to relate the charge that is transfered from one stage to the next one with the voltage drop in order to evaluate the equivalent Thévenin impedance. The charge Q delivered from the capacitors $C_{A_{N-1}}$ or $C_{B_{N-1}}$ to the output capacitors C_{A_N} or C_{B_N}, is the same as the charge Q delivered to the load. Therefore each capacitor C_{A_k} or C_{B_k}, i.e. at the stage k, delivers the charge Q to the stage $k+1$ on each clock phase, in a similar way than the Dickson structure works [1]. The voltage drop ΔV corresponding to Q is $\Delta V = Q/C$, with C the capacitance of all stages. As each stage exhibits a drop of ΔV the output voltage drops by N times ΔV resulting in the following expression:

$$V_{out} = (N+1)V_{DD} - N\Delta V \qquad (4)$$

Figure (3) illustrates the above expression. The voltage across the capacitor U_k at the stage k is represented with bars. In order to remove the increase in voltage, $(k-1) \times V_{DD}$ has been withdrawn. We can notice that the charge Q transfered across the charge pump causes a voltage drop of ΔV from one stage to the next.

Both branches A and B alternately charge the output capacitor. Thus, C_{out} will be charged by Q twice per clock cycle. Defining I_{out} as the average output current and f_{Clk} the clock frequency, it directly results that $I_{out} = 2.f_{Clk}.Q$. Thus, one can relate ΔV with this output current:

$$\Delta V = \frac{I_{out}}{2Cf_{Clk}} \qquad (5)$$

From the two previous expressions, we finally obtain the output voltage V_{out} of the Ker structure with a resistive load:

$$V_{out} = (N+1)V_{DD} - \frac{NI_{out}}{2Cf_{Clk}} \qquad (6)$$

In order to identify the equivalent output resistance of the charge pump, one should note that the output of a Thévenin generator has the following form:

$$V_{out} = V_{th} - R_{th}.I_{out} \qquad (7)$$

Thus we deduce that $V_{th} = (N+1)V_{DD}$ and $R_{th} = \frac{N}{2Cf_{Clk}}$. Similarly to all switched-capacitor circuits, the resistance is inversely proportional to the product $C.f_{Clk}$, meaning that the lower the capacitance the faster the charges transfer has to be, in order to keep the same output voltage. The resistance is also simply proportional to the number of stages.

D. Charge Pump Design

Generally the load resistance and the desired output voltage are imposed by the application. Here, we assume that the clock frequency f_{Clk} and the number of charge pump stages, which will be optimized in the next section, are also imposed. From equation (6), the capacitance C has to be:

$$C = \frac{N.I_{out}}{2((N+1)V_{DD} - V_{out})f_{Clk}} \qquad (8)$$

Furthermore, to provide the output voltage V_{out}, a minimum number of stage N_{min} is required. The charge pump output voltage without load must be higher than the desired output voltage. Therefore N must satisfy the condition $(N+1)\,V_{DD} > V_{out}$. It results in the following expression of N_{min}:

$$N_{min} = \text{INT}\left[\frac{V_{out} - V_{DD}}{V_{DD}}\right] + 1 \qquad (9)$$

Similarly, we can show that there is a maximum for N, N_{max}. This is due to the fact that the drop voltage ΔV should not lead to switch the PMOS transistors off during the charge transfer from cell k to cell $(k+1)$. One can deduce that $\Delta V < V_{DD} - V_t$ where V_t is the maximum between V_{t_n} and V_{t_p},

respectively the threshold voltage of the NMOS and PMOS transistors. It results in:

$$N_{max} = \text{INT}\left[\frac{V_{out} - V_{DD}}{V_t}\right] \qquad (10)$$

The switching transistors should also be sized carefully. There is a minimal ratio W/L for the transistors to make sure the output voltage or the charge pump efficiency are not degraded. In order to charge the load capacitor with at least 99% of their maximum over half a period, the time constant must satisfy the following expression:

$$\frac{T_{Clk}}{2} = 5\,\tau \quad \rightarrow \quad \tau = \frac{1}{10\ f_{Clk}} \qquad (11)$$

Because of the series connection of one NMOS transistor and one PMOS transistor between two load capacitances, the time constant τ is:

$$\tau = (r_{on_n} + r_{on_p}) \times \frac{C}{2} \qquad (12)$$

with the ON resistance of the NMOS and PMOS transistors:

$$r_{on_{n/p}} = \frac{L_{n/p}}{(K_{n/p}W_{n/p}(V_{DD} - \Delta V - Vt_{n/p}))}. \qquad (13)$$

Thus, from equations (11), (12) and (13), with $r_{on_n} = r_{on_p}$, one can deduce that:

$$W_n = \frac{L_n.C}{K_n.(V_{DD} - \Delta V - Vt_n).\tau} \qquad (14)$$

$$W_p = \frac{L_p.C}{K_p.(V_{DD} - \Delta V - Vt_p).\tau} \qquad (15)$$

III. DESIGN OPTIMIZATION

A. Area Optimization

The purpose of this part is to analytically determine the optimal number of charge pump stages N_s in order to minimize the silicon area, while respecting the imposed specifications. We consider that the surface of the charge pump is the area of all the capacitors. This implies that the area of the switching transistors is negligible, which is generally the case ($S_{capa}/S_{TMOS} >> 1000$). Considering that the load output capacitor is sized such as $C_{out} = 2 \times C$, the total area S_{CP} reads:

$$S_{CP} \propto 2(N+1)C \qquad (16)$$

From equation (8), one can express S_{CP} as:

$$S_{CP} \propto \frac{I_{out}}{f_{Clk}} \times \frac{N(N+1)}{(N+1)V_{DD} - V_{out}} \qquad (17)$$

To find the minimum area, the last expression has been derived with respect to N:

$$S_{CP} = S_{CP_{min}} \quad \rightarrow \quad \frac{dS_{CP}}{dN} = 0$$

By introducing $\beta = (V_{out} - V_{DD})$, one can show that the optimal number of stages N_A for minimizing the area is expressed by:

$$N_A = \text{INT}\left[\frac{-\beta}{\beta - \sqrt{\beta V_{out}}}\right] + 1 \qquad (18)$$

B. Efficiency/Area Optimization

The area optimization of the charge pump generally yields in a low capacitance value and a great number of stages. This has two major drawbacks: efficiency reduction and higher ripple in the output voltage. The efficiency is expressed by:

$$\mu = \frac{P_{out}}{P_{in}}$$

where P_{in} is the power entering the charge pump and P_{out} the power delivered by the pump. As mentioned in section II, each stage provides the same charge Q to the next stage. Consequently the charge taken from the input power supply on each clock phase is Q. The drivers that supply the pump with the two clock signals also deliver the charge Q to each negative electrode of the loading capacitors for which the electrical voltage is pushed up to V_{DD}. Since there are N capacitors that are "pushed" on each clock phase, the total quantity of charges that is delivered by the input generator is $(N+1) \times Q$ under the voltage V_{DD}. Therefore, the input power entering the charge pump is simply $I_{out} \times (N+1)V_{DD}$. This input power is also the power delivered by the equivalent Thévenin generator of the charge pump. We can thus deduce the efficiency expression:

$$\mu = \frac{V_{out} \times I_{out}}{V_{th} \times I_{out}} = \frac{V_{out}}{(N+1)V_{DD}} \quad (19)$$

This expression clearly shows that the efficiency decreases with the number of stages. The surface optimization of charge pump does not necessarily minimize the number of stages. Therefore, the trade off between the area and the efficiency can be quantified by the following figure of merit:

$$\gamma = \frac{\mu}{S} \quad (20)$$

From equations (17), (19) and (20), we get the expression of γ with respect to N and constant terms that depend only on the charge pump specifications:

$$\gamma = \frac{f_{Clk}V_{out}}{V_{DD}I_{out}} \times \frac{(N+1)V_{DD} - V_{out}}{N(N+1)^2} \quad (21)$$

The optimal number of stages offering the best compromise between efficiency and area is given by maximizing γ.

$$\gamma = \gamma_{max} \quad \rightarrow \quad \frac{d\left(\dfrac{(N+1)V_{DD} - V_{out}}{N(N+1)^2}\right)}{dN} = 0$$

Finally, it can be shown that the optimal number N_{opti} is obtained by solving a polynomial function of third degree with real coefficients. The solution is deduced from the Cardano-Tartaglia formula:

$$N_{opti} = \text{INT}[u + v - h] + 1 \quad (22)$$

with

$$
\begin{cases}
h &= \frac{1}{2}G - \frac{5}{6} \\[2mm]
u^3 &= \dfrac{q - i^{\left(\frac{1 - signe(\Delta)}{2}\right)}\sqrt{|\Delta|}}{2} \\[2mm]
v^3 &= \dfrac{q + i^{\left(\frac{1 - signe(\Delta)}{2}\right)}\sqrt{|\Delta|}}{2} \\[2mm]
\Delta &= q^2 + \dfrac{4}{27}p^3 \\[2mm]
p &= -\dfrac{3}{4}G^2 + \dfrac{1}{2}G - \dfrac{1}{12} \\[2mm]
q &= \dfrac{1}{4}G^3 - \dfrac{1}{4}G^2 - \dfrac{1}{12}G + \dfrac{1}{108}
\end{cases}
$$

where G represents the voltage gain of the charge pump : $G = V_{out}/V_{DD}$.

IV. SIMULATIONS AND RESULTS

This section presents the validity of the model on which the previous optimization is based. Furthermore we show the benefit of designing a charge pump that has been optimized.

A. Modeling validation

On figure 4, we present the output voltage of the charge pump with respect to the number of stages. The dash line represents the output voltage imposed by the specification, i.e. 8, 15 and 30 V, for a clock frequency set to 10 MHz. For each output voltage, the model has been used to calculate the capacitances and the transistor sizes required to deliver a current of either 10 μA or 100μA, using the High-Voltage AMS 0.35 μm CMOS technology [13]. The corresponding charge pumps were thus simulated with Spectre®. The diamond points and the star points represent the simulation results for an output current of respectively 10 μA or 100μA.

For each output voltage and each output current, the worst case discrepancy between the simulation results and the model is 5%, which corresponds to the maximal number of stages N_{max}. This maximal number of stages is a limit that results in the lowest efficiency of the charge pump. Therefore the accuracy of our model around this point is not relevant.

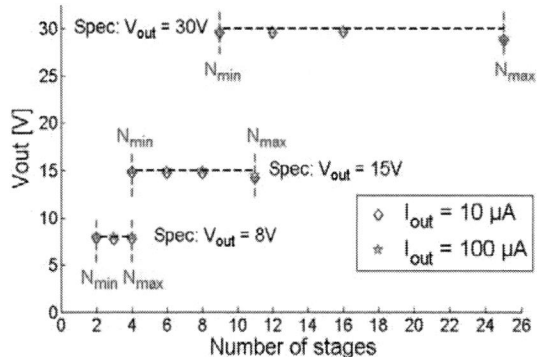

Fig. 4. Simulated output voltage of the Ker charge pump with respect the number of stages for two different output currents at a clock frequency of 10MHz.

Considering only the points where the efficiency is higher than 50%, the error does not exceed 3%.

B. Optimization discussion

In this section, we impose the following specifications to the charge pump: output voltage 30 V, output current 100 μA and clock frequency 10 MHz. Thus, with the previous modeling, we deduce the minimal and maximal number of stages achieving the desired output voltage, respectively N_{min} and N_{max} given by equation (9) and (10). For N between N_{min} and N_{max}, one can deduce the capacitance C of each stage with the expression (8). Figure (5) shows this relationship. One can notice that, as predicted from the model, the capacitance decreases with the number of stages.

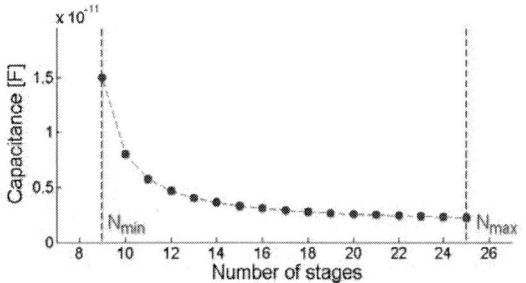

Fig. 5.   Capacitance with respect to the number of stages

From the capacitance, it is easy to evaluate the area of the charge pump given by the equation (17) and represented on figure (6). We note that the minimum area may not be reached for the smallest number of stages or for the smaller capacitance. In this case, equation (18) predicts that the optimum is reached for $N_A = 16$, as we can see on figure (6).

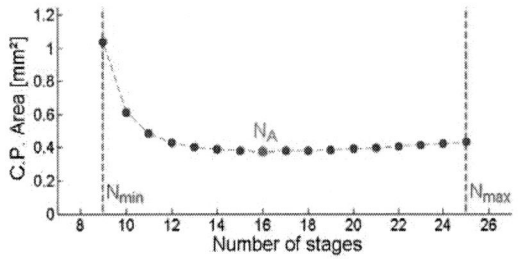

Fig. 6.   Area of the charge pump with respect to the number of stages

While complying with the charge pump specifications, figure (7) shows the reduction of the efficiency with the number of stages. As expected the maximum of efficiency corresponds to the minimum number of stages. However, this design is not interesting because for N_{min}, the capacitance is higher, thus the area is too big.

In this example, on one hand, for the minimum number of stages N_{min} the efficiency is excellent ($\mu = 90.9\%$). However,

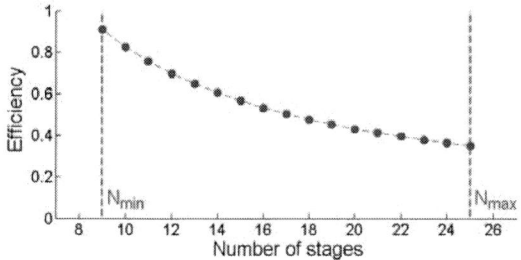

Fig. 7.   Efficiency of the charge pump with respect to the number of stages

the corresponding area is 1.04mm^2, which is about three times the minimum area ($S_{CP_{min}} = 0.38mm^2$). On the other hand, when using this minimum area, the efficiency drops to 53.5% corresponding to $N_A = 16$. While using the figure of merit γ (Eq. (21)), we find an optimal number of stages N_{opti} offering the best compromize between efficiency and area. N_{opti} takes the value of 12 in this case study, as shown on figure (8). Otherwise, it is also possible to propose another figure of merit:

$$\gamma' = \frac{\mu^\alpha}{S^\beta} \qquad (23)$$

with α and $\beta \in \mathbb{N}$. The coefficients α and β offer the possibility to increase respectively the weight of the efficiency or the area. By computation, as opposed to the analytical method used to calculate N_{opti}, it is possible to find the new number of stage N'_{opti}, with respect to the new figure of merit (Eq. 23). Finally, all the possible optimizations with this new figure of merit result in a triplet $\{N'_{opti}; S_{CP}; \mu\}$ in between $\{N_{min}; S_{CP_{max}}; \mu_{max}\}$ and $\{N_A; S_{CP_{min}}; \mu_{N_A}\}$. Therefore the region where N is greater than N_A is irrelevant. The figure of merit guaranties this condition. As previously mentioned, it demonstrates that the region where the model is less accurate (error lower than 5% in terms of the output voltage) is of no interest.

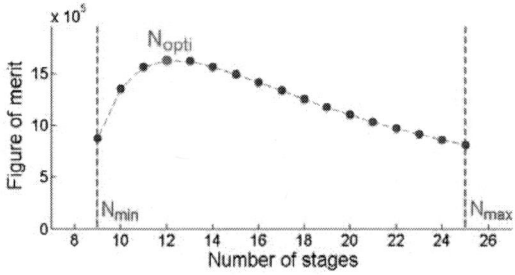

Fig. 8.   Figure of merit with respect to the number of stages

V. CONCLUSION

This paper presents an accurate model of the Ker charge pump with a resistive load. The operating principle was explained and the equations governing the charge pump output voltage were derived. Analytical results are in close agreement with simulations results, validating our model. This model allows the designer to quickly size a charge pump that has to comply with imposed specifications and to find the best trade-off in terms of area or efficiency versus area. This optimization impacts the number of stages, the value of the load capacitors and the size of the switching transistors. Finally, it provides the underlying guidelines that can be used to automatically synthesize a Ker charge pump.

ACKNOWLEDGMENT

The authors would like to acknowledge the Agence Nationale de la Recherche (ANR, France) for its financial support in the framework of the CAPTEX project n° ANR-09-SECU-01-04.

REFERENCES

[1] J.F. Dickson, *"On-chip high voltagegeneration in NMOS integrated circuit using an improved voltage multiplier techniques"*, IEEE J.Solid-state Circuits , vol.11, pp.374-378, June, 1976

[2] G. Di Cataldo and G. Palumbo, *"Double and triple charge pump for power IC: Ideal dynamical models to an optimized design"*, in Proc. Inst/ Elec. Eng. -G. vol. 140, no. 1, pp. 33-38, Feb, 1993

[3] G. Di Cataldo and G. Palumbo, *"Dynamic Analysis of 3 stage Dickson voltage multiplier for an optimized design"*, in Proc. 7th Mediterranean Electrotechnical Conf., pp. 633-636, 1994

[4] T. Tanzawa and T. Tanaka, *"A Dynamic Analysis of the Dickson Charge Pump Circuit"*, IEEE J. Solid-state Circuits, Vol. 32, No. 8, pp. 1231-1240, August, 1997

[5] Favrat et al., *"A High-Efficiency CMOS Voltage Doubler"*, IEEE, vol. 33, No. 3, pp. 410-413, 1998

[6] Nakagome et al., *"An Experimental 1.5-V 64-Mb DRAM"*, IEEE vol. 26, No. 4, pp. 465-471, 1991

[7] J. A. Starzyk, Y-W. Jan, and F. Qiu, *"A DC-DC Charge pump Design Based on Voltage Doublers"*, IEEE Transactions on Circuits and Systems : Fundamental Theory and Applications, vol. 48, No. 3, pp. 350-359, March, 2001

[8] M.S. Makowski, *"Realizability Conditions and Bounds on Synthesis of Switched-Capacitor DC-DC Voltage Multiplier Circuits"*, IEEE Transactions on Circuits and Systems : Fundamental Theory and Applications, vol. 44, No. 8, pp. 684-691, August, 1997

[9] Jongshin Shin, In-Young, Young June Park, and Hong Shick Min, *"A New Charge Pump without Degradation in Threshold Voltage Due to Body Effect"*,, IEEE J. Solid-state Circuits, Vol. 35, No. 8, pp. 1227-1230, August, 2000

[10] J.-T.Wu and K.-L. Chang, *"MOS charge pump for low-voltage operation"*, IEEE Journal of Solid-State Circuits, vol. 33, pp. 592-597, April, 1998

[11] M. Zhang, and N. Llaser, *"Low-Voltage Charge Pump"*, Electronics Letters, Vol.42, February, 2006

[12] Ming-Dou Ker, Shih-Lun Chen, and Chia-Sheng Tsai, *"Design of Charge Pump Circuit With Consideration of Gate-Oxide Reliability in Low-Voltage CMOS Processes"*, IEEE Journal of Solid-State Circuits, Vol.41, No. 5, pp. 1100-1107, May, 2006

[13] http://www.austriamicrosystems.com/Products/Full-Service-Foundry/Process-Technology/High-Voltage

MIXDES 2012, 19th International Conference *"Mixed Design of Integrated Circuits and Systems"*, May 24-26, 2012, Warsaw, Poland

Offset Compensation for Voltage- and Current Amplifiers with CMOS Inverters

Witold Machowski, Jacek Jasielski
Department of Electronics
AGH University of Science and Technology
Kraków, Poland

Abstract—**The paper presents a study of offset reduction techniques dedicated for analog building blocks based on CMOS inverters. Circuit implementations, symbolic analysis as well as SPICE simulation results including global and local variations of model parameters are presented.**

Index Terms—**Analog circuits, CMOS inverter, offset compensation, UMC 180nm**

I. INTRODUCTION

Starting from Nauta's famoust transconductor [1] complimentary push-pull CMOS inverter, invented by Wanlass and dedicated for low power digital logic has been more and more frequently used in analog applications [2-9]. Various authors have different motivations for such approach – one might be expected compatibility of EDA tools for both digital as well as analog CMOS, despite full design automation for analog so far seems to be unreachable goal. Our motivation for interest in inverter based analog circuits is extremely low supply voltage requirement of this block – the minimum one, as long as circuits working in saturation region are under consideration.

From our experience in this area comes out that one of the most important factor limiting the performance of inverter based circuits are offsets. Apparent example of this issue may be illustrated in Fig. 1, where we collected experimental data obtained for inverter-based analog multiplier described in more detail in [8]. The circuit in question is a slight modification of the preceding one, with main idea presented first in [6, 7]. Measurements were made using Agilent/HP 4155A semiconductor parameter analyzer for a lot of six pieces, but in

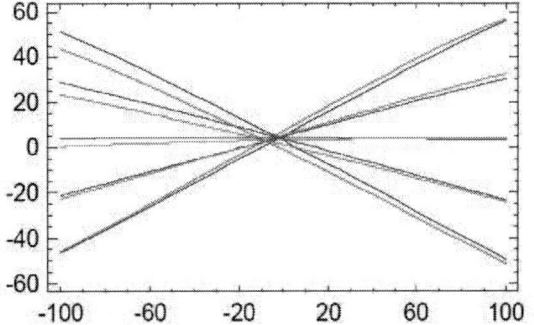

Figure 1. Concatenated measurement results for two specimens of inverter based analog multiplier. X axis is one input voltage in mVolts swept smoothly, while the other is stepped from –100mV up to 100mV by 50mV. The Y axis is the output current in μAmps.

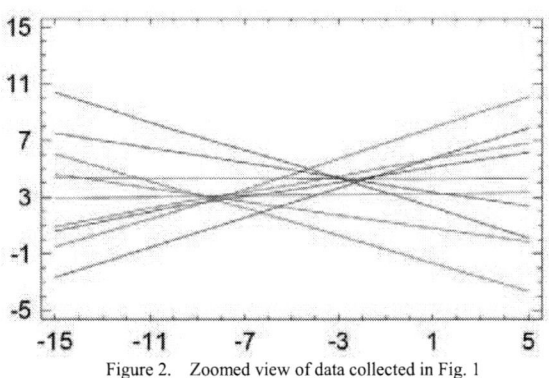

Figure 2.   Zoomed view of data collected in Fig. 1

the sake of clarity in the figure we collected the data for two specimens only, otherwise the collection becomes fuzzy and hard for presentation. Even for two sets it is necessary to use different colors for both curves' families to make them distinguishable. Anyway from Fig. 1 and especially Fig. 2, when the same data are displayed in zoom, comes out that both circuit generate pencils of lines (intersecting more or less accurately in single point – see Fig. 2) with reasonable linearity range, but the vertex of both pencils does not meet with each other. The same apply for another pairs of specimens, the chosen pair represents extreme distance of vertex in the measured lot. Deviation from expected intersection point's coordinates, as well as random character of this shift are clearly visible, anyway the effect is not as much harmful for the circuit as it happened to another topology presented e.g. in [9], where vertex distances exceeded the linear range of operation. While interactive handling of arbitrarily chosen chip from the experimental manufacturing lot makes possible to perform a real four quadrant multiplier operation by appropriate input signal conditioning, the direct implementation of the circuit exhibiting such offsets in real mixed signal VLSI system may be problematic.

The obvious reason for observed discrepancy between simulation results with typical mean condition parameters and actual measurements are components' mismatches. On the other hand, from the manufacturability point of view, global parameters variations should be also taken into consideration, since measurements from one production lot may reflect local variations only. The main question arising here: why this issues had not been taken into consideration before the chip was made, has a simple answer – for the technology used (UMC 180 to be more specific) statistical simulation data were introduced only in 2010 version of Foundry Design Kit, while the chip had been sent for manufacturing a bit earlier. We performed simulations with worst case corners available at time of the design, but from previous experience as well as according to e.g. [10] worst case corners are sometimes exaggerated and may lead to very pessimistic predictions. The same dissatisfaction was attributed to unrealistic (if at all available) Monte Carlo data for many Europractice technologies. Discussion with more experienced designers pointed out, that the aforementioned rules are not necessarily true for more modern nanometer technologies.

This study is dedicated to circuit implementations suitable for offset reduction in CMOS inverter based analog building blocks. Advanced offset reduction techniques [11] use digital calibration (comprising sub-binary DACs), which may be hardly implementable in assumed class of CMOS inverter circuits. Therefore we were seeking the right solution entirely in aforementioned class. In the paper we present some general considerations with particular circuit implementations followed by basic analysis and statistical simulations with data provided by the foundry. Two basic classes of the circuit have been considered so far – complimentary current mirror with simple transistors and high swing cascodes as well as voltage buffer based on CMOS inverters.

II. OFSET REDUCTION FOR CURRENT AMPLIFIERS

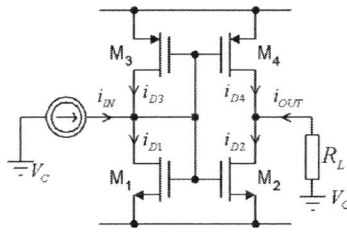

Figure 3.   Complementary current mirror – a current amplifier

In our previous works we often use as a basic building block the current amplifier comprising complementary current mirrors. It is depicted in Fig. 3. The circuit being essentially the superposition of top pMOS sourcing mirror and bottom nMOS sinking mirror, even gained a separated article in 24 volume set of Wiley Encyclopedia [12] under not very relevant title after [13]. However, from another perspective this configuration may be seen as a cascade of two CMOS inverters with the first one having shorted input and output.

Sensitivity analysis for the circuit under consideration may be performed using simplified analytical models (e.g. Shickman-Hodges). Following the way shown in [14] we can estimate the output current variation for simple current mirror. Most important factors determining the drain current and consequently its deviation from nominal are V_T (threshold voltage) and β (transconductance parameter including aspect ratio W/L). For transistors not suffering from body effect (for conventional current mirror as well as for push-pull inverter it

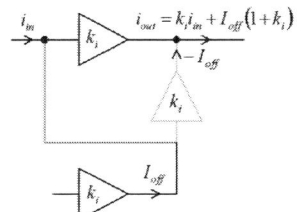

Figure 4.   Concept of current offset compensation

is always the case) γ (the body factor) may be neglected. Moreover due to relationships between electrical characterization and process parameters V_T and β variations are usually considered to be uncorrelated.

The relative variation of output current may be calculated as [14]:

$$\frac{\Delta I_{OUT}}{I_{OUT}} = \frac{2}{V_{GS} - V_T} \Delta V_T + \frac{\Delta \beta}{\beta} \qquad (1)$$

Since for minimizing the first term on the left side of equation (1) the more overdriving voltage the better, stronger inversion is welcome and to minimize the offset small W/L ratio is usually recommended [14].

On the other hand, in this simplified model we did not take into consideration the channel length modulation λ which introduces systematic discrepancy for simple current mirror. For complementary current mirror the position is even a bit worse, since the resulting output current is actually the difference of output currents from both mirror and inherently exhibit bigger offset.

Anyway, performing quantitative analysis involving both global as well as local variations of transistor parameters for specific technology is rather difficult task, since for long time a "communication vehicle" [15] between the designer and the foundry is SPICE model (usually BSIM) without clear values of V_T and β used by simplified model. Good point is, that more recently silicon foundries provide users not only with typical models and corners, but also model sections containing statistical data for both global and local variations.

The concept of method of offset reduction for the circuit from Fig. 3 is depicted in Fig. 4

It implements main idea of correction, assuming that "dummy" bottom structure generates signal which should be subtraced from the output (green signal path). But since we have block performing current inversion, it is beter to utilize it twice (red signal path in Fig. 4). Practical circuit diagram for this architecture is show in Fig. 5.

Let's mention that similar method may be adopted for another inverter based analog block having an output preceded by inverting current amplifier e.g. some circuit variants of already mentioned inverter based analog four quadrant multipliers. From principle this approach should be effective basically for offset resulting from rather global than local variations, but on the other hand assuming un-correlated offsets

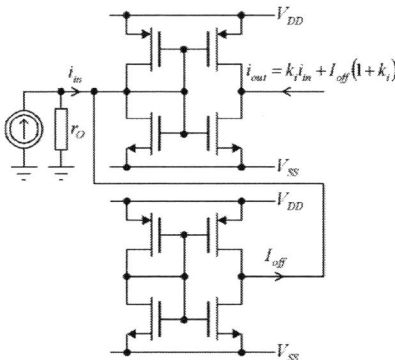

Figure 5. Circuit implementation of scheme from Fig. 4

will average them, thus it should introduce some improvement even in the last case.

III. OFFSET REDUCTION FOR INVERTING VOLTAGE BUFFER

Another basic building block based on inverter used often in our previous design is an inverting voltage buffer shown if Fig. 6. Actually the buffer itself comprise only M1-M4 transistors, while the rest of the figure shows the buffer together with the injector of correction current. Conceptually the voltage buffer (ku block in the bottom part) is composed of transconductor (M1-M2) followed by active load or transimpedance amplifier (M3-M4). Another transconductor (M5-M6) is used for proper output voltage conditioning by injecting to the actual output appropriate current I_{COR} which is set by V_{COR}. One possible implementation The method of making the aforementioned signal and circuit used for this purpose is shown in Fig. 7. Transistors M1-M4 form a "dummy" structure, while M7-M8 compose lame (or "tailless") diffpair with active dynamic load M8-M10 ultimately making desired V_{COR} . This solution is actually more elaborated and symmetrical variant of circuit already presented in [8].

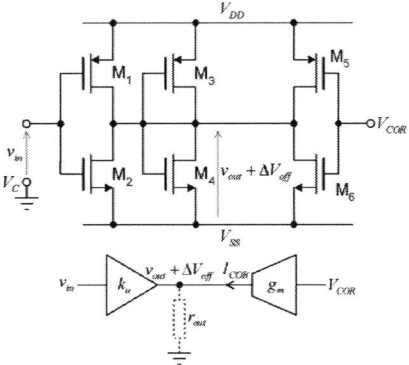

Figure 6. Offset compensation principle for voltage buffer

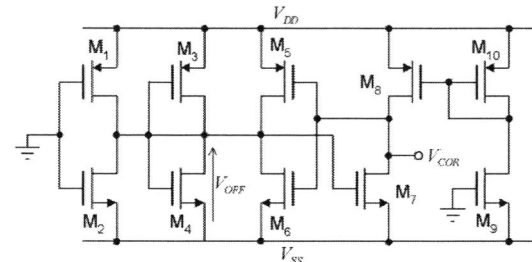

Figure 7. Dummy voltage buffer with error amplifier

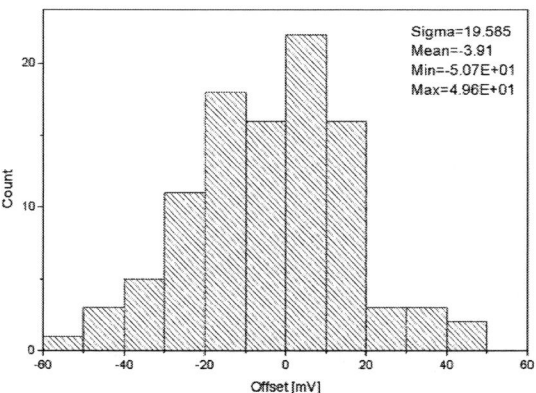

Figure 8. Offset for bare voltage buffer – process varaitions

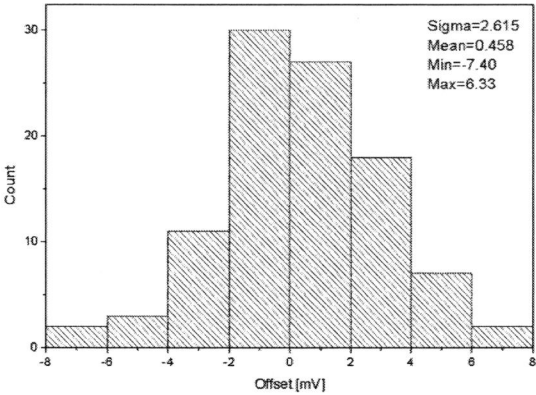

Figure 9. Offset for compensated voltage buffer – process variations

IV. SIMULATION RESULTS

We performed intensive simulation using Cadence Spectre with models provided by the foundry. First the bare circuits were subject of intensive simulations including various configurations of statistical model parameters including *process*, *mismatch* and *process+mismatch* Monte Carlo runs with different correlation coefficients. Actual simulation did

384

include full DC sweeps within ± 100µA or ±100mV respectively for current and voltage buffer and were repeated 300 times. After that the output values corresponding to zero input one were found. The last value represents an output offset, which later was undergoing the statistical visualization with determining most important estimates. Exemplary results for voltage buffer are shown in Figs. 9 and 10. They present histogram of offsets for bare (i.e. without any compensation) or corrected variant.

Complete results comprising essential parameters from statistical runs are summarized in Tables I and II

TABLE I.   OFFSET ESTIMATES FOR CURRENT BUFFER

	Current Mirror			
	Uncorrected		Corrected	
	Process	Process/ Mismatch	Process	Process/ Mismatch
Sigma [µA]	9.55E-2	12.52E-2	3.33E-2	11.04E-2
Mean [µA]	-3.13E-1	-3.27E-1	-2.19E-2	-2.76E-2
Min [µA]	-5.52E-1	-6.08E-1	-8.34E-2	-3.28E-1
Max [µA]	-1.31E-1	-1.88E-2	7.03E-2	1.89E-1

TABLE II.   OFFSET ESTIMATES FOR CURRENT BUFFER

	Voltage Inverter			
	Uncorrected		Corrected	
	Process	Process/ Mismatch	Process	Process/ Mismatch
Sigma [mV]	19.585	16.042	2.615	4.423
Mean [mV]	-3.91	1.74	0.458	-.371
Min [mV]	-5.07E+1	-3.59E+1	-7.40	-12.5
Max [mV]	4.96E+1	3.44E+1	6.33	9.63

V. CONCLUSIONS

We proposed circuits compensating offsets in inverter-based voltage and current buffers suitable for implementation e.g. in various proprietary architectures employing such blocks.

Intensive simulations with statistical parameters confirm functionality of proposed circuit solutions indicating significant reduction of the mean offset as well as its spread. Spectacular results were obtained for process variations, much less, yet observable for total (*process+mismatch*) options.

Frankly speaking pure *mismatch* option shows that some improvement in the circuit performance can be observed only after introducing significantly high correlation factors between instances in Cadence. This is actually expected result, since considered structures seems to be extremely sensitive to mismatch. For obvious reason analog simulation environment does not include real correlation coefficients, which involve layout data. While known from Pelgrom laws [16] that bigger elements match with each other better than smaller ones – and this factor is included in data passed to simulator, the other factor – the distance between elements' pair is not. Pure mismatch model suggest that our bare or corrected circuits should not work at all, which is not the case. So the "virtual prototyping" environment would be desirable for more valuable simulation results. Without any doubt our laying out skills are far from perfection.

To make the considered class of circuits more reliable it is necessary to follow strictly DFM (Design for Manufacturabilty) guidelines for particular technology (if only such a document is provided by the foundry and available for the designer) to improve matching.

ACKNOWLEDGMENTS

The authors express their gratitude to our masters and colleagues prof. St. Kuta and prof. R. Golański for their support and fruitful discussions about various aspect of analog design. Precise measurements were available courtesy of prof. W. Kucewicz, and the access his measurement lab with high class equipment is highly appreciated. Last not least, we thank prof. P. Gryboś and dr. R. Szczygieł for their help with appropriate setup of Cadence™ environment for UMC FDK and discussion about Monte Carlo and worst case corners.

REFERENCES

[1] B. Nauta B., "A CMOS Transconductance-C Filter Technique for Very High Frequencies", IEEE J. of Solid-State Circuits, vol. 27, pp. 142.-153, 1992

[2] H. Barthelemy et al., "CMOS inverters based positive type second generation current Conveyor", Analog Integr. Circ. Sig. Process., vol. 50, pp. 141–146, 2007

[3] H. Barthelemy et al ., "OTA based on CMOS inverters and application in the design of tunable bandpass filter", Analog Integr. Circ. Sig. Process., vol. 57, pp. 169-178, 2008

[4] M. Elnozahi, Y. Massoud "Efficient synthesis methodology for optimal inverter-based transimpedance amplifiers", Analog Integr. Circ. Sig. Process. vol. 50 pp. 205–211, 2007

[5] M. Figueiredo et al., "A Two-Stage Fully Differential Inverter-Based Self-Biased CMOS Amplifier With High Efficiency", IEEE Trans.Circ. Syst. I, vol. 58, 1591-1603, 2011

[6] W. Machowski, St. Kuta, J. Jasielski, "Four-quadrant analog multiplier based on CMOS inverters", MIXDES'06, pp. 290–293,Gdynia, 2006

[7] W. Machowski, St. Kuta, J. Jasielski, "Four quadrant analog multiplier based on CMOS inverters", Analog Integr. Circ. Sig. Proc., vol. 55, pp. 249-259, 2008

[8] W. Machowski, J. Jasielski., "Low Voltage, Low Power Analog Multipliers based on CMOS Inverters", MIXDES'11, pp. 94-97, Gliwice, 2011

[9] W. Machowski, "CMOS inverter based analog multipliers", Przegląd Elektotechniczny, vol. 86 nr 4 s. 209–212, 2010

[10] W. Kuźmicz, "Design for manufacturability in analogue domain" Analogue CMOS IC Design Course, Warsaw UT, 1999

[11] M. Pastre, M. Kayal, "Methodology for the Digital Calibration of Analog Circuits and Systems", Springer, 2006

[12] Newcomb R.W., Sellami L. "Differentiating Circuits" article, in "Wiley Encyclopedia of Electrical and Electronics Engineering", Wiley-Interscience, 1999

[13] E.I. El-Masry, J.W. Gates "A Novel Continuous-Time Current-Mode Differentiator and Its Applications", IEEE Trans. Circ. Syst.-II vol. 43, pp. 56-59, 1996

[14] K.R. Laker, W.M.C. Sansen "Design of Analog Integrated Circuits and Systems", McGraw-Hill, 1994

[15] D. Foty, The SPICE FET Models: Pitfalls and Prospects, Custom Integrated Circuits Conference, Tutorial 1997

[16] M.J.M Pelgrom., C.J. Duinmaijer, A.P.G. Welbers, "Matching Properties of MOS Transistors", IEEE J. of Solid-State Circuits, vol. 24, pp. 1433-1440, 1989

MIXDES 2012, 19ᵗʰ International Conference *"Mixed Design of Integrated Circuits and Systems"*, May 24-26, 2012, Warsaw, Poland

Surface Potential Model of a High-k HfO$_2$-Ta$_2$O$_5$ Capacitor

George Angelov, Nikolay Bonev
Rostislav Rusev, Marin Hristov
Dept. of Microelectronics, ECAD Laboratory, FETT
Technical University of Sofia
8 Kl. Ohridksi str., 1797 Sofia, Bulgaria
gva@ecad.tu-sofia.bg, nbb@ecad.tu-sofia.bg
rusev@ecad.tu-sofia.bg, mhristov@ecad.tu-sofia.bg

Albena Paskaleva
Institute of Solid State Physics
Bulgarian Academy of Sciences
72 Tzarigradsko Chaussee Blvd.
1784 Sofia, Bulgaria
paskaleva@issp.bas.bg

Abstract—**A compact model of a MOS capacitor with high-k HfO$_2$–Ta$_2$O$_5$ mixed layer stack is developed in Matlab. Model equations are based on the surface potential description of PSP model. After fitting the C–V characteristics in Matlab the model is coded in Verilog-A hardware description language to interface with Spectre circuit simulator within Cadence CAD system. The results are validated against experimental measurements of high-k dielectric structure.**

Index Terms—**Device modeling, compact models, PSP, circuit simulation, high-k gate dielectric, Verilog-A, Spectre**

I. INTRODUCTION

The semiconductors industry has been facing new challenges due to CMOS device downsizing. Linear scaling will not be possible in the future unless new materials are introduced in CMOS device structures or unless new device architectures are implemented. The strong association between devices and materials research is the key enabler here. The demand for low voltage, low power and high performance are the great challenges for the engineering of sub 45-nm gate length CMOS devices.

In this context device modeling is the milestone to efficiently implementing design objectives based on the new devices [1]. The scaling of classical bulk Si CMOS transistors approaches its physical limits. The SiO$_2$ gate dielectric thickness of a few atoms raises unwanted quantum mechanical effects such as electron tunneling and gate leakage currents that compromise the classic MOS transistor operation. To maintain the Moore's law progress in microelectronic technologies [2] it is needed to use new materials with higher dielectric constant (high-k materials) to replace the conventional SiO$_2$. The high-k gate dielectrics are also required for ensuring high-performance and low-power CMOS applications in the 45 nm technology node and beyond [3]. The emerging nanoelectronic transistors will rely on non-silicon high-k materials with target effective oxide thickness (EOT) of less than 10 Å to advance beyond the sub-20 nm regime [2], [4].

There are many high-k candidates being studied. Ionic metal oxides, having highly polarized metal-oxygen bonds, would have much larger k values than that of the covalent dielectric materials. Amongst those materials, Hf-based

materials, such Hf silicates, Hf aluminates, have been considered as the most promising materials and have already been used in the state-of-the-art CMOS technology.

Promising high-k candidates for alternative gate dielectric materials are the multicomponent dielectrics based on a multiple metal oxides. Ta$_2$O$_5$ is best high-k candidate for storage capacitors of nanoscale DRAMs; HfO$_2$ appears to be the respective candidate for nano-MOSFETs [4], [5], [6]. The electrical characteristics prove that the structure composed of HfO$_2$–Ta$_2$O$_5$ mixed layer on Si performs as a high-k layer in terms of permittivity, allowable level of leakage current, and appropriate oxide interface properties [7].

II. SURFACE POTENTIAL BASED MODELING

Compact device models need to be physical, simple (compact), accurate, and technology independent. Fitting of device data from different technologies across the industry with high accuracy is the most challenging task. The models are generally coded in circuit simulators using general-purpose languages. Accordingly, they are targeted specifically to the interface and internal data structures of their host simulator, and hence are inherently non-portable. In this context modification and optimization of a given model becomes a time-consuming and error-prone task.

An effective approach to obtain flexible modeling approach is to formulate open source code models in analog hardware description languages (HDLs) such as Verilog-A/AMS or VHDL-AMS. In the recent years Verilog-A has become increasingly viewed as most promising candidate for compact modeling purposes [8].

The basic equations for describing the MOS device characteristics are the Poisson's equation, the continuity equations, and the current-density equations [9]. Historically there are two major approaches to analytically describing device behavior: piece-wise modeling approach (also called regional approach or threshold-voltage based approach) and surface-potential based approach [10].

Piece-wise models describe MOSFET operation in the linear and saturation regions with separate equations. A fundamental problem is the discontinuity of drain current characteristics which is solved by smoothing functions to interpolate the *I-V* characteristic between linear and saturation regions. With surface-potential based approach, on the other hand, model development focuses on surface potential (ϕ_s) formulation. These models allow an inherently single equation and accurate calculation of I_D. From the wide spread models in electronic design automation (EDA) industry the BSIM3/4 models are piece-wise based and PSP – surface-potential based.

In this paper a compact model for circuit simulation of the high-*k* MOS capacitor HfO_2–Ta_2O_5 mixed layer structure presented in [7] is developed. The model is coded in Verilog-A HDL based on the PSP model core. Capacitance–voltage (*C–V*) characteristics are compared to the measurements to validate the model. The BSIM3v3 formulation of the model of this same HfO_2–Ta_2O_5 structure is described in [11].

III. MODEL FORMULATION

The test structures for electrical measurements are MIS capacitors with a back side electrode of ~ 300 nm evaporated Al. The detailed characteristics of the modeled structure can be found in [7]. The capacitors are electrically characterized by means of *C–V* (Fig. 1) curves in the frequency range 50 kHz ÷ 100 kHz for minimizing the effects of parasitic series-parallel circuits [12].

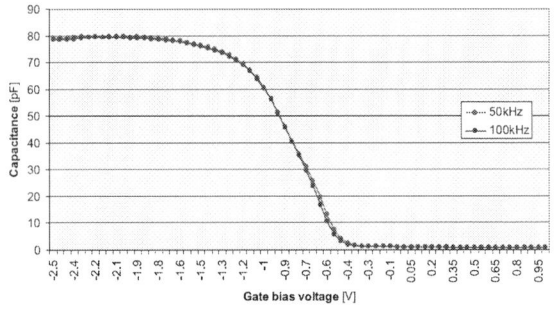

Figure 1. *C–V* characteristics measured by RLC meter versus two frequencies across 10 nm HfO_2-Ta_2O_5 capacitor stack.

By MOS capacitor measurements are obtained the physical properties of the developed technology for fabrication of high-*k* MOS devices. For example the effective dielectric constant ε_{eff} of the films is determined from the capacitance C_0 at an accumulation using ellipsometrically measured values of *d*. The oxide charge Q_f is also evaluated from the *C–V* curves.

A. Parameter extraction

The modeled high-*k* dielectric capacitor stack has parameters which are directly measured after its fabrication – gate area defined by width *W* and length *L*, and type of substrate conductivity (acceptor or donor). These parameters are typical design inputs which can be changed depending on the requirements of the layout.

Other parameters needed for the model are: relative permittivity of the dielectric, dielectric thickness, substrate doping concentration and flat band voltage. These parameters are technology dependant inputs for model adaptation which are not changed during layout design. Their values are summarized in **Table I** and they are determined either by direct measurements or by extraction based on the characterization *C-V* curves.

TABLE I.   SUMMARY OF PRELIMINARY MEASURED OR EXTRACTED TECHNOLOGY PARAMETERS

Technology Parameter	Value	Dimension
Relative dielectric permittivity – ε_{rox}	9	dimensionless
Substrate doping concentration – N_{sub}	$1.25 \cdot 10^{21}$	$[m^{-3}]$
Flat band voltage – V_{FB}	-0.55	[V]
Dielectric thickness – t_{ox}	10	[nm]

B. Model description

The objective of our MOS capacitor compact model is to enable simulation of the high-*k* MOS device using different design and technology parameters keeping model equations as simple as possible while giving highly accurate results. A model meeting the above requirements was already developed in [11] based on the BSIM3v3 core which is one of the most well-known regional models. It describes different operating regions with different equations.

Here we are focused on developing a compact model based on the novel surface-potential approach proceeding from PSP model equations that inherently possess continuous *I-V* characteristics. In this approach the equations are explicit functions of the surface potential ϕ_s instead of the applied voltage and as a result they are continuous for all bias conditions [13]. The main disadvantage is the requirement of separate iterative procedure for calculation of the surface potential ϕ_s as function of the applied voltage. In this model the iterative procedure is replaced with regional approach which is an easy way to incorporate all significant effects into the surface potential description

After simplifying the original PSP equations the intrinsic charge at the gate is expressed with the product of the oxide capacitance C_{ox} and the mid-point voltage V_{oxm}:

$$Q_g = C_{ox} V_{oxm} \tag{1}$$

$$C_{ox} = (\varepsilon_{rox}\varepsilon_o WL)/t_{ox} \tag{2}$$

$$V_{oxm} = \phi_T . x_{gm} \tag{3}$$

where $\phi_T = kT/q$ is the temperature potential and x_{gm} is a variable which depends on the surface potential described by the following expressions:

$$x_{gm} = G.\sqrt{D_m + P_m} \qquad (4)$$

$$D_m = [1/E_s - x_s - 1 - \chi(x_s)]\Delta_{ns} \qquad (5)$$

$$P_m = x_s - 1 + E_s \qquad (6)$$

In equations (5) and (6) x_s is dimensionless potential at the silicon substrate surface which is computed from the temperature and real surface ϕ_s potentials:

$$x_s = \phi_s / \phi_T \qquad (7)$$

The parameter G is calculated from the body effect coefficient γ:

$$G = \gamma / \sqrt{\phi_T} \qquad (8)$$

where $\gamma = \sqrt{2.q.\varepsilon_{Si}.N_{sub}} / C'_{ox}$. The parameter $C'_{ox} = (\varepsilon_{rox}\varepsilon_0) / t_{ox}$ is the oxide capacitance per unit gate area.

The variables E_s and $\chi(x_s)$ are functions of the surface potential:

$$E_s = \exp(-x_s) \qquad (9)$$

$$\chi(x_s) = x_s^2 / (2 + x_s^2) \qquad (10)$$

The variable Δ_{ns} is in exponential dependence from the dimensionless bulk potential deep in the silicon substrate $x_{ns} = \phi_B/\phi_T$:

$$\Delta_{ns} = \exp(-x_{ns}) \qquad (11)$$

Equations (1) ÷ (11) show the modeled charge is explicit function only of the surface potential and it is already described with same formulas for all bias operation regions.

However, the surface potential is not described continuously even in the PSP model. It is split into two regions separated by the marginal dimensionless band bending parameter x_{mrg} calculated from the body effect coefficient:

$$x_{mrg} = 10^{-5}(1 + G/\sqrt{2}) \qquad (12)$$

The dimensionless band bending caused by the bias voltage is:

$$x_g = (V_{GB} - V_{FB} - ST_{V_{FB}}\Delta T) / \phi_T \qquad (13)$$

The parameter ST_{VFB} is the temperature coefficient of the flat band voltage and ΔT is the temperature difference from the nominal temperature (21 °C).

The surface potential is described in two regions:

1) Accumulation and depletion when $x_g < x_{mrg}$

$$x_s = A_1\eta + a.\tau / (a + c) \qquad (14)$$

$$\tau = -A_2\eta + A_3 \ln(a / G^2) \qquad (15)$$

$$a = (-x_g - \eta)^2 \quad \text{and} \quad c = 2.(-x_g - \eta) \qquad (16)$$

$$\eta = \left(z + 10 - \sqrt{(z-6)^2 + 64}\right) / 2 \qquad (17)$$

$$z = -1.25 x_g / \xi \qquad (18)$$

2) Inversion and depletion when $x_g \geq x_{mrg}$

$$x_s = B_1\eta + a.\tau / (a + c) \qquad (19)$$

$$\tau = x_{ns} - \eta + B_2 \ln(a / G^2) \qquad (20)$$

$$\eta = \left(x_g + b_x - \sqrt{(x_g - b_x)^2 + 5}\right) / 2 \qquad (21)$$

$$b_x = x_{ns} + 3 \qquad (22)$$

For the variables a and c are used expressions (16).

IV. Simulation Results and Fitting

The above model equations are coded in Matlab. Equations (14) ÷ (15) are simplified PSP equations in which the fitting non-physical variables A_1, A_2, A_3, B_1 and B_2 are introduced. The need for further fitting of these variables arises after comparing measurements versus simulation results at $A_1 = A_2 = A_3 = B_1 = B_2 = 1$ (cf. Figure 2).

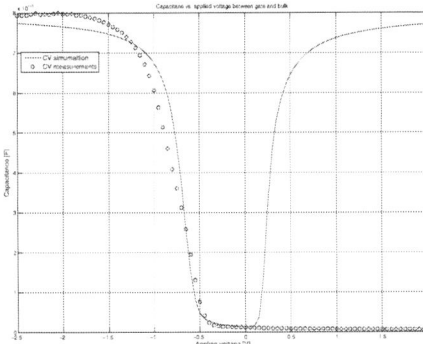

Figure 2. Simulation of *C-V* curve compared to measurements based only on the simplified PSP equations. Further fitting in all bias regions is need.

Both regions are additionally split into smaller pieces where different effects are dominant over different parts of the curve. In Figure 3 and Figure 4 it is observed how the MOS capacitor characteristics are changed after applying the piece-wise fitting of the variables.

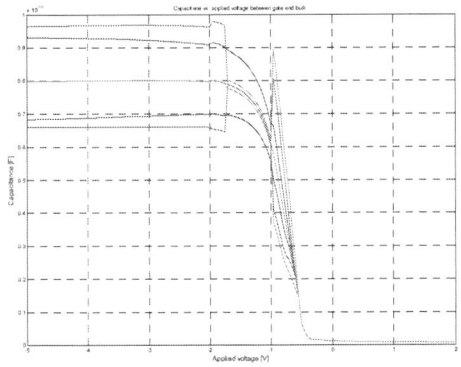

Figure 3. Fitting *C-V* curves by changing variables A1 (green), A2 (red) and A3 (blue) in accumulation and depletion regions.

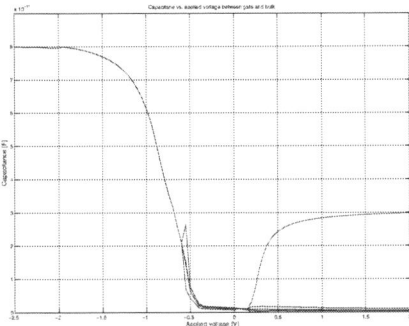

Figure 4. Fitting *C-V* curves by changing variables B_1 (blue) and B_2 (red) in inversion and depletion regions.

The performed additional split enables further easy adjustment of the surface potential so that the integral error of the mismatch between the simulations and measurements curves is calculated to be below a certain maximum of e.g. 3%. The outcomes from fitting of the variables within the entire bias range are given in the lookup Table II. The achieved matching between the model and the experimental data is presented in Figure 5.

TABLE II. LOOKUP TABLE OF THE FITTING VARIABLES A_1, A_2, A_3, B_1 AND B_3

Parameters: Bias Voltage Range [V]	A_1	A_2	A_3	B_1	B_2
$< (V_{FB} - 2.0)$	1	0.955	0.979	1.7	0.5
$(V_{FB} - 2.0) \div (V_{FB} - 1.75)$		1.022	1.02		
$(V_{FB} - 1.75) \div (V_{FB} - 1.0)$	0.04	0.002	0.9875		
$(V_{FB} - 1.0) \div (V_{FB} - 0.65)$	1.15	1.84			
$(V_{FB} - 0.65) \div (V_{FB} - 0.5)$	1.3	1.94	1.5		
$(V_{FB} - 0.5) \div (V_{FB} - 0.45)$				1.5	
$(V_{FB} - 0.45) \div (V_{FB} - 0.4)$				0.6	
$(V_{FB} - 0.4) \div (V_{FB} - 0.3)$				0.5	
$(V_{FB} - 0.3) \div (V_{FB} - 0.2)$				0.6	
$(V_{FB} - 0.2) \div (V_{FB} - 0.0)$				0.7	
$(V_{FB} - 0.0) \div (V_{FB} + 0.15)$				0.9	
$(V_{FB} + 0.15) \div (V_{FB} + 0.25)$				0.938	
$> (V_{FB} + 0.25)$				0.94	

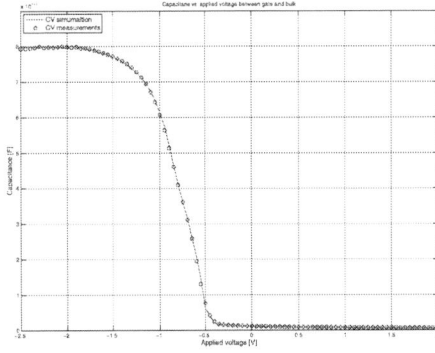

Figure 5. Plot of *C-V* simulation compared to experimental data after fitting. Highly accurate matching is achieved.

Essential part of the model is the surface potential ϕ_s and its function of the applied voltage is plotted in Figure 6. The abstract description of the surface potential is very important because it is developed together with the technology of the researched high-*k* MOS devices.

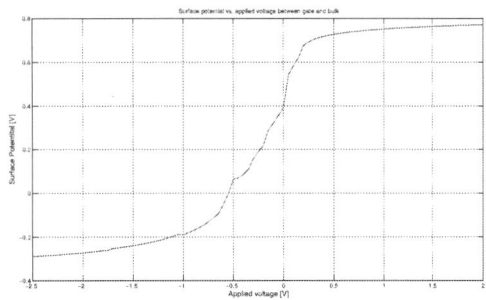

Figure 6. Plot of the surface potential versus the applied voltage. This function is technology dependant.

To perform the circuit simulations, the Matlab code was recoded in Verilog-A in order to input it to Spectre circuit simulator as an external model. The existing MOSFET in Cadence design kit can be simulated as MOS capacitor if the source, bulk, and drain nodes are connected together as described in [13]. The C–V characteristics are simulated in AC mode by plotting the capacitance as a calculation based on the amplitude of the current through the gate node for a frequency of 50 kHz. The input voltage is sinusoidal with fixed small signal amplitude of 10 mV and DC voltage sweep between –5 V ÷ +5 V.

Simulations with the developed model can be run well beyond the bias range (–2.5 ; 2.5) V for which we have experimental data. Outside this range the device behavior follows the natural asymptotic expectations. This is proven with parametric simulations within twice extended range (–5 ; 5) V using the dielectric thickness for parameter. The simulation plots in Figure 7 validate the model by confirming the proper asymptotic behavior.

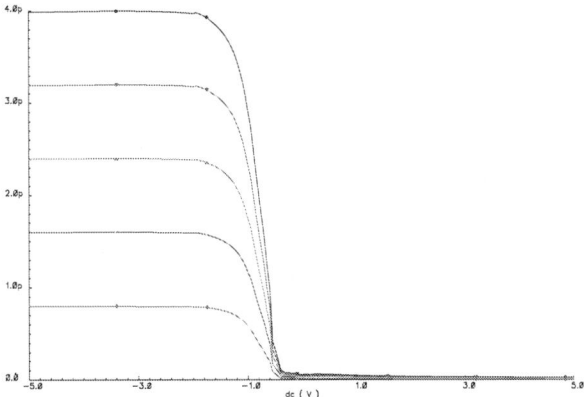

Figure 7. Plots of parametric C-V simulations for different values of dielectric thickness in extended bias range. The model behaves naturally as expected.

V. VERILOG-A CODE

Below we list an excerpt of the Verilog-A code of our model showing the computation of the surface potential and the continuous description of the intrinsic charge.

```
//Surface potential expressions
if (xg < -margin) begin
//accumulation and depletion regions
SP_S_ysub  = -1.25 * xg * inv_xi;
SP_S_eta   = 0.5 * (SP_S_ysub + 10 -
              pow(((SP_S_ysub - 6.0) *
              (SP_S_ysub - 6.0) + 64.0),0.5));
SP_S_temp  = -xg - SP_S_eta;
SP_S_a     = SP_S_temp * SP_S_temp;
SP_S_c     = 2.0 * SP_S_temp;
if(Vgs<(-2.0+VFBO+0.55)) begin
SP_S_tau = -0.955*SP_S_eta +
           0.979*ln(SP_S_a * inv_G02);
end
if((Vgs>=(-2.0+VFBO+0.55))&&(Vgs < (-1.75 + VFBO +
   0.55))) begin
SP_S_tau   = -1.022*SP_S_eta + 1.02 *
             ln(SP_S_a * inv_G02);
end
if((Vgs>=(-1.75+VFBO+0.55))&&(Vgs < (-1.0 + VFBO +
   0.55))) begin
SP_S_tau   = -0.002 * SP_S_eta + 0.9875 *
             ln(SP_S_a * inv_G02);
end
if((Vgs>=(-1.0+VFBO+0.55))&&(Vgs < (-0.65 + VFBO +
   0.55))) begin
SP_S_tau   = -1.84*SP_S_eta + 1.5*ln(SP_S_a *
             inv_G02);
end
if(Vgs>=(-0.65 + VFBO + 0.55)) begin
SP_S_tau   = -1.94*SP_S_eta + 1.5*ln(SP_S_a *
             inv_G02);
end
nu     = SP_S_a + SP_S_c;
x_s    = -(SP_S_eta +SP_S_a * SP_S_tau / nu);
if((Vgs>=(-1.75 + VFBO + 0.55))&&(Vgs < (-1.0 +
   VFBO + 0.55))) begin
x_s    = -(0.04*SP_S_eta + SP_S_a * SP_S_tau/nu);
end
if((Vgs>=(-1.0 + VFBO + 0.55))&&(Vgs < (-0.65 +
   VFBO + 0.55))) begin
x_s    = -(1.15*SP_S_eta + SP_S_a * SP_S_tau/nu);
end
if(Vgs>=(-0.65 + VFBO + 0.55)) begin
x_s    = -(1.3*SP_S_eta + SP_S_a * SP_S_tau/nu);
end
end
else begin
//inversion and depletion regions
SP_S_bx    = xn_s + 3.0;
SP_S_eta = 0.5*(xg + SP_S_bx - pow(((xg -
             SP_S_bx)*(xg - SP_S_bx)+(5.0)),0.5));
SP_S_temp = xg - SP_S_eta;
SP_S_a     = SP_S_temp * SP_S_temp;
SP_S_c     = 2.0 * SP_S_temp;
SP_S_tau = xn_s - SP_S_eta + 0.5*ln(SP_S_a/G02);
nu       = SP_S_a + SP_S_c;
if(Vgs<(-0.5+VFBO+0.55)) begin
x_s    = 1.7*SP_S_eta + SP_S_a * SP_S_tau / nu;
end
if(Vgs>=(-0.5 + VFBO + 0.55)) begin
x_s    = 1.5*SP_S_eta + SP_S_a * SP_S_tau / nu;
end
if(Vgs>(-0.45 + VFBO + 0.55)) begin
x_s    = 0.6*SP_S_eta + SP_S_a * SP_S_tau / nu;
end
```

```
if(Vgs>(-0.4 + VFBO + 0.55)) begin
x_s  =  0.5*SP_S_eta + SP_S_a * SP_S_tau / nu;
end
if(Vgs>(-0.3 + VFBO + 0.55)) begin
x_s  =  0.6*SP_S_eta + SP_S_a * SP_S_tau / nu;
end
if(Vgs>(-0.2 + VFBO + 0.55)) begin
x_s  =  0.7*SP_S_eta + SP_S_a * SP_S_tau / nu;
end
if(Vgs>(VFBO + 0.55)) begin
x_s  =  0.92*SP_S_eta + SP_S_a * SP_S_tau / nu;
end
if(Vgs>(0.15 + VFBO + 0.55)) begin
x_s  =  0.938*SP_S_eta + SP_S_a * SP_S_tau / nu;
end
if(Vgs>(0.25 + VFBO + 0.55)) begin
x_s  =  0.94*SP_S_eta + SP_S_a * SP_S_tau / nu;
end
end
//Calculation of the intrinsic charge//
temp      = 1.0 / (2.0 + x_s * x_s);
xi0s      = x_s * x_s * temp;
delta_1s  = exp(x_s);
Es        = 1.0 / delta_1s;
delta_1s  = delta_ns * delta_1s;
Dm        = delta_1s - delta_ns * (x_s + 1.0 + xi0s);
Pm        = x_s - 1.0 + Es;
Xgm       = G0 * pow((Dm + Pm),0.5);
Voxm      = xgm * phit;
COX       = `EPSO * EPSROXO * W * L / TOXO;
Qg        = Voxm * COX;
```

VI. CONCLUSION

The MOS capacitor behavior of high-k HfO_2-Ta_2O_5 layer stack was studied proceeding from the surface potential description embedded in the intrinsic charge PSP model. The model was coded in Matlab for fitting purposes. The curves are fitted to the experimental data published in [7] with highly accurate matching – below 3% error; the curves also meet the natural asymptotic expectations. The optimized code was then programmed in Verilog-A to integrate with the Spectre simulator of Cadence Design Framework CAD tool.

In addition to the simulation results themselves the model realization represents a straightforward example of an all-purpose methodology for coding compact model equations in a portable, open-source environment applicable to various simulation platforms.

ACKNOWLEDGMENT

This paper is prepared in the framework of Contract No. ДТК – 02/50/17.12.2009.

REFERENCES

[1] G. Angelov, T. Takov, and St. Ristiç "MOSFET Models at the Edge of 100-nm Sizes", *Proc. of the 24th Intl. Conf. on Microelectronics (MIEL 2004)*, Niš, Serbia and Montenegro, Vol. 1, pp. 295-298, 2004.

[2] The International Technology Roadmap for Semiconductors http://www.itrs.net

[3] R. Chau et. al. "Application of High-k Dielectrics and Metal Gate Electrodes to Enable Silicon and Non-Silicon Logic Nanotechnology", *Microelectronic Engineering*, Vol.80, pp. 1-6, 2005.

[4] G. D.Wilk, R. M.Wallace, and J. M. Anthony, "High-k Gate Dielectrics: Current Status and Materials Properties Considerations," *J. Appl. Phys.*, Vol. 89, pp. 5243–5275, 2001.

[5] M. Houssa, ed., "*High-k Gate Dielectrics*", Institute of Physics Publishing, Bristol and Philadelphia, 2004. ISBN 0-7503-0906-7.

[6] E. Atanassova and A. Paskaleva, "Challenges of Ta_2O_5 as high-k dielectric for nanoscale DRAMs", *Microelectronics Reliability* 47(6), pp. 913-923, 2007.

[7] E. Atanassova, M. Georgieva, D. Spassov, and A. Paskaleva, "High-k HfO_2-Ta_2O_5 mixed layers: Electrical characteristics and mechanisms of conductivity", *Microel. Engin.* **87**, pp. 668-676, 2010.

[8] M. Mierzwinski, P. O'Halloran, B. Troyanovsky, R. Dutton, "Changing the paradigm for compact model integration in circuit simulators using Verilog-A", *Technical Proceedings of the 2003 Nanotechnology Conference and Trade Show (Nanotech 2003)*, Vol. 2, February 2003, pp. 376–379.

[9] S.M. Sze, "*Physics of Semiconductor Devices*", Wiley, 1981.

[10] G. Angelov, T. Takov, and St. Ristic "MOSFET Models at the Edge of 100-nm Sizes", *Proc. 24th Intl. Conf. on Microelectronics (MIEL)*, Niš, Serbia & Montenegro, Vol. 1, pp. 295-298, May 2004.

[11] G. Angelov, N. Bonev, R. Rusev, M. Hristov, A. Paskaleva, D. Spassov, "Verilog-A Model of a High-k HfO_2-Ta_2O_5 Capacitor", *Proc. of 18th International Conference Mixed Design of Integrated Circuits and Systems (MIXDES2011)*, pp. 470-475, Gliwice, Poland, June 16-18, 2011. ISBN 978-83-932075-0-3.

[12] K. J. Yang, C. Hu, "MOS Capacitance Measurements for High-Leakage Thin Dielectrics", *IEEE Transactions on Electron Devices*, Vol. 46, No. 7, July 1999.

[13] G. Gildenblat, X. Li, W.Wu, H. Wang, A. Jha, R. van Langevelde, G.D.J. Smit, A.J. Scholten and D.B.M. Klaassen, "PSP: An Advanced Surface-Potential-Based MOSFET Model for Circuit Simulation ", *IEEE Transactions on Electron Devices*, Vol. 53, No. 9, pp. 1979-1993, September 2006.

MIXDES 2012, 19th International Conference *"Mixed Design of Integrated Circuits and Systems"*, May 24-26, 2012, Warsaw, Poland

Symbolic Analysis in Gyrator-Capacitor Filters

Piotr Katarzynski, Michal Melosik, Mariusz Naumowicz, Szymon Szczesny
Chair of Computer Engineering
Poznan University of Technology
Poznan, Poland
piotr.katarzynski@put.poznan.pl

Abstract—**The paper deals with symbolic analysis used for obtaining the transfer functions of filtering structures based on gyrator-capacitor prototype circuits. Following sections discuss the principle of symbolic analysis that utilizes structural numbers to represent data. There are provided algorithms for obtaining the frequency responses of the filter. The work is summarized with an example filter design treated as the case study for the discussed issue.**

Index Terms—**symbolic analysis, gyrator-capacitor, EDA, sensitivity**

I. INTRODUCTION

Filtering is the most common operation applied to electrical signals. It allows to cope with redundant data as well as attenuate the unwanted noise disturbances. Hence the filters are present in a variety of applications covering communication, medicine and industry applications. The still raising demand on mobility, lightweight and low power consumption forces the inclusion of such structures into integrated circuits. Nowadays most of the signals is processed in digital domain. However there is a class of the signals that are produced by sensing devices or those oriented for wireless communication. These signals may be initially processed in the analog domain and filtering is one of the most basic elements in that issue. In terms of compact circuit design, the analog filters should be connected to the remaining digital part of the processing unit preferably in one chip.

The concept of analog signal processing with passive LC filters is well known. Such filtering structures have good selectivity and are organized into simple, ladder based topologies[18]. However, implementing such circuits on chips causes many problems because of the presence of inductive elements. The inductors remain bulky and are hard to design to be implemented as the parts of integrated circuits in classical CMOS technology. Some alternative processes like the *MagnaChip* are nowadays extensively exploited for the use in mixed signal processing [9,10]. The alternative approach in LC filters implementation assumes using the prototype circuits. A floating inductance L may be replaced by two gyrators g and a capacitance C satisfying:

$$L = Cg^2. \tag{1}$$

Gyrator-capacitor modeling is a simple way for representing magnetic components of the circuits and dealing with their imperfections such as core saturation. This approach is used mainly in power electronics however it may also be utilized for filter design issue[11,3]. The gyrator-capacitor circuit may be thus treated as the fully functional equivalent of the corresponding *LC* counterpart. These circuits may be realized in Switched Current (SI), Switched Capacitance (SC) techniques or either by using the operational amplifiers (OTA-C) [17]. All the mentioned techniques are doable in the standard CMOS fabrication process. The design of electric filters by using gyrator-capacitor equivalents is weakly supported by Electronic Design Automation systems. In this paper we propose the g-C circuit analyzer for obtaining the symbolic transfer function of the given multiport gyrator-capacitor structure. The formulation of symbolic transfer function and further analysis is performed in *s* domain which is the most popular approach [14].

The following sections present the basic concept of gyrator-capacitor circuit representation. The symbolic node potential method is introduced as the one applied for the analysis. Finally the circuit simulator is presented. The paper is summarized with some example designs.

II. DESCRIBING THE TOPOLOGY

The gyrator-capacitor prototype circuit may be represented in various ways. The most common assume using the textual files with predefined syntax of *Spice* or *VHDL-AMS* [13,4]. In our approach the prototype circuit is composed of logical blocks. Moreover, we assume, that both the circuits' excitations and responses are voltages. The single, fully equipped g-C block is presented in Fig. 1.

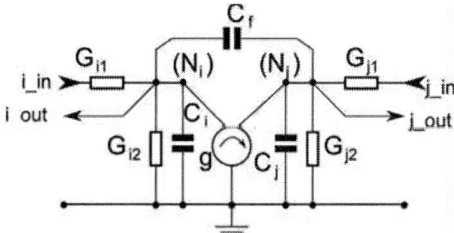

Figure 1. Single gyrator-capacitor block

Each block is spanned between two different nodes of the circuit. The nodes are numbered with integral numbers starting from 1. Besides, each block shares the common reference node which by default has the number 0. The presence of the gyrator g and capacitors C_i C_j C_f in the block is optional and depends on design requirements. The series conductances G_{i1}, G_{j1}

appear only for excitation signals fed by connections *i_in* or *j_in*, whereas the responses may be acquired from parallel conductances G_{i2}, G_{j2} by output connections *i_out, j_out*. In order to provide the agile form of expressing the topology for gyrator-capacitor circuits, the appropriate parsing module was implemented capable of reading the VHDL-AMS description. Fig.2 presents the example topology consisting of three blocks *B1,B2,B3*.

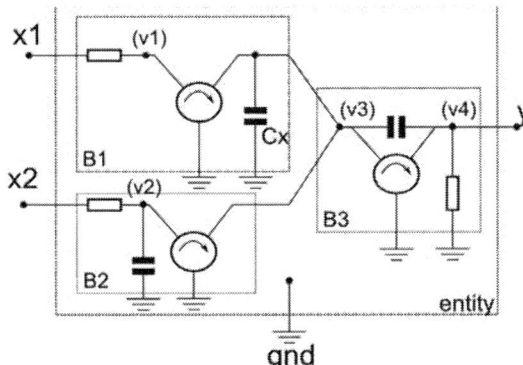

Figure 2. Example gyrator-capacitor circuit with block partitioning within design entity.

Each block is spanned between two internal nodes (marked in brackets) and the common, reference node. The VHDL-AMS description for such formulated topology is presented in Listing 1. The block partitioning may be performed in various ways for the same topology. To give an example, the C_x capacitance may be incorporated as C_j in *B1* or as C_i in *B3*.

```
library VLSI;
use VLSI.SI.all;

entity example_circuit is port (
terminal input x1 : electrical;
terminal input x2 : electrical;
terminal output y : electrical;
terminal ground gnd : electrical);
end entity filter;

architecture example_arch of example_circuit is

variable v1,v2,v3,v4: real;

begin
    B1 : GC_BLOCK generic map(ig=>1,Cj=>1)
    port map(Ni=>v1,Nj=>v3,i_in=>x1);
    B2 : GC_BLOCK generic map(ig=>1,Ci=>1)
    port map(Ni=>v2,Nj=>v3,i_in=>x2);
    B3 : GC_BLOCK generic map(ig=>1,Cf=>1)
    port map(Ni=>v3,Nj=>v4,j_out=>y);
end architecture;
```

Listing 1. Example definition of gyrator-capacitor circuit assuming the VHDL-AMS syntax.

The syntax of the VHDL-AMS model was tailor to meet the block partitioning idea. Hence, there are instances of GC_BLOCK component that represent particular blocks. The exact definition of such component is currently unavailable in terms of VHDL-AMS hence such component may be treated as software dependent. In other words, the script will be positively

validated for occurrences of g-C blocks only with our dedicated software. The generic mapping allows to customize the block's content whereas the port mapping defines the actual interconnections between blocks including internal nodes and external excitation/output signals.

Such formulated file may be then parsed in order to extract the basic information about the elements that appear in the circuit. Each element E is represented by the data structure having the following fields:

$$E=\{type,n,m,symbol\}. \qquad (2)$$

The *type* field identifies the element as one of the following $G_{i1},G_{i2},G_{j1},G_{j2},C_i,C_j,C_f$ and g according to the markings in Fig.1 The fields n and m store integral numbers representing the circuit nodes to which the element is connected. If the element is connected to the reference node or one of the excitation signals, the corresponding element's field is assigned 0. The *symbol* field represents unique numeric identifier for the element. The elements are numbered in order of appearance within blocks defined in VHDL-AMS input model. The list E of elements has N_i input elements which are of type G_{i1} or G_{j1} and N_j number of output elements G_{i2} and G_{j2}

III. SYMBOLIC NODE POTENTIAL METHOD

After obtaining the list of circuit's elements it is possible to prepare the symbolic matrices for the node potential method assuming the general matrix equation:

$$YV=I. \qquad (3)$$

Where Y is the admittance matrix, V is the vector of unknown node potentials and I is the vector holding input excitations expressed by using series conductances as:

$$G_{in}V_{in}=I_{in}. \qquad (4)$$

Where V_{in} is the actual voltage excitation associated with input conductance G_{in} of type G_{i1} or G_{j1}. This approach allows to formulate the transfer functions for the circuit with respect to all its inputs and outputs. The transfer function associated with signal passing from *i-th* input to *j-th* output of the structure is given by the ratio of two determinants:

$$H_{ij}(s)=det(Y_{ij}(s))/det(Y(s)). \qquad (5)$$

Where Y is the admittance matrix and Y_{ij} is the admittance matrix where *j-th* column is replaced by the I vector including only the excitations connected to its *i-th* row. As we express the set of elements in a list with unique symbol name assigned to each element it is then possible to obtain the symbolic form of transfer functions. The major difficulty in such approach is providing the effective algorithms for calculating the determinants for symbolic matrices. In case of matrices having numerical data, the partial results of determinant calculation merge with each other. Hence as we take the determinant for matrix A

$$A = \begin{vmatrix} 1 & 3 \\ 2 & 4 \end{vmatrix}. \qquad (6)$$

The determinant is still a single number equal to *-2*. However the result of determinant calculation for symbolic data in fact represents the history of mathematical operations that were performed. Thus taking the determinant for symbolic matrix B

$$B = \begin{vmatrix} a & b \\ c & d \end{vmatrix}. \tag{7}$$

Results with the determinant describing the sequence of consecutive multiplications and final subtraction with respect to the priority order of calculations

$$det(B)=ad-bc. \tag{8}$$

In order to perform symbolic analysis more efficiently the agile form of representing symbolic matrices must be introduced [15]. In our software we used structural numbers for storing the symbolic expressions. These numbers were introduced originally by Bellert [1,2] and their algebra was proposed in order to formalize the mathematical methods of synthesis and analysis of electrical circuits. Nowadays that approach remains inefficient as computer enhanced methods of analysis are available. It is worth to stress however, that structural numbers may represent symbolic polynomials of given rank. Moreover, their algebra defines multiplication and addition as two major arithmetic operations. It is then possible to represent the matrix elements from (5) as structural numbers. Let us consider the example of symbolic polynomial W defined with respect to Laplace's operator s.

$$W(s)=(x_1x_2+x_3x_4)s^2+(x_1^2+2x_3x_4)s+2x_2x_3+x_1^2. \tag{9}$$

It may be expressed as the structural number assuming the notation presented in Listing 2

```
%! BEGIN SN
%! RANK 2
%! LENTRIES 6
%! COLUMNS 2
%! S 2
    1   1   2
    1   3   4
%! S 1
    1   1   1
    2   3   4
%! S 0
    2   2   3
    1   1   1
```

Listing 2. Example of structural number representing the symbolic polynomial.

The structural number may be represented in textual form. This makes it readable by humans and third party computer software. Each structural number holds several rows of data. The rows starting from '%!' give diagnostic and statistical information. The remaining ones store the series of numbers that refer to the symbolic notation. First number in each row defines the constant multiplier. The remaining ones denote the symbolic variables being multiplied. The rows are grouped with respect to the descending powers of 's'. Representing symbolic variables and expressions with rows of numbers makes it easy to implement such data in PC software.

Let us now consider the algorithm responsible for populating the admittance matrix Y with structural numbers representing the circuit's elements. It uses the list E of recognized elements and the routine *Add(multiplier,rank,index)* which assigns a new row to the given structural number. The

row holds one number *index*, has the associated constant *multiplier* and is assigned the *rank* which denotes the power of Laplace's operator. The appropriate algorithm is presented in Listing 3 Initially it prepares N by N matrix of empty structural numbers. The N parameter denotes the number of internal vertices within the circuit's topology. Then it passes throughout all the elements inside the Y matrix indexed by actual row r and column c. At each iteration the algorithm matches the elements that belong to actual combination of vertices r and c and puts their symbolic representations to the appropriate structural numbers with associated rank and multiplication. The rank is assigned 1 for capacitors thus they appear in structural number as sC expressions. The remaining elements have rank 0. Additional set of conditions solves the proper representation of gyrators with respect to their direction.

```
Y = new SN[N][N]
for r = 1 to N
    for c = 1 to N
        if r == c then
            foreach e in E
                if e.m == r or e.n == c then
                    if e.type in {C_i, C_j, C_f} then
                        Y[r][c].Add(1,1,e.symbol)
                    if e.type in {G_i1, G_j1, G_i2, G_j2} then
                        Y[r][c].Add(1,0,e.symbol)
            else
                if e.n == r and e.m == c then
                    if e.type == C_f then
                        Y[r][c].Add(-1,1,e.symbol)
                    if e.type == g and e.m<e.n then
                        Y[r][c].Add(-1,1,e.symbol)
                    if e.type == g and e.m>e.n then
                        Y[r][c].Add(1,1,e.symbol)
                if e.m == r and e.n == c then
                    if e.type == C_f then
                        Y[r][c].Add(-1,1,e.symbol)
                    if e.type == g and e.m<e then
                        Y[r][c].Add(1,1,e.symbol)
                    if e.type == g and e.m>e.n then
                        Y[r][c].Add(-1,1,e.symbol)
```

Listing 3. The algorithm populating the symbolic admittance matrix

Formulation of the Y_{ij} may be realized by extending the general algorithm for the Y matrix. The actual input excitation applied to the *i-th* ($i=1..N_i$) node of the circuit passes through the *ei* element being either G_{i1} or G_{j1} . The actual output node pointed by *j-th* ($j=1..N_j$) is one of the nodes with G_{i1} or G_{j2} connected.

```
Yij = Y
for r = 1 to N
    Yij[r][j].Clear()
    if ei.m == r or ei.n == r then
        Yij[r][j].Add(1,0,ei.symbol)
```

Listing 4. The algorithm populating the symbolic admittance matrix

Such formulated matrices may be then processed by any algorithm for determinant calculation. It is essential to provide the algorithm that performs multiplications and additions of matrix elements as these operations are doable for elements being structural numbers [1]. The result of determinant calculation is also the structural number representing the polynomials that appear in symbolic transfer function. Two algorithms were considered for calculating determinants. First of them was based on the permutations' method according to the Leibnitz formula [16]. In that case the result is expressed as the sum of all permutations among matrix's elements. This approach was however rejected as it proved to produce a vast number of permutations that reduce with each other. The reduction stage involved high computational power as compared to the second method that was based onto the Laplace's expansion.

IV. SIMULATING THE FREQUENCY RESPONSES

Symbolic form of the transfer function state a decent model for evaluating frequency responses. In terms of electric filters, the gyrator-capacitor prototype circuit shall be investigated for the magnitude and phase response. After calculating the symbolic determinants we obtain two symbolic polynomials represented by structural numbers. These may be further utilized for simulation purposes. Consider vector P storing the actual values for elements that compose the prototype circuit. This vector may be obtained during the circuit synthesis by matching coefficients[6, 5]. The actual realization of the synthesis may involve algebraic, direct pattern search [8, 12] or genetic approach [7] As the structural numbers store unique numeric identifiers for symbolic variables they may be used for indexing the associated values form P. Listing 5 summarizes the issue

```
N = {0,…,NUM.rank}
D = {0,…,DEN.rank}

foreach row in NUM
    tmp = 1
    foreach r in row.data
        tmp = tmp*P[r]
    N[r.rank] = N[r.rank]+tmp

foreach row in DEN
    tmp = 1
    foreach r in row.data
        tmp = tmp*P[r]
    D[r.rank] = D[r.rank]+tmp

complex s = (0,omega)

complex tmp_n = (0,0)
complex tmp_d = (0,0)
complex smp  = (0,0)

for i=0 to 1 NUM.rank
tmp_n = tmp_n + N[i]*s^i

for i=0 to 1 DEN.rank
tmp_d = tmp_d + D[i]*s^i

smp = tmp_n/tmp_d

mag = 20log(abs(smp))
ph = atan2(smp.re,smp.im)
```

Listing 5. The algorithm for calculating the magnitude and phase responses

In order to perform frequency response analysis three data structures are needed. *NUM* and *DEN* are structural numbers representing the determinants obtained in (5) . Vector P stores the values for symbolic variables representing circuit's elements. Numerator and denominator are used to calculate the actual coefficients of polynomials N and D. These are further

used for evaluation at given angular frequency *omega*. Finally the response sample *smp* is obtained by following the equation

$$smp=H(j\omega)=N\,(j\omega)/D(j\omega) \qquad (10)$$

The *smp* value may be then used for calculating the values of magnitude *mag* and phase *ph*.

V. RESULTS

Let us consider the 5-th order elliptic LP filter realized by using the gyrator-capacitor prototype circuit [7].

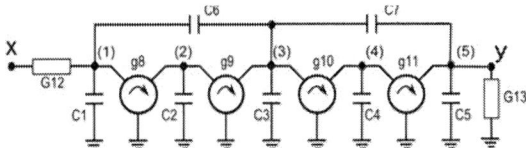

Figure 3. Example gyrator-capacitor circuit with block partitioning within design entity.

The VHDL-AMS model for such structure is presented in listing 6

```
library VLSI;
use VLSI.SI.all;
entity filter is port
(
terminal input x : electrical;
terminal output y : electrical;
terminal ground gnd: electrical);
end entity filter;

architecture aa of filter is
variable v1,v2,v3,v4,v5 : real;
begin
B1 : GC_BLOCK generic map(ig=>1,Ci=>1)
port map (Ni=>v1,Nj=>v2,i_in=>x);
B2 : GC_BLOCK generic map(ig=>1,Ci=>1)
port map (Ni=>v2,Nj=>v3);
B3 : GC_BLOCK generic map(ig=>1,Ci=>1)
port map (Ni=>v3,Nj=>v4);
B4 : GC_BLOCK generic map(ig=>1,Ci=>1,Cj=>1)
port map (Ni=>v4,Nj=>v5, j_out=>y);
B5 : GC_BLOCK generic map(Cf=>1)
port map (Ni=>v1,Nj=>v3);
B6 : GC_BLOCK generic map(Cf=>1)
port map (Ni=>v3,Nj=>v5);
end architecture;
```

Listing 6. VHDL-AMS file representing the topology of the discussed filter.

The file was processed by the *gc_analyser* software written in *C++* language as the console application. The software implements the presented method of symbolic analysis. It produced two structural numbers being the determinants of the transfer function. First determinant refers to the denominator and originates from the admittance matrix Y whereas the second determinant is the numerator for input x to output y. Additionally the *gc_analyser* software produced the matrices formulated in Matlab compatible syntax. Listing 7 presents these symbolic matrices.

```
syms s x1 x2 x3 x4 x5 x6 x7 x8 x9 x10 x11 x12
x13 ;
% ---------------------------------------------
% Admittance matrix
% ---------------------------------------------
Y =[x1*s+x6*s+x12, x8, -x6*s, 0 ,0;
    -x8, x2*s, x9, 0, 0;
    -x6*s, -x9, x3*s+x7*s, x10, -x7*s;
    0, 0, -x10, x4*s, x11;
    0, 0, -x7*s, -x11,x5*s+x7*s+x13 ]
% ---------------------------------------------
% INPUT x versus OUTPUT y
% ---------------------------------------------
x_y =[x1*s+x6*s+x12, x8, -x6*s, 0, x12;
    -x8, x2*s, x9, 0, 0;
    -x6*s, -x9, x3*s+x7*s, x10, 0;
    0, 0, -x10, x4*s, 0;
    0, 0, -x7*s, -x11, 0 ]
% ---------------------------------------------
```

Listing 7. The symbolic matrices exported to Matlab script

The determinants for such matrices may be obtained in Matlab by using the symbolic analysis toolbox and the *det()* function. Listing 8 presents the textual form of the determinant produced in *Matlab* compared to the structural number that was produced in *gc_analyser*. The numerators for transfer function were presented as the symbolic denominator for that case is relatively vast structural number with 64 rows of symbolic data.

```
det(x_y)
ans =

x8*x9*x10*x12*x11+x8*x9*x7*s^2*x12*x4+x2*s^2*x6
*x10*x12*x11+x2*s^4*x6*x7*x12*x4
---------------------------------------------
%! S 4
    1   2   4   6   7  12

%! S 3

%! S 2
    1   2   6  10  11  12
    1   4   7   8   9  12

%! S 1

%! S 0
    1   8   9  10  11  12
```

Listing 8. Symbolic determinant obtained in Matlab and its counterpart produced by *gc_analyser* as the structural number

The results are exactly the same, which proves the reliability of the software. As we consider the time needed for analysis, the gc_analyser makes it at comparable speed especially for more complicated circuits. Table I presents execution times for calculation of the symbolic determinants in *Matlab* and in *gc_analyser*. The time amount measured for gc_analyser involve symbolic matrix formulation as well as the solving routine and stroring the outcome into file. In case of *Matlab* the appropriate times were measured only for the determinant calculation phase.

TABLE I. COMPARISON OF EXECUTION TIMES

Example	Parameters			
	Nodes count	Elements count	gc_analyser	Matlab
5ord	5	13	0.00s	0.01s
11ord	11	28	1.09s	0.35s
7pair	9	30	0.28s	0.28s
multi8	7	38	2.32s	1.58s
multi11	7	42	4.13s	2.50s

Having the circuit's model expressed in one hardware description language makes it possible to easily convert that model into another, recognizable by circuit simulating software. The *gc_analyser* utility was equipped with the option for writing *hSpice* files representing gyrator-capacitor circuits. Listing 9 presents the *hSpice* model of the discussed filter prepared for AC analysis.

```
gc_model
vin 6 0 dc 0 ac 1
C1:  1 0 1.363951
C2:  2 0 0.836127
C3:  3 0 0.889401
C4:  4 0 0.786910
C5:  5 0 0.615968
C6:  1 3 0.160094
C7:  3 5 0.308869
xg8:  1 2 gyrator g=0.538639
xg9:  2 3 gyrator g=0.293381
xg10: 3 4 gyrator g=0.324530
xg11: 4 5 gyrator g=0.404189
R12: 6 1 1.083121
R13: 5 0 4.964158
.subckt gyrator in out g=1
G1 0 out poly(1) in 0 0 g
G2 in 0 poly(1) out 0 0 g
.ends gyrator
.options post
.ac dec 10000 0.01 1
.end
```

Listing 9. The *hSpice* file obtained from initial VHDL-AMS model.

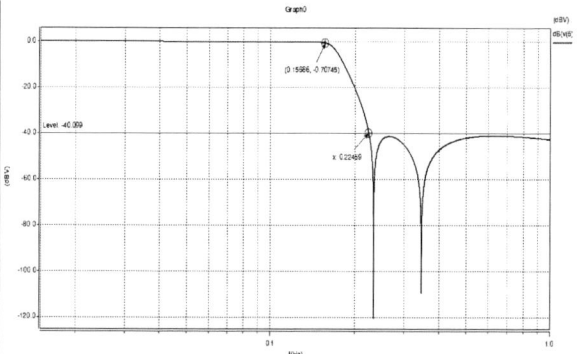

Figure 4. Magnitude response obtained with *hSpice*

The parameters' values were assigned after completing the synthesis by matching the coefficients from algebraic and symbolic transfer functions [5]. Such formulated model was analyzed with *hSpice*. Fig. 4 presents the magnitude response over normalized frequency. The filter design requirements assumed the elliptic LP approximation with attenuation of *40dB* in the stop-band and maximum allowable ripple amount of *0.1dB* in the pass-band. The frequency point at which the transitional band starts was set to *1 rad/s* which equals to *0.159 Hz*.

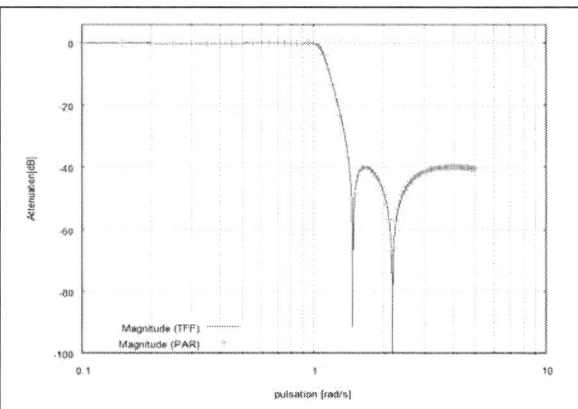

Figure 5. Magnitude response obtained with the discussed software

Fig. 5 presents the magnitude response that was obtained by using symbolic transfer functions expressed as structural numbers. The algorithm of obtaining subsequent samples of frequency response may be then easily expanded to store such samples in data files that are readable by graph plotting software. In case of the discussed filter the *GNUPlot* tool was used for visualization. Comparison between two frequency response curves from Fig. 4 and Fig. 5 claimed the obtained simulation software to be reliable.

VI. SUMMARY

Symbolic analysis is the essential step in computer enhanced design and synthesis of analogue circuit. It states the design endpoint for parameter matching and allows to obtain circuits' simulation. Our studies proved that structural numbers are good form of symbolic data representation. The textual format assumed to store structural numbers makes them readable both for humans and the third party software tools. When we construct the matrices utilized in node potential method as the arrays of structural numbers, the process of obtaining symbolic determinants simplifies substantially. The basic algebraic operations are defined for structural numbers, thus the classical algorithms of determinant computation may be implemented to deal with symbolic data. The algorithm of Laplace's extension proved to be most reliable among those investigated in our studies. When matched with the *Matlab* software the proposed algorithm for circuit analysis has the comparable time complexity. The presented algorithm of symbolic analysis may be also expanded for multiport networks, especially those oriented for preconditioning analogue 2D signals in vision chips. The presented form of

symbolic data representation makes it possible to simulate the circuits' frequency responses at broad range of frequencies without the need of re-launching the topological analysis. This makes the proposed software more favorable in comparison to *hSpice*.

REFERENCES

[1] S. Bellert, "Topological analysis and synthesis of linear systems," Journal of the Franklin Institute, ISSN 0016-0032, vol. 274, no. 6, pp. 425-443, 1962.

[2] G. Bongiovani "A property of structural numbers and cut sets," Calcolo Springer Milan, vol. 16, no. 1, pp. 1-3, 1979.

[3] D. Hamill., "Gyrator-Capacitor Modelling: A Better Way of Understanding Magnetic Components," Conference for Applied Power Electronics, vol. 1,pp. 326-332, 1994.

[4] D.C. Hamill, "Lumped equivalent circuits of magnetic components: the gyrator-capacitor approach," Power Electronics, IEEE Transactions on , vol.8, no.2, pp.97-103, 1993.

[5] A. Handkiewicz, "Mixed-Signal Systems. A Guide to CMOS Circuit Design," Wiley-IEEE Press, ISBN 978-0-471-22853-0, 2002.

[6] A. Handkiewicz.,P. Katarzynski, et al., "Analog filter pair design on the basis of a gyrator-capacitor prototype circuit, " International Journal of Circuit Theory and Applications, DOI: 10.1002/cta.741, 2010.

[7] A. Handkiewicz, P. Katarzynski, et al., "Genetic Algorithms in Gyrator-Capacitor Filters," Elektronika, no. 12, pp. 56-58, 2011.

[8] R. Hooke,T. A. Jeeves T., "Direct Search Solution of Numerical and Statistical Problems," Journal of the ACM, vol. 8, no. 2, pp. 212-229,1961.

[9] Ch. Hwang, "An On-Chip Electromagnetic Bandgap Structure using an On-Chip Inductor and a MOS Capacitor," IEEE Microwave and Wireless Components Letters, vol. 21, no. 8, pp. 439-441, 2011.

[10] G. Kim, S.-B. Park, "CMOS LC-ring oscillator with adaptive purity control," Electronics Letters , vol. 41, no. 10, pp. 569-570, 2005.

[11] Y. Liang, B. Lehman, "Better understanding and synthesis of integrated magnetics with simplified gyrator model method," Power Electronics Specialists Conference, vol. 1, pp. 433-438, 2001.

[12] M. Melosik, M. Naumowicz, „Implementation and comparison for methods of solving non-linear algebraic equations in design of lossless prototype circuits," Engineering Thesis, Poznan University of Technology, 2008.

[13] F. Pecheux, C. Lallement, A. Vachoux, "VHDL-AMS and Verilog-AMS as alternative hardware description languages for efficient modeling of multidiscipline systems," IEEE Transactions on Computer-Aided Design of Integrated Circuits and Systems, vol. 24, no. 2, pp. 204- 225, 2005.

[14] R. Rutenbar, G. Gielen, B. Antao, "Canonical Symbolic Analysis of Large Analog Circuits with Determinant Decision Diagrams", Computer-Aided Design of Analog Integrated Circuits and Systems, Wiley-IEEE Press, pp. 344-361, 2002.

[15] C. Shi, T. Xiang-Dong, "Compact representation and efficient generation of s-expanded symbolic network functions for computer-aided analog circuit design," IEEE Transactions on Computer-Aided Design of Integrated Circuits and Systems, vol. 20, no. 7, pp. 813-827, 2001.

[16] D.W. Shin,"The Permutation Algorithm for Non-Sparse Matrix Determinant in Symbolic Computation," Applied Mathematics and Computation, vol. 192, no. 2, pp. 382-388, 2007.

[17] F. Yuan, "CMOS gyrator-C active transformers," Circuits, Devices & Systems, IET , vol. 1, no. 6, pp. 494-508, 2007.

[18] A. I. Zverev, "Handbook of filter synthesis," J. Wiley & Sons, 1967.

MIXDES 2012, 19th International Conference *"Mixed Design of Integrated Circuits and Systems"*, May 24-26, 2012, Warsaw, Poland

Technology and Device Design and Optimization for the MOSFET Hall Sensor on SOI Structure

Leonid Dolgiy, Ivan Lovshenko,
Vladislav Nelayev, Ibrahim Shelibak
Micro- and Nanoelectronics Department
Belarusian State University
of Informatics and Radioelectronics
Minsk, Belarus
nvv@bsuir.by

Sergey Shvedov,
Arkady Turtsevich
Research and Development Center "BelMicroSystems"
Joint Stock Company "Integral"
Minsk, Belarus
office@bms.by

Abstract—**Results of the magnetosensitive device (a MOSFET Hall sensor on the SOI structure) manufacturing simulation are presented. Electrical features of the device were calculated and the optimization research of the process parameters influence on voltage-current characteristics of the device was made.**

Index Terms—**Semiconductor sensor; Hall effect; SOI structure; technology; simulation, device features.**

I. INTRODUCTION

Hall sensors are the most commonly used converters of a magnetic field into electric signal. They are employed in various science and technology fields (sensors of current, position, flow, rotation angle, vibration, rotational velocity, variable rotational speed drives, motor control and protection circuits, collectorless DC motors, contactless potentiometers, detectors of magnetic elements, tachometers, etc.) [1-5].

Principle of operation of these sensors is based on the Hall effect. Difficulties concerned with poor sensitivity, high noise level, and high bias voltage emerge when integral Hall sensors are manufactured in bulk silicon. Manufacturing a MOSFET Hall sensor on the SOI structure (MHS-SOI) allows these problems to be solved and a number of device parameters to be essentially improved. A sensor body thickness reducing with a reduction in the thickness of the device silicon layer in the SOI structure results in the enhancement of the MHS sensitivity to low magnetic fields. An increase of the sensor series resistance ensured that current flowing through the sensor decreases, with a consequent reduction of power consumption. An increase of the threshold sensitivity results in the dynamic range extension. The high threshold magnetic sensitivity and low input current determine much higher specific magnetic sensitivity, i.e. the transfer function slope of MHS-SOI, which is the ratio of the Hall electromotive force (EMF) value to the values of current and magnetic induction through the sensor. Also, the temperature operating range extends to 200 °C. Moreover, devices manufactured on the SOI substrates are of the enhanced radiation resistance.

The intent of this work was a construction design and technology development for MHS-SOI, a determination and

optimization of the process variables substantially influencing electrical characteristics of the device.

The manufacturing process flow for the MHS-SOI and electrical characteristics of the device were simulated using SILVACO Inc. software package [6].

II. MHS-SOI DESIGN

The MOSFET Hall sensor on the SOI structure comprises a semiconductor resistor, the active thickness of which is modulated by applying suitable potentials to the top and bottom (substrate) gates of the field-effect structure. The induced channel of the MOSFET is used as a conducting layer, while source and drain regions act as current contacts of the sensor. Fig. 1 shows a MHS-SOI construction.

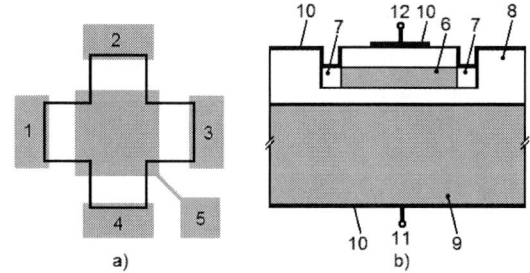

Figure 1. MHS-SOI construction.
(a) top view: 1, 3 – Hall electrodes; 2, 4 – current electrodes; 5 – top gate electrode; (b) cross-sectional view: 6 – P-type operating silicon layer (well in the MOSFET); 7 – MOSFET source and drain (current contacts of the Hall sensor); 8 – insulating silicon oxide of the SOI structure; 9 – silicon substrate; 10 – aluminum metallization layers to source, drain, and gate regions, and to substrate; 11 – contact to substrate; 12 – contact to gate)

Two types of the SOI structures, namely, UNIBOND manufactured by SOITEC Inc. using the Smart-Cut technology [7] and SIMOX [8] have a wide application.

Despite different fabrication methods, UNIBOND and SIMOX SOI structures are the same in design. In such the SOI structures, the silicon layer commonly has a relatively low doping level ($5 \cdot 10^{15}$ cm^{-3}) and can be of both n-type and p-type conductivity. The silicon layer thicknesses are from 0.1 μm (superthin SOIs) to 0.3 μm (thin SOIs). The insulating buried silicon oxide layers are from 0.2 μm to 0.5 μm in thickness.

The manufacturing process for the MHS-SOI has the following distinctive features [9-11]. The "Hall cross" structure formed by silicon strips typically 100 μm in length and 30 μm in width is made in the silicon layer. The source and drain regions (current contacts of the hall sensor), 10 μm in length silicon regions adjacent to the strip ends are formed by the phosphorus doping up to the concentration of $2 \cdot 10^{20}$ cm^{-3}. The surface of the silicon layer is oxidized to form the 40 nm thick gate silicon oxide layer. The aluminum layer is deposited over the gate silicon oxide layer to form a top gate. The silicon substrate with the deposited aluminum film serves as a bottom gate.

III. SIMULATION OF THE MHS-SOI TECHNOLOGY

The manufacturing process flow for the MHS-SOI consists of the following sequence of basic technological operations: a formation of a masking layer for further local oxidation of silicon; the local thermal oxidation of silicon (a formation of the LOCOS insulation); a phosphorus ion implantation into the exposed regions in the photoresistive mask to form source and drain regions of the MOSFET (current contacts of the Hall sensor); post-implantation drive-in; a gate dielectric formation; silicon oxide etching for further formation of contacts; a formation of current contacts by the aluminum deposition followed by aluminum local etching.

Fig. 2 shows a cross-sectional view of the MHS-SOI structure formed.

The boron concentration in the initial substrate (Fig. 2, 6), which is the channel region of the MOSFET, is $5 \cdot 10^{15}$ cm^{-3}, and the phosphorus concentration in the regions of ohmic contacts is $2.5 \cdot 10^{20}$ cm^{-3}.

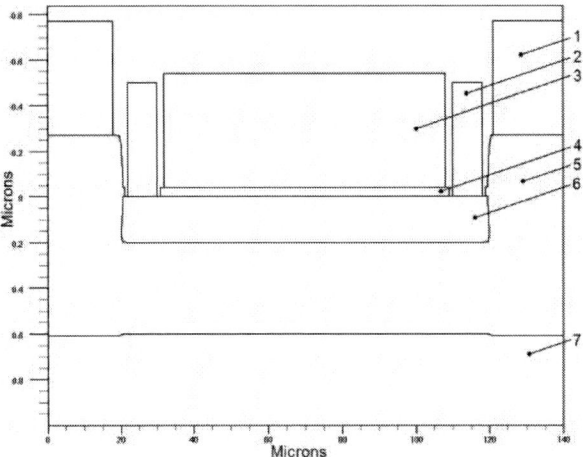

Figure 2.   MHS-SOI cross-sectional view.
1 – metallization; 2 – current contact; 3 – gate; 4 – gate dielectric; 5 – insulating silicon oxide; 6 – silicon of the active sensor region; 7 – substrate.

IV. RESULTS AND ANALISYS OF FIELD HALL EFFECT SENSOR I-V DEPENDENCES SIMULATION

Typical current-voltage characteristics of the MOSFET in the MHS-SOI structure used for further optimization of technological parameters resulted from the simulation.

Fig. 3 demonstrates a family of dependences of drain currents I_D on the gate voltage V_G for the MOSFET in the MHS-SOI structure. Drain voltage V_D is a variable parameter for these dependences. The calculations were made for $V_D = 1, 2$ and 3 V.

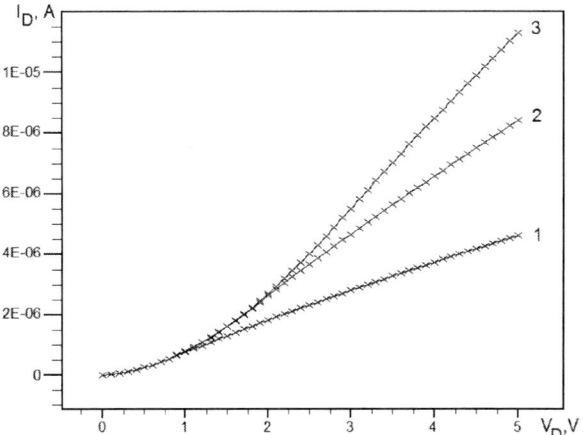

Figure 3.   Dependences of drain current I_D on gate voltage V_G for the MOSFET in the MHS-SOI structure under consideration. Here 1 – V_D =1 V; 2 – V_D = 2 V; 3 – V_D =3 V.

Fig. 3 shows a family of dependences of drain currents I_D on the drain voltage V_D at various gate voltages V_G (1.65; 3.3 and 5 V).

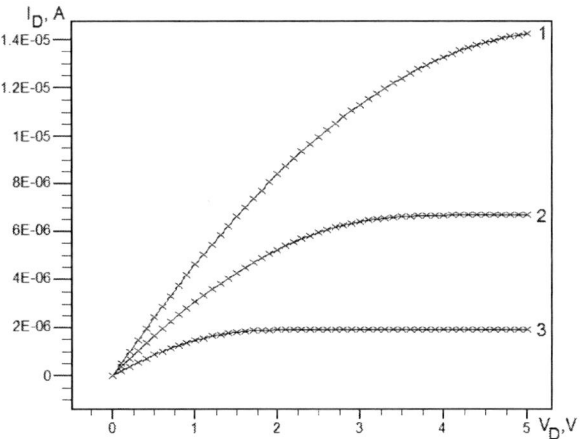

Figure 4.   Dependences of drain current I_D on drain voltage V_D at various gate voltages of the MOSFET in the MHS-SOI structure. 1 – V_G =1,65 V; 2 – V_G =3,3 V; 3 – V_G = 5 V.

As the simulation data indicated, a threshold voltage for the structure based on bulk silicon is higher (about 0.6 V) than for the MHS-SOI (about 0.2 V).

Breakdown voltage is considerably lower for the bulk silicon based structure.

To optimize electrical characteristics of the MOSFET in the MHS-SOI structure and discover ways for the increase in the sensitivity of the magnetic sensor, a dependence of current-voltage characteristics of the MOSFET on the most valuable parameters of the manufacturing process was studied.

Main results of the optimization research are as follow.

(1) Based on the analysis of the dependences of the drain currents I_D on the gate voltage V_G and drain voltage V_D as well as on the impurity concentration in the substrate (transistor channel), it is shown that the drain current increases by a factor of almost three from $5 \cdot 10^{-6}$ A до $1.4 \cdot 10^{-5}$ A as the boron concentration in the channel decreases by one half (from $1 \cdot 10^{16}$ cm^{-3} to $5 \cdot 10^{15}$ cm^{-3}).

(2) The studies of the dependences of the drain currents I_D on the gate voltage V_G and drain voltage V_D at various implantation doses (from $5 \cdot 10^{14}$ to $5 \cdot 10^{16}$ cm^{-2}) and energies (10 – 30 keV) of phosphorus ions implanted into source – drain regions of the MOSFET showed that doping level of these regions practically have no influence on the current-voltage characteristics of the MOSFET.

(3) The analysis of the MOSFET channel length influence on the current-voltage characteristics of the MHS-SOI showed that with the increase in the channel length, operating currents in the MOSFET decrease, causing less increment of the charge carrier concentration in the channel and resulting in the less MHS deterioration. Thus, with the increase in the channel length, MHS output characteristics depend on the gate voltage and drain voltage as well as drain current to a lesser extent. The channel length in these studies was varied from 38 μm (a half of the channel length of the basic structure equal to 76 μm) to 152 μm (doubled channel length of the basic structure).

(4) It is shown that when the thickness of gate dielectric decreases from 50 to 30 nm, electric features of the structure improves as well.

V. Conclusions

The manufacturing technology flow for the MOSFET Hall sensor structure is discussed in the paper. The simulation data of the MOSFET Hall sensor technology with the use of the Silvaco software package are described. The possibilities for the optimization of technological parameters in MOSFET Hall sensor technology are demonstrated. The obtained results were applied as the input parameters to calculate and optimize the electric features of the investigated MOSFET Hall sensor.

Acknowledgments

The work was supported by the Ministry of Education of Belarus, Project "Electronics 1.1.03". The authors would like to thank Dr. V. Bondarenko from the Micro- and Nanoelectronics Department of Belarusian State University of Informatics and Radioelectronics for fruitful discussion and valuable suggestions.

References

[1] J.E. Lenz, "A review of magnetic sensors," Proc. IEEE, Vol. 78, 1990.

[2] J. Heremans, "Solid state magnetic sensors and applications," Phys. D: Appl. Phys. Vol. 26, 1993.

[3] P. Ripka, "Magnetic sensors for industrial and field applications," Sensors and Actuators A, Vol. 41-42, 1994.

[4] H. Baltes, "Magnetic sensors," Semiconductor Sensors, S.M. Sze, New York: John Wiley, 1994.

[5] "HALL EFFECT SENSING AND APPLICATION," Honeywell MICRO SWITCH Sensing and Control, 2010.

[6] http://www.silvaco.com

[7] G. Celler, M. Wolf, "Smart Cut. A guide to the technology, the process, the products," SOITEC, July 2003.

[8] Maria J. Anc, "SIMOX," London, The Institution of Engineering and Technology, 2004.

[9] United States Patent No. 4772929 США, H 01 L 27/22, H 01 L 29/82, H 01 L 29/96, H 01 L 43/00. Hall sensor with integrated pole pieces / Manchester K.E.; Assignee and Inventor: Sprague Electric Company. – No 2059; Prior publication date 09.01.87; Published date 20.09.88.

[10] United States Patent No. H 01 L 27/14, H 01 L 29/82, H 01 L 29/84. Magnetic sensor integrated with CMOS / Berndt D.F., Peczalski A., Vogt E.E., Witcraft W.F.; Assignee and Inventor: Honeywell International Inc. – No 2004/0207031; Prior publication date 21.10.2004; Published date 07.06.2005.

[11] United States Patent No. 4700211 H 01 L 27/22, H 01 L 43/00, H 01 L 29/72, G 01 R 33/02. Sensitive magnetotransistor magnetic field sensor / Popovic R., Baltes H.P.; Assignee and Inventor: LGZ Landis & Gyr Zug AG. – № 514881; Prior publication date 18.07.83; Published date 13.10.87.

Verilog-A Modeling of Electrical Circuit with Adding Element Based on Branched Hydrogen Bonding Network

Elitsa Emilova Gieva, Rostislav Pavlov Rusev, George Vasilev Angelov, Rossen Ivanov Radonov,
Tihomir Borisov Takov, Marin Hristov Hristov

Department of Microelectronics
Technical University of Sofia, ECAD Laboratory
Sofia, Bulgaria
gieva@ecad.tu-sofia.bg, rusev@ecad.tu-sofia.bg, gva@ecad.tu-sofia.bg, radonov@ecad.tu-sofia.bg,
takov@ecad.tu-sofia.bg, mhristov@ecad.tu-sofia.bg http://ecad.tu-sofia.bg

Abstract—A microelectronic circuit emulating the behavior of branched hydrogen bonding network is developed. Each hydrogen bond of the network is described in the electrical circuit by three or four-terminal block-element; the residue, which branches the hydrogen bonding network, is presented as a signal adding element. Each block-element is coded in Verilog-A language and implemented in Cadence. DC and transient analyses are performed and compared to previous results obtained in Matlab. The microelectronic circuit analogous to hydrogen bonding network operates as DC Level Shifter, transistor, amplitude limiter and decoder.

Index Terms—Hydrogen bonding network, behavioral modeling, Verilog-A, proteins

I. INTRODUCTION

In recent years the microelectronics is dynamically progressing and facing physical limitations. For this reason new solutions to keep up to roadmap of development are sought. Emerging new field of electronics that could provide solutions in this context is the bioelectronics [1]. One of its objectives is to use bioorganic compounds for signal processing and transfer. For example interesting opportunities for integration of natural neuron networks in semiconductors microstructures are discovered. On their basis hybrid neuron-semiconductor systems for dynamical memories are developed [2].

One of the structures used in bioelectronics are the proteins. Different proteins with various structures and properties are investigated, but for microelectronics the proteins that can transport charges are most important. For example, the Green Fluorescent Protein [3] and Cytochrome C [4] transfer electrons and bacteriorhodopsin [5] transfers protons. The bacteriorhodopsin is used for optical memory [6], biomolecular devices [7], etc.

The bacteriorhodopsin possesses hydrogen bonding networks that executes the proton transport. This implies that the other proteins with their hydrogen bonding networks (HBN) can process information. Such protein with hydrogen bonding networks is β-lactamase [8]. In this paper a microelectronic circuit analogous to β-lactamase HBN is realized. The proton transfer is investigated by Marcus theory [9]. After that, a block-element is juxtaposed to each hydrogen bond implemented in Cadence [10] by coding it in Verilog-A [11]. The DC and transient analyses are compared to previous results from Matlab [12].

II. MODEL AND EQUATIONS

The branching hydrogen bonding network is taken from [13] and showed on Fig.1. On Figures 2 and 3 are given the analogous microelectronic circuit which block-elements are coded in Matlab [14] and Verilog-A, respectively. Here, block-element T1 is the current source juxtaposed to strong proton donor arginine residue R259. T1 is described by three-terminal block-element. Its input voltage controls its output current and the output voltage is identical to the input voltage.

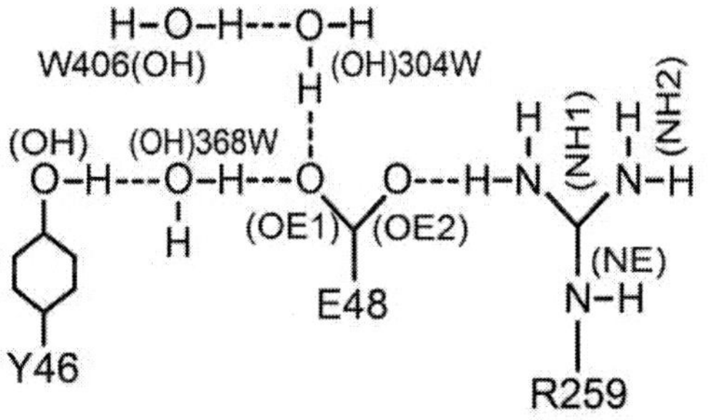

Figure 1. Hydrogen bonding network composed of: NH1, NH2, and NE – nitrogen atoms of Arginine residue R259, OE1 and OE2 – carboxyl oxygen atoms of Glutamic acid residue E48, OH – oxygen atom of Tyrosine residue Y46 and OH are oxygen atoms of water molecules (w304, w368, w406).

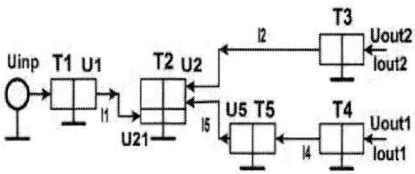

Figure 2. Microelectronic circuit analogous to hydrogen bonding network in Matlab.

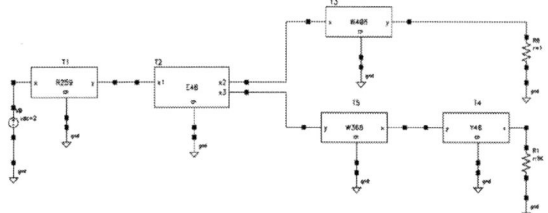

Figure 3. Microelectronic circuit analogous to hydrogen bonding network in Cadence.

The next circuit block-element is T2. It is analogous to the strong proton acceptor glutamic acid residue E48. Since E48 branches the hydrogen bonding network and accepts protons from three donors the analogous block-element T2 has to sum two signals – this is the adding element. These signals are with opposite directions to the input signal. It has to be mentioned that the levels of its output currents/voltages depend only on the input potential.

The other circuit devices are presented as three terminal clock-elements that have identical input-output voltages and different currents.

The model equations that describe the electrical relations of the each block-element are given below.

Equations 1 and 2 describe the T1 block-element.

$$U_1 = U_{in} \qquad (1)$$

$$I_1 = -0.0092 \times U_1^2 + 0.0348 \times U_1 + 0.2152; \qquad (2)$$

Equations 3, 4 and 5 describe the T2 block-element.

$$U_{21} = 0.998 \times U_1 - 1.9408; \qquad (3)$$

$$U_2 = 1.03 \times U_{21} - 0.2327; \qquad (4)$$

$$I_2 = -0.1242 \times U_2^4 + 0.3651 \times U_2^3 + 0.4826 \times U_2^2 \\ -2.3472 \times U_2 + 14.588; \qquad (5)$$

Equations 6 and 7 describe the T3 block-element.

$$U_3 = 1.0347 \times U_2 + 0.1141; \qquad (6)$$

$$I_3 = 4 \times 10^{-5} \times U_3^4 - 0.00013 \times U_3^3 - 8 \times 10^{-5} \times U_3^2 \\ + 0.0012 \times U_3 + 0.0083; \qquad (7)$$

Equations 8 and 9 describe the T4 block-element.

$$U_4 = 0.0378 \times U_1^2 + 0.8262 \times U_1 - 2.2995; \qquad (8)$$

$$I_4 = 0.0023 \times U_4^4 + 0.0029 \times U_4^3 - 0.0385 \times U_4^2 \\ + 0.0618 \times U_4 + 0.641; \qquad (9)$$

Equations 10 and 11 describe the T5 block-element.

$$U_5 = 1.0503 \times U_4 + 8 \times 10^{-5}; \qquad (10)$$

$$I_5 = 0.218 \times U_5^3 - 1.8978 \times U_5^2 + 4.6758 \times U_5 + 162.1; \quad (11)$$

The equations are coded in Cadence with Verilog-A. An example code is given below.

```
// VerilogA for M5, R259, veriloga
`include "constants.h"
`include "discipline.h"
module R259 (x, y, g);
inout x, y, g;
electrical x, y, g;
electrical Vin;
analog

begin
    V(Vin) <+ V(x, g);
    V(y) <+ V(Vin);
    I(x, y) <+ (-0.0092*V(y)*V(y) + 0.0348*V(y)
            + 0.2152)*10e-12;
end
endmodule
```

III. DC ANALYSIS

The DC analysis is performed by feeding input voltage between 200 [mV] and 4.7 [V]. The circuit output voltages are in the interval between -2 [V] and +3 [V] for both simulations Matlab and Cadence (see Figures 4 and 5). All output voltages linearly depend on the input voltage and the level of output 2 is higher than the level of output 1. Therefore the circuit can operate as a DC level shifter.

Despite the negative output voltages the output currents are positive (fig 6-8). The currents in the different outputs are governed by different laws (compared to the input and output voltages) and they have different amplitudes. The current in output 1 is two orders in magnitude larger than the current in output 2.

402

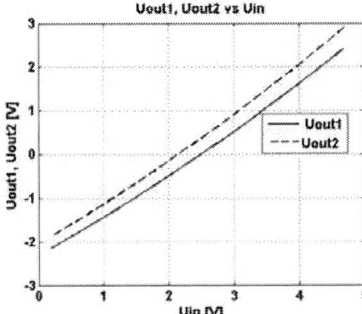

Figure 4. Output voltages (U_{out1} and U_{out2}) versus input voltage (U_{in}) in Matlab.

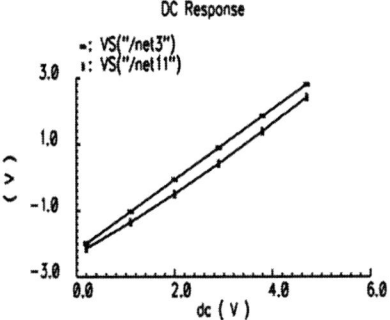

Figure 5. Output voltages (U_{out1} and U_{out2}) versus input voltage (U_{in}) in Cadence.

Analyzing the simulation results from Matalb and Cadence for output 1 (Figs. 6 and 7) we observe that the output curves are identical. The magnitude of the current in Cadence is 10 times higher. The current in input 1 increases linearly (in both simulations) when the voltage U_{out1} is varying between 200 mW and 2 V. When the voltage raises over 2 V the currents saturates. The curve is similar to the output characteristic of a bipolar transistor.

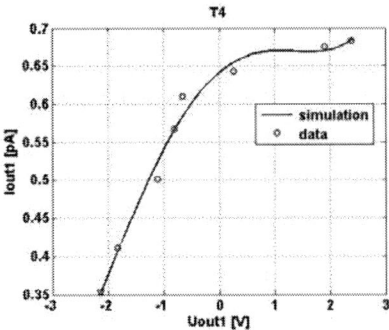

Figure 6. Output current (I_{out1}) versus output voltage (U_{out1}) of first output in Matlab.

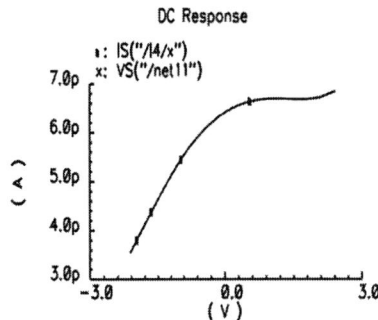

Figure 7. Output current (I_{out1}) versus output voltage (U_{out1}) of first output in Cadence.

On Fig 8 the Cadence output current of I_{out2} is shown (in Matlab it is not shown because it is identical).

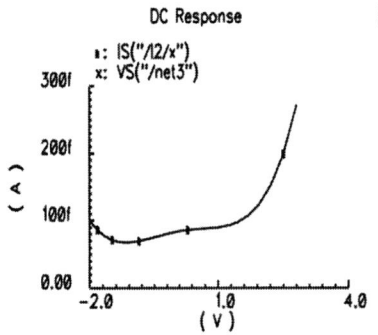

Figure 8. Output current (I_{out2}) versus output voltage (U_{out2}) of second output in Cadence.

Figure 8 shows that at voltages between –2 and 1 V the current in the second channel is almost constant. Above 1 V the currents raises linearly.

IV. TRANSIENT ANALYSIS

Taking into account the specifics of the hydrogen bonding networks and the proton transfer, which takes place for approximately 10^{-11} s, the analogous circuit should transfer signals in the GHz range. On the figures below (Figs. 9-12) the input voltage are given and the output voltages and output currents depends on time. The simulations are made in Cadence (the results from Matlab are identical to those from Cadence and therefore we do not show them).

On Fig. 9 is depicted the dependence of voltage versus time when applying sinusoidal input voltage with amplitude between +0.5 to +1.5 V, the output voltages are negative. Fig. 10 shows the dependence when feeding sinusoidal input voltage with amplitude between +2.6 to +4.6 V; the output voltages are positive.

403

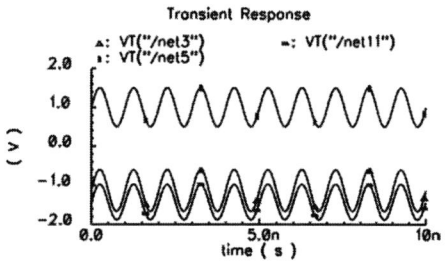

Figure 9. Dependence of voltage versus time in Cadence at input signal with amplitude between +0.5 to +1.5 V.

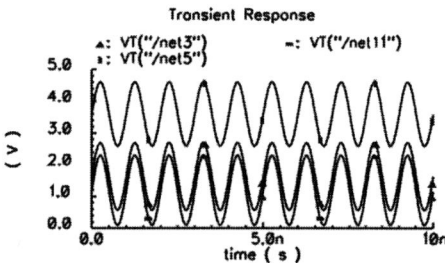

Figure 10. Dependence of voltage versus time in Cadence at input signal with amplitude between +2.6 and +4.6 V.

The output currents I_{out1} and I_{out2} versus time in Cadence are shown in Fig.11 and Fig.12.

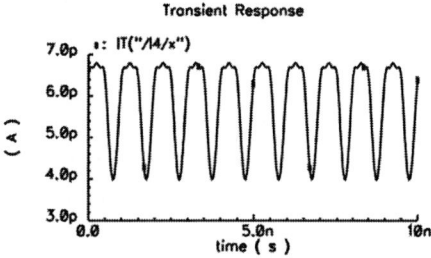

Figure 11. Dependence of I_{out1} versus time.

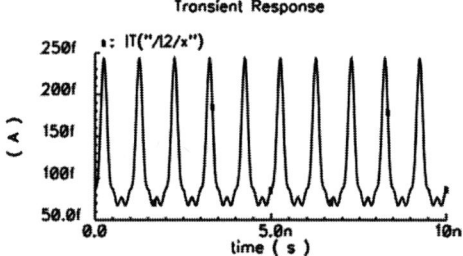

Figure 12. Dependence of I_{out2} versus time.

The output currents are always positive but with different values regardless of the output voltages. At input voltage with amplitude between +0.5 to 4.6 V the output current in the first

output is cut from the top (Fig. 11); the current in the second output is cut from the bottom (Fig. 12).

Therefore the transient analysis proves that the circuit can operate as an amplitude limiter or decoder.

V. CONCLUSION

The developed block-elements of the microelectronic circuit well describe the behavior of the hydrogen bonding network. The DC analysis shows that the circuit can operate as transistor in different amplifying modes or as DC level shifter. From the results of the transient analysis it can be seen that the circuit can operate as amplitude limiter or decoder. The results obtained prove that the modeled circuit can emulate the behavior of hydrogen bonding networks for microelectronics applications.

ACKNOWLEDGMENT

This paper is prepared in the framework of Contract No. D002-126/15.12.2008.

REFERENCES

[1] Wolfgang Göpel, *Bioelectronics and nanotechnologies*, Biosensors & Bioelectronics 13 (1998) 723–728

[2] Prof. Dr. Itamar Willner, Dr. Eugenii Katz, *Bioelectronics: From Theory to Applications*, Chapter 12, Published Online: 23 MAY 2005, DOI:10.1002/352760376X.ch12.

[3] Deborah Stoner-Ma, Andrew A. Jaye, Kate L. Ronayne, Jerome Nappa, Stephen R. Meech, and Peter J. Tonge, *An Alternate Proton Acceptor for Excited-State Proton Transfer in Green Fluorescent Protein: Rewiring GFP*, J. AM. CHEM. SOC. 2008, 130, 1227 1235

[4] M. Fátima Lucas, Denis L. Rousseau, Victor Guallar, *Electron transfer pathways in cytochrome c oxidase*, Biochimica et Biophysica Acta (BBA) - Bioenergetics, Volume 1807, Issue 10, October 2011, Pages 1305-1313

[5] L.E. Petrovskaya, E.P. Lukashev, V.V. Chupin, S.V. Sychev, E.N. Lyukmanova, E.A. Kryukova, R.H. Ziganshin, E.V. Spirina, E.M. Rivkina, R.A. Khatypov, L.G. Erokhina, *Predicted bacteriorhodopsin from Exiguobacterium sibiricum is a functional proton pump*, FEBS Letters, Volume 584, Issue 19, 8 October 2010, Pages 4193-4196.

[6] Jeffrey A. Stuart, Duane L. Marcy, Kevin J. Wise, Robert R. Birge, *Volumetric optical memory based on bacteriorhodopsin*, Synthetic Metals 127 (2002) 3–15.

[7] Stuart, J.A., D.L. Marcy, K.J. Wise, and R.R. Birge, *Biomolecular electronic device applications of bacteriorhodopsin*. In Molecular Electronics: Bio-Sensors and Bio-Computers, L.E.A. Barasanti, Ed., Kluwer Academic Publishers, 2003, 265–299.

[8] Fahd K. Majiduddin, Isabel C. Materon, Timothy G. Palzkill, *Molecular analysis of beta-lactamase structure and function*, International Journal of Medical Microbiology, Volume 292, Issue 2, 2002, Pages 127 137.

[9] Markus A., and V. Helms. 2001. *Compact parameter set for fast estimation of proton transfer rates*. J. Phys. Chem. 114: 3

[10] Website Cadence www.cadence.com

[11] D. Fitzpatrick, I. Miller, *Analog Behavioral Modeling with the Verilog-A Language*, Kluwer Academic Publishers, New York, Boston, Dordrecht, London, Moscow, 2003, pp 41-86, ISBN: 0-7923-8044-4

[12] Rostislav Rusev, Angelov G., Gieva E., Takov T., Hristov M., "Hydrogen Bonding Network as a DC Level Shifter and a Power Amplifier", MIXDES 2010, Poland, June 24-26, 2010

[13] R.Rusev, G. Angelov, T. Takov, B. Atanasov, M. Hristov "Comparison of Branching Hydrogen Bonding Networks with Microelectronic Devices", Annual J. of Electronics, Vol.3, No. 2, pp. 152-154, 2009.

[14] Matlab website http://www.mathworks.com.

Microelectronics Technology and Packaging

Characterization of Test Devices for Development of Nanowire Sensor FETs

Michał Zaborowski, Daniel Tomaszewski, Andrzej Panas, Piotr Grabiec

Institute of Electron Technology
Warsaw, Poland
e-mail: mzab@ite.waw.pl
dtomasz@ite.waw.pl
grabiec@ite.waw.pl

Abstract—Characterization of gate-less nanowire (NW) sensor devices by means of test devices of similar geometry, equipped with metal gate electrode, has been proposed in the paper. Details of NW n-FET technology and microscope verification are presented. $I_{DS}(V_{GS})$ and $I_{DS}(V_{DS})$ DC characteristics have been measured and discussed in relation to a channel implanted phosphorus dose. 2-fin, 8-fin and 32-fin FETs have been compared using a normalization to the single fin device. A spread of NW devices characteristics and an influence of positive or negative incrementation of the voltage have been investigated. Conclusions related to sensor development have been drawn.

Index Terms—nanowires; FinFETs; test structure; chemical sensors

I. INTRODUCTION

Applications of small size FET-type devices for detection and measurement of chemical particles have become standard techniques in biochemistry, environmental research or medicine [1, 2, 3, 4]. Operation of the sensor devices consists in transformation of chemical signals into electrical potential changes, which influence conductivity of the FET channel. A wire or fin shape of the sensor channel produces a large surface area to volume ratio and is expected to be very promising for increasing the sensor sensitivity.

Development of new types of biological sensors has been the aim of the project "Nanoelectronic devices for detection of individual biologically active particles in aqueous solutions", which is presently in progress in Institute of Electron Technology in Warsaw [5]. As an important part of the project technological part a number of test Silicon-On-Insulator (SOI) NMOS transistors and junction-less n-channel devices have been designed for development and verification of the target Si nanowire (NW) sensor devices. The fabrication and characterization of the test devices has been presented and discussed in the paper.

II. EXPERIMENTAL

ISFET-type gate-less nanowire (NW) FETs have been recognized as the sensor devices suitable for the project needs. Dedicated test transistors should be based upon the sensor geometry but in parallel should give an opportunity of measurements of multiple device parameters. Therefore four

types of test FETs with metal gates have been designed. Together with a 4μm wide channel photolithographically defined FET, three FinFET-type devices have been designed. They consist of 2, 8 and 32 fins connected in parallel. They are defined by a smart technique PaDEOx (Pattern Definition by Edge Oxidation), which has been developed in ITE for manufacturing of the NW channels [6] Channel length of all devices is 7μm. Optical images of these transistors are presented in Fig. 1.

(100) oriented Si wafers with 0.4 μm thick buried oxide (BOX) and 0.34 μm thick SOI device layer have been used in the experiment. First, the SOI layer has been thinned down to 285nm using a thermal oxidation process. Next the PaDEOx process has been used. The technique allows for formation of a pairs of channels in one step. The channels width is process-dependent and is constant across the silicon wafer. The width of the NWs is in the range of hundreds of nanometers whereas their length is in the range of up to tens of micrometers.. Taking into account Si consumption during the PaDEOx process the final height of the all NW devices has been

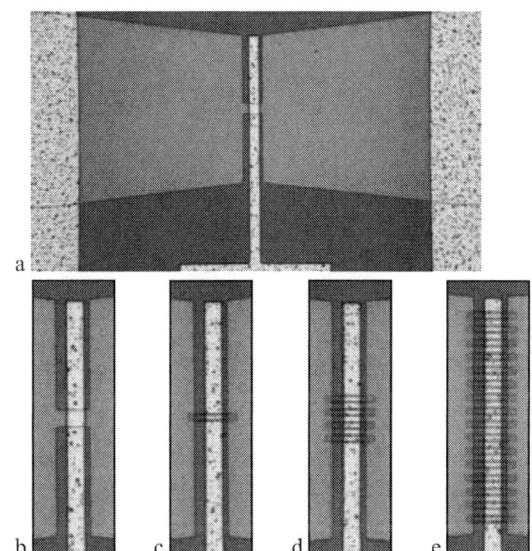

Figure 1. Optical image of the 4μm wide "fin" FET defined by photolithography (a) and magnified images of the four investigated test FETs: 4μm wide "fin" (b); 2 fins (c); 8 fins (d); 32 fins (e).

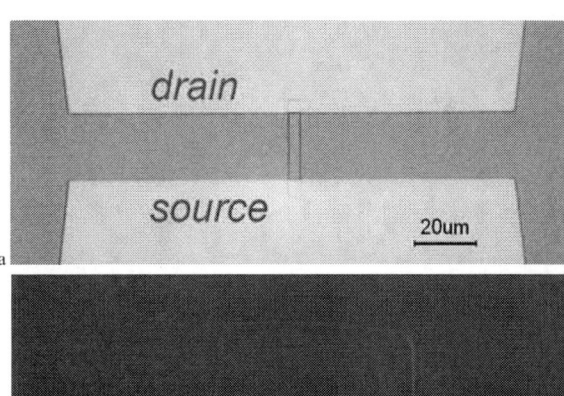

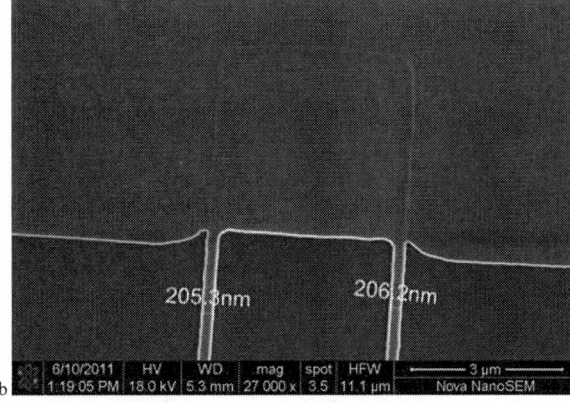

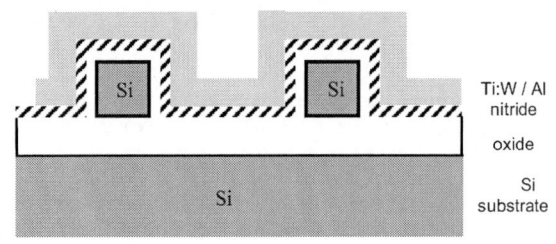

Figure 2. Optical image (a) and SEM micrograph (b) of two Fins of the nanowire FET (prior to gate oxidizing); a cross-section of the final test FET (c).

estimated as 164nm. Final width of the fins is approximately equal to 190nm (205nm prior to gate oxidizing, Fig. 2). Next, channel, source and drain regions have been implanted with phosphorus ions. A 14nm thick thermal silicon oxide layer and a 12nm thick LPCVD stoichiometric silicon nitride layer have been grown as a gate dielectric. Finally, magnetron sputtered Ti10%:W90% alloy layer covered with aluminum layer have been deposited as both the gate and source/drain metallizations.

.Electrical characteristics of the fabricated test transistors have been measured. For this purpose a standard setup of Keithley 2600A SMUs has been used. Based on the measurement results, the device characteristics have been evaluated. The conclusions can be projected for optimization of sensor devices, which can not be directly measured (the have no metallic gates).

III. DISCUSSION OF THE ELECTRICAL TESTING RESULTS

As result of the process described above the accumulation-mode n-MOS transistors have been fabricated. Breakdown voltage of 15V peak to peak was found for the SiO_2/Si_3N_4 gate dielectric stack. Input and output characteristics of the test

devices have been measured using -10V .. +6V V_{GS} range and 0V .. +3V V_{DS} range respectively. Bulk Si substrate has been connected to 0V potential unless it has been stated otherwise.

Selected input characteristics for double-fin NW FET and single 4 μm wide fin FET are compared in Fig.3. It may be easily stated that a control of drain current by the gate voltage is more efficient in the case of the nanowire FETs. A coupling of electric field induced by three gates is responsible for this effect and is widely described in the literature (e.g. [7]. It confirms a recommendation of NW devices for sensor applications.

In order to investigate an effect of the fin doping concentration on the device operation the electrical characteristics of double-fin FETs with different phosphorus implantations have been compared The results are shown in Fig. 4. For the gate voltage negative enough the drain current does not depend on the gate voltage. It corresponds to the residual I_{off} current clearly seen in Fig.3. It may be stated that increasing of the donor concentration increases of I_{off} current

An effect of a back gate should be considered in a majority of SOI devices, particularly for analog or sensor applications. The latter ones may even take advantage of this effect. An efficiency of control of NW FET sensor current by the substrate potential has been verified using double-fin test FETs (Fig. 5). The influence of the back gate bias is distinctly weaker than of the front one, mainly due to large difference of the thicknesses of the box and gate dielectrics. The channel dopant dose also modulates a back gate effect, similarly to the front gate. Thus in our approach a tuning of the device output resistance by means of the substrate potential is available but effective to a small extent only.

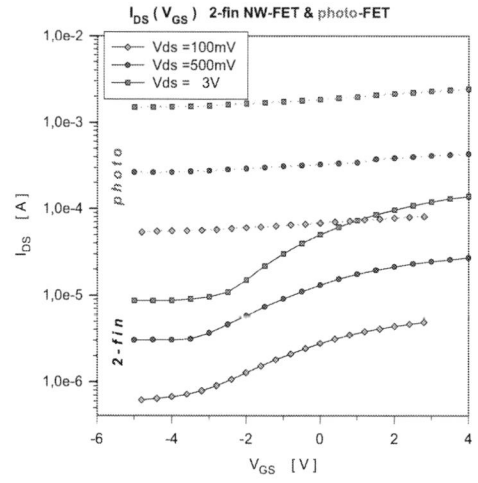

Figure 3. Comparison of transient characteristics for double-Fin FET and 4 μm wide Photo FET in the same test structure.

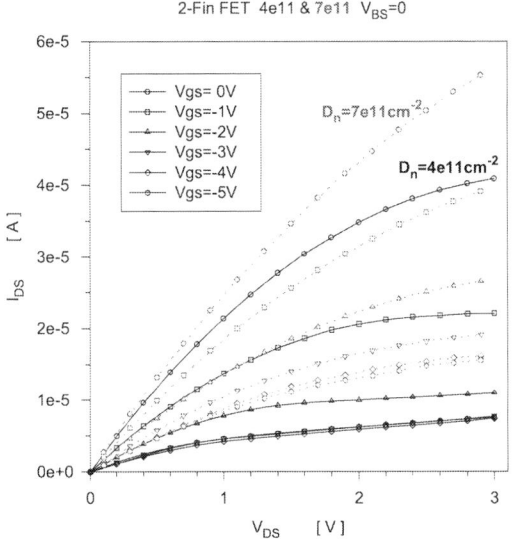

Figure 4. Output characterics dependence on a dose of P⁺-ions implanted in channels of two double-Fin FETs.

As the higher channel doping concentration potentially decreases the device sensitivity to the gate/back gate bias, the devices with phosphorus ion dose $4\times10^{11}cm^{-2}$ have been used in further experiments. A comparison of properties of devices consisting of different numbers of the fins could be easy done by division of measured current by the fin number. Such normalized transient characteristics of 2, 8 and 32-fin FETs are presented in Fig. 6. It may be noticed, that the normalized input I-V curves of multi-fin devices overlap one another in the accumulation range. It means, that all the fins are functional.

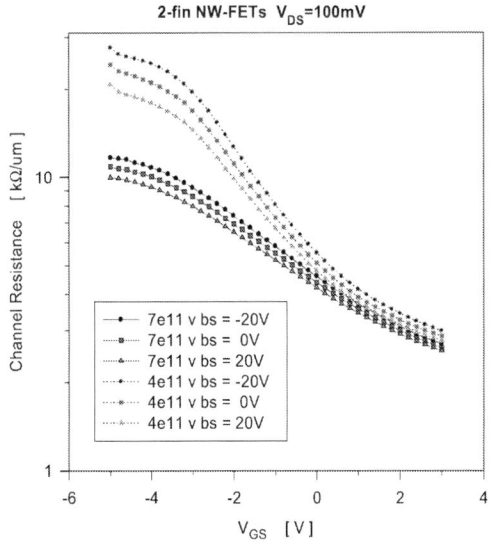

Figure 5. Offseting of channel resistance by front gate and back gate potentials for two two-fin FETs of differently implanted channels.

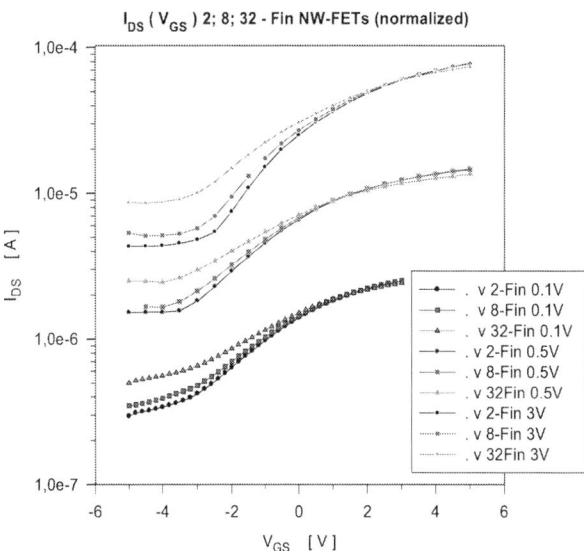

Figure 6. Comparison of input characteristics of three NW FETs, which differ from each other in a number of fins. Results have been divided by the fin number (normalized).

Application of the multi-fin devices is profitable because of their sensitive (gate) area increase without losing of fin advantages, i.e. their higher current sensitivity to gate/back gate voltage.

Input characteristics of the two-fin test transistors placed along the wafer diameter have been measured for three values of the drain-source voltage. Mean values of the $I_{DS}(V_{GS})$ curves and their standard deviations are displayed in Fig. 7. The substantial deviations appear in a depletion region mainly and are depended slightly on the drain-source voltage. Based on the I_D-V_{GS} curve a transconductance characteristic has been

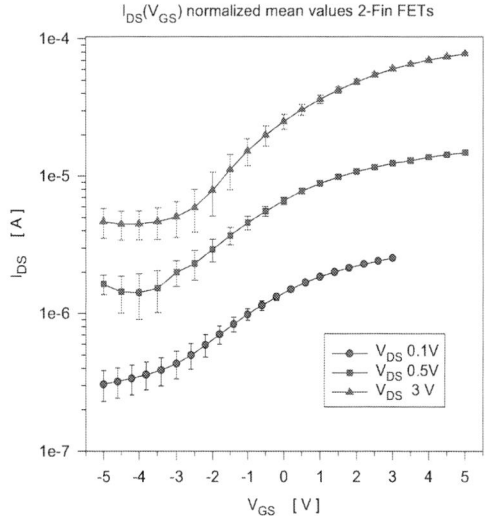

Figure 7. Normalized input characteristics of test two-fin FETs – mean values for a row across the wafer.

409

calculated also – Fig. 8. The curve is smooth, only for far negative gate voltages some sharpenings are visible. The same characteristics describing an individual FET in this row appear in a different manner (Fig. 9). Drain current results depend on a direction of the gate voltage changes. A hysteresis in a region of V_{GS}=-5..-3V for V_{DS}=100mV has been observed for all transistors. The higher spread of the characteristics for this region (Fig. 7) seems to result from this phenomenon. However the appearance of the hysteresis has not been completely studied yet. It is worthwhile to be mentioned that the devices gate double-layer oxide/nitride dielectric stack is similar to the one used in oxide-nitride-oxide ("ONO") stacks for flash memory cells.

Calculation of the mean normalized current or conductance characteristics of fin FETs together with a comparison of characteristics of an individual fin FET seems to be a good method in a technology development to find eventual defected fins.

IV. SUMMARY

Fabrication and electrical measurements of the test n-channel nanowire-type devices has been described in the presented paper. These devices are used as test ones for evaluation of the target nanowire-based devices for sensor applications.

Based on electrical measurements of the test devices optimum operating conditions of the target sensor may be determined. They should correspond to the maximum transconductance. An important constraint is that the operating point should be outside of the gate voltage range corresponding to the hysteresis, which is expected in the devices with silicon oxide/silicon nitride stack. It has been found, that operating point close to V_{GS}=0V is suitable. This solution is advantageous from the eventual application point of view.

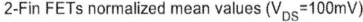

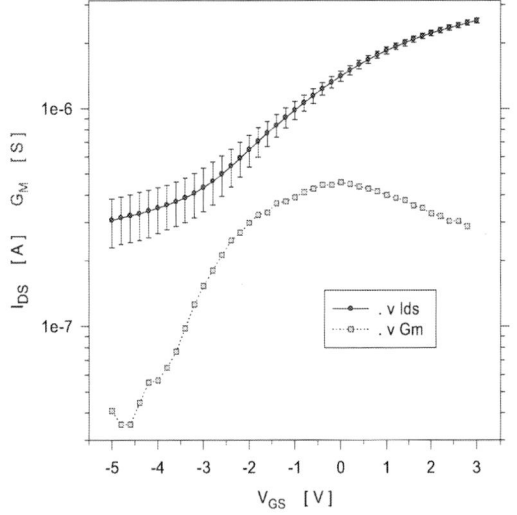

Figure 8. Mean normalized transition characteristics and transconductance characteristics of test two-fin FETs for a full row in the wafer;

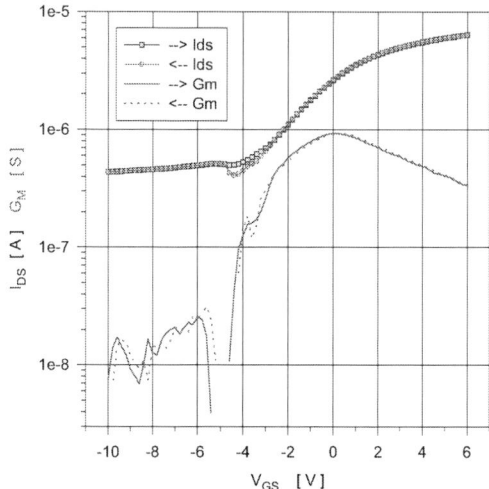

Figure 9. Normalized transfer characteristics and transconductance characteristics for individual two-fin FET, measured with positive and negative increment of the gate voltage; a measurement time intervalis 200ms.

It has been widely demonstrated, that the gate stacks containing silicon nitride layers are suitable for pH measurements. In our investigation it has been shown, that such gate stack works well as the gate dielectric too.

It has been found that I_{ON}/I_{OFF} coefficient in the test devices is low. It does not exceed a value of 10. However it is expected, that using lower channel doping concentration (lower phosphorus implantation doses) results in increase of the device sensitivity. It is an indication for future development of the presented technology.

It has been also demonstrated that the PaDEOx NW multi-fin devices exhibit much better sensitivity of their electrical behavior to the gate voltage (thus environment conditions) than standard devices defined by photolithography steps. Moreover, the multi-fin devices have larger area, than the ones with small number of fins. Thus using the multi-fin devices with the functionalizing layer deposited onto the gate dielectric increases a probability that particles can be detected. It is particularly relevant in case of low concentration of molecules in the solution surrounding the sensor. This issue has to be taken into consideration in further device development towards biochemical application.

ACKNOWLEDGMENT

The work was partially supported by the Ministry of Science and High Education in Poland under grant NRO2 0010 06/2009.

REFERENCES

[1] E. Stern et al., Label-free immunodetection with CMOS-compatible semiconducting nanowires, Nature Vol. 445, 1 Feb. 2007 p.519

[2] N. Elfström, A. Eriksson Karlström, J. Linnros, Nano Letters, Vol. 8, No.3, (2008) pp.945-9

[3] O. Knopfmacher, D. Keller, M. Calame, C. Schönenberger, Procedia Chemistry 1 (2009) pp.678–681

[4] R. Yan et al., Performance Analysis of SOI Junctionless Nanowire Transistors, ELTE2010 Conf. Wroclaw, 22-25 Sept. 2010J. Clerk Maxwell, A Treatise on Electricity and Magnetism, 3rd ed., vol. 2. Oxford: Clarendon, 1892, pp.68–73.

[5] Zaborowski M., Tomaszewski D., Dumania P., Grabiec P.: "Development of Device Technology for Chemical Molecule Detection", Microelectronics Reliability Vol. 51, Issue 7, July 2011 pp.1162-5

[6] M. Zaborowski et al., Acta Physica Polonica A, Vol.116 (2009), pp.139-141

[7] J. P. Colinge, "FinFETs and Other Multi-Gate Transistors", Springer, New York, 2008)

MIXDES 2012, 19th International Conference *"Mixed Design of Integrated Circuits and Systems"*, May 24-26, 2012, Warsaw, Poland

Design Model and Data Management for 3D Integration Technologies

Armin Grünewald, Kai Hahn, Rainer Brück

Institute of Microsystem Engineering
University of Siegen
Hölderlinstr. 3, 57068 Siegen, Germany
armin.gruenewald@uni-siegen.de

Abstract—Developing 3D systems is a highly complex procedure. Next to a huge variation of possibilities on how to vertically integrate two or more dies, a lot of aspects regarding cost, design and application specific selection of technology have to be considered. Therefore a design model, considering the mutual influence of design and process technology during the integration flow development and a data management to create and store process flows, is necessary.

Index Terms—TSV; 3D integration; process flow; die stack; design model; data management

I. INTRODUCTION

Three-dimensional (3D) integration technology is a promising solution for continuing the increase in complexity of integrated circuits described by Moore's law. In the last decades this growth has been achieved by scaling the on-chip feature size. The scaling cannot continue forever, because of physical limits in the shrinking process and increased manufacturing cost. Therefore the stacking of integrated circuits is more and more emerging. During development of a 3D chip, due to the large number of possible processes, it is necessary to take into account restrictions given by fabrication and technology. Additionally, the choice of the integration flow is affected by product specific constraints. This implies the need to explore technology as well as design options, as has already been demanded in [1]. Furthermore [2] points out that a "co-optimization" of the system design and the system hardware is necessary.

This paper will follow these considerations, but will focus on the choice of processes during the development of a 3D integration flow and the management of the required material and process data. Therefore a design model, which makes use of the mutual influence of design and process technology, will be presented. Additionally a prototype of a design flow generator with a dedicated data management is shown. Before this the technology background of 3D integration and 3D integration flows are given. This paper closes with a conclusion and an outline of future work.

II. 3D TECHNOLOGY BACKGROUND

A. Integration Technology

Seen from the process view, 3D integration can be outlined by the following three classes of technology:

1. TSV fabrication (Insertion of vertical connections into the die (TSV = Through Silicon Via))

2. Wafer thinning (Reduction of the thickness of the die to be able to contact the TSVs)

3. Wafer bonding (Connection of two wafers or dies)

Different options exist at which time regarding IC processes and integration process steps the fabrication of TSVs takes place. Creating TSVs before CMOS processing (FEOL = Front-end-of-line) is called "via first", after FEOL but before BEOL (= Back-end-of-line) "via middle". A via formation after the complete IC process (Post-BEOL) is called "via last". The advantage of the latter option is that the IC wafer can be manufactured in a foundry which does not support TSV formation yet. The TSVs can then be added later by another foundry. Creating TSVs after wafer bonding (= via after bonding) is another possibility. Examples of processes are the "Bosch"-process (DRIE = Deep Reactive Ion Etching) to create the trench and Chemical Vapor Deposition (CVD) to fill the trench with copper or tungsten. In order to reduce the thickness of the die it is necessary to perform wafer thinning. Therefore depending on the choice of integration flow two options exist: The IC wafer will be either bonded directly on the 3D IC stack with the backside up and then thinned or temporarily bonded on a wafer handle and thinned.

The two basic factors that have to be considered for the bonding process are the bonding objects and the orientation: The two main approaches to receive a wafer stack are "Wafer-to-Wafer" (W2W) and "Die-to-Wafer" (D2W) bonding. Since W2W stacking means to combine two complete wafers, this method is only suitable if all dies have the same size. Another drawback is the yield of the die stack, because it derives from the yields of the individual dies. If for example two wafers reaching a yield of 90% each, a stacking results in an 81% yield (yield loss through the stacking process is not considered).

D2W stacking facilitates the use of "known good dies" (KGD) so that a higher yield can be achieved. Additionally it is possible to have different die sizes. The orientation of the wafers/dies can be either "Face-to-Face" (F2F), "Back-to-Face" (B2F) or "Back-to-Back" (B2B) [3].

B. Integration Flows

In order to achieve a stacking of only two dies there exist nine possible flows to combine the integration processes with the FEOL and BEOL [3] [4]. Fig.1 shows one example of the "Via last" approach: After the FEOL and BEOL processes the wafer will be temporarily bonded on a wafer handle and then thinned. After that the TSVs will be fabricated from the backside and the wafer will be finally bonded B2F to the other wafer.

With the large number of specific processes (e.g. DRIE or laser ablation for TSV fabrication) and materials to use, there are still a lot of possible variations after selecting one of the integration schemes. This accumulates even more when thinking about heterogeneous integration of several dies having a different technology each. In this case it will be very difficult to determine a process sequence which is fulfilling the initial specifications and is also manufacturable. Product specific constraints are for instance the number of dies, the technology of the dies or the number of TSVs needed. These constraints will affect the selection of processes and TSVs. On the other hand pre- and post-conditions of different process steps, temperature budget and a cost factor need to be considered while developing an integration flow. These aspects can require major changes in the initial design. This train of thought leads to a design model approach, which is described in the following section.

III. METHODOLOGY

The foundation of the design model presented in this paper is the Pretzel Model developed at the University of Siegen (Fig.2) [5]. The Pretzel Model originally designed for MEMS, is now adapted to 3D systems.

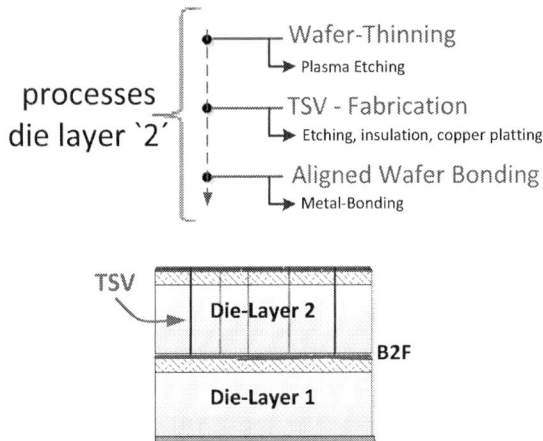

Figure 1. Example process flow.

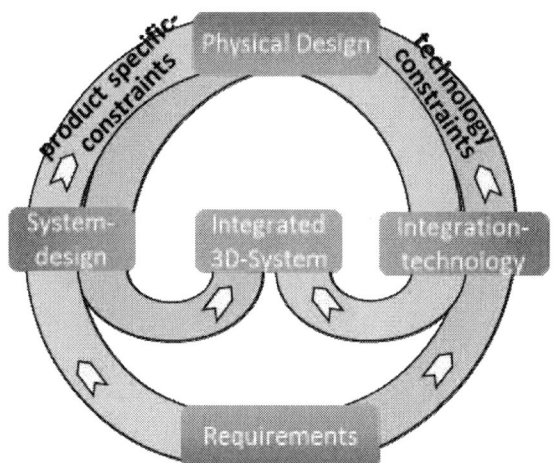

Figure 2. Pretzel Model.

In order to design an integrated 3D system, the bottom-up process design flow (right hand side) has to be taken into account at the same time as the top-down product design flow (left hand side), since with the high amount of dependencies as described before the design cannot be performed detached from the technology. According to this model the designer starts creating a structural description, which is based on the requirements. From this structural description a corresponding physical design of the 3D system can be derived. The next step is to design a process flow for manufacturing the 3D system. At the same time important material and process step data is collected, and the process flow has to be verified with this additional data. The last two steps have to be repeated until the process flow matches the physical design of the 3D system.

In order to be able to develop a process flow and at the same time consider design and technology constraints, a design model has been developed. The design model is divided into four parts (Fig. 3), which are described in the subsequent sections.

A. Building of an integration process step collection

The first step to develop an integration process flow is to build up a collection of available process steps. Each process step needs some parameterization, so that different analysis on process steps and combination of methodology for creating integration flows process steps respectively process flows can be performed. Process parameters basically include information on temperature, time and involved materials as well as other constraints.

One crucial aspect is the cost factor (see section C). Cost is always an important challenge and therefore it is necessary that every process step is parameterized with a cost value, so that not only technical but also economic conditions can be considered during the development of the process flow.

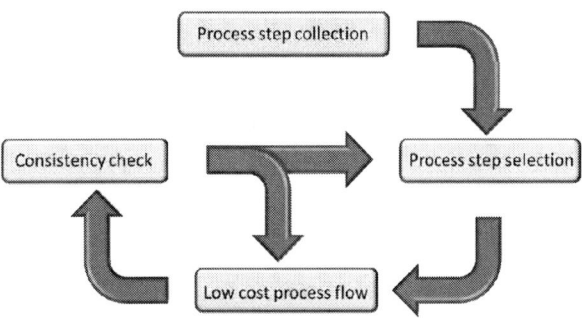

Figure 3. Methodology for creating integration flows.

Process steps require also different temperatures and too high temperatures can damage existing layers on the die. Therefore it is important to know the process step's contribution to the overall temperature budget.

Additionally every process step needs a list of pre- and post-conditions to indicate the dependencies to other process steps. With these conditions it is possible to perform a final check of the flow at the end of the development process to guarantee the manufacturability.

B. Reduction of the number of process steps due to the consideration of product specific constraints

After building an integration process step collection with the mentioned parameters it is necessary to take account of product specific constraints. These constraints give information of the desired product and are derived from the design exploration. They affect the selection of process steps and therefore allow building an integration flow which suits best for the product.

Possible constraints are the number of dies, technology of the dies and number of TSVs. The conditions of the different process steps therefore need to be checked on the individual inputs. In case of violations, the process step has to be removed from the list.

C. Selection of an integration flow with preferably low cost factor

The next step is to build a low cost integration flow from the remaining integration process steps. In recent research papers different approaches have been presented to perform a 3D IC cost analysis. In [6] a cost model for a 3D System-on-Chip is presented where the cost of a 3D process step is determined by the processing time and material consumption per step. Out of these it is possible to derive yearly production cost while considering a yearly target production volume. The cost consists of equipment cost, clean room cost, personnel cost, maintenance cost and material cost. In comparison to this in [7] a cost model is presented where the final 3D chip cost is separated into several different parts: A wafer cost model takes account of the die area and the yield, and a bonding cost model determines the costs of the integration process steps. Both are merged into the overall 3D cost model, which also depends on

additional design options such as Die-to-Wafer/Wafer-to-Wafer bonding or Known-Good-Die cost.

In case it is not possible to develop a low-cost integration flow, e.g. due to lack of available process steps, the building process is canceled and the violating constraints are listed. In this case it is necessary to go back and change the process steps / process parameters or the product specific constraints.

D. Consistency check

The last step of the design model is a consistency check of the chosen process to guarantee manufacturability. The technology constraints of the process flow and the individual process steps are represented by the pre- and post-conditions of each process step, which have been defined in section A. If for example for TSV formation a copper (Cu) deposition process step exists, it may have a deposition of an adhesion layer like TiN as a pre-condition and a thermal treatment as a post-condition [3].

These conditions allow to prove, if all necessary process steps are part of the flow and to directly indicate errors by showing the violating dependency. Another crucial aspect for a sequence of process steps is the temperature budget. Some process steps require a certain temperature limitation for the following process steps or the applied materials can only withstand a certain temperature over a particular period of time. In these cases it is necessary to analyze the different requirements of every process step and every material. In case of violations a return to step B or even to step A (then going along with changes in the design) is necessary, which means that it is also possible that no suitable flow exists.

IV. SYSTEM FOR 3D INTEGRATION FLOW GENERATION

The described design model is currently being implemented in the software ASPIRE ("Application Specific Integration Flow Evolution"). It is realized as a client-server architecture in combination with a PostgreSQL [8] database (Fig. 4).

With the help of the client it is possible to add and edit process steps, materials and TSVs. After creation of this basic data it is possible to develop process flows with a process flow editor. In order to be able to create integration process flows, that allow the assignment of process steps to different dies, a dedicated data model for describing process flows has been developed. With this data model it is possible to combine different process sequences and to pass through a process sequence. This is necessary to be able to perform a consistency check, which will be added soon. Other enhancements will be the implementation of a cost model and a broader consideration of process specific constraints to achieve a partly automatically flow generation.

The technology information about integration process flows, process steps, materials, TSVs and their respective parameters are stored in a database. The process steps are additionally parameterized with rule sets. With the help of these rule sets it is possible to define the pre- and post-conditions described in part D of the previous section. More details on the data management are given in the next section.

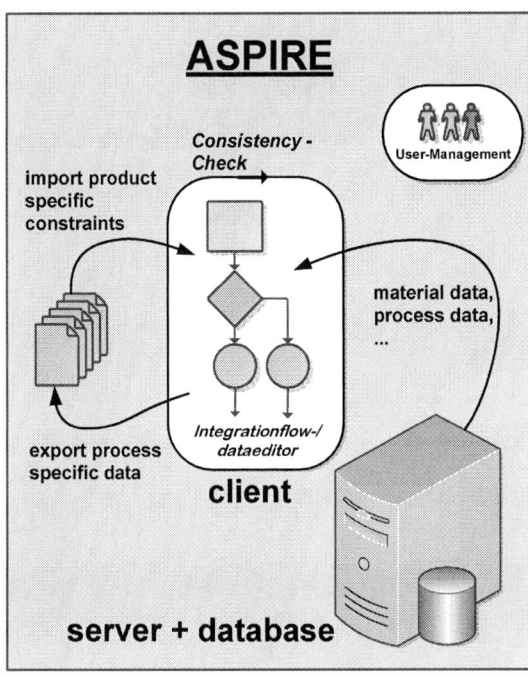

Figure 4. Overview software system.

V. DATA MANAGEMENT FOR PRODUCT SPECIFIC INTEGRATION TECHNOLOGY

Developing 3D systems is a highly complex procedure, since a huge amount of different technology data has to be considered. Therefore it is necessary to be able to create, edit and manage data in a comfortable way and also export the data in a format, so that the information can be easily retrieved by other design tools.

A. Database

The requirement for the database is to store all data that is used for the integration flow generation as well as for other tools regarding the design process of a 3D system. This is achieved by starting with basic information first. To describe technological and physical characteristics, units and parameters are essential components. The data stored in the database to specify a unit are the name, a unique ID, a unit symbol, a short description and information about the author and the creation date. A unit has no dependencies to other entities in the database, so that the data can be retrieved with a simple SQL command "Select * from unit". A parameter is described in a similar way. Additional information are the assigned unit, the type of the value (numerical, textual or material) and attributes. A parameter is initially abstract, which means that it has no attached value. The value will be set when the parameter is assigned to a material, process step, layer or TSV. Attributes adds further information to a parameter, e.g. to specify a location where a certain temperature has to be reached. With the dependencies to attributes and unit it is no longer possible to extract data from the database or insert new data with a simple SQL command. Therefore it is necessary to develop a data editor, which is described in the following section.

The most important data stored in the database are information about materials, TSVs, layer, process steps and process flows. Materials are described with a name, a unique ID, a description, information about the author and creation date, a list of parameters and their values and the material class it belongs to. While TSVs and layers are described in similar way, a process step needs some additional information. The parameters of a process step are divided into process and result parameters in order to give a more accurate specification of a process step. To be able to perform a consistency check over the whole process flow, every process step is parameterized with a rule set, which holds pre- and post-conditions for the process step. The process flows stored in the database are a mapping of the data model mentioned in the previous section.

B. Data editor

As described before, the data structure of the different objects is complex, especially for materials and process steps. To be able to easily add and edit data, the client of the software ASPIRE has different editors with a graphical user interface. Fig. 5 shows a screenshot of the material editor together with the popup for editing a material parameter. In the given example the material copper has two parameters (thermal conductivity and density). The density has the value 8920 kg/m³. Except for the process flow editor all other editors are build up in a similar way. The process flow editor uses a graph model to illustrate a process flow.

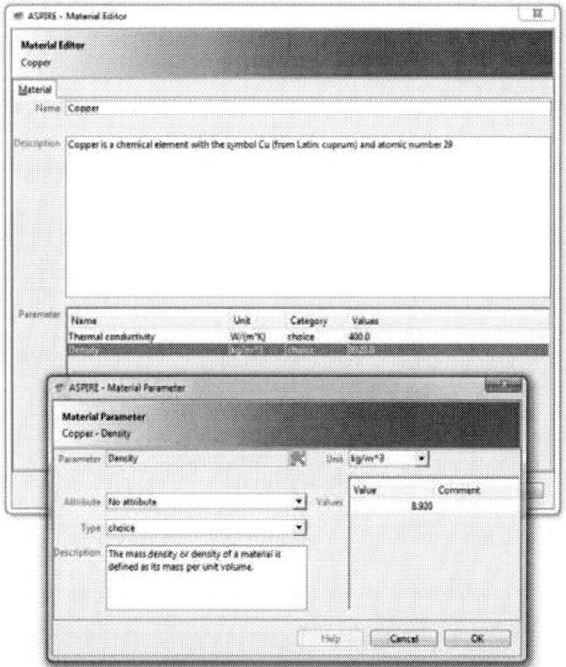

Figure 5. ASPIRE material editor.

C. Exchange format

Next to the functionality to manually add and edit technology data as described in the previous subsection, it is necessary to import and export data in order to communicate with other design tools. As a text format XML (Extensible Markup Language) [9] has been chosen, because of its advantages such as being platform-independent, extendable, flexible and easy to integrate in applications with the help of a parser (e.g. JDOM for JAVA). Fig. 6 exemplary shows the XML export of the material copper (already shown in the editor window in Fig. 5). In the XML-file it is also possible to see the creation date, the ID of the author and the ID of the material.

```xml
<?xml version="1.0" encoding="UTF-8"?>
<database>
 <material>
  <name>Copper</name>
  <materialID>45</materialID>
  <description>Copper is a chemical element with the
        symbol Cu (from Latin:cuprum) and atomic
        number 29.
  </description>
  <authorID>6</authorID>
  <creationDate>2011-12-01T12:21:26Z</creationDate>
  <parameter>
   <name>Density</name>
   <description>The mass density or density of a material
        is defined as its mass per unit volume.
   </description>
   <unit>weight per volume</unit>
   <type>choice</type>
   <value>8920.0</value>
  </parameter>
  <parameter>
   <name>Thermal conductivity</name>
   <description>In physics, thermal conductivity, k, is the
        property of a material's ability to conduct heat.
   </description>
   <unit>Watts per meter kelvin</unit>
   <type>choice</type>
   <value>400.0</value>
  </parameter>
  <materialclass>Metal</materialclass>
 </material>
</database>
```

Figure 6.   XML Description of material copper.

VI. CONCLUSION AND FUTURE WORK

A design model for developing application specific 3D integration flows and a software prototype with a dedicated database were presented in this paper. With the model it is possible to take account of both process and product specific constraints. The software described in the two previous sections will allow the development of manufacturable process flows out of the available integration process steps, in case such a flow exists. Currently editors for materials, TSVs, process steps and process flows exists as well as the connection to a database which allows the storage of all necessary technology data. A connection to other design tools in the project (e.g. 3D floor planner, 3D test) is planned.

ACKNOWLEDGMENT

The work presented in this paper is part of the EDA cluster research project NEEDS (project label 01M3090), which is funded by the German Federal Ministry of Education and Research (BMBF).

REFERENCES

[1] P. Marchal et al., "3-D Technology Assessment: Path-Finding the Technology/Design Sweet-Spot," Proceedings of the IEEE, vol. 97, 2009.

[2] P. Emma, E. Kursun, "3D system design: A case for building customized modular systems in 3D," Interconnect Technology Conference (IITC), 2010.

[3] P. Garrou, C. Bower and P. Ramm, "Handbook of 3D Integration," Wiley-VCH, 2008.

[4] Yole Développement, "Thin wafer handling for 3D TSV," SEMATECH Workshop SEMICON West, 2011.

[5] K. Hahn, A. Wagener, J. Popp and R. Brück, "Process management and design for MEMS and microelectronics technologies," Proceedings of SPIE: Microelectronics: Design, Technology, and Packaging, vol. 5274, 2003.

[6] D. Velenis, M. Stucchi, E.J. Marinissen, B. Swinnen and E. Beyne, "Impact of 3D design choices on manufactering cost," IEEE International Conference on 3D System Integration, 2009.

[7] X. Dong and Y. Xie, "System-Level cost analysis and design exploration for three-dimensional integrated circuits (3D ICs)," Asia and SouthPacific Design Automation Conference, 2009.

[8] PostgreSQL: http://www.postgresql.org/

[9] XML: http://www.w3.org/XML/

Multisensor System for Monitoring Human Psychophysiologic State in Extreme Conditions with the Use of Microwave Sensor

Krzysztof Różanowski[1], Tadeusz Sondej[2], Jarosław Lewandowski[1],
Mariusz Łuszczyk[3] and Zenon Szczepaniak[3]

[1]Aviation Bioengineering Department
Military Institute of Aviation Medicine
Warsaw, Poland

[2]Electronic Department
Military University of Technology
Warsaw, Poland

[3]Strategic Research Office
Przemysłowy Instytut Telekomunikacji S.A.
Warsaw, Poland

krozan@wiml.waw.pl, tsondej@wat.edu.pl, jlewando@wim.waw.pl,
mariusz.luszczyk@pit.edu.pl, z.szczepaniak@pit.edu.pl

Abstract—**The paper concerns recording biomedical signals for the purpose of monitoring psychophysiological parameters in extreme conditions, using a microwave glucose sensor. The intended project involves a modern system for monitoring the psychophysiological activity, integrated with a personal microcomputer system monitoring the human efficiency. The elements of the monitoring system are applicable in the newest trends connected with health care called telemedicine. The elaborated set of micro sensors and algorithms of data processing will allow elaborating proper indicators of the psychomotor state estimation. The interference-resistant technology developed for the military sector enables the proposed solutions to be used also in civilian applications.**

Index Term—**psychophysiology; monitoring; physiological parameter recording; detection algorithms; measuring system**

I. INTRODUCTION

The psychophysiological-state monitoring system presented herein combines technologies of communication between sensor subsystems, technologies of measurement-information transfer, and an IT layer comprising advanced data-processing, decision-making, and management modules.

The implementation uses state-of-the-art technologies in the area of micro-sensors — miniature non-invasive electronic devices which provide digital information about the human body state and location. Thanks to the small size of those sensors and the manner of their application, they do not cause

any restrictions or discomfort in performing everyday tasks by the subjects. The developed set of micro-sensors and the data processing algorithms enable working out certain indicators to assess the psychophysiological state. Such indicators can be transmitted to a remote supervision system (operator console, command center). Two monitoring modes are supported: online and offline (with storing the data to a locally installed memory card).

The system has been designed to be used in the process of selecting subjects by determining their predisposition to combat operations. The requirements for such systems are definitely more stringent than in case of systems designed for use in laboratories with defined device operating conditions. In extreme conditions, most otherwise proven methods of information acquisition may be inapplicable. An important element of the designed system is a non-invasive sensor of glucose level in blood, consisting of a microwave probe with an bio-oscillator and a measuring module.

The works on developing a microwave-based technology to measure changes of glycaemia in humans required research in overlapping areas of such disciplines as microwave electronics, medicine, physiology, chemistry, and signal processing. It is an innovative task. Many years of work resulted in a number or original results. The achievements include a probe which gives distinct resonance in contact with the skin, characterized by a very low level of radiated microwave power, formulas of substances modeling the human tissues, and a measuring module. The design of the mechanical miniature microwave head is protected as an utility model registered in the Polish Patent Office. The techniques of measuring changes in tissue parameters with a bio-oscillator have been patented in Poland.

The general architecture of the psychophysiological fitness monitoring system presented below consists of a hardware Personal Measuring Unit (PMU) and a dedicated software application. The hardware measuring system consists of a Personal Data Server (PDS), a set of external measuring modules, a battery package, and a radio modem.

Such topic is marked in the widely developing subject of the intelligent surrounding of the human, their mutual interactions, in terms of help for the handicapped or ill people (ambient assisted living, intelligent house).

The system is modular, which enables the functional scope to be freely expanded with new modules and measuring methods. It is possible to use contactless sensors which significantly improve the system user's comfort [1, 2]. The system complies with Directive 93/42/EEC in terms of CE certification.

The system has been designed to be used in extreme conditions, with design, hardware, and software solutions resistant to external influence, as well as with contactless technologies for measuring the psychophysiological parameters. Therefore, the technology described herein may be successfully used in civilian applications, such as in the working environments of professional drivers, railway engine drivers, public transport vehicle operators, civil airplane pilots, or trainees in flying clubs, as well as to monitor elderly or ill people at their places of work and living, etc.

II. MEASURING SYSTEM DESIGN

A. System architecture

Developing a life-assistant system requires designing not only the sensor part, but also methods of their integration with available technologies, including data-transmission solutions. Of course, in case of military applications, the data access and data-transmission channels must ensure information security.

Our experience gained during projects conducted by Military Institute of Aviation Medicine suggests that the access to the data and measuring system should be based on a layered model, as presented on Figure 1.

Layer I, the lowest layer of the model, consists of measuring sensors which collect the subject-monitoring data and transmit the data to local centers at Layer II. Layer II may include training centers or decision-making centers. At this layer, the data is analyzed, processed, and stored. The presented location of the Server Radio Data Modem (SRDM) is only an example. The server may be installed in the test area as a base station or may be implemented as a remote unit. It should be noted that the SRDM server communicates with the PMU modules by radio and itself is connected to a computer network. The data and measurements can be accessed through Layer III, the information access layer. Layer III stores the decision-making and data-analysis algorithms, as well as threat definitions. The algorithms will be available to research centers

which can expand the knowledge base through their own research. Layer III includes also servers which analyze signals from the measuring sensors. Threat definitions and decision-making algorithms can be downloaded to and implemented in mobile devices at the sensor level, which enables the solutions to be updated in order to best use the knowledge bases.

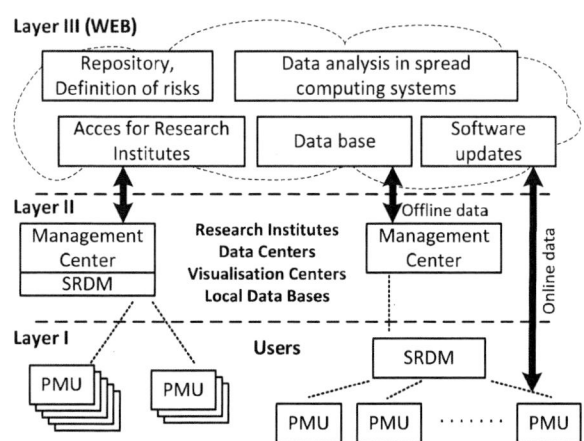

Figure 1. Block diagram architecture of measurement system

The model is universal and in case of military applications should be expanded with a system of data-access authorization.

B. Software for managing the system hardware components

The system software manages the individual hardware components of the system. The developed user interface is clear and intuitive. The read and transmitted data is stored in files in the our SKL format — a universal format for storing in a single file signals sampled both uniformly and non-uniformly. The user interface supports selection of the recorded signal set and of the signal sampling frequency. When a recorder starts, its attributes (signal names and available sampling frequencies) are read in. The user can select relevant signals from the list and select from the sampling frequencies possible for the given signal. Unless set otherwise, the parameters have their default values. The signal viewer supports descriptions of the range marks, in the form of graphic labels above the corresponding marks. The descriptions are entered from the measurement window.

To improve the esthetic quality of the GUI (Graphical User Interface), the application uses the WPF (Windows Presentation Foundation) technology, and in particular the additional libraries WPFtoolkit.dll and WPF.Themes.dll (including the ExpressionDark theme from the latter library).

The interface provides also a map window which presents the current position of the subject, the subject's movement track, and the physiological parameters calculated in real time (displayed in a side panel and directly on the map). The length of the displayed part of the track is configurable (in seconds) in the application settings. A negative value means that the track is recorded infinitely. The settings at the application level enable also configuring the range of parameters displayed in the side channel and on the subject's icon on the map.

III. MICROWAVE BLOOD-GLUCOSE SENSING

As far as extreme intensive physical activity is considered the blood-glucose level becomes very important parameter. The possibility of monitoring glycaemia changes in on-line mode gives very powerful diagnostic tool. From user point of view the most important parameter is non-invasiveness. Therefore, the main designer goal is to provide blood-glucose level sensing without taking a blood drop.

There are several attempts and approaches to solve the problem of non-invasive measurement, e.g. optoelectronic, inverse ionophoresis. Another way is to use microwave techniques. The approach presented here assumes the use of microwave non-invasive probe being in contact with skin surface. The probe is a resonant circuit, which is coupled with an area of tissue (skin) by means of electromagnetic field. A change of blood-glucose level causes change of blood and tissues permittivity thereby causing a change of microwave parameters of the probe [3, 16]. Here, the most important is the resonant frequency of the probe. Further, an electronic circuitry is applied in order to measure the probe parameters. Current design of microwave sensor consists of: microwave head and measuring module (Figure 2).

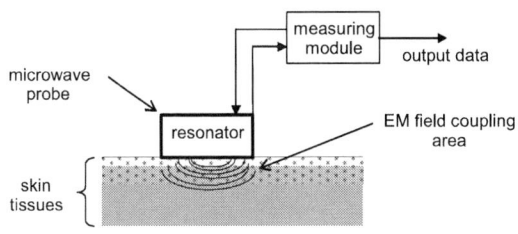

Figure 2.   Concept of microwave glycaemia sensor

The measuring module is integrated with main system computer: power supply is provided by the system and data exchange is accomplished by UART. The probe is planar and comprises Teflon-based substrate with all the metal structure covered by gold. These fulfill the non-irritation and non-inflammatory requirements. The probe must be in tight contact with the skin surface. Depending on the application various setting places may be considered. The most convenient is a wrist band (watch-like) but this position suffers from the artifacts due to limb motions. On the other hand the breast-bone area offers most stable conditions but it may be difficult to use this place, in the case of soldier with bulletproof jacket.

The final features of the developed sensor are:

- non-invasive measurement of glycaemia trends: on-line, update rate - 1 second,

- measurement results sent by common interface: UART, RS-232, etc.

- integration with higher-level system or mobile device i.e. PC tablet,

- emitted microwave power: up to 20 microwatt,

- measurements results comparable with invasive glucometer.

Figure 3 shows a elements of microwave glycaemia sensor.

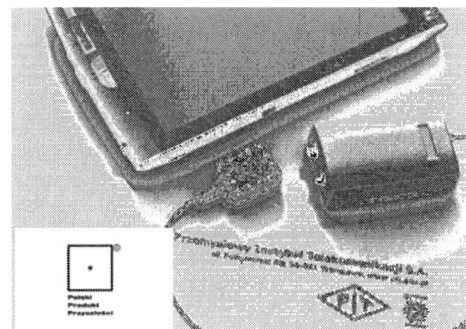

Figure 3.   Microwave glycaemia sensor: microwave head, measuring module and aditional mobile device for data visualisation used in tests

Monitoring of blood-glucose is very important because of the fact that there it may sense and predict of hypoglycaemia state (lowered level of blood-glucose) in a human under exhausting effort (training or real action on battlefield).

During a long-term actitity under exeptional and dynamic effort the drop of bool-glucose level may appear. Then various alert signs appear: blurred vision, fatigue or drowsiness, nausea and vomiting, disturbed memory and headache.

The most serious situation takes place when there is a shortage of glucose in brain tissues: strong headache, disturbed coordination, temporal loss of memory, sudden mood change – e.g. aggression, trembling, lose of consciousness, death. All these symptoms cause disruption of operational ability and threat to health or life. This is the reason to develop multisensor modules to monitor human physiologic parameters [4, 5].

IV. MEASURING EQUIPMENT

Each monitored person (subject) is equipped with a personal measuring unit. PMU is an autonomous system used for reading, recording, and processing the signals, as well as for transmitting them in real time to the management center. Figure 4 shows a block diagram of PMU.

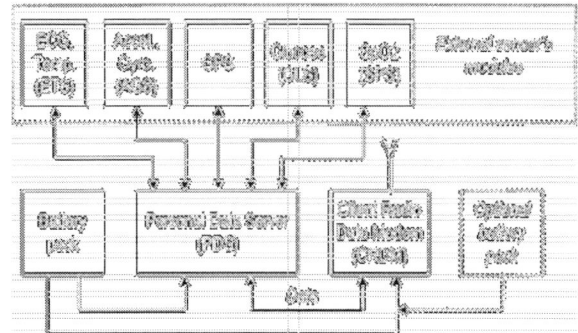

Figure 4.   Block diagram of personal measurement unit

The PMU includes a personal data server (PDS), a set of external measuring modules, a battery package, and a **C**lient **R**adio **D**ata **M**odem (CRDM). The elements are connected by cables with suitable connectors ensuring high mechanical robustness. The external sensors include: the heart rate and temperature measuring module (ETS), accelerometer and gyroscope module (AGS), GPS receiver module, microwave-based glucose measuring module (GLS), and blood oxygenation (SpO2 – saturation of peripheral oxygen) measuring module (SPS). The external sensor modules are powered from the PDS and the two-way data transmission is performed through the UART interface. Each PMU includes a wireless data transmission client (CRDM) used to transmit the data over high distance (at least 0.5 km). The CRDM modem can be powered from the common battery package or from its own battery, depending on the envisaged system operation time.

Each external sensor is implemented as an autonomous microprocessor system, which enables other or additional sensors to be connected to the PDS.

Personal data server, the main component of PMU, manages the whole system, processes the data, and communicates with the computer application. Figure 5 shows a block diagram of PDS [6, 7].

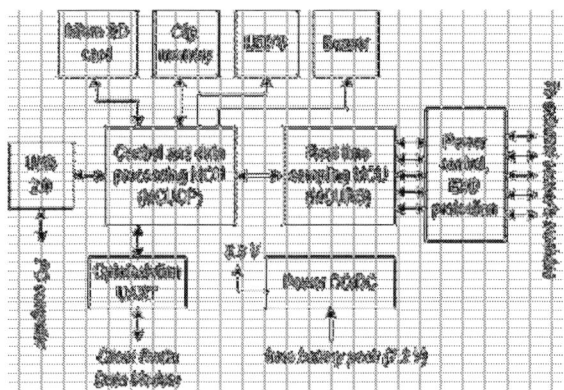

Figure 5.   Block diagram of personal data server

PDS is an autonomous device with STM32-family microcontrollers with the ARM Cortex M3 core. Using those circuits ensures a high computing power with low dissipated power. To support multi-channel synchronous data sampling and processing in real time, two microcontrollers are used. One of them (MCUCP) performs the controlling and data processing functions and the other (MCURS) is responsible for data sampling and pre-processing in real time. They communicate with each other through a UART interface. The MCUCP microcontroller is also responsible for writing and reading the data to/from non-volatile memory (a microSD card) and for controlling the USB interface. The USB connection is used to configure the device and to read the recorded data. Also, MCUCP communicates with the radio modem client through a UART interface with galvanic separation. The operating status of PDS is signaled by LEDs and acoustically. A DC/DC voltage converter provides power to all PDS circuits and external measuring modules [2].

An important element of the monitoring system is the measuring belt which gathers the data about electrical activity of the subject's heart, temperature of the subject's body, and ambient temperature. A series of tests have been conducted to analyze in detail the recorded physiological signals in terms of any irregularities caused by interference. The innovative, proprietary belt, with so-called dry electrodes, has been designed to prevent sliding on the skin surface and to provide sufficient elasticity.

Thanks to the elasticity, some elements of the belt can stretch, but the sensors still operate stably, because the placement of the sensor on the skin surface does not change during body movements. Figure 6 shows schematically the internal front part of the measuring belt.

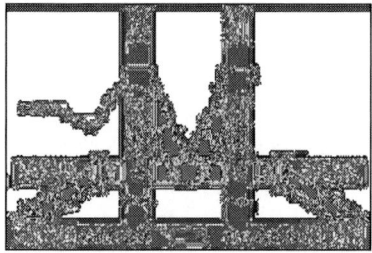

Figure 6.   Real view elements of personal measurement system

The connecting cables are hidden inside. The back part (worn on the back) consists of four elastic adjustable suspenders. They are fastened to the front part by adjustable buckles. The length of all elements is adjustable with self-catching fasteners.

The belt has four ECG electrodes and two temperature sensors mounted inside. The electrodes follow the classic three-point Einthoven system (electrodes LA, RA, LL, RL). The temperature sensors are placed to enable measuring both the body surface temperature and the ambient temperature reflecting the thermal comfort.

A prototype model of the belt has been tested experimentally in terms of the signal quality (Figure 7).

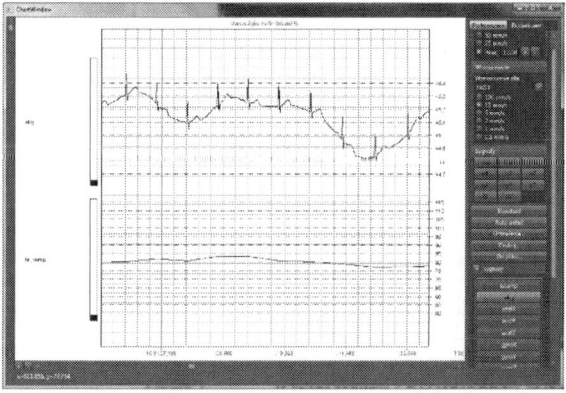

Figure 7.   An ECG signal received from the measuring belt, plotted together with an HR signal

V. SIGNAL PARAMETERIZATION TOOL

The signal analysis program communicates with a database implemented with the PostgreSQL relational data structure management software. The communication with the database uses the Linq technology and the dbLinq relational-object mapping tool. The system database can be installed on a different device than the computer hosting the management application. The digital signal recording sessions are stored as disk files identified by paths kept in the database.

Such files can be loaded asynchronously. The main window layout displays lists of the subjects, test groups, and recording sessions stored in the database. The application includes a graph generator to plot groups of signals. The graph presentation window is very flexible in terms of graph layout and value ranges. Also, it supports picking values by graphic cursors and presents the measuring ranges and position markers of the detected QRS complexes [8].

The ECG signal parameterization software is based on a procedure which involves determining a series of intervals between characteristic moments of the same phases of heart work. Typically, those are intervals between occurrences of the R waves. The determined series of intervals is typically presented as the HR (heart rate) signal expressed in beats per minute.

In order to reduce detection errors, the ECG signal parameterization algorithm analyzes the signal after filtration. The filtration eliminates errors due to pulse interference and due to the signal level fluctuations (slow-changing interference) within the rejection band of the filter. Figure 8 shows an ECG signal recorded with the measuring belt. As can be seen, in highly dynamic measuring conditions, strong interference in the measured signal may be experienced.

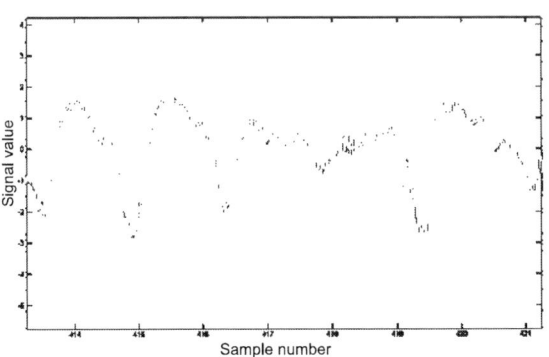

Figure 8. An example of an ECG signal with a strong interference

To determine the optimal filter parameters, frequency analysis has been applied to the ECG signal. Using the Matlab package, waterfall-type graphs were plotted to visualize the spectrum change in the time-frequency domain. In result of such analysis, limit frequencies for the ECG signal filter have been selected. Fourth-order Butterworth band-pass filters have been used. Figure 9 shows a characteristic of this filter.

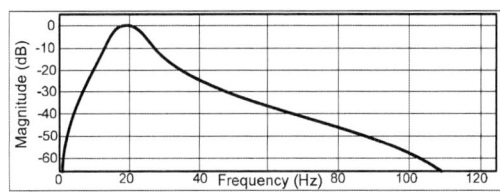

Figure 9.   Frequency characteristic of the filter

Figure 10 shows the effect of filtering with the parameters selected as described above.

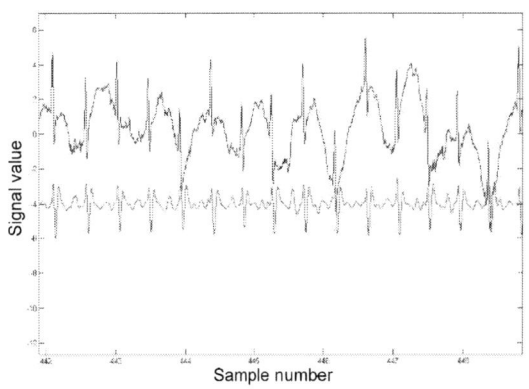

Figure 10. Comparison of corresponding fragments of the ECG signal, without filtration (up) and after filtration (down); the filtered signal is shifted down by the value of 4 for more clarity.

A specific feature of the ECG signal is that characteristic moments of the wave may constitute local maxima or minima (depending on the used lead). To improve the detection quality, the program automatically switches the characteristic extremum detection procedure between detecting the minima and detecting the maxima [9].

Features of the proposed method of signal parameterization:

- The waveform analysis method does not impose any special requirements for the signal, so by selecting a proper pre-filter, the method can be adapted to analyzing signals other than ECG, containing the HR information.

- Possibility to analyze the signals in real time during the tests.

- Independence from the signal sampling method. The analyzing procedure supports also non-uniform signal sampling.

- Small computational complexity of the analysis procedure.

VI. RESULTS OF THE RESEARCH

A. Description of the research method

The project of building a multichannel system for measuring psychophysiological parameters required two stages of laboratory research. The first stage included tests aimed at verification of the project assumptions and at optimization of the system.

The subjects included six males and one female, within the age range from 25 to 34. The subjects performed in total 19 tests, using 5 measuring kits. After each test, they filled a questionnaire, assessing the comfort provided by the system and indicating any events occurring during the tests. The tests were conducted during a time period of 2 weeks [10].

B. Conclusions from the research

It has been established on the basis of optimization tests that the measuring belt, if fitted correctly, performs the ECG signal measurement function as assumed. The belt does not restrict movements and the subjects have not reported any discomfort due to wearing it on their body.

During highly dynamic arm movements, the belt shifts against the skin, resulting in interference in the received ECG signal. Local interference is eliminated by the interference-resistant algorithm of characteristic-parameter detection, enabling the heart rate (HR) and heart rate variability (HRV) parameters to be determined.

Also the temperature sensor installed in the belt properly measured the temperature on the subject's body surface. For the purpose of this project, the measuring range of the sensor has been limited to +20...+40°C.

The quality of the measured ECG signal is also affected by the condition of the Ag/AgCl (silver / silver chloride) coating layer of the dry electrodes, which is prone to rubbing off during intensive use.

The AGS module (Figure 4) installed in the system (a three-axis gyroscope, a three-axis accelerometer, and an ambient temperature sensor) enables the subjects' movements to be efficiently monitored. In particular, the nature of the movements (e.g. regular march, ascending the stairs, small exercises, knee bends, etc.) can be determined.

The test of the GPS module has shown that the module effectively determines the subject's position in open space. The test performed in the Institute's parking lot accurately determined the route walked by the subjects. We cannot publish the visualization of the test due to the fact that military maps have been used in the system.

Figure 11 shows an example set of signals monitored by the proposed system (ekg – one channel of ECG; nX, nY, nZ – signals from three-axis accelerometer; gyroX, gyroY, gyroZ – signals from three-axis gyroscope).

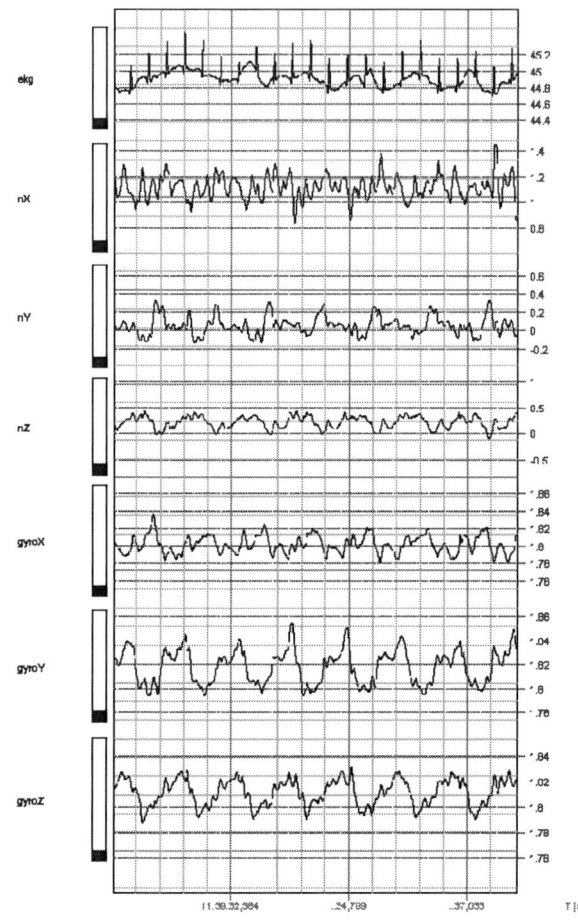

Figure 11. Example of recorded signals. ECG, gyroscope and accelerometer.

C. Tests with participation of an experimental group

To determine the quality and functionality of the proposed system, three-day tests on a group of cadets of the Military University of Technology have been conducted. The tests included several series of effort exercises on various apparatuses, as well as filling out a questionnaire on the comfort of using the system and possibilities of its integration with the uniform. The signals recorded during highly dynamic exercises enabled the system to be tested in conditions close to real.

The average age of the subjects was 28.4±8.1 years, average body mass was 84,8±18.1 kg before entering the thermal chamber and 84.6±18.1 after leaving the chamber, average BMI (body mass index) was 29.4±4.6 kg/m2, and average fatty issue content was 17.8±7.5%. Three persons out of the subjects were overweight and two persons were obese.

TABLE I.   CHARACTERISTICS OF THE SUBJECT GROUP

Subject (person)	Parameters			
	Age	BMI	Fat Mass [%]	HR at rest
MS	35	35.9	27.3	75
BJ	25	22.1	9	65
MD	25	19.8	6.4	82
MŚ	23	20	5.2	104
JL	25	29.1	21.5	87
MZ	23	31.6	28.3	71
PW	22	24.5	18.8	58
GC	24	26.2	17.8	72
JO	49	24	18.1	72
SG	38	24.8	17.6	72
KC	26	25	21.5	80
TŚ	26	26	22.8	75
Avergae	28.42	25.70	17.80	76.08

The tests were conducted using the specialist physical-training device Deptak (manufactured by Military Institute of Aviation Medicine), the muscular strength training device Powertest-M (manufactured by Military Institute of Aviation Medicine), and low-pressure, low-temperature chambers which provided variable climatic conditions. Additionally, anaerobic efficiency and maximal oxygen consumption were measured.

The research aspects related to verification of the hypothesis on diagnostic evaluation of the Mental Effort (ME), Mental Stress (MS), and Fatigue (FQ) indicators during the defined tasks are not discussed in this paper. Those parameters are determined from the ECG signal through frequency spectrum analysis. The tests included also the psychological test DT/S1 from the Vienna Test System. During the tests in the thermal chamber, the volunteers filled the COPE questionnaire (Multidimensional Stress Management Inventory) [11].

The basic objective determining the nature of the conducted tests was to evaluate how the system complied with the functional and non-functional requirements. The most important signal recorded in the system was the ECG signal. Therefore, this paper presents only one descriptive statistic from the time analysis of the ECG signal (the average value — Figure 12) and maintains that the proposed design solution may be successfully used to estimate frequency indicators which reflect the psychophysiological state of the subjects.

The results of our research show that the system may be used also in stationary conditions, e.g. during psychological tests, in order to ensure the required comfort for the subject who quickly forgets about wearing the measuring equipment. The system is very useful due to its non-invasive nature and short preparation time.

Thanks to the scalable and modular architecture, the system functionality can be expanded by adding new signals, such as glucose level, necessary for trustworthy psychophysiological assessment. The system enables accurate measurements regardless of the subject's body mass and build, as confirmed by the tests.

Figure 12 shows a fragment of the recorded data for the conducted tests, namely the average heart rate (aHR) at rest

(light gray), during the psychological test (WTS / black), and under thermal stress (dark gray).

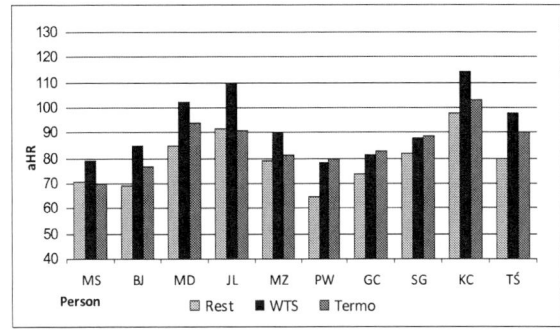

Figure 12. Average heart rate at rest, during the psychological test, and under thermal stress

Both during the psychological test and during thermal stress, the minimal, maximal, and average heart rate increases. The differences between the two stressful factors were higher for the maximal than for the minimal heart rate.

In result of the research, a new carrying system has been developed for ECG, temperature, and (optionally) glucose sensors. The design is the basic component of the Personal Combat System within the national future soldier project. The carrying system designed in Military Institute of Aviation Medicine has been manufactured by MASKPOL, a manufacturer of military uniforms.

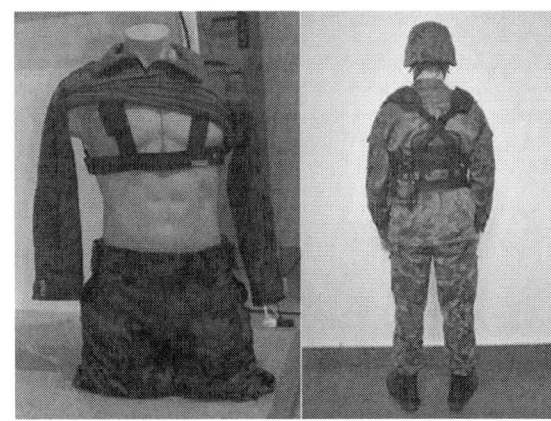

Figure 13. Model of a belt with built-in ECG, temperature sensor, and glucose meter. Picture from Future Soldier Exhibition, Prague 2010

The obtained results have confirmed that the system can be used also in strong-interference conditions.

Simple devices for monitoring the averaged HR value, available on the market, are completely useless for determining the ME, MS, and FQ values. The analysis in the f domain requires information about the heart sinus rhythm identified on the basis of the R wave position determined to an accuracy of 4 ms.

VII. Summary

The authors focus on presenting the technical aspects of the designed system and selected issues of the system testing in simulated conditions. Thanks to the proper design of the measuring part of the system and the interference-resistant algorithms for analyzing signals recorded in extreme conditions, the presented technological solutions can be successfully used in civilian applications, such as monitoring systems for professional drivers, railway engine drivers, civil airplane pilots, or trainees in flying clubs, as well as to monitor elderly people at their places of living, etc.

The psychophysiological signal measuring system described in the paper is currently used to record data in a project involving tests of vehicle drivers. The system is comfortable thanks to its small size and high flexibility. The ECG/temperature measuring module can work both with the sensor carrying belts and with classical adhesive electrodes, as well as with an external wire-based temperature sensor. The GPS module used in the system is capable of determining the device position very fast, while the AGS module enables observation of the ride phases and moments of turning the steering wheel and changing the gears.

Works are under way to develop a miniaturized version of the psychophysiological parameter monitoring system and reduce it to a single integrated circuit.

The system presented here also uses a new concept of microwave technology for non-invasive on-line measurement of human blood-glucose levels. There are advantages of use of this technology in multisensor module for evaluating human physiologic parameters. This evaluation is necessary to monitor and predict soldier health state and efficiency in military application as: candidate selection, military training, battlefield operation and soldier biologic recovery.

Development of on-line and non-invasive sensing of selected physiologic parameters (ECG, heart rate, glycaemia) gives a extremely powerful diagnostic tool for use in military and civil applications.

Acknowledgment

Some of the presented works have been conducted under development project No. O R00 0056 06.

References

[1] M. Życzkowski, B. Uziębło-Życzkowska, K. Różanowski, "Using modalmetric fiber optic sensors to monitor the activity of the heart", 2011, Progress in Biomedical Optics and Imaging - Proceedings of SPIE, 7894, art. no. 789404

[2] K. Różanowski, "Project UDA-POIG.01.03.01-14-136/08-00 — Opracowanie metod monitorowania aktywności psychofizjologicznej z funkcją automatycznego wykrywania zagrożeń (Development of psychophysical activity monitoring methods with an automatic threat detection function)", Innovative Economy Operational Program (Priority I. Research and development of modern technologies; Activity 1.3. Wojskowy Instytut Medycyny Lotniczej, Warszawa, 2011.

[3] Y. Nikawa, T. Michiyama, "Blood-Sugar Monitoring by Reflection of Millimeter Wave", Proceedings of Asia-Pacific Microwave Conference 2007, pp.1581-1584.

[4] Z. Szczepaniak, E. Sędek, M. Łuszczyk, A. Gaździńska, K. Różanowski, "Wybrane problemy nieinwazyjnej techniki mikrofalowej do pomiaru zmian glikemii człowieka w aspekcie monitorowania sprawności psychofizjologicznej (Selected issues of the non-invasive microwave technique of measuring changes of glycaemia in humans for the purpose of monitoring the psychophysiological fitness)". Elektronika konstrukcje, technologie, zastosowania, 2010, 143.

[5] Z. Szczepaniak, E. Sędek, M. Łuszczyk, A. Arvaniti, K. Różanowski, "Wybrane problemy nieinwazyjnych pomiarów poziomu glukozy we krwi człowieka za pomocą technik mikrofalowych w zastosowaniu cywilnym i wojskowym (Selected issues of non-invasive measurements of glucose levels in human blood using microwave techniques in civilian and military applications)". Elektronika konstrukcje, technologie, zastosowania, 2009, 10.

[6] OEM III Module Specification and Technical Information, NONIN® Medical, Inc., 4518-001-10 Rev J, 2007.

[7] Texas Instruments – Medical Application Guide, Medical Instruments, USA 2009.

[8] Z. Piotrowski, K. Różanowski, "Programowy moduł analizatora zmienności rytmu serca (HRV) - opis funkcjonalny oraz algorytmiczny (A software module for Heart Rate Variability (HRV) analysis — functional description and algorithms)". Prace Naukowe. Elektronika. Politechnika Warszawska, 2007, 1.

[9] Z. Piotrowski, K. Różanowski, "Robust Algorithm for Heart Rate (HR) Detection and Heart Rate Variability (HRV) Estimation", 2010, Acta Physica Polonica A, 333.

[10] O. Bar-Or, "The Wingate Anaerobic Test . An update on methodology, reliability and validity". Sports Med. 1987, 4, pp. 381-394.

[11] S. Carver, M. Scheier , J. Weintraub; adapted by: Z. Juczyński, N. Ogińska-Bulik, "Wielowymiarowy Inwentarz do Pomiaru Radzenia Sobie ze Stresem COPE", 2009.

[12] J. Żołądź, "Wydolność fizyczna człowieka (Physical efficiency of a human). in: Fizjologiczne podstawy wysiłku fizycznego (Physiological basis of physical effort)", 2001, Warszawa, PZWL, pp. 456-519

[13] B. Dąbrowska, A. Dąbrowski, "Podręcznik Elektrokardiografii (Electrocardiography manual)". Wydawnictwo lekarskie PZWL, Warszawa 2002, wyd. 4.

[14] Dokumentacja osobistego rejestratora danych (Personal data recorder documentation). Wojskowy Instytut Medycyny Lotniczej, Warszawa, 2009.

[15] K. Różanowski, T. Sondej, T. Radomski, "Wielozadaniowy system monitorowania sygnałów fizjologicznych i środowiskowych (A multi-task system for monitoring physiological and environmental signals)". Elektronika konstrukcje, technologie, zastosowania, 2007, 9, pp. 85-91.

[16] Materiały informacyjne sensora pomiaru glukozy (Glucose level sensor brochure). Przemysłowy Instytut Telekomunikacji, Warszawa, 2009.

Testing and Reliability

426

Determining Effective Testability Degree of Analog Circuits

Zdenek Kincl, Zdenek Kolka

Department of Radio Electronics
Faculty of Electrical Engineering and Communication
Brno University of Technology
Purkynova 118, 612 00 Brno, Czech Republic
Email: xkincl01@stud.feec.vutbr.cz, kolka@feec.vutbr.cz

Abstract—**The paper deals with the parametric fault diagnosis of linear analog circuits in the frequency domain. The testability degree, which is referred to as the total number of testable network parameters, is theoretically independent of nominal values of network parameters, the set of test frequencies and the fault detection method. However, practical results show that the effective testability determined numerically using the Singular Value Decomposition of sensitivity matrix (Jacobian) depends on the normalization of network parameters, frequency response measurement methods and a selected set of test frequencies. The differences between the theoretical and the practical testability degree will be discussed on a practical example of RC phase shifter filter.**

Index Terms—**Parametric fault diagnosis, testability degree, test point selection, analog circuit testing.**

I. INTRODUCTION

The goal of parametric fault diagnosis is to detect fault(s) in analog circuits [1]. It is rather difficult in comparison with digital circuits because analog signals are continuous in time and value. Due to manufacturing process variations, temperature drifts, components ageing and faults the actual values of component parameters are always different from the nominal ones. A component whose parameters are outside allowed tolerances is classified as faulty. The key requirements for a robust parametric fault diagnosis in the frequency domain are appropriate sets of test points and test frequencies with respect to potentially faulty components.

The actual values of some (testable) network parameters are estimated based on the measurements of one or more different network functions on several test frequencies [2]. An arbitrary network function, such as frequency response, input or output impedance function, etc., for a lumped linear time-invariant analog circuit can be written in symbolic form as

$$H(s, \mathbf{p}) = \frac{a_n(\mathbf{p})s^n + \ldots + a_0(\mathbf{p})}{b_m(\mathbf{p})s^m + \ldots + b_0(\mathbf{p})} \qquad (1)$$

where s is the complex frequency, and polynomial coefficients a_i and b_i are nonlinear functions of network parameters $\mathbf{p} = [p_1, \ldots, p_R]^T$. The symbolic expressions for an arbitrary network function can be easily obtained for example using the SNAP program [3].

Since the original nominal values of network parameters are spread over a wide range of magnitudes (resistors - 10^4, capacitors - 10^{-10}, etc.), they should be normalized for correct sensitivity evaluations. Otherwise, the Jacobian matrix contains incomparable columns and its conditionality is poor and the fault diagnosis may return non-relevant results [4]. The value of normalized network parameter $\tilde{p}$ is then

$$\tilde{p} = \frac{p}{p_{nom}} \qquad (2)$$

where p represents the real (actual) value and p_{nom} the nominal value of the network parameter.

The actual values of an a priori selected set of tested network parameters are determined by solving the system of non-linear fault equations

$$H_k(j\omega_{k,i}, \tilde{\mathbf{p}}) = M_{k,i} \qquad (3)$$

where H_k represents the network function of the k-th testpoint in a generalized sense, i.e. it can be a complex network function, its magnitude in decibels or its phase in radians. $M_{k,i}$ are the measurements of the k-th network functions on the i-th test frequency compatible with the definition of H_k.

The Newton-Raphson iteration scheme is considered in the form [5]

$$\tilde{\mathbf{p}}_{n+1} = \tilde{\mathbf{p}}_n - \alpha_n \mathbf{J}(j\omega_i, \tilde{\mathbf{p}}_n)^{-1}(\mathbf{H}(j\omega_i, \tilde{\mathbf{p}}_n) - \mathbf{M}(j\omega_i)) \quad (4)$$

where $\tilde{\mathbf{p}}$ is the vector of unknown network parameters, $\mathbf{H}(j\omega_i, \tilde{\mathbf{p}}_\mathbf{n})$ is the vector of frequency responses evaluated for the actual values of network parameters, $\mathbf{M}(j\omega_i)$ is the vector of measurements, $\mathbf{J}$ is the Jacobi matrix, and α is an adaptively chosen damping parameter which improves the algorithm stability.

II. TESTABILITY ANALYSIS

The total number of testable network parameters with respect to the selected set of test point(s) is referred to as the testability degree T [6]. The testability is associated with the solvability degree of the system of fault equations (3). When the testability degree is lower than the total number of unknown network parameters (potentially faulty components), the system of fault equations (3) does not have a unique solution and some more test points must be chosen.

The theoretical testability of the circuit can be determined directly using symbolic polynomial coefficients a_i and b_i from (1) [7]. The partial derivatives should be evaluated for normalized network parameters.

$$
T = rank \begin{bmatrix}
\frac{\partial a_0}{\partial \tilde{p}_1} & \frac{\partial a_0}{\partial \tilde{p}_2} & \cdots & \frac{\partial a_0}{\partial \tilde{p}_R} \\
\vdots & \vdots & \ddots & \\
\frac{\partial a_n}{\partial \tilde{p}_1} & \frac{\partial a_n}{\partial \tilde{p}_2} & \cdots & \frac{\partial a_n}{\partial \tilde{p}_R} \\
\frac{\partial b_0}{\partial \tilde{p}_1} & \frac{\partial b_0}{\partial \tilde{p}_2} & \cdots & \frac{\partial b_0}{\partial \tilde{p}_R} \\
\vdots & \vdots & \ddots & \vdots \\
\frac{\partial b_m}{\partial \tilde{p}_1} & \frac{\partial b_m}{\partial \tilde{p}_2} & \cdots & \frac{\partial b_m}{\partial \tilde{p}_R}
\end{bmatrix} \quad (5)
$$

The border between a testable and an untestable network parameter is not sharp. It depends on many aspects such as the network parameter normalization process, the difference between the actual and the nominal values of the parameter, the total number of tested parameters, and the set of test frequencies.

The testability degree of the circuit determined using (5) provides only information about the maximal solvability degree of the system of fault equations. For particular values of network parameters the rank of matrix (5) can be lower than the value obtained symbolically. A numerical approach based on the Singular Value Decomposition seems to be optimal for the rank evaluation [8]. The testability degree (5) is valid for complex measurements. A similar form for magnitude-phase measurements is not known yet.

The testability degree of the circuit can be also determined as the rank of the Jacobi matrix of (1), i.e. the total number of linear independent rows or columns of the sensitivity matrix [2]. The rows and columns correspond to the individual test frequencies and tested parameters, respectively.

$$
T = rank \left[\frac{\partial H(j\omega_i)}{\partial \tilde{p}_j} \right] \quad (6)
$$

This testability degree determination provides accurate results in comparison with the method based on (5) and for this reason it is more suitable for test point(s) selection. However, the method requires determining of an appropriate set of test frequencies. An optimum method for the test frequency selection has not been determined yet. A heuristic solution [9], [10] and a procedure based on maximizing the sensitivity in the frequency domain were proposed [1]. Other authors used global stochastic optimization techniques [2], [11].

The linear dependent columns of matrices (5) or (6) determine ambiguity groups. Inside those groups the effects of individual parameters on network functions are indistinguishable from one another. Only some parameters of each group can be tested (the others must be considered fixed) or more test points must be chosen.

The next chapter presents an effective method for determining components of (6). It can be used with both numerical and symbolical calculations.

III. PARAMETER SENSITIVITIES

In this section, the sensitivities of individual network parameters to magnitude and phase frequency characteristics are derived. The original network function (1) can be formulated in the form

$$
H(s, \tilde{\mathbf{p}}) = |H| e^{j\varphi} \quad (7)
$$

where $|H|$ and φ represent magnitude and phase frequency characteristics, respectively.

In the case of complex measurements, entries of (6) are complex as well as equations (3) and (4). However, separate measurements of magnitude in decibels or phase in radians may be available. After some mathematical arrangements of (7) it is possible to derive sensitivities of magnitude and phase frequency response to individual network parameters, which constitute the Jacobi matrix.

$$
S_{\tilde{p}_j}^{|H|_{dB}} = 20 \log(e) \Re \left\{ \frac{\partial H}{\partial \tilde{p}_j} \frac{1}{H} \right\} \quad (8)
$$

$$
S_{\tilde{p}_j}^{\varphi} = \Im \left\{ \frac{\partial H}{\partial \tilde{p}_j} \frac{1}{H} \right\} \quad (9)
$$

For low-complexity circuits, the sensitivities can be evaluated symbolically. However, for moderate and high-complexity circuits the symbolic form of H (1) cannot be even generated. In this case numerical solution should be used.

Any network function H can be obtained as a ratio of two algebraic cofactors of circuit matrix $\mathbf{A}$

$$
H = (-1)^\alpha \frac{det(\mathbf{A}_1)}{det(\mathbf{A}_2)} \quad (10)
$$

Matrices $\mathbf{A}_1$ and $\mathbf{A}_2$ are derived from $\mathbf{A}$ by means of adding and deleting some rows and columns, and α depends on the indices of the rows and columns. Details can be found in classical textbooks of circuit theory, e.g. in [12].

Consider, for example, an admittance-type element g between nodes i and j, which appears in the admittance part of $\mathbf{A}$ as shown in Fig. 1.

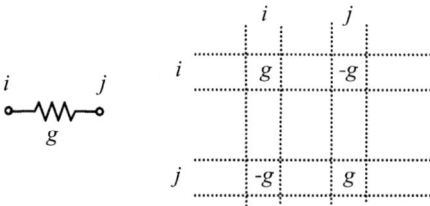

Fig. 1. Floating conductor and its stamp in the hybrid matrix.

Due to the row/column transformation of $\mathbf{A}$ into $\mathbf{A}_1$ and $\mathbf{A}_2$, element g may appear in $\mathbf{A}_1$ or $\mathbf{A}_2$ in the four-position pattern as in Fig. 1, in just one row or column, or in one position. Let us suppose that g appears in both $\mathbf{A}_1$ and $\mathbf{A}_2$ in four positions. Then both determinants in (10) can be expanded to

$$det(\mathbf{A}_1) = det(\mathbf{A}_1') + g\left(\Delta_{i1:i1}^{(A1)} + \Delta_{j1:j1}^{(A1)} - \Delta_{i1:j1}^{(A1)} - \Delta_{j1:i1}^{(A1)}\right) \tag{11}$$

$$det(\mathbf{A}_2) = det(\mathbf{A}_2') + g\left(\Delta_{i2:i2}^{(A2)} + \Delta_{j2:j2}^{(A2)} - \Delta_{i2:j2}^{(A2)} - \Delta_{j2:i2}^{(A2)}\right) \tag{12}$$

where $\mathbf{A}_1'$ and $\mathbf{A}_2'$ are matrices $\mathbf{A}_1$ and $\mathbf{A}_2$ without element g, and $\Delta_{i1:i1}^{(A1)}$ and $\Delta_{i2:i2}^{(A2)}$ are algebraic cofactors of $\mathbf{A}_1$ and $\mathbf{A}_2$, respectively. If element g appeared in $\mathbf{A_1}$ or $\mathbf{A_2}$ in two or one positions, (11) or (12) would contain only the cofactors whose indices correspond to the element coordinates. For impedance-type elements, as well as for controlled sources the formulae will be similar [12].

Thus, for any unique network parameter p, network function (10) can be expressed as

$$H = \frac{ap + b}{cp + d} \tag{13}$$

where a, b, c, and d are complex numbers for a fixed frequency. Then the derivative in (8) and (9) can be expressed as

$$\frac{\partial H(j\omega_i)}{\partial \tilde{p}} \frac{1}{H(j\omega_i)} = \frac{ap_{nom}}{ap_{nom}\tilde{p} + b} - \frac{cp_{nom}}{cp_{nom}\tilde{p} + d} \tag{14}$$

IV. EXAMPLE OF APPLICATION

The differences between theoretical and practical testability degree will be discussed on the example of RC phase shifter filter shown in Fig. 2.

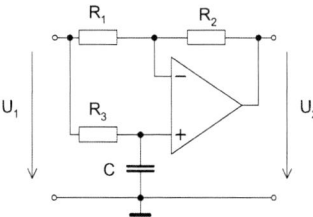

Fig. 2.   RC phase shifter filter.

In the case of R_1 equal to R_2 the circuit behaves as an all-pass phase shifter. In Fig. 3 the frequency responses for nominal values of components $R_1 = R_2 = R_3 = 1 \ \Omega$ and C = 1 F are shown. Only one test point on the output of the filter is considered. The network function is as follows

$$H(s, \tilde{\mathbf{p}}) = \frac{U_2}{U_1} = \frac{R_1 - R_2 R_3 C s}{R_1 + R_1 R_3 C s} \tag{15}$$

According to (5) the theoretical testability degree is 3. Due to linear dependent columns of the matrix (16) the components

R_3 and C belong to an ambiguity group and they cannot be tested simultaneously.

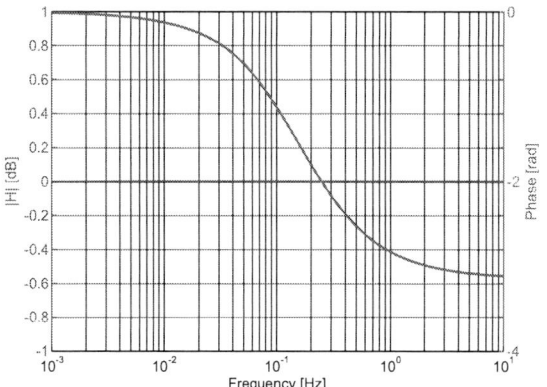

Fig. 3.   Frequency response of the filter.

$$T = rank \begin{pmatrix} & R_1 & R_2 & R_3 & C \\ a_0 & 1 & 0 & 0 & 0 \\ a_1 & 0 & -1 & -1 & -1 \\ b_0 & 1 & 0 & 0 & 0 \\ b_1 & 1 & 0 & 1 & 1 \end{pmatrix} \tag{16}$$

The rank of matrix (16) is evaluated using the Singular Value Decomposition. The singular numbers are shown in (17). As can be seen, there are three nonzero elements, i.e. the theoretical testability degree is equal to three.

$$\mathbf{S} = \begin{pmatrix} 2.3073 & 0 & 0 & 0 \\ 0 & 1.5356 & 0 & 0 \\ 0 & 0 & 0.5645 & 0 \\ 0 & 0 & 0 & 0 \end{pmatrix} \tag{17}$$

In Fig. 4 and Fig. 5 the sensitivities of magnitude and phase frequency responses to network parameters are shown. Because the components R_3 and C belong to the ambiguity group, the curves of their sensitivities coincide. Moreover, these parameters cannot be tested when only magnitude measurement is considered (they have no influence on the magnitude frequency response). The effective testability degree of the circuit is lower than the theoretical testability.

Based on the comparison of theoretical testability degree (5) with the analysis of the Jacobi matrix (6), it is possible to draw some conclusions showing the disadvantage of using (5):

- As can be seen in Fig. 4 and Fig. 5, in the case of the same nominal values $R_1 = R_2$, the curves of individual sensitivities corresponding to R_1 and R_2 are linearly dependent, i.e. there is one more ambiguity group (R_1, R_2). Only one parameter of this group can be tested. The effective testability is two.

- Moreover, in the case of only magnitude measurements, none of the components R_3 and C can be tested (they have no effect on the magnitude frequency response).

429

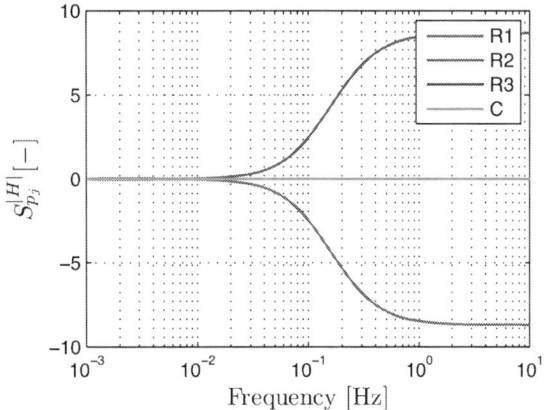

Fig. 4. Sensitivities of network parameters on magnitude frequency response.

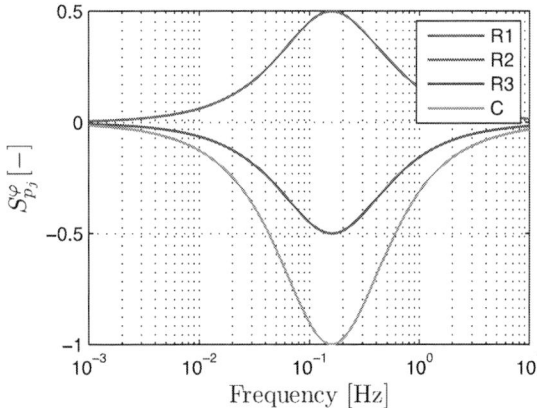

Fig. 5. Sensitivities of network parameters on phase frequency response.

Because R_1 and R_2 form the ambiguity group, only one parameter can be tested. The effective testability is only one.

- In the case of only phase measurements, all of the sensitivities are linearly dependent and all of the network parameters belong to one ambiguity group (R_1, R_2, R_3, C). Only one parameter from the group can be tested. The effective testability is only one.

- For independent testing of two network parameters with respect to the selected test point it is necessary to measure both frequency characteristics (magnitude and phase).

V. CONCLUSION

In the paper the differences between theoretical and practical testability degrees of the RC phase shifter filter are discussed. The same nominal values of some circuit components and the way of frequency response measuring may have an impact on the practical testability degree. The testability degree determined directly using symbolic polynomial coefficients provides only information on the upper bound with respect

to a selected set of test points. But in practice, some of the unknown components may be untestable, i.e. a selected set of test points is not optimal. The testability degree determined as the rank of the Jacobi matrix provides more accurate results, i.e. a set of surely testable components, and for this reason the method is more suitable for a real test point(s) selection. However, it additionally requires a set of test frequencies.

ACKNOWLEDGMENT

The research described in the paper was financially supported by the Czech Science Foundation under grant No. P102/10/1665 and the internal grant BUT No. FEKT-S-11-13. The research is a part of the COST Action IC 0803, which is financially supported by the Czech Ministry of Education under grant No. OC09016. The research was supported by the Czech Ministry of Industry and Trade under grant agreement No. FR-TI2/194. The support of the project CZ.1.07/2.3.00/20.0007 WICOMT, financed from the operational programme Education for competitiveness, is gratefully acknowledged. The described research was performed in laboratories supported by the SIX project; registration number CZ.1.05/2.1.00/03.0072, the operational programme Research and Development for Innovation.

REFERENCES

[1] M. Slamani and B. Kaminska, "Multifrequency Analysis of Faults in Analog Circuits," *IEEE Design and Test of Computers*, 1995, vol. 12, no. 2, p. 70-80.
[2] F. Grasso, A. Luchetta, S. Manetti and M.C. Piccirilli, "A Method for the Automatic Selection of Test Frequencies in Analog Fault Diagnosis," *IEEE Trans. on Instrumentation and Measurement*, 2007, vol. 56, no. 6, p. 2322-2329.
[3] Z. Kolka, "SNAP - Program for Symbolic Analysis," *Radioengineering*, 1999, vol. 8, no. 1, p. 23-24.
[4] Z. Kolka, Z. Kincl, D. Biolek and V. Biolkova, "Parametric Reduction of Jacobian Matrix for Fault Analysis," *In Proc. of the 22nd IEEE International Conference on Microelectronics (ICM 2010)*, 2010, p. 503-506.
[5] C.T. Kelley, "*Solving Nonlinear Equations with Newton's Method*," Philadelphia: Society for Industrial Mathematics (SIAM), 2003.
[6] R. Saeks and N. Sen, "Fault Diagnosis for Linear System via Multifrequency Measurement," *IEEE Trans. on Circuits and Systems*, 1979, vol. 26, no. 7., p. 457-465.
[7] G. Ferdi, S. Manetti, M.C. Piccirilli and J. Starzyk, "Determination of an Optimum Set of Testable Components in the Fault Diagnosis of Analog Linear Circuits," *IEEE Trans. on Circuits and Systems - I*, 1999, vol. 46, no. 7, p. 779-787.
[8] S. Manetti and M.C. Piccirilli, "A Singular Value Decomposition Approach for Ambiguity Group Determination," *Analog Circuits and Systems I: Fundamental Theory and Application*, 2003, vol. 50, no. 4, p. 477-487.
[9] Z. Kincl and Z. Kolka, "Test Frequency Selection for Band-pass Filters," *In Proc. of the 20th International Conference Radioelektronika 2010*, 2010, p. 173-176.
[10] F. Grasso, A. Luchetta, S. Manetti and M.C. Piccirilli, "Symbolic Techniques for the Selection of Test Frequencies in Analog Fault Diagnosis," *Analog Integrated Circuits and Signal Processing*, 2004, vol. 40, no. 3, p. 205-213.
[11] Z. Kincl and Z. Kolka, "Parametric Fault Diagnosis using Overdetermined System of Fault Equations," *In Proc. of the International IEEE Conference on Microwaves, Communications, Antennas and Electronic Systems (COMCAS 2011)*, 2011, p. 1-4.
[12] W.K. Chen, "*Active Network Analysis (Advanced Series in Electrical & Computer Engineering - vol. 2)*," World Scientific Publishing, United Kingdom, 1991.

MIXDES 2012, 19th International Conference *"Mixed Design of Integrated Circuits and Systems"*, May 24-26, 2012, Warsaw, Poland

DFT for Analog and Mixed Signal IC Based on IDDQ Scanning

Badi Guibane

Electronics and microelectronics laboratory
Faculty of Sciences of Monastir
Monastir, Tunisia
guibene.badi@gmail.com

Belgacem Hamdi

Electronic department
Issats
Sousse, Tunisia
Belgacem.hamdi@gmail.com

Abstract—**The cost of integrated circuits increases with the complexity and integration density. This has led designers to consider testing from the design phase; that's what we call DFT (design for testability). In this paper, we propose a DFT solution, based on technique of I_{DDQ} measuring current, by incorporating a Built-In Current sensor, whose function is to detect power consumption of different circuits under test, and by applying an intelligent switching technique, between BICS and the circuits under test. This DFT technique is intended for digital, analog and mixed integrated circuits. The final system represented, by the name of the TEST AND CONTROL UNIT, consists on a test vector generator, an interconnection logic block, a BICS and a diagnostic unit, designed to test all circuits of the wafer by using a single BICS. The aim of system is to reduce the time required for functionality test of each circuit in the mass production. It offers a practical test solution for integrated circuits designers.**

Index Terms—**Buit in current sensor; design for testability; i_{DDQ} testing; analog and mixed circuits; integrated test.**

I. INTRODUCTION

DFT includes all design techniques that aim to facilitate testing of integrated circuits [8] at the last stage of manufacturing as well as in the use of these devices in their applications. Such task is especially difficult in the case of analog and mixed signal integrated circuits. Indeed, techniques applied on analog and mixed circuits are not as developed and mastered as digital ones[9]. The term Design For Testability is used in the design of complex electronic components (Microprocessor, Ram, Operational amplifier....). This is taken into account to:

- Facilitate the analysis, in case of a malfunction of a component. This analysis allows modifying the design, or manufacturing process to obtain functional chips.

- Reducing the time needed for functionality test of each component during mass production.

I_{DDQ} testing can be incorporated into a specific DFT testing technique. Indeed I_{DDQ} testing is a technique based on measuring level of current in an integrated circuit, a functional circuit consumes a well-defined current. A defect is likely to increase the consumption of current in an integrated circuit. This overconsumption is detected by the BICS, which indicate the appearance of a defect. The technique is considered as a valuable complement to other methods for testing circuits, in

the interest that it allows the detection of certain types of defects that could not be found by using the basic techniques. Some defects can cause an excess of current without affecting the performance of the circuit [1], the I_{DDQ} test can detect these defects [2], like the appearance of resistive circuits between two terminals of the transistor quad.

This paper is organized as follows; section 2 presents theoretical background. Section 3 presents types of fault injection at transistor level. Section 4 describes a novel approach for testing integrated circuit by concentrating on the conception and description of control and test unit. Section 5 gives the simulations results. Finally we conclude in the last section.

II. THEORETICAL BACKGROUND

Design of logic integrated circuits in CMOS technology is becoming more and more complex with a rising interest and demand for all sorts of embedded electronics. A common issue to be addressed by integrated circuits designers, manufacturers and users is the testing of these ICs [1].

A. Testing an IC

Testing an integrated circuit can be expressed by checking if its outputs correspond to the inputs applied to it. If the test is positive, then the system is good for use. If the outputs are different than expected, the IC is rejected (Go/No Go test). A diagnosis may be applied to it, in order to point out and identify the problem's causes.

Testing is applied to detect faults after several operations: design, manufacturing, packaging, as illustrated in Figure. 1.

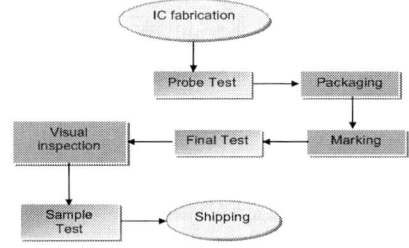

Figure 1.   Typical IC production flow

If a test strategy is considered at IC level, the fault can be detected at early system design stages, located and eliminated at a very low cost (The rule of ten).

B.   Fault Model

A fault is a model that represents the effect of a failure by means of the change that is produced in the system signal.

As a model, the fault does not have to be an exact representation of the defects, but rather, to be useful in detecting the defects. The most common fault model assumes single stuck-at (SSF) lines even though it is clear that this model does not accurately represent all actual physical failures. However, test sets that have been generated for this fault type have been effective in detecting other types of faults.

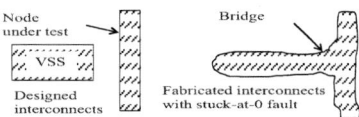

Figure 2.   Physical origin of node fault stuck at 0

Figure 2 illustrates a possible origin for a node stuck at 0 voltage: the implementation is close to a VSS node (here situated close, same layer), and a faulty metal bridge makes a robust connection to the ground.

However, the manufacturing of interconnects may result in interruptions or short-cuts, which may have catastrophic consequences on the behavior of the integrated circuit.

It is possible to mark most commonly used fault models (Table1).

TABLE I.   MOST COMMONLY USED FAULT MODELS

Fault Model	Description
Single stuck-at faults (SSF)	One line takes the value 0 or 1.
Multiple stuck-at faults (MSF)	One, two or more lines have fixed values, not necessarily the same.
Bridging faults	Two or more lines that are normally independent become electrically connected.
Delay faults	A fault is caused by delays in one or more paths in the circuit.
Intermittent faults	Caused by internal parameter degradation. Incorrect signal values occur for some but not all states of the circuit. Degradation is progressive until permanent failure occurs.
Transient faults	Incorrect signal values caused by coupled disturbances. Coupling may be via power bus capacitive or inductive coupling. Includes internal and external sources as well as particle irradiation.

C.   Testing and fault coverage

Testing is the process of determining whether a device functions correctly or not. The question is: How much testing of an IC is enough? The Yield (Y) is defined as the ratio of the number of good dies per wafer to the number of dies per wafer. Fault coverage (FC) is the measure of the ability of a test set T to detect a given set of faults that may occur on the DUT

(Device Under Test). We shall try to achieve FC=1, that is a fault coverage of 100%.

FC= (#detected faults)/(#possible faults)

Defect level (DL) is the fraction of bad parts among the parts that pass all tests.

$$DL= 1 - Y^{(1-FC)}$$

Where FC refers to the real defect coverage (probability that T detects any possible fault in F or not) and DL is the DPM (defects per million). Typical values claimed are less than 200 DPM, or 0.02%.

III.   FAULT INJECTION AT TRANSISTOR LEVEL

Extensive work has been conducted to characterize the fault effects transistor level that typically occur in CMOS ICs [3]. These defects, equally applicable to NMOS and PMOS transistors, consist of short and open circuits as shown in Figure 3 [3]. Three of these defects, the gate–drain short, gate–source short, and gate–channel short, represent gate oxide failures that have been found to exist throughout the life of a CMOS device [3].

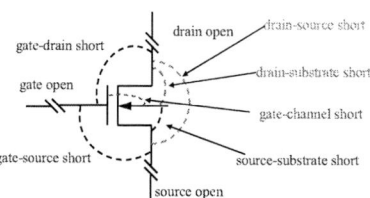

Figure 3.   Nine common defects in transistor level

Bridging faults can appear either at the logical output of a gate or at the transistor nodes internal to a gate. Bridge between the outputs of independent logic gates or an inter-gate bridge can also occur. Bridging fault could happen between the drain and source, drain and gate, source and gate, or bulk and gate nodes.

IV.   CONCEPTION AND SYSTEM DESCRIPTION

A.   Proposed Design

The basic idea is to implement an I_{DDQ} test designed to test several circuits having the same function, using a single built in current sensor. Figure 4 shows our proposed approach. The system consists of five main parts:

- The BICS, it measures I_{DDQ} current and provide pass/fail signal.

- The interconnection logic block : is the block that allows to switch connections between the circuit under test and the BICS.

- The vector test generator

- The diagnostic unit

- Circuits under test

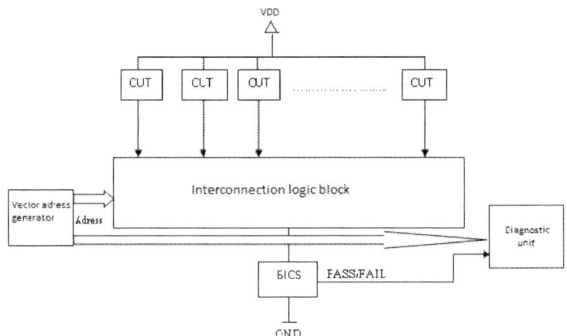

Figure 4. Proposed design

The BICS used is an integrated current sensor.

Each circuit is connected to the BICS via its ground line. Switching connection between the circuits under test and the BICS, is made through the interconnection logic block.

If the measured current exceeds the reference current, the output of the BICS will move to the high state. Otherwise the output remains low, indicating that the circuit is not defected. The voltage output of the BICS can be acquired in a diagnostic unit, placed outside the integrated circuit, allowing the user to verify the validity of each circuit through the address and its adequate associated value [4].

B. System description

The main three parts of the proposed system are:

1) Vector adress generator

The vector address generator is a device used to provide the address of the circuit under test to the interconnection logic block, enabling the binding of selected circuit to the BICS. It generates a vector address for each clock cycle. The test frequency is controlled by the clock generator.

2) Dignostic Unit

The diagnostic unit is used to inform us about the status of each circuit tested by the BICS. Each output value of the BICS and associated CUT address, are treated and stored in the diagnostic unit which is placed outside the integrated circuit.

3) Built in current sensor (BICS)

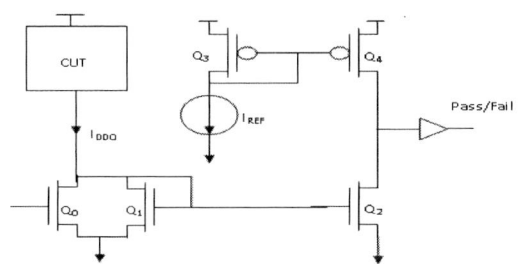

Figure 5. The-built in current sensor (BICS)

One of the most critical issues in an I_{DDQ} test is the BICS that can be used to detect abnormal currents. Figure 5 shows the circuit diagram of the BICS.

This scheme consists of two NMOS transistors forming a current mirror that allows the reflection of the current drawn by the circuit (I_{DDQ}), and also PMOS transistors that form the second current mirror for the reflection of the reference current I_{REF}, an NMOS transistor that allows either isolate or connect the circuit to BICS, a reference current source and an inverter.

In addition, the BICS requires either an external voltage reference or an external power source. Therefore, the BICS requires less space [5]. In test mode, the BICS compares firstly, the quiescent current consumed by the circuit with a reference current I_{REF}.

When the current consumed is greater than the reference, the output signal PASS / FAIL is set to 1, indicating the presence of defects. When the current consumed is less than that value of reference, the output signal PASS / FAIL is set to 0, indicating absence of defects [6].

The NMOS transistor Q0 is used as a switch to isolate or connect the CUT to BICS. The reference current has nearly the same value as the current drawn by the circuit without a fault. By comparing the I_{REF} with I_{DDQ}, The proposed BICS determines whether the circuit under test is functional or defective. The Efficiency of the proposed BICS to test analog and digital circuit is given by [2] [5].

4) Interconnection logic block

The Circuits Under Test (CUT) must be tested by measuring I_{DDQ} current of each circuit. Control unit selects the circuit under test, and activate it to test mode, to finally connect it to the BICS. One BICS measures the current of several circuits. When a particular circuit is under test, its mass line is directed to the BICS to allow observation of IDDQ current. However the activation of one of transistors T1T2 ... Tn led the GND line directly to the circuit bypassing the corresponding BICS [7].

Figure 6 details the different components of the interconnection logic block.

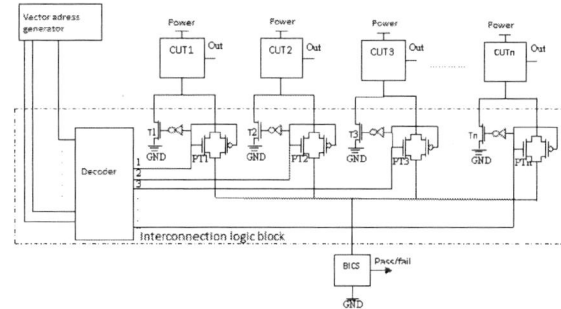

Figure 6. The interconnection logic bloc

I_{DDQ} testing begins by applying the selection vectors at the input of the decoder that enables the transmission gate of desired circuit to connect it to the BICS. Simultaneously, the BICS measures the current, compared it to a reference current, and transmits the test result to the diagnostic unit. During the test operation, the activation of the first transmission gate through the decoder enables the connection of the first circuit under test to the BICS. The other transmission gates are in the off state, so all the other circuits are isolated from BICS and connected directly to the ground line through the activation of the transistors $T_2, T_3 \ldots \ldots T_n$ (All T_i except T_1).

The connection of selected circuit under test to the BICS and the periodic switching of connection of, requires an order made by applying addresses vectors to the input of the decoder generated by the vector address generator at a frequency well defined.

V. SIMULATION AND RESULTS

The block under is composed by three similar operational amplifiers.

To validate the proposed technique SPICE simulations were performed on a device composed by three identical circuits. In addition this technique can be applied to any digital or analog devices . We just have to calculate the reference current I_{REF} to calibrate accordingly the BICS.

First we have made simulation on an operational amplifier to determine its gain-bandwidth.

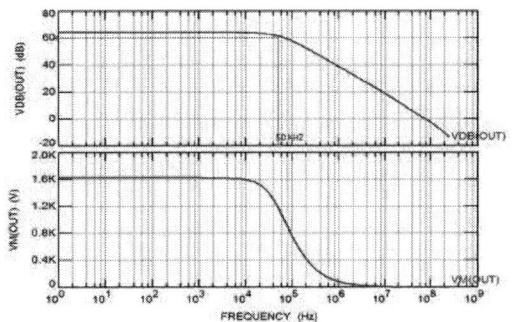

Figure 7. Frequency response curve of the operational amplifier

Accordingly to the figure 7 we can note that the frequency operation of the operational amplifier is working on a range of 50 KHZ (The cut-off frequency is 50 KHZ). For the remaining simulations, we choose an operation frequency of 5KHZ.

This block under test was simulated twice: once with A test frequency of 1 kHz, less than that of operation frequency of the operational amplifier, and then with A test frequency of 50 kHz greater than that of the operating frequency of the operational amplifier. The leakage defect modeled as resistor connected between the drain and the gate of the M6 transistor is introduced, as shown in figure 8.

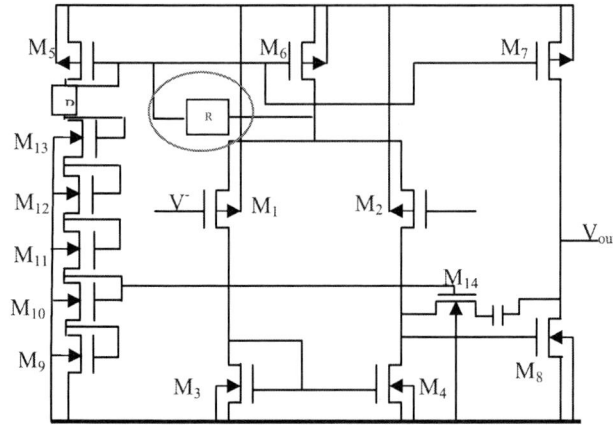

Figure 8. Failing operational amplifier

In the diagram of the results simulation, V (ENABLE), V (ENABLE1) and V (ENABLE2) show the signals for activating the first, second and third transfer gate. The signals V (out), V (out1) and V (out2) respectively correspond the output of the first, second and third amplifier. V (pass / fail) shows the output of the BICS. Injection of default is to insert a resistor (not null and infinite) between the drain and source to model a resistive circuit.

A. Simulation at 1KHZ test frequency

The fault is modeled by a resistor with a value of 200 k Ohms connected between the source and the drain of transistor M6 of the first and third amplifier. Figure 9 provides the simulation of this defect at a test frequency of 1 kHz.

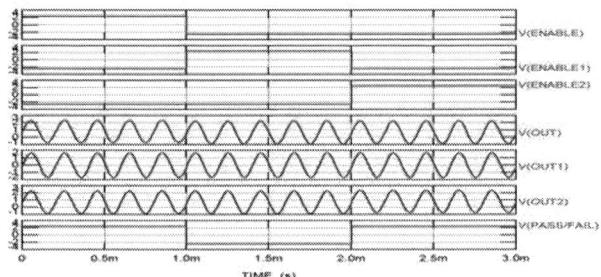

Figure 9. Spice simulation with a fault in the first and third amplifier

We notice that the activation of the default in the first and the second amplifiers involves altering the shape of the output voltage (VOUT and VOUT2) and the decrease of its amplitude which causes a drop in the voltage gain of the operational amplifier. Faults injected at the first and third circuit under test are detected; the output of the BICS goes to the high state when the address selected is that of a faulty circuit. So the test allowed us to distinguish between the functional circuits and the defective circuits.

B. Simulation at 50 KHZ test frequency

The fault is modeled by a resistor with a value of 200 k Ohms connected between the source and the drain of transistor M6 of the second amplifier. Figure 10 provides the simulation of this defect at a test frequency of 50 kHz.

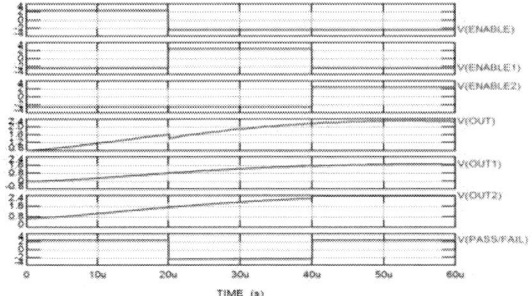

Figure 10. Spice simulation with a fault in the first and third amplifier

According to figure 10 we can notice that simulation results at a frequency of test, greater than the frequency operation of the operational amplifier, were similar to those made at a frequency of test lower than frequency operation of operational amplifier. The defects covered in the first test, at the same type of fault injected, have been detected during the second test which means that the test can be carried out without resorting to consulting output devices to be tested, which reduces significantly test time which has a direct effect on the cost.

VI. CONCLUSION

In this paper we presented a technique for I_{DDQ} test. This off-line testing technique is to implement an I_{DDQ} test of multiple circuits on the same wafer with the same functionality using a single BICS. The effectiveness of the proposed technique is demonstrated by the injection and simulation of faults in the circuits to be tested.

According to the simulation results we can conclude that the proposed test method is efficient for testing the end of production of analog and mixed signal circuits, since the test allowed us to distinguish the functional circuits of the failed circuits. The method can be improved by incorporating this technique in the wafer.

REFERENCES

[1] B. HAMDI, B. GUIBANE, and A. FRADI, "Investigation on fault injection ans analysis in cmos circuits," 2nd International Conference on Computer Modelling and Simulation Brno, Czech Republic, 5 – 7 September 2011 .

[2] B. GUIBANE , "Etude des techniques de détection en ligne et masquage de défauts dans les circuits analogiques CMOS " master of resaerch, National Engeneerig School of Sousse, Tunisia, 2010, pp. 45–52.

[3] D. Shawa, D. Al-Khalilib, C. Rozonb, Fault security analysis of CMOS VLSI circuits using defect-injectable VHDL models, Integration, the VLSI journal, Volume 32 Issues 1-2, November 2002, pp. 77-97.

[4] Miljana Sokolovi, "Design for Testability for SoC Based on IDDQ Scanning", INTERNATIONAL CONFERENCE ON MICROELECTRONICS , NIŠ, SERBIA, MIEL 2008.

[5] Jeong Beom Kim "Design of a Built-In Current Sensor for IDDQ Testing " IEEE JOURNAL OF SOLID- STATE CIRCUITS, VOL. 33, NO. 8, AUGUST 1998

[6] N.ZAIDAN "Conception interface sécurisée pour contrôle-commande de puissances", phd thesis, Institut National Polytechnique de Grenoble. Mai 2002.

[7] Y. Tsiatouhas, Y. Moisiadis, Th. Haniotakis, D. Nikolos, A. Arapoyanni "A new technique for IDDQ testing in nanometer technologies", the VLSI journal 31 (2002) 183–194, May 2002

[8] L. Zaourar, J. Alami Chentoufi, Y. Kieffer, A. Wascrhole "Optimisation du partage de blocs BIST pour le test des mémoires d'un circuit intégré" Laboratory G-SCOP, 2010

[9] M. Larouche, et S. Ethier "Test intégré de circuits analogiques par oscillation avec analyse de signature" Département de génie électrique, École Polytechnique de Montréal, 2007

MIXDES 2012, 19th International Conference *"Mixed Design of Integrated Circuits and Systems"*, May 24-26, 2012, Warsaw, Poland

GNU Radio and USRP2 as a Universal Platform for Verification of Wireless Communication Devices Used in Automotive Applications

Marcin Szelest
Delphi Poland S.A., Technical Center Krakow
Krakow, Poland
Marcin.Szelest@delphi.com

Wojciech Uzdrzychowski, Damian Grzechca
Faculty of Automatic Control, Electronics and Computer Science
Silesian University of Technology
Gliwice, Poland
Wojciech.Uzdrzychowski@gmail.com, Damian.Grzechca@polsl.pl

Abstract—The nature of wireless communication protocols used in automotive industry (eg. TPMS - Tire Pressure Monitoring System, RKE Remote Key-less Entry or PEPS- Passive Entry & Passive Start) and fact, that transmission between remote sensor and main controller has to be received and correctly decoded, require sophisticated receiver systems. Complexity of systems installed in modern vehicles and variety of multiple conditions cause that verification and validation is the most difficult step of whole design process. Especially important part, that must be carefully checked, is behavior of a system in case of incorrect transmission. These false conditions should be repetitive, exact and comprehensive. Usage of Software Defined Radio (SDR) technique makes possible to fulfill all mentioned above requirements. This article describes utilization of SDR modeling with GNURadio to achieve flexible and very powerful Verification Platform for one-way automotive communication protocols. Description is focused on TPMS module however our solution works well for other protocols such as PEPS, RKE and similar.

Index Terms—Software Defined Radio, TPMS, RKE, PEPS, GNU Radio

I. INTRODUCTION

TPMS sensors are encapsulated in irremovable housing, without any external electrical connectors. Packaging technique makes impossible to access to any internal test point, what causes that verification of vehicle's receiver is extremely hard task. To change transmitted frames, sensor must be placed in temperature/pressure chamber. This test requirement is very hard to fulfill and it is almost impossible to simulate moving car. Development of TPMS receiver led us to creating model of TPMS transmitter. The transmitter has been done in SDR technology and together with USRP2 device is able to emulate up to ten TPMS modules. Each of ten modules can work independently and broadcast the data that are necessary from verification point of view. SDR approach guarantee extraordinary flexibility, which allows replacing test cases – usually analyzed in real conditions on the road (drive tests) – by static tests, that can be completed in a laboratory.

In the following chapters we will present our approach and prove that this solution is efficient and extremely flexible.

II. TPMS

Tire Pressure Monitoring System has been defined by automotive industry and is used in modern vehicles to provide information about pressure and temperature inside vehicle's tire [1]. The main purpose of this system is warning driver about tire issues (especially about low pressure). This information is important for safety reasons and also to improve efficiency of fuel consumption [2]. TPMS defines two basic methods of measurement, which are described in the next subsections.

A. Indirect measurement

Indirect measurement bases on data taken from ABS (Anti-Lock Braking System) and derives information from difference of rotation for each wheel. This method reports fault for all tires and is not able to point which specific tire is under fault condition. Indirect measurement method doesn't require additional hardware, however it is not very accurate [1].

B. Direct measurement

This method bases on special sensors equipped with transmitters and placed inside each tire. These sensors broadcast pressure data via a wireless radio link to a central receiver. This type of measurement is very accurate and can detect drop of pressure as small as 1psi (0.069 bar). Unfortunately hardware part is more complex because it requires additional elements in each tire (sensor, transmitter, battery) and dedicated receiver in a vehicle. The most important advantages of the direct measurement method are: high accuracy, compensation with in tire temperature, simultaneous detection of more than one under-inflated tire and reliability under many driving conditions [1]. The article is focused on this measurement method only.

Data link is unidirectional - data is transmitted from each TPMS sensor to vehicle only. Transmission is done in ISM (Industrial, Scientific & Medical) band: at 433.92 MHz in Europe, 315MHz in America. Collected data (pressure and temperature) is fold into frames that contain some additional data as: serial number of sensor, CRC (Cyclic Redundancy Check) checksum and synchronization bits. Fig. 1 contains example structure of TPMS frame. Frames are encoded with Manchester code and modulated in ASK (Amplitude Shift

Keying) or FSK (Frequency Shift Keying) modulator. Bit rate of broadcasted data is constant and equal to 9600 bits per second. Repetition pattern, with delays between each transmision and modulation-switching sequence, is defined in TPMS specification [1].

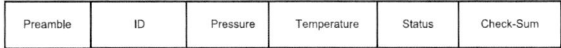

Preamble	ID	Pressure	Temperature	Status	Check-Sum

Fig. 1. Structure of TPMS frame.

III. SOFTWARE DEFINED RADIO

Achievements of modern semiconductor technology makes possible to process quantized analog data with high sampling rate. Digital Signal Processing (DSP) in radio communication area became popular in 1995, when IEEE Communications Magazine published series of articles about Software Defined Radio (SDR) [4].

SDR concept bases on assumption that communication device (receiver or transmitter) is built from two functional parts:

- analog front-end/back-end
  Analog to Digital (ADC) and Digital to Analog (DAC) converters together with linear amplifiers, matching and coupling circuits, etc (RF Hardware).
- digital block
  The most important part of SDR device, usually splitted into two functional parts: sample rate conversion block and DSP processing block. First one is built from DDC (Digital Down Converter) and DUC (Digital Up Coonverter) with CIC (Cascade Integrator–Comb) filters, the second one is specialized DSP processor or FPGA device.

Fig. 2 shows example architecture of the SDR device. Number of DAC and ADC converters depends on required functionality. For example, ADC converters for SDR generators are not necessary, because signals are derived directly from math equations [5]–[7].

One of the biggest advantages of SDR technology is flexibility. As long as core functionality is located in software (hardware acts as universal ADC/DAC board), modifications and improvements can be done without using soldering tools.

However, as long as custom–created code is used for certain implementation, user must be aware that verification of each functional blok is on his side. This is the biggest disadvantage of SDR technology that is not existing in traditional development - where engineer is building his design from ICs tested by manufacturer.

SDR technology does not specify one common processing platform. It is up to the developer to choose architecture which is the most suitable for the implementation. The most common approaches are VHDL/Verilog code for FPGA devices or C code for signal processors. Both of them are good, especially for high-volume devices, however these two methods are not efficient enough for development tasks. Limited number of typical, open–source DSP primitives, extends significantly time necessary for project launch.

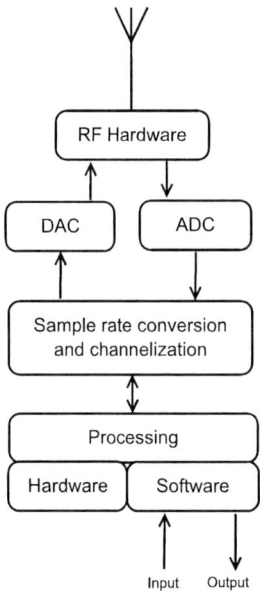

Fig. 2. Typical SDR architecture.

IV. GNU RADIO

GNU Radio is Open Source Software Development Kit (SDK) which provides set of DSP blocks with mechanism of linking them together to build fully functional SDR applications running for instance on desktop computer [7].

Appropriate communication and synchronization protocols are embedded into SDK, so developer does not need to hesitate about interconnection issues. Moreover, SDK contains predefined communication blocks, used for communication with hardware [5]–[7].

V. USRP HARDWARE

GNU Radio is software platform only, so for creation of usable SDR devices hardware platform is necessary. For simple low-bandwidth applications, sound card from desktop computer can be used, but - for modern broadcasting application - commercially manufactured board should be used. One of the most common hardware architecture available currently on market is USRP family. USRP2 device is capable of processing signals up to 100MHz wide in broad frequency spectrum [8].

VI. VERIFICATION PLATFORM - MAIN FEATURES

The Platform can transmit any sequence of ASK/FSK transmission on ten completely independent channels, with different power level for each of them. Number of channels is architecure independent and is limited only by performance of PC (Personal Computer).

Fig. 3 demonstrates Platform architecture: with only one USRP module connected to typical PC, many users can transmit their own frames in few different standards. This architecture is optimal from cost perspective - equipment necessary to create one test bench is reduced to only one PC and one USRP device. Moreover, single test bench is able to transmit three one-way communication protocols (TPMS, PEPS and RKE).

For TPMS standard, Platform can transmit typical TPMS frame or a frame with induced error. Each transmission is repetitive and unequivocal. Thanks to ten independent channels, the system is able to simulate complex test cases (including behavior of a moving vehicle) with almost all possible failures of tire.

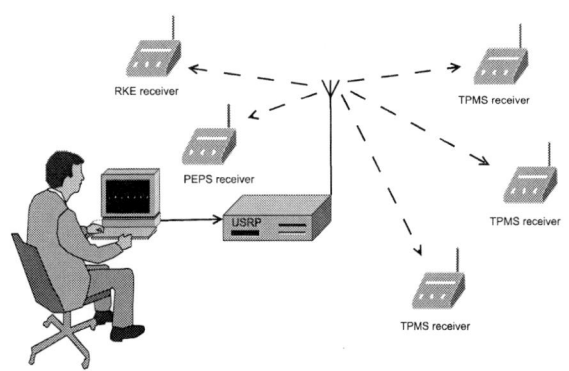

Fig. 3.   Architecture of Verification Platform.

VII. IMPLEMENTATION OF TPMS TESTING SYSTEM

From structural point of view, Verification Platform can be splitted to four layers: hardware layer – Fig. 4, GNU Radio layer – Fig. 5, Test Case layer and Scripting Language layer – Fig. 6.

- Layer 1: USRP2 hardware connected to PC.
  This is the lowest layer, a gateway between software model and physical implementation. At this point software output is transformed to RF (Radio Frequency) signal and transmitted. Output power and frequency are dependent on extension board installed in USRP2 and are out of scope of this article. The extenstion board is responsible for analog conversion from baseband signal to RF signal and amplification.
- Layer 2: TPMS transmitter modeled in GNU Radio.
  This is the most essential part of Verification Platform. The layer is made with use of GNU Radio SDK and is responsible for the whole DSP functionality. Fig. 5 shows block diagram of GNU Radio model. It is top-level view only. In fact, almost each block consists of many smaller DSP primitives (few of them were self-written and added to GNU Radio as extension packages).
  The block is fed with binary data, which is later Manchester encoded and modulated. Used technique enables switching between ASK and FSK without any latency. It means that one bit can be transmitted using ASK modulation then second one with FSK modulation. For this block, user can define: output frequency (limited only by

– mentioned earlier – analog extension boards), deviation (+/- 98kHz), power level (defined as voltage ratio in range between 0 and 1), bitrate (typically 9600 bit/s) and modulation type (ASK, FSK or Continous Wave). This block is duplicated ten times, so the Platform is able to simulate behavior of ten TPMS sensors simultaneously [3].

- Layer 3: Test case interpreter.
  Previous layer reads and transmits raw binary data only. This layer is responsible for analyzing and processing data delivered by test script. It is written as separate C++ application that: translates test scripts to binary data and controls GNU Radio functional block.
- Layer 4: Self-created Scripting Language.
  For ease of use, special scripting language has been defined. Fig. 6 shows example test case described with use of script. All transmission's parameters such as: bitrate (decimal number [bit/s]), output frequency (decimal number [Hz]), total gain and gain per channel (real number between 0 and 1), deviation (decimal number [Hz]), modulation type (enumeric value) and transmitted data (hexadecimal data) can be easily defined and set. Moreover, it is possible to repeat some parts of scripts, because simple loop handling mechanizm has been implemented.

Thanks to splitting whole system to separate layers, big flexibility has been achieved. During development of the TPMS receiver we were asked about support for another automotive communication standard (PEPS) – only few modifications were necessary to do this.

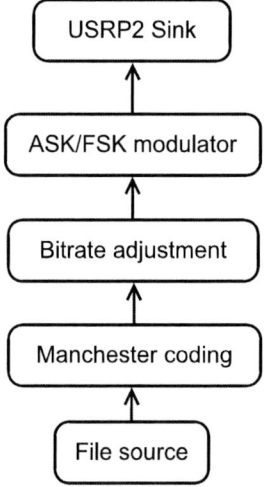

Fig. 5.   Block diagram of TPMS transmitter.

VIII. VERIFICATION OF OUR TEST SYSTEM

Behavior of our transmitter (Fig. 7) has been compared with a commercially available TPMS sensor (Fig. 8). For both cases, the same receiver has been used. We have also manually

438

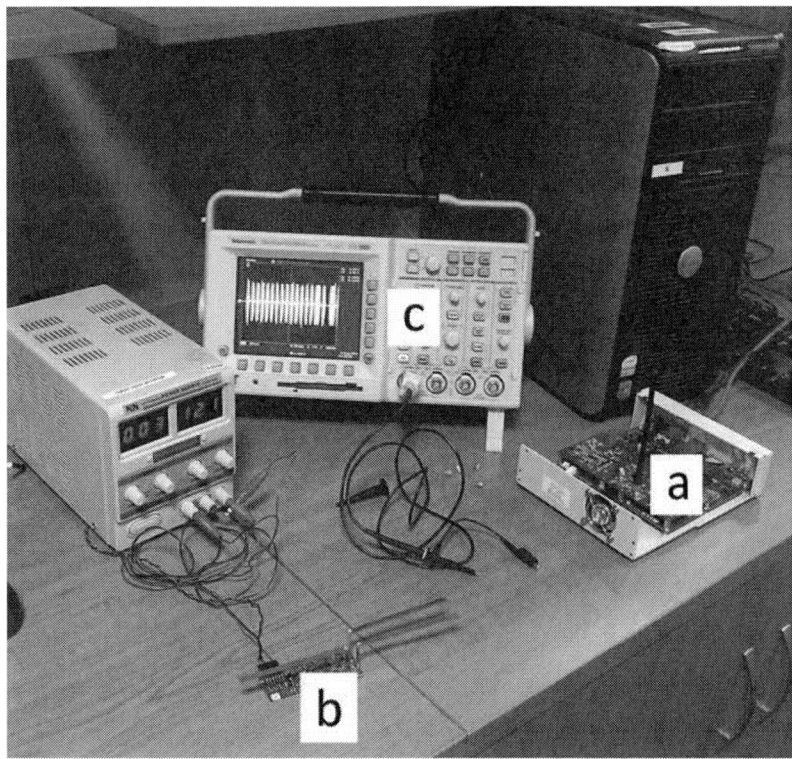

Fig. 4. Verification Platform: a) USRP2, b) TPMS receiver, c) digital oscilloscope with captured transmission.

```
 1 # single frame repeated 4 times - only one sensor
 2 setg bitr 9600          # bitrate
 3 setg freq 433920        # RF freq
 4 setg gain 1.0           # Power setting
 5
 6 set df0 35              # Deviation value
 7 set gain0 0.8           # Power of ch0
 8
 9 label                   # loop starts here
10
11 set mode fsk
12 set id 0x1000001
13 set p 0x3c
14 set t 0x5d
15 set status 0x46
16 crc calc                # calculate CRC for TPMS
17 delay 1056
18
19 repeat 4
20
21 (...)
22 eof
```

Fig. 6. Example test script for TPMS transmission.

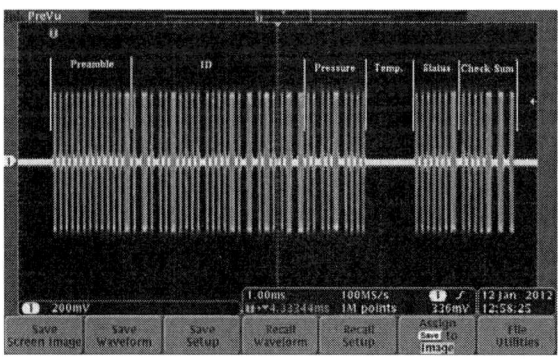

Fig. 7. TPMS transmission (ASK) generated by Verification Platform with induced errors (broken transmission) and captured by digital oscilloscope.

IX. CONCLUSIONS

Our Verification Platform gives us full control of TPMS transmission. It can also generate and transmit fault frames including: incorrect synchronization, CRC errors, erroneous single bit or incorrect intervals between frames. Moreover, the system can transmit signal from up to ten virtual sensors with adjustable power ratio.

Features mentioned above allow us to simulate some conditions that exist in real car during drive cycle, but were

compared transmissions on baseband frequency with use of digital oscilloscope.

For both scenarios observed results were compliant with expectations.

Fig. 8. TPMS transmission (ASK) generated by TPMS sensor and captured by digital oscilloscope.

so far hardly reproductible in the laboratory. We can also analyze cases when sensors are transmitting and interfering with themselves. According to our knowledge the Verification Platform is unique and exceeds other solutions available on the market.

ACKNOWLEDGMENT

We would like to thank our colleagues from Delphi Corporation for many useful and stimulating discussions about technical aspects of the proposed solution.

REFERENCES

[1] NHTSA, *Federal Motor Vehicle Safety Standards,Tire Pressure Monitoring Systems; Controls and Displays*, [Online].
Available: http://www.nhtsa.gov/cars/rules/rulings/tirepresfinal/index.html
[2] S. Velupillai, L. Gven, *Tire Pressure Monitoring*, IEEE Control Systems Magazine, December 2007.
[3] Matthias Fhnle , Markus Hauff, *Analysis of unencrypted and encrypted wireless keyboard transmission implemented in GNU Radio based Software-Defined Radio* , Hochschule Ulm , University of Applied Sciences Institute of Communication Technology , Ulm, Germany, 2011.
[4] M. Szelest, *Cyfrowe przetwarzanie sygnalow w urzadzeniach radiokomunikacyjnych*,Master Thesis, Gliwice, 2004.
[5] E. Blossom, *Exploring GNU Radio*, Linux Journal article, June 2004.
[6] Eric Blossom, *Listening to FM Radio in Software, Step by Step*,Linux Journal article, September 2004.
[7] RadioWare Project, *SDR Documentation*, University of Notre Dame [online]. Available: http://radioware.nd.edu/documentation.
[8] Ettus Research LLC website [online]. Available: www.ettus.com

MIXDES 2012, 19th International Conference *"Mixed Design of Integrated Circuits and Systems"*, May 24-26, 2012, Warsaw, Poland

On-chip Parametric Test of Binary-weighted R-2R Ladder D/A Converter and Its Efficiency

Daniel Arbet, Gábor Gyepes, Juraj Brenkuš, Viera Stopjaková and Jozef Mihálov

Institute of Electronics and Photonics
Faculty of Electrical Engineering and Information Technology
Bratislava, Slovakia
daniel.arbet@stuba.sk

Abstract—**This paper deals with the investigation of the fault detection in separated parts of a mixed-signal integrated circuit example by implementing parametric test methods. The experimental Circuit Under Test (CUT) consisting of an 8-bit binary-weighted R-2R ladder D/A converter and additional on-chip test hardware was designed in a standard 0.35μm CMOS technology. For detection of catastrophic and parametric faults considered in different parts of the CUT, two dedicated parametric test methods: oscillation-based test technique and IDDQ monitoring were used.**

Index Terms—**Fault detection; catastrophic faults; parametric faults; parametric tests; mixed-signal test**

I. INTRODUCTION

The present trends in the development of integrated circuits and new advanced technologies enable integration of complex digital as well as mixed-signal systems on a single chip. These complex systems, know as Systems-on-Chip (SoC), can include digital, analog, and RF circuits as well as MEMS structures, microsensors and another different cores. No doubt, the testability of the respective parts in such systems is greatly decreased [1]. Standard test methods cannot be straightforwardly used to test complex embedded and mixed-signal systems. Therefore, we would need several automatic test equipments (ATE), each dedicated to a particular core integrated in the system. Such approach increases costs of IC production unacceptably, since it requires the expensive and advanced ATE. Due to this reason, test methodology for complex systems becomes the upmost important. Test engineers have been looking for new test methods and approaches, which can assure better testability, higher fault coverage and high quality ICs.

Parametric test methods are most commonly used for testing of analog and mixed-signal ICs. These methods are based on the monitoring of a specific circuit's parameter such as voltage, supply current, frequency, etc. Evaluation of the specific parameter in a complex system is difficult because it requires sophisticated sensing and analyses of the selected parameter in terms of additional hardware needed, setting the Pass/Fail limit, robustness, etc. (in comparison to the simple logic test). On the other hand, parametric test of a part of the complex system may be the only proper test approach in some applications [2,3]. Such approach is based on dividing the complex system into smaller parts, that could be easily tested separately, each part by a proper test method.

In this paper, an 8-bit R-2R ladder D/A Converter (DAC) was used as the test vehicle. Considering the structure of this circuit, it was split into two parts, which have been tested separately using two different parametric test methods. Thus, the control logic, used for selecting the test method and switching the circuit mode (test/functional), has been designed.

II. EXPERIMENTAL CIRCUIT UNDER TEST

Figure 1 shows the block diagram of the experimental mixed-signal circuit designed in 0.35 μm CMOS technology, which consists of the selected circuit under test (CUT) and the necessary on-chip test hardware including the control logic used for switching the circuit into the test mode. As a mixed-signal CUT example, an 8-bit binary-weighted R-2R ladder DAC have been designed and used in our experiment. In the test mode, the DAC circuit is split into two separated parts: a 2-stage operational amplifier (OPAMP) and the R-2R resistor network. Using two additional inputs TEST and MODE, the circuit can be switched in one of two test modes, in order to test each part of the CUT separately employing a proper parametric test method.

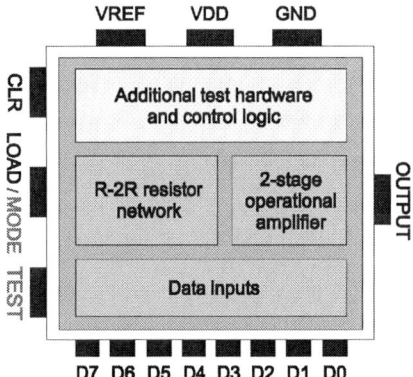

Figure 1. Block diagram of the mixed-signal CUT used

III. PROPOSED TEST STRATEGY

A. Test of the operational amplifier

For the fault detection in the operational amplifier, the oscillation test method, which transforms an analog CUT into an oscillator by inserting the feedback RC network, was used. Using this method, different faults present in the CUT and causing deviation either in the oscillation frequency or in the amplitude of oscillation (exceeding the nominal fault-free tolerance range) can be detected [4].

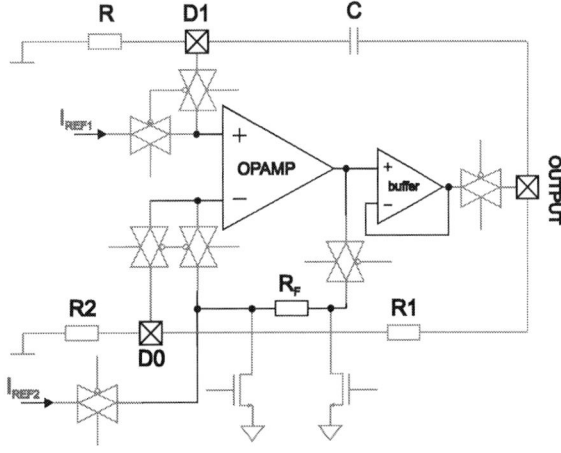

Figure 2.   Circuit diagram of the OPAMP circuit transformed into an oscillator (without control logic)

Components in the feedback network were connected externally (by T-gates) using two additional input pins (D0 and D1) and the output pin OUTPUT. In this configuration, a smaller deviation in oscillation frequency and the amplitude of oscillations were reached. The feedback resistor R_F, which is a part of the DAC, is clamped to ground using two MOS transistors. A buffer is a default part of the converter but in this topology, it was also used to separate the OPAMP from possible outside influences. OPAMP can be tested by applying logic 1 and logic 0 on the pins TEST and MOD, respectively. Then the common control logic (not depicted in Figure 2) generates the control signals for T-gates that for test purposes, connect the feedback network and disconnect the OPAMP from the R-2R resistor network.

T-gates were used for insertion of the feedback. However, they may cause the CUT performance degradation and therefore, their ON- and OFF-resistances must be as low as possible and as high as possible, respectively. Therefore, the ON-resistance of T-gates, which are appearing in the signal path, must be minimized to prevent the undesired performance degradation [5].

The fault-free tolerance bands (representing process variations and temperature influence) for the oscillation frequency and the amplitude of the oscillation frequency were obtained by Monte Carlo analysis, and they are depicted in Figure 3 and Figure 4, respectively. Deviations (obtained by 50

runs of MC analysis) of ±13% in the oscillation frequency and of ±2.85% in the oscillation amplitude were observed.

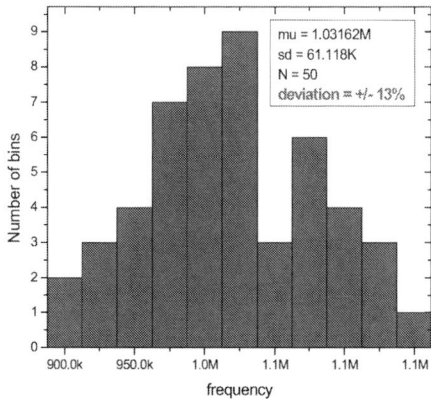

Figure 3.   Nominal deviation in the oscillation frequency

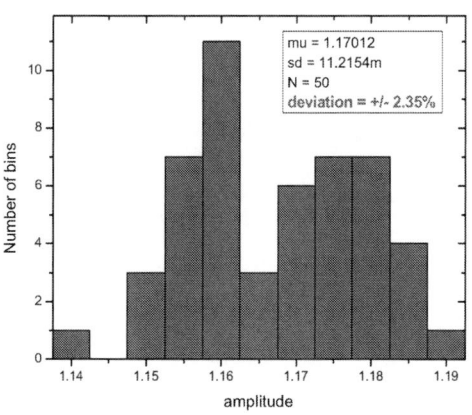

Figure 4.   Fault-free deviation in the amplitude of oscillations

In the proposed test strategy, the feedback network and passive components were realized externally using discrete devices in order to achieve higher accuracy and smaller deviation in the oscillation frequency and the amplitude of oscillations. Taking into account SoC testing requirements, where a sort of BIST strategy to test the mixed-signal cores (e.g. D/A converters) , an on-chip test should be performed Therefore, in case of the on-chip test of the DAC, the OPAMP feedback network must be connected internally, using devices with about 20 % deviation in technology parameters. In [6], such OBIST strategy for testing OPAMP as a part of complex analog and mixed-signal systems is described. To evaluate the efficiency of the proposed test strategy, the circuit oscillation frequency is then compared to the reference frequency given by a Schmitt trigger oscillator, which was used as the on-chip reference to compensate technology variations. Figure 5 shows the circuit diagram of the on-chip oscillator using an on-chip

442

feedback network, transforming the OPAMP circuit into the oscillator.

The oscillation frequency was evaluated by countering a number of oscillation pulses exhibited by the CUT during the time interval generated by the reference oscillator. A more detailed description of the proposed on-chip test strategy and the principle of the oscillation frequency evaluation as well as the PASS/FAIL decision process were presented in [6].

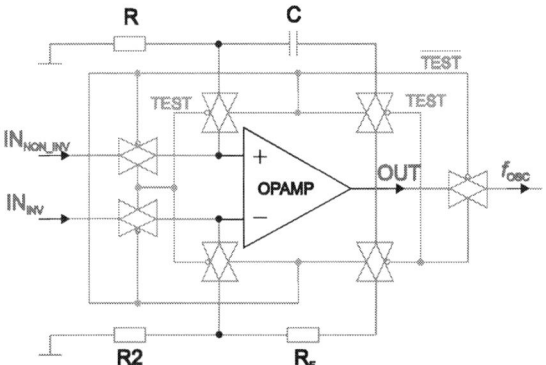

Figure 5.   Circuit diagram of the on-chip oscillator using the internal feedback

B.   Test of R-2R resistor network

The R-2R ladder is a resistor network that uses a cascaded structure of current dividers, which generate binary-weighted currents in the respective branches (Figure 6).

In ideal case, the dividing ratio should be 2:1 but because of

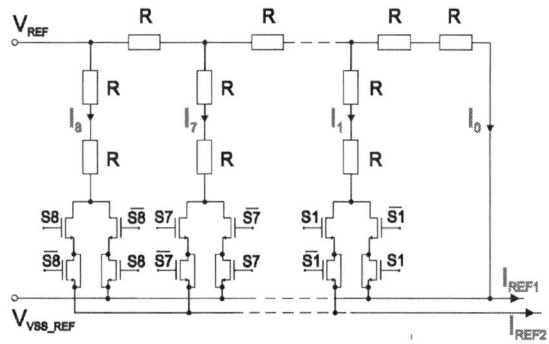

Figure 6.   Circuit diagram of R-2R ladder

resistors mismatch, in reality, the divisions will be imperfect.

The most probable fault in the resistor ladder is that the value of a resistor exceeds its tolerance band (parametric fault). These faults can be detected by the measurement and evaluation of current value in the respective current branches (dividers). The described technique has been used for parametric test of the R-2R ladder, which represent a substantial part of the whole DAC circuit. Modification of this method, used in digital IC test, is known as I_{DDQ} testing/monitoring.

Principle of current testing of the R-2R ladder is as follows: in the first step, current I_8 is compared to the sum of currents I_7 to I_0, then in the second step, the control logic turns-off switches S8 and nS8, and current I_7 is compared to the sum of currents I_6 to I_0. Consequently, control logic turns-off switches S7 and nS7, and another two currents are compared. All switches are controlled by the common control logic.

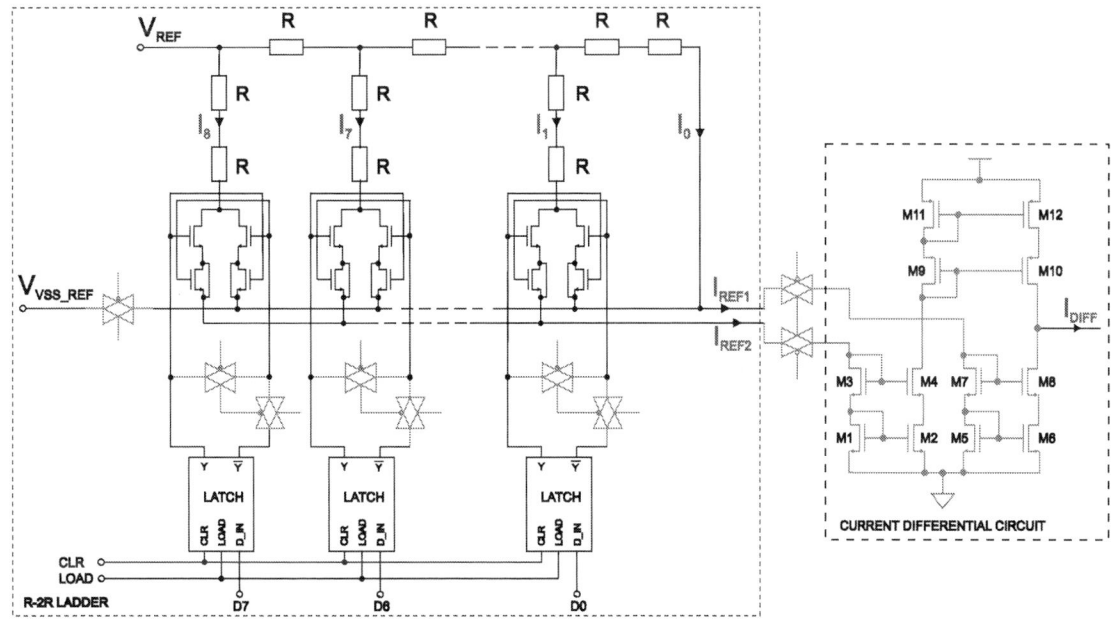

Figure 7.   Circuit diagram of R-2R ladder with additional test hardware (without control logic)

The current difference at each step of the proposed test procedure can be expressed as follows:

$$I_{diff\_N} = I_N - \left(\sum_{i=1}^{N} I_{N-i} \right), \quad (1)$$

where N = 1, 2, 3 ... 8 and I_N is current in the respective branch being sensed.

Circuit diagram of R-2R ladder with the additional test hardware is depicted in Figure 7. In every branch of the resistive network, two T-gates were included to switch-off the corresponding branch. Control signals for the T-gates were generated by the control logic. Circuitry performing the current difference consists of three cascode current mirrors. This circuit also ensures that the differential current (difference of I_{REF1} and I_{REF2}) will flow out to the circuit's output.

This approach makes it possible to test the resistor ladder in total eight steps, by shifting the logic 1 from MBS to LSB. The main problem of this method is that current in the last branch is in order of nA, which is difficult to sense and measure with necessary precision. Therefore, this test technique might be limited to ladders that use resistors with the resistance value smaller than 10 kΩ.

A fault-free tolerance band of the differential current I_{diff} was obtained by Monte Carlo analysis and it will be presented in the following section (Figure 11).

IV. INFLUENCE OF ADDITIONAL TEST HARDWARE

Insertion of the necessary on-chip test hardware might undesirably affect the DAC performance. The main reason is probably the use of T-gates (disconnecting the individual parts), which have some resistance in switch-on state. The most critical are T-gates that are connected in the path leading to the circuit output. Figure 8 shows how the integral and the differential nonlinearity (INL', DNL') of the D/A converter (with the test hardware) depend on the on-resistance of T-gates connected in the path leading to the converter output.

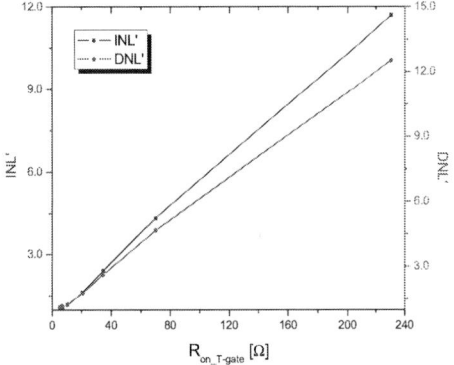

Figure 8. INL' and DNL' versus R_{ON} of T-gate

It can be observed that in order to maintain the original parameters of the circuit, the switch-on resistance of T-gates should be order of tens of Ohms. If the T-gates with the switch-on resistance of about 10 Ω are used, the value of integral and differential nonlinearity will be increased by 18 % and 27 %, respectively. In such case, the additional test hardware will require the area of 0.002 mm², which means that the total area will be enlarged by 13 %. From the aspect of testing it is therefore, necessary to find a good compromise between the CUT performance and the test hardware area overhead.

The figure 9 and 10 show how the main parameters of D/A converter (with the additional test hardware) depend on the switch-on resistance of T-gates connected in path leading to the converter output.

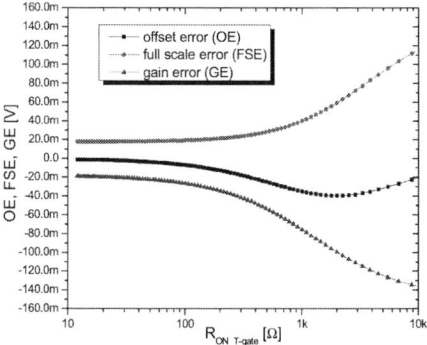

Figure 9. OE, FSE and GE versus R_{ON} of T-gate

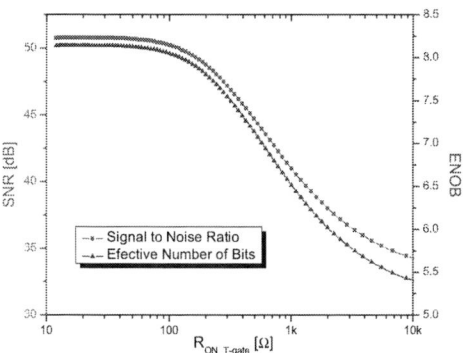

Figure 10. SNR and ENOB versus RON of T-gate

It can be observed that to maintain the original value of the full scale error, SNR and ENOB parameters, it would be sufficient to use the T-gate with switch-on resistance of about 100 Ohms. However, to keep the original value of INL, DNL, offset error and gain error, the switch-on resistance should be about of tens of Ohms. Thus, it can be concluded that to maintain the original value of all parameters of converter, the switch-on resistance of T-gates should be kept in the range from 10 to 100 Ohms. In this case, the total area overhead would be enlarged from 6.5 % to 13 %.

V. ACHIEVED RESULTS

For verification of the efficiency of the proposed test methods in testing parts of D/A converters, four types of catastrophic fault types in the OPAMP, and two parametric faults in the R-2R ladder were considered. Catastrophic faults such as shorts, opens, gate-oxide shorts (GOS) and floating gates (FG) were inserted. Short and open faults were injected in all connection paths, while GOS and FG faults were applied in all transistors forming the OPAMP. Parametric faults, which most commonly arise on passive devices, were modeled and injected in the R-2R resistor network.

A. Results of the operational amplifier test

In the operational amplifier, several catastrophic faults such as opens and shorts as well as floating gates and gate-oxide-shorts, were considered. For their detection, the oscillation-based test method of the OPAMP circuit was employed.

Opens were modeled using a parallel combination of a resistor and a capacitor. Resistor's and capacitor's values depend on the defect location and size. We have considered nine different open faults injected in 24 different locations. From all considered 216 open defects, 209 faults were detected, which means that the fault coverage of 94 % was achieved. However, opens modeled with the resistance value of 1 MΩ represent so call 'hard-detectable' faults, thus, those that have higher resistance would probably not be covered by other test technique either.

Short faults were modeled and simulated using a serial short resistor. Values of resistors considered as short faults are as follows: 500 Ω, 1 kΩ, 10 kΩ, 100 kΩ and 1 MΩ. The achieved fault coverage, presented for different ranges of the short resistance value, is presented in Table 1.

TABLE I. FAULT COVERAGE OF SHORT FAULTS

Short resistance value [Ω]	500 ÷ 10k	500 ÷ 100k	500 ÷ 1M
Fault coverage	100 %	95.58 %	84.7 %

It can be observed that shorts with lower resistance are easier to detect because, in most cases, such shorts lead to loss of the oscillations or significant deviation in the amplitude of oscillations. Shorts with the resistance higher than 100 kΩ usually cause only slight deviation in the oscillation frequency that makes them more difficult to detect.

For floating gates (FG) we used an extended electrical model described in [7], considering also capacity of the break. The model then, include capacitors C_{mp}, C_{pb} and C_{break}, which values were set to 2.82 fF, 3.02 fF and 0.07 fF, respectively. All considered FG faults were easily detected through either loss of the oscillation or a change in the oscillation frequency (one fault).

The worst case of the overall fault coverage (all catastrophic faults), using the external oscillator feedback network, is summarized in Table 2.

TABLE II. WORST CASE OF OFF-CHIP TOTAL FAULT COVERAGE

Faults	Fault coverage
Shorts	84.7 %
Opens	94 %
FGs	100 %
Total	**92.9 %**

The total fault coverage achieved for on-chip realization of the proposed parametric test of OPAMP is summarized in Table 3.

TABLE III. TOTAL FAULT COVERAGE FOR ON-CHIP TEST [6]

Faults	inserted	detected	Fault coverage
Shorts	68	54	79.41 %
Opens	216	180	83.33 %
FGs	9	7	77.77 %
Total	**293**	**241**	**82.25 %**

In case of the off-chip approach, the higher total fault coverage was obtained. However, some of the undetected faults have been manifested by a certain deviation in the amplitude of the oscillations (not evaluated in our experiment). Therefore, the total fault coverage might be increased further by evaluation of this parameter. The obtained results prove that the oscillation-based test approach can be relatively very efficient in detecting different catastrophic faults (including hard-detectable ones) in analog sub-circuits.

B. Results of R-2R ladder test

In contrast to the OPAMP test, parametric faults were considered in the resistor network (R-2R ladder), and the fault coverage by the proposed current test method was investigated. Possible parametric faults in the resistor network were simulated using a resistor with varying resistance value. We considered that the value deviates by ±5 % or ±10 % from its tolerance range. Figure 11 shows the tolerance band and simulated values of the differential current at the output, depending on the test vector being applied.

The simulation results show that almost all parametric faults in the resistor network are detectable. However, parametric fault considered in the last branch is difficult to detect because current is too small to be sensed precisely. Table 4 shows the fault coverage achieved by current monitoring approach for parametric faults considered in the resistor network. It can be observed that ±10 % deviation in resistor's value is fully detectable and 100 % fault coverage is reached. When the resistor's value deviates from its tolerance band by ±5 %, the fault coverage is slightly lowered to 96 % that is still very good result.

The proposed current test method is easy to implement and provides very high parametric fault coverage. However, the method is limited by higher values of the resistance of the resistors used in the ladder, that lead to less current flowing through the branches, which is rather difficult to be measured and processed.

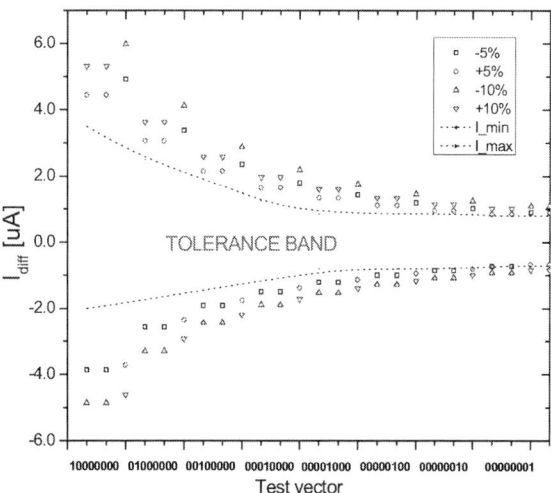

Figure 11. Tolerance band and simulated values of the differential current in R-2R ladder

TABLE IV.   PARAMETRIC FAULTS CONSIDERED IN R-2R LADDER

Parametric fault	±5 % deviation	±10 % deviation
Fault coverage	96 %	100 %

VI.   DISCUSSION & CONCLUSION

Two different parametric test methods have been used for fault detection in the R-2R ladder based D/A converter. For this purpose, an experimental circuit, consisting of the CUT, the additional test hardware, and the control logic for switching the DAC between functional and test modes, has been designed in selected CMOS technology. The control logic was used to split the circuit into two parts, each tested separately by a dedicated method. An operational amplifier was tested by the oscillation-based test strategy, while current monitoring was used to test the resistor ladder.

The crucial point of the used test strategy is that the insertion of the necessary additional test hardware might affect the CUT performance. However, with appropriate setting of ON-resistances of the inserted T-gates, this undesired influence can be minimized. T-gates with ON-resistance of about 10 Ω cause increase in the area of 13 %, and increase of 18 % and 27 % in INL and DNL parameters, respectively.

Catastrophic and parametric faults were considered in the CUT. In the worst case, the total fault coverage of 92.42 % and of 96 % for the OPAMP and the resistor network was achieved in the off-chip realization, respectively. This is an excellent result, especially, taking into account that the fault coverage up to 94.21 % of hard-detectable faults by parametric test of single parts of the DAC has been achieved. In the case of the on-chip test of OPAMP the total fault

coverage of 82.25 % by measuring the CUT oscillation frequency has been achieved. The on-chip total fault coverage might be increased by measuring the amplitude of oscillation. However, for evaluation of the amplitude of oscillation, the additional test hardware will be necessary, which would increase the area overhead.

Parametric test of separated parts is a promising strategy to test complex mixed-signal systems. This approach offers the possibility to identify a defective part and makes test of such systems easier or, in some applications, possible at all. However, insertion of an additional hardware cause the chip area overhead, and in some cases, might disrupt the function of the tested circuit. Therefore, it is necessary to find appropriate method for testing a circuit by splitting it into several parts that helps to maintain its original functionality and specific parameters.

Further research in this field will be focused on realization of BIST structures, based on the oscillation test strategy, which would be generally applicable for D/A converters as cores used in complex SoCs.

ACKNOWLEDGMENT

This work has been supported by the project "Support of the Centre of Excellence for Smart Technology, Systems and Services II", ITMS Code 26240120005, and grant VEGA no. 1/0285/09 founded by Slovak MSVVS, and it has been done with R-DAS research support.

REFERENCES

[1] Jeng-Horng, T.; Ming-Jun Hsiao; Tsin-Yuan Chang; , "An embedded built-in-self-test approach for digital-to-analog converters," Test Symposium, 2001. Proceedings. 10th Asian , vol., no., pp.423-428, 2001

[2] Manhaeve H.: Test Requirements for Today's and Future Circuits: A Perspective, Proc. of Electronic circuits and Systems, 2005, pp. 1-10.

[3] Gurnett, K, Adams, T.: Advanced Devices, Advanced Testing, Semiconductor manufacturing, vol. 6, 2005, pp.27-32.

[4] Arabi, K.; Kaminska, B.; , "Testing analog and mixed-signal integrated circuits using oscillation-test method," Computer-Aided Design of Integrated Circuits and Systems, IEEE Transactions on , vol.16, no.7, pp.745-753, Jul 1997

[5] K. Arabi and B. Kaminska, "Oscillation-Based Test Strategy (OBIST) Scheme for Functional and Structural Tasting of Analog and Mixed-Signal Integrated Circuits", US Patent Application, 1995.

[6] D. Arbet, J. Brenkus, G. Gyepes, V. Stopjakova: Increasing the efficiency of analog OBIST using on-chip compensation of technology variations, Design and Diagnostic of Electronic Circuits and Systems, 13-15 April 2011, pp.71-74.

[7] A.M. Brosa and J. Figueras, "Characterization of Floating Gate Defects in Analog Cells", Journal of Electronic Testing: Theory and Applications, vol 14, 1999, pp. 23-31.

MIXDES 2012, 19ᵗʰ International Conference *"Mixed Design of Integrated Circuits and Systems"*, May 24-26, 2012, Warsaw, Poland

Reliable On-Chip Network Design Using an Agent-based Management Method

Mojtaba Valinataj
Department of Electrical and Computer Engineering
Babol University of Technology
Babol, Iran
m.valinataj@nit.ac.ir

Pasi Liljeberg, Juha Plosila
Department of Information Technology
University of Turku
Turku, Finland
{juplos,pakrli}@utu.fi

Abstract—As the complexity of evolving integrated circuits and the number of cores in each chip increase, reliability aspects are becoming an important issue in complex chip designs. In this paper, we present an on-chip network architecture that incorporates a novel agent-based management method to enhance the reliability and performance of network-based Chip Multi-Processor (CMP) and System-on-Chip (SoC) designs against faulty links and routers. In addition, to utilize the fault information required for the routing process in a scalable manner, we classify the fault information to be exploited in the proposed distributed and hierarchical management structure. The experimental results show that the proposed architecture incurs only a small hardware overhead.

Index Terms—on-chip network; reliability; permanent fault; routing algorith`m;

I. INTRODUCTION

Chip Multi-Processors (CMPs) have been designed to overcome the intrinsic design challenges in order to comply with the increasing processing requirements. CMPs may include hundreds of Intellectual Property (IP) cores, processing elements and embedded memory blocks which communicate with each other [1]. The best scalable interconnection infrastructure for these complex systems is the Network-on-Chip (NoC).

There are reliability, power consumption and thermal issues in NoC-based CMPs that will be more important when the number of nodes increases. In this paper, we concentrate on the reliability aspect in the underlying network. This aspect includes the proposed agent-based management structure and a routing method adapted to exploit this structure to tolerate permanent faults in the nodes and links. We select fault-tolerance against permanent faults since a considerable amount of device failures may occur in both manufacturing and operational phases. To tolerate permanent faults many fault-tolerant routing algorithms have been designed so far. However, because of the size of CMPs, we only consider distributed and scalable routing algorithms such as the methods introduced in [2-4].

The proposed agent-based management architecture consists of distributed agents in a hierarchical structure, which is especially suitable for large CMPs with tens or hundreds of processing elements. In this structure, the agents in each level of hierarchy have the same tasks to gather, manage and distribute the fault information. The previous works related to the hierarchical agents can be found in [5-8]. In [5] and [6] the overall structure for the agent-based management is discussed without any detailed design. In [7] a NoC monitoring scheme based on hierarchical agents is addressed mainly to minimize network power consumption. System level design principles, the basic concepts of the general approach for the hierarchical agent monitoring, and the general tasks of the agents in each level of hierarchy are discussed in [8]. In addition, it includes an approach for DVFS (dynamic voltage and frequency scaling) used in power monitoring. However, it does not present any detailed or low-level design especially for fault-tolerance.

In this paper, a management structure based on hardware agents inside the network components is proposed for the mesh network to optimally utilize the fault information and distribute it among the appropriate nodes. For this purpose, we classify the required fault information for the routing process in detail. The appropriate portions of the fault information will be sent to the direct and indirect neighbor nodes through the hierarchical agents to be used in the routing process. This way, a scalable and fault-aware routing algorithm is achieved with higher performance compared to methods without agent-based management.

The rest of the paper is organized as follows. In Section II the fault information classification needed for the routing process is presented, and in Section III the proposed agent-based management method is explained. The experimental results are presented in Section IV and finally, conclusion and future works are given in Section V.

II. FAULT INFORMATION CLASSIFICATION

In this section we classify the fault information needed for the routing process in the NoC routers. The fault information is provided by the fault detection part. A typical NoC router (Fig. 1) includes a controller, routing unit, crossbar switch as well as input and output ports. The controller mainly includes the switch allocator and virtual channel (VC) allocator if there are virtual channels in the input ports. The input ports include a buffer for each virtual channel and the output ports directly connect to the outgoing links. Based on [9] some test and fault detection circuits can be incorporated in the NoC routers and links to detect the permanent faults in each sub-block with an acceptable hardware overhead. Therefore, we assume that

447

appropriate signals come out from the detection circuits so that we will be aware about the faultiness of five input buffers, four direct links, the routing unit, the controller and the crossbar switch, in each router. In addition, we should also be aware about the faultiness of the other components inside a node that are the Network Interface (NI) and the local core or Processing Element (PE). It is worth mentioning that for simplicity, we assume the links are bidirectional and when any type of permanent fault occurs in any direction, the entire link will be considered faulty.

We say the north direction of a router is faulty or unusable for the routing process if the north link or the north input buffer in the current router or the south input buffer in the north neighbor router is faulty. This condition can be stated by (1) using the appropriate signals from the fault detection circuits:

$$N = Link_N \text{ or } Buf_N^{cur\_router} \text{ or } Buf_S^{N\_router} \qquad (1)$$

In (1) all terms are one-bit status data showing that if any term equals '1' its corresponding component is faulty, otherwise it is healthy. In addition, (1) is a replacement for a common assumption that declares a faulty input buffer can be modeled by its incoming link assumed faulty.

Equation (2) can be generally used for four main directions in each router:

$$X = Link_X \text{ or } Buf_X^{cur\_router} \text{ or } Buf_{(1-X)}^{X\_router} \qquad (2)$$

In (2) X can be N, S, E or W which mean north, south, east or west directions, respectively. In this equation, $Link_X$ means the status of bidirectional link in the X direction of the current router, $Buf_X^{cur\_router}$ stands for the status of the input buffers of all VCs in the X direction of the current router, and $Buf_{(1-X)}^{X\_router}$ stands for the status of the input buffers of all VCs in the opposite direction of X in the neighbor router which is located in the X direction of the current router. In addition, (1-X) stands for the opposite direction of X, which means S, N, W and E for N, S, E and W directions, respectively.

The effect of some faulty components inside a router is that the whole router and as a result the whole node should be considered faulty because the router is unable to perform its main task (sending incoming packets to the correct output ports). For this case, we dedicate a bit called *Node* as a part of fault information regards to this situation based on (3):

$$Node = routing\_unit \text{ or } controllor \text{ or } crossbar \qquad (3)$$

Equation (3) means that we should consider the whole node faulty if the routing unit, the controller or the crossbar switch is faulty.

In NoC-based CMPs, the local cores or the processing elements are connected to the routers via the network interfaces. If a processing element is unusable, the high level system should either migrate its task to other processing elements or perform a remapping process. In a NoC router we assume the local processing element is unusable if it is faulty or its network interface or the buffers located in the local port are faulty based on (4):

$$PE = PE_{local} \text{ or } NI \text{ or } Buf_{local} \qquad (4)$$

In the equation above, *PE* equal to '1' means that the local core or processing element is unusable otherwise it is usable.

The fault information obtained from (2) to (4) should be maintained to be used in the routing process or to be sent to a higher level of the system. This local fault information is stored in a register called the local fault register (LFR). The local fault register also stores the status of four input buffers because the neighbor nodes need them to update their own local fault registers based on (2). This means that the usability of the four output directions in a router depends on the neighbor nodes, too. The local fault register includes 10 bits; four bits for four main directions, four bits for input buffers in the main directions, and two bits for *Node* and *PE* based on (3) and (4) (Fig. 2). It is essential for the routing process that a router be aware about the faultiness of its four main directions. However, it is important that a router be aware about the faultiness of all components inside a small region similar to [3] and [4] because this regional information has a substantial effect on the fault-tolerance capability and the cost of the routing algorithm. We select a region smaller than the one used in [2] which was a 2-hop distance region. This region including all the neighboring links with their names is shown in Fig. 2 in which the central node is the current router. In some manner, the central node should be informed about the faults in this region. Then, the regional fault information is updated and stored in an 8-bit register called the regional fault register (RFR) (Fig. 2).

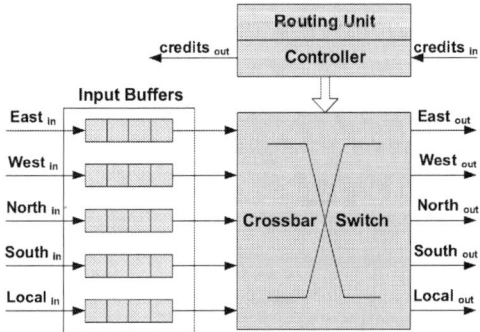

Figure 1.   A typical NoC router architecture

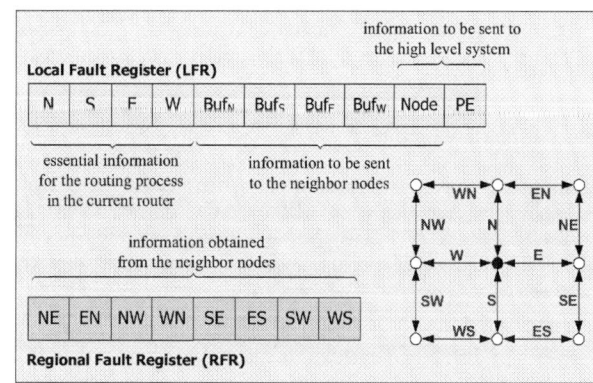

Figure 2.   Fault information registers in each router, and the neighboring area

448

As stated before, faulty input buffers are modeled by their corresponding links assumed faulty and affect the status of the main directions of the routers. For example, based on (2), the status of the NW link (the link situated in the north of west of the central router) shown in Fig. 2 is also affected by the status of the input buffers in its two sides. This way, if its corresponding bit in the regional fault register (NW bit) equals '1', this means that in a minimal routing process the central router should not send any packet from its west direction if the destination node is top-left node.

III. AGENT-BASED MANAGEMENT METHOD

To enhance the performance of a fault-tolerant on-chip network with a large number of components, a scalable management method can be beneficial. Thus, we propose a management method that is agent-based and hierarchical to be more profitable for scalable on-chip networks.

A. Preliminaries

There are two types of agents in the proposed management structure:

- **Cell agent:** Each node or cell includes an agent called the cell agent which collects, manages and distributes the fault information related to the components of its node. In addition, it updates the LFR and RFR.

- **Cluster agent:** Each cluster that includes a number of nodes is controlled by a cluster agent. A cluster agent configures the cell agents inside the cluster by sending the new fault information which is obtained from the other cell agents inside the cluster or other cluster agents.

The incorporated agent hierarchy is shown in Fig. 3. This agent hierarchy differs from that of proposed in the previous works ([5-8]). This is due to the fact that in the proposed structure, for faster reconfiguration the cell agents communicate with their neighbor cell agents even if they are situated inside different clusters. This is a real case because in general, in a CMP, a task may require more than a cluster to be run. On the other hand, the clusters running a common task are not necessarily neighbor clusters. However, the routers should be aware about their neighbors to select the best path for sending the packets to their destinations, and for faster awareness their cell agents should exchange the required fault information.

B. Interconnections for Hierarchical Agents

A small 3×3 network with an agent in each node is shown in Fig. 4. In this figure, R, NI and PE correspond to the router, network interface and core or processing element, respectively. In addition, the agents can be cell or cluster agents but the number of cluster agents is much less than the number of cell agents. For example, in a regular mesh network each 3×3 sub-network can be a cluster with a cluster agent in the center. The proposed agent-based management structure uses two types of communications: a physically separate network for only peer to peer communication between the cell agents, and the baseline data network for control packet communication with a higher

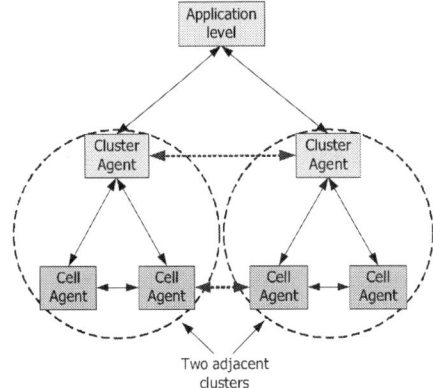

Figure 3. Hierarchical agents in two neighbor clusters

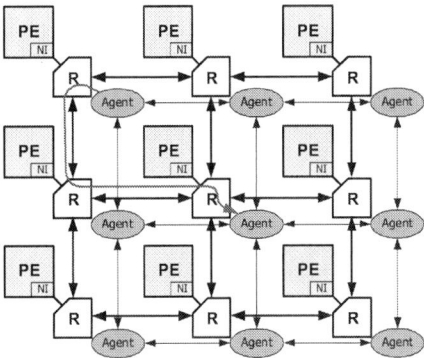

Figure 4. A 3×3 NoC with agents and interconnections

priority compared to data packets. The former is shown in Fig. 4 between the agents and the latter is the network connecting the NoC routers. The control packets including the reconfiguration and fault information are exchanged between the cell agents and the cluster agents in addition to the communications between the cluster agents. In Fig. 4, the shown path is used if at least one of the source or destination agents is a cluster agent. The manner in which these communication networks are utilized is discussed in the next subsection.

C. Agent Tasks

In most previous works it is assumed that for the routing process the NoC routers are aware about the faults or failures occurred in the neighbor nodes in addition to their own components and interconnection links. However, in reality there should be a mechanism to inform the routers about different faults in the network. In the proposed method, this awareness is distributed as fault information by hierarchical agents and their interconnections.

1) Cell agent

In the agent hierarchy shown in Fig. 3, all the fault information needed for the neighbor nodes (the local fault register except its right most bit) can be sent to the appropriate cluster agent, and then the cluster agent can distribute and send it to the appropriate cell agents. However, to minimize the impact of faults on the network performance especially just after fault occurrence, it is better that the cell agents themselves distribute the fault information to their neighbors because this approach is faster. This way, the fault information essential for the routing process is distributed at the lowest level of management hierarchy through the dedicated network between the cell agents. A cell agent sends a portion of LFR to each neighbor cell. In other words, only the fault information required for the routing process in a specific neighbor node is sent to its corresponding cell agent. For example, according to the neighboring area shown in Fig. 2, the cell agent situated in the central node only sends the faulty status of the north direction (N), south direction (S), west input buffer (Buf_w) and whole node (Node) to its west neighbor cell agent (if any bit equals '1'). In the west cell agent, Buf_w is used to update the E bit in LFR; N and S are used to update the NE and SE bits in RFR. The west cell agent does not receive the faulty status of the east direction (E) from the central cell agent because the east link of the central node is situated outside the defined neighboring area for the west node which is smaller than the area obtained by the 2-hop distance introduced in [2]. In addition, Node bit is used to update the NE, SE and E bits simultaneously. In other words, when the west cell agent is notified about the faultiness of the whole central node (its east neighbor node), it will set the NE, SE and E bits to '1'.

To send and receive the essential fault information, the cell agents perform simple encoding and decoding processes. Each fault occurrence should be informed only once. In addition, if more than one fault occurs simultaneously, they can be informed to the neighbors sequentially, in different clock cycles. Thus, we use three bits and as a result a three-bit unidirectional link to encode the status of two directions, one input buffer and the whole node (N, S, Buf_w and Node for the west neighbor) in addition to a non-faulty state to inform each neighbor cell agent. In the neighbor cell agents, each three-bit input is decoded to update the local and regional fault information. Therefore, the width of the links in the dedicated network shown in Fig. 4 will be six bits.

2) Cluster agent

A cluster agent informs the higher level about the critical failures (PE or whole node failures) occurred inside the cluster and receives reconfiguration commands from the higher level about remapping of the tasks on the nodes or task migration. However, it is clear that a cluster agent itself is informed by the cell agents about the critical failures. In addition, for a more effective routing process, a cluster agent should be informed about other faults inside a node such as faulty output directions (which include the effect of faulty input buffers), because it should be able to send different types of fault information to other cluster agents in addition to informing other nodes inside the cluster. When the source or destination agent is a cluster agent the baseline packet-based data network is used. Due to the fact that the number of permanent fault occurrence is low when a NoC-based CMP is running some specific tasks, the amount of fault information that should be distributed is not high. Thus, the number of required packets to carry the fault information is low and as a result, the overhead is negligible. On the other hand, a cluster agent normally has a distance of more than one hop from other cluster agents and from some of

the cell agents inside its cluster. Therefore, the packet-based data network is convenient for cluster agent-based communication.

A cluster agent can be placed in the processing element of a node as a software agent which is implemented entirely in software (SW). It can also be a hardware (HW) component similar to the proposed cell agent or can be implemented in HW/SW co-design manner. In our method a cluster agent is implemented in hardware and it includes the cell agent of the node in which it is located. For packet-based communication and to use existing resources in a node, the cell and cluster agents exploit the local port to send and receive packets. Thus, some extra logic for multiplexing is required to separate data and control packets and then to direct packets to the agents or the local core (PE) or to accept packets from them.

D. Fault-Tolerant Routing Algorithm

The fault-tolerant routing algorithm to incorporate the agent-based method and different types of fault information is a modified version of the method introduced in [4]. This routing algorithm is a low-cost, adaptive and congestion-aware method and since it does not use routing tables and acts in a distributed manner, it is scalable and thus suitable for large NoC-based CMPs. For better utilization of agent-based management method some extra fault information besides the information introduced in Section II is used in the routing algorithm. In each cell agent, this information is obtained from the cluster agent and includes the fault status of the main directions in some nodes inside the cluster. We call it cluster-dependent fault information and it is highly dependent on the size and the shape of a cluster in addition to the type of the routing algorithm.

For a realistic example, we assume that each cluster in the network has a regular structure and includes a 3×3 sub-network similar to the network shown in Fig. 2 in which the central node includes the cluster agent. In such a cluster, if the source and destination nodes are top-right and bottom-left nodes, respectively, the source node should be aware about the status of the links with the labels S, W, SW and WS in Fig. 2 around the destination. These links are located outside of the neighboring area of the top-right node that means this node cannot be aware about the status of the mentioned links from the neighbor cell agents; thus, it should be informed by the cluster agent. The designed routing algorithm acts in such a way that it can correctly deliver all the packets to their destination by selecting the shortest or non-shortest paths if it is not aware about the status of the links with the labels S, W, SW and WS links. However, if it is aware, it definitely selects a shortest path from the source to the destination. This example justifies the usage of the cluster agent that manages and distributes the appropriate fault information related to each node inside the cluster.

IV. EXPERIMENTAL RESULTS

A. Performance Evaluation

To demonstrate the effect of the proposed agent-based management method on the network performance, we simulated 3×3 VHDL-based NoCs with the input buffer size of four flits in each virtual channel and the packet length of 16 flits under the uniform traffic pattern. The number of incorporated virtual channels equals two thus it is minimum. In addition, two different methods are used: The Main-RAFT routing algorithm [4] that does not use any agents, and a modified version of the Main-RAFT that uses the agent-based management method including cluster-dependent fault inform--ation. It is assumed that each cluster is a 3×3 sub-network in which the center node includes the cluster agent. In the 3×3 NoC 25% of the links are faulty (three faulty links) as depicted in Fig. 5a. The average packet latency for each traffic load is measured upon all the packets when each local core generates 2000 packets.

As shown in Fig. 5a, for the source node S and the destination node D there are two different paths P1 and P2 that can be traversed by the packets based on [4]. P1 is a minimal path but P2 is a non-minimal path. However, the routing based on the proposed agent-based management only selects the minimal path P1 for the packets. In addition, similar paths can be obtained for other source-destination pairs in the mentioned 3×3 NoC. The load-latency diagram for this network is shown in Fig. 5b. As shown in this figure, the proposed agent-based routing method has a better performance and a lower saturation point compared to the method introduced in [4].

B. Area Overhead

To evaluate the area overhead of the proposed method we implemented DyXY [10] as the basic adaptive routing algorithm, Main-RAFT [4], and the routing algorithm based on the proposed agent-based management method in a state of the art router architecture using VHDL synthesized with a standard cell library. A medium width of 32 bits has been selected for the links and flits. The size of input buffers is four flits or eight flits. Table I shows the areas for the 5-port routers using the mentioned methods. In addition, this table shows the area overhead of the proposed agent-based router compared to other routers. Based on the obtained results, the area overhead of the proposed router compared to [4] is only 1.1% or 1.6% when the size of the input buffers is eight or four flits, respectively. It is worth mentioning that the DyXY method does not have any means to reliably convey all the packets to their destinations in the fault situations. In addition, the wire overhead of the proposed method (3 bits in each direction) is less than 5% and 10% for 64- and 32-bit flits, respectively.

V. CONCLUSION

In this paper, a scalable agent-based management architecture for fault-tolerant NoC-based CMPs is proposed. To exploit the proposed agent-based architecture for the fault-tolerant routing process, different types of required fault information are classified to be distributed in the network. Each level of agent hierarchy manages and distributes a specific type of fault information. This way, a higher performance can be achieved in the networks of different sizes. The simulation and synthesis results reveal that the proposed architecture improves the network performance with a small hardware overhead. In

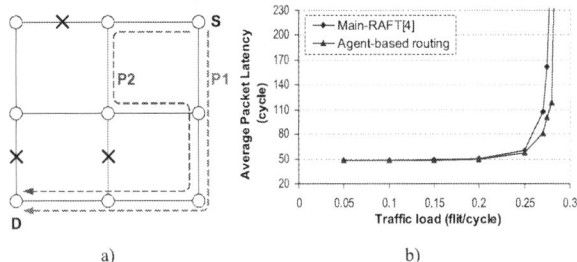

Figure 5.   a) A faulty 3×3 NoC, b) Average packet latencies obtained by two methods

TABLE I.   HARDWARE COST AND OVERHEAD

Routing Method	Area (gate count) for 5-port router		Area Overhead (%)	
	4-flit buf.	8-flit buf.	4-flit buf.	8-flit buf.
DyXY [10]	6889	10450	9.1	6.1
Main-RAFT [4]	7403	10967	1.6	1.1
Agent-based	7519	11083	NA	NA

future, we will investigate the agents in different levels of hierarchy with a more efficient management process. In addition, we will classify and manage the fault information useful for the routing process in the neighbor clusters.

REFERENCES

[1]   O. Cesariow et al., "Multiprocessor SoC platforms: a component-based design approach," IEEE Design and Test of Computers, vol. 19, no. 6, pp. 52–63, 2002.

[2]   C. Feng, Z. Lu, A. Jantsch, J. Li, and M. Zhang, "FoN: Fault-on-Neighbor aware routing algorithm for Networks-on-Chip," Proc. 23th IEEE Int. System-on-Chip Conf. (SOCC), pp. 441–446, 2010.

[3]   M. Valinataj, S. Mohammadi, and S. Safari, "Fault-aware and reconfigurable routing algorithms for Networks-on-Chip," IETE Journal of Research, vol. 57, no. 3, pp. 215–223, 2011.

[4]   M. Valinataj, S. Mohammadi, J. Plosila, P. Liljeberg, and H. Tenhunen, "A reconfigurable and adaptive routing method for fault-tolerant mesh-based networks-on-chip," Elsevier, Int. J. Electronics and Communications (AEÜ), vol. 65, no. 7, pp. 630–640, 2011.

[5]   P. Rantala, J. Isoaho, and H. Tenhunen, "Novel agent-based management for fault-tolerance in network-on-chip," Proc. 10th Euromicro Conf. on Digital System Design (DSD), pp. 551–555, 2007.

[6]   A. W. Yin et al, "Hierarchical agent monitoring NoCs: a design methodology with scalability and variability," Proc. 26th NORCHIP Conf., pp. 202–207, 2008.

[7]   L. Guang, B. Yang, J. Plosila, K. Latif, and H. Tenhunen, "Hierarchical power monitoring on NoC - a case study for hierarchical agent monitoring design approach," Proc. 28th NORCHIP Conf., 2010.

[8]   L. Guang, E. Nigussie, P. Rantala, J. Isoaho, and H. Tenhunen, "Hierarchical agent monitoring design approach towards self-aware parallel systems-on-chip," ACM Trans. on Embedded Computing Systems, vol. 9, no. 3, article 25, 2010.

[9]   A. Kohler, G. Schley, and M. Radetzki, "Fault tolerant network on chip switching with graceful performance degradation," IEEE Trans. on Computer-Aided Design of Integrated Circuits and Systems, vol. 29, no. 6, 2010.

[10] M. Li, Q. Zeng, and W. Jone, "DyXY- a proximity congestion-aware deadlock-free dynamic routing method for Network on Chip, " Proc. 43th Design Automation Conference (DAC), pp. 849–852, 2006.

452

Power Electronics

Survey and Analysis of the Design Issues of a Low Cost Micro Power DC-DC Step Up Converter for Indoor Light Energy Harvesting Applications

Carlos Carvalho

Instituto Superior de Engenharia de Lisboa (ISEL – ADEETC)
Instituto Politécnico de Lisboa (IPL)
Rua Conselheiro Emídio Navarro, n°1
1949-014 Lisboa – Portugal
e-mail: cfc@isel.ipl.pt

João P. Oliveira, Nuno Paulino

UNINOVA/CTS
Departamento de Engenharia Electrotécnica, Faculdade de Ciências e Tecnologia
Universidade Nova de Lisboa
Campus FCT/UNL, 2829-516 Caparica – Portugal
e-mail: nunop@uninova.pt

Abstract—This paper discusses the pertinent issues in designing and developing a DC-DC converter for a low cost, micro power indoor light harvesting system using CMOS technology. The different issues associated to this problem are studied and the relevant literature is analysed. The paper surveys and analyses the design options available for the PV cells, step-up voltage converter circuit architecture, maximum power point tracking (MPPT) methods and energy storage devices. From this analysis a possible solution is discussed.

Index Terms—CMOS, Electronics, Energy harvesting, Energy storage elements, MPPT techniques, Power management circuits, PV cells

I. INTRODUCTION

The ability of circuits to obtain energy from the surrounding environment for self powering is an interesting feature that has gained increasing importance [1]. In opposition to traditional powering methods, like those that involve a cord connection to the power grid, or the use of batteries, obtaining energy from ambient sources promises to take over the powering paradigm for sensor networks [2], [3] and embedded systems [4]. Sensor networks that fully rely on grid powering only have the opportunity to monitor building indoor parameters, as they depend upon the cord connection. If one wants ubiquity and truly pervasive operation, relying on the power grid can be a limiting factor. One step forward, towards unlimited sensor location, could be the use of batteries. Although improving the freedom of sensor distribution, one obstacle remains, related to the batteries themselves. As their stored energy gets depleted, batteries need to be replaced. This can be a problem, if a large number of sensors are deployed, some in places that are difficult to reach. Thus, the trivial operation of battery replacement can result in costs a lot greater than the batteries themselves. The costs with staff and other logistic means may not pay off the utility of the sensor network. To achieve indefinite operation, the sensors must be supplied such that they obtain their power directly from the surrounding environment. This procedure is known as energy harvesting. By not using batteries for system main powering

purposes, the sensor system will not be responsible for contributing to chemical pollution caused by disposing them of, or their manufacturing. In economical terms, not using batteries represents cost reduction in both devices and replacement operation procedures.

The organization of this paper is as follows: Section II presents some considerations about the light energy source. Section III discusses energy conditioning systems to process the energy provided by the harvester. Section IV gives an insight into some devices in which it is possible to store the harvested energy. In Section V, the MPPT aspects are addressed, so as to obtain the maximum efficiency out of PV cells. In the end, Section VI presents the conclusions about this survey, foreseeing general future applications.

II. LIGHT ENERGY HARVESTING

There are different possible energy sources in the environment that can be harvested to power electronic applications [5]. This paper will only be focused on light energy because, comparing all sources, this is the one that shows the highest energy density by volume unity, followed by mechanical and thermal energy respectively [5].

Light energy comes primarily from the Sun, but it also can be obtained indoor. At the maximum of its height and intensity, the Sun can provide as much as $1\,kW/m^2$, but in indoor environments, as the light gets attenuated, the Sun provides about one tenth of this energy density [6]. Relying on artificial illumination can result in only about $10\,W/m^2$ [7]. In some situations it proves useful to have a history of the solar irradiation of the location where the harvester system is to be placed [8], [9].

The key element for this kind of harvesting is the PV cell. This solid state device converts light energy directly to electrical energy without using moving parts. A PV cell is basically a photodiode and it can be manufactured in CMOS technology [10], [11]. However, it is not simple to integrate PV cells with other circuits in the same die to obtain a complete System-on-Chip (SoC) [12], [13]. The photons of the incident light can penetrate the die until they reach the substrate where they can cause a positive charge build up ultimately resulting in latch-up. When a series connection of several PV cells is needed, building the diodes on the same substrate may be limiting [14]. To tackle these limitations, it is necessary to use more expensive technologies, such as Silicon-on-Insulator (SOI), allowing for an arbitrary number of series connected PV cells to obtain a higher voltage value [15].

The efficiency of common PV cells is still relatively low, at about 20% [6]. Some solar cells can reach efficiencies as high as 50%, but involving the use of new layout architectures and less common materials [16], resulting in more expensive systems.

To reduce production costs, it is possible to use PV cell technologies with lower efficiencies and lower manufacturing costs, resulting in a larger area for the PV cell, for the same power level. Using amorphous silicon PV cells is an example of such a trade-off, in which costs are lower, but at the expense of a larger area [17].

Based on an amorphous silicon PV cell that was built and experimentally characterized, an equivalent electrical model was obtained [17], as shown in Figure 1. This model does not intend to be an exact physical representation of the cell itself, but rather to translate its electrical behavior.

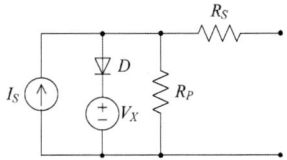

Figure 1.   Equivalent electrical circuit of the amorphous silicon solar cell

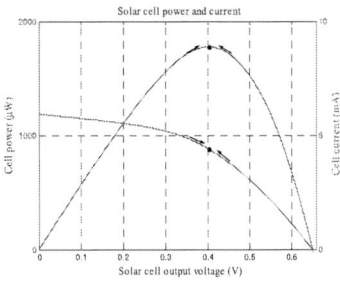

Figure 2.   Power and current curves of the solar cell equivalent circuit model for maximum illumination (AM1).

The power and current curves of the cell for AM1 conditions are depicted in Figure 2. , where the maximum power point (MPP) of the PV cell is shown by the dots. This performance was obtained with a cell having an area of about 1 cm^2. The maximum power (P_{max}) obtained is 1.775 mW, at a voltage of about 400 mV (v_{max}).

If this same cell is to be used indoor, the level of usable power is substantially lower. Also, the voltage at which the power of a single cell is optimal (maximum) is insufficient to power CMOS circuits, which typically require 1.2 V. As such,

it is mandatory to use two PV cells in series and a voltage doubler circuit to step the voltage up. The data in TABLE I. shows some parameters that were obtained by simulating the series of two PV cells using the previously described electrical model. The light intensity is swept from 10% to 100%.

TABLE I.   PERFORMANCE OF A SERIES OF TWO AMORPHOUS SILICON PV CELLS

Light intensity	Maximum power (P_{max})	Open circuit voltage (v_{oc})	Optimal voltage (v_{max})
100%	3.550 mW	1.303 V	803.2 mV
80%	2.830 mW	1.263 V	818.1 mV
60%	2.037 mW	1.210 V	806.5 mV
40%	1.187 mW	1.132 V	774.1 mV
20%	367.4 µW	954.4 mV	591.8 mV
10%	93.49 µW	620.9 mV	316.8 mV

The area of each cell is 1 cm^2. A light intensity of 20% of the maximum is the threshold that allows for an output voltage of about 1.2 V after the step up operation.

III.   ENERGY CONDITIONING SYSTEMS

Since the voltage of a PV cell depends on the light intensity, temperature and load value, it is necessary to use a circuit that adjusts the load seen by the PV cell, to maximize the collected energy and to produce a stabilized output voltage.

As previously explained, the PV cell voltage must be stepped up (boost operation), or stepped down (buck operation). These circuits can be inductor based or switched capacitor based. Either way, the goal is to perform a conversion as efficient as possible. The regulator also plays the role of protecting the energy storage device from overload and when dealing with a PV based system, it sets the output voltage of the PV cell, in order to have optimal power operation conditions.

A.   Inductor Based Converters

The inductance value required by most inductor based converters is outside the range of values that are possible to integrate in CMOS. Therefore, inductor based voltage converters require a discrete inductor placed outside the system. There are numerous examples of energy conditioning systems based on traditional and more elaborated architectures employing inductors, such as in [8], [18], [19], [20].

A comparative study between inductor and SC based conversion technologies is given in [21]. This study concludes that converters based on SC have less losses and that capacitors have a greater energy and power density, when compared to inductors, if small devices are used. Also, in [22] it is shown that for very low power systems (< 1Watt), capacitor based converters lead the performance, in terms of efficiency.

B.   Switched-Capacitor (SC) Based Converters

It is possible to have a SC voltage converter circuit entirely built using standard CMOS technology without using external

456

components. These converters can reach high efficiencies. When dealing with micro-power harvester systems, the limited available energy makes the converter efficiency a critical issue. One of the biggest problems is the bottom plate capacitance associated to every capacitor [18], [23]. If the system has a substantial power (more than 1 Watt) and integration is not a requirement, usually the inductor-based converters can achieve a better efficiency.

Regarding SC topologies, there are some well know, among which is the Series-Parallel, performing an elevation of the input voltage, according to the number of capacitors involved [23]. In general, this topology shows a good performance. A voltage doubler whose base is according to the same principle can be found in [24].

To overcome performance limitations due to the parasitic bottom plate capacitance, [25] proposes some configurations that try to minimize the amount of charge that is lost. Also, in [26] and [27] another technique is employed in order to minimize the bottom plate capacitance loss. For convenience, the circuit topology used in [26] is depicted in Figure 3.

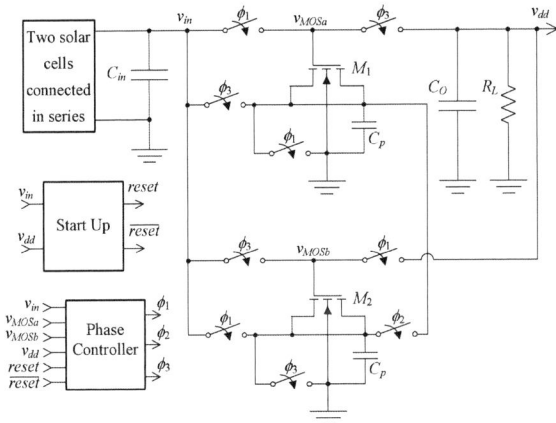

Figure 3. Switched capacitor doubler, with charge reusing.

In general, when dealing with SC converters performance studies, work can be found in [28]–[31]. One interesting technique is used in [32], where a "gearbox", switching between different topologies, is used so as to maximize the overall performance of the converter system. The need for voltage elevation is also related to the number of PV cells connected in series at the input.

IV. ENERGY STORAGE DEVICES

Once the energy has been harvested and conditioned, some means must be used to store that energy, so that it can used at a later time. For a small system, there are two storing devices, batteries and supercapacitors, available to perform this task [4]. Depending on the energy usage profile of the system, any of these devices can be used in accordance. However, to optimize their utility and lifetime, each must fit into the appropriate energy usage regime [9]. If appropriate, both devices can be used in the same application, aiming to extend the lifetime of

each other. Each device requires special attention, as their characteristics involve very specific charging strategies [9].

A. Batteries

Batteries are used when large energy density is required, but their lifetime is seriously affected by the number of charging/discharging cycles. So, trying to minimize the number of these cycles is an important objective. This aspect is related to the amount of time that a battery can remain in operation, so that the stored charge can hold for as long as possible. An example of a work concerning such an issue is presented in [33].

There are several common types of rechargeable batteries, for instance, Li+/Li-polymer (Lithium-ion / Lithium polymer), NiMH (Nickel Metal Hydride) or NiCad (Nickel Cadmium). Typical operating voltages for these kinds of battery technologies can be approximately 1.2 V for the last two types and 3.7 V for the first type [34]. Conventional Li-ion batteries have typical operating voltages that range from 2.7 V to 4.2 V [35]. There are some emerging technologies using materials like $LiCoO_2$ or graphite, that in conjunction with PVDF-Ionic electrolyte, have given promising results, just like technologies based on other types of materials and electrolytes [6]. It is important to note that each type of technology might require a different circuit to control its charging phase.

Besides being electrical charge buffers, batteries serve as voltage stabilizers, providing a constant voltage at the output of the regulator circuit. Examples of systems that use batteries to store harvested energy can be found in [36], [37].

B. Supercapacitors

Supercapacitors exhibit characteristics that make them different from ordinary capacitors. The electrical model of a supercapacitor is not simply a highly valued capacitor, but a set of several branches with different time constants [38], for the case of the double-layer capacitor. Capacitance values can be as high as 1500 F. This type of device can even achieve power densities higher than that of conventional batteries.

These devices can stand a higher number of charge/discharge cycles than batteries, being suitable for applications where this kind of regime is usual. The number of cycles can be as high as a million, leading to an operational lifetime of ten years, until the capacitance value starts to show some degradation [3].

Supercapacitors are cheap, being very appealing to use in opposition to batteries, as these are more expensive. An example of a discrete system that makes use of a supercapacitor to store harvested energy from a solar harvester can be found in [39].

Nonetheless, there are some applications that use both battery and supercapacitor [9], [34]. These act as a primary and secondary energy buffers, respectively.

In the case of a low cost system that is only required to operate when light is available, or a short time after that, the most economical solution is to use a supercacitor. As an example, consider a 1 F supercapacitor was charged to 1.3V and then allowed to power a circuit until its voltage decreases to 1.1V. The energy supplied by this discharging capacitor is $E = \frac{1}{2} \times C \times \Delta V_{out}^2$, in this case 0.02 J. This energy could be used to power a 10 mW circuit for 2 seconds, which is enough time, in most cases, to transmit some bytes of information.

V. MAXIMUM POWER POINT TRACKING TECHNIQUES

There are some limiting factors when building a solar energy powered micro sensor system, such as a low energy budget, due to size limitations. This budget must be enough to enable the controlling circuits to operate and to maintain the interaction between the energy processing system and the harvester at an optimum level [9].

To maximize the energy obtained from the PV cell, there are a set of techniques known as Maximum Power Point Tracking (MPPT), which can be used to reach this goal. For instance, it is possible to manipulate the PV cell orientation, in order to maximize the light intensity on its surface [40], but in small systems this option is not valid because of the low power and low cost budgets. In indoor applications it is possible to place the PV cell as close as possible to a light source, with the correct orientation, in order to maximize the received light energy.

Figure 2. illustrates the MPPT principle. Its algorithm should try to reach the dot position, as the equivalent impedance of both the cell and the converter circuit will vary, according to load, temperature or irradiance changes. The arrows in Figure 2. represent the consequence of the controller action, dealing with an impedance mismatch, vectoring the PV cell to the new MPP. A substantially wide set of MPPT techniques can be found in [41], providing a broad perspective. Most of the ones that are presented in this section were developed for large PV arrays that provide hundreds or thousands of Watts. In some cases, these techniques can be extended to systems with power around μW or mW.

A. True MPPT Techniques

These techniques are concerned with obtaining and tracking the MPP of the PV cell, independently of light and temperature conditions. This accurate estimation is often based on microcontroller computation requiring the use of an ADC. In general, these MPPT techniques do not need to know the PV cell characteristics beforehand, as the converter system adapts itself automatically to the given PV cell. Examples of such techniques include algorithms like the Hill-Climbing, used in [27], [36], [37], [42], [43], and the Ripple Correlation Control (RCC), used in [44]–[47]. Both of these algorithms can also be implemented using analog circuits, thus reducing the power needed to operate. RCC is possibly the best MPPT method, but it requires a multiplication to compute the instantaneous power. Since an analog multiplier is difficult to design and usually dissipates a large power, the RCC method is not suitable for micro power systems, such as for the case of indoor light energy harvesting.

B. Quasi-MPPT Techniques

This type of algorithms cannot reach the true MPP of a PV cell. Since the power value does not change significantly from the maximum value around its vicinity, this is not a big problem. The Fractional Open Circuit Voltage (Fractional V_{OC}) method requires to previously determining the characteristics of the PV cell [48]. The open circuit voltage can be obtained using a pilot PV cell, smaller than the main cell, exposed in the same way as the latter. The circuitry needed to implement the fractional V_{OC} method is very simple and dissipates little power, at the cost of producing only an approximation of the MPP of the PV cell. This trade-off is acceptable for a micro power system. Work based on this MPPT method can be found in [48] and [49].

1) Determination of the Fractional V_{OC} Coefficient (k) for the PV Cell experimentally characterized

In the case of a micro power step-up converter it is possible to tolerate some inaccuracy in the determination of the MPP, in exchange for using a simpler method for determining the MPP of the PV cell that requires less complex circuits and dissipates less power. The Fractional Open Circuit Voltage (Fractional V_{OC}) meets these requirements, because it is a very simple and inexpensive (hardware wise) method.

The Fractional V_{OC} method explores an intrinsic characteristic of PV cells: there is a proportionality factor between their open circuit voltage and the voltage at which the MPP occurs. This factor must be determined beforehand, by studying the solar cell behavior under several conditions of illumination and temperature. The cell whose model is depicted in Figure 1. was simulated and showed the performance depicted in Figure 4. and Figure 5.

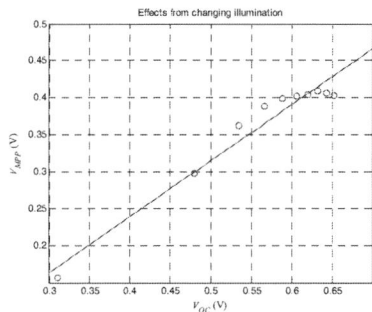

Figure 4.   Fractional open circuit voltage relation between V_{OC} and V_{MPP} under several conditions of illumination

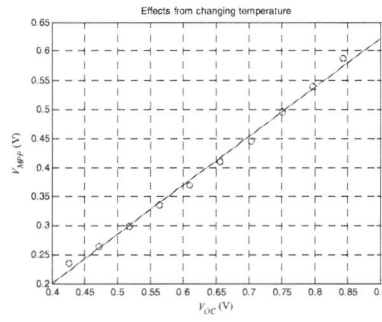

Figure 5.   Fractional open circuit voltage relation between V_{OC} and V_{MPP} under several conditions of temperature

By performing a linear regression over the points plotted on the obtained graphs, the same way as in [49], one can determine the slope of these functions. By sweeping a range of temperatures that spanned from $-55\ ^\circ$C to $+125\ ^\circ$C, the ratio V_{MPP}/V_{OC} was around 0.84. By sweeping illumination from 10% to 100%, the ratio V_{MPP}/V_{OC} was around 0.76. Assuming that illumination has more importance, as it is more likely to vary, a value of 0.77 was selected for k, the Fractional V_{OC} coefficient. This value agrees with the ones stated in [41].

VI. CONCLUSIONS

This paper presented a survey on the relevant issues to take into account when designing a low cost, micro power, light harvesting system for indoor applications. The relevant literature was analyzed, regarding the light energy source, types of energy conditioning systems, energy storage devices and MPPT techniques. Design options about each of these items were presented. From this analysis, several conclusions about the constituting elements of the system were reached. The system should use a low cost PV cell, such as a-Si, the system should use a SC DC-DC converter and it should use supercapacitors as the energy storing devices.

REFERENCES

[1] Paradiso, J.A.; Starner, T.; , "Energy scavenging for mobile and wireless electronics," *Pervasive Computing, IEEE* , vol.4, no.1, pp. 18- 27, Jan.- March 2005

[2] Kansal, A.; Srivastava, M.B.; , "An environmental energy harvesting framework for sensor networks," *ISLPED '03. Proceedings of the International Symposium on Low Power Electronics and Design,* pp. 481- 486, 25-27 Aug. 2003

[3] Chou, P.H.; Chulsung Park; , "Energy-efficient platform designs for real-world wireless sensing applications,", *ICCAD-2005. IEEE/ACM Int. Conf. on Computer-Aided Design*, pp. 913- 920, 6-10 Nov. 2005

[4] Raghunathan, V.; Chou, P.H.; , "Design and Power Management of Energy Harvesting Embedded Systems,", ISLPED'06. Proceedings of the International Symposium on Low Power Electronics and Design, pp.369-374, 4-6 Oct. 2006

[5] Chalasani, S.; Conrad, J.M.; , "A survey of energy harvesting sources for embedded systems," Southeastcon, 2008. IEEE , pp.442-447, 3-6 April 2008

[6] Rabaey, J.; Burghardt, F.; Steingart, D.; Seeman, M.; Wright, P.; , "Energy Harvesting - A Systems Perspective,", IEDM 2007. IEEE International Electron Devices Meeting, pp.363-366, 10-12 Dec. 2007

[7] Hande, A.; Polk, T.; Walker, W.; Bhatia, D., "Indoor solar energy harvesting for sensor network router nodes," Microprocessors and Microsystems, Vol. 31, No. 6. (01 September 2007), pp. 420-432.

[8] Dondi, D.; Bertacchini, A.; Larcher, L.; Pavan, P.; Brunelli, D.; Benini, L.; , "A solar energy harvesting circuit for low power applications," ICSET 2008. IEEE International Conference on Sustainable Energy Technologies, pp.945-949, 24-27 Nov. 2008

[9] Jeong, J.; Jiang, X.; Culler, D.; , "Design and analysis of micro-solar power systems for Wireless Sensor Networks," INSS 2008. 5th Int. Conf. on Networked Sensing Systems, pp.181-188, 17-19 June 2008

[10] Lee, J.S.; Hornsey, R.I.; Renshaw, D.; , "Analysis of CMOS Photodiodes I - Quantum efficiency," IEEE Transactions on Electron Devices, vol.50, no.5, pp. 1233- 1238, May 2003

[11] Lee, J.S.; Hornsey, R.I.; Renshaw, D.; , "Analysis of CMOS Photodiodes II - Lateral photoresponse," IEEE Transactions on Electron Devices, vol.50, no.5, pp. 1239- 1245, May 2003

[12] Guilar, N.J.; Kleeburg, T.J.; Chen, A.; Yankelevich, D.R.; Amirtharajah, R.; , "Integrated Solar Energy Harvesting and Storage," IEEE Transactions on Very Large Scale Integration (VLSI) Systems, vol.17, no.5, pp.627-637, May 2009

[13] Guilar, N.J.; Fong, E.G.; Kleeburg, T.; Yankelevich, D.R.; Amirtharajah, R.; , "Energy harvesting photodiodes with integrated 2D diffractive storage capacitance," ACM/IEEE Int. Symposium on Low Power Electronics and Design (ISLPED), pp.63-68, 11-13 Aug. 2008

[14] Ferri, M.; Pinna, D.; Dallago, E.; Malcovati, P.; , "A 0.35μm CMOS Solar energy scavenger with power storage management system," Ph.D. Research in Microelectronics and Electronics, PRIME 2009, pp.88-91, 12-17 July 2009

[15] Ferri, M.; Pinna, D.; Malcovati, P.; Dallago, E.; Ricotti, G.; , "Integrated stabilized photovoltaic energy harvester," ICECS 2009. 16th IEEE Int. Conf. on Electronics, Circ. and Systems, pp.299-302, 13-16 Dec. 2009

[16] Barnett, A.; Honsberg, C.; Kirkpatrick, D.; Kurtz, S.; Moore, D.; Salzman, D.; Schwartz, R.; Gray, J.; Bowden, S.; Goossen, K.; Haney, M.; Aiken, D.; Wanlass, M.; Emery, K.; , "50% Efficient Solar Cell Architectures and Designs," Conference Record of the IEEE 4th World Conf. on Photovoltaic Energy Conv., vol.2, pp.2560-2564, May 2006

[17] Amaral, A.; Lavareda, G.; Nunes de Carvalho, C.; Brogueira, P.; Gordo, P.M.; Subrahmanyam, V.S.; Lopes Gil, C.; Duarte Naia, V; de Lima, A. P., "Influence of the a-Si:H structural defects studied by positron annihilation on the solar cells characteristics", Thin Solid Films, vol. 403–404, pp. 539–542, 2002.

[18] Richelli, A.; Colalongo, L.; Tonoli, S.; Kovacs, Z., "A 0.2V-1.2V converter for power harvesting applications", Proc. 34th European Solid-State Circuits Conf. ESSCIRC 2008, 2008, pp. 406-409.

[19] Huang, M.-H.; Chen, K.-H.; , "Single-Inductor Multi-Output (SIMO) DC-DC Converters With High Light-Load Efficiency and Minimized Cross-Regulation for Portable Devices," IEEE Journal of Solid-State Circuits, vol.44, no.4, pp.1099-1111, April 2009

[20] Sze, N.-M.; Su, F.; Lam, Y.-H.; Ki, W.-H.; Tsui, C.-Y.; , "Integrated single-inductor dual-input dual-output boost converter for energy harvesting applications," ISCAS 2008. IEEE International Symposium on Circuits and Systems, pp.2218-2221, 18-21 May 2008

[21] Seeman, M.D.; Ng, V.W.; Hanh-Phuc Le; John, M.; Alon, E.; Sanders, S.R.; , "A comparative analysis of Switched-Capacitor and inductor-based DC-DC conversion technologies," IEEE 12th Workshop on Control and Modeling for Power Electronics (COMPEL), pp.1-7, 28-30 June 2010

[22] Pique, G.V.; Bergveld, H.J.; , "State-of-the-art of integrated switching power converters," AACD 2011, April 2011

[23] Seeman, M.D.; Sanders, S.R.; , "Analysis and Optimization of Switched-Capacitor DC–DC Converters," IEEE Transactions on Power Electronics, vol.23, no.2, pp.841-851, March 2008

[24] Su, F.; Ki, W.-H.; Tsui, C.-Y.; ,"Regulated Switched-Capacitor Doubler With Interleaving Control for Continuous Output Regulation," IEEE Journal of Solid-State Circuits, vol.44, no.4, pp.1112-1120, April 2009

[25] Ramadass, Y.K.; Chandrakasan, A.P.; , "Voltage Scalable Switched Capacitor DC-DC Converter for Ultra-Low-Power On-Chip Applications," Power Electronics Specialists Conference, 2007. PESC 2007. IEEE , pp.2353-2359, 17-21 June 2007

[26] Carvalho, C.; Paulino, N.; , "A MOSFET only, step-up DC-DC micro power converter, for solar energy harvesting applications," *International Journal of Microelectronics and Computer Science*, vol.1, no.2, pp.112-119, 2010, ISSN 2080-8755

[27] Carvalho, C.; Lavareda, G.; Lameiro, J.; Paulino, N.; , "A step-up μ-power converter for solar energy harvesting applications, using Hill Climbing maximum power point tracking," IEEE International Symposium on Circuits and Systems ISCAS 2011, pp. 1924–1927, 15-18 May 2011

[28] Ngo, K.D.T.; Webster, R.; , "Steady-state analysis and design of a switched-capacitor DC-DC converter," IEEE Transactions on Aerospace and Electronic Systems, vol.30, no.1, pp.92-101, Jan 1994

[29] Makowski, M.S.; Maksimovic, D.; , "Performance limits of switched-capacitor DC-DC converters," Power Electronics Specialists Conference, PESC '95 Record., 26th Annual IEEE , vol.2, pp.1215-1221 vol.2, 18-22 Jun 1995

[30] Zhu, G.; Ioinovici, A.; , "Switched-capacitor power supplies: DC voltage ratio, efficiency, ripple, regulation," ISCAS '96. 'Connecting the World', IEEE International Symposium on Circuits and Systems, 1996, vol.1, pp.553-556 vol.1, 12-15 May 1996

[31] Pan, Z.; Zhang, F.; Peng, F.Z.; , "Power losses and efficiency analysis of multilevel dc-dc converters," APEC 2005. Twentieth Annual IEEE Applied Power Electronics Conference and Exposition, vol.3, pp.1393-1398 Vol. 3, 6-10 March 2005

[32] Van Breussegem, T.; Steyaert, M.; , "A Fully Integrated Gearbox Capacitive DC/DC-converter in 90 nm CMOS: Optimization, Control and Measurements," Proceedings of COMPEL 2010, 12, Boulder, 2010

[33] Ramadass, Y.K.; Chandrakasan, A.P.; , "Minimum Energy Tracking Loop With Embedded DC–DC Converter Enabling Ultra-Low-Voltage Operation Down to 250 mV in 65 nm CMOS," IEEE Journal of Solid-State Circuits, vol.43, no.1, pp.256-265, Jan. 2008

[34] Jiang, X.; Polastre, J.; Culler, D.; , "Perpetual environmentally powered sensor networks," IPSN 2005. Fourth Int. Symposium on Information Processing in Sensor Networks, pp. 463- 468, 15 April 2005

[35] Torres, E.O.; Rincon-Mora, G.A.; , "Electrostatic Energy-Harvesting and Battery-Charging CMOS System Prototype,", IEEE Trans. on Circ. and Systems I: Regular Papers, vol.56, no.9, pp.1938-1948, Sept. 2009

[36] Hui Shao; Chi-Ying Tsui; Wing-Hung Ki; , "An Inductor-less Micro Solar Power Management System Design for Energy Harvesting Applications," ISCAS 2007. IEEE International Symposium on Circuits and Systems, pp.1353-1356, 27-30 May 2007

[37] Shao, H; Tsui, C.-Y.; Ki, W.-H.; , "The Design of a Micro Power Management System for Applications Using Photovoltaic Cells With the Maximum Output Power Control," IEEE Trans. on Very Large Scale Integrat. (VLSI) Systems, vol.17, no.8, pp.1138-1142, Aug. 2009

[38] Zubieta, L.; Bonert, R.; , "Characterization of double-layer capacitors for power electronics applications," IEEE Transactions on Industry Applications, vol.36, no.1, pp.199-205, Jan/Feb 2000

[39] Simjee, F.; Chou, P.H.; , "Everlast: Long-life, Supercapacitor-operated Wireless Sensor Node," ISLPED'06. Proc. of the 2006 Int. Symposium on Low Power Electronics and Design, pp.197-202, 4-6 Oct. 2006

[40] Mashohor, S.; Samsudin, K.; Noor, A.M.; Rahman, A.R.A.; , "Evaluation of Genetic Algorithm based solar tracking system for Photovoltaic panels," ICSET 2008. IEEE International Conference on Sustainable Energy Technologies, pp.269-273, 24-27 Nov. 2008

[41] Esram, T.; Chapman, P.L.; , "Comparison of Photovoltaic Array Maximum Power Point Tracking Techniques," IEEE Transactions on Energy Conversion, vol.22, no.2, pp.439-449, June 2007

[42] Shao, H.; Tsui, C.-Y.; Ki, W.-H.; , "An inductor-less MPPT design for light energy harvesting systems," ASP-DAC 2009. Asia and South Pacific Design Automation Conference, pp.101-102, 19-22 Jan. 2009

[43] Kim, Y.; Jo, H.; Kim, D.; , "A new peak power tracker for cost-effective photovoltaic power system," IECEC 96. Proceedings of the 31st Intersociety Energy Conversion Engineering Conference, vol.3, pp.1673-1678 vol.3, 11-16 Aug 1996

[44] Midya, P.; Krein, P.T.; Turnbull, R.J.; Reppa, R.; Kimball, J.; , "Dynamic maximum power point tracker for photovoltaic applications ," PESC '96 Record., 27th Annual IEEE Power Electronics Specialists Conference, vol.2, pp.1710-1716 vol.2, 23-27 Jun 1996

[45] Lim, Y.H.; Hamill, D.C.; , "Simple maximum power point tracker for photovoltaic arrays," Electronics Letters , vol.36, no.11, pp. 997- 999, 25 May 2000

[46] Lim, Y.H.; Hamill, D.C.; , "Synthesis, simulation and experimental verification of a maximum power point tracker from nonlinear dynamics," PESC 2001. IEEE 32nd Annual Power Electronics Specialists Conference, vol.1, pp.199-204 vol. 1, 2001

[47] Esram, T.; Kimball, J.W.; Krein, P.T.; Chapman, P.L.; Midya, P.; , "Dynamic Maximum Power Point Tracking of Photovoltaic Arrays Using Ripple Correlation Control," IEEE Transactions on Power Electronics, vol.21, no.5, pp.1282-1291, Sept. 2006

[48] Carvalho C.; Lavareda, G.; Paulino, N.; , "A DC-DC Step-up ⊓ Power Converter for Energy Harvesting Applications, using Maximum Power Point Tracking, Based on Fractional Open Circuit Voltage," Technological Innovation for Sustainability – IFIP Advances in Information and Communication Technology, vol. 349, pp. 510-517, 2011

[49] Brunelli, D.; Moser, C.; Thiele, L.; Benini, L.; , "Design of a Solar-Harvesting Circuit for Batteryless Embedded Systems," IEEE Transactions on Circuits and Systems I: Regular Papers, vol.56, no.11, pp.2519-2528, Nov. 2009

The Effect of Thermal Inertia
in Photovoltaic Module Simulation

Maciej Piotrowicz and Witold Marańda

Department of Microelectronics and Computer Science
Technical University of Lodz
Lodz, Poland
{piotrowi, maranda}@dmcs.p.lodz.pl

Abstract—**Performance of photovoltaic systems greatly depends on climate conditions. Apart from values of solar radiation and ambient air temperature, highly varying radiation can also have an influence on energy yield from a PV-system. In less favourable climate conditions (e.g. in central Europe), highly varying radiation is very frequent and the accurate energy yield predictions from PV-systems require introducing thermal inertia into photovoltaic module thermal models.**

The presented paper demonstrates the simulation of the simple RC-thermal model of photovoltaic module. Long time period data from measurements of field-installed PV-system is utilized in module temperature calculations.

Index Terms—**Photovoltaics, PV-system, Thermal model**

I. Introduction

The efficiency of solar panel strongly depends on the temperature. PV-cells warm up significantly when exposed to sunlight. The efficiency then drops typically 0.4% per 1°C for the typical crystalline-Si panels.

Usually, the efficiency of panels is specified for so called standard test conditions (STC), characterized by air temperature of 25°C and radiation 1000W/m^2 with defined spectrum. However such conditions are hardly ever observed in operation of field-installed PV-systems. Due to higher operating temperatures, the annual energy yield from solar system in climate of central Europe may be lower even by 30% compared to the theoretical yield at STC [1]. Therefore, the realistic radiation, ambient temperature and module temperature conditions should be taken into consideration to increase the accuracy of the simulations of photovoltaic systems.

Thermal modelling of PV devices has been addressed by many authors in recent years and many detailed models were developed, taking into account material properties, various heat transfer mechanisms and weather conditions, e.g. [2, 3]. The correctness of those models was successfully confirmed experimentally. They are, however, unsuitable for PV-application practice.

The industry-standard thermal model on the other hand is sufficient for low variations of solar radiation. This model uses single thermal resistance for PV-cell temperature calculations under given ambient temperature and radiation level. RC

thermal model should be considered in case of highly variable radiation conditions.

The paper demonstrates simulation of simple RC thermal model parameters, that was earlier presented and verified by authors [4]. Both the extraction of parameters and verification were based on the measurements taken during normal operation of photovoltaic system. Unlike laboratory experiment, such approach allows to identify real-life thermal parameter values combining all cooling mechanisms in operating conditions.

The measurements were performed in the photovoltaic system at the Department of Microelectronics and Computer Science (DMCS) of Technical University of Lodz in Poland (N 51° 44' 46", E 19° 27' 20").

II. Thermal Model of PV-module

The simplest approach to PV module thermal modelling is to bind the cell temperature T_c with ambient air temperature T_a and the energy flux from solar radiation G using a single parameter R_{th}. The heat transfer can then be described with the following formula:

$$T_c - T_a = R_{th} G \qquad (1)$$

It must be noticed, that the factor R_{th} represents all the heat transfer mechanisms between module and air, combined with light reflection losses. However, due to its units K/W, it is often called thermal resistance.

In practice, it is easier to characterize the thermal behaviour of the solar panel with the temperature of solar cell under specific test condition. The most popular is the parameter NOCT (Normal Operating Cell Temperature), measured in open circuit, under 800 W/m^2 radiation, 20°C air temperature, 1 m/s wind velocity and open back-side mounting. In most cases, it is the only thermal parameter published in data sheets.

Typical NOCT values range from 30°C to 55°C; the lower the parameter is, the more efficient cooling it represents. This parameter can be easily related with the previously described R_{th} factor, using (1):

$$\frac{NOCT - 20\,^{\circ}C}{800\,W/m^2} = R_{th} \qquad (2)$$

For the locations with favourable climates, typical daily radiation profile consists of single smooth line that represents slow, monotone changes of the insolation value; the exemplary profile is shown in Fig. 1. The dynamics of thermal processes is rather low under such conditions, thus single thermal resistance model can be accurate enough for majority of application.

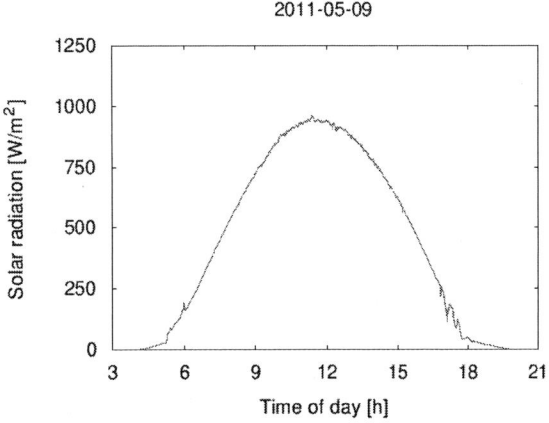

Figure 1. Favourable insolation conditions (clear sky)

Central Europe is an example of the less favourable climate area. In this case, rapid variations of the irradiance, caused by clouds movement, can be observed in daily radiation profiles (a typical one is shown in Fig. 2). The thermal inertia of PV modules should be taken into account under such conditions. However, in many simulation software radiation data have longer time intervals between samples (e.g. 0.5 h) or are averaged. All the dynamic phenomena are hidden in such case, thus the simple thermal resistance model appears to be accurate enough.

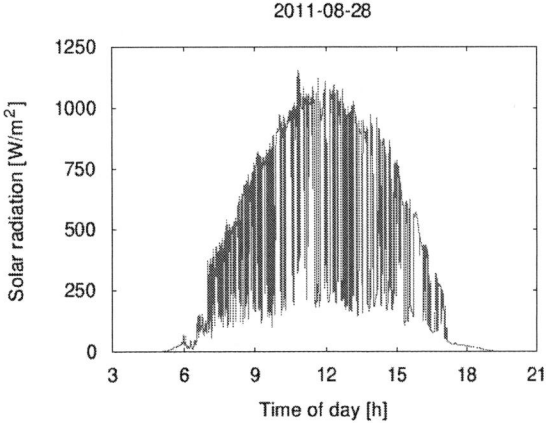

Figure 2. Typical daily radiation profile for Central Europe

As it was presented in [4], the authors elaborated single section RC model of PV module, that is both simple and sufficient for handling variations of solar radiation. Its parameters were extracted from measurements conducted at photovoltaic system working at DMCS. All the data from sensors (especially radiation and temperature values) have been read every 5 seconds – time resolution that is sufficient for dynamic phenomena consideration. Parameters R and C values (thermal resistance and capacitance respectively) fit for the field-installed module are shown in Table I.

TABLE I. Extracted model parameters

Param.	R	C
Value	0.0134	25790
Unit	K/Wm²	J/Km²

The data from Table I allows for calculation of NOCT=32°C (according to Eq. 1 and 2) and the thermal time constant RC=345s (5.75 min) for the examined PV modules. The last factor of almost 6 minutes is longer than the length of majority of short spikes in radiation profiles (as in Fig. 2). Nevertheless highly variable insolation conditions often span long time intervals and deliver considerable amounts of solar energy. The photovoltaic module temperature T_c cannot then be precisely predicted without thermal capacitance taken into account.

It is particularly visible, if graph of the temperature difference $(T_c - T_a)$ vs. solar radiation is analyzed. For favourable insolation conditions the single thermal resistance can easily be identified and sufficient. In other case, the thermal inertia is clearly visible. The two exemplary graphs are presented in Fig. 3 (for clear sky) and Fig. 4 (varying insolation); refer to the profiles from Fig. 1 and 2 respectively.

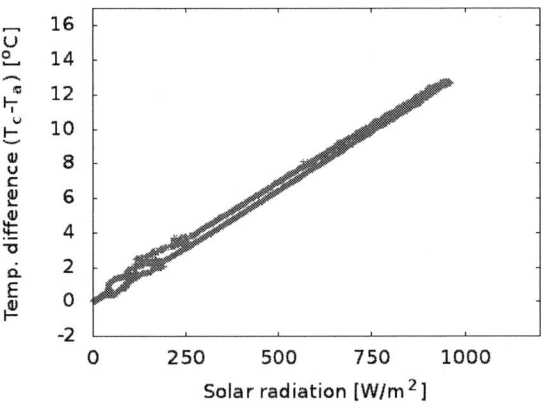

Figure 3. Module excess temperature for profile in Fig. 1.

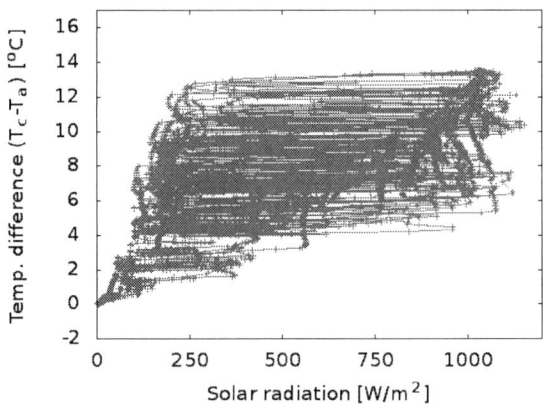

Figure 4. Module excess temperature for profile in Fig. 2.

III. LONG-TERM SIMULATION OF THE MODEL

The RC model parameters mentioned above have been extracted basing on measurements covering the period of over three weeks in September 2009. As monitoring of the PV system at DMCS is being continued, data for a longer time are available, allowing for the testing of the model.

As it was stated earlier, the temperature and radiation data are sampled every 5 seconds. This gives a large sets of values – 17280 samples per 24 hours and about 500000 samples per month (depending on a number of days in particular month).

Therefore, for the presented paper, data range has been limited to 10 months of year 2011 (from January to October).

The module temperature $T_c(t)$ is found by simulation of the RC thermal circuit from Fig. 5 with $G(t)$, $T_a(t)$ as the inputs for all samples for time moments t_i. Calculations were based on the following iterative formula:

$$T_c(t_i) - T_a(t_i) = (T_c(t_{i-1}) - T_a(t_{i-1}))$$
$$+ [\frac{-(T_c(t_{i-1}) - T_a(t_{i-1}))}{RC} + \frac{G(t_i)}{C}] * (t_i - t_{i-1}) \quad (3)$$

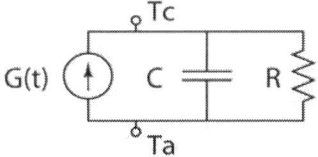

Figure 5. RC thermal circuit used in calculations.

All the necessary calculations were performed with the usage of GNU Octave software. Presented graphs were prepared with gnuplot package.

IV. RESULTS

It was not possible to present the results of the calculations directly on the graphs like Fig. 3 or 4, even for particular months – a large number of data samples made it unreadable. Therefore for each month the temperature difference ($T_c - T_a$)

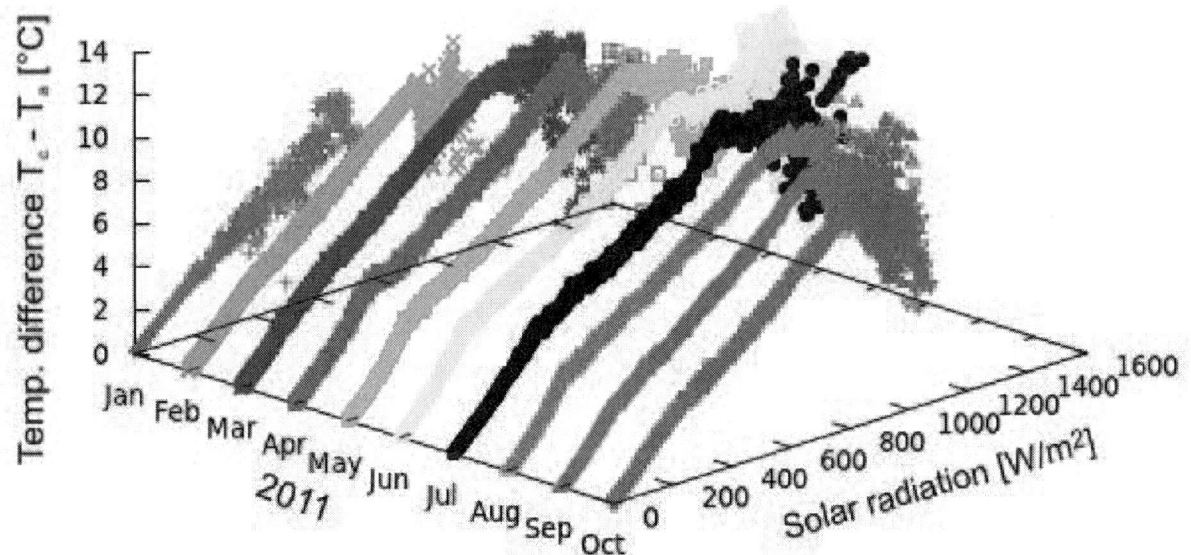

Figure 6. PV module excess temperature averaged for the individual months.

data were averaged for each value of solar radiation. According to [4], such operation does not hide the thermal inertia of the PV-module. Figure 6 presents the results of the calculations for the individual months of the examined period.

The linear dependence $T_c - T_a$ vs G up to around 850 W/m^2 in Fig. 6 reflects the fact that periods of heating and warming conditions compensate and the thermal inertia may be neglected. However, under highly variable insolation, the solar modules temperature is not rising much due to the short time of sunny periods. In effect, the modules appear as operating with lower temperature. The manifestation of a thermal inertia of the module is the negative value of the thermal resistance for this region (radiation values higher than 850 W/m^2).

Figure 7 presents the same dependency, but for the values of $T_c - T_a$ for particular values of radiation averaged for the whole period (10 months). The thermal inertia of the module is visible also in this case, which suggest the importance of that phenomenon.

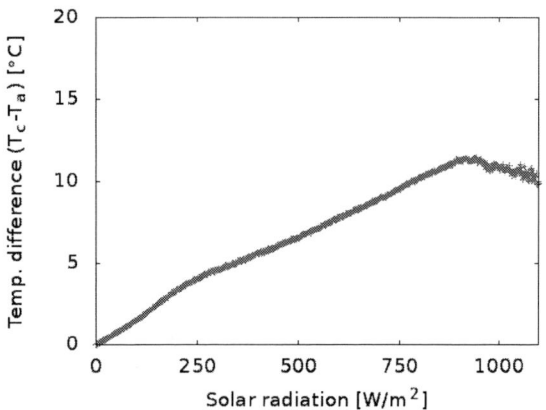

Figure 7. Module excess temperature averaged for 10 months

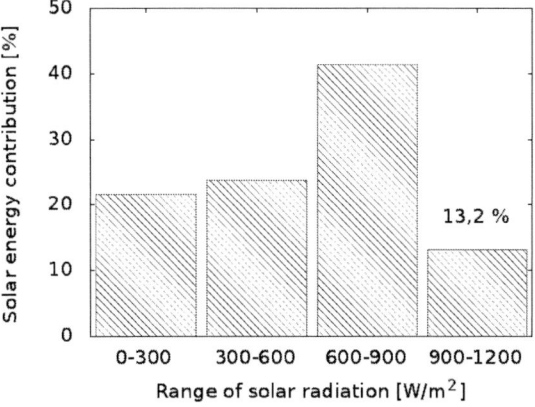

Figure 8. Distribution of received solar energy

The amount of energy delivered by solar radiation at high radiation conditions cannot be neglected. Figure 8 shows that over 13% of total solar power was received in short intervals of highest radiation in examined period. The thermal inertia effect increases the efficiency of photovoltaic modules in such conditions.

V. CONCLUSIONS

The calculation of PV-devices operation temperature is the key element for accurate energy yield predictions for PV-systems. During favourable insolation conditions (slow radiation changes) the simple NOCT-based thermal model is sufficient.

The collected data of solar radiation (with 5 s sampling rate) has revealed common presence of rapid variations in the climate of Central Europe. In such conditions a significant amount of solar energy can be delivered.

The simulation of PV-systems for highly variable solar radiation should take into account the thermal capacitance of PV-modules. The single section RC thermal model has proved to be sufficiently accurate, in contrast to NOCT-based models.

The most pronounced effect of thermal inertia may be observed for peaks of high radiation shorter than this time constant. Such radiation conditions correlate with scattered clouds, strong wind and lower ambient temperature. High solar radiation occurs only as short peaks of duration usually shorter than 1 min.

The operation of PV-systems under highly varying radiation has not drawn much attention so far due to the requirement for high time-resolution monitoring. However, such conditions are very frequent in Central Europe and this can have consequences for modelling and energy yield prediction accuracy for photovoltaic systems.

REFERENCES

[1] T. Kozak, G. De Mey, W. Maranda, and A. Napieralski, "Influence of ambient temperature on the amount of electric energy produced by solar modules", Proceedings of the 16th MIXDES Conference, Lodz 2009, Poland.

[2] A.D. Jones and C.P. Underwood, "A thermal model for photovoltaic systems", Solar Energy, Vol. 70, No. 4, pp. 349-359, 2001, Elsevier Science Ltd.

[3] M.Mattei, G. Notton, C. Cristofari, M. Muselli and P. Poggi, "Calculations ofhte policrystalline PV module temperature using a simple method of energy balance", Renewable Energy, Vol. 31, Issue 4, April 2006, pp. 553-567, Elsevier Science Ltd.

[4] W. Maranda and M. Piotrowicz, "Extraction of thermal model parameters for field-installed photovoltaic module", Proceedings of the 27th International Conference on Microelectronics MIEL2010, Nis 2010, Serbia.

Signal Processing

466

MIXDES 2012, 19th International Conference *"Mixed Design of Integrated Circuits and Systems"*, May 24-26, 2012, Warsaw, Poland

A Hierarchical Algorithm for Moving Vehicle Identification Based on Acoustic Noise Analysis

Sergei Astapov, Andri Riid

Department of Computer Control
Tallinn University of Technology
Ehitajate tee 5, 19086, Tallinn, Estonia
sergei.astapov@dcc.ttu.ee, andri@dcc.ttu.ee

Abstract—**This paper considers a multistage hierarchical algorithm of acoustic signal analysis and pattern recognition for identification of moving vehicles in an open environment. The algorithm applies several standalone techniques to enable complex decision-making during event identification. Computationally inexpensive procedures are specifically chosen in order to provide real-time operation capability. The algorithm is tested on pre-recorded audio signals of passing passenger cars and displays promising classification accuracy.**

Index Terms—**Vehicle identification, audio signal analysis, feature extraction, noise classification, fuzzy logic.**

I. INTRODUCTION

Moving object identification is one of many tasks of environment monitoring systems. The applications of moving motor vehicles identification vary from speed limit control to traffic density analysis and traffic behavior prediction. The most important aspect of such monitoring systems is real-time computation and timely result processing as the nature of the problem most often implies time-critical operation. Most state of the art systems typically rely on single sensor ultrasonic, acoustic, video, infrared, radar, microwave, magnetic, laser, vibration based, etc. signal analysis, otherwise they employ combinational multisensory detectors. The main advantage of acoustic [1]-[5] and video [6] methods lies in the ease of data signal interpretability, i.e., the acquired data is perceptual without additional manipulations.

Video based methods of vehicle identification are generally more effective and robust in changing weather conditions if provided sufficient visibility and illumination. However, the large amounts of video data and significantly more complex pattern search algorithms, if compared to algorithms for one-dimensional data streams, put significant constraints on the possibilities of real-time system implementation. Acoustic systems on the other hand do not rely on visibility factors, yet are sensitive to background acoustic noise variation. Thus the accuracy of acoustic system vehicle identification is directly dependent on its ability to distinguish the sound patterns of passing vehicle noise from a limitless amount of noises occurring in the environment.

Acoustic noise analysis provides the possibility to distinguish well separable classes of motor vehicles, such as different passenger cars and trucks. The harmonic nature of the motor noise is, however, seldom present in the vehicle

sound pattern due to the fact that motor sounds are well dampened in modern cars. This fact complemented by the Doppler Effect renders the spectral analysis based on fundamental frequency detection (e.g. [7]) ineffective.

This paper considers different methods of digital audio signal analysis, namely the estimation of spectral energy levels and energy envelope, the analysis of several frequency spectrum instantaneous features and spectral pattern matching. The proposed algorithm possesses a hierarchical structure, beginning with the detection of signal perturbation and ending with the classification of the detected vehicle. The algorithm is computationally inexpensive and thus is well implementable on an embedded sensor device. The section of the paper, devoted to system testing, proves that the algorithm is well applicable to the task of identifying motorized vehicles under varying weather conditions.

II. METHODS OF SIGNAL ANALYSIS

The audio signal is analyzed in frequency domain. The frequency domain representation of the signal is achieved by applying a temporal signal decomposing operation, namely the Fourier Transform (FT). Frequency features are less affected by noise than temporal ones, also most of the temporal features may be approximated in the frequency domain. Furthermore, the frequency spectrum of a temporal signal frame consists of half as many points as there are in the temporal frame, which is relevant in computation complexity critical systems.

A. The Fast Fourier Transform

The discrete temporal signal is decomposed by the Discrete Fourier Transform (DFT). For a finite duration discrete signal $x(m)$ of length N the DFT function is

$$X(k) = \sum_{m=0}^{N-1} x(m)e^{-j\frac{2\pi}{N}mk}, k = 0, ..., N-1. \quad (1)$$

In this manner the transform is performed along two integer dimensions: m and k, thus having quadratic complexity. In order to reduce its computation the Fast Fourier Transforms were developed. The proposed system applies a specific implementation of the FFT developed by Frigo and Johnson, called FFTW [8].

The resulting frequency spectrum $[X(0), X(1),...,X(N-1)]$ is symmetrically divided into complex conjugate "positive" and "negative" frequencies, the positive ones residing in the interval $[X(0),...X(N/2+1)]$ with $X(0)$ being the signal DC component, which is ignored. In order to obtain the absolute amplitude spectrum, the absolute values of this portion of the spectrum are calculated.

B. Instantaneous Feature Extraction

In order to acquire the specific signal properties several features are extracted from the amplitude frequency spectrum [9]. These are referred to as instantaneous features due to the fact that they are extracted from every single spectral frame independently, not relying on previous information. The list of features is signal-specific and is formed during the process of sample signal analysis in order to distinguish well separable, desirably weakly correlated features, which best indicate the nature of signal fluctuations corresponding to the concerned events. The six spectral features considered in this paper are extracted from the absolute magnitude spectrum frame $X(k)$ of length K.

Root Mean Square (RMS) Energy of the power spectrum conveys the general spectral energy level:

$$X_{RMS} = \sqrt{\frac{1}{K}\sum_{k=1}^{K}|X(k)|^2} \qquad (2)$$

The **band energy** measures the energy of the power spectrum at the ith band and is computed as

$$X_{BE}(i) = \frac{\sum_{l \in S_i}|X_t(l)|^2}{\sum_{k=1}^{K}|X_t(k)|^2}, \qquad (3)$$

where S_i is the set of power spectrum samples belonging to the ith band. The bands are chosen according to the Mel-scale denoted by

$$Mel(f) = 2595 \cdot \log_{10}\left(1 + \frac{f}{700}\right). \qquad (4)$$

The Mel-scale is chosen for its increasing spread towards the higher frequencies which ultimately means that the bands of lower frequencies, where most of the spectral energy resides, are shorter than the bands of low-energy higher frequencies. This allows for better distribution of spectral energy by bands.

The **spectral centroid** represents the first central moment of the magnitude spectrum. It is calculated as the frequency averaged over the absolute magnitude spectrum:

$$X_{SC} = \frac{\sum_{k=1}^{K}k \cdot |X_t(k)|}{\sum_{k=1}^{K}|X_t(k)|}. \qquad (5)$$

Spectral roll-off measures the frequency below which a certain amount of spectral energy resides. This amount denoted by $TH = [0,1]$ is the threshold, which in this case is chosen to be equal to $TH = 0.9$.

$$X_{SR} = \arg\max_{p}\left[\sum_{l=1}^{p}|X_t(l)|^2 \le TH \cdot \sum_{k=1}^{K}|X_t(k)|^2\right] \quad (6)$$

Spectral slope is a measure of spectral energy decrease in the direction of the higher frequencies. It is determined by the gradient and y-intersect parameters of a straight line calculated applying linear regression to the magnitude spectrum frame. Thus for a set of data points $\left(k,|X_t(k)|\right)$, where $k = 1,...,K$ the gradient of the best fitted straight line is denoted as

$$m = \frac{K\sum_{k=1}^{K}k \cdot |X_t(k)| - \sum_{k=1}^{K}k \sum_{k=1}^{K}|X_t(k)|}{K\sum_{k=1}^{K}k^2 - \left(\sum_{k=1}^{K}k\right)^2}, \qquad (7)$$

and the y-intersect is denoted as

$$c = \frac{\sum_{k=1}^{K}|X_t(k)|\sum_{k=1}^{K}k^2 - \sum_{k=1}^{K}k\sum_{k=1}^{K}k \cdot |X_t(k)|}{K\sum_{k=1}^{K}k^2 - \left(\sum_{k=1}^{K}k\right)^2}. \qquad (8)$$

In the proposed algorithm the RMS energy is used independently. On the other hand the rest of the mentioned features are concatenated into a feature vector, which is analyzed during the later stages of classification. The application of instantaneous features is further described in Section III.

C. Attack Sustain Release Envelope

The process of a vehicle passing the measurement point at any given velocity consists of three stages: approach (spectral energy increases), passing (spectral energy remains stable), retreat (spectral energy decreases). This pattern is detected by applying the Attack Sustain Release (ASR) envelope estimation. It is conducted by analyzing the RMS spectral energy denoted by (2).

The amount of deviation of RMS energy of the present frame $X_{RMS}(i)$ is estimated by the difference between it and the mean value of M previous RMS energy readings to account for noise. The parameter $\delta \in [0,1]$ is the lower threshold of energy deviation. Such, RMS energy deviation is coded to three states by the following principle:

$$state_i = \begin{cases} 1, & X_{RMS}(i) > (1+\delta) \cdot mean_{RMS}(i) \\ -1, & X_{RMS}(i) < (1-\delta) \cdot mean_{RMS}(i) , \\ 0, & \text{otherwise} \end{cases} \qquad (9)$$

where 1 denotes energy increase, 0 denotes stable energy levels and -1 denotes energy decrease. The mean of M previous energy levels is calculated by

$$mean_{RMS}(i) = \frac{1}{M}\sum_{M}^{j=i-1} X_{RMS}(j). \qquad (10)$$

Therefore the transitions $1 \rightarrow 0 \rightarrow -1$ and $1 \rightarrow -1$ are suspected for car passing event occurrence and the quantities of -1, 0, and 1 coded frames denote the lengths of attack, sustain, and release components respectively.

III. THE HIERARCHICAL ALGORITHM

The proposed hierarchical algorithm, presented in Fig. 1., consists of two independent stages. The hierarchical decision-making scheme (on the left), firstly differentiates relatively loud sounds from mild background noise, secondly distinguishes vehicle-produced sounds from heavy background noise and lastly estimates the vehicle type from a set of predefined types. This part of the algorithm operates in a frame-by-frame manner, computing a single class label per signal frame. The ASR envelope estimating procedure runs parallel to the decision-making one and complements the past frame classifications with reassurance of positive vehicle-passing event detection. The hierarchy of the algorithm is conditioned by the supremacy of vehicle detection priority over vehicle classification priority, i.e., differentiation between vehicle-produced sound and other types of noise is more important than correct vehicle type estimation. The real-time constraints for the algorithm consist of limiting the processing time of every iteration to the time duration of a single frame.

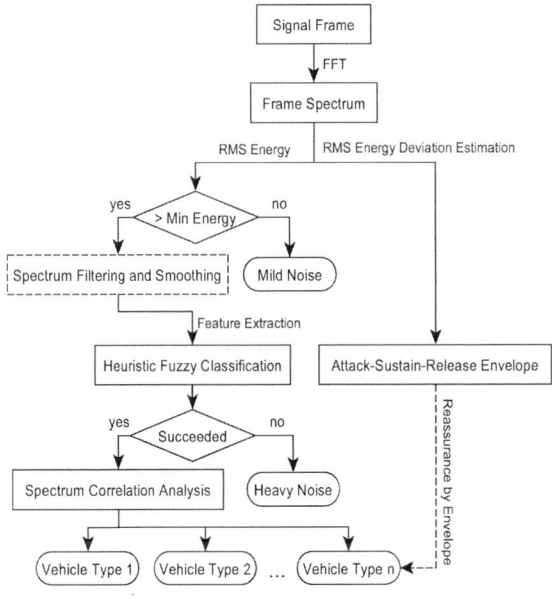

Figure 1. Block diagram of the proposed hierarchical algorithm for vehicle detection and classification.

A. Lower Energy Threshold

The first stage of the hierarchical procedure is the estimation of sufficient signal energy. The energy level of a signal frame is calculated and compared to the lower energy threshold, if the threshold is not exceeded, the procedure terminates and the frame is marked as mild noise. The estimation of the lower energy threshold occurs during algorithm parameter estimation by means of test signal analysis. The initial threshold is chosen as the minimal value of RMS energy of all the frames that correspond to vehicle passing instances.

B. Heuristic Fuzzy Classification

The sound pattern of a moving object passing a measuring device is not consistent. Due to the variance of signal energy and complex spectral shape alternations caused by engine sounds, moving object velocity and the Doppler Effect, and also the influence of background noise, the spectral features of the signal also vary to some extent. Thus the variance of a feature vector of length L produces an L-dimensional feature space, to which all the feature vectors corresponding to an event of the same class label must belong. The extent of this belonging is estimated using a fuzzy inference algorithm derived by a heuristic training procedure [10]. This fuzzy algorithm operates by applying fuzzy inference to an input feature vector. The algorithm is relatively lightweight if L does not exceed 20-30, that being the reason of applying spectral features instead of the whole spectral vector.

For a feature vector $X = [x_1,...,x_L]$ of length L, which is to be classified by assigning one of R different discrete valued labels, the heuristic fuzzy classifier rule-base consists of R rules denoted as

$$\text{Rule}_r: \text{ If } x_1 \text{ is } A_{1r} \text{ and ... and } x_L \text{ is } A_{Lr}, \\ \text{then } y \text{ belongs to class } c_r, (r = 1,...,R) \qquad (11)$$

where A_{ir} is the linguistic term of the ith input (i.e., feature vector element) associated with the rth rule and c_r is the class label assigned by the rth rule. Thus in the inference system each class of R possible classes is represented by a single rule and the area of the L-dimensional feature space corresponding to this class is represented by L linguistic terms, each quantified by a single Membership Function (MF).

The class label is assigned in a winner-takes-all manner by specifying the rule with the highest degree of activation

$$y = c_r, \arg\max_{1 \leq r \leq R}(\tau_r), \qquad (12)$$

where τ_r is the activation degree of the rth rule

$$\tau_r = \bigcap_{i=1}^{L} \mu_{ir}(x_i), \qquad (13)$$

where μ_{ir} is the MF of the linguistic term A_{ir}. The conjunction operator in this specific case is the minimum operation.

Classifier training consists of estimating the parameters of the input MFs. This is done to specify the area of the L-dimensional feature space corresponding to each class. For implementation of the classifier at hand triangle-shaped MFs are used:

$$\mu_{ir}(x_i) = \begin{cases} \dfrac{x_i - a_{ir}}{b_{ir} - a_{ir}}, & a_{ir} \le x_i \le b_{ir} \\ \dfrac{c_{ir} - x_i}{c_{ir} - b_{ir}}, & b_{ir} \le x_i \le c_{ir} \\ 0, & (x_i < a_{ir}) \vee (x_i > c_{ir}) \end{cases} \quad , \quad (14)$$

where a and c locate the base of the triangle and b locates the peak. Triangle-shaped MFs allow to thoroughly restrict the effective feature space by having linear drops to zero membership.

In the proposed hierarchical algorithm the heuristic fuzzy classification may be applied in two different manners. First a general feature space corresponding to the features of all vehicle sounds is estimated and the vehicle produced sound features are differentiated from the noise sound features. Alternatively a separate feature space for the features of each vehicle class is estimated and the features are differentiated not only between vehicle and noise sound, but also between different vehicles classes. In Section V both methods are used during hierarchical algorithm testing.

C. Correlation Coefficient Analysis

The final stage of vehicle identification is correlation analysis between the unknown amplitude spectrum vector and the reference spectrum vectors, each corresponding to a single vehicle type class. For a more rigorous classification several reference vectors per class may also be used. Correlation coefficients are simple and effective metrics for similarity estimation, however this method is very susceptible to noise. Such, a spectrum of loud background noise may correlate to any of the reference spectra enough for it to be incorrectly classified. The application of the fuzzy classifier in the previous stage of the algorithm minimizes this possibility.

During correlation analysis the correlation coefficients between the unlabeled spectrum vector $x = |X_t(k)|$ and C reference vectors of length K, $r_i = [r_i(1),...,r_i(K)]$, $i = 1,...,N$, are calculated using the following equation:

$$\rho_i = \left[K \sum_{k=1}^{K} x(k) \cdot r_i(k) - \sum_{k=1}^{K} x(k) \sum_{k=1}^{K} r_i(k) \right] \Bigg/$$
$$\left[\sqrt{ K \sum_{k=1}^{K} x(k)^2 - \left(\sum_{k=1}^{K} x(k) \right)^2 } \times \right.$$
$$\left. \sqrt{ K \sum_{k=1}^{K} r_i(k)^2 - \left(\sum_{k=1}^{K} r_i(k) \right)^2 } \right] \quad . \quad (15)$$

The correlation is defined on the interval $-1 \le \rho_i \le 1$, -1 meaning total inverse correlation, 0 specifying uncorrelated processes, and 1 meaning total direct correlation. The class label corresponding to the reference vector of maximum

correlation is declared the winner:

$$y = \arg \max_{1 \le i \le C} (\rho_i). \quad (16)$$

D. Reassurance by ASR Envelope

As it was said earlier, the detection of the ASR dynamic of signal energy complements the past identification results. If the ASR pattern is detected, an additional notification is generated and the labels assigned to the frames belonging to the time interval of this pattern are searched for the most frequent one. Additional restrictions may also be applied to the ASR envelope detection. Such if the potential velocity of the moving object is known, the lower and upper bounds for the attack, sustain or release components may be specified, so the detection is invalid if these restrictions are not met.

IV. ALGORITHM COMPLEXITY MINIMIZATION

The most time consuming operations are feature extraction and correlation coefficient calculation due to a large number of lengthy vector summations. To reduce the number of summations several feature extraction techniques were specifically chosen with similar summands. Analyzing equations (2), (3), (5) – (8), the repeating elements are $\sum_{k=1}^{K} |X_t(k)|$, $\sum_{k=1}^{K} |X_t(k)|^2$ and $\sum_{k=1}^{K} k \cdot |X_t(k)|$, the first two of which are also present in the correlation calculating equation (15). Computing these sums only once and minimizing the number of cycles during feature extraction greatly reduces the number of overall operations.

Equations (7) and (8) may be further simplified if k is taken as an integer vector index of the corresponding frequency component. Such the sum of K first successive integers is then equal to

$$\sum_{k=1}^{K} k = \frac{1}{2} K (K+1), \quad (17)$$

and the sum of squares of K first successive integers is

$$\sum_{k=1}^{K} k^2 = \frac{1}{6} K (K+1)(2K+1). \quad (18)$$

Calculating the sums of reference vectors and the sums of squared reference vectors only once during the off-line stage of algorithm parameter specification turns (15) to a lightweight equation with only one specific summation, which must be performed for each correlation coefficient calculation:

$$\sum_{k=1}^{K} x(k) \cdot r_i(k).$$

Using a power of two as the signal frame length also reduces computation complexity due to the fact that in this case many multiplications and divisions are replaced by simpler and faster bitwise arithmetic shifts.

V. ALGORITHM TESTING RESULTS

A. First Test Signal

The first test audio signal was measured using a Shure SM58 microphone and digitized using a Roland Edirol UA-25EX audio signal processor at 44.1 kHz sampling rate in mono channel mode and saved in a 16-bit Waveform Audio File (WAV) format. For the acquisition of the test signal a microphone was placed at an empty parking lot and two cars (Mercedes S320 and Mazda MX-5) were in turn passing the microphone stand at a speed of 35 – 45 km/h at the passing point, starting to accelerate from a distance of approximately 40 meters. Each car has overall passed the microphone three times: the Mercedes first three passes and the Mazda second three passes. The sounds were acquired during summer time in mild weather conditions.

For testing, the frame length of $2^{14} = 16384$ samples is chosen, which corresponds to 0.3715 seconds at a sampling rate of 44.1 kHz. The signal feature vector comprises of eight features: four band energy features (four bands of 1-824, 824-2616, 2616-6514, 6514-15000 Hz), spectral centroid, spectral roll-off and spectral slope. In total 2 class labels are used: 1 for Mercedes and 2 for Mazda. The reference spectral vectors used in correlation analysis are estimated by averaging several spectra of sounds produced by vehicles of the same class, in total one reference vector per class is applied.

The results of algorithm testing are presented in Fig. 2. The general results are satisfying – every vehicle passing instance is detected and successfully classified. As it can be seen on the second and third subplots of Fig. 2., each vehicle passing instance ASR envelope is correctly detected. Though it can be noticed that approximately on the 107th frame the ASR dynamic is falsely detected, however the energy of the signal is below the threshold and the fuzzy classifier fails classification, consequently the detection does not occur. Also approximately on the 145th frame the ASR dynamic is present, but is not detected, as for the known vehicle speed corridor the attack and release components of the envelope are set to be no less than 2 frames in duration for the dynamic to be detected.

The heuristic fuzzy algorithm, trained to identify the general vehicle feature space, succeeds in doing so for the majority of signal frames thus allowing the correlation coefficient calculation procedure to analyze only the frames corresponding to vehicle pass time intervals. The fourth subplot of Fig. 2. shows, that the correlation coefficient values are unreliable during the periods between vehicle passing instances, but during these instances they become more separate, indicating an obvious leader.

B. Second Test Signal

The second test signal was acquired using a miniature condenser microphone Sennheiser KE 4-211-2 and an embedded computing device Gumstix Overo Water (600MHz, 256MB RAM, 4GB microSD). The signal was also sampled at 44.1 kHz mono channel mode and saved to a 16-bit WAV file.

Signal acquisition was conducted at a lively two-lane highway during dense traffic in late fall under heavy wind and light rain.

The frame length was chosen the same as for the first test signal. For the vehicle classes two were chosen: 1 for passenger cars and 2 for trucks and busses. Feature vectors comprise of eight features, which are the same as for the first signal, except the bands for the band energy features are less spread: 220-818, 818-2592, 2592-6438, 6438-14780 Hz. For the acquisition of reference spectral vectors the same technique as before is used.

The results of signal analysis are presented in Fig. 3. As the time intervals between car passes are very short and often non-existent altogether, reference class labels, which are also used during fuzzy algorithm training, are introduced in the first subplot. The results are as follows: out of 46 instances of class 1 vehicles, 37 were successfully detected and classified, 5 were undetected and 4 were confused with class 2; for 11 instances of class 2 vehicles, 9 were correctly classified, 1 was not detected and 1 confused with class 1. Thus the classification accuracy for class 1 vehicles is 80.43% and for class 2 – 81.82%. The vague differences between passenger and heavy cars of some vehicles lower the classification quality. Furthermore a heavy truck can emit a noise loud enough for it to mask the sound of a nearer lighter car thus making it undetectable.

C. General Testing Results

The algorithm operates well in both the cases of motor vehicles passing with a certain time interval between the passes and heavy traffic. Though, if the flow of vehicles is consistent and very dense, the decrease of identification quality is witnessed. The influence of background noise, such as wind, is reduced due to the algorithm's multistage decision-making logic. Thus the algorithm is applicable under different weather conditions.

Providing a variety of tunable parameters, the sensitivity of the algorithm is proven to be well adjustable to the needed extent. This provides the opportunity to apply the algorithm for classification of various types of moving objects not limited to motorized vehicles.

VI. CONCLUSION

In this work we have introduced a hierarchical algorithm for moving vehicle identification by means of acoustic noise analysis. The algorithm is developed specifically for real-time application and is therefore computationally inexpensive and simple in computation. Algorithm testing results indicate the algorithms potency in the task of detecting and classifying motor vehicles.

For future developments algorithm robustness may be increased by applying soft discretization to the transitions of the algorithm decision-making path [11] thus transforming its appearance to a fuzzy tree. The final class label therefore may be decided based on degrees of membership.

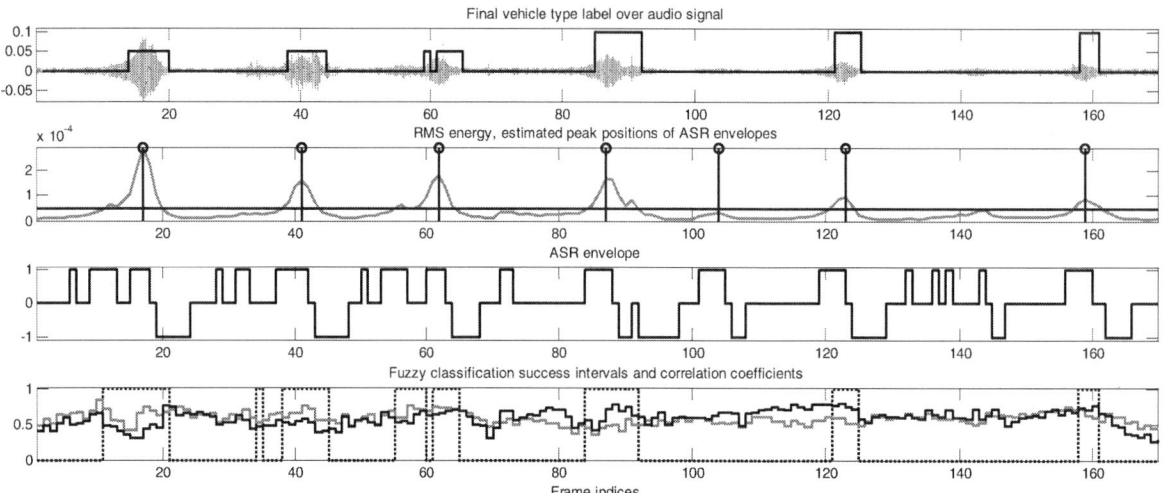

Figure 2. Algorithm testing results, first test signal. From top to bottom, First Subplot: test signal with 6 instances of passing vehicles (grey); final estimated labels with values 0.05 corresponding to class 1 and 0.1 – to class 2 (black); Second Subplot: RMS energy readings per frame (grey); signal energy threshold (black horizontal line); energy peaks approximated by ASR envelope (black stems); Third Subplot: coded RMS energy dynamic of the ASR envelope; Fourth Subplot: intervals of positive fuzzy membership to vehicle feature subspace (dotted vertical lines); coefficients of correlation to the reference spectral vectors (grey – class 1, black – class 2).

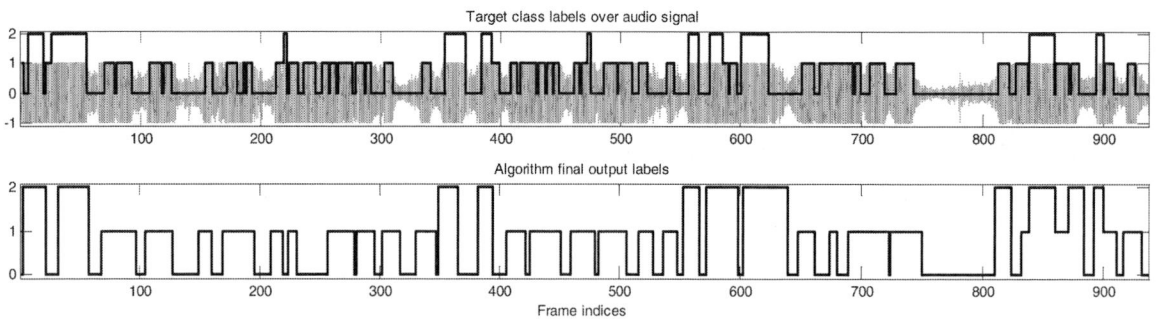

Figure 3. Algorithm testing results: second test signal. From top to bottom, First Subplot: test signal with instances of passing vehicles (grey), reference labels with values 1 corresponding to class 1 and 2 – to class 2 (black); Second Subplot: final estimated labels with values 1 corresponding to class 1 and 2 – to class 2.

REFERENCES

[1] T. Takechi, K. Sugimoto, T. Mandono and H. Sawada, "Automobile identification based on the measurement of car sounds," Proc. 30th Annual Conference of IEEE. IECON 2004, vol. 2, pp. 1784- 1789, November 2004.

[2] A. Starzacher and B. Rinner, "Single Sensor Acoustic Feature Extraction for Embedded Realtime Vehicle Classification," Proc. International Conference on Parallel and Distributed Computing, Applications and Technologies, pp. 378-383, December 2009.

[3] N. A. Rahim, M. P. Paulraj, A. H. Adom, and S. Sundararaj, "Moving vehicle noise classification using backpropagation algorithm," Proc. 6th International Colloquium on Signal Processing and Its Applications (CSPA) 2010, pp. 1-6, May 2010.

[4] S. Maithani and R. Tyagi, "Noise Characterization and Classification for Background Estimation," Proc. International Conference on Signal Processing, Communications and Networking, pp. 208-213, January 2008.

[5] S. S. Yang, Y. G. Kim and H. Choi, "Vehicle identification using wireless sensor networks," Proc. IEEE SoutheastCon, 2007, pp. 41-46, March 2007.

[6] G. Gritsch, N. Donath, B. Kohn and M. Litzenberger, "Night-time vehicle classification with an embedded, vision system," Proc. 12th International IEEE Conference on Intelligent Transportation Systems, 2009. ITSC '09, pp. 1-6, October 2009.

[7] M. Zivanovic, A. Roebel and X. Rodet, "Adaptive threshold determination for spectral peak classification," Proc. 10th International Conference on Digital Audio Effects, September 2007.

[8] M. Frigo and S. G. Johnson, "FFTW: an adaptive software architecture for the FFT," Proc. IEEE International Conference on Acoustics, Speech and Signal Processing, vol. 3, pp. 1381-1384, May 1998.

[9] G. Peeters, "A large set of audio features for sound description (similarity and classification) in the CUIDADO project," CUIDADO I.S.T. Project Report, 2004.

[10] A. Riid and E. Rustern, "An integrated approach for the identification of compact, interpretable and accurate fuzzy rule-based classifiers from data," Proc. 15th IEEE International Conference on Intelligent Engineering Systems (INES), pp. 101-107, June 2011.

[11] Y. Peng and P. Flach, "Soft Discretization to Enhance the Continuous Decision Tree Induction," Proc. Integrating Aspects of Data Mining, Decision Support and Meta-Learning, pp. 109–118, September 2001.

Application of the Newton Method to First-order Implicit Fractional Transfer Function Approximation

Aleksei Tepljakov, Eduard Petlenkov, and Juri Belikov

Department of Computer Control
Tallinn University of Technology
Ehitajate tee 5, 19086, Tallinn, Estonia
{aleksei.tepljakov, eduard.petlenkov, juri.belikov}@dcc.ttu.ee

Abstract—In this paper, a method for approximating a first-order implicit fractional transfer function, that corresponds to a frequency-bounded fractional differentiator or integrator, is presented. The proposed method is based on the well-known Newton's method for iterative root approximation. First-order implicit fractional transfer functions have several applications in modeling and control. This type of transfer function is the basis for the fractional lead-lag compensator. In the following, we provide the description of our algorithm, that enhances the existing technique, and illustrate its use in modeling and control with relevant comments.

Index Terms—fractional calculus, Newton's method, Carlson's method, Matlab, implicit fractional transfer function, fractional power zero-pole

I. INTRODUCTION

Today, fractional-order calculus is a rapidly evolving scientific field. It allows for more accurate modeling of complex systems, such as those that possess memory and hereditary properties [1]. The benefits of using fractional calculus in control are also evident. New types of controllers have been developed [2], [3], [4] based on the added flexibility of the fractional-order models.

However, many problems arise in the implementation of fractional-order controllers. Since fractional models are inherently complex, which follows from the fact that they describe infinite-dimensional systems [5], deducing an effective direct realization method is a difficult task. Therefore, methods for approximating the fractional operators have been developed, including both continuous and discrete approximations.

In this paper, we focus on one particular continuous integer-order approximation, derived from Newton's method. In Section II the underlying method and its modification are summarized and applications to fractional-order modeling are presented. In Section III our method used for implicit first-order fractional transfer function approximation is proposed and discussed. Applications of this method to controller implementation are also presented and a MATLAB realization is described. Illustrative examples follow in Section IV. Some issues and limitations of the proposed method are discussed in Section V. Finally, conclusions are drawn in Section VI.

II. APPLICATION OF NEWTON'S METHOD TO FRACTIONAL CALCULUS

Newton's method, also known as Newton-Rhapson method [6], is a numerical algorithm for finding a real root of a function $f(x)$. It suggests that in order to solve a general nonlinear equation $f(x) = 0$, the following iterative formula can be used, given an initial estimate x_0:

$$x_{k+1} = x_k - \frac{f(x_k)}{f'(x_k)}. \tag{1}$$

A modified algorithm is proposed in [7], [8] such that the convergence of the sequence $\{x_k\}$ is more rapid than that resulting from using formula (1). The corresponding formula is called Halley's formula:

$$x_{k+1} = x_k - \frac{f(x_k)}{f'(x_k) - \frac{f(x_k)f''(x_k)}{2f'(x_k)}}. \tag{2}$$

Consider now a problem of finding an nth root of a real number. The corresponding function is $f(x) = x^n - A$ and using (2) the following particular iteration formula is obtained:

$$x_{k+1} = x_k \cdot \frac{(n-1)(x_k)^n + (n+1)A}{(n+1)(x_k)^n + (n-1)A}. \tag{3}$$

This formula is considered by Carlson [9] and more recently in [5], [10], [11]. In his paper, Carlson has shown, that this formula holds for both even $n = 2m$ and odd $n = 2m + 1$ roots. The method can be applied to approximation of fractional capacitors of the form $(1/s)^{1/n}$ in the following way:

$$G_{k+1}(s) = G_k(s)\frac{(n-1)\left(G_k^n(s)\right) + (n+1)\left(H(s)\right)}{(n+1)\left(G_k^n(s)\right) + (n-1)\left(H(s)\right)}, \tag{4}$$

$$H(s) = \frac{1}{s}, \quad G_0(s) = 1.$$

Since in this case the real variable A is replaced by the transfer function $H(s)$, convergence and rate of convergence cannot be evaluated in the same way as in the case of a real-valued function.

Consider now an example. Using equation (4) we shall obtain an approximation of a fractional capacitor $\sqrt[5]{1/s}$. With

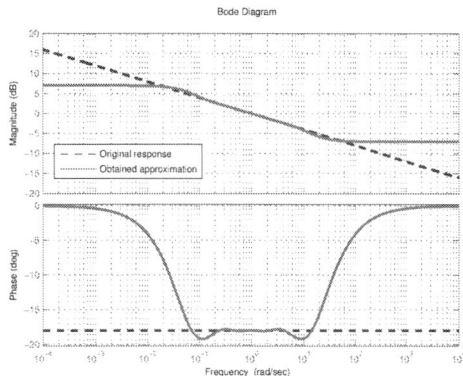

Fig. 1. Frequency response of Carlson's approximation of the fractional capacitor $\sqrt[5]{1/s}$

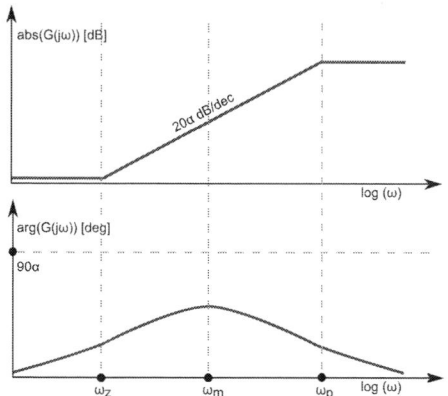

Fig. 2. Bode diagram corresponding to a fractional power pole-zero pair transfer function frequency response with $\alpha > 0$ and $b > a$

two iterations the following transfer functions are obtained:

$$G_0(s) = 1,$$
$$G_1(s) = \frac{0.66667s(s + 1.5)}{s(s + 0.6667)},$$
$$G_2(s) = \frac{G_{21}(s)}{G_{22}(s)},$$

where

$$G_{21}(s) = 0.4444s^7 + 9.062s^6 + 39.47s^5 + 77.81s^4$$
$$+ 82.5s^3 + 47.75s^2 + 13.23s + 1,$$
$$G_{22}(s) = s^7 + 13.23s^6 + 47.75s^5 + 82.5s^4$$
$$+ 77.81s^3 + 39.47s^2 + 9.063s + 0.4444.$$

In Fig. 1 the frequency response of the obtained approximation is shown. The response of the corresponding ideal fractional capacitor is also given for comparison. It can be seen, that the frequency range where the approximation is valid is quite narrow. It is possible to improve this result by increasing the number of formula iterations. However, in this case the order of the obtained rational transfer function may be very high.

Thus, the method may not be very effective for approximating fractional differentiators and integrators on a wide frequency range. However, the method may be used for frequency-bounded implicit fractional transfer function approximation. In this work, we treat the case of a first-order fractional transfer function.

III. APPROXIMATION METHOD FOR FIRST-ORDER IMPLICIT FRACTIONAL TRANSFER FUNCTIONS

A. First-order Implicit Fractional Transfer Function in Modeling and Control

In general, a frequency-bounded non-integer differentiator/integrator may be represented by a first-order implicit fractional transfer function in the form

$$G(s) = \left(\frac{bs + 1}{as + 1}\right)^{\alpha}, \tag{5}$$

where $0 < \alpha < 1$. The frequency of the zero is in this case $\omega_z = 1/b$ and the frequency of the pole is $\omega_p = 1/a$, when $\alpha > 0$. Following the terminology in [12] and since in this case the transfer function has a single fractional power zero and a single fractional power pole, we also refer to this form as a Fractional Power Zero-Pole (FPZP) pair.

The benefits of using fractional calculus in modeling are most evident when analyzing the frequency behavior of the resulting models. A bode diagram corresponding to (5) with $\alpha > 0$ and $b > a$ is given in Fig. 2. It can be seen, that by varying α a magnitude slope of 20α dB/dec and a phase of $90\alpha°$ can be achieved, which allows for more intricate modeling possibilities. This additional freedom is also very important in control design. In particular, the transfer function in (5) corresponds to the fractional part of the fractional lead-lag compensator — a generalization of the conventional controller used in many industrial applications [13], [14]. The fractional lead-lag compensator has the following form [15]:

$$C(s) = K_c x^{\alpha} \left(\frac{\lambda s + 1}{x \lambda s + 1}\right)^{\alpha}, \quad 0 < x < 1, \tag{6}$$

where $\lambda = b$ and $x\lambda = a$ in (5) and $K_c x^{\alpha}$ is the controller gain.

We now describe the algorithm, which can be used to obtain accurate approximations in form of zero-pole distributions for the fractional transfer function in (5).

B. Approximation Algorithm

Based on the previous discussion, several problems of the original algorithm in [9] may be outlined:

- the initial estimate for approximation problem is not addressed,
- the method only allows to obtain approximations for transfer functions of order $1/n$,
- resulting approximations can be of a very high order,
- the limited frequency range where the approximation is valid.

The specific application of Carlson's method could be different. In fact, when applied to the problem of approximating the transfer function in (5) for a limited frequency range, the algorithm provides very accurate results. Further, we describe the refined algorithm, which aims to solve the aforementioned problems.

First, we consider the initial estimate problem. Using the iteration formula (4) results in a recursive distribution of zeros and poles around a central frequency. In case of the fractional power zero-pole pair transfer function, this frequency is the geometric mean computed from the zero and pole frequencies such that

$$\omega_m = \sqrt{\omega_z \omega_p} = \frac{1}{\sqrt{ab}}. \tag{7}$$

It relates to the initial estimate choise through the magnitude of the fractional transfer function obtained at this frequency:

$$G_0(\omega) = |G(j\omega_m)| = \left| \frac{jb\omega_m + 1}{ja\omega_m + 1} \right|^\alpha. \tag{8}$$

When selecting the initial estimate according to (8) the resulting zero-pole distribution is then centered around ω_m ensuring that way the validity of the approximation around this frequency. When the ratio a/b is small, only two iterations are usually required to achieve a good result in the full frequency range.

The problem of approximating transfer functions of arbitrary real order using this method is much more difficult to solve. Here, we must choose a balance between accuracy and efficiency, since in case of order $1/n$ each iteration step involves computing the nth power of a transfer function obtained in the previous step. The order of the approximation grows rapidly. Thus, until a different, more efficient iteration formula is developed, we limit the resolution to $1/10$. This allows to obtain approximations of orders accurate to at least one decimal place. However, there is no reason why a class of arbitrary orders could not be considered as well.

The problem of using the method to obtain an approximation for an arbitrary real α falls under the Egyptian fraction decomposition class of problems, i.e. an order α is decomposed into k simple fractions $1/m_k$:

$$\alpha = \frac{1}{m_1} + \frac{1}{m_2} + \cdots + \frac{1}{m_k}, \tag{9}$$

where $m_k \in \mathbb{N}$. The order decomposition algorithm is depicted in Fig. 3 and is discussed below.

The optimized decomposition is conducted using fractions $1/2$ (most efficient), $1/5$ and $1/10$ (accuracy consideration). The decimal fractions are then decomposed as follows:

$$0.1 = \tfrac{1}{10}, \quad 0.2 = \tfrac{1}{5}, \quad 0.3 = \tfrac{1}{5} + \tfrac{1}{10},$$

$$0.4 = 2 \cdot \tfrac{1}{5}, \quad 0.5 = \tfrac{1}{2}, \quad 0.6 = 3 \cdot \tfrac{1}{5},$$

$$0.7 = \tfrac{1}{2} + \tfrac{1}{5}, \quad 0.8 = 4 \cdot \tfrac{1}{5}, \quad 0.9 = 4 \cdot \tfrac{1}{5} + \tfrac{1}{10}.$$

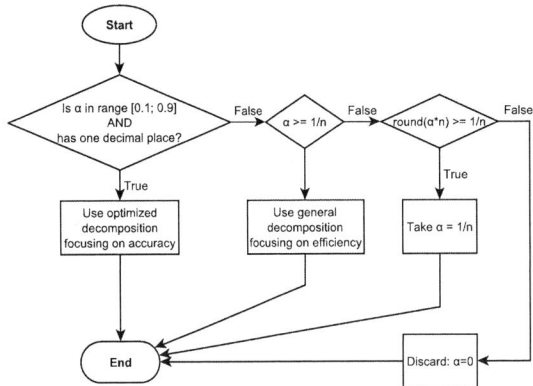

Fig. 3.   Order α decomposition algorithm

for $P = 2$ to M **do**
 if $\alpha \geq (1/P)$ **then**
  $G \leftarrow G \cdot G^{1/P}$
  $\alpha \leftarrow \alpha - (1/P)$
 end if
end for

Fig. 4.   General decomposition algorithm

The fractional transfer function is then approximated as

$$G^\alpha(s) = \prod_{j=1}^{k} G_{base}^{\frac{1}{m_j}}(s), \tag{10}$$

where

$$G_{base}(s) = \frac{bs + 1}{as + 1}. \tag{11}$$

Note, that the initial estimate is computed for every approximation of $G_{base}^{1/m_j}(s)$.

The general decomposition algorithm is given in Fig. 4. In our case $M = 10$ and thus for $0 < \alpha < 1$ a decomposition will always be found, since

$$\sum_{k=2}^{10} \left(\frac{1}{k} \right) > 1.$$

For $\alpha > 1$ the general commutative property of a fractional operator is considered, so the approximation is found such that

$$G^\alpha(s) = G^n(s) \cdot G^\gamma(s), \tag{12}$$

where $n = \alpha - \gamma$ denotes the integer part of α and $G^\gamma(s)$ is obtained using (10). For the case when $\alpha < 0$, the approximation is

$$G^{-\alpha}(s) = \left(\frac{1}{G(s)} \right)^\alpha. \tag{13}$$

Finally, we address the problem of approximation order. We propose two possibilities for order reduction:
1) Reduction of matching zeros and poles;
2) Applying a balancing reduction technique, e.g. [16].

The first method may be invoked on each step of iteration when the order α is small to improve performance. The second method can be applied to the resulting approximation.

We conclude this section by noting the similarities in the approaches to realization of the fractional transfer function in (5) found in this work and in [12], [17]. Also, a similar implementation can be found in [11].

Therefore, it is possible to obtain the fractional differentiator/integrator approximations in the desired frequency range $\omega = [\omega_z; \omega_p]$ by selecting $b = \frac{1}{\omega_z}$, $a = \frac{1}{\omega_p}$ and using the following equation:

$$s^\alpha \approx a^\alpha G(s), \tag{14}$$

where $\alpha > 0$ corresponds to a fractional-order differentiator, $\alpha < 0$ corresponds to a fractional-order integrator and $G(s)$ is the approximation obtained using the above algorithm.

C. Implementation in MATLAB

Hereafter, we provide a description of the function, in which the algorithm is implemented. The function is part of the FOMCON toolbox [18]. The calling sequence is the following:

```
[G,J,err]=fpzp_new(b,a,alpha,N,w,retol)
```

Input arguments:

- b, a, and α — parameters in (5);
- N — desired approximation order (default $N = 2$);
- ω — frequency range in rad/s, used for approximation validation (default $\omega = [0.0001; 10000]$ rad/s);
- $retol$ — matching pole-zero pair reduction tolerance (empty by default).

Output arguments:

- G — the resulting integer-order approximation;
- J — error index, used to assess approximation quality, computed in the following way

$$J = \frac{1}{n_\omega} \sum_{i=1}^{n_\omega} \left| G(j\omega_i) - \hat{G}(j\omega_i) \right|^2,$$

where n_ω is the number of frequencies in ω, $G(j\omega_i)$ is the response of the original fractional transfer function at frequency ω_i and $\hat{G}(j\omega_i)$ is the response of the obtained approximation at the same frequency;

- err — order error, the fraction decomposition residue of the fractional order α.

Consider now two illustrative examples.

IV. EXAMPLES

A. Example 1

We shall obtain an approximation for the following implicit transfer function:

$$G_1(s) = \left(\frac{0.137s + 1}{15.294s + 1} \right)^{-1.115}.$$

In order to do this, the following MATLAB command could be used:

```
[G1,J,err]=fpzp_new(0.137,15.294,-1.115,2)
```

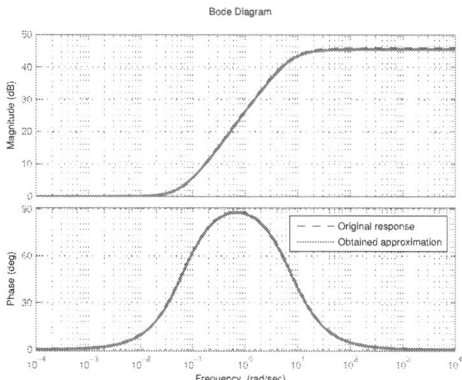

Fig. 5. $\hat{G}_1(s)$ approximation frequency response vs. $G_1(s)$ ideal frequency response

The resulting performance index is $J = 6.0065$ and order error is $\epsilon = 0.0039$. The comparison of the ideal response and the response of the obtained approximation is given in Fig. 5.

B. Example 2

In this example we will implement a fractional lead compensator, discussed in [15]. Consider a transfer function that describes a position servo:

$$G_2(s) = \frac{1.4}{s(0.7s + 1)} e^{-0.05s}.$$

Based on some performance specifications (phase margin $\varphi_m = 80°$ and gain crossover frequency $\omega_{cg} = 2.2$ rad/s), the controller was proposed such that

$$C_1(s) = \left(\frac{2.0161s + 1}{0.0015s + 1} \right)^{0.702}.$$

In order to obtain a rational approximation of this controller, the following MATLAB commands can be employed:

```
C1=fpzp_new(2.0161,0.0015,0.702,3)
```

The resulting performance index $J = 0.8115$ and error is $\epsilon = 0.002$. However, the order of the approximated model is 328. Applying the `minreal()` function results in a system of 56th-order. One may use the balancing reduction technique with the MATLAB function `balred()` of the Control System toolbox in the following manner:

```
C1=balred(C1, 5)
```

The performance index is now $J = 0.8289$, and the reduced model of order 5 is still a very good approximation of the fractional transfer function. The resulting control system open-loop frequency response $C_1(j\omega)G_2(j\omega)$ is shown in Fig. 6. It can be seen, that the design specifications are correctly fulfilled.

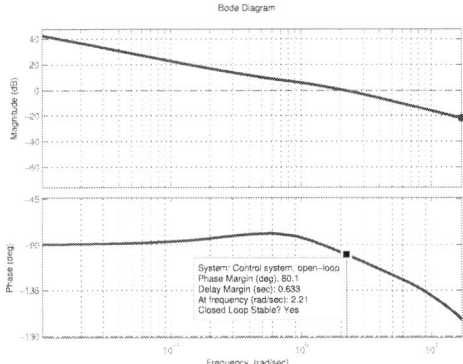

Fig. 6.   Control system open-loop frequency response

V. DISCUSSION

In the following, we list the limitations of the proposed approximation method.

- Limited resolution, currently fixed at $1/10$;
- Unequal distribution of computational complexity with different orders;
- High order of the resulting approximation, which may not be practical in certain situations.

It is also important to consider other existing methods, that can be applied to the same approximation problem, for example [12], [17].

VI. CONCLUSIONS

In this paper, we presented a method, derived from the Newton root-finding method, for approximating a first-order implicit fractional-order transfer function. We have shown, that although the method has limitations, it can be applied to solving a class of modeling and control problems. Further research should be devoted to derivation of an alternative iteration formula so that the expensive operations of taking an nth power for fractional orders $1/n$ could be avoided, thus enhancing the resolution of the approximated order. It is also natural to expect, that this method, given correct treatment, could be applied to approximation of more complex systems.

ACKNOWLEDGMENT

The work was partially supported by the European Union through the European Regional Development Fund and the target project SF0140113As08, the Estonian Science Foundation Grant no. 8738, the Estonian Doctoral School in Information and Communication Technology under interdisciplinary project FOMCON and by European Social Fund's Doctoral Studies and Internationalisation Programme DoRa.

REFERENCES

[1] I. Podlubny, *Fractional differential equations*. ser. Mathematics in science and engineering.   Academic Press, 1999.

[2] I. Podlubny, L. Dorcak, and I. Kostial, "On fractional derivatives, fractional-order dynamic systems and PI$^\lambda$D$^\mu$-controllers," in *Proc. 36th IEEE Conf. Decision and Control*, vol. 5, 1997, pp. 4985–4990.

[3] I. Podlubny, "Fractional-order systems and PI$^\lambda$D$^\mu$-controllers," *IEEE Trans. Autom. Control*, vol. 44, no. 1, pp. 208–214, 1999.

[4] A. Oustaloup, P. Melchior, P. Lanusse, O. Cois, and F. Dancla, "The CRONE toolbox for Matlab," in *Proc. IEEE Int. Symp. Computer-Aided Control System Design CACSD 2000*. 2000, pp. 190–195.

[5] B. M. Vinagre, I. Podlubny, A. Hernández, and V. Feliu, "Some approximations of fractional order operators used in control theory and applications." *Fractional Calculus & Applied Analysis*, vol. 3. pp. 945–950, 2000.

[6] F. Hildebrand, *Introduction to numerical analysis*, ser. International series in pure and applied mathematics.   McGraw-Hill, 1956.

[7] H. S. Wall, "A modification of newton's method," *The American Mathematical Monthly*, vol. 55, no. 2, pp. pp. 90–94, 1948.

[8] J. F. Traub, "Comparison of iterative methods for the calculation of nth roots," *Commun. ACM*, vol. 4, pp. 143–145, March 1961.

[9] G. Carlson and C. Halijak, "Approximation of fractional capacitors $(1/s)^{1/n}$ by a regular newton process," *IEEE Trans. Circuit Theory*, vol. 11, no. 2, pp. 210–213, 1964.

[10] I. Podlubny, I. Petráš, B. M. Vinagre, P. O'Leary, and L. Dorčák. "Analogue realizations of fractional-order controllers." *Nonlinear Dynamics*, vol. 29, pp. 281–296, 2002.

[11] D. Valério. (2005) Toolbox ninteger for MatLab, v. 2.3. [Online]. Available: http://web.ist.utl.pt/duarte.valerio/ninteger/ninteger.htm

[12] A. Charef, H. H. Sun, Y. Y. Tsao, and B. Onaral, "Fractal system as represented by singularity function," *IEEE Trans. Autom. Control*, vol. 37, no. 9, pp. 1465–1470, 1992.

[13] C. A. Monje, B. M. Vinagre, A. J. Calderon, V. Feliu, and Y. Q. Chen, "Auto-tuning of fractional lead-lag compensators," in *Proceedings of the 16th IFAC World Congress*, 2005.

[14] C. Monje, B. Vinagre, V. Feliu, and Y. Chen, "Tuning and auto-tuning of fractional order controllers for industry applications," *Control Engineering Practice*, vol. 16, no. 7, pp. 798–812, 2008.

[15] C. A. Monje, Y. Chen, B. Vinagre, D. Xue, and V. Feliu, *Fractional-order Systems and Controls: Fundamentals and Applications*, ser. Advances in Industrial Control.   Springer Verlag, 2010.

[16] A. Varga, "Balancing free square-root algorithm for computing singular perturbation approximations," in *Proc. 30th IEEE Conf. Decision and Control*, 1991, pp. 1062–1065.

[17] A. Oustaloup, F. Levron, B. Mathieu, and F. M. Nanot, "Frequency-band complex noninteger differentiator: characterization and synthesis," *IEEE Trans. Circuits Syst. I*, vol. 47, no. 1, pp. 25–39, 2000.

[18] A. Tepljakov, E. Petlenkov, and J. Belikov, "FOMCON: Fractional-order modeling and control toolbox for MATLAB," in *Proc. 18th Int Mixed Design of Integrated Circuits and Systems (MIXDES) Conference*, 2011, pp. 684–689.

MIXDES 2012, 19ᵗʰ International Conference *"Mixed Design of Integrated Circuits and Systems"*, May 24-26, 2012, Warsaw, Poland

Combining Sound Source Tracking Algorithms Based on Microphone Array to Improve Real-Time Localization

Christian Ibala
Department of Electronics and computer Eng
University of Limerick
Limerick, Ireland
sibala@acm.org

Julien Vachaudez, Georgios Fourtounis, Paulo Possa,
Carlos Valderrama
Department Electronic and Microelectronics
Polytechnic Faculty of Mons
Mons, Belgium
{julien.vachaudez,carlos.valderrama}@umons.ac.be

Abstract—**In this paper we present a novel approach to reduce the computation burden of a Field Programmable Gate Array based beamforming localization algorithm. This approach combines GCC (Generalize Cross Correlation) and DSB (Delay and Sum Beamforming). In this work, we reduce the position search spectrum by detecting the direction of arrival of the sound source before the computation of its position, this reduce the DSB computation volume by at least 50% without affecting the localization precision in the speech extended bandwidth.**

Index Terms—**DSB, Reconfigurable Platform, Localization, Voice detection**

I. INTRODUCTION

The auditory system of living creatures provides a vast amount of information about the world, such as localization of sound sources [1].

Using several closely positioned microphones, which is called a microphone arrays, allow listening to the sound coming from one direction, while reducing noise and interference sound coming from other directions. This signal processing technique is called beamforming. It can be used to both, locate the sound source and concentrate the sensors beam to that particular direction [2] as it is shown on Figure 1. A sound source in front of microphones 3 and 4 is located at a distance of 7 meters above the linear 8-microphone array with an angle of 98.13 degrees from the center of the array. The Scale at the far right of Figure 1 represents the Steered Response Power (SRP) that will be detailed later.

The sequential implementation of a beamforming algorithm supporting many microphones over a wide band of frequencies presents a significant computational challenge in real-time processing. Our contribution in this work is to combine two well-known algorithms, the GCC (Generalize Cross Correlation) and the DSB (Delay and Sum Beamforming), in order to accelerate DSB computation by restraining the search area.

The remainder of this paper will be organized as follow: In section 2 we present beamforming algorithms. In section 3 we formulate the problem and explain our contribution. In section 4 we present the algorithm hardware and software. In section 5

we discuss results and resources utilization. In section 6 we advices on further work and conclude this work.

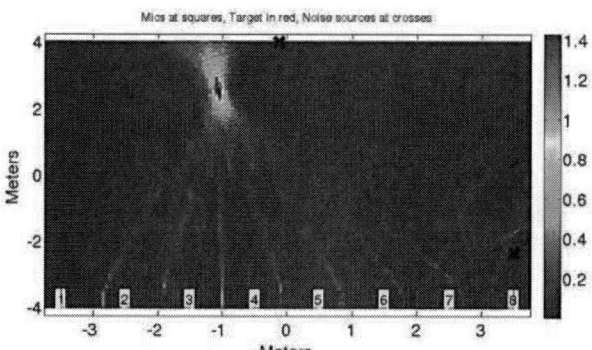

Figure 1. Sound source localization

II. BEAMFORMING RELATED WORK

A. Beamforming principles

Microphone arrays have been studied for more than three decades. One of the most important functionalities of microphone arrays is to extract the speech of interest from its observation corrupted by noise, reverberation, and competing sources. This is done by aiming the beam towards the desired sound source. Beamforming basic idea is to sum up the contribution of each microphone; as a result, signals from this so called "look-direction" are reinforced while signals from all the other directions are attenuated [3].

The response of a linear array of N sensors, with a uniform inter-element spacing d, is known as directivity pattern D. D is a function of the direction φ and frequency f. The far-field directivity pattern is given by equation (1):

$$D(f,\phi) = \sum_{n=0}^{N} w_n(f) \cdot e^{\frac{j2\pi nd \cos \phi}{\lambda}} \qquad (1)$$

- $w_n(f)$ is a complex weight associated to the nth sensor;
- ϕ being the angle measured from the array axis in the horizontal plane;
- λ is the wavelength;

- $r = |r|$ is the radial distance from the sound source to the microphone aperture;
- f is the working frequency;
- d is the distance between two consecutive microphones.

The far-field directivity pattern applies to planar wave fronts respecting the equation (2), with $|r|$ as the sound source minimum distance to the sensors:

$$|r| > \frac{2(Nd)^2}{\lambda} \qquad (2)$$

A simple horizontal directivity for equally weighted sensors $w_n(f) = 1/N$ is shown by the bold line of Figure 2. , illustrating the directional nature of the array response. From the directivity pattern, we see that the sensor array is capable of enhancing a signal arriving from a certain direction with respect to signals arriving from all other directions.

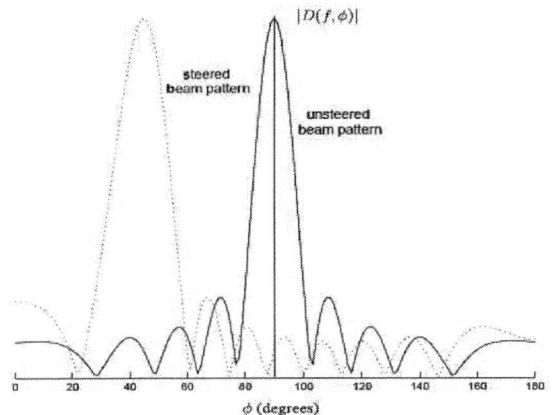

Figure 2. Unsteered and steered ($\phi=45°$) directivity patterns ($f = 1kHz$, N = 10, $d = 0.15m$)[12]

In general, the complex weighting $w_n(f)$ can be expressed by its amplitude and phase as shown by equation (3):

$$w_n(f) = a_n(f) \cdot e^{j\phi_n(f)} \qquad (3)$$

where $a_n(f)$ and $\phi_n(f)$ are frequency dependent amplitude and phase weights respectively. By modifying the amplitude $a_n(f)$ we can modify the shape of the directivity pattern. Similarly, by modifying the phase $\phi_n(f)$, we can control the angular location of the response's main lobe. Thus, the response of the array can be controlled to enhance the signal arriving from a specific direction.

B. Experiment hypotheses

Few hypotheses have been taken into account for this work:

- A far-field approximation is used. (cf equation (2))
- The Microphone array is linear.

As the sound source is far enough from the microphone array, the difference between signal received by the nth microphone x_n and the center of the array is a pure delay [11]. That delay in time units can be expressed as:

$$\tau_n = \frac{f_s \cdot d_n \cos \phi}{c} \qquad (4)$$

- where f_s is the sample frequency, d_n is the extra distance travelled by the wave sound to reach the nth sensor compared to the reference sensor, ϕ is the wave front incident angle, and c being the sound speed.

- No multipath contribution is taken into account.
- The only sources of noise are the microphones themselves and stationary noise.

C. Beamforming algorithms

There are two major groups of microphone-array processing algorithms: time-invariant and adaptive [8]. The first group is fast and simple to get a real-time working implementation. Acoustic adaptive algorithms are able to automatically adapt their response to different weightings or time-delays. However, they require more CPU power and are complex to implement. Thus, only the DSB and GCC will be used for this work.

1) GCC

The main concept of GCC is the estimation of the temporal shift between two microphones that lead to the maximum cross-correlation function. The GCC computation load is associated to the number of cross-correlation it needs to compute. That number is modeled by equation (5):

$$C_N^P = \frac{N!}{P! (N - P)!} \qquad (5)$$

$P = 2$ as cross-correlation is always computed between two signals. With $P = 2$ and N representing the number of microphones in the array = [4, 8, 16, 32, 64]. The number of cross-correlation will vary as follow: C_N^P =[6, 28, 120, 496, 2016]. The CPU power increases with the number of microphones.

2) DSB

The DSB is a beamforming algorithm which uses a predefined Field Of View (FOV) and resolution. A FOV is the region in the space where the sound source is susceptible to be found and the resolution is the measurement of the smallest distinguishable region. A delay and weight is pre-computed for each region. The number of weights and delay computation is function of the number of microphones and the FOV resolution. The incoming signals are combined so that the theoretical delays computed for a particular position are compensated and the signals get added constructively [13].

The FOV size and shape is application dependent. Figure 3. shows an example of a FOV region of size 150x150 cm² split into 9 small squares of size 50x50 cm2. The number of small squares (NoSS) is related to the FOV resolution as defined by the equation (6) below:

$$NoSS = \frac{FOV}{Resolution} = \frac{L \cdot H}{\Delta x \cdot \Delta y} \qquad (6)$$

To locate the source, the DSB must be calculated for each small square.

The DSB algorithm combines accurate sound source localization to the flexibility of having pre-computed coefficients if necessary. Those coefficients (weight and delay) could then be stored in a FPGA BRAM or in an external memory.

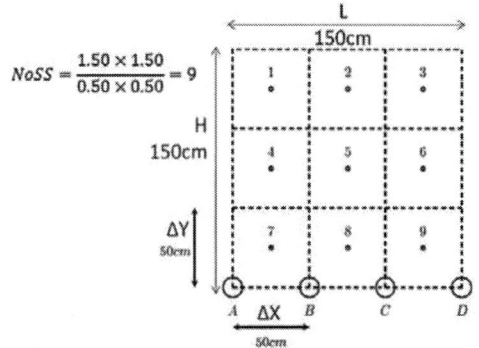

$$NoSS = \frac{1.50 \times 1.50}{0.50 \times 0.50} = 9$$

Figure 3.   2D 3x3 FOV with 150cm x 150cm size and a 50cm x 50cm resolution

3)   Comparison between GCC and DSB

The GCC computation burden only depends on the number of microphones whereas in the DSB, it is dependent on the number of microphones, the FOV size and resolution. Multiple sound source tracking is possible with DSB but not with GCC. Globally we can say that the GCC algorithm is faster than the DSB, but the localization made with the DSB is more precise (GCC just provides the angular direction). A possible way to improve the localization using GCC is to use two sub microphone arrays. But this method is error sensitive, because a small error on the estimation of both angles, results in a false estimation of the sound source position.

D.   Algorithm architecture

Supposing an array of N microphones with outputs denoted as $x_n(k)(n = 1, \cdots, N)$ with k being one of the signal samples, the beamforming output $y(k)$ is achieved by the manipulation of the signals $x_n(k)$ as shown in Figure 4.

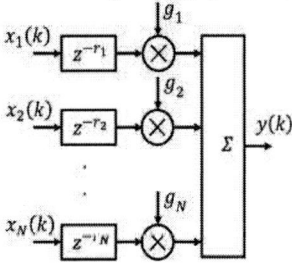

Figure 4.   Structure of the Delay-and-Sum Beamformer [3]

The output of each microphone is delayed (or advanced) by a proper amount of time τ_n, defined in equation (4) so that the signal components from the desired source are synchronized across all the sensors. The resulting signals are then weighted by a factor g_n and summed together. Since the signals are added up coherently, the desired signal components are reinforced. In contrast, other sources and noise are reduced or even eliminated as they are added destructively.

III.   PROBLEM STATEMENT AND CONTRIBUTION

A. Problem statement

Our contribution in this work is to accelerate DSB computation by using GCC and DSB. To illustrate the problem we will use a 4-microphone array as represented in Figure 5. The search region limited by the angles $\phi \pm \varepsilon$ can be obtained by applying first the GCC algorithm, thus reducing the search spectrum and the number of DSB computations for the same resolution.

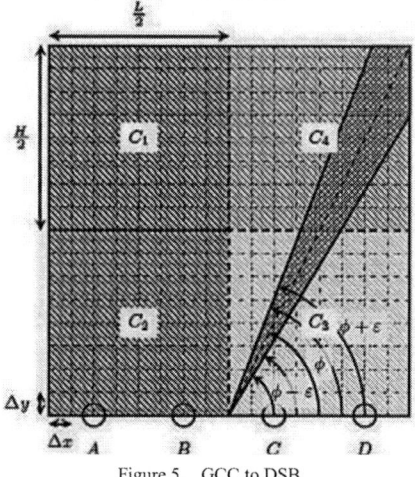

Figure 5.   GCC to DSB

B.   GCC to DSB

The basic idea of this approach is to create a reduced detection zone by drawing two lines one above and the other below the GCC detected angle with an inclination $\pm \varepsilon$ chosen by the user. Figure 5. shows one case of this approach, the remainders are mathematically detailed below. Our algorithm can be described as follows:

For every angle ϕ returned by the GCC, we will assume that:

$$0 \le \phi \le 180$$

There are two special case angles, which are the upper border of the FOV. These two angles denoted δ_1 and δ_2 are defined as follow:

$$\delta_1 = \arctan\left(\frac{H}{L/2}\right)$$

$$\delta_2 = 180° - \arctan\left(\frac{H}{L/2}\right)$$

- **If ($\phi - \varepsilon \le 0$) then the region where we need to compute the DSB is defined by the coordinates:**

$$\{0,0\}; \left\{\left(\frac{L}{2}\right),0\right\}; \left\{\left(\frac{L}{2}\right),\left(\frac{L}{2} \cdot \tan(\phi + \varepsilon)\right)\right\};$$

- **If ($\phi - \varepsilon > 0$ and $\phi + \varepsilon \le \delta_1$) then the region is:**

$$\{0,0\}; \left\{\left(\frac{L}{2}\right),\left(\frac{L}{2} \cdot \tan(\phi - \varepsilon)\right)\right\}; \left\{\left(\frac{L}{2}\right),\left(\frac{L}{2} \cdot \tan(\phi + \varepsilon)\right)\right\}$$

- **If ($\phi - \varepsilon < \delta_1$ and $\phi + \varepsilon > \delta_1$) then the region is:**

$$\{0,0\}; \left\{\left(\frac{L}{2}\right),\left(\frac{L}{2} \cdot \tan(\phi - \varepsilon)\right)\right\}; \left\{\frac{L}{2},H\right\}; \{H \cdot \cot an(\phi + \varepsilon),H\}$$

- **If ($\phi - \varepsilon > \delta_1$ and $\phi + \varepsilon < 90°$) then the region is:**

$\{0,0\}; \{H \cdot \cot(\phi - \varepsilon), H\}; \{H \cdot \cot(\phi + \varepsilon), H\}$

- **If $(\phi - \varepsilon < \delta_2$ and $\phi + \varepsilon > \delta_2)$ then the region is:**

$$\{0,0\}; \{-H \cdot \cot(\phi - \varepsilon), H\}; \left\{\left(\frac{-L}{2}\right); H\right\} \left\{\left(\frac{-L}{2}\right), \left(\frac{-L}{2}\right)\right.$$
$$\left. \cdot \tan(\phi + \varepsilon)\right\}$$

- **If $(\phi - \varepsilon > \delta_2$ and $\phi + \varepsilon < 180°)$ then the region is:**

$$\{0,0\}; \left\{\left(\frac{-L}{2}\right), \left(\frac{-L}{2}\right) \cdot \tan(\phi - \varepsilon)\right\}; \left\{\left(\frac{-L}{2}\right), \left(\frac{-L}{2}\right) \cdot \tan(\phi + \varepsilon)\right\}$$

- **If $\phi - \varepsilon < 180°$ and $(\phi + \varepsilon > 180°)$ then the region is:**

$$\{0,0\}; \left\{\left(\frac{-L}{2}\right), \left(\frac{-L}{2}\right) \cdot \tan(\phi - \varepsilon)\right\}; \left\{\left(\frac{-L}{2}\right), 0\right\}$$

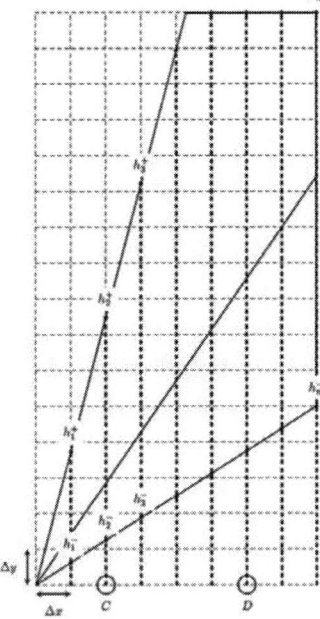

Figure 6. Region search description

C. Memory region access

The FPGA internal memory structure can be read by implementing a simple counter. In GCC to DSB the memory access is more complex as the addresses read are not contiguous but governed by equations (7) and (8).

$$h_n^- = n \cdot \Delta x \cdot \tan(\phi - \varepsilon) \qquad (7)$$
$$h_n^+ = n \cdot \Delta x \cdot \tan(\phi + \varepsilon) \qquad (8)$$

The h_n^- and h_n^+ represents the lowest and highest line in Figure 6. The region between these two points represents the addresses that need to be read by the algorithm. With n varying with a step of Δx. See equation (9).

$$0 < n < \frac{L}{2 \cdot \Delta x} \qquad (9)$$

IV. SYSTEM VERIFICATION HARDWARE AND SOFTWARE TASK PARTITION

A. Task partition

Figure 7. shows the architecture used to implement the GCC and DSB algorithm. Each module will be explained briefly.

1) Microphone array and Sigma-Delta demodulation

MEMS microphones used [6] provide to the FPGA a 2.4MHz Sigma-Delta fourth order digital modulated signal. CIC (Cascaded Integrator Comb) decimates by 56, therefore reduces the signal frequency down to 44.64kHz.

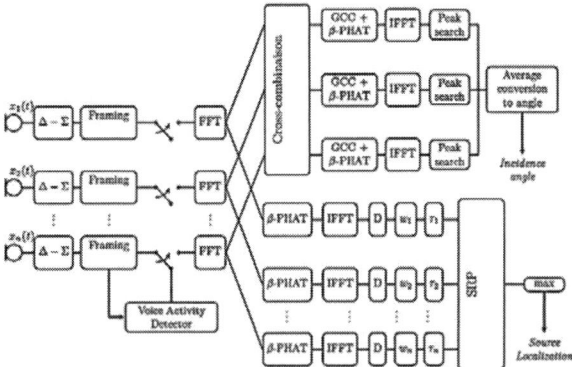

Figure 7. GCC-DSB sound source localization architecture

2) CIC filter

CIC filters achieve sampling rate decrease (decimation) without using multipliers, which make it very suitable for the implementation, as logic is located everywhere in a FPGA. CIC structure is shown in Figure 8.

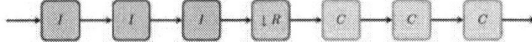

Figure 8. CIC decimation filter (I – Integrator Filter, C – Comb Filter, R – Down sampler)

Equation (10) shows the magnitude response of a N-stage CIC filter at high frequency (f_S) [14]. In this equation M is the differential delay, and is usually limited to 1 or 2.

$$|H(f)| = \left[\frac{\sin(\pi M f)}{\sin\left(\frac{\pi f}{R}\right)}\right]^N \qquad (10)$$

3) Framing and Voice Activity Detector

Framing module cuts the audio signal in frames of (512 samples) each sample being 2 bytes long. The VAD (Voice Activity Detector) module is used to compute noise mean and variance over the 10 first frames. These values are used to determine a threshold equation (13). Mean equation (11) and variance equation (12). Each incoming frame is compared to that threshold. If the computed value is lower than the threshold we assume that we are in presence of a noised frame otherwise there is a sound source. In Figure 7. , the VAD enable the computing of FFT, if we are in presence of a sound source, otherwise that frame is ignored.

$$\mu = \frac{\sum_{i=0}^{L} x_i}{L} \quad (11) \qquad \sigma^2 = \frac{\sum_{i=0}^{L} x_i^2}{L} - \mu^2 \quad (12)$$

where μ is the mean, σ^2 is the variance, L is the number of samples in the frames, x_i is the value of the sample i. The sound detection threshold can be computed as follows (where cst is a constant taken with a value≥ 3):

$$N_{tresh} = \mu + cst \cdot \sigma \qquad (13)$$

4) FFT, IFFT and β-PHAT

After detecting a sound source, the VAD trigger a FFT computation which is modeled by equation (14). The purpose of moving in the Frequency domain is to perform the β-PHAT, which aims to reduce the importance of signal amplitude compared to the phase. The β-PHAT is modeled by equation (16).

$$X(f) = \int_{-\infty}^{+\infty} x(t)e^{-j2\pi ft}\, dt \qquad (14)$$

where x(t) and X(f) are the signal input and output and f the frequency. For computation speed, a Fast Fourier Transform (FFT) which is a fast DFT algorithm that reduces the computing burden from N^2 to $N \cdot \log_2 N$ is used. After the β-PHAT an IFFT is performed to return in time domain to compute signals shift as shown by Figure 4. The IFFT is defined by equation (15).

$$x(n) = \frac{1}{2\pi} \int_{-\infty}^{+\infty} X(f)e^{2\pi ft}\, df \qquad (15)$$

The formula of β-PHAT is given by equation below:

$$W(f) = \frac{X(f)}{|X(f)|^{\beta}} \qquad (16)$$

where X(f) is the signal spectrum, W(f) is the modified spectrum and β is a constant varying between $]0,1[$. If $\beta = 0$, the GCC is a cross correlation (CC), if $\beta = 1$ the magnitude influence is totally removed.

5) Steered Response Power

For a time frame n, the $SRPP_n(\vec{x})$ of the DSB is not affected by a different kind of noise. For example for a single source, the location estimate, $\hat{x}_s^n(1)$ is:

$$\hat{x}_s^n(1) = \underset{\vec{x}}{\operatorname{argmax}}\, P_n(\vec{x}) \qquad (17)$$

$\vec{x}$ is a spatial vector. Given $m_i(t)$ is the signal from the microphone i in a N microphone system, then the SRP for some finite-length frame of length T is defined as:

$$P_n(\vec{x}) \equiv \int_{nT}^{(n+1)T} \left| \sum_{i=1}^{N} w_i m_i(t - \tau(\vec{x}, i)) \right|^2 dt \qquad (18)$$

where w_i is a weight and $\tau(\vec{x}, i)$ is the direct time of travel from the location $\vec{x}$ to microphone i. It has been shown that the SRP may be exactly computed by summing the generalized cross-correlations for all possible pairs of the set of microphones [7].

6) Cross Combination GCC and β-PHAT

With GCC, the main intention is to find the time delay between two microphones that will give the maximum

cross-correlation between them. Using the frequency domain form:

$$R(k) = IFFT[F^*(\omega) \cdot G(\omega)] \qquad (19)$$

with $F^*(\omega)$ and $G(\omega)$ being respectively the Fourier transform of the microphone signals f(t) and g(t), the delay we are looking for is the value of k to which the cross-correlation is maximum.

$$\Delta_{fg} = \underset{k}{\operatorname{argmax}}\, R(k) \qquad (20)$$

In the time domain that corresponds to the number of samples that need to be shifted between f(n) and g(n).

$$\delta_{fg} = \frac{\Delta_{fg}}{f_s} \qquad (21)$$

f_s is the sampling frequency and δ_{fg} is expressed in seconds.

The cross-correlation computation is affected by any change on f(t) or g(t) signals magnitude. To compensate that issue a Generalized Cross Correlation is often used. A normalized coefficient $\varphi(\omega)$ is added to the cross correlation.

$$R(\tau) = \int_{-\infty}^{\infty} \varphi(\omega) F^*(\omega) \cdot G(\omega) e^{j\omega\tau}\, d\omega \qquad (22)$$

$$\varphi(\omega) = \frac{1}{|F^*(\omega) \cdot G(\omega)|^{\beta}} \qquad (23)$$

The GCC reduces the influence of the signal magnitude compared to the phase.

7) Peak search and conversion to angle

A normalize delay factor is computed under the assumption of far field model. The equation (24) expresses the delay:

$$\delta_{ij} = \frac{d_{ij}}{c} \cdot \cos\theta \qquad (24)$$

where c is the sound speed; d_{ij} is the distance between two consecutives microphones and θ is given by equation (25):

$$\theta_{ij} = \arccos\left(\frac{c \cdot \delta_{ij}}{d_{ij}}\right) = \arccos(c \cdot \Delta_{ij}) \qquad (25)$$

With these equations, the normalized delay $\Delta_{ij} = \delta_{ij}/d_{ij}$ can be calculated and also an average delay:

$$\Delta = \frac{1}{C_N^2} \sum_{i=1}^{N} \sum_{j=i+1}^{N} \Delta_{ij} \qquad (26)$$

Therefore

$$\theta = \arccos(c \cdot \Delta) \qquad (27)$$

8) Software

MATLAB and C are used as verification engine and to pre-compute twiddle factor for FFT or weights coefficients for beamforming. The gain of this approach will be presented in section 5. Figure 9. presents the hardware and software interaction. It shows that many pre-compiled coefficients from software can be passed to the hardware to accelerate hardware

482

processing and reduce the logic used. From hardware data can be passed to the software to compare the results against those expected.

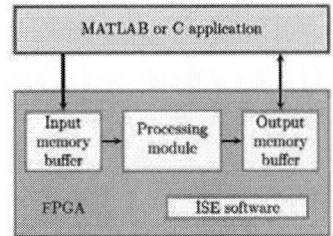

Figure 9.   Hardware/Software systems interaction

V.   RESULTS AND DISCUSSION

This section will present results and approximations made to reduce design resources utilization, improve implementation flexibility and reduced computation points which are useful for real-time applications.

A. Reduction of resources utilization using a software hardware approach

TABLE I. shows the advantage of pre-computing a huge part of the design in software before hardware implementation. It shows the profit of pre-computations of the weight (g_1) and shift (r_1) of Figure 4. above on speed and resources. This approach is applicable when hardware configurations are known in advance.

TABLE I.   WEIGHT AND SHIFT COMPUTATION BOARD ML505

Ressources	Hardware+Sotfware (%)		Hardware (%)	
Slice Registers	8	1%	8192	28%
SLICE LUTs	8	1%	6227	21%
RAM	2	3%	7	11%
DSP48	0	0%	142	9%
Estimate Speed	450	MHz	80.64	MHz

Hardware + Software "Estimate Speed" is much higher because one of the limitation in this approach is the blockRAM speed. For Virtex-5 it is around 550MHz, higher for Virtex-6.

TABLE I. clearly shows that combining software and hardware improve the design speed estimation and reduced logic used. The experiment was done for 9 points FOV and 4 microphones. No multiplications or divisions are necessary in hardware + software co-design, those values are pre-computed in software, Figure 9. explains the zero number of DSP48 used. All the computation was done in software.

B.   GCC to DSB compared with DSB

TABLE II. shows that when comparing the number of small square FOV, Our (GCC to DSB) approach reduces the number of computation points compared to a simple DSB by more than 80%. It also shows that for small angle the computation point's decrease which will be the opposite if the FOV was rectangular. This last result is FOV geometry dependent. TABLE II. computation takes into account the FOV NoSS, the GCC angle with his ε approximation. TABLE II. results were computed for a 16x16 FOV with $\varepsilon = 10°$.

VI.   FURTHER WORK

The present implementation can benefit from dynamic partial reconfiguration and Network on Chip. Indeed, those techniques allow a balance between shared/parallel and reuse of resources.

TABLE II.   DIFFERENT ALGORITHMS COMPUTATION NUMBER FOR A 16X16 FOV

Algorithm	DSB	GCC to DSB	$\phi + \varepsilon$	$\phi - \varepsilon$
Region	256 pts	32 pts	63.5°	43.5°
Region	256 pts	25 pts	53.5°	33.5°
Region	256 pts	18 pts	45°	25°
Region	256 pts	12 pts	20°	0°
Region	256pts	12 pts	< 20°	0°

REFERENCES

[1] Jean-Marc Valin, François Michaud, and Jean Rouat, "Robust Localization and Tracking of Simultaneous Moving Sound Sources Using Beamforming and Particle Filtering," Robotics and Autonomous Systems, vol. 55, no. 3, pp. 216-228, 2007.

[2] Jacek Dmochowski, Jacob Benesty, and Sofiène Affes, "A Generalized Steered Response Power Method for Computationally Viable Source Localization," Audio, Speech, and Language Processing, vol. 15, no. 8, pp. 2510 - 2526, Nov 2007.

[3] Jacob Benesty, Jingdong Chen, Yiteng Huang, and Jacek Dmochowski, "On Microphone Array Beamforming From a MIMO Acoustic Signal Processing Perspective," Audio, Speech, and Language Processing, vol. 15, no. 3, pp. 1053 - 1065, March 2007.

[4] Iain McCowan, Ivan Himawan, and Mike Lincoln, "Microphone Array Shape Calibration in Diffuse Noise Fields," Audio, Speech, and Language Processing, vol. 16, no. 3, pp. 666 - 670, March 2008.

[5] Daniel Jackson Alfred, "Evaluation and Comparison of a Beamforming Algorithm for Microphone Array Speech Processing".

[6] Analog Device, "Application Note 1003,".

[7] Hoang Do, Harvey Silverman, and Ying Yu, "A Real-Time SRP-PHAT Source Location Implementation Using Stochastic Region Contraction (SRC) on a Large-aperture Microphone Array," in IEEE International Conference on Acoustics, Speech and Signal Processing, 2007, pp. I-121 - I-124.

[8] Ivan Tashev, Sound Capture and Processing, Wiley, Ed., 2009.

[9] Kim Keonwook, George Alan, and Sinha Priyabrata, "Experimental Analysis of Parallel Beamforming Algorithms on a Cluster of Personnal Computers," HCS Research Lab, Electrical and Computer Engineering Department, University of Florida, 2000.

[10] Qi Ziming, "Real-Time Adaptive Noise cancellation for Automatic Speech Recognition in a Car Environment," Massey University, School of Engineering and Advanced Technology, Auckland, New Zealand, PhD Thesis 2008.

[11] Qi Li, Manli Zhu, and Whei Li, "A portable USB-Based Microphone Array Device For Robust Speech Recognition," in IEEE International Conference on Acoustics, Speech and Signal Processing, Taipei, 2009, pp. 1301 - 1304.

[12] Ian McCowan, "Robust Speech Recognition using Microphone Arrays", PhD Thesis, Queensland University of Technology, Australia, 2001.

[13] Krishnaraj Varma, "Time-Delay-Estimate Based Direction-of-Arrival Estimation", Virginia Polytechnic Institute and State University, Blacksburg, Master Thesis 2002.

[14] Altera, "Understanding CIC Compensation Filters,"

MIXDES 2012, 19th International Conference *"Mixed Design of Integrated Circuits and Systems"*, May 24-26, 2012, Warsaw, Poland

Design and Implemetation of
a Monopulse Radar Signal Processor

Bo Liu, Wenge Chang, Xiangyang Li
School of Electronic Science and Engineering
National University of Defense Technology
Changsha, 410073, P.R.China
liubo19830120@163.com, changwenge@nudt.edu.cn, lxyniu@sina.com

Abstract—**Signal processor is a key part for radar seeker. In this paper, a missile-borne monopulse radar signal processor is porposed and described. According to the requirement of monopules radar, a signal processor with better commonality and versatility is designed with a FPGA&DSP structure. Firstly, some related signal processing theories are reviewed and the implementation techniques in engineering are analyzed. Secondly, the hardware realization of the signal processor is presented. Finally, some measurements and compensation for the signal processor are implemented, and the results indicate that the design of the signal processor is successful.**

Index Terms—**monopulse radar; signal processor; pulse compression; DSP; FPGA**

I. Introduction

Monopulse radar has being played an important role in missile guidance for its high-precision angular resolution. A monopulse radar signal processor presented in this paper is a key part of a MMW radar seeker. The major functions performed by the signal processor are to search, detect and track target, anti-jamming, data fusion and radar system control.

Real-time signal processing and high reliability are the principal requirements of the signal processor. A real-time pulse compression of 16K points is accomplished by FPGA with FFT-IP core technology; high speed DSP signal processing technology is adopted to complete the follow-up real-time radar signal processing; and a FPGA&DSP digital signal processing structure is designed to improve the re-configurability and versatility of the signal processor.

The paper is organized as follows. Section II reviews the related signal processing theories and analyses their implementation techniques in engineering. Section III presents the hardware design of the signal processor, with the main emphasis on the implementation of pulse compression by FPGA. Section IV gives some system testing results and some analysis. Finally,some conclusions are made in section V.

II. Related Signal Processing Theorys

A. Bandpass sampling, digital quadrature demodulation and signal extraction

The three-channel input signals are intermediate frequency (IF) signals for the monopulse radar signal processor. The band-pass sampling theorem is adopted in data sampling for the purpose of reducing the sampling rate. For a band-pass signal with bandwidth $B = f_H - f_L$, the band-pass sampling theorem requires the sampling frequency f_s meet [1]:

$$\begin{cases} 2B < f_s < 2f_H \\ 2f_H/n \le f_s < 2f_L/(n-1) \end{cases} \quad (1)$$

Where f_H is the upper cut-off frequency, f_L is the lower cut-off frequency, n is an integer determined by (2), int represents rounding off:

$$1 < n \le \text{int}[f_H / B] \quad (2)$$

In digital signal processing, digital quadrature demodulator converts a digital real signal centered at an intermediate frequency to a baseband complex signal centered at zero frequency. In addition, digital quadrature demodulator can lower sampling rate by digital down-conversion (DDC). A digital quadrature demodulator consists of three subcomponents: a direct digital synthesizer (DDS), a low-pass filter (LPF), and a DDC (which may be integrated into the low-pass filter). Digital quadrature demodulation and DDC are accomplished by FPGA IP-corn in the signal processor [2], the realization structure is shown in Fig. 1.

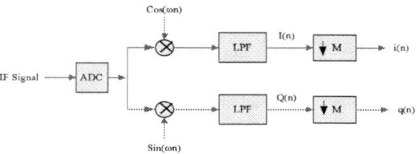

Figure 1. Digital quadrature demodulator and DDC

Supposing IF sampled signal is:

$$x_{IF}(n) = Aa(n)\cos[\omega_0 n + \theta(n) + \varphi] \quad (3)$$

Where A is the amplitude, $a(n)$ is the amplitude modulation, ω_0 is the digital IF carrier, $\theta(n)$ is the phase modulation and φ is a stochastic initial phase. After digital quadrature demodulation and low-pass filtering, the baseband complex signal (I\Q signal) can be expressed as:

484

$$\begin{cases} I(n)=\dfrac{A}{2}a(n)\cos[\theta(n)+\varphi] \\ Q(n)=-\dfrac{A}{2}a(n)\sin[\theta(n)+\varphi] \end{cases} \qquad (4)$$

If the sampling rate is lowered from f_s to f_s/M, then the extraction factor is M. After extracted, the output of DDC is:

$$\begin{cases} i(n)=\dfrac{A}{2}a(Mn)\cos[\theta(Mn)+\varphi] \\ q(n)=-\dfrac{A}{2}a(Mn)\sin[\theta(Mn)+\varphi] \end{cases} \qquad (5)$$

In order to avoid the frequency spectrum aliasing, the sampling frequency of extracted signal should meet the Nyquist sampling theorem.

B. Pulse compression of LFM signal

LFM pulse signal is adopted in the monopulse radar, the transmitting signal can be expressed as follows:

$$s(t)=A_s\cdot rect\left[\frac{t}{T_A}\right]\exp[j(2\pi f_0 t+\pi Kt^2+\varphi)] \qquad (6)$$

Where A_s is the signal's amplitude, T_A is the pulse width, f_0 is carrier frequency, $K=B/T_A$, B is the bandwidth, φ is a stochastic initial phase.

If there is a point-target at the range of R, then the echo delay $\tau=2R/c$, c is the velocity of light. The echo of the sum channel's baseband digital signal is:

$$u_\Sigma(n)=A_\Sigma\cdot rect\left[\frac{n\Delta t-\tau}{T_A}\right]\exp[j(\pi K(n\Delta t-\tau)^2+\varphi-2\pi f_0\tau] \qquad (7)$$

Where A_Σ is the amplitude of sum channel signal, $\Delta t=1/f_{sm}$, f_{sm} is the sampling rate of baseband signal. Pulse compression is accomplished by digital matched filter in frequency domain:

$$x_\Sigma(n)=IFFT\{FFT[u_\Sigma(n),N]\cdot FFT[s_0(n),N]\} \qquad (8)$$

Where $s_0(n)$ is the reference signal, N is the length of $u_\Sigma(n)$. The implementation structure of digital matched filtering in frequency domain is shown in Fig. 2.

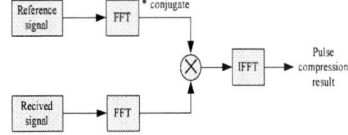

Figure 2. Digital matched filtering in frequency domain

The result of sum channel pulse compression is:

$$x_\Sigma(n)=D_\Sigma\cdot rect\left[\frac{n\Delta t-T_d}{T_g-T_A}\right]Sinc[\pi B(n\Delta t-\tau)]$$
$$\cdot\exp[j(\varphi_\Sigma-2\pi(f_0\tau+\frac{1}{2}K(n\Delta t-\tau)^2))] \qquad (9)$$

Where D_Σ is the amplitude gain of pulse compression, T_d is the time delay of range tracking gate, T_g is the width of

range tracking gate, φ_Σ is the phase factor which is decided by the performance of sum channel.

Identically, the results of two difference channels' pulse compression results can be expressed as:

$$x_{\Delta a}(n)=D_{\Delta a}\cdot rect\left[\frac{n\Delta t-T_d}{T_g-T_A}\right]Sinc[\pi B(n\Delta t-\tau)]$$
$$\cdot\exp[j(\varphi_{\Delta a}-2\pi(f_0\tau+\frac{1}{2}K(n\Delta t-\tau)^2))] \qquad (10)$$

$$x_{\Delta e}(n)=D_{\Delta e}\cdot rect\left[\frac{n\Delta t-T_d}{T_g-T_A}\right]Sinc[\pi B(n\Delta t-\tau)]$$
$$\cdot\exp[j(\varphi_{\Delta e}-2\pi(f_0\tau+\frac{1}{2}K(n\Delta t-\tau)^2))] \qquad (11)$$

Where $x_{\Delta a}(n)$ is the azimuth difference channel pulse compression result, $D_{\Delta a}$ is the azimuth amplitude gain, $\varphi_{\Delta a}$ is azimuth phase factor. $x_{\Delta e}(n)$ is elevation difference channel pulse compression result, $D_{\Delta e}$ the elevation amplitude gain, $\varphi_{\Delta e}$ is elevation phase factor.

C. Coherent integration

In frequency domain, coherent integration is a matched filtering process for a coherent pulses train [3]. It can be accomplished by discrete Fourier transform (DFT) in the signal processor.

Supposing the target movement obeys uniform linear motion, the relative radial velocity is v. Sum channel pulse compression results with L successive echoes are ready for coherent integration. And assuming that $\tau_i=\tau-i\Delta\tau$ is the echo delay, $i=0,1,...L-1$, $\Delta\tau=2vT_p/c$, T_p is the pulse repetition interval, Doppler frequency $f_d=2v/\lambda_0$, λ_0 is the wavelength of the carrier frequency, then the pulse compression results of the pulse train with L pulses in sum channel is expressed as:

$$x_{i\Sigma}(n)=D_{i\Sigma}\cdot rect\left[\frac{n\Delta t-T_d-iT_p}{T_g-T_A}\right]Sinc[\pi B(n\Delta t-iT_p-\tau_i-Kf_d)]$$
$$\cdot\exp[j(\varphi_\Sigma-2\pi(f_0+f_d)\tau_i-\frac{1}{2}K(n\Delta t-iT_p-\tau_i)^2)]$$
$$, \quad i=0,1,...L-1. \qquad (12)$$

If target movement during the interval of the pulse train is no more than a range resolution cell, the result of DFT at $iT_p+\tau_i+Kf_d$ is expressed as follows:

$$X_\Sigma(k)\approx\sum_{i=0}^{L-1}D_{i\Sigma}\exp[j(\varphi_\Sigma-2\pi(f_0+f_d)\tau)]\exp(j2\pi if_dT_p)\exp(-j2\pi ik/L)$$
$$=\exp[j(\varphi_\Sigma-2\pi(f_0+f_d)\tau)]\sum_{i=0}^{L-1}D_{i\Sigma}\exp[j2\pi iT_p(f_d-\frac{kf_r}{L})]$$
$$, \quad k=0,1,...L-1. \qquad (13)$$

Where f_r is the pulse repetition frequency, $f_r=1/T_p$, $D_{i\Sigma}$ is the amplitude of the pulse compression result of the i th echo.

485

It can be seen, when $f_d - \dfrac{kf_r}{L} = mf_r$, or $f_d = mf_r + \dfrac{kf_r}{L}$, $m = 0,1,2...$, the result of coherent integration has the maximum value:

$$Max\{|X_\Sigma(k)|\} = \sum_{i=0}^{L-1} D_{i\Sigma} \qquad (14)$$

So, the echo of sum channel after coherent integration is:

$$X_\Sigma(k) = \exp[j(\varphi_\Sigma - 2\pi(f_0 + f_d)\tau)]\sum_{i=0}^{L-1} D_{i\Sigma} \qquad (15)$$

Similarly, the echoes of two difference channels after coherent integration are:

$$X_A(k) = \exp[j(\varphi_{\Delta a} - 2\pi(f_0 + f_d)\tau)]\sum_{i=0}^{L-1} D_{i\Delta a} \qquad (16)$$

$$X_E(k) = \exp[j(\varphi_{\Delta e} - 2\pi(f_0 + f_d)\tau)]\sum_{i=0}^{L-1} D_{i\Delta e} \qquad (17)$$

Where $X_A(k)$ is the coherent integration result of azimuth difference channel, $X_E(k)$ is the coherent integration result of elevation difference channel.

D. Amplitude-comparison monopulse angle measurement

According to the principle of amplitude-comparison monopulse radar [4], the angle-tracking-error is:

$$E(\varepsilon) = \dfrac{A_\Delta}{A_\Sigma}\cos\theta = k_m\dfrac{\varepsilon}{\theta_{0.5}} \qquad (18)$$

Where A_Δ is the amplitude of difference signal, A_Σ is the amplitude of sum signal, θ is the phase angle between sum and difference signals which is either $0°$ or $180°$ [4] when the monopulse radar system is properly adjusted , k_m is the normalized angle-error slop of monopulse radar, ε is the angle deviation of target from the antenna axis, $\theta_{0.5}$ is the principle half-power beam width. k_m can be obtained by measuring S-curve, angle-tracking-error can be measured by sum-difference signals, so ε can be obtained by a simple calculation:

$$\varepsilon = \dfrac{k_m}{\theta_{0.5}}\cdot\dfrac{A_\Delta}{A_\Sigma}\cos\theta \qquad (19)$$

The signal processor calculates angle-tracking-error based on the coherent integration results of sum-difference three-channel. According to (14) and (15), the azimuth angle-tracking-error is:

$$E_A(\varepsilon) = \dfrac{\sum\limits^{L-1} D_{i\Delta a}}{\sum\limits^{L-1}_{i=0} D_{i\Sigma}}\cos(\varphi_{\Delta a} - \varphi_\Sigma) \qquad (20)$$

Where $\varphi_{\Delta a} - \varphi_\Sigma$ is the phase angle between sum and difference signals of the monopulse radar. Normally, the phase angle is either $0°$ or $180°$. In fact an error is inevitable, however as

long as the error do not change the sign of $\cos(\varphi_{\Delta a} - \varphi_\Sigma)$, the azimuth angle-tracking-error can be calculated as follows:

$$E_A(\varepsilon) = \dfrac{\sum\limits_{i=0}^{L-1} D_{i\Delta a}}{\sum\limits^{L-1} D_{i\Sigma}}\text{sgn}[\cos(\Phi_{\Delta a} + \varphi_{\Delta a} - \varphi_\Sigma)] \qquad (21)$$

Where sgn[.] is the sign function, $\Phi_{\Delta a}$ is the phase angle error between sum and azimuth difference channels. Generally, it can be controlled less than $20°$ by compensation.

Similarly, the elevation angle-tracking-error is:

$$E_E(\varepsilon) = \dfrac{\sum\limits_{i=0}^{L-1} D_{i\Delta e}}{\sum\limits_{i=0}^{L-1} D_{i\Sigma}}\text{sgn}[\cos(\Phi_{\Delta e} + \varphi_{\Delta e} - \varphi_\Sigma)] \qquad (22)$$

$\Phi_{\Delta e}$ is the phase angle error between sum and elevation difference channels of the monopulse radar.

III. HARDWARE REALIZATION

A. Mapping processing tasks into the signal processor

To map the processing tasks into the hardware of the signal processor, the radar signal processor can be functionally divided into six modules: data sampling, time sequence control, data preprocessing, signal processing, radar system control and power supply. In all of them, the data preprocessing module and the signal processing module are the key modules. Fig. 3 is the functional structure of the signal processor.

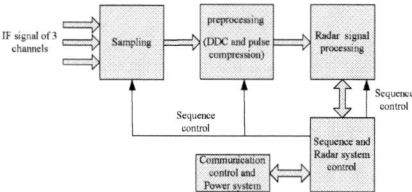

Figure 3. Functional structure of the signal processor

The input of the signal processor is three-channel IF sampled signals for the monopulse radar. The data preprocessing module accomplishes digital quadrature demodulation, DDC and pulse compression for three-channel signals. As monopulse radar requires very high amplitude and phase consistency for three-channel signals, so the structure design and PCB layout should be as symmetrical as possible. Taking into account the commonality and scalability of the signal processor, three FPGAs are used to accomplish the data preprocessing of three-channel signals respectively.

The functions accomplished by the signal processing module are to detect, search, and track targets, anti-jamming and data fusion. There are the characters including frequent data interchange, complex flow for controlling and real-time processing for this module, therefore it is reasonable to use DSP to implement signal processing module. In monopulse radar signal processing, target detection and anti-jamming are accomplished based on the sum channel data, which involves a large amount of data and computation consumption. So two DSP chips are employed in the signal processing module where one DSP implements coherent integration of sum channel signal, target detection and anti-jamming; the other one implements target searching, tracking and data fusion

B. Structure design

Three high performance FPGA (XC4VSX55) produced by Xilinx Company are adopted to complete data preprocessing of three-channel of monopulse radar. TS201 of ADI Company is a high performance floating point DSP chip. Its high-speed computing ability and abundant interfaces make it very suitable for radar signal processing. Therefore two TS201 chips are interconnected through high-speed link ports to form a loosely coupled high-speed signal processing unit [5]. The design of this loosely coupled architecture has advantages: programming is more flexible without regarding to conflict of external bus, and the PCB layout is easier.

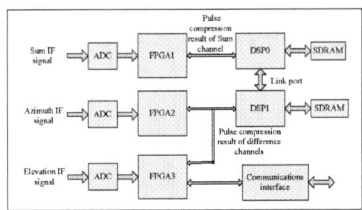

Figure 4.   Hardware architecture of signal processor

The hardware architecture of the signal processor is shown in Fig. 4.

C. Implementation of pulse compression by FPGA

Monopulse radar generally adopts LFM pulse signal. For a long period LFM pulse signal, real-time pulse compression is a problem. We carried out digital pulse compression in frequency domain for 16K complex data with FPGA FFT-IP core technology when the pulse repetition frequency is 1 KHz. The implementation procedure of digital pulse compression in frequency domain is shown in Fig. 2, the matched function can be calculated in advance and stored in the RAM of FPGA. The implementation structure is concise, and the key processes are FFT and IFFT which are implemented with FFT-IP core provided by Xilinx Company [6].

Fig. 5 shows the pulse compression results implemented by FPGA and MATLAB respectively. Because FPGA adopted fixed point calculation, the result involves quantization error. Nonetheless the precision has met the system demand.

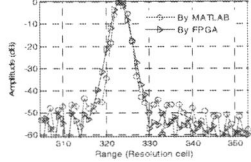

Figure 5.   Comparision of pulse compression in FPGA and MATLAB

The FPGA spend about 615us for implementing pulse compression of 16K complex data, while the pulse repetition interval is 1 ms, so the real-time requirement is fulfilled.

IV. MEASUREMENT

A. Consistency test and compensation for three-channel pulse compression

The monopulse radar system has a high demand on the three-channel consistency. For the sum difference three channels, signal processing tasks include: ADC sampling, DDC, signal extraction and pulse compression. In order to ensure the consistency of three-channel signals processing, the signal processor adopts the same approach and symmetrical hardware circuit to implement parallel processing. However, the clock signal and the data bus in PCB layout can not be exactly identical for these three-channel signals, and the delays of the chips used in each channel are also different, which lead amplitude and phase inconsistency for three-channel pulse compression. Nevertheless, these inconsistencies are system error, which can be eliminated greatly by compensation in signal processor.

To test amplitude and phase consistency among three-channel, an IF signal is split into three-way by a power splitter to feed into sum, azimuth and elevation difference channels respectively. The pulse compression results of three-channel are shown in Fig. 6, where Fig. 6 (a), Fig. 6 (b), and Fig. 6 (c) are the complex pulse compression results of sum channel, azimuth difference channel and elevation difference channel respectively. Fig. 6 (d), Fig. 6 (e) and Fig. 6 (f) are the envelopes of pulse compression results of three-channel respectively. It can be seen that the amplitude consistency is fine form Fig. 6 (d), Fig. 6 (e) and Fig. 6 (f), it can also be seen that the phase consistency is poor from Fig. 6 (a), Fig. 6 (b), and Fig. 6 (c).

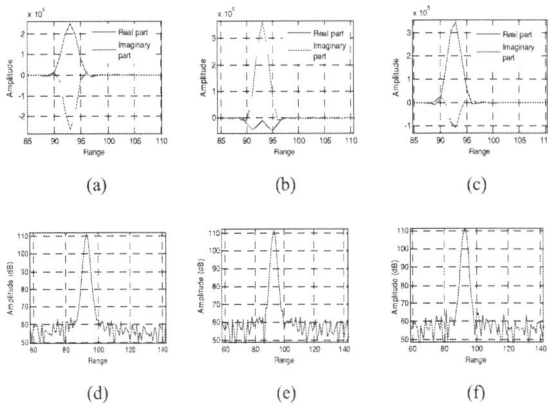

Figure 6.   Pulse compression results of there channels before compensation. (a), (b), (c) are the complex pulse compression results of sum, azimuth deference and elevation difference channels respectively; (d), (e), (f) are the envelopes of pulse compression results of sum, azimuth difference and elevation difference channels respectively.

Sum channel pulse compression result is taken as a benchmark, amplitude and phase compensation of two difference channels are accomplished by DSP1(shown in Fig. 4) reading pulse compression results from FPGA2 and FPGA3 (shown in Fig. 4). Fig. 7 is the pulse compression results of three-channel after compensation. It can be seen that the phase consistency of three-channel signal is good after compensation.

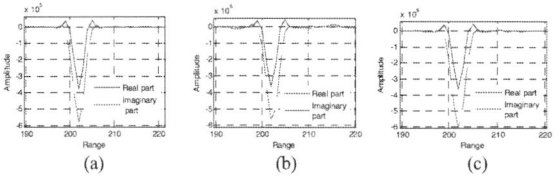

(a)   (b)   (c)

Figure 7. Pulse compression results of there-channel after compensation. (a), (b) and (c) are the complex pulse compression results of sum, azimuth deference and elevation difference channels respectively after compensation

B. S-curve test

It is necessary to test the angle error slope of a monopulse radar before angle tracking, namely to test $\frac{k_m}{\theta_{0.5}}$ in (18), which is normally called S-curve test. There are two reasons for S-curve test: firstly, measure the actual angle error slope k_m, and secondly, test and compensate the phase error generated in radio frequency (RF) end.

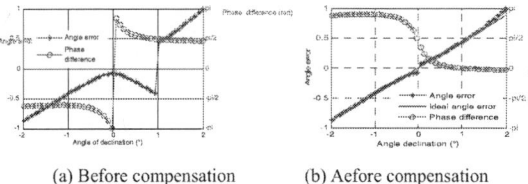

(a) Before compensation   (b) Aefore compensation

Figure 8. Azimuth S-curve

Fig. 8 is an azimuth S-curve test result, where Fig. 8 (a) is the S-curve for the azimuth difference receiver channel before phase compensation, Fig. 8 (b) is the azimuth S-curve for the azimuth difference receiver channel after compensation. The horizontal axis in the figure is the angle deviation of target from the antenna axis. The left ordinate is the angle error $E(\varepsilon)$ while the right ordinate is the phase angle difference between azimuth and sum signals. It can be seen from Fig. 8 (a) that, because of the phase error generated by the monopule radar receiver channels, S-curve is nonlinear in main-beam, which does not meet the principle of amplitude-comparison monopulse radar. So it is necessary to compensate for the phase angle error. It can be seen from Fig. 8 (b) that S-curve is linear in main-beam, but there is an apparent distortion in the vicinity of $0°$, caused by the null of the monopulse radar antenna. The null is defined as the ratio of the corresponding difference beam and sum beam when sum beam is the maximum value, and is very important to the angle tracking precision. The main factor affecting the null of monopulse radar is the antenna feed and the phase error of "sum-difference comparator". In engineering design, the null is required less than -35 dB [7].

V. CONCLUSION

Radar signal processor is a key part for a radar seeker. A missile-borne monopulse radar signal processor has been designed and implemented in this paper. According to the requirement of monopulse radar seeker, the signal processor with FPGA&DSP structure has presented to fulfill commonality and versatility. The key problem of the real-time signal processing in the signal processor is pulse compression. A digital pulse compression in frequency domain has been accomplished for 16K complex data with FPGA FFT-IP core technology when the pulse repetition frequency is 1 KHz. Finally, the amplitude and phase consistency of the system has been tested and compensated in the signal processor, and S-curve of the amplitude-comparison monopulse radar has been measured. These results indicate the signal processor is excellent

REFERENCES

[1] A.V. Oppenheim, R.W. Schafer, "Discrete-Time Signal Processing". (3rd Edition) Prentice-Hall, Inc, 2009.

[2] Tao Guan, Yunhang Zhu, Wenge Chang, Xiangyang Li, "Design and Implementation of a DDC and pulse compression system", Radar Science and Technology. 2010.4.

[3] J. Hoffman, "Numerical Methods for Engineers and Scientists". NewYork: Marcel Dekker, 2001.

[4] D. R. Rhodes, "Introduction to Monopulse". New York: McGraw-Hill, 1959.

[5] Analog Devices, Inc. "ADSP-TS201 Tiger SHARC Processor Hardwaie Reference, Revision 1.1", December 2004.

[6] Xilinx, Inc. Product Specification: "Fast Fourier Transform V4.1", DS260 February 15, 2007.

[7] Huai Huang, Rundong Qi, Shuliang Wen, "Guidance radar technology", Beijing: Electronic Industry Press, 2006, pp.68–100.

MIXDES 2012, 19th International Conference *"Mixed Design of Integrated Circuits and Systems"*, May 24-26, 2012, Warsaw, Poland

Fractional Delay Filter Design with Extracted Window Offsetting

Marek Blok

Faculty of Electronics, Telecommunications and Informatics
Gdansk University of Technology
Gdańsk, Poland
mblok@eti.pg.gda.pl

Abstract—**This paper presents the concept of fractional delay (FD) filter design with window method. In proposed approach we use window extracted from impulse response of the optimal filter which is offset by the fractional delay of the designed filter. The FD filter designed using this approach is not optimal if a single FD filter is considered but it offers better performance when applied to sampling rate conversion (SRC). The main advantage is the lack of large lobes in the stopband of overall interpolation filter specific to the use of optimal filters in SRC. Additionally we attain the ability to adjust the position of transition band of overall interpolation filter.**

Index Terms—**Sampling rate conversion; fractional delay filter; interpolation filter; extracted window; offset window**

I. INTRODUCTION

In digital signal processing the delay operator is one of basic building blocks. Such an element can be readily implemented as it simply stores samples for few processing cycles so they can be used in farther processing without any quality loss. However, its capabilities to delay digital signal are limited to the integer multiple of the sampling period. On the other hand many applications need to delay signal for a fraction of the sampling period and although this can be achieved with digital filter, the task is not simple and signal quality must be compromised to some extent. However, with careful selection of specifications, optimal filters can be used. High performance of the optimal FD filter is, however, paid for with complex design procedure which excludes direct utilization of optimal methods in applications where variable fractional delay (VFD) filter is required, such as synchronization in digital modems [1], modeling of instruments [2] and sampling rate conversion [3, 4].

In this paper we will consider the last of aforementioned applications, since it requires an FD filter with different fractional delay for each output sample and the required fractional delay changes in the whole range from 0 to 1. With Farrow structure [5, 8] or extracted window method optimal FD filters might be applied offering the best performance with the lowest order. The use of adjustable fractional delay allows for change between any two arbitrary selected sampling rates. However, the resampled signal must be bandlimited which

means that additional lowpass prefilter is required. We will, however, demonstrate in this paper that the prefilter is not needed if instead of optimal filters we use the offset window method [6]. The FD filter design in this method is quite simple, though the proper window selection and its offsetting is problematic. In this paper we propose to use the window extracted from optimal filter like in extracted window method and offset it with short maximally flat (MF) FD filter.

II. FRACTINAL DELAY FILTER

The ideal FD filter has frequency response [7]

$$H_{id}(f) = \exp(-j2\pi f \tau_d), \quad (1)$$

where τ_d is the total delay and $f \in [-0.5, 0.5]$ is the normalized frequency. As we can see from (1), FD filters are characterized by constant magnitude response and constant group delay τ_d. In time domain impulse response of the FD filter approximates the ideal infinite impulse response

$$h_{id}[n] = \text{sinc}(n - \tau_d). \quad (2)$$

Because of the causality requirement, good FD filters are characterized with nonzero integer delay $D = \text{floor}(\tau_d)$, which for FIR filters is usually selected close to the bulk delay $\tau_N = (N-1)/2$. With those two delays defined, we come to the following formula for the total delay

$$\tau_d = D + d = \tau_N + \varepsilon \quad (3)$$

where $d \in [0, 1)$ is the fractional delay and ε is the net delay.

Three main optimal FD filter design methods based on different optimality criteria are maximally flat (MF filter) approach, minimization of least squared error (LS filter) and minimization of peak error (minimax filter). In the MF FD filter design a maximal flatness of error frequency response means that the error function

$$E(f) = H_N(f) - H_{id}(f), \quad (4)$$

where

This work was supported by the Polish Ministry of Science and Higher Education under the research project financed from the state budget designated for science in the years 2010-2012.

$$H_N(f) = \sum_{n=0}^{N-1} h[n]\exp(-j2\pi f n) \qquad (5)$$

is the frequency response of an FIR filter of the length N, has to be zero together with its $N-1$ consecutive derivatives.

In case of LS FD filter [7] we minimize error

$$E_{LS}(f_a) = 2\int_0^{f_a} |E(f)|^2 \, df \qquad (6)$$

in the approximation band $f \in [0, f_a]$ while designing minimax (Chebyshev) FD filter [7, 8] we minimize peak error (PE)

$$E_{PE}(f_a) = \max_{f \in [0, f_a]} |E(f)| \qquad (7)$$

also in the approximation band.

In all three cases, MF, LS and minimax, column vector of impulse response $\mathbf{h}$ of the optimal filter with fractional delay d requires solving matrix equation [7]

$$\mathbf{Ph} = \mathbf{p} \qquad (8)$$

for each fractional delay d which can be converted into the following formula for computation of impulse response

$$\mathbf{h} = [h[0], \quad h[1], \quad \ldots, \quad h[N-1]]^T = \mathbf{P}^{-1}\mathbf{p} \qquad (9)$$

where ^{T} denotes transposition and matrix $\mathbf{P}^{-1}$ is inverse of matrix P.

The coefficients of matrix $\mathbf{P}$ and vector $\mathbf{p}$ depend on optimization criteria. For MF filter matrix $\mathbf{P}$ is a Vandermonde matrix

$$P_{k+1,n+1} = n^k \qquad (10)$$

and vector $\mathbf{p}$ has elements

$$p_{1,k+1} = \tau_d^k \qquad (11)$$

In case of LS filter coefficients of matrix $\mathbf{P}$ and column vector $\mathbf{p}_d$ can be computed straightforward from formulas

$$P_{k,n} = f_a \operatorname{sinc}(f_a(k-n)) \qquad (12)$$

and

$$p_{1,k} = f_a \operatorname{sinc}(f_a(k-\tau_d)) \qquad (13)$$

where $k, n = 1, 2, \ldots, N$.

For minimax filters a set of $N+1$ frequency points, called extremal points, must be found using recursive complex Remez algorithm [15]. With those frequencies computed, coefficients of matrix $\mathbf{P}$ and vector $\mathbf{p}_\varepsilon$ can be computed from

$$P_{k+1,n+1} = \cos(2\pi f_k n) - \sin(2\pi f_k n) \text{ and } P_{k,N+1} = (-1)^k \quad (14)$$

and

$$p_{1,k+1} = \cos(2\pi f_k \tau_d) - \sin(2\pi f_k \tau_d) \qquad (15)$$

where $k = 1, 2, \ldots, N+1$ and $n = 1, 2, \ldots, N$. In this case vector obtained from (9) has one additional element with magnitude equal to peak approximation error (7). For FSD filters with real impulse coefficients, extremal points are symmetrically placed on frequency axis and complex matrix equation defined by (14) and (15) can be readily converted into to two sets of real matrix equations, separate for even and odd length filters.

III. SAMPLING RATE CONVERSION WITH FD FILTERS

Digital signal resampling algorithm has to find a new set of samples of the analog signal $x_a(t)$ on the basis of sequence of samples previously acquired with different sampling rate. This means that starting from sequence

$$x[n] = x_a(nT_{s1}); \quad n = -\infty, \ldots, -1, 0, 1, \ldots + \infty \qquad (16)$$

where $T_{s1} = 1/F_{s1} > 0$ denotes sampling interval (reciprocal of input sampling rate F_{s1}), we need to compute sequence

$$y[m] = x_a(mT_{s2}) \qquad (17)$$

sampled with sampling interval T_{s2} (output sampling rate $F_{s2} = 1/T_{s2}$).

Classic sampling rate converter by an arbitrary rational factor L/M is presented in Fig. 1 where the integer factors L and M are related to the input/output sampling rates

$$L = F_{s2}/\gcd(F_{s1}, F_{s2}), \quad M = F_{s1}/\gcd(F_{s1}, F_{s2}) \qquad (18)$$

where $\gcd(x, y)$ denotes the greatest common divisor of x and y.

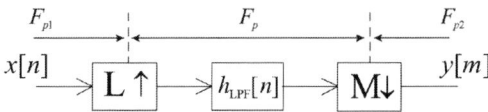

Figure 1. Classic sampling rate converter.

In the classic approach, the input signal $x[n]$ is upsampled by the integer factor L by means of zeroinserting which causes spectrum replication. Therefore the interpolation filter with normalized cutoff frequency at $0.5/L$ is needed to remove replicas. The M-fold downsampler following lowpass filter H keeps every Mth sample of the filtered sequence. As the downsampling operation can result in aliasing for $L < M$, the interpolation filter H in such a case should have cutoff frequency changed to $0.5/M$.

Classic sampling rate conversion algorithm described above has simple interpretation and implementation but it has many disadvantages. The transitional sampling rate is L times higher than input sampling rate, thus, for example in case of CD to DAT conversion ($L = 160$) it is as high as 7.056 MHz. Additionally for large L the interpolation filter bandwidth is very narrow which means that very long impulse response is needed. This makes it's design extremely difficult and in more demanding cases even impossible. The whole impulse response must be computed beforehand and stored in memory. Moreover, we cannot implement incommensurate rate conversions since L and M should be relatively small integer numbers.

The alternative approach based on FD filters [3, 4] is more universal since it allows for any resampling ratio (Fig. 2). The long interpolation filter is replaced with a group of much shorter FD filters that can be designed on demand in real time with computations performed at output sampling rate. Let us notice that the current output sample $y[m]$ is closest to the input sample $x[n[m]]$, where

$$n[m] = \text{floor}\left(mT_{p2}/T_{p1}\right) \tag{19}$$

with the distance between those two samples defined as fractional delay is following

$$d[m] = \left(mT_{p2} - n[m]T_{p1}\right)/T_{p1} = m\frac{T_{p2}}{T_{p1}} - n[m] \in \langle 0,1) \tag{20}$$

The resampling algorithm is a simple recipe: take $\Delta n[m]$ new input samples and compute m-th output sample using FD filter with delay $d[m]$, where algorithm parameters are updated with following formulas [14]

$$\Delta n[m] = \text{floor}(d[m-1]+\frac{T_{p2}}{T_{p1}}) \in \{0,1,2,3,...\} \tag{21}$$

and

$$d[m] = d[m-1]+\frac{T_{p2}}{T_{p1}} - \Delta n[m] \in [0,1) \tag{22}$$

Apart its easy implementation the algorithm described above has one disadvantage: its performance evaluation is very difficult. The problem is that for each output sample we use different FD filter and performance of such filter depends on its delay. However, for commensurate sampling rates a sequence of fractional delays computed with (16) is periodic with period equal to L. In such cases we only need L FD filters to compose an overall filter $h_{\text{all}}[\cdot]$, which is equivalent to interpolation filter

used in the classic approach. This overall filter is composed of all FD filters $h_{d[m]}[n]$ used in the resampling

$$h_{\text{all}}[nL + (L-1) - m] = h_{d[m]}[n]; \quad m = 0, 1, ..., L-1 \tag{23}$$

with $d[m]$ computed with the assumption

$$T_{p2}/T_{p1} = 1/L \tag{24}$$

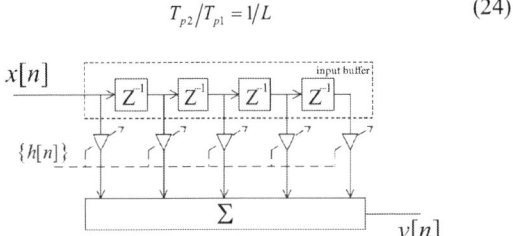

Figure 2.   Sampling rate converter based on adjustable FD filter.

IV. FD FILTER DESING WITH WINDOW METHOD

Let us now investigate consequences of the use of FD filters designed with window method in resampling algorithm. In case of FD filter there are two approaches to the window method. First uses symmetric window with gain correction dependent on fractional delay [9, 10]. This method allows for nearly optimal filter design with window extracted from optimal filter. Second method uses offset windows [6]. This approach has not been widely used because of problems with window selection and its offsetting. In this paper we, however, propose to use window extracted form optimal filter, like in the first approach and offsetting implemented using short FD filter.

A. Design with Symmetric Window

With single symmetric window $w[n]$ we can design a family of FD filters with different fractional delays. We simply multiply N samples of the ideal impulse response (2) by the window and additionally must correct filter gain.

$$h_{N,d}[n] = \alpha(d)w[n]h_{\text{id},d}[n] \tag{25}$$

Approximation band and approximation error of the designed FD filter depends on a selected window and thus this choice is crucial. The best performance is achieved if we take window extracted from impulse response of the optimal FD filter $h_{\text{opt},d}[n]$ with parameters satisfying our needs.

$$w_{\text{ext}}[n] = h_{\text{opt},d}[n]/h_{\text{id},d}[n] \tag{26}$$

The window extraction can be done for any arbitrary selected fractional delay but when the extracted window has non-zero odd part it is best to use only its even part

$$w_{\text{ext},e}[n] = (w_{\text{ext}}[n] + w_{\text{ext}}[N-1-n])/h_{\text{id},d}[n] \tag{27}$$

Fig. 3 presents example of the extracted window together with its magnitude response. With extracted window we can design FD filters which are practically optimal [9, 10, 11]. We only need to properly correct gain of the designed filter. The correction factor can be computed using the following formula [12]

$$\alpha(d) = 1 \bigg/ \left(\sum_{n=0}^{N-1} \mathrm{sinc}(2f_a(n-\tau_d)) w[n] h_{id}[n] \right) \quad (28)$$

In practice it is best to compute the gain correction curve beforehand and approximate it with a low order polynomial which can be used at runtime [11].

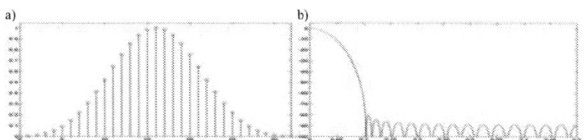

Figure 3.   Window extracted from minimax FD filter with impulse response length $N = 33$ and approximation band $f_a = 0.4$ (a) and its normalized magniture response (b).

The advantage of the approach described above is that it results in the best performance of each FD filter in a given approximation band. However, when adjustable filter designed with single extracted window is used in SRC a problem surfaces. In Fig. 4 and 5 we can see high lobes in the stopband of overall filter. Location of these lobes corresponds to images of input signal components located above approximation band of the FD filter used in resampling. This means that if the processed signal spectrum is limited then those lobes can be ignored. In other cases lowpass filter is required before SRC. The problem is, that such a filter needs to have very narrow transition and stop bands and therefore has long impulse response which increases numerical costs. Additionally width of overall transition band is a sum of widths of transition bands of prefilter and SRC overall filter. This means that when prefilter is used approximation band of FD filter must be increased and thus the length of impulse response of FD filters must be increased which leads to additional increase in numerical costs.

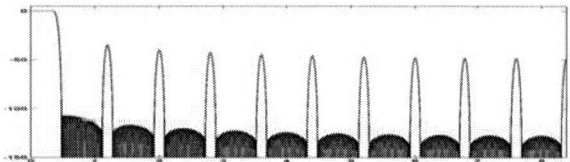

Figure 4.   Magnitude response of overall filter composed of FD filters designed with window from Fig. 3. Interpolation factor $L = 21$.

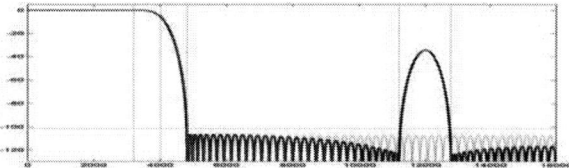

Figure 5.   Zoomed magnitude response of overall filter from Fig. 4 and magniture response of minimax interpolation filter (dotted line).

The source of problems lies in the use of symmetric window which results in stair-like overall window (Fig. 6a) which in turn leads to magnitude response of overall window with undesirable lobes (Fig. 6b).

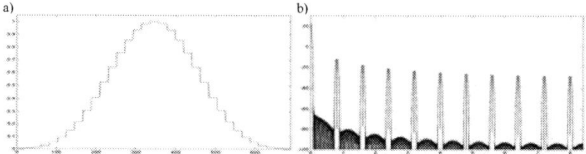

Figure 6.   Overall window (a) and its magnitude response (b) corresponding to overall filter from Fig. 4.

B. Design with Offset Window

The second approach to the design of FD filter with window method uses slightly different window for each fractional delay. The procedure is called window offsetting [6] since we shift the window prototype by the fractional delay we require in the designed filter. The concept can be readily implemented with cosine windows [6, 16]. However, a set of windows defined with closed form formulas is limited and performance of the overall filters is significantly worse that of optimal ones. There is however one advantage that makes this method interesting.

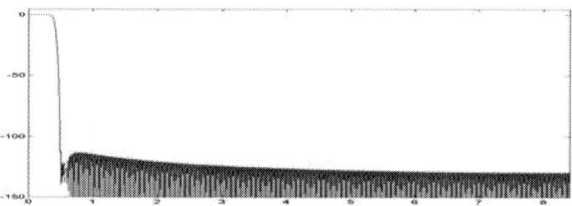

Figure 7.   Magnitude response of overall filter composed of FD filters designed with Balckman-Nutall window of the lengh $N = 33$. Interpolation factor $L = 21$.

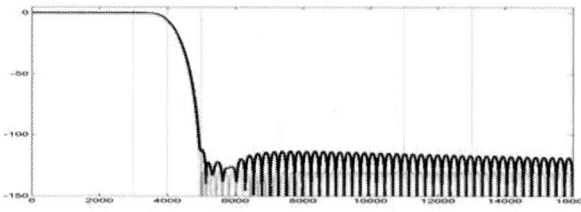

Figure 8.   Zoomed magnitude response of overall filter from Fig. 7 and magniture response of minimax interpolation filter (dotted line).

Let us notice that the obtained overall window is simply the window of same type as the prototype window but just L times longer. Such a window by definition is smooth (Fig. 9a) and in consequence large lobes in overall window magnitude response (Fig. 9b) and in the stopband of the overall filter disappear (Fig. 7 and 8). Additional problem is that the control over the width of approximation band is limited since its change for given window can be only achieved by change of the window length. For example Blackman-Nutall window of the length $N = 33$ has approximation band width $f_a = 0.3754$.

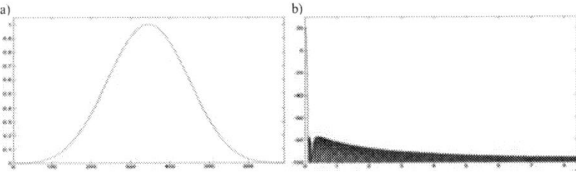

Figure 9. Overall window corresponding to overall filter from Fig. 7 (a) and its magnitude response (b).

C. Offseting Extracted Window

To improve the performance of SRC algorithm using FD filters designed with offset windows we propose to use window extracted from the optimal FD filter and offset it simply by filtering it with FD filter.

Since the main energy of the window is located at low frequencies (Fig. 3) the FD filter used for window offsetting does not need to have a wide approximation band. This means that low order filters can be used. We propose to use MF FD filter as it offers the best approximation around zero frequency. Our investigations showed that FD filters with order from 5 or 7 give good results and can be used if we need SRC algorithm with stopband attenuation of the overall filter around 100 dB (see Fig. 10, 11 and 12).

The filters used for offsetting are very short and perform well but the impulse response of the designed filter is lengthened by $N_{off}-1$ samples, where N_{off} denotes length of the offsetting filter, with overall filter impulse response lengthened by $(N_{off}-1)L$ samples. While performance of overall filter is close to the performance of optimal LS or minimax filter of the length $N_{prot}L$, where N_{prot} is the length of the prototype window, it is worse by about 15 dB than the performance of optimal filters of length NL. This performance loss might be, however, neglected in case of incommensurate SRC where we cannot design and use interpolation filters designed using optimal methods.

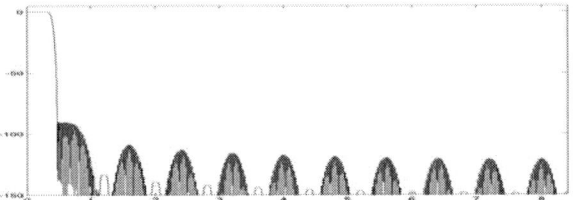

Figure 10. Magnitude response of overall filter composed of FD filters designed with window extracted from minimax FD filter of the length $N = 33$ and approximation band $f_a = 0.4$ (Fig. 3) offset with MF FD filter length 7. Interpolation factor $L = 21$.

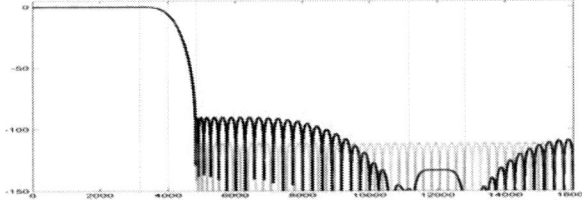

Figure 11. Zoomed magnitude response of overall filter from Fig. 10 and magnitude response of minimax interpolation filter (dotted line).

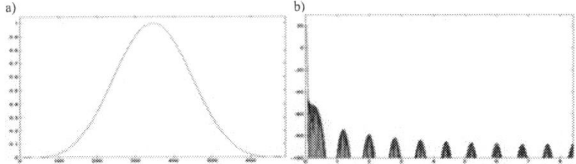

Figure 12. Overall window corresponding to overall filter from Fig. 10 (a) and its magnitude response (b).

However, by analogy to cosine windows we can treat a extracted prototype window as samples of a periodic function. In such case in window offsetting we can replace linear convolution with circular convolution and get offset window of the same length as prototype window. By this modification we obtain performance improvement which can be seen in Fig. 13, 14 and 15.

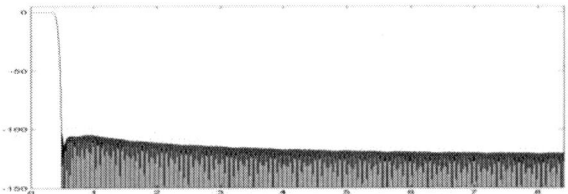

Figure 13. Magnitude response of overall filter composed from FD filters designed with window from Fig. 3. Interpolation factor $L = 21$.

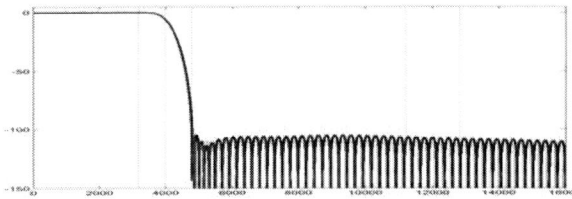

Figure 14. Zoomed magnitude response of overall filter from Fig. 13 and magniture response of minimax interpolation filter (dotted line).

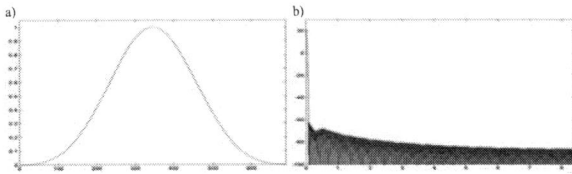

Figure 15. Overall window (a) and its magnitude response (b) corresponding to overall filter from Fig. 13.

With extracted window method we had no control over the position of transition band but now, with offset windows we can readily change the position of transition band by using ideal response of band limited FD instead of fullband one (2)

$$h_{id,B}[n] = 2B\,\text{sinc}(2B(n-\tau_d)) \qquad (29)$$

where B is the bandwidth of the ideal FD filter. This is important when SRC has to decrease sampling rate and the filter should have narrower bandwidth. In Fig. 16 we see effects of using such prototype impulse response for both methods with only offset window method giving satisfactory results.

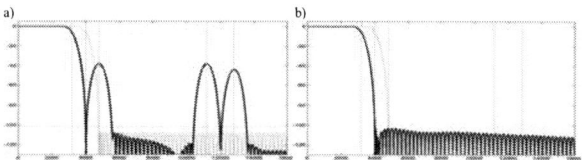

Figure 16. Zoomed magnitude response of overall filter composed of FD filters designed using lowpass ideal FD filter and (a) symetric window method, (b) offset window. Dotted line - magniture response of minimax interpolation filter without shifted transision band.

V. PRACTICAL EXAMPLE

Let us now consider a practical case of SRC from DAT (F_{s1} = 48000 Sa/s) to CD (F_{s2} = 44100 Sa/s) standard (L = 147, M = 160). Assume that we want approximation band up to 20 kHz and do not allow for alias above $F_{s2}/2$ = 22.05 kHz with stopband attenuation A_{dB} = 100 dB. From this specification we obtain approximation band of FD filters which is equal to half of the width of transition band (22050 Hz – 20000 Hz) f_a = 0.5 – 1025/F_{s1} = 0.4686. On this basis we can estimate impulse response length of FD filters offering errors smaller then A_{dB} with the following formula [13]

$$N = \text{ceil}\big(0.125(A_{dB}/(2-4B)+3)\big) = 147 \cdot \qquad (30)$$

Finally, bandwidth of lowpass ideal FD filter B = 20000/F_{s1} = 0.4671.

Overall filter has length NL-1 = 21608 and it is worth noting that direct design of the optimal interpolation filter fulfilling our requirements using firpm function in MATLAB fails. Extracted window method with FD offsetting is, however, successful (Fig. 17 and 18). Using this method offers additional advantage since we can readily change the SRC ratio as we can design FD filter on demand for any delay.

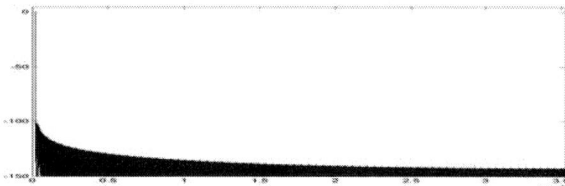

Figure 17. Magnitude response of overall filter composed from FD filters designed using offset window for DAT to CD conversion.

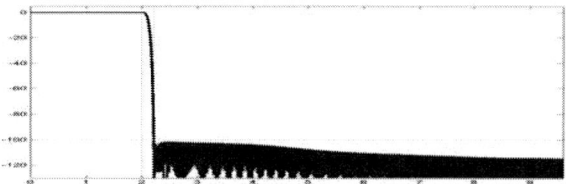

Figure 18. Zoomed magnitude response of overall filter from Fig. 17.

VI. CONCLUSIONS

In the paper problems of FD filter design for the use in SRC algorithm have been discussed. As it has been presented here FD filters designed using offset window should be used instead of optimal ones when they are intended to be used in SRC. Therefore a solution based on offset window method has been proposed. In order to achieve the performance close to optimal solutions the approach based on a window extracted from optimal filter has been proposed. Offsetting extracted window is done with the use of a short MF FD filter. With the proposed method a designer has control over the location of transition band and the attenuation in stopband of overall interpolation filter without any limitations on resampling ratio. The efficiency of this approach has been illustrated with example of DAT to CD conversion.

REFERENCES

[1] M. Makundi, T. I. Laakso, and A. Hjørungnes: "Generalized symbol synchronization using variable IIR and FIR fractional-delay filters with arbitrary oversampling ratios," IEEE Conference on Acoustics, Speech & Signal Processing ICASSP 2004.

[2] T. Tolonen, V. Valimaki and M. Karjalainen: "Modeling of tension modulation nonlinearity in plucked strings," IEEE Trans. on Speech & Audio Processing, vol. 8, no. 3, pp. 300-310, 2000.

[3] K. Rajamani, Y. S. Lai and C. W. Farrow: "An efficient algorithm for sample rate conversion from CD to DAT," IEEE Signal Processing Letters, vol. 7 (No. 10), pp. 288-290, 2000.

[4] E. Hermanowicz, R. Rojewski and M. Blok: "A sample rate converter based on a fractional delay filter bank," ICSPAT 2000, Dallas, Tx, USA, 16-19 October 2000.

[5] C. W. Farrow: "A continuously variable digital delay element," IEEE Proc. Int. Symp. Circuits and Systems, Espoo, Finland, pp. 2641-2645, June 1988.

[6] A. Yardim, G. D. Cain and A. Lavergne: "Performance of fractional-delay filters using optimal offset windows," ICASSP'97, vol. 3, pp. 2233-2236, 21-24 April 1997.

[7] T. I. Laakso, V. Valimaki, M. Karjalainen and U. K.. Lain: "Splitting the unit delay," IEEE Signal Processing Magazine, vol.13 (No.1), pp. 30-60, January 1996.

[8] M. Blok: "Farrow structure implementation of fractional delay filter optimal in Chebyshev sense," Proceedings of SPIE, vol. 6159, 61594K, Wilga, Poland 30 May-2 June 2005.

[9] E. Hermanowicz: "A nearly optimal variable fractional delay filter with extracted Chebyshev window," IEEE Int. Conf. on Electr., Circuits and Systems, vol. 2, pp. 401-404, Lisboa, Portugal, 7-10 September 1998.

[10] M. Blok: "Gain Correction for Nearly Optimal Variable Fractional Sample Delay Filter Design," SPA 2011, Poznań, Poland, 29-30 September 2011.

[11] M. Blok: "On practical aspects of optimal FSD filter design using extracted window method," ECCTD 2011, Linkoping, Sweden, 28-30 August 2011.

[12] M. Blok: "Versatile Structure for Variable Fractional Delay Filter Based on Extracted Window Method," PWT 2011, Poznań, Poland, 9 December 2011

[13] M. Blok: "Length Estimation of Fractional Delay FIR Filter," in polish, KST'2000, Bydgoszcz, vol A, pp. 325-332, 2000.

[14] M. Blok: "Collective filter evaluation of an FSD filter-based resampling algorithm," OSEE 2002, 15 September 2002.

[15] M. Blok: *"Optimal fractional sample delay filter with variable delay,"* OSEE 2002, 18 March 2002.

[16] H. H. Albrecht: *"A family of cosine-sum windows for high-resolution measurements,"* ICASSP'01), vol. 5, pp. 3081-3084, 2001.

Gap in pagination due to withheld paper.

Pages 495-506

Measurement of Settling Time of High-speed D/A Converters

Rokas Kvedaras, Vygaudas Kvedaras, Tomas Ustinavičius

Electronics Faculty

Vilnius Gediminas Technical University,

Vilnius Lithuania

vygaudas.kvedaras@el.vgtu.lt

Abstract—Investigation results of sampling device with peak detecting and algorithm of digital signal processing developed by authors for automated measurement of settling times of fast D/A converters are presented. The analysis of influence of different time scale transformation methods, internal noise and other noises of sampling device with peak detecting to the measurement results is presented. A method of reduction of noises of the measurement signal by using an ideal comb filter in the frequency domain and averaging of filtered signals in time domain has been investigated. It is shown that such method ensures significant reduction of white and $1/f$ noise in the measurement signal. Results of the analysis of the developed digital signal processing algorithm are submitted.

Index Terms—dynamic parameters of high-speed D/A converters, automated measurement of high-speed D/A converters settling time, implementation of digital comb filter

I. INTRODUCTION

Digital-to-Analog Converters (DAC) are gaining significant value in today's digital world. With the development of the consumer electronics, measurement devices, industry and manufacturing equipment and other fields the needs for higher speed and accuracy (resolution) of DACs increases. Therefore manufacturers of DAC ICs are forced to satisfy such needs. In the production of high-speed and high resolution DACs the main problem remains measurement of AC parameters and especially measurement of settling times of the produced DAC ICs in order to classify these according to speed and accuracy and also to identify faulty ICs that do not correspond to the required parameters [1].

In order to measure settling time of high-speed (with settling time measured in nanosecond range) DACs with resolution of 12 bit or more it is necessary to gain a very high accuracy of measurement equipment that is commonly designed using sampling converters [2–4] and additional digital signal processing means [5–8].

Requirements to the measurement equipment are so high that it is necessary to gain highest possible result in each step of signal processing step.

Some of aforementioned problems are solved in earlier works [2–8] but these are not summarized and require further investigations.

II. SAMPLING CONVERTERS

Sampling converters are commonly used in settling time measurement equipment as: these are high-speed enough to investigate settling times of nanosecond and subnanosecond range; protects further measurement equipment from overloads and amplitude distortions [3]; ensure change of time scale to

operate with non-high-speed accurate ADCs, FIFOs, etc. without influencing shape changes of measurement signal. Nevertheless requirements for parameters of such sampling converters are very strict to meet the needs for measurement of settling times.

Peak detecting sampling converters with two quartz oscillators [5] or Numerically Controlled Oscillators (NCOs) [6, 7] are commonly used.

Such converters ensure comparably simple circuitry and high transfer coefficient ($K = 0.5$–0.9) that is much higher compared to converters with feedback.

Peak detecting converter because of the simple circuitry and small dimensions can be placed directly on the DAC output pin (or measurement head pin). In this case signal transmission channel is excluded from measurement device and noise interferences and distortions of it are excluded as well.

High transfer coefficient ensures that internal noise of the converter is low. Sample step formation circuit designed to use two quartz oscillators with automatic frequency tuning ensures stable and accurate transformation factor of the converter.

Nevertheless peak detecting converters are distorting measurement signals during conversion process. It happens because of: finite speed of the converter; non equable the flat part of transient characteristic; unlinearity of amplitude characteristic.

A. Backward Sampling method

Sampling converter with two quartz oscillators can be easily set for forward sampling ($f_{IN} > f_S$, where f_{IN} – frequency of the input signal; f_S – frequency of the sampling pulses) and for backward sampling ($f_{IN} < f_S$).

With backward sampling (Fig. 1) measurement signal with transformed time scale is reversed (starting from falling edge and moving to the rising edge) compared to the input signal. It is very convenient for settling time measurement as: a) simplifies settling time measurement (it is not necessary to find last shot of the measurement signal to the readout levels); b) protects further circuitry from overloads and distortions of the measurement signal caused by them; c) finite rise time of the

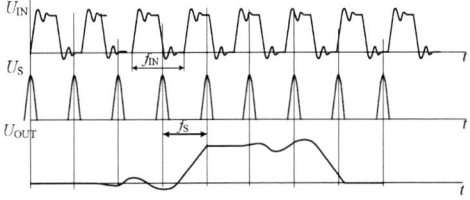

Fig. 1. Backward sampling in sampling converter

converter transient characteristic is less influencing measurement errors.

Modelling of the influence of the sampling converter transient characteristic rise time to settling time measurement in case of forward and backward sampling has been performed. It has been shown that in case backward sampling is used the requirements for rise time of transient characteristic of sampling converter are 8–10 times lower compared to requirements for forward sampling (Fig. 2). Therefore it is possible to use converter with significantly narrower band and reduce white noise of the converter.

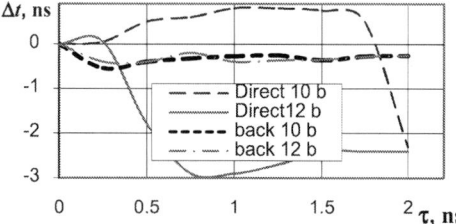

Fig. 2. Dependence of DAC settling time measurement errors to the rise time of the transient characteristic of the sampling converter

B. Trancient Characteristic

Memory capacitor of the peak detecting converter is charged through small resistance of the open diode during short sampling pulse. During the pause between two sampling pulses memory capacitor is discharged through relatively high load resistance R.

With each sampling pulse memory capacitor is charged more than it is discharged during the pause. In case sampling pulse has shape of triangle then after some time capacitor load comes into balance (because effective duration of the sampling pulse becomes shorter) and it means that capacitor additional charge during sampling pulse is equal to capacitor discharge during the pause. Mixer diode is opened by sampling pulse of relatively small amplitude and duration and it is assumed that characteristic of the mixer diode is exponential.

With these assumptions expression of the voltage on the memory capacitor of the peak detector sampling converter after i-th sampling pulse has been obtained:

$$ u_{ki} \approx \left\{ u_{k(i-1)} + \frac{1}{\lambda} \left[\exp(-2v_i) + \frac{2v_i}{z} \exp z_i \right] \exp(-\beta_i) \right\}; (1) $$

where $u_{k(i-1)}$ – capacitor voltage after $(i-1)$ sampling pulses $z_i = S\lambda\tau_i ; v_i = \lambda\tau_i i_0 / C$, S and τ_i – slope of i–th sampling pulse and its effective duration; λ, i_0 – parameters of exponential approximation ($i_d = i_0(\exp(\lambda u_d)-1)$) of current-voltage characteristic of the mixer diode; u_d – voltage drop on mixer diode; $\beta_i = t_S / \tau_s$ – relative time constant of the memory capacitance discharge; $t_s = 1/f_S$ – period of repeating of sampling pulses; τ_s – time constant of memory capacitance.

For sampling peak detectind converters it is assumed that number of sampling pulses on one point of input signal is

$N = \Delta t_r / \Delta t_N$, where $\Delta t_r = F_D / f_{IN}^2$ – sampling step of the converter, $F_D = f_S - f_{IN}$, f_{IN}, f_S – frequencies of the input and sampling pulses, $\Delta t_r = 1/2 f_a$ – sampling step set according to Nyquist-Kotelnikov-Shannon theorem, f_a – highest frequency of the measurement signal.

Results of the analysis are displayed on Fig. 3.

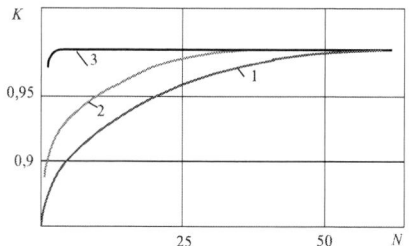

Fig. 3. Dependence of the relative transfer coefficient of the converter from number of sampling pulses on one point when relative time constants of memory capacitance discharge are $1 - \beta_i = 2 \cdot 10^{-3}; 2 - \beta_i = 0.01; 3 - \beta_i = 0.2$

It is obvious that discharge time constants should be selected according to the sampling step (transformation coefficient). In real sampling converters N is usually in range of 10-30 and βi is selected in range 0.1–0.2 accordingly. In this case absolute transfer coefficient of the peak detecting sampling converter is reduced from 0.9-0.95 to 0.5. Output signal of such converter usually has relatively high instabilities of memory capacitor charge-discharge that are not acceptable for high accuracy measurements. Frequency of such instabilities is f_S and the highest frequency of the measurement signal with transformed time scale is

$$ F_A = \frac{0.45}{t_f q}, \qquad (2) $$

where t_f – duration of the rising edge of the measurement signal at the input of the converter, q – time transformation coefficient of the converter.

Therefore the instabilities of the charge-discharge are filtered by low-pass filter on the output of the sampling converter.

Usually low-pass filter on the output is filtering instabilities of charge-discharge without influencing measurement signal shape when $F_A \ll f_S$ and other parameters of the converter are set right.

Analysis of the amplitude characteristic nonlinearity has shown that nonlinearity of the amplitude characteristic does not exceed ± 0.3 % even if amplitude of the sampling pulse is exceeding amplitude of the measurement signal by 3-4 times. In addition to that usually settling time measurement equipment performs measurement signal amplitude normalization and therefore influence of this nonlinearity is relatively small and can be neglected.

C. Circuit of Compensating Peak Detecting Sampling Converter

Measurement of DACs settling time is related to very small voltage readout levels (hundreds of microvolts – related to resolution of the DAC) that are set from the measurement signal settled value. Therefore internal noise of the converter are highly influencing measurement results.

Converter noise can be separated to measurement signal tract internal noise and noise-like disturbances influenced by the sampling pulse shaper.

Noise of the measurement signal change and processing tract are common noise of the electronic circuits – $1/f$ type and white noise.

It has been established that the highest influence (70–80 %) to the peak detecting sampling converter noise have amplitude instability and noise of the sampling pulse that are generated in sampling pulse shaper. These disturbances are by its character additive noise and multiplication noise.

Analytical analysis and investigation of a digital converter model has shown that low frequency components of the additive noise of sampling pulse are effectively filtered in sampling pulse shaper and high frequency components are filtered in amplification circuit of measurement signal with transformed time scale and are effectively suppressed by 50–60 dB in case parameters of these circuits are selected in the right way and do not influence measurement results.

Carried out investigations of the converter model has shown that multiplication noise of the sampling pulse amplitude are directly transferred to the output of the converter and the spectrum of these noise are partially or fully overlapping with the spectrum of the measurement signal and causes significant errors of the settling time measurements.

In order to solve this issue the compensating peak detecting sampling converter has been developed (Fig. 4). This converter consists of two mixers ($D1$, $C1$, $R2$, $R4$ and $D2$, $C2$, $R3$, $R5$,

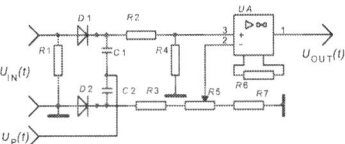

Fig.4. The circuit of compensating peak detecting sampling converter

$R7$) fed by the same sampling pulses $U_P(t)$. Input of the first mixer is measurement signal from the DAC under test and the output is the total voltage of the measurement signal $U_{IN}(t)$, sampling pulse $U_P(t)$ and possible disturbances with transformed time scale. Input of the second mixer is 0V, and the output is the total voltage of the sampling pulse $U_P(t)$ and possible disturbances with transformed time scale. The output of the measurement amplifier UA shall be a voltage of the measurement signal and the difference of the sampling pulse $U_P(t)$ and possible disturbances (anticipated by differences of the parameters of the mixers) with transformed time scale.

Investigation of the model of the compensating peak detecting sampling converter has shown that it effectively reduces (up to 1700 times) (Fig. 5) multiplication noise of the sampling pulses and drifts of the converter. Experimental investigation of the compensating converters has shown that internal noise of such converters are 40–60 µV and voltage of the disturbances has been reduced by 2-3 times.

Nevertheless it has been shown [5] that even by using compensating converters it is not possible to obtain accurate

measures of settling times of DACs with resolution exceeding 12 bit because of internal noise of the sampling converter.

Therefore the output signals of the converter are processed by digital signal processing algorithms for further processing and measurement of the settling time [6–8].

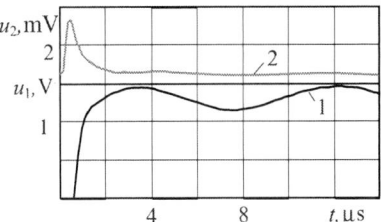

Fig. 5. Internal noises of the sampling converter influenced by multiplication instabilities of the amplitude of the sampling pulse: 1 – peak detecting sampling converter (u_1); 2 – compensating sampling converter (u_2)

III. DIGITAL SIGNAL PROCESSING

Results of the investigations have shown that the residual internal noise of the converter is sum of the $1/f$ type noise and white noise.

In order to reduce these noise N realizations of a measurement signal is combined to pseudo-periodic sequence of the measurement pulses which is then filtered by the proposed comb filter [6–8]. Filter is blocking spectral components that are not corresponding to periodic signal and does not influence spectral components of the measurement signal. In this way the influence of the converter internal noise is highly reduced. Nevertheless the spectral components of the noise signal that is equal to the periodic signal spectral components are not removed. In order to improve filtering results it has been proposed to use filtering and averaging algorithm. Investigation of the proposed algorithm by using LabView® has shown that the white noise is decreased up to 100 times. This improvement allows test DACs with 6-7 bits higher resolution compared to the test equipment without the proposed digital signal processing algorithm. The results are obtained by using pseudo-periodic sequence of 40 signal realizations and averaging 10 such filtered sequences.

Signal model with influence of $1/f$ type of noise has been developed (Fig. 6). In this case it is seen that the signal is distorted so that it is out of readout levels and therefore settling time measurement cannot be done. After application of the proposed algorithm influence of the $1/f$ type of noise is significantly lower (Fig. 7).

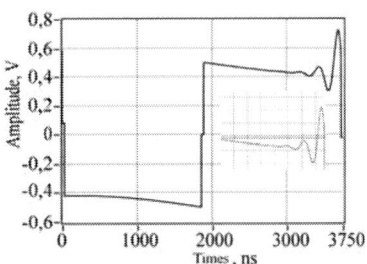

Fig. 6. Realization of the measurement signal with $1/f$ type of noise

Investigation of the digital processing algorithm has shown that the most effective reduction of the noise (up to 53 times) if 100 periods of the signal is used for pseudo-periodic sequence and 100 filtered signals are used for averaging (Fig. 8).

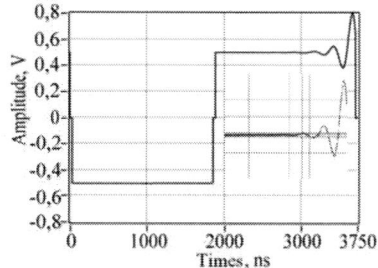

Fig. 7. Measurement signal after digital processing

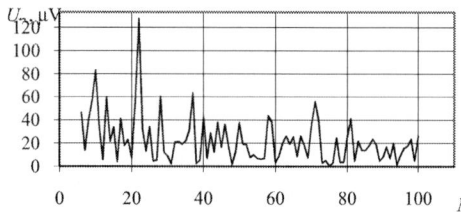

Fig.8. Noise level after digital processing of an investigated signal

Obtained results show that the algorithm is more effectively reducing white noise (1.5–2 times) than $1/f$ type of noise.

IV. CONCLUSIONS

1. Analysis of the performance of the peak detecting sampling converters has shown that it is recommended to use backward sampling for settling time measurement of DACs as it allows to: a) reduce errors of settling time measurements caused by performance of the converter by 8-10 times; b) reduce errors of the measurement caused by overload of the converter amplifier; c) simplifies equipment or algorithm of settling time measurement.

2. Analysis of the flat part of the transient characteristic of the converter has shown that by the selection of time constant of the memory capacitance discharge of the peak detecting sampling converter it is possible to obtain equable flat part of transfer characteristic and avoid measurement errors.

3. Analysis and modelling of the internal noise of the sampling converters has shown that the amplitude multiplicative noise-like disturbances of the sampling pulse has the highest influence to the internal noise of the sampling converters. Compensating peak detecting sampling converter is designed that ensures suppression of these disturbances by 50-60 dB.

4. Analysis of the digital signal processing algorithm using digital comb filter and averaging has shown that white noise is reduced by 100 times and $1/f$ type noise is reduced by 50 times. These results ensure that by using the same measurement equipment it is possible to measure settling time of DACs with 5-6 bit higher resolution.

REFERENCES

[1] E. Balestrieri, "Some critical Notes on D/A CONVERTER Time Domain Specifications," Proceedings of Instrumentation and Measurement Technology Conference, IMTC 2006. Sorrento, Italy, 24-27 April 2006, pp. 930–935.

[2] E.-A. Bagdanskis, V. Kvedaras, "Measurement of the dynamic characteristics of digital-to-analogue converters", High Optoelectronics, Instrumentation and Data Processing No. 5, 1985, pp. 101-105.

[3] J. Williams, "Precisely measure settling time to 1 ppm", EDN, March 4, 2010, pp. 20-24.

[4] J. Williams, "Measuring wideband-amplifier settling time", EDN, August 12, 2010.

[5] R. Kvedaras, V. Kvedaras, "The measurements of dynamic parameters of high-speed multi-bit DACs", Electronics and electrical engineering, No. 3 (83), 2008. p.11-14.

[6] R. Kvedaras, V. Kvedaras, T. Ustinavicius, "Method of Dynamic Parameters Measurement of High-speed D/A Converters", Proc. 18th Int. Conf. on Microwave Radar and Wireless Communications (MIKON 2010), Vilnius (Lithuania) 2010, p. 48.

[7] R. Kvedaras, V. Kvedaras, T. Ustinavicius, "Settling Time Testing of Fast DACs", Acta Physica Polonica A, No 4, 119, 2011, p. 521–527.

[8] T. Ustinavičius, R. Kvedaras, V. Kvedaras, Z. Jankauskas, "Initial Digital Filtering of Testing Signals of High-Speed DAC's", Electronics and Electrical Engineering, No 7(95), 2009, p. 62-67.

MIXDES 2012, 19th International Conference *"Mixed Design of Integrated Circuits and Systems"*, May 24-26, 2012, Warsaw, Poland

Realtime Physics Engine for Robots Movement

Enhanced Virtual Environment

Aleksandra Burdziuk, Janusz Pochmara, Krzysztof Łakomy, Piotr Szablata, Radosław Koppa

Poznan University of Technology

Poznan, Poland

Abstract—In our investigation we focus on the A* Algorithm, which is widely used in the video games design theories. A* algorithm is the most common choice for solving the path-finding problems, because it's fairly flexible and can be used in a wide range of contexts. The main problem of A* Algorithm is the finite computer memory. When in need of finding a path on considerably large map, computer have to remember complex list of examined and open nodes, which can occupy most free space in computer memory. Nonetheless, this solution shows the best results and it is worth analyzing as the algorithm for the intelligent robot movements.

Index Terms—**A*, A-Star, Microsoft Robotics, virtual simulation, optimal paths, Dijkstra algorithm, Gready Best-First-Search algorithm, AI, intelligent automatics**

I. INTRODUCTION

Movement for a single object seems easy, but path finding might become a very complex problem[1]. Given map can be considered as a graph with each tile being a vertex and edges drawn between tiles adjacent to each other. Figure a) presents typical map, where obstacles are depicted as black color, starting point is a light gray color cell, and ending point is a dark gray cell. The map is two dimensional XY. Moving object doesn't know a topography of the map[2].

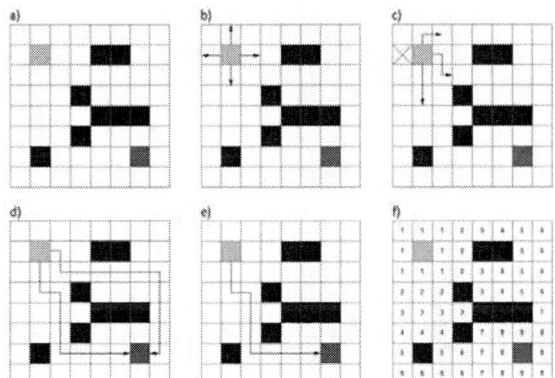

Figure 1. Basic concepts of path-finding algorithm based on the AI, a) topology of the map, b) starting movements, c) probably routing, d) two best ways to reach target cell, e) the low cost optimal route from starting to ending point, f) cost function as a way from cell to cell (map with attached values of cost function needed to reach the cell)

The main goal of the object is to calculate the best way from the starting to the ending points with as low cost as possible. Despite the fact that the information about topography is not located on the map, we can simply represent the land features by assigning to each edge the coefficient, which will define the difficulty of covering the route from point A to B. Determining the coefficient makes it possible to consider not only the slope, but also many more features, such as type of the ground[3].

II. A-STAR ALGORYTHM

A* Algorithm estimates the value of $f(u)$ function (which is returning the approximated distance from the actual point (u) to the end point). Function f(u) is defined as [4]:

$$f(u) = g(u) + h(u) \qquad (1)$$

where $g(u)$ is the weight of the current and optimal path from s to u, and h(u) is an estimate (lower bound) of the remaining costs from u to a goal, called the heuristic function. The heuristic function h(u), which is returning the linear distance between the actual and the ending cell. Heuristic function have to be admissible. We can insert it into the inequality:

$$h(s) \le c(s,s') + h(s') \qquad (2)$$

where s and s' stands as actual and next vertex and c(s,s') is the distance between s and s'. This formula is true for any node s different than the goal point. If the vertex s is equal to the end point then h(s) = 0. Inequality guarantees the admissibility – it is impossible that h(s) would be larger than the real value of cost function reaching the goal point. If such situation would arise and the h(s) would be greater than the real path, algorithm might oversee the optimal path[5]. Due to some limitations, several variants of A* might be distinguished:

- HPA* (Hierarchical Path-Finding A*) - Divides the map to clusters of cells. Then it is finding the route in two steps, first at local level – it finds the optimal route in single cluster and then globally it finds the best way possible by treating clusters as a single units[4]

- IDA* (iterative-deeping A*) – if the value of cost function exceeds a limit value of T (if f(u) > T) the algorithm goes back to another cells and tries to find the route, which length would be shorter than the limit

- SMA* (Simplified memory-bounded A*) - if the algorithm has no more memory available to remember checked cells, it removes memory from bad-promising cell to make place for new ones

The A* Algorithm is based on the Dijkstra's Algorithm, which also can be used to find the shortest path. Dijkstra's algorithm solves the Single Source Shortest Path problem if all edge weights are greater than or equal to zero. Without worsening the runtime complexity, this algorithm can compute the shortest paths from a given start point s to all other nodes [4]. Dijkstra's algorithm works by visiting vertices in the graph, starting with the object's starting point.

The algorithm checks all the edges connected with the actually checked node to search the optimal way to the vertex located across the proper edge. It then repeatedly examines the closest not-yet-examined vertex, adding its vertices to the set of vertices to be examined. it expands outwards from the starting point until it reaches the goal. With usage of Dijkstra's algorithm it is guaranteed to find a shortest path from the starting point to the goal, as long as none of the edges have a negative value of cost function [6].

The main difference between these two algorithms is the appearance of heuristic in the A* Algorithm. If good heuristic is implemented in A*, our object will move exactly to the goal point. It'll bring benefit if comparing this algorithm to spreading-in-every-direction Dijkstra Algorithm. In conclusion, the Dijkstra Algorithm will go through the larger amount of cells.

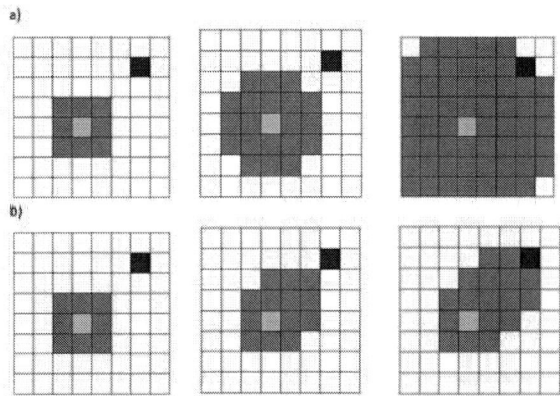

Figure 2.   Cells searched by a) Dijkstra Algorithm, b) A* Algorithm

Algorithm A* can be casted as Dijkstra's Algorithm in a reweighted graph, where we incorporate the heuristic into the weight function as equation:

$$wd(u, v) = w(u, v) - h(u) + h(v) \qquad (3)$$

Figure 1c) presents the process of reweighted edges. All idea hidden behind A* Algorithm can be formalized as lemma [4]:

Let G be a weighted problem graph and h: $V \Rightarrow R$. Define the modified weight $w_d(u, v)$ as $w(u, v) - h(u) + h(v)$. Let $\delta(s, t)$ be the length of the shortest path from s to t in the original graph and $\delta d(s, t)$ be the corresponding value in the reweighted graph.

1. For a path p, we have $w(p) = \delta(s, t)$, if and only if $w_d(p) = \delta_d(s, t)$
2. Moreover, G has no negatively weighted cycles with respect to w if and only if it has none with respect to w_d.

Dijkstra's algorithm works by visiting vertices in the graph commencing with the object's starting point. It then repeatedly examines the closest not-yet-examined vertex, adding its vertices to the set of vertices to be examined. it expands outwards from the starting point until reaching the goal. Dijkstra's algorithm is guaranteed to find a shortest path from the starting point to the goal, as long as none of the edges have a negative cost.

The Greedy Best-First-Search algorithm works in a similar way, except that it has some estimate (called a heuristic) of how far from the goal any vertex is located. Instead of selecting the vertex closest to the starting point, it selects the vertex closest to the goal. (Greedy) Best-First-Search does not guarantee to find the shortest path. However, it runs much quicker than Dijkstra's algorithm, because it uses the heuristic function of guiding its way towards the goal very quickly.

The main advantage of the A* Algorithm is that it combines the pieces of information of the Dijkstra's algorithm (favoring vertices that are close to the starting point) and uses data of favoring vertices that are close to the goal[7].

III.   THE MAZE SIMULATOR

Windows-based environment for the robot control simulation - Microsoft Robotics Developer Studio - gives us great opportunity to analyze virtual machines behavior.

MRDS includes built-in support for packages, which makes it possible to add other services to the suite. One of them is a community-developed Maze Simulator, program that allows to easily build an environment with various types of walls, that can be explored by a virtual robot[8].

The problem we are trying to solve is to get an object from the starting point to the goal through the virtual maze in shortest, optimal time.

IV. PRACTICAL APPLY – THE LABIRYNT PROJECT

Microsoft Robotics Studio[9] is the programmatic environment, which gives us wide range of opportunities in simulating robot movements. The possibility of testing the robot behavior in virtually prepared situation is an important advantage. It's also the easiest way to test our algorithms, it also requires less financial outlay[10].

The physical engine (Nvidia PhysX[11]) expresses the real objects behavior and the forces between them. That's why the program possibilities are virtually limitless. Unfortunately, the understanding of how the multithreaded application works may be hard for the beginners. The biggest advantages of Microsoft Robotics Studio are located in CCR (Concurrency and Coordination Runtime), which is responsible for concurrency and management of asynchronous operations and DSS (Decentralized Software Services), which makes programming efficient programs easier.

The aim of this project was to test and use the A-star algorithm in a model labyrinth. The robot equipped only with the simple touch sensor (bumper), digital compass and CMOS camera. It had to create a map of the area, reaching the aim as quick as it is possible. The results of these tests are presented below[12].

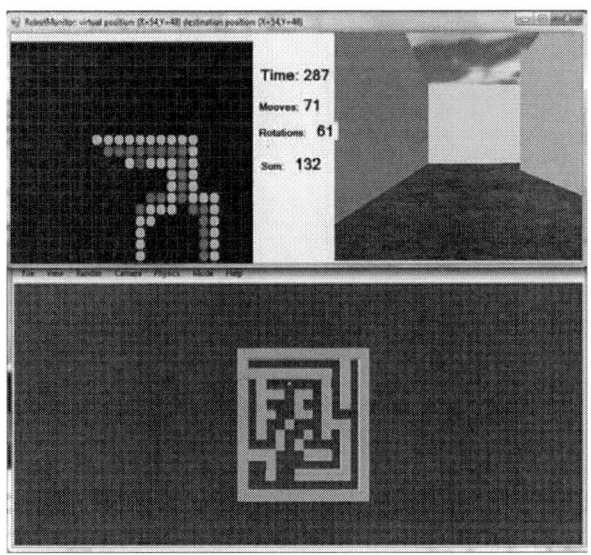

Figure 3. Default settings – first test run

Often occurring collisions forced the usage of proven materials, which had to withstand random errors in relocations and most of all – possibility of turning over. Flat iRobot model was chosen as the test object. Its design stands out from the

competitors with its well-placed center of gravity, which positively affects the amount of generated errors.

The robot is using simplified linear type of movement, which enables the options of learning the terrain and calculating most optimal way of relocation[13]. Installed compass makes it possible to show detailed orientation in defined space and robots current location[14].

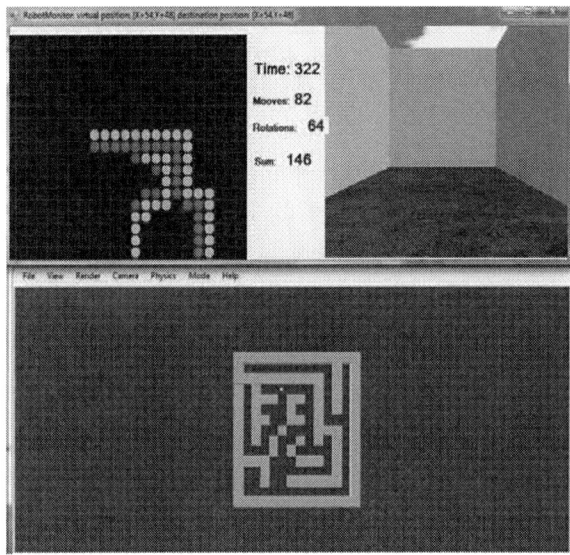

Figure 4. Third serie – the worst result

Presented object requires considerable amount of time to rotate itself and the best effects are shown for the method with the highest power cost (regulations and eventual corrects included). It was proven with the formula specifying cost for the single node (hX stands for the heuristic cost):

$$TotalCost = parentCost + moovingCost + rotationCost + hX \qquad (4)$$

Testing process started in point P(50,50), being the center of virtual system, divided into rectangles with size of 1 meter. General single block relocation cost was set at 3 seconds. Rotation time totaled at 2 seconds for full angle and 1 second for half full angle. Speed of the object was chosen with consideration for low slip and wall bounce.

First test finished at 287 seconds with 71 turns and 61 rotations. Object got to the finish point in 132 operations.

Second run was conducted with higher heuristic cost (times 100), which didn't show noticeable impact on the results. Robot made to the ending location with one turn less and one rotation added.

The third test used modified rotation cost with the same heuristic cost as in the second run. Results were non-satisfactory with the time rising from 287 to 322 second (35 seconds of difference). These analyzes shown that the biggest time costs are generated by rotations algorithm, which stimulates excessive movements in path finding system implemented in this project.

Last test was conducted with higher rotation cost (times 100), which shown the best results for the whole project.

TABLE I.   TEST RESULTS

Serie	Total time	Moves	Rotations	Sum
1	287	71	61	132
2	287	70	62	132
3	322	82	64	146
4	275	69	57	126

Grouping results in series allowed to show the big picture for the conducted tests. Runs presented with first two lines of the table were using default or slightly changed parameters and they managed to produce solid but probable sum. The worst result was observed for the third test, which generated 10% more moves and rotations than algorithms presented on default settings. The last run ended with notable result of 126 sums, which was the best score seen within this testing session.

The biggest losses were discovered in excessive movements, occurring because of wrongly adjusted rotation parameters (mostly at the beginning of the route). Same impact was shown with wall collisions, which generated considerable amount of extra time in the relocation process.

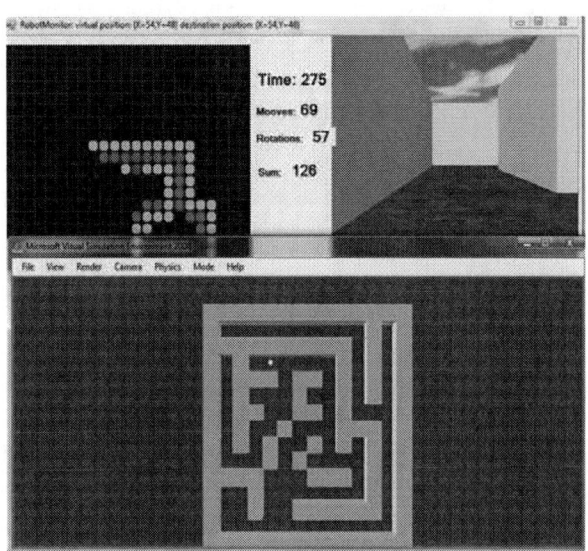

Figure 5.   Best results with highest rotation costs

V.   CONCLUSION

The Impossible Maze Problem project was conducted to show the connection between theoretical algorithms and real-life/virtual simulation intelligent robot solution movements. Analyze results from the conducted test showed the importance of factors included in Dijkstra's and The Greedy Best-First-Search algorithms formulas[15]. Even the smallest changes in values of implemented variables had impact on generated robot intelligence and outcome of the relocation process realization. Details were the most important part of data in this project.

Microsoft Robotics Studio helped to show live results in understandable format and effective presentations form. Virtual simulation preview made it possible to notice the problems the robot had while finding its way through the complicated maze prepared specially for this test session. Solutions changing the amount of preferences like cost of rotations and turns were made in respond to these observations.

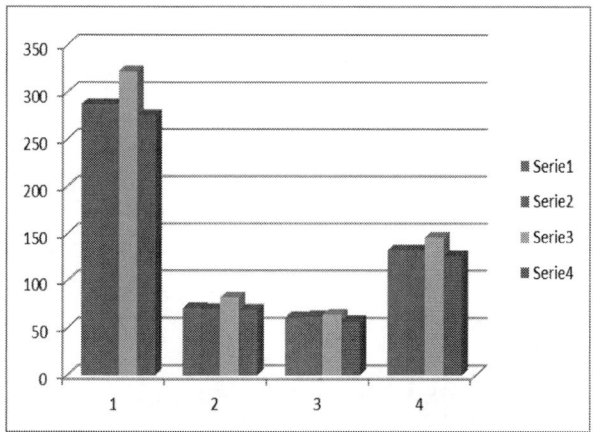

Figure 6.   Results grouped in four series

Project has proven that it is possible to create an effective artificial intelligence[16,17] from simple formulas and documented assumptions with simple tools and low cost robot vehicle[18].

Efficient AI has its place in modern robots and finding out how to implement it in most optimal way is the subject that can result in great power savings in embedded designs. Presented analyzes took us one step closer to revealing the best algorithm for intelligent robot path finding navigation system.

ACKNOWLEDGMENT

The authors would like to acknowledge the contribution of S.K.I.M. department at Poznan University of Technology in making valuable suggestions in improving the content of this paper.

REFERENCES

[1] R. Murphy, Introduction to AI robotics, "Navigation", MIT Press, 2000

[2] P. Sheu,Q. Xue, Intelligent robotic planning systems, Robot Path Planning, World Scientific Publishing Co., 1993

[3] G. Cook, Mobile Robots: Navigation, Control and Remote SensingRobot Navigation, John Wiley & Sons, Inc., 2011

[4] S. Edelkamp, S. Schrodl, Heuristic Search, Theory and Applications, Elsevier Morgan Kaufmann, Inc., 2011.

[5] A. Botea, M. Muller, J. Schaeffer, Near Optimal Hierarchical Path-Finding, Department of Computing Science, University of Alberta Edmonton

[6] G. Gordon, S. Thrun, M. Likhachev, ARA*: Anytime A* with Provable Bounds on Sub-Optimality, pp. 8, School of Computer Science Carnegie Mellon University Pittsburgh

[7] D. Floreano, C. Mattiussi, Bio-Inspired Artificial Intelligence: Theories, Methods, and Technologies, The MIT Press, 2008

[8] S. Russell, P. Norvig, Artificial Intelligence: A Modern Approach, pp. 97–104, Prentice Hall, 2003

[9] K. Johns,T. Taylor, Professional Microsoft Robotics Developer Studio, Concurrency and Coordination Runtim", Decentralized Software Services, Wiley Publishing, Inc.,2008

[10] Toshinori Munakata. Fundamental of the New Artificial Intelligence, Springer, 2008

[11] P.Levi, Symbiotic Multi-Robot Organisms: Reliability, Adaptability, Evolution,Simulation Environments, Springer, 2010

[12] C. Hurtado, A. Valerio, L. Sánchez, Virtual Reality Robotics System for Education and Training, Mechatron. Automotive Res. Center (CIMA), ITESM Toluca, Toluca, Mexico

[13] J. Pochmara,A. Rybarczyk, J. Wencel, The evolutionary approach to control mobile robot, the 4th International Conference on Cybernetics and Information Technologies, Systems and Applications (CITSA 2007), Orlando, Florida, USA

[14] M.Buckland, Programming game AI by example, Vectors, Wordware Publishing, Inc., 2005

[15] D, Bourg,G, Seemann, AI for game developers, A* Pathfinding, O'Reilly Media, Inc., 2004

[16] I. Millington, J. Funge, Artificial Intelligence for Games, 2.2.2 Heuristics, Evlsevier Morgan Kaufmann, Inc.,2009

[17] M. Jones - Artificial Intelligence - A Systems Approach, Jones & Bartlett Publisher, 2008

[18] K. Allen, Launching New Ventures, iRobot: Robots for the home, Houghton Mifflin Company, 2009

Embedded Systems

Digital Hardware for Prime Numbers Generation

Przemysław M. Szecówka, Wojciech Buszko[*]

Faculty of Microsystem Electronics and Photonics
Wrocław University of Technology
Wrocław, Poland
przemyslaw.szecowka@pwr.wroc.pl
[*]now with IBM, Wrocław, Poland

Abstract—**Prime numbers testing algorithm was implemented in specialized digital hardware. The design was coded in VHDL, verified, synthesized and loaded into FPGA. In-house developed Ethernet interface was integrated to provide external control of the computing machine from a desktop computer. Address Resolution Protocol (ARP) was added to extend this connection beyond the local area network. The device may test a single number or generate a series of prime numbers on request. The prototype was constructed and experimentally tested. Prime numbers testing unit may be replicated for higher performance and used for various cryptographic tasks.**

Index Terms—**prime number; digital; hardware; VHDL; FPGA; Ethernet; TCP/IP**

I. INTRODUCTION

Prime numbers have attracted interest of mathematicians and philosophers for hundreds of years. This interest invoked continuous seek for the consecutive prime numbers and development of algorithms for search and check, including several simplified methods delivering subsets of prime numbers fulfilling special conditions. In the last century it was revealed that prime numbers may bring great contribution to cryptography. This induced development of new efficient algorithms followed by the appropriate software. But there is also another approach to the problem - development of dedicated digital hardware [1,2]. For this approach the classic knowledge about algorithms efficiency may need revision, due to specific features of hardware, e.g. natural concurrency (i.e. the *poor* methods may outperform the *smart* ones). Shall be expected that decomposition of prime numbers searching process to consecutive ranges processed in parallel by separate digital blocks will lead to ultimate decrease of search time.

This paper shows preliminary study of contemporary programmable logic circuits application for construction of computing machine dedicated to search for prime numbers. We propose the architecture and implement it in FPGA. The key assumptions were to enable replication of a module to provide concurrent operation. Another issue is user interface and/or communication with software (e.g. cryptographic) utilizing the prime numbers. The authors decided to use LAN connection and hence the appropriate interface was implemented. TCP/IP protocol was coded in hardware description language - VHDL, embedded in FPGA circuit and integrated with prime numbers algorithm. For testing purposes the appropriate software was prepared to provide graphic user interface, control and communication of prime numbers circuitry with host PC computer.

II. PRIME NUMBER TEST ALGORITHM

A. Algorithm

Prime number is, by definition an integer which cannot be divided by another integer (excluding 1 and the number itself) to produce integer result. There are numerous algorithms to check this condition, starting from primitive exhaustive search for divider in the full range, via skipping even numbers, shrinking the search space to square root of the number etc. The authors decided to apply a method based on operations on integer numbers exclusively, due to the nature of digital hardware providing much higher speed for fixed point arithmetic. The algorithm is presented in Fig. 1. After checking parity of the number, variables a and b are preset to 3 and series of iterations takes place. Then b is increased by 2, with a following its previous value until the square of b exceeds the input number. If the equality betweens multiplication result and the input number is not found in all these iterations, another series of steps take place. This time a is decreased by 2 until it reaches 3. If equality is not found at any step, the number is proved to be prime.

B. Implementation

The algorithm was transformed to digital architecture, providing operation in two modes. In the first mode the machine checks whether the number provided on the input is prime. In the second mode it sends to output all prime numbers found in a specified range defined on input. The architecture is full synchronous, with single clock and common asynchronous reset. Series of registers store actual values of a and b, actual number checked, the last prime number found, iteration counter and series of control signals. Combinatorial logic provides increment/decrement arithmetic and multiplying. The design was implemented in hardware description language, VHDL [3]

TABLE I. SYNTHESIS RESULTS FOR PRIME NUMBERS GENERATOR (XILINX SPARTAN 3E, 500 K GATES)

	Used	% of all
Slices	528	11
Slice Registers	261	2
Slice LUTs	939	10
18 x 18 multipliers	8	40
Clock frequency	54 MHz	

as a single entity, verified and synthesized for an FPGA using Xilinx ISE tools [4]. Results of synthesis are summarized in Table I. Logic complexity of prime number generator (single unit) may be estimated to 50 000 gates. The module may be replicated for concurrency in two ways. The trivial method is cloning of the structure to provide simultaneous testing of various numbers. Another approach is variation of each unit by specific start values and algorithm stop conditions defined for a and b variables. For this approach the search space may be decomposed to series of subsets, with specialized unit allotted to each of them. In such case all the units may process a single number in parallel, decreasing the response latency.

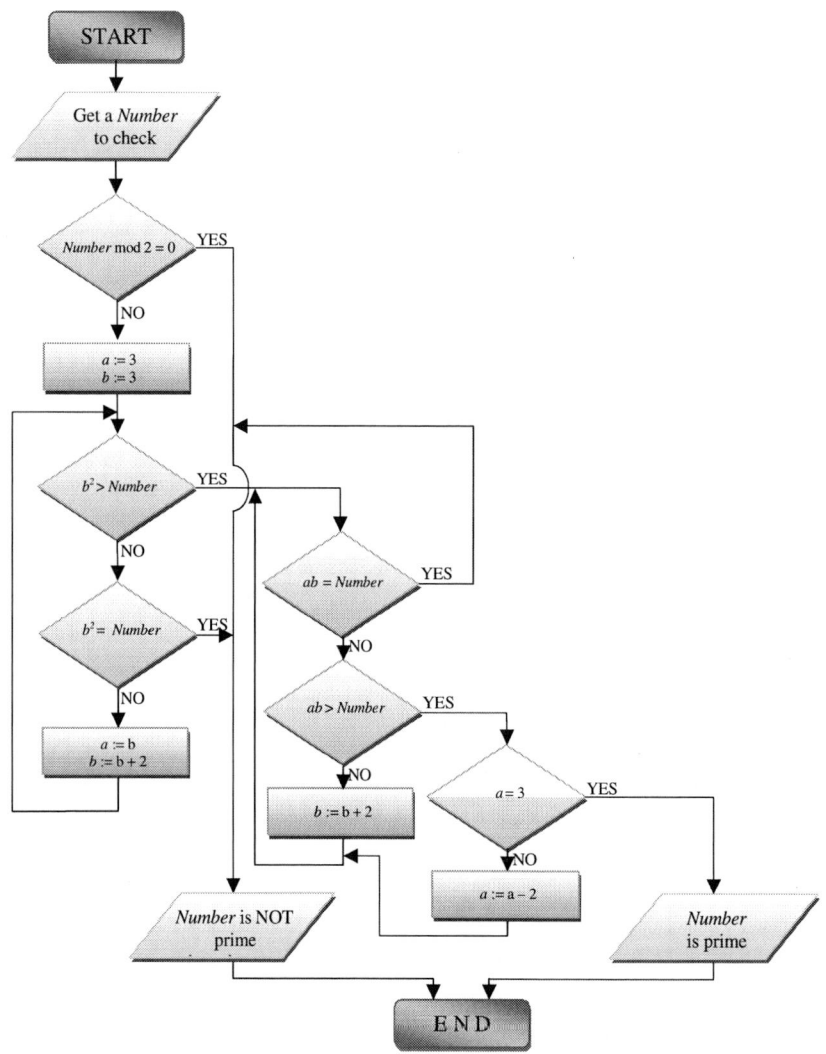

Figure 1. Prime number checking algorithm applied for the design.

III. ETHERNET INTERFACE

Ethernet interface for prime numbers generator was based on original in-house developed architecture. The concept is presented in Fig. 2. It consists of classic modules providing sending and receiving data and more advanced IP&ARP sender block. ARP provides translation of IP addresses understood by the internet to physical addresses of Ethernet nodes – MAC [5].

It means that besides a local cable connection of a PC Ethernet card and the FPGA board (just like in popular Wi-Fi routers, configurable via Internet Explorer), this device shall be accessible from any computer in the world, connected to the internet. All the protocol was implemented directly in hardware, with no microprocessor involved.

Roughly speaking, internet communication consists in series of encapsulations of data structures into bigger ones on the transmitter side and then the reverse process of extraction on the receiver side. Depending on position of the device in transmission mechanism, more or less detailed extraction takes place. Each level of encapsulation may have its own mechanisms of addressing and control.

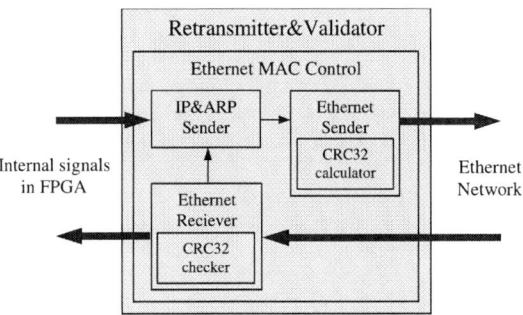

Figure 2. Ethernet interface schematic.

Receiver part is responsible for collection of data coming from the internet. The process is controlled by the FSM outlined in Fig. 3. After awaking, the incoming bitstream is analyzed to extract frames, IP datagrams, UDP packets and eventually the data is collected. For IP datagrams classic version 4 of IP protocol was applied. At any stage of analysis if the header doesn't match the defined one, the process of receiving is stopped.

Transmitter block is responsible for construction of frames transmitted over Ethernet. This process is controlled by relatively simple FSM presented in Fig. 4. Most of the states are reached in a plain schedule, except No. 4 and 5, dedicated for data byte transmission. Two states are dedicated for each byte because only 4 bits are transferred in a clock cycle.

Advanced functionality of the interface is secured by the IP&ARP transmitter block. It supports operation of both receiver and transmitter parts, as shown in Fig. 2. It is controlled by another FSM, shown in Fig. 5. In the begin the ARP messages are exchanged. Then if transmission is enabled, in consecutive states the appropriate headers are transmitted .

TABLE II. SYNTHESIS RESULTS FOR PRIME NUMBES GENERATOR INTEGRATED WITH COMMUNICATION MODULE

	Used	% of all
Slices	2474	53
Slice Registers	2760	29
Slice LUTs	3069	32
18 x 18 multipliers	8	40
Max clock frequency	54 MHz	

The transmission is protected by standard 32-bit CRC. It was implemented as a shift register with XOR gates (Fig. 6). Same structure was replicated for receive and send blocks.

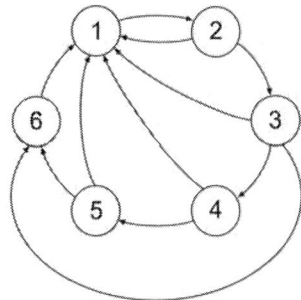

Figure 3. Finite State Machine for Ethernet receiver. 1 – idle state, 2 – skip preamble, 3 – frame header analysis, 4 - IP header analysis, 5 – UDP header analysis, 6 – payload extraction.

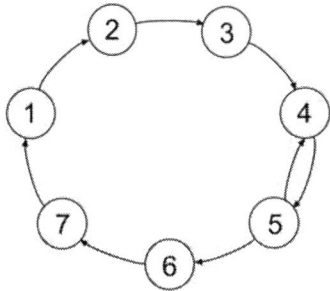

Figure 4. Finite State Machine for Ethernet transmitter. 1 – idle state, 2 – send preamble, 3 – send frame header, 4 – send lower part of byte, 5 – send higher part of byte, 6 – send CRC, 7 – sending blocked.

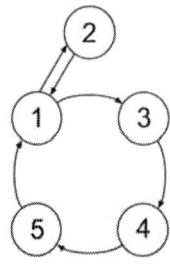

Figure 5. Finite State Machine for IP&ARP transmitter. 1 – idle state, 2 – send ARP response, 3 – send IP header, 4 – send UDP header, 5 – send payload

Ethernet interface was implemented in VHDL, verified and synthesized. This part of device required substantially higher design effort and eventually used much more FPGA resources than single prime numbers generator. The results of synthesis of the whole design (prime numbers generator and the interface) are shown in TABLE II. Similar maximum clock frequency was obtained thus it shall be presumed that Ethernet communication module could operate with its own clock, faster than prime numbers checker. It may be considered to apply two separate clocks for generator and interface. For the design with multiple prime number generators, the interface could be a bottleneck for operation. In the reference design there was only one generator implemented and a single clock was applied for the whole device to avoid synchronization problems.

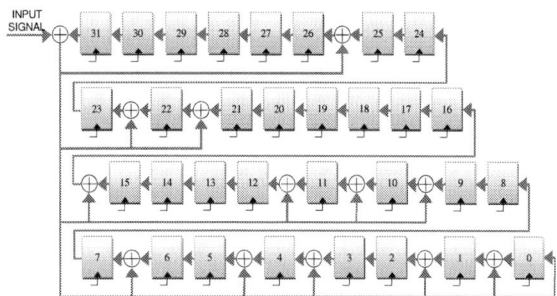

Figure 6. CRC32 control of frame.

IV. PROTOTPE AND EXPERIMENTS

Prime numbers generator together with Ethernet interface was physically implemented on a prototyping board with Xilinx Spartan 3E FPGA [6]. Complementary software was written for PC to provide user interface for control of prime numbers generation and communication with the FPGA. Program window is shown in Fig. 7. It presents current results of search for primary numbers in range from 100 to 200 millions. Consecutive lines present prime numbers and their serial numbers. Additional information, in brackets, is number of clock cycles used to find particular numbers.

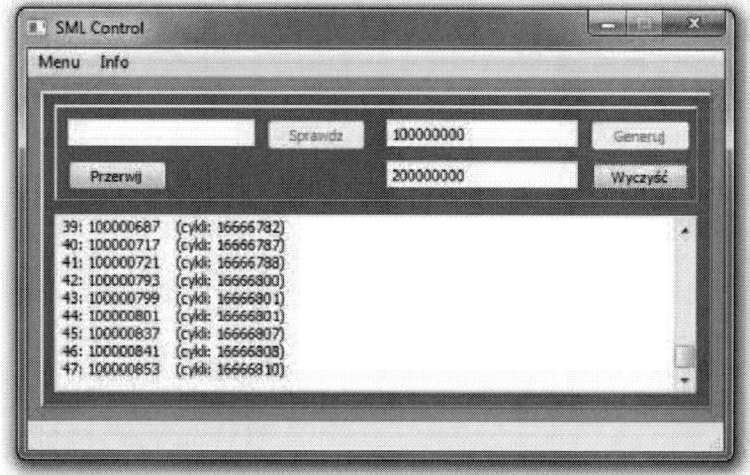

Figure 7. User interface of computer software controlling prime numbers generator implememnted in hardware.

V. CONCLUSIONS

Specialized computing machine for prime numbers testing was proposed. The device provides two modes of operation – generation of all prime numbers from a given range and checking whether a specific number is prime. The experiments were performed on a singular structure. It may be replicated to provide concurrent testing of several numbers as well as for allocation of several units to the shared task of testing a single number.

Special part of experiment was implementation of in-house developed Ethernet interface. This module provided communication with user interface software run on a PC computer. Additional feature was Address Resolution Protocol (ARP), which provides remote access to the device from internet. It was shown that compact version of computing machine, performing tasks defined by any user logged into the internet, may be constructed of digital hardware only, with no microprocessors and no software involved.

Prime number algorithm and communication module were integrated in a single programming device (FPGA). Working

prototype of device was constructed, using universal board delivered by Xilinx. Experiments have shown that the device works properly. For small numbers to check the hardware appears to be relatively slow, with most of the time consumed by communication. Then, for increasing values the growing superiority of hardware over a desktop computer is visible.

Further research will focus on application of prime number checker for real cryptographic tasks and on well balanced decomposition of search process to appropriate ranges.

REFERENCES

[1] M.H. Rais, S.M. Qasim, FPGA implementation of Rijndael algorithm using reduced residue of prime numbers, Proc. 4th International Design and Test Workshop (IDT), Nov. 2009, pp.1-4.

[2] R.C.C Cheung, A. Brown, W. Luk, P.Y.K. Cheung, A scalable hardware architecture for prime number validation, Proc. IEEE International Conference on Field-Programmable Technology, 2004, pp. 177- 184.

[3] VHDL, IEEE Std No. 1076, 2000.

[4] Xilinx ISE Web Pack, www.xilinx.com, 2011.

[5] RFC 826:1982, An Ethernet Address Resolution Protocol.

[6] Spartan 3-E FPGA Starter Kit Board User Guide [online], Xilinx, Inc., 2008.

MIXDES 2012, 19th International Conference *"Mixed Design of Integrated Circuits and Systems"*, May 24-26, 2012, Warsaw, Poland

Hierarchical UML Activity Diagrams into Control Interpreted Petri Nets Transformation

Michał Grobelny*, Iwona Grobelna*, Marian Adamski

Institute of Computer Eng. And Electronics
University of Zielona Góra
Zielona Góra, Poland
m.grobelny@weit.uz.zgora.pl, {i.grobelna, m.adamski}@iie.uz.zgora.pl

Abstract — **The paper presents bi-directional transformation of two hierarchical hardware behavior description diagrams - UML Activity Diagrams and Control Interpreted Petri Nets. The transformation rules covers representation of UML complex activities by means of Petri Net macroplaces and macrotransitions. Bi-directional transformation ensures fully efficient cooperation of designers teams working with both mentioned technologies. Additionally, transformation of UML Activity Diagrams enables use of various analysis, verification and synthesis techniques available for Petri Nets, which is especially important considering safe reconfigurable logic controllers.**

Index Terms — **UML Activity Diagrams, Control Interpreted Petri Nets, logic controller, hierarchical specification techniques**

I. INTRODUCTION

Logic controller design phase is responsible for the shape and functionality of final project. Considering safe reconfigurable logic controllers, the importance of the phase increases significantly. Behavior modeling is a part of the phase, which describes the functionality rules of desired product. Behavior specification can be realized using various forms. Mostly known are Sequential Function Charts, Algorithmic State Machines and Petri Nets [2, 3]. On the other hand Unified Modeling Language (UML) [7], which was designed to visualize software specification, appears in miscellaneous new domains. Moreover, usage of UML in logic controller design becomes every day more popular. One of the UML diagrams, exactly Activity Diagrams [4, 7], are very useful in behavior specification. Although there is a lack of analysis, verification and synthesis techniques dedicated for logic controller design with use of Activity Diagrams. The gap can be filled with use of transformation into Petri Nets. Petri Nets have enormous amount of desired quality improvement techniques. Control process specification can be formally verified against behavioral properties using model checking techniques [4, 5]. Transformation of Activity Diagrams into Petri Nets with ensured system behavior consistency will

enable usage of Petri Net techniques to designers familiar with UML. Logic controller design may vary from simple couple actions diagram to very complicated and multi-diagram specification. The design complexity often enforces usage of hierarchical specification techniques. Working with complex design hierarchically organized is more efficient and it improves readability and understanding of designed functionalities. Therefore transformation method of hierarchical Activity Diagrams of UML into hierarchical Control Interpreted Petri Nets [2] increases design efficiency parallel with decrease of possible errors ratio. Proposed novel approach extends available transformation techniques [1, 6] making them available in hierarchical control process description. It is based on dedicated subset of elements of both diagrams. Subsets fulfill all logic controller design requirements and in the same time ensure bidirectional transformation [4].

II. UML ACTIVITY DIAGRAMS IN LOGIC CONTROLLER DESIGN

UML [7] in last couple of years is rapidly gaining importance in software and hardware development domains. Originally it was dedicated to model system architecture in software engineering. However, with technology progress UML has appeared in slightly different regions than previously desired. Moreover, it simplifies information flow between team members and enables easy understanding of system design and functionality for non-experienced staff. Actually, UML is present in such areas as business modeling, workflow specification and last but not least in discrete controller design. Multiple diagram types can be exploit in the design of different aspect of the controller. Behavioral embedded system specification can also be prepared by using some types of UML diagrams, like Activity Diagrams, state machines or sequence diagrams. The paper concentrates on hardware behavioral modeling acquired with Activity Diagrams of UML, which are dedicated to describe system dynamic.

Formally, UML Activity Diagram dedicated to describe logic controller specification or control process can be defined

  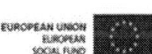

* Authors are scholars within Sub-measure 8.2.2 Regional Innovation Strategies, Measure 8.2 Transfer of knowledge,
Priority VIII Regional human resources for the economy Human Capital Operational Programme co-financed by European Social Fund and state budget

as a 7-tuple AD = (A, T, G, F, S, E, Z), where A: set of actions/activities; T: set of transitions; G: set of guard conditions corresponding to transitions (input signals); F: set of flow relation between the activities and transitions; S: an initial state (or set of states); E: a final state (or set of states); Z: set of output signals.

System design decomposition and hierarchical representation of systems are important issues in large control systems. Usually a behavioral specification cannot be presented in one diagram with enough readability maintained. This aspect enforces decomposition of complex designs into hierarchical structures concerning autonomous segments of the design. In UML hierarchy of activities is reached be usage of complex activities, which contain sub-diagrams specifying sub-processes.

III. CONTROL INTERPRETED PETRI NETS

Petri Nets [2, 3] were firstly introduced as a general purpose mathematical model for describing relations between conditions and events. Graphic representation of Petri Net can be understood even for non-technical staff. It allows e.g. to specify such behaviors as parallelism and concurrency, choice, synchronization, memorizing, reading or resources sharing [2].

Formally, a Petri Net can be defined [6] as a 3-tuple PN = (P, T, F), where: P: set of places; T: set of transitions; F: set of flows between elements of P and T or elements of T and P (it is forbidden to create connections between two elements of the same type e.g. T and T or P and P).

Control Interpreted Petri Nets [2] specify and model the behavior of concurrent logic controllers and take into account properties of controlled objects. Local states may change after firing of transitions, if some events occur. Transition guards are associated with input signals of controller, while places are associated with its output signals. Global state of logic controller is built of simultaneously holding local states.

Formally, Control Interpreted Petri Net can be defined [6] as a 6-tuple PNIO = (PN, X, Y, ρ, λ, γ), where PN: alive and safe Petri Net; X: set of input states; Y: set of output states; ρ: T $\rightarrow$ 2X is a function, that each transition assigns the subset of input states X(T); 2X states for the set of all possible subsets of X; λ: M $\rightarrow$ Y is a function of Moore outputs, that each marking M assigns the subset of output states Y(M); γ: (M x X) $\rightarrow$ Y is a function of Mealy outputs, that each marking M and input states X assigns the subset of output states Y.

IV. TRANSFORMATION RULES OF HIERARCHICAL DIAGRAMS

Usually in software or hardware projects there is one defined and imposed specification type. However, technology progress enforces the growth of design size with the growth of systems complexity. Therefore, systems specification and design can be simultaneously developed by more than one project team. Decomposition of design is then necessary. Moreover, two different teams cooperating can prefer different specification techniques. Therefore, usage of both UML Activity Diagrams and Control Interpreted Petri Nets in one

big project is probable. Furthermore, specification is usually discussed with the sponsor or client. Not always it is highly professional with knowledge about extensive design and specification techniques. Usage of UML Activity Diagrams seems to be justified, taking into account better readability and understanding of UML for non-professionals. Transformation between discussed diagram types can be a bridge connecting two cultures and enabling efficient communication and cooperation during project lifetime.

A. Specified controll process

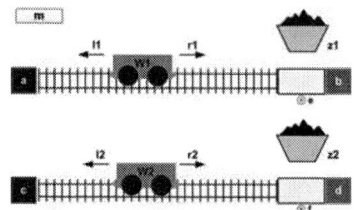

Figure 1. Real model of specified process

A simple process for transport of friable goods with two carriages is considered (presented in Fig. 1). It is realized as follows. Initially, the carriages are in starting points a and c. The process is started after pressing the m button. Firstly, carriages move to the right simoultanously. The movement lasts so long, until the carriages reach destination points, b in case of carriage W1 and d in case of W2. When carriage reaches own destination point, the adequete chute is opened (z1 for W1 carriage, and z2 for W2) and as the result friable goods flows into the carriage, until it is full (indicated by e signal in case of W1 and f signal in case of W2). Then the chute is closed and the carriage W1 begins to move left (signal l1). The movement lasts so long, until the carriage reaches its staring point a. Then, carriage W2 moves left (signal l2) until it reaches own starting point c.

B. UML Activity Diagram into Control Interpreted Petri Net transformation with macrotransitions

Fig. 2a presents specification of discussed process with use of UML Activity Diagrams. Diagram is presented in decomposed form with simple actions. There are activities marked with rectangles, accordingly ACT1, ACT2, ACT3 and ACT4. ACT1 and ACT2 describe two concurrent processes realizing carriages movement to the right and feeling with friable goods. ACT3 defines sequential process of carriages return trips. ACT3 and ACT4 are the top most activities (see Fig. 2e). Moreover ACT4 can be decomposed to a middle level with ACT1 and ACT2 (see Fig. 2c). The transformation of UML Activity Diagram into Control Interpreted Petri Net is done accordingly at all levels of abstraction. Rules of transformation without hierarchy can be found in [4]. Transformation of hierarchical structures is strictly to the rule: activities are transformed into macrotransitions. Equivalent diagrams of both specification techniques are Fig. 2a with Fig. 2b, Fig. 2c with Fig. 2d and Fig. 2e with Fig. 2f. Moreover, it is a simple extending of rule defining, that actions are transformed into transitions of Petri net.

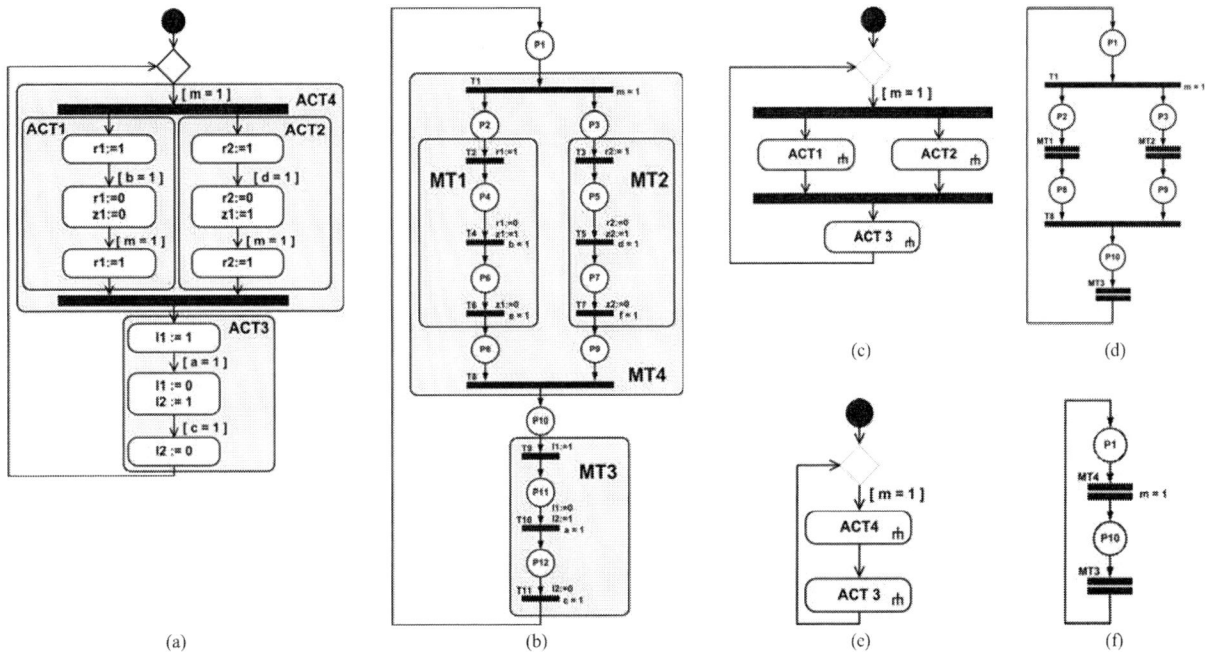

(a)   (b)   (c)   (d)   (c)   (f)

Figure 2.   Specification of presented process with use of hierarchical UML Activity Diagram (a), (c), (e) and hierachical Control Interpreted Petri Net (b), (d), (f).

Thus, activity ACT1 of UML Activity Diagram is transformed into macrotransition MT1. Furthermore, activities ACT2, ACT3 and ACT4 correspond to macrotransitions MT2, MT3 and MT4.

C. Control Interpreted Petri Net into UML Activity Diagram transformation with macroplaces

Due to ensure fully compatibility between two graphical behavioral logic controller specification techniques backward transformation provide. In the previous section discussed process is defined with use of Mealy outputs and activities. It is enforced by the specific construction of UML Activity Diagrams. This technique does not contain any static elements as places of Petri Nets. Therefore, there is no possibility to present connection of output to a state of the process. Single actions present the dynamic of the process, which means that one action can activate a signal and the other one can deactivate it. However designer using Control Interpreted Petri Nets as a major technique can specify processes with Moore outputs and moreover can use macroplaces (see Fig. 3a). Both elements are not available in Activity Diagrams. Therefore, during the transformation of the specification source diagram (Control Interpreted Petri Net) has to be partially modified.

Firstly, Moore outputs have to be swapped into Mealy outputs and secondly, macroplaces have to be changed into macrotransitions. Further processes are equivalent to backward transformation presented in previous section B. In details process is described further in current section. The first necessary step is to exchange Moore outputs into Mealy outputs. Therefore, signals assigned to places have to be removed. In order that processes have to be functionally equal, the signal is swapped into activation and deactivation in

surrounding transitions. In some cases, there have to be additional place and transition inserted to a process. For instance to place P2 with signal r1 assigned from Fig. 3a an extra transition have to be added. This is necessary in order to preserve the process behavior unchanged and to change Mealy into Moore output. Then, transition T2 from Fig. 3a becomes transition T4 in Fig. 3b and place P2 is changed into place P4 without any signal assigned. Additional new place P2 and transition T2 are created. In place of assigned to place P2 signal r1 there is an activation of signal r1:=1 in transition T2 and deactivation of signal r1:=0 in transition T4 (Fig. 3b). The above process presents swapping of Moore output into Mealy output, with addition of extra place and transition. In discussed control process specification three similar places are present when expansion of net is necessary. It is the case by signal r1, r2 and l1. Other Moore signals into Mealy signals modifications do not enforce growth of the diagram. For example, transformation of signal z1 assigned to place P4 is less complex. Activation of the signal (z1:=1) is inserted to transition T4 and deactivation is inserted to transition T6. Thus, no extra added elements are necessary and existing ones are sufficient to fulfill the requirements of Moore into Mealy output transformation.

Further step of whole UML Activity Diagram into Control Interpreted Petri Net transformation process is swapping of macroplaces into macrotransitions. In the case of presented control process there are three macroplaces MP1, MP2 and MP3. The macroplaces are accordingly transformed into macrotransitions MT1, MT2 and MT3 as follows. Macroplace MP1 contains places P2 and P8 which do not have any functionality and can be omitted without changing the process behaviour.

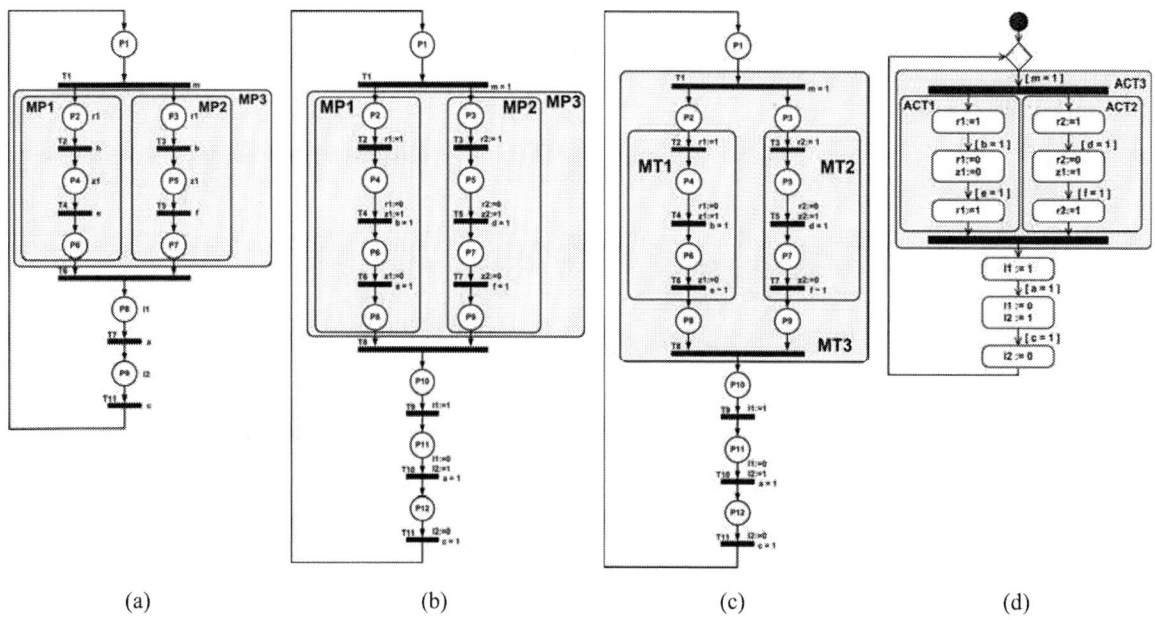

Figure 3. Four steps of transformation from Control Interpreted Petri Net into UML Activity Diagram, Control Interptered Petri Net with macroplaces and Moore outputs (a), macroplaces and Mealy outputs (b), macrotransitions (c) and UML Activity Diagram with activities (d)

Thus, macrotransition MT1 is created. Accordingly macroplace MP2 is transformed into macrotransition MT2. The next step during whole transformation process is to exchange macroplace MP3 into macrotransition MT3. Ought to create a macrotransition surrounding transitions are combined together with macroplace MP3.This scenario is alternative to additional transition creation. It is possible due to the existence of surrounding transitions and the fact that such modification does not change the process functionality. Generally meaning the swapping of macroplace into macrotransition process can be described as change between "the process is in the x-th state" into "the process is executing the x-th activity". Last step in this transformation is the change of macrotransitions into activities. The transformation of simple elements as places and transitions is not discussed as it is out of scope of the paper. Rules of simple transformation of UML Activity Diagrams into Control Interpreted Petri Nets (and reverse transformation) can be found in [4].

V. CONCLUSIONS

Behavioral logic controller specification can be realized with use of several graphical specification techniques. UML Activity Diagrams and Control Interpreted Petri Nets are good representatives. Both techniques are well known and popular. Transformation described in the paper is a bridge between both specification techniques. Moreover, use of hierarchical structures and decomposition can be helpful in large and complex designs. On the other hand, decomposition can be very helpful in processes embedded in reconfigurable logic

controllers. There is a possibility to describe atomic processes, which can be easily changed during the partial reconfiguration even during employment. Bi-directional characteristic of the transformation enable existence of both specification techniques in one project. Furthermore, designers working with both technologies can communicate and exchange own parts of the design without any delays.

REFERENCES

[1] F. Basile, P. Chiachio, D. Del Grosso, "Modelling automation systems by UML and Petri Nets", Proceedings of the 9th International Workshop on Discreet Event Systems, pp. 308 – 313, 2008.

[2] R. David, H. Alla, "Discrete, Continuous, and Hybrid Petri Nets", Berlin Heidelberg: Springer Verlag, 2010.

[3] C. Girault, R. Valk, "Petri Nets for Systems Engineering. A Guide to Modeling, Verification, and Applications", Berlin Heidelberg: Springer Verlag, 2003.

[4] I. Grobelna, M. Grobelny, M. Adamski, "Petri Nets and activity diagrams in logic controller specification - transformation and verification", Mixed Design of Integrated Circuits and Systems (MIXDES), Proceedings of the 17th International Conference, 2010, pp. 607 – 612.

[5] I. Grobelna, M. Adamski, „Model Checking of Control Interpreted Petri Nets", Mixed Design of Integrated Circuits and Systems (MIXDES), Proceedings of the 18th International Conference, 2011, pp. 621 – 626.

[6] T.S. Staines "Intuitive Mapping of UML 2 Activity Diagrams into Fundamental Modeling Concept Petri Net Diagrams and Colored Petri Nets", 15th Annual IEEE International Conference and Workshop on the Engineering of Computer Based Systems, pp. 191 – 200, 2008.

[7] Unified Modelling Language website: http://www.uml.org

MIXDES 2012, 19ᵗʰ International Conference *"Mixed Design of Integrated Circuits and Systems"*, May 24-26, 2012, Warsaw, Poland

Wireless Communication Solutions for Distributed Strain Measure Systems in Mechanical Structures

Artur Andrzejczak, Piotr Pietrzak, Maciej Makowski, Andrzej Napieralski

Department of Microelectronics and Computer Science

Technical University of Lodz

Lodz, Poland

andrzejczak@dmcs.pl

Abstract—**This paper describes considerations associated with design of wireless distributed strain measure system of mechanical structures with particular emphasis on selection of wireless communication technology, which will be applied in TULCOEMPA project. Problems associated with specific wireless communication standards, the reasons of their use, benefits and problems that come with them. Wi-Fi, Bluetooth and IEEE 802.15.4 networks are being presented and compared with each other. The summary of the features allows to identify the areas of applications, in which those communication protocols are suitable, primarily in context of proposed structure of wireless strain measure system.**

Index Terms—**strain, stress, humidity, temperature measure, wireless communication, Wi-Fi, Bluetooth, IEEE 802.15.4, ZigBee, distributed measure systems**

I. INTRODUCTION

Strain monitoring systems allow to control and maintain condition of various mechanical structures, like bridges, halls, buildings and dams. The main reasons for applying this kind of solutions are safety issues. Overload of mentioned structures may cause substantial costs. What is more, damage or even collapse comes with major threat for their users. There are many factors, which can cause stress overload of structures, for example weather conditions. Strong wind, heavy snowfalls may exert major impact on the exposed structure. Ageing of materials has also significant influence on object's preservation. Because of that, popularity of strain monitoring systems is permanently rising. Constant measure of strain or/and stress allows to take appropriate action in a short time, providing safety and cost efficient operation of monitored structures.

In particular cases it is worth to consider the wireless communication, which can come with several benefits. The most significant advantage is avoidance of structural cabling. There are many situations, where providing wire connection between measure nodes comes with unacceptable costs. Obstacles can be both physical or formal. In first case, high temperature or humidity may lead to faster degradation of insulation and connectors. Those conditions can cause higher costs, which are connected with installation and proper wire quality. Formal obstacles often comes with additional time and expenses. Mentioned issues become increasingly important when measure system is being installed for short, specified amount of time, usually for diagnostics purpose. In this kind of situations, in spite of mains availability, wireless communication is still efficient solution.

This article focuses on choosing proper wireless communication standard for strain measure system intended for TULCOEMPA project. Its aim is to develop strengthening technology for concrete bridges the information about strain will provide the feedback about the efficiency of the applied enhancements, which will allow to improve them.

II. ISM BANDS

Most of currently applied wireless communication solutions for measure devices use the ISM (Industrial Scientific Medical) band. It is a set of bands, which doesn't require any licence or permission to use with limited transmit power. Primarily ISM frequencies were intended for industrial, scientific and medical purposes, other than communication, for example, microwave ovens. Today, in spite of original destination, they are generally used for short range radio communication, often accompanied by creation of small network infrastructures.

TABLE I. EXAMPLE ISM BANDS

Example ISM bands	
6.765 MHz	7.000 MHz
13.553 MHz	13.567 MHz
26.957 MHz	27.283 MHz
40.660 MHz	40.700 MHz
433.050 MHz	433.920 MHz
868.000 MHz	870.000 MHz
2.400 GHz	2.500 GHz
5.725 GHz	5.875 GHz
24.000 GHz	24.250 GHz
61.000 GHz	61.500 GHz
122.000 GHz	123.000 GHz
244.000 GHz	246.000 GHz

Due to initial use and wide applicability of this kind of equipment, transceivers working on ISM band should be able to deal with various interferences. Scopes of ISM frequencies are defined by Radiocommunication Sector of International

Telecommunication Union, however they may vary in different countries, because of local regulations. (Table I) presents example ISM frequencies.

III. COMPARISON OF TECHNOLOGIES USING ISM BANDS

For strain measure system three most popular standards are taken under consideration (Table II), WiFi, Bluetooth in currently most popular version 3, and ZigBee, which is based on IEEE 802.15.4.

Significant differences between those technologies result in high importance of matching area of application to the merits of various standards. Mismatch of chosen protocol and needs of a particular solution can lead to unexpected downtimes and additional costs. Hence, it is essential to understand features and destination of mentioned standards.

A. ZigBee

ZigBee is suitable for solutions requiring high power efficiency, wireless nodes are often battery powered. Because of that, low hardware and protocol complexity and as short as possible network joining time are expected. Physical layer, medium access control layer and logical link control sublayer for ZigBee are provided by IEEE 802.15.4. Using only those three first layers gives the possibility to lessen transceiver complexity and power consumption. The mentioned characteristics in conjunction with the possibility of large (2^{64}) number of clients in one networks and link security, suits to the concept of distributed wireless sensor networks, which makes ZigBee accurate solution for industrial electronics. The most significant drawback of IEEE 802.15.4 and ZigBee networks is low bandwidth, which restricts considered technology to applications with low demand for link data rate.

B. Wi-Fi

WiFi networks due to the complexity protocols, and as a result high computational requirements, are not able to operate for longer time periods on battery power, which restricts them to the mains supply. In comparison to IEEE 802.15.4 and ZigBee networks, bandwidth is a substantial advantage, which opens up possibilities for high data rate demanding applications, like video broadcast or measure systems with high frequency dynamics. Despite this essential difference, the two of mentioned technologies can complement each other. As an example may serve wireless measurement network. Nodes send acquired data to the local coordinator using IEEE 802.15.4, which forwards results to data centre using Wireless LAN.

C. Bluetooth

When considering the Bluetooth, its major advantage is its presence in most portable consumer electronics, for example smart phones, tablets and notebooks. Therefore it is proper solution for wireless configuration interfaces, when the operator uses e.g. netbook to set up or read machine's

parameters. High complexity of the protocol connected with significant restrictions tightens the set of possible applications.

TABLE II. COMPARSION BETWEEN WIRELESS STANDARDS

	ZigBee	Bluetooth	WiFi
Throughput	250 kbit/s	24 Mbit/s	600 Mbit/s
Estimated range	From 10 m to few kilometers	From 10 m to few hundred meters	From 50 m to few kilometers
Available Bands	868, 928, 2400 MHz	2400 MHz	2400, 5000 MHz
Protocol complexity	Low	High	High
Power consumption	Low	Medium	High
Network join time	Below 30 milliseconds	Few seconds	Few seconds
Topologies	ad-hoc, star, mesh	ad-hoc, star	ad-hoc, star
Maximum number of clients	Very high (2^{64})	Little (7)	High, but bandwidth shared between clients
Security	High – AES	Medium – SAFER	High – AES
Example applications	Wireless, distributed controls systems like industrial automation and home control	Mainly consumer electronics, usually ad-hoc networks in order to exchange data	Wireless networks with TCP/IP protocol, complex control and diagnostics systems with high data exchange
Popularity	Average in industry and consumer electronics	High in consumer electronics, low in industry	High in both consumer and industry electronics

Medium power consumption and relatively long network join time induces to consideration other standards. As a confirmation of outlined opinion might be a new Bluetooth standard, version 4. It divides Bluetooth to low energy version using low-complexity, with IEEE 802.15.4-like bandwidth, protocol and high speed version which is based on WiFi. In the future, Bluetooth v4 Low Energy might compete with IEEE 802.15.4 and ZigBee networks.

D. Characteristic of IEEE 802.15.4

IEEE 802.15.4 defines the two first layers of Open Systems Interconnection model – physical (PHY) and medium access control (MAC). This protocol provides reliability and throughput acceptable for low-frequency measure and control systems, while ensuring low power consumption and complexity of transceiver. When more sophisticated communication features are required, additional higher-layer protocols like ZigBee or 6LoWPAN might be applied.

Major application field for solutions using IEEE 802.15.4 are wireless sensor networks. They consist of many small, similar to each other, nodes, deployed in particular area. Typically physical quantities like temperature, humidity or light intensity are being measured. As previously mentioned, considered measurement system analyses strain. Wireless sensor network applied in this particular solution allows to continuous monitoring strain and stress of mechanical structures and ambient conditions around it. Gathered quantities from many points lead to more accurate assessment of the state of the analyzed object.

There are three available frequency groups (Table III), which can be used by IEEE 802.15.4. Depending on selected frequency and modulation, different throughput and range might be available.

TABLE III.   AVALIBLE FREQUENCIES AND MODULATION FOR IEEE 802.15.4

Band	Number of channels	Throughput	Modulations	Area
868.0 868.6 MHz	1	20 kbit/s	BPSK	Europe
		100 kbit/s	O-QPSK	
		250 kbit/s	ASK	
902.0 928.0 MHz	30	40 kbit/s	BPSK	North America
		250 kbit/s	ASK	
		250 kbit/s	O-QPSK	
2400.0 2483.5 MHz	16	250 kbit/s	O-QPSK	World

IV. RELATIONSHIP BETWEEN BAND AND COVERAGE

2.4 GHz band is a popular choice because of the worldwide availability. In earlier specification of IEEE 802.15.4, 860 MHz and 900 MHz band offered lower throughput (20 and 40 kbit/s), which made 2.4 GHz even more appealing.

The distance over which connection can be set up depends on the band, in which transceiver operates. When the antennas have no visual contact, communication takes place through the obstacles, reflecting from them and bending. What is more, attenuation of materials depends on frequency of electromagnetic wave passing through. Attenuation is rising with frequency of the wave.

Attenuation is caused by resonance of particles in the atmosphere, mainly water and oxygen. As a result of this phenomenon, the range decreases. Another relevant effect is reflection of the electromagnetic wave. Similarly to the attenuation, higher frequency of the wave, more energy is absorbed by the object, which reduces energy reflected from obstacle, and as a final result – decreases range. Next phenomenon is diffraction. When the wave encounters an obstacle, it bends around it. The angle of deflection depends on the length of the wave and the relative size of the object. The lower frequency, the greater deflection angle. When dealing with large obstacles, for example mountains, electromagnetic wave with lower frequency might be more efficient with reaching client behind the obstacle.

Another important issue is also the amount of the interference in selected band. In 2.4 GHz band requires most concern. This frequency is used by many ISM devices, like microwaves, WiFi networks, Bluetooth adapters, analog surveillance cameras, etc. Accumulation of large number of mentioned devices in a limited space, causing interference, which will lead to decreased IEEE 802.15.4 network coverage and capacity.

Presented factors make 860 MHz band an interesting choice in Europe because of its lessen popularity, hence, decreased chance for interference. The main drawback of this frequency range is availability of only one channel, which might be insufficient in particular cases. 2.4 GHz band have 16 channels available, 2 MHz widespread. On both bands, many IEEE 802.15.4 transceivers have output power above 100 mW, which makes their range competitive to WiFi networks. In most cases, IEEE 802.15.4 uses DSSS spread spectrum technique.

V. NETWORK TOPOLOGY FOR STRAIN MEASUREMENT

IEEE 802.15.4 defines two types of network devices. FFD (Full Function Device) has full protocol functionality, offering connectivity to any client of the network and as a result is able to work as network coordinator, network node or end device. Second type is RFD (Reduced Function Device), often with reduced computing and energy resources, being able to communicate only with FFD, hence, being unable to route traffic from other nodes. Picture below presents two most popular network topologies, star and peer to peer.

IEEE 802.15.4 doesn't define routing, which can be implemented on higher OSI layers. One of the most common standard basing on IEEE 802.15.4 is ZigBee, relation between them is being shown on figure 1.

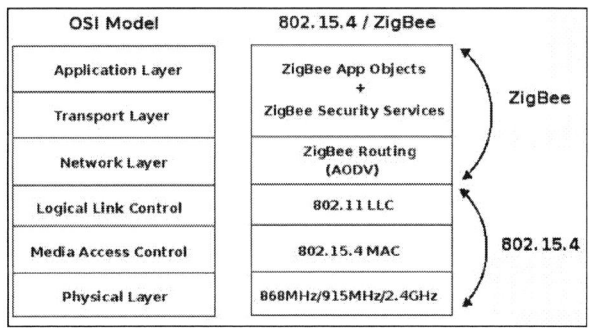

Figure 1. ZigBee and IEEE 802.15.4 OSI layers (libelium.com)

ZigBee was developed by ZigBee Alliance, which associates many well-known companies from elecronics industry, for example Atmel, Analog Devices, Freescale, Texas Instruments and many others . ZigBee provides mesh topology and appropriate to it routing algorithms. It also defines ways of joining and authorizing devices into the network.

ZigBee Protocol defines three types of network nodes – coordinator, router and end device. Coordinator is gate of the network, allowing to e.g. store and analyze acquired data externally. Routers are responsible for forwarding data packets from source to destination. Coordinator also has a functionality of the router, as well as some end devices, provided that they have sufficient power resources. If not, they do not participate in routing, confining themselves to end device functionality – commonly measurement.

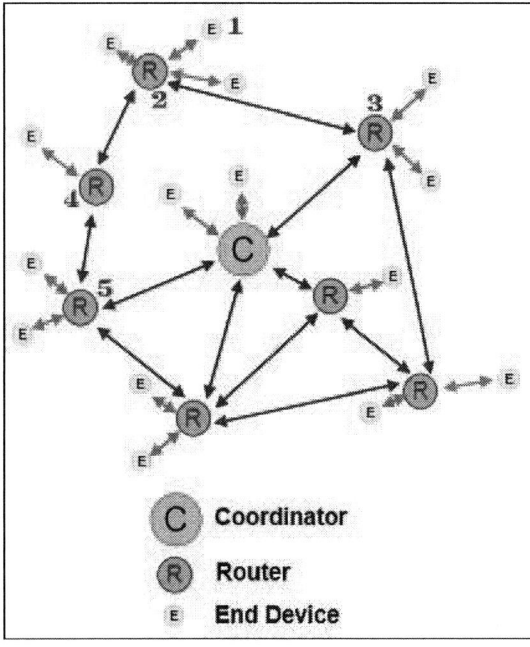

Figure 2. ZigBee Mesh network topology

Due to mesh network topology, when sending data, source node do not require direct radio connection with recipient (Fig. 2). Packets between them can be passed through one or more routers. For example, when node (1) wants to send a message to coordinator, it forwards it to router (2). (2) may choose a way to coordinator, for example by node (3). If there is no connection with device (3), caused by its malfunction or interference, data can go through (4) and (5). ZigBee uses AODV (Ad hoc On-Demand Distance Vector) routing. Connection pathway is collated only on request. Described mechanisms applied in ZigBee standard provide reliable, tolerant to faults and unexpected changes, data exchange.

VI. STRUCTURE OF MEASUREMENT SYSTEM

Considered system should be able to measure strain in mechanical structures and send results wirelessly to a Linux computer, which could allow further analysis of the data. Furthermore, humidity and temperature will be acquired to assess environmental conditions, which might have impact of obtained results. Exchange of data between measure network and PC will be carried out by IEEE 802.15.4, possibly enhanced with ZigBee standard, which ensure, that presented in this article issues will be efficiently solved.

The project focuses on developing a measurement module, which will be able to operate in distributed strain measure network. It is expected to be power efficient, while providing high measurement accuracy. Two basic modes of operation will be implemented. Continuous high speed measure and low power, periodical measure, will allow both to precisely asses behaviour of construction and monitoring its condition over a longer period of time. This makes several demands to proposed device, like high performance microcontroller and ability to power down currently unused parts of the system. Schematic diagram of considered solution is being presented on figure 3.

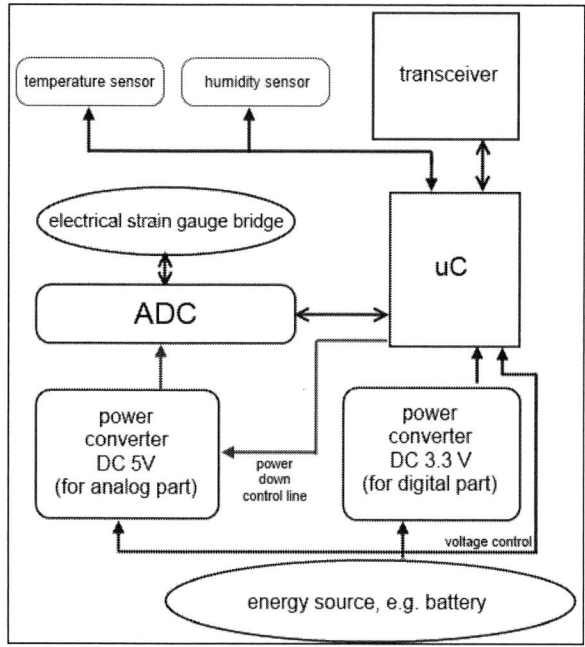

Figure 3. Proposed system structure

Analog-digital converter reads voltage from electrical strain gauge bridge, and sends data to the microcontroller, which would be also responsible for acquiring data from temperature and humidity sensor. When measure values are ready to send, microcontroller uses transceiver to pass through results to computer. Using additional power converter for analog circuit, primarily electrical strain gauge bridge comes with two substantial benefits. Analog supply voltage can be filtered out of distortions, which significantly increases accuracy of strain measure results. Moreover, microcontroller is able to power down analog circuit, when no measurement is performed, which essentially reduces power consumption in case of periodical strain monitoring.

Presented aspects of wireless communication for strain measurement systems lead to conclusion, that due to IEEE 802.15.4 reliability and minor hardware requirements, it seems to be the best available solution. The proposed system structure is initial state of development of more advanced strain measurement system, which is being elaborated for TULCOEMPA project.

VII. SUMMARY

The article has described considerations associated with design of wireless distributed measure system in context of bridge strain monitoring related to TULCOEMPA project. Problems associated with specific wireless communication standards, the reasons of their use, benefits and problems that come with them have been discussed. Wi-Fi, Bluetooth and IEEE 802.15.4 networks features were presented and compared. The conducted analysis led to the conclusion, that in this particular project, IEEE 802.15.4 is most suitable solution. What is more, initial structure of strain measurement system has been proposed, which is a basis for further development.

REFERENCES

[1] Shahin Farahani, "ZigBee Wireless Networks and Transceivers", Newnes, 2008.

[2] Pejman Roshan, Jonathan Leary, "802.11 Wireless LAN Fundamentals", Cisco Press, 2004.

[3] David Kammer, Gordon McNutt, Brian Senese, "Bluetooth, Application Developer's Guide: The Short Range Interconnect Solution", Syngress

Medical Applications

MIXDES 2012, 19th International Conference *"Mixed Design of Integrated Circuits and Systems"*, May 24-26, 2012, Warsaw, Poland

Automatisation of Computer-aided Burn Wounds Evaluation

Wojciech Tylman, Marcin Janicki, Andrzej Napieralski
Department of Microelectronics and Computer Science
Technical University of Łódź
Łódź, Poland
e-mail: tyl@dmcs.p.lodz.pl

Abstract—**This paper presents work on a PC-based software solution for evaluation of burn wounds. Improvements leading to automatisation of important elements of the evaluation procedure, together with other enhancements requested by the physicians are presented.**

Index Terms—**thermography, medical imaging, medical diagnosis, image processing**

I. INTRODUCTION

The ability to perform temperature measurements of human body areas is important in some types of medical diagnosis. In particular, it has been shown that precise temperature measurements carry valuable indications for the treatment of burn wounds. Such measurements may be used, for example, when the physician has to determine whether the particular area of the wound may be expected to heal by itself, without surgical intervention.

One possible approach to such measurements is utilisation of thermovision measurements, in particular the solutions where matrix infrared sensors are used. Such sensors are capable of capturing the temperature of many (hundreds of thousands) points at the same time. The resulting temperature map – a thermovision image – is ideally suited to the purpose of burn wound evaluation, as the burn wound itself is very often a large object, requiring multi-point measurements.

Once the thermal image is captured, a detailed analysis is required in order to extract useful informations. This analysis should be carried out by the physician in the hospital environment. Precision of results, short analysis time and ease of use are important here. For this purpose a dedicated software solution has been proposed, described in [1]. This paper presents further work aiming at improving and enhancing the software.

The remainder of this paper is organised as follows: Section II introduces BurnDiag, the discussed software solution. Section III presents enhancements for automatic image registration, while Section IV describes other improvements requested by the physicians. Section V outlines future research directions and provides closing remarks.

The research presented in this paper was supported by the grant of the Polish Ministry of Scientific Research and Higher Education No. N515 2423 37.

II. BURNDIAG – A PC SOLUTION FOR ANALYSIS OF BURN WOUNDS

This section briefly introduces BurnDiag software. For details consult [1].

A. Input data and user interface

The BurnDiag software is intended to work with two types of images: thermovision and visible light. The main input is the thermovision image, the visible light one is supplemental, providing means to map burn wound areas to the body areas as seen by the human eye. The user interface is able to display both images side-by-side, making their comparison straightforward. Apart from the image, the program main window contains controls (buttons, edit boxes, etc.) required for performing analyses and displaying their results.

The interface can also be switched to text view, displaying textual information about the patient and the wound. These data are stored in a database using DICOM format [2].

B. Tools and analyses available in the original version

The software, as presented in [1], supports following operations on the thermographic image:

- presenting image in user-selected false colour palette,
- adjusting lower and upper temperature limit,
- freehand drawing of a closed path in order to delimit the wound area,
- setting the point of (optional) biopsy and reading its temperature,
- displaying an isotherm and computing burn area having temperature below and above the isotherm.

The software is able to read thermographic images in FLIR jpg format and also visible light images in standard jpg format. The information about the patient and wound are stored in a database.

The current user interface view (image mode) is presented in Fig. 1.

III. AUTOMATIC IMAGE REGISTRATION

As already mentioned, BurnDiag software is able to display thermographic and visible light images side-by-side.

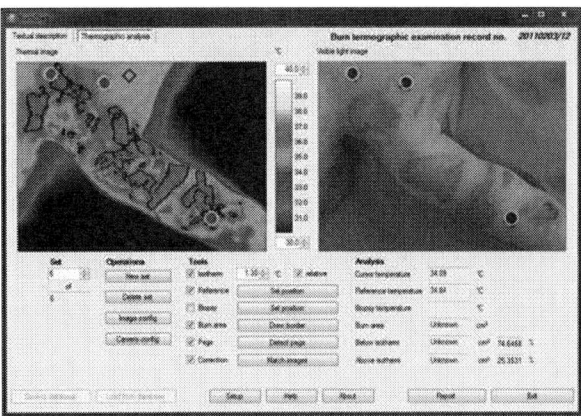

Figure 1. BurnDiag user interface

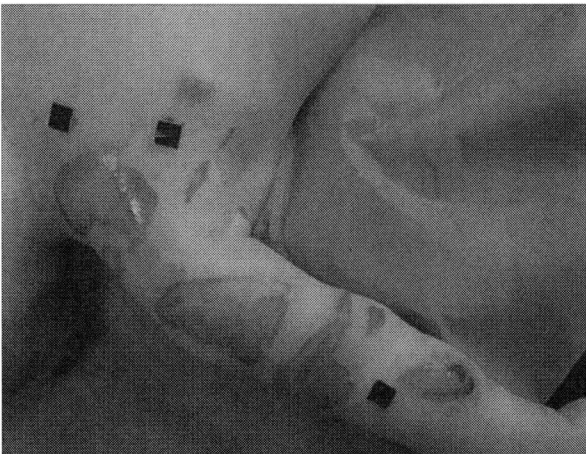

Figure 3. Visible light image of the wound – uncorrected. Compare with Fig. 2 the position of two uppermost reference squares.

Unfortunately, usefulness of this feature can be seriously crippled if these images do not present exactly the same view of the wound. Consequently, both images should be taken from the same point in space and using cameras with exactly the same angle of view.

In practice, these conditions are never met. Even when thermovision – visible light combo camera is used (such as for example FLIR ThermaCam P660 [3] utilised in the development of the software), the lenses of the two cameras are separate and offset by several centimetres, they angles of view are also vastly different (the angle of view of the visible light camera being usually much wider). This problem is illustrated in Figs. 2, 3, 4 and 5, which present (in pairs) the thermographic and visible light images of the same object. When separate thermovision and visible light cameras are used, the problem becomes even more obtrusive, due to possible rotation of the cameras.

For this reason, correction has to be applied to at least one image in order to match it (register) it with the other one. This requires two distinct steps:

a) determining corresponding points in the images,

b) applying suitable transformations.

These steps will now be covered in details.

A. *Determining Position of Reference Points*

If a number of point pairs in the images being matched can be identified that correspond to the same objects in the recorded scene, then it is possible to compute image transformation coefficients. The more pairs are identified, the more parameters can be used in the transformation process and therefore more complicated deformations can be corrected. BurnDiag employs three pairs of points, which allows to perform affine transform (see section III.B).

The points in question can be selected in various ways, ranging from manual identification to fully automatic approach that includes discovery of the features of the image that can be treated as reference points. For the purpose of BurnDiag software, a solution has been proposed that performs automatic

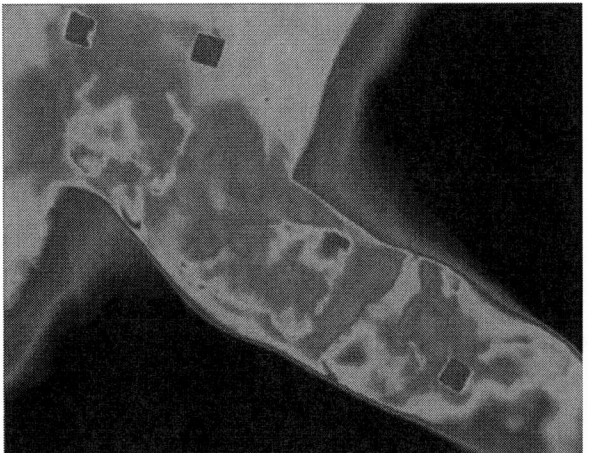

Figure 2. Thermographic image of a wound. The blue rectangles are the reference squares.

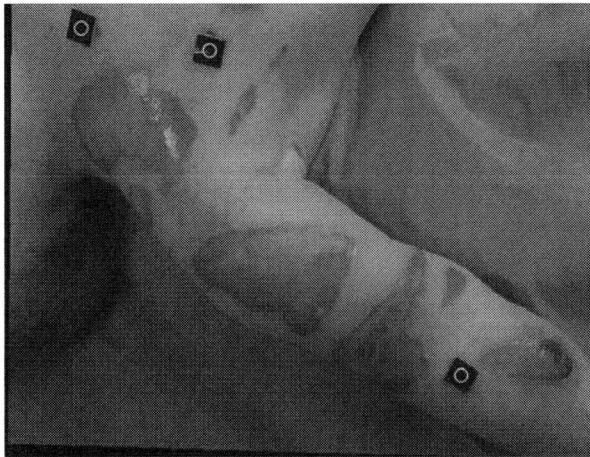

Figure 4. Visible light image of the wound – corrected. The centres of the green circles over reference squares denote the reference points, as detected by the algorithm

536

detection of reference points in the images, but the points themselves are marked by placing predefined shapes on the investigated object (human body in this case). This offers more robust performance and less chance of errors, which in case of medical diagnosis are particularly unwanted. The mentioned predefined objects are black squares with area of 1 cm² each.

The detection procedure consists of several steps and is similar for both the thermographic and visible light image. However, the former requires additional considerations, which will be covered towards the end of this subsection.

The first step of the procedure is conversion of the image to a grayscale equivalent. Next, an edge detection procedure is applied, allowing to disclose objects which differ significantly in brightness from the surrounding – in this case the reference squares. This is followed by a sequence of dilatation and erosion filters – the purpose is to expose distinct objects in the image and increase the chance they will be outlined by a closed path. The dilatation filter grows the objects in the image, consequently filling the gaps that may appear in the path outlining the reference squares. The erosion filter removes small stand-alone objects, which reduces noise and smooths the outline of the squares.

The resulting image is passed to blob detector that labels regions corresponding to distinct objects larger than predefined threshold.

At this point the algorithm has a list of objects; among them the three squares of similar area have to be found. The next step is therefore removing the objects which shapes significantly differ from squares. For this purpose every object is tested for a) being a quadrilateral, and b) having sides and angles of similar values. The objects that pass the test are sorted by their area. The final step of the algorithm searches through the list for three adjacent (in terms of position on the list) objects having similar areas. If such triple is found, the algorithm finishes successfully, otherwise it fails.

The performed tests showed the algorithm usually produces good results, but its optimal parameters (e.g., number of passes of the dilatation and erosion filters, minimum blob size, tolerances for side length and angle values of quadrilateral etc.)

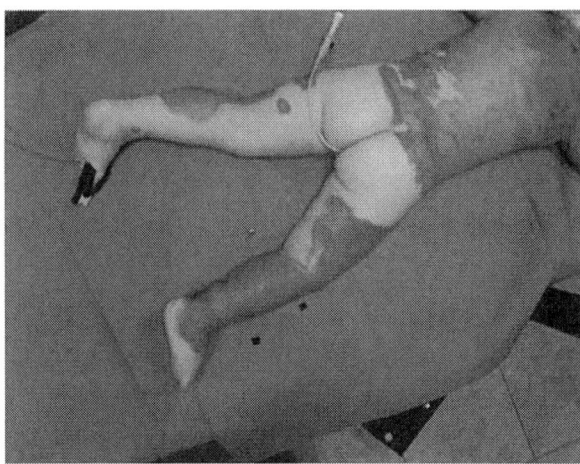

Figure 6. Visible light image of the wound – uncorrected. Note vastly different angle of view. The reference squares are very small.

may vary depending on the image analysed (due to, for example, different distance of the camera from the object). For this reason several sets of parameters has been proposed; if the algorithm fails using one set, it is repeated using other sets. This increased robustness of the solution.

As already mentioned, processing of thermographic image requires additional steps. Unlike the visible light image, the thermographic one is in fact a pseudo-image, created by application of false-colour palette to a set of numerical measurements. The chosen palette, as well as the temperatures corresponding to its lower and upper boundaries, can be arbitrarily set and this choice significantly impacts the performance of the algorithm. During tests it has been shown that the settings producing in most cases the best results are using a grayscale palette with lower and upper boundary set to 30°C and 40°C, respectively. For some images, a narrower span of 32°C to 38°C produced better results. These values are of course mainly determined by the temperature of the human body.

Figure 5. Thermographic image of a wound. The reference squares are hardly discernible (greenish ovals above and below patient's leg).

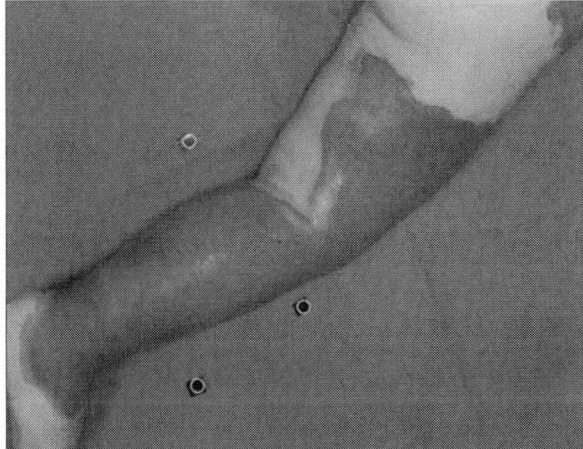

Figure 7. Visible light image of the wound – corrected using manual selection of reference points.

Another problem with the thermographic images are their very low spatial resolution (compared to the typical visible light image). At 640×480 pixels the reference squares may be to small for reliable detection. It has also been observed that the depth of field for the thermographic camera was small, which led to the squares (placed outside of the wound, and as a result, often at different distance from the camera than the fragment on which the camera focused) being out of focus. To handle such situations, the BurnDiag software allows also manual (by mouse click) marking of the reference squares in the images.

Before actual correction can be performed, one additional step is needed: the reference points in the visible light image have to be matched to the corresponding points in the thermographic image. For three reference squares there are six possible ways of joining the detected points in pairs. In order to select the correct one, the Euclidean distances between points in the thermographic image and the visible light image are calculated. Next, for each way of joining them in pairs, the sum of distances is computed and the solution that gives the lowest sum is selected. This approach gives correct results provided the rotation of one image against the other one is not extreme.

B. Correction of the Image

The three pairs of points discovered by the algorithm described above allow to perform an affine transform, i.e., a transform that can map a parallelogram onto square [4]. The transform of point *(x, y)* into corrected point *(u, v)* is given by (1) and the computation of coefficients a_0-a_2 and b_0-b_2 is straightforward. Tests showed that this kind of transform is sufficient for the discussed application. It should be noted that the corrected image is the visible light image, as the thermographic one should not be subject to any unnecessary processing.

$$u = a_0 + a_1 x + a_2 y$$
$$v = b_0 + b_1 x + b_2 y$$
(1)

The result of the fully automatic correction procedure is presented in Fig. 4, which presents the image from Fig. 3, after correction.

As already mentioned, in some circumstances the algorithm fails do detect the reference squares. In such case (or when the reference squares were not used during examination) it is still possible to perform image registration using the manual selection of reference points. This situation is presented in Figs. 5, 6 and 7.

IV. OTHER IMPROVEMENTS

This section describe new or improved functionality, added after initial tests of the software.

A. Relative Temperature Isotherm

The temperature value at which the isotherm is drawn can now be expressed relatively to an user-selected point of the image. This is in accordance with the procedure of wound evaluation, where the optimal treatment of the wound is chosen based on its temperature with relation to the healthy skin.

B. Improved Temperature Readout

The temperature of the point under cursor is continually displayed in the user interface window. A secondary cursor is displayed over the visible light image and points to the same location (provided the image registration procedure described in Section II has been performed). A precise temperature scale is displayed next to the palette, facilitating analysis of the image using pseudo-colour approach.

C. Reporting

In order to provide a way to print the results of the examination or incorporate it into other documents, reporting capability has been added to the software. The report is automatically generated at a click of the button; it contains textual description of the wound and patient data (As retrieved from the database) and also both images of the examination. The report can be generated as Microsoft Word, Microsoft Excel or Adobe PDF file.

V. CONCLUSIONS AND FUTURE RESEARCH

This paper presented new tools and improvements in BurnDiag – a PC program burn wounds evaluation. In particular, a procedure for automatic image registration has been described. In the current state of development, BurnDiag is already a fully-equipped tool for thermographic analysis of burn wounds. Addition of reporting capability allows to incorporate results of the analysis in documents used in hospital practice.

Tests in hospital environment will determine the need for subsequent improvements of the software.

REFERENCES

[1] W. Tylman, M. Janicki, and A. Napieralski., " Computer-aided approach to evaluation of burn wounds", Proceedings of the 18th International Conference on Mixed Design of Integrated Circuits and Systems MIXDES 2011, pp..653–656, Łódź, 2011

[2] DICOM specification, ftp://medical.nema.org/medical/dicom/2009/

[3] FLIR P-Series Infrared Cameras,
http://www.flir.com/thermography/eurasia/en/content/?id=31667

[4] B. Zitová and J. Flusser, "Image registration methods: a survey", Image and Vision Computing, vol. 21, No. 11, pp. 977-1000 , October 2003

MIXDES 2012, 19ᵗʰ International Conference "*Mixed Design of Integrated Circuits and Systems*", May 24-26, 2012, Warsaw, Poland

Low Power System for Measurement of Skin Conductance and Temperature of Patient Body

Mateusz Majchrzycki, Igor Karoń, Krzysztof Kolanowski, Andrzej Rybarczyk
Chair of Computer Engineering
Poznań University of Technology
Poznań, Poland

Abstract—In the article we present a system for measurements of skin conductance and temperature of patient body. Presented system is a part of the recorder module of the MOnOff system. We propose an improved algorithm to measure of skin conductance and temperature, which reduces the power consumption of the measuring circuit. Construction of measurement circuits and practical experiments are presented.

Index Terms—skin conductance, temperature measurement, low power, MOnOff

I. INTRODUCTION

Complex, small and standalone systems are necessary for chronic disease where patient condition needs to be evaluated all the time to improve course of a treatment. Those systems have to be from one side simple enough that patient can use it on their own while in the same time should give accurate results of the measurement. Due to a wide range of diseases and symptoms in each of them, data measurement systems must be designed for each of them individually.

In treatment, skin conductance is treated as a factor of electrophysiological evaluation of autonomic nervous system (ANS) health condition. The study of sympathetic skin potentials is a method involving registration with the electrode surface, and then analysis of the synchronous activity of sweat glands induced stimulation of sympathetic fibers by unexpected stimulation [1]. Large extent of previous work on the study of skin conductance focused on exploring properties of the skin without concerning clinical implications of obtained measurements [2], [3], [4], [5] or as a additional method to improve other measurements [6]. Long term monitoring of the skin conductance in order to determine the state of the ANS in a natural conditions without the use of external stimuli can be a source of additional information for Parkinson's disease. Due to a certain influence of noise signals caused by the same properties in both skin like time dependency, random variation, non-reproducibility and non-linearity [4], [7] and complexity of the human body measurement of skin conductance is insufficient.

While the conductance of the skin is treated as an information of the ANS state, body temperature may be treated as a simplified information of the internal state of the patient body. Combining these two measurements it is possible to obtain a fuller picture of the state of the patient body with a minimum noise in the acquired measurements.

Presented paper focuses on measurements of both factors in low-power long term monitoring system.

II. SYSTEM OVERVIEW

Described in this paper measurements of skin conductance and temperature are an integral part of the MOnOff system. Presented system is used to monitor the movement and outside the movement disorders in patients with the extrapyramidal syndromes. To require the differentiation of the extrapyramidal syndromes in clinical practice include Parkinson's disease, genetically determined parkinsonism, dementia with Lewy bodies, progressive supranuclear palsy, multisystemic atrophy, corticobasal degeneration, and degeneration of the fronto-temporal dementia with parkinsonism linked to chromosome 17th.

The main symptoms of the extrapyramidal syndromes are associated with the sphere of motion, however, increasing attention is paid to the signs outside the movement. These include: cognitive impairment, neuropsychiatric symptoms, Dysautonomia, sleep disturbances, impaired sense of smell, pain. The task of the system is to monitor the movement disorders by recording each limb tremor in three axes and outside the movement disorders such as skin moisture, body temperature and blood pressure. All measurements are non-invasive and do not require prior calibration by the end user.

The data collected by the recorder module are used to analyse the effects of patients treatment. The analysis is supported by the artificial neural network which is used to recognize and classify of the characteristic stages of patient activity. This information are used by doctors to select the correct doses of drugs individually for each patient. It allows faster diagnosis and the appointment of the current stage of the disease.

During the design of the MOnOff system, the most important were the two requirements. The first one was a low power consumption by all the components. The second was a maximal reducing of the size of the recorder module. All components used in the presented circuits meet both requirements, as described below.

III. METHODOLOGY

Measurement of skin conductance is to provide feedback on a person's physiological state changes. Conductance of the skin may be subject to change due to the presence and activity of sweat glands secreting sweat which constitutes a weak electrolyte can increase the conductance of the skin. The body's physiological changes are monitored by a suitable device, such as measuring computer system. This method is used, inter alia, in psychology, medicine, but also in sport or business. Enables the measurement of physiological parameters, the patient consciously uncontrolled skin conductance, such as a skin conductance and body temperature [8]. Electrodermal measurement can be done using one of two methods.

The first method is to record changes in skin electrical conductance during the flow of a weak electric current [9]. The conductance increases with increase in the sweating of the skin, which in turn is associated with activation of the autonomic nervous system in the body. Activation of the autonomic nervous system is coupled with physiological arousal, and therefore changes in skin electrical conductance can mean experiencing emotions (e.g. fear, anger, stress, disease attack, etc.).

The second method is to record the skin potential produced by the body [9]. The skin, like other organs (e.g. brain - EEG, heart - ECG and other muscles - EMG) produces its own electric potential which can be measured. This method involves the precise measurement of voltage relative to a known reference point on the body.

Of the two methods the preferred is measurement of skin conductance [10]. This method is unambiguous and less complicated to implement than measuring the electrical potential of the skin [9]. In the presented system, skin conductance measurement method is used.

The modules of the MOnOff system are placed on the wrists and above the ankle joints both legs. Because of this, the temperature measure could be in one of this four places. All of them are non-standard for temperature measuring, however for purposes of diagnosis the outside the movement disorders they are sufficient. In the system assumptions, it was important to investigate the correlation between body temperature and others measured factors. Finally the temperature is measured on the right wrist of the patient body. This place was chosen because of integration of the measurement of skin conductance and temperature in the one recorder module.

Presented system introduces few innovations in the scope of skin conductance measurement and the body temperature measurement. It uses components with very low power consumption. Further the power consumption is reduced with use of the proposed algorithm of the measurements.

IV. SKIN CONDUCTANCE MEASUREMENT

In the measurement circuit the Wheatstone bridge is used. In one of the legs of the bridge is placed skin conductance. Schematic of the measurement circuit is shown on Fig. 1. Resistors R_1, R_2 and R_3 are known, there is also dependence $R_1 = R_2$. To measure the offset voltage of the bridge the differential amplifier is used. If the bridge is balanced, the voltage across the resistor R3 and skin conductance is equal. Thus the output voltage of the differential amplifier is zero. Change of skin conductance influences the imbalance of

the bridge, then the differential amplifier output voltage also changes. To prevent the load of the measurement bridge were placed unity buffer amplifiers between the bridge and inputs of the differential amplifier.

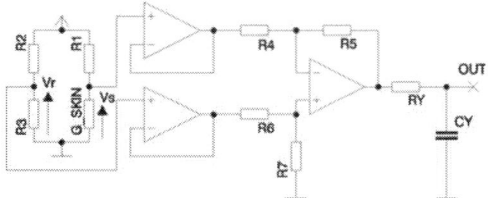

Fig. 1. Schematic of skin conductance measurement unit

The output voltage is calculated using the equation proper for differential amplifier [11] assuming that $R_4 = R_6$ and $R_5 = R_7$. Finally the equation (1) is obtained.

$$V_{out} = \frac{R_4}{R_5}(V_r - V_s) \tag{1}$$

The value of the voltage V_r is calculated as shown in (2).

$$V_r = V_{cc}\frac{R_3}{R_2 + R_3} \tag{2}$$

The value of voltage drop obtained on skin conductance is shown in (3).

$$V_s = V_{cc}\frac{1}{G_{skin}R_1 + 1} \tag{3}$$

As a result is obtained the equation (4), on assumption $R_1 = R_2$. The equality of resistors R_1 and R_2 ensures the ability to adjustment measurement range. It is possible by change the value of the resistor R_3, which determines the minimal possible to measure value of the skin conductance.

$$V_{out} = \beta \frac{R_1(R_3 G_{skin} - 1)}{(R_1 + R_3)(R_1 G_{skin} + 1)} \tag{4}$$

where

$$\beta = \frac{R_4}{R_5}V_{cc} \tag{5}$$

Using equation (4) is possible to determine the values of the resistors in the bridge and differential amplifier on Fig. 1. Calculated resistor values should reflect the value of supply voltage V_{cc} which influences gain factor β. It should also be taken into account the range and the resolution of the ADC which is used to measure the voltage V_{out}. In the presented system the ADC uses reference voltage of $2.56V$ and it has 10-bits resolution. The last factor affecting the value of the resistors in the measurement system is the scope of skin conductance changes which the system should measured. The expected value of the skin conductance is different for each patient, therefore the system should perform measurement in range $5\mu S$ to $100\mu S$[9].

Transforming equation (4), we obtain the relationship between voltage V_{out} and the value of skin conductance as shown in equation (6).

$$G_{skin} = \frac{\beta R_1 + V_{out}(R_1 + R_3)}{\beta R_1 R_3 - V_{out}(R_1^2 + R_1 R_3)} \tag{6}$$

The equation (6) is used by the analysis software to convert the value from ADC to value of skin conductance.

As previously mentioned, the output voltage from the measurement circuit is converted to a digital form. The ADC integrated in the microcontroller of the MOnOff recorder module is used. In the firmware of the MOnOff recorder's module has been implemented the algorithm for measurement of the skin conductance value. The measuring cycle takes 7.5 seconds and consists of four ADC readings at intervals of 2.5 seconds. The cycle is repeated every minute. The samples acquired from ADC during the measurement cycle are averaged using the arithmetic mean. This technique allows to reduce the noise in the measuring circuit.

The measurement algorithm reduces power consumption by measuring circuit. This is achieved by controlling the power supply of the measuring system. This technique uses the digital output of microcontroller to turn on and off the power supply. At the beginning of the measurement cycle, microcontroller turns on the power supply for measurement circuit. It is turned on when sampling the output voltage of the differential amplifier by 7.5 seconds. After the measurement the measuring circuit supply voltage is switched off. This way of operating reduces power consumption for 87.5% compared to operation in continuous mode. It is significantly important to ensure the long time of the operation of the recorder module.

Although the algorithm supports energy efficiency in the measurement circuit, all the components used in the system should meet the requirement for low power consumption. Because of this, as a operational amplifier is used the MCP6034 [12]. It is a quad amplifier with very low power consumption. In the presented circuit there are three main current collectors. The first of these are the operational amplifiers. They consume up to $1.35\mu A$ of current. The second one is the Wheatstone bridge, which consumes $3.91\mu A$ of current. This value is achieved for $R_1 = R_2 = R_3 = 1.2M\Omega$, $G_{skin} = 10\mu S$ and $V_{cc} = 3.3V$. The rest of the current is consumed by resistors in differential amplifier circuit. This circuit consumes approximately $2\mu A$ of current. This value depends on voltage value on differential amplifier inputs. Total power consumption by the measuring circuit is approximately $7.26\mu A$. Laboratory measurements have shown that the real power consumption in continuous mode is $7.35\mu A$. The measuring algorithm further reduces power consumption by 87.5%. The measuring circuit is enabled by a 7.5 seconds during a 60-seconds cycle. Therefore the circuit consumes the current for 12.5% of work time.

V. TEMPERATURE MEASUREMENT

The measurement circuit of the temperature is very similar to the measurement circuit of the skin conductance which is shown on the Fig. 1. The difference consists in replacing the skin conductance with the use of the thermistor. Complete schematic of the temperature measurement circuit is shown on Fig. 2. The values of measurement circuit resistors are different than used in the skin conductance measurement circuit. This is due to the type of the thermistor used for measurement and its

nominal resistance. In the presented system the MC65F103B thermistor is used [13]. It is the NTC thermistor with the nominal resistance at $25°C$ equal to $10k\Omega$.

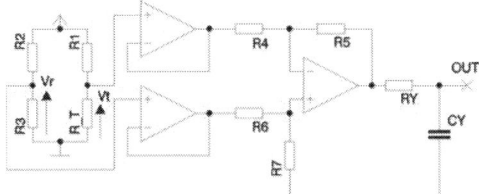

Fig. 2. Schematic of temperature measurement unit

As described in section III, the temperature of the patient body is measured on the wrist. Therefore, we assume that the temperature of the thermistor can be changed within $33°C$ to $42°C$. It is a sufficient measurement range to be able to register all changes of the the patient body temperature.

The formula for the output voltage in the temperature measurement circuit is identical to the equation (4). In this equation the V_s should be replaced with V_t. Using equations (7), (8) and (4) we obtain the relationship for the actual thermistor value in function of the V_{out} voltage described by the equation (9).

$$V_r = V_{cc}\frac{R_3}{R_2 + R_3} \tag{7}$$

$$V_t = V_{cc}\frac{R_t}{R_t + R_3} \tag{8}$$

$$R_t = \frac{\beta R_1 R_3 - V_{out}\left(R_1^2 + R_1 R_3\right)}{\beta R_1 + V_{out}(R_1 + R_3)} \tag{9}$$

Calculated value of the thermistor resistance is used to calculate the value of the temperature of the patient body. For this purpose, the Steinhart-Hart equation (10) is used.

$$\frac{1}{T} = a + b\left(\ln\frac{R_t}{R_{25}}\right) + c\left(\ln\frac{R_t}{R_{25}}\right)^2 + d\left(\ln\frac{R_t}{R_{25}}\right)^3 \tag{10}$$

In the equation (10), T is the temperature in K, R_t is the actual thermistor resistance and R_{25} is the nominal thermistor resistance at $25°C$. The value of coefficients a, b, c and d can be found in the manual of the material of the thermistor which is used [14].

The algorithm for temperature measure is similar to algorithm for skin conductance measure. It consist of measurement cycle which takes 7.5 seconds and is repeated every minute. During the measurement cycle, the output voltage of the measuring circuit is sampled by ADC. Sampling is performed four times at intervals of 2.5 seconds. Every four samples acquired during the measurement cycle are averaged using the arithmetic mean. It allows to reduce the noise in the measuring circuit.

Compared to the skin conductance measurement circuit, the temperature measurement circuit consumes more power.

The main reason is power consumption by the Wheatstone bridge which consumes $102\mu A$ of current. This is due to the nominal thermistor resistance and the need of selection the values of the remaining resistors in the bridge. The values of these resistors are on the order of tens kiloohms. The operational amplifier consumes $1.35\mu A$ of current. Resistors on inputs of the differential amplifier consume approximately $2\mu A$ of current. The total calculated power consumption of the temperature measurement circuit is $105.35\mu A$ of current. The supply voltage of the measurement circuit is $3.3V$. The measurements have shown that the real power consumption is $107.4\mu A$.

The algorithm of temperature measurement saves the energy consumed by the measuring circuit using the technique described in IV. This allows the system saves 87.5% of energy.

VI. PRACTICAL EXPERIMENTS

The presented system was tested on several people. Tests helped to improve measurement algorithm and consequently to increase energy efficiency. The measuring circuits described above have been integrated into one of the MOnOff recorder's module. Ready measurement set is presented in Fig. 3.

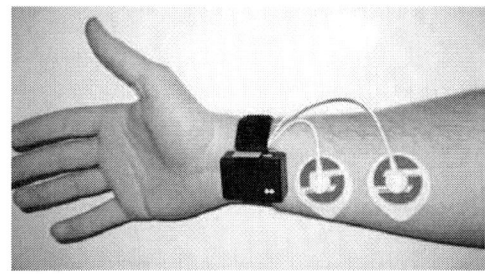

Fig. 3. The MOnOff system founded on the patient's hand.

Data acquired during one of the tests are presented in Fig. 4. They present the activity of one person for over 12 hours. Noticeable is large fluctuation of the temperature measure data. This is a result of the mounting method of the thermistor and construction of the MOnOff recorder module. However the data obtained from the temperature measurement circuit can be processed to obtain a trend line which is clearer.

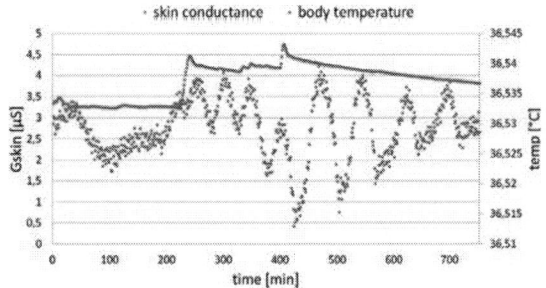

Fig. 4. The acquired results of temperature and skin conductance measurements.

VII. CONCLUSION

The proposed system for measuring of the skin conductance and body temperature satisfies the initial assumption. Accuracy and sample rate collection of samples are sufficient to carry out treatment and to provide necessary information for medics. Moreover, miniaturization and low energy consumption of the presented system allows normal life of the patient without necessity of the medical supervisor. Research proves the reduction of power consumption by 87.5% by use the proposed measurement algorithm compared to continuous mode of operation.

REFERENCES

[1] B. Zakrzewska-Pniewska, "Elektrofizjologiczne metody oceny dysautonomii w chorobach ukladu nerwowego." *Polski Przeglad Neurologiczny,* vol. 4, no. 2, pp. 58–64, 2008.
[2] S. Kim, R. F. Yazicioglu, T. Torfs, B. Dilpreet, P. Julien, and C. Van Hoof, "A 2.4ua continuous-time electrode-skin impedance measurement circuit for motion artifact monitoring in ecg acquisition systems," in *Proc. IEEE Symp. VLSI Circuits (VLSIC),* 2010, pp. 219–220.
[3] T. Yamamoto and Y. Yamamoto, "Analysis for the change of skin impedance." *Med Biol Eng Comput,* vol. 15, no. 3, pp. 219–227, May 1977.
[4] G. Lee and J. Choi, "Differential method against motion artefacts in skin conductance response measurement," *Electronics Letters,* vol. 43, no. 7, pp. 375–377, 2007.
[5] A. L. Toazza, F. Mendes de Azevedo, and J. M. Neto, "Microcontrolled system for measuring skin/electrode impedance in bioelectrical recordings," in *Proc. Second IEEE Int Devices, Circuits and Systems Caracas Conf,* 1998, pp. 278–281.
[6] T. Degen and H. Jackel, "Continuous monitoring of electrode–skin impedance mismatch during bioelectric recordings," *IEEE Transactions on Biomedical Engineering,* vol. 55, no. 6, pp. 1711–1715, 2008.
[7] Y. Yamamoto and T. Yamamoto, "Dynamic system for the measurement of electrical skin impedance." *Med Biol Eng Comput,* vol. 17, no. 1, pp. 135–137, Jan 1979.
[8] J. Malmivuo and R. Plonsey, "The elektrodermal response." *Bioelectromagnetism - Oxford University Press,* pp. 428–434, 1995.
[9] D. C. Fowles, M. J. Christie, R. Edelberg, W. W. Grings, D. T. Lykken, and P. H. Venables, "Committee report. publication recommendations for electrodermal measurements." *Psychophysiology,* vol. 18, no. 3, pp. 232–239, May 1981.
[10] D. T. Lykken and P. H. Venables, "Direct measurement of skin conductance: a proposal for standardization." *Psychophysiology,* vol. 8, no. 5, pp. 656–672, Sep 1971.
[11] G. Rutkowski, *Operational amplifiers: integrated and hybrid circuits,* ser. A Wiley-Interscience publication. Wiley, 1993.
[12] *MCP6031/2/3/4: 0.9µA, High Precision Op Amps,* http://ww1.microchip.com/downloads/en/DeviceDoc/22041b.pdf, Microchip Technology Inc.
[13] *NTC Thermistors: type MC65,* http://www.ge-mcs.com/download/temperature/920_306a.pdf, GE Sensing.
[14] *Material type: F,* http://www.thermometrics.com/assets/images/f.pdf, GE Sensing.

MIXDES 2012, 19ᵗʰ International Conference *"Mixed Design of Integrated Circuits and Systems"*, May 24-26, 2012, Warsaw, Poland

Prediction of Fatigue and Sleep Onset Using HRV Analysis

Małgorzata Szypulska, Zbigniew Piotrowski

Faculty of Electronics
Military University of Technology
Warsaw, Poland
goskaszyp@wp.pl, zpiotrowski@wat.edu.pl

Abstract—**In the paper objective syndromes associated with sleep onset and fatigue based on the analysis of heart rate variability (HRV) have been presented. An algorithm for detection of the moment of sleep onset and fatigue has been described. It is based on the determination of the LF/HF ratio on the basis of an RR tachogram and assigning its value to three basic states: activity, drowsiness and sleep.**

Index Terms—**Heart Rate Variability, HRV, tachogram, sleep onset, fatigue, sleep onset detection**

I. INTRODUCTION

Research on sleep onset and fatigue is conducted in many scientific centres around the world by many research teams. The standard definition of the moment of sleep onset is based on changes in EEG, EOG, ECG and EMG, and usually reliably detects changes within a few seconds. When in an alert state, as long as a particular person has their eyes closed and at the moment of sleep onset, the EEG record shows a domination of alpha waves. However, the detection of SO (*sleep onset*) is much more complicated, for example due to the fact that the subjective perception of falling asleep does not always coincide with that specified and defined. During sleep onset a number of different parameters change in specific moments, but they do not coincide in time. The conclusion from this is that sleep onset is not a discrete occurrence but a continuous process.

It has been observed that before and during sleep onset the power level in the bands VLF and LF decreases, which reflects a decrease in sympathetic system activity. However, an increase in the power of the HF band can be seen, which means movement toward parasympathetic domination [2, 5]. These trends persist along with the deepening of NREM sleep [4]. A significant increase in the value of RR intervals during sleep has also been noticed, although their variability decreased. The mean RR interval (RRI) increased (heart rate decreased) during SO. A significant change in the level of RR intervals occurred 30s after sleep onset compared to the average RRI level for 7 - 9 minutes before SO [2]. Another indicator of sleepiness is a drop in the value of the LF/HF ratio. During sleep, it has minimum values [1].

II. HRV ANALYSIS

The RR interval represents the length of the ventricular cardiac cycle, it is measured between two successive R waves and serves as an indicator of ventricular rate. It is also called the inter-beat interval. Heart rate variability is the degree of fluctuation in the length of intervals between heart beats which are presented as R waves. HRV is a mirror image of heart beat regularity: the greater the regularity, the lower the HRV and vice versa. The length of RR sections is dependent on external regulation of heart rate. HRV is perceived as a reflection of the adaptability of the heart to changing conditions through the detection and rapid response to unexpected stimuli. Normal heart rate variability stems from autonomic control of the heart and the circulatory system. Sustained functioning of the sympathetic nervous system (SNS) and the parasympathetic nervous system (PNS) which are branches of the autonomous nervous system and control it. Increased SNS activity or diminished PNS activity result in an acceleration of heart rate. The opposite is true in case of low SNS activity or high PNS activity, which results in a slowing of the heart rate. The degree of variation in the heart rate provides information about the functioning of the nervous system when the pulse is taken and the ability of the heart to respond to it. HRV analysis is divided into: temporal and frequency.

A. Temporal Analysis

Temporal analysis consists of the measurement of RR interval variability over a specific time period and a calculation of the mean and variability. The following temporal analysis parameters are distinguished: Mean RR (mean heart rate) SDNN (standard deviation of all RR sinus rhythm intervals), RMSSD (square root of the mean of the sum of the squared differences between successive RR sinus rhythm intervals) and p50NN (percentage of differences between successive RR sinus rhythm intervals beyond 50 ms) [7].

B. Frequency Analysis

Frequency analysis is a representation that uses the sum of a finite number of harmonic waves with different wave Frequency analysis is a representation that uses the sum of a finite number of harmonic waves with different wave frequencies for determining heart rate variability. Analysis of frequency consisting in setting parameters for the power

543

spectral density estimate describes periodic oscillations in heart rate signal spread over different frequencies and amplitudes, and provides information about the relative intensity of the heart's sinoatrial rhythm. Spectral analysis can be performed in two ways: using nonparametric methods, i.e. the fast Fourier transform (FFT), which is characterized by discrete peaks for the selected couple of frequency components, and parametric methods describing the autoregressive model, resulting in a continuous, smoothed down activity spectrum. The following frequency indicators are distinguished:

- HF (*high frequencies*) – from 0.15 to 0.4 Hz – associated with breathing and parasympathetic system activity;

- LF (*low frequencies*) – from 0.04 to 0.15 Hz – associated with the activity of the sympathetic and parasympathetic systems; dependent on changes in blood pressure and baroreceptor reflex oscillation.

- VLF (*very low frequencies*) – from 0.0033 to 0.04 Hz – variability modulated by chemoreceptor activity, dependent on vascular - movement and thermoregulatory reflexes.

- TP (*total power*) – up to 0.4 Hz, total heart rate variability.

A coefficient is also distinguished that is equal to the quotient of power in the LF band and power in the HF band recorded as LF/HF.

III. DETECTION OF SLEEP ONSET

In accordance with a simple behavioural definition – sleep is a reversible state of detachment from reality and lack of response to the surrounding environment. Sleep is also a complex combination of physiological and behavioural processes. The precise determination of sleep onset has been the subject of many studies, mainly due to the fact that there is no single method from among such examinations as: EEG, EMG and EOG which would make it possible to unambiguously state that at a particular moment of time a person is certainly asleep. For example, a change in the EEG pattern is not always associated with the subjective perception of sleep. Even when a research subject claims that he or she is still fully conscious, distinct changes in indicators of sleep onset may be observed. Falling asleep (sleep onset, SO) begins, in normal circumstances in the case of adults who do not show any disturbance of sleep, with NREM phases [6]. During sleep onset, changes in particular parts of polysomnographic studies are as follows:

- Electromyogram (EMG) – a gradual diminution of muscle tonus as sleep approaches, but discrete changes in EMG indicating the moment of sleep onset are rare. Moreover, the level of EMG just before sleep onset can be, in principle and when a person is relaxed, completely indistinguishable from the sleep state.

- Electrooculogram (EOG) – when sleep is coming, EOG exhibits slow changes potentially resulting from

asynchronous eye-ball movements which usually disappear within a few minutes after the appearance of EEG changes suggestive of SO. An occasional start of these free eye movements is associated with the start of the sleep onset process of a particular person. More often however, research subjects report that they are still conscious (aware).

- Electroencephalogram (EEG) – the process of sleep onset begins with phase 1 of NREM sleep and is associated with changes in EEG, i.e. the appearance of alpha waves in the occipital region of the skull. Such changes usually appear in the EEG after a few seconds, up to a minute after the start of free eye movements. The beginning of phase 1 in the EEG pattern may or may not coincide with sleep onset. For this reason it was concluded that changes in the EEG indicating the start of this process will be K-complexes and sleep spindles, characteristic for phase 2 NREM sleep. Sleep onset is not a unitary event, various functions such as sensory awareness, memory, self-awareness, continuity of logical thought, delayed response to stimuli, and changes in brain potentials indicate the start of this process, but their time of occurrence does not always coincide with SO.

In addition to information about SO provided by a polysomnographic examination, information gained through observation on the basis of behaviours associated with falling asleep is also important. Responses to visual, auditory stimulation are weakened, and the sense of smell diminished.

The variety of EEG rhythms is significant and depends on many different factors, including a person's mental state, e.g. degree of awareness, sleepiness. Rhythms are conventionally characterized by their frequency band and relative amplitude, and are divided into:

- Delta rhythm, < 4Hz, usually seen during deep sleep, with a large amplitude;

- Theta rhythm, 4 - 7 Hz, occurs during certain stages of drowsiness and sleep;

- Alpha rhythm, 8 - 13Hz, the rhythm is significant in relaxed people with eyes closed but remaining conscious; alpha activity is suppressed immediately after eyes are opened. The amplitude of the alpha rhythm is higher in the occipital regions;

- Beta rhythm, 14 - 30 Hz, this is a fast rhythm with low amplitude, associated with the activation of the cerebral cortex and can be observed, for example, during certain sleep phases. The beta rhythm occurs mainly in the frontal and central areas of the skull.

- Gamma rhythm, > 30 Hz, this rhythm is associated with active information processing in the cerebral cortex.

Fig. 1 presents the results of all examinations and analysis performed by Israeli researchers [2] for a person who fell asleep without any problems, in a normal and correct way. These results contain a hipnogram (diagram of sleep phases), the power level of alpha and delta waves, the EMG signal, a tachogram, power levels in VLF, LF and HF bands, and the LF/HF ratio. Changes in brain wave activity during sleep onset are clearly visible, where the activity of the alpha band is replaced by delta band activity. Parameters of spectral analysis also change: VLF and LF range power falls, and a few minutes after sleep onset HF increases.

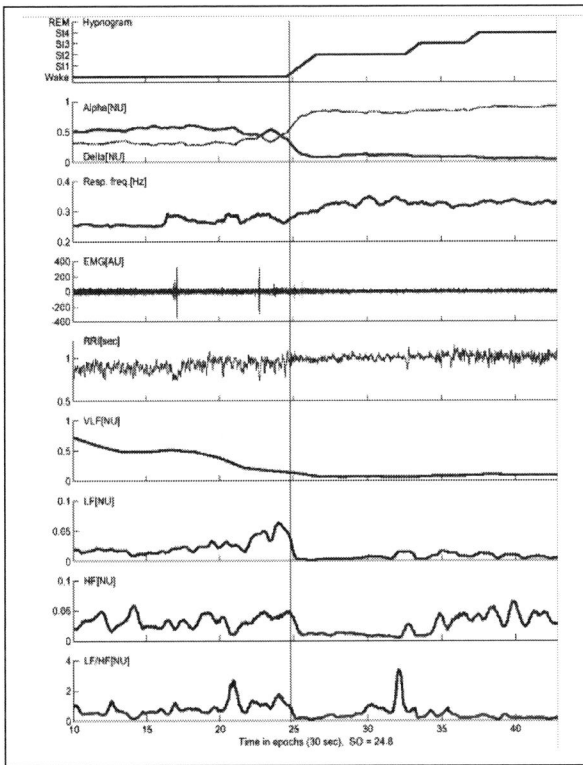

Figure 1. The results of analysis conducted for a person without problems with falling asleep [2].

IV. FATIGUE DETECTION ALGORITHM

The developed algorithm is a simple detector for the state of activity of a person over periods of 30 seconds. Similar changes were observed during analysis of the relative values of LF/HF for all research subjects, on the basis of which the state of the subject can be described. The time interval, the so-called analysis window, of 30 seconds, determines the possibility of detecting whether the person is in phase 1 of sleep. Unfortunately, it is not possible to use it to distinguish phase 2 sleep, which is the first real sleep phase and not of sleep onset, so a phase when momentary sleep onset may occur.

The algorithm is composed of the following steps: preprocessing the ECG signal which consists in, in very simplified terms, filtering out the noise in the electrical network band (50 Hz) or of muscle disturbances (35 Hz) using a digital filter, trend removal and amplitude correction.

The next step is to detect, using the selected standard algorithm, R-waves, or more precisely the whole QRS complex. Then frequency analysis is performed in which the power level is determined for individual bands: VLF, LF and HF, and the value of LF/HF is calculated based on the value of LF and HF components. The next step is to check the value of the LF/HF coefficient and, on the basis of a fixed threshold value, assigning the coefficient to certain states of consciousness. The last stage is imaging the state of consciousness of the examined person in 30-second periods. If the LF/HF coefficient is greater than 2, it indicates a state of activity, while if the value of LF/HF is in the range of 1 to 2, it indicates drowsiness (phase 1). When the value of LF/HF is lower than 1, the person's condition is classified as sleep. Level values have been determined on the basis of spectral analysis and are identical for each tachogram.

V. SPECTRAL ANALYSIS RESULTS

During the study, 22 recordings from six different people aged 20 to 53 years were collected, but only 5 were selected for further analysis, 3 belonging to a 20-year old man and two to a 22-year old woman. They were recorded during sleep onset and sleep in the evening. None of the subjects had any dysfunctions in ECG graphs and was observed with the aim of recording the approximate time of sleep onset. The recognition of this moment is possible on the basis of certain behaviours, such as: arm and leg spasms, slow, steady breath. The ECG examination was conducted with 5 leads: 4 limb and 1 thoracic. The basis for determining the LF/HR coefficient were tachogram charts that were a recording of RR intervals as a function of the next R wave sequence number. A sample tachogram is shown in Fig. 2. It belongs to the 20-year old man. The stages of sleep and the moment of sleep onset have been marked.

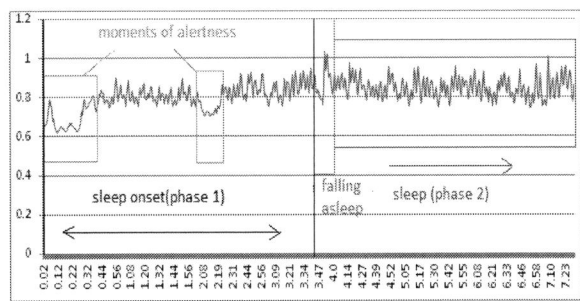

Figure 2. ECG tachogram.

Five tachograms have been subjected to the spectral analysis. Fig. 3 shows the percentage of individual frequency components for the examination performed on the 22-year old woman. The power level of the LF component fell during sleep onset and afterwards it reached a minimum value, while in the case of the HF component it was the opposite - its power increased during sleep onset and was at a high level during sleep. This flow of power from one band to the other is associated with the activity of the autonomic nervous system. Prior to sleep, the sympathetic nervous system and its related

LF range dominated. During sleep, the parasympathetic system and its associated HF component dominated. A correlation was also observed between the VLF band and the LF band, but this fact cannot be completely linked to the process of falling asleep, as the VLF band is not related in part to sympathetic activity, but concerns mainly thermoregulation and humoral regulation. Similar conclusions also apply to the other four graphs regarding frequency components. However, such significant changes in LF and HF power levels were observed only in one other case. In the other three, the HF power level after sleep onset was slightly higher than the LF.

Fig. 4 illustrates the total power level (0-0.4 Hz) for the ECG of 4 December performed on a 20-year old man. An area (between the green lines) of sleep onset was marked, during which there was a visible increase in total power. This observation did, however, concern only two tachograms coming from the same person and has not been used because of this for the design of an algorithm for detecting the state of consciousness.

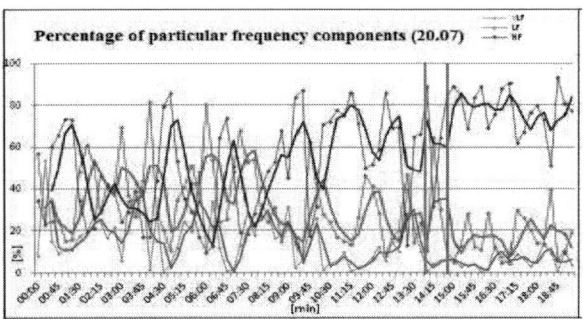

Figure 3.   Chart of particular frequency components.

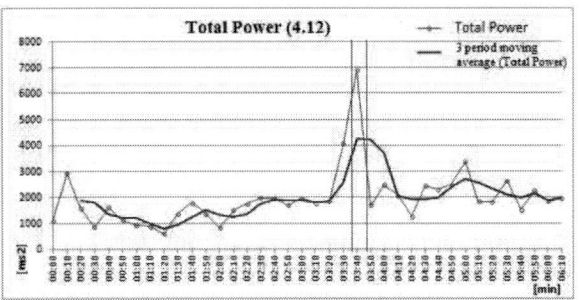

Figure 4.   Chart of total power level.

VI. CONCLUSIONS

The examinations performed should be seen as the beginning for subsequent ones which would be aimed at conducting the experiment again with a group of people larger than five (of which two were qualified for final analysis) and in a wider age range. Records should also be carried out over a longer period of time, including a longer activity time of the examined persons. They should also be carried out at the beginning in reproducible conditions for several days in order to observe trends in HRV analysis parameters that would be characteristic for the examined person or, preferably, a larger part of the examined group. Later, if other fundamental changes in HRV indicators are confirmed and identified, the conditions for registration of the biological signal can be adapted in order to determine characteristic changes to these indicators during sleep onset. The results obtained in this paper should therefore be treated as guidelines for further research and the developed algorithm as the basis for the creation of a next version covering a greater number of variables.

REFERENCES

[1]   E. Michail, A. Kokonozi, I. Chouvarda, N. Maglaveras, "EEG and HRV markers of sleepiness and loss of control during car driving." 30th Annual International IEEE EMBS Conference, Canada, 2008.

[2]   Z. Shinar, S. Akselrod, Y. Dagan, A. Baharav, "Autonomic changes during wake-sleep transition: a heart rate variability based approach", Israel, Auton Neurosci 2006 Dec 30;130(1-2):, pp.17-27.

[3]   Z Shinar, A Baharav, S Akselrod, "Changes in Autonomic Nervous System Activity and in Electro-Cortical Activity during Sleep Onset", Tel-Aviv University, Tel-Aviv, Israel, Computers in Cardiology, 2003, pp. 303-306

[4]   Hélène Otzenberger, Claude Gronfier, Chantal Simon, Anne Charloux Jean Ehrhart, François Piquard and Gabrielle Brandenberger, "Dynamic heart rate variability: a tool for exploring sympathovagal balance continuously during sleep in men", Am J Physiol Heart Circ Physiol 275:H946-H950, 1998.

[5]   G Dorfman Furman, A Baharav, C Cahan, S Akselrod, "Early Detection of Falling Asleep at the Wheel: A Heart Rate Variability Approach", Tel Aviv, Israel, Computers in Cardiology 2008; pp.1109−1112.

[6]   R. D. Ogilvie, "The process of falling asleep", Department of Psychology, Canada, 2001, Sleep Medicine Reviews, Vol.5, No.3 pp. 247 − 270.

[7]   U. Rajendra Acharya, K. P. Joseph, N. Kannathal, Ch. Min Lim, J. S. Suri, "Heart rate variability: a review", International Federation for Medical and Biomedical Engineering, 2006.

[8]   P. Bušek, J. Vaňková, J. Salinger, J. Opavský, S. Nevšímalov "Spectral Analysis of Heart Rate Variability in Sleep", Palacky University, Physiol. Res. 54. 2005, pp.369-376.

Student Projects

MIXDES 2012, 19th International Conference *"Mixed Design of Integrated Circuits and Systems"*, May 24-26, 2012, Warsaw, Poland

A 6-bit 122 MS/s Digital-to-Analog Converter for Contactless Applications in CMOS 90 nm Technology

Michał Brzeziński
Institute of Microelectronics and Optoelectronics
Warsaw University of Technology
Warsaw, Poland
e-mail: M.Brzezinski.1@stud.elka.pw.edu.pl

Tomasz Pomorski
INSIDE Secure Poland
Warsaw, Poland
e-mail: tpomorski@insidefr.com

Witold A. Pleskacz
Institute of Microelectronics and Optoelectronics
Warsaw University of Technology
Warsaw, Poland
e-mail: W.Pleskacz@imio.pw.edu.pl

Abstract—**This paper describes a 6-bit 122 MS/s CMOS current steering digital-to-analog converter (DAC) for contactless applications. Main design requirement for proposed circuit was the ability to generate ISO 14443 and ISO 15693 compliant signal for 13.56 MHz contactless communication while maintaining minimum area and power consumption. Presented DAC architecture consists of globally biased matrix of unary current sources. Each source is accompanied by a latch holding present state of the cell and AOI gate, which controls on/off state based on input signals from external binary to thermometer code decoder. The design was implemented in UMC CMOS 90 nm technology process and resulted in complete layout of circuit along with sets of post-layout and mismatch simulations.**

Index Terms—**DAC; digital-to-analog converter; CMOS IC; contactless communication; NFC; current steering**

I. INTRODUCTION

Digital-to-analog converters are one of the most widely used integrated circuits (IC). Their popularity is increasing constantly. With digital processors available so cheaply and freely it became a habit to utilize them for tasks, which previously required specialized analog circuits. This approach depends on the ability to convert signal from digital to analog domain and therefore a large selection of DACs have been designed to fulfill this purpose. They differ largely from the basic concept to the targeted applications. Majority of design architectures employ interpolation and oversampling techniques (sigma-delta DACs), pulse width modulation techniques (PWM DACs) or summing currents techniques (current steering DACs). Choice of the proper architecture depends mostly on application within, which such DAC should be used, as there is no universal solution and most of those circuits have significant tradeoffs in some and other areas.

Proposed circuit has been designed in cooperation with INSIDE Secure Poland as a part of signal generator for RF contactless applications. Therefore it has to comply with sets of ISO standards for such application. In the following paragraphs we will present design requirements and chosen architecture with obtained results.

II. DESIGN REQUIREMENTS AND LIMITATIONS

This digital-to-analog converter is a part of a signal generation path for contactless 13.56 MHz transmitters. Digital logic drives the DAC with digital values and DAC directly feeds power amplifier connected to the antenna. Because of this fact, signal generated by the converter had to be compliant with ISO 14443 [1] and ISO 15693 standards [2]. These norms define strict policies on the field's envelope shape. Both standards allow for 100% modulation index in one of the approach and another modulation with index between 8% and 14% for ISO 14443 and between 10% and 30% for ISO 15693. Overshoot voltage could not exceed 10% of difference between amplitudes of unmodulated and modulated field. Carrier signal has to be within ±7 kHz range around 13.56 MHz. This concludes most vital requirements imposed by ISO standards. Further limitations were based on external causes. DAC could not contain any down-pass filters at the output and power amplifier had to suffice in cutting off higher frequencies. Targeted platform will be an embedded device and therefore space and power consumed by DAC needed to be relatively low. Within whole chip there are two separate voltage domains: 1.0 volt for digital and 1.8 volt for analog part. This enforced a usage of both thick gate and thin gate MOS transistors.

III. DIGITAL-TO-ANALOG CONVERTERS

A. Current Steering Digital-to-Analog Converters

As it has been mentioned before, different architectures of DACs have significant tradeoffs associated with them. The current steering DACs utilize the idea of creating a group of matched current sources connected to the output node. Output value of DAC is the sum of the currents of presently enabled sources. Example of such circuit for 3-bit DAC is presented in Figure 1.

Current summing does not require any specific circuitry and allows for relatively simple architectures, which results in smaller area and power consumption of complete device. Furthermore, in CMOS technology it is really simple to create closely matched sources from current mirrors with common bias network. Transistors also take less space than complementary resistors (resistor ladders DACs) or capacitors (charge division DACs). Hence, this architecture is more applicable when small size of a circuit is crucial. Problem with this family of DACs comes when voltage output is needed instead of current. In this case I-V conversion is needed. One of the common solutions of this problem is the usage of operational amplifier (OPAMP) or operational transconductance amplifier (OTA) for this task. Both circuits increase complexity of the design and introduce their own sets of problems, as they are usually the slowest elements of the DACs.

B. Binary and Unary-Weighted Architectures

Simplest form of current steering digital-to-analog converter consists of binary weighted sources. In that way each current source corresponds to appropriate bit of input word and no decoding is necessary. This simplifies overall design but can generate large glitches during transitions. An example shown in Figure 1 belongs to this category. One can imagine the following situation. Current state of DAC corresponds to "011" and next input word is going to be "100". Difference between values is equal to the least significant bit (LSB) but transition requires switching the states of all sources and they will not occur at exactly the same moment in time. As a result two the worst-case scenarios may happen. In the first one, the most significant bit is turned on first and two following bits are turned off next. In the second one, most significant bit is turned on after the remaining bits were already turned off. Former

event generates brief output spike equal to "111" of input word and latter event sets output to zero before setting it again to appropriate value. This illustrates how glitches in this architecture can be of the same magnitude as a desired output value itself.

To reduce this effect, instead of binary weighted current sources, set of unary-weighted sources can be used. Whenever transition occurs glitch is always proportional to the amount of switched sources and not to their absolute value. This comes at the expense of size and simplicity of the design, but guarantee monotonicity of DAC. For n-bit resolution, 2^n current sources have to be used, so complexity of the design grows exponentially and practically DACs with resolutions higher than 6-bits are formed from combining both approaches. Least significant bits are formed from unary weighed current sources and most significant bits are binary weighted.

IV. PROPOSED CIRCUIT

Because of requirements explained in Section II, 6-bit unary current steering digital-to-analog architecture have been used. Signal with 100% modulation is easy to obtain, because it can be created by setting constant value at the output of the DAC. Modulation with the index value at 8% requires the ability to produce stable signal (in frequency domain) with amplitude smaller than amplitude of unmodulated signal by approx. 15%, which translates to over 9 values of least significant bit for a 6-bit DAC (peak-to-peak). As our simulations proved, this is enough to be ISO compliant and still keep complexity of the design on affordable level. Converter block has access to 122.04 MHz (9*13.56 MHz) frequency oscillator, which allows meeting strict frequency requirements for generated carrier signal. Power amplifier, at the output of the DAC, has to be voltage driven, so I-V conversion needs to be done.

A. Global Overview

Fig. 2 shows global overview of proposed circuit. Input word is divided in two 3-bit chunks, which are being sent to "Row decoder" and "Column decoder" blocks. Those are two identical binary-to-thermometer code decoders. Their outputs are connected to the matrix of 64 identical cells containing current sources and small digital logic. This digital logic, based on inputs from both binary-to-thermometer decoders, decides about the on/off state of the cell. All cells are connected to common node where summing of currents occurs. After that I-V conversion is achieved on with simple resistor. Two things are worth noting here. First, connected CMOS power amplifier is acting as a capacitive load and also require DC biasing of its input stage at 1.35 volts (3/4 of supply voltage for analog part). Second, in RF transmission, absolute amplitudes of generated carrier are not as important as the shape of the signal. Therefore global mismatch will not pose a problem in this case. If we set output range in upper half of analog supply voltage (which is 0.9 V – 1.8 V), resistor can guarantee high linearity and monotonicity of the conversion with additional benefits of low complexity and power consumption.

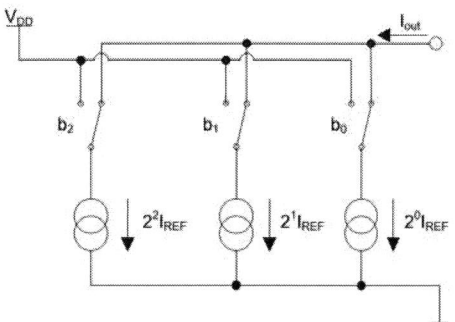

Figure 1.   Example of current steering DAC.

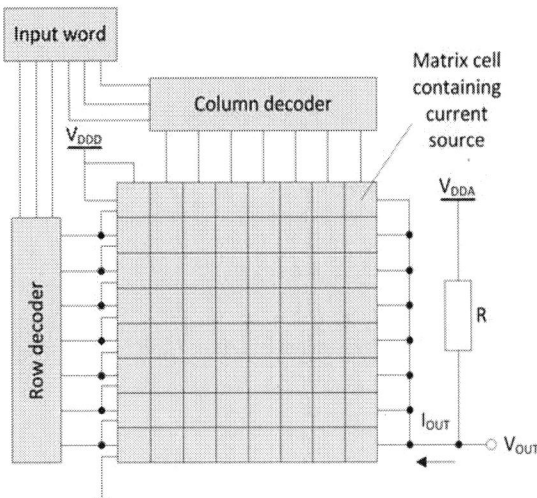

Figure 3. Global overview of proposed DAC.

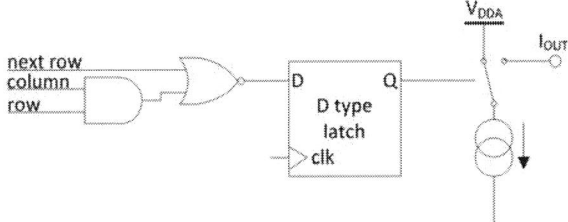

Figure 2. Matrix cell overview.

V_T of available transistors was as high as 0.55 V and, with typical cascode biasing, minimal output voltage was to close to lower border of operation region. To reduce minimum operational output voltage, the high-swing biasing has been used. By properly sizing NM0 transistor and forcing I_{REF} current through its drain one can lower minimum allowable output voltage of the current mirror to

$$V_{out,min} = 2\sqrt{\frac{2I_{REF}}{\beta}}, \qquad (3)$$

thus eliminates influence of high V_T and reduces chance to fall out of operational range even with the worst-case process corners.

Transistors NM4 and NM5 along with inverter created from transistors PM2 and NM3 are used for turning DAC off during power down state. When input "On" is in low state, active loads formed from transistors PM0 and PM1 act as open circuits. At the same time transistors NM4 and NM5 discharge remaining voltage from VB1 and VB2 nodes, which otherwise would remain on "diode" transistors.

The devices NM8 and NM9 are being driven by D-latch outputs and direct current to either power supply node or output node, while transistors NM6 and NM7 help in canceling charge gathered on the drain of NM10 transistor during switching [3]. Without those devices, charge would propagate to the output, generating a significant spike.

After examination of mismatch data for technology process, sizes of cascode transistors have been selected to obtain standard deviation (for local mismatch) of drain current below 0.8 %, which is enough to guarantee precision higher than 0.5 LSB.

B. Matrix Cells

Each matrix cell consists of and-or-invert gate driven by the outputs of the row and column decoders. Output of the gate sets the state of source cell and is held in D type latch to separate current source from transition states of the gate. The latch has its Q output connected to a switch directing current of the source between output node and analog power supply node. Schematic of matrix cell is presented in Fig. 3.

C. Current Mirrors

All of the current mirrors work in high-swing cascade configuration and are globally biased from one circuit. This has been illustrated in Fig. 4 with transistor sizes given in Table 1. Output resistance of such source is given by the equation

$$r_{out} = g_{m10} / g_{ds10} * g_{ds11}, \qquad (1)$$

where g_m is transconductance, and g_{ds} is common-source output conductance of respective transistors. Based on mismatch data for technology process, reference current, flowing through the drains of NM0, NM1 and NM2 transistors, has been set to $I_{REF} = 10$ μA, and is the same for bias circuit and current mirrors.

During design phase of DAC, technology provider offered only two types of thick gate oxide transistors for voltage supply of 2.5 V and 3.3 V, and for analog part former type has been selected. Downsize, caused by the lack of transistors prepared to work with specific supply voltage, was the necessity to use remaining transistor types with higher threshold voltage.

With typical cascode biasing, minimal allowable output voltage, which keeps output transistor out of triode region, equals

$$V_{out,min} = 2\sqrt{\frac{2I_{REF}}{\beta}} + V_T, \qquad (2)$$

with β being transconductance parameter and V_T being threshold voltage of transistors in diode configuration [2].

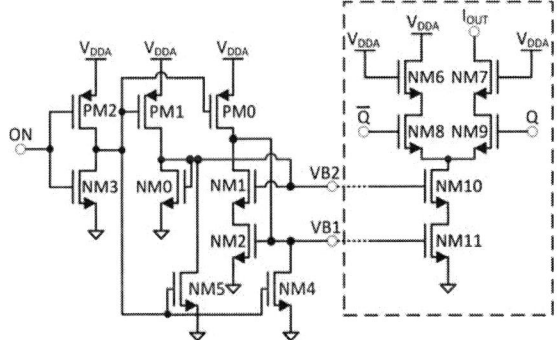

Figure 4. High-swing cascode current mirrors (dashed line surrounds part replicated in matrix cells).

551

TABLE I. TRANSISTORS USED IN HIGH-SWING CURRENT MIRROR

Name	Transistors parameters		
	Transistor type	*Width [μm]*	*Length [μm]*
NMO	NMOS 2.5 V	1	6
NM1-NM2	NMOS 2.5 V	2.3	1.4
NM3-NM5	NMOS 2.5 V	0.36	0.24
NM6-NM7	NMOS 2.5 V	1	0.24
NM8-NM9	NMOS 2.5 V	2	0.24
NM10-NM11	NMOS 2.5 V	2.3	1.4
PM0	PMOS 2.5 V	0.9	3.2
PM1	PMOS 2.5 V	1.2	2.4
PM3	PMOS 2.5 V	0.9	0.24

D. D Type Latch

D type latch used in the matrix cells is a typical design based on transmission gates, however two things are worth mentioning here. The latch is a part of digital domain and has been implemented with 1.0 V, thin gate oxide transistors. Simulations confirmed that voltage of digital power supply was enough to fully open thick oxide NMOS transistors. Changes inflicted by process variations are not significant enough to justify a switch to thick gate oxide transistors or usage of level shifters. Even with largely disturbed process output signal of DAC maintains proper shape. Another important fact regards switching characteristic of latch outputs. Normally, during transition change, voltage at inverted and noninverted outputs should be symmetrical and cross paths in the middle between supply voltage and ground. However, this can result in brief moment when both NMOS switches are turned off at the same time, which would lead to generation of voltage spike at the output and longer setting time [4, 5]. To avoid this situation n-type transistors of output inverters have significantly longer channels, so one NMOS switches can turn completely on before second one will turn off.

E. Additional Circuits

Targeted platform for proposed digital-to-analog converter will be embedded device and therefore requires quick and easy way to shutdown converter block completely. The part of the solution has been described in Sec. IV.C, with introduction of "On" input in bias block. This input operates in analog power domain and logic translation is needed to drive it directly from the digital part. For this purpose conventional high-level shifter circuits with cross-coupled PMOS transistors have been used. Digital part of DAC can be turned off by clock gating (for quick and temporary power downstate, when only leakage current flow through transistors) or by disconnecting digital supply (which requires longer setting time after switching on).

V. LAYOUT

The complete layout of the proposed DAC occupies 5565 μm^2 of chip area and can be seen in Fig. 5. This does not include binary-to-thermometer code decoders. Those modules were written in Verliog code and were synthesized as a part of digital core. The cells containing current mirror, D-type latch and AOI gate are physically organized in matrix to guarantee shortest connections between cells. This is significant because the furthest cells have higher nets capacity to charge when switched and their setting time increases accordingly. Each current mirror is surrounded by the guard ring to separate it from the influence of switching latches located in its immediate proximity.

On the left side of the design the global bias circuitry and the polysilicon resistor (made from highly resistive Poly) are clearly visible.

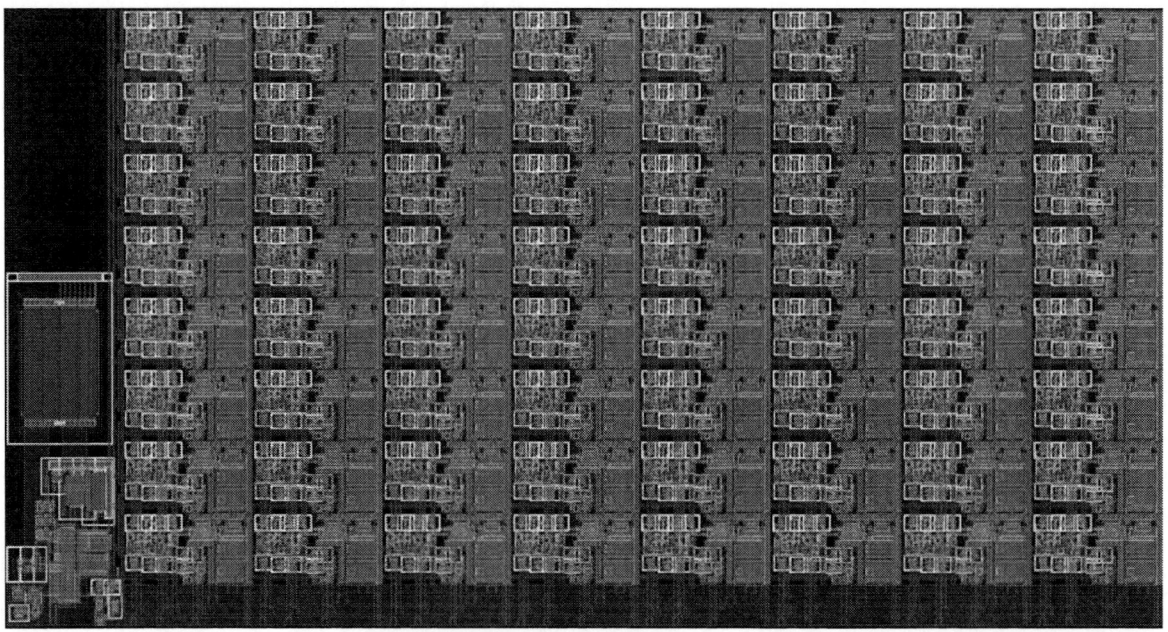

Figure 5.   Layout of proposed DAC (without binary-to-thermometer decoders).

VI. SIMULATION RESULTS

Final simulations have been done after post-layout parasitic extraction. Generated carrier signal at the output of extracted circuit is presented in Fig. 6.a. It represents the synthesized 6-bit 13.56 MHz sinus signal with 122 MS/s sampling rate. Small crosstalk from digital part is visible at the output. This should be expected from such simplistic architecture. One could also observe how nets capacitance prolongs setting time while more sources are being connected to the output at the same time. However, both of the previously described effects do not pose a problem, as the circuit has been designed for usage with specific power amplifier as the receiver. This allowed taking advantage of cutoff frequency and limited slew rate of said amplifier and using it as output filter for generated carrier. The results of Monte Carlo analysis are shown in Fig. 6.b (for mismatch of DACs output voltage) and in Fig. 6.c (for mismatch of DAC with connected power amplifier). As it is shown, only the amplitude of generated signal has changed. This has been caused mainly by process variation and therefore change of the current of every source was more or less the same. Because of that, change of the amplitude was linear and shape of a waveform remained the same. Discrete Fourier transform of generated carrier is presented in Fig. 6.d. Spurious free dynamic ratio equals 25.6 dBc.

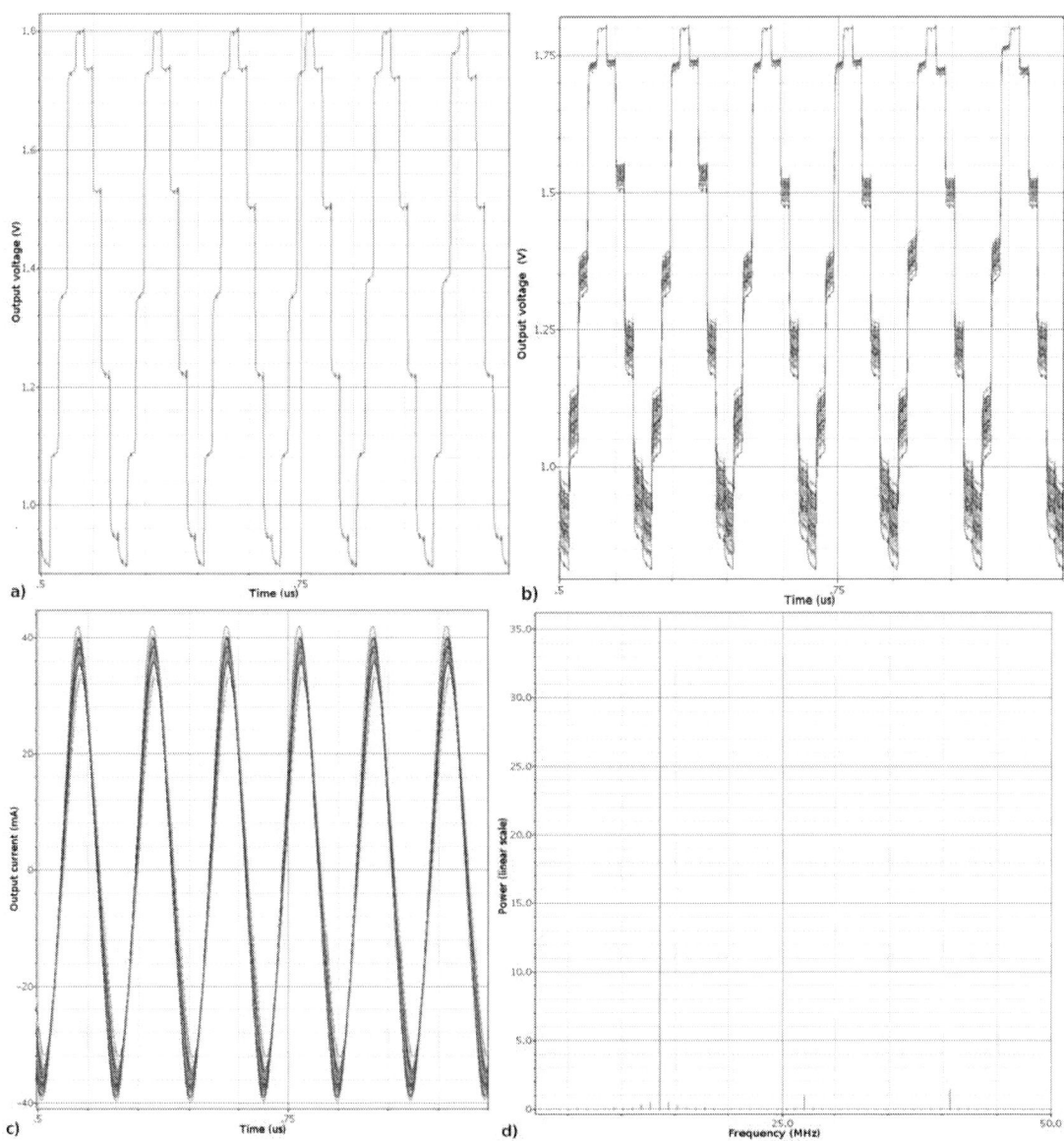

Figure 6. Post-layout simulation results for proposed DAC: a) Output voltage, b) Monte Carlo analysis of output voltage, c) Output current of fed power amplifier, d) Fourier transform of generated carrier.

TABLE II. PERFORMANCE SUMMARY OF PROPOSED DAC AND COMPARISON BETWEEN OTHER WORKS

Parameter	Proposed DAC	Work [4]	Work [7]	Work [8]	Work [9]	Work [10]	Work [11]	Work [12]
Technology [nm]	90	180	180	180	130	130	130	90
Resolution [bit]	6	8	6	6	6	6	6	10
Sampling rate [MS/s]	122.04	500	100	1000	1000	2700	3000	1000
Power consumption [mW]	1.23	7.88	15.8	24	7.5	28	29	23
Power supply [V]	1.8 / 1.0	1.8	1.8	1.8	1.5	1.2	1.2	1
Die area [mm^2]	0.006	- (*)	0.66	- (*)	0.3	0.76	0.2	0.47
SFDR [dBc]	25.6 @ 13.56 MHz	44.83 @ 10 MHz	35 @ 50 MHz	35 @ 5 MHz	37 @ 25 MHz	41 @ 300 MHz	47 @ 30 MHz	64.7 @ 50 MHz

(*) Electrical schematic was the final stage of the design described in the paper therefore no die area has been given.

VII. CONCLUSIONS

The comparison between proposed and other DACs has been presented in Table 2. Simplicity of the proposed DAC allowed us to achieve main design goals. Power consumption is sufficient for use in embedded devices, and chip area is much smaller than area used in most of the other 6-bit DACs. This architecture has some visible disadvantages but the knowledge of the specifics of operating conditions and connected amplifier gave us freedom in some of design aspects without the significant drawback in performance. At the time of the conference only post layout simulations have been done, however chip is ready to be manufactured in MPW service to test its final efficiency.

REFERENCES

[1] ANSI Committee, "ISO/IEC FDIS 14443-2. Identification cards - Contactless integrated circuit(s) cards - Proximity cards - Part 2: Radio frequency power and signal interface," 2000.

[2] ANSI Committee, "ISO/IEC FDSI 15693-2. Identification cards - Contactless integrated circuit(s) cards - Vicinity cards - Part 2: Air interface and initialization," 2000.

[3] Chen Wai-Kai, "The VLSI Handbook" - 2nd ed., Chicago: CRC Press, 2007, pp. 557-559.

[4] S. Sarkar S. Banerjee, "An 8-bit 1.8 V 500 MSPS CMOS segmented current steering DAC," Tampa: IEEE Computer Society Annual Symposium on VLSI, 2009, pp 3.

[5] G. A. M. V. der P. J. Vandenbussche, W. Sansen, M. S. J. Steyaert, G. G. E. Gielen, "A 14-bit intrinsic accuracy Q^2 random walk CMOS DAC," Santa Clara: IEEE Journal of Solid-State Circuits, 1999, vol. 34, pp. 212-213.

[6] H. K. Y. Nakamura, A. Kondo, H. Amishiro, T. Miki, K. Okada, "A 350-MS/s 3.3-V 8-bit CMOS D/A Converter using a delayed driving scheme," Santa Clara: Custom Integrated Circuits Conference 1995, 1995, pp. 1713.

[7] J. Moon, S. Hwang, D. Kim, H. Kang, S. Yeo, D. Lee, M. Song, "Design of a Current Steering CMOS D/A Converter with an Adaptive Control Switch and a Novel Layout Technique," Grenoble: International Conference on Integrated Circuit Design and Technology, 2008, pp. 29-32.

[8] K. Farzan, D.A. Johns, "A power-efficient architecture for high-speed D/A converters," Bangkok: International Symposium on Circuits and Systems, 2003, pp. 89-900, vol.1.

[9] S. Spiridon, F. Op't Eynde, "A 6 bit resolution, 1 gsamples/sec digital to analog converter," Sinaia: International Semiconductor Conference, 2005, pp. 455-458 vol. 2.

[10] J. Jae-Jin, P. Bong-Hyuck, C. Sang-Seong, L. Shin-Il, K. Suki, "A 6-bit 2.704Gsps DAC for DS-CDMA UWB," Singapore: IEEE Asia Pacific Conference on Circuits and Systems, 2006, pp. 347-350.

[11] P. Palmers, Xu Wu, M. Steyaert, "A 130 nm CMOS 6-bit Full Nyquist 3GS/s DAC," Juju: IEEE Asian Solid-State Circuits Conference, 2007, pp. 348-351.

[12] Y. Chueh-Hao, H. Ching-Hsuan, S. Tim-Kuei, C. Wen-Tzao, "A 90nm 10-Bit 1GS/s Current-Steering DAC with 1-V Supply Voltage," Hsinchu: IEEE International Symposium on VLSI Design, Automation and Test, 2008, pp. 255-258.

MIXDES 2012, 19th International Conference *"Mixed Design of Integrated Circuits and Systems"*, May 24-26, 2012, Warsaw, Poland

Application of the DLFSR Generators in Spread Spectrum Communication

Rafał Stępień, Janusz Walczak

Faculty of Electrical Engineering
Silesian University of Technology
Gliwice, Poland
rafal.stepien@polsl.pl, janusz.walczak@polsl.pl

Abstract—**The following article provides a description of an application of the pseudo random signals generators, based on dynamic linear feedback shift registers, used in the DSSS spread spectrum communication. Due the changes of generator's feedback loop in time, the length of pseudo random sequence is much longer. An implementation of the dynamic linear feedback shift register generator (DLFSR) in VHDL language was created. Exemplary communication chain with FPGA was designed and constructed. Some signals measurements were done. Based on carried out measurements some conclusions about the use of the DLFSR generators were drawn.**

Index Terms—**DLFSR; random number generators; DSSS; LFSR**

I. THE DYNAMIC LINEAR FEEDBACK SHIFT REGISTER GENERATOR -DLFSR

The pseudo random number generators are widely used in a number of areas of science and industry [1], [2]. One of the basic pseudo random generator is the Linear Feedback Shift Register (LFSR) generator [3]. These generators have static in time topological structure of the feedback loop. LFSR generators are well known and described in literature, for instance in: [3], [4], [5], [6]. The general idea of the LFSR generator is shown on figure 1.

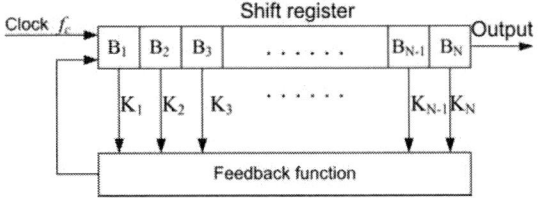

Figure 1.   The general LFSR generator

Selected register's taps $(K_1...K_N)$ are forwarded into the feedback function block which is an multiple input XOR or NXOR gate. Selected taps are written as a polynomial. Additional information about creating the polynomial description of the feedback loop can be found in [5]. When the LFSR generator is turned on, the set of active taps does not change – the register is described by the one feedback polynomial. Introducing an additional block that enables the change of feedback loop polynomial (the set of taps) while

generator is turned on, leads into an DLFSR generator [7]. The general idea of DLFSR generator is shown on figure 2.

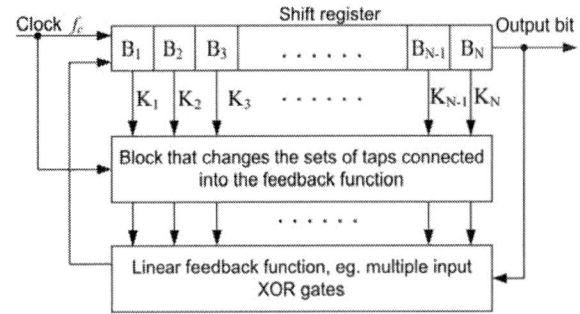

Figure 2.   A DLFSR generator

The feedback taps change block uses the clocking signal. This block counts the cycles of clocking signal and when the number of counted cycles is between the set range, it changes the structure of the feedback loop (the feedback loop polynomial is changed). The selection of feedback loop polynomials and the moments of their switch are dependent of the generator's construction. If the selection of feedback loop polynomials and the switching criteria are proper, the length of the output sequence can be increased. Additional information about description of the DLFSR generators can be found in literature [8].

II. THE DSSS SPREAD SPECTRUM COMMUNICATION.

The DSSS is one of the spread spectrum communication method [2]. It has a significant position in telecommunication systems and it is a base of UMTS, GPS, and WLAN networks [9]. The basic DSSS communication system is shown on figure 3. The information signal $m(t)$ is multiplied by the spreading sequence $r(t)$. The output signal $s(t)$ is wideband signal. In the receiver the wideband signal $s(t)$ is again multiplied by the spreading sequence $p(t)$. If both of the spreading sequences are exactly the same, the receiver's output signal will be the same as the information signal $m(t)$. In order to ensure the correct work of the system shown on figure 3 is to provide a proper synchronization between the transmitter and the receiver. This problem can be solved in different ways [10], [11], [12]. The synchronization mechanism was not the main goal of this article, however it will be considered in future Authors' research. Additional information about the DSSS system can be found in literature [9], [13].

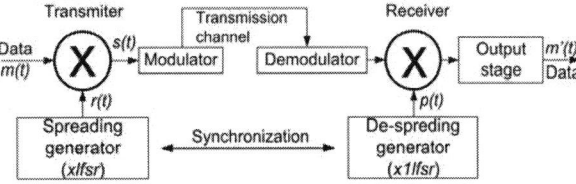

Figure 3.   The Basic DSSS communication system

III.   THE CONSTRUCTED DSSS SYSTEM

The proposed solution of telecommunication system with spread spectrum technique is shown on figure 4. Proposed circuit consists of the following blocks:

a) two DLFSR generators used as a spreading sequence generator (*xlfsr*) and despreading sequence generator (*x1lfsr*),

b) two XOR gates which work as a modulator (*XOR1*) and demodulator (*XOR2*),

c) clock input signal path which is synchronizing both *xlfsr* and *x1lfsr* generators,

d) 27MHz clocking source, which is connected with *p1* input.

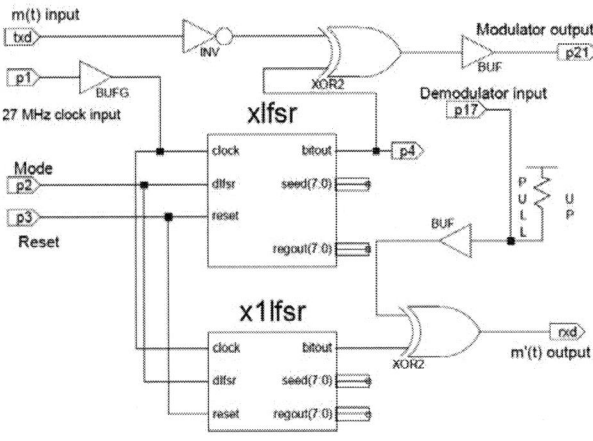

Figure 4.   Scheme of the constructed DSSS system

The DLFSR generator is made of 8 bits shift register, the reset input, the clocking signal, the dynamic feedback loop change enable signal and the 8 bits internal bus that sets the initial condition. Every input is synchronized with the clock signal. Active state of all inputs (except the *reset* input, which is active in a low state) is high state. Both DLFSR generators have 8 bits output bus and 1 bit output line. Low state on the *dlfsr* input switch the generator into a LFSR generator (only one polynomial is used). In an active state of the *dlfsr* input the generator is switched into dynamic feedback loop and it works with the following polynomials [14]:

$$y = x^8 + x^7 + x^6 + x^1 + 1 \qquad (1)$$

$$y = x^8 + x^7 + x^3 + x^2 + 1 \qquad (2)$$

$$y = x^8 + x^7 + x^6 + x^5 + x^4 + x^2 + 1 \qquad (3)$$

Each polynomial (1), (2), (3) is used as a feedback function according to the following algorithm. If number of the counted clock cycles modulo 256 is between 64 and 128 then polynomial (2) is set. If polynomial (2) is set between 64 and 128 times then polynomial (3) is set. In any different case the polynomial (1) is set. The number of clocking cycles is counted by the variable called *licznik*. Variable *licznik1* counts the sets times of polynomial (2).

The VHDL code of proposed DLFSR generator is shown on listing 1.

```
architecture Behavioral of xlfsr is
begin
process(clock, reset,dlfsr)
variable licznik: integer range 0 to 255;
variable licznik1: integer range 0 to 255;
begin
seed<="11111111";          --Initial condition
if rising_edge(clock) then  --Synchronous with rising edge
    if (reset='0') then      --if reset is active
    --write initial condition into the register
        regout(7 downto 0)<=seed(7 downto 0);
    else                     --if reset is not active
    licznik:=licznik+1;      --increase the clock cycle counter
    regout(7 downto 1)<=regout(6 downto 0); --shift bits left
                          -- (1) feedback polynomial
    regout(0)<=regout(7) xor regout(6) xor regout(5) xor regout(0);
                          --if dlfsr input is active
    if ((dlfsr='1') and (licznik>64) and (licznik<128))then
    licznik1:=licznik1+1;    --increase the polynomial (2) counter
                          |--(2) feedback polynomial
    regout(0)<=regout(7) xor regout(6) xor regout(2) xor regout(1);
    if (licznik1>64) and (licznik1<128)then
                          --(3) feedback polynomial
    regout(0)<=regout(7) xor regout(6) xor regout(5)
    xor regout(4) xor regout(3) xor regout(1);
    end if;
    end if;
    bitout<=regout(7);      --output bit
    end if;
end if;
end process;
end Behavioral;
```

Listing 1. The DLFSR code written in VHDL

Currently, as far as it is known, methods of creating switching algorithms have not been designed yet. For this reason the switching algorithm shown on listing 1, which concerns three polynomials, was set by experiments. An important problem in proposed solution was determination of the switching conditions (see variables *licznik* and *licznik1*) The constant values that were used to compare with variables *licznik* and *licznik1* were set by the experiment in order to

556

achieve the longest period of pseudo random sequence (measured by spectrum analysis of the output signal) . Above remarks determine the generator's structure that is being considered in this article.

An initial state of both DLFSR generators is important. Initial state can be determined trough 8 bit input bus or directly in the VHDL code. In the following article a direct method was used, and the initial condition is set in VHDL code. If both generators *xlfsr* and *x1lfsr* work with the same initial condition, sharing the same clocking signal and the same set of feedback polynomials as well as the algorithm of its changes, the transmitted input signal m(t) will be correctly received. Every difference between initial states, sets of feedback polynomials and the algorithm of its changes and different clocking frequencies of both generators leads to communication errors. In this situation receiver will generate a high frequency square waveform on its output that does not have any correlation with information signal *m(t)*. The scheme shown on figure 4 was implemented in the Spartan III FPGA [15], [16] family which is a part of development board. The transmission channel that connects the transmitter and the receiver was a 0,2m of wire. The influence of ground circuit, due its short length could be omitted. In constructed circuit, the influence of external noise that could be feed trough into channel was not researched.

IV. EXPERIMENTAL RESULTS

A. DLFSR generator output signal spectrum

Goal of the first measurements was to show that constructed pseudorandom generator (spreading signal generator), build as a 8 bit DLFSR generator, has wide and dense bandwidth. In this purpose an FFT of the output signal that is generated on pin *p4* (figure 4), was done. The output spectrum is shown on figure 5. Clocking signal of the DLFSR generator is equal to 27MHz and the -3dB bandwidth of the spectrum analyzer is 70MHz.

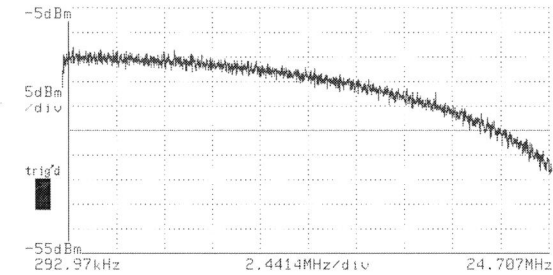

Figure 5. DLFSR output signal spectrum

The obtained spectrum (figure 5) is not flat and falls down with the Sinc envelope [4]. The -3dB point is located at 11,5MHz.

B. Spectrum of the modulator's signal.

The modulator's output signal and its spectrum are shown on figure 6. In addition the *m(t)* signal was shown. Signal *m(t)* with TTL level was generated by the laboratory generator. Its frequency is set to 2MHz. In the modulator's output signal the

phase changes are observed. Phase change occurs on the edges of information signal m(t). The output signal looks like noise signal, and no signs of the *m(t)* signal can be found in its spectrum.

C. Transmission test with the same initial conditions.

The transmission test was carried out. Information signal *m(t)* has a 2MHz square wave shape with TTL levels. Initial conditions of both DLFSR generators (figure 4) were the same. On figure 7 are shown:

a) upper waveform – output of the demodulator *m'(t)*,

b) middle waveform – information signal *m(t)*,

c) lower waveform – the *rxd* signal (figure 4).

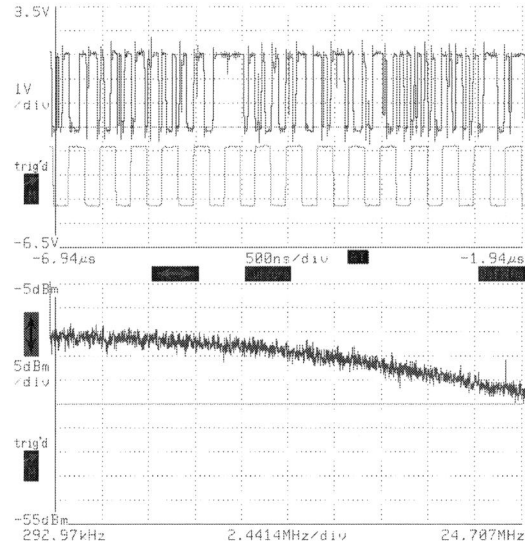

Figure 6. Output of the modulator, its spectrum and information signal *m(t)*

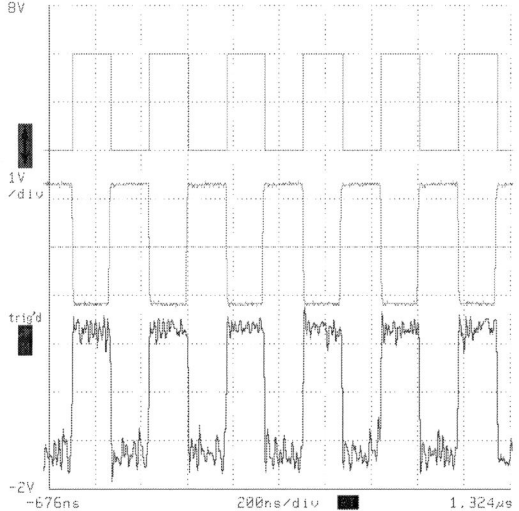

Figure 7. Received signals

The demodulator's output signal was computed digitally in the oscilloscope. A mathematical algorithm checked the sign of the rxd signal and returned +1V when rxd was positive and -1V when rxd was negative. Signal rxd is negation of m(t). This is caused by the INV inverter on the m(t) input. This inverter was used to provide the right phase relations of the system when the input signal was a RS232 waveform.

D. Transmission test with different initial conditions.

The transmission test was carried out. Information signal m(t) has a 2MHz square wave shape with TTL levels.

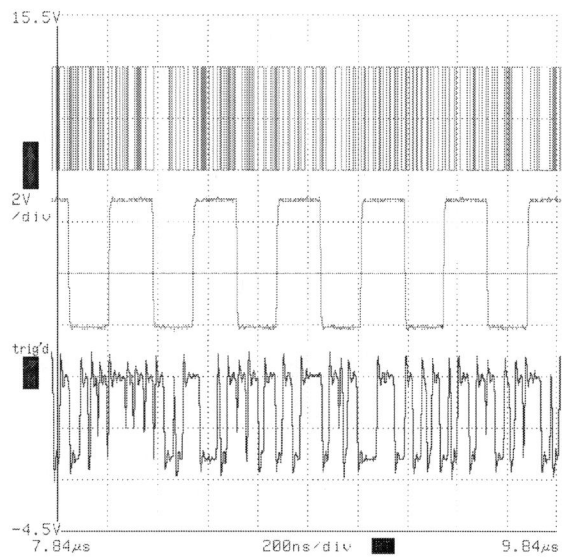

Figure 8. Received signals

Initial conditions of both DLFSR generators (figure 4) were as follows: xlfsr: 0x01, x1lfsr: 0x41. On figure 8 are shown:

a) upper waveform – output of the demodulator m'(t),

b) middle waveform – information signal m(t),

c) lower waveform – the rxd signal

Signal received at demodulator's output is a high frequency square waveform. This waveform does not have any correlation with the input signal m(t). Its impossible to reconstruct the correct information that was send.

V. SUMMARY

This article provides a description of exemplary communication system that uses DSSS technique. The spreading and despreading generators are built as a 8-bit DLFSR generators. An adequate switching method of selected feedback polynomials results in much longer output sequence (and much dense output spectrum) in comparison to standard LFSR generator. Research that was made proves that DLFSR generators can work in DSSS communication systems.

REFERENCES

[1] Kotulski Z.: Generatory liczb losowych: algorytmy, testowanie, zastosowania, Matematyka Stosowana 2,2001.

[2] Golomb S. W.: Shift-Register Sequences And Spread-Spectrum Communications, IEEE Third International Symposium on Spread Spectrum Techniques & Applications Oulu, Finland, July 4 - 6, 1994

[3] Golomb S. W.: Shift Register Sequences. Laguna Hills, C A Aegean. Park Press, 1982

[4] Mutagi R.N.: Pseudo noise sequences for engineers, Electronics & Communication Engineering Journal,Vol.8 Issue 2, April 1996, pp:79-87

[5] Schneier B.: Kryptografia dla praktyków, Vol. 2, WNT, Warszawa 2002.

[6] Massey, J. L. (1969), Shift-register synthesis and BCH decoding, IEEE Trans. Information Theory IT-15 (1):,pp:122–127.

[7] R. Mita, G. Palumbo, S. Pennisi and M. Poli. Pseudorandom bit generator based on dynamic linear feedback topology. Electronic Letters, Vol. 28, No. 19, pp. 1097–1098, 2002.

[8] Walczak J., Stępień R.: Shift registers with dynamic feedback loop. Proceedings of XXXIV conference IC-SPETO May, 2011, ss:125-126

[9] Wesołowski K.: Podstawy cyfrowych systemów telekomunikacyjnych, Wydawnictwo Komunikacji i Łączności, Warszawa 2006, ss:348-366.

[10] Stephens J.P, Norman D.M.: Direct-Sequence Spread Spectrum System, Aerospace and Electronics Conference, 1991. NAECON 1991., Proceedings of the IEEE 1991 National Aerospace and Electronics Conference, pp:462-366

[11] Kaage, U. Kolble, E.: Efficient DSSS burst synchronization methods, 1996, IEEE 4th International Symposium on Spread Spectrum Techniques and Applications Proceedings, pp:1017-1023

[12] Mangalvedhe, N.R., Reed, J.H.: An eigenstructure technique for soft synchronization of DSSS signals, Acoustics, Speech, and Signal Processing, 1996. ICASSP-96. Conference Proceedings, pp:1751-1754

[13] Wesołowski K.: Systemy radiokomunikacji ruchomej, Wydawnictwo Komunikacji i Łączności, Warszawa 2006.

[14] Ward R, Molteno T, Department of Physics, University of Otago: Table of Linear Feedback Shift Registers, 2007.

[15] Xilinx corporation.: Spartan III family data scheet, http://www.xilinx.com/support/documentation/data_sheets/ds099.pdf

[16] Alfke P.: Efficient Shift Registers, LFSR Counters, and Long Pseudo-Random Sequence Generators, Xilinx application note, July 7,1996, Vol 1.1.

MIXDES 2012, 19th International Conference *"Mixed Design of Integrated Circuits and Systems"*, May 24-26, 2012, Warsaw, Poland

FPGA Implementation of an Evolutionary Algorithm Based Charge Management for Electric Vehicles

Matthias Mielke, Simon Hardt, Armin Grünewald, Rainer Brück

Institute of Microsystem Engineering
University of Siegen
Siegen, Germany
matthias.mielke@uni-siegen.de

Abstract—Electric vehicles today are getting more usable. Due to the progress in accumulator technology and power electronics, modern electric vehicles offer performances allowing using them in daily life. It is forecasted that the number of battery electric vehicles (BEV) and plug-in hybrid electric vehicles (PHET) will increase to 5 million vehicles by the year 2020. The increase of electric vehicles will have an impact on the existing power infrastructure, especially at specific times of day. In this paper a decentralized approach to calculate a charging schedule for electric vehicles is presented. Each electric vehicle in a so called consumer grid is equipped with a power controlling unit (PCU). The PCUs are connected to each other and communicate their individual demands for power. By using an evolutionary algorithm, a schedule is calculated that ensures that all electric vehicles are charged and that a given maximum peak power is not exceeded. The proposed approach was implemented in a Java program and its performance was evaluated for different scenarios. Afterwards, a VHDL implementation was created and verified using simulation and FPGAs.

Index Terms—evolutionary algorithm; FPGA; charge management; renewable energy; electric vehicle

I. INTRODUCTION

Electric vehicles are becoming increasingly attractive. In the last years, several car manufacturers presented new electric vehicles with operating distances beyond 100 km per charge. They are regarded as an environmental friendly alternative to gasoline vehicles. The International Energy Agency predicts that a total number of 5 million battery electric vehicles (BEV) and plug-in hybrid electric vehicles (PHEV) will be used in the year 2020 [1]. The influence of the electric vehicles on existing power infrastructure was subject on different studies.

According to [2] the number of electric vehicles in Germany will raise to 1 million vehicles in the year 2020 and 6 million vehicles in 2030. Such a high number of new electric power consumers will have an impact on the power grid. Today, the impact on the current power grid is still low, because of the low number of electric vehicles. Additionally, current battery electric vehicles are charged using one phase of the power grid only, limiting the charge power to 3.7 kW (with 220 V and 16 A). Depending on the construction of the charging device and its connection to the power grid, the maximum charging power can be up to 22 kW per vehicle (400 V, three phases, 32 A fuse) [3]. With a charging power of 3.7 kW and an estimated daily need of 6 kWh per day and vehicle,

the impact of electric vehicles on the German power grid will become visible when 1 million electric vehicles will be used. The vehicles' influence depends also on the overall charging concept and the use of the vehicles. Reference [3] considers different scenarios when and where electric vehicles will be charged. It is shown that the power consumption will rise at different times of day, which can lead to overload of regional power grids.

The authors of [4] examined the impact of the increasing number of electric vehicles on a power grid in the Danish island Bornholm. Using intelligent charging schemes, the share of electric vehicles in the total number of vehicles can be up to 40%; without intelligent charging schemes, this share can only be 10%. Another case-study of the influences on a small electrical grid was published in [5]. It was shown that the existing load will increase by 18% in 2040 in the examined power grid, due to the power demand of plug-in electric vehicles. The intelligent shift of the charging processes is suggested. The calculations show that the peak power can be reduced by 7.5% by shifting the charging processes into times of day when the demand for power is low, like in the night time.

To handle the presented problems above, different approaches were already proposed in literature to reduce peak power load and prevent overload situations in power grids. An approach specifically developed for controlling the charging of electric vehicles, was presented in [6]. It was tailored to charge a fleet of electric vehicles. Each of the vehicles is equipped with a controllable charging unit. The grid operator identifies the power demand by monitoring the power consumption of the fleet. If the power consumption exceeds a certain limit, the grid operator scales the charging power of the complete fleet. The authors of [7] presented an application of evolutionary algorithms to load management. Evolutionary algorithms are used to design and select load management actions and are evaluated for different scenarios.

In this paper a system for charge management, based on an evolutionary algorithm, is proposed. The approach was implemented in a Java program to show feasibility. After verifying the functionality, a prototype was implemented in VHDL and realized using FPGAs. In the following section, the scenario and setup are presented. The evolutionary algorithm is explained in section III, followed by presentation of simulation results in section IV. The FPGA implementation is introduced in section V and a conclusion is presented in section VI.

II. SCENARIO

The presented system was developed with a specific scenario in mind: a distribution center for mail operates a number of battery power electric vehicles (BEV). The idea of the scenario was influenced by the project ELectric vehicle CIty DIStribution systems (ELCIDIS). In the project a new organization of distribution centers, in combination with using electric vehicles, was examined for mail and goods distribution. The ELCIDIS project was run from March 1998 to July 2002 in seven European cities. The results of the project are described in [8].

The basic idea of the proposed system is that the battery electric vehicles are connected in a grid, the so called consumer grid, which provides the power and a way of communication between the single BEVs and the power supplier. The power supplier assigns a power limit, the so called power budget, to the consumer grid. This power is guaranteed to the consumer grid for a specific time. After that time, the power supplier provides a new power budget to the grid. Besides this constraint, every BEV knows the state of its own battery and the point in time when the battery must be charged (the deadline). From this information, the BEV can calculate its demand for power. The point in time when the battery must be charged can be assigned by the user of the individual vehicle.

To calculate and execute a charging schedule for the BEVs in the consumer grid, each BEV is equipped with a so called power controlling unit (PCU). The PCUs communicate with each other to exchange their demand for power and the deadlines when the charging has to be finished. This information is used to determine a schedule for activation and deactivation of the BEVs' battery chargers. The setup of the scenario is illustrated in Fig. 1.

III. EVOLUTIONARY ALGORITHM

Finding a good or optimal scheduling for charging the battery electric vehicles in a consumer grid is a multi-dimensional optimization problem. The target of the optimization is to guarantee that, in the best case, all vehicles are charged before their individual deadline is reached and that the given power budget is not exceeded, which are orthogonal targets. Finding an optimal schedule for a given consumer grid can be traced back to the Knapsack problem, which is NP complete [9]. In the specific case, the problem is a two-dimensional optimization problem. An evolutionary algorithm can be used to find a good solution for the optimization problem.

As described before, each PCU has knowledge about the power budget for the consumer grid, the BEV's own power consumption, the point in time when the charging process has to be finished, the time needed to fully charge the battery, and a flag which designates if the charging process may be interrupted. To perform the calculations of the evolutionary algorithm each PCU uses three lists. A list, the BEV list, is used to store which BEVs are present in the consumer grid and

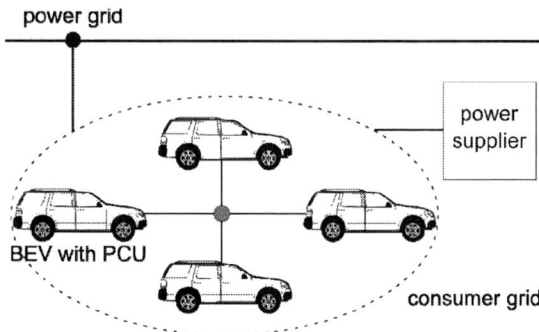

Figure 1. Setup of the scenario.

their individual parameters. A second list contains the schedules with best fitness values. It is called the best list. Its size depends on the number of PCUs in the consumer grid. Best mutation results are stored in a third list (mutation list).

The algorithm (a more detailed description is presented in [10]) starts with an initialization phase. In this phase, it calculates a first schedule with the parameters of the BEVs in the consumer grid. The quality of each schedule is rated by evaluation of a fitness function for it and the best schedule is sent out to the other PCUs. The assessment of the calculated schedules is called selection and is the second phase in the evolutionary algorithm. With help of the list containing the best up-to-date schedules, new possible solutions are generated in the mutation phase. The algorithm processes each schedule sequentially and decides if another PCU is added to the schedule. An interruptible PCU is added with a probability of 50%. Afterwards the algorithm randomly exchanges between 1 and n PCUs in the schedule; n is the number of PCUs present in the consumer grid. PCUs in the schedule are removed and PCUs, which are present in the consumer grid and which are not already in the examined schedule, are added. The resulting schedules are again assessed by calculating the fitness function. The best calculated schedule is distributed to the other PCUs.

The fitness function can be tuned by a parameter, called deadline-factor, to adapt the algorithm to different usage scenarios. With the deadline-factor the algorithm can be tuned whether to prefer power budget utilization or to guarantee that all BEVs finish the charging before deadline.

IV. SIMULATION

Initially, the algorithm was implemented in Java-program and simulated for different scenarios. The program is capable of simulating a single scenario of multiple PCUs or multiple scenarios in a batch mode. Fig. 2 shows the consumption during one charging period of a single scenario. In this scenario there are three types of BEVs. The selection of BEVs bases on the ELCIDIS-project [8]. The first type of a BEV is a scooter, Modenas Ctric. It has a range of 30-50 km, needs 3.5h for charging and has a capacity of 1.2 kWh. This is a typical scooter for short ranges, as for example for a post delivery service. The second BEV is a Renault Berlingo First Electric. It has a range of 120 km, needs 7h to charge (5h for 80% of battery capacity) and has a capacity of 23.5 kWh. This is a typical example for medium ranges, as for example for package

560

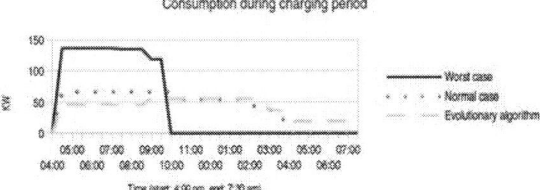

Figure 2. Power consumption of the compared charging schemes.

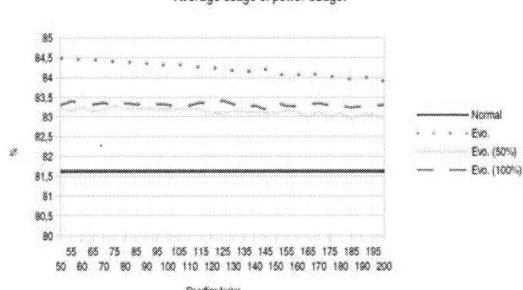

Figure 3. Power utilization dependent on deadline-factor.

delivery service. The third BEV is a Renault ElektroMidlum. It has a range of 100 km, needs 8h for charging (6h for 80% of battery capacity) and has a capacity of 150 kWh. This is a typical example of a vehicle for transporting goods between delivery stations.

A delivery station is used for the simulation scenario. In this scenario, three scooters, eight Berlingos and five ElektroMidlums are used. The scooter needs 0.36 kW, the Berlingo needs 3.52 kW and the Renault needs 18.7 kW for charging. To get real boundary conditions, the beginning of work of the delivery service was set to 8:00 am and ends at 4:00 pm. In the worst case no charging management is used and all vehicles of the station are loaded at the same time (4:30 pm). The power consumption of the worst case scenario is plotted in Fig. 2. The loading period is finished after 6h, and the maximum peak is between 4:30 pm and 9:00 pm with a consumption of 137.04 kW. In the time frame between 5:00 pm and 8:00 pm power consumption has its peak [3].

For the application of intelligent charging scheme, it is assumed that the power supplier sets a power budget to 66.33 kW, equivalent the consumption of three ElektroMidlums and three Berlingos. These two BEVs need a long time for charging, so it could be critical to get them fully charged. The power consumption of the so called normal case in Fig. 2, was calculated as a reference algorithm. This algorithm orders the PCUs in ascending order of their deadlines minus their durations. Then the PCUs are activated in this order until the power budget is reached. The maximum peak of this algorithm is between 4:30 pm and 10:30 pm, with a consumption of 66.33 kW. A negative aspect here is that the loading period ends at 4:00 am. This algorithm doesn't make full use of the available time.

In the third case, the proposed evolutionary algorithm is used (see evolutionary algorithm in Fig. 2). The peak power consumption is between 9:00 pm and 2:30 am, with 56.25 kW. In this time it's recommended to load BEVs, see [3]. The charging period of the evolutionary algorithm ends at 07:30 am, just in time before the work day starts. In this scenario the developed algorithm has an optimal power consumption over a given time period.

In order to proof that the algorithm is also capable of optimizing the power consumption of multiple scenarios, many simulation runs were done. For the simulation 400 randomly generated scenarios were used, which have been simulated ten times each. In summary 4000 simulations have been done. For the simulation of the evolutionary algorithm three types of cases have been used. In the first case no PCU was

interruptible. In the second case half of the PCUs were interruptible. And in the third case every PCU was interruptible. The results are shown in Fig. 3. With a low deadline-factor the Evolutionary algorithm with no interruptible PCU has an average power budget utilization of 84.5%, while the normal algorithm (described above) reaches 81.5%. So the evolutionary algorithm is capable to reach a 3% higher power budget utilization. With a higher deadline-factor the average value lowers to 84%, because the algorithm now selects the PCUs with a stronger focus on the deadline and does not prefer a higher maximal consumption. The evolutionary algorithm with 50% and 100% interruptible PCUs is constant between 83% and 83.5%. This result is about 2% better than the result of the normal algorithm. In summary it is possible to raise the average consumption for a randomly chosen scenario at about 3% with the developed algorithm.

V. IMPLEMENTATION

The presented approach was laid out to be simple in computation and implementation, so that it can be integrated easily in a wide variety of technologies and devices. In the following a prototype implementation of the PCU using VHDL and Field Programmable Gate Arrays (FPGA) will be presented. A block diagram of the realized design is shown in Fig. 4. The calculation of the evolutionary algorithm described above can be done for an arbitrary number of BEVs. For implementation of the algorithm in a FPGA, and later in an application specific integrated circuit (ASIC), the maximum number of PCUs in the consumer grid was restricted to 16 PCUs. The best list contains five schedules.

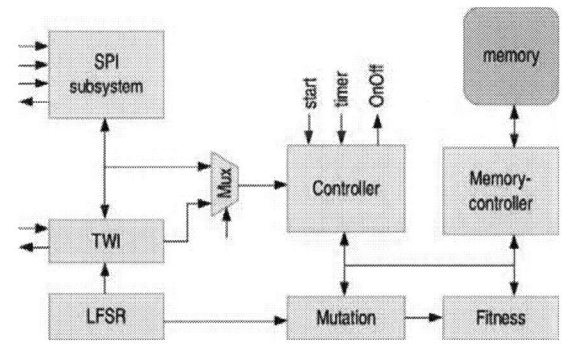

Figure 4. Block diagram of the PCU.

561

Each PCU has information about the power budget, the own power demand, the point of time when the charging has to be finished and a flag which indicates if the charging process is interruptible. These parameters are stored in a SRAM memory and are accessed via the Memory-controller (see below). The available power budget is saved in a register. Beside the parameters directly needed for execution of the evolutionary algorithm, some additional parameters are set to configure the PCU. Each PCU is assigned a unique ID stored in a register. The parameter deadline-factor (see section III) is stored in a register.

Data flow in the PCU is managed by the module named Controller. The controller reacts on the messages from the communication interfaces (see below) and control signals from outside. It is realized by a finite state machine. The operation of the PCU is started by the input signal start. When the signal is asserted, the Controller transmits its own parameters to the other PCUs via the two wire interface. The Controller handles all incoming data and prepares the own data for transmission. When a schedule is negotiated between the PCUs, the Controller executes the activation and deactivation of the BEV's charging process according to that schedule by assertion and de-assertion of the signal OnOff. A global time signal, connected to each PCU, is used by the Controller for execution of the schedule. In the prototype setup, a low frequency clock, connected to all FPGAs boards, is used for the synchronization between the PCUs.

The evolutionary algorithm described in section III is realized by the two modules Fitness and Mutation. The Mutation module generates new schedules basing on the schedules in the best list and passes them to the Fitness module for assessment. Communication between the two modules is managed by a handshake mechanism. The Mutation module accesses the entries in the best list consecutively and adds a new PCU to the list or exchanges PCUs in the list randomly. If the best list is empty, i.e. no schedules were generated before, the Mutation module generates new schedule by randomly selecting PCUs present in the consumer grid. When the Mutation module has modified a schedule, the Fitness module is started to calculate the fitness for the generated list. After all five schedules in the best list are modified by the Mutation module and the fitness for the schedules are calculated, the Fitness module asserts a flag to indicate that calculation has finished, writes the best schedule to memory and outputs the fitness value of this schedule. Now the Controller reads the schedule with best fitness and distributes it via the two wire interface to the other PCUs in the consumer grid. If a received schedule's fitness is better than a fitness of a schedule in the best list, the schedule in the best list with lowest fitness is exchanged with the received schedule. When all PCUs have finished their calculation and exchanged their best results, the next iteration is started. 50 iterations are done until a new best list is fixed and the charging processes are started.

As described above, the single PCUs are connected to a communication bus for sharing their parameters and the calculation result. As stated earlier different bus systems can be used to implement the communication. For the prototype implementation a two wire interface (TWI), similar to the I²C [11], was implemented for direct connection between the single devices. A data frame consists of 40 bit and starts with a 3 bit header, identifying the type of frame. Two frame types are defined; a frame for transmission of PCU parameter and a frame for transmission of schedules with their associated fitness. Since a collision is possible when two PCUs want to access the bus, collision detection and a backoff-strategy are implemented. When a collision occurs, a jam-signal is transmitted via the TWI. The PCUs receive the jam-signal and stop transmission. After a random waiting period, a single PCU checks if the bus is free and – if it's free – tries to restart the transmission. The backoff-time depends on the number of collisions occurred before.

In addition to the two wire interface, a SPI slave interface is implemented in the PCU. It is possible to access all data with the SPI interface, which is used to set the PCU's parameters and debug the PCU. The PCU can be used as a coprocessor for a microcontroller. It is connected to the microcontroller with the SPI interface. When used as a coprocessor, it can be chosen if the TWI module is used for communication between the PCUs or another bus which is implemented with the microcontroller. A single SPI frame consists of 48 bit. The highest bit is a parity bit, followed by 5 bit operation code and 42 bit of data. To decouple the SPI communication from the main functionality of the PCU, the control of the SPI communication is handled by a dedicated controller, the SPI subsystem. It handles all SPI access to the PCU and is directly connected to the registers saving the BEV's power consumption, its deadline-factor and its ID. These parameters are read by the Controller and by the Fitness module. For debugging purposes and for usage as a coprocessor it is possible to access the SRAM memory via the Memory-controller for write and read operations.

The central data storage of the design is realized by a 128 by 8 bit SRAM memory. Access to the memory is controlled by a Memory-controller. All modules that need to access the memory are connected to the Memory-controller, which handles the write and read operations and hides the organization of the data in the SRAM. Following data are stored in the memory:

- The BEV list, which contains the parameters of all BEVs in the consumer grid. Each entry consists of 4 Byte.

- The best list containing the five best schedules with their associated fitness. Each schedule occupies 4 Bytes in the memory.

- The mutation list with the five new schedules – without their fitness – generated by the Mutation module. Each schedule consists of 2 Byte.

- The best schedule calculated in the current iteration. It is stored with its fitness, occupying 4 Byte in memory.

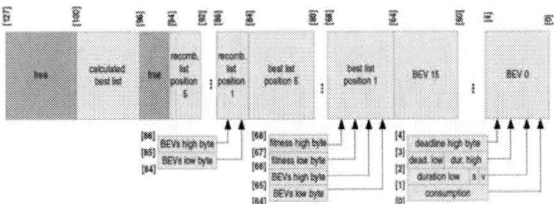

Figure 5. Organization of the data in the memory.

The organization of the data in memory is illustrated in Fig. 5. Access to memory is prioritized to guarantee that the most urgent access is handled first. The highest priority is given to the SPI-controller. Second highest priority is given to the Controller for writing data. The modules for calculation of the schedules, Fitness and Mutation, have lower priorities. The lowest priority is given to the Controller for read access.

The evolutionary algorithm and the TWI module use random numbers for operation. In the implementation, a 32 bit linear feedback shift register (LFSR) is used to generate a pseudo random number. The LFSR is initialized with the value of a counter when the start pin is asserted. The counter starts incrementing immediately after reset is released.

VI. CONCLUSION AND FUTURE WORK

In this paper the authors presented a concept and a prototype implementation of a decentralized, evolutionary algorithm based, system for scheduling the charge process of BEV in a small scale power grid. In contrast to the approach proposed by [6], the proposed approach considers the energy demand of each electric vehicle connected to the power grid directly to calculate a charging schedule which minimizes the peak power load. The system was implemented in a Java program and its performance evaluated by simulation for different scenarios, basing on a distribution center for mail. The results of the simulation were compared to a simple reference algorithm. The utilization of the power grid can be up to 84.5% when using the evolutionary algorithm compared to an 81.5% utilization when using the reference algorithm.

After implementing a proof-of-concept in Java, a VHDL implementation of the algorithm was created. It supports up to 16 devices in a consumer grid and employs a simple two-wire interface for communication between the single BEV. The description was verified using simulation and a setup for demonstration was created using Xilinx's Spartan3 FPGA. The VHDL design of the PCU is equipped with a two-wire interface for use as common bus between the PCU. For later implementation, the individual clocks of the PCUs can be synchronized by using the Precision Time Protocol, described in the IEEE 1588 standard [12]. To load the parameters, it has a SPI interface. A microcontroller can access all data in the PCU via the SPI interface and use the PCU as a coprocessor for schedule calculation.

Even though, the system was developed with an industrial scenario in mind, it would be beneficial to equip private BEVs with a PCU. The power savings for a single household are rather low, whereas usage of the system in a neighborhood can noticeably reduce peaks at the energy grid. One further level above the neighborhood, for example on the level of a district, the peak power can be reduced even better. In conclusion it is possible to reduce the number of peak power by using more PCUs. Furthermore, it seems to be feasible to use this approach in a smart home for load regulation as well. The approach can be used in a small grid with a limited number of electric load, e.g. in a single household / company or a small number of households/ companies. In order to increase the effectiveness of the system, machine learning techniques could be implemented. The evolutionary algorithm would then be able to improve its behavior by gaining experience from different scenarios and usage.

The presented FPGA implementation serves as a prototype for an ASIC implementation in a 0.35 μm CMOS process from austriamicrosystems.

ACKNOWLEDGMENT

Special thanks go to our highly motivated students Jan-Christopher Barczak, Björn Brachthäuser, Simon Hardt, Kai-Uwe Müller and Jens Schmidt for their time and effort developing and testing the system.

REFERENCES

[1] International Energy Agency Report, "Technology road map: Electric and plug-in hybrid electric vehicles," June 2011, http://www.iea.org/papers/2011/EV_PHEV_Roadmap.pdf.

[2] Informationtechnische Gesellschaft im VDE, "VDE Position Paper Energy Information Networks and Systems Taking Stock and Development Trends," Verband der Elektrotechnik Elektronik Informationstechnik e.V., Frankfurt a.M., December 2010.

[3] The Power Engineering Society, "VDE-Studie: Elektrofahrzeuge Bedeutung, Stand der Technik, Handlungsbedarf," Verband der Elektrotechnik Elektronik Informationstechnik e.V., Frankfurt a.M., April 2010.

[4] J.R. Pillai, B. Bak-Jensen., "Impacts of electric vehicle loads on power distribution systems," Vehicle Power and Propulsion Conference (VPPC), 2010 IEEE , vol., no., pp.1-6, 1-3 Sept. 2010

[5] J. Schlee, A. Mousseau, J. Eggebraaten, B. Johnson, H. Hess, "The effects of plug-in electric vehicles on a small distribution grid," North American Power Symposium (NAPS), pp.1-6, 4-6 Oct. 2009

[6] S. Bashash, H.K. Fathy, "Robust demand-side plug-in electric vehicle load control for renewable energy management," American Control Conference (ACC), pp.929-934, June 29 2011-July 1 2011.

[7] A. Gomes, C.H. Antunes, A.G. Martins, "A multiple objective evolutionary approach for the design and selection of load control strategies," Power Systems, IEEE Transactions on, vol.19, no.2, pp. 1173- 1180, May 2004.

[8] T. Vermie, "ELCIDIS FINAL REPORT," European Commision, 2002, http://elcidis.org/elcidisfinal.pdf.

[9] R. M. Karp, "Reducibility among combinatorial problems," Complexity of Computer Computations: Proc. of a Symp. on the Complexity of Computer Computations, R. E. Miller and J. W. Thatcher, Eds., The IBM Research Symposia Series, New York: Plenum Press, pp. 85-103, 1972.

[10] A. Grünewald, S. Hardt, M. Mielke, R. Brück, "A decentralized charge management for electric vehicles using a genetic algorithm," in press.

[11] NXP Semiconductors, "UM10204 I²C-bus specification and user-manual," NXP B.V., 19. June 2007.

[12] IEEE, "IEEE Standard for a Precision Clock Synchronization Protocol for Networked Measurement and Control Systems," IEEE Std 1588-2008 (Revision of IEEE Std 1588-2002), pp.c1-269, July 24 2008.

MIXDES 2012, 19th International Conference *"Mixed Design of Integrated Circuits and Systems"*, May 24-26, 2012, Warsaw, Poland

Programmable Gain Amplifier for 13.56 MHz Radio Receiver in CMOS 90 nm Technology

Mateusz Teodorowski, Witold A. Pleskacz

Institute of Microelectronics & Optoelectronics
Warsaw University of Technology
Warsaw, POLAND
e-mail: M.Teodorowski@stud.elka.pw.edu.pl
W.Pleskacz@imio.pw.edu.pl

Tony Takeshian, Tomasz Pomorski

INSIDE Secure
Sophia Antipolis, FRANCE; Warsaw, POLAND
e-mail: {ttakeshian, tpomorski}@insidefr.com

Abstract—In this paper, a PGA (Programmable Gain Amplifier) for 13.56 MHz radio receivers will be presented. The gain of the simulated PGA, based on the circuit presented in [1], is controlled by 4-bit digital word and is linear in decibels using a pseudo-exponential function. The circuit achieves a gain range of 22.91 dB (0.53–23.44 dB) with an absolute gain error not higher than 0.74 dB. The maximum gain step and minimum 3 dB bandwidth are equal to 2.02 dB and 125.8 MHz, respectively. The amplifier consumes 1.52 mW in active and 1.50 nW in standby mode from 2.5 V single power supply.

Index Terms—dB-linear gain characteristic, digitally-controlled analog amplifier, Automatic Gain Control (AGC), Common Mode Feedback (CMFB), CMOS, Programmable Gain Amplifier (PGA), Variable Gain Amplifier (VGA)

I. INTRODUCTION

Variable gain amplifiers (VGAs) are commonly used in many devices such as communication systems, disk drives, image sensors and medical equipment (e.g. hearing aids) [2]. They are often used in Automatic Gain Control loops (AGC), which have to provide a constant amplitude output signal while the input amplitude of the signal can vary greatly [2]. In an AGC circuit it is desirable to have a VGA which has a gain that is an exponential function of the control signal (linear-in-decibels gain characteristic) in order to achieve a constant loop settling time [3]. The time to adjust the gain in response to a change in the amplitude of the input signal remains constant and is independent of the level.

In bipolar transistors the exponential function, which makes possible the dB-linear gain characteristic, is built-in. Unfortunately, parameters of available lateral and vertical bipolar transistors in a typical CMOS technology process are often not satisfying while BiCMOS technology is not a cost-effective solution. For a direct exponential function implementation, MOS transistors can work in the weak inversion region. Such a usage is usually not possible, because of high frequencies used in the signal path [4] and significant device noise [5]. MOS transistors can provide useful square or linear characteristics. In order to realize the dB-linear gain characteristic using MOS transistors it is necessary to use pseudo-exponential or Taylor series approximation functions [2], [4], [6].

The gain of the amplifier can be controlled in a digital or analog fashion. VGA controlled in the digital domain is referred to as a Programmable Gain Amplifier or PGA. In this work, a variable gain amplifier that is controlled by analog signal is called VGA, while one controlled by a digital signal is called a PGA.

Disadvantage of using a PGA comparing to a VGA is the need to use many control bits in order to minimize the gain step. For applications that require a smooth gain transition, VGAs are preferred [2]. Moreover, a gain of a PGA varies as a discrete function of the control signal which can lead to discontinuous signal phases [2]. An advantage of a PGA is the fact that it can be controlled directly by the digital section of the system where complex signal processing is performed which is the case in modern communication systems [7]. In case of using an analog control signal, an ADC (Analog-to-Digital Converter) is required. Furthermore, the digital nature of the control signal exhibits all its advantages comparing to analog one: the control signal is more immune to noise and temperature variations. Conversely, when the gain control signal is digital, a DAC (Digital-to-Analog Converter) is needed for a VGA. However, in many cases, beside an implementation of a decision circuit that yields the gain control voltage, an appropriate gain control circuit is needed that transforms the gain control signal changes into proper variations of the circuit's gain [2], [8].

Regardless of the chosen approach, it is very challenging to obtain a wide gain control range with a small gain error while implementing a dB-linear gain characteristic. Implementation of a wide dB-linear gain control characteristic in a single VGA or PGA stage reduces power consumption and die area, because additional gain stages are not needed.

The paper is organized as follows: Section II presents the purpose of the project and a part of the receiver where the PGA is used. In Section III, possible solutions are briefly discussed and one of them is selected. Section IV provides a detailed discussion and a design process of the proposed PGA, followed up by simulation results in Section V. Section VI concludes the paper.

II. Purpose and Aims

The purpose of this project was to design a variable gain amplifier for a 13.56 MHz radio receiver. In order to create the AGC loop shown in Fig. 1, a variable gain amplifier is necessary. The input signal amplitude varies with the changing distance between transmitter and receiver. The aim of the VGA is to maximize the signal swing at the input of the ADC. The output from the ADC is transmitted to the digital section, where the complex signal processing is performed and the decision is made as to what the VGA gain should be in order to avoid exceeding the input dynamic range of the ADC. Since the gain control signal is digital, the natural solution is to use a PGA. A VGA can also be used, but doing so would require adding a DAC. It is worth noting that the higher the output dynamic range of the PGA, the lower the ADC resolution can be.

The section of the radio receiver shown in Fig. 1 is an AGC circuit. As mentioned in section I., it is advantageous to design a PGA circuit which has a dB-linear gain control characteristic and the widest possible gain range. For this project, no attenuation was required so the value of the minimum gain was set to ~0 dB.

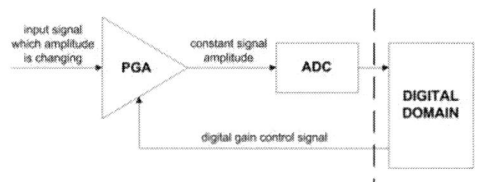

Figure 1. The part of the receiver which realizes AGC function where a PGA is needed

III. Overview of Possible Solutions

In this section, different PGA and VGA solutions are briefly discussed. VGA circuits which can function as a PGA without the need for implementing a DAC are presented. Additionally, circuits where dB-linear gain characteristic can be easily obtained without the need of implementing special exponential circuits are taken into account. Other possible solutions, especially for VGA circuits, can be found in [7]-[10].

A. High-gain amplifier with resistor-network feedback

Fig. 2a shows a high-gain amplifier with resistor-network feedback. The gain varies by changing the ratios of resistors. If the gain of the operational amplifier is not high enough, the variation of the feedback factor results in variations of the bandwidth and hence increases the total harmonic distortion (THD) [7]. The power consumption is not optimized when the circuit is designed to cover the worst-case scenario [7]. The dB-linear characteristic is obtained by switching in corresponding amount of resistance.

The resistors can be realized as poly resistors in series with MOSFET switches as shown in Fig. 2b. According to [7], in order to obtain an optimal tradeoff between the overall linearity and available bandwidth, accurate prediction of the nonlinear

effects of the MOSFET switch and the switched resistor is essential. The MOSFET switch of the variable resistor should always be connected to inputs of the operational amplifier in order for voltage variation across the switches to be minimized. In this way, the non-linearity of the switches has a minimum impact on the output signal values.

For lower power consumption (lower currents), larger resistors have to be implemented. This leads to a larger die area, a higher cost and a more demanding layout. For lower gain step, more resistors are needed and again the layout becomes more challenging.

Large resistor values can be replaced by switched capacitor circuits [11]. Moreover, capacitors exhibit lower temperature and voltage coefficients than resistors [11], [12].

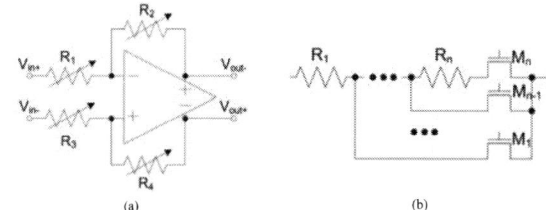

Figure 2. (a) Operational amplifier which gain is controlled by varying the ratio of feedback resistors; (b) Realization of the variable resistor

B. R-r attenuator

Fig. 3a shows a differential pair whose load resistance is varied by turning on and off the MOS transistors which simulate resistors. To achieve acceptable THD, the MOS transistor must be kept in the linear region mode of operation. The dB-linear gain characteristic can be obtained by switching in corresponding amounts of resistance. The resistor/switch pair shown in Fig. 2b can be used instead of MOS transistors.

C. Differential pair with source degeneration

Fig. 3b shows a differential pair whose transconductance is varied by changing the resistance of the degeneration resistor R_{diff}. Again the dB-linear gain characteristic can be achieved by switching in corresponding amount of resistance when using the digital control signal. Resistor R_{diff} can be implemented as shown in Fig. 2b or in Fig. 3a. An advantage of this topology is that it can achieve constant signal-to-noise-and-distortion ratio for the fixed output level regardless of the gain settings [7]. This is achieved by using a small R_{diff} when the input signal is weak in order to obtain high gain and low noise. A large R_{diff} is used when the input signal is large resulting in low gain and high linearity.

Since all above presented solutions use resistors as the main element that determine gain, it is important to be aware of problems than can arise while using resistors.

Generally speaking, integrated circuits technologies concentrate on manufacturing transistors. According to [13], there are no special layers dedicated to serving as resistors, but every free-floating layer in an integrated circuit can, when properly patterned, become a resistor. Processes where excellent resistors are available exist, but they are not universally available and the additional process steps increase die cost significantly [12]. The choice of a layer for an implementation of a resistor should be preceded by analysis of the following parameters: sheet resistance ($\Omega/\square$) which determines the dimensions of the resistor, tolerance, temperature coefficient, voltage coefficient and parasitic capacitances.

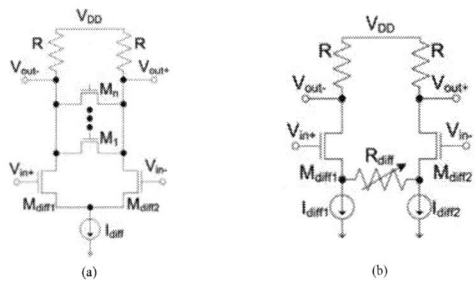

(a)   (b)

Figure 3. (a) R-r attenuation circuit; (b) Differential pair which gain is controlled by changing values of the common source resistor

It is highly desirable to design integrated circuits so that they are independent from values of passive elements which can vary greatly. The biggest advantage of any integrated technology is its ability to match well both active and passive elements [13]. Hence, it is advantageous for parameters of an integrated circuit to be defined by the ratio of two elements. Moreover, this ratio should be independent from the temperature coefficient [11].

For optimum matching of two different resistors, both should be constructed using a unit resistor and a proper layout technique (e.g. interdigitated layout [11], common-centroid layout) should be applied. Matching of all devices depends strongly on the dimensions used [13]. The bigger the devices, the higher the chance of better matching.

In the solutions presented above, the R-r attenuation circuit (Fig. 3a) is dependent on values of the type of resistor used, while the two other solutions (Fig. 2a and 3b) are dependent on the ratio.

D. Transconductance ratio

Fig. 4 shows two differential pairs. The M_1/M_2 transistors constitute the input pair, while M_3/M_4 are diode-connected transistors that work as load for the input pair. The differential gain is equal to the transconductance of the input differential pair divided by the transconductance of the load differential pair. The gain varies by varying the transistor's size and bias currents simultaneously by the same ratio. Thus, the gain follows a pseudo-exponential function given as $e^x \approx (1+x)/(1-x)$. According to [14] the circuit can achieve more than 20 dB gain range with a gain error less than $\pm$ 0.5 dB.

In this work the topology depicted in Fig. 5 and described in [1] was selected, because of the following advantages:

- Lack of resistors should allow to use this topology in any IC technology.

- Simple gain control using a digital control signal, which turns on and off proper differential pairs.

- Relatively wide gain range (20 dB) while achieving a small gain error ($\pm$ 0.5 dB). According to [2] gain range around 15 dB and a gain error of less than $\pm$ 0.5 dB when using simple pseudo-exponential and Taylor series approximation function can be achieved.

- Possibility of doubling the gain range easily [15].

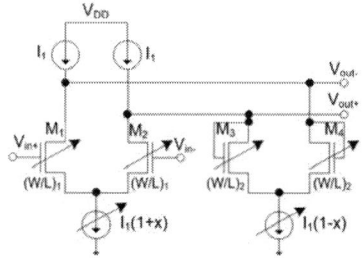

Figure 4. PGA consisting of input differential pairs and output diode-connected differential pairs. Original schematic presented in [14]

IV. PGA DESIGN

The proposed circuit was designed and simulated in UMC (United Microelectronics Corporation) CMOS 90 nm technology using the Cadence environment. A single 2.5 V power supply was used.

A. Detailed analysis of the gain control function

A detailed schematic for the selected PGA topology, presented in [1], is shown in Fig. 5 (the body terminals of NMOS and PMOS transistors are tied to ground and power supply voltage, respectively). It consists of two differential pairs and a common-mode feedback circuit (CMFB) that sets the output DC voltage. The first differential pair is the input of the PGA while the second differential pair works as the load for the first pair. Transistors in the second pair are diode-connected. The gain of a simple differential amplifier is equal to the transconductance of the input differential pair multiplied by its load impedance. For the circuit of Fig. 5 the gain is equal to the transconductance of the input differential pair divided by the transconductance of the load differential pair or $g_{m\ input\text{-}pair}/g_{m\ load\text{-}pair}$. The ratios $(W/L)_{input\text{-}pair}$, $(W/L)_{load\text{-}pair}$, $I_{input\text{-}pair}$ and $I_{load\text{-}pair}$ (the aspect ratios and bias currents of the input and load transistors, respectively) are controlled by 4 bits, binary weighted, transistor arrays $(A_0 A_1 A_2 A_3)$ and mathematically expressed as [1]:

$$(W/L)_{input-pair} = (W/L)_1 (2^0 A_0 + 2^1 A_1 + 2^2 A_2 + 2^3 A_3 + k), \quad (1)$$

$$(W/L)_{load-pair} = (W/L)_2 (2^0 \bar{A}_0 + 2^1 \bar{A}_1 + 2^2 \bar{A}_2 + 2^3 \bar{A}_3 + k), \quad (2)$$

$$I_{input-pair} = I_1 (2^0 A_0 + 2^1 A_1 + 2^2 A_2 + 2^3 A_3 + k), \quad (3)$$

$$I_{load-pair} = I_2 (2^0 \bar{A}_0 + 2^1 \bar{A}_1 + 2^2 \bar{A}_2 + 2^3 \bar{A}_3 + k), \quad (4)$$

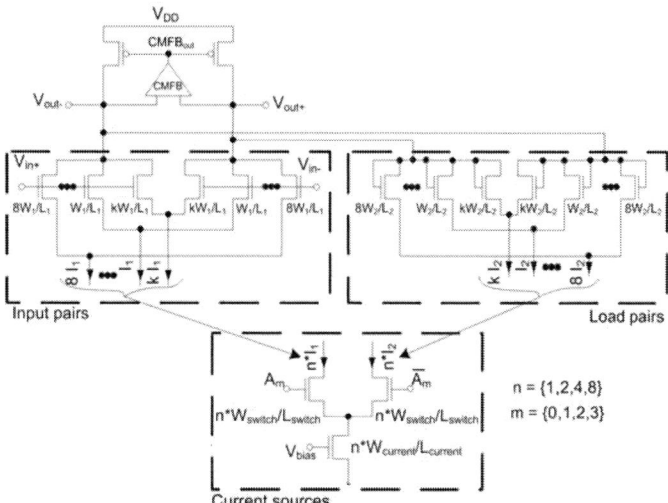

Figure 5. Schematic of the simulated PGA. Original schematic was presented in [1]

where k is a chosen positive rational constant number for adjusting the gain range. Using equations (1) thru (4) the gain of the PGA is equal to:

$$A_v = \frac{g_{m \text{ input-pair}}}{g_{m \text{ load-pair}}} = \sqrt{\frac{(W/L)_{\text{input-pair}}}{(W/L)_{\text{load-pair}}} \frac{I_{\text{input-pair}}}{I_{\text{load-pair}}}} = a\frac{k+d_1}{k+d_2} \quad (5)$$

Where a, d_1 and d_2 are equal to $\sqrt{\frac{(W/L)_1}{(W/L)_2}\frac{I_1}{I_2}}$, $2^0 A_0 + 2^1 A_1 + 2^2 A_2 + 2^3 A_3$ and $2^0 \bar{A}_0 + 2^1 \bar{A}_1 + 2^2 \bar{A}_2 + 2^3 \bar{A}_3$, respectively.

In analog version of the proposed circuit, dimensions of differential pairs are fixed and only the bias current is varied. In the digital version, twice the gain range and improved linearity can be achieved by controlling both the sizes and currents of input and diode-connected loads. The current densities of the input and diode-connected loads are kept constant for the different gain settings.

Referring to (5), increasing k will result in a more linear function while reducing the gain range. The parameter a allows to set the minimum gain value. Smaller gain steps are possible by increasing the number of binary weighted transistor arrays.

B. Calculating parameter values for the PGA

For this work the a and k parameters in equation (5) were experimentally chosen to be equal to 4 and 5, respectively. For these values of a and k simulation predicts the gain control range, maximum gain error and maximum gain step (a difference between two adjacent gain values) to be 24.08 dB (0-24.08 dB), 0.58 dB and 2.03 dB, respectively.

The maximum possible output signal amplitude is determined by the acceptable linearity zone of the output differential pair. The gain can vary with the applied input signal amplitude due to nonlinearities of the input and load pairs. The lower the $(W/L)_2$ ratio and/or higher I_{02} current are,

the wider the linearity zone of the load differential pair is. Large values of I_{02} current should not be used to increase the linearity zone, because it leads to higher power consumption and more importantly, it necessitates larger PMOS transistors (see Fig. 5). This results in greater parasitic capacitance values on the output nodes, which reduces the PGA bandwidth. To maximize the bandwidth the smallest transistor size possible should be used. The MOS transistor f_T is inversely proportional to the square of its length[1]. All mentioned demands are of course in a conflict and a compromise must be found. For the proposed PGA, the dimensions of W_2/L_2 and the current I_{02} are set to 0.36/1.1 µm and 22 µA, respectively, in order to ensure a linearity zone equal to 650 mV$_{\text{pk-pk}}$. Once the values for W_2, L_2 and I_{02} are chosen, the remaining parameters W_1, L_1 and I_{01} are defined using the constant a=4 in equation (5).

The dimensions of PMOS transistors have to be chosen such that they ensure that for their highest chosen absolute gate-source voltage value, these PMOS transistors are able to source the maximum current required by the PGA circuit. A relation $I_{01} = I_{02}$ ensures a constant overall current sourced by PMOS transistors for different gain settings. Thus, the gate voltage value of PMOS transistors does not change. If current I_{01} is smaller than current I_{02}, larger $(W/L)_1$ ratio would be needed in order to keep the a parameter value what, because of Miller effect, can lead to reducing the circuit's bandwidth

For the proposed PGA circuit the following relations were assumed:

$$I_1 = I_2 \, , \, L_1 = 2 \cdot L_{min} = 0.48 \text{ µm} \, , \, (W/L)_1 = a^2 \cdot (W/L)_2 \, ,$$

giving $(W/L)_1$ = 2.515/0.48 µm and I_1 = 22 µA. For these values of parameters the input linearity zone is equal to 100 mV$_{\text{pk-pk}}$. The smaller L_1 is, the larger is the effect of the output signal on the input differential pairs current, which

---

[1] For transistors in saturation that are not limited by effects of velocity saturation ($V_{DS} \ll E_{crit}$). For higher V_{DS}, f_T is inversely proportional to L.

results in reducing the gain. Cascoding the input pairs can help to stabilize the gain, but this requires more voltage head-room.

Values of currents for the input and output differential pairs are equal. When the current for a selected input pair is on the current for a corresponding load pair is off. Only one current source is required which operates for both the input or load pair depending on the control digital word. The switches consists of single NMOS transistors which are sized for small drain-source voltage values (wide transistors) when on. Similar to the input/load differential pairs, the switches are binary weighted to keep the voltage drop the same across the array.

For the lowest possible parasitic capacitance of the PMOS transistors (the smallest dimensions) the highest absolute value of PMOS gate-source voltage should be used. This higher value must allow the PMOS transistors to work in saturation for the required output swing. The lower the length of PMOS transistors the lower parasitic capacities on output nodes. The currents of the PMOS transistors are more dependent on the output signal levels, which decrease the gain of the PGA circuit. If there is no need to maximize the output signal swing, cascodes can be used for PMOS transistors.

C. CMFB circuit

Fig. 6 shows the implemented CMFB circuit. It consists of two differential pairs and a current mirror as an active load. In equilibrium, when no signal is applied to the input differential pair of the PGA circuit, the PMOS transistors from Fig. 5 source the current required by the input and load arrays. In this situation the output common mode DC voltages are equal to V_{ref}. When an input signal is applied, V_{out+} and V_{out-} move in the opposite directions and currents in the current mirror M_5-M_6 do not change. When I_{D1} increase I_{D4} decrease by the same amount, simultaneously.

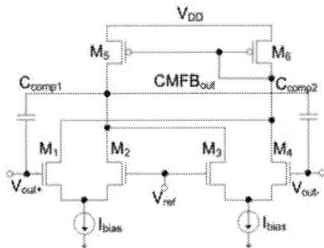

Figure 6. Circuit schematic of the CMFB block

The CMFB circuit is a negative feedback loop. Proper phase and gain margin for unity gain frequency are ensured by the use of capacitors C_{comp1} and C_{comp2} of 82 fF each, in a Miller topology. Using a single capacitor placed at the output of the CMFB circuit instead of using the Miller technique leads to a large value for the compensation capacitor. This results in long settling times when the gain of the circuit is changed. Two compensation capacitors are used in order to load the output nodes of the PGA equally. The phase margin of the CMFB circuit is equal to 60°. While ensuring proper phase margin the bond wire inductance of the power supply wires were taken into consideration.

V. SIMULATION RESULTS

A received gain control function is shown in Fig. 7. The gain control range, maximum gain error and maximum gain step are equal to 22.91 dB (0.53–23.44 dB), 0.74 dB and 2.02 dB, respectively. Simulated values in section IV.B. are equal to 24.08 dB (0-24.08 dB), 0.58 dB and 2.03 dB, respectively. As predicted earlier the gain of the circuit varies little with a value of input signal amplitude. For two input values of amplitudes equal to 20 mV$_{pk-pk}$ and 42.2 mV$_{pk-pk}$, the biggest difference between their gains is for the maximum gain setting and is equal to 0.30 dB.

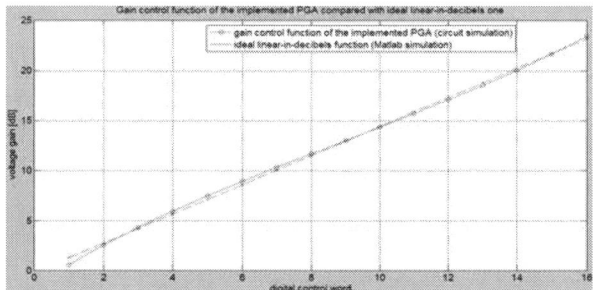

Figure 7. Gain control function compared to ideal line. Applied input sinusoidal signal: 20 mV$_{pk-pk}$, f = 13.56 MHz

In order to estimate the time needed for the outputs to achieve the proper level while the gain of the circuit is changing, output signal values between different gain settings were compared with output signal values for one gain setting. Signals on single outputs V_{out+} and V_{out-} need time equal to ~22 ns after the control bits achieve values equal to 10 % or 90 % of the power supply. Proper differential signal value is set faster within ~3 ns, because the outputs suffer for the similar distortions while changing the gain settings. Rise and fall time of the control bits were equal to 3 ns.

The bandwidth of the circuit varies with gain, due to the Miller effect. The input capacitance is proportional to the amplifier gain, which leads to lowering the bandwidth. Furthermore the higher the gain, the more active transistors are in parallel at the input and hence the higher the input capacitance is. The lower the gain, the lower the THD value. Its highest and lowest values for a sinusoidal differential input signal of 20 mV$_{pp}$ (f = 13.56 MHz) are equal to 1.22% and 0.25%, respectively.

The proposed PGA cannot be loaded by a resistive load directly and when capacitive loads are used, the higher the capacitance the lower the bandwidth. A buffer for these situations is needed. For example simple source followers can be used.

The overall PGA performance is summarized in TABLE II. and its performance is compared in TABLE I. .

TABLE I. PERFORMANCE COMPARISON OF THE PROPOSED AND PREVIOUSLY REPORTED PGAs

Ref.	CMOS Technology	3 dB bandwidth [MHz]	Output load	Power consumption	Gain variation (gain range)/Number of stages	Gain variation in one stage [dB]	Size of gain step [dB]	Maximum absolute gain error [dB]	Input-referred noise [nV/√Hz]	Method of dB-linear gain control	Active area [mm²]	Year
[16]	0.8 μm	(ᵃ)	7 pF	25mW/5V	14 (-2 to 12) / 1	14	2	0.05ʲ	12ᶠ	(ᵍ)	0.175	1996
[7]	0.35 μm	125ᵇ	2 pFᵉ	21mW/3.3V	19 (0 to 19) / 1	19	2	1	8.63ᶜ	(ᵍ)	0.18	2003
[17]	0.25 μm	100	62.5 kΩ	6.75mW/2.5V	{5.8, 12.5, 17.0} / 1	11	{6.7, 4.5}	(*)	16.75ᶠ	(ᵍ)	0.038	2001
[1]	0.13 μm	200	(*)	7.80mW/1.5V	84 (-32 to 52) / 4	21	(*)	< 1	(*)	(ʰ)	0.46	2008
[16]	0.18 μm	84ᶜ	(*)	1.37mW/1.8V	42 (-21 to 21) / 1	42	1.3	0.55	(*)	(ʰ)	0.05	2008
[15]	0.18 μm	60	(*)	3.15mW/1.5V	42 (-21 to 21) / 1	42	1.3	0.54	(*)	(ʰ)	0.078	2009
This work	0.09 μm	125.8ᶜ	none	1.52mW/2.5V	23 (0.5 to 23.5) / 1	23	1.3 – 2	0.74	3.50ᶜ	(ʰ)	(ⁱ)	2012

(*) not found; a. 15 MHz unity gain bandwidth (> 60 MHz for low-ohmic termination); b. constant bandwidth over the entire gain range; c. for the maximum gain setting; d. for the minimum gain setting; e. on each output; f. source: [7]; g. variable resistors; h. pseudo-exponential function; i. simulation; j standard deviation σ

TABLE II. OVERALL PERFORMANCE SUMMARY OF THE PROPOSED PGA

Technology:	UMC CMOS 90 nm
Supply Voltage:	V_{DD}=2.5 V, V_{SS}=0 V
Voltage Gain Range:	22.91 dB (0.53–23.44)
Maximum Absolute Gain Error/Step:	0.74 dB / 2.02 dB
3 dB Bandwidth for the Maximum/Minimum Gain Setting:	125.8 MHz/398.9 MHz
Input Referred Noise:	
for the maximum gain setting:	$3.50 \, nV/\sqrt{(Hz)}$
for the minimum gain setting:	$40.97 \, nV/\sqrt{(Hz)}$
Current consumption for on/off circuit:	606.49 μA / 160.42 pA
Assumed useful input/output differential signal amplitudes:	100 mV$_{pk-pk}$/650 mV$_{pk-pk}$

VI. CONCLUSION

In this work the PGA controlled by 4 bits was simulated in UMC 90 nm CMOS technology. The advantage of this PGA architecture is its simplicity. Equation (5) shows the gain to be independent of process. The gain control range, maximum gain error and maximum gain step, can be determined by the same equation. These values are close to the values achieved by the simulated circuit. The limitation of the circuit is a fact that values of maximum input signal amplitude, maximum output signal amplitude and values of achieved gains are exchangeable between each other. It means that after setting values of two of three above parameters (in the proposed circuit values of constant a and maximum output signal amplitude was set), the value of the third parameter is set automatically. The next limitation is a fact that the maximum output signal amplitude is limited by the width of the linear zone of the output differential pair. In this work a maximum allowable output signal swing was tried to be achieved. When using a single power supply of 2.5 V the maximum output differential signal swing is equal to 650 mV$_{pk-pk}$.

REFERENCES

[1] H.-H. Nguyen, Q.-H. Duong and S.-G. Lee, "84 dB 5.2 mA digitally-controlled variable gain amplifier," vol. 44, no. 5, Electronics Letters 28th Feb. 2008.

[2] Quoc-Hoang Duong, Quan Le, Chang-Wan Kim, Sang-Gug Lee, "A 95-dB Linear Low-Power Variable Gain Amplifier," IEEE Transactions on Circuits and Systems, vol. 53, no. 8, Aug. 2006, pp. 1648–1657.

[3] E. J. Tacconi, C. F. Christiansen, "A Wide Range and High Speed Automatic Gain Control," IEEE 1993, pp. 2139–2141.

[4] R. Harjani, "A low-power CMOS VGA for 50 Mb/s disk drive read channels," IEEE Transactions on Circuits and Systems, vol. 42, no. 6, Jun. 1995, pp. 370–376.

[5] Yuanjin Zheng, Jiangnan Yan and Yong Ping Xu, "A CMOS dB-linear VGA with pre-distortion compensation for wireless communication applications," ISCAS 2004, pp. I-813 – I-816.

[6] I.-Hsin Wang, Shen-Iuan Liu, "A 0.18-μm CMOS 1.25-Gbps Automatic-Gain-Control Amplifier," IEEE Transactions on Circuits and Systems, vol. 55, no. 2, Feb. 2008, pp. 136-140.

[7] Cheng-Chung Hsu, Jieh-Tsorng Wu, "A Highly Linear 125-MHz CMOS Switched-Resistor Programmable-Gain Amplifier," IEEE J. Solid-State Circuits, vol. 38, no. 10, Oct. 2003, pp. 1663-1670.

[8] Louis Fan Fei, "CMOS AGC Design Strategies", Microwave Journal, Feb. 2008 pp. 156-162.

[9] H. Elwan, A. Tekin, K. Pedrotti, "A Differential-Ramp Based 65 dB-Linear VGA," IEEE J. Solid-State Circuits, vol. 44, no. 9, Sep. 2009, pp 2503-2514.

[10] Hui Dong Lee, Kyung Ai Lee, Songcheol Hong, "A Wideband CMOS Variable Gain Amplifier With an Exponential Gain Control," IEEE Transactions on Microwave Theory and Techniques, vol. 55, no. 6, Jun. 2007.

[11] R. J. Baker, "CMOS – Circuit Design, Layout, and Simulation," Third Edition, IEEE Press, Wiley, ch. 4, 5, 25.

[12] T. H. Lee, "The Design of CMOS Radio-Frequency Integrated Circuits", Second Edition 2004, ch. 4.

[13] H. Camenzind, "Designing Analog Chips", February 2005, ch. 1.

[14] H.-H. Nguyen, Q.-H. Duong, H.-B. Le, J.-S. Lee, S.-G. Lee, "Low-power 42 dB-linear single-stage digitally-controlled variable gain amplifier," vol. 44, no. 13, Electronics Letters 19th Jun. 2008.

[15] H.-H. Nguyen, H.-N. Nguyen, J.-S. Lee, S.-G. Lee, "A Binary-Weighted Switching and Reconfiguration-Based Programmable Gain Amplifier", Transactions on Circuits and Systems, vol. 56, no. 9, Sep. 2009, pp. 699-703.

[16] J. J. F. Rijns, "CMOS low-distortion high-frequency variable-gain amplifier," IEEE J. Solid-State Circuits, vol. 31, pp. 1029–1034, July 1996.

[17] K. Philips and E. C. Dijkmans, "A variable-gain IF amplifier with -67dBc IM3-distortion at 1.4 V$_{pp}$ output in 0.25 μm CMOS," in Symp. VLSI Circuits Dig. Tech. Papers, 2001, pp. 81–82.

MIXDES 2012, 19th International Conference *"Mixed Design of Integrated Circuits and Systems"*, May 24-26, 2012, Warsaw, Poland

Switched Capacitor Low Noise Voltage Converter Design Strategies in 90 nm CMOS Process

Andrzej Grodzicki

Institute of Microelectronics & Optoelectronics
Warsaw University of Technology
ul. Koszykowa 75, 00-662 Warszawa, Poland
A.Grodzicki@imio.pw.edu.pl

Abstract—**Switched capacitor voltage converters (also known as charge pumps) are used for on–chip internal voltage generation. There are several techniques known in the art for designing low noise charge pumps. This paper reveals pros and cons of low noise charge pump different design techniques. Simulation results in 90 nm CMOS technology are presented.**

Index Terms—**Switched capacitor; charge pump; voltage converter; low noise; voltage doubler**

I. INTRODUCTION

Switched capacitor voltage converters are commonly used in integrated circuits for internal voltage generation. Modern switched capacitor voltage converters originate from Dickson [1] charge pump. One of well known issues needed to be addressed in charge pump design is noise. There are two major noise types that designer need to deal with.

First is a switching noise. Switched capacitor voltage converters operate on principle of charge transferring (switching) through capacitors. There are current/voltage spikes when capacitor matrix is switched. Typically frequency of the switching noise correspond to frequency of oscillator that is used for the converter clocking. The switching noise is responsible for spikes in the converter current consumption. It also creates an output voltage ripple.

Another problem is a regulation noise. The regulation noise is a result of regulation loop imbalance (for example due to immediate increase of load current consumption). The imbalance leads to the converter output voltage overshoots and undershoots. The regulation noise has not been considered in this paper.

There are several techniques known in the art to reduce the charge pump switching noise. This paper gives answers which design technique is the most suitable for modern CMOS technology. The design methods were compared by simulating several different charge pump architectures. Motivation for this work was that although many results were published, but usually for different technology and it is difficult to decide

which approach is better. Additionally switching noise reduction methods are not explored enough.

II. METHODOLOGY

In this paper the switching noise in voltage converters has been analyzed based on design examples. UMC CMOS 90 nm technology was used. Different architectures of charge pump were designed and simulated. Additionally to pump designs, other system elements were designed and simulated as well. The other system elements are: power up signal generator, reference voltage and current generator, ring oscillator. In power efficiency measurements oscillator power consumption was taken into consideration. Power up and current reference generator was not taken into account. No layout preparation or post-layout simulations were done.

A. Design Goal

The design goal was a switched capacitor voltage converter with minimized switching noise, input voltage: 1.5 V, output voltage: 2 V, current delivered to output: 100 nA.

As "core" transistors operate at 1.0 V and 1.2 V only higher voltage devices operating at 2.5 V maximum were used. For implementing charge transfer and noise filtering function a N-channel capacitors were used (gate capacitance).

B. Simulations and results preparation

For each of considered architectures simulations were done using UMC 90 nm CMOS technology. In order to neglect the regulation noise the regulation loop was removed (regulation was done manually setting frequency and capacitor area parameters). The load was an ideal 2 KΩ resistor.

III. BASIC SWITCHED–CAPACITOR CONVERTER

The first analyzed architecture was a basic Switched–Capacitor converter without any noise reducing improvements. This pump is presented in Fig. 1 and is well known in the art.

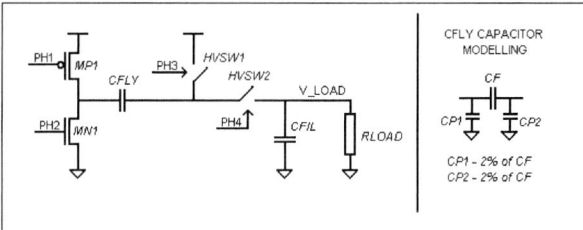

Figure 1. Basic Switched–Capacitor architecture.

Non–overlapping clock phases PH1, PH2, PH3, PH4 are controlling switching transistors MP1, MN1 and high voltage switches HVSW1 and HVSW2. The high voltage switch HVSW1 consists of NMOS transistor and local clock boost circuitry (the transistor gate voltage need to be threshold voltage greater than the supply to turn the transistor on). Consequently the HVSW2 switch consists of PMOS transistor and local clock booster. Charge transferring capacitor (hereinafter called "flying capacitor") CFLY modelling is presented in Fig. 1. According to [7] parasitic capacitances in switched-capacitor converters reach 1% to 10% of the flying capacitor value. In this paper flying capacitor (in this and in next architectures) is modeled with main capacitance CF and parasitic capacitances CP1, CP2 being 2% of CF. All three capacitances are NMOS type. Only CF area is later presented in tables. Charge transferring capacitor CFLY is charged and discharged to load sequentially. Additional filtering capacitor CFIL reduces output voltage (V_LOAD) ripple. A conventional pump would have a regulation loop for controlling output voltage level. In this paper in order to neglect regulation noise there is no regulation loop. Output voltage is regulated by frequency and flying capacitor capacitance ratios.

Design considerations for this architecture can be substantially formed in following sentence. In order to reduce output voltage ripple filtering capacitor CFIL capacitance should be significantly larger than flying capacitor CFLY. Table I presents ways of achieving output voltage ripple reduction. The first parameters set in the table represents a voltage converter which have relatively large output voltage ripple (133 mV) and need to be optimized. One of the options for minimizing output voltage ripple is increasing filtering capacitor capacitance. This approach is presented in second position of the Table I. Increasing of the filtering capacitance approach is chip area ineffective. Additionally it leads to output voltage setup time increase. Third parameter and measurements set in the table represent frequency increase approach. Although chip area is being kept low, overall efficiency drops because of power loses increase.

TABLE I. BASIC SWITCHED-CAPACITOR ARCHITECTURE SIMULATION RESULTS

Id	C_{FLY} [um²]	C_{FIL} [um²]	F_{OSC} [MHz]	V_{OUT} [V]	V_{RIPP} [mV]	IS_{MAX} [mA]	T_{SETUP} [us]	P_{EFF} [%]
1	510	2000	55.5	2.011	132	1.27	0.43	46.1
2	510	4000	55.5	2.024	66	0.92	0.65	45.8
3	375	2000	77	1.99	101	0.95	0.45	42.4

Description of parameters used in Table I.

- Id – identification number,
- C_{FLY} – flying (NMOS) capacitor total area,
- C_{FIL} – filter (NMOS) capacitor total area,
- F_{OSC} – oscillator frequency,
- V_{OUT} – output voltage middle level (this and further measurements except T_{SETUP} are taken in stable operating condition, when output voltage reached desired level),
- V_{RIPP} – output voltage ripple amplitude,
- IS_{MAX} – maximum current consumption peak from supply voltage source,
- T_{SETUP} – time required for output voltage to reach 90% of its nominal value at powering up,
- P_{EFF} – power efficiency, power dissipated in load resistor versus power consumed from supply voltage source. Oscillator power consumption is included.

Although presented architecture is simple there is one design challenge related. A typical output voltage (V_LOAD) plot is presented in Fig. 2. The plot was simulated for first design case in Table I.

Fig. 3 presents a zoom in at final period of Fig. 2 (steady state operation). There are several markers attached. Marker M0 indicates beginning of charge transfer from flying capacitor to load. At time indicated by marker M1 the charge transfer process is complete. A discharge from filtering capacitor to load begins. However between M1 and M2 the filtering capacitor discharge is steeper than between M2 and M3. It is because reverse charge transfer through high–voltage switch HVSW2. The problem of reverse charge transfer is well known in the art. Reverse charge transfer can be caused by clocking inaccuracy, or by high voltage switches isolation issues. As it is presented in Fig. 3 the reverse charge transfer not only decreases power performance, but increases switching noise as well. In this particular example working out the reverse charge transfer issue would result in approximately 20 mV ripple voltage reduction.

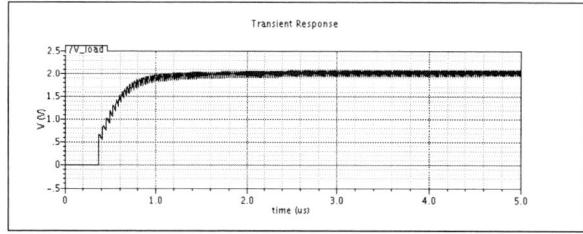

Figure 2. Basic Switched-Capacitor output voltage.

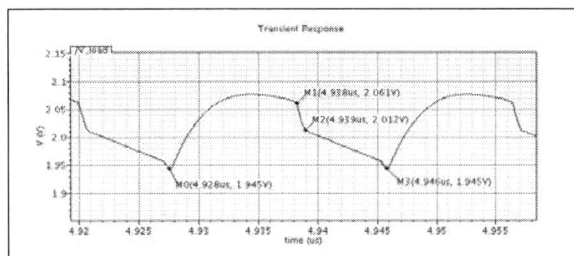

Figure 3. Basic Switched-Capacitor output voltage. Zoom in.

IV. INTERLEAVING SCHEME

A switching noise reduction technique known as interleaving scheme ([2], [3]) is multiplying pump instances and clocking them with shifted in phase clocks. The idea is presented in Fig. 4. Pump cells are the same as previously presented in Fig. 1. The main advantage of the interleaving approach is that smaller pieces of charge are transferred to load at higher ratio. To some extend the technique is similar with frequency increasing. In this particular example there are 5 converters used. Therefore effective frequency the load is charged is 5 times higher than if individual pump instance was used. There are advantages over the simple frequency increase. Oscillator frequency is not being increased (current consumption issue). The simulation results are gathered in Table II.

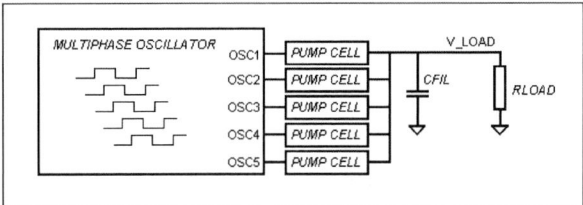

Figure 4. Interleaving scheme.

TABLE II. INTERLEAVING ARCHITECTURE RESULTS

Id	C_{FLY} [um²]	C_{FIL} [um²]	F_{OSC} [MHz]	V_{OUT} [V]	V_{RIPP} [mV]	IS_{MAX} [mA]	T_{SETUP} [us]	P_{EFF} [%]
4	700	2000	55.5	2.0	35	1.1	0.3	27.4

Surprisingly power efficiency dropped significantly comparing to previous design examples. It is because individual pump instances contain not only switches, but other circuitry as well. In particular there is a non–overlapping clock generator. Conclusion is that interleaving scheme is promising in noise reduction. But designer should keep in mind power dissipation. Circuitry like non-overlapping clock generators are repeated in every pump instance. The additional circuitry power dissipation should be carefully considered.

V. CURRENT LIMITING AND CURRENT MODE CONVERTER

Another noise reduction technique is limiting switches current. There are several possible aspects of the technique. A basic approach is to properly size flying capacitor drivers [4].

The drivers should be no stronger than required. As the drive capability is reduced, the immediate charge injection from flying capacitor to filtering capacitor will be reduced as well. First design example from Table I. was reexamined. Drive capability of transistor MP1 was reduced by 30%. These results are presented in Table III.

TABLE III. TRANSISTOR MP1 DRIVE CAPABILITY REDUCTION

Id	C_{FLY} [um²]	C_{FIL} [um²]	F_{OSC} [MHz]	V_{OUT} [V]	V_{RIPP} [mV]	IS_{MAX} [mA]	T_{SETUP} [us]	P_{EFF} [%]
5	510	2000	55.5	1.99	123	0.95	0.45	46.2

More sophisticated aspect of current limiting technique is using current limiters and current sources instead of simple switches ([5], [6]).

In this paper a design attempt that goes even further was simulated. Filtering capacitor was removed. Load was supplied with constant current directly through flying capacitor. An architecture for such attempt is presented in Fig. 5. Assuming ideal components (current source, flying capacitor, resistor) the output voltage would be constant. However, as already mentioned, a real 90nm technology components (2.5V transistors with their respective models) were used in this work. It was key important to properly design a current source. A simple current mirror is not acceptable. A precise current source as illustrated in Fig. 6 was used.

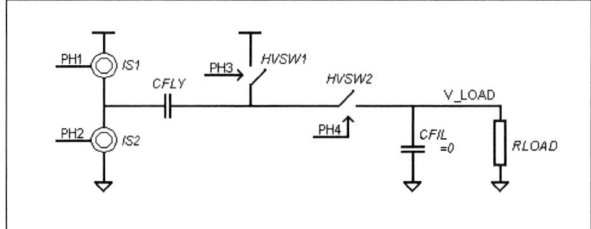

Figure 5. Current Mode converter without filtering capacitor.

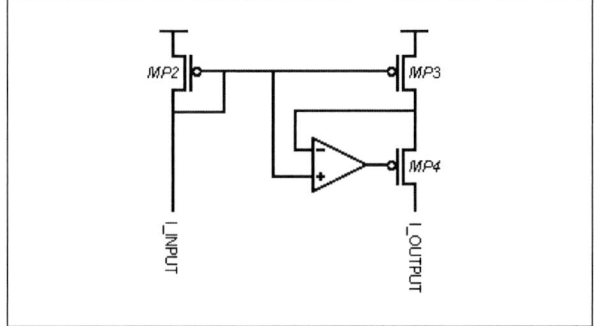

Figure 6. Precise Current Source IS1.

The principle of operation is as follows. Two pumps of architecture as presented in Fig. 5 are used in interleaved scheme. It means if first pump is recharging its flying capacitor, the second pump instance is in an opposite state -

charge is being transferred to filtering capacitor. Considering now a phase of charge transfer to load. That is when only current source IS1 and switch HVSW2 are active. The precise current source IS1 is supplying flying capacitor with constant current. Initially the flying capacitor is charged, but now, because of the constant current flowing through, the capacitor is gradually discharging. Assuming ideal elements: current source IS1, flying capacitor CFLY and load resistor RLOAD the output voltage remains constant. However flying capacitor capacitance is limited and that is why two pumps of the above architecture are required in interleaved scheme. Simulation reveals switching noise in the proposed architecture is difficult to be avoided. However, it is possible to achieve an extended time intervals when output voltage has low ripple. The price is a very large flying capacitor to be used. Simulation plot of output voltage is presented in Fig. 7 and Fig. 8. Simulation results are gathered in Table IV.

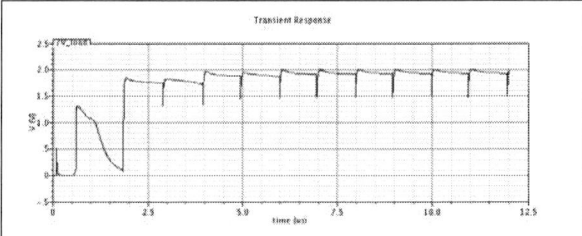

Figure 7. Current Mode converter output voltage.

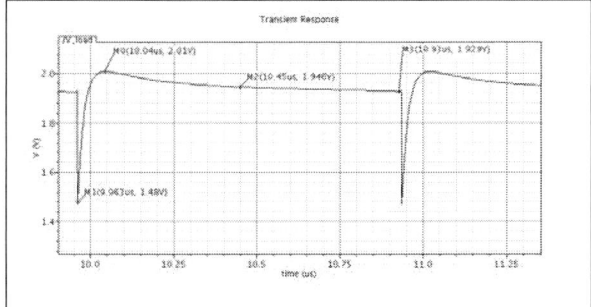

Figure 8. Current Mode converter output voltage. Zoom in.

TABLE IV. CURRENT MODE CONVERTER RESULTS

Id	C_{FLY} [um²]	C_{FIL} [um²]	F_{OSC} [MHz]	V_{OUT} [V]	V_{RIPP} [mV]	IS_{MAX} [mA]	T_{SETUP} [us]	P_{EFF} [%]
6	88000	0	0.5	1.74 1.94(*)	530 17(*)	1	1.5	58.5

(*) - measured assuming half period is useful (between markers M2, M3)

Table IV contains two values marked with star (*). As it can be observed in Fig. 8 there is an almost flat time interval between markers M2 and M3. The additional values in Table IV characterize time interval between the M2 and M3 markers.

The presented architecture is promising, but there are still design challenges to be solved. The first of design challenges is fast switching precise current source. The precise current

source as presented in Fig. 6 has slow response for turning on. It is basically because of parasitic gate-drain capacitance of the MP3 transistor (Fig. 6). When MP3 drain voltage changes rapidly during turning the current source on, the MP3 gate voltage changes too. The output current of the precise current source is highly dependent on MP3 gate voltage fluctuations. In particular referring now to Fig. 5 and Fig. 8, the IS1 current source is initially disabled (before time M1) and is then enabled (at time M1). There is an output voltage overshoot at time M0. Other precise current source, or improvements like a different clocking scheme is required.

Another difficulty is non-linearity of flying capacitor. Flying capacitor is a NMOS type capacitor using gate capacitance. With gate–source voltage change the capacitance is being modified as well. This phenomena could be neglected in previously presented architectures, but it plays significant role in current mode pump. Fortunately the capacitor capacitance versus voltage changes very little in higher voltages range. A conclusion is that NMOS flying capacitor should not be discharged too much.

The last challenge is parasitic capacitance. As already mentioned parasitic capacitance may have values in range 1% to 10 % of flying capacitor. A 2% value was picked in this work. Current from the precise current source will flow not only to the flying capacitor, but to parasitic capacitance as well. As a result converter output voltage will be affected. The parasitic capacitance is main reason for voltage drop at time moment marked as M1 in Fig. 8.

VI. CONCLUSIONS

Several design techniques has been examined for designing low noise switched capacitor voltage converters. Simulations and analysis were focused on deep sub–micron technology usage. Results were put in tables and can be compared for choosing the best design strategy. Within conventional techniques interleaving is promising. However a precise and low power consuming clock distribution is required. Current mode converters are interesting as well. An experiment of removing filtering capacitance in current mode converter revealed design challenges still remain.

REFERENCES

[1] J. Dickson, "On-chip High-Voltage Generation in NMOS Integrated Circuits Using an Improved Voltage Multiplier Technique," IEEE J. Solid-State Circuits, vol. 11, no. 6, pp. 374-378, June 1976.

[2] Dung Q. Tran,"Low Ripple Bias Voltage Generator," USP 5,036,229, July 1991.

[3] T. Kohama and R. Tsunesada, "Design Guidelines for Low-Ripple Paralleled Converter System," PESDS2009, pp.1131-1136.

[4] J. Y. Lee, S. E. Kim, S. J. Song, J. K. Kim, S. Kim, H. J. Yoo, "A Regulated Charge Pump With Small Ripple Voltage and Fast Start-Up," IEEE J. Solid-State Circuits, vol. 41, no. 2, pp. 425-432, Feb. 2006.

[5] T. Das, P. Mandal, "Switched-Capacitor based Buck Converter Design using Current Limiter for better Efficiency and Output Ripple," 2009 22nd International Conference on VLSI Design, pp.181 - 186.

[6] M. J. Kobayashi, "Low Ripple Sclable DC-DC Converter Circuit," USP 6,617,832, Sept. 2003.

[7] R. T. Burt, H. Zhang, T. L. Botker, V. V. Ivanov, "Low Ripple Charge Pump For Charging Parasitic Capacitances," USP 6,794,923, Sept.2004.

Index of Authors

ADAMSKI M.	523	DOLGIY L.	398
AHMED F.	495	DOROSZ P.	300
ALIZADEH B.	192	DUMAS N.	376
ALMUDÉVER C.G.	120, 124	DUSBIN F.	45
ALVARADO J.	74	ELHADI B.R.	495
AMAT E.	115, 120	ESKANDARI S.	360
ANDRZEJCZAK A.	527	FARINE P.-A.	227
ANGELOV G.	386, 401	FEIZY H.	243
ARBET D.	441	FINO H.	277
ARORA V.K.	17	FOURTOUNIS G.	478
ASTAPOV S.	467	FOUTRIS N.	109
AUDZEYEU M.	239	FRANKIEWICZ M.	309, 323
AVOLIO G.	84	FRICK V.	376
AXELOS N.	109	GAYDAZHIEV D.	370
AYMERICH N.	120	GIEVA E.	401
BALTUS P.G.M.	133	GŁĄB S.	300
BASZCZYK M.	300	GŁUSZKO G.	60
BELIKOV J.	473	GOES J.	178
BLALOCK B.J.	208	GOŁDA A.	189, 309
BLOK M.	489	GÓRECKI K.	304, 313
BODEA M.	356	GRABIEC P.	407
BONEV N.	386	GRAUPNER A.	364
BRENKUŠ J.	441	GROBELNA I.	523
BRINSON M.E.	94	GROBELNY M.	523
BRITTON, JR. C.L.	208	GRODZICKI A.	570
BRÜCK R.	412, 559	GRÜNEWALD A.	412, 559
BRZEZIŃSKI M.	549	GRYBOŚ P.	219, 223, 273
BURDZIUK A.	511	GRZECHCA D.	436
BUSCH L.	183	GUIBANE B.	431
BUSZKO W.	519	GYEPES G.	441
CADDEMI A.	84	HABEKOTTÉ E.	139
CANAL R.	103, 120	HABIB S.B.	147
CARRETERO J.	109	HABIBI H.	133
CARVALHO C.	455	HADIDI K.	243
CARVALHO J.	178	HAENZSCHE S.	283
CEDERSTRÖM L.	364	HAFIDA M.	495
CERDEIRA A.	74	HAHN K.	412
CHANG W.	484	HAMDI B.	431
CHATZIATHANASIOU V.	328	HAMEDANI F.T.	360
CHEN H.	198	HARDT S.	559
CHENG Y.	169, 174	HARTMANN S.	283
CHERALATHAN M.	74	HATZOPOULOS A.A.	249
CHETVERIKOV G.	502	HEINEN S.	183
CHEVILLON N.	78	HEITZ J.	376
CLAEYS C.	37	HERRERO E.	109
CLAUDIUS D.	356	HÉBRARD L.	376
CONTRERAS E.	74	HOFMANN K.	265
CREOSTEANU A.	346	HOLLEMAN J.	208
CREOSTEANU L.	346	HOLTIJ T.	88
CRUPI G.	84	HÖPPNER S.	283
DADASHI A.	192, 235, 255	HRISTOV M.	386, 401
DĄBAL P.	260	IBALA C.	478
DE MEY G.	328	IÑÍGUEZ B.	25, 74, 88
DENG S.	269	IONESCU A.M.	55
DŁUGOSZ R.	227	ISKANDER R.	45

JABŁOŃSKI G.	156	MIHÁLOV J.	441
JAKŠIĆ Z.	103	MIK Ł.	300
JAKUBOWSKI A.	51	MILOSEVIC D.	133
JANICKI M.	319, 535	MILOVANOVIC V.	198
JANSSEN E.J.G.	133	MORRISON A.P.	269
JASIELSKI J.	382	MUELLER J.H.	183
JAVID F.	45	NAE B.	25
KAŁUŻA M.	328	NAPIERALSKI A.	152, 156, 161, 319, 527, 535
KAROŃ I.	539	NAUMOWICZ M.	392
KATARZYŃSKI P.	392	NELAYEV V.	398
KAUTH C.	289	NENZI P.	69
KAYAL M.	289, 335	NOACK M.	214
KHOEI A.	243	OLIVEIRA J.P.	178, 455
KINCL Z.	427	OLIVEIRA L.B.	178
KLOES A.	88	OLIVIERI M.	69
KŁECZEK R.	219	ORLIKOWSKI M.	161
KMON P.	223, 273	OTFINOWSKI P.	219, 273
KOEHLER C.	340	PANAS A.	407
KOLANOWSKI K.	539	PAPAGIANNOPOULOS I.	328
KOLKA Z.	427	PAPPAS I.	249
KOPPA R.	511	PARTZSCH J.	214
KOROLCZUK S.	147	PASHEV A.	370
KOS A.	189, 309, 323	PASKALEVA A.	386
KOUZEHKANAN M.K.	235, 255	PASTRE M.	289
KOZAK T.	152	PAULINO N.	455
KUCEWICZ W.	300	PAUN M.-A.	335
KUCHARSKI K.	60	PEDRYCZ W.	227
KVEDARAS R.	507	PEŁKA R.	260
KVEDARAS V.	507	PEREIRA P.	277
LALLEMENT C.	78, 376	PEREK P.	156, 161
LANNUTTI F.	69	PETLENKOV E.	473
LAOPOULOS T.	203	PFLANZL W.C.	297
LESHCHYNSKYI V.	502	PIETRZAK P.	527
LEWANDOWSKI J.	417	PIOTROWICZ M.	461
LI X.	484	PIOTROWSKI Z.	543
LILJEBERG P.	447	PLESKACZ W.A.	549, 564
LIME F.	25	PLOSILA J.	447
LIU B.	484	POCHMARA J.	511
LOUËRAT M.-M.	45	POMORSKI T.	549, 564
LOVSHENKO I.	398	POSSA P.	478
ŁAKOMY K.	511	POUYAN P.	115
ŁUKASIAK L.	51	PRÉGALDINY F.	78
ŁUSZCZYK M.	417	PUKNEVA D.	370
MACHOWSKI W.	382	RADONOV R.	401
MAHMOUDI A.	192, 235	RAFFO A.	84
MAJCHRZYCKI M.	539	RAMAN A.	350
MAKOWSKI D.	152, 156, 161	RAMÍREZ T.	109
MAKOWSKI M.	527	RAUZA J.	273
MALESIŃSKA J.	60	RIID A.	467
MANOLOV E.	370	RITZENTHALER R.	25
MARAŃDA W.	461	RÓŻANOWSKI K.	417
MARGRAF M.	94	RUBIO A.	115, 120, 124
MASOUMI S.	235, 255	RUSEV R.	386, 401
MAYER A.	340	RUSU A.	55
MAYR C.	214	RYBARCZYK A.	539
MELOSIK M.	392	SALLESE J.-M.	78, 335
MERABET A.	495	SAŁEK P.	51
MIELCZAREK A.	156	SANCHEZ D.	109
MIELKE M.	559	SAPOR M.	300

SCHIEFER S.	283
SCHREURS D.	84
SCHULTZ M.	214
SCHÜFFNY R.	214, 283
SCHWARZ M.	88
SEEBACHER E.	297
SHELIBAK I.	398
SHEN L.	265
SHVEDOV S.	398
SIKORA D.	147
SILVA M.M.	178
SISKOS S.	203, 249
SIWIEC K.	231
SONDEJ T.	417
SPERANSKY D.	66
SPIRIDON S.	356
SRIVASTAVA A.	30
STEPANETS U.	239
STĘPIEŃ R.	555
STOPJAKOVÁ V.	441
STRACHE S.	183
SZABLATA P.	511
SZCZEPANIAK Z.	417
SZCZĘSNY S.	392
SZCZYGIEŁ R.	223
SZECÓWKA P.M.	519
SZELEST M.	436
SZERMER M.	319
SZEWIŃSKI J.	147
SZYPULSKA M.	543
TAKESHIAN T.	564
TAKOV T.	401
TALAŚKA T.	227
TEODOROWSKI M.	564
TEPLJAKOV A.	473

TEYMOURI M.	192, 255
TOMASZEWSKI D.	60, 407
TRUNG T.T.	66
TUROWSKI M.	350
TURTSEVICH A.	398
TYLMAN W.	535
ULAGANATHAN C.	208
USTINAVIČIUS T.	507
UZDRZYCHOWSKI W.	436
UZUNOV I.	370
VACHAUDEZ J.	478
VALDERRAMA C.	478
VALINATAJ M.	447
VAN DER WILT F.	139
VAN ROERMUND A.H.M.	133
VANNINI G.	84
VECHIRSKA I.	502
VENTIM-NEVES M.	277
VERA X.	109
VOULKIDOU A.	203
WALCZAK J.	555
WANG H.	169, 169, 174, 174
WANG T.	169, 174
WIĘCEK B.	328
WUNDERLICH R.	183
WURM M.	340
WYCHOWANIAK J.	161
YAO Y.	169, 174
YAZDANJOUEI H.	243
ZABOROWSKI M.	407
ZAITSAY V.	239
ZAJĄC P.	319
ZARĘBSKI J.	304, 313
ZIMMERMANN H.	198
ŻOŁĄDŹ M.	223, 273

IEEE
445 Hoes Lane
Piscataway, NJ 08854-4141

ISBN 978-1-4577-2092-5